SIA

Surface and Interface Analysis

ECASIA 85

Surface and Interface Analysis

(*SIA. Surf. interface anal.*)

AN INTERNATIONAL JOURNAL DEVOTED TO THE DEVELOPMENT AND APPLICATION OF TECHNIQUES FOR THE ANALYSIS OF SURFACES, INTERFACES AND THIN FILMS

Surface and Interface Analysis (ISSN 0142-2421) is published by John Wiley & Sons Limited, Baffins Lane, Chichester, Sussex, England. Orders should be addressed to: Subscriptions Department, John Wiley & Sons Limited, Baffins Lane, Chichester, Sussex, PO19 1UD, England. Copyright © 1986 by John Wiley & Sons Ltd. Printed and bound in Great Britain by J. W. Arrowsmith Ltd, Bristol.

USA Postmasters: Send address changes to *Surface and Interface Analysis*, c/o Publications Expediting Services Inc., 200 Meacham Avenue, Elmont, NY 11003.

SIA

Surface and Interface Analysis

ECASIA 85

Proceedings of the European Conference on Applications of Surface and Interface Analysis

14–18 October 1985, Veldhoven, The Netherlands

1986 **Volume 9**

Wiley

SIANDQ 9(1–6)(1–522)
ISSN 0142–2421

Surface and Interface Analysis
(*SIA. Surf. interface anal.*)

VOLUME 9 ISSUES 1-6 JULY 1986

Proceedings of the European Conference on
Applications of Surface and Interface Analysis
14–18 October 1985, Veldhoven, The Netherlands

CONTENTS

Preface/Editorial i

Opening Speeches iii

SECTION 1. TECHNIQUES

Practical Surface Analysis: State of the Art and Recent Developments in AES, XPS, ISS and SIMS. Invited paper 3
S. HOFMANN

Quantitative Surface Analysis by Glow Discharge Optical Spectrometry 21
J. PONS-CORBEAU, J. P. CAZET, J. P. MOREAU, R. BERNERON AND J. C. CHARBONNIER

X-ray Diffraction Investigations on Ultra-thin Gold Films 27
K. HÄUPL, M. LANG AND P. WISSMANN

The Application of Laser Ionization Mass Spectrometry to the Study of Thin Films and Near-surface Layers 31
N. S. CLARKE, J. C. RUCKMAN AND A. R. DAVEY

Auger Electron Spectroscopy Depth Profiling During Sample Rotation 41
A. ZALAR

Scanning Acoustic Microscopy. Invited paper 47
L. J. BALK

Crack Imaging in Alumina: a Scanning Acoustic Microscope Study 55
G. C. SMITH AND M. G. GEE

Extended Abstracts

Laser Ionization Mass Analysis: its Potential as a Quantitative Microanalytical Technique 65
T. DINGLE AND B. W. GRIFFITHS

On the Influence of Crystallographic Orientation on the Shape of Depth Profiles 66
M. KÜHLEIN

Laser Probe Microanalysis: Performance Assessment and Applications 67
M. J. SOUTHON, A. HARRIS, V. KOHLER, S. J. MULLOCK AND E. R. WALLACH

LIMA Analysis of Ga(1 − *x*)Al(*x*) As Alloys: the Effect of Alloy Composition on Secondary Ion Yields 68
G. J. EDGE AND E. A. CLARK

A New Equipment for Angle Resolved Electron Spectroscopy of Gases and Solid Surfaces 69
L. KÖVÉR, Á. KÖVÉR, I. CSERNY, T. VESZPRÉMI AND GY. ZSOMBOK

Comparative Studies of Contrast Mechanisms in Electron and Ion Induced Secondary Electron Images 70
A. BROWN, D. BRIGGS, S. Y. LAI AND J. C. VICKERMAN

Data Acquisition and Control System for Quadrupole SIMS Instruments 71
W. HILGERS AND J. HERION

Short Abstracts 72

SECTION 2. QUANTIFICATION

Surface Chemical Analysis—Report on the VAMAS Project 79
C. J. POWELL AND M. P. SEAH

Data Compilations: their Use to Improve Measurement Certainty in Surface Analysis by AES and XPS 85
M. P. SEAH

Prediction of Auger Backscattering Factors from Measurements of Spectral Backgrounds 99
M. PRUTTON AND M. E. EL GOMATI

Distribution of Hydrogen in the NPL Standard Ta_2O_5 Films 105
H. E. BISHOP

Characterization of the Imaging Properties of a Double-pass Cylindrical-mirror Analyzer 111
N. E. ERICKSON AND C. J. POWELL

On the Use of the Quartz Crystal Microbalance for Surface Analysis in Combination with Surface Sensitive Spectroscopies 119
J. ONSGAARD, W. HUANG AND E. TAGLAUER

Measurement of the Sputtering Yield with a 3D Stylus Surface Mapping System: Application to Copper and α-Brass 125
J. C. QUANTIN, E. DARQUE-CERETTI, R. COMBARIEU AND F. DELAMARE

Extended Abstracts

The Backscattering Contribution to Auger Line Intensity 129
L. MEDA AND G. QUEIROLO

Elastic and Inelastic Scattering of Electrons Emitted from Solids: Effects on Energy Spectra and Depth Profiling in XPS/AES 130
A. L. TOFTERUP, S. TOUGAARD AND P. SIGMUND

Positive Secondary Ion Yields of Elements in Amorphous and Crystalline Silicon 131
J. HERION, W. BEYER, J.-R. CHEN, B. STRITZKER AND H. WAGNER

Monochromator Versus Deconvolution in XPS: Signal-to-Noise Considerations 132
M. F. KOENIG AND J. T. GRANT

Electromigration and Charging Effects During Auger and XPS Analysis of Insulators 133
J. CAZAUX

Short Abstracts 135

SECTION 3. METALLURGY

Grazing Incidence X-ray Diffraction of an AISI 1006 Nitrogen Implanted Steel 139
N. MONCOFFRE, M. BRUNEL, P. DEYDIER AND J. TOUSSET

Wetting and Spreading of Liquid Metals: the Rôle of Surface Composition 145
G. C. SMITH AND C. LEA

Surface Segregation in Fe–18Cr–3Mo Single Crystal 151
B. SCHIFFMANN AND M. POLAK

Simultaneous Analysis of Fracture Surfaces by Auger and Energy Dispersive X-ray Mapping 157
A. WIRTH, I. ANDREONI AND G. GREGORY

On the Use of Markers and Tracers to Establish the Growth Mechanism of Alumina Scales During High Temperature Oxidation 163
E. W. A. YOUNG, H. E. BISHOP AND J. H. W. DE WIT

An AES(SAM), ESCA and ISS Study of the Oxidation of Pt_3/Sn. 1: Low Pressure Exposure to O_2 169
G. B. HOFLUND, D. A. ASBURY, P. KIRSZENSZTEJN AND H. A. LAITINEN

Analysis by Scanning Electron Microscopy and Structure Determination by High Resolution Electron Microscopy of Ni/Ni_3B Eutectic 175
J. DESSAUX-THIBAULT AND S. HAMAR-THIBAULT

Extended Abstracts

Auger Lines Relative Intensities as a Tool for the Characterizing of Oxides Formed During the Oxidation of Single Crystals of Dilute Cu–Mn Alloys 181
G. VAN STEENE, M. JARDINIER-OFFERGELD AND F. BOUILLON

Characterization of Buried Interfaces in Metallic Film-structures by SIMS 182
P. WILHARTITZ, M. GRASSERBAUER, H. G. HIESBÖCK AND C. BARTL

The Bonding State of Segregated Elements on α-Iron Surfaces and in their Iron Compounds Studied by Electron Spectroscopy 184
W. DIEKMANN, G. PANZNER AND H. J. GRABKE

Microchemical Characteristics of Metal Surface Oxidation Related to Quantitative EMPA Profiles in the Depletion Zone of High-temperature Alloys 185
H. GRÜBMEIER, L. MALDONADO, A. NAOUMIDIS AND H. NICKEL

Combined Use of EPMA, XPS, GDOS for the Characterization of Steel Surfaces 186
P. DE GREGORIO, S. FIORAVANTI AND G. M. INGO

Preliminary Study of the Formation and Characterization of Rh Alloys with an Element of the 1B Group: Copper 187
M. P. DELPLANCKE, M. JARDINIER-OFFERGELD AND F. BOUILLON

Automatic Quantitative Measurement of Carbon Profiles in Case-hardened Steels 188
M. LAMOTHE AND F. CONVERT

Short Abstracts 189

SECTION 4. CATALYSIS

Surface Analysis and the Complexity of Real Catalysts. Invited paper 195
B. DELMON

AEM, XPS and ISS Characterization of Catalyst Modifications During Propene Oxidation Over a Supported Mixed Oxide Catalyst 207
G. KREMEŃIC, V. CORTÉS CORBERÁN AND J. M. D. TASĆON

AES and SIMS of Carbonaceous Adsorbate Layers on Pt, Rh, and PtRh 215
A. D. VAN LANGEVELD AND J. W. NIEMANTSVERDRIET

Characterization of Supported Bimetallic FeIr/SiO_2 Catalysts by Mössbauer Spectroscopy, Temperature-programmed Reduction and X-ray Photoelectron Spectroscopy 221
J. W. NIEMANTSVERDRIET AND A. M. VAN DER KRAAN

XPS and AEM Studies of Carbon-supported HDS Catalysts 227
J. S. BRINEN, L. F. ALLARD AND F. P. DALY

Electronic State and Catalytic Activity of SiO_2 Doped γ-Al_2O_3 237
I. BERTÓTI, G. MINK, T. SZÉKELY, G. VARSÁNYI AND F. RÉTI

Extended Abstracts

Surface Microstructure Investigation of Calcined θ-Al_2O_3-supported Molybdena Catalyst 243
M. I. ZAKI, H. M. ISMAIL AND R. B. FAHIM

XPS Study of Rhenium on Silica 245
A. CIMINO, D. GAZZOLI, M. INVERSI AND M. VALIGI

Surface Characterization of CuO–ZnO–Al_2O_3 Methanol-synthesis Catalysts by XPS 246
G. MORETTI, M. LO JACONO, G. FIERRO AND G. MINELLI

Surface Investigation, by XPS, ISS and IR Spectroscopy of Absorbed NO, of the Structural Changes Occurring during the Reduction of Co-Mo/γAl_2O_3 Catalysts 247
R. PRADA SILVY, J. M. BEUKEN, J. L. G. FIERRO, P. BERTRAND AND B. DELMON

XPS Study of Irradiated Polycrystalline TiO_2 248
G. MUNUERA, A. R. GONZÁLEZ-ELIPE, J. P. ESPINÓS, A. NAVÍO AND A. FERNÁNDEZ

The Influence of Reductive Pretreatments on CuO/MnO_2 Catalytic Activity: an XPS Study 249
V. DI CASTRO, C. FURLANI, C. FRAGALE, M. GARGANO AND M. ROSSI

Surface Analysis of Pt-black by XPS 251
L. KÖVÉR, J. TÓTH AND Z. PAÁL

Reactions on Thin Films of Cobalt–Molybdenum Oxides: I. Characterization of Molybdenum and Cobalt Oxide Species by High Resolution XPS 253
N. S. MCINTYRE, D. JOHNSTON, L. L. COATSWORTH AND J. R. BROWN

Reactions on Thin Films of Cobalt–Molybdenum Oxides: II. Reduction, Sulphidation, and Hydrodesulphurization (HDS) Activities 255
J. R. BROWN, N. S. MCINTYRE, D. JOHNSTON AND L. L. COATSWORTH

An ISS Study of Platinized Titania 258
G. B. HOFLUND, A. L. GROGAN, JR. AND D. A. ASBURY

XPS Studies on Doped Vanadia Catalysts in CO Oxidation 260
G. MINK, I. BERTÓTI AND T. SZÉKELY

Effect of the Pretreatment on the Surface Structure of Amorphous $Fe_{80-x}Ni_xB_{20}$ Alloys and their Catalytic Activity and Selectivity in the $CO+H_2$ Reaction 261
L. GUCZI, G. KISFALUDI, Z. SCHAY AND A. LOVAS

Short Abstracts 262

SECTION 5. ELECTRONICS, SEMICONDUCTORS AND THIN FILMS

Perspectives in Surface and Interface Analysis for Electronic Devices and Circuits. Invited paper 265
R. HECKINGBOTTOM

Study of Near-surface Disorder and Surface Residues after Reactive Ion Etching of Silicon 275
G. S. OEHRLEIN, G. J. COYLE, J. G. CLABES AND Y. H. LEE

Applications of SIMS, SAES and XPS to Problems in the Semiconductor Industry. Invited paper 283
P. ALNOT, A. M. HUBER AND J. OLIVIER

Analysis of Plasma Enhanced Chemical Vapour Deposition-grown Borophosphosilicate Glass with a Combination of Rutherford Backscattering and Elastic Recoil Detection 293
J. B. OUDE ELFERINK, K. DUNSELMAN, E. J. EVERS, F. H. P. M. HABRAKEN, W. F. VAN DER WEG AND J. HOLSBRINK

Depth Profiling of Carbonaceous Films, Produced *in situ* in the Tokamak TEXTOR 297
K. G. TSCHERSICH, U. LITTMARK AND H. P. FLEISCHHAUER

SIMS Analysis of $Al_xGa_{1-x}As/GaAs$ Layered Stuctures Grown by Metal–Organic Vapour Phase Epitaxy 303
P. R. BOUDEWIJN, M. R. LEYS AND F. ROOZEBOOM

The Application of Secondary Ion Mass Spectrometry to Surface Analysis of Semiconductor Substrates and Devices 309
A. BROWN, J. A. VAN DEN BERG AND J. C. VICKERMAN

A Characterization Study of the Native Oxide Layer Formed on Chemically Etched InP (111) 319
G. B. HOFLUND AND C. F. CORALLO

Extended Abstracts

Investigation of Surface Contamination on Silicon Wafers with SIMS 325
G. STINGEDER, M. GRUNDNER AND M. GRASSERBAUER

AES and SIMS Investigation of Different Cu_xS/CdS Solar Cells 327
A. HUBIN, H. TERRYN, J. VEREECKEN, M. MOENS AND F. ADAMS

Direct TEM Study of the Roughening of SiO_2/Si Interface Induced during AES Depth Profiling 328
P. B. BARNA, L. GOSZTOLA, A. ZALAR AND M. RASIGNI

Low Temperature Oxidation of Titanium and Titanium Silicide 329
G. QUEIROLO, G. DE SANTI AND G. OTTAVIANI

AES, SEM and EDX Study of Diffusion Phenomena in Reed Switches 330
R. VANDEN BERGHE, R. VLAEMINCK, M. SAILLE AND R. VAN PAEMEL

An AES Study of the Electron-induced Dissociation of SiO_2 331
P. JULIET AND B. BLANCHARD

AES Study of Thermal Behaviour of Sb-doped Sn-oxide Coatings 332
P. H. DUVIGNEAUD, P. F. HENNAUT, A. LAUDET, H. TERRYN AND J. VEREECKEN

An Electron Spectroscopy Study of Sputtered Nickel Ferrite Films 333
A. G. FITZGERALD AND G. MUIR

The Diffusion of Copper and Silver in Amorphous Chalcogenide Thin Films 334
A. G. FITZGERALD AND C. P. MCHARDY

Effects of a Caesium Ion Beam on GaAs 335
K. MIETHE, W. H. GRIES AND A. PÖCKER

Photoemission Study of Cu/Al Interface using Synchrotron Radiation 336
V. DI CASTRO AND G. POLZONETTI

ISS-study of Dopant Redistribution in $TiSi_2$ Layers during Silicidation by Rapid Thermal Processing 338
H. TOLLET, P. BIELEN, C. CREEMERS, H. VAN HOVE, A. NEYENS, K. MAEX AND R. F. DE KEERSMAECKER

Short Abstracts 340

SECTION 6. CORROSION

The Role of Electron Spectroscopy in Corrosion Science. Invited paper 345
J. E. CASTLE

XPS and ISS Examinations of Electrode Surfaces and Passive Layers with a Specimen Transfer in a Closed System 357
S. HAUPT, C. CALINSKI, U. COLLISI, H. W. HOPPE, H.-D. SPECKMANN AND H.-H. STREHBLOW

AES/EDX Studies of Reactions in Phosphate Layers During Corrosion 367
W. J. VAN OOIJ

Application of GDOS (Glow Discharge Optical Spectroscopy) and SIMS to the Analysis of Surface Corrosion Products on Stainless Steel After Exposure to Liquid Alkali Metals 371
M. G. BARKER AND I. E. SCHREINLECHNER

ESCA Study of the Oxidation of a Nickel–Molybdenum Alloy 377
P. MARCUS AND I. OLEFJORD

Extended Abstracts

Surface EXAFS and Magic Angle Spinning NMR Studies of Anodically Formed Oxide Films on Aluminium 383
A. J. FORTY, S. M. EL-MASHRI, R. G. JONES, R. DUPREE AND I. FARNAN

Quantitative Analysis of Conversion Layers on Aluminium 384
H. BUBERT, R. P. H. GARTEN, H. PUDERBACH AND S. STORP

XPS Study of Thin Oxide Films Formation on Iron Alloys 385
A. G. AKIMOV AND V. YU. DEMIN

Surface Layers and Failure Areas on Lacquered Pretreated Steel Sheets 386
B. MAURICKX, J. M. ANNE AND J. FOCT

Short Abstracts 388

SECTION 7. POLYMERS

SIMS for the Study of Polymer Surfaces: a Review. Invited paper 391
D. BRIGGS

The $\pi \rightarrow \pi^*$ Shake-up Phenomena in Polyesters Containing Backbone Aromatic Groups 405
L. C. LOPEZ, D. W. DWIGHT AND M. B. POLK

Extended Abstracts

Ion Beam Damage Effects in Poly(Methyl Methacrylate) 411
M. J. HEARN AND D. BRIGGS

Application of XPS to the Characterization of Polymer-loaded Drugs 412
F. GARBASSI AND F. CARLI

XPS Investigation of Fluoropolymers Thin Films Deposited in Glow Discharges 414
E. DESIMONI, C. MALITESTA, L. SABBATINI AND P. G. ZAMBONIN

An ESCA Study of a Contact Lens Co-polymer RF Plasma-discharge Treated at Various Power Levels 416
D. FAKES, M. NEWTON, J. F. WATTS AND M. J. EDGELL

XPS Study on the Thermal Degradation of Poly-N,N′/4,4′-diphenylether/pyromellitimide 417
A. TÓTH, I. BERTÓTI, T. SZEKELY, T. A. ANTONOVA AND J. N. SAZANOV

Short Abstracts 418

SECTION 8. ADHESION

Use of Surface Behavior Diagrams to Study Surface and Interfacial Reactions. Invited paper 421
G. D. DAVIS

XPS Analysis of Barium Labelled Carbon Fibres and Interfacial Effects in Fibre-Epoxy Composites 431
P. DENISON, F. R. JONES AND J. F. WATTS

The Adhesion of Evaporated Metal Films to Polyethylene Substrates 437
P. BODÖ AND J.-E. SUNDGREN

Kinetics of Bromine Migration and Copper Reactions at Copper–Epoxy Resin Interfaces 441
A. TORRISI AND S. PIGNATARO

Extended Abstracts

SIMS Study of Alcohol-amines: model Compounds for Amine Cured Epoxy Resins 445
S. AFFROSSMAN AND R. A. PETHRICK

Surface Chemistry of Aminoalkylated Polystyrene Resins and Adhesion of Yeast Cells 446
V. DOURTOGLOU, W. E. E. STONE, M. GENET AND P. G. ROUXHET

The Catalysis of Electroless Copper Deposition on Fibreglass–Epoxy Laminates 447
C. LEA, M. T. ANTHONY AND G. C. SMITH

Short Abstracts 449

SECTION 9. CERAMICS

Results of Microanalytical Investigations on SiC-Ceramics and their Impact on Materials Development. Invited paper 453
F. THUMMLER, R. HAMMINGER AND G. GRATHWOHL

Combined TEM and AES Study of CVD TiC and Al_2O_3/TiC Coatings on Cemented Carbides 463
E. MINNI AND S. VUORINEN

Study of the Thermal Behaviour of $^{56}Fe^+$ Implanted Yttria Stabilized Zirconia 467
D. SCHOLTEN AND A. J BURGGRAAF

Extended Abstract

SIMS Analysis of Garnet Epilayers 473
H. J. TOLLE AND P. WILLICH

Short Abstract 474

SECTION 10. TRIBOLOGY AND ENVIRONMENTAL

Electron Spectroscopy of Reaction Layers Formed Under Different Wear Test Conditions. Invited paper 477
H. J. MATHIEU, D. LANDOLT AND R. SCHUMACHER

The Influence of Surface Films on the Protection of Metal Surfaces Under Boundary Lubrication Conditions 493
J. L. SULLIVAN AND L. F. WONG

XPS Studies of Xanthate Adsorption on Metals and Sulphides 501
L.-S. JOHANSSON, J. JUHANOJA, K. LAAJALEHTO, E. SUONINEN AND J. MIELCZARSKI

An AES Depth Profiling Study of the Deposits Formed on Aluminium During the Jet Fuel Thermal Oxidation Test 507
L. B. HAZELL, C. BAKER, P. DAVID AND A. D. FACKERELL

Extended Abstracts

Studies of Ion Exchange Reactions on Naturally Weathered Medieval Glass Surfaces 515
M. SCHREINER, G. STINGEDER AND M. GRASSERBAUER

Trace Element Determination by a Combined AES–Chemical Deposition Procedure 516
K. MACHO, R. P. H. GARTEN, H. BUBERT, J. AUFFARTH, D. KLOCKOW AND G. TÖLG

Rutherford Backscattering and Channelling Studies of Laser Irradiated Stainless Steel 517
T. LAURSEN, J. L. WHITTON, J. NILSON, W. NIP, I. V. MITCHELL, H. H. PLATTNER AND M. L. SWANSON

Aluminium Mirror Coatings in Space: a Study of the Decrease in Ultraviolet Normal Incidence Reflectance Produced by Controlled Oxidation of Evaporated Aluminium Mirror Surfaces 518
M. GRANDE AND W. M BURTON

Short Abstracts 519

Author Index 521

SIANDQ 9(1–6) 1–522 (1986)
ISSN 0142-2421

Preface/Editorial

ECASIA 85 was born during informal discussions between members of the European Surface Analysis User Groups SCADEG and UKESCAUG in 1982 — the result of a common feeling that the applied surface analysis community needed an opportunity to meet at the international level. It was a privilege for me to announce formally in SIA, in February 1983, that ECASIA 85 was on the road to reality.

Despite several problems *en route* the Conference took place and can be counted as a major success. The maximum possible number of papers was presented, over 300 delegates attended, with many non-Europeans amongst them, and a major instrument exhibition was hosted including five large working systems. This success owes everything to the hard work of the organizing committees and the support of many organizations, noted overleaf.

These proceedings have been organized as follows. In order to keep the size within reasonable bounds and to ensure rapid publication, poster papers appear as extended abstracts. Papers from the parallel oral sessions are published in full. In those cases where either no manuscript was submitted or the referees recommended extensive revision the original conference booklet abstract has been included. Papers of all types have been subdivided into sections which mirror the parallel sessions subject areas.

A large number of the authors of poster papers have indicated that full versions of their papers will be submitted for normal publication in SIA. In fact many are currently 'in press' and they will be designated as papers presented at ECASIA 85.

D. BRIGGS
Editor-in-Chief

ORGANIZING COMMITTEE:

H.W. Werner, Philips, Eindhoven, Chairman
J. Kragten, University of Amsterdam, Treasurer
W.J. van Ooij, Akzo, Arnhem, Secretary
A.W. Witmer, retd. from Philips, Eindhoven, Secretary.

SCIENTIFIC COMMITTEE:

A. Benninghoven, University of Münster
D. Briggs, ICI, Wilton-Middlesbrough
J.E. Castle, Surrey University, Guildford
J.C. Rivière, UKAERE, Harwell
M.P. Seah, NPL, Teddington.

LOCAL COMMITTEE:

M.J.A. van Craen, Bell Telephone, Gent
C. la Lau, retd. from Shell, Amsterdam
A.P. von Rosenstiel, TNO, Apeldoorn.

CONFERENCE BUREAU:

QLT Convention Services
Keizersgracht 792
1017 EC Amsterdam, The Netherlands.

SPONSORS AND SUPPORTING ORGANISATIONS:

Royal Netherlands Chemical Society, KNCV
Royal Society of Chemistry (UK)
Royal Netherlands Academy of Sciences, KNAW
Flemish Chemical Society, VCV
German Chemical Society, GDCh
UK SIMS Users Forum
Thin Films and Surfaces Group of the Institute of Physics (UK)
Société Française du Vide
Société Française du Métallurgie
Austrian Society for Micro- and Analytical Chemistry, ÖGMAC
Swedish Vacuum Society
Danish Physical Society
Belgian Society for Surface Treatment of Metals, VOM
VG Scientific (UK)
Akzo (NL)
Philips NPB (NL)
Section for Analytical Chemistry KNCV (NL)
Section for Inorganic and Physical Chemistry KNCV (NL)
German Vacuum Society, DAGV
ATOMIKA (FRG)
Perkin-Elmer (FRG)
Leybold-Heraeus (FRG)
JEOL (Europe) b.v. (NL)
CAMECA (F)
KRATOS (UK)
Imperial Chemical Industries PLC (UK).

Opening Speeches

Welcome to ECASIA 85

Dear Participants,

It is an honour and pleasure to welcome you to our country and to ECASIA 85, the first of a long series — we hope — of congresses on surface and interface analysis. In our circulars the aim and scope of this meeting have been amply explained and somehow we seem to have convinced you, since you have come in such large numbers, from a variety of countries.

Your arrival marks the end of the preparative work and also the point where the responsibility for the success of the congress is no longer borne by the organizing committee but is shared by every participant. i.e. by you. Your contribution to the scientific program and discussion will determine that success.

From our side we have done our utmost to create the right atmosphere for this congress and we shall stay vigilant to maintain optimum conditions. The combination of these conditions and your good efforts will guarantee the success of this colloquium.

We wish you a fruitful and pleasant congress.

Dr. H.W. Werner
President of the Organizing Committee

ECASIA

A conference like the ECASIA conference would not have been possible 30 years ago. In the period 1955–1960 Lander of Bell Labs had just published a paper on the possible use of Auger electrons to study surfaces, Farnsworth at Brown University was just obtaining the first LEED data and Werner, the present chairman of this conference, was among the first to develop a surface sensitive method, which was coined later as SIMS. In that time there would have not been enough people involved in surface studies to fill the first row of this conference room.

A conference like this one would not have been possible 15 years ago. At that time a large number of people all over the world were studying surfaces, even interfaces were probed. This was the stage of new techniques, new instruments, new three and four letter combinations. Research was being done to make AES and SIMS quantitative. There were many conferences held and the subjects ranged from hydrogen on nickel to heated discussions about the structural models of the 7 × 7 super-structure of Si(III). All this work was published in a journal called *Surface Science*.

Today is the first day of the first ECASIA conference, European Conference on the Application of Surface and Interface Analysis. This is a clear proof that surface and interface science has matured to the stage of useful application. The time-span of 30 years is short enough that a few people here in the audience have been involved from the very beginning of modern surface science and that many others will have known the pioneers in the field from the first conferences in the sixties. I suppose that you can draw the line between young and old in the field of surface science there, where people believe that Pierre Auger discovered the method of Auger

Electron Spectroscopy and went on to develop the super-SAM (scanning Auger microscope).

In the early days things were different. For instance vacuum was a difficulty. To obtain 10^{-9} torr took days of hard labour, whereas nowadays it takes hard cash. The problem has shifted from the scientist to his boss.

Surface science has grown in two directions:

– In the direction of interface science. A surface is a very special interface, actually the solid–vacuum interface is meant. This type of interface is hardly encountered in daily practice.

– That brings me automatically to the second direction. Surface and interface science has developed from a phenomenological to a fundamental science and from there to an applied science. The methods are being used for practical applications.

There is still a lot of room to do new and exciting studies, both in the field of applications and in the field of new tools. For instance it is still very difficult, or impossible even, to obtain in a non-destructive way the chemical composition and the chemical structure at interfaces. I know many problems that could be solved by such methods.

In summary: surface and interface studies have a past of about 30 years, a present, which is the ECASIA conference today and a future where many scientific and applied problems can be, and will be, solved.

I wish you a very fruitful conference.

F. Meijer
Philips Research Laboratories

Section 1.

Techniques

SURFACE AND INTERFACE ANALYSIS, VOL. 9, 3–20 (1986)

Practical Surface Analysis: State of the Art and Recent Developments in AES, XPS, ISS and SIMS

Siegfried Hofmann
Max-Planck-Institut für Metallforschung, Institut für Werkstoffwissenschaften, Seestr. 92, D-7000 Stuttgart 1, FRG

This review considers the well established and most frequently applied surface analysis techniques: AES, XPS, ISS and SIMS in a comparative manner, confronting fundamental principles and present instrumental performance with the requests of the scientist. The primary goal of surface analysis is the determination of the elemental composition of the outermost atomic layer of a solid. In addition, characterization of binding states, of homogeneity and of structural details will often be required. Whereas the information gathered generally contains also contributions from deeper layers, the inherent high in depth resolution of the techniques allows thin film analysis with high precision. Thus, surface analysis includes the analysis of interfaces and thin films by depth profiling. Recent progress has led to improvements e.g. in sensitivity, spectral resolution and data acquisition and handling. With the trend to high spatial resolution, surface analysis is becoming a special branch of microlocal analysis. Quantification is of primary interest. However, the variety of approaches and their dependence on instrumental parameters hamper interlaboratory comparisons, pointing out the need of standardized methods and correction procedures, of well characterized test samples and of accurate data banks. Advance in precise quantification will depend on future progress in our understanding of matrix effects and beam-sample interactions.

INTRODUCTION

The importance of surface and thin film analysis for the study of the properties as well as for the development of new materials results from the direct dependence of their properties on the chemical composition on a microscopic scale particularly in thin films and structural discontinuities, i.e. at external and internal interfaces. Properties and processes such as contamination, adsorption, oxidation, friction and wear, segregation induced embrittlement, adhesion, interdiffusion, passivation, coating, epitaxy, doping and crystal growth can only be controlled by surface, interface and thin film characterization with high spatial resolution.

The development of all-metal ultra-high vacuum chambers and powerful electronic detection systems in the mid sixties has led to the commercial availability of surface analysis equipment in the early seventies. Since then, a variety of about 50 different methods has evolved and it would be fruitless to review them all. For practical surface analysis, four of those techniques have been steadily improved by several companies during the past fifteen years, and have now matured in instruments being reliable analysis tools. These techniques are Auger electron spectroscopy (AES), x-ray photoelectron spectroscopy (XPS or electron spectroscopy for chemical analysis, ESCA), secondary ion mass spectrometry (SIMS) and ion scattering spectroscopy (ISS) which are described in several review articles.[1–12] Their common feature is the capability of probing the third, in-depth dimension perpendicular to a surface from the nanometer range down to monoatomic layers, which has enhanced their nowadays widespread use in many areas of materials research and fabrication, as compiled in Table 1.

Table 1. Some of the main application fields of surface analysis in materials research

Surfaces Direct	Interfaces Via fracture and/or profiling	Thin films Depth profiling
Segregation	Segregation	Interdiffusion
Diffusion	Diffusion	Ion implantation
Contamination	Embrittlement	Reaction layers
Adsorption	Intercrystalline corrosion	(Oxides, passivation layers etc.)
Oxidation	Sintering	Evaporation layers
Catalysis	Adhesion	Protective coatings
Friction and wear	Composites	Microelectronic devices

The age of practical surface analysis is precisely 16 ± 1 years. This statement is supported by a compilation of publications per year, obtained by a computer research based on metal abstracts and chemical abstracts and shown in Fig. 1. It can easily be recognized, that the most popular techniques AES and XPS have their point of take-off in 1969. The usage of SIMS and ISS starts in around 1973 and 1974, respectively. Furthermore, a comparison of Fig. 1(a) with Fig. 1(b) discloses AES to be more suitable for metallurgical application, whereas XPS is the preferred technique for chemists. SIMS and particularly ISS are less frequently used than the electron spectroscopies, probably due to their relative insensitivity to matrix effects and therefore easier quantification, combined with good elemental specifity and principally nondestructive detection capability. It is interesting to note that SIMS makes a leap forward from 1983 to 1984 in Fig. 1(b), which may be due to the recent advances in SIMS of organic molecules.[13,14] Fig 1(a) indicates that a satura-

0142–2421/86/120003–18 $09.00

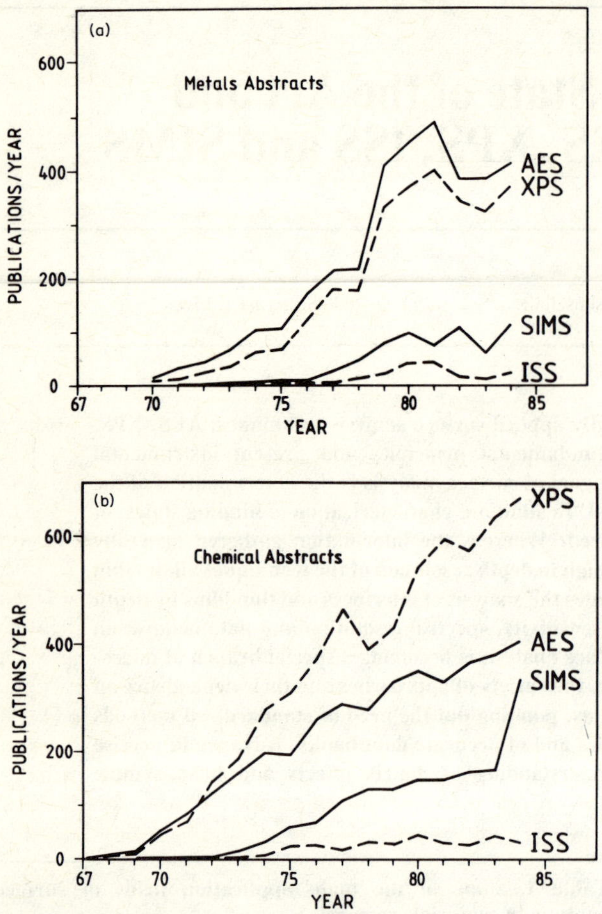

Figure 1. Publications per year from 1967 to 1984 in the field of AES, XPS, SIMS and ISS, based on keyword computer research in (a) metals abstracts; (b) chemical abstracts.

tion was attained about five years ago, a characteristic of a mature, commonly used analysis method particularly for industrial research and process control.

The question 'what is a surface' can be answered in many different ways, depending on the specific structure, property or application of interest. For the two dimensional vacuum/solid interface, the 'physical' solid surface can be defined as the outmost atomic layer. In practice, this layer is not stable with time as it is subject to various physico-chemical reactions with the surrounding gas phase. Thus, a surface 'from the real world' would consist of a contaminated reaction layer (most often oxide) of typically about 5–10 nm thickness with more or less pronounced roughness. This layer is relatively stable at room temperature and is commonly called 'practical surface', which already is a thin film when considered from the physical definition given above. Although — with the possible exception of ISS — all surface analysis techniques have a certain information depth larger than a monolayer thickness, the relative contribution of the first topmost surface layer is always superior to that of the following, deeper layers. Analysis of the first atomic layer, the 'physical surface', requires some kind of deconvolution of the measured data with respect to the sampling or information depth and involves a knowledge of the composition of the layers below. Therefore, surface analysis and thin film analysis are principally interlinked.

A detailed analysis requires not only a high resolution perpendicular to the surface, but its combination with high lateral resolution to obtain a three dimensional characterization. Thus, surface analysis is becoming a most powerful branch of microlocal analysis.[5,6]

Figure 2 shows a comparative diagram of different analysis methods with respect to lateral resolution and in depth resolution (or depth of information). For comparison, electron probe micro-analysis (EPMA), a typical bulk analysis method, is also included, as well as field ion microscopy (FIM) in the form of the atom proble (A–P). FIM is at present the most powerful method for micro-local analysis at an atomic scale, but requires careful and cumbersome sample preparation which limits its application range.

Another method shown in Fig. 2 is analytical transmission electron microscopy (TEM), either as energy dispersive x-ray analysis (EDX) or as transmitted electron energy loss spectroscopy (TEELS). Due to the thickness of the sample of typically 50–100 nm, this method is restricted to edge-on interface analysis, where it approaches similar lateral resolutions as obtained for in depth resolution in surface analysis. Therefore, modern STEM instruments, capable of operation under ultra high vacuum and equipped with energy analyzers for secondary and transmitted electrons, offer a unique combination of surface and interface analysis with high spatial resolution.

The aim of this review is to briefly summarize the physical principles and special features of the most popular surface analysis techniques: AES, XPS, ISS and SIMS in order to elucidate their practical application capabilities and their limitations in present instrumentation.

GENERAL CONSIDERATIONS

Instrumentation principles

A schematic block diagram of different surface analysis methods is shown in Fig. 3.[15] It consists of a vacuum chamber, which contains the excitation source (x-rays, electrons, or ions), the sample and the analyzer/detector system (energy and/or mass filter). Data aquisition and handling is a major part of the equipment. An

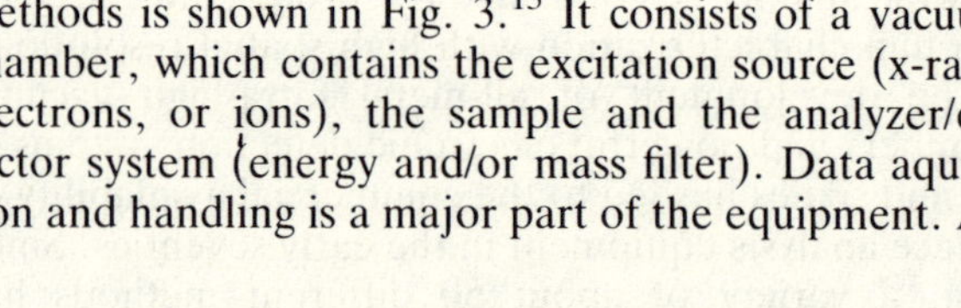

Figure 2. Typical in-depth and lateral resolution of the most popular surface and micro-local analysis techniques. For details see text.

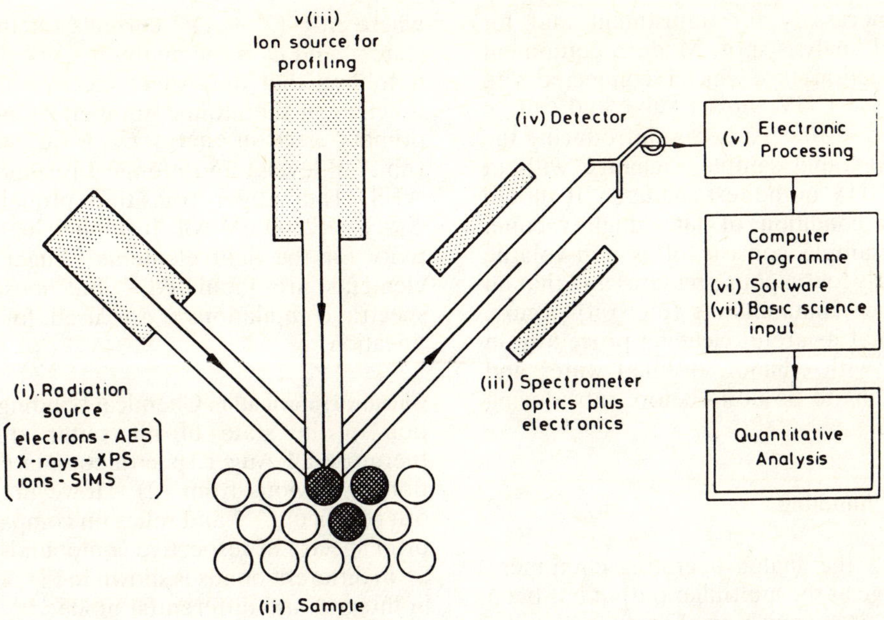

Figure 3. Principal instrumentation for surface analysis techniques, according to Seah[15]

auxiliary ion gun for the electron spectroscopies is a necessity. Sample manipulation and introduction is done by air-lock stages of various sophistication. The common structure of all surface analysis methods enhances their combination in one instrument. Today, almost any manufacturer offers basic systems with add-on capability of the three remaining methods.

Vacuum requirements

Any surface under study should be free of contamination from the ambience. This means that the residual partial pressure of reactive gases such as CO, H_2O. C_xH_y etc. should be low enough to ensure an undisturbed detection during the analysis time. Assuming a sticking probability of one, a simple calculation shows that about one monolayer coverage in one second is obtained at a pressure of 10^{-4} Pa. If, as a moderate requirement, this characteristic time is supposed to be > 1000 s, the pressure should be $< 10^{-7}$ Pa ($< 10^{-9}$ Torr). i.e. in the UHV range. During ion sputtering, the above condition is somewhat relaxed due to sputter desorption. The stationary surface coverage θ_s (for $\theta_s << 1$) is in this case given by:[16,17]

$$\theta_s \approx 10^3 \,.\, p\,(\text{Pa})/\dot{z}(\text{nm s}^{-1}) \qquad (1)$$

where z is the sputtering rate, p is the partial pressure and θ_s is the ratio of adsorbed atoms/matrix atoms. Although θ_s may be below the detection limits for electron spectroscopies, the resolution of depth profiling can be distorted.[18] In dynamic SIMS with its high sensitivity, (1) defines the detection limit given by the background due to adsorption. This was shown by Magee and Botnick[17] for hydrogen analysis in silicon as a function of residual water vapour pressure down to 10^{-8} Pa.[17,19]

In any case, UHV ($< 10^{-7}$ Pa) in the ambience of the sample is a necessary condition for reliable surface analysis. It can only be cautiously relaxed for depth profiling with high sputtering rates. To achieve UHV, the general instrumentation is based on a bakeable stainless steel chamber with attached ion getter-Ti sublimation-, turbomolecular-, cryo- or diffusion pumps (with cold traps) or a suitable combination.[7]

Ion source and sputtering

Most auxiliary ion sources for AES, XPS and ISS use electron impact ionization for the generation of noble gas ions (generally high purity argon), electrostatic focusing of the beam of 0.1 . . . 1 mm diameter, at 0.5–5 kV acceleration voltage and a x-y-rastering facility up to 10×10 mm^2. Whereas in older equipment, backfilling of the whole chamber with argon at typically 3×10^{-3} Pa (5×10^{-5} Torr) provides the necessary argon pressure in the ion formation chamber, in modern ion sources argon is directly supplied through a needle valve and the gun is differentially pumped by an auxiliary turbomolecular or diffusion pump. This avoids build-up of contaminant gases desorbed from the walls of the chamber when using argon backfilling, and allows higher pressures in the formation chamber to yield typical ion current densities of 10^2–10^3 μA cm^{-2}.

For SIMS, the ion source has to be more elaborate to avoid impurities in the beam and to achieve higher sputter rates. Most often duoplasmatron sources are used, with mass/energy filtering and a beam bend to separate neutrals.[20] Facilities to switch between negative and positive ions, and to use caesium or liquid metal sources are often available. The higher standard for SIMS ion sources is necessary because the mass analysis detects impurities down to the ppm level or below and is directly coupled with ion sputtering.

Sample stage

The sample is generally mounted on a manipulator which allows x, y, z movements and sometimes rotation. This is necessary for adjustment and for locating the desired analysis spot. Modern equipment has a sample introduction stage which is connected with the test chamber by a UHV shutter valve and can be evacuated at least to $< 10^{-1}$ Pa before introducing the sample. This allows rapid sample exchange without breakdown of the UHV in the test chamber. It should be noted that the condition of ultra high vacuum requires that the sample consists of a non-volatile substance (sufficiently low vapour pressure) and that no desorption of surface contaminants (e.g. oil) occurs. This often means that a careful cleaning procedure in an ultrasonic bath with ethanol, distilled water and subsequent drying has to be done before introducing the sample.

Data aquisition and handling

During recent years, the analog operating equipment measuring the voltage at the multiplier output has been outdated by digital data aquisition. For example, the common lock-in amplifier with high frequency modulation yielding directly the derivative spectrum in AES is replaced by a voltage to frequency converter to obtain a digital output. For lower currents, pulse counting is directly applicable and gives, together with the voltage ramp steering of the analyzer, $N(E)$ spectra in AES, XPS and ISS, or $N(m/z)$ spectra in SIMS. The digital data are stored in a microcomputer memory. They can be displayed in real time on a CRT monitor and subsequent data manipulation like smoothing, differentiation, background subtraction, peak fitting, quantitation calculations etc. can be easily performed. A hard copy printer or plotter completes the data output.

In modern equipment, the steering of all analyzer and excitation source functions is done by the firmware/software of the computer. Therefore the software package becomes of increasing importance as an integral part of the instrument.

SURVEY OF TECHNIQUES

A comparison of the four main surface analysis techniques is based on the principal features of each of them, and a brief survey is given in the following. for more detailed information, the reader is referred to numerous recent review articles[1–12] and the references given therein. Table 2 shows an intercomparison of the main features of the techniques.

Auger Electron Spectroscopy (AES)

Principle and spectral identification Auger electron emission occurs when an inner electron shell of an atom is ionized by photon, electron or ion excitation. AES specifically refers to electron impact excitation. The kinetic energy of an Auger electron E_{WXY} for an ionized level of binding energy E_W and the participating outer levels E_X and E_Y is given by:[1–12]

$$E_{WXY} = E_W(Z) - E_X(Z) - E_Y(Z + \Delta) - \phi_s \quad (2)$$

where $\Delta = (Z + 1)/2$ corrects for the doubly ionized state.[21] and ϕ_s is the analyzer work function. From (2) it follows that the Auger electron energy E_{WXY} is a function of the atomic number Z, independent of the primary electron energy $E_p > E_W$, and the H and He (no X,Y-levels) and atomic Li cannot be detected by AES. The Auger transition probability is high for $E_{WXY} < 2500$ eV which implies high detection sensitivity for the light elements. Auger energies for the elements are tabulated,[21] but most often standard spectra compilations[22] are used for elemental identification.

Chemical information Chemical bonding means an alteration in the state of the outer atomic levels, and therefore all Auger spectra contain chemical information as obvious from (2). However, interpretation is often difficult[23,24] and relies on comparison with fingerprint spectra of respective compounds. An example for Sn in different bonds is shown in Fig. 4, with the spectra in integral and differential mode.[12]

Additional information The secondary electron spectrum contains fine structure details which are — enhanced by high resolution digital data aquisition and storage — more and more exploited for additional information: The fraction of elastically backscattered electrons (BSE) increases monotonously with the atomic number[25,46] and can be used for elemental identification. This is particularly advantageous in elemental mapping with high lateral resolution and frame time restrictions,[26] due to the high signal to background ratio compared to Auger peaks. Spectroscopy of electron energy loss features at the low energy side of an Auger or elastic peak (electron energy loss spectroscopy, ELS or EELS) can also be used for additional information. This comprises primarily plasmon losses and ionization losses,[27–31] the energy of which depends on the chemical bonding status. Oxide formation[27] and depth profiling[28,2] have been studied by EELS. Recently, Bevolo *et al.* used plasmon losses for chemical mapping of rare earth hydrides[30] in SAM and coined the term 'reflection electron energy loss microscopy' (REELM).

The dependence of the bulk plasmon peak energy of Sn on temperature and bond strength was recently demonstrated by Seah.[31] It provides an elegant method

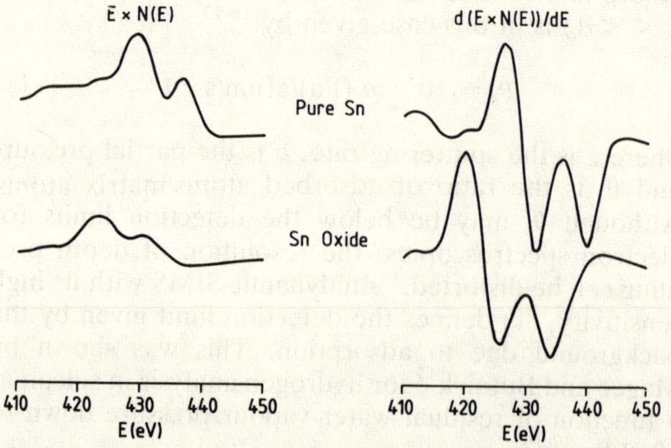

Figure 4. Chemical effects in AES: comparison of the $Sn(M_5N_{4,5}N_{4,5})$ and $(M_4N_{4,5}N_{4,5})$ peaks in the normal ($E \,.\, N(E)$ and derivative ($d(E\,.\,N))/dE$) modes for pure Sn and SnO_2, showing a shift of −6 eV and shape alterations.[12]

Table 2. Comparison of some characteristics of surface analysis techniques

Techniques	AES	XPS	ISS	SIMS Static	SIMS Dynamic
Principle: Excitation	electrons	photons	ions	ions	ions
emission	electrons	electrons	ions	ions	ions
analysis	energy	energy	energy	mass	mass
Basic information	elements	elements chemical bonding	elements	elements isotopes	elements isotopes
Additional information	imaging chem. bonding plasmon losses structure	depth distrib. (> resolved) valence band structure	isotopes structure (single cryst.) TOF–MS	compounds	imaging compounds
Typical depth of information	1 nm	1 nm	0.3 nm	0.6 nm	10 nm
Lateral resolution	1 μm–0.1 μm	1 mm–100 μm	1 mm	1 mm	0.5 μm–0.05 μm
Sensitivity (atomic fractions)	$10^{-3}-10^{-2}$	$10^{-3}-10^{-2}$	10^{-2}	10^{-5}	10^{-6}
Ease	+ + + + +	+ + + + +	+ + +	+	+ +
Quantification accuracy	+ +	+ + + + +	+ +	+	+ + + +
Elements not detected	H, He	H, He	H	—	—
Ease of use	+ + + +	+ + + +	+ + +	+ +	+ + + +
Versatility	+ + +	+ + + +	+ + +	+ +	+ + +
Extend of support data base	+ + + +	+ + + +	+ +	+ +	+ + + +
Overall percentage of use	40	40	5	5	10
Main application fields: materials:	Metals Semiconductors	Metals Semiconductors Ceramics Organics	Metals Semiconductors	Metals Semiconductors Organics	Semiconductors Metals Ceramics
Surface analysis	+ + +	+ + + +	+ + + + +	+ + + +	+
thin film	+ + + + +	+ + +	+ +	+	+ + + + +
Present main hardware developments	e⁻-gun (F.E.) Spatial resolution Energy resolution	Multichan. Detec. Energy resolution (Monochromatic) Small spot	Mass resolution (∢ – dep.) TOF–MS	Mass resol. sens. (TOF-MS) FAB	small spot ion source Mass resolution

to measure *in situ* the temperature rise due to the primary electron beam.

The low energy side modulation of ionization loss peaks contains structural information analogous to EXAFS. De Crescenzi *et al.* were able to determine the nearest neighbour distance in Ni and NiO by this method of 'surface extended electron energy loss fine structure' (SEELFS).[32]

In-depth resolution (AES and XPS) In electron spectroscopy, the depth of information is given by the yield decay with distance z from the surface and is described by an exponential function $\exp(-z/\lambda_e)$. The decay parameters λ_e is the mean escape depth (or effective escape depth) of the Auger- or photoelectrons and depends on the geometry between analyzer and sample surface. With φ denoting the angle between the normal to the surface and the emission angle of the detected electrons, $\lambda_e = \lambda \cdot \overline{\cos\varphi}$. The fundamental parameters is λ = inelastic mean free path (imfp) of the electrons. The factor $\overline{\cos\varphi}$ is the weighted mean of all contributing emission angles. For the geometry of the CMA, the latter is depicted in Fig. 5.[33] Note that beyond the critical angle 47.8° a part of the CMA acceptance cone is shadowed ($I/I_o < 1$) which is responsible for the hump in the curve. It can be seen that angle dependent AES with a CMA can only give a variation of the effective escape depth of 43% which limits its use for non-destructive depth profiling (see fifth section).

The inelastic mean free path (imfp) λ of the Auger- or photoelectrons depends on both their energy and on the matrix. There are several theoretical, semi-empirical and empirical approaches to predict λ in the above terms.[34–40] In general, $\lambda \alpha E^n$ with $0.5 < n < 0.8$ is found for kinetic energies $E > 150$ eV. According to

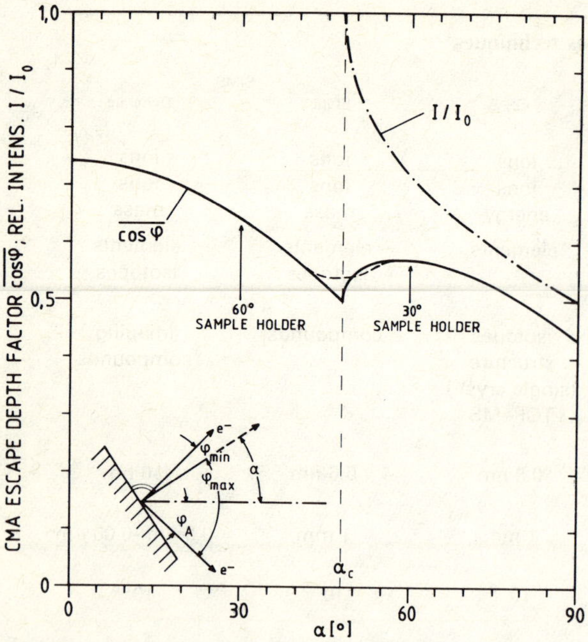

Figure 5. The relative effective mean escape depth of Auger- or photoelectrons $\overline{\cos\varphi} = \lambda_e/\lambda$ as a function of the angle α between the normal to the sample surface and the axis of a cylindrical mirror analyzer (CMA).[33]

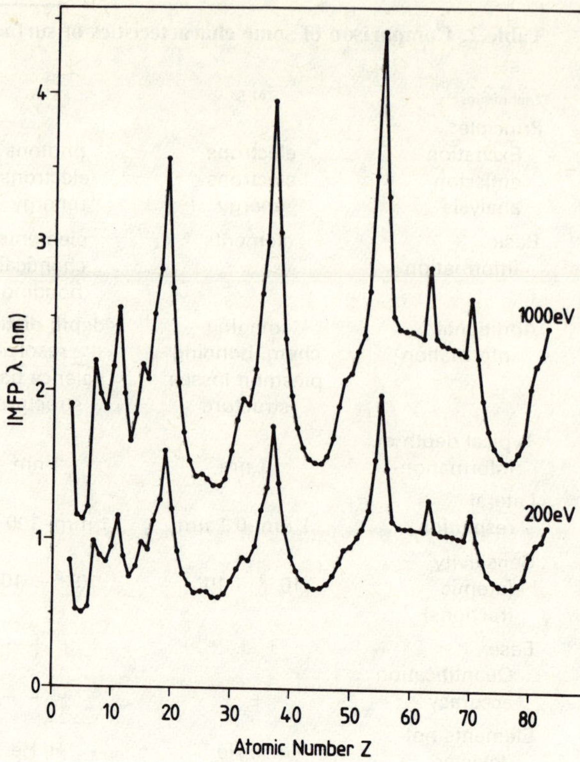

Figure 7. Inelastic mean free path λ for pure elements as a function of atomic number z according to the model of Seah and Dench[35] for E = 200 eV and E = 1000 eV kinetic energy.

Powell,[39] the exponent n as well as the matrix factor depend mainly on the electronic structure of the valence band.[40] In Fig. 6 three popular models are compared with experimental data for SiO_2,[38] showing at least that neither approach is completely wrong.

This unpleasant situation is presently the reason for the most severe limitation in quantification of AES and XPS (see third section and Ref. 41). To date, the most practical prediction is that of Seah and Dench[36] which gives for elements (and $E > 150$ eV):

$$\lambda = 0.41 \cdot a^{3/2} \cdot E^{1/2} \tag{3}$$

with λ and the mean atomic diameter a in nm and the kinetic energy (E) in eV. Figure 7 shows a plot of λ vs. the atomic number Z of the elements for E = 200 eV and E = 1000 eV according to (3). Seah and Dench[36] have also proposed λ calculations for inorganic compounds (e.g. oxides) and organic compounds. For oxides it is often felt that their model gives somewhat too high values and may better be replaced by adding 10% to the value from (3).[41]

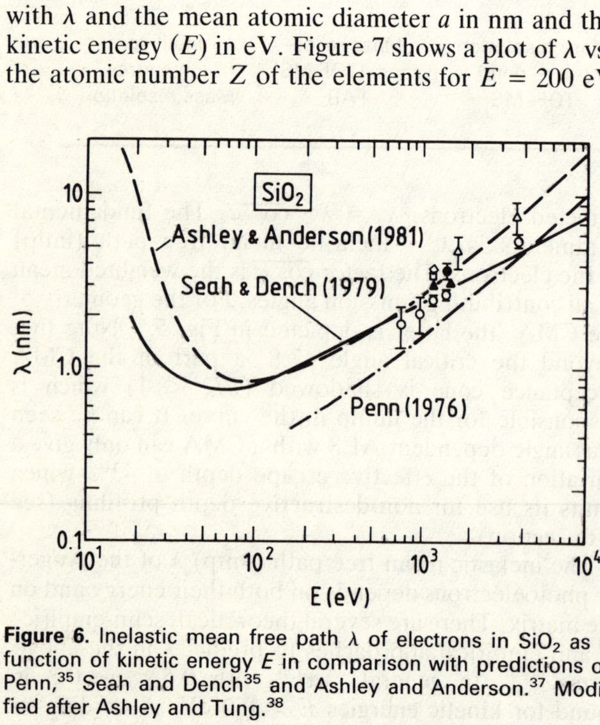

Figure 6. Inelastic mean free path λ of electrons in SiO_2 as a function of kinetic energy E in comparison with predictions of Penn,[35] Seah and Dench[35] and Ashley and Anderson.[37] Modified after Ashley and Tung.[38]

Quantification of AES For a more comprehensive discussion, the reader is referred to recent reviews.[7,11,15,42] In the case of constant concentration with depth, the general expression for the atomic fraction X_A of an element A in a substitutional compound or alloy with componenets j is:[42]

$$X_A = \frac{I_j/I_j^0}{\sum_j F_{jA}^A \cdot I_A/I_j^0} \tag{4}$$

Where I_j is the measured Auger signal intensity and I_j the intensity of a pure element standard measured under identical analysis conditions. It is customary to use the Auger peak to peak height (APPH) in the derivative mode as a measure of Auger intensity (see e.g. Refs. 42–44 for a discussion whether this mode or the integral peak area should be used), and to set $I_j = S_{X,Ag}$, the relative elemental sensitivity factors with respect to pure silver, which can be adapted from a popular handbook.[22] For the latter procedure, care must be taken to ensure identical primary electron energy and analyzer transmission function and a sufficiently low modulation voltage or point smoothing routine.[42,43] The Auger matrix correction factor F_{jA} is set to unity in a first order approximation, which is only valid for similar elements and Auger energies. As first shown by Hall and Morabito[45] this factor is in general between 0.5 and 2 for binary alloys and is approximately given by[42]:

$$F^A_{BA} = \left[\frac{1+r_B(E_A)}{1+r_A(E_A)}\right] \cdot \left[\frac{\lambda_B(E_A)}{\lambda_A(E_A)}\right] \cdot \left[\frac{a_A}{a_B}\right]^3 \qquad (5)$$

The first term refers to the energy and matrix dependent backscattering factor $(1 + r_j)$,[45,46] the second to the different imfp values[36] and the third term is a density correction. Introduction of (5) in (4) reduces the possible error, e.g. in Fe–C from about 30% ($F_{FeC} = 1$) to less than 5% ($F_{FeC} = 1.28$).[7]

It should be noted that in many practical cases (e.g. segregation or oxide layers) steep concentration gradients within a few λ exist, and therefore the full integral solution must be applied[7,11,42] which is straight forward for layered structures of known thickness. Only when the in-depth distribution of composition is known, a correct quantification can be performed. This is most important for sputtered alloy or compound surfaces prone to preferential sputtering of one component.[47,48]

Lateral resolution and detection limit A focused electron beam of a diameter between 1μm and 25 nm is generally available in modern instrumentation. The detection limit c_{min} ($= 3 \times S/N$) as defined by the signal to noise (S/N) ratio depends on the square root of the product of primary current i_p and measurement time t:[6,7]

$$S/N = k \cdot (i_p \cdot t)^{1/2} \qquad (6)$$

where $k = I_j$ * spectrometer function. Assuming a constant brightness of the electron source, i_p = const. * d^2 (beam diameter) is limited by the beam diameter. Werner[49] and more recently Cazaux[50] have discussed those relations and Fig. 8 shows a plot of the detection limit c_{min} versus beam current i_p or diameter d for AES. It is obvious that high lateral resolution can only be obtained by sacrificing detection sensitivity, thus limiting the useful lateral resolution.

The lateral resolution limit in AES is mainly due to the influence of backscattered electrons on Auger emission, as shown by El Gomati *et al.*[52] and by Dudek,[53] resulting in a primary beam excitation broadening depicted in Fig. 9.

Electron beam effects in AES Care must be taken to avoid electron beam effects which are generally proportional to the primary current density.[54–57] Electron beam induced temperature increase is most detrimental for thin films on substrates with low thermal conductivity.[56,57] Electron stimulated desorption is often encountered in studies of ionic compounds,[58] e.g. in SiO_2.[59] Charging of insulators is sometimes difficult to avoid[7,9] and can even lead to induced electromigration.[60] In general all those effects are minimized for low primary current density and short sampling times.

Instrumentation of AES Most up to date electron sources use a high brightness LaB_6 cathode, an electrostatic ($d \geqq 200$ nm) or magnetic ($d \geq 25$ nm) focussing and a scanning facility (SAM) to produce secondary electron images, point or line Auger analysis and elemental mapping. The analyzer is generally a CMA with an energy resolution $\Delta E/E = 0.3 \ldots 2\%$. The advantage of high point transmission ($\approx 10\%$ of the half space) of the CMA leads to reduced sampling times. However, CHA analyzers with multichannel detection seem to compete favourably in this respect.

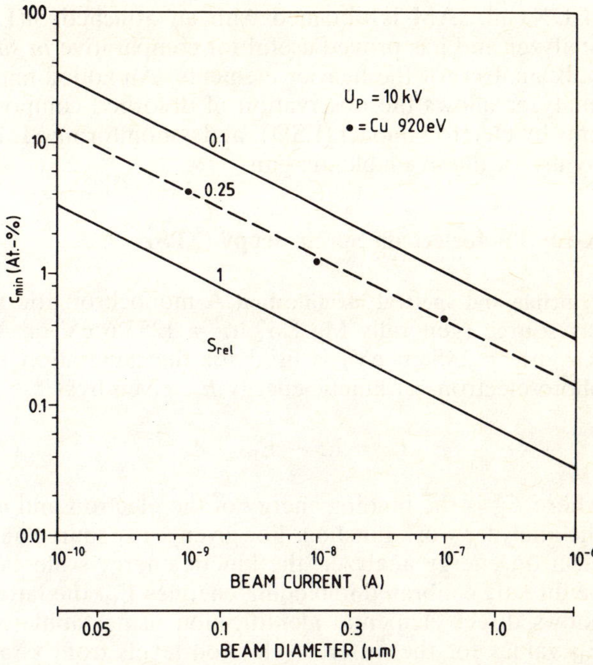

Figure 8. Typical relation between minimum detectable concentration c_{min} (at %) and primary beam current and diameter at 10kV for AES. Measured data are for the Cu 920 eV peak, t = 1s/eV. S_{rel} are relative elemental sensitivities.[12]

Combinations with other techniques A straight forward change of the analyzer voltage allows the performance of ISS by use of the auxiliary ion gun, if a double path analyzer with a rotable slit[62] or a CHA is used. XPS can be readily performed with similar analyzers by adding an x-ray source. Electron dispersive x-ray analysis

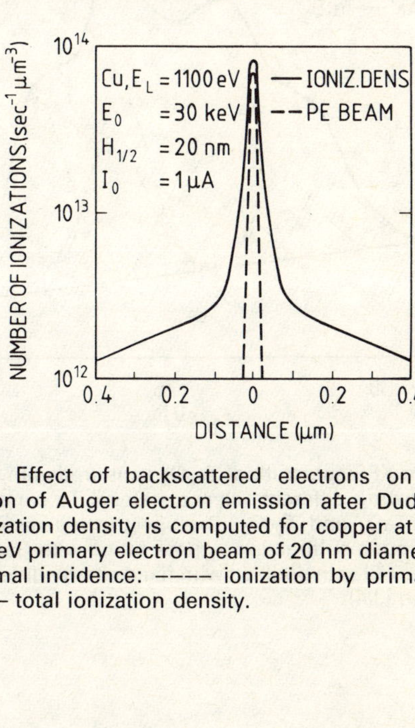

Figure 9. Effect of backscattered electrons on the lateral distribution of Auger electron emission after Dudek.[53] The L-shell ionization density is computed for copper at the surface. for a 30 keV primary electron beam of 20 nm diameter (FWHM) with normal incidence: ---- ionization by primary electron beam —— total ionization density.

(EDX) in SAM is obtained with an attached Si(Li) analyzer and has proved useful for comparative *in situ* bulk analysis of the heavier elements. An added mass analyzer allows the observation of desorbed components by electron impact (ESD), and a monitoring SIMS by use of the available ion gun.[62]

X-ray Photoelectron Spectroscopy (XPS)

Principle and spectral identification A monochromatic x-ray source (generally Mg Kα, hν = 1253.6 eV or Al Kα, hν = 1486.6 eV) is used for the generation of photo-electrons of kinetic energy E_k, given by:[7,11]

$$E_k = h\nu - E_B - \phi_s \qquad (7)$$

where E_B is the binding energy of the electron and ϕ_s the analyzer work function. For given x-ray source and electron energy analyzer, the kinetic energy scale can be directly calibrated in binding energies E_B. the latter allows direct elemental identification using tabulated E_B values for the different electron levels from x-ray fluorescence tables.[71] Photoionization cross sections for H and He are too low for detection.

Chemical information from XPS Because the binding energy is influenced by the chemical bonding state, its detection is the main application field of XPS. Different valuence states in compounds are recognized by specific energy shifts. An example is shown in Fig. 10 for Ar^+ sputtered Ta_2O_5 which is decomposed into different suboxides and elemental Ta.[63] Binding energies for elements in many compounds have been compiled.[7,4] Holm and Storp have shown that in general a correlation between formation enthalpies of oxides and E_B shift exists.[65]

Since Auger electrons are also detected, the difference between the Auger and photoelectron energy ($E_A - E_k$), the Auger parameter after Wagner[64] provides a fundamental definition of the binding energy independent of total energy scale shifts, e.g. by wrong calibration or due to charging effects.

The valence band structure can also be detected by XPS if the energy resolution is sufficient, as for monochromatized x-rays.[66]

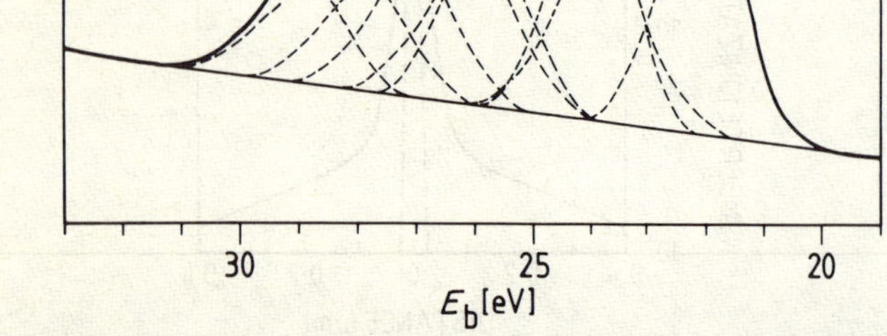

Figure 10. An XPS spectrum of Ta_2O_5 sputtered with 3 keV Ar^+ ions,[47,63] showing different valence statesof Ta corresponding to Ta, TaO, TaO_2 and Ta_2O_5. The spectrum is resolved after linear background subtraction by peak fitting with Gaussians of the Ta 4f (5/2, 7/2) spin doublett with fixed energy difference and intensity ratio (3/4).

Additional informations in XPS For thin amorphous layers, (< 5 nm thickness) the dependence of the effective escape depth of the photoelectrons on the emission angle (as discussed above) can be used to reveal the in-depth distribution.[63,67–72] Recently, Tougaard and Ignatiev[73] have shown that the relative intensity of the background intensity provides quantitative layer thicknesses determination.

For single crystalline layers, angle resolved XPS gives structural information.[74,75]

In-depth resolution in XPS Of course the interaction of photoelectrons with matter follows the same physical laws as for Auger electrons. Therefore, for a given kinetic energy, the inelastic mean free path λ is identical and the remarks about it above are also valid for XPS.

Quantification of XPS Quantification in XPS is also analogous to AES.[42,43] However, it is more reliable because of the following reasons: (a) photoionization cross sections according to Scofield[76] are well known; (b) the peak intensities are not influenced by backscattered electrons as in AES; (c) the background is lower and easier to subtract, i.e. peak intensities are obtained by integration over $N(E)$ after background subtraction (for a discussion of the latter, see e.g. Refs 7 and 43).

Equation (4) can also be used for XPS quantification if the matrix correction factor F_{jA}^{A} is replaced by the XPS factor F_{jA}^{X}, which is defined for binary alloys analogous to (5) as:[42]

$$F_{jA}^{X} = [\frac{\lambda_B(E_A)}{\lambda_A(E_A)}] \cdot [\frac{a_A}{a_B}]^3 \qquad (8)$$

For thin layers with steep concentration gradients, the effect of a finite λ has to be corrected as in AES.[7,42,70]

Lateral resolution in XPS At present most instruments have an area of analysis in dimensions of several mm diameter. However, small area XPS is getting more widespread in use, and modern instruments allow the detection of about 100 μm diameter.[77] Photoelectron spectro-microscopy (PESM) can be performed with about 10 μm resolution using a magnetic lens for imaging via photoelectrons.[78] The different modes of small area XPS and its limitations have recently been discussed by Cazaux[79] and Drummond *et al.*[80]

Instrumentation (XPS) Most XPS instruments use Mg Kα as the standard x-ray source. The anode is excited by electrons of 10–15 keV and 300 to 600 W electric power. Different anode material (e.g. Al, Si, Ti, Ag) provides, according to (7), higher $h\nu$ and therefore higher E_k values which enlarges the depth of information. Besides double-pass CMA,[61,67] the most frequently used analyzer is based on concentric hemispheres (CHA) with pre-retardation lens system, allowing operation with constant, pre-selected energy resolution E independent of kinetic energy E_k. for systems without monochromators $E_{min} \approx 0.75$ eV and is limited by the natural line width of the Mg Kα radiation. Monochromators using natural quartz crystals mounted on a Rowland circle through x-ray source and sample surface can achieve 0.4 eV resolution. The best value reported so far is E ="0.23 eV of Gelius *et al.*[66] using double-focusing of Al Kα x-rays from a 40 kW rotating anode source.

Modern instruments use the energy dispersion of the CHA at the output slit for position sensitive detection with multichannel detectors, as depicted in Fig. 11.[81] These are either discrete channeltrons or a twin channel plate arrangement above a resistance plate to locate the pulse position, or a luminescence screen with video camera recording.[66] Small area XPS is available as discussed in above.

Combinations with other methods Changing the voltages of retarding lenses of the CHA allows energy analysis of positive ions and therefore ISS with the ion gun used as excitation source.[61,82,83] Addition of an electron gun is necessary to obtain AES analysis, with the advantage of higher energy resolution as compared to the usual CMA detection. If an energy dispersive x-ray analyzer is attached, x-ray fluorescence analysis (XRF) is possible giving useful *in situ* bulk information as shown by Castle.[84] X-ray stimulated desorption (XSD) can be observed by detection of ions with an attached quadrupole mass analyzer or — more elegantly — with time-of-flight mass spectrometry in the analyzer.[61,85] Mass analysis in conjunction with the available ion gun allows a monitoring SIMS operation.

Particularly for valence band studies, an ultra violet (HeII) radiation source can be added to perform UPS.[7]

Ion Scattering Spectrometry (ISS)

Principle and spectral identification Often more precisely named LEIS (low energy ion scattering spectrometry), this technique is based on elastic scattering of the incident primary ions. Their relative energy loss depends on the scattering angle and on the difference between the primary ion mass M_1 and the target atom mass M_2. For ions incident with energy E_1^0. the kinetic energy E_1 of the scattered ions for 90° scattering angle is.[2–,86]

$$E_1 = E_1{}^0 \,.\, (M_2 - M_1)/(M_2 + M_1) \qquad (9)$$

Equation (9) shows that protons (H) cannot be detected, as well as He for the generally used He^+ ions of 0.3–2 keV energy. It is obvious that all higher masses can be determined, the strong nonlinear decrease of mass resolution for higher masses limits the spectral identification to masses below 100.[87,88] Use of higher scattering angles and of Ne^+ or Ar^+ increases the resolution for higher masses, but the isotopic selectivity which is, in principle possible, is restricted to lower masses.[88] A typical ISS spectrum of a Mo surface contaminated with O, F, Na and K is shown in Fig. 12.[85] Note the nonlinearity between $E_1/E_1{}^0$ and atomic masses according to (9). Whereas atomic collisional process contains no specific chemical information, the latter can be gleaned from yield-energy relations.[87,89]

Additional information from ISS The structure of ordered atomic positions on single crystalline surfaces can be detected by angle resolved ISS due to shadowing effects.[90,91] Some information of the electronic structure may be obtained by its influence on the neutralization probability.[87–89] Increased primary ion energy (> 2 keV) gives additional information on the composition of deeper layers.[92]

In-depth resolution of ISS For sufficiently low energy, the steeply increasing neutralization probability with depth confines the measured backscattered ions to those

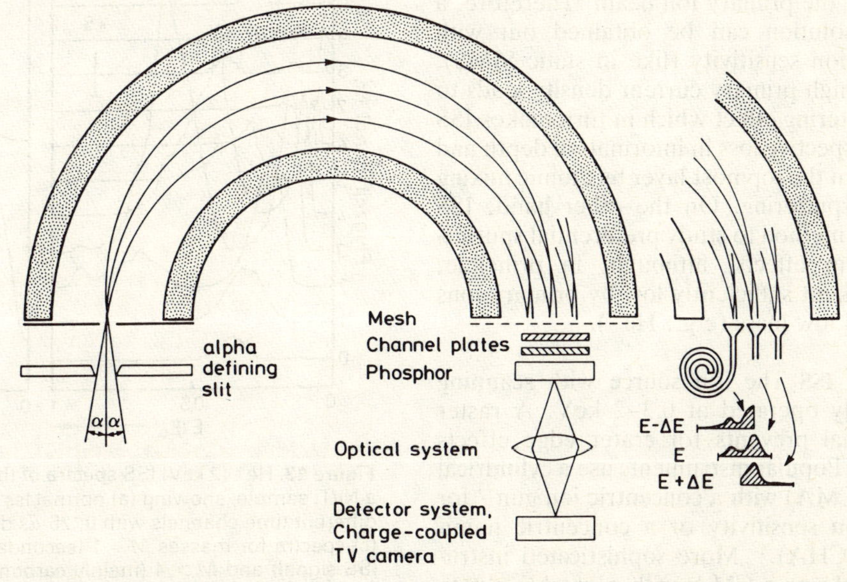

Figure 11. Multichannel detection in XPS at the output slit of a concentric hemispheres analyzer (CHA), after Seah.[81]

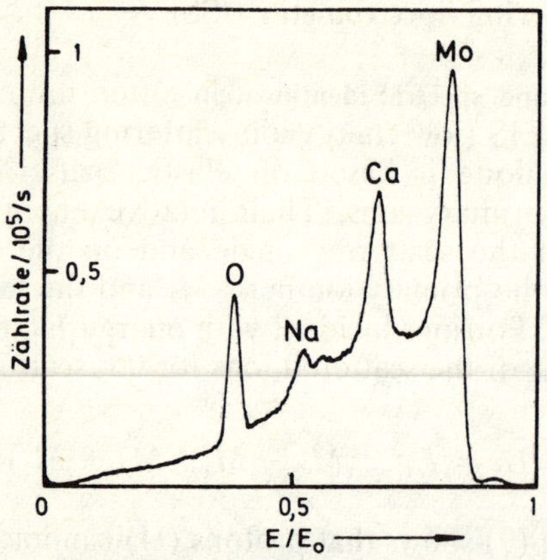

Figure 12. Typical ISS spectrum (taken with CMA with concentric ion gun, 1 keV He^+) of a partially contaminated Mo surface, from Hauser.[85] Note that there are 24 atomic mass units between O and Ca but 56 between Ca and Mo.

which have interacted only with the topmost atomic layer. This means extreme surface selectivity which is particularly useful for segregation studies.[94] On the other hand, a slight contamination may considerably distort the signal from the solid surface.[85]

Quantification of ISS The principle of quantification can be understood from the relation between the number of detected primary ions N_{12} per time, which are backscattered by the areal density N_2 of the considered target atoms:[85,94]

$$N_{12} = \text{const.}\ Y_{12} \cdot (1 - P_n) \cdot N_2 \qquad (10)$$

where the constant contains the total transmission function, Y_{12} is the scattering cross section dependent on Z_2, and P_n is the neutralization probability. The latter quantity is difficult to predict and is the main problem in quantification of ISS.

Lateral resolution of ISS Typically, the lateral resolution is of order of some millimeters, but depends principally on the focusing of the primary ion beam. Therefore, a better spatial resolution can be obtained but with sacrifice of detection sensitivity (like in static SIMS). Moreover, using high primary current density leads to an increasing sputtering effect which in turn makes ISS dynamic with a respective loss in information depth and a coupled change in the topmost layer by atomic mixing and preferential sputtering. On the other hand, ISS provides a unique method to study preferential sputtering.[93,95] Ion beam effects, although, in principle, unavoidable, are kept sufficiently low by primary ions of low energy and low mass (e.g.: He^+).

Instrumentation for ISS The ion source with scanning facility is generally operated at 0.3–2 keV. A raster gated output signal prevents for crater edge effects (similar to SIMS). Popular instruments use a cylindrical mirror analyzer (CMA) with a concentric ion gun[94] for optimum detection sensitivity or a concentric hemisphere analyzer (CHA).[82] More sophisticated instruments use a double pass CMA with a rotable slit to allow variation of the scattering angle.[61,90] In this way, an adjustment of the optimum mass resolution for a certain range is facilitated.

Combinations with other techniques Time-of-flight mass spectrometry is easily performed by a pulsed ion source and time resolved detection.[61,85] For example, this enables hydrogen determination. Figure 13 shows an example from Hauser.[85]

ISS/XPS is an easy combination which is provided by almost any XPS equipment (see above). Addition of an electron gun and changing the analyzer potential gives access to AES.[85] Conventional static SIMS can be performed by adding a quadrupole mass filter.[61] Likewise combination with an optical spectrometer can detect the ion induced radiation for analysis.

Secondary ion mass spectrometry (SIMS)

Principle of SIMS Under energetic ion or atom bombardment, any solid responds by emission of a variety of

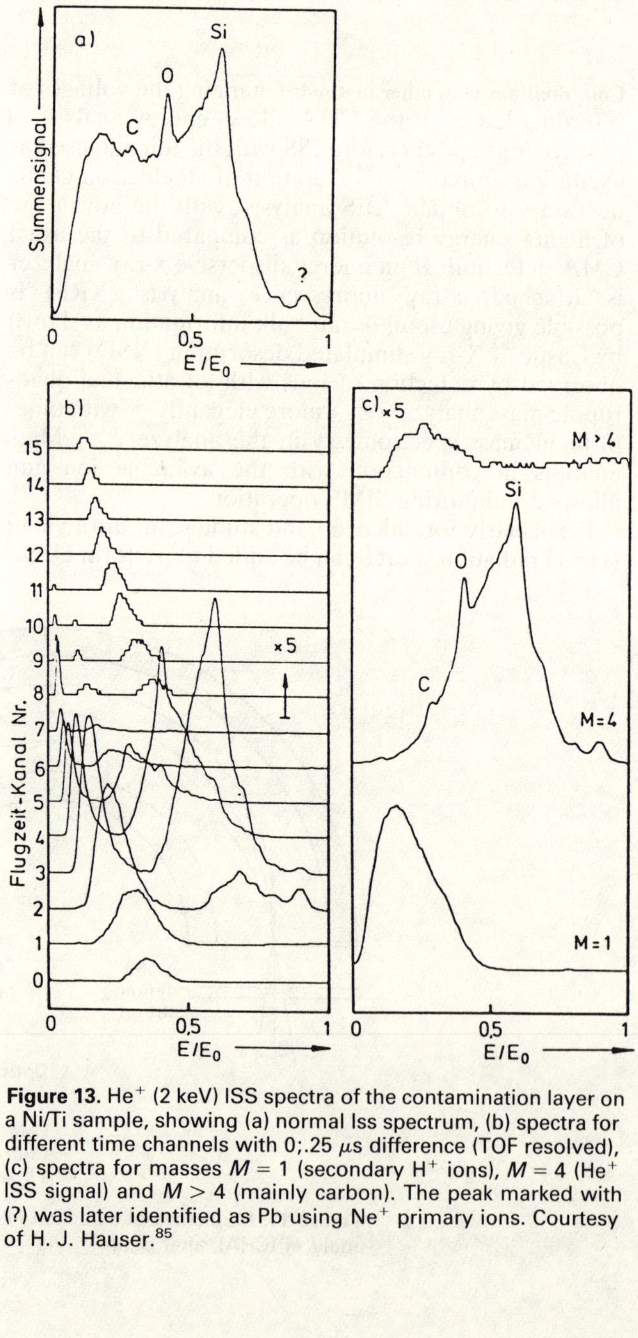

Figure 13. He^+ (2 keV) ISS spectra of the contamination layer on a Ni/Ti sample, showing (a) normal Iss spectrum, (b) spectra for different time channels with 0;.25 μs difference (TOF resolved), (c) spectra for masses $M = 1$ (secondary H^+ ions), $M = 4$ (He^+ ISS signal) and $M > 4$ (mainly carbon). The peak marked with (?) was later identified as Pb using Ne^+ primary ions. Courtesy of H. J. Hauser.[85]

secondary particles which constitute the sample surface: neutrals, positive and negative ions, partly in different clusters and multiply charged. The emitted ions are collected by a mass spectrometer which gives a readout in terms of mass/electric charge (*m/z*) ratio.[1–6,49] All elements including H can be detected if the mass range extends to 300. A unique feature is the capability to detect isotopes which allows tracer studies.

The elemental specifity of SIMS depends mainly on mass resolution and, to a lesser extent, on the possibility of energy resolved detection.[96,97]

Two methods of SIMS can be differentiated:[1–6,98] static and dynamic SIMS. According to Benninghoven,[13] static SIMS means that only a small fraction of a monolayer is used for analysis (low primary current density), whereas dynamic SIMS means relatively fast sputtering rate typically for obtaining depth profiles.

Chemical information of SIMS Cluster ions and molecule fragments can be used to trace back the chemical environment of the target atoms.[13,98] It was shown by Barber *et al.*[99] and by Benninghoven[13] that bombardment with low energy ions or atoms ('fast atom bombardment', FAB) allows the direct determination of large organic molecules immersed on a surface, since often only one proton is stripped. For general purpose, the chemical information is more indirect and may be blurred by high energy primary ions at high current density typical for dynamic SIMS.

In-depth resolution of SIMS; The mean escape depth of secondary ions is about 0.4 nm for Ar^+ primary ions on Ni between 0.5 keV and 10 keV energy, as recently shown by Monte Carlo calculations of Eckstein and Biersack[100] and depicted in Fig. 14. Whereas this figure may apply for static SIMS (primary ion current $j_p < 10^{-9}$ Acm^{-2}), the depth resolution for thin film profiling with dynamic SIMS ($j_p > 10^{-8}$ Acm^{-2}) is limited by atomic mixing within the collisional cascade

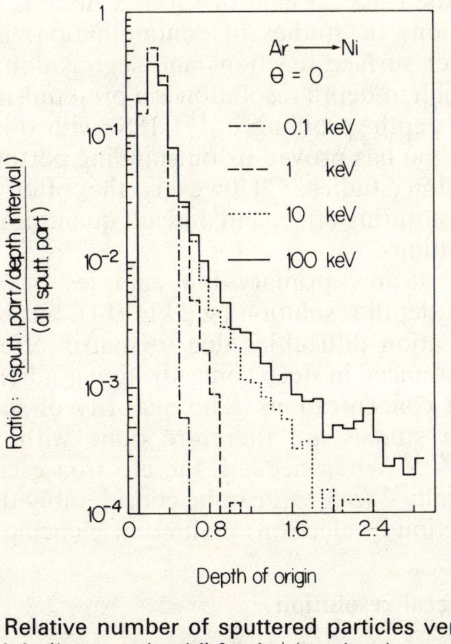

Figure 14. Relative number of sputtered particles versus their depth of origin ('escape depth') for Ar^+ bombardment of Ni with different energies at normal incidence. After Monte Carlo calculations of Eckstein and Biersack.[97] Depth of origin in nm.

induced by the primary ions.[101,102] This effect blurrs sharp interface profiles within a depth of the order of the range of the primary ions, i.e. about 1 nm for 1 keV Ar^+ ions in Si. therefore, the depth resolution deteriorates with increasing ion energy.[101] A further limitation in dynamic SIMS is the sampling time per measured point, which can be adjusted in fast recording modern instruments.

At very low ion energy, the depth resolution is expected to be optimum. As shown by Oechsner[103] a depth resolution below 3 nm can be obtained in SNMS (secondary neutrals mass spectrometry) by high frequency plasma excitation with typically 100 eV primary ion energy.

Lateral resolution of SIMS Static SIMS needs low ion current density and is therefore confined to a relatively large analyzed area (typically 10 nm diameter) to get sufficient detection sensitivity.

Dynamic SIMS also looses some detection sensitivity when using spatially resolved detection. However, elemental mapping with high ion current densities is possible by two different approaches: Either by a fine focussed rastered primary ion beam (typically 1 μm diameter),[104] or by a secondary ion image performed through an electron optical system tuned to a specific mass (typically 0.5 μm resolution).[98] Usage of high brightness liquid metal sources (Ga^+) allows secondary ion imaging with a rastered beam down below 40 nm resolution.[105]

Quantification of SIMS The basic equation for quantitative SIMS can be written:[1–6,98]

$$N_A = \text{const. } S_A \cdot \alpha^{\pm} c_A \qquad (11)$$

with N_A the current of detected particles of mass A, S_A the respective sputtering rate, c_A the concentration of component A. The constant contains primary current and apparatus transmission. The most important factor in (15) is $\alpha^{\pm}$, the ionization probability, which is strongly dependent on the sputtered species and on the work function at the surface and may vary for different atoms and surface conditions between 10^{-5} and 10^{-1}. (The change of S_A is generally within one order of magnitude.[106]) This is a strong limitation for quantitative SIMS. The spread in $\alpha^{\pm}$ can be minimized by using oxygen primary ions or an oxygen jet at the sample surfaces, which increases the ionization probability for positive secondary ions and stabilizes the surface state.[107]

Only a small fraction of the sputtered particles is ionized, most of them are neutrals. The detection of these neutrals would mean ($\alpha^{\pm} \rightarrow 1$) an increased detection sensitivity and less scatter between relative output intensity and concentration. This goal is approached by post-ionization techniques, like GDMS (glow discharge mass spectrometry).[108] or SNMS (secondary neutrals mass spectrometry),[103] and TEMS (thermal equilibrium mass spectrometry).[109] The beneficial effect of post-ionization on quantification has been demonstrated by Lipinsky *et al.*[107] in a comparison of SNMS and SIMS results of a sputtered silicon surface under increasing oxygen exposure as shown in Fig. 15.

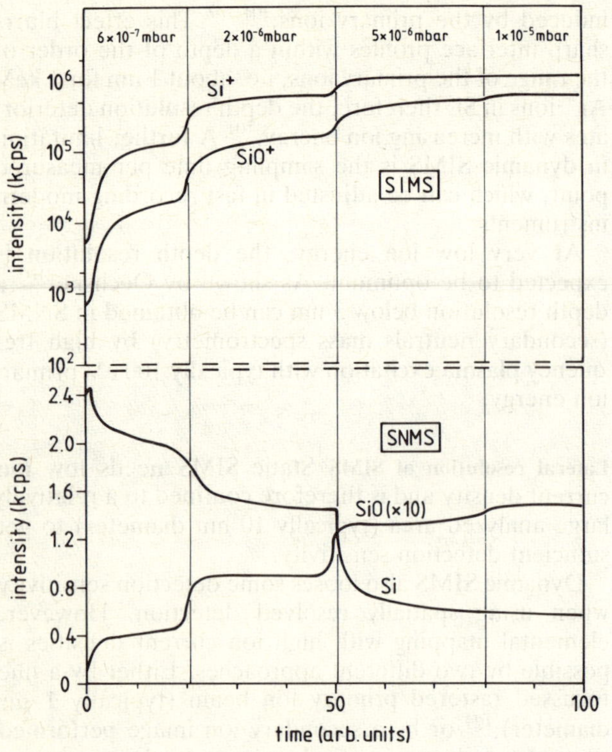

Figure 15. Comparison of SIMS and SNMS results of a Si(100) surface under simultaneous 3 keV Ar^+ bombardment and increasing oxygen pressure. Note that only SNMS shows the inverse correlation of the Si and SiO signal intensities as expected for the onset of oxidation with increasing oxygen exposure. After Lipinsky *et al.*[107]

Instrumentation of SIMS The direct dependence of the output signal on the incident primary ions as well as the high dynamic range (10^4 . . . 10^8) requires a more sophisticated ion gun equipment than the more generally used for sputter cleaning and/or depth profiling in electron spectroscopies. Up to date equipment for dynamic SIMS consists of a duoplasmatron source, ion extraction and focusing (often a Wien filter[20] of a parallel magnetic and electric field is used to select the transmission of one m/z only), and a slight electrostatic deflection of the beam to exclude primary neutrals, which do not undergo the x–y scanning via four deflection plates. A variety of ion sources (gas bottles), including Cs^{+}[110] and liquid metal (Ga^+) can be attached for the ionization chamber.[111]

In front of the analyzer, an energy filter[20] is generally used to preselect for monoatomic or cluster particles according to their different emission energy distributions.[97] For quadrupole mass analyzers, the resolution is limited to $m/\Delta m > 2000$.[98] Combined electrostatic and magnetic sector analyzers allow a better $m/\Delta m$ up to about 10000. Recently developed time-of-flight mass spectrometers achieve high mass resolution ($m/\Delta m = 4650$[112]) with the advantage of highest transmission ($\approx 50\%$), independent of mass range and unlimited mass number.[112,113]

Combinations with other techniques Modification of the sample chamber allows SNMS by HF plasma discharges (glow discharge mass spectrometry, GDMS) or by electron beam post-ionization. Addition of an optical spectrometer results in the complementary method of glow discharge optical spectrometry (GDOS) which is often used as a separate technique for depth profiling of thicker layers without the need of UHV.[114] A source for neutrals with relatively low energy (< 1 keV) (FAB) is often provided for analysis of insulators without charging and for organic materials analysis.[115]

Addition of an electron energy analyzer and an x-ray source of electron gun results in combined XPS or AES techniques. As pointed out above, SIMS can also be performed by using a pulsed primary ion source and an energy analyzer in the TOF detection mode. It should be noted, however, that any combination of techniques — particularly one including SIMS — suffers from a loss in optimized performance as compared to an instrument dedicated to one technique alone.

COMPARISON OF SURFACE ANALYSIS TECHNIQUES

Application of an appropriate technique to a specific problem should depend on the respective capabilities and limitations, as discussed above. A comparison of the most important parameters and characteristic features is presented in Table 2, which may serve as a guideline. It should be noted that such tables contract any information into a keyword or number, which often means to choose those as what is felt as 'typical', i.e. there may always be outstanding characteristics surpassing the figures in Table 2.

A problem-oriented selection of an analysis technique is fundamental for any successful application. The next section will try to pinpoint some criteria by which a proper choice may be enhanced.

OPTIMIZATION AND APPLICATIONS TO DIFFERENT PROBLEMS

High in-depth resolution at the surface

This is the classical field of surface analysis with main applications or studies of contamination, adsorption and other surface reactions and segregation. Furthermore, high in-depth resolution is a prerequisite for high quality depth profiling.[5,7,116] ISS with low energy primary ion has proved its outstanding performance in segregation studies.[93] However, the often not negligible sputtering effect and difficult quantification show its limitations.

SIMS at low primary ion energies has 1–2 ML inherent depth resolution (see Fig. 14), but is prone to quantification difficulties due to matrix effects which are pronounced in steep concentration gradients at high absolute concentrations. The majority of surface and interface studies are therefore done with AES and XPS.[7–12,117] When needed, the electron escape depth of generally 2 to 5 ML can be considerably diminished by detection of electrons emitted at glancing angles.[70]

High lateral resolution

At present, AES with a focused primary electron beam (SAM) of about 30 nm diameter has the highest lateral resolving power.[12,52] A focused primary electron beam, as used with high brightness liquid metal sources (Ga^+) in dynamic imaging SIMS, has recently shown to yield 40 nm of resolution.[118] Because principally SIMS does not suffer from distortional background effects, this value is still not the limit, which is given by the lateral momentum transfer spreading of roughly twice the range of the primary ions. Imaging SIMS by SI optics is prone to ion optical aberrations and is presently limited to about 0.2 μm.[119]

XPS of the order of 100 μm lateral resolution is now possible with many instruments (down to 10 μm in PESM[72]), but will certainly not be regarded as a high resolution technique. The case is even worse for ISS, where high density of the primary ion current is precluded by the increased surface damage due to sputtering.

Chemical bonding detection

Being the classical field of XPS, the detection of chemical bonding is enhanced by the use of monochromatized x-rays which are a prerequisite for obtaining energy resolutions of the order of 0.2 eV.[66] On the other hand, chemical effects in AES are in general larger than those in XPS.[23] However, detrimental effects of the electron beam as well as analog data acquisition and difficult interpretation impede a clear chemical effect detection in many cases. Due to the availability of digitized $N(E)$ data aquisition, this situation is now rapidly improving, and the data base is increasing.

Static SIMS, particularly with special low energy ions (or atoms) techniques has proved as an outstanding tool to determine organic molecules.[13] In both, static and dynamic SIMS, clusters and molecular fragments are used for monitoring the chemical composition.[120]

Interface analysis

The analysis of internal interfaces with surface analysis techniques is only possible if those are transformed into a surface. This can be done either by (a) cleavage or fracture *in situ* or (b) by depth profiling with ion sputtering. Method (a) is most often used in studies of grain boundary segregation,[117] thus keeping the depth resolution power of direct surface analysis. When (a) is not possible, the universally applicable method (b) is used for interface studies, with some inevitable loss of depth resolution.

Thin film depth profiling

In the majority of applications, practical surface analysis includes thin film analysis by depth profiling. Only the electron spectroscopies (AES & XPS) are capable of none destructive depth profiling by variation of the emission angle (i.e. the effective escape depth) up to about 5 nm total thickness.[67–72] Ion sputtering is a destructive method, but has principally no limitations with respect to the depth scale. SAM offers the possibility of crater edge profiling with the advantage of versatile interface studies at different locations after

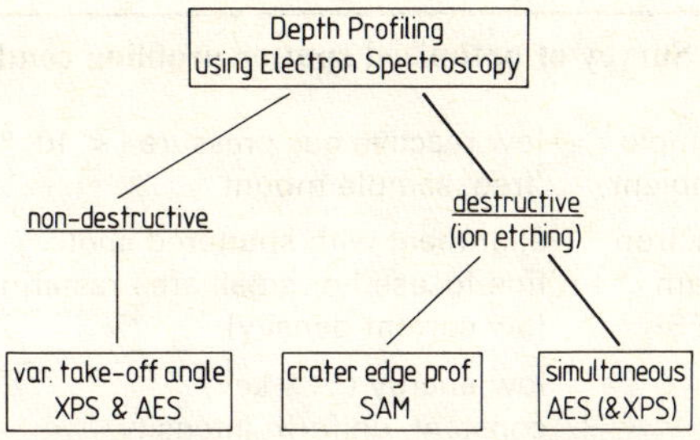

Figure 16. Scheme of the different thin film analysis methods with electron spectroscopy techniques. ISS and SIMS are confined to ion sputtering.

sputtering. A survey of the different techniques is shown in Fig. 16.

Whereas dynamic SIMS is inherently a depth profiling method, AES and XPS use an auxiliary ion gun with the advantage that sputtering and analysis are not coupled.

The depth resolution is of primary concern in depth profiling with ion sputtering, since it determines the matching between a measured profile and the true, original in-depth distribution of composition ('real profile'). There are numerous papers about this topic.[7,116,123–125] Therefore it should suffice here to summarize in Table 3 the main causes of distortional effects encountered in sputter profiling, which affect not only depth resolution, but may also lead to serious failures in thin film analysis (e.g., preferential sputtering, enhanced diffusion and segregation). Instrumental effects can be readily minimized, whereas ion beam induced effects are of more fundamental nature,[126] and of course the analyst has to cope with the special characteristics of the sample. A summary of optimized conditions for sputter profiling, which include optimum depth resolution, is given in Table 4.

Magee and Honig[110] have demonstrated how depth resolution and dynamic range is gradually improved in dynamic SIMS when going from static ion beam to rastering, raster plus lens, raster plus electronic gating (to remove secondary ions from crater walls), and finally ion beam raster plus electronic gate plus lens.

The advantage of using low energy primary ions is

Table 3. Compilation of the main distortional effects in sputter profiling

Instrumental factors	Adsorption from residual gas
	Redeposition of sputtered matter
	Impurities in Ion Beam
	non-uniform Ion Beam Density
	Information depth of analysis method
Sample characteristics	surface roughness
	crystallinity and structural defects
	alloys, compounds, 3nd phases
Ion beam induced effects	erosion statistics
	primary ion implantation
	atomic mixing
	enhanced diffusion and segregation
	induced roughness
	preferential sputtering
	decomposition of compounds

Table 4. Survey of optimized sputter profiling conditions

Sample ambient	low reactive gas pressure (< 10^{-8} Pa) 'free' sample mount
Electron beam (AES)	alignment with sputtered spot fine focused or small area rastering (ow current density)
Ion beam	low energy (≤ 1 keV) constant, uniform intensity scanning (raster gating in SIMS) large incidence angle (e.g. 75° for NiCr) reactive gas ion beam (O_2 in SIMS)
Sample	smoothed, polished surface non-crystalline, no 2nd phases (oxides, semiconductors) components with similar sputter rate good electrical and thermal conditions

demonstrated in Fig. 17, which shows the ion energy dependence of the depth resolution Δz (i.e., measured interface width) for Ni/Cr layer structures.[127,128] Another important parameter is the ion incidence angle θ.[127–130] Fig. 18 shows Δz as a function of θ for 1 keV AR^+ ions at two different sputter depth in Ni/Cr multilayer structures.[127] The worst case (maximum Δz) is obtained at $\theta \approx 30°$, which ironically is near the most frequently used angle in commercial equipment. Optimum depth resolution is obtained for large incidence angles (> 60°) as recently shown by Zalar[132] (Fig. 19a). A compilation of recent measurements of Δz as a function of sputter depth with optimized conditions in Ni/Cr multilayers by Fine *et al.*[128] and Zalar[127,131,132] is given in Fig. 19b in comparison to Ta_2O_5/Ta interfaces.[125] (Both materials can be obtained as standard references from the NPL (Ta_2O_5/Ta) and the NBS (Ni/Cr)). Oxides or semiconductors not only show a smaller Δz than metallic materials,[125] but also a different behaviour with respect to the ion incidence angle, as seen in Fig. 20. This difference can be qualitatively explained by the predominant effect of sputtering induced roughness for Ni/Cr, whereas in Ta_2O_5 the original surface roughness is responsible for increased Δz at $\theta < 60°$. Both effects are superimposed in the profiling of rough Ni/Cr surfaces[130] as seen in Fig. 21. We may conclude from Fig. 18–21, that larger ion incidence angles are only favourable for smooth surfaces of metallic materials, whereas for rough surfaces, smaller incidence angles are preferable. Of course, further research is needed to establish a predictive model for angular dependence in sputter profiling.

For a more detailed discussion of the various factors affecting depth resolution and true profile evaluation, the reader is referred to review articles.[110,116,123–126,133]

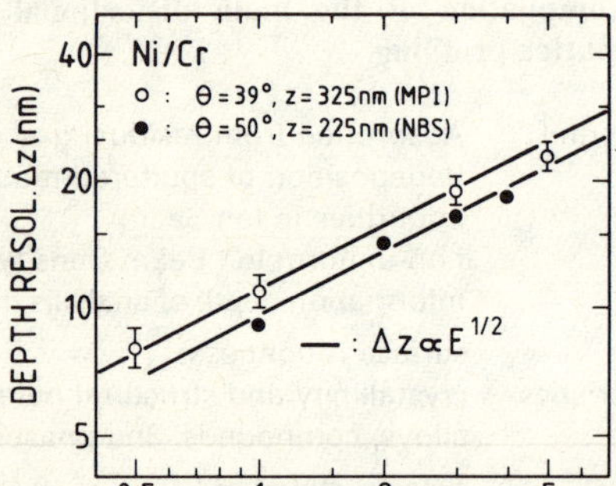

Figure 17. Dependence of depth resolution Δz on Ar^+ ion energy at 39° and 50° incidence angle and 325 nm and 225 nm sputtered depth, respectively, measured with AES on Ni/Cr multilayers. Data are from Zalar *et al.* (MPI and IEVT)[127,130–132] and from Fine *et al.* (NBS).[128]

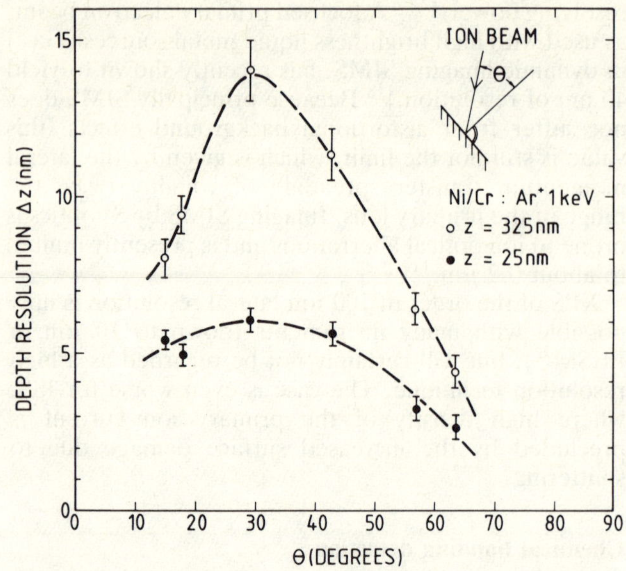

Figure 18. Dependence of the depth resolution Δz on the angle of incidence θ, from AES depth profiling of Ni/Cr multilayers deposited on a smooth Si(111) surface, sputtered with 1 keV Ar^+ ions; at $z = 25$ nm and $z = 325$ nm sputtered depth.[127,130–132] Similar results were recently obtained by Fine *et al.*[128]

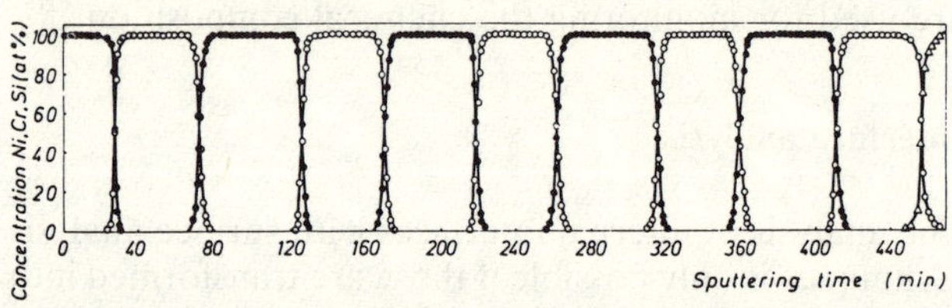

Figure 19. (a) Optimized AES composition-depth profile of a Ni/Cr multilayer structure (50 nm layer thickness, top Ni layer: 25 nm) on a smooth Si (111) substrate, obtained by sputtering with 1 keV Ar^+ ions at 63° incidence angle. From Zalar.[127,132]

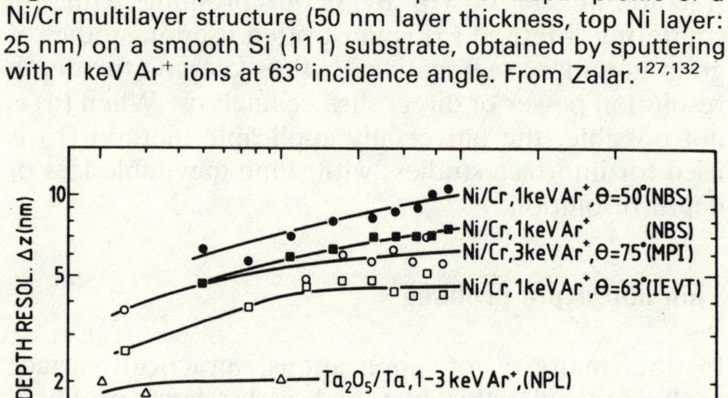

Figure 19. (b) Depth resolution Δz as a function of sputtered depth z obtained on smooth Ni/Cr multilayer structures from different laboratories under high incidence angles θ (NBS,[128] MPI,[12] IEVT (Fig. 19)). for comparison, the corresponding plot for Ta_2O_5/Ta (NPL[133]) is also shown.

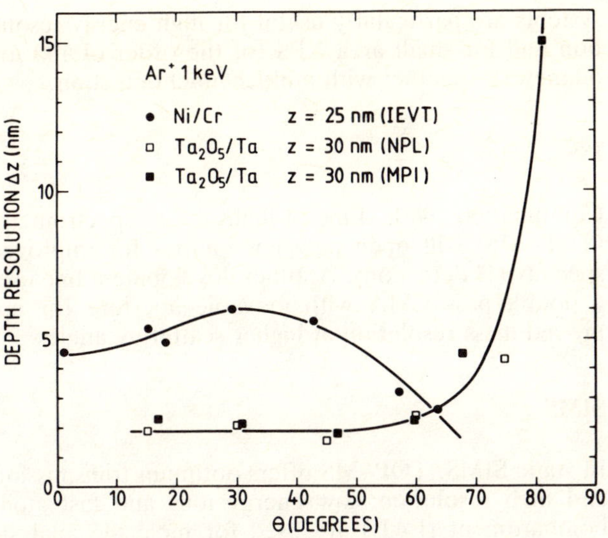

Figure 20. Comparison of the depth resolution Δz (AES profiling with 1 keV Ar^+ ions) as a function of the ion incidence angle of a Ni/Cr multilayer with smooth surface at $z = 25$ nm sputtered depth (IEVT[127]) and for Ta_2O_5/Ta anodic layers ($z = 30$ nm) (NPL[133]), (MPI[131]).

Evaluation of Sputtering Profiles The first step to evaluate 'real' composition/depth-profile ($c = f(z)$) from the measured intensity/sputter time profile is the calibration of time in terms of depth and that of intensity in terms of concentration.[123]

Depth scale calibration requires evaluation of the sputtering rate $\dot{z}$. Measurement of the primary ion current density and the use of tabulated sputter yield values[106] is generally less accurate than direct comparison with standard samples of known thickness (e.g., Ni/Cr or Ta_2O_5/Ta) profiled under identical conditions. Whereas these procedures can only give an average value for $\dot{z}$, *in situ* measurements with x-ray spectrometry,[134] laser interferometry[135] or quartz micro-

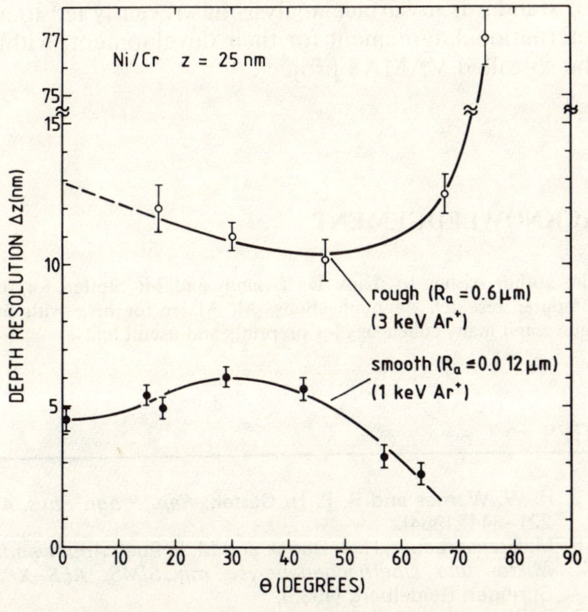

Figure 21. Comparison of the depth resolution Δz as a function of the ion incidence angle for smooth (roughness $R_a = 0.012$ μm) and rough ($R_a = 0.6$ μm) Ni/Cr multilayers at $z = 25$ nm; 1 keV and 3 keV Ar^+ ion bombardment, respectively.[131,132]

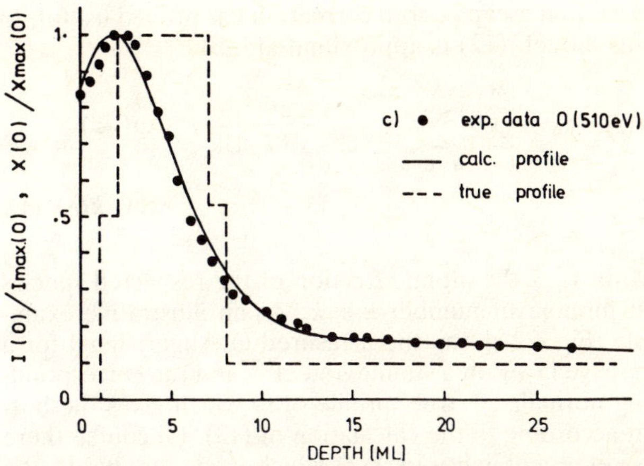

Figure 22. Fitting of the measured O 510 eV Auger intensity sputtering profile (points, full drawn line) of a passive layer on a stainless steel by means of (13) using the assumed true profile (dashed line) for X (Oxygen).[140] The non-vanishing oxygen content for larger depth can be explained by ion beam shadowing effects due to sample roughness.[130–132]

balance[136] can provide instantaneous values for $\dot{z}$, which is particularly important for layered stuctures of materials with different $\dot{z}$.[123] Known implantation profiles can also be used for this purpose.[110]

The concentration scale calibration is done by application of the quantitation algorithms for the specific analysis method used (see third section).

The then obtained 'real' profile is generally not yet the 'true', original composition — depth distribution which is wanted as the result of depth profiling. Even if there is no preferential sputtering, the latter is convoluted by a resolution function.[133,137]

Determination of the true composition-depth profile For a given true concentration $c(z')$ at depth z', the measured signal intensity is described by the convolution integral:[133,137]

$$I(z) = \int_{-\infty}^{+\infty} c(z') \cdot g(z-z')\mathrm{d}z \qquad (12)$$

where $g(z - z')$ is an appropriate resolution function This may be, e.g., an exponential function $\exp(-z'/\lambda)$ in AES assuming ideal layer by layer sputtering. More generally, it can be approximated by a Gaussian function which includes such effects as atomic mixing and/or sputtering induced roughening. For a known resolution function and $I(z)$, $c(z')$ can be evaluated from (12) by iterative procedures, e.g. the Van Cittert method.[137,138] In practice, a 'convolution' or fitting procedure is easier to perform. This means: assumption of a true profile $c(z')$, calculation of $I(z)$ and comparison with the measured intensity profile after normalization. Thus, by trial and error, the assumed profile which gives the best fit is obtained as the true profile.[137–140]

For AES depth profiling of very thin layers, the use of the so-called SLS model,[9,123] which assumes a monoatomic layer structure and a Poisson distribution as resolution function, in combination with the exponential escape depth correction has proved useful. In this model, (12) is approximated by:[9,139,140]

$$I(z) = I_0 \cdot \sum_{k=0}^{M} \sum_{n=0}^{N} X_{n+k} \cdot \frac{z^{(n-1)}}{(n-1)!} \cdot \exp(-z) \cdot \exp(-k/\lambda) \quad (13)$$

with X_{n+k} the atomic fraction of the respected species at monolayer number $n + k$. As an illustrative example, Fig. 22 shows the measured O Auger signal for a passive layer on a stainless steel[140] and the corresponding normalized 'true' profile of O which gives the best fit according to the calculation of (13). Of course there is some ambiguity in $X(z)$ which is mainly due to the error in the measured $I(z)$.[140]

In principle, using appropriate resolution functions, the general formalism of (12) and (13) can be used for any other method (XPS, ISS, SIMS) applied to depth profiling. This points out the importance of a well known resolution function, whose evaluation is the main task in fundamental profiling studies.

CONCLUSIONS AND OUTLOOK

Today the four main surface analysis techniques AES, XPS, ISS and SIMS have reached a mature state with respect to instrumentation. Nevertheless, the development of both hard- and software is going on steadily. Whereas some trends are clearly recognized, future developments are difficult to predict and may be biased by personal experience. The author's view of the main present trends in hardware development is briefly summarized:

AES

High spatial resolution, although almost at its physical limit, will be improved for special cases by field emission cathodes. Due to multichannel detection, concentric hemisphere anlyzers with preretardation may become more popular for AES since they allow the recording of a better resolved spectrum for fine features evaluation.

XPS

High power x-ray sources (with a variety of anodes), improved x-ray monochromators and analyzer lens systems are particularly useful for high energy resolution and for small area XPS (of the order of 100 μm diameter), together with multichannel detection.

ISS

Combination with time-of-flight-mass spectrometry (TOF–MS) will open new possibilities for improved specifity (H detection). Another development line uses a double-pass CMA with movable aperture for improved mass resolution at higher scattering angles.

SIMS

In static SIMS, TOP–MS offers optimum transmission and high resolution; low energy ions and fast atom bombardment (FAB) are used for molecule analysis and insulator studies. In dynamic SIMS, post ionization techniques (SNMS) with mass and optical emission spectrometry improve quantitative profiling; liquid metal sources provide high resolution imaging SIMS.

In summary, besides sensitivity and specifity, optimized micro-local analysis in all three dimensions is the primary aim.

For any of the above techniques, software developments are characterized by increased sophistication, particularly with fully automized computer steering of all hardware components, including sample manipulation. Complete measurement and data processing routines are becoming available, and image processing for AES and SIMS. Data banks and calculation routines will be included for on line quantification.

Thus, in the future, we will see more massaged and quantitative data being reported. However, we may face an increasing danger of results, the origin of which cannot be traced back in detail to their physical generation. therefore, standard reference materials and test routines together with carefully established data bases will become of increasing importance. The need for standards in surface analysis has recently led to an international agreement for their development, within the so-called VAMAS project.[141]

ACKNOWLEDGEMENT

The author wishes to thank Dr Gramm and Mr Steffen for the computer research on publications, Mr Maier for help with the figures and many colleagues for preprints and useful hints.

REFERENCES

1. A. W. Czanderna, (ed.), *Methods of Surface Analysis*, Elsevier, Amsterdam (1975.
2. R. Holm and S. Storp, in: Hrsg. Ullmans Encycl.d.techn. Chemie, Vol. 5, pp 519–579, ed. by H. Kelker Verlag Chemie, Weinheim (1980).
3. S. Hofmann, in: *Analytiker-Taschenbuch*, pp. 287–313 Springer-Verlag, (1979).
4. H. J. Grabke, (ed.), *Oberflächenanalytik in der Metallkunde*, Deuitsche Gesellschaft für Metallkunde, Oberursel (1983).
5. H. W. Werner and R. P. H. Garten, *Rep. Progr. Phys.* **47**, 221–344 (1984).
6. M. Grasserbauer, H. J. Dudek and M. F. Ebel, *Angewandte Mikro- und Oberflächenanalyse mit SIMS, AES XPS*, Springer, Heidelberg (1985).
7. D. Briggs and M. P. Seah, (eds), *Practical Surface Analysis by Auger and Photoelectron Spectroscopy*, Wiley,. Chichester (1983).

8 .P. H. Holloway, *Adv. in: Electronics and Electron Physics*, Vol. 54, pp. 241–298 Academic Press (1980).
9. S. Hofmann, in: *Wilson & Wilson's Comprehensive Analytical Chemistry*, Vol. IX, chapter 2, pp. 89–172, ed. by C. Svehla, Elsevier, Amsterdam (1979).
10. J. C. Riviere, *The Analyst*, **108**, 649 (1983).
11. M. P. Seah, *Vacuum*, **34**, 463 (1984).
12. S. Hofmann, in: *Proc. 6th Int. Symp. High Purity Materials in Science Technology, Akad. d. Wiss. d. DDR, Dresden 1985*, Vol. II, pp. 149169 A. Drescher, ed.
13. A. Benninghoven, *J. Vac. Sci. Technol.* **A3**, 451 (1985).
14. D. Briggs, M. J. Hearn and B. D. Ratner, *Surf. Interface Anal.* **6**, 184 (1984).
15. M. P. Seah, *J. Vac. Sci. Technol.* **A3**, 1330 (1985).
16. P. H. Holloway and H. Stein, *J. Electrochem. Soc.* **123**, 723 (1979).
17. C. W. Magee and E. M. Botnick, *J. Vac. Sci. Technol.* **19**, 47 (1981).
18. H. J. Mathieu and D. Landolt, *J. Microsc. Spectr. Electron.* **3**, 113 (1978).
19. S. Hofmann, *Proc. 3rd Int. Conf. on SIMSS (III), Budapest (1981)*, pp. 186–200 eds: A. Benninghoven, J. Laszlo, M. Riedel, H. W. Werner, Springer-Verlag (1982).
20. C. W. Magee, W. L. Harrington and R. E. Honig, *Rev. Sci. Instr*, **49**, 477 (1978).
21. W. A. Coghlan and R. E. Clausing, *Atomic Data*, **5**, (1973), 317.
22. L. A. Davis, N. C. MacDonald, P. W. Palmberg, G. E. Riach and R. E. Weber: *Handbook of AES*, 2nd ed., Perkin-Elmer, Phys, El. Div., Eden Prairie (1979).
23. H. H. Madden, *J. Vac. Sci. Technol.* **18**, 677 (1981).
24. R. Weismann and K. Müller, *Surf. Sci. Reports*, **105**, 251 (1981).
25. H. Seiler, *Ultramicroscopy*, **17**, 1 (1985).
26. J. H. Thomas III, *J. Vac. Sci. Technol.* **A2**, 84 (1984).
27. J. M. Sanz and S. Hofmann, *J. Electron Spectr. Rel. Phen*, **34**, 149 (1984).
28. J. H. Thomas III and S. Hofmann, *Surf. Interface Anal.* **4**, 156 (1982).
29. J. M. Sanz and S. Hofmann, *Thin Solid Films*, **120**, 185 (1984).
30. A. M. Bevolo, J. D. Verhoeven and M. Noack, *Surf. Sci.* **134**, 499 (1983).
31. M. P. Seah, to be publ. in *J. Mat. Sci.*
32. M. De Creszenzi, F. Antonangeli, C. Bellini, A. Rosei, *Phys. Rev. Lett.* **50**, 1949 (1983).
33. S. Hofmann, *Depth Profiling*, Chap. 7, pp. 141–180 in Ref. 7.
34. K. Berresheim, *Surf. Interface Anal.* **9**, (1986).
35. D. R. Penn. *J. Electr. Spectr.* **9**, 29 (1976).
36. M. P. Seah and W. Dench, *Surf. Interface Anal.* **1**, 2 (1979).
37. J. C. Ashley and V. E. Anderson, *J. Electron Spectr.* **24**, 127 (1981).
38 J. C. Ashley and J. C. Tung, *Surf. Interface Anal.* **4**, 52 (1982).
39. J. C. Powell, *J. Vac. Sci. Technol.* **A3**, 1338 (1985).
40. C. D. Wagner, L. E. Davis and W. M. Riggs, *Surf. Interface Anal.* **2**, (1980).
41. M. P. Seah, Private communication and this conference.
42. M. P. Seah, *Scanning Electron Microscopy 1983/II*, p. 521, SEM Inc.
43. M. P. Seah, J. L. De Segovia, ed., *Proc. IX Int. Vac. Congr. and Conf. Sol. Surf., ASEVA, Madrid 1983*, pp. 63–75.
44. J. P. Langeron, L. Minel, J. L. Vignes, S. Bouquet, F. Pellerin, G. Lorang, P. Ailloud and J. C. LeHericy, *Surf. Sci.* **138**, 610 (1984).
45. P. M. Hall and J. M. Morabito, *Surf. Sci.* **83**, 391 (1979).
46. R. Shimizu and S. Ichimura, *Quantitative Analysis by AES*, Toyota Foundation Res. Rep., I–006 No. 76–9175, Osaka (1981).
47. S. Hofmann and J. M. Sanz, *Fresenius Z. Anal. Chem.* **314**, 215–219 (1983).
48. J. M. Sanz and S. Hofmann, *Surf. Interface Anal.* **8**, (1979).
50. J. Cazaux, *Surf. Sci.* **140**, (1984).
51. C. LeGressus, D. Massignon, R. Sopizet, *Surf. Sci.* **68**, 338 (1977).
52. M. M. El Gomati, A. P. Janssen, M. Prutton and J. A. Venables, *Surf. Sci.* **85**, 309 (1979).
53. H. J. Dudek, in D. A. Degras and M. Costa, eds.: *Proc. 4th Int. Conf. Sol. Surf. a. 3rd ECOSS, Cannes 1980*, p. 1365.
54. H. F. Bauer and H. Seiler, *Electron Microsc.* **3**, 214 (1980).
55. J. Cazaux, *Ultramicroscopy*, **12**, 83 (1983).
56. S. Hofmann, A. Zalar, *Thin Solid Films*, **56**, 337 (1979).
57. K. Röll, *Appl. surf. Sci.* **5**, 388 (1980).
58. C. G. Pantano and T. E. Madey, *Apl. Surf. Sci.* **7**, 115 (1981).
59. S. Thomas, *J. Appl. Phys.* **45**, 161 (1974).
60. F. Ohuchi, M. Ogino, P. H. Holloway and C. G. Pantano, *Surf. Interface Anal.* **2**, 85 (1980).
61. G. B. Hoflund, D. A. Asbury, P. Kirszenstejn and H. A. Laitinen, *Surf. Sci.* **161**, L583 (1985) and G. B. Hoflund, *Surf. Interface Anal.* **9**, (1986).
62. G. B. Hoflund, *Scanning Electron Microscopy*, Vol. II, 560 SEM Inc. (1985).
63. S. Hofmann and J. M. Sanz, *J. Trace Microprobe Techn.* **1**, 213 (1982–83).
64. C. D. Wagner, W. M. Riggs, L. E. Davis, J. F. Moulder and G. E. Muilenberg, *Handbook XPS*, Perkin-Elmer Corp., Eden Prairie (1979).
65. R. Holm and S. Storp, *Appl. Phys.* **9**, 217 (1976).
66. U. Gelius, L. Asplund, E. Basilier, S. Hedman, K. Helenelund and K. Siegbahn, *Nucl. Inst. a. Meth.* **B1**, 85 (1984).
67. C. S. Fadley, *Prog. Solid State Chem.* **11**, 265 (1976).
68. M. F. Ebel, *Surf. Interface Anal.* **3**, 191 (1981).
69. M. Pijolat and G. Hollinger, *Analusis*, **10**, 8 (1982).
70. S. Hofmann and J. M. Sanz, *Surf. Interface Anal.* **6**, 75 (1984).
71. H. J. Mathieu and D. Landolt, *Surf. Interface Anal.* **6**, 82 (1984).
72. T. D. Bussing and P. H. Holloway, *J. Va. Sci. Technol.* **A3**, 1973 (1985).
73. S. Tougaard and A. Ignatiev, *Surf. Sci.* **129** 355 (1983) and S. Tougaard and P. Sigmund, *Surf. Interface Anal.* **9**, 1986).
74. D. Briggs, R. A. Marbrow and R. M. Lambert, *Sol. State Comm.* **26**, 1 (1978).
75. S. Evans and M. D. Scott, *Surf. Interface Anal.* **3**, 269 (1981).
76. J. H. Scofield, *J. Electron Spectrosc.* **8**, 129 (1976).
77. N. Gurker, M. F. Ebel and H. Ebel, *Surf. Interface Anal.* **5**, 13 (1983).
78. G. Beamson, H. Q. Porter and D. W. Turner, *Nature*, **290**, 556 (1981).
79. J. Cazaux, *Scanning electron Microscopy*, Vol. III, p. 1193, SEM Inc (1984).
80. I. W. Drummond, T. A. Cooper and F. J. Street, *Spectrochim. Acta*, **40B** 809 (1985).
81. M. P. Seah, to be published in: J. M. Walls, ed., *Methods of Surface Analysis: Techniques and Applications*, Cambridge University Press (1986).
82. S. Storp, K. Berresheim and M. Wilmers, *Surf. Interface Anal.* **1**, 96 (1979).
83. H. H. Streblow, *Surf. Interface Anal.* **9**, (1986).
85. H. J. Hauser, Thesis, Univ. of Hohenheim (1983).
86. E. Taglauer and W. Heiland, *Appl. Phys.* **9**, 216 (1976).
87. W. L. Baun, *Surf. Interface Anal.* **3**, 243 (1981).
88. A. W. Czanderna, A. C. Miller, H. H. G. Jellinek and H. Kachi, *J. Vac. Sci. Technol.* **14**, 227 (1977).
89. R. L. Erickson and D. P. Smith, *Phys. Rev. Lett.* **34**, 297 (1975).
90. H. Niehus and E. Bauer, *Surf. Sci.* **47**, 222 (1975).
91 W. Englert, E. Taglauer and W. Heiland, *Surf. Sci.* **117**, 124 (1982).
92. D. P. Jackson, T. M. Buck and G. H. Wheatley, *Nucl. Instr. Meth.* **B2**, 440 (1984).
93. S. Swartzfager, S. V. Ziemacki and M. J. Kelley, *J. Vac. Sci. Technol.* **19**, 185 (1981).
94. E. N. Haeussler, *Surf. Interface Anal.* **2**, 134 (1980).
95. P. Varga and E. Taglauer, *Nucl. Instr. Meth.* **B2**, 800 (1984).
96. F. Degrève, R. Figaret and P. Laty, *Inst. J. Mass Spectr. Ion Phys.* **27**, 262 (1978).
97. J. B. Clegg, *Surf. Interface Anal.* **2**, 91 (1980).
98. H. W. Werner, *Surf. Interface Anal.* **2**, 56 (1980).
99. M Barber, R. S. Bordoli, R. D. Sedwick and A. N. Tyler, *Nature*, **293**, 270 (1981).
100. W. Eckstein and J. Biersack, *Nucl. Instr. Meth.* **B2**, 550 (1984).
101. H. H. Andersen, *Appl. Phys.* **18**, 131 (1979).

102. U. Littmark and W. O. Hofer, *Nucl. Instr. Meth.* **168**, 329 (1980).
103. H. Oechsner, *Fresenius Z. Anal. Chem.* **314**, 211 (1983).
104. H. Liebl, *Scanning*, **3**, 79 (1980).
105. G. L. R. Mair and T. Mulvey, *Scanning Electron Microscopy*, Vol. III, p. 959, SEM Inc. (1985).
106. M. P. Seah, *Thin Solid Films*, **81**, 279 (1981).
107. D. Lipinsky, R. Jede. O. Ganschow and A. Benninghoven, *J. Vac. Sci. Technol.* **A3**, 2007 (1985).
108. J. W. Coburn and E. Kay, *Appl. Phys. Lett.* **19**, 350 (1971).
109. G. Blaise, *Scanning Electron Microscopy*, Vol. I, p. 31, SEM Inc. (1985).
110. C. W. Magee and R. E. Honig, *Surf. Interface Anal.* **4**, 35 (1982).
111. A. R. Bayly, A. R. Wangh and P. Vohralik, *Spectrochim. Acta*, **40B**, 717 (1985).
112. R. B. Opsal, K. B. Owens and J. P. Reilly, *Anal. Chem.* **57**, 1884 (1985).
113. A. Benninghoven, Private communication.
114. R. Berneron and J. C. Charbonnier, *Surf. Interface Anal.* **3**, 134 (1981).
115. F. Degreve and J. B. Lang, *Surf. Interface anal.* **7**, 177 (1985).
116. H. Oechsner, ed., *Thin Film and Depth Profile Analysis, Current Topics in Physics*, Vol. 37, Springer-Verlag, (1984).
117. M. P. Seah, *AES in Metallurgy* Chapter 7 in Ref. 7.
118. R. Levi-Setti, G. Crow and Y. L. Wang, *Scanning Electron Microscopy*, Vol. II, p. 53, SEM Inc. (1985).
119. F. G. Rüdenauer, in: A. Benninghoven, J. Giber, J. Laszlo, M. Riedel and H. W. Werner, eds., *Secondary Ion Mass Spectrometry SIMS III*, p. 2 Springer-Verlag, (1982).
120. A. Benninghoven, *Surf. Sci.* **53**, 596 (1975).
121. A. Zalar and S. Hofmann, *Surf. Interface Anal.* **2**, 183 (1980).
122. L. L. Levenson, T. P. Massopust, J. Dick, M. C. Jaehnig and D. Griffith, *Scanning Electron Microscopy, Vol. III*, p. 1087, SEM Inc. (1985).
123. S. Hofmann, *Surf. Interface Anal.* **2**, 148 (1980).
124. E. Zinner, *J. Electrochem. Soc.* **130**, 199 (1983).
125. M. P. Seah and C. P. Hunt, *Surf. Interface Anal.* **5**, 33 (1983).
126. K. Wittmaack, *Vacuum*, **34**, 119 (1984).
127. A. Zalar, Thesis, Univ. of Ljubljana (1986).
128. J. Fine, P. A. Lindfors, M. F. Gorman, R. L. Gerlach, B. Navinsek, D. F. Mitchell and G. P. Chambers, *J. Vac. Sci. Technol.* **A3**, 1413 (1985).
129. S. S. Makh, R. Smith and J. M. Walls, *Surf. Interface Anal.* **2**, 115 (1980).
130. A. Zalar, *Three country vacuum conference, Debrecen* (1985).
131. A. Zalar and S. Hofmann, to be published.
132. A. Zalar, *Surf. Interface Anal*, **9**, 41 (1986).
133. S. Hofmann, *Depth Profiling*, Chap 4 in Ref 7.
134. J. Kirschner and H. W. Etzkorn, *Appl. Surf. Sci.* **3**, 251 (1979).
135. J. Kempf, *Surf. Interface Anal.* **4**, 116 (1982).
136. B. Navinsek, P. Panjan, A. Zabkar and J. Fine, *J. Vac. Sci. Technol.* **A3**, 671 (1985).
137. S. Hofmann and J. M. Sanz, Chap 7 in Ref. 116.
138. S. Hofmann and J. M. Sanz, F. Abels and M. Croset, eds., *Proc. 8th Int. Vac. Congr., Cannes 198*, Vol. I, p. 90, Paris (1980).
139. S. Hofmann and J. M. Sanz, *Surf. Interface Anal.* **6**, 78 (1984).
140. J. M. Sanz and S. Hofmann, to be publ. in *Surf. Interface Anal.*
141. M. P. Seah and C. J. Powell, *The Coordinated Development of Standards for Surface Chemical Analysis*, US Dept. of Commerce, NBSIR 85–3120, Gaithersburg (1985).

SURFACE AND INTERFACE ANALYSIS, VOL. 9, 21–25 (1986)

Quantitative Surface Analysis by Glow Discharge Optical Spectrometry

J. Pons-Corbeau, J. P. Cazet, J. P. Moreau, R. Berneron and J. C. Charbonnier
Institut de Recherches de la Sidérurgie Française (IRSID), 185, rue du Président Roosevelt, 78105 Saint-Germain-en-Laye Cedex, France

In an analysis by glow discharge optical spectrometry the relation between the intensity of an optical emission line of an element and its concentration on the surface is exploited for quantitative determination of variations of composition in galvanized coatings and for an estimation of the enrichments on the surface of annealed extra mild steel sheets.

PRINCIPLE

In previous study[1,2] it was shown how the different parameters vary which connect the intensity $J^{A\,(\lambda)}$ of a spectral line of an element A located on the surface of a specimen submitted to an ionic erosion in GDOS with the atomic concentration of this element C^A, at sputtering steady state : $J^{A(\lambda)} = k^{A(\lambda)} R^A Y^M I_G C^A$, where $k^{A(\lambda)}$ is the collection efficiency of the line λ (A). It is an instrumental factor which depends upon the operating conditions. R^A is the emission yield of the element A for the line λ ; it does not depend upon the matrix in which A is present; Y^M is the total sputtering yield of the matrix. It depends upon the nature and the composition of the matrix; I_G is the current intensity in the discharge.

Let us consider a matrix which is composed of several constituents A, B, C and so on and the depth profiles of the lines intensities J of the elements A and B for instance (Figure 1).

At any time t_1 it may be written :

$$J^A = k^A R^A Y^{M(t_1)} I_G^{t1} C^{A(t_1)} = K^A Y^{M(t_1)} I_G^{(t_1)} C_A^{(t_1)} \tag{1}$$

$$J^B = k^B R^B Y^{M(t_1)} I_G^{t_1} C^{B(t_1)} = K^B Y^{M(t_1)} I_G^{(t1)} C_B^{(t_1)}$$

The sputtering yields and the current intensities, which are a function of the erosion time are the same in both expressions. On the other hand : $k^A R^A = K^A$ and $k^B R^B = K^B$ do not vary with the depth because the collection conditions of photons remain the same and, as it was previously shown, the emission yields are independent of the matrix.[1]

So it is possible to estimate the atomic fraction of the elements such as A, x^A_{at} at any erosion depth z (t_1) and for the area of the specimen submitted to the discharge, that is to say a few mm[2,3,4] through the equation :

$$x^A_{at} = \frac{J^A/K^A}{J^A/K^A + J^B/K^B + J^C/K^C + \ldots}$$

Practically we use the ratio of the coefficient K for every element towards that one of the major element (B for example). So we obtain the atomic fractions of the elements such as A :

$$x^A_{at} = \frac{J^A . K^B/K^A}{J^B + J^A\, . \,K^B/K^A + J^C . K^B/K^C + \ldots}$$

Consequently it is sufficient to know beforehand the ratios K^B/K^A, K^B/K^C . . .

They may be purely and simply determined with the aid of a glow discharge on any binary system B–A, B–C, and so on. . . of known composition or in default of this on any kind of systems containing B–A, B–C, and other elements.

The examples presented below meet the different needs of quantification in surface analysis, for instance : (a) determination of the chemical composition of various galvanized coatings; (b) quantification of enrichments occuring during batch annealing of low carbon steel sheets.

The equipment used for the measurements has already been described in various publications.[3,4] The

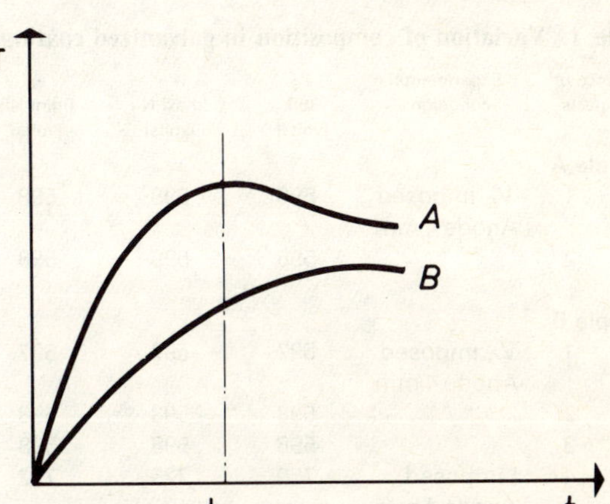

Figure 1. Spectral intensities J versus erosion time t.

0142–2421/86/120021–05 $05.00

erosion rates have been determined with a surface profilometer; a computer gives the average depth from 2500 measurements of the crater depth made along a diameter.

APPLICATION TO GALVANIZED COATINGS

Taking into account their interest and their relative simplicity we have chosen to consider Zn–Ni alloy in this paper. The profiles of the chosen line for Ni (λ_{Ni} = 341,48 nm) often show sensitive variations of depth intensity. We have proposed to estimate the corresponding variations of composition. We did not have Zn–Ni standard available of known composition, so we used the average composition of the electroplated coating given by the manufacturer and controlled by chemical analysis : $\bar{x}_{at}^{Zn} = 0.90$ $\bar{x}_{at}^{Ni} = 0.10$, for determining the ratio K^{Zn}/K^{Ni} and we took the average value of the intensities $\bar{J}^{Zn}$ and $\bar{J}^{Ni}$ on the layer depth.

So we obtain : $0.90 : \bar{x}_{at}^{Zn} = \dfrac{\bar{J}^{Zn}}{\bar{J}^{Zn} + \bar{J}^{Ni} \,.K^{Zn}/K^{Ni}}$

Henceforth the ratio K^{Zn}/K^{Ni} being known, we calculate the zinc amount with the aid of this relation and consequently that of nickel at any depth.

The composition is determined at the very depths where the nickel profile shows respectively a minimum and a maximum. The corresponding spectral variation for zinc is of course very weak due to the large prevalence of this element.

The measurements were carried out on two specimens A and B which differentiate each other by the place of sampling on the sheet.

Different discharge conditions were used : imposed current intensity (I_G) or imposed lamp voltage (V_G), anode diameter 4 or 8 mm.

The results are shown in Table 1. An example of intensity profiles is given Fig. 2.

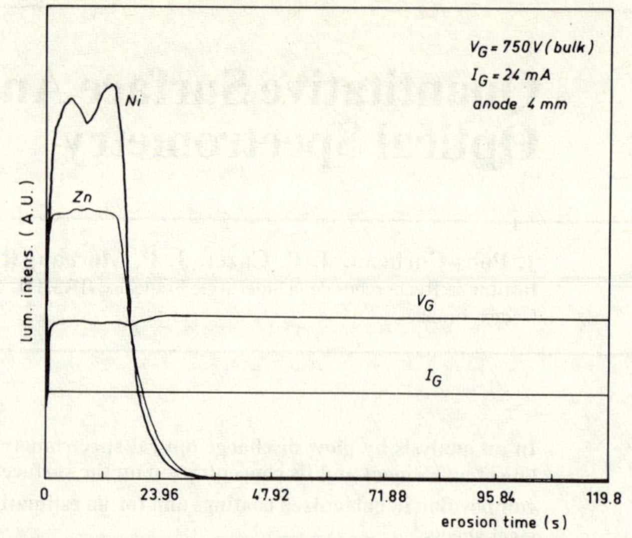

Figure 2. Depth profiles of Zn and Ni in galvanized sheets V_G: voltage, I_G: current in the lamp, anode 4 mm.

Comments on the results

The measurements performed on both specimens in an interval of a few weeks lead to very similar results for a given specimen whatever may be the operating conditions : I_G or V_G imposed, anode diameter.

The values of the ratios K^{Zn}/K^{Ni} shown in Table 1 vary with time. It is not surprising since they use the photon collection efficiency which depends upon the operating conditions such as the cleaness of the window of the discharge lamp. It has no incidence on the investigated values of concentrations if one takes care of determining these ratios for every series of measurements. This is very quickly made.

The calculations of concentation performed as mentioned above lead to the following results : (a) for the specimen A there is no noticeable difference in the zinc amount through the whole thickness of the layer ; (b) for the specimen B the zinc amount is very slightly different (about 1%) and this for all the impacts,

Table 1. Variation of composition in galvanized coatings.

Reference of the impacts	Experimental conditions	V_G Bulk (volts)	V_G (maxi Ni) (volts)	V_G (mini Ni) (volts)	I_G mA	I_G (maxi Ni) mA	I_G (mini Ni) mA	K_{Zn}/K_{Ni}	x_{Zn} at % (maxi Ni)	x_{Zn} at (mini N
Sample A										
1	V_G imposed Anode 4 mm	598	598	598	19.3	19.3	19.3	0.118	0.899	0.90
2	"	598	598	598	19.2	19.2	19.2	0.119	No noticeable difference	
Sample B										
1	V_G imposed Anode 4 mm	597	597	597	19.2	19.3	19.2	0.120	0.895	0.904
2	"	598	598	598	18.9	18.7	18.9	0.118	0.896	0.903
3	"	598	598	598	19.1	19	19	0.123	0.895	0.905
4	I imposed Anode 4 mm	750	725	722	24.2	24.2	24.3	0.078	0.896	0.906
5	I imposed Anode 8 mm	1000	817	846	100	100	100	0.127	0.890	0.902

between the depth where the nickel profile shows a maximum and that where this profile exhibits a minimum. As to the amount of nickel it varies within 10% since there is in average 10 times less nickel than zinc in the layer.

Finally let us note that for a voltage $V_G = 600$ volts and a current intensity of 20 mA the erosion rate in the coating is of the order of 200 nm s^{-1}

APPLICATION TO ELEMENT ENRICHMENTS ON THE SURFACE OF ANNEALED SHEETS

Two sorts of extra mild steel sheets having different compositions were studied : (a) sheet A (Laboratory prepared); composition (wt %) : Mn 0.349, C 0.051; (b) sheet B (industrial); composition (wt %) : Mn 0.191, C 0.049, Al 0.052, Si 0.022, Cr 0.024, P 0.010, S 0.018, Cu 0.026, Ni 0.027. These sheets were annealed at 700°C during various times.

As it is now known selective oxidation phenomena of elements occuring during batch annealing of mild steel sheets play an important role in surface reactivity. This is the reason why we tried to find out the quantities of elements segregated in the outermost layers.

We took interest in the manganese enrichment on the surface of the sheet A and in the manganese – phosphorous – chromium – silicon and aluminium enrichment on the surface of the sheet B. In both cases the largest prevalent element is iron and the following approached relation is sufficient for determining the atomic fraction of an element i :

$$x^i_{at} = \frac{J^i . K^{Fe}/K^i}{J^{FE}}$$

When it is not the case we make use of the complete formula (2) after having determined the coefficients K^{Fe}/K^i with the aid of standards.

Sheet A

We use a non annealed specimen to determine the ratio K^{Mn}/K^{Fe} when knowing its manganese content from the formula :

$$x^{Mn}_{at} = \frac{J^{Mn} . K^{Fe}/K^{Mn}}{J^{Fe}} \tag{3}$$

Two series of measurement have led to the following results :

$$K^{Mn}/K^{Fe} = 56 \qquad K^{Mn}/K^{Fe} = 57$$

They were carried out in the same discharge conditions as for the annealed specimen.

The spectral background of manganese (that of iron is negligible compared to the signal) is determined with the aid of a very pure iron specimen and then subtracted from the total spectral intensity.

The measurements were performed with a 4 mm anode diameter and an imposed tension of 1000 V. The composition of the matrix being not very different on the surface and in the bulk, the current intensity does suffer very low variations (in the order of 1/100). Under these discharge conditions (1000 V, 25 mA), the sputtering rate is 70 nm $s.^{-1}$.

We calculate the average atomic fraction of manganese on the surface from the average value of the iron and manganese light signals and from the (3). We then do infer the ratio between the manganese average concentration $\overline{C}_s$ on the surface and the bulk concentration C_c as well as the segregated manganese content on the surface in μg cm^{-2}.

The thickness of the layers concerned by the enrichment are estimated on one hand from the intersecting points of profiles of the manganese light signals in the annealed specimen and in the reference (normalized with the global spectrum intensity *FI* which has been collected) and on the other hand from an erosion rate of 70 nm s^{-1} for the admitted discharge conditions. The results are mentioned in Table 2 and an example of profile is given in Fig. 3.

As regarding the specimen annealed during 16 and 8 hours the manganese profile shows a peak on the surface and then a 'shoulder'. We have calculated the amount of segregated manganese only corresponding to this peak and then the whole segregated quantity.

Finally let us remark that the largest prevalent element of the matrix is evidently iron. Consequently if even there is a maganese enrichment the total sputtering yield of the matrix is noticeably the same on the surface and in the bulk (that is to say equally in the non-annealed reference specimen). In this case we can use a more direct method for interpreting the measurements : it suffices to do the ratio between the

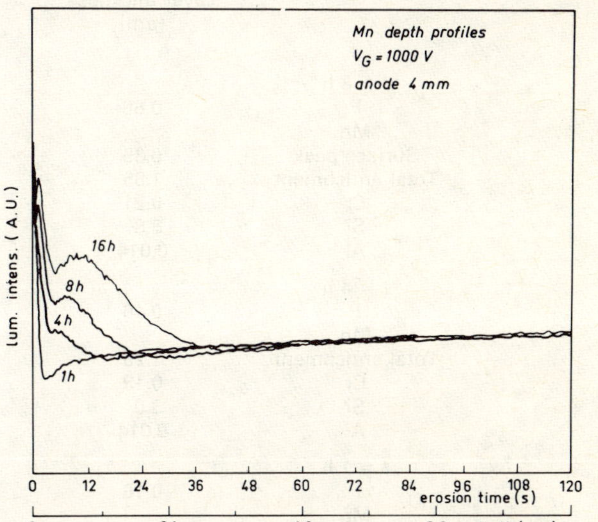

Figure 3. Mn intensity depth profiles in 1 → 16 h annealed steel sheets (A).

Table 2. Evaluation of the enrichment of Mn in sheets A; annealing conditions : 700°C; dew point : −10°C; time *t* variable.

	Layer thickness (μm)	$\frac{\bar{C}_s}{C_c}$	Mn segregated (μm cm^{-2})	Method 2 $\frac{\bar{C}_s}{C_c}$
t = 16 h				
Total enrichment	2.0	2.0 − 1.8	5.5 '− 4.5	1.85
Surface peak	0.36	3.1 − 2.9	2.0 −1.8	
t = 8 h				
Total enrichment	1.13	1.9 − 1.8	2.6 − 2.4	1.8
Surface peak	0.31	2.8 − 2.7	1.55 − 1.45	
t = 4 h				
total enrichment	0.64	1.9 − 1.7	1.65 −1.25	1.8
t = 1 h				
Total enrichment	0.117	3.6 − 3.5	0.83 − 0.81	3.6

spectral intensities of manganese in the annealed specimen and the spectral intensities in the reference specimen for the same depth interval. So we obtain the corresponding concentrations ratio :

$$\frac{J^{\text{spec}}}{J_{\text{ref}}} = \frac{I_G^{\text{spec}}}{I_{\text{Gref}}} \times \frac{C^{\text{spec}}}{C_{\text{ref}}}$$

(the spectral intensities being normalized with regard to the entire collected light *FI*).

This method was employed (method 2 in Table 2) for calculating again the average segregations on the surface. A similarity between the results issued from both methods has been ascertained.

Sheet B

The problem is to estimate the enrichment on surface with the following elements : manganese – phosphorus – chromium – silicon – aluminium. The sheets were annealed at 700°C during times varying between 8 to 1 h. The ratio K^{Fe}/K^i for each element *i* was determined from standard specimen containing more than 95% iron, the *i* element, and others which are often negligible for the calculation. Two or three standards were used for each element *i*. The results obtained are as follows:

K^{Fe}/K^{Mn}	K^{Fe}/K^{P}	K^{Fe}/K^{Cr}	K^{Fe}/K^{Si}	K^{Fe}/K^{Al}
1.49×10^{-2}	1.43×10^{-2}	1.56×10^{-2}	4.7×10^{-2}	2.76×10^{-2}

Table 3. Evaluation of the enrichment of P, Mn, Cr, Si, Al in sheets B: annealing conditions : 700°C; dew point : −10°C; time *t* variable.

	Layer thickness (μm)	$\frac{\bar{C}_s}{C_c}$	quantity segregated μ cm^{-2}	Method 2 $\frac{\bar{C}_s}{C_c}$
t = 8 h				
P	0.66	18	0.90	18
Mn				
Surface peak	0.35	5.7	2.5	6.2
Total enrichment	1.85	2.2	3.4	2.3
Cr	0.21	7.0	0.24	
Si	3.8	1.3	0.19	
Al	0.014	8.0	0.040	4.9
t = 4 h				
P	0.46	17	0.58	16
Mn				
Total enrichment	1.25	1.95	1.7	2.0
Cr	0.13	10	0.22	
Si	3.0	1.3	0.17	
Al	0.014	16	0.090	9
t = 1 h				
P	0.16	11	0.13	8.6
Mn				
Total enrichment	0.45	1.7	0.50	1.8
Cr	0.092	8.8	0.14	
Al	0.014	15	0.080	9

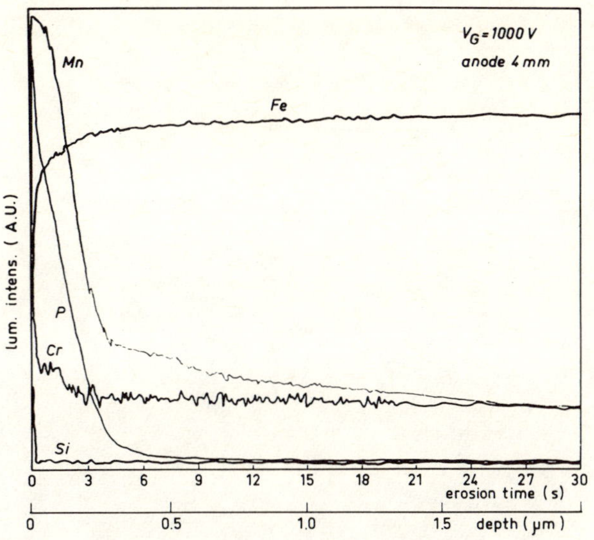

Figure 4. Intensity depth profiles of Mn, P, Cr, Si, in an eight-hour annealed steel sheet (B).

The discharge conditions are the same as previously seen except for the current which is somewhat lower. These conditions led to an erosion rate of 66 nm s^{-1}.

The spectral backgrounds of the different elements, the thicknesses of the enriched layers are estimated as for the sheets A and the results are given in Table 3. The profiles of the 8 h annealed specimen are shown in Fig. 4 for example. In this case again the enrichment corresponding to the only one peak of the manganese profile in surface and the whole enrichment were separately calculated.

The direct method (method 2) could not be used for chromium and silicon, the spectral intensities of these elements not being strong enough in the non annealed standard specimen. For the other elements it gives results close to those issued from the general method except for aluminium where the enrichment is distributed within a particularly short depth. Perhaps the reason of this must be found in the sputtering phenomenon, the sputtering steady state not having been reached yet.

At last, in the present study we were not particularly interested in the composition of the specimen at every depth under the surface, but evidently the method allows to determine it; as an example, the curves

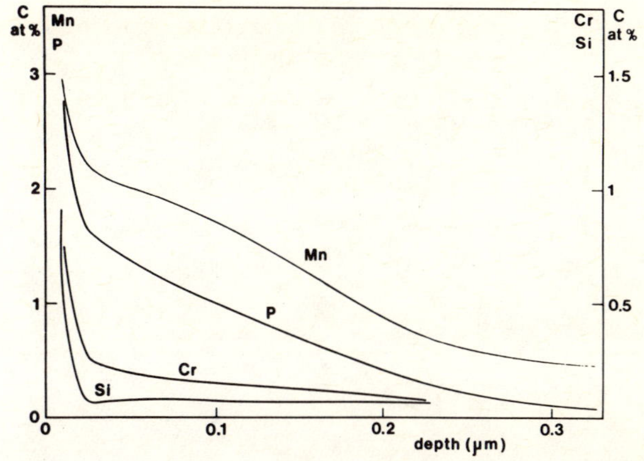

Figure 5. Concentration depth profiles of Mn, P, Cr, Si, in an eight-hour annealed steel sheet (B).

representing the concentrations of the elements versus depth have been drawn for the eight-hour annealed specimen (Fig. 5).

CONCLUSION

The studies presented hereabove concerning electrogalvanized coatings and the enrichments in annealed sheets allowed to show how the relation between the intensity of an emission line of an element and its concentration could easily be used for quantifying the GDOS surface analysis.

This method is of general interest and does not raise any particular operating problem. Of course it requires, as for all spectroscopic methods, the use of standards but we have seen through the studied examples that their choice does not prove very demanding.

At last the results obtained require for being rigorous that the steady state of sputtering be reached but this problem is inherent to all methods using ionic erosion.

REFERENCES

1. J. Pons-Corbeau, *Surf. Interface Anal.* **7**, 169 (1985).
2. J. Takadoum, J. C. Pivin, J. Pons-Corbeau, R. Berneron, J. C. Charbonnier, *Surf. Interface Anal.* **6**, 174 (1984).
3. R. Berneron, *Spectro Chimica Acta* **33B**, 665 (1978).
4. R. Berneron, J. C. Charbonnier, *Surf. Interface Anal.* **3**, 134 (1981).

SURFACE AND INTERFACE ANALYSIS, VOL. 9, 27–30 (1986)

X-ray Diffraction Investigations on Ultra-thin Gold Films

K. Häupl†, M. Lang and P. Wissmann
Institut für Physikalische und Theoretische Chemie der Universität, Egerlandstraße 3. 8520 Erlangen, FRG

The structure of epitaxially grown ultra-thin gold films is investigated by x-ray diffraction, texture analysis and LEED/Auger techniques. High sensitivity of the x-ray diffraction analyzer has been achieved by means of a step scanning unit and an impulse height discriminator. The resolving power is then sufficient for the detection of the reflections of films as thin as 1 nm. It is concluded that the particle size broadening is the limiting factor for a further extension of the investigation range towards smaller thicknesses.

INTRODUCTION

Usually a minimum thickness of several hundreds of Angstroms is postulated in the literature for the applicability of x-ray diffraction analysis for studying the structural properties of thin films.[1] In a recent paper we showed however, that this under limit can be lowered drastically if well-oriented films are investigated.[2]

The present studies are concerned with gold films epitaxially grown on Cu(111). This system is known to provide particularly monocrystal-like film structures without a disturbing amount of built-in lattice faults or grain boundaries.[3] Moreover, gold is a very suitable cover metal because it permits the x-ray diffraction measurements to be performed in air at atmospheric pressure. The thickness was decreased as much as possible in order to elucidate the structural peculiarities in the ultra-thin film region. Special emphasis was given to the evaluation of the lattice parameter.

EXPERIMENTAL

Weakly boron doped (111) oriented silicon single-crystals (Wacker, Burghausen) were used as substrates. After a suitable pretreatment (etching, glowing under UHV conditions and predoping with gold[3]) of the silicon a 10 nm thick copper film was deposited which showed a substantial (111) single-crystal structure. The gold film was then deposited by heating the evaporation source, a tungsten helix with spectrally pure gold riders fixed on it. The film thickness was recorded *in situ* with the help of a vibrating quartz thickness monitor (Kronos FTT 300). After completion of the evaporation process the sample was removed from the vacuum apparatus and put into the holder of an x-ray diffraction analyser. For reasons of clarity it should be pointed out that the normal Bragg-Brentano focusing arrangement was used instead of the particularly surface sensitive small angle scattering.[4,5]

Details of the detector system are described elsewhere,[6] the most important item being the step-scanning unit. It provides a decisive enhancement of the signal/noise ratio by counting single points of the reflection for a long time. A Co-Kα fine-focus tube served as radiation source ($\lambda = 0.17902$ nm).

Figure 1 shows a typical example of a spectrum of a 1.7 nm thick gold film. The gold and copper peaks are sufficiently separated, maxima and minima in the overlapping region can be easily traced back to the extremely smooth surface of the copper film serving as substrate.[3] For very thin films ($d \leq 5$ nm) only the (111) peaks have sufficient intensity to be detected against the back ground noise, for thicker films also the (222) peaks can be recorded. No other reflections are seen in the spectra. The evaluation of the exact angular position 2θ of the peak maxima according to Bragg's equation

$$a_{hkl} = \frac{\lambda}{2} \frac{\sqrt{h^2 + k^2 + l^2}}{\sin\theta} \tag{1}$$

allows the determination of the lattice parameters a_{hkl}. Here h, k, l are Miller's indices and λ is the wavelength of the Co-Kα mixed radiation. Lattice parameters with different indices normally differ from each other because of unavoidable adjustment errors. These errors, however, can be eliminated by plotting a_{hhl} versus $\cos^2\theta/\sin\theta$ and extrapolating to $\cos^2\theta/\sin\theta \rightarrow 0$.[6] The characteristic constant C in the equation

$$\frac{\Delta a_{hkl}}{a} = C \frac{\cos^2\theta}{\sin\theta} \tag{2}$$

is derived from calibration measurements with a bulk (111)-oriented single-crystal gold standard; in the present case we obtain $C = 1{,}2 \times 10^{-3}$. Since this value is independent of the special sample under investigation the extrapolation can be performed even for such films which do not reflect sufficient intensity for the detection of the (222) peak.

† Present address: Semikron GmbH, 8500 Nürnberg, FRG.

0142–2421/86/120027–04 $05.00

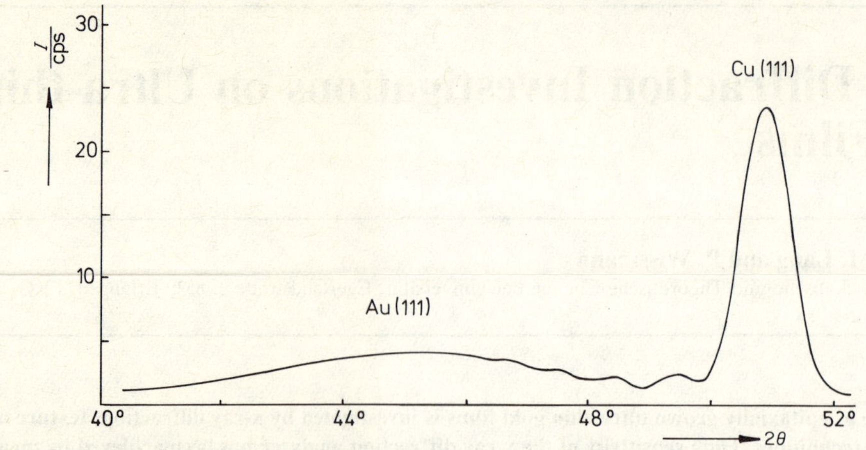

Figure 1. Typical example of the x-ray diffraction spectrum obtained for a 1.7 nm thick gold film.

X-RAY DIFFRACTION MEASUREMENTS

The analysis of the line widths and intensities leads to the conclusion that the films consist of (111)-oriented crystallites the dimensions of which perpendicular to the film plane correspond to film thickness.[6] The dimensions of the crystallites in the film plane are expected to be remarkably larger than film thickness.[7]

A plot of the lattice parameters a_{hkl} calculated according to (1) vs. $\cos^2\theta/\sin\theta$ is shown in Fig. 2. Obviously the extrapolated value initially increases and then decreases with decreasing thickness. This dependence becomes even more distinct if the relative deviation of the lattice parameter from the bulk value is plotted vs. thickness. Such a plot is presented in Fig. 3. The reason for the initial enhancement of the lattice parameter is probably a net expansion of the gold lattice plane distances due to surface effects.[8] The distinct drop of the lattice parameter for thicknesses below 5 nm seems to indicate that the gold lattice at the copper/gold interface is compressed to some extent.[9]

Interdiffusion phenomena were considered not to play an important role at room temperature.[10,11] Nevertheless we have tried to get additional information on this subject by performing LEED and Auger measurements on our films. These results are reported in the next section.

LEED AND AUGER MEASUREMENTS

The samples have been prepared in a combined LEED/Auger apparatus[12,13] keeping all parameters as comparable as possible to the evaporation conditions described above. A typical Auger spectrum obtained for a 1.9 nm thick gold overlayer is shown in Fig. 4. All peaks can be attributed to the gold while no copper components are detected. Obviously an island formation can be excluded at least at this thickness. If an Au/Cu alloy would have been formed, then the mixing area must be restricted to the direct Au/Cu interface while the gold surface is copper-free. It should be noted that no changes in the spectra occur if the thickness is further increased. Moreover, no traces of other elements like silicon or oxygen appear in the spectra.

If the thickness of the gold overlayer is lowered to about 1 nm, then indeed the copper 60 eV peak appears in the spectra. Future work must decide whether this peak has to be attributed to gold island formation[9] or to alloying. The well-known escape depth of the relatively slow 60 eV electrons does not suffice for an explanation countrary to the case of highly energetic electrons which would show a distinctly higher escape depth and hence a remarkable contribution to the copper signal even at 1 nm overlayer thickness. In each case final

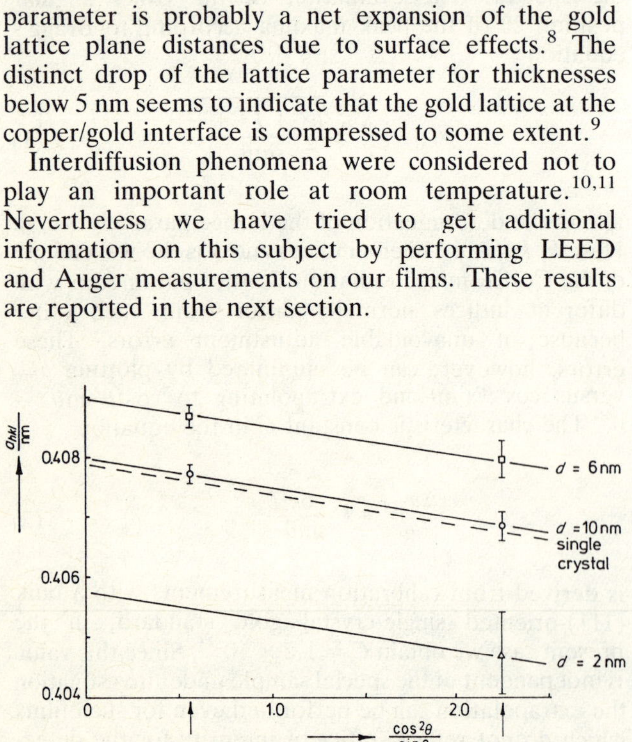

Figure 2. Plot of the lattice parameter a_{hkl} vs. $\cos^2\theta/\sin\theta$ in order to eliminate the influence of eccentric sample arrangement.

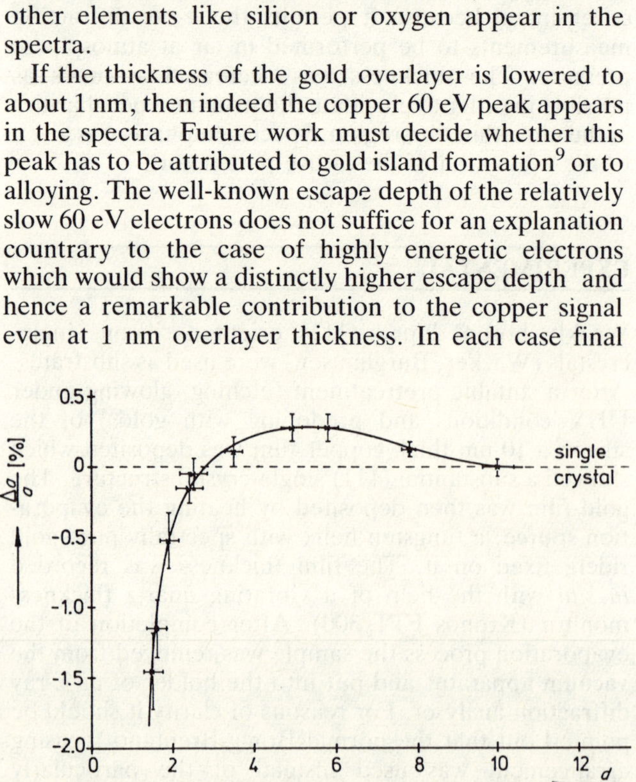

Figure 3. Relative change of the extrapolated lattice parameters as a function of gold overlayer thickness.

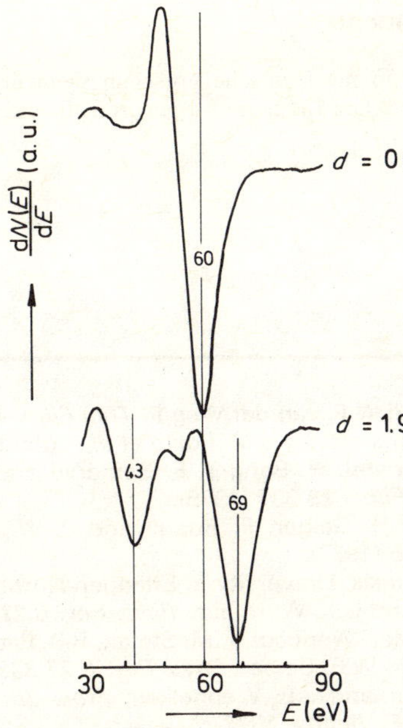

Figure 4. Auger spectra of the clean copper (111) substrate (d_{Au} = 0 nm) and of a d_{Au} = 1.9 nm thick gold overlayer.

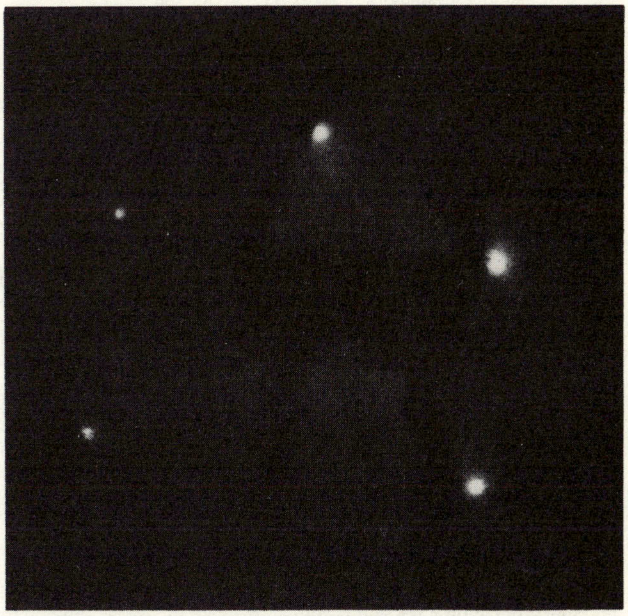

Figure 5. LEED pattern of an 8 nm thick gold overlayer (U = 110 V).

statements of the behaviour of the Au/Cu system should take care of the exact preparation conditions of the films in the ulta-thin thickness range. It would also be helpful to apply modern high-resolution XPS techniques which allow differentiation between various surface states of the gold and copper, respectively.[14.15]

The single-crystalline structure of the gold films is reflected by the LEED pattern of Fig. 5. One easily recognizes the typical sixfold symmetry of the diffraction peaks. Characteristic superstructures as reported by other authors[16,17] are missing in the present case.

TEXTURE ANALYSIS

Another direct insight into the single-crystalline structure of the gold films is provided by texture analysis. Details of the experimental procedure are described elsewhwere.[12] A typical (111) pole figure obtained for a 3 nm thick gold overlayer is shown in Fig. 6. The contours represent levels of constant intensity, β is the inclination angle of the (111) planes against the film plane, and θ is the azimuth. The profile (Fig. 6 (b)) clearly illustrates the main peak at $\beta = 0$ and the side maxima at $\beta = 70.5°$ which are due to the fact that at this angle tilted (111) planes arrive at the reflecting position. These features do not change drastically if the thickness of the film is less than 2 nm, but then the intensity of the side maxima is insufficient for a secured detection.

The pole figure provides a direct insight into the reciprocal lattice of the films similar to the LEED pattern in Fig. 5. The half width of the spots in the azimuthal direction is about 4° while the corresponding value taken in the radial direction is about 1°. Since both values are distinctly larger than the experimental resolution power (about 0.3°), the films evidently exhibit a substantial mosaic structure.[18]

CONCLUSIONS

The lower limit of the thickness necessary for the detection of x-ray diffraction data can be reduced by at least one order of magnitude if single-crystalline films are studied. The reason for this behaviour is obviously the excellent (111)-alignment of the films which allows extraordinarily high reflected radiation intensity to pass through the entrance slit of the counter tube detector. Only for film thickness smaller than 1 nm the influence of particle broadening becomes so strong that no further reasonable information can be derived from the peaks.

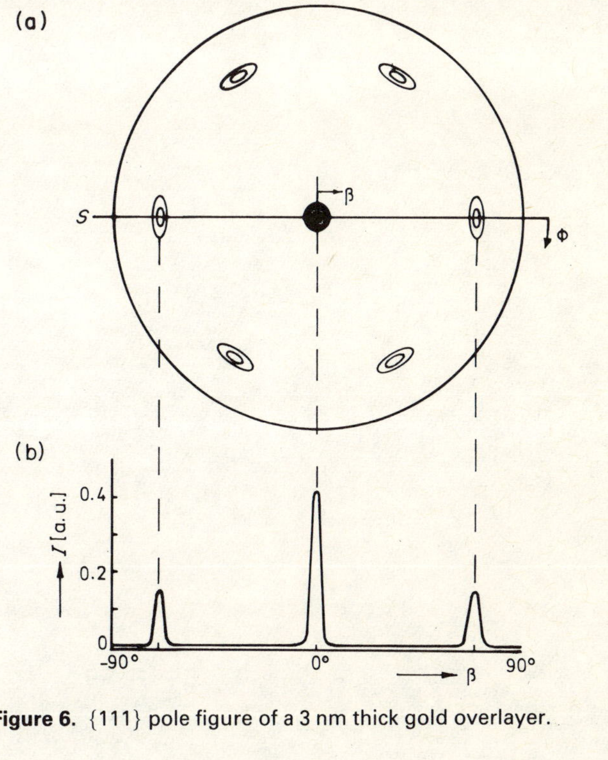

Figure 6. {111} pole figure of a 3 nm thick gold overlayer.

The sensitivity is shown to be sufficient for the unambiguous detection of a weak increase and a subsequent strong decrease of the lattice parameter of thin epitaxially grown gold films with decreasing film thickness. Similar effects were recently observed by Macur *et al.*[9] with the help of high energy electron diffraction measurements.

Acknowledgements

Thanks are due to the Deutsche Forschungsgemeinschaft and the Fonds der Chemischen Industrie for generous financial support.

REFERENCES

1. R.W. Vook in *Epitaxial Growth*, Part A, p. 339, ed. by J.W. Matthews, Academic Press (1975).
2. W. Fischer and P. Wißmann, *Appl. Surface Sci.* **11**, 109 (1982).
3. K. Häupl and P. Wißmann, *Fresenius Z. Anal. Chem.* **314** 337 (1983).
4. K. Kiessig, *Ann. Physik* **10** 51 (1931).
5. P. Croce, G. Devant, M.G. Sere and M.F. Verhaeghe, *Surf. Sci.* **22** 173 (1970).
6. W. Fishcher and P. Wißmann *Z. Naturforsch.* ***31a*** *183 (1976).*
7. H. Geiger, K. Häupl, P. Wißmann and E. Wittmann, *Vakuum-Technik* **34** 135 (1985).
8. R.W. Vook, S. Ouyang and M.A. Otooni, *Surf. Sci.* **29** 277 (1972).
9. J.E. Macur and R.W. Vook, *Thin Solid Films* **66** 371 (1980).
10. S.S. Lau and W.F. van der Weg in *Thin Films-Interdiffusion and Reactions,* ed. by J.M. Poate *et al.,* John Wiley (1978).
11. A. Wagendristel, H. Banger, E. Semerad and P. Skalicky, *Thin Solid Films* **28** 337 (1975).
12. W. Fischer, H. Geiger, P. Rudolf and P. Wiβmann, *Appl. Phys.* **13** 245 (1977).
13. M. Lang, Thesis, University of Erlangen-Nürnberg (1986).
14. P.H. Citrin and G.K. Wertheim, *Phys. Rev. B* **27** 3176 (1983).
15. D.L. Weissman-Wenocur, P.M. Stefan, B.B. Pate, M.L. Shek, I. Lindau and W.E. Spicer, *Phys. Rev. B* **27** 3303 (1983).
16. D.M. Zehner and J.F. Wendelken, *Proc. 7th Intern. Vac. Congr.* p.517, Vienna, 1977.
17. H. Melle and E. Menzel, *Z. Naturforsch.* **33a** 282 (1978).
18. H.G. van Bueren, *Imperfections in Crystals,* pp. 322 (1961).

SURFACE AND INTERFACE ANALYSIS, VOL. 9, 31–40 (1986)

The Application of Laser Ionization Mass Spectrometry to the Study of Thin Films and Near-surface Layers

N S Clarke, J C Ruckman, A R Davey
MOD (PE), Metallurgy Division, AWRE Aldermaston, Reading, UK

The laser ionization mass analyser, LIMA, was originally developed as a bulk microprobe instrument giving elemental information similar to that obtained from the conventional electron probe microanalyser, but with the advantage of full periodic table coverage. The LIMA is therefore capable of detecting all elements including hydrogen, helium, lithium, and beryllium, which are inaccessible to EPMA. Other advantages of LIMA over EPMA are isotope discrimination and the ability to obtain chemical information from materials as a result of the emission of complex ionic fragments from surfaces under laser irradiation. As experience grew with the use of the prototype LIMA at Aldermaston, it became clear that the technique was a particularly versatile tool which could not only perform well as a bulk microprobe, but with suitable experimental procedures could also be used as a sensitive depth profiling instrument with moderate depth resolution (>0.01 μm). It became apparent that in its performance parameters, and in many of its applications in bulk and surface analysis, LIMA was directly comparable with dynamic SIMS. This paper is, therefore, concerned with the applications of LIMA when operated in a surface sensitive mode. In general, this involves using low power densities, and a defocused laser — the defocusing means that some sacrifice in lateral resolution must be made. Nonetheless, it is shown in this paper that the LIMA has a genuine contribution to make in depth profiling and thin film analysis.

INTRODUCTION

The aim of this paper is a description of the technique of laser ionization mass spectrometry applied to solids and solid surfaces. Before proceeding further it will be profitable to define certain terms used frequently in this branch of science. LIMA (laser ionization mass analyser) and LAMMA (laser microprobe mass analyser) are trade names of instruments manufactured by Cambridge Mass Spectrometry Limited[1] and Leybold Heraeus Gmbh[2] respectively. These instruments embody a technique which may be termed 'laser-solid ionization mass spectrometry' which employs high laser power densities to produce high temperature microplasmas from bulk or thin solids resulting in significant numbers of atomic ions and extensive fragmentation of molecules.[3] This technique should be recognised as being distinct from all others grouped under the generic term 'laser ionization mass spectrometry' for example the gas phase resonant ionization mass spectrometry (RIMS).[4]

The so called 'laser desorption mass spectrometry' (LDMS) involves low laser power densities and the evaporation of the top few atomic or molecular layers with subsequent ionization by purely thermal mechanisms with minimal molecular fragmentation.[5,6] It might be prudent to recognise this latter technique as a laser induced *evaporation* process as opposed to the true *desorption* of surface species. It is hoped to make this distinction clear in this paper. It is useful to note that both LIMA and LAMMA instruments can be made to operate in either the 'hard' laser ionization mode or the 'soft' laser desorption mode by suitable adjustment of the laser focus.[7]

In this paper, only a brief description of the prototype LIMA[8] will be given, the experimental methods employed in thin film, and depth profile analyses will be described in greater detail together with some applications and results using LIMA in the study of chemisorption will be presented.

A semiquantitative comparison of LIMA with other surface sensitive and surface specific techniques is shown in Table 1 and in Fig. 1 (Ref. 9). We consider that the information available from LIMA bridges the gap between the bulk elemental analysis generated by EPMA on the one hand, and the surface profiling/bulk information given by dynamic SIMS on the other, with of course the advantages that any mass spectrometric technique has over an electron probe i.e. isotopic analysis and full periodic table coverage.

EXPERIMENTAL

The design and construction of the AWRE LIMA has been discussed extensively in the literature[10,11] and consequently only a cursory description will be presented here. Schematic diagrams of the instrument are shown in Figs 2(a) and (b). The basic principle of operation is that a finely focused pulsed laser beam is used to generate ions from a bulk target situated in a UHV chamber. The resultant ions are separated by Time-of-Flight (TOF) mass spectrometry and detected by an electron multiplier. The detector signal is digitized by a fast waveform analyser and the raw data manipulated by desk-top computer.

0142–2421/86/120031–10 $05.00

Table 1. Depth profile characteristics of surface and near surface physico chemical analysis techniques

	LIMA		Dynamic SIMS	Auger	ESCA
Depth profiling mode	'Self' photon probe		'Self' ion probe	Auxillary ion etch	Auxillary ion etch
Profiling rate	0.1 – 1.0 μm per shot		50 Å sec^{-1}	10 Å. min^{-1}	10 Å. min^{-1}
Lateral	10μm	interdependent	0.5 μm	500 Å	> 1mm
Depth resolutions	0.1–1.0 μm		1 nm	2 nm	2 nm
Detection limit	1 ppm atomic		20 ppb atomic	1000 ppm atomic	5000 ppm atomic
Element range	H–U 0<m/z<1000		H–U 0<m/z<1000	Be–U	Li–U
Additional information	Isotopic chemical (imaging)		Isotopic chemical imaging	Imaging	Chemical (imaging)

Notes

The profiling rates quoted here are merely order of magnitude values intended to establish relative profiling capabilities of the various techniques. It should be noted that compositional and topographical alterations occur during ion etching, static SIMS and dynamic SIMS experiments due to removal of material, collisional mixing, and differential elemental sputter yields, resulting in artefactual analytical information. During pulsed laser irradiation, relative compositional changes of the sample should be much less than during ion bombardment because primarily there is no collisional mixing problem, i.e. the photon probe is inert. The spatial extent of any chemical alteration by melting, thermal diffusion or laser induced shock wave process is small compared with the overall crater dimensions because the probe time is so short (20 ns). It is however conceivable that trace impurities may be concentrated in the melt zone around the crater. Most importantly the laser does produce gross, on a micron scale, surface deformation i.e. craters which can affect the optical absorption properties of the sample at the irradiation point. This is why LIMA can only be described as 'quasi non destructive'.

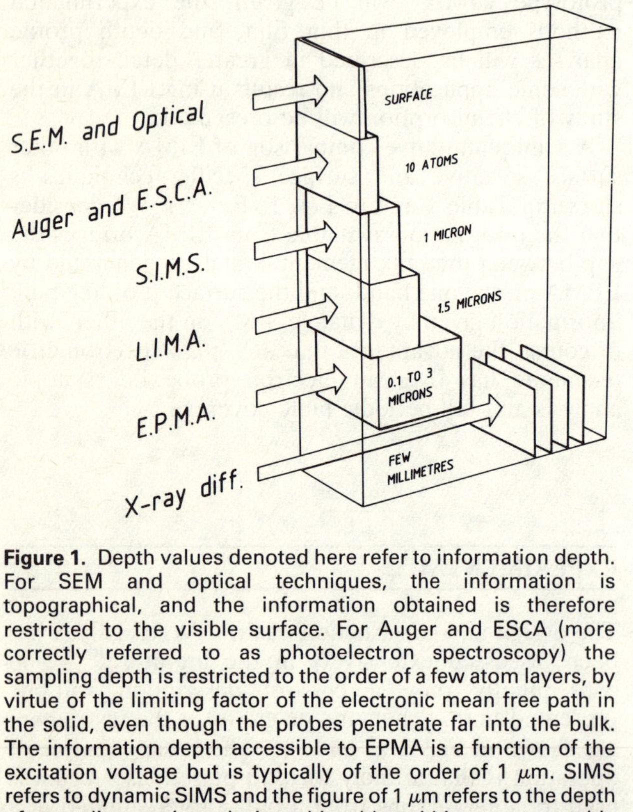

Figure 1. Depth values denoted here refer to information depth. For SEM and optical techniques, the information is topographical, and the information obtained is therefore restricted to the visible surface. For Auger and ESCA (more correctly referred to as photoelectron spectroscopy) the sampling depth is restricted to the order of a few atom layers, by virtue of the limiting factor of the electronic mean free path in the solid, even though the probes penetrate far into the bulk. The information depth accessible to EPMA is a function of the excitation voltage but is typically of the order of 1 μm. SIMS refers to dynamic SIMS and the figure of 1 μm refers to the depth of sampling and analysis achievable within an acceptable experimental period, say 1 h, using a typical erosion rate of, say, 3 Å s^{-1}. The figure of 1.5 μm quoted for LIMA refers to the depth of a typical single shot crater made using a focused high power (10^9 W) laser pulse. The 'few millimetres' quoted for x-ray diffraction serves to illustrate that this latter is a true bulk technique.

Depth Profile and Thin Film Analyses

LIMA depth profiling involves irradiation of a given sample spot with successive laser pulses. In order to establish a strategy for depth analysis for a given system, account must be taken not only of the analytical requirements, but also the limitations of the LIMA technique. Such considerations will include the shot-to-shot reproducibility of the laser output, the minimum detection limit of the species of interest, the depth resolution and the total depth of analysis, as well as matrix effects. One significant factor which may adversely affect analyses is the instability of the sample stage (VG HPT) and the concomitant difficulty in focusing the laser reliably onto the same spot each time. Tests have shown that our sample stage is adversely affected by vibration and mechanical shock, but measures to damp the resultant stage oscillations are to be taken.

The laser focus *per se* can affect the absolute sensitivity of the analysis because of the different power densities available with a given laser energy output at different degrees of focus. For example, a fully focused laser pulse will have a high power density because of the small area into which the energy is deposited. However, at the same nominal energy output, a defocused pulse will deposit less power per unit area because the spot diameter will be greater. The highly focused pulse will be able to vaporize matrices with high specific heats (C_p) and enthalpies of evaporation (ΔH_{vap}) and will be able to ionize elements with high ionization energies. The defocused pulse will, however only vaporize those solids with low C_p and ΔH_{vap}

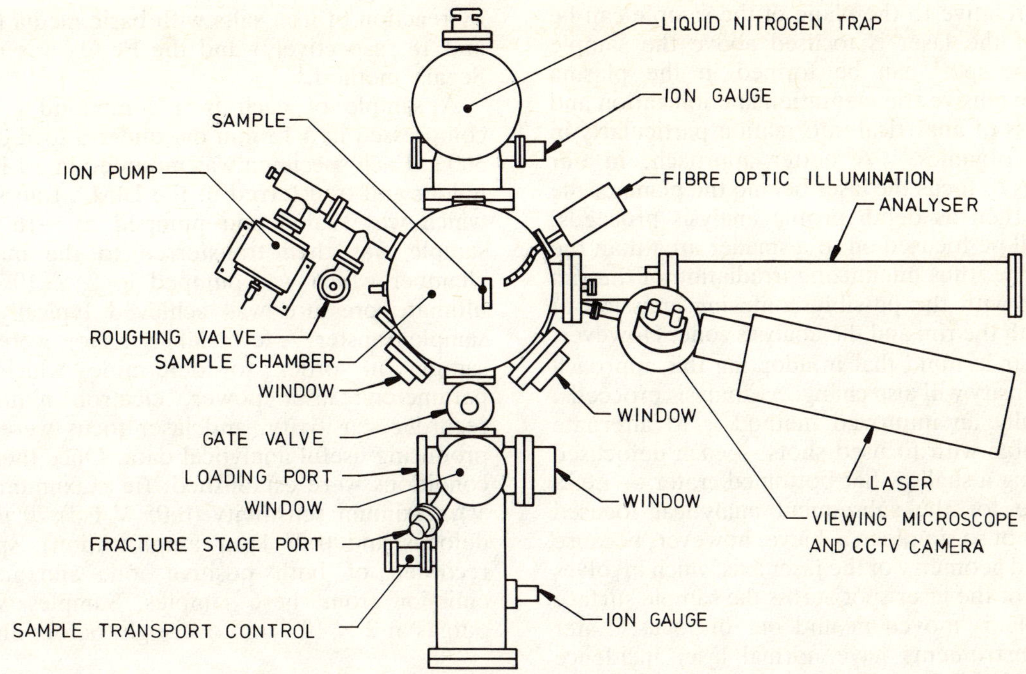

Figure 2 (a). Schematic plan view of Lima.

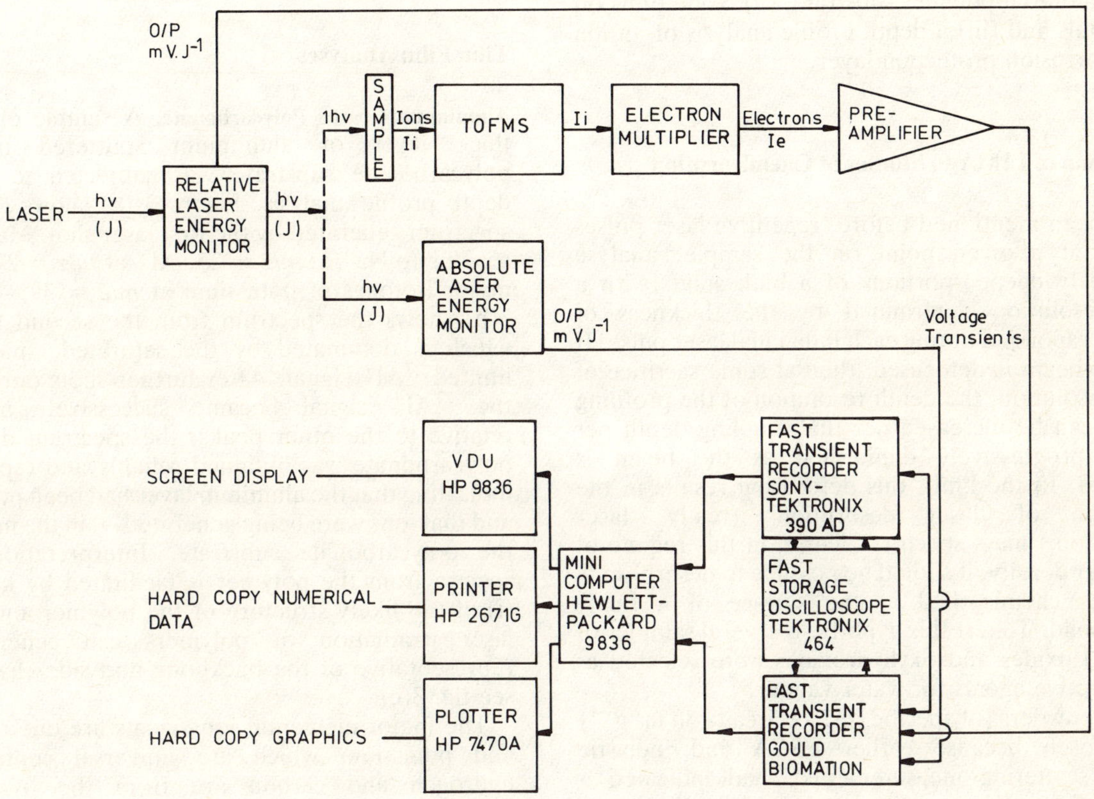

Figure 2 (b). Block diagram AWRE Lima.

values, and will only ionize those species with low ionization energies. Sensitivity, particularly for low ionization energy species at low concentration may possibly be achieved using the defocused laser beam, because this will irradiate a relatively large area of the sample with sufficient energy to excite and ionize the species of interest. The foregoing discussion assumes a constant total laser energy output for both focused and defocused cases. One notable point concerning the degree of laser focus is that in general, the more finely focused the beam, the deeper the penetration per shot, and *vice versa*.

The degree of laser focus, and consequently the power density change as a depth profile analysis proceeds; this is a direct result of the 'conical' focus arrangement. Consequently the position of the laser focal plane, relative to the plane of the sample can be important. If the laser is focused above the sample plane, a 'hot spot' can be formed in the plasma resulting in extensive fragmentation and ionization and possibly a loss of analytical information particularly in the case of organics.[11] A better approach, in our experience, is to focus the laser behind the plane of the sample and then as depth profile analysis proceeds, each shot will be focused on to a smaller area than the proceeding one, thus minimizing irradiation of the rim of the crater with the possibly confusing emission of ions from both the rim and the analysis zone. However, one must bear in mind that in adopting this approach the power density will also change as analysis proceeds.

Theoretically an improved method is to alternate defocused shots with focused shots — each defocused shot producing a shallow flat-bottomed crater — i.e. a clean surface for the subsequent analytical focused shot. This is not so simple to achieve, however, because of the inclined geometry of the laser axis which involves a translation of the laser spot across the sample surface as the sample is moved in and out of focus. Later generation instruments have normal laser incidence and do not suffer from this limitation.

The results[12] presented here will be concerned with (i) a depth profile analysis of a thin metal layer sputtered onto a polymer substrate, (ii) oxide films on bulk metals and (iii) a depth profile analysis of an ion plated corrosion protection layer.

Application of LIMA to Studies of Chemisorption

As has been mentioned before, repetitive laser pulses directed at a given point on the sample, analyse successively deeper portions of a bulk solid, with a depth resolution determined by the thickness of material vaporized during each individual laser pulse. If the laser beam is defocused, then at some sacrifice of lateral resolution, the depth resolution of the profiling analysis can be increased i.e. the sampling depth per shot is progressively diminished as the beam is defocused. In the limit, this defocusing results in the generation of 'laser desorption' (really 'laser evaporation') mass spectra. Clearly, in this regime of low power density, it should be possible to desorb small molecules chemisorbed at the surface of a finely divided solid. To test this hypothesis several samples of iron (III) oxides and oxyhydroxides were selected as chemisorptive agents for water vapour.

These powders (of specific surface area $> 30\ m^2\ g^{-1}$) were chosen because earlier ESCA and inelastic neutron scattering measurements[13] had indicated a residual, finite concentration of chemisorbed water at the surface of the oxides even after prolonged outgassing at room temperature, whereas infra red measurements had indicated the complete removal of molecular water by room temperature evacuation.[14] It was hoped that LIMA could help to resolve the dilemma.

The samples chosen for study were; α–Fe_2O_3, γ–Fe_2O_3, α–FeOOH, β–FeOOH, γ–FeOOH, and Fe_2O_3 was prepared by thermal decomposition of α FeOOH[15] the $\alpha-$, $\beta-$, and $\gamma-$ FeOOH were prepared by reaction of iron salts with basic media (Refs 16, 17 and 18 respectively) and the Fe_3O_4 was prepared by Segal's method.[19]

A sample of each iron compound ($\sim$ 5 g) was compressed in a 13 mm die under a load of 80 kN for 300 s. Each specimen was mounted in a LIMA sample holder and transferred to the LIMA transfer chamber which was sealed and pumped to $\sim 16^{-6}$ mb. The sample was then transferred to the main analysis chamber which was pumped to 2×10^{-9} mb. This ultimate pressure was achieved typically 1 h after sample transfer. A few trial laser shots were fired at the sample in order to determine which operating parameters: laser power, electron multiplier gain, recorder sensitivity, and laser focus were optimal in producing useful analytical data. Once these optimum conditions were established, (ie maximum gain (3500 V),maximum sensitivity (0.05 V fsd) 20 μm negative defocus and 0 V laser amplification), spectra were recorded of both positive ion, and negative ion emission from these samples. Samples were left to outgas at 2×10^{-9} mb overnight before analysis.

RESULTS AND DISCUSSION

Thin Film Analyses

Aluminium Coated Polycarbonate. A sample of a 12 nm thick layer of aluminium sputtered on to a polycarbonate substrate was subjected to a LIMA depth profile analysis. Figure 3 (a) shows the LIMA spectrum generated by the first laser shot. Major peaks are due to Na^+ at $m/z = 23$, Al^+ at $m/z = 27$, and the major isotopes of potassium at $m/e = 39, 41$. Figure 3(b) shows the spectrum from the second laser shot which is dominated by the saturated, space-charge limited $^{27}Al^+$ signal. After further shots during which the $^{27}Al^+$ signal became successively attenuated relative to the other peaks, the spectrum due to the polycarbonate was obtained reliably and reproducibly indicating that the aluminim layer had been penetrated, and that ions were being generated — in the main, from the polycarbonate substrate. Interpretation of the spectra from the polymer is facilitated by knowledge (i) of the likely structure of the polymer and (ii) that laser irradiation of polymers can generate ions representative of the backbone and side chains[20,21,22] see fig. 3(c).

The major elemental ion signals are due to sodium and potassium which are universal contaminants, hydrogen and carbon are from the pyrolysis of hydrocarbon and/or polymer, the residual aluminium signal will originate from irradiation of the 'lip' of the laser crater. The remaining peaks can generally be assigned to molecular ions such as carbon clusters, saturated alkyl cations ($C_nH_{2n+1}^+$), other hydrocarbon fragments, and fragments from the polymer backbone and side chains.

Uranium Oxide Films This example describes some preliminary results from the analysis of oxide films on uranium in particular, and forms part of a more

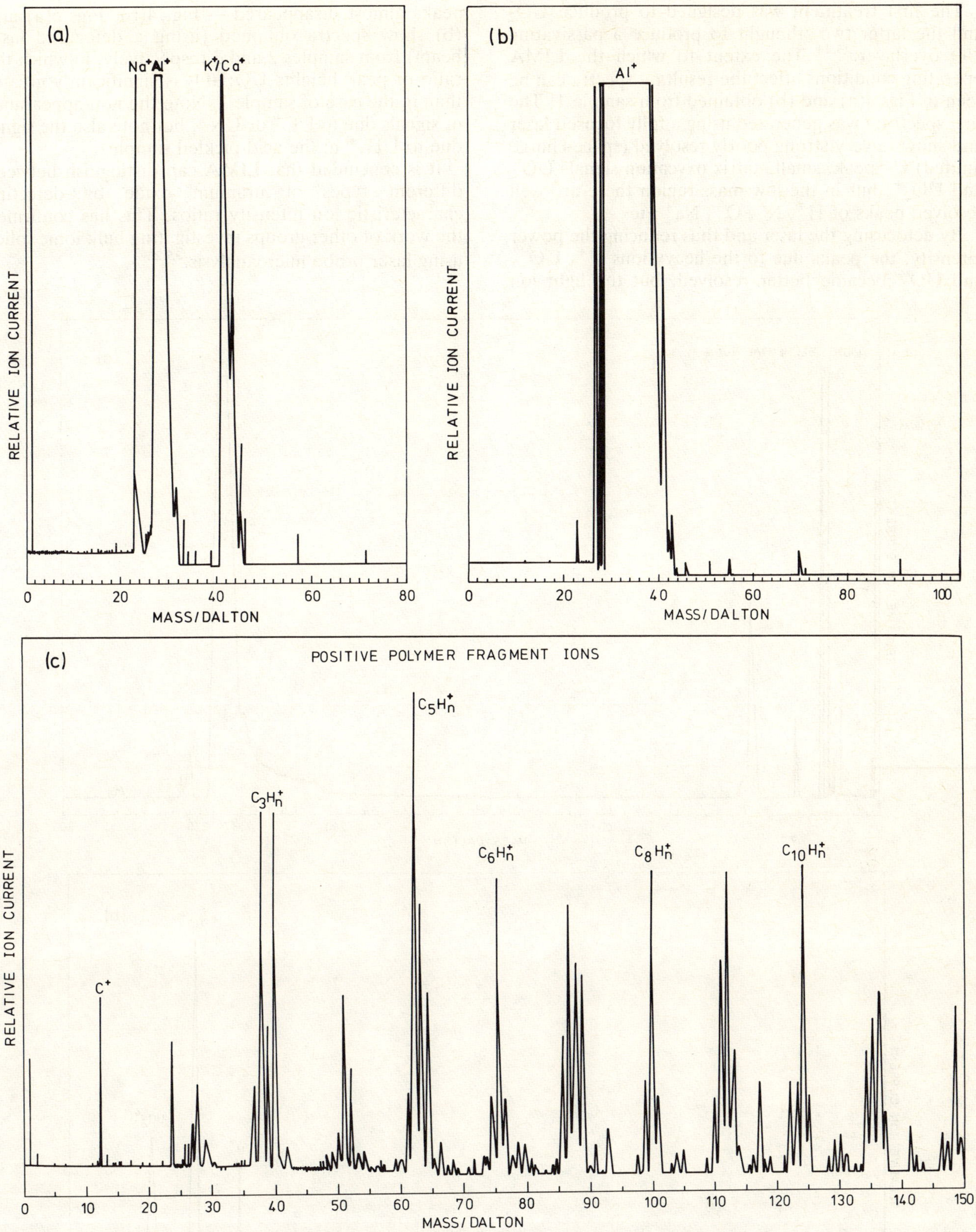

Figure 3. (a) Aluminium sputtered polycarbonate surface contamination.
(b) Aluminium sputtered polycarbonate aluminium overlayer 12 nm thick.
(c) Polycarbonate substrate.

extensive programme of work to investigate the characteristic fragmentation of different bulk and surface oxides under laser irradiation.

Three 12 mm × 12 mm coupons of as-machined depleted uranium were given different oxidizing treatments as follows:

1. 4 days at 80°C, 75% H_2O(g) in air.
2. 1 hour at 1000°C *in vacuo*.
3. Acid pickled, 50% HNO_3, 50°C followed by 1 h at 1000°C *in vacuo*.

The first treatment was designed to produce UO_2 and the latter two, thought to produce a passivating UO overlayer.[23,24] The extent to which the LIMA operating conditions affect the resultant spectra can be seen in Figs 4(a) and (b) obtained from sample 1. The first spectrum was generated using a fully focused laser and shows a very strong poorly resolved (space-charge limited) U^+ peak, small matrix oxycation signals UO^+ and ${UO_2}^+$, but in the low mass region there are well resolved peaks of H^+, N^+, O^+, Na^+ etc.

By defocusing the laser and thus reducing the power intensity, the peaks due to the heavy ions U^+, UO^+, and UO_2^+ became better resolved, but the light ion peaks almost disappeared — Fig. 4(b). Figs 5(a) and (b) show spectra obtained (using a defocused laser beam) from samples 2 and 3 respectively, in which the ratio of peak heights $UO:UO_2$ is significantly greater than in the case of sample 1. Note the non-appearance of signals due to UC^+ or UN^+, but note also the signal due to ${UH_3}^+$ in the acid pickled sample.

It is concluded that LIMA can distinguish between different types of uranium oxide by detecting characteristic ion intensity ratios. This has confirmed the work of other groups investigating bulk ionic solids using laser probe microanalysis.[25,26]

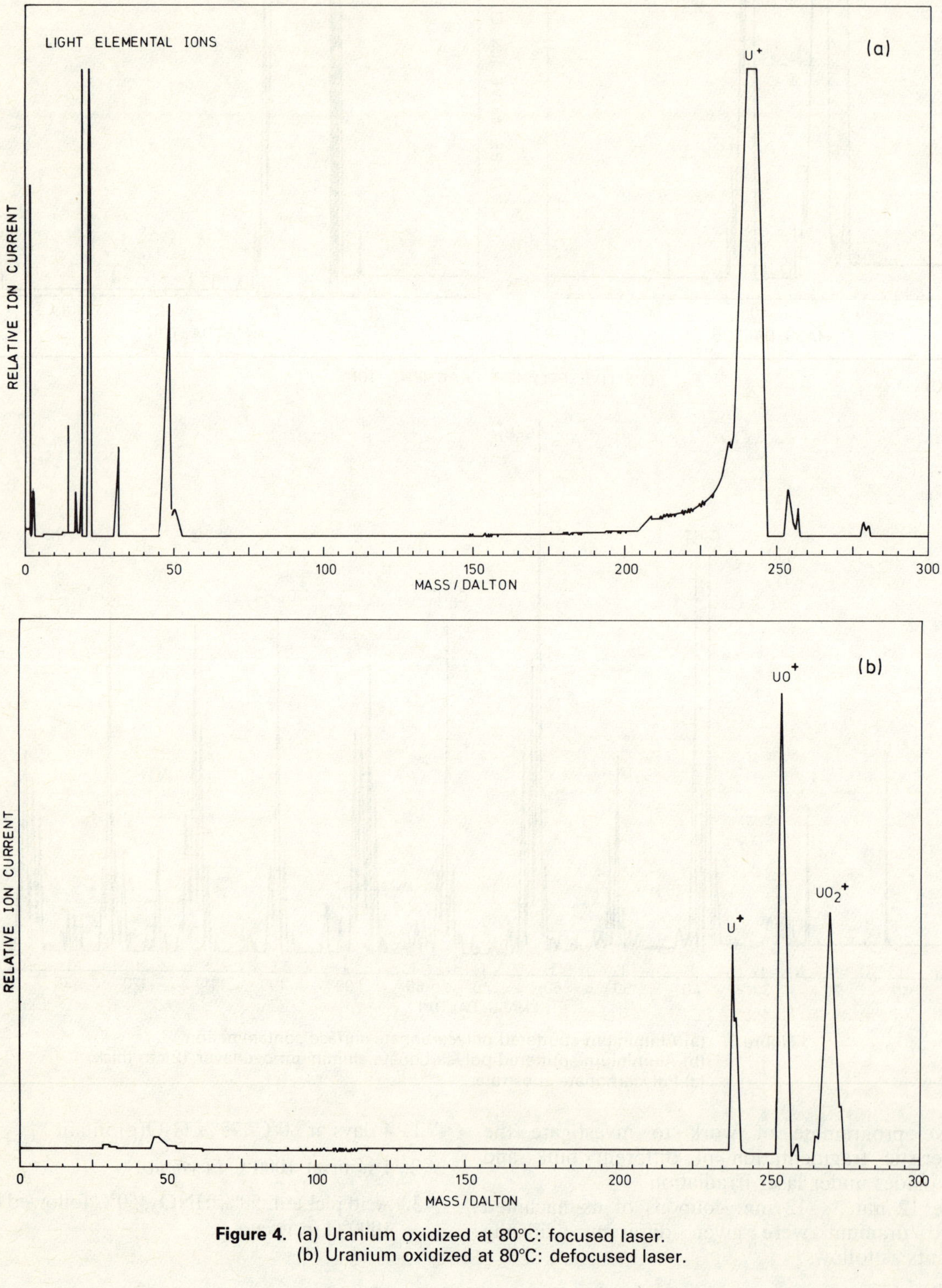

Figure 4. (a) Uranium oxidized at 80°C: focused laser.
(b) Uranium oxidized at 80°C: defocused laser.

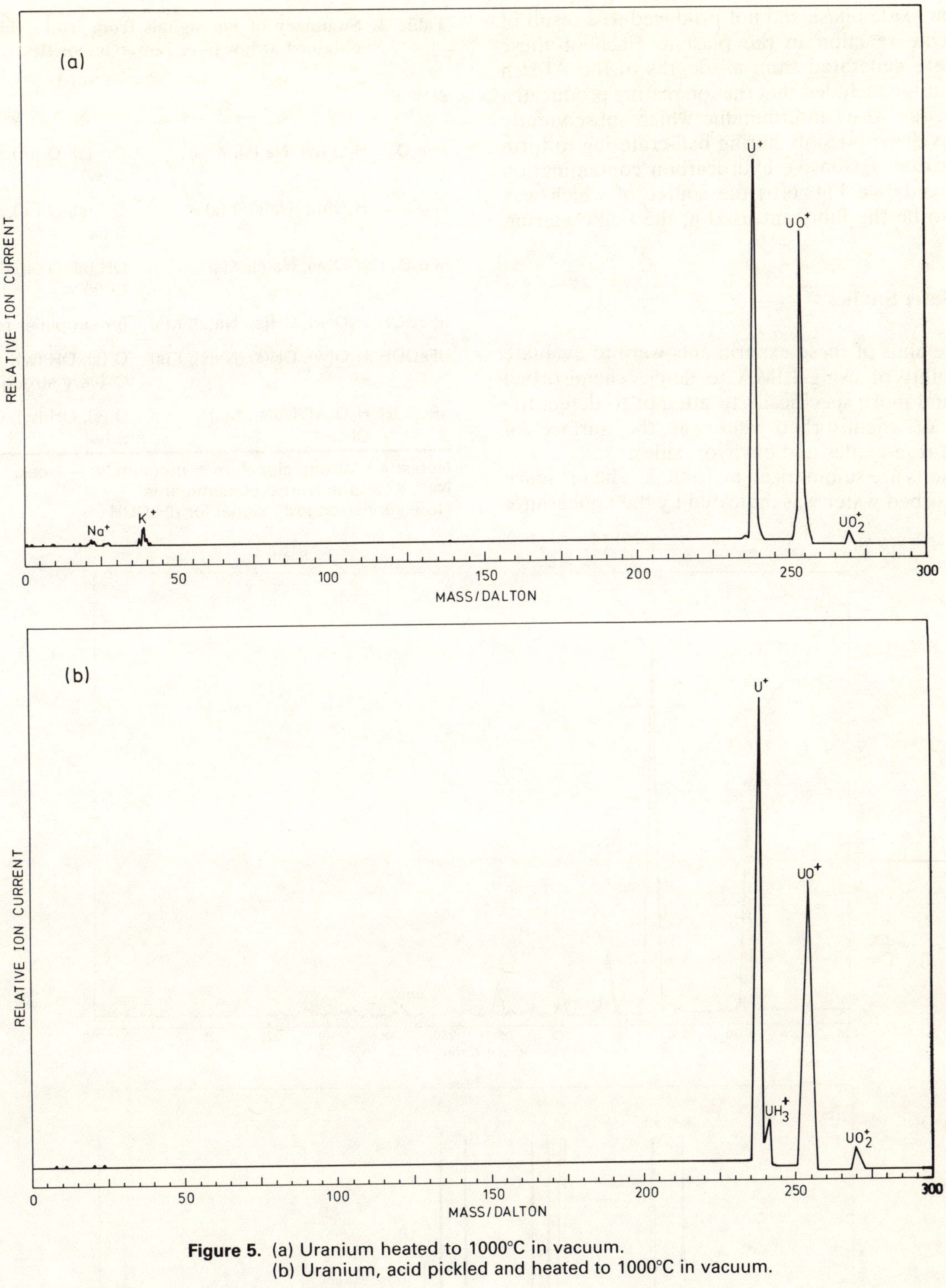

Figure 5. (a) Uranium heated to 1000°C in vacuum.
(b) Uranium, acid pickled and heated to 1000°C in vacuum.

Aluminium Ion Plated Uranium In order to facilitate depth profile analyses of films more than a few tens of μm in thickness, where 'self' profiling techniques lose resolution and/or sensitivity and where ion etching would take inordinately long, it is necessary to resort to some mechanical means of abrasion to expose deep sub-surface layers. Such methods include sectioning, taper sectioning and ball cratering.[27,28] We have used LIMA in conjunction with the latter technique to investigate interphases in a sample of aluminium ion plated uranium. The nominal thickness of the aluminium layer was 30 μm.[29] A ball-crater approximately 1200 μm in diameter and 6 μm in depth[28] was produced in this aluminium layer, LIMA analyses were taken across this crater, a typical mass spectrum is shown in Fig 6(a), and shots from the whole region showed signals due to Al^+ (not shown in the figure) and U^+ as would be expected. Signals due to matrix oxycations $Al_xO_y{}^+$ (not shown in the figure) and $U_xO_y{}^+$ (x=0,1; y = 1,2) were also seen in addition to UAl O^+ and $UAlO_2{}^+$. These latter two signals we assume to be generated from a mixed uranium-aluminium oxide phase and not produced as a result of ion-molecule reaction in the plasma. Each of these signals were generated from all depths of the Al-rich layer and it is concluded that the ion plating produced a homogeneous Al–U intermetallic which subsequently became oxidized possibly during ball-cratering to form a mixed oxide. Extensive hydrocarbon contamination was observed, see Fig. 6(b) the source of which was believed to be the lubricant used in the ball cratering process.

Sorbed Water Studies

The prime aims of these experiments were to evaluate the feasibility of using LIMA to detect chemisorbed species, and more specifically to attempt to detect the presence of chemisorbed water at the surface of outgassed iron oxides and oxyhydroxides.

The results are summarized in Table 2. The presence of chemisorbed water was indicated by the appearance

Table 2. Summary of ion signals from iron oxide surfaces obtained at low laser power intensities.

SAMPLE	IONS DETECTED +ve	IONS DETECTED −ve
αFe_2O_3	H_2O (w), Na (s), K (s)	OH (s), O (m), Cl (w), S (w)
Fe_3O_4	H_2O(m), Na(s), K(s)	OH (s), O (m), Cl (s), S (w)
γFe_2O_3	H_2O(w), Na(s), K(s)	OH (s), O (s), S (w), Cl (s)
αFeOOH	H_2O(w), OH(s), Na(s), K(s)	(preamplifier failed)
βFeOOH	H_2O(w), OH(s), Na(s), K(s)	O (s), OH (w), Cl (very strong)
γFeOOH	H_2O(w), Na(s), K(s), OH(m)	O (s), OH (w), Cl (s), S (w)

Notes: s = strong signal, m = medium, w = weak.
Na^+, K^+ are universal contaminants.
Note very strong Cl^- signal for βFeOOH

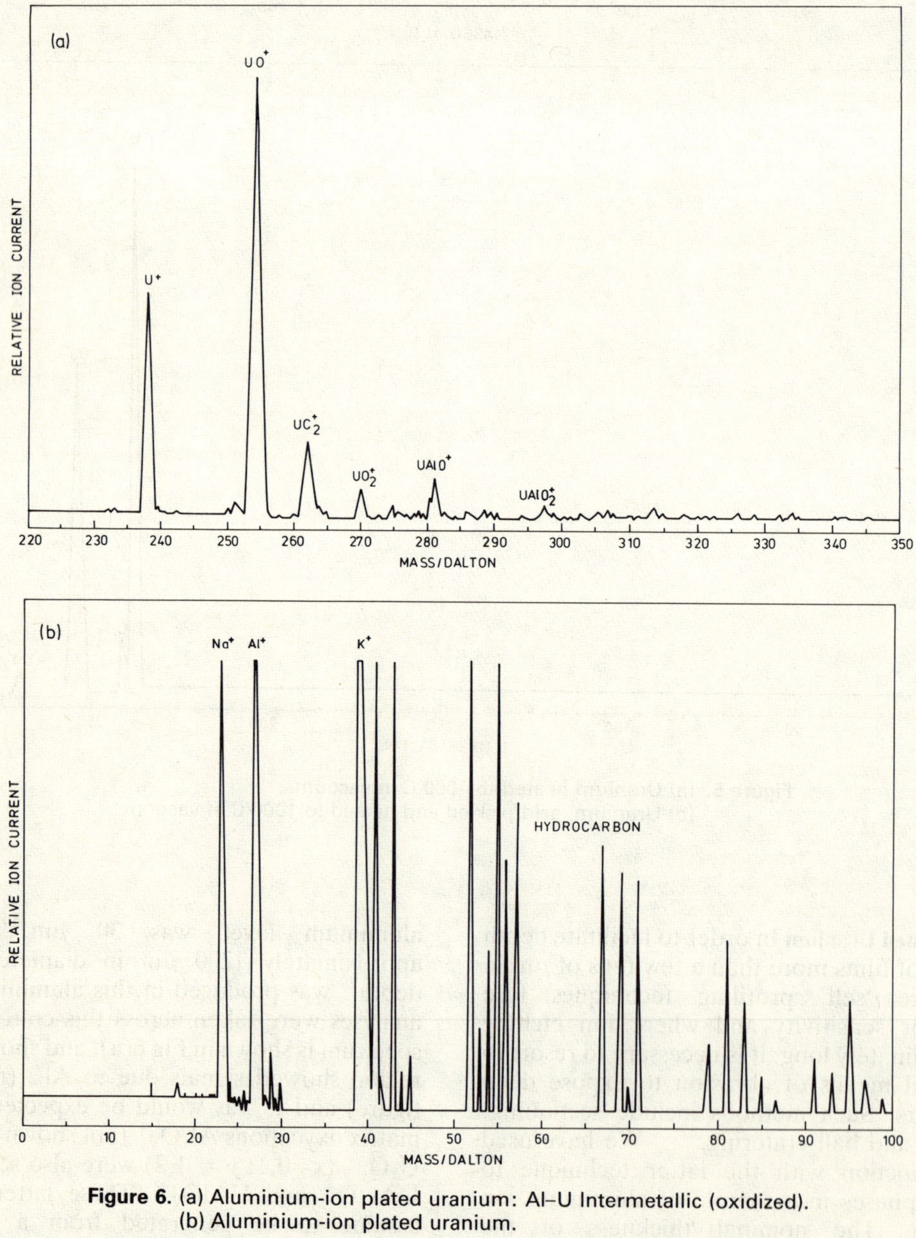

Figure 6. (a) Aluminium-ion plated uranium: Al–U Intermetallic (oxidized). (b) Aluminium-ion plated uranium.

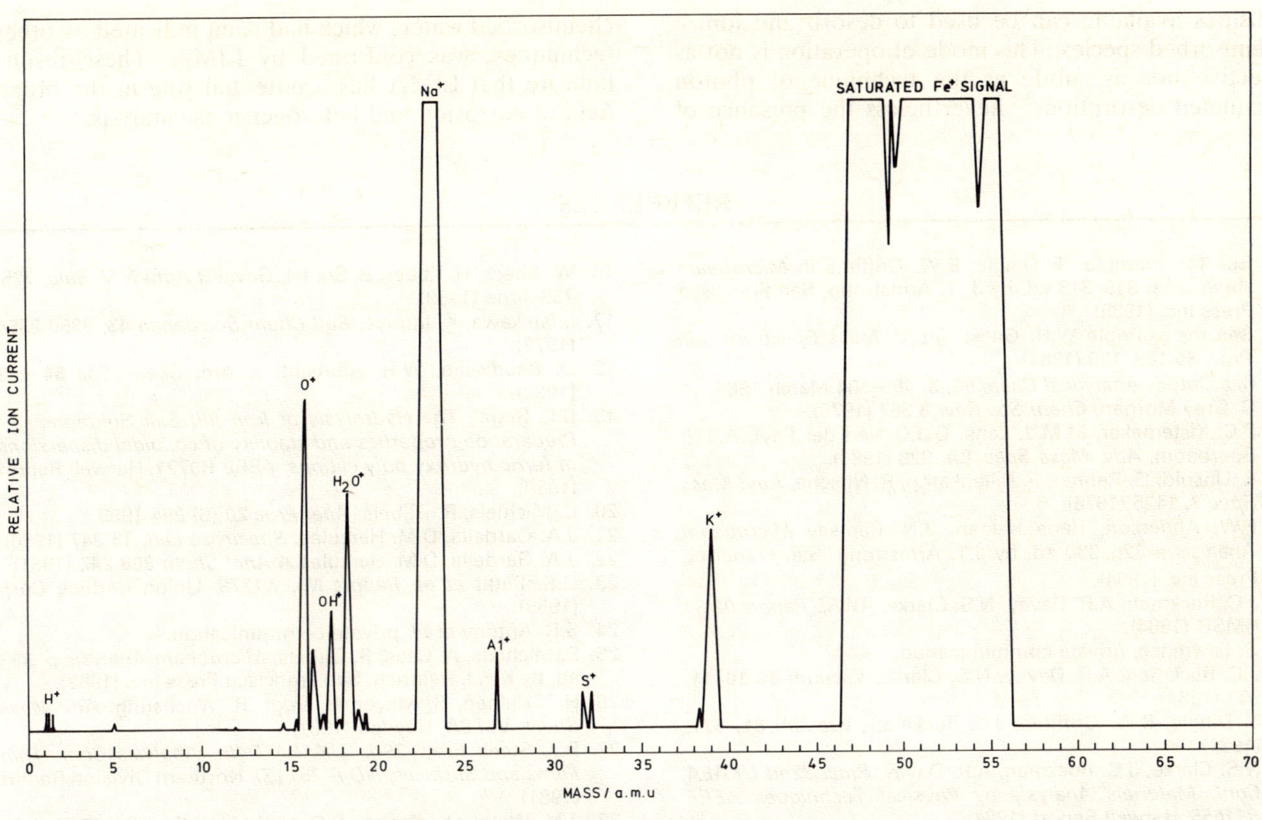

Figure 7. Positive ion spectrum from magnetite.

of the H_2O^+ plasma phase cation signal at m/z = 18 Dalton, which was obtained reliably and reproducibly at low power densities. A typical mass spectrum is shown in Fig. 7.

The mass range 0 to 60 Dalton was deemed to be the most important since information on the sorbed surface species was unlikely to be contained in the region of the mass spectrum $m/z < 60$ Dalton where the complex 'lattice' ions are detected.[20]

The table shows the intensities of the positive and negative ions signals obtained by laser irradiation of these iron-oxygen compounds. The signal intensity scale (I) has been arbitrarily divided into three: weak (0–30% fsd), medium (30–60% fsd) and strong (60–100% fsd). In all cases the Fe^+ signal was saturated reflecting the high gain setting of the electron multiplier and the high sensitivity setting of the transient recorder.

The ion signal due to molecular water was only obtained in the positive ion detection mode. In all cases this signal was only obtained at weak or moderate intensity. The intensity of this signal is correlated with the specific surface area of the oxides since for α, γ Fe_2O_3 and α, β, γ FeOOHS ~ 30 m^2 g^{-1}, I = 0–30% fsd, however for Fe_3O_4, S = 70 m^2g^{-1}, I = 30–60% fsd.

The OH^+ signal was only obtained for the oxhydroxides but was however, detected at moderate to strong intensity. This perhaps reflects the structure and stoichiometry of these compounds with −OH coordinated to the central Fe, and linked together to various extents by hydrogen bonds.

OH^- was obtained for all samples in moderate or strong intensities. The presence of this ion merely reflects the method of preparation of these materials. A significant observation is the detection of a strong Cl^- signal from β FeOOH. The chloride ion is thought to stabilize the open tunnel-like structure of β FeOOH.

CONCLUSIONS

Although LIMA was not designed to be a surface sensitive instrument, it can be used in thin film applications. Thin films have been investigated and these have ranged in thickness from a few tens of μm to ~0.01 μm. The major advantage of LIMA in this thickness range is its rapidity of analysis compared with techniques using auxiliary ion etching and the 'self' profiling ability of dynamic SIMS.

This paper has described three discrete varieties of sample which have been subjected to LIMA depth profiling analyses. Two types of uranium oxide film have been analysed. Semiquantitative discrimination could be made between the two on the basis of the different relative intensities of the UO^+ and UO_2^+ signals from each sample. In an aluminium ion-plated uranium specimen, evidence was found for a 30 μm thick Al–U intermetallic overlayer which had become oxidezed during ball cratering. Results from an aluminium layer sputtered on to a polycarbonate substrate were interesting, because they showed that LIMA is capable of analysing films as thin as 12 nm and also the complex spectra from polymers can be interpreted in terms of characteristic degradations of the polymer backbone and fission of the side chains.

An area of analysis which appears promising is that of true laser desorption whereby the low laser power densities available can be used to desorb and ionize chemisorbed species. This mode of operation is not as selective nor as subtle as the technique of photon stimulated desorption,[30] nevertheless the presence of chemisorbed water, which had been indicated by other techniques, was confirmed by LIMA. These results indicate that LIMA has a potential role in the broad field of corrosion and heterogeneous catalysis.

REFERENCES

1. See for example: T. Dingle, B.W. Griffiths in *Microbeam Analysis,* p 315–318 ed. by J. T. Armstrong, San Francisco Press Inc. (1985)
2. See for example W.H. Guest. *Int. J. Mass Spectrom. Ion Proc.* **60** 189–199 (1984).
3. R.J. Cotter, *Analytical Chem* **56**, 3, 485–504 March 1984.
4. C. Grey Morgan, *Chem Soc Rev,* **8** 367 (1979).
5. P.C. Kistemaker, M.M.J. Lens, G.J.Q. van der Peyl, A.J.H. Boerboom. *Adv. Mass Spec.* **8A,** 928 (1981).
6. E. Unsold, G. Renner, F. Hillenkamp, R. Nitsche, *Adv. Mass Spec* **7,** 1425 (1978).
7. F.W. Anderson, Hans Heinen, J.N. Ramsay *Microbeam Analysis* p 325–330 ed. by J.T. Armstrong, San Francisco Press Inc. (1984).
8. J.C. Ruckman, A.R. Davey, N.S. Clarke, *AWRE Report 02/84* HMSO (1984).
9. G. Lawrence, private communication.
10. J.C. Ruckman, A.R. Davey, N.S. Clarke, *Vacuum* **34 10–11,** 911 (1984).
11. T. Dingle, B.W. Griffiths, J.C. Ruckman, *Vacuum* **31,** 571, (1981).
12. N.S. Clarke, J.C. Ruckman, A.R. Davey, *Proc. 32nd UKAEA Conf. Materials Analysis by Physical Techniques AERE R11655,* Harwell Report (1984).
13. N.S. Clarke, P.G. Hall, C.J. Wright, unpublished work.
14. C.H. Rochester, S.A. Topham, *JCS Far I* **75 (5),** 1073 (1979).
15. T. Ishikawa, K. Inouye, *Bull Chem Soc Jap* **46,** 2665–2668 (1973).
16. W. Abeck, H. Kober, B. Seidel, *Gevaert Agfa N.V. Belg,* 725, 759 June (1969).
17. I. Ishikawa, K. Inouye, *Bull Chem Soc Japan* **43,** 2350-2354 (1972).
18. O. Baudische, W.H. Albrecht, *J. Am. Chem Soc* **54** 943 (1932).
19. D.L. Segal. *The Hydrolysis of Iron (III) Salt Solutions I — Preparation properties and stability of colloidal dispersions of ferric hydroxy poly cations,* AERE R9721, Harwell Report (1980).
20. C. Michiels, R. Gijbels. *Anal Proc* **20** (6) 294 1983.
21. J.A. Gardella, D.M. Hercules. *Spectrosc Lett.* **13** 347 (1980).
22. J.A. Gardella, D.M. Hercules *A Anal Chem* **308** 242 (1981).
23. G.S. Pettit *et al. Report No. K1778,* Union Carbide Corp (1969).
24. S.R. Antony *et al.* private communication.
25. E. Michiels, A. Celis, R. Gijbels, *Microbeam Analysis* p. 383 ed. by K.F.J. Heinrich, San Francisco Press Inc. (1982).
26. H.J. Heinen, S. Meyer, H. Vogt, R. Wechsung. *Adv Mass Spect, Vol* **8A,** Heyden (1981).
27. R.J. Steel. *Proc. 29th. UKAEA Diffn. Analysis Conf. Thin Films and Surfaces, ND-R-751 (S),* Northern Division Report (1981).
28. J.M. Walls, I.K. Brown, D.D. Hall, *Appl Surf Sci* **15,** 1–4 93 (1983).
29. D.J. Prince, R.C. Greenwood, private communication.
30. M.L. Knotek *Nature* **291** 452 (1981).

SURFACE AND INTERFACE ANALYSIS, VOL. 9, 41–46 (1986)

Auger Electron Spectroscopy Depth Profiling During Sample Rotation

A. Zalar
Inštitut za elektroniko in vakuumsko tehniko, Teslova 30, 61000 Ljubljana, Yugoslavia

There are many factors that affect depth resolution of AES compositional depth profiles; quite often, it is degraded mostly by the sample surface roughness. We differentiate between the initial roughness of the sample surface and the roughness induced by ion sputtering during depth profiling. The influence of the ion beam incidence angle and of the ion energy on the depth resolution of multilayer Cr/Ni thin film structures was initially investigated on the samples that were stationary during AES depth profiling. The sputtering induced surface topography was investigated by transmission electron microscopy. In order to diminish topographical effects, we applied sample rotation during AES depth profiling. The sample holder of a commercial AES instrument was equipped with a mechanism that enabled sample rotation (1 rpm) during the simultaneous ion sputtering and AES analysis. The influence of sample rotation on depth resolution was studied on multilayer Cr/Ni thin film structures deposited on smooth substrate and on SiO_2/Si samples. In all the cases studied, sample rotation during depth profiling improved depth resolution. This improvement was in some cases surprisingly large, whereas in other cases it was less substantial, depending on the type of sample and its original surface topography, on the ion beam incidence angle, on the ion beam energy, and on the sputer depth. The main reasons for improved depth resolution resulting from sample rotation during depth profiling are the decrease of the angular dependence of the sputtering yield for variously oriented microplanes with respect to the ion beam direction, and the decrease of the roughness induced during ion sputtering.

INTRODUCTION

The accuracy of AES sputter depth profiling of thin film structures depends on the obtained depth resolution. Although in the past years much work has been done on the determination of factors affecting depth resolution,[1,2,3] this area nevertheless requires further research. New analysis instruments with a rastered ion beam or with two ion guns and with a small diameter of the primary electron beam, substantially diminish the instrumental effects on depth resolution that were and still are very noticeable when older instruments are used. The nature of the sample and of the ion beam are such that we still see sample effects and ion beam effects on depth resolution. Sample effects are strongly dependent on composition, crystal structure, and surface topography. There is no perfectly smooth solid surface, because in discussing polished samples we still refer to the atomic scale topography (< 10.0 nm).[4] For this reason, surface roughness topography always exerts some influence on depth resolution, and frequently represents the effect that degrades it the most.

The effects of ion beam on depth resolution, such as sputter-induced surface roughness, atomic mixing, radiation-induced diffusion and segregation, are more noticeable and easier to measure on the samples with a smooth surface. The influence of sputter-induced surface roughness is less noticeable at smaller sputtering depths and at a higher degree of the original roughness of samples. We can assume that the analysis point consists of a larger number of microplanes that are differently inclined with respect to the ion beam direction, which results in local differences in sputtering yields.[5]

In order to diminish some sample effects and some ion beam effects on depth resolution Δz, we introduced sample rotation during AES sputter depth profiling.[6] In the present work, the influence of sample rotation during AES depth profiling on depth resolution was studied on a multilayer Cr/Ni thin film structures and on SiO_2/Si samples. We compare the results of the analyses carried out using the rotating samples with those using stationary samples; likewise, the influence of the ion beam incidence angle and of the energy of ions on depth resolution of Cr/Ni samples was observed.

EXPERIMENTAL DETAILS

Depth resolution of AES sputter depth profiles was measured on two kinds of samples: the multilayer Cr/Ni thin film structures with a multicrystalline structure and the samples with an amorphous SiO_2 layer formed on a Si substrate. The Cr/Ni structure was deposited on Si substrates each 0.3 mm thick with different roughness degrees. In the present paper, however, we will report on our analyses on smooth substrates only. These had centre-line average value $R_a = 0.012$ μm (measured by the Talysurf method). The deposition of individual layers of Cr and Ni making up the multilayer system was made according to the present author's (AZ) design at the Institut Jožef Stefan, Ljubljana, in 1982. Since it has been described elsewhere,[5,7,8] this paper provides only the basic data. The multilayer structure

0142–2421/86/120041–06 $05.00

consists of five cycles of chromium and nickel thin films each about 50 nm thick, except that the top nickel protective layer was about 25 nm thick. The total thickness of multilayer structure, measured by optical interferometry, is 485 ± 10 nm.

The SiO_2/Si samples were prepared by means of the thermal growth of SiO_2 on a smoothly polished monocrystal of Si. We analyzed the samples with two different thickness degrees of the SiO_2 layer, 80.0 nm and 0.5 μm.

All the samples were analyzed in the system Physical Electronics Industries SAM 545 A. The intention of using sample rotation during the simultaneous AES analysis and ion sputtering was to obtain the optimum depth resolution Δz which depends, among other factors, on the ion beam incidence angle and on the energy of ions. Therefore the stationary Cr/Ni samples were first examined with respect to the two mentioned effects on Δz. Different ion beam incidence angles were achieved by placing the samples on metal bases attached to the original sample holder. The angles between the normal on the average sample surface and the ion beam direction were 0°, 15.0°, 17.8°, 29.1°, 42.6°, 56.8°, and 63.5°. At these incidence angles of ion beam the samples were sputter profiled with a rastered beam of 1 keV Ar^+ ions, ion beam current being constant. It was measured at the ion beam incidence angle of 42.6° and was 6.8×10^{-9} A, which corresponds to ion current density at this angle of 0.138 Am^{-2}.

At the ion beam incidence angle of 42.6° the influence of the energy of ions on the depth resolution of Cr/Ni samples was studied using beams of 0.5 keV, 1 keV, 3 keV, and 5 keV Ar^+ ions. In these investigations, the ion current density was also maintained constant at 0.138 Am^{-2}, with beam rastering on the surface which always exceeded 6 mm × 6 mm. The pressure in the chamber during depth profiling was about 7.3×10^{-3} Pa, base pressure $< 2 \times 10^{-7}$ Pa. We made use of a static primary electron beam of 3 keV, 1.0 μA and diameter $d_{eb} \approx 45$ μm. The angle betwen the ion beam direction and the direction of the primary electron beam was in all the cases about 66°.

The sample holder was equipped with the rotation mechanism driven by a mechanical spring, which enables sample rotation (1 rev min^{-1}) during the simultaneous AES analysis and ion etching. The analysis point was invariably located in the center of the rotation axis. The early experimental insights into sample rotation during AES depth profiling and details of the rotation mechanism are given elsewhere.[6]

During the simultaneous AES analysis and ion etching the peaks of Cr (529 eV) and Ni (848 eV) were recorded at intervals that were as short as possible During the analysis of the SiO_2/Si samples, the peaks of Si (78 eV) and O (507 eV) were measured.

After the AES depth profiling some samples were investigated by transmission electron microscopy in order to ascertain the change in their surface topography induced by the ion sputtering of the samples. The surfaces were investigated by the transmission electron microscope Philips EM 301 by using platinum preshadowed carbon replica technique.

RESULTS AND DISCUSSION

The concentration of elements, presented in depth profiles, was calculated on the basis of the relative elemental sensitivity factors $S_{Cr} = 0.34$, $S_{Ni} = 0.28$, and $S_{Si} = 0.34$. The depth resolution Δz was defined as the depth interval between 16% and 84% concentration change at a single interface.

Figure 1 shows the depth resolution Δz as a function of sputtering depth z for multilayer Cr/Ni thin film structures depth profiled with 1 keV Ar^+ at various ion beam incidence angles: at 0°, 15.0°, 17.8°, 29.1°, 42.6°, 56.8°, and 63.5°. The best depth resolution was determined on the samples which were sputtered at grazing ion beam incidence angles of 63.5° and 56.8°; it became increasingly poorer in the following order with respect to the ion beam incidence angles: 0°, 15.0°, 17.8°, 42.6°, and 29.1°. At this point, we must stress that ion current was kept constant during all the measurements, whereas ion current density was changing with the cosine of the ion beam incidence angle. We can see that the change in Δz is not proportional to the change of ion beam incidence angle over the entire measured range, from 0° to 63.5°. Therefore we conclude that Δz depends strongly on the crystallographic orientation of the multicrystalline Cr/Ni samples during ion sputtering which induces the surface roughness. It is interesting to note that one of the poorest characteristics of Δz is obtained at the ion beam incidence angle of 42.6°, obtained on the 60° sample holder, which most producers still build into their instruments. The sputtering rates of Cr, Ni and Cr/Ni structures depending on the ion beam incidence angles are given in Table 1. The sputtering rates are calculated on the basis of the sputtering time, the thickness of 5 Ni layers (225 nm), and 5 Cr layers (250 nm). It is necessary to point out that the sputtering rate is dependant except on the ion current density and ion energy also strongly on the precise focusing of the ion beam.

Figure 2 shows the depth resolution Δz as a function of sputtering depth z for the Cr/Ni samples depth

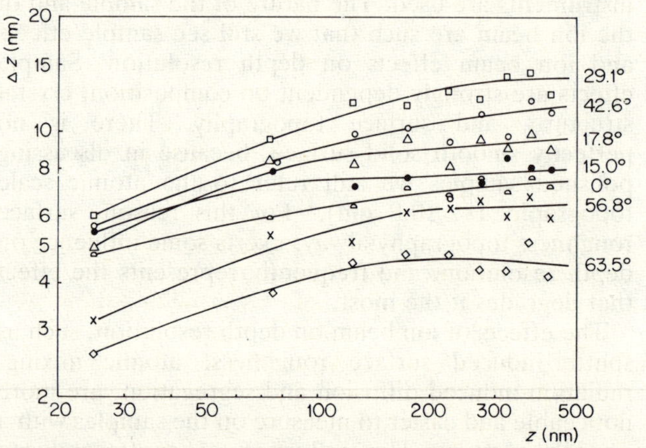

Figure 1. Depth resolution of multilayer Cr/Ni thin film structures as a function of sputtered depth, depth profiled at different ion beam incidence angles: 0°, 15.0°, 17.8°, 29.1°, 42.6°, 56.8°, and 63.5°. The samples were sputter profiled with a rastered beam of 1 keV Ar^+ ions, ion current density 0.138 Am^{-2}, measured at 42.6°.

Table 1. The sputtering rates of Cr, Ni and Cr/Ni structure obtained by ion sputtering of multilayer Cr/Ni thin film structures at different ion beam incidence angles: 0°, 15.0°, 17.8°, 29.1°, 42.6°, 56.8°, and 63.5°. The samples were ion sputtered with rastered beam of 1 keV Ar^+ ions at ion current of 6.8 × 10^{-9} A measured with a Faraday cup of aperature diameter 0.25 mm. The ion current density of 0.138 Am^{-2} is valid for ion beam incidence angle of 42.6°

Incidence angle α (°)	Sputtering rate Z (nm min^{-1}) Cr	Ni	Cr/Ni
0	1.10	1.29	1.19
15.0	1.46	1.41	1.43
17.8	1.55	1.50	1.52
29.1	1.63	1.46	1.54
42.6	1.79	1.62	1.70
56.8	1.55	1.33	1.44
63.5	1.13	0.94	1.03

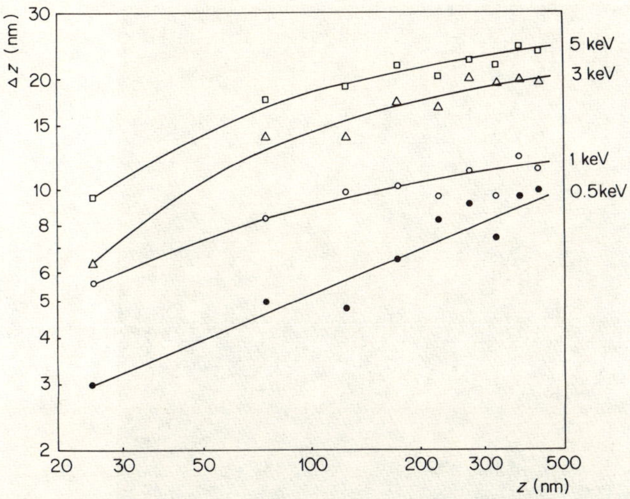

Figure 2. Depth resolution of multilayer Cr/Ni thin film structures as a function of sputtered depth z. Depth profiling was done with rastered beams of 0.5 keV, 1 keV, 3 keV and 5 keV Ar^+ ions, at the ion beam incidence angle of 42.6°, ion beam current 6.8 × 10^{-9} A, ion current density 0.138 Am^{-2}.

profiled with 0.5 keV, 1 keV, 3 keV and 5 keV rastered argon ion beams at an incident angle of 42.6°. In all cases ion current density was 0.138 Am^{-2}. Figure 2 clearly indicates that the depth resolution Δz degrades with the increasing energy of Ar^+ ions, and that $z \propto \sqrt{E}$. The sputtering rates of Cr, Ni and Cr/Ni structures given in Table 2 show the dependence on the ion beam energy.

On the basis of the above measurements, we decided to carry out the analysis of multilayer Cr/Ni thin film structures during sample rotation, at the ion beam incidence angles of 42.6° and 66.4°, with 1 keV Ar^+ ions and at 42.6° with 5 keV Ar^+ ions. The diagram in Figure 3 enables us to compare the depth resolution of the Cr/Ni structures analyzed on rotating samples with those analyzed on stationary samples. Except for the ion beam incidence angle, other analysis parameters were not changed unless specifically indicated otherwise. If we make first a comparison of the results obtained at the ion beam incidence angle of 42.6°, we can state that sample rotation contributed to the essential improvement in the depth resolution, in the

Table 2. The sputtering rates of Cr, Ni and Cr/Ni structure obtained by ion sputtering of multilayer Cr/Ni thin film structures with rastered beam of Ar^+ ions with different energy: 0.5 keV, 1 keV, 3 keV and 5 keV. The incidence angle of ion beam and the ion current density were in all cases the same: 42.6° and 0.138 Am^{-2} respectively. The ion current measured with a Faraday cup of aperture diameter 0.25 mm was 6.8 × 10^{-9} A

Ion energy E (keV)	Sputtering rate Z (nm min^{-1}) Cr	Ni	Cr/Ni
0.5	1.49	1.27	1.38
1	1.79	1.62	1.70
3	2.57	2.68	2.62
5	2.93	3.07	3.00

cases of ion sputtering with both 1 keV Ar^+ ions and 5 keV Ar^+ ions

The comparison of the depth resolution of stationary with rotating samples (Fig. 3, curves (a) and (b)) at the depths of 25 nm and 425 nm shows that in sputtering with 1 keV Ar^+ ions on the stationary sample, Δz is greater than on the rotating sample by a factor of about 1.4. A monotonous increase of the depth resolution with the increase in sputter depth is observed in the case of both samples.

At the same ion beam incidence angle but with the use of the ion beam with a higher energy of 5 keV Ar^+, the depth resolution improved even more: at the depth of 25 nm the depth resolution of the stationary sample is higher by a factor of 1.6, and at the depth of 425 nm even by a factor of 3.6 with respect to the rotated sample (Fig. 3, curves (c) and (d)). Under these conditions of ion sputtering, almost constant Δz is

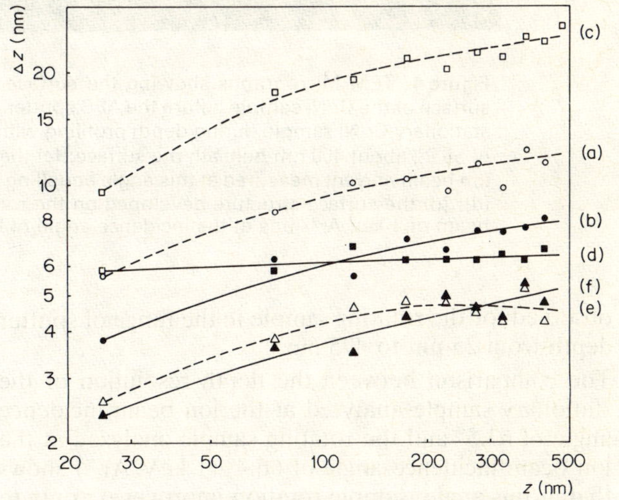

Figure 3. Depth resolution of multilayer Cr/Ni thin film structures as a function of sputtered depth: curve (a), for a stationary sample and rastered beam of 1 keV Ar^+ ions at the incidence angle of 42.6°, curve (b), the same as curve (a) but the sample rotated at 1 rev min^{-1}, curve (c), for a stationary sample and rastered beam of 5 keV Ar^+ ions at the incidence angle of 42.6°, curve (d), the same as curve (c) but the sample rotated at 1 rev min^{-1}, curve (e), for a stationary sample and rastered beam of 1 keV Ar^+ ions at the incidence angle of 63.5°, curve (f), for a rotated sample and rastered beam of 1 keV Ar^+ ions at the incidence angle of 66.4°. The ion beam current was in all cases the same: 6.8 × 10^{-9} A, measured at 42.6°.

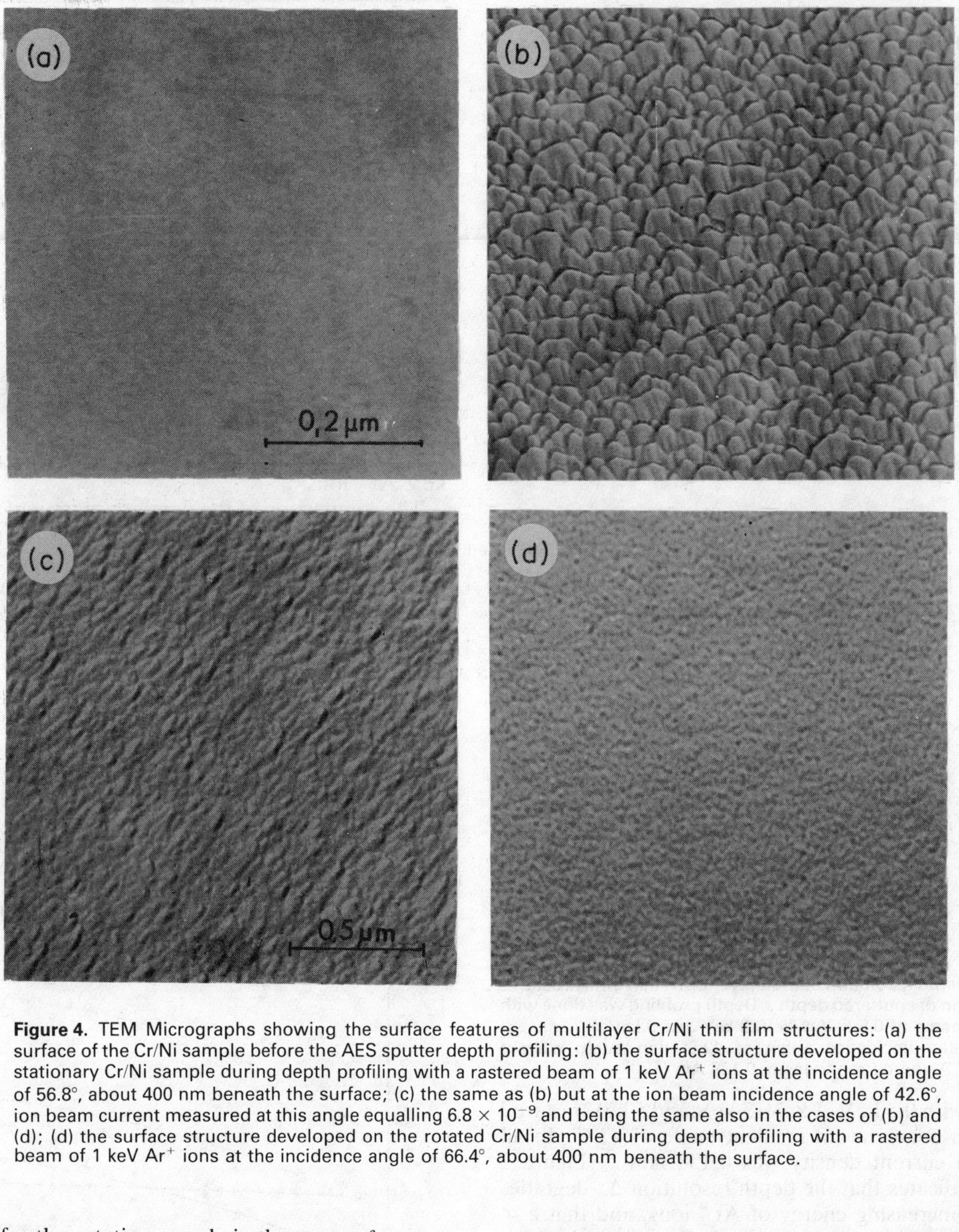

Figure 4. TEM Micrographs showing the surface features of multilayer Cr/Ni thin film structures: (a) the surface of the Cr/Ni sample before the AES sputter depth profiling: (b) the surface structure developed on the stationary Cr/Ni sample during depth profiling with a rastered beam of 1 keV Ar^+ ions at the incidence angle of 56.8°, about 400 nm beneath the surface; (c) the same as (b) but at the ion beam incidence angle of 42.6°, ion beam current measured at this angle equalling 6.8×10^{-9} and being the same also in the cases of (b) and (d); (d) the surface structure developed on the rotated Cr/Ni sample during depth profiling with a rastered beam of 1 keV Ar^+ ions at the incidence angle of 66.4°, about 400 nm beneath the surface.

observed for the rotating sample in the range of sputter depth from 25 nm to 475 nm.

The comparison between the depth resolution of the stationary sample analyzed at the ion beam incidence angle of 63.5° and the rotating sample analyzed at the ion beam incidence angle of 66.4° (1 keV Ar^+) shows that at this angle, sample rotation improves Δz only to the depth of up to about 175 nm, while at greater depths the value of Δz is approximately of the same order of magnitude for both samples (Fig. 3, curves (e) and (f)). The influence of the small difference in the angle of incidence between the rotated and the stationary sample, which in our case equaled 2.9°, upon Δz is supposed to be negligible. In any case, we must point out that on the rotating sample a very good depth resolution of 2.4 nm was attained at the depth of 25 nm; at greater depths, however, it increased monotonously.

The depth resolution of multilayer Cr/Ni thin film structures is strongly dependent on the ion beam incidence angle and on the energy of ions. Both these effects exert an influence on the roughness of the sample surface induced by ion sputtering during the AES depth profiling, and on the depth of the atomic mixing zone. The greatest depth of the atomic mixing zone — viewed in the direction perpendicular to the surface — can be expected at the perpendicular incidence of ions on the sample (ion beam incidence angle being 0°); it should diminish with the increase of the ion beam incidence angle. The atomic mixing zone increases with the energy of ions. The surface roughness induced by ion sputtering depends — among other factors — on the orientation of grains and defects in the metal structure. The sample surfaces investigated by the transmission electron microscopy (TEM) after

AES sputter depth profiling, show great differences in the topography of the surfaces that were ion sputtered under various conditions (Fig. 4 (a)–(d)).

Although TEM was used for investigating a number of samples depth profiled under various conditions, this paper presents only four micrographs (Fig. 4 (a)–(d)). Figure 4(a) shows the surface of a multilayer Cr/Ni thin film structure, i.e. the surface of the top Ni layer, before the AES depth profiling. Figures 4(b) and 4(c) present the topography of the samples that were stationary during AES depth profiling and sputtered with a rastered beam of 1 keV Ar^+ ions, the one at the ion beam incidence angle of 56.8° (Fig. 4(b)) and the other at the angle of 42.6° (Fig. 4(c)). The depth resolution of both samples is shown in Fig. 1. The micrographs were made about 400 nm beneath the surface on the walls of the craters formed on the samples because of the sputtering by the rastered ion beam. At approximately the same depth the micrograph of the topography of the sample was taken, which during AES depth profiling rotated at the rate of 1 rev min^{-1} (Fig. 4(d)). The incidence angle of rastered beam 1 keV Ar^+ ions was 66.4°. The depth resolution of this sample is shown as the lowest curve in Fig. 3. The measurements of roughness by the Talysurf method yielded the roughness factor of the topography in Fig. 4(a) $R_a = 0.012$ μm. The surface topographies formed after AES depth profiling show an essential increase in roughness with respect to the original sample surface (note the much larger increase in Fig. 4(a) than those of 4(b)–(d). The quantitative estimation of the amplitude of the roughness of samples after ion sputtering (Fig. 4(b)–(d) has yet to be made, but the micrographs allow a qualitative statement: the roughness is greatest on the stationary sample sputtered at 42.6° ion incidence (Fig. 4(c)); it is followed by the stationary sample sputtered at 56.8°

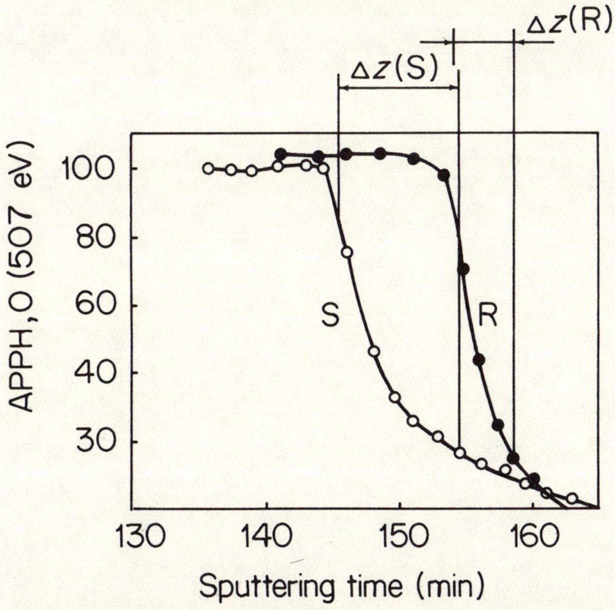

Figure 5. Auger peak-to-peak heights of O (507 eV) as a function of sputtering time for the SiO_2/Si samples with the SiO_2 layer 0.5 μ thick: (S) for the stationary sample and a rastered beam of 3 KeV Ar^+ ions at an incidence angle of 42.6°, ion beam current 5.6×10^{-9} A, ion current density 0.114 Am^{-2}; (R) for the rotating sample and a rastered beam of 3 keV Ar^+ ions at an incidence angle of 66.4°, ion beam current 5.6×10^{-9} A.

(Fig. 4(b)). The lowest roughness amplitude and the most favourable angular distribution of microplanes was ascertained for the sample which during AES depth profiling rotated and was sputtered at 66.4° angle of ion incidence. It is not known whether the micrographs in Figs 4(b)–(d) were taken on the surface of Ni or Cr layers; nevertheless, the investigations of the topography of sputtered Cr/Ni samples we have so far performed do not point to the existence of the difference in the roughness degree between two adjacent layers on crater walls.

The rotation of samples during ion sputtering diminished the effects of the angular dependence of sputtering yield as observed on differently inclined microplanes, while at the same time it also reduced the roughness induced by ion etching (Fig. 4(d)). The influence of rotation upon the depth of the atomic mixing zone has not been established yet, but it is possible that it is also diminished because of the incidence of ions on the sample from different directions. By introducing sample rotation, at 42.6° angle of incidence (5 keV Ar^+) we attained an almost constant depth resolution in the range from 25 nm to 475 nm. At the same angle of incidence but with the use of the ion beam of 1 keV Ar^+, however, the depth resolution increases monotonously with the depth of ion sputtering in spite of sample roation. Since we believe that during depth profiling both the depth of the atomic mixing zone and ion current density are unchanged, we hold that the main reason for the increase of the dpeth resolution with sputtering depth is the formation of surface structure induced by ion etching. It may well be that some other rate of sample rotation, which has yet to be ascertained, would enable us to obtain also in this position of the sample and with 1 keV Ar^+ ion sputtering, a favourable depth resolution that did not change with sputtering depth. The analysis of the multilayer Cr/Ni thin film structure at 63.5°angle of ion incidence and with 1 keV Ar^+ shows that there are such favourable conditions for achieving a good depth resolution already on the stationary sample that sample rotation at this angle contributes only to a small improvement in the surface section of a multilayer structure, up to 175 nm, while at greater depths no such improvement occurs any more.

The SiO_2/Si samples with two different thicknesses of SiO_2 layers were ion sputtered by 3 keV Ar^+ ion beam on the surface over an area of 6 mm × 6 mm at constant ion current. The stationary and rotating samples with 80.0 nm thick layer of SiO_2 were ion sputtered at 42.6° ion incidence angle. The ion current measured at this angle was 5.6×10^{-9} A, which corresponds to ion current density equal to 0.114 Am^{-2}. The comparison between the two samples shows that the improvement of depth resolution obtained by sample rotation at a depth of 80.0 nm equals 15%. The samples with a 0.5 μm SiO_2 layer on Si substrate were ion sputtered at two different ion incidence angles. The stationary sample was ion sputtered with rastered beam of 3 keV Ar^+ ions at 42.6° angle of incidence and the rotating sample at 66.4° (ion current was 5.6×10^{-9}A). Figure 5 shows Auger peak-to-peak heights of O (507 eV) as a function of sputtering time for a stationary and for a rotating sample. The relative depth resolution at a depth of 0.5

μm on the stationary SiO_2/Si sample is 6.1% and on the rotating sample it is 2.7%. Both the sample rotation during depth profiling and gracing incidence angle of Ar^+ ion beam contributed to the improvement of depth resolution. The development of the surface topography on the amorphous SiO_2 layer is more noticeable after prolonged ion sputtering at greater depths. That is why the difference between the stationary sample and the rotating sample is more pronounced on the sample with SiO_2 layer 0.5 μm thick than on the sample with SiO_2 layer 80.0 nm thick.

CONCLUSION

The depth resolution of multilayer Cr/Ni thin film structures on smooth (as well as rough substrates) is strongly dependent on the ion beam incidence angle and on the energy of ions. The surface topography induced during ion sputtering and the depth of the atomic mixing zone also depend on both of these parameters.

The rotation of samples during depth profiling diminished the influence of the angular dependence of sputtering yield observed on differently inclined microplanes, and the roughness induced by ion sputtering. The influence of rotation on the depth of the atomic mixing zone has not yet been investigated. In the case of multilayer Cr/Ni thin film structures, the rotation of samples at 42.6° angle of ion incidence substantially improved the depth resolution during sputtering with the ion beam energy of 1 keV and 5 keV; by use of a 5 keV ion beam resulted in an almost constant depth resolution in the range from 25 nm to 475 nm. In order that we might attain constant depth resolution while using the ion beam energy of 1 keV, the optimum rate of sample rotation during depth profiling should be first established. In the case of the samples with the amorphous SiO_2 layer on Si, the improvement of depth resolution because of sample rotation becomes more noticeable at greater depths only (> 80.0 nm).

Acknowledgements

The author wishes to thank Dr. B. Navinšek and Mr. A Žabkar for the deposition of the multilayer Cr/Ni thin film structures, and Dr. V. Marinkovič for taking TEM micrographs; all of the Institut Jožef Stefan, Ljubljana, He is also grateful to Dr. S. Hofmann, Max-Planck-Institut für Metallforschung Stuttgart, for valuable discussions. This work is part of the doctoral thesis (forthcoming) by A. Zalar, Univerza Edvarda Kardelja, Ljubljana, 1985.

REFERENCES

1. S. Hofmann, *Surf. Interface Anal.* **2** 148 (1980).
2. M. P. Seah and C. Lea, *Thin Solid Films*, 5781, 257 (1981).
3. H. Oechsner, *Thin Film and Depth Profile Analysis,* (ed.) Springer Verlag, Berlin (1984).
4. G. Carter, B. Navinšek and J. L. Whitton in *Sputtering by Particle Bombardment 11*, ed. by R. Behrisch, Chap. 6, p. 240, Springer Verleg (1983).
5. A. Zalar, S. Hofmann and A. Žabkar, *Thin Solid Films*, **131**, 141 (1985).
6. A. Zalar, *Thin Solid Films*, **124**, 223 (1985).
7. A. Zalar, in J.L. de Segovia (ed.), *Extended Abstracts in Proc. 9th Int. Vac. Congr. and 5th Int. Conf. Sol. Surf., A.S.E.V.A.* Madrid, 1983, p.134.
8. A. Zalar and B. Navinšek, ibid.

SURFACE AND INTERFACE ANALYSIS, VOL. 9, 47–54 (1986).

Scanning Acoustic Microscopy

Ludwig Josef Balk
Universität Duisburg, Fachgebiet Werkstoffe der Elektrotechnik, Leiter: Prof. Dr.-Ing E. Kubalek, Kommandantenstraße 60. D4100 Duisburg 1, FRG

Generation and propagation of sound and ultrasound in materials are strongly affected by various quantities such as elastic and thermal, electronic and magnetic properties. Accordingly, local changes of one of these quantities can give information on the physical and chemical nature of that region which the acoustic signal is generated in or in which the propagation occurs. Based on this, different techniques of 'scanning acoustic microscopy' have been developed, important methods in this respect being conventional scanning acoustic microscopy (CSAM), scanning photo- or laser acoustic microscopy (SPAM) and scanning electron acoustic microscopy (SEAM). In CSAM ultrasound is focused onto the specimen surface, then its transmission or reflexion within the specimen is determined. Thus propagative properties and acoustic scattering are dominant in CSAM. In SPAM and SEAM a focused laser or electron beam, mostly intensity-modulated with time, is used to generate an acoustic signal within the probed surface region. Recently these two techniques have been accompanied by the introduction of scanning ion acoustic microscopy (SIAM), which uses a focused ion beam for sound generation. In these three techniques the sound generation mechanism is the main source of information, as typical operation frequencies are low compared to CSAM and quite often very long acoustic wavelengths are resulting. However, for a detailed analysis of the acoustic image contrast, contributions of acoustic propagation to the signal have to be taken into account, too. Due to the different mechanisms involved in signal formation, the mentioned techniques allow determination of a large variety of material parameters, and they are not only competing methods, but add up to a complete set of techniques for material microcharacterization. The applicability of these methods, their advantages and disadvantages are discussed.

INTRODUCTION

Generation and propagation of acoustic waves, both in the sound and in the ultrasound regimes, are strongly affected by various material parameters, such as elastic, thermal, electronic and magnetic properties. Accordingly, local variations of each of these quantities can give information on the physical or chemical nature of the region which the acoustic wave is generated in or in which its propagation occurs. If furthermore the acoustic wave is focused onto or within the specimen and if the generation of acoustic wave is carried out by focused particle beams or wave probes, determination of material parameters based on changed acoustic generation or propagation can be combined with a high spatial resolution. Based on this concept a large variety of so-called scanning acoustic microscopes was developed or — which is true for most of these instruments — is still in the development stage. In order to limit the number of microscopical techniques to be described there, only those methods will be discussed which are based on the general scheme of a scanning microscope. As shown in Fig. 1, such a microscopical system involves the interaction of a probe with a specimen, which leads to some interaction products. These must be detected and afterwards signal-processed. The two-dimensional imaging is realized by variation of the probe position with respect to the specimen, which can be done both by scanning the probe or the specimen. Simultaneously to this scanning the temporal variation of the interaction

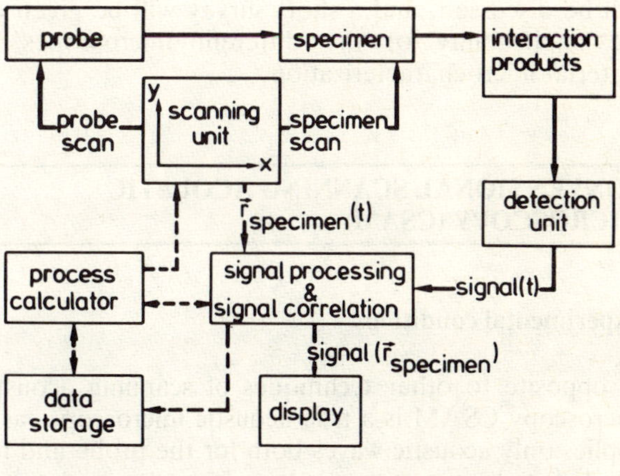

Figure 1. Scheme of a scanning microscope.

products has to be recorded as a function of time and finally correlated by means of a signal correlator to the probe location. By means of video displays and photographic systems or by process calculators micrographs can be obtained both as half-tone (z-modulation) or Y-modulation images. In this respect two different groups of scanning acoustic microscopes have to be mentioned. One uses focused acoustic waves as a probe and the propagation (and by this adsorption and reflexion) of the acoustic wave within the specimen (in the following denoted as conventional scanning acoustic microscopy). The other group mainly relies on the generation of sound or ultrasound within a small region probed by a particle beam or a

0142–2421/86/120047–08 $05.00

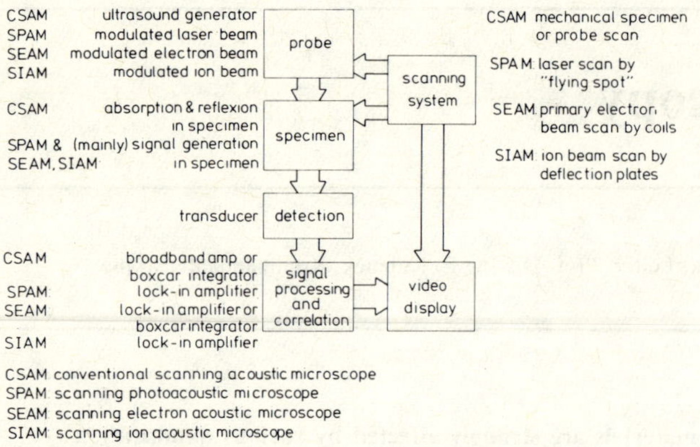

Figure 2. Schemes of scanning acoustic microscopes.

focused electromagnetic wave. At present three different types are realized (compare Fig. 2): scanning photoacoustic microscopy using a laser beam as a source for sound generation, scanning electron acoustic microscopy, which usually uses the primary electron beam of a scanning electron microscope as a sound generation source, and finally scanning ion acoustic microscopy using an ion source of secondary ion mass spectrometer. Though these three types of scanning acoustic microscopy are quite similar concerning their instrumentation, their capability for revealing material parameters is quite different due to the different physical nature of the three probes.

In the following the principle experimental conditions for the different techniques will be given, the operation conditions and amplification techniques will be discussed, and a short survey will be given on the applicability of the different microscopes to material microcharacterization.

CONVENTIONAL SCANNING ACOUSTIC MICROSCOPY (CSAM)

Experimental conditions

In opposite to other techniques of scanning acoustic microscopy CSAM is a true acoustic microscope, as it applies only acoustic waves both for the probe and for the detected interaction product. It can be used in two principally different modes, in reflection and transmission.[1–4] Figure 3 shows a simplified set-up for a transmission mode CSAM. The acoustic waves are generated by a piezoelectric transducer, typically a polycrystalline zinc oxide film, and then focused by a sapphire lens onto the specimen. To obtain a sufficient acoustic coupling and to minimize aberration effects, water is used as a coupling fluid between the sapphire lens and the specimen. The transmitted wave intensity is then detected by an identical transducer-lens-water system, in which the acoustic signal, which can be amplified either by a broadband amplifier or, in more critical signal conditions, by a boxcar integrator. Scanning can be done only mechanically, either by moving the specimen stage[5] or by scanning the whole transducer-lens-system parallel to the specimen surface.[6] As the spatial resolution of this system is largely determined by the diffraction limit, it correlates directly to the acoustic wavelength used. Therefore high frequencies have to be used to get high spatial resolution; for 375 MHz about 1.6 μm and less than 0.5 μm for low GHZ frequencies.[2,5] Unfortunately, the absorption losses increase significantly with higher frequency, therefore it is necessary to minimize these losses and furthermore to enable measurement of samples which are not thinned down to a thickness of some acoustic wavelengths. To do this, the reflexion mode of CSAM was developed.[1,6,7] In this mode the acoustic waves are reflected by the specimen back to the same transducer-lens system, which emitted the primary acoustic wave front. An electronic gating circuitry separates the emitted and the detected acoustic signal and allows switching of the electronic from a transmitting to a receiving operation condition.

Operation conditions and contrast mechanisms

To discuss possible contrast mechanisms for CSAM, different operation conditions have to be considered. These are continuous wave and pulsed operations and high excitation conditions. Due to the high frequencies used mostly, magnitude (or amplitude) images are taken. A phase sensitive detection of the relative phase angle between primary wave front and detected signal is not yet realized. For the transmission mode CSAM the contrasts are discussed by Wickramasinghe,[3] one possible mechanism is demonstrated for magnitude images in Fig. 4. The CSAM acts in this mode as a phase contrast microscope. A phase gradient in the specimen tilts the acoustic wave front during its passage

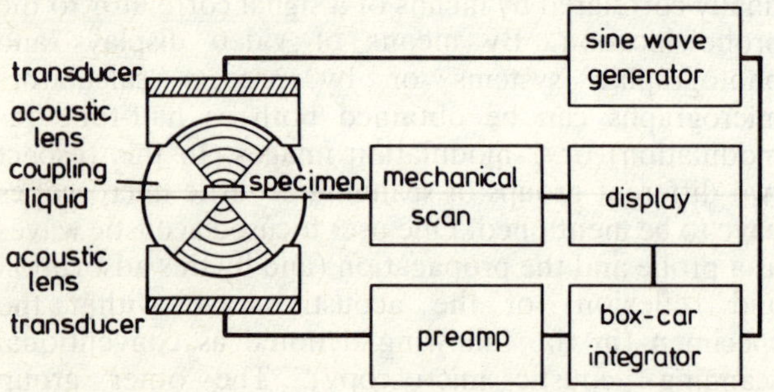

Figure 3. Schematic experimental set-up for transmission CSAM.

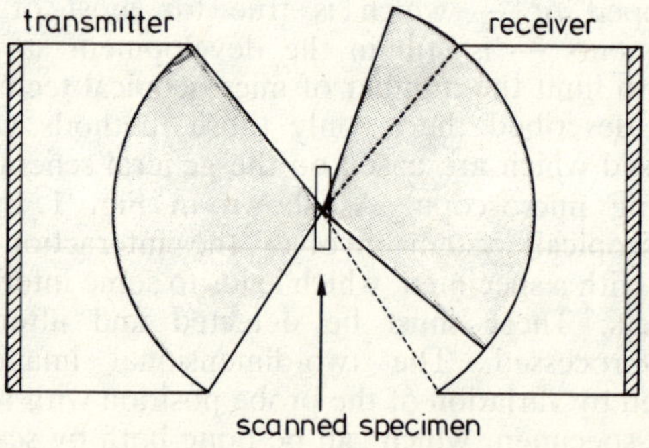

Figure 4. One possible contrast mechanism in transmission CSAM (after H.K. Wickramasinghe, Stanford University, 1979).

through the material and thereby reduces the acoustic power detected by the receiver. Furthermore signal attenuation is an important contrast mechanism in transmission CSAM,[8] which strongly correlates to the acoustic impedance. As this is the product of mass density and sound velocity, it gives rise to a material contrast.[9]

In the reflexion mode CSAM contrasts are more difficult, as the reflected waves may originate from various specimen depths. As these waves interfere, a high (i.e. bright) signal region does not necessarily indicate a higher reflexion coefficient due to a higher density or stiffness of the material. The actual signal depends on these interferences and thereby on the focus depth of the primary wave front. This is expressed in terms of so-called $V(z)$-curves, with z the depth coordinate.[10] This $V(z)$-function depends on the pupil function of the acoustic lens system and on the reflectance function of the specimen. When varying the acoustic focus depth, this function shows a oscillatory behaviour in a manner, which is typical for the material investigated.[9] Similarly a grain boundary contrast arises from boundaries intersecting the surface. Under defocus operation condition (compare Fig. 5) the directly reflected waves interfere with waves which travel along the surface and reentry into the vacuum again. These so-called leaky waves, which are typically Raleigh waves, are significantly disturbed by any imperfection, which leads to a contrast in the received signal.[6,11] As is indicated in Fig. 5, the directly reflected wave and the leaky wave arrive at different times (t_1 and t_2) at the acoustic lens. By use of a pulsed operation and short enough primary pulses, it is possible in principle to separate these two waves by sampling techniques, as has been already done for low frequency (100 MHz) CSAM.[12] Then the information gained is actually a transmission mode result — due to the transmission of the leaky waves through the specimen surface — though being fulfilled in an reflexion mode experiment. Further, by such an arrangement one possibly could overcome the interpretation problems of the $V(z)$-curves. The experimental arrangement of Fig. 5 was discussed up to now mainly in the leaky-wave model. For layered structures, however, this same experimental condition gives rise to surface acoustic waves.[13] Again this SAW contribution to the signal may be separated by ultrashort pulses and sampling techniques from the other signals, which should at least

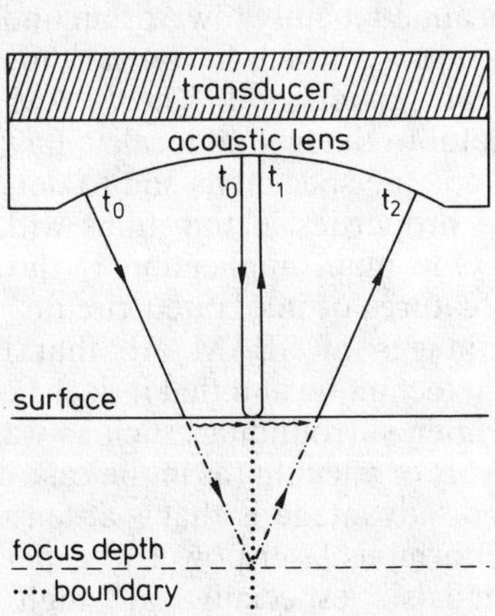

Figure 5. Leaky-wave model for grain boundary contrast in CSAM.

in principle enable a microscopy of surface and interface structures.

Finally it should be mentioned that by high excitation conditions nonlinear signal generation can occur within the specimen. This is due to a breakdown of the assumption that the density of the material remains unchanged by the propagating wave. This nonlinearity gives rise to an improved spatial resolution, though the signal is still detected at the fundamental, i.e. the primary wave frequency. This can be understood in that manner that the 2f wave recovers to the fundamental frequency due to a decreased intensity after being reflected at the specimen surface.[8,14]

Applications of CSAM

CSAM, preferably in the reflexion mode, has been applied up to date to many different materials. It can show crystalline structure without etching, reveal grain boundaries and surface imperfection, image subsurface delamination of passivation layers. These and further applications are reported by Hoppe *et al.*[9] and Briggs[8]. One main advantage of CSAM is that it can analyse materials without the need for vacuum, it even can analyse live biological material. Further, as only propagation properties are involved, the theoretical understanding is quite sufficient, especially compared to the other scanning acoustic microscopes. In principle it can be used for nondestructive depth profiling, though at present the results have to be improved to a depth resolution in the order of below 1 μm. This could be done by image processing or time-resolved techniques. One of its disadvantages is that it needs sufficient specimen polishing to avoid an image intrusion due to surface roughness.

SCANNING PHOTOACOUSTIC MICROSCOPY (SPAM)

Experimental conditions

A typical experimental set-up for SPAM is shown in Fig. 6[15] A laser beam is chopped mechanically, scanned by means of movable mirrors (flying spot scan) and then focused onto the specimen surface. Due to the periodic energy deposition an acoustic wave is generated within the specimen. One main mechanism for this acoustic signal is a primary production of

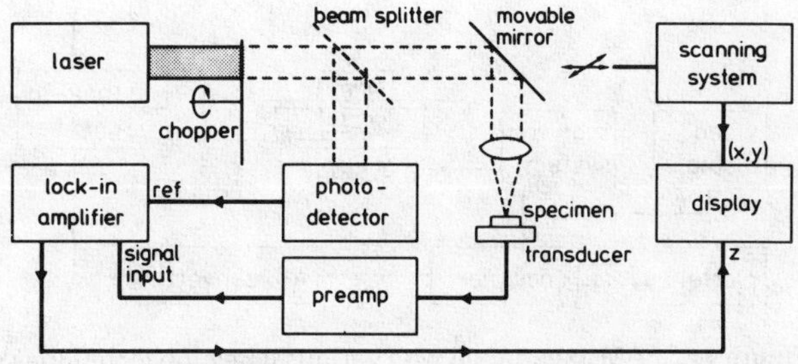

Figure 6. Schematic of a SPAM set-up.

thermal waves due to local heating and a subsequent conversion of the thermal wave into elastic via the thermal expansion coefficient.[16] The acoustic waves propagate to the bottom of the specimen and are detected by a transducer being in tight contact with it. Typically piezoelectric lead zirconate titanate (PZT) ceramics are used, though a pyroelectric detection is possible, too.[17,18] Further detection techniques imply noncontacting detection such as photo-acoustic beam deflection techniques (compare Fig. 7). Here the deflection of a laser with glancing incidence monitors the surface deformation due to the sound wave at the surface by a changed deflection angle.[19] A method which is very suitable to the investigation of thin films is demonstrated in Fig. 8. The thin film to be investigated is evaporated on a piezoelectric substrate, which at the same time carries an interdigital transducer. This transducer can directly detect the acoustic waves generated in the thin film.[20]

Typically a lock-in amplifier is used for amplification of the detected signal. This is possible, because the frequency range used at present is typically below 50 MHz, sometimes even down to some hundred Hz. Therefore both magnitude and phase images can be obtained, where phase means the relative phase change of the signal with respect to the primary beam modulation.

Besides of the generation of acoustic waves by a square wave modulation of a laser beam, other techniques have been applied, too. Even without a temporal modulation of the primary laser beam an acoustic signal can be generated. One possibility is use of a scan modulation by superposing a sine wave to the saw tooth for the scan movement. By doing this, the first derivative with respect to the modulated coordinate can be gained.[21] In an even simpler experiment such a signal generation was done by suitable adjustment of the scan speed.[21] Up to date no attempts have been reported on the use of short laser pulses and a subsequent signal amplification by means of a boxcar integrator, though this technique has proven to be very useful in photoacoustics.[23]

Operation conditions and contrast mechanisms

Depending on the modulation frequency different signal generation and by this contrast mechanisms can occur. For low frequencies generally a thermal wave

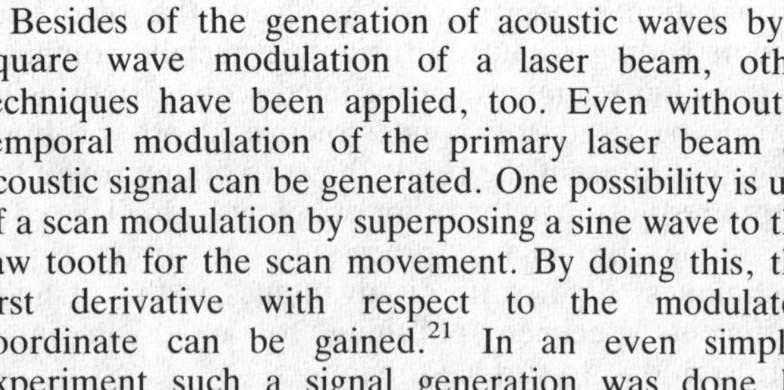

Figure 7. SPAM experiment with various detection techniques[19] (after M. Kasai *et al.* University of Tokyo, 1985).

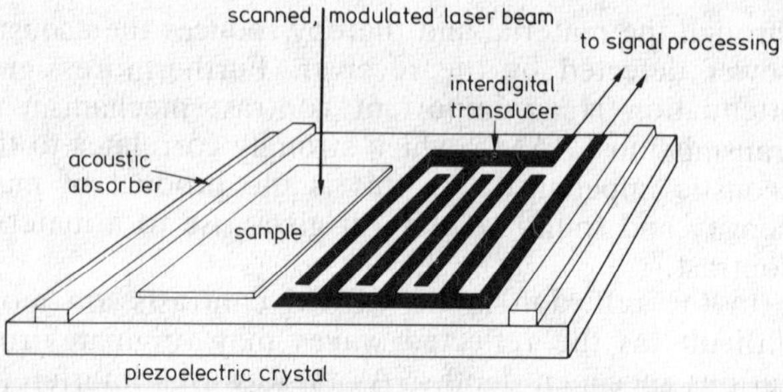

Figure 8. SPAM for thin film investigations (after G. Veith, Universität Wien, 1982).

production should be the main process for the signal generation. As already mentioned, the generation of thermal waves is the primary interaction. These thermal waves are strongly damped and decay within a few wavelengths. Thus the thermal wave or decay length determines the volume in which a conversion of thermal into acoustic waves occurs. Therefore it limits the spatial resolution in a much larger extend than the diameter of the laser spot. This thermal decay length is proportional to the square root of the reciprocal modulation frequency. Thus the attained spatial resolution is frequency dependent and can vary from several hundred μm up to a few μm. For higher operation frequencies and especially for semiconducting materials other signal generation mechanisms have to be considered, such as the sound coupling to transport properties of the excess carriers generated by the laser beam.[24,25]

Both magnitude and phase of the signal can be used to obtain acoustic images. Whereas the magnitude is affected both by optical and by elastic-thermal properties, the phase signal is sensitive only to variations in thermal or elastic parameters.[26] Furthermore it is assumed that the phase signal has a higher capability of monitoring specimen features from deeper locations.[27]

Application of SPAM

As the origin of photoacoustic signal is determined by the thermal decay length, it is possible to detect subsurface structures. By recording of a sequence of micrographs with various modulation frequencies, a depth profiling seems to be possible. Up to now the applications aimed mainly towards an understanding of the arising contrast, which in contrary to CSAM is still not completely understood. The SPAM technique is certainly useful to detect and localize flaws and defects in optically opaque specimens and to determine elastic and thermal properties in thin films with high spatial resolution. One main application is the detection of subsurface features of integrated circuits.

The advantages of SPAM are that it is an easy experimental technique and that it does not require any special specimen surroundings such as water (as in the case of CSAM) or vacuum (as in the case of SEAM and SIAM). Its disadvantage is that – at least at present – one needs enormous laser power to generate sufficient acoustic signals, especially at high modulation frequencies. This disables the use of SPAM for thermally sensitive materials.

SCANNING ELECTRON ACOUSTIC MICROSCOPY (SEAM)

Experimental conditions

In SEAM an intensity-modulated primary electron beam of a scanning electron microsope is used to generate acoustic waves in a manner which is very similar to SPAM. The signal generated within the probed specimen region penetrates the sample and is detected by a piezoelectric transducer at the bottom specimen surface. As in SPAM PZT-ceramics are used for detection mainly. To enable lock-in amplification the modulation waveform is a precise square wave. This is chosen instead of a sine wave in order to avoid a degradation of spatial resolution due to the beam chopping.[20] Amplification of the acoustic signal at the chopping (fundamental) frequency yields the so-called linear SEAM images.[30–32] Again both magnitude and phase of the signal can be monitored separately,[32] though some instruments only deliver a mixed information as in-phase ($A\cos\phi$) or out-of-phase ($A\sin\phi$) signal, with A the amplitude and ϕ the phase relative to the exciting wave from.[33] By tuning the lock-in amplifier to even harmonics of the fundamental, mostly to second harmonic excitation, so-called nonlinear SEAM images can be gained,[32] which deliver additional information on local variations of material parameters. Figure 9 shows a set-up, in which both linear and nonlinear SEAM experiments can be done at one time.

Besides of a harmonic beam modulation pulsed excitation can be used, too. In this case a boxcar-integrator is used to amplify the SEAM signal[34] enabling the recording of time-resolved SEAM micrographs (Fig. 10). Finally, in a similar manner as in SPAM first experiments are carried out to generate a SEAM image without beam intensity modulation but with a modified scanning procedure.[35]

Operation conditions and contrast mechanisms

Again, similarly to SPAM, the operation conditions are important for the resulting contrast mechanism. In SEAM this situation is even worse, as besides of the usual linear image nonlinear and time-resolved experiments can be carried out. Furthermore, the

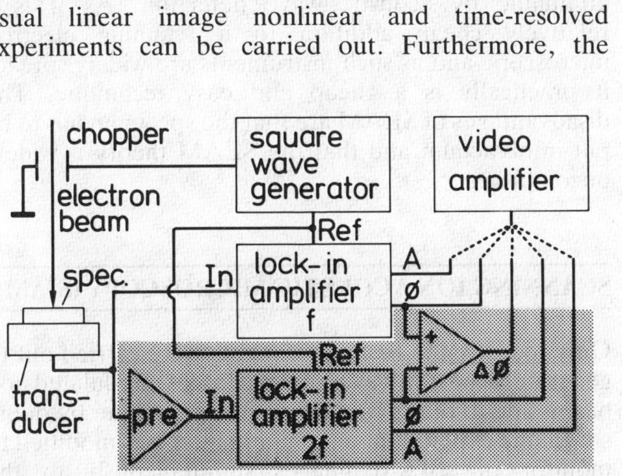

Figure 9. SEAM experimental set-up for simultaneous recording of linear and nonlinear images.

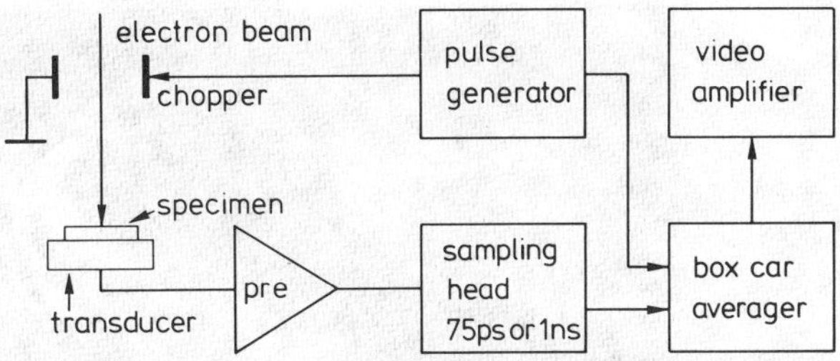

Figure 10. Set-up for time-resolved SEAM experiments.

interaction of the primary electron beam with the specimen, leading to a final generation of acoustic waves, is more complicated than is the case for a laser beam. Besides of a thermal wave coupling other mechanisms can occur due to the fact that the electron beam both deposits electric charges into the sample and carries a magnetic field with it. In this context theoretical models are discussed attributing the acoustic signal generation to piezoelectric, electro- and magnetostrictive coupling.[36] The latter two coupling mechanisms are inherently nonlinear due to a square law between deposited charge rsp. influenced magnetic field to an elongation of the sample. For low operation frequencies and non-semiconducting materials the thermal wave model is valid for the signal generation, and by this thermal properties determine the overall magnitude of the signal. Nevertheless, the contrast can be determined by elastic properties.[37,38] The spatial resolution is determined by either the primary electron dissipation volume or the region of decaying thermal waves whichever is larger, as sketched in Fig. 11. Variation of the primary beam modulation only gives a reasonable improvement in spatial resolution up to, say, 10 to 20 MHz. For the example of Cu–Zn–Al alloy the attained spatial resolution could be varied from 300 μm at a modulation frequency of 50 Hz up to 3 μm for 500 kHz, both for linear detection.[38]. Further increase of the modulation frequency can cause a strong intrusion of the micrographs by superposed vibrational mode patterns,[38] especially in metals. The contrast mechanisms, even in the simple thermal wave model, are not clear at all at present. Many different possibilities have to be considered, such as scattering of the thermal wave, the conversion of thermal into acoustic waves, acoustic scattering, especially for high frequencies, and a thermal to acoustic conversion at a

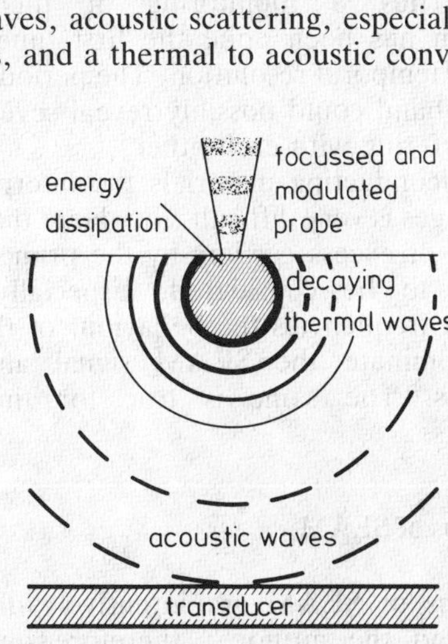

Figure 11. Thermal wave model for electron beam excitation.

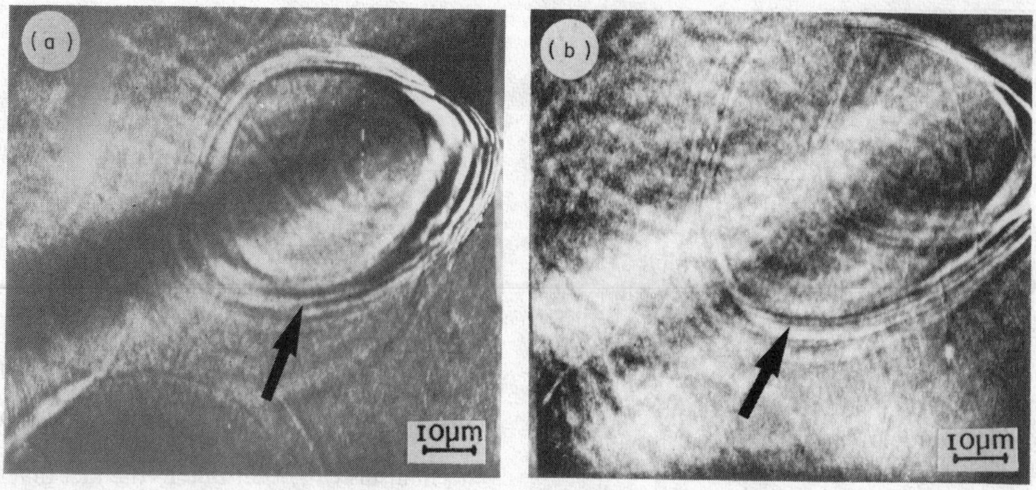

Figure 12. Time-resolved SEAM micrographs around a martensite in Cu–Zn–Al repetition rate: 180 kHz, primary electron energy: 30 keV.

scattering site.[39] This situation becomes important in the case of time-resolved experiments, as the different mechanisms are associated with different time delays. This may be explained for the example of monoclinic martensite laths in cubic Cu–Zn–Al alloy. At the end of such martensites oval rings could be detected by means of nonlinear SEAM.[38,40] Davies attributed their signal origin to a hysteretic shrinkage and growth of the martensites due to thermal influence,[41] a contrast due to mode conversion at a scatterer site. In order to prove this, time-resolved SEAM micrographs were recorded at such locations. Under the experimental condition chosen, the velocity for the thermal wave should be in the order of about 1 m/s. Time-resolved SEAM images should show ring like structures around such martensites which become larger in diameter with increased delay time, as the travel distance of the locally generated heat towards the martensite becomes greater, too. This can be seen in Fig. 12, in which two SEAM micrographs of 10 μs time delay difference are compared. The ring like band around the martensite, which penetrates from the bottom left corner to the top right corner of the images, is broadened just by about 20 μm, which correlates to the above mentioned velocity. Thus a monitoring of thermal wave propagation has been done the first time with high spatial and temporal resolution. The periodic structure within the band could possibly reveal several thermal waves interfering with each other.

For semiconducting materials the interpretation of SEAM images is very difficult, too. Here the enormous production of excess carriers by the primary electron beam has to be considered. Especially for high modulation frequencies the behaviour of these excess carriers dominate the SEAM signal and contrast mechanisms. The same is true for time-resolved experiments.

Applications of SEAM

Applications of SEAM are intensively discussed by Cargill[42] and the author,[35] therefore only a short summary shall be given here. Due to the complexity of sound generation by a high energetic electron beam, many material parameters can be monitored with SEAM and nearly any material can be investigated. Applications to metals, ceramics, semiconductors, ferroelectrics and ferromagnets are reported. In spite of the great problems in interpreting SEAM micrographs, for many cases an idenfitication of signal origin and by this a determination of the imaged structure was possible. These are grain contrast, grain boundaries, ferroelectric and ferromagnetic domain contrast, imaging of regions of changed material stiffness and, finally, determination of doping structures and pn-junctions in semiconducting devices. As SPAM the SEAM technique can visualize subsurface structures. As the penetration depth of the primary electron beam can be varied over a wide range, the depth of information can be optimized to the actual investigation problem.

The advantages of SEAM are: that it can deliver micrographs of micron and quite often submicron sized spatial resolution, even at moderate modulation frequencies and thus at moderate experimental expenses; it can image the largest variety of material parameters; linear, nonlinear and time-resolved techniques are possible and complete the information attainable by acoustic wave detection. As it is a relatively cheap addition to a scanning electron microscope and as such instruments are widely spread, it practically is a cheap and easy technique. The disadvantages of SEAM are that the specimen has to be put into vacuum and that the SEAM theory is widely uncomplete.

SCANNING ION ACOUSTIC MICROSCOPY (SIAM)

Only recently the first experiments were carried out to generate acoustic waves by means of modulated ion beams and to realize a microscopic technique by doing so. In one experiment an ion implanter was modified to monitor the SIAM images simultaneously to the implantation process,[43] whereas in the other case a secondary ion mass spectrometer was modified to a

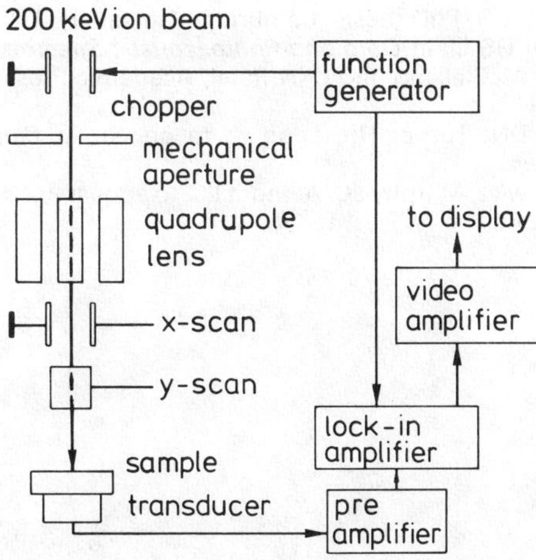

Figure 13. Experimental set-up for SIAM[43] (after D.N. Rose, US Army TACOM, Warren, 1985).

SIAM set-up.[44] The general set-up is very similar to SPAM and SEAM, as can be seen from Fig. 13. One aim of the experiments was to show that for SIAM a direct momentum transfer from the ions into the solid is the major signal generation contribution, an effect not observed until now for SPAM and SEAM. Up to now, however, this direct momentum contribution was not clearly identified.[44] As SIAM is the youngest of the discussed techniques, both theory and applications are at a very low level. One advantage will certainly be the fact that this additional sound source of different particle properties may help to analyse the signal generation and contrast mechanisms involved with the different techniques of scanning acoustic microscopy. One disadvantage of SIAM is similar to SEAM that the specimen has to be brought into vacuum, a further disadvantage is that it is the only destructive technique of the methods discussed in this paper.

CONCLUSIONS

All four techniques of scanning acoustic microscopy discussed in this paper are useful techniques for a microcharacterization of materials, both for surface and subsurface structures. The different quality in theoretical understanding and the amount of applications reported do not necessarily correspond with the quality of the method, but — to the authors point of view — clearly are a measure of the actual history and development stage of each of these techniques. CSAM as the oldest one is well interpreted. whereas for SIAM as the youngest about no theoretical background is given. It seems worthwhile to quantitatively compare the methods by means of calibration samples in order to obtain a maximum of information out of scanning acoustic microscopy.

Acknowledgements

The author would like to thank Prof. E. Kubalek for many discussions and encouragements, Mr. N. Kultscher for his assistance in experiments and in the development of theories, Dr. D.G. Davies from Cambridge University helped significantly to understand many SEAM experiments. The micrographs shown are taken from a bachelor's thesis of Mr. N. Nöthen.

REFERENCES

1. Lemons RA, Quate CF, in *Physical Acoustics,* Vol 14, pp 1–92, ed. by W.P. Mason and R. N. Thurston, Academic Press, London (1979).
2. Weglein RD, Wilson RG, Bonnell DM, *15th Annual Proc. Reliability Physics, Las Vegas, 1977,* 37–43.
3. Wickramasinghe HK, *J. Appl. Phys.* **50** (2), 664–672 (1979).
4. Hildebrand JA, Lam LK, *Appl. Phys. Lett.* **42,** 413–415 (1983).
5. Weglein RD, Wilson RG, *Appl. Phys. Lett.* **31** (12), 793–796 (1977).
6. Thaer A, Hoppe M, Patzelt WJ, *Leitz-Mitt. Wiss. u. Techn.* **VIII** Nr3/4, 61–67
7. Hollis RL, Hammer RV, in *Scanned Image Microscopy Scanned Image Microscopy* pp. 155–164, ed. by EA Ash, Academic Press, London. (1980).
8. Briggs GAD, *J. Scanning Electron Microscopy* **III,** 1041–1052 (1984).
9. Hoppe M *et al, Leitz-Mitt. Wis.u.Techn.* **VIII** Nr.5, 125–131 (1983).
10. Quate CF, Atalar A, Wickramasinghe HK, *Proc. IEEE* **67,** 1092–1114 (1979).
11. Adler L, deBilly M, Quentin GJ, *J. Appl. Phys.* **53** (12), 8756–8758 (1982).
12. Gilmore RS, to be published.
13. Smith IR, *et al, Appl. Phys. Lett.* **42,** (5), 411–413 (1983).
14. Rugar D, *J. Appl. Phys.* **56** (5), 1338–1346 (1984).
15. Busse G, *Proc. IIIrd Int. Conf. Photoacoustics, Paris (1983).*
16. Rosencwaig A and Gersho A, *J. Appl. Phys.* **47,** (I), 64–69 (1976).
17. Hadni A *et al, Applied Optics* **15** (9), 2150–2158 (1976).
18. Coufal H, *Appl. Phys. Lett.* **44** (I), 59–61 (1984).
19. Kasai M *et al,* to appear in *Canadian J. Physics* (presented at *4th Int. Topical Meeting on Photoacoustic, Thermal and Related Sciences, Montreal)*
20. Veith G, *Appl. Phys. Lett* **41** (II), 1045–1046 (1982).
21. Busse G, Orgabeck AS, *J. Appl. Phys.* **51** (7) 3576–3578 (1980).
22. Quimby RS, *Appl. Phys. Lett.* **45,** 1037 (1984).
23. Tam AC, Coufal H, *J. de Physique,* Colloque **C6** Suppl. 10, **44,** C6–9 — C6–20 (1983).
24. Sablikov VA, Sandomirskii VB, *Sov. Phys. Semicond.* **17** (I), 50–53 (1983).
25. Skumanich A *et al.* to appear in *Canadian J. Physics.*
26. Thomas RL *et al, IEEE Ultrasonics Symposium,* 586–590 (1982).
27. Rosencwaig A, Busse G, *Appl. Phys. Lett.* **36** (9), 725–727 (1980).
28. Sugitani Y, Akinori U, to appear in *Canadian J. Physics.*
29. Menzel E, Kubalek E, *J. Scanning Electron Microscopy Part I,* 305–318 (1979).
30. Cargill GS III, in *Scanned Image Microscopy,* pp 319–330, ed. by EA Ash, Academic Press, London, (1980).
31. Brandis E, Rosencwaig A, *Appl. Phys. Lett.* **37,** 98–100 (1980).
32. Balk LJ, Kultscher N, *Inst. Phys. Conf. Ser.* **67,** 387–392 (1983).
33. Rosencwaig A, *J. Scanning Electron Microscopy Part IV,* 1611–1628 (1984).
34. Balk LJ, Kultscher N, *J. de Physique,* Colloque **C2** Suppl.2, **45,** C2-873 — C2-876 (1984).
35. Balk LJ, to appear in *Canadian J. Physics.*

36. Kultscher N, Balk LJ, to appear in *Scanning Electron Microscopy* (1986).
37. Holstein WL, Begnoche BC, *J. Scanning Electron Microscopy Part III,* 1033–1040.
38. Balk LJ, Davies DG, Kultscher N, *Phys. Stat. Sol* (a) **82** (23), 23–33 (1984).
39. Favro LD, Kuo PK, Thomas RL, to appear in *Canadian J. Physics.*
40. Balk LJ, Davies DG, Kultscher N, *J. Scanning Electron Microscopy Part IV,* 1601–1610 (1984).
41. Davies DG, PhD thesis, Cambridge University, UK (1985).
42. Cargill GS III, in *Opto and Photoacoustic Spectroscopy* ed. by JF McClelland and CKN Patel, Academic Press, London (1984).
43. Rose DN, Turner HR, Legg K, to appear in *Canadian J. Physics.*
44. Satkiewicz, Murphy JC, Aamodt LC, to appear in *Canadian J. Physics.*

SURFACE AND INTERFACE ANALYSIS, VOL. 9, 55–63 (1986)

Crack Imaging in Alumina: a Scanning Acoustic Microscope Study

G.C. Smith and M.G. Gee
Division of Materials Applications, National Physical Laboratory, Teddington, Middlesex TW11 0LW, UK

A scanning acoustic microscope (SAM) operating at 750 MHz has been used to image artificially induced cracks in polycrystalline and single crystal specimens of alumina. Although the resolution of the instrument used in this work, at 1.6μm, was not as good as that of an optical microscope, significantly more information about the crack characteristics was revealed acoustically. Cracks of finite width act as scattering points for acoustic waves and thus sources of contrast in the SAM image, even when their width is less than the theoretical spatial resolution. In polycrystalline samples it was found that defocussing the SAM both above and below the focal plane, utilising different regions of the *V*(*z*) response, enabled contrast due to grain boundaries, cracks, and individual grains to be distinguished. It was also possible to identify different modes of cracking. For the single crystal samples, distinctive contrast was obtained from both horizontal and vertical sub-surface cracks within one Rayleigh wavelength of the surface. In most materials, features such as these are not readily detectable by other techniques.

INTRODUCTION

Reflection scanning acoustic microscopy (SAM) is rapidly becoming established as a method of non-destructive evaluation (NDE) particularly suited to the detection of surface or near-surface inhomogenieties and discontinuities on a microscopic scale.[1] Since the contrast in the images is a function of the elastic properties of the specimen quite different information from that given by optical or electron microscopy is obtained. Interaction of acoustic waves with the sample proceeds primarily via either direct reflection of longitudinal waves back into the coupling medium giving contrast due to variations in the reflection coefficient across the sample, or by mode conversion of the incident acoustic radiation to surface acoustic waves which propagate on the surface of the sample and scatter, giving rise to the distinctive contrast modes associated with the SAM.

Engineering ceramics are brittle materials that fail from flaws without the occurrence of gross plastic flow. These flaws are the result of manufacturing defects or surface damage produced during component preparation or service. Cracks associated with this surface damage often propagate in aggressive environments to form defects which significantly weaken the material. In order to improve our understanding of microcrack propagation and evaluate the suitability of the SAM as a NDE technique for these materials, a range of polycrystalline and single crystal alumina specimens which contain artificially induced microcracks has been studied. Alumina was chosen as the model material for this study since it is widely used industrially and is elastically and acoustically well characterised.[2,3]

0142–2421/86/120055–09 $05.00

OPERATION OF THE REFLECTION SAM

The reflection SAM uses the large velocity mismatch between acoustic waves in sapphire and water to focus a beam of ultrasound to the diffraction limit using a spherical sector lens.[4] The lens is mechanically scanned in a raster across the specimen surface and the image built up by detection of the reflected ultrasound intensity for appropriate points of the scan (Fig. 1). This echo intensity can be used immediately to modulate the brightness of a CRT rastered synchronously with the lens, or digitised and read into a frame-store for subsequent display and image processing.

The instrument used in this study[5] employs ultrasound of frequency 750 MHz giving a spatial resolution on the sample surface at focus of 1.6μm. The

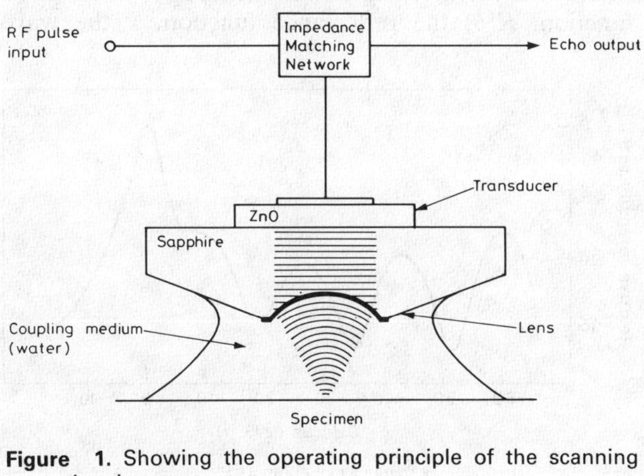

Figure 1. Showing the operating principle of the scanning acoustic microscope.

lens is mounted in a miniature scanner which mechanically scans the x-direction of the raster at approximately 30Hz. Data are acquired digitally and stored as 256 × 256 pixel images with 8-bit precision. Thus a scan time of about 10 s per frame is achieved, and the scan size is variable up to 400 μm square.

CONTRAST IN THE REFLECTION SAM

The $V(z)$ effect

It has been known for some time that the transducer output voltage of the SAM shows a periodic variation as a function of the lens to specimen spacing.[6] This has been described either as the $V(z)$ effect[7] or acoustic material signature.[8] A $V(z)$ curve obtained from the surface of sapphire using a cylindrical acoustic lens is shown in Fig. 2.[3] The cylindrical lens gives a better response than the conventional spherical lens in the presence of elastic anisotropy. Similar curves are obtained from isotropic materials using spherical lenses. A spherical lens must, of course, be used for imaging. The periodicity in the $V(z)$ curve is dependent upon the elastic properties of the specimen and is an important source of contrast in acoustic microscopy. The phenomenon may be understood either within the framework of an approximate ray model[9] or by the use of Fourier optics.[10]

The ray diagram of Fig. 3 shows the normal ray and a ray incident at the critical angle for generation of surface acoustic waves (Rayleigh waves) for an acoustic lens displaced a distance z from the focus towards the specimen (conventionally, a defocus of $-z$). The normal ray is directly reflected from the surface, whereas the ray incident at the critical angle is mode-converted to a propagating Rayleigh wave which 'leaks' acoustic energy back into the coupling medium at the critical angle. These re-radiated waves are focused and collected by the lens and interfere with the reflected normal ray. It is this interference that causes the observed periodicity in the $V(z)$ response.

Alternatively, the $V(z)$ response may be written in terms of Fourier optics as[10]

$$V(z) = \int_0^{\alpha} P^2(\theta)R(\theta) \exp(2jkz\cos\theta) \sin\theta \cos\theta \, d\theta$$

where α is the half-angle of the lens, $P(\theta)$ is the pupil function, $R(\theta)$ the reflectance function, k the wave

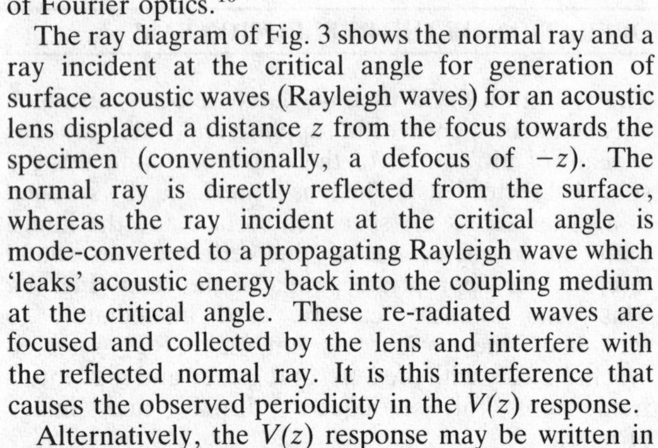

Figure 2. $V(z)$ of the sapphire surface.[3]

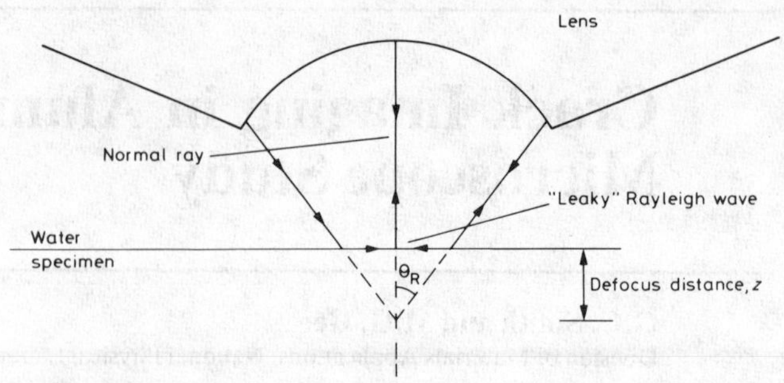

Figure 3. Ray diagram for a scanning acoustic microscope lens defocused towards the specimen.

number in the coupling medium and z the defocus distance. The dominant feature in $R(\theta)$, calculated from the acoustic Fresnel equations,[11] is the change in phase around the critical angle θ_R for generation of Rayleigh waves, giving rise to a periodicity $\Delta_{kz} = \pi/(1-\cos\theta_R)$ in the Fourier transform, which may be re-expressed as $\Delta z = \lambda_w/2(1-\cos\theta_R)$. Thus, the periodicity in $V(z)$ may be used as a measure of the Rayleigh critical angle θ_R, and hence the Rayleigh wavelength, λ, given by $\sin\theta_R = (V_W/V_R)$ where V_W is the velocity of longitudinal waves in the coupling medium (water) and V_R is the Rayleigh wave velocity.

Grain Contrast

An important aspect of imaging in the reflection SAM is the ability to detect contrast due to the random orientation of grains in a material and thus to visualise the grain structure at a polished but un-etched surface. This effect has recently been discussed by Somekh *et al.*[12] and shown to arise as a result of the elastic anistropy of individual grains of the material. For an anistropic material the reflectance function is dependent on ϕ, the azimuthal angle. Somekh and co-workers[12] include this in the expression for $V(z)$ by writing

$$V_{hkl}(z) = \int_0^{\alpha} R'_{hkl}(\theta) \, P^2(\theta) \exp(2jkz\cos\theta) \sin\theta \cos\theta d\theta$$

where $R'_{hkl}(\theta)$ is the complex mean reflectance function for the (hkl) plane and is given by

$$R'_{hkl}(\theta) = \frac{1}{2\pi} \int_0^{2\pi} R_{hkl}(\theta,\phi) d\theta$$

It is this variation of $R(\theta)$ with geometry that results in each grain orientation having its own characteristic $V(z)$ response and hence an image taken with a fixed value of lens defocus shows contrast from grain to grain depending on orientation.

Response in the Presence of a Discontinuity

Fig. 4 shows a ray diagram for an acoustic microscope lens defocused towards the surface of a sample containing a reflective discontinuity, for example a narrow crack, oriented parallel to the axial ray. In this case the 'leaky' Rayleigh wave is reflected from the discontinuity and interferes with the normal Rayleigh

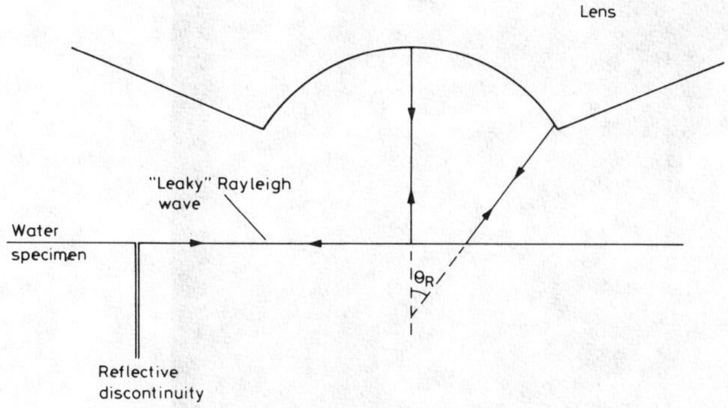

Figure 4. Ray diagram for a scanning acoustic microscope lens defocused towards a specimen containing a narrow reflective defect.

wave contribution in a manner which produces characteristic fringes of spacing $\lambda_R/2$ in the image.[13,14] The modification of the $V(z)$ curve in the presence of a discontinuity has been the subject of recent theoretical effort.[15,16,17] Somekh and co-workers[15,16] follow Bertoni and Tamir[18] in dividing the reflectance function for an isotropic material into two parts: a geometrical reflection coefficient and a Rayleigh wave reflection coefficient to account for the excitation and re-radiation of Rayleigh waves. In the case of a narrow discontinuity the geometrical part is not affected and contrast occurs as a result of the interaction of the Rayleigh waves with the discontinuity. Using a two-dimensional Green's function incorporating the Rayleigh wave reflectance leads to modified forms of the usual $V(z)$ expression for the presence of the discontinuity.

Within this framework, it is found that the modification of the $V(z)$ curve due to the discontinuity is a function of its Rayleigh wave transmission and reflection coefficients. Features in the sample surface may be classified as 'lossy' (zero reflection and finite transmission coefficients) or reflective (zero transmission and finite reflection coefficients) and it is found that reflective features are sources of contrast when the lens is defocused in both the positive and negative z directions, whereas absorbers of acoustic energy (the 'lossy' features) affect only the negative z region of the curve. The characteristic ripple of periodicity $\lambda_R/2$ is seen at the edge of a reflective discontinuity only, and is not produced by a 'lossy' feature. The theory also predicts contrast associated with the boundary between two regions of different Rayleigh wave velocity.

It is significant that detection of the discontinuity in the image depends on its ability to interact with Rayleigh waves propagating in the surface plane. Thus features such as grain boundaries or narrow microcracks can act as sources of contrast in the acoustic image even though their width may be much less than the theoretical resolution of the technique and hence their contribution to the geometrical part of the reflection function is negligible.

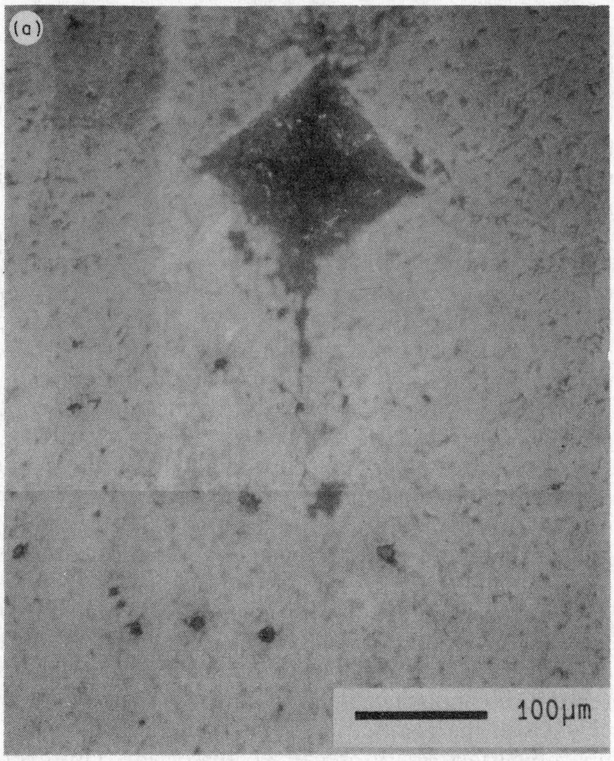

Figure 5 (a). Optical micrograph of the corner of a 196N microhardness indent in polycrystalline alumina.

RESULTS AND DISCUSSION

Polycrystalline Alumina

Alumina is widely used as an industrial ceramic but, due to its partly translucent character, it is difficult to examine for cracks using conventional optical microscopy without coating the specimen, or using other specialised techniques. Scanning electron microscopy suffers from the disadvantage that, for useful contrast to be obtained from narrow cracks, the field of view is rather small. Fig. 5 shows the optical and acoustic images of a crack propagating from the corner of a 196 N microhardness indentation in a polished alumina surface. It is clear that, although the resolution of the optical microscope is the higher, when observing a relatively large area (low magnification) the detectability of the crack is greater in the acoustic microscope. At this magnification the optical microscope, with the sample illuminated from above in the normal reflection contrast mode, shows the crack extending approximately 150 μm from the corner of the indentation whereas acoustically it is revealed extending to at least 200 μm. Full details of the crack could be brought out optically by using a series of high magnification micrographs, but this is inconvenient and makes direct measurement of the crack length difficult. Comparison of SAM and SEM measurements show that microcracks of width around 100 nm are detectable in the acoustic microscope.

The surface finish on the sample of Fig. 5 was 0.08μm centre line average (CLA). Fig. 6 shows two acoustic images taken with different amounts of defocus from a sample of the same material polished to 0.04μm CLA. The improved surface polish dramatically reduces Rayleigh wave scattering from the surface roughness and allows direct imaging of the

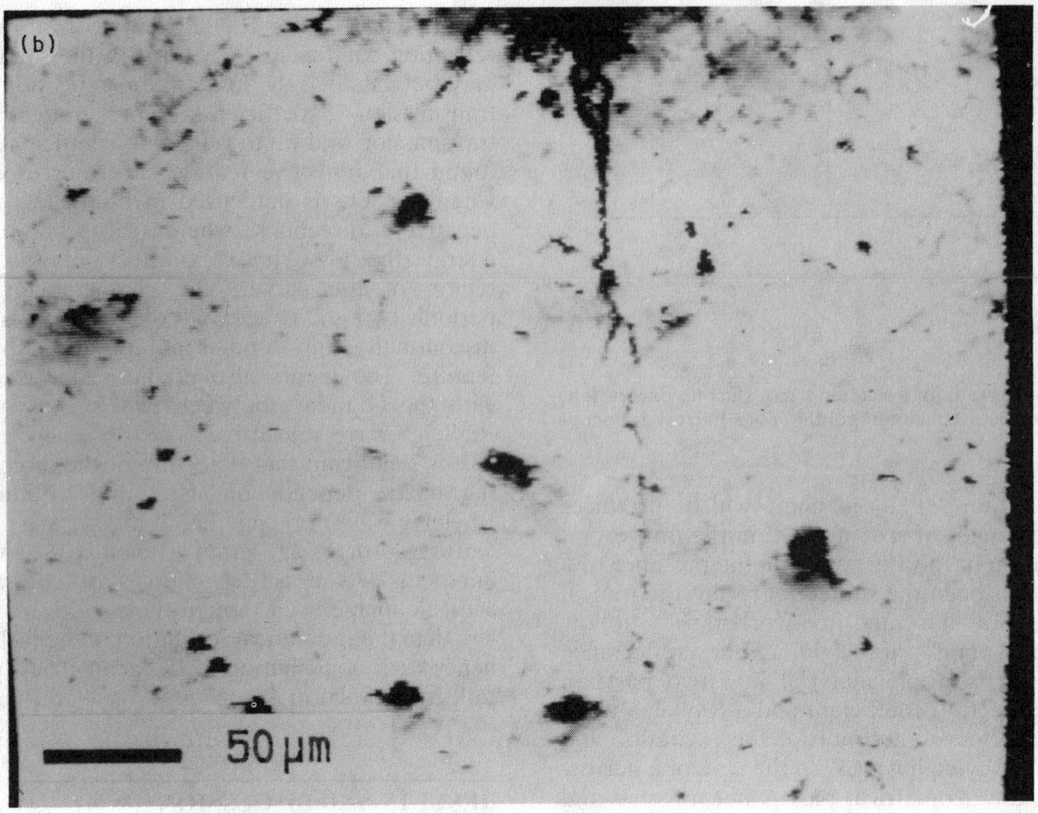

Figure 5 (b). acoustic image of part of the area shown optically in Figure 5(a).

grain structure of the material.[19] Also visible are a large number of pores, characteristic of the sintering process by which the material is manufactured. It is of interest to note that small changes in the defocus distance z can cause contrast reversal in the images. The effect is well known and the contrast mechanism is as discussed in the previous section. Its significance in the context of the present work is that, in the presence of microcracking, the crack pattern observed in the acoustic microscope can immediately be related to the grain structure.

The specimen of Fig. 6 was a pin of 99.5 wt% alumina of length 20 mm and diameter 6 mm which was prepared as part of an investigation into the wear properties of this material. During the wear test, the pin was held against a wheel of the same alumina for 180 min with a force of 120 N while the wheel was rotated to give a relative speed of 0.37 ms^{-1}. The sample was then sectioned longitudinally, mounted, and polished to a good optical finish using diamond paste and silica suspension to reveal the crack pattern formed as a result of the wear test. Fig. 6 shows an undamaged part of the specimen. Fig. 7(a) is an optical micrograph of an area near to the wear surface showing a fine crack which can just be discerned passing approximately horizontally across the centre of the field of view. The corresponding acoustic image taken with the lens just on the negative side of the focus (Fig. 7(b)) reveals the crack quite clearly and, in addition, shows a further network of fine lines. These correspond to the edges of the grains which are shown by moving to the positive z side of focus (Fig. 7(c)). The fine lines in Fig. 7(b) are clearly grain boundaries. This image shows the crack to follow partially grain boundaries and to be partially transgranular; information not readily available from other non-destructive techniques.

Single Crystal Alumina (Sapphire)

As a model material for the study of crack propagation, the single crystal form of alumina, sapphire, is convenient. The interaction of the cracks with the grain structure does not need to be considered. The material also has the advantage of transparency, enabling sub-surface features to be examined in the optical microscope.

Two optical micrographs of the impression made by a Vickers indenter with a load of 4.9 N in a polished sapphire surface are shown in Fig. 8. Figure 8(a) was taken with the optical microscope focused on the surface and shows the characteristic symmetric crack pattern produced at the corners of the indentation. Figure 8(b) was taken with the optical microscope moved from the surface focus by 5 μm towards the surface of the sample and shows, in addition to the out of focus surface crack pattern, a sub-surface crack at an angle to the main left hand radial crack. The refractive index of sapphire is approximately 1.6 so this feature is approximately 3 μm below the surface.

The acoustic image of this indentation obtained with the lens at focus is shown in Fig. 9(a). The surface crack-pattern seen in the optical micrograph is reproduced and, since image contrast at focus is due to changes in the geometrical reflection coefficient only, no sub-surface features can be discerned. However, when the acoustic microscope is defocused by 32 μm towards the specimen surface two additional effects

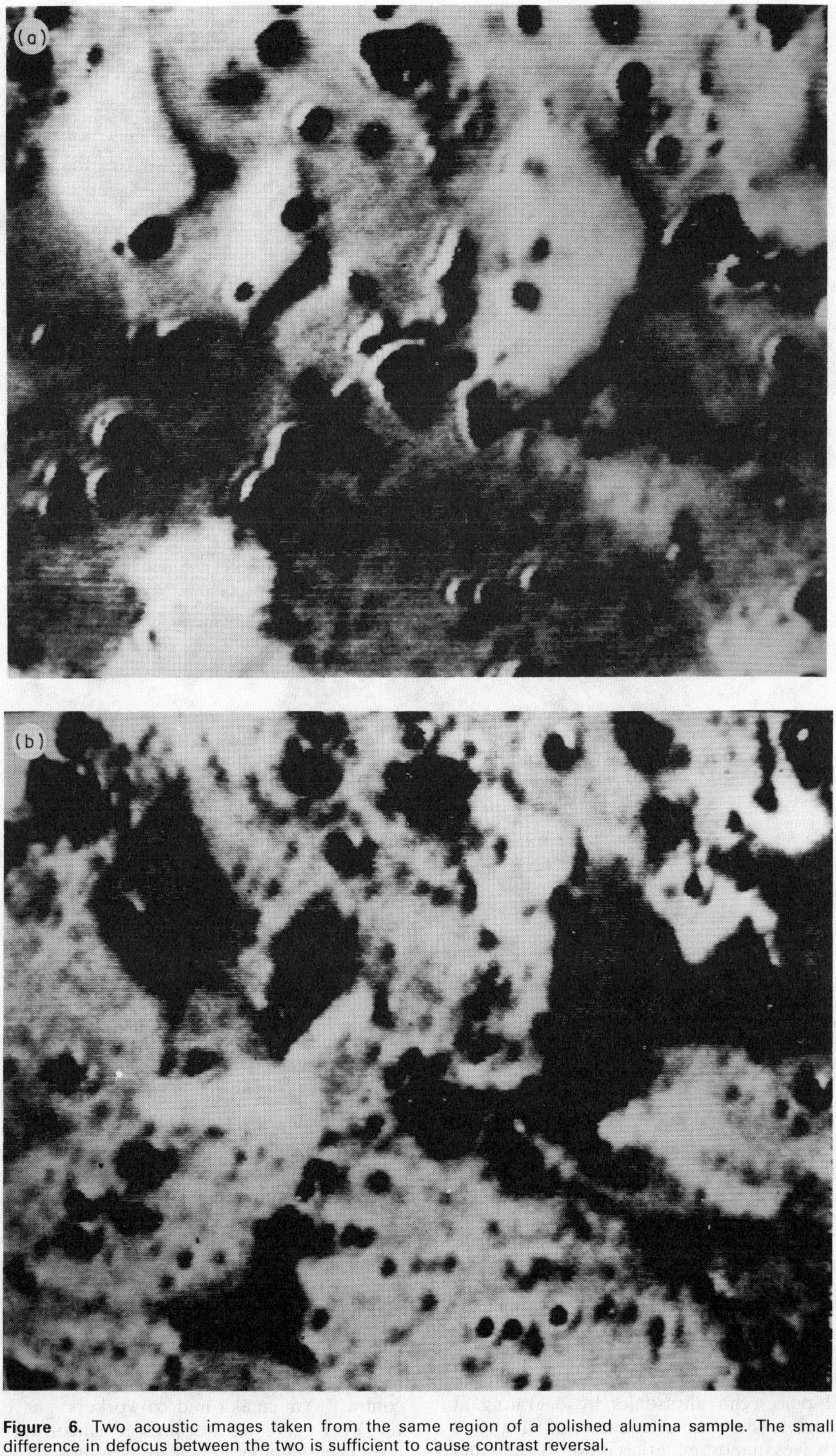

Figure 6. Two acoustic images taken from the same region of a polished alumina sample. The small difference in defocus between the two is sufficient to cause contrast reversal.

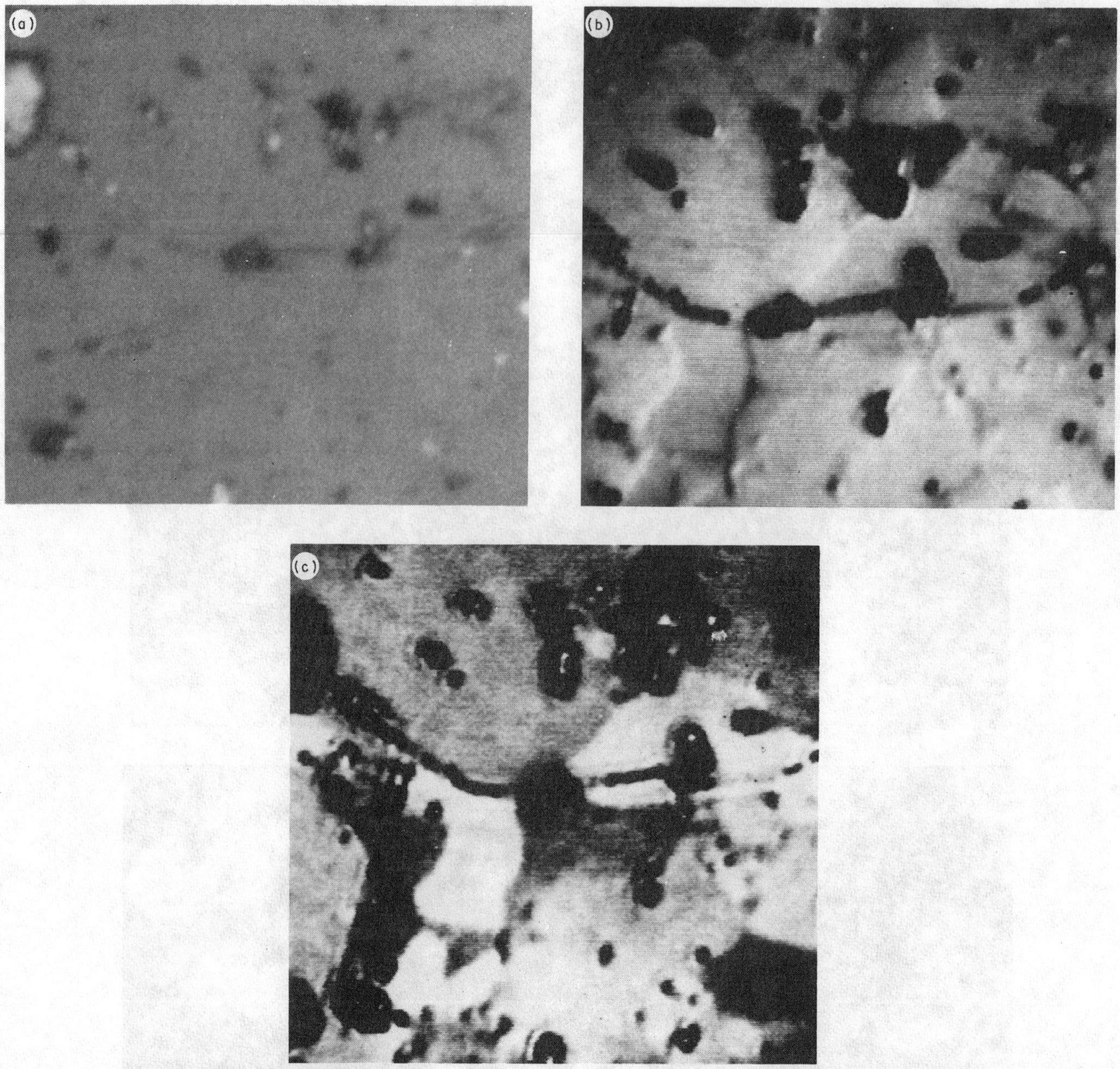

Figure 7. (a) Optical micrograph of a polished alumina surface containing a fine microcrack; (b) acoustic image of the area shown in (a), taken with lens displaced slightly towards the sample (negative *z* defocus); (c) acoustic image of the area shown in (a), taken with lens displaced away from the sample (positive *z* defocus).

become apparent (Fig. 9(b)). Characteristic interference fringes of spacing $\lambda_R/2$ are seen around the surface-breaking features and, in addition, interference fringes can be observed at an angle to the surface-breaking crack in a position corresponding to the sub-surface crack seen optically in Fig. 8(b). It is important to note that this sub-surface defect causes contrast in the acoustic image by its interaction with the surface acoustic waves generated by the defocused lens. It is not correct to think of the lens as 'focusing below the surface' in any way. However, it is possible to detect sub-surface defects in the SAM in the manner of conventional pulse-echo ultrasonics by operating at lower frequencies where the attenuation of shear and longitudinal waves in most materials is not so great and significant depth penetration can be achieved.[20] In the present case, the optical transparency of the sample enabled identification of the near-surface buried defect. For opaque materials the SAM shows a unique ability to non-destructively detect sub-surface features lying within approximately one Rayleigh wavelength of the surface.

The idealised crack pattern produced by the microhardness indentation tool is shown schematically in Fig. 10.[22] As well as the radial cracks that can be viewed from above, a horizontal lateral crack is formed. Using the arguments given above, if this lateral crack is within λ_R of the surface then the $V(z)$ response will be affected and it will act as a source of contrast. Yamanaka and co-workers[21] have modelled the $V(z)$ curve for hot-pressed silicon nitride in the presence of a lateral crack at various depths and shown that this is indeed the case. Moving the lens slightly towards the sample from focus is predicted to give contrast from the crack region. This also appears to be the case for sapphire as illustrated in the image shown in Fig. 11 taken of a 4.98 N microhardness indent in the (0001) basal plane of sapphire.

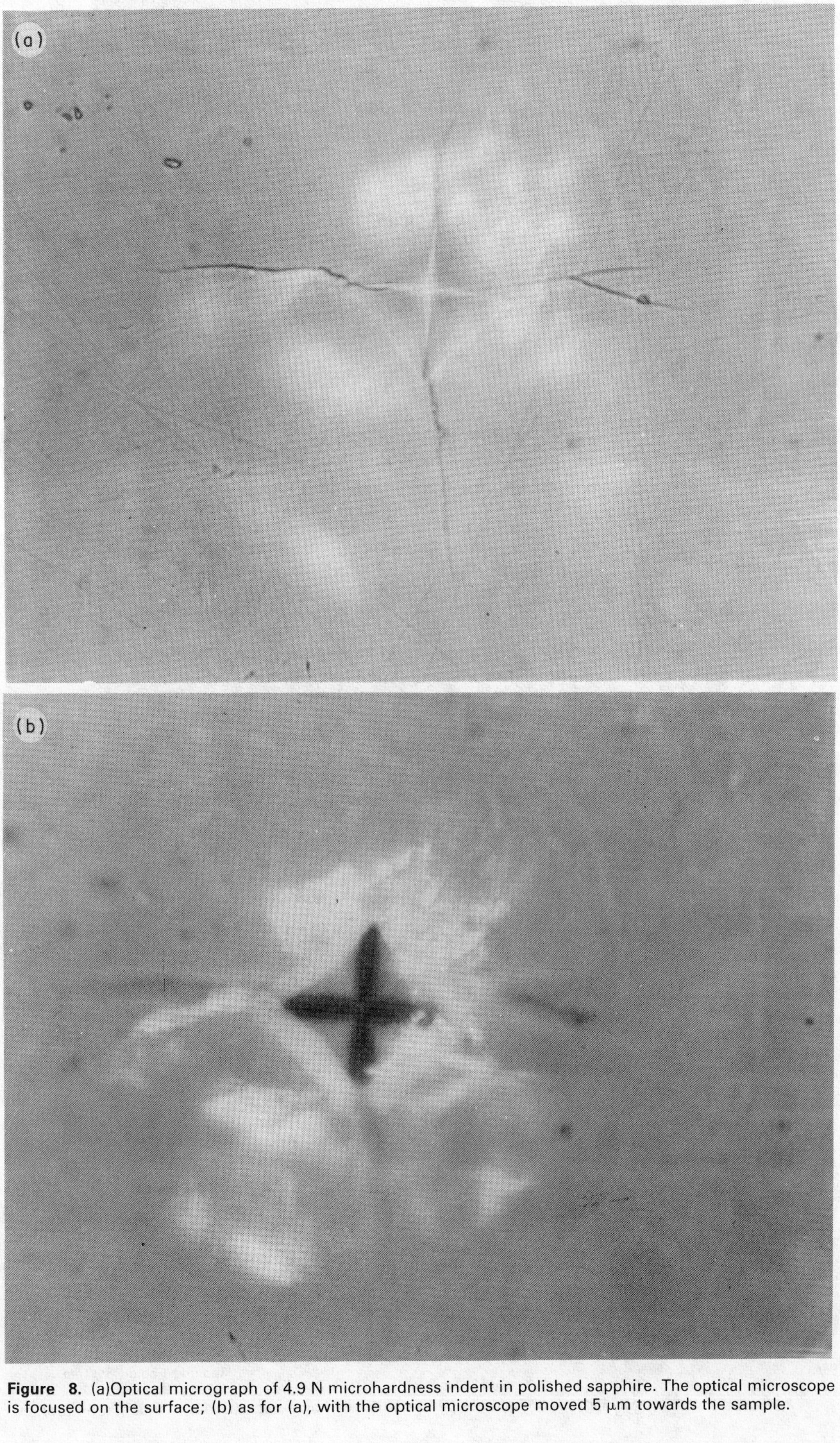

Figure 8. (a)Optical micrograph of 4.9 N microhardness indent in polished sapphire. The optical microscope is focused on the surface; (b) as for (a), with the optical microscope moved 5 μm towards the sample.

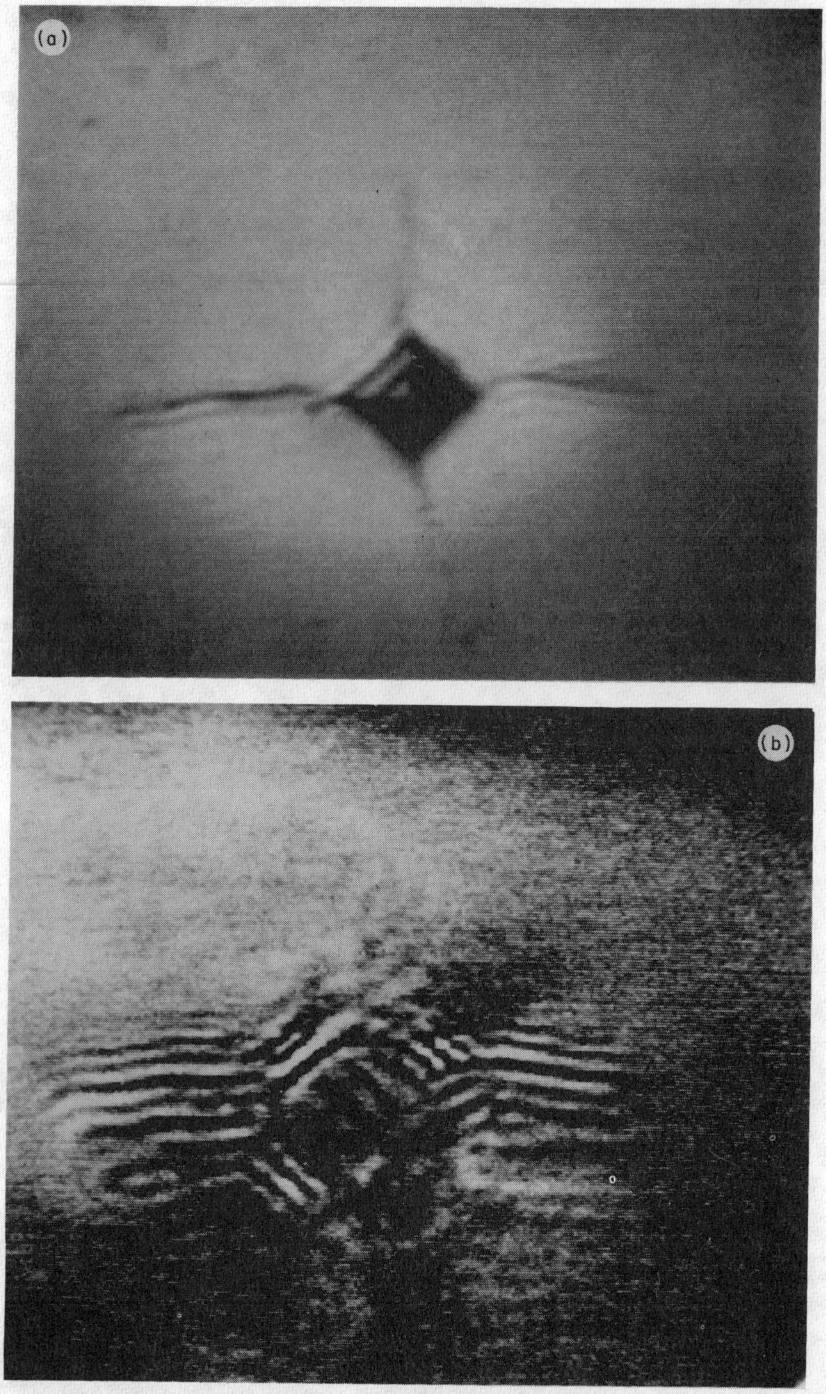

Figure 9. (a)Acoustic image corresponding to Fig. 8(a). Acoustic lens focused on sample surface; (b) as for (a), with the acoustic lens defocused by — 32 μm (towards the sample surface).

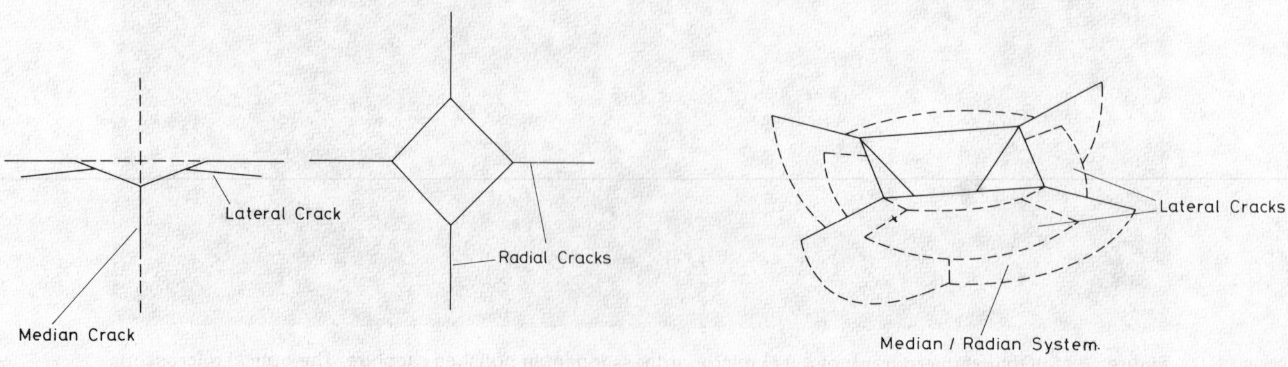

Figure 10. Schematic diagram of the crack pattern produced by a microhardness indentation tool.

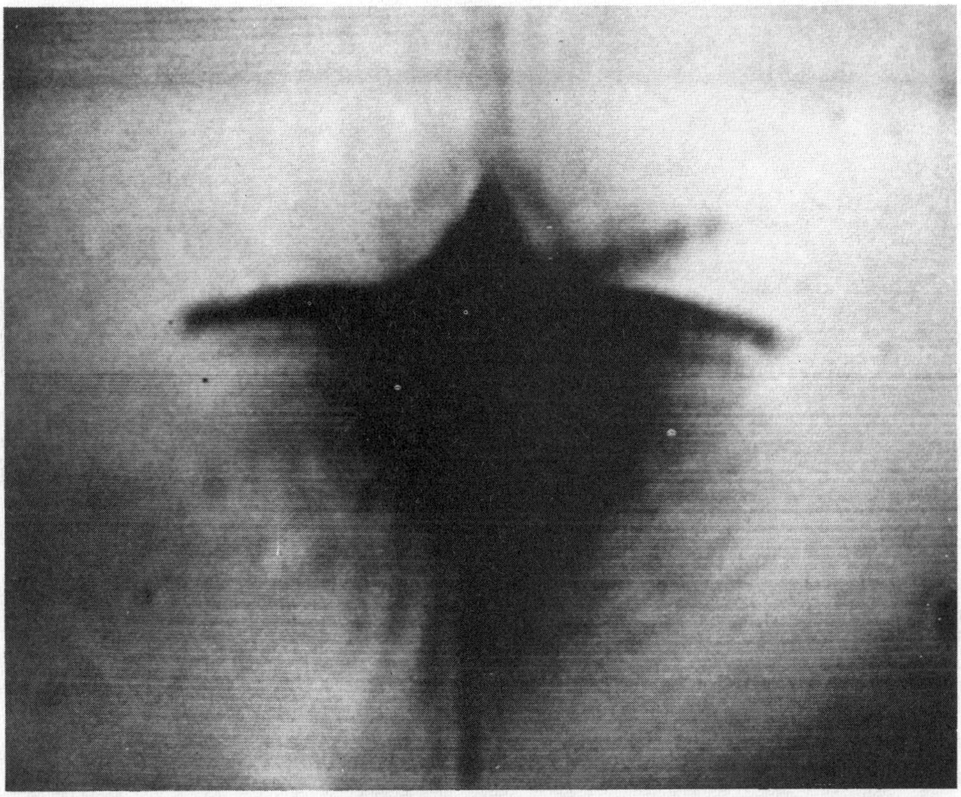

Figure 11. SAM image of a 4.9 N microhardness indent in sapphire with the lens displaced slightly from the focus towards the specimen, showing enhanced contrast from the region of the lateral crack.

CONCLUSIONS

The SAM shows several clear advantages over other techniques for the non-destructive examination of cracks in alumina. When viewing a large area at low magnification cracks not readily apparent in the optical microscope become visible acoustically. The contrast mechanism is Rayleigh wave reflection and, as a consequence, narrow cracks of width much less than the intrinsic resolution of the instrument can be readily detected. These cracks have a negligible effect on the geometric reflection function. For samples with suitably polished surfaces it is possible to reveal the grain structure by utilising different parts of the $V(z)$ curve, and hence to study the interaction of the crack pattern with the microstructure of the material.

However, the most promising area for the application of the SAM to the study of defects in materials appears to be for the detection of features buried within approximately one Rayleigh wavelength of the surface. Such features affect the $V(z)$ response in a manner which can be modelled theoretically, and, for opaque materials, are difficult to reveal non-destructively by any other technique.

REFERENCES

1. M. Nikoonahad, *Contemp. Phys.* **25,** 129 (1984).
2. J.B. Wachtman Jr, W.E. Tefft, D.G. Lam Jr and R.P. Stinchfield, *J. Res. Nat. Bureau Standards* **64A,** 213 (1960).
3. J. Kushibiki, A. Ohkubo and N. Chubachi, *Electronics Letts.* **17,** 534 (1981).
4. C.F. Quate, A. Atalar and H.K. Wickramasinghe, *Proc. IEEE* **67,** 192 (1979).
5. I.R. Smith, R.A. Harvey and D.J. Fathers, *IEEE Trans.* **SU–32,** 274 (1985).
6. A. Atalar, *J. Appl. Phys.* **49,** 5130 (1978).
7. H.K. Wickramasinghe, *Electronics. Letts.* **14,** 305 (1978).
8. R.D. Weglein and R.G. Wilson, *Electronics. Letts.* **14,** 352 (1978).
9. W. Parmon and H.L. Bertoni, *Electronics. Letts.* **15,** 684 (1979).
10. C.J.R. Sheppard and T. Wilson, *Appl. Phys. Letts.* **38,** 858 (1981).
11. L.M. Brekhovshikh, *Waves in Layered Media,* 2nd Ed., Academic Press, New York (1980).
12. M.G. Somekh, G.A.D. Briggs, and C. Ilett, *Philos. Mag.* **A49,** 179 (1984).
13. C. Ilett, M.G. Somekh and G.A.D. Briggs, *Proc. Roy. Soc. Lond.* **A393,** 171 (1984).
14. K. Yamanaka and Y. Enomoto, *J. Appl. Phys.* **53,** 846 (1982).
15. M.G. Somekh, H.L. Bertoni and G.A.D. Briggs, *Proc. IEEE. Symp. Sonics & Ultrasonics, Dallas* (1984) published by IEEE.
16. M.G. Somekh, H.L. Bertoni, G.A.D. Briggs and N.J. Burton, *Proc. Roy. Soc. Lond.* **A401,** 29(1985).
17. B.N. Cox and R.C. Addison in *Rev. Prog. in Quant. N.D.E.,* Vol. 3B, pp 1173–1184, ed. by D.O. Thompson and D.E. Chimenti, Plenum (1984).
18. H.L. Bertoni and T. Tamir, *Appl. Phys.* **2,** 157 (1973).
19. G.C. Smith, E.A. Almond, M.G. Gee and M. Nikoonahad, *Phil. Trans. R. Soc. Lond.* in press (1986).
20. A. Atalar, *IEEE. Trans.* **SU-32,** 164 (1985).
21. K. Yamanaka, Y. Enomoto and Y. Tsuya, *IEEE. Trans.* **SU-32,** 313 (1985).
22. B. Lawn and R. Wilshaw, *J. Mater. Sci.* **10,** 1049 (1975).

SURFACE AND INTERFACE ANALYSIS, VOL. 9, 65–71 (1986), EXTENDED ABSTRACTS

Laser Ionization Mass Analysis: its Potential as a Quantitative Microanalytical Technique

T. Dingle and B. W. Griffiths
Cambridge Mass Spectrometry Ltd, Cambridge Science Park, Milton Road, Cambridge CB4 4FX, UK

In general the laser microprobe is considered to be a non quantitative technique. Two routes are available for quantification, theoretical prediction of ion yields from known or measured experimental parameters, and the empirical approach of using fixed analytical conditions chosen to minimize variations in ion yields, and subsequent comparison of data from known and unknown samples. This paper describes some work we have performed recently using the second of these methods.

The main experimental variable is the incident laser power density. This can be controlled at will, but in practical situations is not known to be better than $\pm$ 10%. It is clear that ion yields of various elements will vary since the energy of the exciting photons is $\approx$ 4 eV whilst the ionization potentials of the elements vary from 4 eV to 25 eV. Experimentally if sufficient photons are present, i.e. high incident laser power, it is possible to keep the variation in yield to less than an order of magnitude. At low laser powers the yield variations are more analagous to those in SIMS (4–5 orders of magnitude). The use of high laser power also reduces the effect of the uncontrolled shot to shot variation in absorbed power.

By using high irradiances and averaging the results of 5 shots we have analysed a series of silicate glasses and corrected the results using calibration coefficients derived from analysis of an NBS standard glass. The results obtained were accurate to $\pm$ 10% on the bulk constituents of the glass and to $\pm$ 30% on the trace elements at levels down as far as a few tens of ppm.

If the calibration coefficients obtained by this method are graphed against ionization potential a clear correlation is present, although anomalies do exist and it seems that other effects, both instrumental and physical are at work. We have, however, by judicious extrapolation from this graph been able to estimate successfully concentrations of elements for which no calibration coefficient was known. Comparison of the calibration coefficients generated from the glass system, with those from a single crystal inorganic, $CaCO_3$, have shown variations of $\pm$ 15% suggesting that an all embracing system may be possible.

We have shown that by empirical means the laser microprobe may be used as a semi-quantitative tool in cases where control samples are available, and that promise exists for the more general extension of this form of system.

On the Influence of Crystallographic Orientation on the Shape of Depth Profiles*

Maria Kühlein
Erich-Schmid-Institut für Festkörperphysik der Österreichischen Akademie der Wissenschaften, 8700 Leoben, Austria

Depth profiles of silver layers on polycrystalline copper substrates were investigated in an ultrahigh vacuum unit equipped with an Auger electron spectrometer (AES) and an ion-bombardment gun. Depth profiles were established by argon-ion bombardment ($p = 6 \times 10^{-5}$ Torr; $j_p = 38$ μA cm^{-2}) and simultaneous AES analysis. Further experimental details are described in a prior paper.[1]

Silver layers of approximately 1500 nm thickness were vapour-deposited on recrystallized copper substrates. A strong scatter of the depth resolution $\Delta z = |z_{0.84} - z_{0.16}|$ was observed. Δz varied from 294 nm to 796 nm. Investigations by scanning-electron microscopy (SEM) of sputtered specimens show copper grains without silver and grains covered with silver cones. Different orientation of the copper grains causes different growth of the silver layer. The sputtering yield depends on the orientation of the grains.[1,2] These two effects contribute to the strong scatter of depth resolution.

Heavily cold-worked copper substrates (true strain $\epsilon_t = 3.0$) with a pronounced $<111> - <100>$ fibre texture show a drastically reduced scatter of depth resolution which varied from 301 nm to 417 nm.

To investigate the growth of the silver layer further experiments were performed with silver layers of 610 nm thickness deposited on recrystallized copper substrates. These layers were sputtered off stepwise (40–100%) by argon-ion bombardment. In SEM investigations the silver layers reflect the grain structure of the copper substrate already after sputtering off 40% of the initial thickness.

* Sponsored by "Fonds zur Förderung der wissenshaftlichen Forschung", Vienna, Austria, under contract No. P 5649.

REFERENCES

1. M. Kühlein and I. Jäger, *Surf. Interface Anal.* **6,** 129 (1984).
2. G. Carter, A. Gras-Marti and M. J. Nobes, *Rad. Eff.* **62,** 119 (1982).

Laser Probe Microanalysis: Performance Assessment and Applications

M. J. Southon, A. Harris, V. Kohler, S. J. Mullock and E. R. Wallach
Department of Metallurgy and Materials Science, Cambridge University, CB2 3QZ, UK

The LIMA 2A instrument used for this work has been described in detail elsewhere.[1] A focused frequency-quadrupled Nd-YAG laser incident on the surface of a solid sample, via Cassegrain optics, provides a burst of ions for analysis in a time-of-flight mass spectrometer. This consists of a folded field-free flight tube with an electrostatic reflector, designed to compensate for the initial spread in ion energies,[2] and a modified venetian-blind electron multiplier for ion detection. The current transient at the output of the electron multiplier is buffered and stored in a fast transient recorder for further processing. Material is sampled from a crater 2–3 μm in diameter and ~ 0.1 μm in depth with the option of a crude depth profile provided by repeated shots at the same position. As yet, in common with other laser microprobes,[3] the resulting mass-peak areas give only semi-quantitative information on the relative abundances of elements in a sample. For many problems this information is sufficient and LIMA has proved a valuable tool in the investigation of the surface and near-surface conditions in a variety of samples, due to the rapid rate at which data is obtained together with the minimal specimen preparation required; some specific examples are detailed later. Efforts have also been made to identify factors responsible for the variable elemental sensitivities and the scatter in experimental results.

QUANTIFICATION

The measured peak area of a particular element depends primarily on its concentration, the ionization potential, the absorbed laser energy and a machine transmission/detection efficiency.[3] To obtain quantitative analyses these dependences must first be estimated either from theoretical models or empirically, but in either case, consistent experimental results are required. The dependence on absorbed laser energy is strong and although the incident laser energy at a constant setting has a shot-to-shot variation of only 3% (one standard deviation, 1 σ), the measured scatter in absolute peak areas for Si ranges from about 130% (1 σ) at low laser-power to about 10% at about double this power. Most of this scatter is probably due to variations in the efficiency with which the laser energy is absorbed. The scatter is reduced if peak-area ratios are used, but differences in the dependence of ion yield on absorbed power from element to element and the presence of non-linearities in the detection system mean that the scatter is still large, especially at low laser-power.

One approach to quantification is to use a high laser-power, where the relative sensitivities appear to saturate, and an empirical relationship for sensitivity vs ionization potential.[4] However, the penalty is a loss in spatial resolution with craters as big as 20 μm in diameter. If two different elemental ratios, measured at various laser powers on a homogeneous specimen, are plotted against each other a curve is obtained which suggests that the measured ratio of two elements of known concentration could be used to predict a correction factor for other pairs of elements at all powers.

Isotope ratios have also been investigated, as in this case the dependence of ionization efficiency on laser power should be the same for each component. The results rêveal a serious nonlinearity in the response of the electron multiplier, as found elsewhere.[5] This can be described by a single curve of correct vs measured peak area, valid for a range of different gain settings and for both the elements investigated so far, Si and Ca: hence a relatively simple correction may be made. It should be noted that this non-linearity causes the scatter in absolute peak areas to be reflected in the peak-area ratios.

Finally, preliminary calculations have been made to predict the temperature and electron density in the laser-induced plasma assuming the LTE model.[6] These values, 12000–17000 K and $\sim 10^{18}$cm^{-3}, are consistent with the establishment of LTE and agree with those found by other workers.[7]

APPLICATIONS

Applications in materials science include analysis of metal surfaces and interfaces with particular interest in the detection of impurities and light elements (e.g. carbon) which can be more difficult to assess by other means. The wear rate of Ti has been found to depend strongly on the atmosphere present, but the reason for this is not known. LIMA has been used to look for the presence of oxygen and carbon in the surface which may affect the hardness or inhibit local welding. Perhaps surprisingly, similar levels were found in specimens worn both in air and in argon. A similar search was made on the fracture surface of a specimen that had undergone a fatigue-crack growth-rate threshold test, showing a segregation of oxygen to the threshold region. The lateral resolution of LIMA has also allowed the investigation of interfaces. for example, the intercellular regions in a sample of gas-atomized Al–Li powder have been studied in order to assess the segregation of alloy impurities.[8] The low-mass coverage of LIMA has also allowed a measurement of variations in the Li concentration to be made.

Relatively little sample preparation is required for LIMA and the technique does not suffer from the charging problems associated with a charged probe. For example, a sample of red mud containing low concentrations of gold was simply smeared onto double-sided sticky tape. The particles could be roughly sorted by eye in the viewing optics and a LIMA survey made of gold content. It was possible to show the presence of high concentrations of gold in certain types of particle even though the overall concentration was only ~ 1 ppm. The rest of the mass spectrum for the gold-bearing particles was then examined to choose a leaching process for extracting the gold.

Finally, some preliminary efforts have been made to discover whether there is a high-mass limit to the sensitivity of the machine. A sample of oxidized insulin, mixed with CsI as a mass marker and glycerol to promote ionization, has given a mass peak at 3519 amu.

REFERENCES

1. M. J. Southon *et al. Vacuum*, **34**, 23 (1984).
2. B. A. Mamyrin *et al. Sov. Phys. JETP*, **37**, 45 (1973).
3. T. Mauney, *Microbeam Anal.* 299 (1985).
4. T. Dingle and B. W. Griffiths, *Surf. Interface Anal.* **9**, 65 (1986).
5. D. S. Simons, *Int. J. Mass Spec. Ion Process*, **55**, 15 (1983).
6. P. K. Carroll and E. T. Kennedy, *Contemp. Phys.* **22**, 61 (1981).
7. R. Kaufmann, *Microbeam Anal.* 341 (1982).
8. R. A. Ricks *et al. Proc. 3rd Int. Al-Li Conf.* (1985).

LIMA Analysis of Ga(1 − x)Al(x)As Alloys: the Effect of Alloy Composition on Secondary Ion Yields

G. J. Edge and E. A. Clark
Department of Physics, Loughborough University of Technology, UK

Laser-induced ion mass analysis (LIMA) is a relatively new technique which owing to an incomplete understanding of the processes involved in the interaction between the laser beam and the sample is at present limited in its ability to perform fully quantitative analysis. Because LIMA is a microprobe technique many materials which are suitable standards for other techniques are often inappropriate for providing calibration of LIMA analyses due to localized inhomogenity.

However, Ga(1 − x)Al(x)As alloys are available in a range of known compositions and these were chosen for this investigation. Since there is considerable interest in the relationship between LIMA and SIMS techniques it was appropriate to compare the data from LIMA analysis with that for SIMS analysis. The SIMS analysis is also important in its own right since such information is necessary for the analysis of SIMS of super-lattice structures made from these alloys. For efficient LIMA analysis it is necessary to determine the optimum ionization conditions for the sample. This is a complex function involving laser power density and ionization cross-section. As a first approximation the degree of ionization of an element can be related to its ionization potential. In Ga(1 − x)Al(x)As alloys two elements, Ga and Al, have virtually identical ionization potentials (Ga I.P. = 6.00 Al I.P = 5.98) whereas that of As is significantly higher at 9.81. To find the optimum ionization conditions for As a sample of a GaAs wafer was analysed to determine the relationship between signal intensity and laser power. As would be predicted from ionization potentials Ga was found to ionize at lower laswer power than As, but it was found that the maximum signal for both Ga and As was obtained under similar values of laser power.

The conditions of high ionization efficiency determined for GaAs were subsequently used in the analysis of the Ga(1 − x)Al(x)As alloys. It was found that under these conditions the Al : Ga ratio increased with increasing Al concentration, but that the measured ratio was approximately twice that expected from the sample stoichiometry and significant scatter in the results was also observed. No variation in As signal intensity was observed over the range of Al concentrations analysed. However, the As signal intensity was found to be only 0.6 of the value obtained with GaAs. This would indicate that the introduction of Al into the system inhibited the ionization of As.

The analyses were then repeated under conditions of lower power density in order to suppress As ionization. Under these conditions it was found that the Al : Ga ratio obtained was in good agreement with the sample stoichiometry and that significantly improved reproducibility of results was achieved.

For SIMS analysis, depth profiles on each of the samples were performed using 15 keV/O_2^+ ion bombardment. The Al : Ga and Ga : As secondary ion yield ratios were calculated once equilibrium conditions had been achieved.

It was found that the Al : Ga ratio increased linearly with increasing Al concentration (x) over the range of compositions examined, $x = 0.21$ to $x = 0.53$.

The Ga : As ratio as a function of (x) over the range $x = 0.21$ to $x = 0.53$ was found to be non-linear and showed more scatter in the results than did the Al : Ga ratio. It was observed that the As signal was enhanced by the presence of Al in contrast to the results obtained in the LIMA analysis. The Al : As ratio showed little variation with composition.

In conclusion it was found that both LIMA and SIMS show an increase in Al : Ga ratio as a function of increasing Al concentrations. Further, suppression of As ionization in the LIMA analysis resulted in an Al : Ga ratio which agreed well with the stoichiometry of the sample. The significant difference between the two techniques was in the As signal, for SIMS this was found to increase with increasing Al concentration whereas in LIMA it was found to be constant but significantly lower than that for GaAs.

A New Equipment for Angle Resolved Electron Spectroscopy of Gases and Solid Surfaces

L. Kövér, Á. Kövér and I. Cserny
Institute of Nuclear Research of the Hungarian Acad. of Sciences, H–4001 Debrecen, PO Box 51, Hungary

T. Veszprémi and Gy. Zsombok
Technical University of Budapest, H-1521 Hungary

A new UHV instrument and measuring control system has been designed and built for angle resolved electron spectroscopy of gaseous and solid samples. Figure 1 shows the scheme of the analyser part of the equipment. The novel, 'distorted field' electrostratic analyser[1,2] built into the system, is applicable for the analysis of energy and angular distribution of UV excited photoelectrons, electron beam excited Auger-electrons and electrons coming from interactions between molecules and metastable atoms, in a wide electron energy range (up to 15 keV). The modular construction of the spectrometer makes possible its applicability also for coincidence (e, 2e) electron spectroscopy.

Angular distributions of electrons emitted in the 0°–133° angular range (related to the direction of the exciting beam) can be detected by using two overlapping angular regions corresponding to two different directions for excitation (Fig. 1.). For the excitation of photoelectrons, a DC capillary discharge UV source[3] has been developed. The energy resolution of the spectrometer obtained without retardation is demonstrated on Fig. 2 for the case of HeI excited Ar 3p photoelectron lines. The vacuum system is pumped by three diffusion pumps ensuring $3000\ \mathrm{l\ s^{-1}}$ total pumping speed. The attainable vacuum level is 10^{-9} mbar without baking.

For the acquisition and handling of the measured data, as well as for the control of the measurements a microprocessor based modular system[4,5] with a software package written in FORTH has been constructed and built.

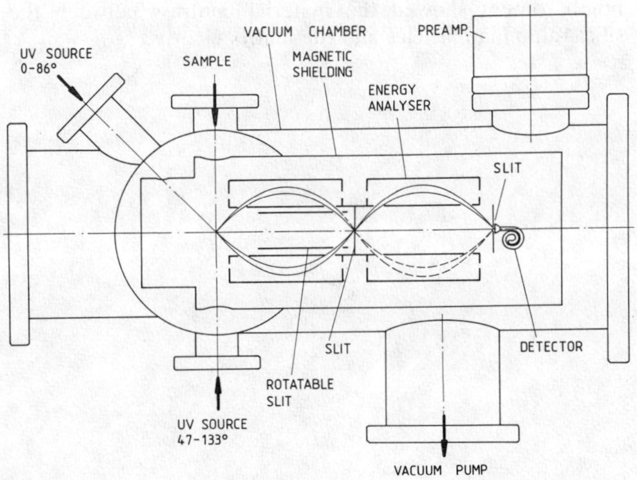

Figure 1. Schematic diagram of the analyser part of the equipment.

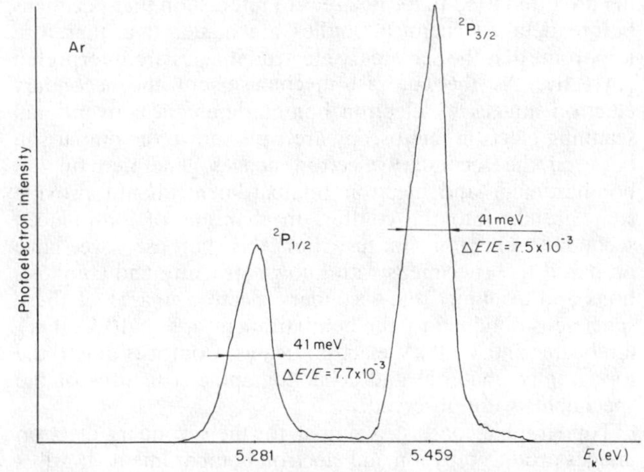

Figure 2. Ar 3p spectrum obtained by HeI excitation.

REFERENCES

1. D. Varga, Á. Kövér, L. Kövér, L. Redler, *Nuclear Instrum Meth Physics Res.* **A238**, 393 (1985).
2. Á. Kövér, D. Varga, E. Szmola, J. Herbák. L. Kövér, I. Cserny, *Proceedings of the 5th Seminar on Electron Spectroscopy of Socialist Countries, Aug. 26–29, 1984, Dresden*, p. 69.
3. Y. Harada, personal communication.
4. I. Cserny, T. Lakatos, J. Molnár, G. Pintér, S. Fekete, *Proceedings of the 5th Seminar on Electron Spectroscopy of Socialist Countries, Aug. 26–29, 1984, Dresden*, p. 29.
5. I. Cserny, *Nucl. Instrum. Meth. Physics Res.*, to be published.

Comparative Studies of Constrast Mechanisms in Electron and Ion Induced Secondary Electron Images

A. Brown†, D. Briggs*, S. Y. Lai† and J. C. Vickerman†
† Department of Chemistry, UMIST, PO Box 88, Manchester, M60 1QD, UK
* ICI Petrochemicals and Plastics Division, PO Box 90, Wilton, Middlesbrough, Cleveland TS6 8JE, UK

The secondary electron images obtained during SIMS imaging are often used to locate areas of interest on the specimens before detailed chemical studies are made. It is therefore important that the secondary electron images are interpreted correctly. As the contrast mechanisms of the secondary electron images by electron bombardment in conventional scanning electron microscopy are well known, a comparison between the secondary electron images generated on ion bombardment and electron bombardment should provide some insight into the contrast mechanisms of ion-induced secondary electron images. For this purpose, we have prepared test specimens with known structure and compositions and examined the secondary electron images of these specimens obtained by the bombardment of 5 or 10 kV Ga^+ ion beams and 4–30 kV electron beams. Contrasts due to the topography, material and crystallographic structures of the speciments were observed.

Topographic contrast is similar for the secondary electron images produced by ion and electron bombardment. It arises out of the differing inclination of different parts of the specimen surface with respect to the direction of the primary beam and the position of the detector. Also regions that are sheltered by the surrounding areas of the specimen, e.g. small hollows tend to appear dark due to the absorption of the emitted secondary electrons by the adjacent parts of the specimen.

Material contrast depends on the nature of the primary beam. For the tin–zinc alloy and tin–copper alloy studied, the grains with different compositions have different secondary electron yields. In both cases, the contrasts between the phases in the Ga^+ ion-induced images is opposite to that of the electron-induced images. The presence of a thin layer of surface contaminants on a specimen alters the material contrast.

Crystallographic contrast arises out of the channelling of the primary beams when the primary beam is aligned with a low index direction of a crystalline specimen. On a clean polycrystalline brass specimen, different grains show different brightness depending on their crystallographic orientations. The contrast between the grains can be changed by varying the specimen tilt angle. The presence of a thin layer of surface contaminant removes crystallographic contrasts.

Due to the much smaller range of an ion beam compared with an electron beam of similar energy, the ion probe is much more surface sensitive. The secondary electron image of a specimen of silica–alumina particles embedded in a plastic matrix and covered with ca. 20 nm of carbon produced by 10 kV Ga^+ bombardment showed only the surface topography whereas the image produced by 10 kV electron bombardment showed the material contrast between the silica–alumina particles and the matrix clearly.

Data Acquisition and Control System for Quadrupole SIMS Instruments

W. Hilgers and J. Herion
Institut für Grenzflächenforschung und Vakuumphysik, Kernforschungsanlage Jülich, Postfach 1913, D-5170 Jülich, FRG

If SIMS is used as a research method both high performance and versatility of the data acquisition and control system are required. These properties tend to be in contradiction with easy and fast operation and, therefore, are lacking in many commercial instruments. In this paper, we describe a system which combines the above properties with relative low costs. In our case, the system acts as an interface between a HP 200 series desktop computer and ATOMIKA SIMS components. Due to the use of standard techniques, it can be used also with other SIMS components and desktop computers.

Most important in SIMS is the achievement of sputtering craters with flat bottoms and the elimination of signals from the crater edges. These requirements are met in our case by scanning the focused primary beam over a rectangular area and gating the signal in such a way that the counting electronics is activated only if the ion beam is within the flat part of the sputtering crater (electronic window).

A schematic diagram of the data-acquisition and control system is given in Fig. 1. The flow of information in the system is controlled by the scan control unit which provides the voltage ramps for the beam scanning, the gate signals for the electronic window and the 'beam reset' pulses which are generated when the ion beam, after completing one frame, returns to its starting position. The secondary ion pulses originating from the channeltron of the mass spectrometer are fed via a preamplifier and an amplifer/discriminator to the counting electronics. The counting results corresponding to one frame of the ion beam are transferred to the computer by an IEEE 488 interface. The interface system contains also digital-to-analog converter modules which provide the voltages for the setting of the mass of the mass spectrometer and for the target potential. A switching module is used for switching the ion beam, the high voltage supply of the channeltron and the rf power of the quadrupole mass spectrometer.

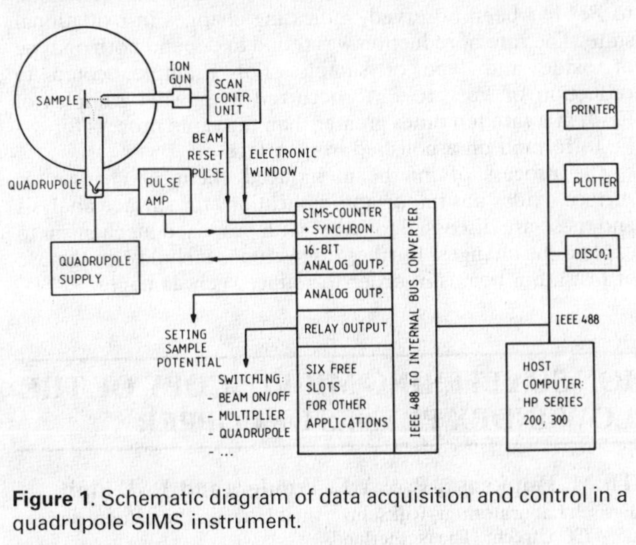

Figure 1. Schematic diagram of data acquisition and control in a quadrupole SIMS instrument.

Figure 2 gives more insight to the interfacing electronics. The 19-inch main frame contains the power supply for the eurocard modules. + / − 5 volt for logic components and + / − 15 volt for analog devices are available.

The conversion between the IEEE 488 bus and the internal data bus is achieved by a microprocessor interface, which consists of an 8085 processor, the GPIB controller, input/output drivers and the EPROM resident software. The internal bus splits up into 24 input lines, 16 output lines, 4 address lines and 4 lines for other purposes. The IEEE address which is assigned to the interface allows to connect up to 15 modules. Various input/output modules are available. Each contains an address logic and a data storage.

One of the modules with analog output consists of a 16 bit digital-to-analog converter with min. monotonicity of 14 bit and, optional, 0 . . . + 10 V, 0 . . . − 10 V or − 10 V . . . + 10 V range. The total acquisition time of this module with the driver program on a HP series 200 computer is 15 ms. The SIMS counter consits of a very fast input amplifier with a dual pulse resolution of 4 ns followed by a 24 bit counter. The input is optional ECL, NIM or TTL level. The levels for gate and trigger inputs are TTL, normal or inverted. In addition to its counting function this module delivers an interrupt pulse via IEEE bus to the host computer in order to synchronize the computer with the scan control unit. The total time for the data transfer between counter and computer is 24 ms.

Much convenience and flexibility in operation and in data manipulation is due to the fact that the programming language is high-performance BASIC. The results obtained during two years of operation show that the data acquisition and control system permits the best use of the possibilities of the instrument.

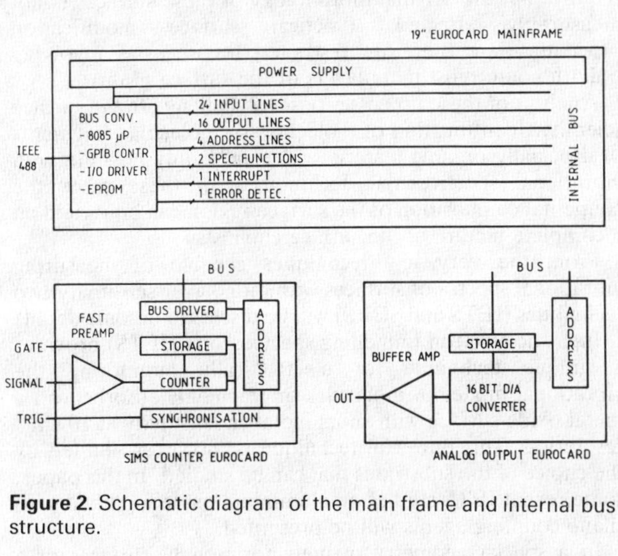

Figure 2. Schematic diagram of the main frame and internal bus structure.

this mode (SERS) the Raman technique can be used to study monolayers as will be illustrated by some examples.

CHARACTERIZATION OF THIN FILMS AND INTERFACES BY MEANS OF ANALYTICAL TRANSMISSION ELECTRON MICROSCOPY

M.P.A. Viegers
Philips Research Laboratories, PO Box 80.000, 5600 JA Eindhoven, The Netherlands

A survey is given of the characteristic properties of an analytical transmission electron microscope (TEM) with respect to microstructural features of thin layers and interfaces. It is illustrated that the technique is capable of providing chemical and crystallographic information on a scale close to the atomic level.

Using phase contrast and structure factor contrast small structural (monolayer) or compositional (less than 1%) features can be distinguished. Scanning transmission electron microscopy (STEM) can be applied to the detection of compositional fluctuations via monitoring of transmitted electrons, back scattered electrons, secondary electrons or x-rays generated when the incident electrons interact with specimens. These capabilities are shown to be indispensable in various fields of materials research. Factors affecting spatial resolution and detectable concentrations are discussed. Most of the examples given will be related to semiconductor research and technology.

VIBRATIONAL SPECTROSCOPIC ANALYSES OF CHEMICAL SURFACE MODIFICATION

P.N.T. van Velzen
Philips Research Laboratories, PO 80.000, 5600 JA Eindhoven, The Netherlands

Surface chemical properties of materials are largely determnied by the particular functional groups present at the surface. A controlled change of these chemical structures is a versatile and successful means of modifying surface properties in a well-defined manner, i.e. for improved adhesion.

Modern electronic and electro-optic applications usually require that the original morphology of these surfaces is not measurably affected. Practical surface modification procedures are therefore restricted to processes involving only the outermost monolayers of the surface (1nm).

Analysis of these processes poses a challenge to the surface scientist. Identification of molecules or molecular fragments of especially organic materials is usually only possible with vibrational spectroscopic techniques. In these cases the elemental composition of the surface region can only yield an incomplete picture of the surface chemistry.

From the very few techniques capable of measuring vibrational spectra of surfaces with monolayer sensitivity two techniques (IETS and SERS) will be discussed in some detail.

Inelastic electron tunnelling spectroscopy (IETS) provides a unique technique for electronically monitoring the vibrational modes of molecules or fragments adsorbed on a metal oxide surface with submonolayer sensitivity at 10 cm^{-1} resolution. This powerful technique is, however, limited by the choice of the substrates that can be studied. In this paper, some recent IETS studies on aluminium oxide modified with silane coupling agents will be presented.

In a special geometry, namely for 'rough' surfaces of a select number of metals, the Raman signal of molecules adsorbed onto these surfaces shows enhancements of 10^6. In

POSITRON ANNIHILATION SPECTROSCOPY FOR CHEMICAL ANALYSIS (PASCA) AS A NON-DESTRUCTIVE AND IN-SITU SURFACE TECHNIQUE

Y.C. Jean and K.L. Cheng
Departments of Chemistry and Physics, University of Missouri-Kansas City, Kansas City, Missouri 64110–2499, USA

Positron annihilation spectroscopy for chemical analysis (PASCA) has been utilized as a new tool for the surface analysis. Unlike conventional analytical methods, PASCA possesses several unique advantages in the surface characterization. It is considered as a non-destructive *in situ* technique which can analyse the surface reactions under both high vacuum and high pressure conditions. It has been used for determining the total surface area of highly porous materials by measuring the positron-electron annihilation lifetime. Most surface techniques are generally semi-quantitative because of many parameters involved. The PASCA has demonstrated excellent quantitative results. Detailed chemisorption and desorption reactions of NO_2, SO_2, Na^+, K^+, Cu^{2+}, Ni^{2+}, Fe^{3+}, Cr^{3+}, etc. with zeolites and γ-alumina will be presented. Mechanisms of adsorption and the determination of active sites on zeolites and γ-alumina will be discussed.

THE REDUCTION OF SURFACE OXIDES OF IRON BY ION BEAM BOMBARDMENT

J.L. Sullivan
Department of Mathematics and Physics, University of Aston, Birmingham, UK

Changes in chemical state of the oxides of iron when bombarded with argon ions in the energy range 0.5 keV to 2 keV have been investigated through a study of shifts in the energies of core electron levels in iron by means of x-ray photoelectron spectroscopy.

Rapid reduction of Fe^{3+} ions to Fe^{2+} and further reduction to Fe^0 has been observed, indicating changes in oxidational state. The rate of reduction was found to depend both on type of oxide and type of sample. For example, complete reduction of Fe^{3+} to Fe^{2+} occurred in powder samples of Fe_30_4 at a rate ten times greater than the same process for $\alpha-Fe_20_3$ formed on a polished iron surface.

The process of ion beam induced chemical changes in surface oxides has important implications for surface analysis and these are discussed together with proposed mechanism to explain the changes. Further discussion is included on the use of fast atom beams in order to reduce such damage.

ION SCATTERING MICROSCOPY OF THE LOW-INDEX PLANES OF COPPER

Th.M. Hupkens*, P.A.A.F. Wouters and J.M. Fluit
Fysisch Laboratorium, Rijksuniversiteit Utrecht, Princetonplein 5, 3584 CC Utrecht, The Netherlands

Low energy ion scattering (LEIS) can be used to determine relative positions of atoms in the first few surface layers of single crystals. Hence LEIS is well suited to study *ordering* effects in adsorbed layers as well as *reconstruction* of surface layers. The maximum information depth ranges from one layer to about four layers depending on the type of projectile, energy and scattering angle. The information depth also depends on the interatomic distance within the surface layer and the interlayer spacing. This will be illustrated on the basis of so-called 'photograms' — shadow patterns deduced from singular reflected projectile intensities at varied polar and azimuthal incidence directions — measured for large-angle scattering of Ne^+. In these photograms, atoms of the second layer of Cu{111} are only faintly visible, whereas the second layer atoms of Cu{100} and Cu{110} are clearly visible. If H_20 scattering combined with measuring the O^- dissociation product after scattering is used, a much larger information depth can be obtained. Then the 120° symmetry (!) of the first and second Cu{111} surface layers shows up clearly in the photogram, even if small scattering angles are used. Some interpretation problems that sometimes arise in LEIS due to focusing and neutralization effects will be discussed on the basis of angular distributions of bombarded clean and oxygen-covered Cu surfaces.

*Present address: Kamerlingh Onnes Laboratorium, Rijksuniversiteit Leiden. Nieuwesteeg 18, 2311 SB Leiden, The Netherlands

HYDROGEN BACKSCATTERING BY A SMALL NUMBER OF SURFACE LAYERS

J.M. van Zoest, J.M. Fluit and A. Niehaus
Fysisch Laboratorium, Rijksuniversiteit Utrecht, Princetonplein 5, 3584 CC Utrecht, The Netherlands

In low energy noble gas ion scattering (LEIS), depth information is restricted to the first two surface layers. Therefore also the *in situ* control of surface quality (*Nucl. Instr. Methods B*, Jan. 1985) in LEIS studies usually is restricted to the first two layers.

In this work we show that information on the quality of deeper layers can be obtained by means of keV proton scattering. We use H_2^+ and H^+ projectiles and obtain information by comparing backscattered intensities of H^+, H^0, and H^-, whereby the energy spectra for the ions are measured by an electrostatic analyser, and those for the neutrals by the time-of-flight method. All spectra show the importance of electronic stopping and of multiple scattering. Information on deeper layers is obtained from the low energy parts of the spectra and is demonstrated in photograms. Structures in photograms due to planar channelling are used to determine the subsurface quality of the crystal. In addition the phenomena of charge exchange and of dissociation will be discussed.

ELECTRON BEAM IMPACT POSTIONIZATION OF SPUTTERED NEUTRALS

D. Lipinsky, R. Jede, J.Tümpner, O. Ganschow, A. Benninghoven
Physikalisches Institut der Universität, Domagkstr. 75, D-4400 Münster, W Germany

Postionization of sputtered neutrals for mass analysis with a quadrupole mass filter can be achieved with good analytical performance if one succeeds in suppression of secondary ion and residual gas signals. Using this approach to SNMS, typically 3×10^4 cps per μA of 3 keV KAr^+ at 55° angle of incidence can be achieved for most metals. Signal-to-background ratios then exceed 10^4. Sensitivity factors for different elements vary by about one decade with those instruments, irrespective of the matrix material. First analytical applications of our system included bulk analysis of ultrapure steels, where we tracked the carbon concentration down to 8 ppm quantitatively, Se, Sn, In redistribution over a few hundred angstroms in the 10^{18} cm^{-3} concentration range, and elimination of preferential sputtering artefacts in anodic oxide layers on $Cd_xHg_{1-x}Te$. Gnaser *et al*[1] applied their system to the analysis of He implantation in the first wall of a tokamak.

Our instrumental approach allows rapid switching between the positive and negative SIMS modes, SNMS and residual gas analysis in a true UHV environment. In this way, mutual calibration of SIMS and SNMS becomes possible, so that experiments addressing the SIMS matrix effect in detail become feasible. Thus we were able to quantify secondary ion and sputtered neutral emission from oxygen covered Si and Fe surfaces in terms of oxygen concentration. The main result of these experiments concerning SNMS is, that as a result of the SIMS matrix effect, secondary ion yields from these surfaces in general cannot be neglected for SNMS quantification. This is due to an exponential increase of the ionization probability of sputtered metal atoms with oxygen concentration, in accordance with out previous SIMS/XPS data. From the same experiments, one can deduce cluster formation probabilities as a function of oxygen concentration, which allow us to discuss model prediction in molecular secondary ion and neutral emission. Thus electron beam impact postionization of sputtered neutrals increases the potential of mass spectrometric surface and thin film analysis and adds to our understanding of sputtering and secondary ion emission.

Reference

1. H. Gnaser *et al., Appl. Phys. A,* in press.

ENHANCEMENT OF DETECTION LIMITS OF XPS WITH MONOCHROMATIZED ESCA COMBINED WITH MULTICHANNEL DETECTION: APPLICATION OF HIGH PERFORMANCE MXPS FOR ROUTINE ANALYTICAL WORK AT LOW CONCENTRATION LIMITS

J. Krizek, G. Panzner, J. Hochkirchen, D. Schellenberger, K. Bergert, M. Wilmers.
Leybold-Heraeus GmbH, BonnerStrasse 504, D-5000, Köln, FRG

High background in ESCA spectra combined with a number of satellite structures has been the main cause of limitation of this method for quantitative detection of small concentrations. For monochromatized XPS, the background satellite structure is naturally reduced but the method is limited by signal to noise ratio due to small intensity of MXPS lines.

Further problems come from spurious secondary electrons, both in the electron optics and in the detector, which contribute to the background, reduce overall resolution and therefore the intensity yield of the instrument and reduce the detection dynamics of the XPS spectrometer.

The problems are especially critical for high energy resolution methods such as MXPS, as some of the effects are energy dependent and mix up spurious ghost peaks with XPS spectral lines. In this paper we present a new design of an electron spectrometer which is optimized for high resolution XPS, MXPS work and at the same time capable of count rates approaching the current limits of MCP.

The spectrometer uses a MCP system with a parallel detection. The ghost peaks background is limited to 1% over all energy ranges of the analyser which makes it possible to take meaningful results from a MCD–MXPS/XPS system. As an example the systems Si; Si:P; Si:As are studied with various doping concentrations.

The spectra show that concentrations below 0.1% can be readily detected with this spectrometer. Using this, ESCA opens new opportunities in the field of quantitative analysis at small concentrations. The system is microcomputer-controlled and can be connected to any minicomputer for automated data acquisition and evaluation.

MULTIPHOTON, 'MONOCHROMATIC' X-RAY SOURCE FOR XPS

David W. Dwight
Center for Adhesion Science, Virginia Tech, Blacksburg, VA 24061, USA

A design is proposed for modifying the secondary fluorescence approach used by Henke *et al.* (*J. Appl. Phys.* **52,** 1509 (1981) in soft x-ray spectrometry in which 18 x-ray energies ranging from 109 eV to 7469 eV have been demonstrated. Applied to XPS, this excitation source promises to provide many advantages over conventional sources. Similarly to crystal monochromatization, the reduced Bremsstrahlung acts to narrow photoelectron peak widths, enhancing resolution of small chemical shifts, as well as improving the signal-to-noise ratio for quantitative analysis. Photoemission cross sections can be varied or enhanced for specific orbitals, and energy-dependent depth profiles can be obtained (especially useful on rough, fibrous or powder specimens) by the use of numerous x-ray energies. Also the higher x-ray energies can extend the analysis depth to 50 nm or greater, and access additional Auger transitions.

ION MILLING PROCEDURES FOR CROSS-SECTIONAL SPECIMENS FOR TRANSMISSION ELECTRON MICROSCOPY

C.W.T. Bulle-Lieuwma
Philips Research Laboratories, PO Box 80.000, 5600 JA Eindhoven, The Netherlands

Major efforts in material science and technology have yielded significant advancements in the preparation of ultra-thin layers and hyperabrupt interface regions.

Consequently the demands on the characterization facilities are continuously increasing. In particular XTEM (cross-sectional transmission electron microscopy) has proven to be most valuable in this field. Its major drawback was the time consuming and elaborate way of specimen preparation. Here we report upon a method, which we recently developed to obtain a considerable improvement of the quality and the throughput of specimens of crystalline semiconducting materials.

It will be shown that a high sputter rate can be obtained using 5 kV Ar ion bombardment, without introducing the well-known surface roughness. The apparent disagreement with previous studies on the development of surface roughness and the disagreement with the models proposed to explain these effects, will be discussed.

A NEW INSTRUMENT FOR SURFACE AND SUB SURFACE IMAGING

E.M. Cashell
Cork Regional Technical College, Bishopstown, Cork, Ireland
L. McDonnell
Tekscan Ltd., RTC Campus, Bishopstown, Cork, Ireland

A novel UHV instrument incorporating Auger electron spectroscopy (AES), scanning electron microscopy (SEM), scanning Auger microscopy (SAM) and scanning thermal wave microscopy (STWM) is described. This instrument provides, within a single system, the ability to correlate sub-surface information such as grain structure, cracks and other features with surface topography and chemical composition. Preliminary applications of the instrument in ceramics, metallic alloys and microelectronic devices are presented and used to illustrate the advantages of this experimental approach.

SURFACE FREE ENERGIES OF GLASS AFTER ADSORPTION OF HYDROCARBONS WITH A VARIABLE CHAINLENGTH

W.J. Postma, H.M.W. Uyen, H.J. Busscher and J. Arends
University of Groningen, Laboratory for Materia Technica, Ant. Deusinglaan 1, 9713 AV Groningen. The Netherlands

Surface properties of glass can be influenced by the use of surfactants. In this study α–Cn–DMEA–HCL ($8 \leqq n \leqq 16$) molecules have been adsorbed on glass slides during 2 min. at various surfactant concentrations (0.02 – 10 mM). Surface free energy changes have been determined from contact angle measurements using the concept of polar and dispersion components. Surface free energies gradually decreased as a function of concentration for all surfactants up till 1.5 mM. For concentrations higher than 1.5 mM no distinct relation between surface free energy and concentration was found. This probably indicates that at 1.5 mM the glass surface is completely covered with a monolayer of surfactant molecules. Good linear relations (correlation coefficients 0.96) were observed between the hydrocarbon chainlength n and the surface free energy at surfactant concentrations below 1.5 mM. The lack of correlation above 1.5 mM is most likely caused by the formation of non-oriented, adsorbed multilayers. From the results of this study it can be seen that minor changes at a surface, such as the chain length of adsorbed hydrocarbons, can be determined from contact angles. Contact angle measurements can thus be considered as a highly sensitive technique for studying surfaces.

SMALL SPOT ESCA APPLIED TO ELECTRONIC MATERIALS PROBLEM SOLVING

J.S. Hammond
Perkin–Elmer Physical Electronics Division, 5609 Flying Cloud Drive, Eden Prairie, Minnesota 55344, USA
H.Goretzki
Perkin–Elmer Verkauf GmbH, Physical Electronics Division, Bahnhofstr. 30. D–8011 Vaterstetten, W-Germany

In the area of electronic materials ESCA has become an established tool for non-destructive chemical characterization of surface residue and oxides which impede material performance. However, for many electron materials problems, sample dimensions less than 1 mm limit the application of traditional ESCA techniques. With a newly designed instrument which permits lens-defined image selection down to less than 200 micron in diameter, techniques such as Angle resolved Esca, photon dependent ESCA and Auger parameter analysis are still possible. Examples will be shown of chip packaging contamination, bond failure on leads and oxide formation on silicon nitride dielectric layers where this unique lens design provides the only viable approach for problem solving with small spot ESCA.

RAPID DEPTH PROFILING WITH ESCA

W.T. Jansen, C.E. Bryson, and D.K. Bakale
Surface Science Labs., 465 National Avenue, Mt. View, CA 94043, USA

Whilst a very important technique in Auger spectroscopy, rapid depth profiling with ESCA has had limited practical use, despite its large potential and applicability to insulators. We discuss a number of constraints which have restricted the use of ion etching with ESCA, and report on recent work to remove some of them. In particular, we demonstrate rapid profiling (60 Å min^{-1}) with good interface resolution (5 Å) using small area analysis and rapid spectral acquisition (5–10 s per spectrum) using a high resolution imaging detector.

Section 2.

Quantification

SURFACE AND INTERFACE ANALYSIS, VOL. 9, 79–83 (1986)

Surface Chemical Analysis — Report on the VAMAS project

C.J. Powell
Surface Science Division, National Bureau of Standards, Gaithersburg, MD 20899, USA
M.P. Seah
Division of Materials Applications, National Physical Laboratory, Teddington, Middlesex, TW11 0LW, UK

The VAMAS project on surface chemical analysis is a multi-national cooperation for the provision of standards data and materials for surface chemical analysis measurement and for the provision of the basic understanding necessary for these activities. This project is one of a rapidly growing suite of projects initiated as a result of decisions following the 1982 Versailles Summit Meeting of the Heads of State or Government of Canada, France, Germany, Italy, Japan, UK, USA and the representatives of the Commission of the European Communities. During the past year national representatives to this VAMAS project have been appointed and national committees established. This article summarizes the philosophy, scope, and organization of the project and describes specific activities that have been initiated. Information is given on how individuals, both within and outside the group of member states, may participate.

INTRODUCTION

The Versailles Project on Advanced Materials and Standards (VAMAS) was established following the Economic Summit meeting at Versailles, France in 1982 of the Heads of State or Government of Canada, France, the Federal Republic of Germany, Italy, Japan, the United Kingdom, and the United States of America and representatives of the Commission of the European Communities (CEC). This project is a scheme for stimulating the introduction of advanced materials into high technology products and engineering structures. The overall aim of encouraging international trade is intended to result from international agreement on codes of practice and performance standards and from multilateral research aimed at providing the scientific and metrological base necessary to achieve agreement on standards.[1]

Advanced materials configurations are being increasingly developed in thin-film or modified-surface form. Such configurations are integral to established technologies such as microelectronics and are of growing importance in fields such as corrosion, catalysis, and biomaterials. Coating technology, vacuum metallurgy, ion implantation, and plasma processing are but some of the many surface-science-related sectors which have acknowledged high growth potential for manufacturing industry. The development of very novel technologies in these areas depends on an ability to measure the chemical state of deposited films or of treated surfaces. The reliability of these films or surfaces, and the assurances that can be made about them, depend on an ability to measure and verify surface compositions. The ubiquity of surface analysis through all sectors of high-technology industry has been clearly shown in recent surveys.[2]

Owing in part to the recent and rapid utilization of surface analysis in modern technologies and in part to the diversity of the applications, the development of relevant standards has been slow. The problem has been recognized in the USA, where the focus has been the E–42 Committe on Surface Analysis of the American Society for Testing and Materials (ASTM), and in Europe through the Community Reference Bureau. These programs have no formal relation with one another but attempt, as far as is practicable, to be complementary. Many other countries have projects which add to this body of work in an ad-hoc manner as demonstrated, for example, at the third International Conference on Quantitative Surface Analysis at the UK National Physical Laboratory in November, 1983 and the first European Conference on Applications of Surface and Interface Analysis in October, 1985. International collaboration is urgently needed to make the most constructive use of this effort and to halt divergencies of approach.

The VAMAS Surface Chemical Analysis (SCA) project was approved by the VAMAS Steering Committee in June, 1984. The main objective of this project is to produce, by coordinated effort, the reference procedures, reference data, and reference materials necessary to establish standards for surface chemical analysis. This effort is expected to lead to the international coordination of standards-related activities before national standards are adopted and to the development of needed standards on an international level. Information on the rationale and planned scope of this VAMAS project is contained in two reports available from the authors.[3,4] We note here that this VAMAS project is one of eleven projects that include wear test methods, polymer blends, polymer composites, ceramics, hot salt corrosion resistance, weld characteristics, bio-engineering materials, creep crack growth, material databanks, and superconducting and cryogenic structural materials.[1]

0142–2421/86/120079–05 $05.00

This article describes the scope and initial objectives of the VAMAS SCA project and the nature of current activities. Information is given on how interested persons may participate in projected activities.

SCOPE OF THE VAMAS SCA PROJECT

The most urgent needs for surface-analysis standards pertain to the use of Auger-electron spectroscopy (AES), x-ray photoelectron spectroscopy (XPS), secondary-ion mass spectroscopy (SIMS) and sputter-depth profiling (SDP).[3,4] As projects in these areas become established and as other techniques come into greater use, it is expected that the program will be broadened to include, for example, ion-scattering spectroscopy, Rutherford backscattering spectroscopy, and high-resolution electron energy-loss spectroscopy.

Qualitative and quantitative surface analyses require consideration of the following three items:

(i) the physical principles of the measurement methods;
(ii) the behaviour of the measuring instrument; and
(iii) the interpretation and expression of the final result.

The basic principles of the physics of the various techniques and the selected analytical methods are vital since they lead to the working formulae which are used for quantitative analysis. Agreement must be obtained on these formulae and their regimes of applicability when working to certain accuracies for particular types of specimens and particular techniques. These formulae involve the measurement of intensities by the instrument which are then related to the composition at the surface by functions of the specimen material and the particular measurement process and by functions of the measuring instrument. Tabulations are needed of particular material properties relevant to the various techniques and information is needed on instrument properties; the latter information should be established by the user or be provided by the manufacturer. Additionally, surface analyses can give chemical state information, the interpretation of which requires a similar hierarchy of understanding, data, and reference procedures. Therefore, the analyst requires:

(a) an agreed base of principles, definitions of terms, and equations to relate the techniques and procedures to the measurement;
(b) reference procedures specifying how to measure intensities and spectral features;
(c) reference procedures, reference data, and reference materials for calibrating instruments, determining instrument functions, and inter-relating instruments of different designs;
(d) reference data of elemental sensitivity factors, matrix terms, information depths, libraries of chemical-state spectra, information on ion-sputtering rates, radiation damage, etc.; and
(e) standard methods of specifying an analysis.

Details of specific standards needs for AES, XPS, SIMS, and SDP are given elsewhere.[3,4] Activities of this VAMAS project (summarized below) to meet these needs can be of several different types. First, such activities may involve work by one or more individuals in a particular nation on a given topic. Examples of this type may be found in current activities of th ASTM E—42 Committee on Surface Analysis in the USA.[5] Second, projects may involve work on a given topic by scientists in two or more nations. Examples of such projects are interlaboratory comparisons of measurements of the same type, for example, that published by Clegg *et al.*[6] Third, projects may involve coordination with other appropriate national and international organizations, for example the International Union of Pure and Applied Chemistry, the Community Bureau of Reference of the CEC, the ASTM E–42 Committee, etc. Finally, workshops may be organized to discuss particular problem areas and to assist in the planning of new cooperative projects.

ORGANIZATION OF THE VAMAS SCA PROJECT

National representatives have been appointed to this VAMAS project as shown in Table 1. These representatives lead national activities in their countries, often with the aid of a national committee, and develop bi-national or multi-national activities where these are considered desirable. It is expected that the national representatives will meet as a group approximately annually to discuss plans, progress, and opportunities.

It should be made clear that VAMAS is a cooperative venture among the member states. Although it is intended that the reference data,

Table 1. National Representatives to the VAMAS Surface Chemical Analysis Project

Canada	N.S. McIntyre, Surface Science Western, University of Western Ontario, London, Ontario, Canada N6A 5B7 (519–679–6373)
FRG	S. Hofmann, Max-Plank-institut für Metallforschung, Institut für Werkstoffwissenschaften, Seestrasse 92, D 7000 Stuttgart 1, FRG (0711–2095–320)
	H. Hantsche, Bundesanstalt für Materialprüfung, Unter den Eichen 87, D–1000 Berlin 45, FRG (030–8104–1)
France	C. Le Gressus, Department de Physico Chimie, Section des Solides et des Surfaces, CEN Saclay, 91191 Gif-Sur-Yvette Cedex, France (6–908–4762)
Italy	F. Garbassi, Istituto Guido Donegani SpA, Centro Ricerche Novara, Via G. Fauser 4, 2810 Novara, Italy (0321–24701)
	G. Queirolo, SGS/ATES Componenti Eletronici SpA, Via C. Olivetti 2, 20041 Agrate Brianza (MI), Italy (6555–594)
Japan	R. Shimizu, Department of Applied Physics, Faculty of Engineering, Osaka University, Suita, Osaka 565, Japan (06–877–5111, ext. 4658)
UK	M.P. Seah (Vice-Chairman), Division of Materials Applications, National Physical Laboratory, Teddington, Middlesex, TW11 0LW, UK (01–977–3222, ext. 3634)
USA	C.J. Powell (Chairman), Surface Science Division, National Bureau of Standards, Gaithersburg, MD 20899, USA (301–921–2188)
CEC	D. Gould, Community Bureau of Reference, Commission of the European Communities, Rue de la Loi 200, 1049 Brussels, Belgium (02–235–9313)

reference procedures, and standards resulting from the VAMAS SCA project will be widely disseminated, participation in current project activities is limited to persons of the member states and others who receive invitations to join. Scientists interested in participation should contact a national representative if they are from a member-state institute or otherwise contact the Chairman or Vice-Chairman (Table 1).

No central funding exists for VAMAS projects. Each participating nation in VAMAS decides whether it wishes to join in a particular project authorized by the Steering Committee. Those nations which agree to cooperate in a project such as SCA are expected to make use of appropriate national organizations, funds, or funding mechanisms. The details of any funding will therefore be different in each state. All nations participating in VAMAS have, in fact, agreed to take part in the SCA project.

The function of the CEC (Table 1) is *not* to represent those countries of the CEC which are not individual participants in VAMAS. Instead, the purpose of CEC representation in VAMAS is to represent the CEC as a whole in those fields (such as SCA) where there exists a related community-funded activity for the benefit of all CEC states.

It is expected that the various VAMAS cooperative projects will lead to the development of pertinent standards. (The term standards is used here in a broad context to include reference or recommended procedures, reference data, and reference materials.) Each VAMAS project should stimulate and expedite the development of national standards in each state; because of VAMAS, the national standards in the cooperating states should be compatible with each other. These national standards could then be established as international standards through the International Organization for Standards.

Each VAMAS technical project reports to the VAMAS Steering Committee which in turn reports to the Versailles Working Group on Technology, Growth, and Employment.[1] Reports on progress are also presented by the national representatives to VAMAS projects to appropriate national government agencies.

ACTIVITIES OF THE VAMAS SCA PROJECT

Three types of activities have been initiated to date: (1) multinational activities, (2) national activities, and (3) workshops on surface analysis. These activities will now be described in turn.

The first meeting of the national representatives to the VAMAS SCA Working Party took place in October, 1985 during the ECASIA Conference in The Netherlands. A number of possibilities for cooperative projects were discussed and it was agreed that the following multinational activities should be implemented.

(a) Development of thin oxide films as reference materials. It is planned to produce thin films of transition-metal oxides (Fe_2O_3, NiO and Cr_2O_3) with thickness of about 25nm on metal substrates in Canada. Films of similar thickness of SiO_2 on silicon will be produced in Italy. These films will be characterized by sputter-depth profiling at a number of laboratories in Canada and Italy. The ion sputter yields of these oxide films under common conditions of sputter-depth profiling will be compared with similar measurements on existing Ta_2O_5 and Ni/Cr reference materials at the UK National Physical Laboratory (NPL) and the USA National Bureau of Standards (NBS).

(b) Development of calibration data for the energy scales of Auger-electron spectrometers. Current projects exist at both NBS and NPL for the independent measurement of the kinetic energies of the principal Auger-electron transitions of copper, silver, and gold to be used as calibration data in the energy range 50–2000 eV. After intercomparison, the calibration data will be utilized in an interlaboratory comparison of Auger-electron measurements to be organized by NPL.

(c) Procedures for quantitative XPS. It is planned to investigate at NBS the suitability of recently fabricated Ta_2O_5 films on silicon and, if necessary, the oxide films prepared in project (a) as test materials for an interlaboratory comparison of angle-dependent XPS measurements. If these films can be cleaned adequately with minimal ion sputtering, NBS will organize the comparison of XPS measurements to give information on film thickness and film stoichiometry.

Two additonal cooperative projects were discussed and it is likely that these will be initiated following the preparation of more definite plans. The first such project is the development of alumina reference materials with known concentrations of oxygen vacancies. These materials will be used as reference materials for problems in materials science such as wetting, adhesion, and electrical breakdown as well as for determining conditions for radiation damage. The second project is the development of reference procedures for high-resolution scanning Auger-electron microscopy so that the effects of topography, specimen crystallinity, and angle of emission can be minimized.

A large number of national SCA activities have been initiated during 1985 and these are summarized in Table 2. Some of these activities represent contributions to the multinational activities described above and some are part of the ongoing programs of institutions (such as NBS or NPL) or of existing organizations (such as the ASTM E–42 Committee) that have agreed to cooperate with the VAMAS SCA Project.

The projects listed in Table 2 should be regarded only as initial plans. It is expected that as projects mature, cooperations between scientists in two or more VAMAS states will occur in areas of mutual interest.

Plans are being developed for two workshops on surface chemical analysis for which VAMAS will be a co-sponsor. The first workshop is planned for October 24, 1986 at NBS and will precede the Tenth International Vacuum Congress and Sixth International Conference on Solid Surfaces to be held in Baltimore, October 27–31, 1986. The second workshop is planned for November 17, 1986 at NPL and will precede the fourth Conference on Quantitative Surface Analysis at NPL on November 18–20, 1986. These workshops will publicize VAMAS activities, disseminate results of recent work on quantitative surface analysis, and promote the development of new cooperative VAMAS projects. Specific attention will be given at both workshops to needs for reference data, needs for reference materials, methods of instrument calibration, and analytical methodology.

Status reports on the VAMAS SCA project are expected to be presented to national and international scientific meetings. Announcements and reports will also be sent to appropriate journals.

Table 2. Summary of technical activities contributing to the VAMAS Surface Chemical Analysis project together with the participating scientists or organizations where known. Further information on these activities can be obtained from the national representatives listed in Table 1

(a) Canada
- Development of thin film oxides for energy and sputter-rate calibration (National Research Council, Ottawa)
- Measurements of secondary ion yields in silicon (University of Western Ontario)
- Development of an improved gold standard for energy referencing of electron spectrometers (University of Western Ontario)

(b) Federal Republic of Germany

(i) Activities at Max-Planck Institut für Metallforschung, Stuttgart
- Quantitative depth profiling by AES and ion sputtering
- Dependence of depth resolution in depth profiling of metallic thin films on sample and ion beam parameters

(ii) Round robin in eight European laboratories on combined electrochemical and surface-analysis (AES, XPS) characterizations of passive films on Fe–Cr and Fe–Cr–Mo alloys

(c) France

(i) Physical principles of the measurement methods
- Study of the shape function of the background in AES
- Deconvolution of instrument and inelastic scattering effects in AES
- Simulation of AES background by Monte Carlo calculations (Cailler, Langeron)
- Calculation of the shapes of Auger peaks for aluminum and for carbon in different carbides (Darque–Ceretti)
- Measurement of electron attenuation lengths by two methods (Gillet, Vergand, Dufour)
- Comparison of optical and electron-spectroscopy methods for surface analysis (Remond, Paulmier)
- Data reduction: use of spline functions and calculation of plasmon profiles (Boiziau, Stefani)

(ii) Instrument behaviour and characterization of standards
- Instrument calibration through use of the background, loss function, and secondary emission
- Use of polyacrylonitrile, Al, Al_2O_3, and AlN as standards for electron spectroscopy (Le Gressus, DeMoncy)
- Surface preparation by laser evaporation (Stefani)

(d) Italy

(i) Activities at Agrate Brianza
- Measurement of backscattering factors for metal/silicon sandwiches at different angles of incidence
- Qualitative analyses of transition–metal and refractory–metal silicides
- Quantitative depth profiles for As and BF_2 in silicon
- Oxidation kinetics of polycrystalline silicon films and SiO_2/Si interface width measurement

(ii) Group Activities
- Comparison of multi-laboratory, XPS measurements of an Al film on polyethylene
- Compilation and evaluation of sputtering yields and AES sensitivity factors
- Measurement of backscattering coefficients for AES

(e) Japan
- Round robin of AES measurements on Al–Cu alloys (Shimizu)
- Monte Carlo calculation of backscattering coefficients for AES (Shimizu)

(f) United Kingdom
- Provision of reference SIMS spectra for polymers and organic coatings (Briggs)
- Provision of chemical-state data for selected systems for XPS (Chadwick)
- Development of a standard data transfer format (Hazell)
- Comparison of NPL and NBS sputtering reference materials (Seah)
- Provision of traceable calibration line positions for AES (Seah, Smith)
- Calibration of spectrometer transmission functions (Seah, Smith)

(g) United States of America

(i) Activities of the ASTM E–42 Committee
- Development of additional definitions of terms for AES, XPS, SIMS, and Energetic Ion Analysis
- Development of the following documentary standards:
  - — Guide to Depth Profiling in Auger Electron Spectroscopy
  - — Guide for Specimen Handling in AES and XPS
  - — Practice for Reporting Spectra in XPS
  - — Proposed Method for Calculating Data-Gathering Speed in XPS
  - — Proposed Guide to Steady-State Charge-Referencing Techniques in XPS
  - — Practice for Reporting Sputter Depth Profile Data in SIMS
  - — Proposed Guide to Determine the Depth Resolution in SIMS
  - — Proposed Guide for Specification of Ion Beams for Use in Surface Analysis
  - — Procedure for Sample Positioning with Cylindrical Mirror Analyzers in ISS
  - — Procedure for Approximate Energy Scale Calibration with Cylindrical Mirror Analysers Used in ISS
  - — Proposed Guide for Data Reporting in ISS
- Report of an ISS round robin has been submitted for publication
- Survey of the E–42 membership is to be undertaken shortly to determine needs for new types of reference materials
- The Standard Reference Data Subcommittee is conducting a survey of needs for reference data and of format requirements

(ii) Activities at the National Bureau of Standards
- Characterization of the imaging properties of a double-pass cylindrical mirror analyser (Erickson, Powell)
- Comparison of core-level excitation energies by three different spectroscopic excitation energies (Erickson, Powell)
- High-accuracy measurements of reference calibration data for AES and XPS energy scales (Erickson, Powell)
- Calculations of electron inelastic mean free paths from experimental dielectric data and analysis of the energy dependence of experimental electron attenuation-length data (Powell)
- Measurement of electron attenuation lengths in condensed molecular films (Kurtz, Stockbauer, Madey)
- Compilation of XPS chemical shift data (Wagner)
- Compilation of ion sputtering-yield data (Fine)
- Development of nickel-chromium thin-film depth-profile standard reference material SRM 2135 (Fine)

SUMMARY

The formation of VAMAS has come at an opportune time for the development of standards for surface chemical analysis on an international basis. Although groups such as the ASTM E–42 Committee on Surface Analysis and the CEC Community Reference Bureau have been productive, the needs for standards in this field are greater than can be filled by groups such as these in a desirable period of time.[3,4]

The VAMAS SCA project has aroused a considerable amount of interest and enthusiasm in many of the VAMAS states, as exemplified by the activity listing in Table 2. Many scientists believe that VAMAS has already provided and will continue to provide a considerable stimulus to the development of standards for surface chemical analysis in each state. Furthermore, through cooperation via VAMAS, national standards will be developed more efficiently and, it is hoped, without conflicts. Areas where international cooperation through VAMAS will be most effective are believed to be: (1) the development of data banks and preferred data formats; (2) the development of reference materials; (3) the development of documentary standards for instrument operation and calibration; and (4) the development of recommended procedures for data acquisition, data reduction, and expression of final results.

The availability of needed standards (including reference procedures, reference data, and reference materials) should enable more accurate and reliable surface analyses to be made, should permit meaningful transfer of information from one laboratory to another, and should stimulate and make more efficient the development of advanced materials in new technologies. Scientists interested in participation in VAMAS SCA activities should contact a national representative if they are from a member-state institute (Table 1) or otherwise contact one of the authors of this article.

REFERENCES

1. Further information on VAMAS and its Technical Working Areas can be obtained from the Secretary of the VAMAS Steering Committee, Dr. T.I. Barry, National Physical Laboratory, Teddington, Middlesex, TW11 0LW, UK. Information on areas of cooperation resulting from the Versailles Economic Summit is summarized by D. Dickson, *Science* **224,** 1317 (1984); *ibid. 228,* 825 (1985).
2. C.J. Powell, *Appl. Surf. Science* **1** 143 (1978); M.P. Seah, *Surf. Interface Anal.* **2** 222 (1980).
3. M.P. Seah and C.J. Powell, *A Proposal for the Co-ordinated Development of Surface Chemical Analysis,* UK National Physical Laboratory Report DMA (A) 89, September 1984 (available from M.P. Seah).
4. M.P. Seah and C.J. Powell, *The Coordinated Development of Standards for Surface Chemical Analysis,* USA National Bureau of Standards Report NBSIR 85–3120, March 1985 (available from C.J. Powell).
5. P.H. Holloway, *J. Vac. Sci. Tech. A* **1,** 1570 (1983); G.C. Nelson, *Surf. Interface Anal.* **6,** 144 (1984); P.H. Holloway, *Surf. Interface Anal.* **7** 204 (1985).
6. J.B. Clegg, A.E. Morgan, H.A.M. de Grefte, F. Simondet, A. Huber, G. Blackmore, M.G. Dowsett, D.E. Sykes, C.W. Magee, and V.R. Deline, *Surf. Interface Anal.* **6,** 162 (1984).

SURFACE AND INTERFACE ANALYSIS, VOL. 9, 85–98 (1986)

Data Compilations: their Use to Improve Measurement Certainty in Surface Analysis by AES and XPS

M.P. Seah
Division of Materials Applications, National Physical Laboratory, Teddington, Middlesex TW11 0LW, UK

Data compilations are an integral and vital part of the development of the measurement scheme for surface analysis. They have three main functions that will be considered in this survey: (i) the provision of numerical values for use in calculations of quantitative surface analysis, (ii) the provision of unbiased data for the development of theory, and (iii) the establishment of a benchmark to judge the value of current research. A brief review will be made of practical data relating to AES and XPS instruments; electron guns, x-ray sources, spectrometers and detectors. Next, data compilations of experimental measurements of relative sensitivity factors, attenuation lengths and backscattering factors will be contrasted. Finally, theoretically derived data banks of cross sections, inelastic mean free paths, asymmetry parameters, etc will be considered and conclusions drawn of the need for further work.

INTRODUCTION

Data compilations are an integral and vital part of the measurement infrastructure for all analytical techniques. Surface analysis by Auger electron spectroscopy (AES) and x-ray photoelectron spectroscopy (XPS) are no exception. The first compilations date from 1967[1] and continue at an impressive rate to the present day. In this paper we briefly review some of these data compilations and consider their strengths and weaknesses in the context of present day surface analysis. At the time that these compilations were made it was often not clear what directions would be taken by future developments but now we can consider these aspects more clearly. At times, certain standard conditions were chosen for the work as these then seemed most sensible but subsequent developments in, for instance, instrumentation or procedure dictates that, today, different conditions are adopted in practice. These differences will be noted and appropriate corrections indicated.

In all areas we see compilations originating on the one hand with experimental data and, on the other, with theory. The first correlation between the two is often not good but with refinement and focused effort the two converge to an extent which, in some areas, is still poor. Until this convergence is completed, however, it will be difficult to have complete confidence in the full analytical power and generality of the techniques. Thus, in this survey we shall also cover some of the theoretically based data sets to see how far they have progressed even though they may not, at this stage, be used directly in improving the measurement certainty in surface analysis.

Fig. 1 shows a schematic measurement chain for surface analysis. Compiled data exists for the three main aspects involved: (i) instrumental, (ii) physics of the process, and (iii) the sample configuration. We must understand the current status of all of these in order to obtain the most reliable analysis of our samples. Apart from the most obvious of these data sets to provide numerical values for the use in quantitative surface analysis, they are also necessary to aid the development of the theory and also to provide a base of experience upon which we may judge the value of current research and instrumentation. Data sets exist in each of these areas.

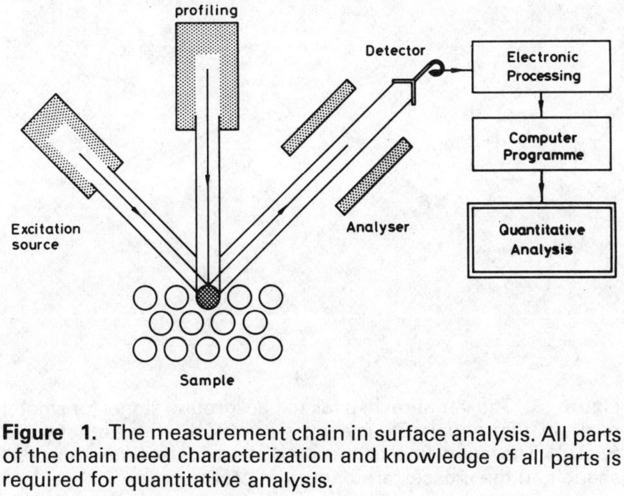

Figure 1. The measurement chain in surface analysis. All parts of the chain need characterization and knowledge of all parts is required for quantitative analysis.

INSTRUMENTAL DATA SETS

In order to provide a logical order to this survey we shall organise the data sets starting from the source of radiation on the left hand side in Fig. 1 and follow

0142–2421/86/120085–14 $07.00

events through to the final analysis. We shall not, of course, cover everything!

X-ray Sources for XPS

Characteristic data for the radiation sources in XPS are not much in dispute. Data for relative intensities and energies of the satellite lines for Al and Mg anodes are given by Carlson[2] whereas the relative heights of the satellites have been determined by Wagner *et al.*[3] No data exist for any changes that may result from contamination of the anodes. Data for other higher energy sources are given by Barrie.[4] Certain of the higher energy sources may be used in monochromators and their optimum resolutions, using the quartz diffracting crystals, are given by Siegbahn.[5] One remaining piece of data that is useful in XPS studies is the position of ghost lines from contaminants of the x-ray sources. These contaminants are usually Mg, Al, Cu and O and ghost positions are tabled by Wagner *et al.*[3] and by Briggs and Riviere.[6]

Electron sources

Characteristic data for the electron sources for AES are also not in dispute but rarely appear in a general form. In Fig. 2 we collate data for the most common commerical instruments[7–10] and we see a general convergence in the behaviour of the probe size as a function of the beam current close to the theoretical optimum both for the conventional and field emission (*FE*) sources. We may note several aspects from Fig. 2 which give a clearer idea of behaviour. Firstly, the Helmholtz-Lagrange law indicates that the beam diameter will be proportional to $E_p^{-0.5}$ where E_p is the electron beam energy. Thus, the 10 keV diameters would be reduced to 70% at 20 keV. Secondly, the brightness of LaB_6 sources, at around 10^6 A/cm²/steradian, is an order of magnitude higher than that of conventional tungsten filaments and so should give an order of magnitude more beam current for equivalent conditions. The single example of a field emission (*FE*) Auger microscope, the VG HB50, shows a somewhat different behaviour which makes use of the very high brightness (25.10^6 A/cm²/steradian) at small beam currents but aberrations in the system become progressively more important as higher beam currents are utilised.

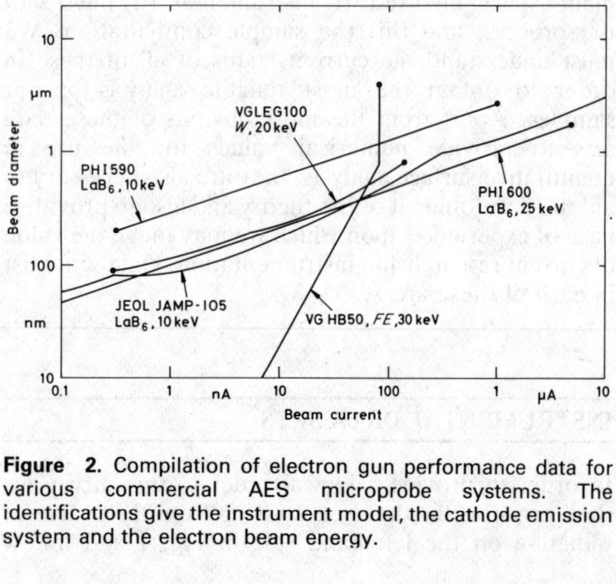

Figure 2. Compilation of electron gun performance data for various commercial AES microprobe systems. The identifications give the instrument model, the cathode emission system and the electron beam energy.

Electron Spectrometers and Detectors

Electron spectrometers in commercial AES and XPS systems largely belong to the cylindrical mirror analyser (CMA) or hemispherical sector analyser (HSA) classes. For AES the spectrometers are usually operated in the constant $\Delta E/E$ mode, where ΔE is the energy resolution at an energy E, since this is convenient. The convenience lies in the fact that the higher energy Auger electron peaks tend to be weak but the increasing ΔE offsets this so that the variation in sensitivity to different elements is reduced. The increasing ΔE broadens the high energy peaks but, since AES is not usually used for chemical state identification, this is no problem. In the constant $\Delta E/E$ mode the transmission of the spectrometer, irrespective of type, is usually constant in solid angle but increases linearly in ΔE with energy E. Thus, the transmission in terms of peak areas in the direct energy spectrum, is usually taken as proportional to E.

In practice the situation is a little more complex since the focal point of the spectrometer or the area from which electrons are selected,[11,12] may move across the sample, and therefore also the point analysed, as the energy E is scanned. This may arise through stray magnetic fields[13] or manufacturing tolerances and leads to losses of transmission, especially at low energies. Two further complicating factors exist; the effect of modulation for derivative spectra and the effect of the channel electron muliplier. The effect of modulation has been discussed in detail by Anthony and Seah[14] and others previously. There it is shown quite clearly that accurately reproducible derivative spectra may only be obtained if the spectra are measured with exactly the same modulation amplitude and spectrometer energy resolution. It is not the purpose here to discuss that problem in detail but to see how it affects published data sets. In Fig. 3 we reproduce the analysis from Anthony and Seah[14] of the effect of a modulation energy V eV on the intensity of the derivative peak with width W eV contributed by the quadrature sum of the

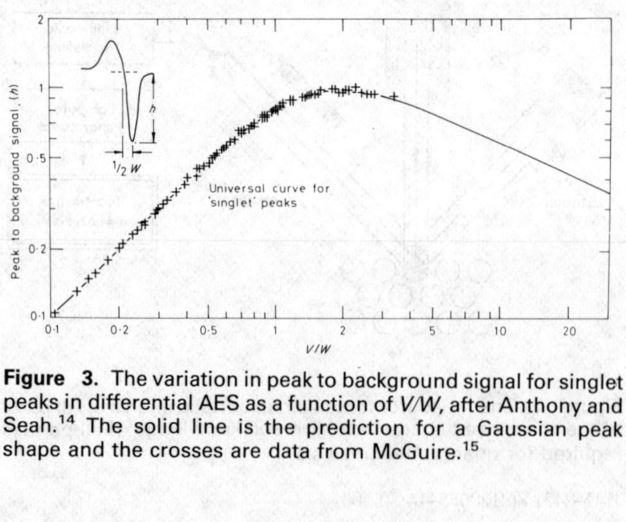

Figure 3. The variation in peak to background signal for singlet peaks in differential AES as a function of *V/W*, after Anthony and Seah.[14] The solid line is the prediction for a Gaussian peak shape and the crosses are data from McGuire.[15]

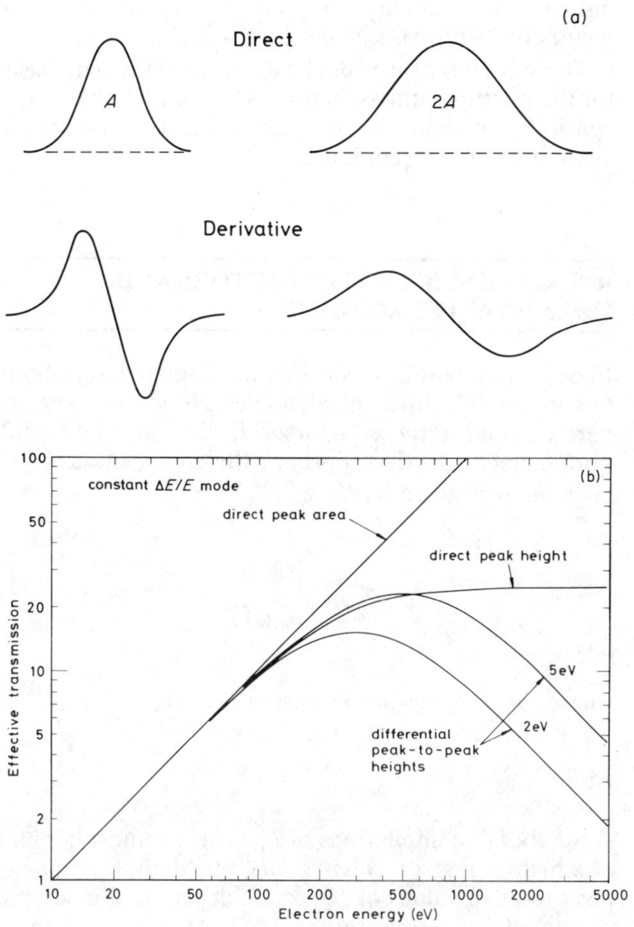

Figure 4. (a)Schematic diagram of a given intensity occurring at energies E and $2E$ in the direct and differential spectra. Note that although the area doubles, the differential signal is halved; (b) The transmission functions for a spectrometer operated in the constant $\Delta E/E$ mode for measurements of direct spectra peak areas, direct spectra peak heights, and differential spectra peak-to-peak heights at 2 eV and 5 eV peak-to-peak modulations respectively.

intrinsic processes and the analyser resolution. In general the most intense peaks which are used for AES are the sharpest and have an intrinsic width of about 1.5 eV. Thus, in a spectrometer with 0.6% resolution, as used in the *Handbook of Auger Electron Spectroscopy*,[16] the value of W will be 1.5 eV at low energies, increasing to 2 eV at 220 eV, 5 eV at 800 eV and 10 eV at 1650 eV. In terms of Fig. 3, for a 2 eV constant modulation V through the spectrum, we see that we move from the peak at $V/W = 1.3$ at low energies down to the lower part of the linear region with $V/W = 0.2$ at 1650 eV. The decline in intensity, due to peak broadening at energies above 300 eV, is more or less as E^{-1}. In the constant $\Delta E/E$ mode, used for AES noted above, we find the spectrometer transmission is proportional to E. However, this is only for peak area measurement in the direct spectrum mode. If we refer to Fig. 4(a) which shows a peak at area A at low energy and an equivalent peak at twice the energy, and therefore area 2A, we find that measurements of peak height do not scale with energy and measurements of peak gradient (i.e. the intensity in the differential spectrum) actually fall as E^{-1}. The intrinsic peak widths, as discussed above, affect these simple dependencies but in the energy range where the spectrometer resolution dominates we may summarise the spectrometer transmissions for AES in the constant $\Delta E/E$ mode as proportional to: (i) E, for peak area measurement; (ii) constant, for peak height measurement; (iii) E^{-1}, for differential peak height measurement.

For a peak of intrinsic width 1.5 eV the effective transmission functions are shown more fully in Fig. 4(b) for these three cases. If the intrinsic width is increased, the departures from the early linearity occur at proportionately higher energies.

The actual intensities measured in practice are the product of the above spectrometer transmission functions and the channel electron multiplier efficiency at each energy. In many commercial spectrometers the front of the multiplier is kept at earth potential[16] so that the electrons are incident at the analysed energy E. In spectrometers designed for combined AES and XPS analysis the electrons are often retarded from their initial energy to a lower energy for dispersion in the deflection part of the spectrometer. This retardation ratio, R, also applies at the electron multiplier and then electrons are incident there at the reduced energy E/R.

Figure 5 shows compiled data of channel electron multiplier efficiencies in the counting mode accumulated from the published literature[17]. It is not known whether the scatter reflects the true variability of the efficiencies of new multipliers from one manufacturer or of various types from different manufacturers or after different degrees of ageing. Additionally, even for the use of one multiplier in one instrument we expect differences between the analogue and counting modes as well as differences in one instrument with variations in the retardation ratio.[18]

The above terms all add to the variability of the instrument response in AES and before we can transfer published data banks to be suitable for our own use we must be able to remove this variability by a logical and systematic process. For instance, if we consider the relative intensity of the copper 60 and 920 eV peaks in the derivative mode, using CMA analysers and a 5 keV electron beam, we find ratios of 1.04[16], 0.12[15], 0.76[19] and 0.97[20], respectively, in the different reference handbooks. This variability would cause a similar

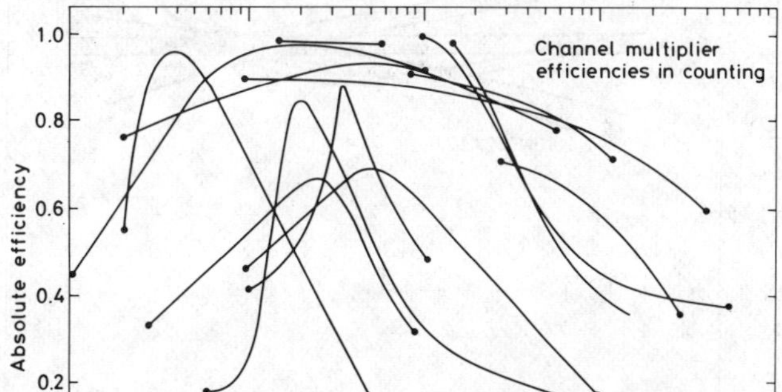

Figure 5. A compilation of channel electron multiplier efficiencies in the counting mode, as a function of the incident electron energy, after Seah.[17]

scatter in the quantification of, say, Pb or Bi in Cu. This variability is also reflected in the compiled data from the ASTM E–42 sponsored interlaboratory survey where a value of $0.75 \overset{\times}{\div} 2.5$ was found for the above ratio. The scatters have been expressed as $\overset{\times}{\div}2.5$ arising from the standard deviation evaluated from the log intensity scales. Thus, before we can use a reference data bank for quantification we need a reference procedure to establish the relative intensity–energy response function of the measuring instrument to that used to set up the data bank.

The situation is much better for XPS. Spectra are only measured in the counting mode and electrons generally strike the front of the channeltron at a fixed energy. Thus, the electron multiplier can generally be ignored. Additionally, spectra are generally measured as direct spectra in the constant ΔE mode of the spectrometer so that the resolution is constant and there are no modulation problems. Thus, for XPS, the only remaining problem is that of the varying acceptance solid angle of the spectrometer as the energy scale is scanned. This variation arises as a result of the Helmholtz–Lagrange law mentioned earlier and the precise behaviour depends on the apertures of the electron lenses, internal slits and the choice of spectrometer pass energy.[21,22] It is not the purpose here to discuss how these differences arise but merely to note that they do and that whilst particular manufacturers' instruments of a given model all appear to behave similarly, the differences between models and manufacturers are quite marked. Interlaboratory tests of a number of instruments give the relative intensity-energy response functions shown in Fig. 6,[23] normalized to the VG Scientific ESCA3 Mk II. Correction with these functions reduces the inter-instrument variability to some 4% rather than the factors of 250% we saw earlier for AES.

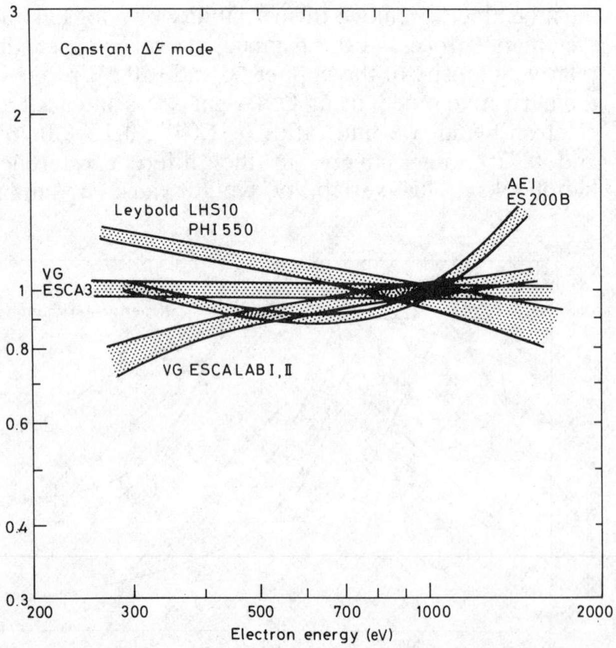

Figure 6. Compilation of commercial XPS intensity-energy response functions relative to the VG Scientific ESCA3 Mk II for the constant ΔE mode.[23] These functions may be determined by the use of NPL reference materials SCAA83.[23]

These brief passages deal with some of the data bases for the instrumental aspects of AES and XPS. We now consider the data sets that are used for quantitative analysis by both techniques.

RELATIVE SENSITIVITY FACTORS AND QUANTITATIVE ANALYSIS

In order to quantify spectra we need relative sensitivity factors for the different elements. Notations vary but here we refer to these factors as I_i^∞, the intensity for the pure element. In this approach the composition X_A of element A is given by:[18,24,25,26]

$$X_A = \frac{I_A/I_A^\infty}{\sum_i F_{iA}\, I_i/I_i^\infty} \quad (1)$$

where the F_{iA} are matrix factors for AES or XPS.

AES

A number of compilations of I_i^∞ exist, the most popular of which is that of Davis *et al.*[16] In their *Handbook*, peak to peak values of I_i^∞ for the differential mode are presented for the primary electron beam set at 3, 5 or 10 keV along the surface normal and measured with a CMA with a resolution, $\Delta E/E$, of 0.6%. An example of their sensitivity factors for the 5 keV electron beam is shown in Fig. 7. The original presentation has been modified to emphasise the different transitions that are chosen as the atomic number increases. According to the introduction to the *Handbook*, the data have been smoothed and in some cases interpolated or averaged from many samples.

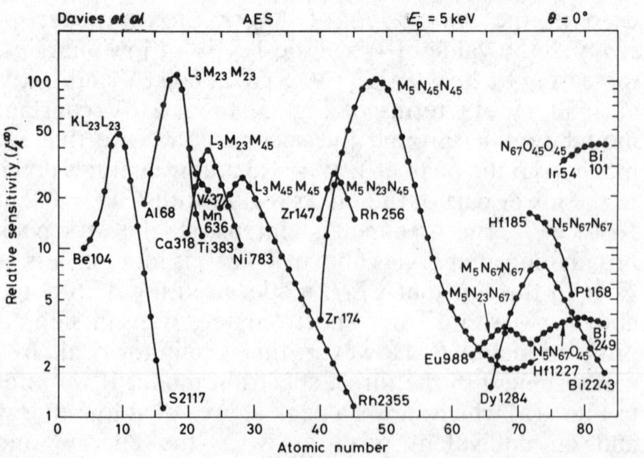

Figure 7. AES peak to peak relative sensitivity factors, I_A^∞, derived experimentally. The values are for 5 keV normally incident electrons and are compilted from the data of Davis *et al.*[16]

We note now that the Auger matrix factors F_{iA}^{A} are approximately given by:[26]

$$F_{iA}^{A} = \left\{ \frac{1 + r_i(E_A)}{1 + r_A(E_A)} \right\} \frac{\lambda_i(E_A)}{\lambda_A(E_A)} \frac{a_A{}^3}{a_i{}^3} \qquad (2)$$

where $\lambda_i(E_A)$ is the electron attenuation length of electrons of energy E_A in the matrix i, $r_i(E_A)$ is the fractional contribution of Auger electron intensity arising from the backscattered electrons compared to those generated by the primary electron beam and a_i^3 is the total volume occupied by one i atom. We shall return to these terms later on but here we should note a possible simplification to (2). In 1979 Seah and Dench,[27] analysing all of the previous experimental data of electron attenuation lengths showed that, for elements at electron energies, E_A, above 60 eV,

$$\lambda_i(E_A) = 0.41\, a_i^{3/2} E_A{}^{1/2}$$

where a_i and λ_i are in nm and E_A is in eV. Thus,[26],

$$F_{iA}^{A} = \left\{ \frac{1 + r_i(E_A)}{1 + r_A(E_A)} \right\} \left[\frac{a_A}{a_i} \right]^{3/2}$$

If we now consider new relative sensitivity factors for (1) defined as $I_i^\infty(a_i/a_{Ag})^{3/2}$, where a_{Ag} is the atom size of silver, we may rewrite (1) with the new factors replacing the old but with the advantage that the matrix factors become simplified by removal of the $(a_A/a_i)^{3/2}$ term. This approach was first suggested by Payling[28] who lists both the original data of Davis *et al.* at 5 keV and the modified factors multiplying by $(a_i/a_{Ag})^{3/2}$. This exercise highlights the point that the values of the matrix factors are not solely an intrinsic property of the system but arise through the way we choose to express the quantitative relationship. We should also observe that, whilst Payling's approach is a correct simplification of the situation if true I_i values are available, in practice it may not improve matters. The smoothing between elements and the averaging by analysing several samples by Davis *et al*[16] may already have gone some way towards producing a set of $I_i^\infty(a_i/a_{Ag})^{3/2}$ sensitivity factors in the *Handbook of Auger Electron Spectroscopy* in the first place!

Similar data sets to that of Davis *et al.*[16] have been produced by McGuire[15], the Anelva Corporation[19] and by Sekine *et al.*[20] McGuire's *Reference Manual of Auger Electron Spectroscopy*[20] is particularly useful since it provides detailed spectra on an expanded energy scale together with data for modulation energies between 1.6 eV and 16 eV[15,29] McGuire's *Manual* also allows I_i^∞ values to be determined which may be contrasted with those of Davis *et al.*,[16] both compilers using a PHI CMA of 0.6% resolution and a 5 keV electron beam. One difference which we should note is the use of a normal incidence beam by Davis *et al.*[16] and one at 30° away from the normal by McGuire. Figure 8(a) shows the ratio of these sensitivity factors as a function of atomic number and Fig. 8(b) as a function of the Auger electron energy. There appear to be no systematic variations and the standard deviation of $\overset{\times}{\div}$ 1.56 would appear to arise from the scatter of measurement in one or other data set. Figure 8(c) shows McGuire's data plotted in the format of Fig. 7.

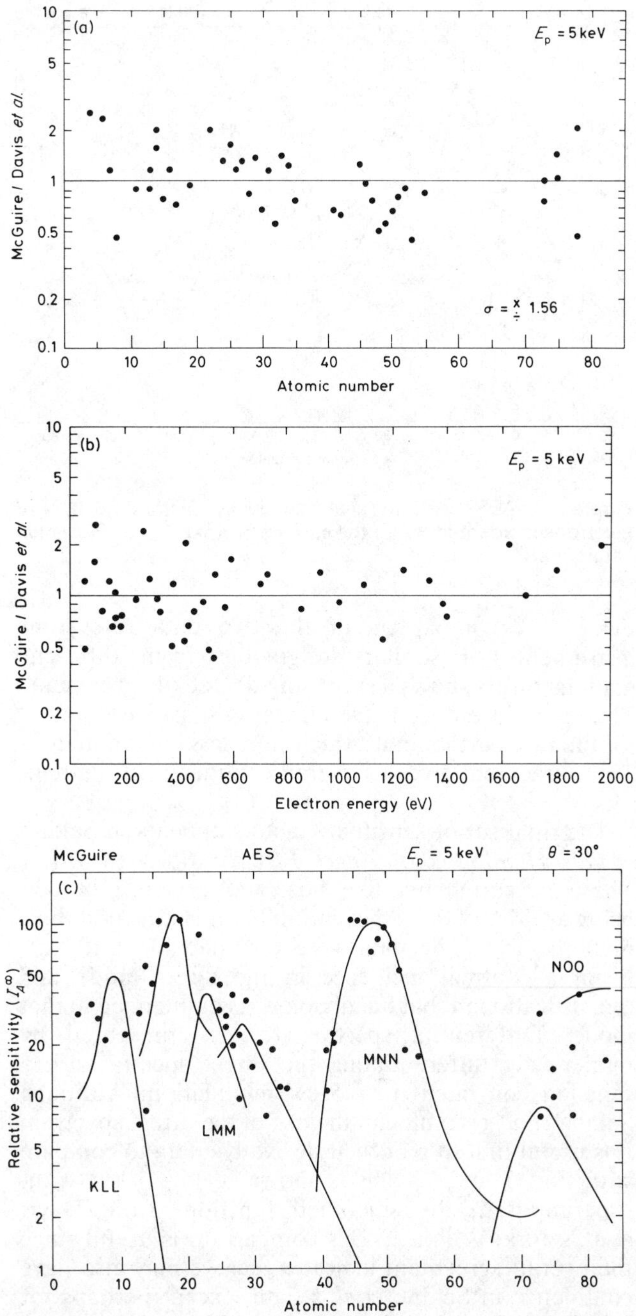

Figure 8. A comparison between the peak to peak relative sensitivity factors for AES presented by McGuire[15] and Davis *et al.*,[16] (a) as a function of atomic number, and (b) as a function of kinetic energy. Figure 8(c) shows McGuire's data for 5 eV peak to peak modulation in the format of Fig. 7.

The Anelva Corporation catalogue[19] provides another data set which uses a CMA but here with a fixed 2 eV modulation from 0 to 1000 eV and 5 eV from 1000 eV upwards. Again the samples are at 30° to the electron beam and two beam energies, 3 keV and 5 keV, are used. The spectra for many reactive elements such as the lanthanide rare earth metals show less contamination than in Davis *et al.*[16] and so many give more reliable sensitivity factors. Instead of analysing this data as extensively as we did with McGuire, we reproduce the data in Fig. 9 simply using the format of

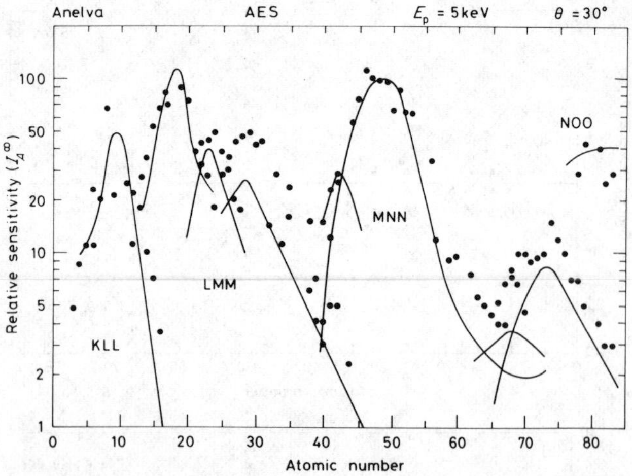

Figure 9. AES peak to peak sensitivity factors for 5 keV electrons incident at 30° to the surface normal, from the Anelva catalogue.[19]

Fig. 7. A comparison of the two data sets from instruments of similar design but from different manufacturers shows a surprising degree of agreement. The main difference is the increased sensitivity factors for the rare earth metals which were less contaminated. This agreement gives us overall confidence in the data sets.

The final set of sensitivity factors appears in Sekine *et al.*'s *Handbook of Auger Electron Spectroscopy.*[20] These measurements, like those above, were recorded using a CMA of 0.5% resolution but this time of a third manufacturer. The data were recorded for a 10 keV beam at normal incidence in the direct mode and quantification in that handbook is recommended in that mode. Differential spectra are also presented by numerically differentiating the direct spectra with a function equivalent to a 5 eV modulation. Although Sekine *et al.* recommend the use of the direct spectrum it is useful first to plot their derivative data to compare with Figs 7 and 9. This is shown in Fig. 10(a) with, superimposed, the smoothed functions from Davis *et al.*'s 10 keV data.[16] This comparison is useful since the overall agreement indicates that we may also have confidence in Sekine *et al.*'s data, except perhaps for Be, Y and the low energy peaks of Mg and Al.

Before considering Sekine *et al.*'s data for the direct mode it is worth returning to Fig. 4 to consider the effects of analyser resolution or chemical state which alter the observed widths of peaks, but not their area, in the direct spectrum. In the traditional derivative spectrum for high energy peaks where $W > V$ in the spirit of the section on electron spectrometers and detectors, we find that the signal falls by the factor x^2 when the width increases by the factor x. For lower energy peaks where $W < V$ this sensitivity is progressively reduced. In the direct spectrum if we measured the peak area there would of course, never be any problem with either the analyser resolution or the chemical state. At the present time, however, we cannot do this as we do not have an agreed method for removing the background necessary to define the area. An alternative which is less accurate is to measure the peak height in the direct spectrum. This is better than the derivative spectrum since the signal only falls by the factor x when the width increases by that factor. In Fig. 10(b) we present Sekine *et al's* peak height data set. This data comes from the same spectra as Fig. 10(a) and therefore attracts the same confidence except perhaps for Be, Y and the low energy peaks of Mg and Al. Note that here Y seems not to be out of place and so it may be that the smoothing in Davis *et al's* data is incorrect. Note also, for the reasons discussed above, that the intensities do not fall as rapidly in the direct spectra as in the derivative as the atomic number, and hence the peak energy, increases.

The above data sets for AES are largely determined using metal foils, sputter cleaned in UHV, or simple binary compounds. This approach appears to be less successful for XPS, as discussed below.

XPS

Two types of data set exist for XPS, one concerned with intensities in quantification and the other with peak position for chemical state identification. In this survey we shall be limited to the former type although, for XPS, the latter may be considered the more

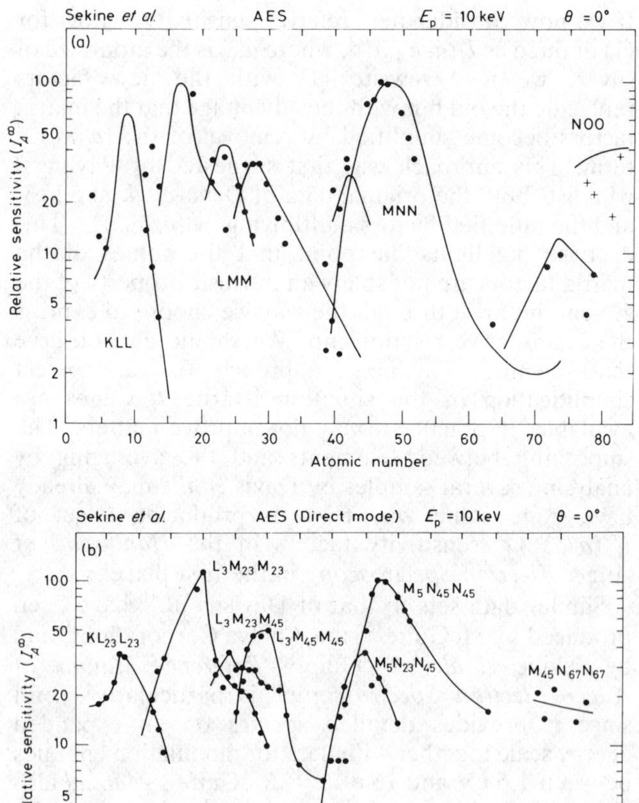

Figure 10. AES sensitivity factors for 10 keV electrons incident along the surface normal, from Sekine *et al's*[20] *Handbook*, (a) in the differential peak to peak mode with Davis *et al's*[16] curves for comparison, and (b) for peak heights in the direct mode.

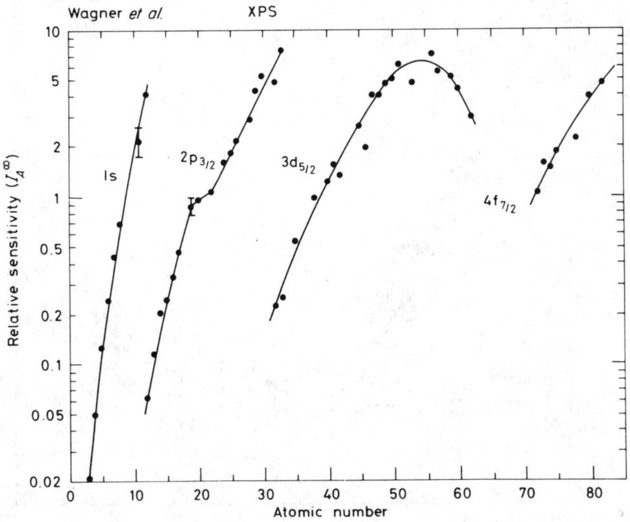

Figure 11. Peak area sensitivity factors for XPS, after Wagner *et al.*

important. The data sets for XPS intensities have many differences from those for AES. The intensities are usually measured as peak areas in the direct spectrum but only with a simple straight line to define the background to be subtracted. As discussed before, the different manufacturers' spectrometers have a noticeable effect on relative peak areas.

Two data sets, by Brillson and Ceasar[31] and by Vulli, Starke and Janghorbani,[32–34] exist which use the AES method of separate sputter cleaned foils or evaporated films. For some reason this does not appear to be very satisfactory and the earlier method of using freshly crushed powders gives more consistent results. We shall not analyse all of the data sets[35–44] since this is done elsewhere[45,46] but merely consider the final and most up-to-date set by Wagner.[46] For consistency these data are plotted as the AES data previously, in Fig. 11. We shall not comment further at this stage but note that these data relate to instruments with an intensity-energy transmission function equivalent to the Perkin–Elmer PHI 550 and transference to other instruments requires the use of Fig. 6.

A Broader Quantification

For quantitative analysis we use (1) with the XPS matrix factor, F^x_{iA}, given by the analogue to (2):

$$F^x_{iA} = \frac{\lambda_i(E_A)}{\lambda_A(E_A)} \frac{a_A^3}{a_i^3} \tag{4}$$

If, as discussed at the beginning of the third section of this paper, we use (3) for $\lambda_i(E_A)$ and renormalize the pure element sensitivity factors to be J^∞_i equal to $I^\infty_i(a_i/a_{Ag})^{3/2}$ then we find (1) may be rewritten:

$$X_A = \frac{I_A/J_A^\infty}{\Sigma_i\, I_i/J_i^\infty} \tag{5}$$

Thus, if (3) is valid, we may renormalize the pure element sensitivity factors so that the matrix factors all become unity! Conversely, if we all determine sensitivity factors from an equation of the form of (5) from many compounds we do, in fact, determine the J^∞_i. This is, in fact, what is done for the data sets.[35–46] Thus, in XPS we should, perhaps, use (5) with matrix factors of unity.

Of course, (1) is only the crudest level of quantification in which the sample is assumed to be homogeneous over the surface region to a depth of, say, 2 $\lambda_i(E_A)$. There are many other types of sample and these may be found in the relevant texts.[18,26,47,48] As an illustration we shall consider just one of these models, that of a monolayer adsorbate, A, on a homogeneous single element substrate, B. We then find the equivalent of (1) becomes:

$$X_A^S = \frac{I_A/I_A^\infty}{Q_{BA}\, I_B/I_B^\infty} \tag{6}$$

where X_A^S is the fraction monolayer coverage of A and the matrix factors Q^A_{BA} and Q^X_{BA} for AES and XPS respectively, are given by:

$$Q^A_{BA} = \left\{ \frac{1 + r_B(E_A)}{1 + r_A(E_A)} \right\} \frac{1}{\lambda'_A(E_A)\cos\theta} \tag{7}$$

$$Q^X_{BA} = \frac{1}{\lambda'_A(E_A)\cos\theta} \tag{8}$$

Here $\lambda'_A(E_A)$ is the attenuation length of electrons of energy E_A in the material A in units of monolayers of A and θ is the angle of collection of the electrons from the surface normal. It is clear that, for quantification of these situations, we need data on both the attenuation lengths of electrons and the matrix sensitivity of backscattering.

The Attenuation Lengths

In our earliest works in this area[49] the term inelastic mean free path (λ) was used as the distance over which the characteristic electron intensity fell by l/e. In this sense it was a description of the observed events. Calculations of inelastic mean free paths were subsequently made which we shall discuss below, but there was an intrinsic difference. The calculations focused on the total or gross track length which may or may not have been diverted from a straight line by the effect of elastic scattering events whereas the experimental observations were of the net track length. To clarify and remove this difference Powell[51] has retained the use of inelastic mean free path for the total lengths which are those directly calculated and redefined the net lengths, usually those derived from experiment, as attenuation lengths. The experimental observations actually measure the electron escape depth which may be defined[50] as the distance normal to the surface at which the probability of an electron escaping without significant energy loss due to inelastic scattering processes drops to l/e of its original value. It is then assumed that the escape depth is equal to the attenuation length multiplied by cos θ.

It is not the purpose here to review the state of knowledge in this field; this is done thoroughly by Powell.[51] We wish to comment on the available data sets. Various compilations of attenuation lengths have been published in early reviews but the first most systematic assessment was that of Powell.[52] Seah and Dench[53] extended this data base to include later publications and found, by including all of the data without selection, that certain empirical equations could be written to describe the attenuation length, λ, for different classes of material. The resulting relations and classes of material were:

for elements:

$$\lambda = \frac{538}{E^2} + 0.41\,(aE)^{1/2} \text{ monolayers } [1.36] \qquad (9)$$

for inorganic compounds:

$$\lambda = \frac{2170}{E^2} + 0.72\,(aE)^{1/2} \text{ monolayers } [1.38] \qquad (10)$$

and for organic compounds:

$$\lambda = \frac{49}{E^2} + 0.11\,(E)^{1/2} \text{ mgm m}^{-2}\,[2.10] \qquad (11)$$

where a, the atom size in nm is derived from the relation

$$\rho N n a^3 = 10^{24} A_0 \qquad (12)$$

where A_0 is the atomic or molecular weight, n is the number of atoms in the molecule, N is Avogradro's number, and ρ is the bulk density in kg/m^{-3}. Additionally, the number in square brackets represents the factor of uncertainty ($\overset{\times}{\div}$) at one standard deviation on the log λ vs log E plots, one example of which is shown in Fig. 12. This level of uncertainty is a major problem for quantification in surface analysis. Effectively 32% of λ values diverge from the above equations by more than a factor of 1.36 for elements etc. At the present time no clear experimental method has been evolved which is generally applicable and allows easy and accurate measurements of λ.

In considering the above equations for λ we should note that the equation for inorganic compounds is dominated by data for the alkali halides. If we separate the alkali halides from the oxides, for data above 100 eV we find:

Alkali Halides

$$\lambda = 0.89\,(aE)^{1/2} \text{ monolayers } [1.19] \qquad (13)$$

Oxides

$$\lambda = 0.55\,(aE)^{1/2} \text{ monolayers } [1.37] \qquad (14)$$

This value for the oxides is much closer to that for elements and is better than using the overall average for inorganic compounds.

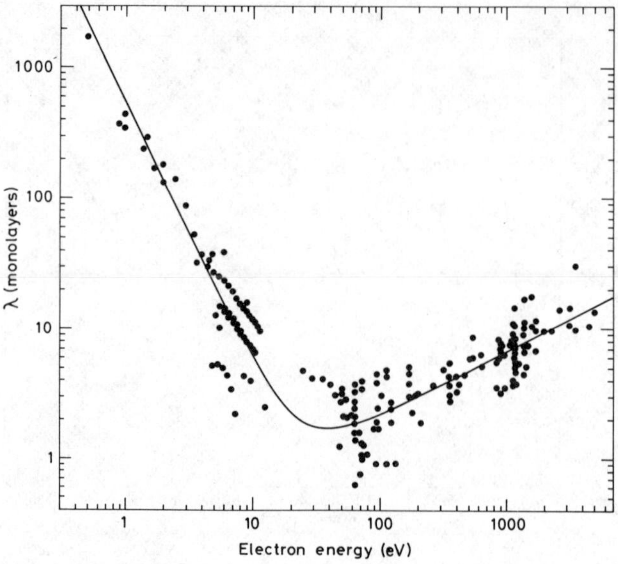

Figure 12. Compilation of the electron attenuation lengths in elemental solids as a function of the electron energy, after Seah and Dench.[53]

A more recent analysis of specific sets of measurements has been made by Wagner *et al.*[53] who find that, for sets of measurements by individual workers, the energy dependence above 400 eV is in the range $E^{0.54}$ to $E^{0.81}$ and that the average power is 0.66 $\pm$ 0.10 rather than 0.5 as shown in (9), (10), (11) etc. It is clear however, that a power law dependence could have been taken for energies above 100 eV instead of selectively excluding some low energy points. If this is done in Wagner *et al.*'s[53] compilation we find that the average power dependence falls to 0.56 $\pm$ 0.14, really not very different from the empirical values of (9) to (14).

The above data sets for λ apply equally well for AES as for XPS. This is evident because the characteristic electrons undergo the same inelastic scattering in each case. Actually one can envisage small differences but these are usually ignored.

The remaining data set noted above in the previous section is that of the backscattering factors. These are needed for AES but not XPS and are discussed below.

Backscattering Factors

Compiled data sets of measurements of backscattering are very limited. Two elegant types of experiment allow precise measures of the backscattering factors to be determined, both arising from the study of microelectronic circuits. In the first, Kirschner[54] studied devices in which a thin film of Al (a metal interconnect) overlaps thick regions of Au and Si. The ratio of the Al intensity between the two regions is simply $\{1 + r_{Au}\,(E_{Al})\}/\{1 + r_{Si}\,(E_{Al})\}$ for the appropriate electron beam energy. In the second, Meda and Queirolo[55] analyse sputter depth profiles of thick (1 μm) films of Si on Ta. The ratio of the silicon intensity in the layer to that just before the interface gives the ratio $\{1 + r_{Si}\,(E_{Si})\}/\{1 + r_{Ta}\,(E_{Si})\}$, similar to Kirschner's determination. Both types of method allow the determination of the backscattering from many substrates, as a function of both the Auger electron energy and the primary beam energy, with relative ease and precision. In Meda and Queirolo's[55] study the experimental value for the ratio was found to agree with the predictions of Shimizu[56] from Monte Carlo calculations to about 1.3%. With this level of agreement a full experimental data bank would seem unnecessary and it merely remains to test Shimizu's prediction with a few further points to see if this accuracy is confirmed.

THEORETICAL DATA BANKS FOR QUANTITATIVE ANALYSIS

In the above we have covered, somewhat sketchily, the basic experimentally determined data bases relating to quantitative surface analysis by AES and XPS. It is impossible to list all of the caveats and comments that should go with them but we have tried to list the most important and some of those often ignored.

General Considerations for AES

For AES we may write the Auger electron intensity from a pure element standard as:[26]

$$I_A^{\infty} = I_0\, \sigma_x(E_p)\, \gamma_{xyz}\, [1 + r_A(E_A, \alpha, E_p)]\; N_A\, \lambda_A(E_A) \cos\theta\, T(E_A) D(E_A) \qquad (15)$$

Where I_0 is the primary electron beam current, $\sigma_x(E_p)$ is the cross section for ionization of the core level x, γ_{xyz} is the decay probability that leads to ejection of an Auger electron of energy E_A involving the three core levels x, y and z, $r_A(E_A,\alpha,E_p)$ is the backscattering factor with its dependence on the primary beam angle of incidence, α, N_A is the density of A atoms, $T(E_A)$ is the transmission of the spectrometer and $D(E_A)$ is the efficiency of the detector. For the relative sensitivity factors both I_0 and $\cos\theta$ cancel between two intensities and so may be ignored, N_A is readily available[57] and both $T(E_A)$ and $D(E_A)$ are as discussed in the second section.

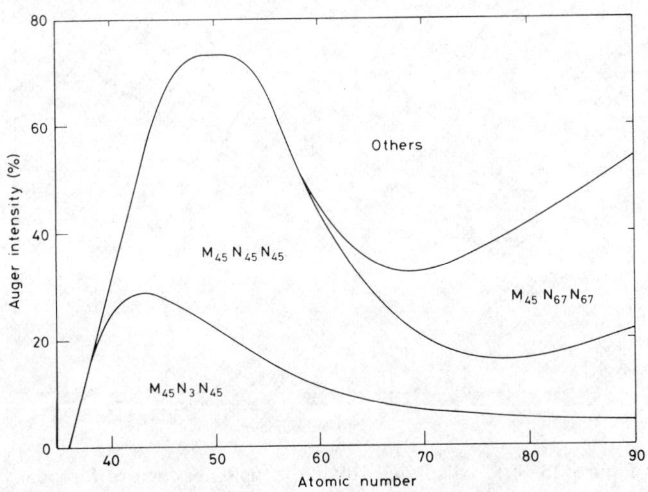

Figure 13. The fractional intensity in different Auger transitions originating in the M_4 and M_5 shells of elements between Kr and Th, after McGuire.[61]

Amongst those who calculate intensities from first principles, the cross section $\sigma_x(E_p)$ is usually calculated according to Gryzinski's expression.[58] The various expressions that have been developed over the years and their agreement with one another and with experimental data is considered in some detail in the review of Powell[59] and will not be considered here. One point that should be added occurs where the level x is $L_{2,3}$ or, say, $M_{4,5}$. As can be seen from Fig. 7 these are the most important peaks for elements above Mg. For these shells extra ionisation occurs due to Coster–Kronig transitions to deeper shells and therefore the effective cross section $\sigma_x(E_p)$ should be increased by values of up to 30%.[60,61] The term γ_{xyx} is the product of two further terms, the probability of an Auger electron being emitted and the probability that, of the possible Auger transitions originating from level x, the decay xyz will occur. The first term is usually that of Burhop[62] but the second factor is often ignored. The relative intensities in the different Auger electron peaks have been calculated by many authors and these may be found inthe excellent review by Weightman.[63] As an example, Fig. 13 shows the results of calculations by McGuire[61] for the relative intensities of some of the more intense peaks for elements between Kr and Th arising from ionizations in the M_4 and M_5 shells. The remaining two terms, the backscattering factor and attenuation lengths will be considered below.

The backscattering factor. The earliest backscattering factors used in quantitative AES were those of Bishop[65] and Reuter.[66] More recently Jablonski[67] has provided much data and provided a general expression[68] for r at normal incidence. However, by far the most general calculations are those provided by Ichimura and Shimizu.[69,70] Their Monte Carlo calculations of the scattering of the primary electrons with both eleastic and inelastic scattering in the solid provide an enormous data set which, fortunately, has been distilled into three equations[56] depending on the angle of incidence, α, of the primary electron beam:

$$r = (2.34 - 2.10\, Z^{0.14})\, U^{-0.35} + (2.58\, Z^{0.14} - 2.98), \quad \alpha = 0^\circ \qquad (16)$$

$$= (0.462 - 0.777\, Z^{0.2})\, U^{-0.32} + (1.15\, Z^{0.2} - 1.05), \quad \alpha = 30^\circ \qquad (17)$$

$$= (1.21 - 1.39\, Z^{0.13})\, U^{-0.33} + (1.94\, Z^{0.13} - 1.88), \quad \alpha = 45^\circ \qquad (18)$$

where Z is the atomic number of the matrix, $U = E_p/E_x$ and where, in turn, E_x is the binding energy of the core level x. These equations are simply the best fits to describe the data generated by the Monte Carlo calculations and are valid for $4 < Z < 80$, for $100 < E_x < 2000$ and for 3 keV $< E_p <$ 10 keV which ranges encompass most experimental situations. The experimental data of Meda and Queirolo[55] support these equations as discussed earlier.

The inelastic mean free path and attenuation length. Various calculations of inelastic mean free paths have been published, the most general of which is that of Penn.[71]

For $E_A > 200$ eV the energy dependence is of the form:

$$\lambda = \frac{E}{a(\ln E + b)} \qquad (19)$$

and, as analysed by Wagner *et al.*,[53] this gives a dependence close to $kE^{0.77}$ for a range of materials. Penn provides data for all elements except the rare gases and for energies, E, between 200 and 2400 eV. This, therefore, represents a very complete data set. As discussed by Powell,[51] there are better theoretical expressions than that giving Penn's data set but few provide the basis of generating a full data set without some effort on the part of the analyst.

A very useful derivation has been made more recently by Szajman *et al.*[72] This extends Penn's[71] model to inorganic solids and especially to insulators. They find for valence electron scattering

$$\lambda_v(E_A) = (2.12\ E_A\ \bar{E}/E_{pl}^2)\ /\ [\ln(\alpha E_A/\bar{E}) + \phi]$$

where

$$\alpha = 8Q/[(\bar{E}\ (4.18\ Q^2 + 0.68)/E_{pl})^{1/2} + 0.66\ Q^2 + 0.18],$$

$$\phi = 3E_{pl}(Q + 0.11\ E_{pl}^2/\bar{E}^2)/4Q^2\bar{E},$$

$$Q = E_F/E_{pl},$$

and where, in turn, E_{pl} and E_F are the free-electron values of the plasma and Fermi energies respectively and $\bar{E}$ is the energy centroid of the loss function, all in electron volts. Some simplifications were made to make calculation easy for most solids. Thus,

$$\bar{E} \simeq E_{pl} + E_g$$

where E_g is the band gap (zero in metals) and, if additionally E_{pl} is taken to be in the range 5 to 30 eV and the core electron scattering is 15% of that of the valence electrons, they derive:

$$\lambda \simeq 0.1\ \frac{8\bar{E}\ E_A^{3/4}}{E_{pl}^{\ 2}}\ \text{nm}$$

although a power dependence of $E_A^{0.6}$ *to* $E_A^{0.7}$ would seem more appropriate. With rather more complex calculations, which it is not appropriate to present here, they show a universal plot which combines the elemental and inorganic compound data of Seah and Dench's compilation[27] into one plot which, by excluding 9 points with gross deviations, gives a correlation with an rms scatter of 1.32. Although the correlation is little better than Seah and Dench's equations, this approach has the merit of being physically based. The poor quality of the correlation probably has more to do with the poor quality of the data than with the theory. One aspect that is generally missed in the theories is that of the elastic scattering. Jablonski *et al.*[73] and Ferron *et al.*[74] are now including this but general data tables and the effect on the predicted energy dependence are not yet available. Limited calculations by Dwyer and Matthew[75] indicate a 20% reduction in the net path length and very minor alterations to the energy dependencies.

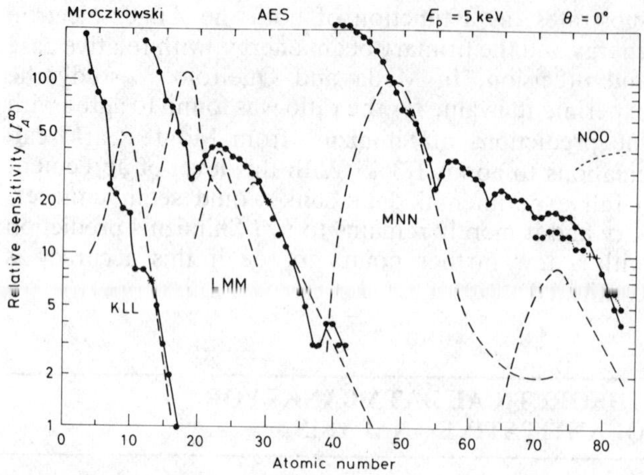

Figure 14. The calculated relative sensitivity factors for peak to peak AES, after Mroczkowski and Lichtman.[75,76] The points marked * are used for fitting each set of transitions to the dotted lines representing the data of Davis *et al.*[16]

The Relative Sensitivity Factors. The only attempt to provide a full 'pseudo-first principles' calculation of Auger electron relative sensitivity factors is that of Mroczkowski and Lichtman.[76] Their analysis, based on (15) uses Gryzinski's cross section;[58] Burhop's probability of an Auger process,[62] Reuter's backstattering factor[65] and Seah and Dench's[27] prediction for λ. It is further assumed that $T(E_A)D(E_A)$ is constant above 100 eV although, as discussed above, we see that, in the derivative mode the effective function may vary by a factor of 3 over the AES energy range.

However, using the above approach Mroczkowski and Lichtman[76] calculate the relative sensitivity factors for the elements to compare with the factors of Davis *et al.*,[16] a somewhat extended version of which is shown in Fig. 7. Their results for a 5 keV beam energy are shown in Fig. 14 together with Davis *et al.'s* original presentation. Because the peak shapes change somewhat from element to element and more particularly from transition group to transition group

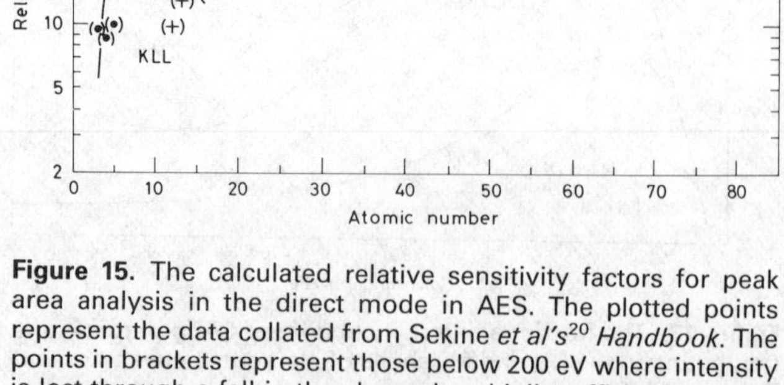

Figure 15. The calculated relative sensitivity factors for peak area analysis in the direct mode in AES. The plotted points represent the data collated from Sekine *et al's*[20] *Handbook.* The points in brackets represent those below 200 eV where intensity is lost through a fall in the channel multiplier efficiency.

they have simply normalized the KLL data to be correct for Al, the LMM for Zn and the MNN for Sb. Similar plots are also presented at 3 and 10 keV. This work is interesting since it provides, for the first time, a serious attempt to initiate a convergence between theory and experiment in this difficult area. The details of the analysis have been criticised by Ferron *et al.*[77] Two criticisms are really important. The first is that the use of only Burhop's term for γ_{xyz} means that the total Auger yield for the level x is calculated. This is represented by the sum of the areas under the relevant peaks in the direct spectrum, rather than the steepest gradient on one of the peaks. Secondly, the spectrometer transmission function must be included. If, however, we consider the peak areas in the direct spectrum then the transmission function of the spectrometer becomes proportional to E rather than being complex. Thus, it seems sensible to use Mroczkowski and Lichtman's approach and compare the results with the peak areas estimated from Sekine *et al's* direct spectra.[20] Figure 15 shows the results of such calculations, using (15), with $\sigma_x(E_p)$, γ_{xyz} and $\lambda_A(E_A)$ calculated using Gryzinski's cross section,[58] Burhop's factor[62] and Seah and Dench's equation for λ for elements.[27] For the backscattering factor we have used (16) from Shimizu[56] and $T(E_A)$ has been assumed to be proportional to E_A. No relative adjustment has been made between the KLL, L_{23}MM and M_{45}NN intensities, as was done by Mroczkowski and Lichtman.[75,76] The peak areas to be derived from Sekine *et al.*'s data[20] represent rather more of a problem. We could simply draw a straight line background under the appropriate peaks but a simple observation of the peaks for C, Mg, Ge or Ag shows that this would include many inelastically scattered electrons. Alternatively we could use a Shirley background[78] subtraction but this would be both difficult with the data as presented and also lead to similar, but smaller, errors as the straight line.[79] Bearing in mind that Fig. 10(a) shows we may be dealing with errors of up to a factor of two anyway, the following simple procedure was adopted and applied equally to all peaks. The

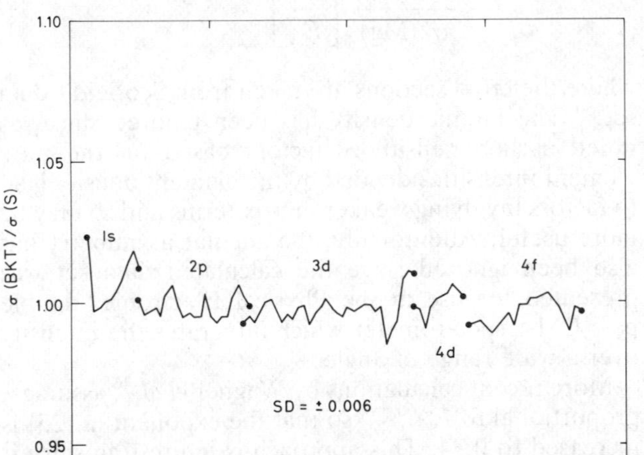

Figure 16. The ratio of Band *et al.*'s[81] photoelectron cross section for Al Kα radiation, to that of Scofield's,[80] for the most intense transitions as a function of atomic number.

intensity was measured by summing all of the peak heights of the appropriate transition, for instance all of the peaks between 1400 and 2200 eV for the M_{45}NN transitions of Au, and by multiplying this by a notional peak width. The peak width was estimated from the quadrature sum of the analyser resolution and an intrinsic width of 1.5 eV, i.e. $(1.5^2 + (0.005\ E_A)^2)^{1/2}$. This whole procedure is not recommended for quantitative analysis but was simply adopted to see if the criticisms of Mroczkowski and Lichtman's analysis could be implemented and if the agreement with experiment is thus improved. A simple comparison of Fig. 14 and 15 shows that, even without separate adjustment for the three sets of transition an order of magnitude improvement has been effected for the rare earth region and the low atomic number ends of the series.

General Considerations for XPS

For XPS we may write the photoelectron intensity from a pure elemental standard as:

$$I_A^\infty = J_0\ \sigma_x(h\nu)\ L_A(\gamma)\ N_A\ \lambda_A(E_A)\cos\theta\ T(E_A)\ D(E_A) \quad (21)$$

where, of the new terms, J_0 is the x-ray flux density, $\sigma_x(h\nu)$ is the cross section for ejection of a photoelectron from the level x by photons of energy $h\nu$ and $L_A(\gamma)$ is an asymmetry term describing the angular intensity of the emission as a function of the angle γ between the x-ray source and the direction of the emitted electrons.

The Photoelectron Cross-section. Most popularly, Scofield's[80] Hartree–Slater calculations, which cover all subshells of all elements for both Mg and Al Kα x-rays, are used. In his conclusions Scofield notes that experimental measurments of ratios of cross sections diverge as much as a factor of two from his calculations. It is likely that a large portion of this error arises from the difficulty of the experimental measurements. Other calculations of photoionisation cross sections have appeared since Scofield's. For instance Band *et al.*,[81] using a relativisitc self-consistent Dirac–slater potential, calculate the subshell cross sections for many x-ray sources for all elements. They note that for Al and Mg Kα the results practically coincide with Scofield's. Figure 16 shows the ratio of Band *et al.*'s calculations for Al Kα to that of Scofield for some of the more important lines. More recently, Yeh and Lindau[82] have provided further data, extending the photon energy range down to 10.2 eV. They note that the work of Band *et al.*[81] takes account of the *j*-splitting of subshells which is not available within their framework and also that Band *et al.*, including all of the multipoles of the radiation field, provide a more advanced treatment than theirs. Figure 17 shows the equivalent plot to Fig. 16 for Yeh and Lindau's calculations and we now see systematic divergences, especially for the 4f subshell. Such divergence is not seen in Goldberg *et al.*'s calculations for a limited set of elements.[64]

It seems reasonable to conclude that for consistency, in comparisons of experiment with theory. the experimentalists should continue to use Scofield's[80]

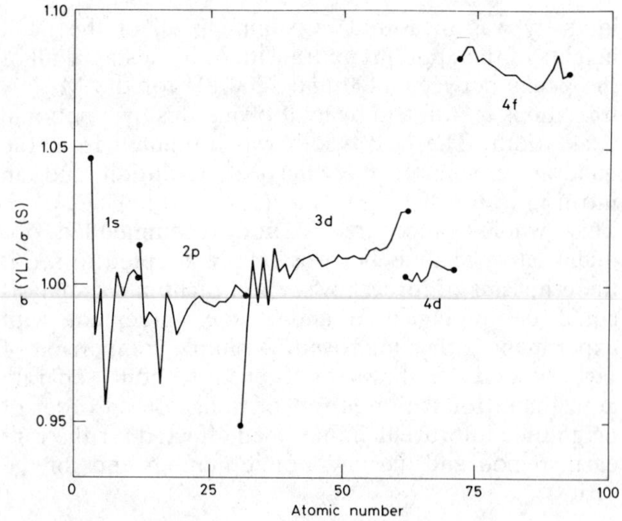

Figure 17. The ratio of Yeh and Lindau's[82] photoelectron cross sections for Al Kα radiation, to that of Scofield's[80], for the most intense transitions, as a function of atomic number.

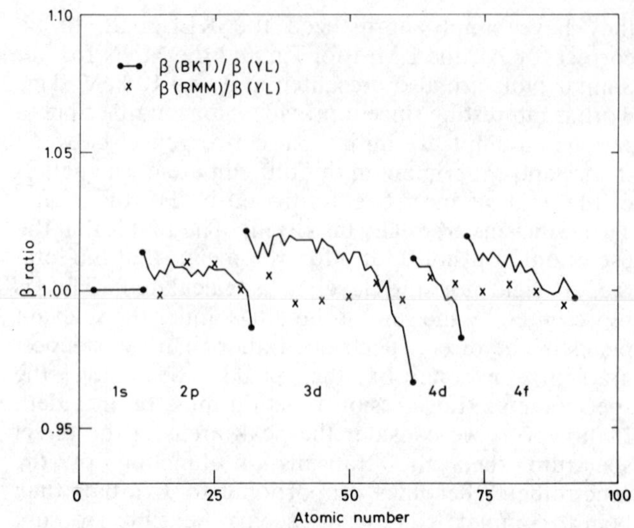

Figure 18. The ratio of Band *et al.*'s[81] β value to those of Yeh and Lindau[82] and of Reilman *et al.*'s[83] also to those of Yeh and Lindau, as a function of atomic number.

data as they have in the past. It would seem reasonable also to hope that those calculating cross sections in future would indicate where they diverge significantly from previous calculations!

The Angular Asymmetry Parameter. The angular asymmetry term which describes the intensity at an angle γ to the incoming x-ray flux direction is given by:

$$L_A(\gamma) = 1 + \frac{1}{2}\beta_A\left[\frac{3}{2}\sin^2\gamma - 1\right] \qquad (22)$$

where β_A, the angular asymmetry parameter is given in tables and depends on the photon energy, the atomic number and the subshell of the atom. Traditionally, most experimentalists use the calculations of Reilman *et al.*[83] which cover both Mg and Al radiations, and all subshells but only for atoms at 5 amu intervals. The β values vary slowly and regularly and so interpolation to all Z values is relatively simple. No figure will be given here as a visual presentation is given elsewhere.[26,45] Similar calculations appear in the tables of Band *et al.*[81] and of Yeh and Lindau.[82] Figure 18 shows how agreement between Reilman *et al.* and Yeh and Lindau exhibits an rms scatter of around 0.5% leading to a scatter in $L_A(\gamma)$ of around only 0.1%. A similar correspondence occurs in Goldberg *et al.*'s calculations.[64] These are negligible divergencies in terms of the experimental errors and so, again, experimentalists should continue to use Reilman *et al.'s* data.[83] One proviso, however, is necessary. Traditionally, (22) has been used for solids although it is actually derived for gases. In solids elastic scattering alters angular distribution[84,85] and, as a result of the analysis for many elements Ebel *et al.*[85] show that the effective value of β in solids β^*, is given by:

$$\beta^* = (1.0688 - 0.0235\,Z + 0.000188\,Z^2)\beta \qquad (23)$$

for photoelectrons in the kinetic energy range 1000 to 1500 eV and for elements up to $Z = 46$. Bearing in mind this large modification to β, any differences in the data sources for β, as shown in Fig. 18, are fairly trivial.

The Relative Sensitivity Factors. Many publications contain calculations of the elemental intensities in their comparisions with experimental data. However, few of these provide a data set for use directly for quantitative analysis. One such set does exist in the *Handbook of X-Ray Photoelectron Spectroscopy*.[3] It is not clear from the *Handbook* exactly how the calculations have been performed but it appears that the sensitivity factors, I_A^∞, are derived from

$$\frac{I_A^\infty}{I_F^\infty} = \frac{\sigma_A(\mathrm{Mg})}{\sigma_F(\mathrm{Mg})}\;\frac{\lambda_A(E_A)}{\lambda_F(E_F)}\;\frac{T(E_A)}{T(E_F)} \qquad (24)$$

where the normalization is to the flourine 1s peak. Furthermore $\lambda_A(E_A)/\lambda_F(E_F)$ is taken as $(E_A/E_F)^{0.75}$ and $T(E_A)/T(E_F)$ as E_F/E_A. Thus:

$$\frac{I_A^\infty}{I_F^\infty} = \frac{\sigma_A(\mathrm{Mg})}{\sigma_F(\mathrm{Mg})}\left[\frac{E_F}{E_A}\right]^{0.25} \qquad 25$$

where the cross sections are taken from Scofield's data set.[80] The atomic density has been ignored since, as noted earlier, sensitivity factors based on the pure element intensities divided by the element density lead to factors involving weaker matrix terms and so may be more useful. Additionally, the angular asymmetry has also been ignored since the calculated data set was presented for use in the Physical Electronics double pass CMA spectrometer which integrates the intensity over a wide range of angles.

More recent calculations by Wagner *et al.*[46] assume λ proportional to $E^{0.66}$ [44] so that the exponent in (25) is increased to 0.34. This approach is interesting since it allows us to compare the experimental data set of Wagner *et al.*[46] with theoretical predictions calculated as discussed above but with the new energy dependence for λ. If we consider this ratio of experiment to theory we notice that the experimental intensities for the low kinetic energy peaks are too weak. This arises because the powdered samples used by Wagner *et al.* are contaminated by hydrocarbons.[44] If we assume that a monolayer of hydrocarbons attenuate the peaks according to the factor

$$\exp(-25E^{-0.66})$$

then we may derive contamination-corrected experimental data from their Table 2 as:

$$I_A^\infty \text{(expt)} = 0.54\, I_A^\infty \text{(Wagner)} \exp(25E^{-0.66}) \qquad (26)$$

and the theoretical predictions as:

$$I_A^\infty \text{(theory)} = \frac{\sigma_A}{\sigma_F}\left[\frac{E_F}{E_A}\right]^{0.34} \qquad (27)$$

The ratio I_A^∞(expt)/I_A^∞ (theory) is shown in Fig. 19. From this it is clear that the overall intensities of the 1s, 2p, 3d and 4f peaks commonly used for analysis are all in reasonable order but that there are experimental scatters of ± 20%. In particular we do not see any overall loss of intensity at low or high energies and nor are there any systematic changes with core level. Figure 19 is quite different from the analysis of apparent differences between empirical and theoretical sensitivity factors shown in Fig. 12 of the paper by Wagner *et al.*[46] In their paper they take the original experimental data and compare it with the theoretical predictions on a log intensity versus log binding energy plot. From this comparision they bring all of the measurements for a particular transition into line so that the scatter between adjacent elements is eliminated. This approach leads to sensitivity factors which are a hybrid between experiment and theory. As far as their use in quantitative analysis is concerned, this hybrid approach seems good commonsense. This hybrid data appears as Table 3 in the paper by Wagner *et al.*[46] and in Appendix 5 of *Practical Surface Analysis by Auger and X-ray Photoelectron Spectroscopy.*[87] Of course, for the present analysis in Fig. 19, we return to the primary source of data since inevitably any

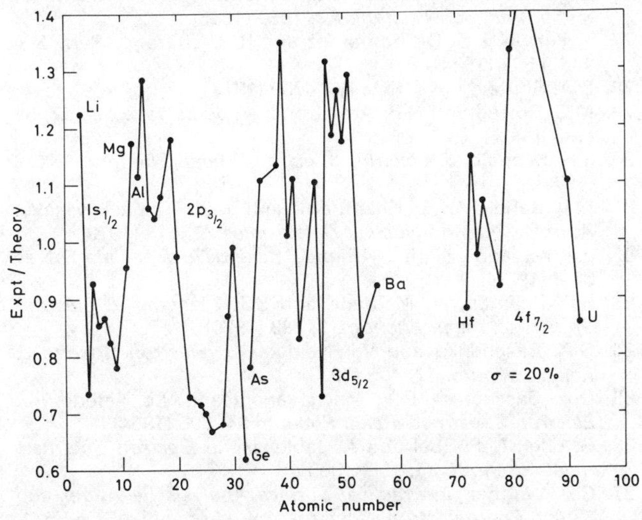

Figure 19. The ratio of Wagner *et al's*[46] measurements of peak area corrected for contamination and the predictions according to (27).

homogenisation with theory must lead to a better correlation with that theory. One final point that is worth noting is that the average numerical value of 1.0 on the ordinate of Fig. 19 is, of course, quite arbitrary and meaningless. An inspection of the deconvoluted no-loss spectrum[88] shows that a value nearer 0.5 would be more appropriate.

In the comparisons between experiment and theory it appears that the shifts have been less with the theoreticians and that, when we know how to handle shake-up events and background subtraction, the final data set will probably be very similar to that predicted as discussed above.

CONCLUSIONS

Data compilations exist for many aspects of the use of surface analysis instruments as well as their calibration and the interpretation of results. Their importance should not be underestimated. The book *Practical Surface Analysis by Auger and X-ray Photoelectron Spectroscopy*[88] contains 100 pages of both tabulated data and procedures and a number of journals now publish data banks. In this survey we have tried to bring together the different data banks and procedures relevant to the quantification of surface analysis by AES and XPS.

The data banks relating to the instrumentation appear to be in very good shape except those for the channel electron multipliers. The data bank of experimental relative sensitivity factors are at a more interim stage and cannot progress further until we have a proper method of background subtraction for the direct spectrum in both AES and XPS. When this is done we should see a convergence between theory and experiment. At the present time peak to peak height factors are available for analysis by differential spectrum AES and these are broadly compatible with a similar data bank for peak height factors in direct spectrum AES. The latter data are, in turn, broadly compatible with the theoretical predictions. For XPS, again, peak area data are available which, when corrected for contamination, are also broadly compatible with the theoretical predictions. On the theoretical side we need a data bank of electron attenuation lengths which include elastic scattering and additionally we need further work on the photoelectron asymmetry parameters also including elastic scattering. Both parameters need rather more accurate experimental measurements than have been available hitherto. On the other hand the calculations of the backscattering factors and of the cross sections in both electron and x-ray ionisation appear to have sufficient accuracy for the current requirements.

REFERENCES

1. K. Siegbahn, G. Nordling, A. Fahlman, R. Nordberg, K. Hamrin, J. Hedman, G. Johansson, T. Bergmark, S.-E. Karlsson, I. Lendgren and B. Lindberg, ESCA, *Atomic Molecular and Solid State Structure Studied by Means of Electron Spectroscopy,* Almqvist and Wiksells BAB, Uppsala (1967).

2. T.A. Carlson, *Photoelectron and Auger Spectroscopy,* Plenum, New York (1975) p 24.
3. C.D. Wagner, W.M. Riggs, L.E. Davis, J.F. Moulder and G.E. Muilenberg, *Handbook of X-ray Photoelectron Spectroscopy*, Perkin-Elmer Corp., Minnesota (1978)
4. A. Barrie in *Handbook of X-ray and Ultraviolet Photoelectron Spectroscopy,* Chapter 2, p. 79, ed. by D. Briggs, Heyden (1977).
5. K. Sigbahn, *J. Electron Spectrosc. Relat. Phenom* **36**, 113 (1985).
6. D. Briggs and J.C. Riviere in *Practical Surface Analysis by Auger and X-ray Photoelectron Spectroscopy,* Chapter 3, p. 87, ed. by D. Briggs and M.P. Seah, Wiley (1983).
7. *JEOL JAMP-1OS brochure no. 1203H 469,* JEOL Ltd. Tokyo.
8. *PHI Multiprobe 600, PHI Data Sheet 1064 8-82 13M,* Perkin–Elmer, Minesota.
9. C.P. Hunt, private communication (1985).
10. J.A. Venables, A.P. Janssen, C.J. Harland and B.A. Joyce, *Philos Mag.* **34** 495 (1976).
11. D.C. Peacock, M. Prutton and R. Roberts, *Vacuum,* **34**, 497 (1984).
12. M.P. Seah and H.J. Mathieu, *Rev. Sci. Instrum.* **56,** 703 (1985).
13. P.H. Holloway and D.M. Holloway, *Surf. Sci.* **66,** 635 (1977).
14. M.T. Anthony and M.P. Seah, *J. Electron Spectrosc. Relat. Phenom.* **32,** 73 (1983).
15. G.E. McGuire, *Auger Electron Spectroscopy Reference Manual*, Plenum (1979).
16. L.E. Davis, N.C. MacDonald, P.W. Palmberg, G.E. Riach and R.E. Weber, *Handbook of Auger Electron Spectroscopy*, 2nd Ed., Physical Electronics Industries Inc., Minnesota (1976).
17. M.P. Seah, in *Methods of Surface Analysis: Techniques and Applications*, Chapter 3, ed. by J.M. Walls, Cambridge University Press (1986).
18. M.P. Seah, *Vacuum*, to be published.
19. *Auger Electron Spectra Catalogue — A Data Collection of Elements*, Anelva Corporation, Tokyo (1979).
20. T. Sekine, Y. Nagasawa, M. Kudoh, Y. Sakai, A.S. Parkes, J.D. Geller, A. Mogami and K. Hirata, *Handbook of Auger Electron Spectroscopy,* JEOL, Tokyo (1982).
21. M.P. Seah, *Surf. Interface Anal.* **2**, 222 (1980).
22. M.P. Seah and M.T. Anthony, *Surf. Interface Anal.* **6**, 230 (1984).
23. M.P. Seah, M.E. Jones and M.T. Anthony, *Surf. Interface Anal.* **6,** 242 (1984).
24. P.H. Holloway, *Surf. Sci.* **66**, 479 (1977).
25. P.M. Hall and J.M. Morabito, *Surf. Sci* **83** 391 (1979).
26. M.P. Seah, in *Practical Surface Analysis by Auger and X-Ray Photoelectron Spectroscopy,* Chapter 5, p. 181, ed. by D. Briggs and M.P. Seah, Wiley (1983).
27. M.P. Seah and W.A. Dench, *Surf. Interface Anal.* **1,** 2 (1979).
28. R. Payling, *J. Electron Spectrosc. Relat. Phenom.* **36**, 99 (1985).
29. M.P. Seah and M.T. Anthony, *J. Elec. Spectrosc. Relat. Phenom.* **32**, 87 (1983).
30. M.P. Seah, M.T. Anthony and W.A. Dench, *J. Phys. E,* **16**, 848 (1983).
31. L.J. Brillson and G.P. Ceasar, *Surf. Sci.* **58**, 457 (1976).
32. M. Janghorbani, M. Vulli and K. Starke, *Anal. Chem.* **47**, 2200 (1975).
33. M. Vulli, M. Janghorbani and K. Starke, *Anal. Chim Acta,* **82**, 121 (1976).
34. M. Vulli and K. Starke, *J. Micros. Spectrosc. Electron.* **2**, 57 (1977).
35. C.D. Wagner, *Anal. Chem.* **44,** 1050 (1972).
36. C.K. Jorgensen and H. Berthou, *Faraday Discuss. Chem. Soc.* **54,** 269 (1972).
37. V.I. Nefedov, N.P. Sergushin, I.M. Band and M.B. Trzhakovskaya, *J. Electron Spectrosc. Relat. Phenom,* **2**, 383 (1973).
38. V.I. Nefedov, N.P. Sergushin, Y.V. Salyn, I.M. Band and M.B. Trzhakovskaya, *J. Electron Spectrosc. Relat. Phenom.* **7**, 175 (1975).
39. H. Berthou and C.K. Jorgensen, *Anal. Chem.* **47**, 482 (1975).
40. C.D. Wagner, *Anal. Chem.* **49**, 1282 (1977).
41. J.M. Adams, S. Evans, P.I. Reid, J.M. Thomas and M.J. Walters, *Anal. Chem.* **48**, 2001 (1977).
42. S. Evans, R.G. Pritchard and J.M. Thomas, *J. Phys. C.* **10**, 2483 (1977).
43. S. Evans, R.G. Pritchard and J.M. Thomas, *J. Electron Spectrosc. Relat. Phenom.* **14**, 341 (1978).
44. C.D. Wagner, L.E. Davis and W.M. Riggs, *Surf. Interface Anal.* **2**, 53 (1980).
45. M.P. Seah, *Surf. Interface Anal.* **2**, 222 (1980).
46. C.D. Wagner, L.E. Davis, M.V. Zeller, J.A. Taylor, R.M. Raymond and L.H. Gale, *Surf. Interface Anal.* **3**, 211 (1981).
47. L.V. Phillips, L. Salvati, W.J. Carter and D.M. Hercules, in *'Quantitative Surface Analysis of Materials'*, p. 47 ASTM STP 643, ed. by N.S. McIntyre, ASTM, Philadelphia (1978).
48. M. Prutton, *SEM I*, 83 (1982).
49. M.P. Seah, *Surf. Sci.* **32**, 703 (1972).
50. *1985 Annual Book of ASTM Standards, Volume 03.06, Standard E 673–85*, p. 304 ASTM, Philadelphia (1985).
51. C.J. Powell, *SEM IV,* 1649 (1984).
52. C.J. Powell, *Surf. Sci.* **44**, 29 (1974).
53. C.D. Wagner, L.E. Davies and W.M. Riggs, *Surf. Interface Anal.* **2**, 53 (1980).
54. J. Kirschner, *SEM I*, 215 (1976).
55. L. Meda and G. Queirolo, *Surf. Sci.* **147**, 576 (1984).
56. R. Shimizu, *Japan J. Appl. Phys.* **22**, 1631 (1983).
57. R.C. Weast, ed. *Handbook of Chemistry and Physics, 58th Edn.*, CRC Press (1977–78).
58. M. Gryzinski, *Phys. Rev.* **138A**, 336 (1965).
59. C.J. Powell, *Revs. Mod. Phys.* **48**, 33 (1976).
60. E.J. McGuire, *Phys. Rev.* **A3**, 587 (1971).
61. E.J. McGuire, *Phys. Rev.* **A5**, 1043 (1972).
62. E.H.S. Burhop, *The Auger Effect and Other Radiationless Transitions,* Cambridge University Press (1952).
63. P. Weightman, *Repts. Prog. Phys.* **45**, 753 (1982).
64. S.M. Goldberg, C.S. Fadley and S. Kono, *J. Electron Spectrosc.* **21**, 285 (1981).
65. H.E. Bishop, *Brit. J. Appl. Phys.* **18**, 703 (1967).
66. W. Reuter in *Proc. 6th Intern. Conf. on X-ray Optics and Microanalysis*, p. 121. ed. by G. Shinoda, K. Kohra and T. Ichinokawa, Univ. of Tokyo Press (1972).
67. A. Jablonski, *Surf. Interface Anal.* **2**, 39 (1980).
68. A. Jablonski, *Surf. Interface Anal,* **4**, 135 (1982).
69. R. Shimizu and S. Ichimura, *Quantitative Analysis by Auger Electron Spectroscopy*, Toyota Foundation Research Report I–006 No. 76–0175, Osaka (1981).
70. S. Ichimura and R. Shimizu, *Surf. Sci.* **112**, 386 (1981).
71. D.R. Penn, *J. Electron Spectrosc. Relat. Phenom.* **9**, 29 (1976).
72. J. Szajman, J. Liesegang, J.G. Jenkin and R.C.G. Leckey, *J. Electron Spectrosc. Relat. Phenom.* **23**, 97 (1981).
73. A. Jablonski, P. Mrozek, G. Gergely, M. Menyhard and A. Sulyok, *Surf. Interface Anal* **6**, 291 (1984).
74. J. Ferron, E.C. Goldberg, L.S. De Bernardez and R.H. Buitrago, *Surf. Sci.* **123**, 239 (1982).
75. V.M. Dwyer and J.A.D. Matthew, *Vacuum* **33**, 767 (1983).
76. S. Mroczkowski and D. Lichtman, *Surf. Sci.* **131** 159 (1983) and *J. Vac. Sci. Technol.* **A3**, 1860 (1985).
77. J. Ferron, L.S. De Bernardez and R.H. Buitrago, *Surf. Sci.* **145**, L501 (1984).
78. D.A. Shirley, *Phys. Rev.* **B5**, 4709 (1972).
79. M.C. Burrell and N.R. Armstrong, *Appl. Surface Sci.* **17**, 53 (1983).
80. J.H. Scofield, *J. Electron Spectrosc. Relat. Phenom.* **8**, 129 (1976).
81. I.M. Band, Yu. I. Kharitonov and M.B. Trzhaskovskaya, *Atomic Data and Nuclear Data Tables* **23**, 443 (1979).
82. J.J. Yeh and I. Lindau, *Atomic Data and Nuclear Data Tables* **32**, I (1985).
83. R.F. Reilman, A. Msezane and S.T. Manson, *J. Electron Spectrosc. Relat. Phenom.* **8**, 389 (1970).
84. O.A. Baschenko and V.I. Nefedov, *J. Electron Spectrosc. Relat. Phenom.* **17**, 405 (1979).
85. O.A. Baschenko, G.V. Machavariani and V.I. Nefedov, *J. Electron Spectrosc. Relat. Phenom* **34**, 305 (1984).
86. H. Ebel, M.F. Ebel and A. Jablonski, *J. Electron Spetrosc. Relat. Phenom.* **35**, 155 (1985).
87. C.D. Wagner, in *Practical Surface Analysis by Auger and X-Ray Photoelectron Spectroscopy,* Appendix 5, p. 511, Wiley, (1983).
88. S. Tougaard and B. Jorgensen, *Surf. Sci.* **143**, 482 (1984).

SURFACE AND INTERFACE ANALYSIS, VOL. 9, 99–103 (1986)

Prediction of Auger Backscattering Factors From Measurements of Spectral Backgrounds

M. Prutton and M.M.El Gomati
Department of Physics, University of York, Heslington, York YO1 5DD, England

The shape of $N(E)$ has been obtained from a carefully characterized electron spectrometer for 24 different elements and compounds. The secondary electron cascade has been found to obey a form $N(E) = AE^{-m}$. A and m have been measured and used together with a model of the energy distribution of rediffused primary electrons to calculate an Auger backscattering factor for each material when the primary energy is 20 keV. These calculations have been compared with backscattering coefficients measured at the carbon 1s binding energy in the same instrument on the same samples. The results show the backscattering factor to be linear in atomic number. The measurement of A and m from a spectrum can be used to estimate the backscattering factor within an estimated error of about ±0.05.

INTRODUCTION

Knowledge of the value of the Auger backscattering factor, r, is essential for quantitative Auger electron spectroscopy. Experimental methods specially designed to measure this parameter have been described by a number of authors.[1–6] and theoretical estimates have been given by others.[7–11] All of these approaches require a purpose-designed experiment on the material of interest or a calculation based upon some (usually approximate) description of the electron scattering processes in the solid. This paper describes another approach in which the same spectrum, $N(E)$, from which Auger peak heights or areas are measured is also used for the estimation of r. If such an approach can yield values of r of acceptable accuracy then it may be useful for the surface analyst and would avoid the uncertainties associated with empirical approaches to estimating r such as are described by Reuter[12] and Ichimura *et al.*[13]

METHOD

In order to use $N(E)$ to obtain r use is made of the experimental observation, first noted by Sickafus,[14] that the slowly varying background $B(E)$ upon which Auger peaks are superimposed can be described by the equation:

$$B(E) = A'\,E^{-m} \tag{1}$$

In this equation A' and m are constants of the material, the primary beam current, i_b, and energy E_p and the geometry of the experiment. This background shape has been found experimentally to describe the spectra $N(E)$ of a large number of materials using the York scanning Auger microscope with E_p of above 20 keV for kinetic energies E below 2 keV after correction for the transmission function of the electron spectrometer.[15–17] As E_p is reduced from 20 keV the component of the background which rises towards the elastic peak begins to modify the form of $B(E)$ in the range below 2 keV. This component, due to rediffused primary electrons, becomes significant for E_p below ca 15 keV and a minimum in $N(E)$ at some energy E_{cp} can be detected. Peacock and Duraud[17] have shown that the experimental background shapes for Al and Si and the primary energy dependence of E_{cp} can be described by a function of the form:

$$B(E) = A\,E_p^{-n}\,E^{-m} + B\,E_p^{-(1+q)}E^{q} \tag{2}$$

here, the first term describes the cascade of secondary electrons (cf (1)). The second term is an approximation to the contribution arising from 'backscattered' (rediffused) primary electrons. An adequate fit for $0.3 < E/E_p < 0.9$ was obtained with $q = 2$ for elemental Ag and Cu. The backscattering factor is related to $B(E)$ by:

$$r = 1 + \frac{1}{\phi(E_p)} \int_{E_A}^{E_p} \phi(E)B(E)\,\mathrm{d}E, \tag{3}$$

where $\phi(E)$ is the ionization cross-section for the substrate material. The simplest evaluation of r can be made using a Born cross-section for $\phi(E)$ and the form of $B(E)$ given in (2). For fixed E_p and gun/specimen/analyser geometry (3) becomes

$$r = 1 - \frac{\alpha A}{E_p^{m-1}\log \epsilon_A} \int_{\epsilon_A}^{1} \log\left(\frac{\epsilon}{\epsilon_a}\right) \epsilon^{-(m+1)}\,\mathrm{d}\epsilon - \frac{\beta Z}{\log \epsilon_A}\left(\frac{\epsilon_A}{2} - \log \epsilon_A - 1\right) \tag{4}$$

In this equation ϵ and ϵ_A are the reduced energies E/E_p and E_A/E_p respectively; E_A is the binding energy of the electron ionized to launch the Auger process in

0142–2421/86/120099–05 $05.00

question and Z is the atomic number of the substrate. In deriving (4) the coefficient B in (2) is assumed[7] to be linear in atomic number Z and the constants α and β are treated as properties of a particular measuring instrument used at a fixed primary energy. Inspection of (4) reveals that E_A, E_p and Z are known, and A and m can be measured by fitting (2) to the experimental data $N(E)$ above the Auger peak in question. The integral is easily evaluated. As a result r can be determined from measurable quantities once the coefficients α and β are known for the user's apparatus. In this work, α and β for the York SAM at $E_p = 20$ keV have been determined by linear regression between measured values of the backscattering coefficient, r_x, and calculated values, r_b; r_x has been obtained using the method described by El Gomati[5] but outlined below for clarity. r_b has been determined from measurement of $N(E)$ and the use of (4). A related approach to the determinatin of r has been described by Wang *et al.*[18] Their approach is confined to low primary energies, where ionization is dominated by the rediffused primaries and the secondary electron cascade can be neglected. They present data for a Au–Cu and a Ag–Cu alloy.

EXPERIMENTAL TECHNIQUE

The measurements were made in a scanning Auger microscope which has been described previously.[16] The samples were measured in 3 sets, each set being covered with a different film thickness of carbon.

The carbon was evaporated in a conventional vacuum system onto NaCl substrates. It was then floated off of the NaCl in distilled water and picked up onto either electron microscope grids or small solid samples made of the materials for which r_x, A and m were to be measured. Thus, the value of r_x at the C KVV energy (272 eV) could be measured and compared with r_b calculated using (4) and the values of A and m measured for the same sample as r_x. The first set of samples were of C, Al, Cu, Ag, W and Pt covered with a film of C 4 ± 1 nm thick; the second were of Al, Si, Mo, Gd, Ta, NiO and InP covered with a film of C 25 ± 3 nm thick and the third were of C, Si, Cr, Gd, Ta, Pb, InP and NiO covered with a film of C 10 ± 2 nm thick. These C thicknesses were measured using a 'Talystep' instrument after exposure of the samples to the air. Of course, this method of preparation results in contamination of the substrate/C interface before the sample is installed in the SAM. However, it has been observed that only C Auger peaks (plus signs of substrate peaks for the first set of samples) are present in the spectra. Thus, C is the dominant material present on top of the substrates and has the thickness noted above. All spectra were collected with a primary energy of 20 keV and a beam current in the region of 5 nA which was measured for each spectrum with a Faraday cup. The angle of incidence was 40° as was the take-off angle to the electron spectrometer. This spectrometer was a concentric hemispherical analyser collecting 1.7% of the 2π steradians of emitted electrons and operating in constant pass energy so as to produce an output count giving $N(E)$ with a constant energy window of 6 eV from energies of 50 eV to 2000 eV. The known transmission function $T(E)$[15] was applied to the raw spectra to obtain a spectrum characteristic of the specimens themselves. All spectra were normalized to 1 nA beam current so that absolute values of A could be compared. The signal to noise in the spectra was improved by smoothing them with the optimal Wiener filter as described by Kosarev and Pantos.[19] This process reduced the number of Fourier components from 1024 in the raw data to about 120 in the data used for measuring r, A and m.

Normalization of the raw data were carried out for two factors. In the first place all the raw data were normalized to the same incident beam current (1 nA). The second, more difficult, correction was for variations from time to time in the gain of the channeltron used to detect electrons passing through the electron spectrometer. It is well known that the gain of a channeltron varies both with the total dose of electrons to which it has been exposed and with its exposure to air or other gases.[23] Although the effects of such gain changes can be reduced by careful choice of the discriminator amplitude window within which electrons are counted they cannot be eliminated altogether because the tails of the pulse height distribution cut the discriminator levels. These gain changes were estimated by integrating the number of counts under the spectra between 1800 and 1900 eV after normalization to 1 nA beam current. By comparing these areas for the same materials measured in each set of observations allowances for the channeltron gain changes could be made. The largest such effect seen was a 40% increase in channeltron gain after a series of down-to-air and bakeout procedures in a short period of time.

The values of A and m were obtained by least squares fitting a power law cascade as in (1) to the normalized, corrected, spectra above the C KVV peak in an energy range depending upon the specimen and the presence of Auger peaks from the substrate above the C KVV peak. The region chosen for this fit was always free of substrate Auger peaks or even changes in the magnitude of m. An example of such a fit is shown in Fig. 1. The electron cascade above C KVV was then subtracted from the data to allow estimation of the area under the C peak so as to measure r_x. The remaining background stripping was carried out by fitting a new secondary cascade of the form in (1) to the region between 60 and 100 eV and then matching that cascade to the zero in signal just above the C KVV peak at 290 eV using the modification of Shirley's procedure which is described by Peacock.[20] One iteration of Shirley's procedure was used and a typical result is shown in Fig. 1. This background was subtracted from the data and the area under the peak remaining was used to estimate the C KVV Auger yield.

In each set of samples an internal 'standard' was included. This 'standard' was a piece of the same C film used to cover the samples. It was mounted on a standard electron microscope grid and the SAM was used to produce a spectrum on an unsupported region of this film. The Auger yield was estimated using the same procedure as described above. The value of r_x was then derived by taking the ratio of the area under the C KVV peaks on the sample being measured and of the unsupported film of the same thickness.

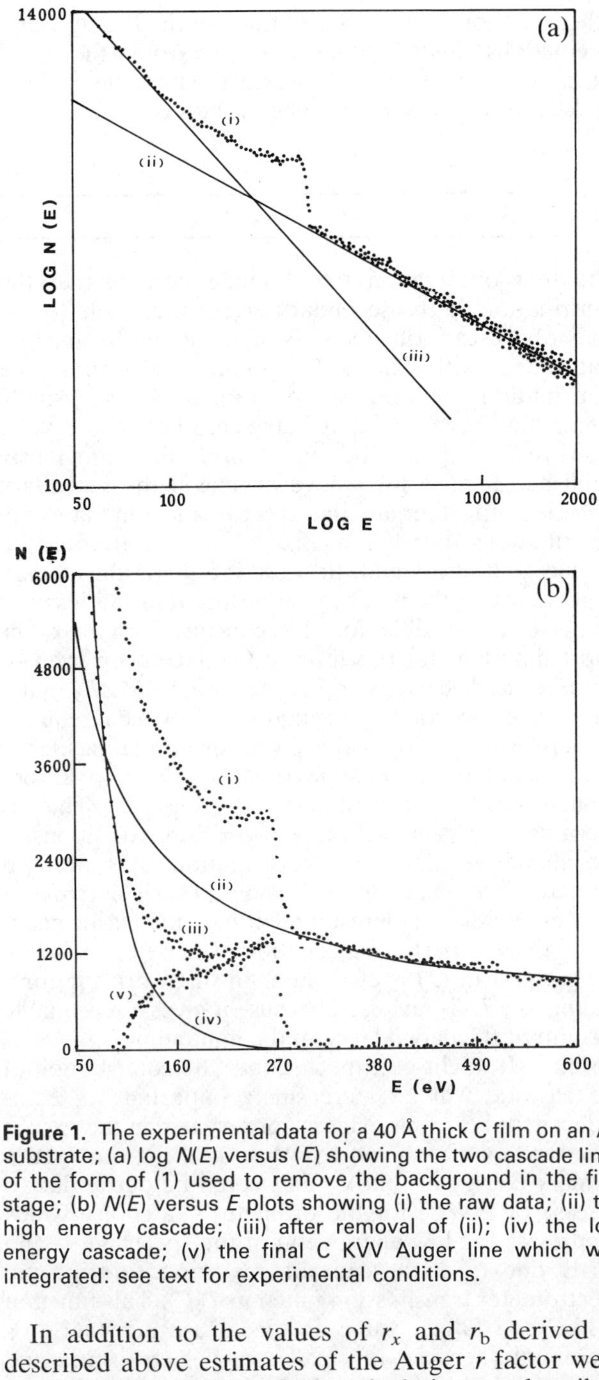

Figure 1. The experimental data for a 40 Å thick C film on an Ag substrate; (a) log $N(E)$ versus (E) showing the two cascade lines of the form of (1) used to remove the background in the first stage; (b) $N(E)$ versus E plots showing (i) the raw data; (ii) the high energy cascade; (iii) after removal of (ii); (iv) the low energy cascade; (v) the final C KVV Auger line which was integrated: see text for experimental conditions.

In addition to the values of r_x and r_b derived as described above estimates of the Auger r factor were made using the Monte-Carlo calculations as described by El Gomati and Prutton[21] and El Gomati *et al.*[22] This calculation used a Rutherford scattering cross-section and Bethe continuous energy loss to model the electron interaction with the solid. It tends to underestimate the Auger backscattering coefficient because it neglects the ionization due to the true secondary cascade and the Rutherford cross-section is increasingly too small as Z increases. However, it does serve as an additional check of the validity of the value of r_b obtained from (4). The Monte-Carlo estimate of r is referred to here as r_m.

RESULTS

Figure 2 displays the results of comparing 17 values of r_x with the corresponding calculated values r_b. The

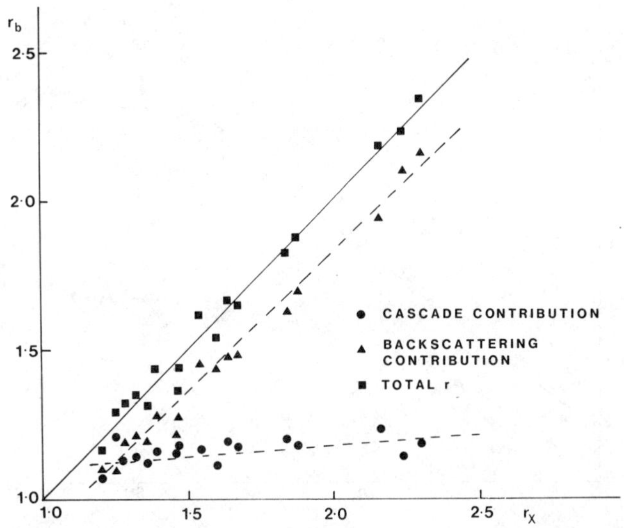

Figure 2. The correlation of calculated values, r_b, and measured values, r_x, of the Auger backscattering factor 272 eV for various materials. E_p = 20 kV, angle of incidence 45° ■ total r_b; ● Secondary casade contribution to r_b; ▲ Rediffused primary contribution to r_b. The coefficients α and β (4) are 7.9×10^{-8} and 1.9×10^{-2} respectively.

correlation is good — the standard deviation of r_x from r_b being 0.045. In the same figure the contributions of the cascade (second term in (4)) and the rediffused primaries (last term in (4)) are shown separately. The point where these two contributions to r_b are equal occurs near Z = 14. For higher Z the rediffused primary contribution becomes increasingly important.

One interesting feature of the data in Fig. 2 is that it includes four compounds. The two materials InP and NiO were used as substrates directly but, in addition, one of the Gd samples and the Zn sample were heavily oxidized before the C film was floated on to them. Accordingly the value of the average atomic number, Z, used for that Gd sample was 30.4 (corresponding to Gd_2O_3) and for the Zn sample was 19.0 (corresponding to ZnO). The data for the two Ta samples and the Pb sample are not included in Fig. 2 because they too were oxidized and r_x measured on them was low. If they had been completely oxidized then an average value for Z could have been estimated and the data then used in (4) and Fig. 2. However, if the surface is inhomogeneously oxidized it is difficult to estimate a value Z. All the values of r_x measured are presented in Fig. 3 where it can be seen that the Ta and Pb samples fall below the general trend of $r_x(Z)$ if they are treated as elements and above the general trend if they are treated as stoichiometric oxides.

The straight line fitting the data points in Fig. 3 is given by:

$$r = 1 + 0.15 + 0.0152\ Z.$$

This is consistent with (4) if A and m are independent of Z. The raw data for A and m were examined for systematic behaviour with Z but none could be found. m was in the range 0.5 to 1.0 but otherwise appeared to be random. A was also apparently randomly scattered with values between 5×10^3 (eV)^m and 10^6 (eV)^m. The physics of the origin of the form AE^{-m} for the

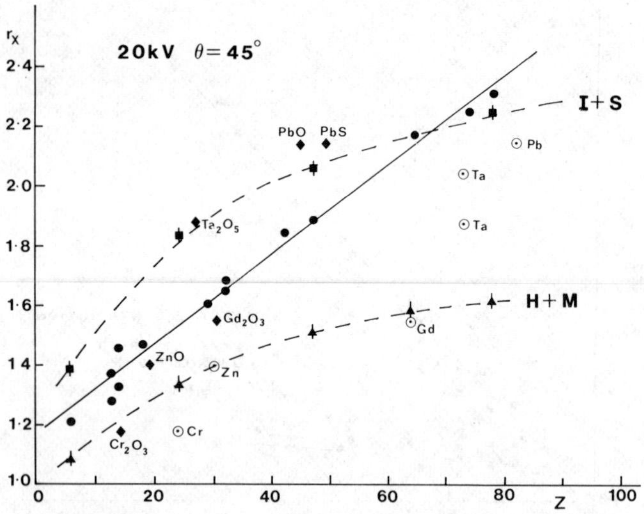

Figure 3. Dependence of measured r_x versus atomic number Z. r_x measured at 272 eV; $E_p = 20$ keV; angle of incidence 45°. The straight line is a least square fit to all the points *except* those labelled (Ta and its oxide, Pb and its oxide or sulphide) — see text. Odd results, due to surface reactions (see text), are indicated by ○. Compounds of these materials are given the same value of r_x and are indicated by ◆. The theoretical prediction of Ichimura and Shimizu[11] (IS) and Hall and Morabito[24] (HM) are indicated by the broken lines and markers ■ and ▲ respectively.

background cascade needs to be more completely understood in order to comment upon this behaviour.

A blind test of the applicability of (5) was conducted by using (4) to calculate r_c for 7 polycrystalline clean elemental standards (C, A1, Cr, Ni, Zr, Ag, La) for which all the corrections described above could be made. The resulting values of r_b are plotted in Fig. 4 where the line drawn is the same as that fitting the measurements r_x in Fig. 3. The agreement is very good indeed.

The Monte-Carlo results of Ichimura and Shimizu[11] and of Hall and Morabito[24] are plotted as lines in Fig. 3 and do not fit the data points r_x. The same shape of $r_m(Z)$ was found in the present Monte-Carlo calculations but with values falling between those of the

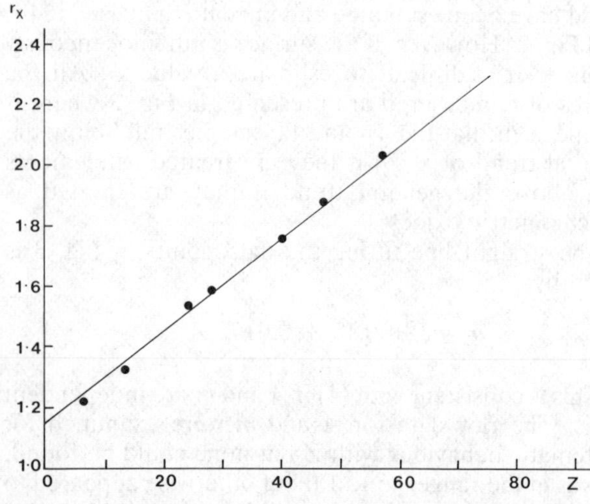

Figure 4. Backscattering factors calculated using clean elemental standards with measured A and m and (4). The solid line is the best fit line to the data of Fig. 3.

other authors. It appears that both Monte-Carlo calculations and estimates of r from the total backscattering factor[24] underestimate the Auger backscattering factor for higher values of Z.

DISCUSSION

The first observation to be made here is that the contribution of the secondary electron cascade to the Auger backscattering factor is small but non-negligible. Naturally, with 20 keV primary electrons the contribution of rediffused primaries is usually dominating because E_{cp} is between 2 keV and 3 keV. However, it should be noted both that previously published models for r have neglected the secondary cascades altogether and that this cascade dominates the contributions to r below Z of about 15. It could be argued that the simplistic description of the energy dependence of the number of rediffused primaries used in (4) is responsible for this conclusion. Two other approximations for this term in (4) have been tried — the Everhardt background (which scales as ZE) and a more realistic Niedrig background[25] (which requires numerical evaluation and is given for normal incidence only). Both of these approximations lead to a very similar cascade contribution in Fig. 2. This is presumably because all the analytic approximations to the shape of the energy distribution of rediffused primaries over-estimate the number of such electrons at low energies and underestimate it near the elastic peak. Thus, there is partial cancellation of these errors when the integral in (3) is evaluated. In this work the form scaling as ZE^2 is favoured because it gives a reasonable account of the dependence of the minimum E_{cp} as E_p is varied.[17] It might be expected that the contribution of the cascade will be increasingly important as E_p is reduced and E_{cp} falls. This is not revealed in the $EN(E)$ spectrum for Cu shown by Wang *et al.*[18] who used a primary energy of 2 keV. This could be partly due to the use of $EN(E)$ for the ordinate — a choice which suppresses the low energy end of the spectrum — and partly due to an absence of any correction for the spectrometer transmission function $T(E)$. This function is likely to fall in value as the measuring energy is decreased (because of small stray electro-static and magnetic fields) and so will have attenuated the low energy end of the spectrum.

The work of Peacock and Duraud[17] demonstrates that the essential shape of the spectral background for analysis energies in the range 50–2000 eV is of the form of (2) for primary energies between 2 keV and 20 keV This suggests that the procedures described here will be applicable to primary energies lower than the 20 keV results described in the present work. Since the primary energies used in conventional Auger spectrometers tend to be in the range 5–10 keV, it will be interesting to carry out similar experiments in this range to test the usefulness of this general approach.

The results presented here are all for r evaluated at the C Auger energy. Equation (4) can be evaluated at other core level binding energies E_A to explore the variation of r with this energy. More experimental work with adlayers other than C is required to test the validity of (4) for this variation. Nevertheless, (4) can be evaluated, for example, for Cu using the experimental values of A and m and r_b is found to fall smoothly from 1.69 at $E_A = 100$ eV to 1.34 at $E_A = 2$ KeV.

The precision of results calculated using (4) can probably be improved by using a more realistic ionization cross-section and by using a more realistic description of the energy distribution of rediffused primaries, e.g. an extension on Niedrig's analytic expression[25] to the angle of incidence of 45° actually used in the experiment to determine r_x. Contamination effects in the calibration experiments required to determine the constants α and β in (4) could be reduced by using *in situ* deposition of C on the clean substrates.

Nevertheless, without going to these lengths, it is concluded that a once and for all calibration experiment on a particular instrument with a particular sample configuration will provide the instrumental constants α and β from which (4) can be used. Thus, a value for r can be derived from a measurement of A and m using the spectrum that has to be collected if quantitative Auger analysis is to be carried out. A simple computer programme will evaluate (4) quickly once A, m, Z, and i_b are known.

Acknowledgements

The authors are pleased to be able to thank the Science and Engineering Research Council (UK) for their financial assistance with this work. Also they wish to thank C G Walker and D C Peacock for frequent useful discussions and help whilst the data presented here was being collected.

REFERENCES

1. T.E. Gallon, *J. Phys. D: Appl. Phys.* **5**, 822 (1982).
2. F. Meyer and J.J. Vrakking, *Surf. Sci.* **32**, 21 (1972).
3. R.L. Gerlach and A.R. Ducharm, *Phys. Rev.* **A6**, 1892 (1972).
4. K. Goto, T. Ishikawa, T. Koshikawa and R. Shimizu, *Surf. Sci.* **47**, 477 (1975).
5. M.M. El Gomati, *Surf, Sci.* **152/153**, 877 (1985).
6. D. Gramari and J. Cazaux, *Surf. Sci.* **136**, 296 (1984).
7. T.E. Everhart, *J. Appl. Phys.* **31**, 1483 (1960).
8. H.E. Bishop and J.C. Riviere, *J. Appl. Phy.* **40**, 1970 (1969).
9. D. J. Fathers and P. Rez, *Scanning Electron Microsc.* **I**, 55 (1979).
10. A. Jablonski, *Surf. Inteface Anal.* **1**, 122 (1979).
11. S. Ichimura and R. Shimizu, *Surf. Sci.* **112**, 386 (1981).
12. W. Reuter in: *Proc 6th Int. Conf. on x-ray Optics and micro-analysis*, p 121 ed. by G. Shinoda, K. Kohra and T. Ichinokawa, Univ. Tokyo Press (1972).
13. S. Ichimura, R. Shimizu and J.P. Langeron, *Surf. Sci.* **124**, L49 (1983).
14. E.N. Sickafus, *Phys. Rev.* **B16**, 1436 (1977).
15. D.C. Peacock, M. Prutton and R. Roberts, *Vacuum,* **34**, 497 (1984).
16. M. Prutton, R. Borwning, M.M. El Gomati and D.C. Peacock, *Vacuum,* **32**, 351 (1982).
17. D.C. Peacock and J.P. Duraud, *Surf Interface Anal.* **8**, 1 (1986).
18. Q. Wang, Q.J. Zhang and Z.Y. Hua, *Applic. Surf. Sci.* **22/23**, 160 (985).
19. A. Kosarev and E. Pantos, *J. Phys. E.* **16**, 537 (1983).
20. D.C. Peacock, *Surf. Sci.* **152/153**, 895 (1985).
21. M.M. El Gomati and M. Prutton, *Surf. Sci.* **72**, 486 (1979).
22. M.M. El Gomati, J.P. Broomfield, J.A.D. Matthew and M. Prutton, *Scanning Electron Microscopy,* **III**, 1101 (1981).
23. P. LeComte and V. Perez-Mendez, *IEEE Trans. Nuclear Science.* **25**, 964 (1978).
24. P.M. Hall and J.M. Morabito, *Surf. Sci.* **83**, 391 (1979).
25. H. Niedrig, in *Electron Beam Interactions with Solids*, p. 51, SEM Inc. Chicago (1983).

SURFACE AND INTERFACE ANALYSIS, VOL. 9, 105–109 (1986)

Distribution of Hydrogen in the NPL Standard Ta_2O_5 Films

H.E. Bishop
Materials Development Division, AERE Harwell, Didcot, Oxon OX11 0RA, UK

The NPL has produced well characterized tantalum pentoxide anodized films for use as a thickness and depth resolution standard for profiling in surface analytical techniques. Very sharp interfaces for these films have been demonstrated using AES. Studying these films by dynamic SIMS in the CAMECA IMS 3F, well defined profiles with a sharp interface were obtained with an oxygen primary beam. More complex profiles were obtained with argon and caesium primary beams, apparently showing the presence of hydroxide at the metal/oxide interface and a high concentration of hydrogen in the metal near the surface. However an argon primary beam with an oxygen flood gave the same type of profile as that obtained with an oxygen primary beam. Elastic Recoil Detection Analysis confirmed that there was no high concentration of hydrogen at the metal-oxide interface and showed that hydrogen in the tantalum was present at 1 at %.

INTRODUCTION

The National Physical Laboratory (NPL) have produced well characterized tantalum pentoxide films on tantalum to serve as calibrants for depth resolution and etch rate in compositional depth profiling in Auger electron spectroscopy (AES) and x-ray photoelectron spectroscopy (XPS).[1,2,3,4] In the course of commissioning a Cameca IMS-3F ion microscope in our laboratory, it was decided that a study of these films would provide both a check on the depth resolution that could be achieved with SIMS (Secondary Ion Mass Spectroscopy) depth profiles and a useful subject for operator training. When they characterized these films,[3] Hunt and Seah employed AES and obtained profiles with a well defined interface for both the tantalum and oxygen signals. In this investigation it has been found that SIMS profiles are more complex than those measured with AES. Many ion species show anomalous increases or decreases in the vicinity of the metal oxide interface and the hydrogen profiles obtained with caesium or argon primary beams indicate a high hydrogen concentration in the tantalum metal within a few tens of nm of the interface.

In this paper we will describe how the hydrogen distribution in the NPL tantalum oxide standard films, as measured by SIMS, varies with experimental conditions and compare the results with the hydrogen distribution measured by Elastic Recoil Detection Analysis (ERDA),[5] a scattering technique that is not subject to the artefacts induced by SIMS. In addition, interface widths, measured from the oxygen profiles obtained with both caesium and oxygen primary beams for a range of primary beam energies, are reported.

EXPERIMENTAL

All The measurements were made in a CAMECA IMS–3F secondary ion microscope. In this instrument a primary ion beam is directed on a specimen that is held at either +4.5 or −4.5 keV, depending on the polarity of the secondary ions under investigation. Secondary ions are extracted through an aperture in an earthed plate into a mass spectrometer. The secondary ion optics retain the spatial distribution of the ions leaving the specimen and form an image of the selected ion species at the output of the mass spectrometer. A field aperture may be used to select which region of the specimen contributes to the output signal. For the work reported here the primary ion beam was focused into a probe some 50–100 μm diameter and rastered over a 250 μm square to ensure uniform erosion of the specimen over a substantial area. For all measurements of interface width the field aperture selected was an 8 μm diameter area of the specimen, but in other cases a 64 μm diameter area was selected where a higher signal was desirable.

Primary beam energies in the range 5 to 10 keV were used. However because of the 4.5 keV specimen potential, the primary ions are either accelerated or retarded onto the specimen, with the result that the actual incident energy of the primary beam is either increased or decreased by 4.5 keV and the incident angle of the beam will vary with primary energy and secondary ion polarity. The energies quoted in the text are the actual incident energies and the corresponding incident angles are given in Table 1.

Table 1. Measured interface widths

Ion energy (keV)	Secondary polarity	Incident angle	Interface width (nm) 30 nm	100 nm
Cs^+ primary beam				
14.5	−ve	66°	2.3	4.0
9.5	−ve	69°	2.1	2.5
5.5	+ve	51°	1.5	2.1
3.0	+ve	44°	1.5	2.1
O_2^+ primary beam				
5.5	+ve	51°	2.9	–
3.0	+ve	44°	2.0	–
2.0	+ve	37°	1.2	1.6

0142–2421/86/120105–05 $05.00

The specimen chamber of the IMS 3F is pumped by a helium cryopump to a pressure of better than 10^{-8} torr, measured near the pump. When making hydrogen measurements in SIMS the background due to hydrogen reaching the surface from the residual atmosphere has to be considered. However as the hydrogen levels in the samples proved to be relatively high no special precautions were taken to reduce the hydrogen background.

Two standard tantalum pentoxide films of nominal thickness 30 and 100 nm from the NPL were examined. These were grown anodically under carefully controlled conditions and are certified to be 28.4 ± 1.7 nm, and 96.3 ± 2.8 nm respectively. The interface widths measured by AES were found to be 1.5 and 1.7 nm.[3]

HYDROGEN DISTRIBUTION

The initial measurements on the films used a caesium primary beam and negative secondary ions. Although a well defined interface was found, the profiles for O^- and TaO^- showed an unexpected peak at the interface. Further investigation showed that this effect could be associated with hydrogen; there was a sharp peak at the interface corresponding to $TaOH^-$ and the hydrogen signal increased by two orders of magnitude from the oxide to the metal, Fig. 1.

The shape of the hydrogen profile was both unusual and unexpected. It was also found to be dependent on the sputtering rate. Figure 2 shows how the hydrogen profile varied with erosion rate. At the highest erosion rate, 1.6 nm sec^{-1}, the hydrogen profile showed a sharp peak at the interface. At lower erosion rates a progressively broader peak was observed until at the lowest rate recorded, 0.27 nm sec^{-1} the hydrogen signal was virtually constant for the first 250 nm into the metal before starting to drop away slowly. (The depth scale in the figures is not corrected for any difference between the sputtering rate in tantalum oxide and in the metal.)

Similar hydrogen profiles were found with a ceasium primary beam and positive secondary ions and with an argon primary beam, although the latter case was not investigated as fully. The situation with an oxygen primary beam proved to be quite different, no sputter rate dependence was observed over the range of rates in Fig. 2 and the hydrogen signal within the metal was constant. Although the hydrogen signal from the oxide showed a marked difference between positive and negative ions (Fig. 3), it was substantially higher relative to the signal from the metal than was the case for caesium or argon primary beams.

The measured hydrogen profile was found to be altered by the use of an oxygen flood. For an argon primary beam (and by inference caesium as well) the influence of the oxygen flood is shown in Fig. 4. With the oxygen flood, the sputtering rate of the oxide film was reduced, the relative hydrogen signal from the oxide was very substantially increased and that from the metal was constant, as for an oxygen primary beam. The profile obtained with an oxygen primary beam

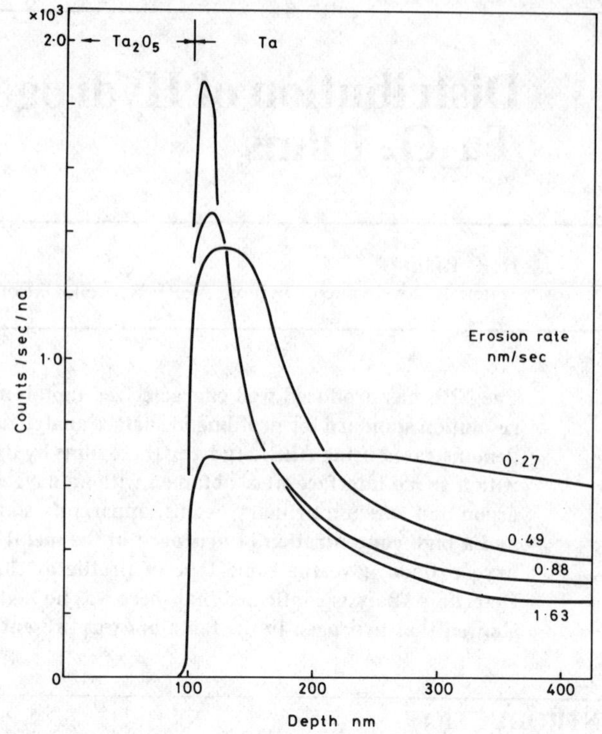

Figure 3. Hydrogen profiles with O_2^+ primary beam, beam energy 5.5 keV for H^+ and 14.5 keV for H^-.

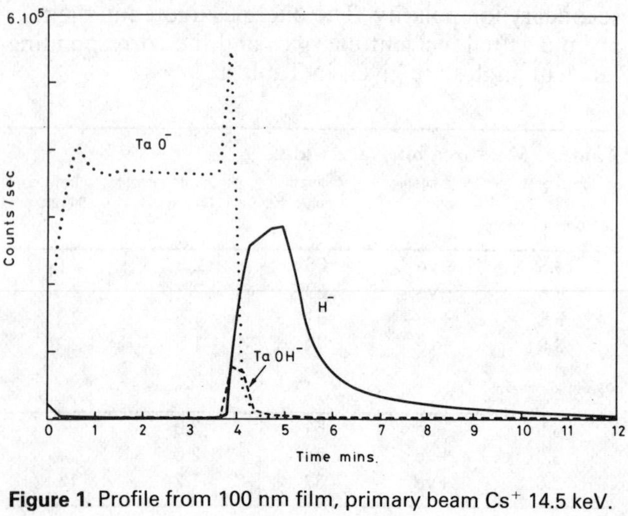

Figure 1. Profile from 100 nm film, primary beam Cs^+ 14.5 keV.

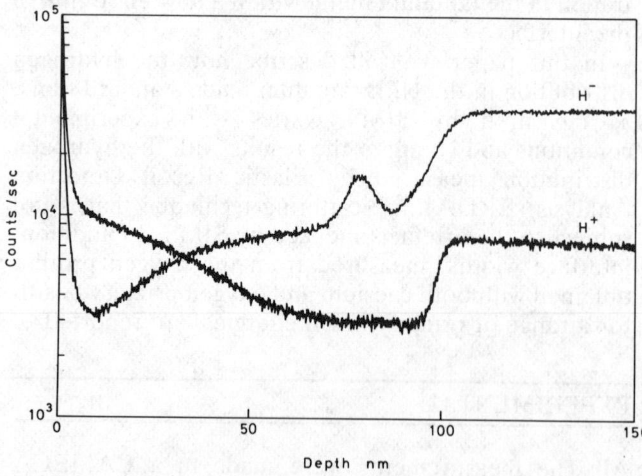

Figure 2. Variation of H^- profile from 100 nm film with sputtering rate; primary beam Cs^+ 9.5 keV.

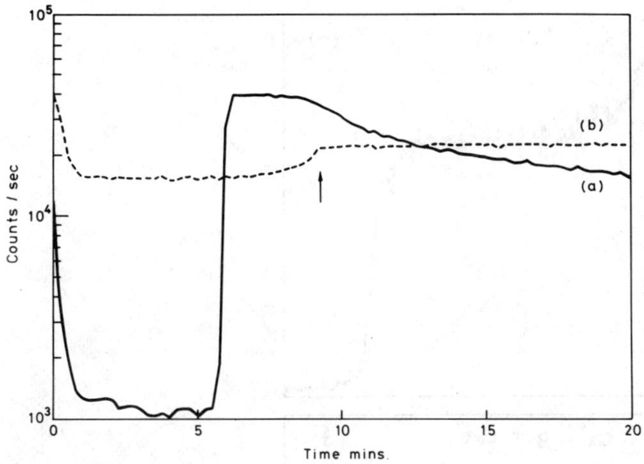

Figure 4. H^+ profile for argon primary beam, energy 5.5 keV, (a) without oxygen flood, (b) with oxygen flood; ↑ indicates metal oxide interface for (b).

(positive secondary ions) was also changed by an oxygen flood. Again the sputtering rate was reduced and the hydrogen signal from both the oxide and the metal was increased, by about 30% in the case of the metal. The signal from the oxide was increased by more than that from the metal so that there was practically no change in the hydrogen signal across the interface.

In view of the considerable variations in the hydrogen profiles obtained by SIMS under different experimental conditions it was decided to try an alternative method of measuring the hydrogen distribution within the samples. ERDA was chosen because, although it does not have the depth resolution of SIMS, it is based on very well understood scattering processes and is much less susceptible to artefact than SIMS. In ERDA a 30 MeV chlorine ion beam is directed at the sample at a glancing angle. Light elements are forward scattered out of the sample by Rutherford scattering and are measured by an energy sensitive detector. The scattered hydrogen is preferentially selected by placing a suitable absorber in front of the detector. The distribution obtained from the 100 nm film is shown in Fig. 5. This shows a uniform hydrogen concentration of about 1 at % within the bulk of the film. The surface peak, due to surface contaminants is seen by this technique on the majority of samples. There was a

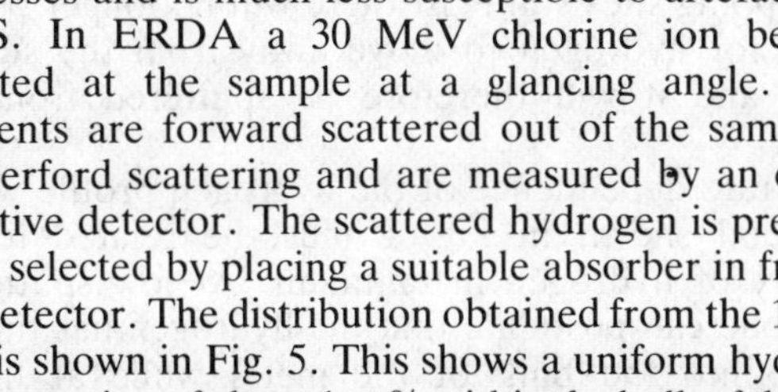

Figure 5. ERDA profile for hydrogen from the 100 nm film.

dip between the surface peak and the uniform bulk signal suggesting that the hydrogen content of the oxide film is rather lower than that of the underlying metal. Given the poorer depth resolution of ERDA this profile is consistent with that shown in Fig. 4 with the oxygen flood.

INTERFACE WIDTH

Hunt and Seah[3] adopted the distance, Δz, between the 84% and 16% intensity levels as their measure of interface width. Such a definition presents no problems with the simple profiles obtained from AES. However, with the more complex profiles obtained from SIMS, as in Fig. 1, where the ion yield for a particular ion species is often influenced by the chemical state of the surface with the result that its sensitivity may vary through the interface, it is by no means clear how to define the true interface width. For the purpose of this study, it was decided, rather arbitrarily, to use the O^+ or O^- profiles for measurements of interface width. Although the oxygen profiles are far from ideal the interface effects were less dramatic than for Ta or TaO species. A selection of profiles for oxygen and caesium primary beams is shown in Fig. 6. At the start of the profile there is an initial period during which equilibrium is set up. A similar effect is seen in the AES profiles and is ascribed to a reduction of the Ta_2O_5 by the ion bombardment. Unlike the Auger profiles, the O^+ profiles show a further drop starting some 5–10 nm from the interface with a small increase in intensity just before the interface itself. The negative ion profile, for O^- with caesium, shows an increase rather than a decrease in signal near the interface. For the purpose of measuring the interface width the maximum just before the interface was taken as the 100% level and the 84 and 16% levels were then determined. The resultant widths are given in Table 1. This approach worked well where the maximum was well defined but for broader interfaces it was difficult to determine a consistent value for the 100% level — as in Fig. 6(b). This uncertainty in the 100% level may be partly responsible for some of the higher values in Table 1.

DISCUSSION — HYDROGEN DISTRIBUTION

The SIMS measurments described above show that the sputtering process gives rise to complexities in the secondary ion yield and to mass transport of hydrogen in the films. Indeed the profiles probably tell us more about the nature of the sputtering process than about the composition of the films themselves. It is well known[2–4] that tantalum pentoxide is reduced under inert ion bombardment and this effect is seen clearly in both the AES and SIMS profiles. The drop in O^+ signal in SIMS as one approaches the interface, Fig. 6, is more difficult to explain as neither AES nor the O^- profile show this effect. This drop does not represent a true decrease in surface oxygen concentration, as a similar drop would appear in the AES profile, and it must therefore reveal some more subtle change in the

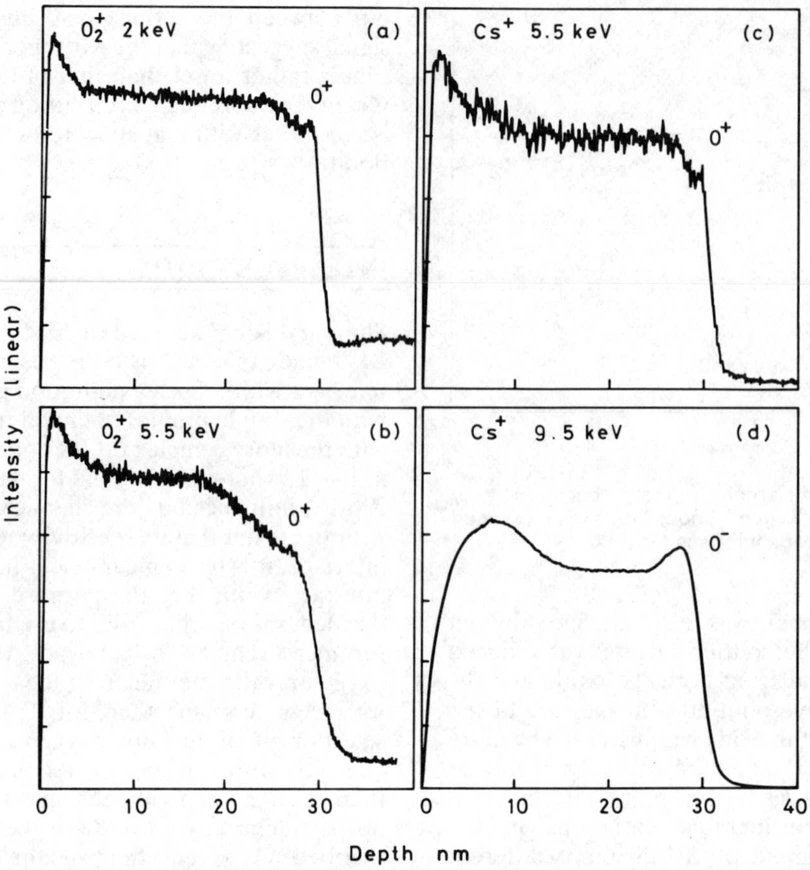

Figure 6. Oxygen profiles under different conditions: (a) and (b) O_2^+ primary beam and positive secondaries, (c) and (d) caesium primary beam.

electrical or chemical state of the surface that influences the ion yield.

The behaviour of the hydrogen in the films as they are sputtered is particularly intriguing. With an oxygen primary beam, or with argon with an oxygen flood, the uniform hydrogen signal from the metal is consistent with the ERDA measurements. However, in view of the differences in hydrogen profiles obtained from the film shown in Figs. 2 and 3 it is difficult to see how SIMS could be used to provide a reliable measurment of hydrogen distribution in this type of sample.

Sputtering in the absence of oxygen was found to give a very low hydrogen signal from the film coupled with anomalously high signals from the metal near the interface. The hydrogen in the oxide must be driven ahead of the sputtering front until it meets the metal–oxide interface. The behaviour in the oxide may be explained if the sputtering process produces a very high density of traps for hydrogen some distance below the surface, but with many fewer such traps in the surface region itself. As sputtering proceeds the trapped hydrogen moves forward into new traps as they are formed leaving the actual surface that is being sputtered relatively free of hydrogen.

When oxygen is present either as the primary ion beam or as a flood, the hydrogen no longer appears to be driven through the film. This could result from two different mechanisms; either the oxygen in some way inhibits the formation of, or poisons, the subsurface traps, or, by increasing the oxygen content of the immediate surface region, it increases the trap density in the near surface region. In the latter case although the trapping sites still exist there will no longer be any reason for hydrogen to move away from the surface region and it will therefore be sputtered from the surface.

The rate dependence of the hydrogen profile within the metal shown in Fig. 2 must be related to the mobility of hydrogen in tantalum. At low sputtering rates, one can postulate that the hydrogen has time to diffuse into the bulk of the metal while at higher sputtering rates the hydrogen has no time to escape and is released almost immediately. However, as the diffusion coefficient in tantalum at room temperature is about 10^{-6} cm^2 sec^{-1},[6] i.e. hydrogen can diffuse 10 μm into tantalum in a second, such an explanation must be an oversimplification as the sputtering rates are three orders of magnitude less than the diffusion rate. As the bulk hydrogen concentration measured by ERDA is 1 at % the local concentrations of hydrogen near the surface in Fig. 2 must be substantially higher than this. At such high concentrations the effective diffusion coefficient may be reduced by the high concentration itself and by hydride formation, but not sufficiently to explain the discrepancy found here. It is therefore necessary again to predict that the sputtering process produces a high density of traps for hydrogen as in the oxide and that the mobility of hydrogen in the region of the traps is greatly reduced. With such traps the hydrogen cannot escape ahead of the eroded surface at a high sputtering rate but at lower rates the hydrogen can diffuse across the whole defected region, reducing the surface concentration of hydrogen and hence the rate of hydrogen loss. Given the high mobility of hydrogen within tantalum the presence of such surface traps could cause misleading hydrogen depth profiles from tantalum specimens, particularly thin films, even if there is no accumulation of hydrogen from overlayers. As the sputtering front enters the tantalum, hydrogen from the bulk would rapidly be attracted into the surface traps giving an anomalously high hydrogen signal. For a thin film, the hydrogen may be depleted rapidly leading to a drop in hydrogen signal across the film which would not represent the true hydrogen distribution in the film.

DISCUSSION — INTERFACE RESOLUTION

In AES, atomic mixing and the electron escape depth influence on the interface width, and their relative contributions, have been assessed quantitatively by Seah and Hunt.[4] With SIMS the escape depth is less significant but this is replaced by the more serious problem of changes in ionization efficiency. At the tantalum oxide–tantalum interface the latter effects are severe and a full analysis of them is well beyond the scope of the present investigation. The interface widths reported in Table 1 provide a useful test of instrumental factors, such as uniformity of erosion and the influence of surface topography, but there is some uncertainty about the physical significance of the measurement as we do not fully understand the physical processes influencing the ion yield in the vicinity of the interface.

The measurements given in Table 1 show the expected trends with primary beam energy and film thickness. The best resolution was obtained with a low energy oxygen primary beam where the interface widths are less than those reported by Seah and Hunt.[3] While such a result is not physically improbable, it is likely that the SIMS results are an underestimate of the true interface width. The small peak just before the interface suggests that the oxygen ionization efficiency may be greater at the interface and that the peak intensity is an underestimate of the value that should be used in identifying the 84 and 16% points in the determination of Δz. In view of the uncertainties associated with the measurements of Δz, it is not safe to attach too much significance to the absolute values or to the variations with beam energy. It is, however, interesting to note that the CAMECA IMS–3F, using either an oxygen or a caesium primary ion beam, is capable of producing depth profiles with depth resolutions comparable to those obtained with AES.

CONCLUSIONS

The reduction of the surface of tantalum oxide under ion bombardment, together with the formation of subsurface defects, leads to complex effects in the SIMS profiles obtained from this material. The behaviour of hydrogen is particularly strongly influenced by the presence of traps induced in the subsurface region of both the oxide and the underlying metal of anodized tantalum foils. The standard tantalum pentoxide on tantalum foils provided by the National Physical Laboratory provide a useful check on the depth profiling performance of a SIMS instrument. However ion beam induced processes make the detailed interpretation of the profiles difficult and the foils are therefore not ideal for systematic studies of the effects of beam energy and ion species on depth resolution in SIMS.

The author would like to thank D.M. Poole and C.K. Jackson for helpful suggestions, P.M. Read for the ERDA measurements and A.D. LeClaire for advice on the diffusion of hydrogen.

REFERENCES

1. M. Seah, *Tantalum Pentoxide on Tantalum Foil BCR No. 261,* EUR 8839 EN, (1983).
2. M.P. Seah and C.P. Hunt, *Surf. Interface Anal.* **5** 33–37, (1983).
3. C.P. Hunt and M.P. Seah, *Surf. Interface Anal.* **5** 199–209, (1983).
4. M.P. Seah and C.P. Hunt, *J. Appl. Phys.* **56** 2106–2113 (1984).
5. J. L'Ecuyer, C. Brassard, C. Cardinal, J. Chabbel, L. Deschenes, J.P. Lebrie, B. Terrecult, J.C. Martel and R. St. Jacques, *J. Appl. Phys.* **47,** 381 (1976).
6. J. Volkl and G. Alefeld, *Diffusion in Solids: Recent Developments,* p. 231, ed. by A.S. Nowick and J.J. Burton, Academic (1975).

SURFACE AND INTERFACE ANALYSIS, VOL. 9, 111–117 (1986)

Characterization of the Imaging Properties of a Double-pass Cylindrical-mirror Analyzer

N.E. Erickson and C.J. Powell
Surface Science Division, National Bureau of Standards, Gaithersburg, MD 20899, USA

The imaging properties of a double-pass cylindrical-mirror analyzer have been investigated using extensions of the technique recently described by Seah and Mathieu. Elastic-peak intensity data obtained as an electron beam was rastered across a test specimen, was recorded in digital form and later used to generate 'three-dimensional' images. Images have been obtained for a range of operating conditions and for displacements of the specimen from its optimum position. Measured image widths were in semi-quantitative agreement with those expected from a simple analysis although deviations attributed to instrument imperfections were found. Elastic-peak images provide a convenient and quantitative means of assessing instrument performance and defining the specimen area being analyzed for the selected conditions.

INTRODUCTION

Seah and Mathieu[1] have recently described a method to determine the active analysis area of an x-ray photoelectron spectrometer and have reported area measurements on a particular instrument as a function of specimen alignment and various instrumental parameters. The method, based on a technique used by others,[2,3] consists of rastering an electron beam of a selected energy across a test specimen and of measuring the current of scattered electrons in the spectrometer at selected energies in phase with the position of the incident beam on the specimen. Seah and Mathieu used the signals for the elastic peak and for selected lower energies to modulate the brightness of an oscilloscope, rastered in synchronism with the beam on the specimen, to create images representing the specimen area contributing to the elastic peak. Their instrument, however, only allowed three levels of brightness so that only semiquantitative information on the dependence of the measured intensity on electron-beam position on the specimen could be obtained.

We have extended the Seah–Mathieu technique in the following ways. First, the elastic-peak intensity was stored in a computer (as a function of beam position on the specimen). The intensity data were later plotted to provide quantitative 'images' representing the imaging properties of the spectrometer, in our case a double-pass cylindrical mirror analyzer, as a function of various experimental parameters. Second, a test specimen in a normal operating configuration was used, in our case at an angle of 60° to the analyzer axis. This orientation provides a characterization of the system as it is often used; Seah and Mathieu placed their specimen normal to the analyzer axis in order to better compare their spectrometer performance with predictions based on design data. Finally, images were obtained from measurements of the absorbed current to the specimen, i.e., scanning electron micrographs. These images have been useful in determining the deflection-voltage limit at which beam current is reduced and in assessing the roughness of the specimen in order to ensure that structures found in the elastic-peak images were not associated with specimen topography or other artifacts.

We report here elastic-peak and absorbed-current images that have been recorded using a Perkin–Elmer–Physical Electronics Model 549 XPS/AES system equipped with a Model 15–255G double-pass cylindrical-mirror analyzer (CMA).[4] In the XPS mode of operation, electrons are retarded to a selected analyzer pass energy by a pair of spherical grids before energy analysis in the CMA while in the AES mode there is no retardation.[2] Our instrument is similar in design to the instrument tested by Seah and Mathieu[1] but does not have an aperture on a rotating drum to select desired azimuths.

Elastic-peak images have been obtained for selected combinations of the following experimental conditions: (a) with the specimen aligned using the conventional elastic-peak method and with the specimen displaced from that position by up to ± 2 mm along the analyzer axis; (b) for electron beam energies of 100, 500, and 1000 eV in both the XPS and AES modes of operation; (c) for analyzer pass energies of 25, 50, and 100 eV in the XPS mode of operation (d) the selection of small or large analyzer apertures intended for AES or XPS measurements, respectively.[2] Comparisons have been made of measured image widths with values expected from design considerations. A separate paper gives elastic-peak images for changes of the analysis energy about the incident-electron energy; in addition, energy offsets are reported in the position of the elastic peak as the incident beam is deflected to different specimen regions in the field-of-view of the analyzer.[5]

EXPERIMENT

The voltages applied to the deflection plates of the electron gun on the axis of the analyzer were supplied by programmable power supplies. At each electron energy, the maximum deflection voltage was set so that

0142–2421/86/120111–07 $05.00

the electron beam was rastered over horizontal and vertical ranges each of 13 mm; since the specimen surface was at an angle of 60° with respect to the analyzer axis, the specimen was scanned over an area of 13 mm (horizontal) by 15 mm (vertical) in all the experiments reported here.

The test specimen for the data reported here was a polycrystalline silver foil. The foil had been cleaned by ion sputtering at an earlier time but had been left in uhv for several months without further cleaning. This particular specimen had only small amounts of surface roughness as revealed by the absorbed-current images (see below).

The elastic-peak count-rate and absorbed-current data were acquired as the incident electron beam was stepped horizontaly at a given vertical position on the specimen, the beam stepped vertically prior to another horizontal sweep, and so on. These measurements were made at 51 equally-spaced points for each horizontal sweep and for 50 equally-spaced horizontal lines within the vertical sweep range. Digital data were thus acquired at intervals of about 0.25 mm horizontally and 0.3 mm vertically. The full-width at half-maximum (FWHM) intensity of the incident electron beam was less than 0.25 mm.

Software has been written so that 'three-dimensional' images could be plotted from the current or count-rate data stored in the computer. These images could appear as being 'seen' from selected viewing points. All images presented here have been normalized to approximately the same 'height', although there were, in general, variations of intensity from one set of conditions to another.

RESULTS

Fig. 1 shows an absorbed-current image recorded with an incident electron energy of 100 eV and no bias applied to the specimen. Similar images although with different contrasts were found at other energies and with biases applied to the specimen. The large, circular plateau in the centre defines the specimen region receiving the same beam current; outside this region, the absorbed-current falls off due to interception of the beam by the drift tube at the exit of the gun. Small features in the central plateau may be due to specimen topography; these features move as the specimen is translated. The spike in the lower-right region of the image, is due to the edge of the specimen-mounting clip. Similar absorbed-current images have been recorded at electron energies of 500 and 1000 eV.

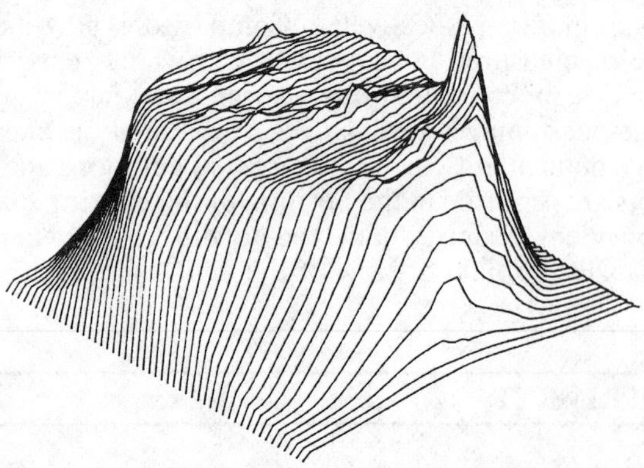

Figure 1. Absorbed-current image recorded with an incident electron energy of 100 eV and no bias applied to the specimen. The horizontal (bottom left to right) distance scanned by the electron beam is 13 mm and the vertical distance 15 mm.

Fig. 2(a)–(c) show sets of elastic-peak images for an analyzer pass energy of 50 eV, use of the AES aperture, and for incident electron energies of 100, 500, and 1000 eV, respectively. The central image, designated $\Delta z = 0$ in each figure, corresponds to the optimum specimen position on the analyzer axis, as found using the conventional method of maximizing the elastic peak intensity at an energy of 2000 eV. We note here that the optimum analyzer focus (minimum image width) occurs at the $\Delta z = 0$ position as expected from the analyzer design.[2] Seah and Mathieu, however, found in their instrument that there was a 1.5 mm difference in the axial positions for specimen alignment by the elastic-peak method and for best analyzer focus; this difference was attributed to a mechanical misplacement of the retarding grids at the entrance to their double-pass CMA.

The $\Delta z = 0$ images in Fig. 2 are generally symmetrical about the peak maxima although shoulders are apparent in the upper left portions of Figs 2(b) and 2(c) and there are low-intensity ridges in all images. As the specimen is translated by up to $\pm$ 2 mm along the analyzer axis, the images in Fig. 2 show the expected changes from an approximately circular spot for $\Delta z = 0$ to an approximately circular halo. The intensity distributions in the halos, however, are far from uniform. These intensity distributions are associated, in part, with the specimen not being normal to the analyzer axis. In addition, there are localized structures, particularly sharp for the 100 eV incident energy in Fig. 2(a), that may be associated with Moire transmission effects in the grids or with the complex relationship between source position of the specimen, angle of emission, and electron energy that defines the analyzer transmission function.[6] The latter relationship also has to be modified in practice by the blocking effects of the grid support bars.

Fig. 3 presents the elastic-peak images for $\Delta z = 0$ at incident energies of 100, 500, and 1000 eV from Fig. 2. This comparison clearly shows a substantial broadening of the active specimen area contributing to the elastic-peak intensity as the electron energy is increased. This result, which is counter to that expected from design considerations,[1,2] is discussed further below.

Fig. 4 shows elastic-peak images for an incident electron energy of 500 eV, the AES aperture, and analyzer pass energies of 25, 50, and 100 eV; these data were obtained with the specimen at its optimum position. These images become broader (and somewhat more asymmetrical) as the pass energy is reduced, a result also counter to design data and to be discussed later.

We have performed similar measurements to those shown in Figs. 3 and 4 with the XPS aperture instead of the AES aperture in the CMA. Fig. 5 shows images corresponding to Fig. 3 (pass energy = 50 eV; incident electron energies of 100, 500, and 1000 eV); the images in Fig. 5 are broader than those in Fig. 3, as expected, but there is now a minimum in the image width for the 500 eV incident energy. Fig. 6 shows images corresponding to Fig. 4 (incident energy = 500 eV; pass energies of 25, 50, and 1000 eV); the images in Fig. 6 are also broader than those in Fig. 4 but there is a slight minimum in the image width for the 50 eV pass energy. Figs. 5 and 6 thus show that the specimen area contributing to the XPS signal changes significantly as the electron energy for analysis and the analyzer pass energy are varied in typical ranges for XPS.

Finally, we have obtained elastic-peak images with the instrument operated in the AES mode (i.e., no retardation by the grids) and with the AES and XPS apertures. Fig. 7 shows elastic-peak images in the AES mode, with the AES aperture, for incident electron energies of 100, 500, and 1000 eV. Similar images recorded with the use of the XPS aperture are shown in Fig. 8. As expected, there is not a significant change in image width with electron energy; there is, however, a noticeable difference in the shapes of the images in Figs. 7 and 8.

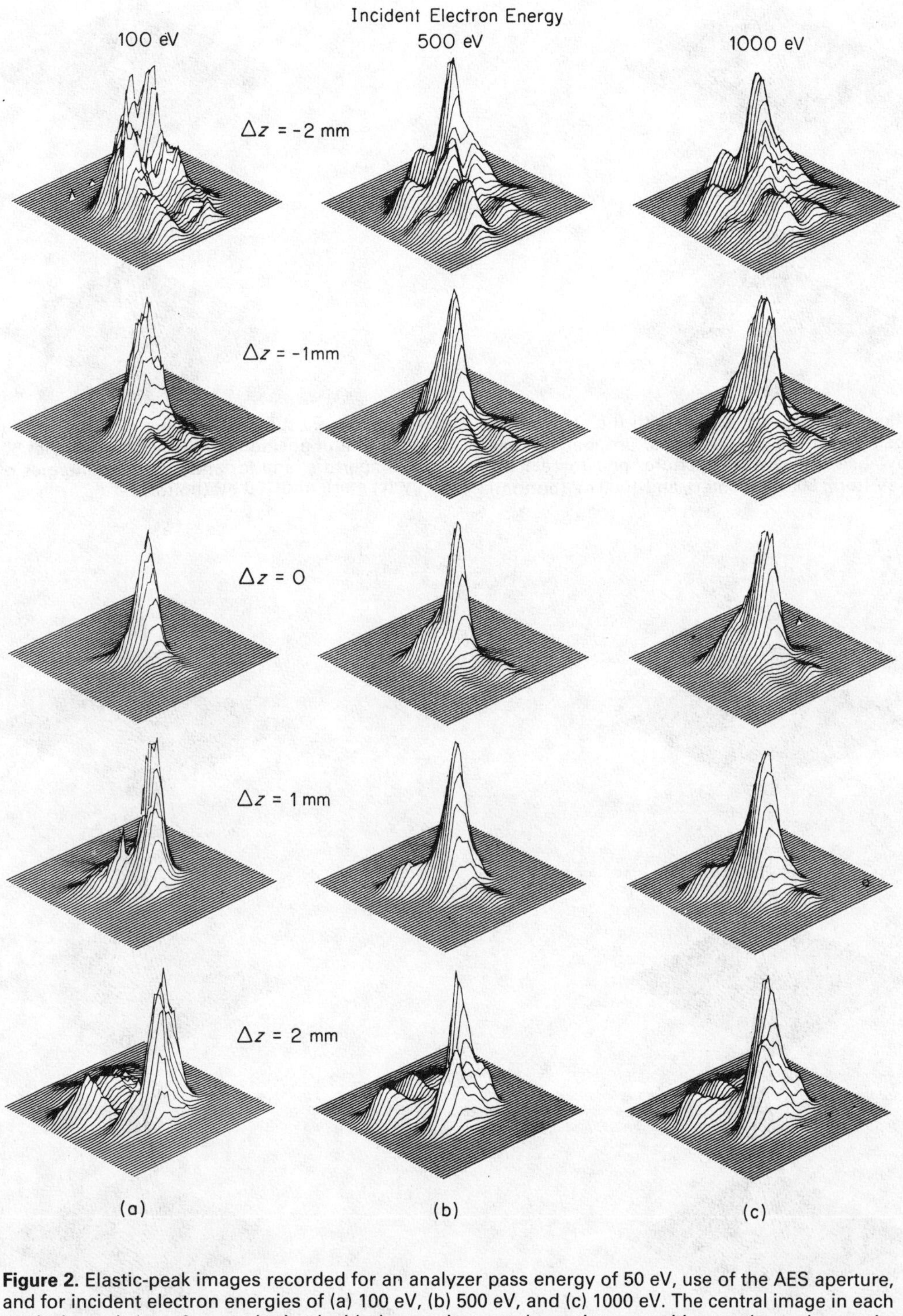

Figure 2. Elastic-peak images recorded for an analyzer pass energy of 50 eV, use of the AES aperture, and for incident electron energies of (a) 100 eV, (b) 500 eV, and (c) 1000 eV. The central image in each vertical panel, $\Delta z = 0$, was obtained with the specimen at the optimum position on the analyzer axis. The other images were recorded after the specimen was translated in steps of 1 mm along the analyzer axis away from (top) and towards (bottom) the analyzer.

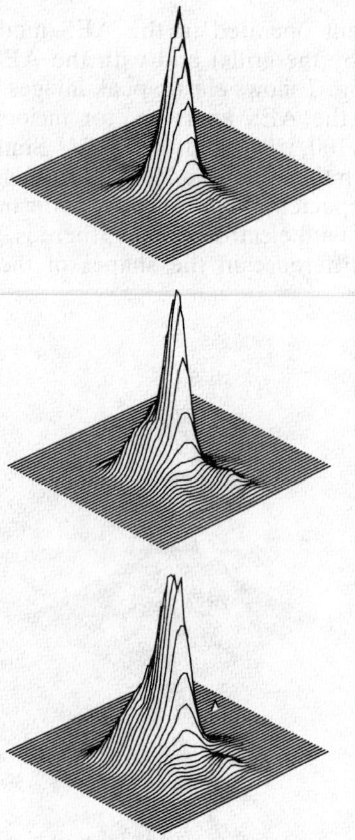

Figure 3. Elastic-peak images recorded with the specimen at its optimum position on the analyzer axis, an analyzer pass energy of 50 eV, use of the AES aperture, and for electron energies of 100 eV (top), 500 eV (center), and 1000 eV (bottom).

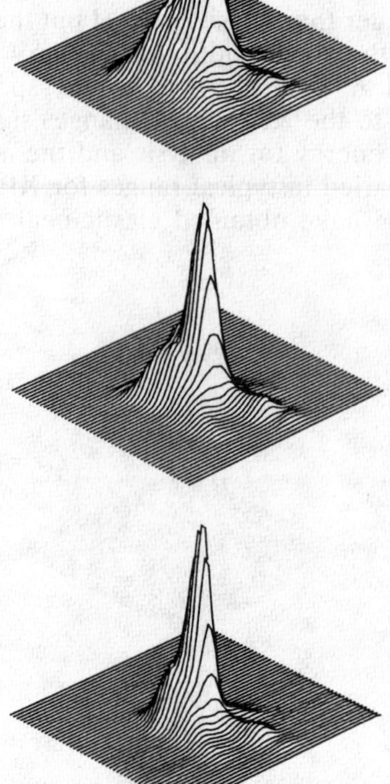

Figure 4. Elastic-peak images recorded with the specimen at its optimum position, and incident energy of 500 eV, use of the AES aperture, and for analyzer pass energies of 25 eV (top), 50 eV (center), and 100 eV (bottom).

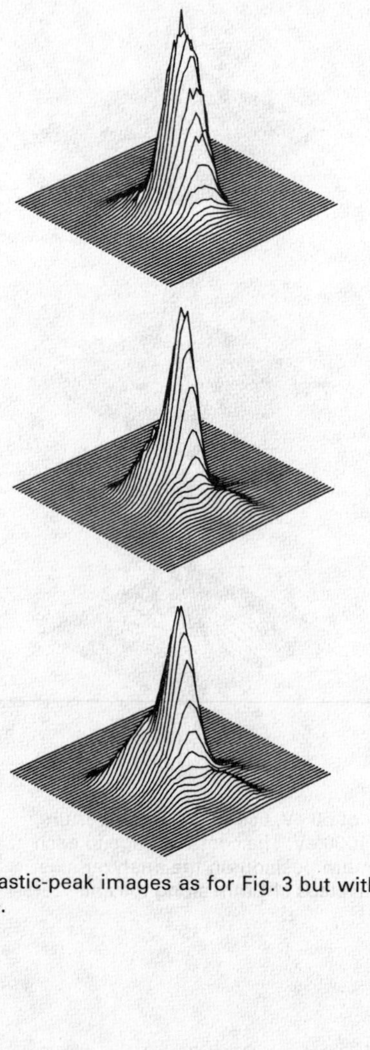

Figure 5. Elastic-peak images as for Fig. 3 but with use of the XPS aperture.

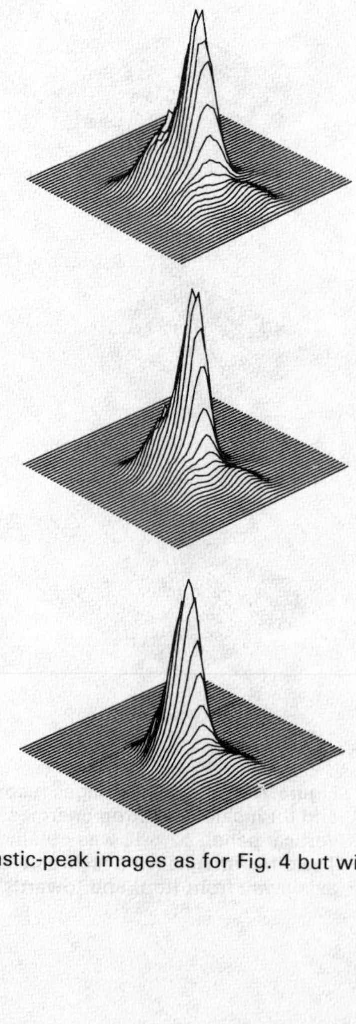

Figure 6. Elastic-peak images as for Fig. 4 but with use of the XPS aperture.

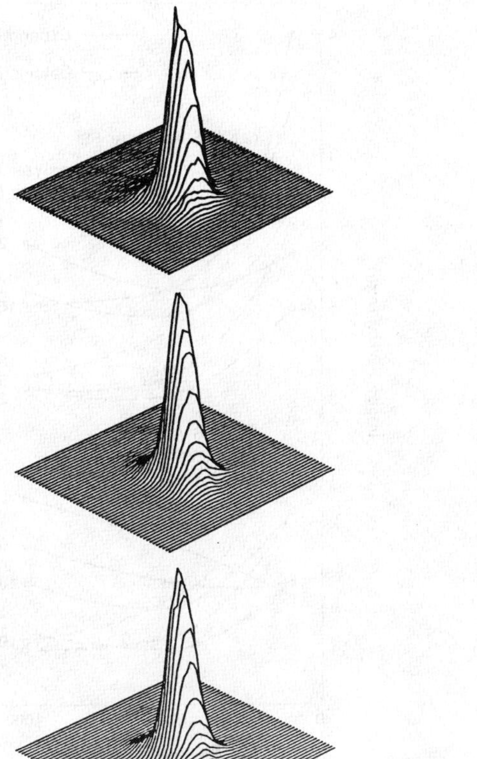

Figure 7. Elastic-peak images as for Fig. 3 but with no electron retardation before energy analysis (i.e. the AES mode).

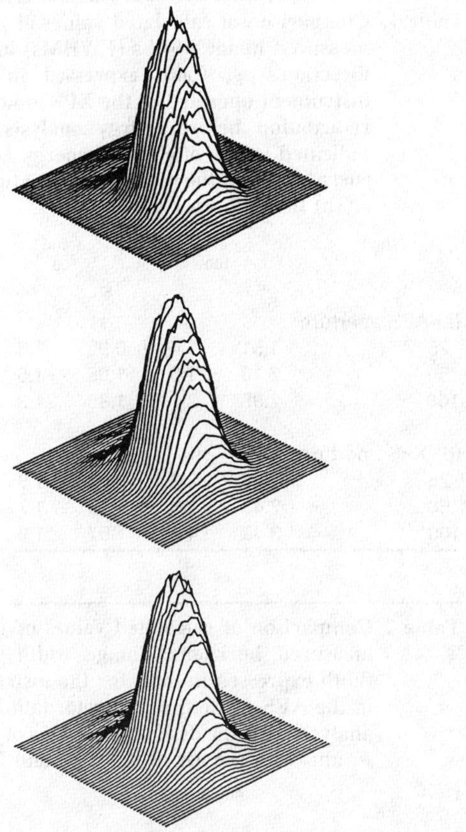

Figure 8. Elastic-peak images as for Fig. 7 but with the XPS aperture instead of the AES aperture.

DISCUSSION

Eagen and Sickafus[6] have analyzed the transmission properties of a CMA in some detail and have pointed out the generally complex relationship of the transmission with source size, the polar angle of electron emission, and the electron energy. The calculations of transmission functions for particular configurations are complicated, however, and deviations from the calculations are to be expected for non-ideal conditions (e.g., lack of perfect grid sphericity due, for instance, to their fabrication from woven mesh) which are difficult to handle in the analysis. We therefore make use of the simplified analysis of Seah and Mathieu[1] who give formulas for the several different contributions to the image widths in a CMA. Their formulas give the base widths of the images and we make the assumption here that the FWHMs of the images should be one half of the calculated base widths.

Seah and Mathieu[1] derive the following formulas for the image radii ρ_R, ρ_G, and ρ_A to give the separate contributions due to pre-retardation (if used), grid refraction, and aperture size, respectively:

$$\rho_R = 0.105\, R\, (E_p/E)^{1/2} \quad (1)$$

$$\rho_G = (aR/8S)\, (E/E_p)^{1/2} \quad (2)$$

$$\rho_A = (d/2)\, (E_p/E)^{1/2} \quad (3)$$

where E is the initial electron energy, E_p is the pass energy in the analyzer, R is the mean radius of the retarding grids, a is the size of the square holes in the grids, S is the separation of the grids, and d is the diameter of the selected CMA aperture. For our instrument, R = 27.04 mm (the two grids have radii of 24.51 mm and 29.57 mm), a = 0.23 mm, S = 5.06 mm, and d = 1.58 mm (AES aperture) or d = 3.81 mm (XPS aperture).[7] Seah and Mathieu write the total image radius ρ_T as the quadrature sum of the separate components:

$$\rho_T = (\rho_R^2 + \rho_G^2 + \rho_A^2)^{1/2} \quad (4)$$

There will, in practice, be additional contributions to ρ_T associated with, for example, non-sphericity of the grid elements and mechanical imperfections of critical components, but these are difficult to quantify.

Tables 1 and 2 show values of ρ_T calculated from (4) for the instrument operated in the XPS and AES modes respectively (i.e., with and without retardation of the electrons before energy analysis). Calculations have been performed for use of the AES and the XPS apertures and for electron energies of 100, 500, and 1000 eV; Table 1 shows ρ_T values calculated for analyzer pass energies of 25, 50 and 100 eV. Each table also shows measured values of image widths (FWHMs) in the horizontal direction, ρ_E, for the corresponding conditions of instrument operation.

There is a rough correspondence between the calculated values of ρ_T and the experimental values of ρ_E in Tables 1 and 2. Close agreement is not expected because of the complex analyzer lineshapes expected[6] (and which vary with changes in operating parameters), because the calculations are based on a specimen

Table 1. Comparison of calculated values of ρ_T from (4) and measured image widths (FWHMs) in the horizontal direction, ρ_E, (both expressed in mm) for the instrument operated in the XPS mode (i.e., electron retardation before energy analysis and with the indicated values of electron energy E, analyzer pass energy E_p, and the use of either (a) the AES aperture or (b) the XPS aperture.

E_p(eV)	E(eV) 100		500		1000	
	ρ_T	ρ_E	ρ_T	ρ_E	ρ_T	ρ_E
(a) AES aperture						
25	1.51	1.4	0.95	1.8	1.07	2.2
50	2.10	1.5	1.05	1.5	0.95	1.8
100	2.95	2.1	1.36	1.3	1.05	1.5
(b) XPS aperture						
25	1.74	2.2	1.03	1.9	1.11	2.3
50	2.43	2.6	1.19	1.7	1.03	1.9
100	3.43	4.0	1.57	1.9	1.19	1.8

Table 2 Comparison of calculated values of ρ_T from (4) and measured horizontal image width (FWHMs), ρ_E, (both expressed in mm) for the instrument operated in the AES mode (i.e., no retardation before energy analysis), for the designated values of electron energy E, and with the use of the AES and XPS apertures

E(eV)	AES aperture ρ_T	ρ_E	XPS aperture ρ_T	ρ_E
100	2.95	1.9	3.42	3.9
500	2.95	1.9	3.42	4.1
1000	2.95	1.8	3.42	4.0

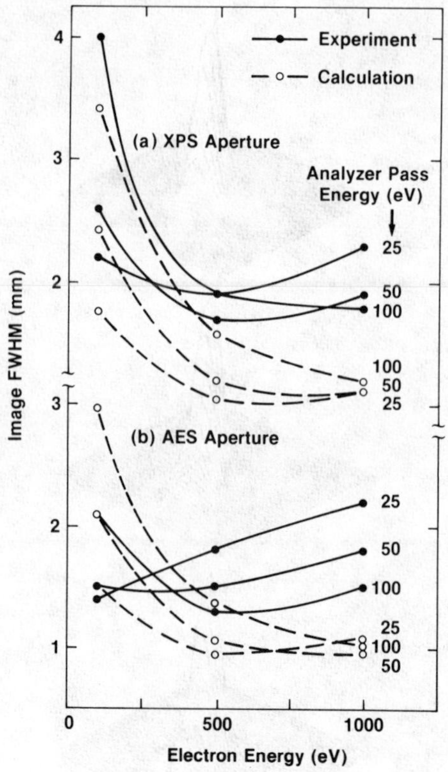

Figure 9. Plot of the calculated values of ρ_T from (4), open circles and dashed lines, and measured image widths (FWHMs) in the horizontal direction, solid circles and solid lines, as a function of electron energy for the designated analyzer pass energies and use of (a) the XPS aperture and (b) the AES aperture.

placed normal to the analyzer axis,[1,7] and because substantial asymmetries are observed in some of the experimental elastic-peak images (e.g., Figs. 3 and 4). Nevertheless, it is instructive to examine trends in the plot of the data of Table 1 in Fig. 9. For a pass energy of 100 eV, there is reasonable agreement between the experimental and calculated image widths. For the lower pass energies, there is reasonable agreement at a electron energy of 100 eV but appreciable deviations occur for higher electron energies (i.e., for higher retardation ratios). In the AES mode where there is no retardation, Table 2 indicates no significant changes in the ρ_E values with electron energy, the result expected from (4).

Rusch[7] has pointed out that image widths in instruments of the present type can be modified appreciably by wrinkles in the grids covering the apertures in the inner cylinder of the CMA and by deviations of the retarding grids, at the entrance to the CMA, from their ideal spherical shape. Measurements in his laboratory of similar type to those reported here show image widths and trends similar to those in Table 1. We therefore believe that differences between the calculated values of ρ_T and the measured ρ_E are due in part to the imperfections of our instrument and in part to the limitations of (4) in predicting image widths.

Although our instrument behaves semi-quantitatively as expected from (4), it is clear from asymmetries in the measured images, particularly for high retardation ratios, that our instrument has some specific imperfections. Our test specimen was at 60° with respect to the analyzer axis, but our images do not show the asymmetry expected from this geometrical effect alone. (We note that a contour plot of the intensity data would be helpful in analyzing image symmetry.) The various observed asymmetrics could be due to the instrumental imperfections noted above or, possibly, to the presence of varying residual magnetic fields in the specimen-grid volume.

SUMMARY

We have described an extension of the method of Seah and Mathieu[1] for characterizing the imaging properties of an electron spectrometer used for AES and XPS. Digital elastic-peak intensity data have been acquired and stored in a minicomputer as an electron beam was rastered across a test specimen. These intensity data were later used to generate 'three-dimensional' images that showed clearly the imaging properties of a double-pass CMA. Elastic-peak images have been obtained for a test specimen in a typical operating position (in our case at an angle of 60° to the CMA axis) and for typical measurement conditions (electron energies of 100, 500, and 1000 eV; analyzer pass energies of 25, 50, and 100 eV in the XPS mode of operation; and selection of either the AES or XPS apertures). We have investigated how the elastic-peak images were modified as the sample was translated along the analyzer axis about its optimum position. Finally, we have obtained absorbed-current images to ensure that the test specimen was sufficiently smooth (or the roughness on a much smaller scale than our electron beam width) to prevent significant artifacts occurring in the elastic-peak images.

We have measured image widths and compared these with values expected from a simple analysis. Semi-quantitative agreement has been obtained although there are differences in trends (as a function of electron energy or analyzer pass energy) that can be associated with non-uniformities of grids on the inner-cylinder apertures, with lack of sphericity of the retarding grids, and with simplifications of the analysis. Asymmetries in the measured images, for example at high retardation ratios or when the specimen is moved from its optimum position, are due to specific imperfections in our instrument.

The specimen area giving rise to detected electrons in the CMA changes with electron energy and pass energy in the XPS mode (with electron retardation prior to energy analysis). These changes, particularly with electron energy, need to be known if lateral inhomogeneities in the specimen can affect the analysis; this would be the case if specimens with non-ideal geometries (e.g., particles) were being examined. In the AES mode of operation (no retardation), no change was found in the active specimen area with electron energy, as expected from design considerations.

The techniques of Ref. 1 and this paper provide a direct means of determining the imaging properties of any electron spectrometer. These techniques should also be useful in optimizing the alignment and rastering of a sputter-ion gun with the specimen area viewed by the electron spectrometer. Finally, similar techniques could be useful in ISS for properly aligning the specimen.

Acknowledgement

The authors are grateful to Dr. T.W. Rusch of Perkin–Elmer–Physical Electronics Division for helpful discussions and for communicating instrumental design data.

REFERENCES

1. M.P. Seah and H.J. Mathieu, *Rev.Sci.Instr.* **56**, 703 (1985).
2. P.W. Palmberg, *J. Electron Spectrosc.* **5**, 691 (1974).
3. D.C. Peacock, M. Prutton, and R. Roberts, *Vacuum.* **34**, 497 (1984).
4. Commercial equipment has been identified to specify experimental conditions and does not imply endorsement by the National Bureau of Standards.
5. N.E. Erickson and C.J. Powell, *J. Vac. Sci. Tech* (in press).
6. C.F. Eagen and E.N. Sickafus, *Rev. Sci. Instr* **48**, 1269 (1977).
7. T.W. Rusch (private communication).

SURFACE AND INTERFACE ANALYSIS, VOL. 9. 119–123 (1986)

On the Use of the Quartz Crystal Microbalance for Surface Analysis in Combination with Surface Sensitive Spectroscopies

J. Onsgaard and W. Huang
Fysisk Institut, Odense Universitet, DK-5230 Odense M, Denmark
E. Taglauer
Max-Planck-Institut für Plasmaphysik, Euroatom-Association, D-8046 Garching bei München, West Germany

The quartz crystal microbalance (QCM) has a wide range of applications in surface physics and surface chemistry. We present results obtained by combining the high areal density sensitivity of the QCM with surface sensitive spectroscopies like Auger electron spectroscopy (AES) and ion surface scattering (ISS). One type of experiments deals with measurements of sputtering yield variations during ion profiling of metal–metal interfaces, situated at the surface of the QCM, as a function of varying surface composition. A new method for online measurements of angular distributions of sputtered particles from a solid surface, by utilizing the QCM's as collectors of the ejected particles, is the subject of the other type of experiment.

INTRODUCTION

The quartz crystal microbalance with its capacity to determine AREAL DENSITY has a wide spectrum of applications in surface science. This spectrum covers areas such as adsorption kinetics, chemical processes occurring on the surface of the QCM, plasma-assisted etching studies and monitoring of thicknesses of evaporated films. A review of applications of the QCM in surface science is given by L.L. Levenson in Ref. 1.

A QCM is a thin piezoelectric quartz crystal plate with a specific crystallographic orientation with respect to the natural crystal structure. When an external, periodic varying electric potential, via electrodes fixed to the crystal plate is applied to the crystal then the induced mechanical stress will vibrate with the frequency of the voltage source. A mode of vibration of the QCM very often used is the thickness–shear vibration with a high frequency. The fundamental vibration mode and the orientation of the so-called AT cut quartz plate with respect to the crystal axes are shown in Fig. 1(a) and Fig. 1(b) respectively.

The basic relationship[2,3] between mass change and the corresponding change in frequency Δf is expressed by

$$\frac{\Delta f}{f_q} = -\frac{m_d}{m_q} \tag{1}$$

where Δf is the frequency shift induced by the addition of material with an areal density m_d, f_q is the original resonance frequency, and m_q is the areal density of the quartz plate. This linear relationship represents according to Miller and Bolef[4] an approximation which can be used if m_d is less than 1% of m_q.

In the fundamental mode the relations between the wavelength λ_q of the wave in the thickness direction,

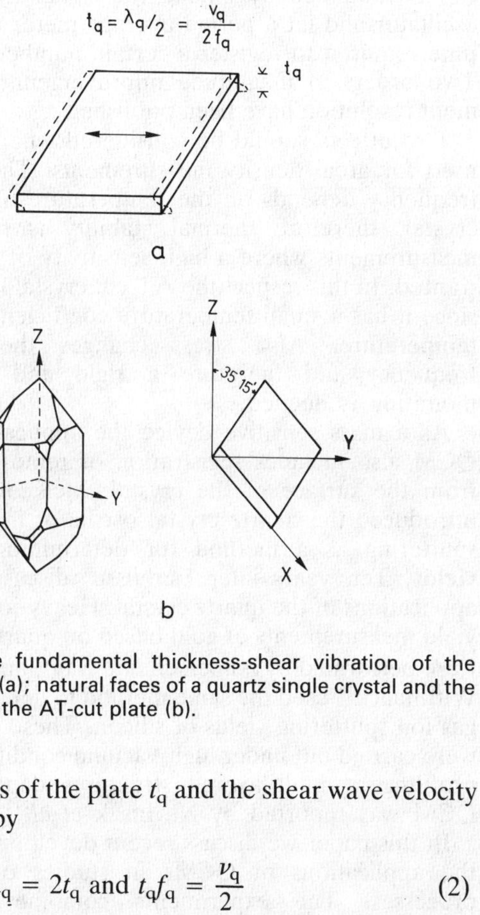

Figure 1. The fundamental thickness-shear vibration of the quartz crystal (a); natural faces of a quartz single crystal and the orientation of the AT-cut plate (b).

the thickness of the plate t_q and the shear wave velocity v_q is given by

$$\lambda_q = 2t_q \text{ and } t_q f_q = \frac{v_q}{2} \tag{2}$$

The mass sensitivity $[\mathrm{Hz \cdot cm^2\, g^{-1}}]$ is introduced by

$$C_d = \frac{2f_q^2}{\rho_q v_q} \tag{3}$$

0142–2421/86/120119–05 $05.00

where ρ_q is the density of the quartz. Substitution of (2) and (3) into (1) yields the simple expression

$$\Delta f = C_d m_d \quad (4)$$

For a 5 MHz quartz crystal with a 4 mm aperture practically the whole sensitivity region is covered. Therefore, C_d is not influenced by the particular shape of the differential sensitivity distribution. For a quartz plate with a resonance frequency of 5 MHz, $\rho_q = 2.65$ g cm^{-3}, and $\nu_q = 3340$ m s^{-1}, a mass sensitivity $C_d = 5.65 \times 10^7$ Hz cm^2 g^{-1} is found. This means that deposition of a film with an areal density of 17.7 ng cm^{-2} results in a frequency change of 1 Hz. Typically the density of one monolayer corresponds to 10^{15} atoms cm^{-2} resulting in a frequency change of 10 Hz for an element with mass number equal to one hundred.

Usually the QCM is a part of an oscillator circuit and the output of the oscillator is coupled to a frequency meter or a circuit for determining periods. If the measurement of the frequency is based on counting the number of cycles within a period, then the uncertainty will be ±1 Hz. This inaccuracy means that the resolution for a 10 MHz crystal, with a 1 s measurement time will be 1 part in 10^7, which is equivalent to an areal density of ~5 ng cm^{-2}.

A higher resolution in the frequency determination can be obtained by mixing the signal from a stable oscillator and let a period average meter measure the time required to register a certain number of periods. Two orders of magnitude improvement in measurement resolution have been published.[5,6]

Two effects should be considered when QCMs are used for areal density measurements. The resonance frequency depends on the temperature of the quartz crystal, therefore thermal stability is required for measurements where a high sensitivity of the QCM is wanted. In this respect the AT-cut crystal is well suited since it has a small temperature coefficient near room temperature. Also stress changes the resonance frequency and therefore a rigid and strain free mounting is needed.

As a mass sensitive device the applicability of the QCM also includes registration of removal of atoms from the surface of the crystal. McKeown[7] in 1961 introduced the quartz crystal oscillator in the field of sputtering as a method for determining sputtering yields. Ten years later EerNisse[8] discussed vacuum applications of the quartz crystal. Heavy-ion sputtering yield measurements of gold based on quartz oscillators were performed by Andersen and Bay[9], and Blank and Wittmaack[10] used the same method to determine noble gas ion sputtering yields of silicon. These experiments were carried out under high-vacuum conditions. Depth profiling of multilayered structures deposited on a QCM was reported by Navinsek *et al.*[11]

In this paper we discuss recent developments[12–13] in the applications of QCMs in studies of sputtering processes. The experiments combine the QCM measurements with results from surface sensitive spectroscopies and all measurements are carried out under UHV conditions. Auger electron spectroscopy and ion surface scattering were utilized for determination of the actual composition of the surfaces and for observation of the surface cleanliness.

In the first type of experiment the sputtering process is performed on the QCM itself, thus making a direct determination of the sputtering yield possible. However, this information can be extended to include a dynamical observation of the variation of the sputtering yield, Y, when a depth profiling of a metal–metal interface takes place.

A new function of the QCM is presented in the other type of experiments where the crystal operates as a collector of atoms sputtered from a solid surface. With a set of QCMs the angular distribution of sputtered particles, $I(\theta)$, can be determined in on-line measurements.

EXPERIMENTAL

The basic pressure in the UHV-vessels was below 1×10^{-10} mbar. The quartz crystal microbalances used in the present experiments were shaped like discs with 14 mm diameter, AT-cut, and operated at a resonance frequency around 5 MHz. Mountings of the QCMs were either in a special holder or suspension at the periphery in steel springs. Apertures with a 4 mm hole defined a central irradiation area of the QCM with a constant mass sensitivity. Sputtering was performed with noble gas ions from guns operating between 1 and 5 keV. In most of the experiments the beams were rastered in order to obtain a homogenous irradiated area on the target.

Depth Profiling of Interfaces

A sketch of the experimental setup for depth profiling of thin films of noble metals evaporated onto a Mo substrate deposited on a QCM, is shown in Fig. 2. The deposition of overlayers on the QCMs were carried out *in situ* via evaporation from helical tungsten filaments. Calibration of the QCM was done by Rutherford backscattering spectroscopy. A 1 keV Ne^+ ion beam was used for the erosion process. During bombardment the frequency rate was followed, meaning that changes in areal density are determined as a function of fluence. Thus a sputtering yield characteristic of the actual

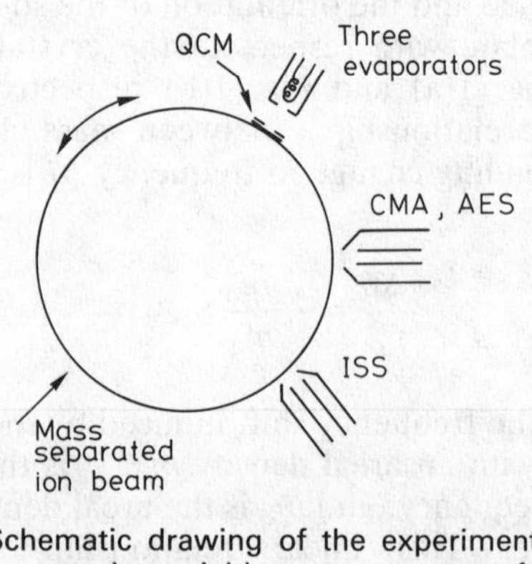

Figure 2. Schematic drawing of the experimental set up for dynamical sputtering yield measurments of metal–metal interfaces. Metal overlayers are evaporated onto a QCM which can be rotated to ISS and AES ports for composition measurements. Erosion is carried out by an ion beam directed along the normal of the target.

composition of the system is measured. Auger electron spectroscopy and ion surface scattering, see Fig. 2, were used to determine the concentrations of adsorbate and substrate during the profiling. These two spectroscopies provide different information depths which cover the average projected range of 20 Å for 1 keV Ne^+ ions.

Angular Distribution of Sputtered Particles

The measuring principle in the $I(\theta)$ experiment is shown in Fig. 3. With the sample mounted in the center of the chamber it is possible to position several QCMs, in the present case eight, in a circular arrangement with the normal of the QCM pointing towards the centre. Furthermore, the crystals were suspended such that they could be rotated 180° for the purpose of analyzing the surface of the QCMs by means of AES or ISS.

The important parameter to be measured in the angular distribution experiment is the frequency change of the quartz crystals as a function of fluence. A direct determination of the deposited areal density is carried out and the frequencies of the different crystals are read succesively via a relay and a frequency counter. With AES the amount of deposited atoms can be controlled and determined in the range, 1/100 of 1 ML to 5–6 ML, depending on the energy of the Auger transition considered. The Auger electron analysis was carried out by excitation of the quartz surface with 2 keV electrons and energy analysis of the emitted electrons. A modulation voltage with a 5 V peak-to-peak amplitude and a 10 μV sensitivity were the basic parameters in the lock-in technique based detection of the Auger electrons.

RESULTS AND DISCUSSION

Depth Profiling of Interfaces

Ion bombardment of an interface consisting of a Au film evaporated on a QCM with a Mo overlayer resulted in the depth profile shown in Fig. 4. The escape depths for the 69 eV Auger line of Au and the 221 eV auger line of Mo are close lying. A comparison

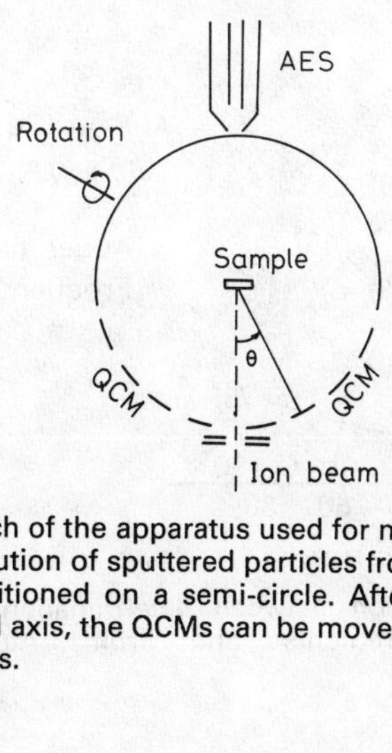

Figure 3. Sketch of the apparatus used for measurement of the angular distribution of sputtered particles from a copper target. QCMs are positioned on a semi-circle. After an 180° rotation about a vertical axis, the QCMs can be moved to a CMA port for surface analysis.

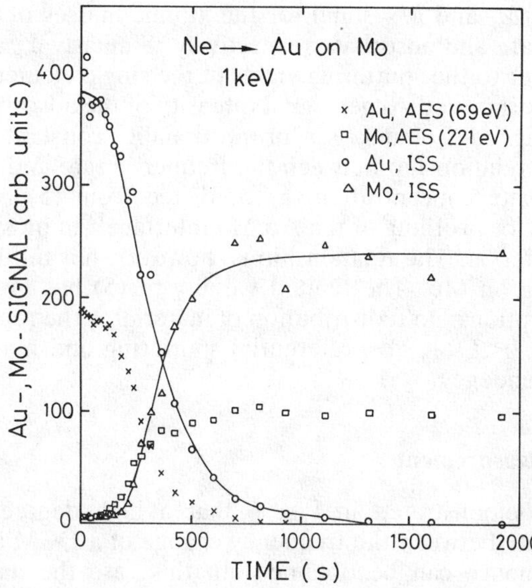

Figure 4. AES and ISS intensity profiles of a Au/Mo interface.

of the intensity decays of the Au ISS signal and the Au Auger line shows a parallel behaviour during the profiling. The substrate ISS and Auger signals also follow each other but there is a delay in the onset of the Mo ISS signal. An effect which can be explained by the high surface sensitivity of ISS.

A factor of 4.5 between the frequency rates of Au and Mo is directly determined from Fig. 5, which shows $\Delta\nu/\Delta t$ as a function of bombardment time. Thus the sputtering yield ratio is 2.2, to be compared with the value 2.6 determined from the graphs published by Matsunami *et al.*[14] Also it is demonstrated in the graph that sufficiently many frequency rates can be determined during the profiling, allowing concentration measurements to be performed in combination with $\Delta\nu/\Delta t$ readings.

In Ref. 12 it was shown that $\Delta\nu/\Delta t$ can be expressed by

$$\frac{\Delta\nu}{\Delta t} = k\left\{\left(\frac{M_s Y_s}{M_A Y_b} - 1\right) I_s + 1\right\} \tag{5}$$

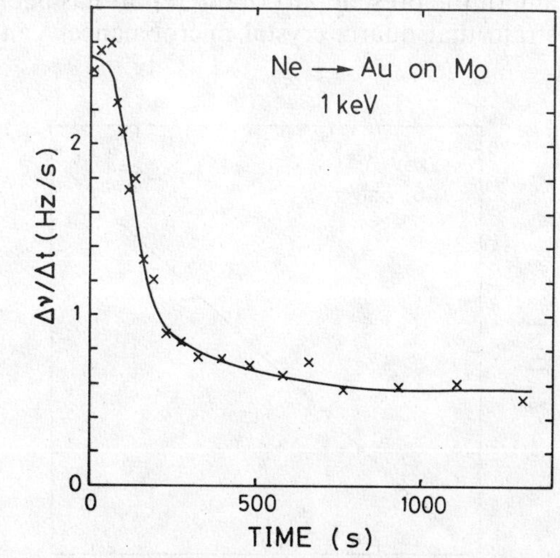

Figure 5. Change in the frequency rate as a function of bombardment time for a Au/Mo interface with a 80Å thick Au layer.

where M_s and M_A stand for the atomic masses of the substrate and adsorbate respectively. Similarly Y_s and Y_b refer to the sputtering yields of the single elements. The normalized Auger signal intensity of the adsorbate is designed I_s and k is a proportionality constant. A linear relationship between the frequency rate and the adsorbate concentration, as predicted from (5), was found for profiling of the Au/Mo interface[13] in most of the interval. The relationship is, however, not fulfilled for Ag on Mo. The limited validity of (5) lies in the assumptions: no redistribution of material in the region where $0<I_s<1$, no preferential sputtering and matrix independency.

$I(\theta)$ Measurements

It is demonstrated in Fig. 6 that a high degree of linearity between the frequency change of a QCM and the fluence can be obtained. In this case the angle between the incoming 2 keV Ar^+ ion beam direction and the direction of the QCM was 13°. Linearity is also found in the low-fluence regime, below a few times 10^{17} ions cm^{-2}, where ion beam induced texture effects in the Cu sample are small. For the QCM positioned 13° relative to the sample normal it is seen that 2×10^{17} ions cm^{-2} results in a 9 Hz frequency shift which corresponds to a Cu deposition of 0.8 ML. Of course, larger angles cause smaller frequency shifts but this can be overcome by counting for a longer period or change the reading technique to measurements of periods.

An intensity distribution $I(\theta)$ corresponding to a fluence of 4.8×10^{17} ions cm^{-2} is shown in Fig. 7. The experimental points have been normalized in order to give the same integrated yield, $2\pi \int Y(\theta) \sin\theta d\theta$, as a cosine distribution, and the results are plotted as a function of $\cos\theta$. Deviation from a cosine-like distribution is observed.

A comparison of $I(\theta)$ as measured with frequency changes and as determined from Auger electron concentration measurements $I(\theta)$ is shown in Fig. 8 where a normalization of the two types of measurements at $\theta = 43°$ is carried out. A reasonable agreement between the two types of measurments is found.

The aim of the present part of the report has been to demonstrate that quartz crystal microbalances can be used for measurements of angular distributions of sputtered particles in on-line measurements under UHV-conditions. Furthermore, a surface sensitive spectroscopy serves to control the method and to monitor the surface cleanliness. Earlier Besocke *et al.*[15] determined angular differential sputtering yield of silver by moving a single QCM around the sample in a circular movement. Quartz crystals as collectors of sputtered Au atoms from a Au plate irradiated with high energy Ar^+ ions under a pressure of 2×10^{-6} torr were used by Mannami *et al*[16]

On-line measurements of $I(\theta)$ mean that deviations from linearity between fluence and $I(\theta)$ can be followed. Non-linearity can be due to changes in the sticking coefficient of the sputtered particles to the QCM or due to irradiation effects on the sample surface. An advantage of the QCM-method compared to an $I(\theta)$ method based on a surface sensitive analysis is that the former method has no constraints concerning information depth and homogenous distribution of the deposited particles. A drawback of the QCM set-up is that the number of crystals in a semicircular arrangement is rather limited. However, it is possible to accomplish an $I(\theta)$ determination with twice as many measuring points as presented here by reducing the diameter of the quartz crystals considerably. The gain obtained by the 180° rotation of the quartz disc is partly that the adsorbate concentration can be measured and partly that the quartz surfaces can be cleaned in situ by ion bombardment.

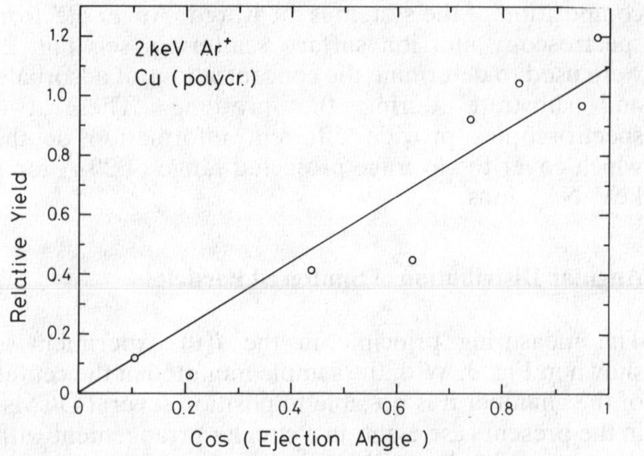

Figure 7. Angular distribution $I(\theta)$ plotted as a function of $\cos\theta$.

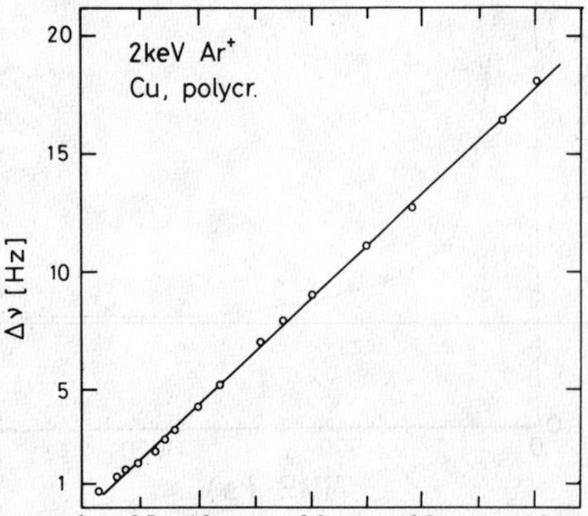

Figure 6. Frequency change of a QCM as a function of fluence.

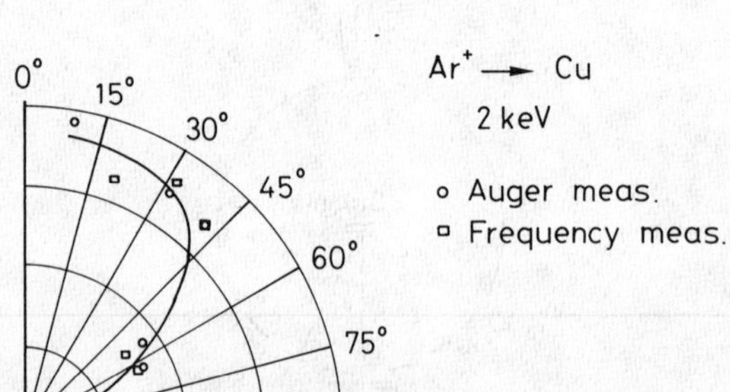

Figure 8. Comparison between determination of $I(\theta)$ from frequency measurements and from Auger electron spectroscopy.

CONCLUSION

The basic idea underlying this work has been to utilize the high areal density sensitivity of the QCM in combination with surface sensitive spectroscopies. It has been shown that online angular distributions of sputtered particles can be measured by a circular arrangements of quartz crystals where the crystals function as collectors. The method can be extended by introducing surface sensitive methods to monitor the surface cleanliness and eventually control the amounts of deposited material on the quartz crystals. It has also been demonstrated that the mass change per time unit during depth profiling of an interface at the QCM surface can be combined with concentration determinations based on ISS and AES. Thus a dynamical measurement of the sputtering yield as a function of the interface composition is performed.

REFERENCES

1. *Applications of piezoelectric quartz crystal microbalances,* vol. 7 of 'Methods and Phenomena: Their Applications in Science and Technology, eds. C. Lu and A.W. Czanderna, Elsevier, Amsterdam (1984), and references therein.
2. G. Sauerbrey, *Phys. Verh.,* **8,** 113 (1957).
3. G. Sauerbrey, Z. Phys. **155,** 206 (1959).
4. J.G. Miller and D.I. Bolef, *J. Appl. Phys.* **39**, 5815 (1968).
5. C. Benndorf, G. Keller, H. Seidel and F. Thieme, *J. Vac. Sci. Technol.* **14,** 819 (1977).
6. B. Kasemo and E. Törnqvist, *Surf. Sci.* **77,** 209 (1978).
7. D. McKeown, *Rev. Sci. Instrum.* **32**, 133 (1961).
8. E.P. EerNisse, *J.Appl. Phys.* **42,** 480 (1971).
9. H.H. Andersen and H.L. Bay, *J. Appl. Phys.* **46,** 1919 (1975).
10. P. Blank and K. Wittmaack, *J. Appl. Phys.* **50**, 1519 (1979).
11. B. Navinsek, P. Panjan and A. Zabkar, *Proc. 'Contributed Papers': The Physics of Ionized Gases,* SPI G–82, Dubrovnik, ed. by G. Pichler, Institute of Physics, Zagreb (1980).
12. J. Onsgaard and O. Ellegaard, *Nucl. Instrum. Meth.* **B2,** 666 (1984).
13. J. Onsgaard and E. Taglauer, *Vacuum* **34,** 831 (1984).
14. N. Matsunami, Y. Yamamura, Y. Itikawa, N. Itoh, Y. Kazumata, S. Miyagawa, K. Morita, R. Shimizu and H. Tawara, *Atomic Data and Nuclear Data Tables* **31**, (1984).
15. K. Besocke, S. Berger, W.O. Hofer and U. Littmark, *Rad. Effects* **66,** 35 (1982).
16. M. Mannami, K. Kinnera and A. Kyoshima, *Nucl. Instr. Meth.* **185,** 533 (1981).

SURFACE AND INTERFACE ANALYSIS, VOL. 9. 125–128 (1986)

Measurement of the Sputtering Yield with a 3D Stylus Surface Mapping System: Application to Copper and α-Brass

J. C. Quantin, E. Darque-Ceretti, R. Combarieu and F. Delamare
Ecole des Mines de Paris, CEMEF (CNRS UA 852), Sophia-Antipolis, 06565 Valbonne Cedex, France

This paper presents a method for determining the sputtering yield of a material bombarded by an ion beam, using a 3D stylus surface mapping system. Data obtained of copper are discussed. The influences on the sputtering yield of progressive addition of zinc is observed. The results are in good agreement with previous theoretical results.

INTRODUCTION

Many techniques of surface analysis use ion beams (ESCA, Auger, SIMS) to provide concentration gradients. They all need prior knowledge of the sputtering yield of the bombarded material in order to convert erosion time into depth.

Many sputtering yield measurements have been performed and a vast amount of data has been published.[1–3] Data is obtained by measurements either on the sputtered material (dynamic techniques, using for example a time-of-flight mass spectrometer, or analysis of the collected sputtered material) or directly on the target. Thus, several methods can be used, such as change of mass, detection of a change of nature of a material of a known thickness by surface analysis or thickness change. Some authors measure the step between the original surface and the bottom of the crater by 2D (two-dimensional stylus profilometer).[4,5]

We have used in the same manner a 3D (tri-dimensional) stylus surface mapping system designed by one of the authors.[6] In this paper, we present the advantages and disadvantages of this method. Experiments conducted with copper will enable us to compare our results with those already published in the literature. The influence of an alloy element (Zn) in a solid solution (α–brass) will then be studied.

PRINCIPLE OF THIS METHOD

It is extremely simple. The sputtering (S_T) yield of a material is defined as the average number of atoms sputtered for incident ion:

$$S_T = N.e. \frac{d}{M} \cdot \frac{V}{It}$$

where N is Avogardro's number; e = electron charge; d = density of the material; M = atomic mass of the material or average atomic mass for an alloy; V = volume of the crater; I = ion current intensity; t = sputtering time.

0142–2421/86/120125/04 $05.00

For a given ion beam (i.e., of a known material and ion charge and predetermined energy and incidence angle (θ)), the problem comes down to measuring the ion current and the volume of the resulting crater.

The ion current is measured globally with a double walled Faraday cage, the hole of which is slightly larger than the size of the beam. One should check that it remains constant during the whole experiment. The volume of the crater is measured via 3D profilometry.

MEASURING THE VOLUME OF THE CRATER

This is done using a digital map of the topography of the area under study, obtained with a 3D stylus surface mapping device, the general diagram of which can be found in Fig. 1.

Once the size of the area to be studied has been selected, it is scanned line by line with a mechanical finger (diamond with a radius of 5 μm). The coordinates of each point are recorded step by step. The current maximum accuracy is of 1 μm for the X and Y coordinates, and 60 nm for the heights.[6] We thus have a digital image of the sputtered surface, which can be used either for visualizing the surface or for computing the volumes of sputtered matter. The

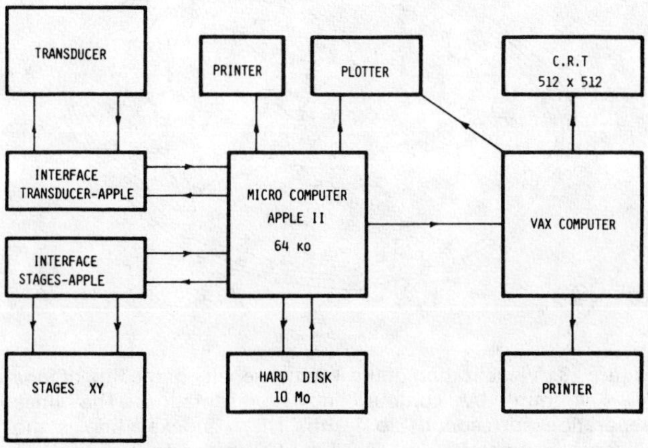

Figure 1. Schematic diagram of our device.

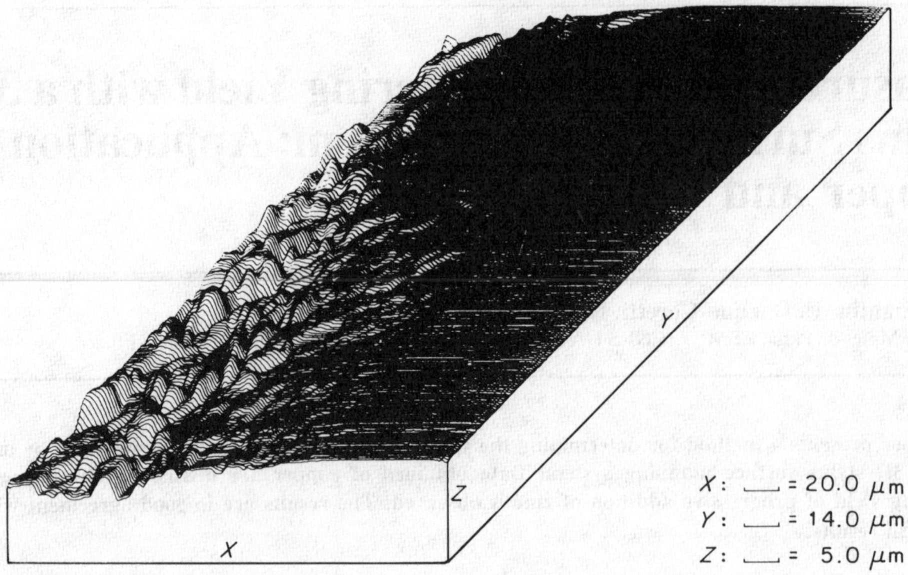

Figure 2. A 3D isometric representation of an edge of a crater showing the increase of roughness due to sputtering. α–brass (C_{Zn} = 30 %) sputtered during 2 h by 15 keV Ar^+ ions.

surface can be visualized in isometric perspective as in Fig. 2 (45° projection angle and removal of hidden parts), or by drawing contour lines (Fig. 3 and 4).

If we select a large scale, we can get a global view of the geometry of the crater and control how homogeneously distributed the flux of incident ions is over the sample (Fig. 3). On a small scale, we can see the details of the microtopography of the sputtered surface. Figures 4 and 5 illustrate very well how the sputtering speed depends on the crystal orientation.

The computation of the sputtered volume is carried out over a rectangle, between the bottom of the crater and a reference plane representing the original surface. In order to minimize the consequences of an inhomogeneous flux of the incident ions or of sputtered atoms redeposition, we choose to carry out the computation over the whole crater. The main difficulty in this measurement lies in determining the reference plane. Indeed, it is quite rare that metallographic polishing yields a surface planar enough. Automatic polishing machines supply more or less spherical surfaces. One needs hand polishing to obtain surfaces which, although not planar, are at least cylindrical. We accept them only if the rise is less than 4 μm for an 8 mm distance.

The reference plane is then replaced by a series of planar strips parallel to the axis of the cylinder and equal in width (Fig. 6). The volume of the crater is then computed as the sum of the various volumes measured under each strip. Experiments show that increasing the width of the strips from 30 to 300 μm, we get the same volume of the crater at 0.5 % only.

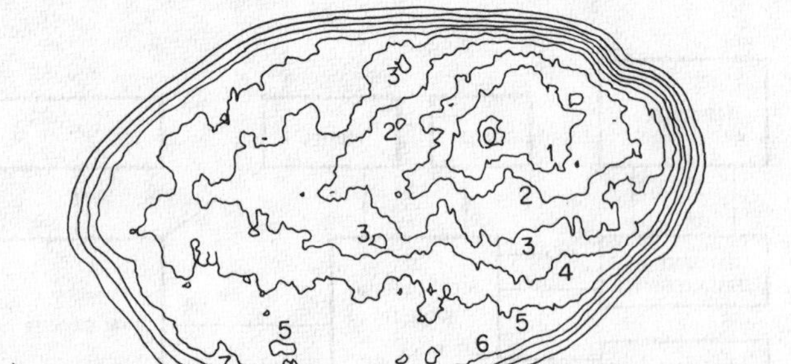

Figure 3. Visualization of the inhomogeneity of the flux of ions (∅ = 4 mm) by contour line representation. The lines separation corresponds to 1 μm. The 0 indexed line is the deepest. α–brass (C_{Zn} = 24 %) sputtered during 2.16 h by 15 keV Ar^+ ions.

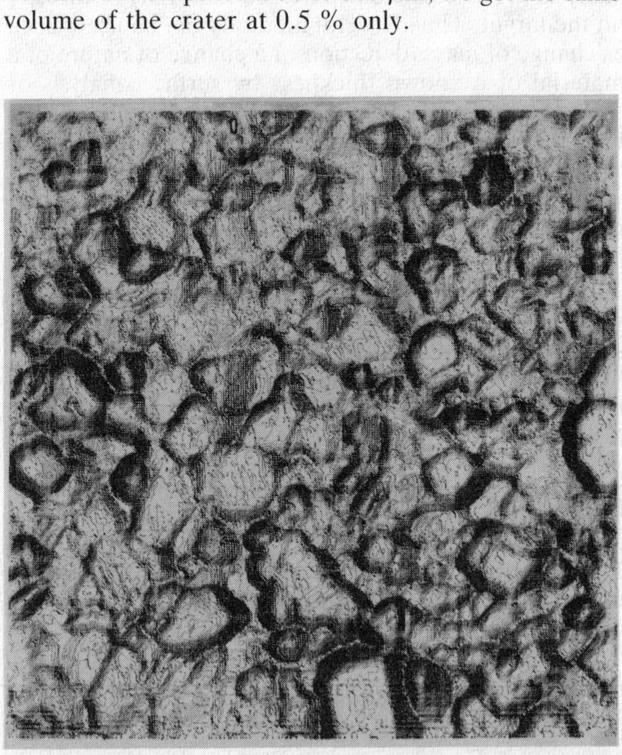

Figure 4. Contour line representation with higher magnification on the bottom of the crater shown in Fig. 3, (256 μm × 256 μm) gives some details on the surface roughness. Because of the dependence of sputtering yield with respect to the crystalline orientation, the grains appear. The lines separation corresponds to 0.1 μm.

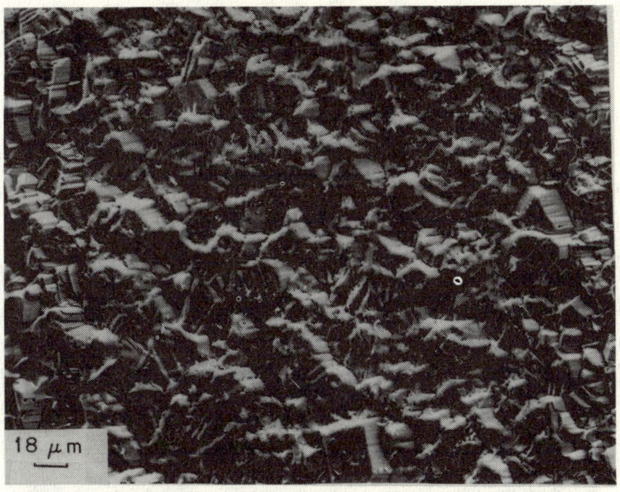

Figure 5. A secondary electron micrograph. Same area as in Figure 4. There is a good agreement for the shape and the dimensions of the grains.

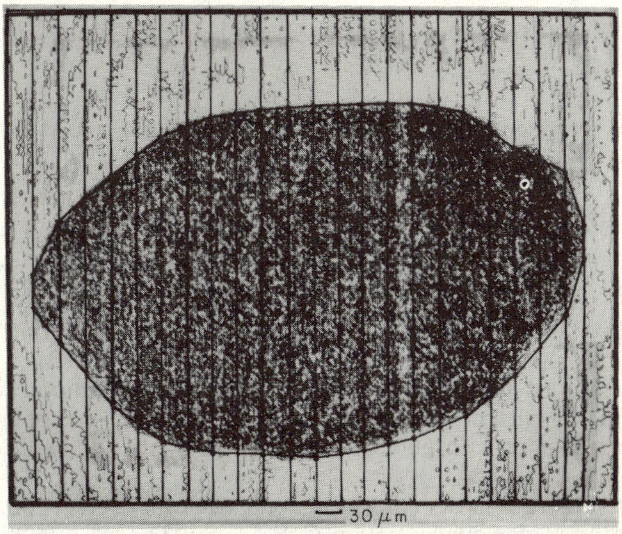

Figure 6. Computation of the sputtered volume on a cylindrical surface. The reference plane is replaced by a series of planar strips parallel to the axis of the cylinder and equal in width (here: 300 μm).

DETERMINING THE SPUTTERING YIELD OF COPPER

Ultra purified polycrystalline copper was sputtered by Ar^+ ions in three different systems. Each sample was polished with progressively finer abrasives, down to diamond paste (1 μm). One of them underwent a further electrolytic polishing in a phosphoric acid bath.

Table 1 summarizes the results we obtained. The yields obtained with a positive incidence angle were revaluated for a zero incidence angle to the formula

$$S_T(0) = S_T(\theta) \cdot \cos\theta$$

Notice that the yield increases with the energy of the ions, and that the sample polishing method has no influence.

Other than that obtained directly under normal influence, the values we get are lower than those published : at 3 keV, $S_T(0) = 5$ according to Ref. 7, at 15 keV, $S_T(0) = 6$ or 6.5 or 7.3 according to Refs. 8, 9 or 10. Although the relative errors for these results are not given, let us try to evaluate ours.

Relative errors on the measurement of volume come from the curvature of the diamond stylus (< 1 %), resolutions in all three spatial directors and the chosen sampling step (< 2 %) and the determination of the reference plane (< 2 %) thanks to the method previously described). Although, this amounts therefore to less than 5 %. There remains only the relative error due to the intensity and stability of the ion current. The values given in Table 1 correspond to applying a zero voltage to the double walls of the Faraday cage. They are systematically underestimated, which can explain our values lower than these given in the literature. On the other hand, stability of the beam is better than 4 %.

We can only conclude that the most important relative error is probably due to the ion current measurement.

SPUTTERING OF α–BRASS

The SM2 device was used for a comparative study of copper and pure α–brass samples containing 5, 15, 20, 24, 30 and 36 wt % of zinc and pure zinc.

All the samples were polycrystalline, mechanically polished down to the 1 μm diamond paste, and bombarded by a 15 keV Ar^+ ion beam at $\theta = 60°$. The

Table 1.

System	STIMS	SM4			SM2
Polishing	Diamond	Diamond		Electrolytic	Diamond
Ion energy (keV)	3.0	15.0	15.0	15.0	15.0
Ion current (μA)	8.69	3.75	3.86	3.76	11.80
θ incidence angle of the ion beam (°)	45	0	60	60	60
Duration (h)	2.00	3.00	4.00	3.66	3.00
Volume of the crater (mm³)	0.0238	0.0189	0.0408	0.0363	0.0921
$S_T(\theta)$	5.37	6.01	9.86	9.83	9.79
$S_T(0)$	3.8	6.0	4.9	4.9	4.9

Table 2.

Sample	Cu	α-brass						Zn
		5 %	15 %	20 %	24 %	30 %	36 %	
Ion current (μA)	11.8	11.8	11.8	13.1	12.5	12.6	13.0	13.0
Duration (h)	3	2	2	2	2	2	2	2
Crater volume	0. 092	0.0589	0.0654	0.0755	0.0759	0.0880	0.0907	0.371
S_T (60)	9.80	9.2	10.1	10.1	10.8	12.3	11.7	42.4

pression measured in the chamber was 10^{-7} during irradiation. The results are given in Table 2.

Let us notice that for pure zinc, the normalization of the results at $\theta = 0°$ gives $S_T(0) = 20.7$, in good agreement with published data ($S_T(0) = 21$ following Ref. 11).

Figure 7 shows a linear dependance of S_T versus the atomic zinc concentration C_{Zn}, $S_T = 6.37\ C_{Zn} + 9.38$ i.e. $S_T = 9.38\ (1 + 0.68\ C_{Zn})$ to be compared to : $S_T = S_{Cu}\ (1 + 0.64\ C_{Zn})$ as predicted by Darque-Ceretti *et al.*[12]

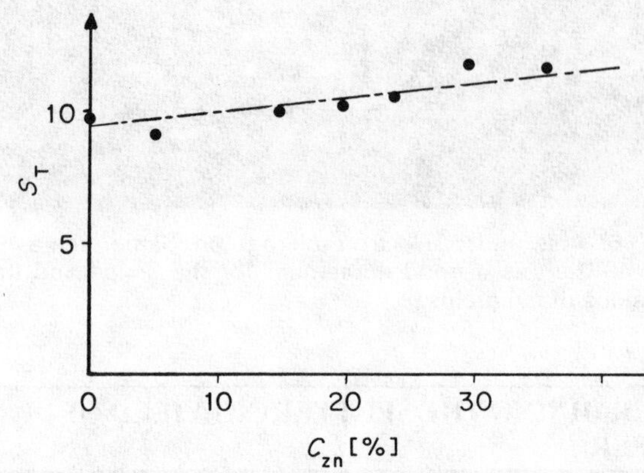

Figure 7. Variation of the sputtering yield for the α-brass versus the zinc concentration. Good agreement between theory and experiments.

CONCLUSIONS

Using a 3D stylus surface mapping device leads to a good understanding of the microtopography of a surface sputtered by an ion beam. Studying the whole crater makes it possible to characterize the homogeneity of the flux of ions coming from the gun, and to better understand the origin of the ions extracted from the surface.

On a finer scale, the microtopography of the crater bottoms is very well described by this type of technique, which yields standardized images which, as opposed to those coming from a scanning electron microscope, can be interpreted quantitatively. As for measuring sputtering yields, the method is particularly attractive for its very simplicity. It nevertheless requires a suitable accuracy of the measurements of the crater volume and ion current. This is what we are currently putting the emphasis on in our work.

Acknowledgements

For the use of mass spectrometer facilities, we are greatly indebted to Dr Devienne (Lab. de Physique Moléculaire des Hautes Energies, Peymeinade, SM2 and SM4) and Prof. Blaise (Lab. Physique du Solide, Orsay, STIMS).

REFERENCES

1. R. Behrish (ed), *Sputtering by particle bombardment I*, Topics in Applied Physics Vol. 47, Springer (1981).
2. G. Betz, *Surf, Sci.* **92**, 283 (1980).
3. M. Szymonski, *Appl. Phys.* **23**, 89 (1980).
4. F. Schulz, K. Wittmaack, J. Maul, *Radiat. Effect* **18**, 211 (1973).
5. T. Kimura, J. Kobayashi, S. Okuda, H. Akimune, *Jpn. J. Appl. Phys.* **15**, 2479 (1976).
6. J. C. Quantin, *Application de la rugosimétrie tridimensionnelle à l'étude des surfaces*, Thesis, Ecole des Mines de Paris, (1986).
7. A. Southern, W. R. Willis, M. T. Robinson, *J. Appl. Phys.* **34**, 153 (1963).
8. C. Fert, N. Colombie, B. Fagot, Phan Van Chuong, in *Le bombardement ionique*, CNRS (1961).
9. O. Almen, G. Bruce, *Nucl. Instrum, Methods,* **11**, 257 and 279 (1961).
10. C. Abatino, G. Luzzi, L. Papagno, *Thin Solid Films*, **56**, 291 (1979).
11. K. H. Krebs, in *Atomic and Molecular data for fusion*, p. 185 IAEA, Wien (1977).
12. E. Darque-Ceretti, F. Delamare, G. Blaise, *Surf. Interface Anal.* **7**, 141 (1985).

The Backscattering Contribution to Auger Line Intensity

L. Meda and G. Queirolo
SGS-Microelettronica, 20041 Agrate Brianza (MI), Italy

Quantitative Auger analysis can be performed with both the derivative $E\,dN(E)/E$ or the direct $E\,N(E)$ electron distribution, using the so-called elemental sensitivity factors, as reported, for instance, by Davis[1] or given by the spectrometer suppliers. In a multicomponent system the elemental sensitivity factors can change, due to a number of matrix effects, like: variations in the atomic density and in the inelastic mean free path for the Auger electrons which change the number of analysed atoms; variations in the backscattering coefficient, which change the effective primary current for the generation of the Auger electrons. The knowledge of these effects is necessary in order to perform reliable quantitative analysis and to correctly interpreting the variations in the Auger current taking place in a multilayered sample.

In this work we perform depth profiles[2] in sandwiched layers to separate the effect of the variations in the backscattering coefficient from the other matrix effect; the values obtained for R are compared with those computed with the available Monte Carlo computations.

The samples studied were Si/Me/Si multilayers deposited on an oxidized silicon substrate; the metal Me can be Ge, Mo, Ta. The thickness of the metal and of the amorphous silicon layers was of the order of magnitude of the range of the primary electrons.

When these samples are sputtering depth profiled, the Si signal in the outer Si layer is first constant, but increases steadily as the Si/Me interface is approached. The distance from the interface at which the effect is first noticed is approximately equal to $L/2$ $(1/\sin\theta)$, where L is the range of the primary electrons in silicon, and θ the incidence angle. This increase is higher, the greater is the atomic number of the metal Me; it is caused by the increase in the effective primary current, due to the increase in the current of electrons backscattered from the high Z layer. Very near to the Si/Me interface, i.e. at a distance of the order of the escape depth of the Auger electrons considered, the Auger current reaches its maximum value. This value, and that 'far' from the interface (i.e. at a distance greater than $L/2$ $(1/\sin\theta)$) contains information on the values of the backscattering coefficients for the Auger peak used, in the two matrix. The ratio of the two signal amplitudes will give the ratio of the corresponding backscattering coefficients. The measured values agree very well with those obtained by Shimizu[3] using Monte Carlo methods.

We can obtain the same evaluation of the backscattering factors using another, less time consuming method. We can in fact stop the ion etching as soon as the Me is reached, and then move the sample under the beam, while measuring the Auger signal for the chosen peak on the crater slope. The results compare well with those obtained with the previous method; in this case it is nevertheless possible to evaluate the backscattering factors ratio for different primary energy with only one sputtering experiment.

REFERENCES

1. L. E. Davis, N. C. McDonald, P. W. Palmberg, G E. Riach and R. E. Weber, *Handbook of Auger Electron Spectroscopy*, Physical Electronics Industries, Eden Prairies, MN, (1976).
2. L. Meda and G. Queirolo, *Surf. Sci.* **147**, 576, (1984).
3. R. Shimizu, *Japan. J. Appl. Phys.* **2**, 1631 (1983).

Elastic and Inelastic Scattering of Electrons Emitted from Solids: Effects on Energy Spectra and Depth Profiling in XPS/AES

A. L. Tofterup, S. Tougaard, and P. Sigmund
Fysisk Institut, Odense Universitet, DK-5230 Odense M, Denmark

OBJECTIVES

The energy spectrum of electrons emitted from a solid surface depends on the depth of the source. Therefore, the intensity and shape of photoelectron and Auger lines contain information about depth profiles. Photoelectron and Auger lines split and broaden by inelastic scattering of electrons travelling from the point of origin to the surface. The number of inelastic scatterings increases with increasing pathlength. The path-length is larger than the depth of origin because of angular deflection mainly by elastic scattering.

The possibility of extracting in-depth composition information through an analysis of XPS lineshapes was previously studied experimentally.[1]

The present work contributes to the theoretical foundation of depth profiling by lineshape analysis. It is based on a general outline given by two of us.[2] A detailed account is being published.[3]

CALCULATIONAL METHOD

The calculational scheme is based on decoupling angular deflection, caused mainly by elastic scattering, from inelastic events, mainly plasmon excitation.[2] The main effort is devoted to an adequate description of the pathlength distribution determined by elastic scattering. In analogy to earlier work on reflected-electron spectra,[4] the P_i approximation of transport theory is utilized to solve the fundamental transport equation.[2] Landau's formula is employed to evaluate energy spectra versus path length for multiple inelastic scattering events.[2] Spectra are evaluated for simple models of the source distribution.

RESULTS AND CONCLUSIONS

Pathlength distributions of emerging electrons as determined by elastic scattering turn out to be close to exponential both for well-defined source depth and for spatially homogeneous sources. The P_1 approximation is asymptotically exact at large pathlengths and yields a 7 pct discrepancy compared with the exact result in the limit of zero pathlength. The latter can be reduced to 2 pct by going to the P_3 approximation.

The intensity of the elastic peak is attenuated approximately exponentially with a characteristic length Λ given by

$$\Lambda = \frac{\lambda}{[3(1+\lambda/\lambda_1)]^{1/2}}, \tag{1}$$

where λ is the inelastic mean free path and λ_1 the transport mean free path for elastic scattering. (1) approaches the result given by the diffusion approximation[2] in the limit of $\lambda_1 \ll \lambda$, i.e., pronounced elastic scattering.

The intensity ratio between the elastic peak and the associated first plasmon peak depends sensitively on the depth of the source. The same holds for ratios involving higher plasmon peaks. Extraction of depth information from peak ratios implies some knowledge of λ and λ_1. Neglecting angular scattering[5,6] is a good starting point for qualitative considerations but insufficient for detailed estimates.

ACKNOWLEDGEMENT

This work has been supported by the Ingeborg and Leo Dannin Foundation for Scientific Research.

REFERENCES

1. S. Tougaard and A. Ignatiev, *Surf. Sci.* **129,** 355 (1983).
2. S. Tougaard and P. Sigmund, *Phys. Rev. B.* **25,** 4452 (1982).
3. A. L. Tofterup, *Surf. Sci.* **167**, 70 (1986).
4. A. L. Tofterup, *Phys. Rev. B.* **32,** 2808 (1985).
5. S. Tougaard, *Surf. Interface Anal.* **9**, (1986).
6. S. Tougaard, *Surf. Sci.* **162**, 875 (1985).

Positive Secondary Ion Yields of Elements in Amorphous and Crystalline Silicon

J. Herion, W. Beyer, J.-R. Chen,* B. Stritzker and H. Wagner**
Institut für Grenzflächenforschung und Vakuumphysik, Kernforschungsanlage Jülich, Postfach 1913, D-5170 Jülich, FRG

Despite the great technological importance of crystalline and hydrogenated amorphous silicon (*c*–Si and *a*–Si : H) there is little information on the secondary ion yields of foreign elements in these materials. Besides the practical interest in quantitative analyses by SIMS a comprehensive compilation of SI yields could provide insight into the physical processes of SI generation which are not yet understood very well.

While SIMS investigations are relatively straightforward in the case of moderately doped crystalline silicon, charging effects may influence the SI yields of *a*–Si : H which is generally much more resisitive than *c*–Si. The problem can be often overcome by illuminating the samples with white light e.g. from a tungsten–halogen lamp. The resulting photoconductivity depends on the concentration of defects, however, and may vary with depth e.g. in a sample which has been subjected to ion implantation.

Charging effects were monitored by measuring the ion yield as a function of the target potential. Substantial changes of this dependence were observed during depth-profiling of *a*–Si:H samples prepared by the glow discharge method and containing implanted C1 atoms.

With the instrument used in this work (based on ATOMIKA components) the ion yields depend strongly on the geometry of the sample and on the exact position of the sample holder. To correct for these effects, after positioning a new sample the potentials influencing the ion collection have to be readjusted. We found that these adjustments did not markedly change the ratios of the ion yields: four samples of *c*–Si containing known amounts of implanted As gave relative sensitivity factors for As^+ ions which agreed within 20%. (Throughout this work, 9 keV O_2^+ primary ions at normal incidence were used. The relative sensitivity factors were obtained by integrating the depth profiles of the implanted elements with respect to depth and normalizing to the $^{16}O^+$ intensity. The erosion rates were determined by measurements of the crater depths by a SLOAN profilometer.)

Ion yields were also measured for various samples of *c*–Si (100 oriented, ~ 10 Ω cm) and *a*–Si : H (prepared by glow discharge deposition) containing known amounts of implanted Na, Cr, Fe, Ti, As, P, and C. Implantation energies were between 100 and 300 keV, and ion doses between 0.1 and 2 mC cm^{-2}. Furthermore, silicon wafers containing 7×10^{18} B atoms cm^{-3} as well as *a*–Si : F and *a*–Si : D films containing 7% F and 10% D, respectively, were analysed.

The relative sensitivity factors of the various elements are shown in Fig. 1. On the horizontal axis the values of $I_{eff} = E_{ox}^{1/3}\, I^{2/3}$ are given where E_{ox} is the binding energy between oxygen and the element under consideration and I is the ionization energy (both energies are taken from Ref. 1). It is interesting to note that the data fit a universal curve

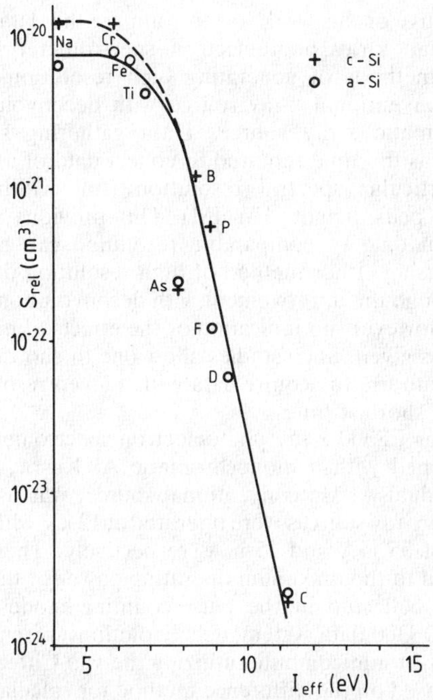

Figure 1. Relative sensitivity factor vs. 'effective' ionization energy.

much better if I_{eff} is used rather than I. As is the only exception. The authors are not aware of any theoretical model that could explain the observed dependence on I_{eff}. The results may qualitatively support the hypothesis that the secondary ions X^+ are produced by dissociation of excited molecules of the type O–X.

Acknowledgements

The authors acknowledge the skilful assistance of F. Birmans in the SIMS measurement and they are grateful to M. Gebauer for performing the ion implantations and to A. Marino for providing the *a*–Si : F samples.

REFERENCE

1. R. C. Weast (ed.), *CRC Handbook of Chemistry and Physics*, CRC Press, Inc. (1978).

* Permanent address: Department of Materials Science and Engineering, National Tsing Hua University, 855 Kuang Fu Road, Hsinchu, Taiwan 300, Republic of China.
** Institut für Festkörperforschung, Kernforschungsanlage Jülich, Postfach 1913, D-5170 Jülich, FRG

Monochromator Versus Deconvolution in XPS: Signal-to-Noise Considerations†

M. F. Koenig and J. T. Grant
Research Institute, University of Dayton, Dayton, Ohio 45469 USA

The objective of this work was to compare the data gathering speed of an x-ray photoelectron spectrometer using two different methods of generating high resolution data, i.e. using a conventional x-ray source with deconvolution or a monochromatic x-ray source. Data gathering speed was measured as the time required to collect data of a given *S*/*N* for a particular spectral resolution (full width at half-maximum peak height, FWHM). Our previous work[1] has shown that data of comparable resolution can be reliably achieved using either method of high resolution data collection, although the improvement with deconvolution is limited to 30%. However, no indication of the effects of noise in the spectra was given. Such studies allow one to choose which x-ray source to use to acquire data with a given resolution and *S*/*N* in the shortest time.

A Kratos ES300 x-ray photoelectron spectrometer, which was equipped with a monochromatic Al Kα-ray source as well as a dual Al/Mg conventional source, was used in this study. The x-ray sources were operated at 12 kV with emission currents of 35 mA and 25 mA, respectively. These settings correspond to the maximum operating power of the sources. Data was collected in the pulse-counting mode using the standard DS300 data system; deconvolutions were performed on the LSI-11 minicomputer utilizing the van Cittert iterative algorith based on the difference method for calculation.

A variety of specimens were used to determine the noise characteristics of the spectrometer: WO_3, MoS_2, Au, and a graphite crystal. Noise measurements were made at three different kinetic energies (*KE*) on each specimen where the backgrounds were of different intensities and relatively flat. Each noise measurement consisted of 50 data points collected at one particular *KE*; the standard deviation (S_n), mean, and peak-to-peak (ptp) noise were then calculated. At least two measurements were performed at three different dwell times at each kinetic energy. A total of 84 noise measurements were made on the four different specimens. The values of $S_n/\text{mean}^{1/2}$ and ptp/S_n were found to be 1.0 ± 0.2 and 4.6 ± 0.6, respectively, with the errors given as standard deviations. These are the values expected for Poisson-distributed noise from the counting of random events.

The noise increase due to deconvolution was characterized by storing the intermediate results (i.e. the results after each iteration) up to and including five iterations. The signal, background, and ptp noise for each spectrum were then measured and compared. Five iterations result in a noise of six times the starting value,so the total accumulated counts must be increased by a factor of 36 to retain the original *S*/*N* value.

Incorporating noise measurements at the peak maximum and background (*B*) on its high kinetic energy side leads to equations relating *S*/*N*, *S*/*B*, and $S + B$ (the total count rate). These noise relationships were used to calculate the data acquisition times needed to acquire data of the same *S*/*N* using monochromatic Al Kα or conventional Mg Kα x-ray sources. It was found that if deconvolution is not necessary to achieve the required resolution, the conventional source is about an order of magnitude faster than the monochromatic source. If deconvolution is needed with the conventional source, it was calculated to be superior when $S/B > 5$, but when $S/B < 1$ the monochromatic source should be superior. This was examined experimentally using S 2p peaks from MoS_2 and Ni 2p peaks from oxidized stainless steel.

These results are totally instrument-dependent since there are many variables which may affect count rate and background levels in XPS spectra. The methods used in this analysis, however, are completely independent of instrumentation. If the only source of noise is that due to counting statistics, the graphs and equations can be used without change for any spectrometer. A more complete presentation of these results can be found in Refs 2 and 3.

REFERENCES

1. M. F. Koenig and J. T. Grant, *J. Electron. Spectrosc. Relat. Phenom.* **33,** 9 (1984).
2. M.F. Koenig and J. T. Grant, *J. Electron. Spectrosc. Relat. Phenom.* **36,** 213 (1985).
3. M. F. Koenig and J. T. Grant, *Surf. Interface Anal.* **7,** 217 (1985).

† This work was sponsored by the Air Force Wright Aeronautical Laboratories, Materials Laboratory, Air Force Systems Command, United States Air Force, Wright-Patterson Air Force Base, Ohio 45433, USA.

Electromigration and Charging Effects During Auger and XPS Analysis of Insulators

J. Cazaux
Laboratoire de Spectroscopie des électrons, UER des Sciences 51062, Reims Cédex, France

The aim of the present paper is to elucidate some conflicting results obtained during Auger analysis of insulators such as glasses. These results concern the decay of the signal from ions of opposite charges, Na^+ [1–4] and Cl^- [1] while the signal of ions of the same charge, K^+ and Ca^+ is sometimes found to increase.[2,4] With slight modifications the results can also be applied to XPS.

To reach this goal, the components of the electric field and the electric potential induced by the trapped charges are evaluated by using simple electrostatic arguments but taking into account image effects. For a semi-infinite sample (dielectric constant ϵ) bounded by a vacuum (ϵ_o) bombarded by a stationary beam at normal incidence, it is assumed that the positive charge distribution ρ_+ due to the emission δ of secondary electrons is uniform inside a cylinder of radius a (incident beam radius) and height d_s, while the non-backscattered electrons $(1 - \eta)$ with density ρ_- are uniformly distributed inside a cylinder of radius a and height d_p.

From classical textbooks of electrostatics,[5] it can be established that the axial, $F_z(o, z)$, and radial $F_r(r, o)$, components of the electric field inside the positively charged layer ($o < z < d_s$; $o < r < a$) are given by (defocusing conditions: $a > > d_s, d_p$):

$$F_z(0\ z) \simeq \frac{\rho_+ + \rho_-}{\epsilon} z - \frac{\epsilon_0}{\epsilon a} V(0,0) \tag{1}$$

$$F_r(r,0) \simeq \frac{2V(0,0)}{\Pi r} \{J_1 (\frac{r}{a}) - J_2 (\frac{r}{a})\} \tag{2}$$

$$V(r,0) \simeq \frac{2}{\Pi} V(0,0)\, J_2 (\frac{r}{a}) \tag{3}$$

where J_1 and J_2 are elliptic integrals and $V(0,0) \simeq$

$$\frac{\rho_+ d_s + \rho_- d_p}{\epsilon + \epsilon_0} a$$

When the sample in the form of a foil, (thickness $d_f < d_s$) is bounded by a vacuum on one side ($K = (\epsilon - \epsilon_o)/(\epsilon - \epsilon_o)$) and a substrate ($\epsilon_1$) on the other side ($K_1 = (\epsilon - \epsilon_1)/(\epsilon + \epsilon_1)$, $F_z(o, z)$ is given by:[6]

$$F_z\ (0,z) \simeq \frac{\rho_+ + \rho_-}{2\epsilon} 2z + (\frac{K - K_1}{1 - KK_1} - 1)\, d_f \tag{4}$$

where it is assumed that there are no trapped charges in the substrate because of their weak time constant, τ_1 ($\tau_1 = \epsilon_1/\gamma_1$; γ_1: substrate electrical conductivity). For metallic substrates $\epsilon_l = \infty$ and $K_1 = -1$.

The above expressions hold as well for XPS (by setting, obviously, $\rho_- = 0$) as for AES (whether or not the minority charges, ρ_-, recombine quickly).

The most important results are:

1. When the surface potential, $V(o, o)$ is positive ($\delta > 1 - \eta$), the F_z component always changes sign inside the positive surface layer and close to the surface, giving rise to opposite directions of migration for ions of a given charge leading to a depletion (if +) or an accumulation (if −) region around the inversion point P_o. The position, d_o, of this inversion point is obtained by setting $F_z(o, d_o) = 0$ in (1) and (4):

$$d_0 = \frac{\epsilon_0\ V(0,0)}{a(\rho_+ + \rho_-)} \tag{5}$$

$$d_0 = (1 + \frac{K_1 - K}{1 - K_1 K}) \frac{d_f}{2} \tag{6}$$

The d_o depends on the electrical parameters values of the sample (ϵ, γ) but its range, and the information depth , δ_0,($\delta_o = \lambda\cos\theta$) of Auger (and XPS) electrons have the same order of magnitude (δ_0, $d_0 < d_s/(\epsilon_r + 1) < 10$ nm).

2. Relative to the Auger signal (S_o) before electromigration, the asymptotic value of the Auger signal after electromigration along the z axis (but before electron stimulated desorption), S_1, is given by:

$$S_1(+)/S_0 \simeq d_0/\delta_0 \text{ (for + ions)};\ S_1(-)/S_0 \simeq \frac{d_1}{\delta_0}\ e^{-d_0/\delta_0} \text{ (for −ions)}.$$

(d_1 = position of a 2nd inversion point — see Fig. 1). Consequently the Auger (and XPS) signals of positive ions are expected to increase (decrease) when $d_0 > \lambda\cos\theta$ ($d_0 < \lambda\cos\theta$). This analysis may explain the increase of the K and Ca signals:[2,4] $E_k(K) = 252$ eV; $E_k(Ca) = 294$ eV, and the decrease of the Na signal, $E_k = 990$ eV, if d_o obeys to (see insert): $\delta_1(K)$, $\delta_1(Ca) < d_o < \delta_2(Na)$. But these increases can sometimes be counterbalanced by the electron-stimulated-desorption of the species (the decrease of the K signal in Ref. 3?).

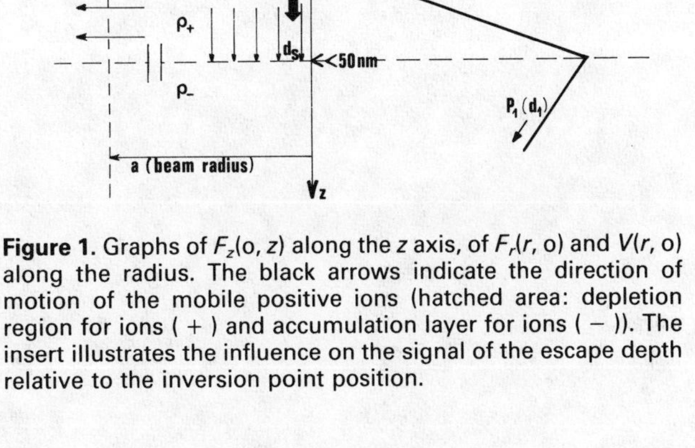

Figure 1. Graphs of $F_z(o, z)$ along the z axis, of $F_r(r, o)$ and $V(r, o)$ along the radius. The black arrows indicate the direction of motion of the mobile positive ions (hatched area: depletion region for ions (+) and accumulation layer for ions (−)). The insert illustrates the influence on the signal of the escape depth relative to the inversion point position.

3. The radial component of the electric field $F_r(o, o)$ takes a maximum value at $r = a$ where its strength is greater than $F_z(o, o)$. Thus a radial migration of ions is expected (around $r = a$) but the Auger and XPS response functions are less sensitive to the radial than to the axial motion of ions (for comparable paths). Consequently when the incident beam is stationary, the electron (or photon) stimulated desorption concerns only a thin region of d_o around the point 0(o, o).

4. The axial change, $V(o, z)$ of the electrical potential can be neglected inside the information depth of electrons of interest (because of the weakness of δ_{ij}). Its radial change (when going from $r = o$ to $r = a$) can be very important:

$V(a,0) = \frac{2}{\Pi}V(0,0)$ and can cause shift and broadening of the Auger and XPS lines.

5. When the defocusing conditions are applied, $F_z(o, z)$ is proportional to the charge densities which are in turn, proportional to J_o (Current density), and is independent of the spot radius a. But despite this fact, these electric effects cannot be be defined by a critical dose because of the influence of $F_r(o, r)$ and overall due to the non-linear variation of ρ_+ as a function of time (influence of $V(o, o)$ on the recombination rate of the majority charges, ρ_+).

6. Except the specific values of δ and d_s for photons, the electromigration and charging effects of XPS are expected to be quite the same as for AES (when $\delta > 1 - \eta$) if the incident particles' densities and spot sizes are assumed to be the same (for instance in high resolution XPS). These effects are reduced when a increases (J_o and ρ decrease) while the incident beam intensity remains constant.

7. Further, for more accurate evaluations the need is to indicate the electrical parameters' values (ϵ and γ) of the investigated samples.

REFERENCES

1. C. G. Pantano and T. E. Madey, *Appl. Surf. Sci*, **7**, 115 (1981).
2. P. T. Dawson, O. S. Heavens and A. M. Pollard, *J. Phys.* **11**, 2183 (1978).
3. R. G. Gossink, H. Van Doveren and J. A. T. Verhoeven. *J. Non-Cryst. Sol.* **37**, 111 (1980).
4. M. Sfar, J. P. Lacharme, P. Lepeut and P. Champion. *J. Phys. C2*, **45**, 375 (1984).
5. E. Durand in *Electrostatique: I les distributions*, Masson Edition, Paris (1964).
6. J. Cazaux, *J. Appl. Physics*, **59**, 1418 (1986).

IN-DEPTH COMPOSITION INFORMATION FROM THE INELASTIC BACKGROUND SIGNAL IN XPS

S. Tougaard
Fysisk Institut, Odense Universitet, DK–5230 Odense M, Denmark

Depth distributions of elements near solid surfaces are determined through a time consuming procedure. Alternative ways of obtaining depth composition profiles are therefore of current high interest.

Based on experimental data a new non-destructive and extremely fast method was previously proposed.[1] Here a quantity $D = A_p/B$ which is essentially the ratio of the XPS peak area to the increase in background signal associated with the peak was defined.

In the present work D is studied theoretically for homogeneous as well as inhomogeneous transition metals. For homogeneous metals D is found to be a weak function of both the XPS peak energy and the metal. This is in agreement with experimental results.[1] For inhomogeneous metals D is independent of the element concentration as well as of peak shape. Furthermore D is a strong function of the in depth composition and for a given composition only a weak function of the host metal.

With these properties D is potentially a good parameter for fast non-destructive extraction of information on depth composition.

There are however two major problems in the practical use of D. Thus it is shown that D, for inhomogeneous samples, is a strong, but well-behaved, function of the XPS peak energy and that the depth composition is not a unique function of D.

The theoretical reason for the above mentioned properties of D and the limitations in using D for depth profiling will be discussed.

Reference

1. S. Tougaard and A. Ignatiev, *Surf. Sci.* **129** 355 (1983)

INELASTIC BACKGROUND REMOVAL IN XPS

S. Tougaard
Fysisk Institut, Odense Universitet, DK–5230 Odense M, Denmark

The inelastic background signal in XPS spectra constitutes a major problem in quantitative surface analysis. Theoretical formulas for background subtraction were previously given.[1,2] Central in the formulas is a detailed knowledge of the cross section for inelastic electron scattering.

In the present work, experimental spectra from Al, Cu, Ag and Au are analysed on this basis. Cross sections are evaluated within dielectric response theory.

For all peaks studied, the primary spectrum after deconvolution of inelastic background signal exhibits a tail on the low energy side which extends ~ 50 eV below the peak energy.

In the use of XPS as a routine tool, the time spent on analysis is an important factor. Therefore it is highly important to look for approximative methods for background subtraction. Such a method should be simple, of reasonable general validity, and yet give a fair description of the physical situation.

The inelastic scattering cross section varies considerably both with primary energy and metal. However it is found that the product of inelastic mean free path and cross section to a good approximation is a 'universal function' for transition metals at all energies of relevance in XPS.

Since this product is the only determining function in the deconvolution formulas,[1,2] a general and very simple method for background subtraction can be set up for this class of metals. This simple method is discussed in relation to the result of alternative more elaborate methods.

References

1. S. Tougaard and P. Sigmund, *Phys. Rev. B* **25,** 4452 (1982).
2. S. Tougaard, *Surf. Sci.* **139,** 208 (1984).

CHARACTERIZATION OF NBS STANDARD REFERENCE MATERIAL 2135 FOR SPUTTER DEPTH PROFILE ANALYSIS

Joseph Fine
National Bureau of Standards, Gaithersburg, MD 20899, USA
Boris Navinšek
Jozef Stefan Institute, Ljubljana, Yugoslavia

A Ni/Cr multilayered thin-film standard reference material (SRM) for sputter depth profile calibration has been developed jointly by the National Bureau of Standards, the Jozef Stefan Institute, and the E–42 Surface Analysis Committee of the American Society for Testing and Materials. This periodically modulated structure consists of 5 layers of Cr (each about 53 nm) and 4 layers of Ni (each about 66 nm) which have been sputter deposited onto smooth silicon substrates. This material can be effectively used to calibrate sputter erosion rates and depth of erosion scales in surface analysis as well as to monitor ion beam stability and to optimize sputtering conditions so as to achieve maximum interface resolution. Characterization results obtained on this first SRM for surface analysis to be issued by NBS indicate that the accuracy of its structure is known to better than 5% and that its sputter profiles are well-defined and reproducible.

Results of the calibration and compositional analysis of this SRM are presented regarding uniformity and periodicity of thin film layers, absolute film thickness, sputtered interface depth-resolution, and relative Ni/Cr sputtering rates and yields. Measurement methods used to characterize this thin-film structure include *EN(E)* Auger sputter depth profiling, Rutherford backscattering spectrometry, and neutron activation analysis.

ON THE MATERIAL DEPENDENCE OF THE INELASTIC MEAN FREE PATH IN ELECTRON SPECTROSCOPY

K. Berresheim and M. Wilmers
Leybold-Heraeus GmbH, Bonner Str 498, 5000 Cologne 51, W. Germany

Electron spectroscopy is used for quantitative chemical analysis as well as for non destructive determination of the thickness of surface layers. For XPS the quantification problem has been solved by supplying the analyst with sufficiently accurate sensitivity factors.[1-4] The determination of the thickness of a thin layer by electron spectroscopy needs the knowledge of the inelastic mean free path (IMFP). Whilst the energy dependence of the IMFP seems to be approximated by a simple equation,[5-7] little is known with respect to its material dependence. This is shown by the large scatter for IMFP values at a fixed energy.[7]

In this work we show a way to determine the IMFP without thin layer measurements. Solid sample measurements together with one single thickness standard are sufficient for all materials examined so far. Plane samples of compounds containing F, Na, Cl and O were prepared and measured under UHV conditions. The results are compared with the predictions of Powell,[6] Penn[8] and Seah.[7]

Reference

1. H. Berthou and C.K. Jørgensen, *Anal. Chem.* **47**, 482 (1975).
2. S. Evans, R.G. Pritchard and J.M. Thomas, *J. Electron Spectrosc.* **14**, 341 (1978).
3. C.D. Wagner, L.E. Davis, M.V. Zeller, J.A. Taylor, R.H. Raymond and L.H. Gale, *Surf. Interface Anal.* **3**, 211 (1981).
4. K. Berresheim, *Surf. Interface Anal.* (to be published).
5. I. Lindau and W.E. Spicer, *J. Electron Spectrosc.* **3**, 409 (1974).
6. C.J. Powell, *Surf. Sci.* **44**, 29 (1974).
7. M.P. Seah and W.A. Dench, *Surf. Interface Anal.* **1**, 2 (1979).
8. D.R. Penn, *J. Electron Spectrosc.* **9**, 29 (1976).

DETERIORATION OF DEPTH RESOLUTION IN SPUTTER DEPTH PROFILING BY RASTER SCANNING AN ION BEAM AT OBLIQUE INCIDENCE AND CONSTANT SLEW RATE

U. Kaiser, R. Jede, P. Sander, J.H. Schmidt, O. Ganschow and A. Benninghoven

Physikalisches Institut der Universität Münster, Domagkstr. 75, D-4400 Münster, W. Germany

Raster scanning of the primary ion beam is used in sputter depth profiling for improving depth resolution by elimination of crater wall effects. In most thin film analysis instruments, in particular in SIMS equipment, the primary ion beam includes a non-zero angle with the target normal. When scanning with constant voltage slew rate applied to the deflection capacitor, the ion fluence will thus decrease with the beam moving away from the gun because of widening of the beam projection and increasing the projected scan rate. As a consequence, in typical geometries the eroded depth varies substantially over the sample. The corresponding contribution Δz to depth resolution in a concentration profile varies linearly with the removed layer thickness. The effect decreases with increasing deflector-to-sample distance and vanishes exactly at normal incidence. We present numerical values for practical situations together with a method for electronic compensation. The resulting improvement is shown for combined SIMS-AES-XPS depth profiles of anodic oxides on $Cd_xHg_{1-x}Te$ and on Ta.

OPERATION OF SCANNING TUNNELLING MICROSCOPY AT AIR AMBIENT PRESSURE TO STUDY TECHNOLOGICALLY RELEVANT MATERIALS

A.M. Baró, N. García and R. Miranda

Depto. de Fisica Fundamental, C-III, Universidad Antónoma de Madrid, Cantoblanco, 28049-Madrid, Spain

We have been operating the scanning tunnelling microscope (STM) [1] at air ambient pressure. In these conditions we obtain a resolution which is better than 10 Å in the lateral and better than 1 Å in the vertical dimensions. This allows to study the microtopography of surfaces in real conditions. Applications of STM in metallurgy, electronics, surface roughness, growth phenomena, industrial finishing, biomaterials, etc., will be shown. This makes STM a very interesting technique for applied surface studies on industrially relevant materials.

Reference

1. G. Binning, H. Rohrer, Ch. Gerber and E. Weibel, *Phys. Rev. Lett.* **50**, 120 (1983).

Section 3.

Metallurgy

SURFACE AND INTERFACE ANALYSIS, VOL. 9, 139–143 (1986)

Grazing Incidence X-ray Diffraction of an AISI 1006 Nitrogen Implanted Steel

N. Moncoffre, M. Brunel†, P. Deydier and J. Tousset
Institut de Physique Nucléaire (and IN2P3), Université Claude Bernard Lyon-1, 43, Bd du 11 Novembre 1918, 69622 Villeurbanne Cedex, France

Nitrogen implantations were performed on a very low alloyed steel XC06, AISI 1006 at various doses (10^{17}, 2×10^{17}, 4×10^{17} ions cm^{-2}) and various temperatures (20, 100, 150, 200°C). A study of the nitride phase evolution as a function of the two parameters (dose and temperature) is performed. The nitrogen distribution profiles are obtained using the $^{15}N(p,\alpha\gamma)^{12}C$ nuclear reaction whereas the iron nitride phases are followed by means of glancing angle RX diffraction. (N)-martensite, ϵ-nitride and ϵ-carbonitride phases of different nitrogen and carbon composition are identified. These results are correlated with Mössbauer results that put in evidence a carbonitride formation depending on implantation temperature and with recent Rutherford backscattering results that quantify the carbon contamination depending on implantation conditions. Moreover a preferential orientation of these phases depending on temperature is noticed.

INTRODUCTION

Nitrogen implantation into steels has been at the origin of a great number of works in recent years because of various fields of application, in particular concerning the improvement of surface tribological properties. A precise characterization of the implanted layers is possible thanks to several surface techniques such as XPS[1,2], AES,[3] nuclear reaction spectroscopy (NRS)[4–6], conversion electron Mössbauer spectroscopy (CEMS)[6–8] or transmission electron microscopy (TEM).[9,10]

In this paper we present a grazing incidence x-ray diffraction study of a very low alloyed XC06 steel, nitrogen implanted. The surface composition analysis is performed as a function of fluence and temperature. One of the main interests of this technique is the possibility to identify the various phases getting informations versus depth by simply increasing the glancing angle between the x-ray beam and the surface. The results thus obtained can be correlated with those of Mössbauer experiments which also allow the identification of iron chemical phases formed during implantation. Carbucicchio *et al.*[11] using depth selective CEMS have elaborated a qualitative picture of the implanted layer. Moreover grazing incidence x-ray diffraction is a good tool to put in evidence a possible texture of the implanted layer phases.

EXPERIMENTAL PROCEDURE

Samples of a XC06 steel (0.05 C; 0.205 Si; 0.375 Mn; 0.07 Cr) (AISI 1006) were mechanically polished with 1 μm diamond paste. Using the isotope separator of the Institut de Physique Nucléaire of Lyon, they were then implanted with 40 keV$^{15}N^+$ at various doses (10^{17}, 2×10^{17}, 4×10^{17} ions cm^{-2}) and various temperatures (20°C, 100°C, 150°C, 200°C).

The sample temperature was kept constant thanks to an air-cooling system surrounding the target holder. The special device used to implant at different temperatures was previously described.[5] The current density was maintained equal to 10μA cm^{-2}. Nitrogen profiles were obtained using the $^{15}N(p, \alpha\gamma)^{12}C$ nuclear reaction presenting a very narrow resonance at 429 keV. The nitrogen distributions corresponding to 10^{17} and 2×10^{17} N^+cm^{-2} fluences at the four temperatures mentioned above were recently published.[5,6] Figure 1 shows the profiles obtained for a 2×10^{17} N^+cm^{-2} implantation fluence at 20, 100, 150, 200°C. The chemical phases formed during implantation were studied by means of grazing incidence x-ray diffraction technique using the x-radiation Cu Kα (wavelength λ = 0.154 nm). The incident angle α between the beam and the sample surface was varying from 0.2 to 1°. For very small glancing angles inferior to the critical angle for total reflexion α_c (for iron), the major part of the beam is entirely reflected. The propagation in the material is thus evanescent in a direction almost perpendicular to the surface. For each incident angle, a penetration depth can be calculated using the Fresnel formula. These penetration depths (τ nm) are very small in the case of iron samples because x wave is much absorbed. Figure 2 represents their evolution versus α and the experimental set-up. Spectra are performed at a constant α angle whereas the detector is moving. The use of Sollers slits on the diffracted beam path reduced the Bragg peak widths. It allows a study of about 1 cm^2 of the implanted layer. The iron fluorescence is suppressed by using a silicium-lithium detector.

The diffraction technique using x-ray beam at grazing incidence is a non destructive method allowing the

† Laboratoire de Cristallographie, CNRS Associé à l'USMG, 166 X, 38042 Grenoble Cedex, France.

0142–2421/86/120139–05 $05.00

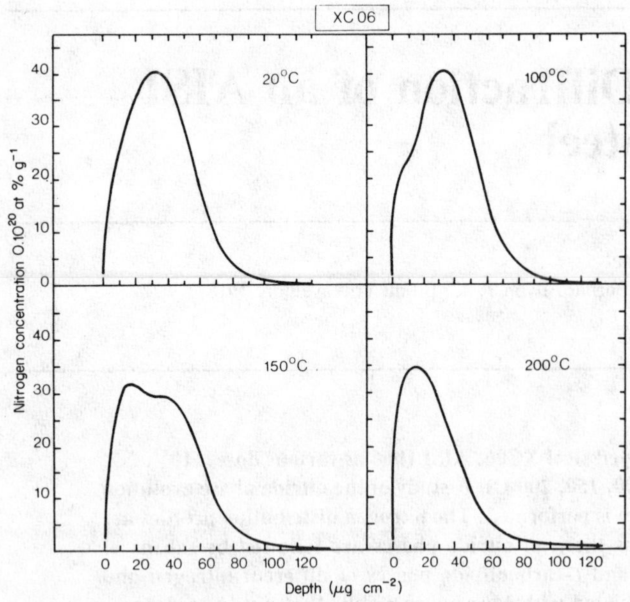

Figure 1. Nitrogen distribution profiles of a 2×10^{17} ions cm^{-2} implanted XC06 steel at different implantation temperatures : 20, 100, 150, 200°C (10μg cm^{-2} roughly correspond to 13 nm).

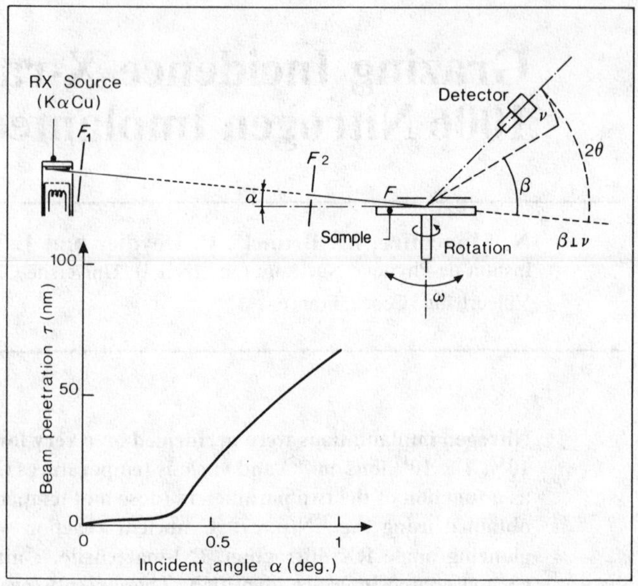

Figure 2. Grazing incidence diffraction experimental set up. Evolution of the penetration depth τ as a function of the incident α angle.

study of depths of the same order of magnitude as the implanted layer (≃ 100 nm). Increasing the incident angle α we can progressively explore depths varying from 5 nm to about 100 nm.

RESULTS

(N)-martensite

At a 10^{17} N^+cm^{-2} fluence and 20°C, (N)-martensite is identified by an iron Bragg reflexion splitting corresponding to a quadratic deformation. The intensity ratio of one of the most intensive martensite reflexions: 101 (θ_{Bragg} around 21°60) with the main α-iron 110 reflexion (θ_{Bragg} around 22°40), rapidly decreases with increasing α angles (Fig. 3). This indicates the presence of (N)-martensite essentially in the near surface region. In the same implantation conditions, but at a grazing incidence of 0.65° (Fig. 4) the 101 reflexion gives rise to a broadened peak indicating variations in the martensite composition. However, the $\alpha''-Fe_{16}N_2$ phase is also probably present (as shown by Mössbauer results.[5] Its more intense reflexions occur at θ_{Bragg}: 22°40(220) and at θ_{Bragg}: 21°37(202).

It is possible to calculate the nitrogen concentration in the (N)-martensite, knowing the unit cell dimensions a and c, from linear relations:[12]

$$a = (2.865 - 0.0019 \times X_N)\ \text{Å}$$
$$c = (2.868 - 0.024 \times X_N)\ \text{Å}$$

where X_N is the nitrogen atom number per 100 iron atoms. It is found to vary between 6 and a little bit more that 10%. When implanting at 100°C (Fig. 4), this composition range narrows significantly.

At 2×10^{17} ions cm^{-2} and 20°C, (N)-martensite is still detected.

Nitride phases

Figure 6 shows the phase evolution as a function of fluence at 20°C (α = 0.5°). ε–phases are identified at 10^{17} N^+ cm^{-2} from 100°C and at $2 \times 10^{17} N^+cm^{-2}$ from 20°C. The unit cell dimensions a and c are given on the basis of the close-packed hexagonal unit of the iron-atom arrangement ($a \simeq 2.71$ Å and $c \simeq 4.38$ Å). Figures 4 and 5 give the diffraction spectra obtained at these two doses versus temperature. A composition of these nitride phases can be calculated. Jack published the unit cell dimension evolution a, c, and c/a as a function of nitrogen concentration in the case of an $\epsilon-Fe_{2+x}N$ structure.[13] The various reflexions obtained from our experimental spectra do not fit with Jack's data, especially at high doses. Considering first Mössbauer results showing an ε-carbonitride formation, and secondly recent RBS experiments proving the existence of carbon near the surface,[14] we can presume that the phases formed correspond essentially to ε-carbonitride. Table 1 gives the various ratios C/N calculated in the case of implantations at different doses and temperatures. It is possible to estimate the carbonitride composition range thanks to empirical relations[15] that establish linear variations of a and c parameters versus a global atomic percentage of carbon and nitrogen. These relations are:

$$a\sqrt{3} = (4.3535 + 0.01337\ (C + N)\ \text{at}\ \%)\ \text{Å}$$
$$c = (4.2332 + 0.00578\ (C + N)\ \text{at}\ \%)\ \text{Å}$$

Mean phase compositions thus obtained are given in Table 2. In this table, the results obtained for a 2×10^{17} N^+ cm^{-2} implantation at 20°C are not mentioned.

For a small incident α angle (0.25°), the 100 reflexion leads to a well defined and rather narrow peak corresponding to an $\epsilon-Fe_{3.3}(C,N)$ carbonitride composition ($a = 2.694$). As α is increased the peak broadens so that for a 1° grazing incidence (τ ≃ 70 nm) a large composition range between $Fe_2(C,N)$ and

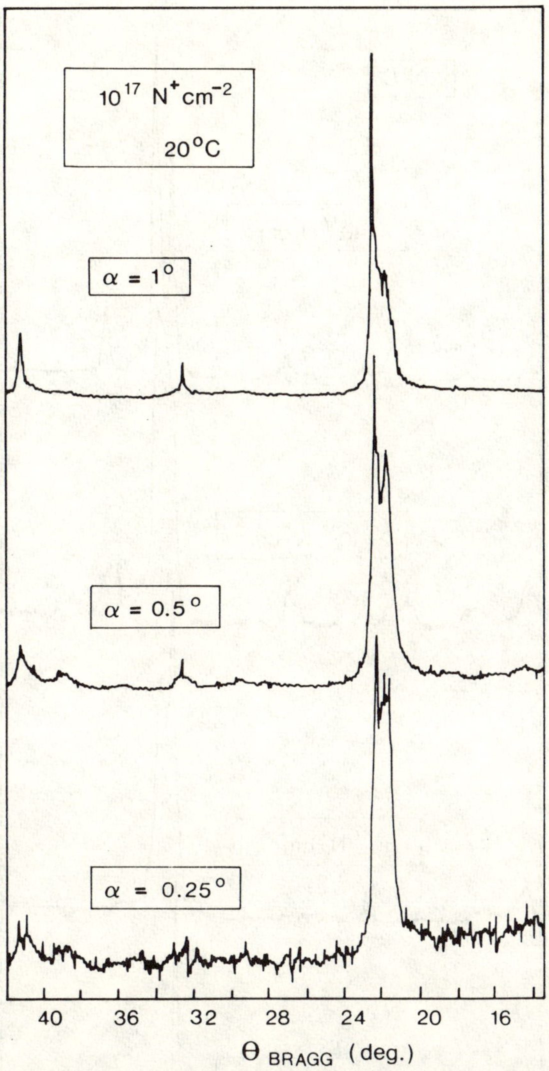

Figure 3. X-ray diffraction spectra of a 10^{17} N^+ cm^{-2}, 20°C implanted XC06 steel at different glancing angles α: 0.25°, 0.5°, 1°.

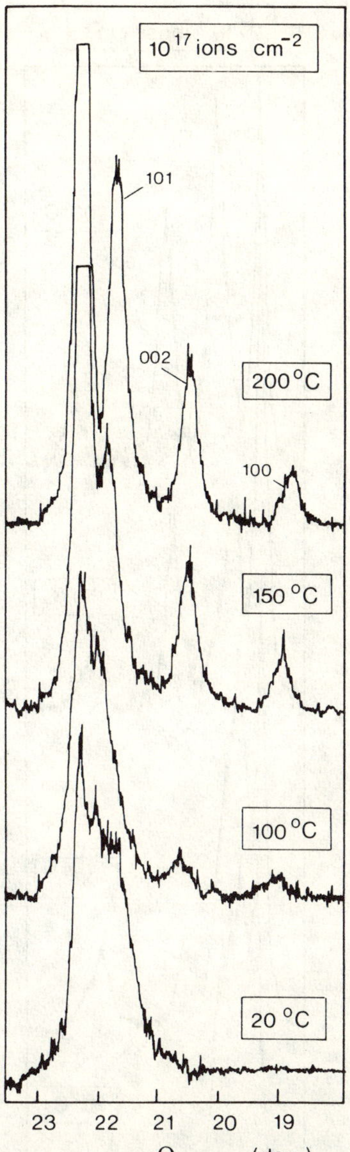

Figure 4. X-ray diffraction spectra of a 10^{17} ions cm^{-2} nitrogen implanted XC06 steel obtained at various implantation temperatures: 20, 100, 150, 200°C (α = 0.65°).

$Fe_{3.3}$(C,N) is identified confirming the presence of richer nitride phases deeper in the implanted layer, probably at the maximum of the distribution.

Texture

A preferential orientation of the ε phases is observed considering the reflexion intensity ratios: / 100 / 002 / 101 that are different from those normally obtained. This orientation is temperature dependent whatever the dose. From a 100°C implantation temperature, the 002 reflexion (not visible at 20°C) appears for a diffracted beam in the incidence plane whereas its intensity decreases for a diffracted beam in the surface plane. For a 20°C implanted sample, a maximum intensity of this reflexion is found in the surface plane. From these observations we conclude that at 20°C the *c* axis of the grains is preferentially oriented in the surface plane thus explaining the total disapperance of the 002 reflexion in certain conditions.

Always using grazing incidence (but with the x-radiation Co Kα preventing from iron fluorescence) an observation of photographic films shows additionally an intensification of some points of the diffusion ring.

Table 1. Quantification of the C/N ratios calculated in the case of a nitrogen implanted XC06 steel at various doses and implantation temperatures.

Nitrogen fluence ions cm^{-2}	Implantation temperature (°C)	Remaining nitrogen ions cm^{-2}	Carbon concentration ions cm^{-2}	*C/N* ratio
10^{17}	20	10^{17}	5×10^{15} to 10^{16}	5×10^{-2} to 0.1
	150	10^{17}	1 to 2×10^{16}	0.1 to 0.2
2×10^{17}	20	2×10^{17}	10^{16}	5×10^{-2}
	150	2×10^{17}	2×10^{16}	0.1

DISCUSSION

On the chemical phases

Figure 3 puts forward at 10^{17} ions cm^{-2} and 20°C a rapid decrease of the (N)-martensite contribution from the surface. Moreover, Fig. 4 shows some variation in

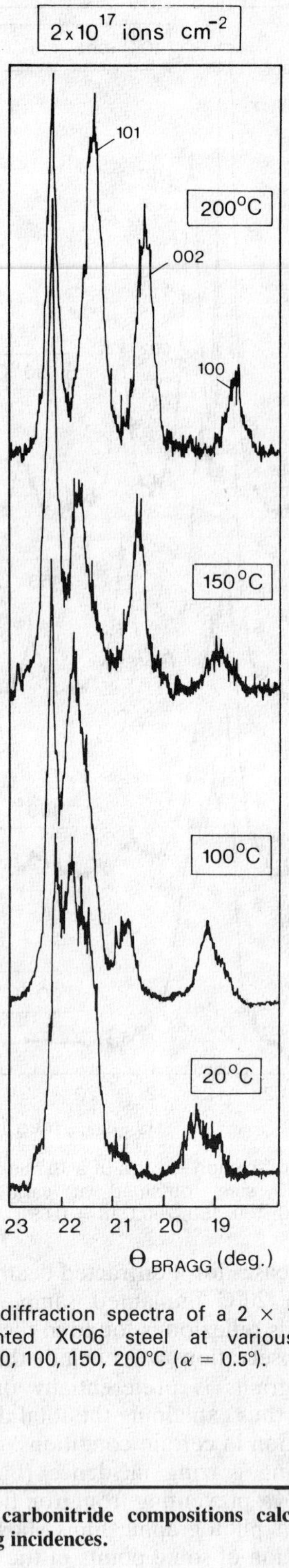

Figure 5. X-ray diffraction spectra of a 2×10^{17} ions cm^{-2} nitrogen implanted XC06 steel at various implantation temperatures : 20, 100, 150, 200°C (α = 0.5°).

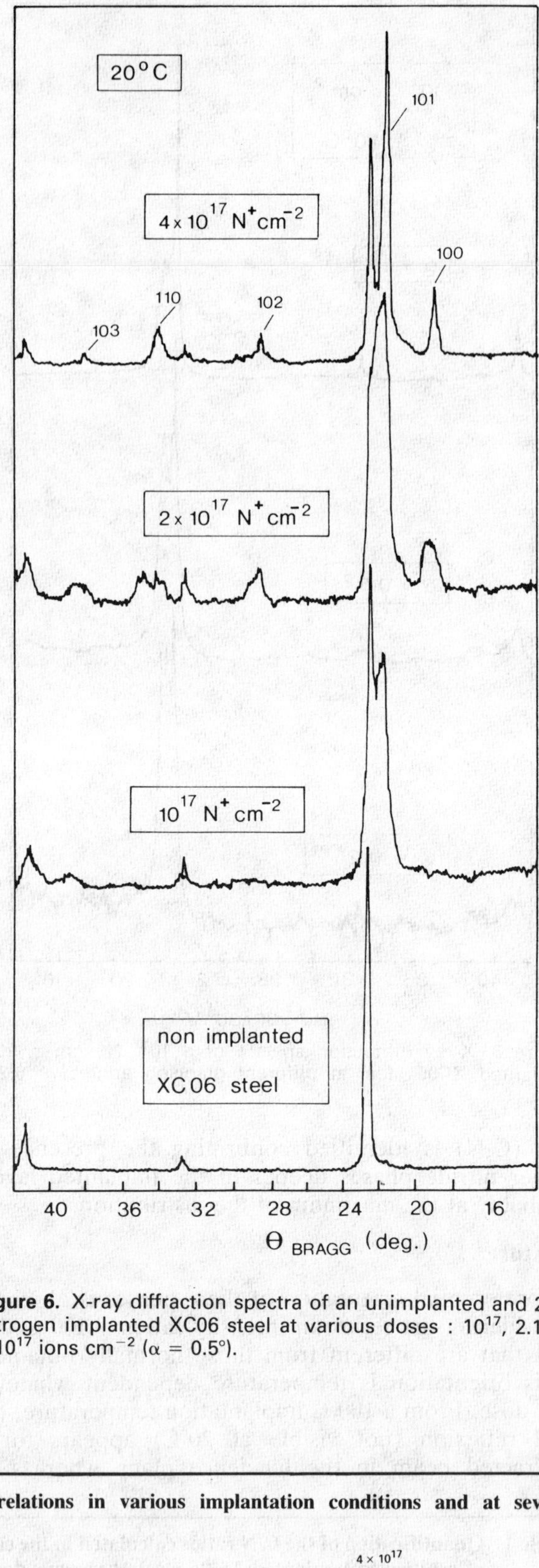

Figure 6. X-ray diffraction spectra of an unimplanted and 20°C nitrogen implanted XC06 steel at various doses : 10^{17}, 2.10^{17}, 4.10^{17} ions cm^{-2} (α = 0.5°).

Table 2. Mean carbonitride compositions calculated from empirical relations in various implantation conditions and at several grazing incidences.

Dose (N^+ cm^{-2})	10^{17}			2×10^{17}									4×10^{17}		
Temperature (°C)	100	150	200	100			150			200			20		
Incident α angle (deg)	1	1	1	0.25	0.5	1.0	0.25	0.5	1.0	0.25	0.5	1.0	0.25	0.5	1.0
a(Å)	2.708	2.715	2.731	2.717	2.718	2.730	2.726	2.730	2.732	2.763	2.764	2.769	2.764	2.767	2.770
c(Å)	4.371	4.374	4.386	4.342	4.358	4.374	4.380	4.378	4.374	4.384	4.400	4.404	4.410	4.410	4.416
c/*a*	1.614	1.611	1.606	1.602	1.603	1.602	1.606	1.604	1.601	1.587	1.592	1.590	1.595	1.593	1.594
Mean carbonitride composition	$Fe_{3.1(1)}$(C,N)	$Fe_{3.0(2)}$(C,N)	$Fe_{2.7(1)}$(C,N)	$Fe_{3.4(6)}$(C,N)	$Fe_{3.2(4)}$(C,N)	$Fe_{2.9(1)}$(C,N)	$Fe_{2.8(2)}$(C,N)	$Fe_{2.8(2)}$(C,N)	$Fe_{2.8(2)}$(C,N)	$Fe_{2.4(2)}$(C,N)	$Fe_{2.3(2)}$(C,N)	$Fe_{2.2(1)}$(C,N)	$Fe_{2.2(1)}$(C,N)	$Fe_{2.2(1)}$(C,N)	$Fe_{2.7(1)}$(C,N)

the range composition of this martensite. Now nuclear distribution profiles prove that the maximum of the nitrogen concentration occurs around 40 nm. The question is to know in which chemical form nitrogen is at this depth. An answer is given by conversion electron Mössbauer spectroscopy (CEMS) that indicates the formation of ε-carbonitride (as magnetic components) and of $\alpha''Fe_{16}N_2$ phases.[5] At this low dose, these phases are probably present as very dispersed and fine microcrystallites or even crystalline seeds not yet detectable by x diffraction.

At a higher dose of 2×10^{17} ions cm^{-2}, right from 20°C, RX spectra show the formation of nitrogen richer and richer carbonitride phases as depth is increasing (from ε-$Fe_{3.3}$(C,N) to ε-Fe_2(C,N): these observations are in agreement with Mössbauer results proving the simultaneous existence of ε-Fe_{2+x}(C,N) (as magnetic components) and of ε-Fe_2N phases (as non magnetic components characterized by a quadrupole doublet).

Carbonitride compositions calculated from the empirical relations of Firrao *et al.* [15] are in fact average compositions deduced from slightly different percentages that depend on *a* and *c* parameters. Rather wide diffraction peaks due to crystallites of very small sizes, variable compositions, presenting moreover numerous defects perhaps as well as a nitrogen over saturation, could explain these differences in the calculated compositions. Nevertheless it can be noticed that for a $2 \times 10^{17} N^+ cm^{-2}$ fluence, except at 150°C, *a* and *c* parameters increase with depth showing a constant enrichment of nitrogen inside the implanted layer as confirmed by nuclear profiles. Table 2 points out that this nitrogen enrichment of carbonitrides is also favored by high implantation temperatures (to the detriment of N-martensite and $\alpha''Fe_{16}N_2$), whereas nuclear analysis indicates a nitrogen removal inside the implanted layer at 200°C.

On the texture

This texture phenomenon has already been mentioned in the case of implanted[16–18] or mixed layers.[19] Van Wyk[16,17] gives an explanation based on channeling mechanisms in which the beam direction becomes a symmetry axis. In our case, in order to check this hypothesis, a 45° implantation was performed at 20°C. The resulting spectra do not differ from those obtained at normal incidence and this whatever the RX incident beam direction compared with the implantation axis. The *c* axis is thus in each case preferentially parallel to the surface at 20°C. This orientation is changing with the implantation temperature. Indeed, from 150°C and progressively it changes so that finally the *c* axis is found to be partially perpendicular to the surface. These orientations linked to the surface and not to the beam direction could come from a distribution of nitrogen isoconcentrated planes and strong resulting strains.

As a conclusion

Nuclear reaction analysis clearly displays that as the implantation temperature is increasing, more and more nitrogen is moving toward the surface[5] when in the same time carbon concentration in the near surface region grows up. Mössbauer spectroscopy also indicates that ε-Fe_2N concentration decreases as a function of temperature.[6] From these results we conclude that more and more carbonitride phases are formed at the surface when the implantation temperature increases. It is possible that the nitride and carbonitride orientation observed at 20°C results in isoconcentration stresses as a consequence of the implantation process. It seems also possible that a change of this orientation at high temperature indicates a new orientation of the surface carbonitrides which progressively replace the deeper nitrides and carbonitrides in thc implanted layer. Experiments are now in progress which will allow to know whether the orientation of these superficial carbonitrides at higher temperature is linked to surface stresses or to some other cause such as the beam direction.

Acknowledgements

The authors would like to thank A. Plantier who performed the implantations, S. Fayeuelle and G. Marest for their very helpful discussions and criticisms. This work was supported by a Ministère de l'Industrie et de la Recherche contract, in collaboration with UNIREC (Usinor Group).

REFERENCES

1. H.G. Feller, R. Klinger, W. Benecke, *Mat. Sci. Eng.* **69**, 173–180 (1985).
2. D.C. Kothari, M.R. Nair, A.A. Rangwala, K.B. Lal, P.D. Prabhawalkar, P.M. Raole, *Nucl. Instr. Meth. in Phys. Res.* **B7/8**, 235–241 (1985).
3. I.L. Singer, *Vacuum,* **34 (10/11)**, 853–859 (1984).
4. G.K. Hubler, *Nuclear Instr. Meth. in Phys. Res.* **191**, 101–113, (1981).
5. N. Moncoffre, G. Hollinger, H. Jaffrezic, G. Marest, J. Tousset, *Nucl, Instr. Meth. in Phys. Res.* **B7/8**, 177–183 (1985).
6. N. Moncoffre, G. Marest, S. Hiadsi, J. Tousset, *7th International Conference on Ion Beam Analysis, Berlin, (RFA) 7–12 Juillet 1985,* to be published in *Nucl. Instr. Meth. Phys. Res.*
7. C.A. dos Santos, B.A.S. de Barros Jr., J.P. de Souza, J.R. Baumvol, *Appl. Phys. Lett.* **41(3)**, 237 (1982).
8. G. Principi, S. Lo Russo, C. Tosello, *Intern. Symp. on the Industrial Appl. of the Mössbauer Effect, Honolulu, 1984.*
9. B. Rauschenbach, A. Kolitsch, *Phys. Stat. Sol.* **A80**, 211, (1983).
10. S. Fayeulle, D. Tréheux, C. Esnouf, *Nucl. Instr. Meth. Phys. Res.* **B7/8**, 171–176 (1985).
11. M. Carbucicchio, L. Bardani, S. Tosto, *J. Appl. Phys.* **52(7)**, 4589 (1981).
12. J. Foct, thesis, Nancy (1973).
13. K.H. Jack, *Acta Cryst.* **5**, 404 (1952).
14. N. Moncoffre, H. Jaffrezic, J. Tousset, A. Chevarier, N. Chevarier, M. Stern, *7th. International Conference on Ion Beam Analysis, Berlin (RFA), 7–12 Juillet 1985,* to be published in *Nucl. Inst. Meth. in Phys. Res.*
15. D. Firrao, B. Debenedetti and Rosso, *Metall. Ital.* **71**, 373 (1979).
16. G.N. Van Wyk, *Rad. Eff. Lett.* **57**, 45–50 (1980).
17. G.N. Van Wyk, *Rad. Eff. Lett.* **57**, 187–192 (1981).
18. D. Dobrev, *Thin Solid Films,* **92**, 41–53 (1982).
19. B.Y. Tsaur, S.S. Lau, J.W. Mayer, *Appl. Phys. Lett.* **36(10)**, 823–826 (1980).

SURFACE AND INTERFACE ANALYSIS, VOL. 9, 145–150 (1986)

Wetting and Spreading of Liquid Metals: the Rôle of Surface Composition

Graham C. Smith and Colin Lea
Division of Materials Applications, National Physical Laboratory, Teddington, UK.

Quantitative measurements of the surface compositional changes which occur during the wetting of copper substrates in vacuum by liquid SnPb solder alloys have been made *in situ*, by Auger electron spectroscopy. Five different spatial regions associated with the wetting front have been identified. Away from the spreading front, the liquid solder shows a surface enhancement of lead. Approaching the front, adjacent bands of Cu_3Sn intermetallic compound, and tin-depleted liquid occur, both of width $\sim 20\ \mu m$. The lead-enriched surface of the solder extends over these two regions, and out over the copper to form approximately 50% monolayer 'halo' of $\sim 80\ \mu m$ width. Beyond the halo, in the fifth region, very small amounts of tin and lead were detected. The lead halo forms rapidly by surface diffusion, and precedes wetting. In the absence of other limitations, it is the reduction of the copper substrate surface energy by the lead halo that restricts the spread of the molten alloy. Intermetallic compound formation is relatively slow and only takes place once spreading has ceased and a quasi-equilibrium reached.

INTRODUCTION

The ability of a liquid metal to wet and subsequently spread over a different solid metal is the basis of soldering and brazing. The efficiency of these processes is of immense technological significance, notably in the electronic assembly industry in which over one million soldered joints are being made daily in the UK alone,[1] and the reflow of electrodeposited solder coatings on copper is a basic process in the manufacture of printed circuit boards. In practical soldering, the wetting of the materials to be joined is governed by the surface energy of the liquid tin-lead alloy and the interfacial tension between it and the substrate. These properties are a function of the local microchemistry. One property of the flux usually used in soldering, besides removing corrosion films and preventing oxidation, is to reduce the surface energy of the liquid solder and so aid spreading. Thus the behaviour of the spreading solder is determined by the local micro-composition of both the free surface and the liquid–substrate interface.[2] These compositions are time-dependent and explain why the rate of spreading varies and is eventually reduced to zero.

In general an alloy, be it solid or liquid, will exhibit the phenomenon of surface segregation in which the surface layer becomes enriched in one of the components[3] such that, at equilibrium, the total energy of the system is minimized. For a binary alloy where the two components form a regular solution with small interaction, the surface concentration of the lower surface energy component will be higher than its bulk composition. In the case of the liquid tin–lead system there is a significant difference between the surface energies of the pure component materials and a segregation of lead to the free surface is expected.

Tin is a material which readily forms intermetallic compounds with many metals — distinguishable homogeneous phases having relatively narrow ranges of composition with simple stoichiometric proportions. With copper, the formation of Cu_3Sn and Cu_6Sn_5 can proceed quite rapidly when the tin or solder is liquid at the temperatures used in soldering[4] and these compounds may well solidify at the copper–liquid interface and affect the spreading of the still molten solder.

Once this quasi-equilibrium has been reached and the liquid tin-lead alloy has ceased to spread, these intermetallic compounds will continue to grow at the liquid–solid interface, and indeed at the solid–solid interface, albeit at a very much reduced rate.

The rôles played by the liquid surface composition and the intermetallic compound formation at the interface, and their relationship to the wetting of the substrate and the spreading of the liquid metal alloy, form the subject of this paper.

SURFACE SEGREGATION AND THE SURFACE ENERGY OF LIQUID TIN–LEAD ALLOYS

Although surface segregation is a well known phenomenon[5,6] and has been the subject of considerable theoretical and experimental effort by many workers, relatively few reports of electron spectroscopic investigations of liquid metal alloy surfaces exist. This is at least in part due to the experimental difficulty of handling such materials in ultra-high vacuum (UHV). Results have been reported for certain low melting point and low vapour pressure alloys.[7,8,9,10] Our present interest is in liquid tin–lead alloys for which the vapour pressure of lead is relatively

0142–2421/86/120145–06 $05.00

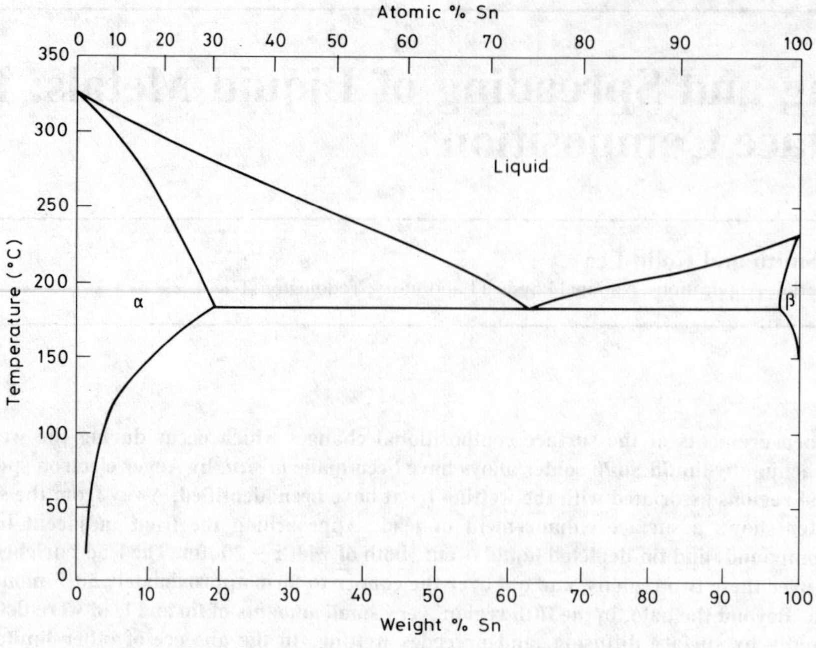

Figure 1. The SnPb phase diagram.

high such that, for UHV operation, the temperature must be kept below about 300°C. The phase diagram of the PbSn system is shown in Fig. 1. It can be seen that restricting the maximum temperature to 275°C allows measurements to be made on alloys with an atomic percentage (at.%) of lead up to about 70 at.%. Practical soldering with SnPb alloys covers the full composition range from pure tin to pure lead, but for soldering of electronics a composition near the eutectic is used normally.[11] The eutectic has a SnPb composition 73.4/26.6 at.% and a melting point of 183°C.

The results of an AES investigation of the surface composition of the SnPb liquid alloy at 275°C are shown in Fig. 2. These measurements were made in UHV with the liquid surface free of all detectable contamination. The details of the experiment and of the quantification procedure used to obtain these results are discussed elsewhere.[12] The measurements reveal a significant surface enrichment of lead across the whole composition range. Furthermore, the experimental points relating the bulk atomic fraction X^b and the surface atomic fraction X^s show a close fit to the theoretical curve obtained from the quasi-chemical bond energy model of Wynblatt and Ku[13], and also to the semi–empirical relationship of Seah.[14] In the case of a liquid alloy, the contribution to the free energy of segregation from the strain energy may be neglected, since a liquid cannot support a lattice strain, and as a result comparison between theory and experiment is reported to be more precise than the corresponding solid state case. The important point to note from Fig. 2 is that, at the eutectic composition of $X^b_{Pb} = 26.6$ at.%, the surface atomic layer has a measured lead composition $X^s_{Pb} = 67$ at.%.

The relationship between the surface energy of a binary mixture and the surface energies of its components has been the subject of much investigation.[15] For an ideal solution (complete mixing, no interaction, no segregation) the surface energy of the alloy is simply equal to the sum of the surface energies of the components added in proportion to their bulk concentration, i.e. for components A and B:

$$\gamma^s_{AB} = \gamma^s_A X^b_A + \gamma^s_B X^b_B \tag{1}$$

where γ^s_m is the surface energy of alloy component m, and X^b_m is its bulk atomic fraction. In the case of a regular solution with relatively small numerical values of the regular solution parameter, Ω, the surface energy is found, to a good approximation, to be the sum of the surface energies of the pure components added in proportion to their surface concentration, i.e.

$$\gamma^s_{AB} = \gamma^s_A X^s_A + \gamma^s_B X^s_B \tag{2}$$

where X^s_m is the surface atomic fraction of alloy component m.

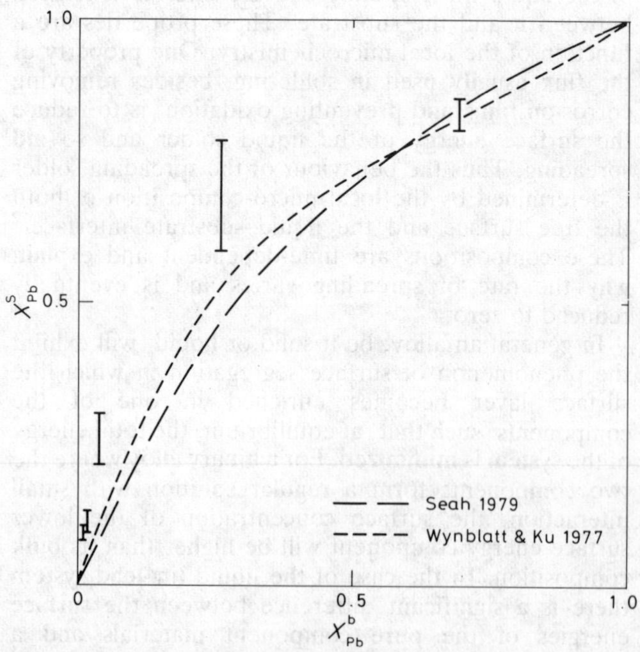

Figure 2. Surface segregation of SnPb in the liquid state.

Recent approaches to the problem which treat the subject thermodynamically, implicitly include the surface segregation.[16]

In the case of liquid SnPb considerable surface tension data exist in the literature[17,18,19] and the above relationship (2) is found to hold to within a few percent across the whole of the composition range.

Using values of γ for the pure components, and their temperature coefficients available in the literature, a surface energy $\gamma^{s}_{SnPb} = 481\ mJm^{-2}$ is obtained for the eutectic composition SnPb alloy at 275°C. This corresponds to the measured surface composition of 67 at.% Pb and 33 at.%Sn.

THEORY OF WETTING

Recent theoretical work has concentrated on treating the phenomenon of wetting as a phase transition of the first order within the framework of a mean field (van der Waals) approximation[20,21], although the details of the theory are a matter of some discussion.[22] However, such theories refer to model systems and, at the moment, are difficult to relate to experiments on 'real' specimens. Historically, wetting has been viewed in the context of the Young equation,[33] which invokes the concept of a contact angle θ defined by the balance of the three interfacial tensions as shown in Fig. 3. For a small change of contact area, ΔA, the energy change is given by

$$\Delta G^{s} = \Delta A(\gamma_{SL} - \gamma_{SV}) + \Delta A\ \gamma_{LV}\cos(\theta - \Delta\theta)$$

where the subscripts S, L and V refer to the solid, liquid and vapour (or flux, or vacuum) respectively. In the limit $\Delta A \to 0$, $\Delta G^{s}/\Delta A \to 0$ and $\Delta\theta \to 0$. Hence

$$\gamma_{SL} - \gamma_{SV} + \gamma_{LV}\cos\theta = 0.$$

This analysis ignores the vertical component $\gamma_{LV}\sin\theta$, which will tend to induce surface stresses. It has been suggested that the contact angle is a result of the balance of surface stresses, but the surface energy or tension balance approach is found to correspond adequately with the experimental observations.

In cases where $(\gamma_{SV} - \gamma_{SL})/\gamma_{LV} > 1$, spontaneous spreading occurs and a spreading coefficient $S_{L/S}$ can be defined:

$$S_{L/S} = \gamma_{SV} - \gamma_{LV} - \gamma_{SL}$$

representing the free energy change for spreading of the liquid on the solid substrate.

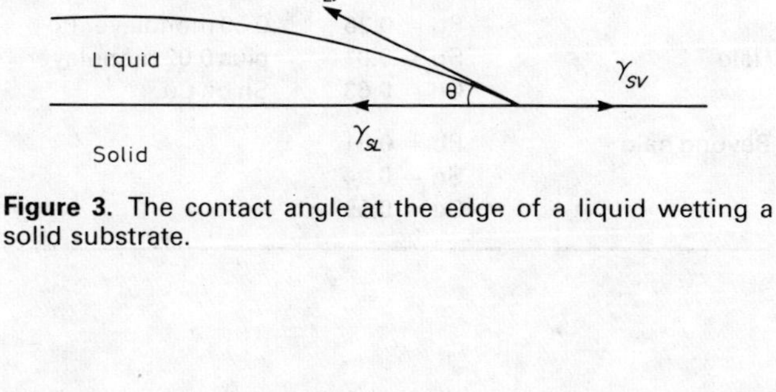

Figure 3. The contact angle at the edge of a liquid wetting a solid substrate.

EXPERIMENT

Auger electron spectroscopy (AES) has been used to monitor the microchemistry of a liquid SnPb solder alloy at the soldering temperature while spreading across a copper substrate. Experiments were performed using a Varian CMA with an axial electron gun in front of which were arranged the solder bath and the copper substrate. The solder bath was constructed from tantalum sheet directly heated from below by a non-inductively wound tungsten resistance heater and fixed on a sample manipulator with X, Y, Z movement, plus rotation and tilt about the vertical axis of the system. A thermocouple monitored the temperature of the solder bath. Copper samples were mounted on a separate $XYZ\Phi$ manipulator mounted horizontally and arranged as shown in Fig. 4. These degrees of freedom in sample manipulation enabled either the copper samples or the molten solder to be cleaned by argon ion bombardment and their surface compositions monitored by AES. In addition, with a copper sample in the analysis position, the bath of molten solder could be raised up from below until partially immersing the copper, initiating wetting and spreading. AES spectra of the molten liquid in the region of the spreading front were then obtained. The system also provided a scanning electron microscopy (SEM) facility, with spatial resolution limited by the integral CMA electron gun to ~ 5 μm. All Auger electron spectra were taken in the differential mode using a beam energy of 5 kV and modulation of 5 V. Quantification proceeded via the sensitivity factor method of Davis *et al.*;[24] the Auger transitions used together with the sensitivity factors and mean free paths are shown in Table 1.

RESULTS AND DISCUSSION

In the SEM image of the molten SnPb alloy wetting the Cu substrate, five distinct regions were visible. The region of molten solder showed high brightness. At the edge of the molten region a dark band of width ~20 μm could be distinguished, followed by a bright band of similar width. Beyond this, a region of reduced

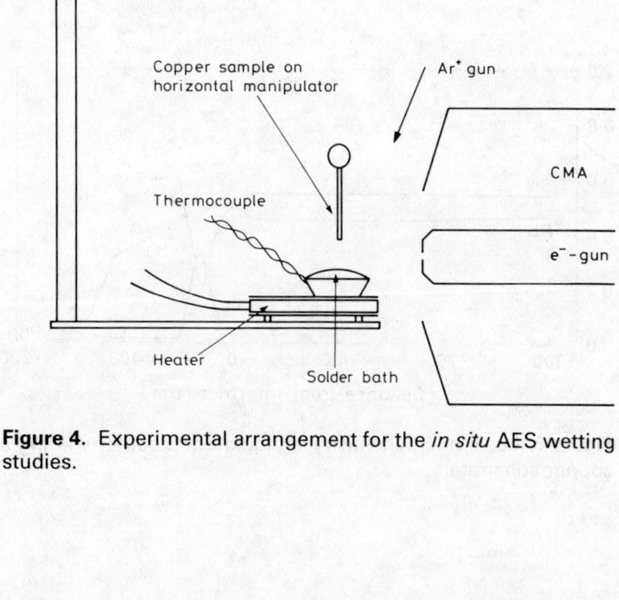

Figure 4. Experimental arrangement for the *in situ* AES wetting studies.

Table 1. The Auger transitions, sensitivity factors and inelastic mean free paths[31] for the materials of this study.

Element	Auger energy (eV)	Sensitivity factor	Inelastic mean free path (monolayers) Sn Pb solder	Cu	Cu_3Sn intermetallic
Cu	60	0.11	1.91	1.67	1.73
Pb	92	0.46	2.25	1.94	2.02
Sn	430	0.90	4.72	4.06	4.23
Cu	920	0.24	6.90	5.94	6.18

brightness extended for a further 80 μm before the original faint contrast of the copper substrate could be observed. Auger electron spectra were taken at points along a line normal to the edge of the molten solder in order to determine the local surface microchemistry of these five regions of differing contrast. Spectra were quantified using the Sn 430 eV, Pb 92 eV and Cu 60 eV peaks to give the atomic fraction composition in each region. The results of the analysis are shown in Fig. 5, together with a model for the structure of the interface.

According to the model, on the left hand side of the interface the compositon detected is that of bulk molten SnPb solder with Pb enrichment in the outermost surface monolayer, in agreement with the results of a study of segregation in the liquid metal alloy.[12] At the interface the Sn component in the spectra begins to decline and, in the dark band seen in the SEM image, some contribution to the spectra from Cu becomes apparent. These data were taken between 10 and 30 mins after wetting, giving time for the formation of an intermetallic compound layer ~ 1 μm thick.[25] The Auger electron spectra show the dark band in the SEM image to be depleted of Sn and the region following this to consist of mainly the intermetallic compound Cu_3Sn. The surface enrichment of Pb extends from the liquid SnPb alloy over both the Sn-depleted region and the region of intermetallic compound to the copper substrate. The 80 μm wide region with bright contrast in the SEM image is shown by the Auger electron spectra to consist of a 'halo' of approximately 50% monolayer Pb on the copper substrate. Beyond the halo the spectra show copper, with Pb and Sn at approximately 1% and 4% monolayer respectively. The mean compositions of these various regions near the interface are summarized in Table 2.

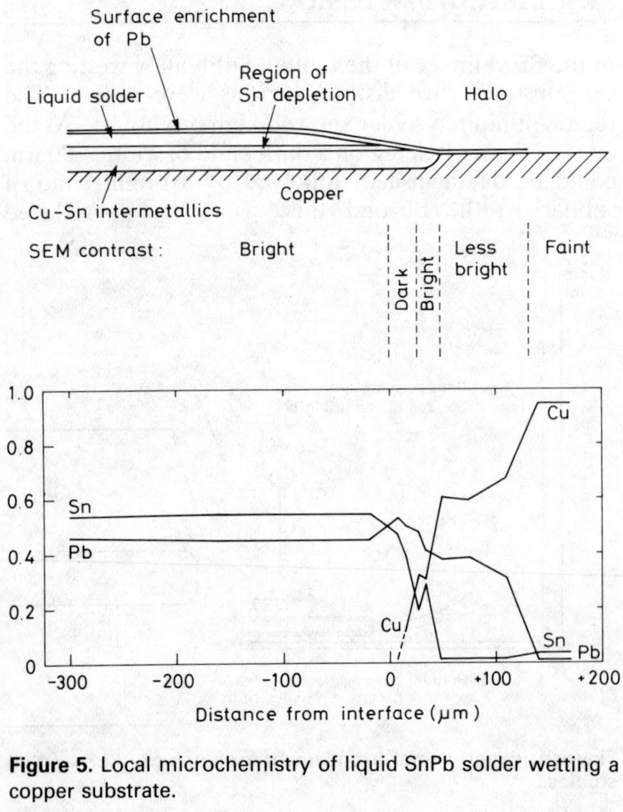

Figure 5. Local microchemistry of liquid SnPb solder wetting a copper substrate.

The compositions in Table 2 are those after spreading has ceased and a quasi-equilibrium reached. Attempts were made to monitor the composition of the wetting front during spreading by tuning the analyser to a particular Auger energy and, with a stationary electron beam, monitoring the intensities of peaks as a function of time as the wetting front swept across the point on the sample at which the incident beam was focused. Traces for the Pb 92 eV and Sn 430 eV peaks are shown in Fig. 6. The high noise level in these traces is a result of operating the spectrometer with a low (10 ms) time constant to allow the system to respond to the rapid changes that took place. The two traces were recorded in separate experiments so it is probably not possible to compare the time on an absolute scale. However, several significant differences can be seen. The Pb trace shows a very sudden and rapid rise, reaching an intensity very near the maximum in a period of about 100 milliseconds, after which it approaches gradually a saturation value over the next 12–15 seconds. The tin trace, however, exhibits a much less sudden onset before the rapid rise to a near saturation level; the total rise time in this case being about 600 milliseconds. Following this there is a further slight increase to a peak value after 2 seconds, followed by a slight drop during a further 2 seconds. The signal then shows a slow rise over the next 10 – 15 seconds.

These results are consistent with the quasi-equilibrium model of Fig. 5. We suggest that formation of the Pb halo is due to fast surface diffusion processes

Table 2. Mean compositions of the various regions identified in Fig. 5.

Region	Mean Atomic Composition	Interpretation
Solder	Pb = 0.45 Sn = 0.55	Bulk solder (Pb 0.27, Sn 0.73) plus 0.58 monolayer Pb
Sn-depleted region	Pb = 0.51 Sn = 0.43 Cu = 0.06	
Intermetallic region	Pb = 0.45 Sn = 0.24 Cu = 0.31	Mainly Cu_3Sn with 0.58 monolayer Pb, and some solder
Halo	Pb = 0.36 Sn = 0.01 Cu = 0.63	0.50 monolayer Pb plus 0.02 monolayer Sn on Cu.
Beyond halo	Pb = 0.01 Sn = 0.04 Cu = 0.95	

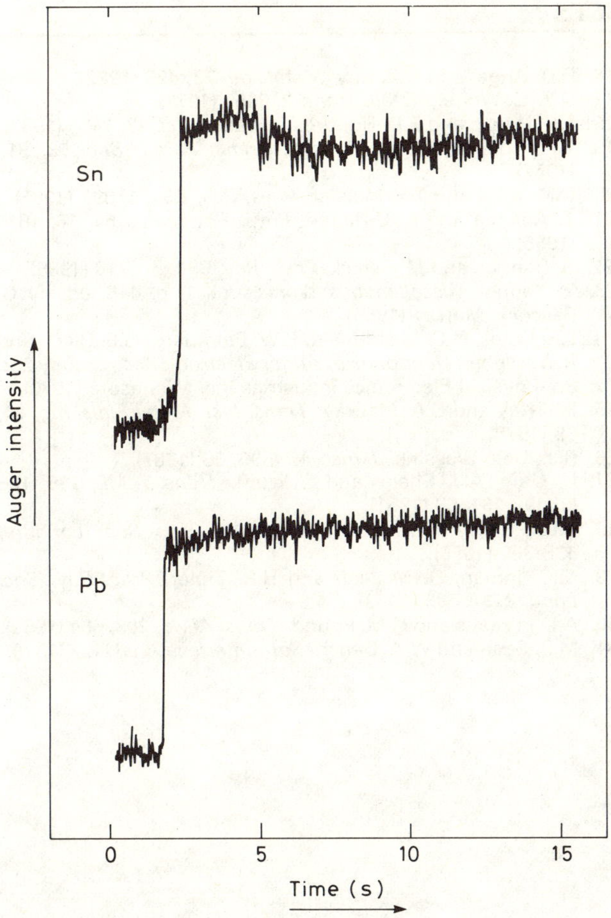

Figure 6. Development of Auger electron intensities vs. time, for the tin and the lead.

taking place from the Pb-enriched outer layer of the molten solder. This accounts for the sudden onset of the Pb signal in the trace of Fig. 5. Actual wetting, i.e. movement of the molten solder, takes place over a Cu surface already partially covered by the Pb halo and is much less rapid. The small changes in the Auger electron intensities after wetting can be attributed to the initiation of Cu_3Sn intermetallic compound growth. Cu_3Sn formation at the solder – copper interface follows a $(\text{time})^{-1/2}$ law.[25] The growth rate at 275°C is about 30 nm $s^{-1/2}$,[25] corresponding to the formation of one monolayer in 80 milliseconds.

It is of interest to apply the Young equation to the situation which exists prior to the formation of any intermetallics. From the second section above, we have taken the surface energy of the solder, γ_{LV}, to be 0.48 Jm^{-2} for the temperature and composition used in the present work. The interfacial tension between the solid copper and liquid solder is estimated using the approximate method of Klein Wassink[26] to be $\gamma_{SL} \sim 0.09\ Jm^{-2}$, and a mean of the surface tension values for solid copper given in the literature is $\gamma_{SV}^{Cu} = 1.73\ Jm^{-2}$.[27,28,29,30] Thus, for solder wetting copper we obtain a contact angle with $\cos\theta = (\gamma_{SV} - \gamma_{SL})/\gamma_{LV} > 1$ and a spreading coefficient $S_{L/S} = \gamma_{SL} - \gamma_{LV} - \gamma_{SL} = 1.16\ Jm^{-2}$, implying spontaneous spreading. To include the effect of the Pb halo, which precedes spreading, we may approximate

$\gamma_{SV} = x\,\gamma_{SV}^{Pb} + (1-x)\gamma_{SV}^{Cu}$,

when x is the fractional coverage of the Pb monolayer. γ_{SV}^{Pb} is the surface tension of solid Pb, and is approximately 0.45 Jm^{-2}.[17]

The spreading coefficient can then be written

$$S_{L/S} = x\,\gamma_{SV}^{Pb} + (1-x)\gamma_{SV}^{Cu} - \gamma_{LV} - \gamma_{SL}$$

$$= 1.16 - 1.28x\ Jm^{-2}$$

Thus for a value of $x = 0.9$ monolayer the spreading coefficient is zero, the contact angle is zero and equilibrium has been reached. This is in qualitative agreement with the experimental observation of an average 0.5 monolayer of lead over the halo region. The extent of the halo is limited by temperature gradients in the copper substrate and by evaporation of lead, leading to a higher lead concentration near the source of lead, i.e., just immediately in front of the advancing molten solder.

This analysis ignores the effect of surface roughness on the wetting front. This is difficult to include quantitatively but in general results in a change in the contact angle to an apparent value ϕ, given by $\cos\phi = f\cos\theta$ when f is the so-called roughness factor, the ratio of real to apparent area.

CONCLUSIONS

Auger electron spectroscopy has been used for *in situ* measurements of the local microchemistry of a molten SnPb solder alloy as it wets a copper substrate. It appears that the surface enrichment of lead, observed in the isolated liquid alloy, continues over the wetting front and spreads across the copper ahead of the solder forming a 'halo' region consisting of a 50% coverage of a single atom layer of lead. As the electron beam is scanned from the molten solder towards the spreading interface, a reduction in the tin concentration is noted, followed by the appearance of copper. This corresponds to intermetallic compound formation between copper and tin of mainly Cu_3Sn. The lead surface enrichment of the SnPb liquid alloy continues over both the tin-depletion region and the intermetallic region and extends to form the halo on the copper substrate.

Consideration of the kinetics lead us to conclude that the lead halo, formed by a rapid surface diffusion process, precedes the wetting front, and that formation of the intermetallic layer does not begin until spreading has ceased and a quasi-equilibrium reached. In UHV, and in the absence of any other restriction (e.g. amount of material available, etc) the solder will spread until sufficient Pb has diffused ahead of it to reduce the solid-vapour surface tension of the substrate sufficiently to give a zero spreading coefficient.

REFERENCES

1. C. Lea, *Brazing and Soldering*, **2**, 20 (1982).
2. C. Lea. Proc 8th Intern. Vac. Congr., Cannes, Sept. 22–26 (1980), Suppl. *Revue Le Vide, Les Couches Minces*, **201**, 467 (1980).
3. E.D. Hondros and M.P. Seah in *Physical Metallurgy 3rd Ed.,* ed. by R.W. Cahn and P. Haasen, Chap. 13, pp. 856–931, Elsevier (1983).
4. P.E. Davis, M.E. Warwick, and P.J. Kay, *Plating and Surf. Finishing*, **69,** 72 (1982).
5. E.D. Hondros and M.P. Seah, *Intern. Metall Rev.* **222,** 262 (1977).
6. M.P. Seah in *Practical Surface Analysis* ed. by D. Briggs and M.P. Seah, Chap 7, pp. 247–281, Wiley (1983).
7. S. Hardy and J. Fine, in *Materials Processing in the Reduced Gravity Environment of Space,* ed. by G.E. Rindone, pp. 503–511, Elsevier (1982).
8. P. Laty, J.C. Joud. and P. Desre, *Surf. Sci.* **104,** 105 (1981).
9. S. Berglund and G.A. Somarjai, *J. Chem. Phys.* **59**, 5537 (1973).
10. L. Goumiri and J.C. Joud, *Acta Metall.* **30**, 1397 (1982).
11. C.J. Thwaites, *Intern. Metall. Rev.* **29**, 2, 45 (1984).
12. G.C. Smith, M.P. Seah and C. Lea, (to be published).
13. P. Wynblatt and R.C. Ku, *Surf. Sci.* **65**, 511 (1977).
14. M.P. Seah, *J. Catal.* **57**, 450 (1979).
15. N. Eustathopoulos, *Intern. Metall. Rev.* **28**, 4, 189 (1983).
16. R.D. Angal and D.Z. Roy, *Metallkde.* **73,** 428 (1982).
17. D.W.G. White, *Metall. Trans.* **2**, 3067 (1971).
18. F.H. Howie and E.D. Hondros, *J. Mater. Sci.* **17**, 1434 (1982).
19. T.P. Hoar and D.A. Melford, *Trans. Farad. Soc.* **53**, 315 (1957).
20. D.M. Kroll and T.F. Meister, *Phys. Rev.* **B31,** 1, 392 (1985).
21. T. Aukrust and E.H. Hauge, *Phys. Rev. Letts.* **54,** 16, 1814 (1985).
22. S. Dietrich and M. Schick, *Phys. Rev.* **B31**, 7, 4718 (1985).
23. T. Young, *Miscellaneous Works,* vol. 1, p. 418, ed. by G. Peacock, Murray (1855).
24. L.E. Davis, N.C. Macdonald, P.W. Palmberg, G.E. Riach and R.W. Weber, *Handbook of Auger Electron Spectroscopy* 2nd Ed, Physical Electronics Industries Inc. Minnesota (1976).
25. P.J. Kay and C.A. Mackay, *Trans. Inst. Metal Finishing,* **54,** 68 (1976).
26. R.J. Klein Wassink, *J. Inst. Met.* **95,** 38 (1967).
27. H. Udin, A.J. Shaler and J. Wulff, *Trans. AIME,* **185,** 186 (1949); **191**, 63 (1951).
28. C.L.J. Bailey and H.C. Watkins, *Proc. Phys. Soc. (London),* **63B** 350 (1950).
29. M.C. Inman, D. McLean and H.R. Tipler, *Proc. Roy. Soc. Lond.* **273A** 538 (1963).
30. A.L. Pranatis and G.M. Pound, *Trans. AIME,* **203,** 664 (1955).
31. M.P. Seah and W.A. Dench, *Surf. Interface Anal.* **1,** 2 (1979).

SURFACE AND INTERFACE ANALYSIS, VOL. 9, 151–155 (1986)

Surface Segregation in Fe–18Cr–3Mo Single Crystal

B. Schiffman and M. Polak*
Department of Materials Engineering, Ben-Gurion University of Negev, Beer-Sheva 84105, Israel

The (100) surface region of the Fe–18Cr–3Mo ternary alloy was investigated during and after argon sputtering at ambient and elevated temperatures by means of Auger electron spectroscopy. Low and high-energy lines, corresponding to different sampling depths, were monitored, and the quantitative analysis was based on relative intensities from an *in situ* scribed crystal, representing the bulk composition. The Cr:Fe steady-state ratio measured during sputtering reflected significant subsurface Fe depletion which reached a maximum at ~800 K and then decreased. This trend could be inverted (i.e. inducing Cr depletion) by Ar sputtering under some residual CO. Similarly, compositional changes observed after sputtering were critically dependent on the cleanliness of the surface. The results are discussed in terms of possible (Gibbsian) surface segregation of Fe vs. chemically-induced segregation of Cr and Mo.

INTRODUCTION

The occurrence of surface segregation (Gibbsian segregation) of a constituent in a metallic alloy, has been observed in most of the systems studied. The segregation trend, i.e. which element enriches the external clean surface of a binary alloy under thermal equilibrium conditions, often correlates with predictions based on simple thermodynamic arguments.[1,2] Thus, according to the bond-breaking theory within the ideal (or regular) solution model, the element which has weaker interatomic bonds is expected to preferentially populate the surface. A second driving-force for segregation is the release of strain-energy which is likely to occur when a solute atom of different size leaves the crystal bulk to occupy a surface site.

Experimental determinations of surface segregation were not always conclusive due to the presence of surface contamination, originating from residual gases in the test chamber, or from bulk impurities. In both cases, the segregation level can be enchanced, or alternatively, its trend inverted as a result of chemical interactions with the foreign species.

Quite recently, it has been suggested that Gibbsian segregation plays a major role in the so-called 'preferential sputtering' often observed in alloys.[3] Although analysis of ion bombardment induced compositional changes can be complicated by contributions from other mechanisms, a sputtering experiment may furnish significant information concerning Gibbsian surface segregation in alloys, especially in cases when impurities limit the time available for measuring the segregation directly.

Surface segregation in several iron-chromium alloys has been studied previously based on quantitative analysis of high-energy Auger[4,5] and XPS[6] lines. Leygraf *et al* reported Cr-enrichment in the (100) and (110) surface regions of Fe-16Cr single crystals. However, the effect was reported to be temperature-independent between 675 K and 1175 K, and moreover it was accompanied by appreciable sulphur enrichment at the surface. (Indeed, more recently co-segregation of Cr and S in stainless steel was reported.[7]) Dowben *et al.*[6] studied surface segregation in Fe–28Cr(110) at 750 K and concluded that Cr segregation occurred, although the surface was not entirely free from contamination. Dumoulin and Guttmann,[5] on the other hand, reported no segregation of Cr in polycrystalline Fe–2Cr between 500°C–700°C. Also no Mo segregation was noted[5] for Fe–2Mo between 425°C–900°C. (Addition of Mo is known to improve the corrosion resistance of stainless steel.)

We have studied the surface and subsurface compositions of Fe–18Cr–3Mo (100) under various UHV conditions over a wide range of temperatures. Both low and high-energy Auger transitions were analyzed, with reference spectrum from an *in situ* scribed sample used for quantification. Data were recorded during Ar sputtering until steady-state conditions were established in the composition, and during subsequent annealing at the same temperature. The segregation phenomena in both experiments critically depended on the state of cleanliness of the surface. The data for the clean surface suggest the possibility that iron (and not chromium) is the segregant in this alloy.

EXPERIMENTAL

The single crystal (79 wt%Fe, 18 wt%Cr, 3 wt%Mo according to EDAX) was mechanically polished to 1/4 micron using diamond paste, while keeping the orientation to less than 1° off the (100) plane as checked by Laue x-ray diffraction. The crystal was heated *in situ* using a tungsten filament mounted behind it, with a stabilized DC power supply. The temperature was monitored with a chromel–alumel thermocouple spot-welded to the sample. The crystal cleaning was done with high-temperature Ar ion sputtering cycles.

All Auger measurements were performed using the

*To whom correspondence may be addressed

0142–2421/86/120151–05 $05.00

SAM unit of PHI 549 system. The chamber was baked at 200°C giving a base-pressure of 3×10^{-10} torr. Sputtering was usually done at 1 kV (4kV in few experiments) with 2×10^{-5} torr argon in the chamber. A multiplex unit was used to monitor line intensities during kinetic measurements.

RESULTS

Steady-State Composition under Ar bombardment

The steady-state Cr–Fe relative concentration established during Ar sputtering showed significant temperature dependence above ~500 K (Fig. 1). As reflected by both low and high-energy Auger transitions, the composition of the surface region quite closely resembles the bulk value at room temperature and at 500 K. At higher temperatures up to ~800 K, the high-energy lines (HE) showed gradual Cr-enrichment, while the low-energy lines (LE) indicate near stoichiometric composition. Since the Cr(529) and Fe(703) lines correspond to an effective mean-free path of ~9.5 and ~11 Å, respectively, and the Cr(36) and Fe(47) lines correspond to λ_{eff} ~3.3 Å,[8] it can be concluded that the subsurface became depleted in iron, while the outmost surface layer stayed closer to the stoichiometric composition. (A typical quantitative in-depth composition profile of the surface region will be given below.) After reaching a maximum around 800 K, the Cr/Fe HE line ratio decreased with increasing temperature, and the LE line ratio first increased somewhat, and then at the highest temperature it decreased significantly as did the HE ratio. (This temperature could only be roughly estimated as being between 1100–1300 K.) 4kV Ar bombardment resulted in qualitatively similar behavior of the steady-state compositions. However, the Cr-enrichment effect in the intermediate temperature range was clearly weaker. Figure 2 shows the variation with temperature of the steady-state concentration of all three constituents of the alloy (relative to the bulk values) under 1 kV Ar sputtering. As can be seen, unlike Cr, the Mo enriched the surface even at room temperature. At higher temperatures this effect increased and at the highest temperature the Mo concentration decreased close to the bulk value.

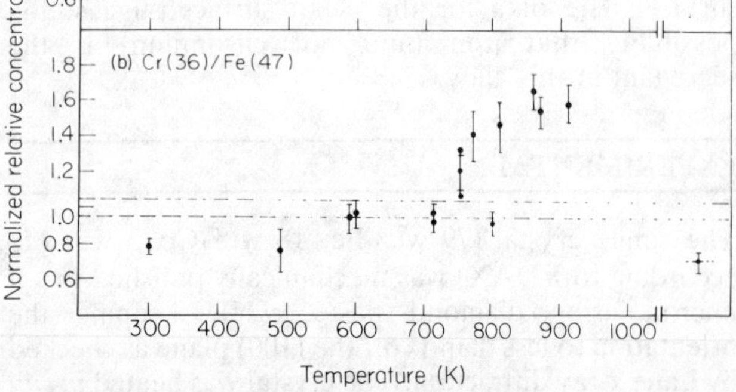

Figure 1. The temperature dependence of the steady-state Cr/Fe normalized relative concentration measured during 1 kV Ar sputtering of Fe–18Cr–3Mo(100). Quantification was done with bulk intensities of the low and high-energy Auger transitions measured for an *in situ* scribed sample. (the dashed lines around 1.0 denote experimental uncerainties in these reference intensities).

Post-bombardment annealing experiments

Drastic changes in the Cr and Fe line intensities occurred at the intermediate temperature range after turning off the argon ion beam. As an example, Fig. 3(top) shows a typical time evolution of the AES signals measured at 910 K. Note the sharp initial *increase* in Fe and the concomitant Cr *decrease*. (At lower temperatures this effect was smaller.) After the Fe/Cr increase, the trend was inverted and the Fe signal gradually *decreased* while that of Cr *increased*. This second stage is also characterized by the appearance of small quantities of carbon and even smaller oxygen. Qualitatively, the low-energy Cr and Fe lines exhibited similar behaviour as function of annealing time. The Mo behaviour during the initial stage is less clear due to the relatively poor signal to noise ratio, but clearly at the second stage, the Mo signal went significantly up, reaching after ~8 min more than 100% enrichment (which is much more than the Cr).

A completely different picture emerged (Fig. 3, bottom) when the annealing was performed under some residual O-containing species (probably CO). The following features should be noted in comparison to the previous experiment (Fig. 3, top):

(i) The initial Fe increase and Cr decrease were minor, and soon the Cr rose to a relatively high level in parallel to the increase in oxygen surface concentration; (ii) The increase in Mo was faster and

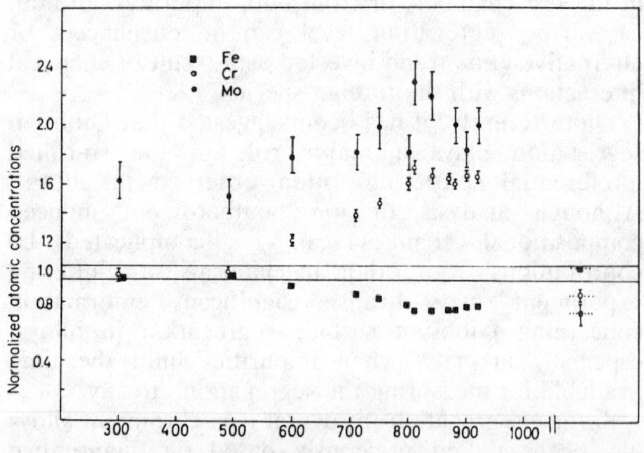

Figure 2. Normalized atomic concentrations during 1 kV sputtering vs. temperature, evaluated from the Mo(186), Cr(529) and Fe(703) line intensities and the corresponding sensitivity factors. The normalization was done with respect to evaluated concentrations for the scribed sample, so that each data point denotes the relative enrichment (or depletion) for each element.

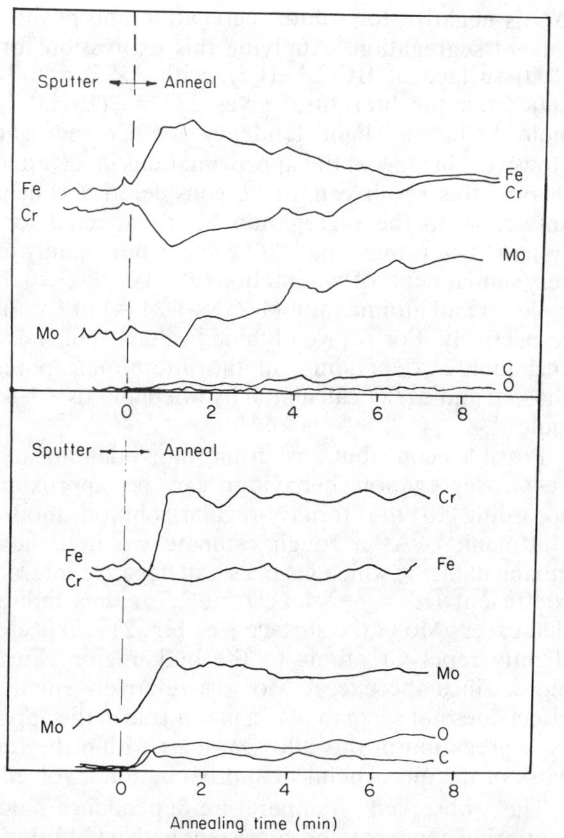

Figure 3. Auger line intensities (Mo × 5) measured at 910 K as function of time after stopping 1 kV Ar sputtering at this temperature: top: annealing under relatively clean residual-gas background; bottom: annealing under residual O-containing species in the chamber.

followed the rise in carbon, which quickly levelled off to a smaller value than the C final level in the previous experiment. (The Mo level was smaller, too.) Clearly, a correlation exists between Mo and C, and between Cr and O.

Ar sputtering under residual CO

The trend described above, namely, subsurface Cr-enrichment during Ar sputtering, could be chemically altered by performing Ar sputtering under small amounts of CO in the chamber. In Fig. 4(a), we present a spectrum recorded during sputtering with relatively clean residual gas background (only some CH_4 released by the ion pumps was present). Fig. 4(b) shows the spectrum recorded under the residual CO at the same temperature (780 K). Compared to the clean surface in the previous experiment (Fig. 4(a), small carbon and oxygen signals could be detected during sputtering. As can be seen, it had an overwhelming effect on the Cr/Fe intensity ratio. Compared to Cr subsurface *enrichment* and near-stoichiometric surface composition, under these conditions chromium *depletion* occurred at the entire surface region. Based on the LE and HE Cr/Fe intensity ratios relative to the reference (scribed sample) ratios, we fitted near-surface composition profiles of an exponential shape[6] to the data. For the first case (Fig. 4(a)) a stoichiometric top surface layer was assumed, while in the second (Fig. 4(b)) only a continuous exponential distribution starting from the outermost layer could fit the data. The profiles con-

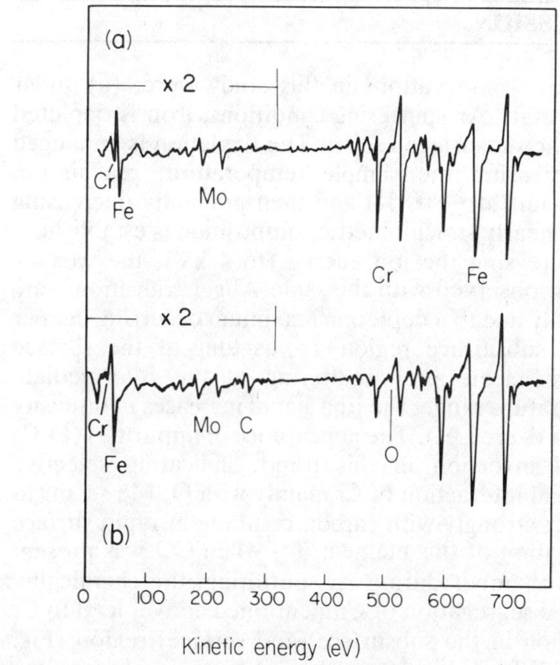

Figure 4. Auger spectra of Fe–18Cr–3Mo recorded during 1 kV Ar sputtering at 780 K: (a) under relatively clean residual gas background; (b) under residual CO background (primary e-beam – 3 kV).

structed in the way (Fig. 5) clearly demonstrate the drastic effect of the residual CO. Besides differences in the first atomic layer and in the basis trend (Cr enrichment vs. Cr depletion), the clean-surface composition profile (Fig. 5(a)) extends deeper into the bulk of the crystal.

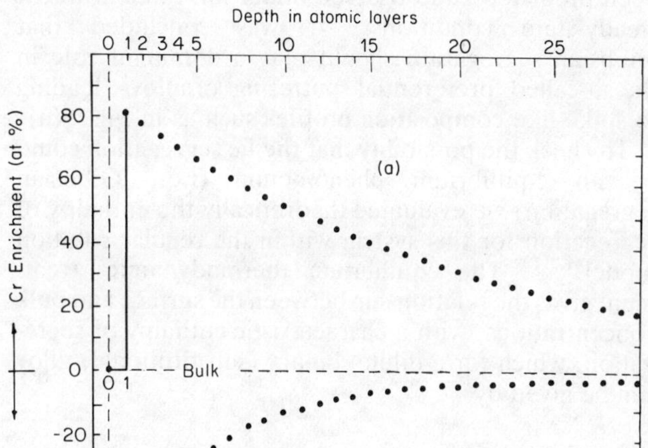

Figure 5. Chromium concentration depth-profiles (a), (b) constructed by fitting exponential-function parameters to the (normalized) experimental Cr/Fe line-intensity ratios (LE, HE) from Fig. 4(a) and 4(b), respectively. Calculated attenuated intensities (with characteristic λ's) from atomic layers parallel to the (100) plane were summed up to give the total intensity.

DISCUSSION

The main observations in this study were: (a) under steady-state Ar sputtering conditions, iron is depleted from the subsurface region. The depletion is enhanced by increasing the sample temperature, reaching a maximum (at ~800 K) and then gradually decreasing until a nearly stoichiometric composition is established. On increasing the ion energy (to 4 kV), the weaker effects, observed with the same Auger transitions, are probably due to a depletion maximum occurring deeper in the subsurface region; (b) as long as the surface remains clean after sputtering at the intermediate temperature range, the iron signal increases in intensity (the Cr decreases). The appearance of impurities (O,C) caused inversion in this trend, indicating selective chemical interaction of Cr mainly with O. Mo seems to interact strongly with carbon resulting in rapid surface segregation of this element; (c) when CO was present in the chamber during Ar sputtering, the chemically-induced segregation of Cr mentioned above, lead to Cr depletion in the subsurface (and surface) region (Fig. 5(b)). When this latter driving-force is absent, the experimental data mentioned above suggest that iron (and not chromium) is the element that segregates to the clean surface of this alloy. As such, it is depleted in the subsurface layers during sputtering, whereas the outermost surface layer remains nearly stoichiometric (Fig. 5(a)) due to the very similar sputtering yields of Fe and Cr.[9] (When chemically-induced Cr segregation occurred, the depletion zone extended to the outmost surface layer (Fig. 5(b)) probably due to the change in the Cr sputtering yield when bonded to the impurity.) Subsurface depletion of the segregating element has been predicted and observed under ion bombardment steady-state conditions.[3] It was concluded that Gibbsian segregation should play a dominant role in the so-called 'preferential sputtering' of alloys, leading to spike-like composition profiles such as in Fig. 5(a)

To check the possibility that the Fe segregation could be an equilibrium phenomenon (i.e., Gibbsian segregation) we evaluated theoretically the enthalpy of segregation for this system within the regular solution model.[1,2,13] The equilibrium thermodynamics treatment gives the relationship between the surface and bulk concentrations, with a characteristic enthalpy of segregation, which for a dilute, binary (substitutional) alloy can be given by

$$\Delta Q = \frac{\Delta Z}{Z}\,(H_2^{\text{sub}} - H_1^{\text{sub}} - \Delta H_{12}^{\text{mix}}) - E_{\text{el}}$$

In the first term, which depends on nearest-neighbour bond-strengths, ΔZ is the difference between the coordination number (Z) of a bulk atomic site and of a surface site, H_1^{sub} and H_2^{sub} are the sublimation heats of the pure solvent and pure solute elements, respectively, and $\Delta H_{12}^{\text{mix}}$ is the heat of mixing of the two components. E_{el} denotes the contribution due to the release of bulk elastic strain energy when a solute atom from the bulk exchanges positions with a solvent atom at a surface site, and it depends on the size difference of the solute and solvent atoms. It can be roughly approximated using continuum elasticity theory.[1] As defined above, ΔQ is negative for solute segregation and positive for solvent segregation. Applying this expression for the (100) surface of BCC Fe(Cr), with $\Delta Z/Z = 1/2$ and data from the literature, gives $\Delta Q = -(1.6\pm0.7)$kcal mole^{-1}, i.e. a slight tendency for Cr segregation. However, in view of the approximations involved in the theory, this result cannot be considered a conclusive answer as to the segregation trend expected for this system. The rather small ΔQ value is due mainly to the very similar heats of sublimation (95±1vs. 99.3±0.3 kcal mole^{-1}) and atomic radii (1.25 vs 1.24 Å) of Cr and Fe respectively. For E_{el} we obtained values of 0.14 or 0.26 kcal mole^{-1} depending on the interatomic potential chosen, and $\Delta H_{12}^{\text{mix}}$ calculated by Miedema[2] is -1.4 kcal mole^{-1}.

Possible contributions from molybdenum to the Fe–Cr segregation behaviour can be approximated according to the ternary-regular-solution model of Guttmann[11]. As a rough estimate we used heat of mixing data,[2,12] which gave a small negative interaction coefficient ($\alpha' = -2.4$ kcal mole^{-1}), thus indicating that excess Mo at the surface (see Fig. 2) is expected to slightly repel Cr atoms to the bulk region. Furthermore, since the excess Mo was relatively small, this effect does not seem to play a major role in the apparent Fe segregation in this alloy, at least within the framework of the quasichemical approximation involved.

The observed temperature-dependence during sputtering appears to be consistent with a Gibbsian-like segregation model. Thus, at ambient temperatures diffusion is too slow to cause any measurable effect, whereas at elevated temperatures thermally-activated diffusion occurs (possibly enhanced by the ion bombardment) by which Fe atoms segregate at the surface layer, and due to the combined effect of sputtering and diffusion, are depleted from the subsurface. The decreasing effect after ~800 K may be indicative of thermal equilibrium conditions so that the level of segregation is reduced by further increasing the temperature. (The slight Cr depletion observed at the highest temperature of this study, Fig. 1, may be due to an evaporation effect.) It should be noted that nearly no change was seen in the composition upon stopping the ion bombardment at this temperature, unlike the behavior observed at lower temperatures after similar annealing times (e.g., Fig. 3(a)).

To conclude this point, it should be noted that theoretically predicted segregation enthalpies for several binary systems were systematically more negative than the experimental values.[13] Thus, it is possible, as suggested by the data reported here, that ΔQ for Fe(Cr) is slightly positive, namely that there is a small tendency for iron segregation in this alloy, although the above calculation predicts small Cr segregation (ΔQ slightly negative). A phenomenon that we would like to mention at this point is diffusion of carbon from the surface inside, as has been recently observed following halocarbon adsorption on Fe (100).[10] Although we can see no evidence for a similar process in our data, additional experiments are planned in order to further establish the above conclusion concerning Fe Gibbsian segregation.

As to the Mo behaviour during sputtering, only the Mo(186 eV) line was monitored (λ_{eff}~5Å), and it reflected Mo enrichment even at room temperature (see Fig. 2). Its variation with temperature might indicate some segregation contributions in addition to mass-related effects (Mo sputter-yield relative to that of Fe is 0.64[9]). Numerical evaluation of ΔQ based on the expression given above shows that the two driving-forces, expressed by the bond-breaking term and strain-release term, are in an opposite sense for both the Fe(Mo) and Cr(Mo) systems. In particular, the considerably higher Mo–Mo bond-strength should contribute to Fe and Cr segregation, whereas the larger size of Mo (solute) should contribute to Mo segregation. Using continuum elasticity theory gives positive or negative ΔQ values, depending on the interatomic potential chosen. Non-linear elasticity theory[1] (plus bond-breaking) predicts solvent (Fe,Cr) segregation. On turning off the ion beam, no clear change could be noted in the Mo intensity until carbon appeared at the surface inducing remarkable Mo enrichment (Fig. 3). This result agrees with the strong synergistic segregation effect observed for Mo-C in a polycrystalline Fe–2Mo binary alloy[5] (estimated surface interaction coefficient ~7 kcal mole^{-1}). We expect no significant difference whether the carbon originates from the residual gas in the chamber, or co-segregates (with Mo) from the bulk, leading possibly to carbide formation at the surface. As to the Cr–C interaction, according to the present data they are attractive, but clearly weaker than the Mo–C interaction as well as the Cr–O interaction.

Acknowledgement

The authors are grateful to C. Leygraf, G. Hultquist and P.A. Dowben for providing the crystal.

REFERENCES

1. F.F. Abraham and C.R. Brundle, *J. Vac. Sci. Technol.* **18**, 506 (1981).
2. A.R. Miedema, Z. Metalle. **69** 455 (1978).
3. R. Kelly, *Surf. Interface Anal.* **7**, 1 (1985); R. Kelly, *Materials Sci. Eng.* **69**, 449 (1985).
4. C. Leygraf, G. Hultquist, S. Ekelund, J.C. Eriksson, *Surf. Sci.* **46**, 157 (1974).
5. H. Dumoulin and M. Guttmann, *Materials Sci. Eng.* **42**, 249 (1980).
6. P.A. Dowben, M. Grunze and D. Wright, *Surf. Sci.* **134**, L524 (1983).
7. C.L. Briant and R.A. Mulford, *Metall. Trans. A.* **13A**, 745 (1982).
8. M.P. Seah and W.A. Dench, *Surf. Interface Anal.* **1**, 2 (1979).
9. H.J. Mathieu and D. Landolt, *Appl. Surface Sci.* **3**, 348 (1979).
10. P.A. Dowben and M. Grunze, *J. Electron. Specrosc. Relat. Phenom.* **28**, 249 (1983).
11. M. Guttmann and D. McLean, in *Interfacial Segregation*, pp. 261–348, ed. by W.C. Johnson and J.M. Blakely, ASM, Metals Park, OH (1979).
12. A.R. Miedma, *Phillips Tech. Rev.* **36**, 217 (1976).
13. P. Wynblatt and R.C. Ku, as Ref. 11, pp. 115–136.

SURFACE AND INTERFACE ANALYSIS, VOL. 9, 157–161 (1986)

Simultaneous Analysis of Fracture Surfaces by Auger and Energy Dispersive X-ray Mapping

A. Wirth, I. Andreoni and G. Gregory
Metals and Materials Engineering Department, Sheffield City Polytechnic, Pond Street, Sheffield S1 1WB, UK

The grain boundary segregation of the metalloid impurities antimony and phosphorus to the prior austenite grain boundaries of an Fe–3.5 Ni–1Cr alloy during embrittlement at 430°C has been studied by simultaneous Auger and energy dispersive x-ray analysis. The segregating species were found to be related to the tempering characteristics of the unembrittled microstructures. Initial intergranular embrittlement in a quenched and tempered martensite microstructure was associated with the prescence of phosphorus whereas the initial embrittlement in a bainitic microstructure was associated with the segregation of antimony. The difference in behaviour between the two microstructures has been explained in terms of site competition at the grain boundaries. In the case of the tempered martensite microstructure the carbon activity is reduced as a consequence of the formation of chromium rich precipitates at the grain boundary. Carbon activity is reduced as a consequence of the formation of chromium rich precipitates at the grain boundary. The reduction in carbon activity allowed phosphorus migration to the boundary producing intergranular embrittlement. In bainites the predominant precipitate M_3C increased the carbon activity with a resultant decrease in phosphorus concentration the grain boundary and the absence of intergranular failure in the early stages of embrittlement. Prolonged embrittlement of the bainites produced a low energy intergranular failure. Increased nickel and antimony concentrations at the grain boundaries were associated with the formation of a fine grain boundary precipitate low in chromium. The increased carbon activity continued to prevent appreciable phosphorus segregation but was not sufficient to inhibit the cosegregation of nickel and antimony.

INTRODUCTION

Much attention has been focused in recent years on the understanding of the relationship between microstructure and grain boundary segregation characteristics in a wide range of steels which fail prematurely in an intergranular manner. Correlation of the information obtained from microstructural and surface analytical studies has often proved extremely difficult in the past mainly because of the problems of carrying out surface analysis in one electron optical system and then microstructural analysis of the fracture surface in a different electron optical system with the resultant repositional problems that specimen transfer involved. Examining the composition and distribution of grain boundary precipitates by energy dispersive x-ray mapping techniques simultaneously with Auger mapping of the same region during depth profiling has provided the authors with valuable information which is helping towards further improving their understanding of the relationship between precipitate composition, grain boundary segregated species and intergranular fracture characteristics of both ferritic and austenitic steels.

The aim of this paper is to relate in more detail the relationship between segregating species and grain boundary precipitate composition in a low alloy steel susceptible to intergranular failure during tempering at 430°C.

EXPERIMENTAL PROCEDURE

The composition of the vacuum melted material studied in this investigation was Fe 3.56 at %Ni, 1.04 at %Cr, 0.36 at %C, 0.24 at %Mn, 0.10 at %Si, 0.06 at %Sb, 0.002 at %S, <10 ppm As, <10 ppm Sn.

Auger fracture specimens 30 mm in length and 5 mm in diameter and notched 15 mm from one end were manufactured from the remains of izod samples previously fractured at ambient temperature. The izod samples had been heat treated to produce either tempered martensite or bainite and then embrittled alongside each other for up to 2 years at 430°C.

Auger and energy dispersive x-ray analysis were carried out simultaneously on the freshly broken Auger samples in a VG Microlab 500 Simslab fitted with a Link Systems energy dispersive x-ray analysis system.

Auger spectra were obtained using 3–20 kV electrons and specimen currents of 5–20 nA. Auger and x-ray mapping used a 10–20 kV electron beam energy and 10 nA specimen current.

Auger maps were processed using the ratio $\frac{P-B}{B}$ to minimize the effects of topography on the results.

RESULTS AND DISCUSSION

Previous examination of low alloy steels susceptible to reversible temper embrittlement (RTE) has highlighted the different low energy fracture morphologies

0142–2421/86/120157–05 $05.00

observed in embrittled quenched and tempered martensites and bainites.[1,2]

For example, embrittled, quenched and tempered martensite invariably fractures to produce a fracture morphology of intergranular, ductile intergranular or ductile dimpling. The amount of each type of morphology depending upon the temperature at which fracture occurs. Embrittled bainites on the other hand rarely contain intergranular or ductile intergranular modes of failure unless the alloy contains large amounts of metalloid impurity and has been embrittled for periods of time of the order of 1000 hours. Figures 2(a) and 3(a) show typical examples of the different fracture characteristics of embrittled bainites and tempered martensites.

Extensive transmission microscopical studies of the different embrittled microstructures, reported elsewhere,[3] showed significantly different prior austenite grain boundary precipitate types, distribution and morphologies. The authors suggested that changes in the precipitation sequence during the tempering of tempered martensite and bainite was responsible for a difference in the rate of diffusion of antimony to the tempered martensite and bainite prior austenite grain boundaries.

Clearly the presence of antimony after long tempering times would then promote the tendency to intergranular failure, hence the difference between the fracture morphologies of tempered martensites and bainites in the early stages of the embrittlement process. Similarly Erhart and Grabke[4] showed recently that the grain boundary phosphorus concentration was

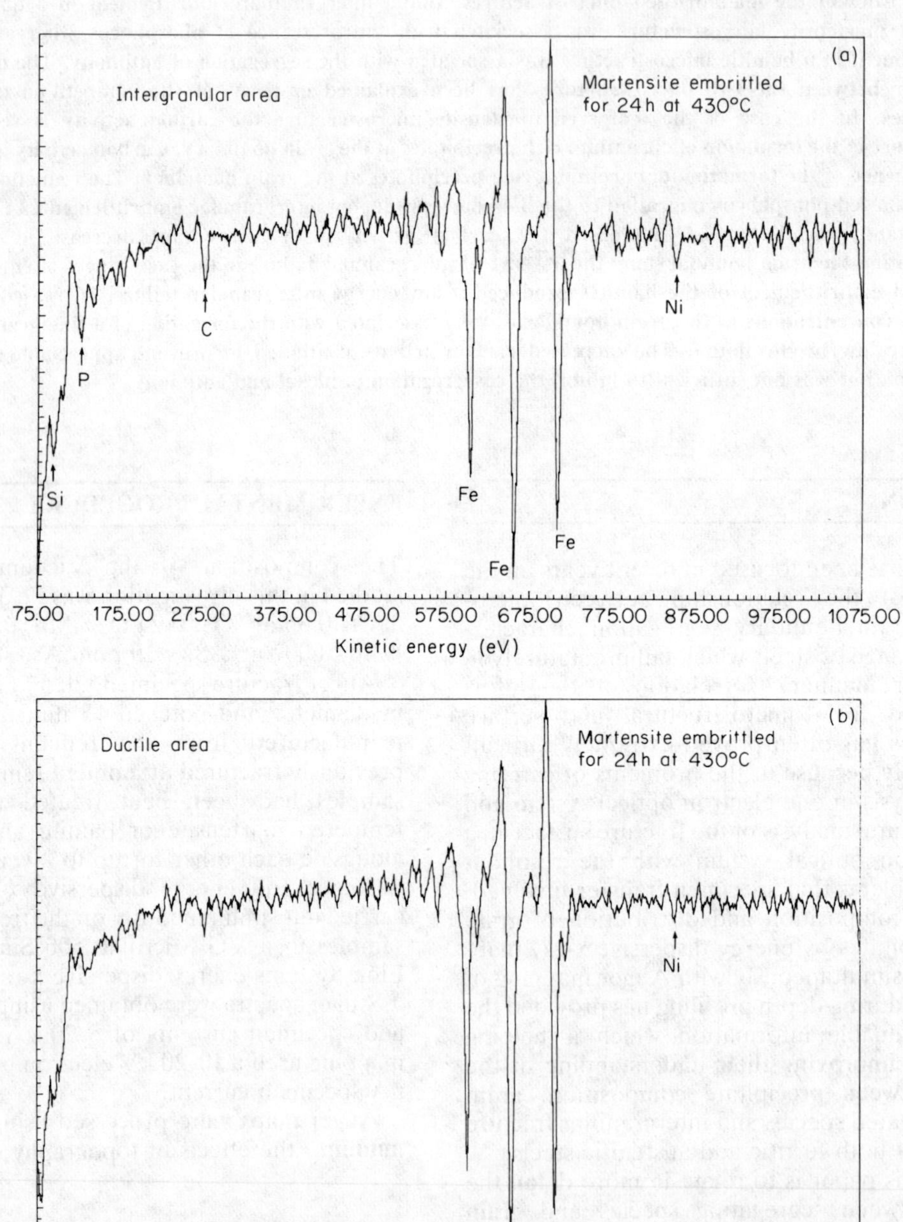

Figure 1. (a) Auger spectrum from intergranular area shown in Fig. 2(a); (b) Auger spectrum from ductile area shown in Fig. 2(a).

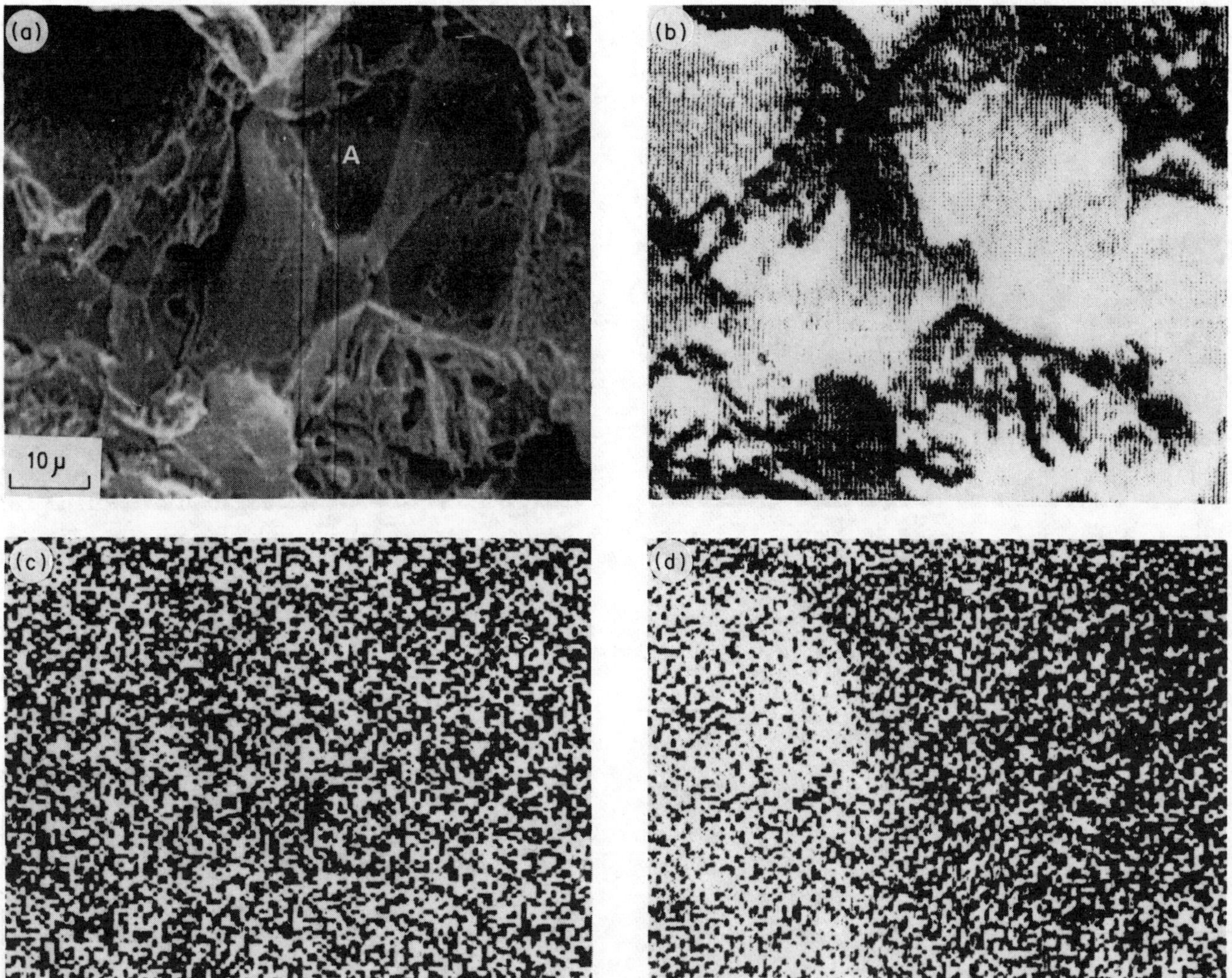

Figure 2. (a) Secondary electron image from fracture surface of embrittled martensitic material; (b) Auger map of background signal for area shown in Fig. 2(a); (c) Auger map showing nickel distribution on fracture surface shown in Fig. 2(a); (d) Auger map showing phosphorus distribution on fracture surface shown in Fig. 2(a); (e) energy dispersive x-ray analysis of particle A on intergranular fracture surface shown in Fig. 2(a); (f) Cr K x-ray map for area shown in Fig. 2(a).

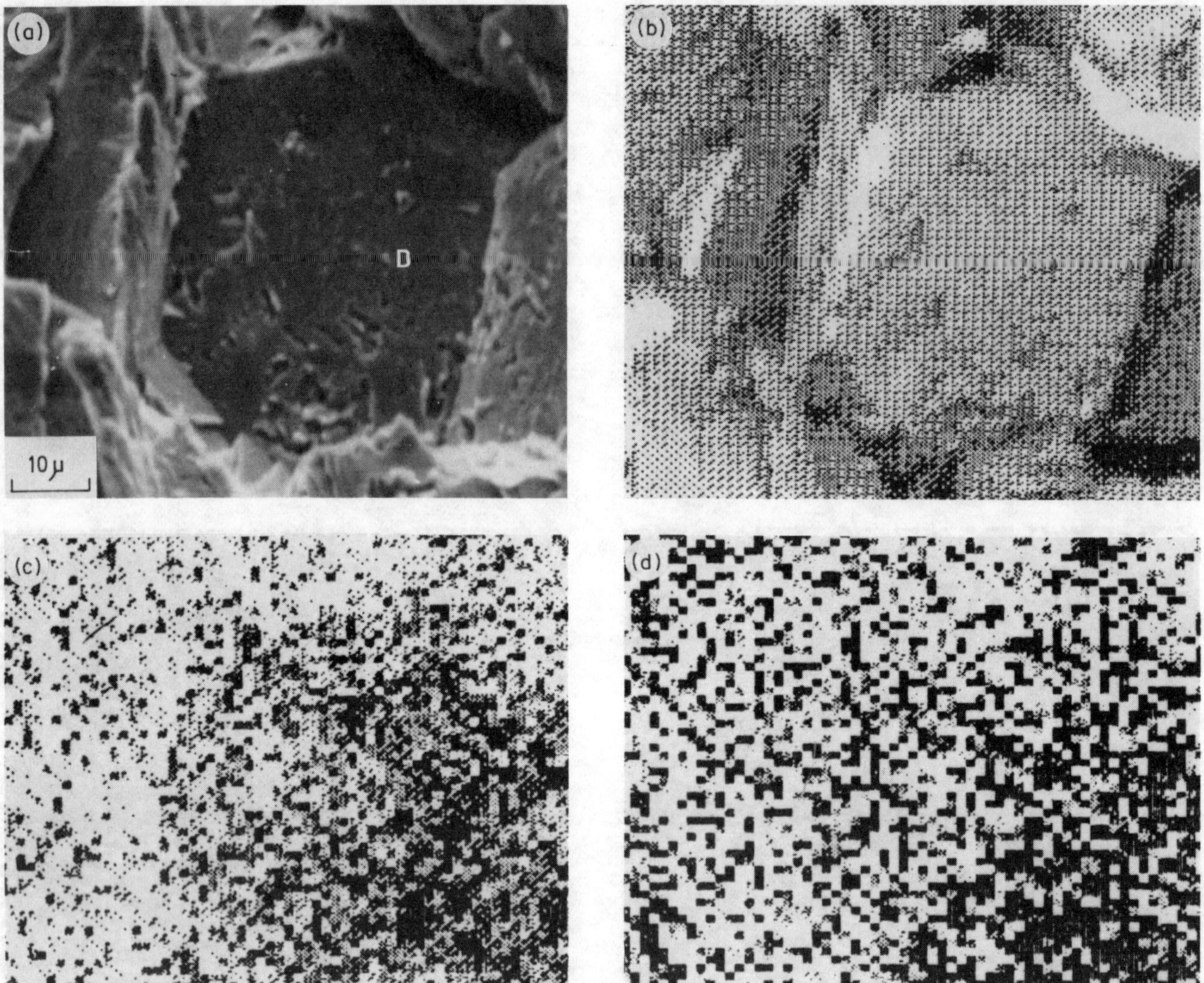

Figure 3. (a) Secondary electron image from fracture surface of embrittled bainitic material; (b) Auger map of background signal for area shown in Fig. 3(a); (c) Auger map showing nickel distribution on fracture surface shown in Fig. 3(a); (d) Auger map showing antimony distribution on fracture surface shown in Fig. 3(a); (e) energy dispersive x-ray analysis of particle on intergranular surface shown in Fig. 3(a); (f) Cr K x-ray map for area shown in Fig. 3(a).

dependent upon the carbide type present at the grain boundary. They reported that competition between carbon and phosphorus for grain boundary sites was related to the bulk concentration of free carbon. Using this argument Erhart and Grabke concluded that a strong carbide former, chromium reduces the carbon activity at grain boundaries in an Fe–C–P alloy allowing increased phosphorus segregations and intergranular failure. Conversely alloying elements e.g. nickel which increase the carbon activity ought to reduce grain boundary phosphorus segregation although they had no evidence available when going to press to confirm their view.

Examination of the Auger spectra shown in Fig. 1(a) and 1(b) from the intergranular and ductile areas shown in Fig. 2(a) indicates that interpretation becomes more complex when both chromium and nickel are present because the phosphorus and carbon concentrations are significantly higher on the intergranular facets compared with the ductile dimplex region. Furthermore, examination of the nickel and phosphorus Auger maps Fig. 2(c) and 2(d) suggests that the distribution of phosphorus is not significantly related to the distribution of nickel even though the nickel concentration in the intergranular spectrum is approximately twice the value in the ductile spectrum (compare Fig. 1(a) and 1(b)). What may be of greater significance, and in agreement with the results of Ehart and Grabke is that the chromium rich carbides are concentrated in the same area where the phosphorus concentration is greatest (compare Fig. 2(d) and 2(f)).

In the case of embrittled bainites the situation becomes even more complex when the low energy fracture changes to the intergranular type because the antimony metalloid impurity rather than phosphorus is associated with an increase in the nickel concentration on the intergranular facets compared with the cleavage region (see Fig. 3). The predominant precipitate associated with the intergranular facets contains much less chromium (see Fig. 3(e)) than the precipitate observed at the grain boundaries in the quenched and tempered material although the Auger peak height ratio P_c/P_{Fe} is similar for both bainite and tempered martensite intergranular facets.

It could be argued that the bainitic grain boundary precipitate which is low in chromium acts in a similar manner to Fe_3C in Fe–C–P alloy[4] in as much as the high carbon activity prevented the phosphorus migrating to the bainitic prior austenite grain boundaries during the early stages of tempering at 430°C. Antimony on the other hand because of its ability to form nickel–antimony precipitates and has a bulk concentration in the alloy of some thirty times greater than the phosphorus content may explain why antimony segregates to the bainitic prior austenite grain boundaries to promote intergranular failure after extensive tempering. It should also be remembered that the relatively clean prior austenite boundaries observed after short embrittling times contain M_3C precipitates after long tempering times,[3] although there is no reason to suppose the carbon activity will have changed significantly with the precipitation of the M_3C at the grain boundaries. Consequently embrittled bainites will always fail with a low energy cleavage mode in the early stages of tempering because the prior austenite grain boundaries are relatively free from precipitates and segregated impurities and the interface between the large matrix M_3C and the ferrite is a more probable area for inducing failure by cleavage.

Clearly there is a need for a great deal of further careful work to be carried out in order to resolve the factors relating grain boundary precipitate composition with the amount and composition of embrittling grain boundary species if intergranular embrittlement is to be understood in multiphase systems.

CONCLUSIONS

Simultaneous Auger and x-ray mapping of fracture surfaces produced *in situ* showed that the segregated species responsible for intergranular embrittlement in an Fe–3½%Ni–1 at%Cr steel tempered at 430°C was related to the precipitate composition at the prior austenite grain boundaries. Increasing the chromium content of the grain boundary precipitate resulted in a decrease in grain boundary phosphorus concentration and an increase in the antimony concentration.

Acknowledgement

One of us, I. Andreoni would like to thank CNPQ-MIC-STI (Brazil) for their continued support.

REFERENCES

1. A. Wirth and B. Clarke, *Advances in Physical Metallurgy and Applications of Steels,* University of Liverpool, September 1981.
2. R. Viswanathan and R. Joshi, *Met. Trans.* **6A,** 2289–2296 (1975).
3. A. Wirth, I. Andreoni and J. Titchmarsh, *Inst. Phys. Conf. Ser. No. 68,* 297–300 (1983).
4. H. Erhart and H.J. Grabke, *Met. Sci.* **15,** 401–408 (1981).

SURFACE AND INTERFACE ANALYSIS, VOL. 9, 163–168 (1986)

On the Use of Markers and Tracers to Establish the Growth Mechanism of Alumina Scales During High Temperature Oxidation

E.W.A. Young, H.E. Bishop,* J.H.W. de Wit†
Inorganic Chemistry Department, State University Utrecht, Croesestraat 77A, 3522 AD Utrecht, The Netherlands
*Materials Development Division, AERE Harwell, Didcot, Oxon, OX11 ORA, UK
†Lab. for Metallurgy and Materials Science, Delft University of Technology, Rotterdamseweg 137, 2628 AL Delft, The Netherlands

Various 'marker methods' have been applied to study the oxidation mechanism of NiAl and Ni–15Cr–24Al–.3Y from 1170 K to 1370 K. Experiments were performed using noble metal markers. The behaviour of both vacuum deposited noble metals as well as implanted noble metals was studied. Also ^{18}O tracers were used. RBS, $^{18}O(p, \alpha)^{15}N$ resonance spectrometry and SIMS analysis were performed to investigate the alumina scale composition and the position of the marker after oxidation of alloys. The results show that (at 1170 K) NiAl oxidizes mainly by outward diffusion of Al through the oxide scale. It is also shown that noble metal markers may give misleading results. The oxidation mechanism of Ni–15Cr–24Al–.3Y is much more complex. The oxide scale formed on this alloy contains Ni, Cr and Y–rich oxide regions. These regions contribute to the apparent overall oxidation mechanism of the alloy.

INTRODUCTION

The study of the transport properties of protective alumina scales on high temperature alloys occupies a position of prime importance in high temperature corrosion research.[1,2] Many investigations concern the relation between the morphology of the alumina scales and the transport properties.[2,3] There is however a need for more direct methods to establish the scale growth mechanism. Often inert metal markers are applied for this purpose. A thin wire of (usually) Pt is positioned on the surface of the alloy or on an existing alumina scale. The position of the wire is observed after subsequent oxidation. Also very thin, vacuum deposited, layers of noble metal have been used.[4] The use of thin layers is thought to avoid the possible perturbation of the scale growth process by the relatively thick Pt wires. The marker is supposed to be immobile with respect to the alumina. Thus, if the scale grows by an inside growth mechanism (oxygen mobile in the scale), the marker will remain at the scale/gas interface. In the case of outside growth (Al mobile), the marker will be incorporated in the scale at or near the metal/oxide interface. All metal marker studies reported in literature for alumina scale forming alloys suggest inside growth, oxygen being the mobile species.[4,5,6,7] Recently however we reported that inert markers can give misleading results. ^{18}O tracer experiments showed that O and Al both contributed to the growth of alumina scales on NiAl.[8] In the present study various marker methods were investigated systematically for the oxidation of NiAl at 1170 K. Experiments were performed using noble metal markers applied either by vacuum deposition of the metal on preoxidized NiAl or by implantation of the noble metal into the scale of preoxidized NiAl. The position of the markers in the scale after subsequent oxidation was determined by Rutherford back-scattering spectrometry (RBS) of 2 MeV $^4He^+$. Also ^{18}O tracer experiments were performed. The depth profile of the ^{18}O tracer was obtained from proton activation measurements using the reaction $^{18}O(p, \alpha)^{15}N$. The single spectrum method of 800 keV protons was used.[9]

The tested marker methods were then applied to study the oxidation mechanism of Ni–15Cr–24Al–.3Y (at%). Scanning electron microscopy (SEM) and Scanning auger microscopy (SAM) measurements were used to determine the composition of the complex scales on this alloy. In addition to the proton activation measurements secondary ion mass spectroscopy was applied to measure the ^{18}O distribution in the scale.

RESULTS

Marker and Tracer Experiments on NiAl

The preparation of the alloy has been described in detail before.[8,10] Oxidation of polished specimens in '^{16}O' was performed in a tube furnace in dried flowing natural oxygen. Oxidation in '^{18}O' was performed in a sealed quartz ampoule, filled with a 50% ^{18}O enriched oxygen gas. The samples were slowly reheated and cooled after each oxidation treatment. A smooth alumina scale is formed on NiAl during oxidation at 1170 K. A micrograph of such a scale is given in Fig. 1.

As reported previously, noble metals deposited on the surface of preoxidized NiAl remain at the surface

0142–2421/86/120163–06 $05.00

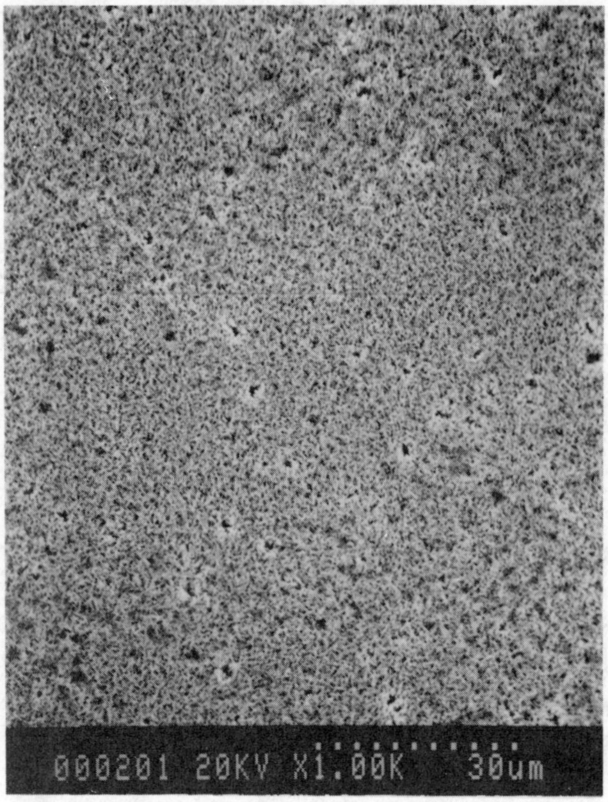

Figure 1. SEM image of NiAl, oxidized 88 h, T = 1170 K, O_2.

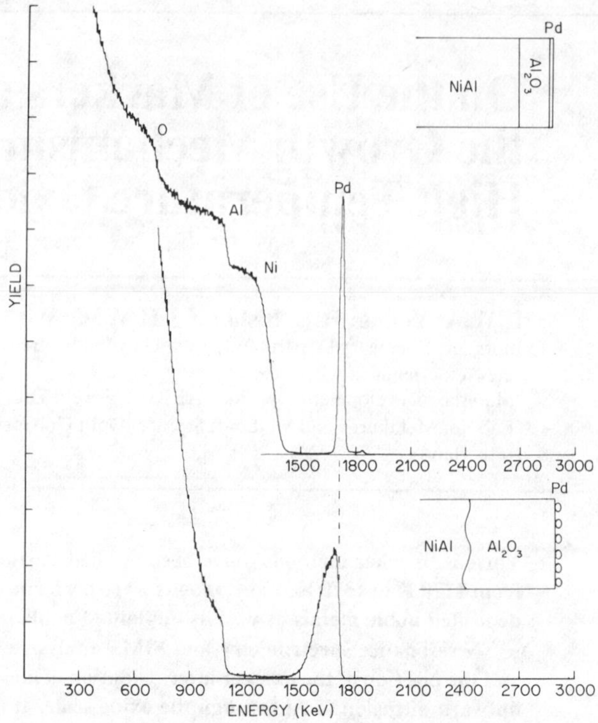

Figure 2. 2 MeV $^4He^+$ RBS: (top) NiAl preoxidized 45 min, 1170 K, O_2 100 nm Pd deposited; (bottom) NiAl preoxidized 45 min, 1170 K, O_2 100 nm Pd deposited, oxidized 16 h, 1170 K, O_2.

during oxidation.[8,10] In Fig. 2 an example of such a surface marker experiment is given. The 2 MeV $^4He^+$ RBS spectrum of preoxidized NiAl (45 min, O_2, 1170 K) covered with 10 nm of vacuum deposited Pd is shown. The sharp peak at 1723 keV represents the Pd layer. From the position of the peak in the spectrum it can be deduced that the Pd is at the specimen surface. The lower spectrum was obtained after the specimen had been given an extra oxidation treatment (16 h). From the shift of the Ni edge in this spectrum to lower energies it can be concluded that the thickness of the alumina scale increased (from ~300 nm to ~800 nm). The Pd however is still present at the specimen surface (Pd at 1722 keV, the surface value). The broadening of the peak can be attributed to the transformation of the Pd layer into small islands.[11] Such islands of about 0.5 μm in diameter have been observed with SEM/EDX analysis of the specimen surface.

If similar experiments are performed with implanted noble metal the marker material is covered by newly formed alumina, as can be seen in Fig. 3. The upper spectrum is for the preoxidized NiAl, implanted with a dose of 1×10^{16} at cm^{-2} Pt. The position of the Pt peak indicates that the Pt is confined to the first few hundred Å of the surface of the alumina scale. After further oxidation of this specimen for 5 h the Pt is no longer just below the surface, but is covered by a layer of 180 nm of alumina. This is reflected in the shift of the Pt peak to lower energies (see lower spectrum of Fig. 3). In total 240 nm of alumina has been formed during this second oxidation treatment (calculated from the shift of the Ni edge in the spectrum).

Figure 4 shows the position of the (implanted) Pt marker in the scale after several subsequent oxidation treatments. The total oxide thickness and the thickness of the newly formed scale after implantation, both derived from the shift of the Ni edge in the RBS spectra, are also given. The figure shows that 2/3 of the newly formed alumina is on top of the Pt marker. Whereas the surface marker experiment suggests an

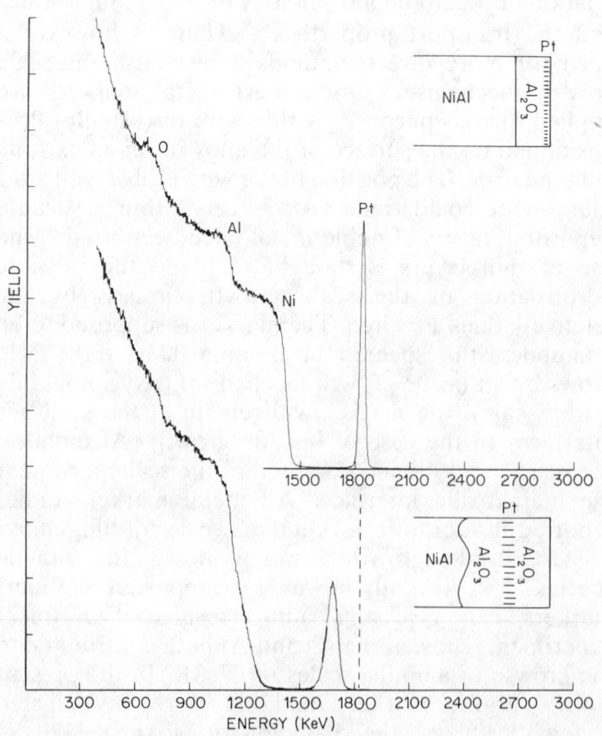

Figure 3. 2 MeV $^4He^+$ RBS: (top) NiAl preoxidized 30 min, 1170 K, O_2, Pt implanted; (bottom) NiAl preoxidized 30 min, 1170 K, O_2, Pt implanted, oxidized 5 h, 1170 K, O_2.

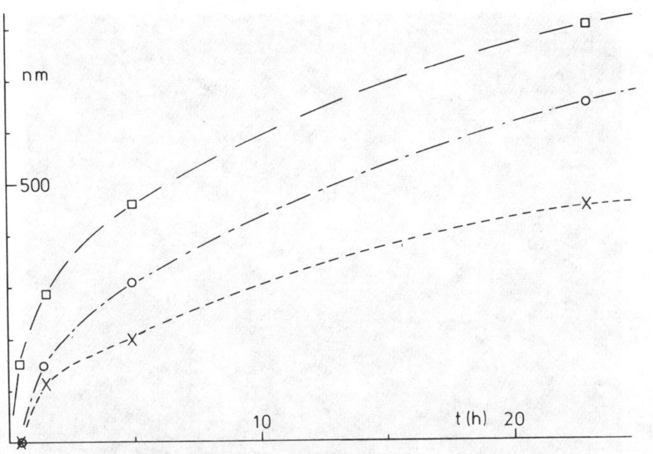

Figure 4. Pt implanted NiAl oxidized T = 1170 K, O_2.
□ thickness of the oxide scale
o thickness of the newly formed oxide
x thickness of the oxide on top of the implanted Pt

inside growth mechanism, the experiment with implanted Pt clearly indicates 66% outside growth.

Figure 5 shows the result of a proton activation analysis of a specimen that has been oxidized for 2 h in natural (^{16}O) oxygen and for 8 h in ^{18}O enriched oxygen gas. As outlined previously,[8] this spectrum in combination with RBS data allows a depth profile to be calculated. A schematical representation of the depth profile of the alumina scale is given in the upper right corner of Fig. 5. (50% ^{18}O enriched oxide formed in the 50/50 % oxygen gas mixture, is referred to as '$Al_2{}^{18}O_3$'.) The ^{18}O distribution confirms the outside growth mechanism due to Al diffusion through the scale. Most of the new ^{18}O rich oxide is situated above the original oxide/gas interface, but some ^{18}O penetrated into the existing ($Al_2{}^{16}O_3$) scale. Roughly the ratio between penetrated and non-penetrated oxygen is 1/2. Evidently the penetration of oxygen below the original oxide/gas interface will cause this interface to move relatively outwards. This explains the observed position of the implanted Pt marker, which is found to be covered with only 2/3 of the newly formed oxide. It can be concluded therefore that the original oxide/gas interface is correctly indicated by the implanted Pt marker.

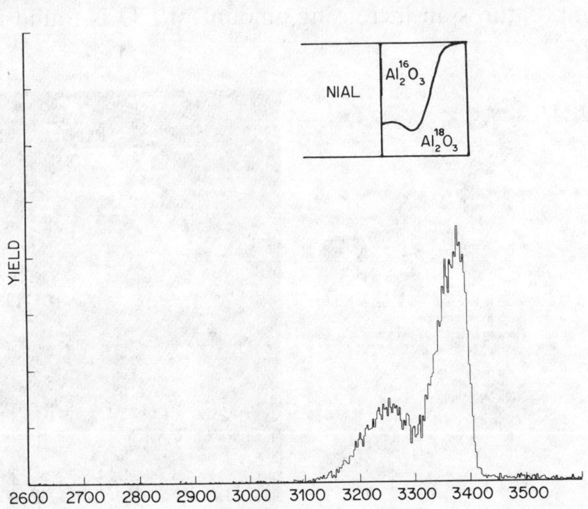

Figure 5. α spectrum (800 keV protons) of NiAl oxidized, 1170 K, first 2 h ^{16}O, later 8 h ^{18}O).

The result of the ^{18}O tracer experiments and the measurements on implanted markers clearly show that 'surface noble metal markers' cannot be used to establish the growth mechanism of alumina scales. These surface markers of noble metals from islands which probably flow over the alumina surface. The mechanical adherence of the metal to the scale presumably is very small. Therefore Al ions will easily reach the surface where they will react with oxygen without interference by Pd. Previously we reported on the behaviour of Y layers deposited on preoxidized NiAl. After oxidation, 2/3 of the newly formed oxide was found on top of the layer, just as is observed in the case of the implanted Pt marker.[10] This can easily be understood considering the very fast oxidation of the Y layer. The adhesion of Y_2O_3 to alumina will probably be excellent because both oxides have a very similar crystal structure. Therefore, if the material does not act as a barrier for either O or Al transport, it will effectively act as a marker like implanted Pt. The present data fully support the results of a previous ^{18}O tracer study.[8] For a more detailed discussion on the actual diffusion mechanism i.e. short circuit vs bulk diffusion we would like to refer to this paper here.

Summarizing we can state that ^{18}O tracers are reliable markers to establish the growth mechanism of scales. They provide direct information on the transport of oxygen. If noble metal markers are to be used, implantation evidently is the better choice.

Oxidation of Ni–15Cr–24A1–.3Y

Buttons of Ni–15Cr–24Al–.3Y were prepared from pressed tablets of an appropriate mixture of Ni, Cr, Al and Y metal powder. The tablets were melted in argon, using induction heating. Samples of about 5 × 5 × 1 mm^3 were cut from the buttons with a slow speed diamond saw. Prior to oxidation one side of the specimen was carefully polished down to 1 μm with diamond paste. Oxidation was performed as described above for NiAl. The NiCrAlY alloy obtained this way contains three different phases. The major phase is found in big grains, (see Fig. 6, phase A). Between the grains there are two other discrete phases. The nominal composition of the phases was measured by wavelength dispersive electron microprobe analysis. The composition of the phases is given in Table 1.

Oxidation of this alloy at 1170 K does not result in a smooth alumina scale as observed on NiAl. The oxide that has grown on the intergranular phases B and C has a convoluted structure (see Fig. 7). Auger mapping of the overall oxide has developed on one of these phases. Al and Ni maps of an oxide surface comparable with that shown in Fig. 7 are presented in Fig. 8. In order to establish the transport properties of these scales again oxidation in alternating ^{16}O and ^{18}O was performed. In addition to proton activation analysis, SIMS profiling of oxygen was applied for these scales. The depth distribution of various other elements can be determined simultaneously, using this method. A CAMECA IMS 3F secondary ion microscope was used. The oxide scales were eroded by a Cs^+ primary beam of about 200 nA. A beam raster 250 μm square was used and the field aperture in the secondary ion optics was set to select an area of 60 μm diamter, in the

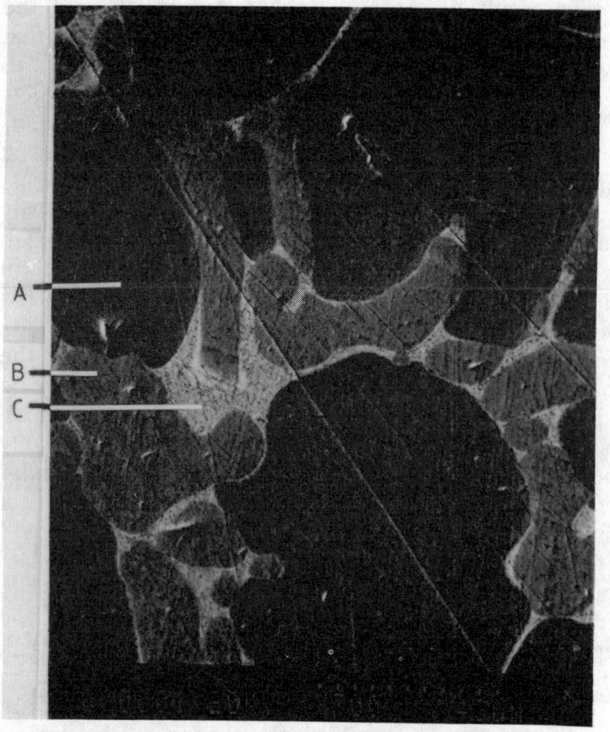

Figure 6. BEI of Ni-15Cr-24Al-.3Y (polished surface).

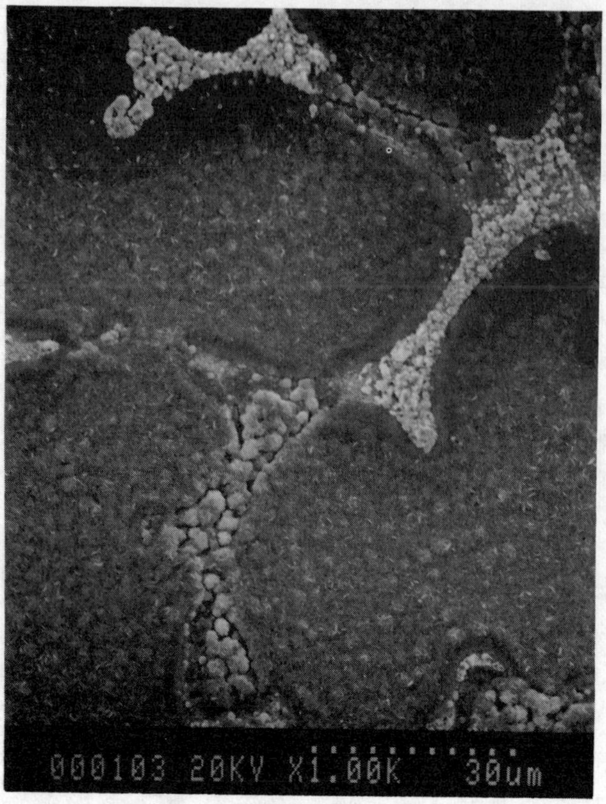

Figure 7. SEM of Ni-15Cr-14Al-.3Y oxidized 18 h, 1170 K, O_2.

Table 1. Composition of the phases in Ni–15Cr–24Al–.3Y

At%	Phase A	Phase B	Phase C
Ni	60	63	65
Cr	9	22	13
Al	31	15	16
Y	–	–	6

middle of the sputtered area, for analysis. To avoid charging of the specimen during the analysis, the specimen was gold coated and an electron flood gun was used. The ^{16}O and ^{18}O depth profile is shown in Fig. 9. The depth scale was deducted from proton activation analysis of the same scale. (The ^{16}O and ^{18}O data were processed to correct for the use of only 50% enriched oxygen gas, the plotted $^{18}O = 2 \times {}^{18}O$ measured, ^{16}O plotted $= {}^{16}O - {}^{18}O$ measured.)

The ^{18}O rich oxide again is found at the specimen surface. The original $Al_2\,{}^{16}O_3$ layer formed during the first oxidation treatment (in ^{16}O) is found near the alloy/scale interface. There is a very good agreement between the SIMS ^{18}O depth profile and the $^{18}O(p, \alpha)$ ^{15}N in depth analysis of the ^{18}O distribution (not shown here). Unlike the scales on NiAl, the outer scale on this alloy also contains a substantial amount of ^{16}O. The distribution of Al, Ni and Cr in the scale, measured in the same run as the O profiles, is shown in Fig. 10. Ni is mainly located near the surface of the scale. SIMS imaging of the scale shows that Cr is present together with some Y in the oxide that has grown on the intergranular metal phases (B and C). SIMS images of Cr and Y in the scale are shown in Fig. 11. Note that the amount of Cr and Y in the scale is rather small, below the detection limit of AES, ~5%.

The different transport properties of the Ni-rich oxide are probably responsible for the different oxygen isotope distribution in this scale compared with the pure alumina scales of NiAl. The most likely explanation is that growth of the alumina scale on the Al rich phase in the NiCrAl alloy takes place primarily through outside growth, as was observed on NiAl. In contrast the scales on alloy phases low in Al grow chiefly by inward diffusion of oxygen. This explains why ^{16}O is still found in the oxide near the surface after oxidation first in ^{16}O and subsequently in ^{18}O. A schematic representation of this scale growth model is given in Fig. 12.

Oxidation of the same alloy at higher temperatures (1270–1370 K) was also studied. At these higher temperatures an increasing amount of ^{18}O is found in

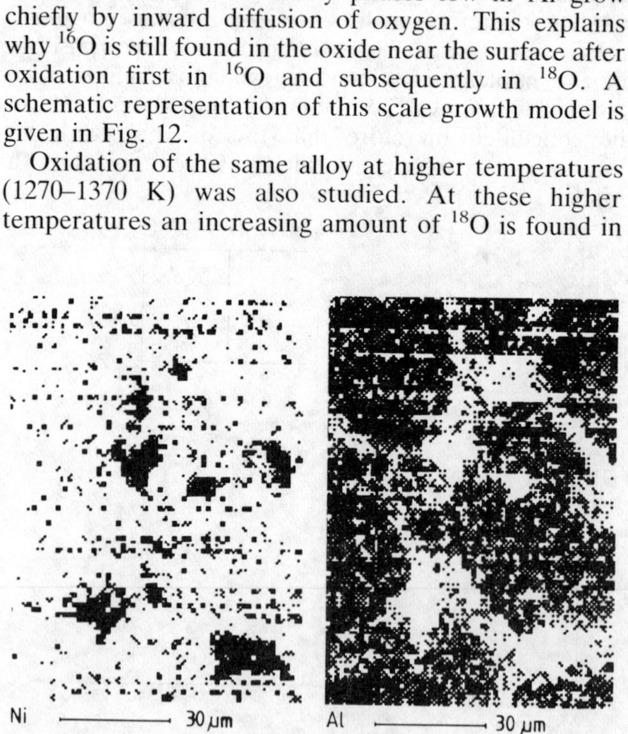

Figure 8. Auger maps of Ni–15C-24Al-.3Y oxidized 10 h, 1170 K, O_2.

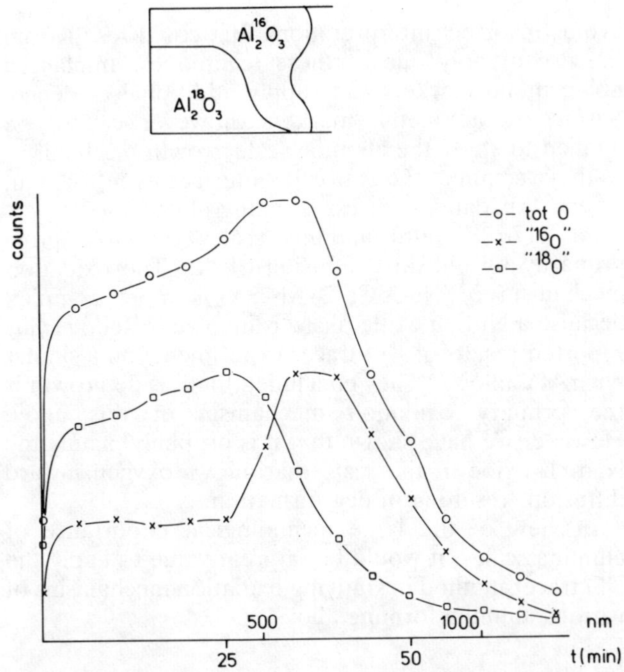

Figure 9. SIMS profile (prim beam Cs, pos. ions detected), Ni-15Cr-24Al-.3Y, oxidized 1170 K, first 2 h ^{16}O, later 16 h ^{18}O.

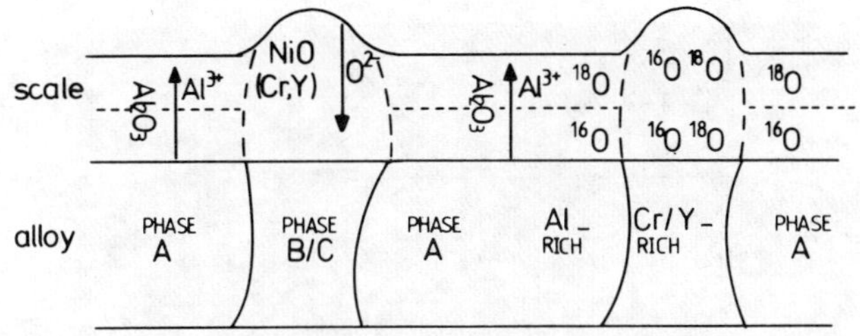

Figure 12. Schematical representation of the scale formation on Ni-15Cr-24Al-.3Y at 1170 K.

the inner region of the scale after the second oxidation treatment in ^{18}O, although, (especially at 1270 K), much of the ^{18}O is still found in the outer surface of the scale. Ni, Cr and Y are again detected in the scale.

A SIMS depth profile of a scale formed after oxidation first in ^{16}O for 2 h, then in ^{18}O for 6 h at 1370 K is shown in Fig. 13 (corrected for the use of a 50% ^{18}O enriched gas mixture again). ^{18}O is found all through the scale, but also a large amount of it is found under the scale. As can be seen in Fig. 14, pegs of oxide have developed under the scale. (The dark area represents the oxide.) Note that the depth profile of only ^{18}O could easily have been interpreted as an indication of primarily overall inside scale growth. Whereas peg formation under the scale seems to be the genuine effect here.

Oxidation experiments with NiAl (50/50%) show that at these temperatures alumina scales still partially grow by an outside growth mechanism.[8] Probably the scale formation on the Al–rich phase of this NiCrAlY alloy is also partly by outside growth. The scale on the alloy phases low in Al seems to be less protective which suggests that there are diffusion paths in this Ni (Cr and Y)–rich oxide for oxygen, supplying oxygen for internal oxidation.

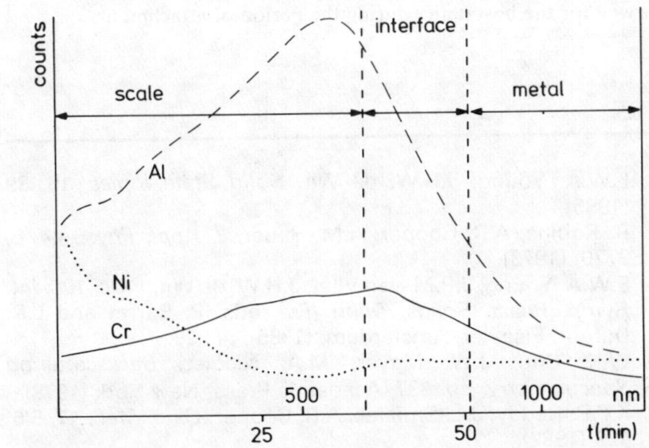

Figure 10. SIMS profiles (prim. beam Cs^+, pos. ions detected), Ni-15Cr-24Al-.3Y, oxidized 1170 K, first 2 h ^{16}O, later 16 h ^{18}O.

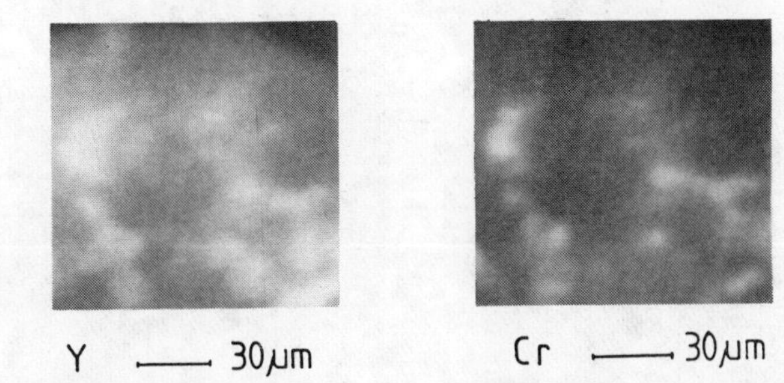

Figure 11. SIMS images (prim. beam Cs^+, pos. ions detected), Ni-15Cr-24Al-.3Y oxidized 1170 K, 18 h O_2.

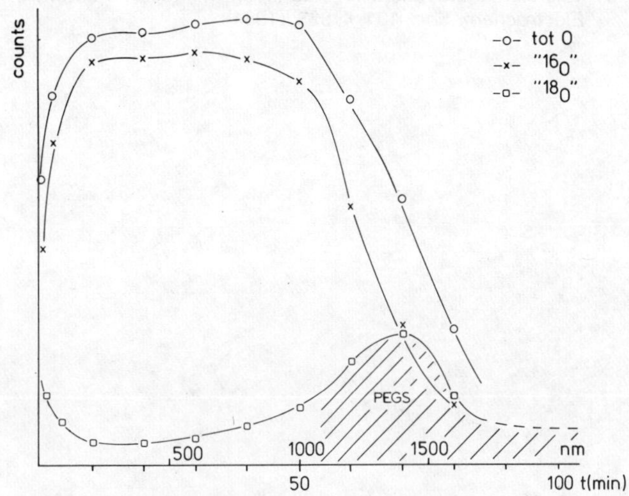

Figure 13. SIMS profile (prim. Cs, pos. ions detected) Ni-15Cr-24Al-.3Y oxidized 1370 K, first 2 h ^{16}O, later 6 h ^{18}O.

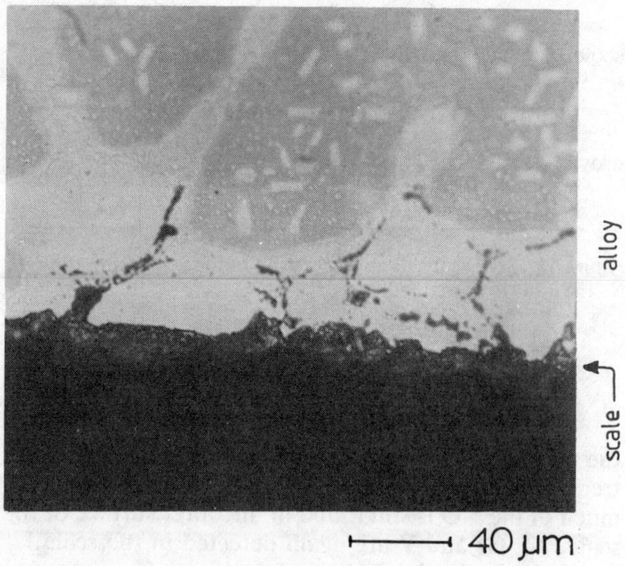

Figure 14. Optical micrograph Ni-15Cr-24Al-.3Y oxidized 1370 K, 10 h, O_2 (polished at 30° angled).

DISCUSSION AND CONCLUSIONS

^{18}O tracer studies proved to be a useful tool in determining the primary transport mechanism in oxide scales on NiAl and NiCrAlY. In particular the combination of proton activation, SIMS analysis and optical data produces reliable and complementary results and avoids incorrect interpretations that could result from the use of only one of these techniques. Implanted noble metal markers can supply additional evidence. Surface noble metal 'markers' however cannot be applied to study the alumina scale growth mechanism, as they remain at the gas/oxide interface even when an external oxidation process is taking place.

At 1170 K pure alumina scales on NiAl grow primarily by outward diffusion of Al. The oxidation mechanism of Ni–15Cr–24Al–.3Y is more complex because a Ni-rich oxide phase is involved. Reddy *et al.* reported results of ^{18}O tracer experiments on a similar NiCrAlY alloy.[12] They concluded that inside growth is the primary oxidation mechanism of this alloy. However we have shown that it is probably mainly the Ni–rich oxide in the scale that allows oxygen inward diffusion, resulting in peg formation.

In view of the large technological importance of alumina scales, it would be of great value to apply the ^{18}O tracer method in studying oxidation mechanisms of various alumina forming alloys.

Acknowledgements

The authors are grateful to the Physics Department of the University of Utrecht for providing 3 MV facilities. They also wish to acknowledge that the implantation of Pt was performed by D. Scholten and R.W. Hawes did the electron microprobe analysis. One of the authors (EY) would like to thank all members of the Surface Anal. Section at Harwell for the hospitality during the period of attachment.

REFERENCES

1. J.L. Smialek, R. Gibala, High Temp. *Corr.*, NACE **6**, 274, (1983).
2. D.P. Whittle, J. Stringer, *Phil. Trans. R. Soc. Lond. A,* **295,** 309, (1980).
3. F.A. Golightly, F.H. Stott, G.C. Wood, *J. Electrochem. Soc.,* **126,** 6, 1035, (1979).
4. H.M. Hindam, W.W. Smeltzer, *Oxid. Met.,* **14,** 4, 337, (1980).
5. H.M. Hindam, W.W. Smeltzer, *J. Electrochem. Soc.,* **127,** 7, 1630, (1980).
6. J.K. Tien, F.S. Petit, *Met. Trans.,* **3,** 1587, (1972).
7. T.A. Ramanarayanan, M. Raghavan, R. Petkovic-Luton, *J. Electrochem. Soc.,* **131,** 4, 923, (1984).
8. E.W.A. Young, J.H.W. de Wit, *Solid State Ionics,* **16,** 39 (1985).
9. R. Robins, A.R. Cooper, A.M. Heuer, *J. Appl. Phys.,* **44,** 8, 3770, (1973).
10. E.W.A. Young, J.P.M. van Vliet, J.H.W. de Wit, *Proc. 10th Int. Symp. React. Solids, Dijon (Fr.),* eds. P. Barret and L.C. Dufour, Elsevier, Amsterdam, (1985), p. 229.
11. W.K. Chu, J.W. Mayer, M.A. Nicolet, *Backscattering Spectrometry,* pp. 337, Academic Press, New York (1978).
12. K.P.R. Reddy, J.L. Smialek, A.R. Cooper, *Oxid. Met.,* **17,** 5/6, 429 (1982).

SURFACE AND INTERFACE ANALYSIS, VOL. 9, 169–173 (1986)

An AES (SAM), ESCA and ISS Study of the Oxidation of Pt_3/Sn

1: Low Pressure Exposure to O_2

Gar B. Hoflund and Douglas A. Asbury
Department of Chemical Engineering, University of Florida, Gainesville, Florida 32611, USA

Piotr Kirszensztejn† and Herbert A. Laitinen
Department of Chemistry, University of Florida, Gainesville, Florida 32611, USA

ESCA, ISS, AES and scanning Auger microscopy have been used to examine a sputter-cleaned Pt_3Sn alloy surface before and after exposure to oxygen at 10^{-6} Torr. It is found that the amount of oxygen adsorbed on a sputtered surface is a strong function of sample temperature with a maximum at about 200°C. The clean surface consists of regions of varying Pt-to-Sn concentration ratios. Regions near grain boundaries are tin rich, and the tin concentration decreases in moving away from a grain boundary until a platinum-rich region is reached. ESCA shows that an exposure to oxygen causes oxidation of the tin but no change in the Pt core-level features. ISS shows that segregation of tin to the surface occurs and that the surface Pt concentration decreases due to coverage by the tin.

INTRODUCTION

The use of bimetallic catalysts in hydrocarbon reforming processes is a major improvement over alumina-supported platinum crystallites. These bimetallic catalysts generally exhibit higher activities for longer operating periods and allow for greater control of the selectivities toward the many reactions which occur during reforming. Bimetallic catalysts are of the form Pt–X/Al_2O_3 where X can be Rh, Pb, Ir, Ge or Sn. Of these Pt–Ir/Al_2O_3 and Pt–Sn/Al_2O_3 are used most widely in industrial processes.

Reforming catalysts are bifunctional in that certain reactions such as hydrogenation and dehydrogenation are catalyzed by metallic sites and other reactions such as isomerization are catalyzed by the acidic sites of the support. It is well known that the presence of tin affects the catalytic behavior of the platinum. However, the form of the tin and the way in which it interacts with the platinum are less completely understood. Previous studies have suggested that it is possible to reduce the tin to metallic form in these catalysts and that alloy formation occurs,[1–5] particularly at high loading. Other studies claim that tin is not fully reduced to the metallic state in these catalysts and that modification of the catalytic properties of the platinum occurs through the influence of SnO, SnO_2, or other oxygen-containing species.[6–11]

It is anticipated that a study of the interaction between platinum and tin in alloys before and after various treatments will be useful in understanding the platinum–tin interaction in Pt–Sn/Al_2O_3. Several previous studies of platinum–tin alloys appear in the literature. It is interesting to note that chemisorption and reaction studies over Pt–Sn alloys demonstrate that these surfaces become less active as the surface tin concentration increases.[12–14] ESCA and AES studies of Pt–Sn alloys[15–17] show that the surface Pt–Sn ratio can vary over a wide range depending upon the surface pretreatment. Sputtering or exposure to hydrogen enriches the surface region in platinum while annealing or exposure to oxygen enriches the surface region in tin.

The study of the oxidation of the surface of a Pt–Sn alloy provides a method of probing the Pt–Sn interaction. In this part of the study, a Pt_3Sn surface is cleaned by sputtering. The changes in the composition of the surface and the oxidation states of the surface tin and platinum are investigated using several UHV techniques including AES, SAM, ISS and ESCA. The next part of this study examines the same surface after an exposure to air with ESCA and angle-resolved ESCA.

Neither AES nor ESCA are highly surface sensitive unless performed in an angle-resolved manner. However, they do indicate trends in the outermost surface-region composition as changes occur. This point is illustrated in this study through the comparison of AES and ESCA spectra with ISS and angle-resolved ESCA spectra (next part of this study).

EXPERIMENTAL METHODS

The Pt_3Sn sample was prepared by electrodepositing a stoichiometric amount of metallic tin onto a platinum

† Permanent address: Department of Chemistry, University of A. Mickiewicz, Poznan, Poland.

0142–2421/86/120169–05 $05.00

cathode followed by annealing under flowing hydrogen. Complete details of the preparation are given in a previous study.[18]

The data shown in this study were taken using a PHI Model 560 ESCA/SAM system manufactured by Perken–Elmer Physical Electronics. The background pressure in the vacuum system was about 1×10^{-10} Torr. AES and SAM spectra were taken using a primary electron beam energy of 5 keV and a beam of current of 50 nA over a spot size approximately 1μm in diameter. ESCA spectra were taken in the retarding mode with an energy resolution of approximately 0.9 eV using Mg X-rays. ISS spectra were collected in the nonretarding mode using a primary beam of 1 keV helium ions and a scattering angle of 132°. The use of a 90° rotating aperture and a smaller scattering angle limits the mass resolution of the spectra as can be seen by comparing the spectra of this study with the spectra in the previous study,[18] which was taken with a 12° aperture. The AES data shown in Fig. 1 was taken in another UHV system with a base pressure of 2×10^{-10} Torr. The AES data were taken using a double-pass CMA (PHI model 15–255GAR) with a primary beam energy of 3 keV.

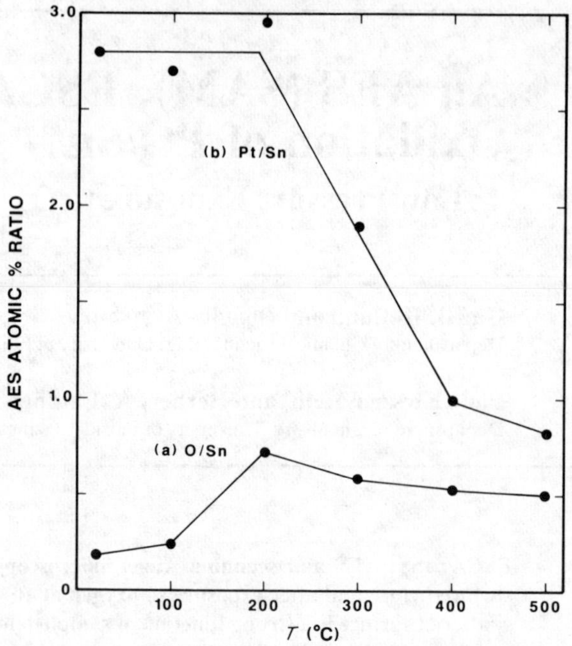

Figure 1. (a) Oxygen–tin ratio determined by AES after a saturation exposure to O_2 at 10^{-6} Torr as a function of sample temperature. Atomic sensitivity factors have been taken into account, but the Sn peak shape changes upon oxidation have been neglected in calculating the O–Sn ratios; (b) Platinum–tin ratio determined by AES after a saturation exposure to O_2 at 10^{-6} Torr as a function of sample temperature. The sample was sputtered before each data point was taken until the platinum–tin ratio no longer changed.

RESULTS AND DISCUSSION

In order to put the sample into a clean and well-defined condition, it was inserted into the vacuum system and sputtered with 1 keV argon ions until the AES signals for oxygen and carbon contamination were no longer detectable and the Pt-to-Sn ratio reached a constant value. The resulting AES spectrum (not shown) is quite similar to that shown in Fig. 3(a) of the previous study except the Pt-to-Sn ratio is somewhat smaller for this sample. As mentioned in the previous study, treatments involving vacuum annealing, oxidation and sputtering cause a gradual reduction in the bulk tin content of Pt-Sn alloys. This sample had oniy been sputtered once while the sample used in the previous study had been through many repeated treatments.

The sample was then exposed to 10^{-6} Torr of oxygen at room temperature for about 16 min (1000 L). This is referred to as a 'saturation exposure' because longer exposures (<10^5L) do not result in the adsorption of noticeably more oxygen. It has been found that the most important variable with regard to the amount of oxygen adsorbed during a saturation exposure is the temperature. The temperature dependence is shown in Fig 1(a) in which the atomic ratio of O–to–Sn is plotted after a saturation exposure to oxygen as a function of the sample temperature during dosing. The sample was sputter cleaned before each data point was taken. The amount of adsorbed oxygen goes through a distinct maximum at about 200°C. This behaviour is not fully understood, but several points are of interest in attempting an explanation. Firstly, below 200°C the amount of Sn at the surface strongly influences the amount of oxygen adsorbed. This suggests that the oxygen is interacting primarily with the tin. Secondly, the maximum occurs at approximately the melting point of pure tin metal. Thirdly, tin segregates more readily to the surface at higher temperatures (>200°C) giving larger Sn–Pt ratios as is shown in Fig. 1(b). Fourthly, it has been shown in previous studies[19–22] that SnO_2 decomposes to SnO by annealing in vacuum or exposure to an electron or ion beam. Fifthly, an equilibrium thermodynamic calculation assuming ideal SnO and SnO_2 phases predicts that SnO should form when SnO_2 is annealed in vacuum. As stated above, reduction does occur but at a higher temperature than predicted by the calculation at a given oxygen partial pressure suggesting that the decomposition reaction is activated.[24] Thus, many processes could influence the shape of the curve shown in Fig. 1(a), but it is possible to formulate an explanation of the observed behaviour. In the temperature range from 20 to 200°C, extensive segregation of tin to the surface does not occur as shown by Fig. 1(b). However, the tin at the surface becomes oxidized to a greater extent as the temperature increases. This is most likely due to the formation of a thicker oxide layer or to more complete oxidation of surface tin by overcoming the activation barrier. In the temperature range from 200 to 400°C, tin segregates to the surface more rapidly as the temperature increases, but the oxidation rate and, hence, the amount of adsorbed oxygen does not increase as rapidly. This results in more tin at the surface, but less of it is oxidized. It also may be that SnO forms preferentially to SnO_2 at higher temperatures, but this is difficult to determine as discussed below. There is very little change in either the Pt-to-Sn ratio or in the O-to-Sn ratio in the 400 to 500°C temperature range suggesting that little change in either the rate of tin segregation or the rate of oxidation occurs in this range.

Unfortunately, distinguishing between SnO_2 and SnO

is difficult. Generally, ESCA does not provide a good means of distinguishing between SnO and SnO_2 because the Sn 3d peaks of both oxides have the same binding energy and peak shape. However, it may be possible to determine the composition of the oxide at the surface using peak area ratios under the O 1s peak and the oxidic portion of the Sn 3d peaks provided that oxygen is present only as tin oxide. Also, valence-band photoemission[23] and energy loss spectroscopy (ELS)[21,22] can be used to distinguish between SnO and SnO_2, but these are complicated by the presence of metallic tin and platinum.

In this study the amount of oxygen adsorbed corresponds to the point at 20°C in Fig. 1(a). It is of interest to determine how this adsorption occurs. It could adsorb weakly or strongly as an oxide and to either the tin or platinum or to both. ISS, SAM and ESCA prove to be a useful combination of techniques in attempting to understand how a sputtered Pt_3Sn surfaces responds to and interacts with oxygen.

Figure 2(a) shows a secondary electron micrograph of the alloy surface after sputter cleaning. The features are magnified by a factor of 250 in the photograph. The notable features in this photograph are A: grain boundaries, B: very flat, smooth regions, C: rough regions and D: bulk surface deposits. A scanning Auger micrograph of tin is shown in Fig. 2(b). The surface concentration of tin is greater in the lighter regions. From this micrograph it can be seen that the regions of greatest tin concentration are those at the grain boundaries (A in Fig. 2(a)). The rough areas (C in Fig. 2(a)) are also tin rich but not as much so as the grain boundary regions. Both the smooth regions and the regions with surface deposits (B and D in Fig. 2(a) respectively) have lower tin concentrations. A scanning Auger micrograph of platinum (not shown) is just the inverse of the tin micrograph indicating that the tin-rich regions are platinum depleted regions. An exception is the region marked with a D in Fig. 2(a) which shows bulk-like deposits. This region is both tin depleted and platinum depleted. Fig. 2(a) and 2(b) show that there is a definite correlation between the geometrical structure of the surface and tin concentration. The grain boundaries have the highest tin concentrations, and the tin concentration decreases in moving away from a grain boundary until a platinum-rich region is reached near the centre of a crystallite. This suggests that tin segregates to the surface by diffusion at grain boundaries and then spreads out over alloy crystallites at the surface. The thickness or concentration of the tin layer is greatest near grain boundaries and decreases toward the centre of the crystallites. Thus, sputtering removes the tin located at the thinner layers selectively which explains the patterns seen in Fig. 2(a) and 2(b).

Figure 2(c) is a scanning Auger micrograph of oxygen taken after the room-temperature, saturation oxygen exposure. The important point demonstrated by Fig. 2(c) is that oxygen adsorbs at regions which are tin rich and, therefore, particularly at grain boundaries. The smooth, platinum-rich regions show very little oxygen adsorption. The region marked with a D in Fig. 2(a) had a high concentration of oxygen even before the oxygen exposure, and its carbon concentration was high both before and after the oxygen exposure. The contaminating deposits, therefore, contain both carbon and oxygen, but the presence of hydrogen cannot be ruled out in this region or others. One possibility is that the contaminated region consists of a surface oxide which is covered with the carbonaceous residue seen in Fig. 2(a). Sputtering would not remove the oxide until

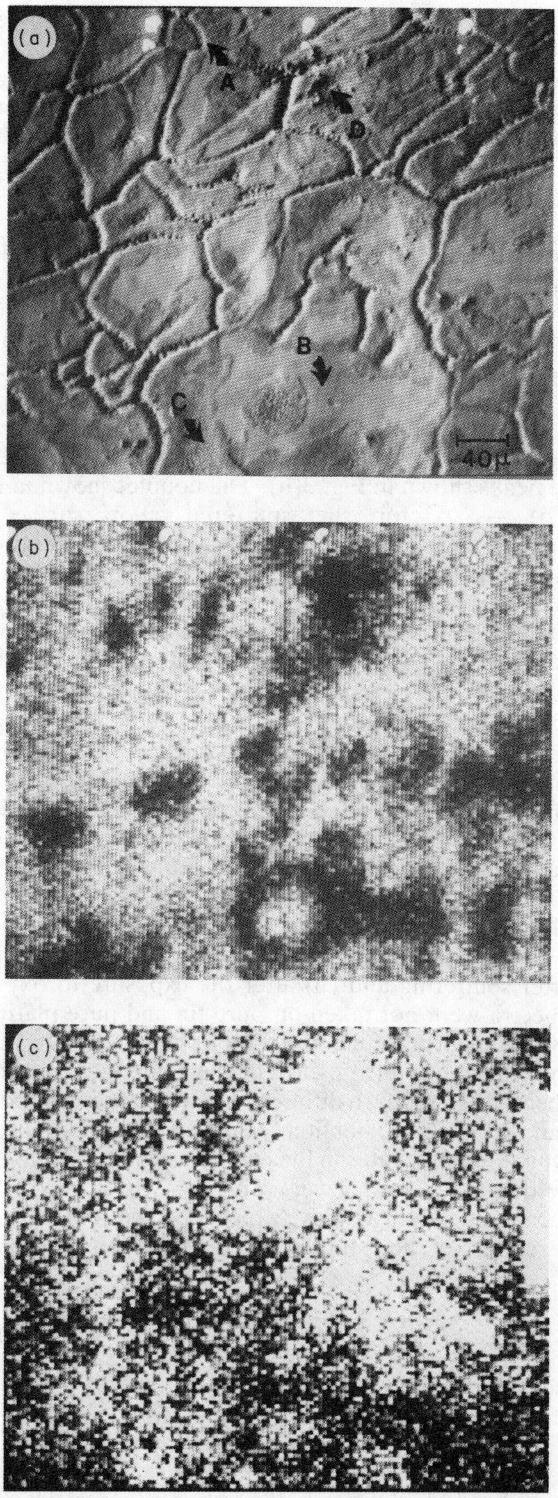

Figure 2. (a) Secondary electron micrograph of the alloy surface after cleaning; the magnification is 180; (b) Scanning Auger micrograph of tin (same region shown in (a)). The lighter regions are tin rich; (c) Scanning Auger micrograph of oxygen (same region shown in (a)) after a saturation exposure to oxygen at 10^{-6} Torr and room temperature. The lighter regions are oxygen rich.

the carbonaceous material is removed. However, continued sputtering would remove these deposits and the oxide. The conclusions drawn from Fig. 2 agree with that drawn from Fig. 1 in supporting the premise that oxygen interacts with the tin and not the platinum after a room-temperature, low-pressure exposure to oxygen.

ESCA spectra taken before and after the oxygen exposure are shown in Fig. 3 and 4. Figure 3(a) and (b) show the Pt 4f and Sn 3d peaks respectively. The Pt $4f_{7/2}$ peak has a binding energy of 70.9 eV both before and after the oxygen exposure which is characteristic of metallic platinum. By comparing A in Figs 3(a) and 2(b), it can be seen that there are no changes in the position or peak shape of the Pt 4f doublet before and after exposure to oxygen. This is not the case for the Sn 3d peaks shown in Fig. 3(b). The doublet shown in Fig. 3(b) — A for the sputtered alloy surface is characteristic of elemental tin. The binding energy of the Sn $3d_{5/2}$ peak is 484.6 eV. After oxidation the Sn 3d peaks shift by 0.2 or 0.3 eV to a higher binding energy, and the peaks become more asymetrical. This is particularly noticeable at the high binding energy portion of the bases of the B peaks shown in Fig. 3(b). The Sn $3d_{5/2}$ peaks for both SnO and SnO_2 have a binding energy of 286.4 eV, which is about 2.0 eV higher than that of metallic tin. The broadening of the Sn 3d peaks seen at B in Fig. 3(b) indicates that tin oxide is formed during the oxygen exposure, but efforts have not been made to determine if the oxide is SnO_2 or SnO or a mixture. It is also clear from Fig. 3(a) that no platinum oxides form.

Figure 4 shows valence-band ESCA spectra taken (a) after sputtering and (b) after the exposure to oxygen. Spectra were not taken on pure tin and pure platinum using this instrument so it is not possible to determine the contribution from each of the components to these spectra. However, a difference between the two spectra can be seen. Oxidation causes an increase in the emission from states in the energy region from 2 to 6 eV below the Fermi level. This is attributed to the formation of tin oxides, which is consistent with the results of earlier studies of the oxidation of metallic tin.[23,25] The difference between the spectra shown in (a) and (b) is small. This is mostly due to the fact that in valence-band ESCA the photoemitted valence electrons leave the surface with very large kinetic energies (1250 eV). Thus, most of the electrons originate from layers beneath the surface which are metallic. UPS or angle-resolved, valence-band ESCA provides more surface sensitive means for examining this oxidation process as is demonstrated in the next part of this study. It is interesting to note that the shape or position of the spectra do not vary near E_F. This is due to the fact that oxidation of tin does not result in a change in work function as shown in the study of Powell and Spicer.[25]

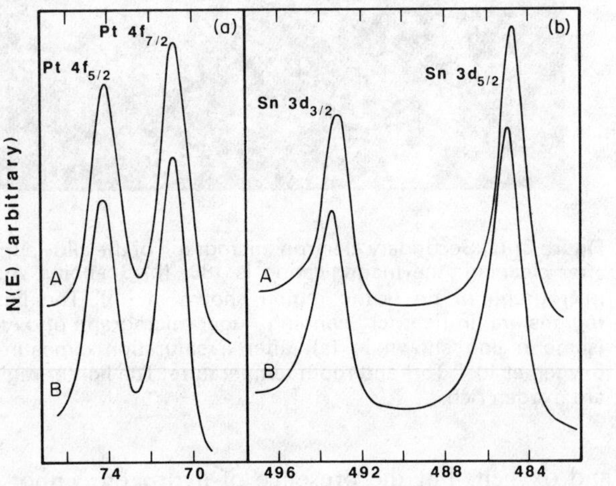

Figure 3. ESCA spectra showing Pt 4f peaks (A) and Sn 3d peaks (B), (a) before and (b) after a room temperature, saturation exposure to oxygen at 10^{-6} Torr.

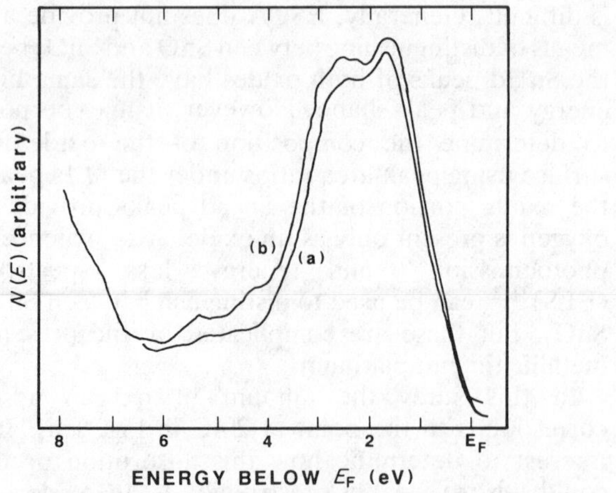

Figure 4. Valence-band ESCA spectrum of the alloy surface taken (a) after sputtering and (b) after the exposure to oxygen.

The usefulness of ISS for examining surfaces of alloys is discussed below and in the preliminary study.[18] ISS is unique in that it is essentially sensitive only to the outermost layer of atoms. In ISS inert-gas ions of a given primary energy are elastically scattered by atoms in the solid. Thus, energy analysis of the scattered ions yields a mass spectrum of the outermost layer of atoms. Its high surface sensitivity is the result of two factors. The first factor is that ions which penetrate beneath the outermost layer of atoms have a very high probability of being neutralized and, therefore, are not energy analysed. The second factor is that even if neutralization does not occur, the ions which penetrate beneath the outermost surface layer have a very high probability of losing energy in the inelastic process. Thus, it is extremely unlikely that a primary ion scatters elastically from a subsurface atom and is counted in the peak of elastically scattered ions.

In the preliminary study,[18] ISS was performed before and after an oxygen exposure of a tin-depleted Pt_3Sn alloy to oxygen. The surface concentration of tin remained unchanged while the concentration of surface platinum decreased by about one-third. In terms of a simple site-blocking model, this suggests that the oxygen adsorbs at platinum atoms and does not interact with the tin. In this study the results of similar experiments of a Pt_3Sn sample which has not been depleted in tin are described. The ISS results are shown in Fig. 5(a) for the sputtered sample and in Fig. 5(b) for the oxygen-exposed sample. The concentration of tin at the surface is much higher than in the previous study after sputtering, which is expected. By comparing Fig. 5(a) and (b), it can be seen that the relative magnitudes of the backgrounds consisting of inelastically scattered ions which escape without being neutralized are different. This background is lower for the metallic alloy surface than for the oxygen-exposed surface. This is due to the fact that the metal is more effective in neutralizing ions than the oxidized surface because the conduction electrons of the metal are free to move throughout the surface region and respond to the presence of the ion very rapidly.

The most important difference between the ISS results of the previous study and this study is how the heights of the Sn and Pt peaks change after exposure to oxygen. In this case the height of the Sn peak actually increases after the oxygen exposure, but the Pt peak still decreases in size as it did previously. This implies that segregation of tin to the surface occurs even at an oxygen exposure pressure of 10^{-6} Torr and that this tin covers surface platinum. As discussed above, tin segregates from the bulk to the surface at the grain boundaries. Then it spreads out over the smooth platinum-rich regions and is oxidized. Similar observations have been made and quantified by Opila[26] in a SAM study of tin-lead alloys. Then, even with a coverage effect by the oxygen, more tin is observed with ISS at the surface due to the presence of an increased amount of surface tin. The decrease in the Pt peak observed in ISS is due to coverage of the platinum by tin and not oxygen. With this understanding it is possible to accurately interpret the ISS results of the previous study. In that case segregation of tin to the surface, coverage of the platinum by the tin and oxidation of the tin all occurred, but on the tin-depleted sample, the amount of tin which segregates to the surface compensates for the amount of tin which is covered by oxygen. Therefore, the net effect appears to be a reduction in the amount of surface platinum due to coverage by oxygen while the actual process is considerably more complicated. A simple site-blocking interpretation leads to an incorrect conclusion.

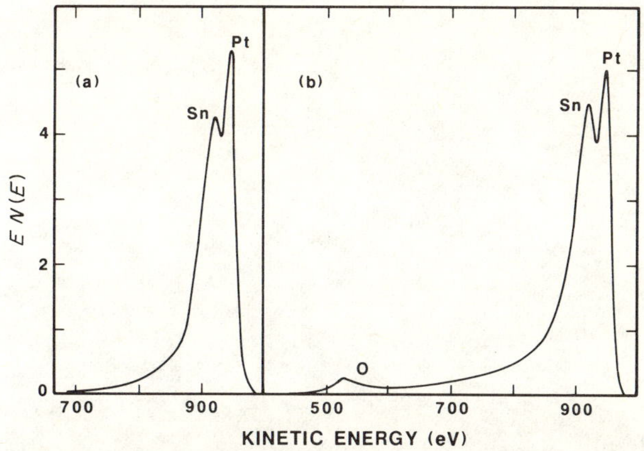

Figure 5. ISS spectra of the alloy surface (a) after sputtering and (b) after exposure to oxygen using a 1 keV beam of He^+ and a scattering angle of 132°.

CONCLUSION

It is shown that a sputter-cleaned Pt–Sn alloy surface is composed of regions which vary in concentration. Grain boundaries are tin rich, and the surface concentration of tin decreases in moving away from a grain boundary until a platinum-rich region is reached near the centres of crystallites. It is shown that exposure to oxygen causes tin to segregate to the surface by grain boundary diffusion. The tin is oxidized while the platinum is not. ESCA, ISS and oxygen uptake experiments are in agreement with this interpretation of the SAM results.

Acknowledgements

The authors appreciate the use of the vacuum system by Perkin-Elmer Physical Electronics and the efforts of John Moulder in data collection. Financial support of this research was received from NSF under grants #CBT–8416380 and #CHE–8309445 and the donors of the Petroleum Research Fund administered by the ACS under PRF #16995–AC5–C.

REFERENCES

1. F.M. Dautzenberg, J.N. Helle, P. Biloen and W.M.H. Sachtler, *J. Catal.* **63**, 119 (1980).
2. R. Bacaud, P. Bussiere, F. Figueras and J.P. Mathieu in *Preparation of Catalysts* p. 509 (ed. by B. Delmon, P.A. Jacobs and J. Poncelet) Elsevier (1976).
3. H. Lieske and J. Volter, *J. Catal.* **90**, 96 (1984).
4. R. Bacaud, P. Bussiere, F. Figueras and J.P. Mathieu, *Compt. Rend.* **C281**, 159 (1975).
5. R. Bacaud, P. Bussiere and F. Figueras, *J. Catal.* **69**, 399 (1981).
6. R. Burch, *J. Catal.* **71**, 348 (1981).
7. A.C. Muller, P.A. Engelhard and E.J. Weisang, *J. Catal* **56**, 65 (1979).
8. S.R. Adkins and B.H. Davis, *J. Catal.* **89**, 466 (1984).
9. B.A. Sexton, A.E. Hughes and K. Foger, *J. Catal.* **88**, 466 (1984).
10. D.F. Cox, G.B. Hoflund and H.A. Laitinen, *Langmuir* **1**, 269 (1985).
11. G.B. Hoflund, D.A. Asbury and R.E. Gilbert, *Thin Solid Films* **129**, 139 (1985).
12. Z. Karpinsky and J.K.A. Clarke, *J. Chem. Soc. Faraday Trans.* **I71**, 893 (1975).
13. H. Verbeek and W.M.H. Sachtler, *J. Catal.* **42**, 257 (1976).
14. A.G.T.M. Bastein, F.G.C.M. Toolenaar and V. Ponec, *J. Catal.* **90**, 88 (1984).
15. R. Bouwman, L.H. Toneman and A.A. Holscher, *Surf. Sci.* **35**, 8 (1973).
16. R. Bouwman and P. Biloen, *Surf. Sci.* **41**, 348 (1974).
17. R. Bouwman and P. Biloen, *Anal. Chem.* **46**, 136 (1974).
18. G.B. Hoflund, D.A. Asbury, P. Kirszensztejn and H.A. Laitinen, *Surf. Sci.* **161**, L583 (1985).
19. G.B. Hoflund, G.L. Woodson, D.F. Cox and H.A. Laitinen, *Thin Solid Films* **83**, 261 (1981).
20. G.B. Hoflund, D.F. Cox, F. Ohuchi, P.H. Holloway and H.A. Laitinen, *Appl. Surf. Sci.* **14**, 281 (1983).
21. D.F. Cox and G.B. Hoflund, *Surf. Sci.* **151**, 202 (1985).
22. D.F. Cox, G.B. Hoflund and H.A. Laitinen, *Appl. Surf. Sci.* **20**, 30 (1984).
23. C.L. Lau and G.K. Wertheim, *J. Vac. Sci. Technol.* **15**, 622 (1978).
24. Unpublished results.
25. R.A. Powell and W.E. Spicer, *Surf. Sci.* **55**, 681 (1976).
26. R.L. Opila, *J. Vac. Sci. Technol.* **A3**, 1298 (1985).

SURFACE AND INTERFACE ANALYSIS, VOL. 9, 175–180 (1986)

Analysis by Scanning Electron Microscopy and Structure Determination by High Resolution Electron Microscopy of Ni/Ni_3B Eutectic

J. Desseaux-Thibault* and S. Hamar-Thibault†
*CEN Grenoble, Department de Recherches Fondamentales, Service de Physique, 85X–38041 Grenoble Cedex, France
† INP Grenoble, ENSEEG - LTPCM (UA29), BP75 Domaine Universitaire, 38402 Saint Martin D'Heres, France

The eutectic Ni/Ni_3B has been found in a number of special hardfacing alloys such as Ni–Cr–B–Si. However, this eutectic has not been studied. This eutectic is a good model. Ni_3B has the same crystallographic structure as Fe_3C. The combination of HREM with EDX and EELS is a means to obtain together data on composition and morphology of defects and grain boundaries. Here an attempt is made to determine the structure by HREM and to analyse by EELS both Ni and Ni_3B. Images of low indexed planes of Ni_3B are obtained and interface observed.

INTRODUCTION

Hardfacing is an important feature for all problems of wear and corrosion. In many uses, the alloys must show good wear and corrosion resistant properties. The Ni–base hardfacing alloys are characterized by a dispersion of hard phases of several kinds (carbides M_7C_3, borides Ni_3B, CrB, and silicides Ni_5Si_2, Ni_3Si). As previously shown,[1,2] the common characteristic of these alloys is the presence of binary eutectic $Ni(\alpha)$–Ni_3B. The structure of Ni_3B is reported to be orthorhombic[3] and isostructural with Fe_3C. Therefore, this eutectic is analogous to the one between austenite and cementite. Furthermore, Ni–B system is not susceptible to another competing eutectic reaction as occurs in Fe–C alloys. It is why Ni–Ni_3B eutectic is a good model. The study is undertaken by means of high resolution electron microscopy (HREM) and electron energy loss spectroscopy (EELS). Both technics are complementary.[4] HREM has been already extensively used to determine the atomic structure of the defects such as dislocations and grain boundaries[5,6] and EELS to chemical analysed precipitates and defects at the same atomic level.[7,8]

EXPERIMENTAL PROCEDURES

Sample preparation

Hardfacing is usually made with techniques involving fast solidification. Mixtures of pure metal (Ni) and nickel boride (NiB) are melted in an arc-furnace under argon atmosphere. The chemical composition of the eutectic occurs at 3.23 wt%B. The specimens are cut and thinned with ion-milling in order to obtain thin foils suitable for HREM and EELS observations.

Structural studies by HREM

The contrast variations on the image are related to the projection of the crystal structure under restricted conditions : the incident electron beam must be aligned with low indexed atomic rows in the crystal (due to resolution limitation) and the thickness of the crystal must be not too large (<200Å). However, the contrast is dependent on the crystal thickness and on the microscope focalization.[5] It is why the complete determination of atomic position requires the comparison between micrographs and computer images (in the same experimental conditions). On one hand the interaction electron-material is computed by the multislice method.[9] The theoretical model of the structure to be tested is a data of computation. The image is then calculated by introducing microscope experimental conditions. On the other hand, it must be pointed out that in the microscope, the information about the crystal structure is contained in the diffracted beams and in diffused scattered beams (when defects are present). The objective aperature must select the appropriate beams to get the most faithful image of the structure having regard to the resolution limitation. HREM observations are made on a JEM200CX (CENG). Images of some low indexed planes in Ni_3B are obtained and an attempt is made to observe the interface.

Chemical analysis by STEM

The chemical information is contained either in the edges in EELS or in the spectra lines in EDX. EELS is particularly suited to study of the edges in the energy loss domain (100 to 2000 eV). This domain contains K edges for low Z elements (Z<15) and is suitable to

0142–2421/86/120175–06 $05.00

detect B. On the other hand, EDX is not able to detect lines below 1000 eV, thus preventing the analysis of light elements. EELS is a more sensitive technique in term of minimum mass detection but EDX is better for minimum fraction detection. High special resolution analysis is required for interface studies. Therefore two parameters have to be taken into account : the first one is the volume to be analysed and the second one is the signal to noise ratio. This last point is the most important for boron detection.

EDX and EEL spectra are obtained with a VG electron spectrometer mounted on a VG–HB5–STEM (LTPCM). The EEL spectra (pulse counting mode) are recorded using a TRACOR TN2000 multichannel system, which is programmed for quantitative light elements analysis. The analysis of Ni and B is performed in either Ni or Ni_3B. The finest probe size leads to a lateral resolution of 1.5 nm at the sample level. The collection semi-angle for EELS set by a 500 μm diameter objective aperture is 12 mrad.

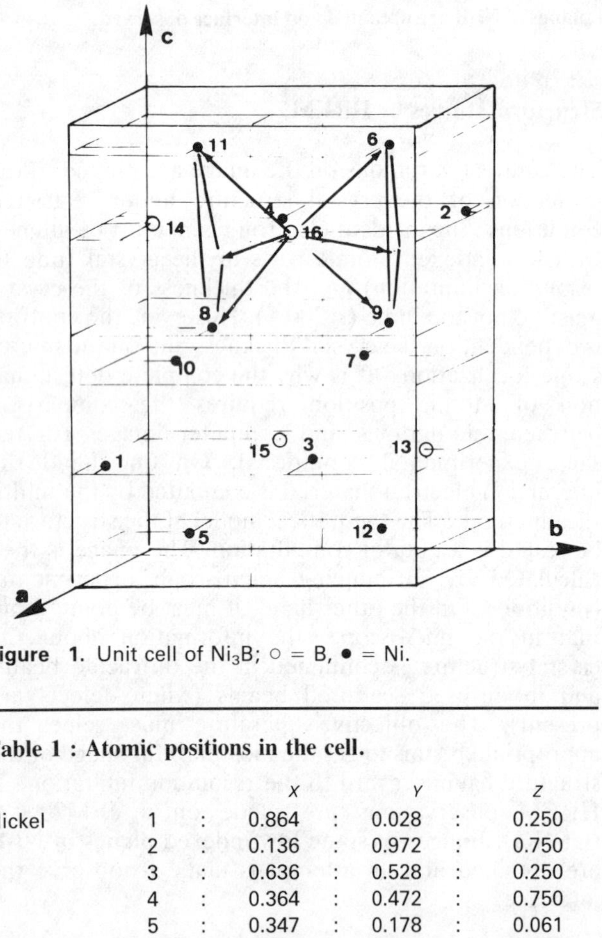

Figure 1. Unit cell of Ni_3B; ○ = B, ● = Ni.

Table 1 : Atomic positions in the cell.

		X	Y	Z
Nickel	1	0.864	0.028	0.250
	2	0.136	0.972	0.750
	3	0.636	0.528	0.250
	4	0.364	0.472	0.750
	5	0.347	0.178	0.061
	6	0.653	0.822	0.939
	7	0.153	0.678	0.439
	8	0.847	0.322	0.561
	9	0.653	0.822	0.561
	10	0.347	0.178	0.439
	11	0.847	0.322	0.939
	12	0.145	0.678	0.061
Boron	13	0.433	0.889	0.250
	14	0.567	0.111	0.750
	15	0.067	0.389	0.250
	16	0.933	0.611	0.750

RESULTS AND DISCUSSION

Structures of Ni and Ni_3B

Ni_3B crystallizes in an orthorhombic structure. The unit cell is shown in Fig. 1. Its parameters are :

$a = 4.399$ Å, $b = 5.211$ Å, $c = 6.619$ Å

These values are close to the lattice parameters of the cementite Fe_3C:[10]

$a = 4.516$ Å, $b = 5.077$ Å, $c = 6.727$ Å

According to Rundqvist,[11] the structure belongs to the space group Pbnm. The unit cell contains 4 Ni_3B groups and the coordinates of the atoms are given in the Table 1. On the same figure, the prisms Ni_6B are represented as often shown for the cementite. This places the boron atom at the centre of a nearly trigonal prism of Ni atoms in a structure in which, Ni–Ni lies between 2.5 and 2.64 Å and Ni–B between 1.99 and 2.08 Å. As seen in Fig. 1, the structure is related to the fcc structure and may be described as an almost close packing of Ni atoms with boron in the interstices as in the Fe–Fe_3C system.[10]

HREM Observations of Ni_3B along some low indexed directions

Figure 2 (a) shows a HRE micrograph of an edge of Ni_3B (001) crystal. The thickness of the foil is increasing from the left to the right. The measured focalization is about −1100 Å. The contrast variation in the micrograph is well reproduced by computed images (Fig. 2(d)) for different thicknesses (20 kV, −1100 Å focalization, defocus spread 200 Å, divergence angle of the electron beam 0.005 rd and objective aperture diameter 0.46 Å^{-1}). The arrows on the images indicate the areas where the thickness is about respectively 33, 99 and 230 Å. Figure 2(b) shows the projection of the atomic potential distribution on the (001) plane in Ni_3B. Figure 2(c) shows the intensity of the electron wave function just behind the crystal. In Fig. 2(c) and 2(d), two cells are imaged. It is noticed that in these experimental conditions, the correlation between the

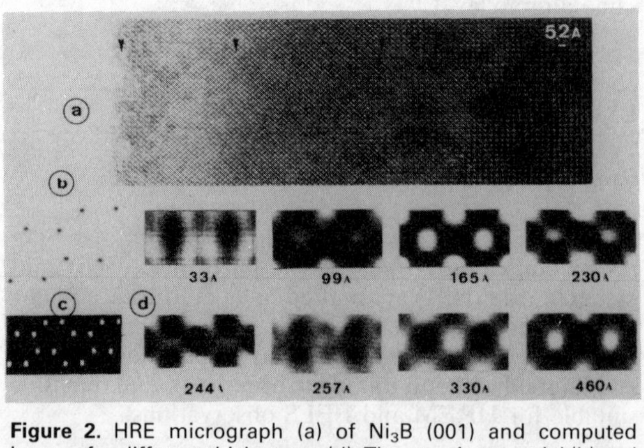

Figure 2. HRE micrograph (a) of Ni_3B (001) and computed images for different thicknesses (d). The atomic potential (b) and the intensity distribution (c) are also presented.

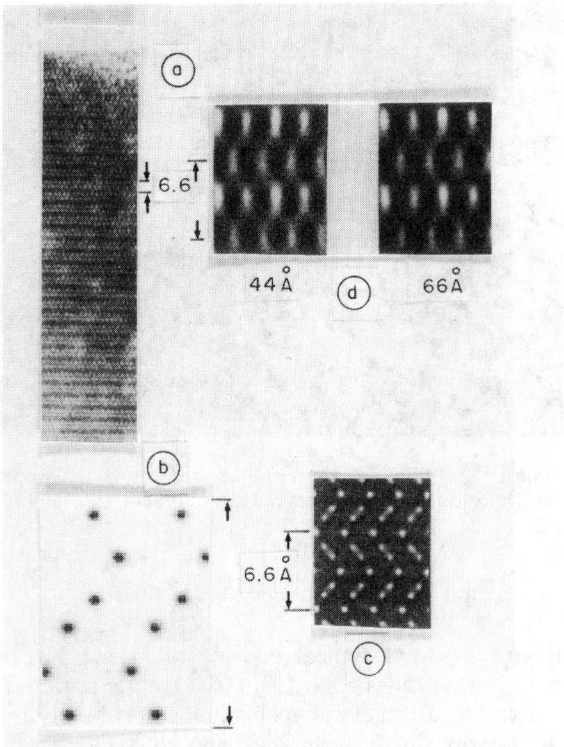

Figure 3. HRE micrograph (a) of Ni_3B (100) and computed images for different thicknesses (d). The atomic potential (b) and the intensity distribution (c) are also presented.

structure and the micrograph is not direct, even if some symmetry elements are observed. The interpretation must be made very carefully and the atomic positions cannot be deduced straightforwardly.

Figure 3 shows HRE micrograph of an edge of Ni_3B (100) crystal. The measured focalization is almost −640 Å and the computed images are performed with the same experimental conditions as for Fig. 2. The computed images (Fig. 3(d)) for the very thin parts of the foil (44 and 66 Å) are close to experimental micrographs (Fig. 3(a)) but once more they are not directly related to the atomic structure of Ni_3B (Fig. 3(b) – (c)). Figure 4 shows a HRE micrograph of Ni_3B (111) crystal. The pseudo-ternary symmetry of the crystal is well evidenced.

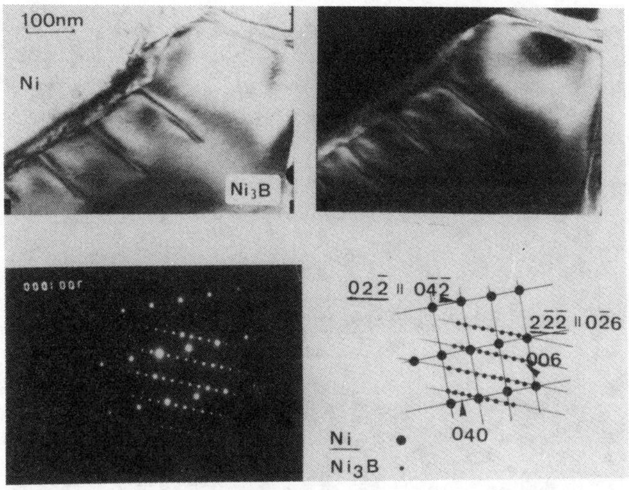

Figure 5. Transmission electron micrograph of one interface between Ni and Ni_3B. Superimposed diffraction patterns and key diagrams.

Structural observations of some interfaces.

Conventional transmission electron micrographs and superimposed diffraction patterns of both compounds at the interface are reported in Fig. 5. From the key diffraction pattern, it is noticed that the (100) plane in Ni_3B is parallel to the (211) plane in Ni. Moreover, the plane $(0\bar{2}6)$ of Ni_3B is parallel to the Ni $(2\bar{2}\bar{2})$ plane. The orientation relationship between the two crystals is represented with the superimposed stereograms reported in Fig. 6. The superimposed stereograms are presented with the (100) plane of Ni_3B parallel to the (211) plane of Ni.

According to previous studies[12,13] the relative orientation relationship between the two crystals is described as :

$$(100)Ni_3B \,//\, (211)Ni$$
$$(01\bar{3})Ni_3B \,//\, (\bar{1}11)Ni$$
and
$$(04\bar{2})Ni_3B \,//\, (02\bar{2})Ni$$

The plane of the interface is :

$$(1\bar{1}\bar{4})Ni_3B \,//\, (130)Ni$$

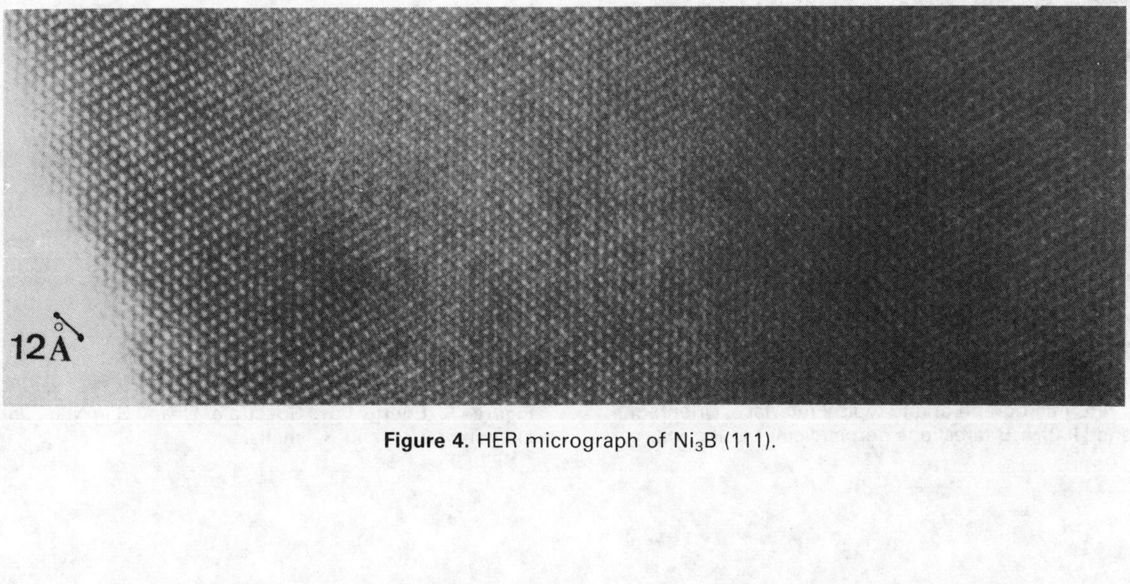

Figure 4. HER micrograph of Ni_3B (111).

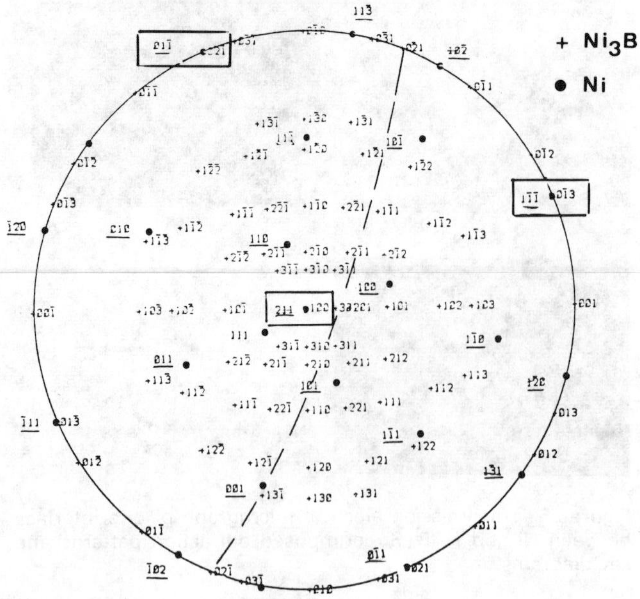

Figure 6. Superimposed stereograms of the two phases.

Figures 7 and 8 show HRE micrographs of two interfaces with different orientations of the sample with respect to the incident beam. The (211) Ni plane (Fig. 7) and the (110) Ni (Fig. 8) plane are respectively parallel to the foil. The interfaces were only found and observed inclined with respect to the foil surface. The moire contrast is due to the superimposition of both Ni and Ni_3B crystals. The periodic disappearance of the moire reveals the crystals misfit. However the interfaces are found to be well defined and located. In these two cases, the intersection of the interface with the foil surface has long straight parts (part A). The change in direction seems to be accommodated by faceted parts (part B). The difference in local configurations is certainly due to energetical states of the boundary in each direction.

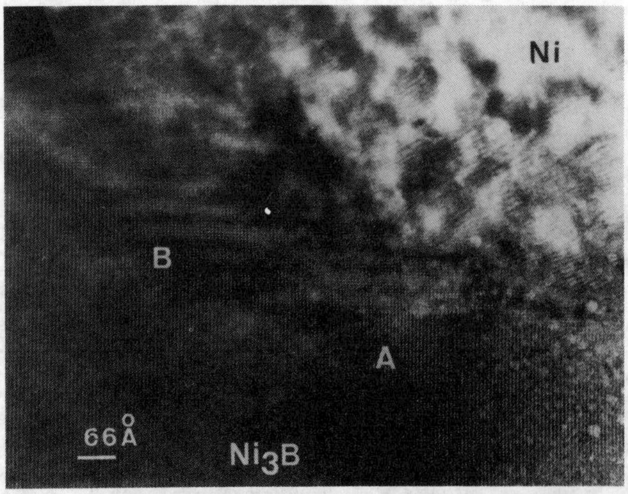

Figure 7. HRE micrograph of one typical interface. Orientation : [211]Ni and [100]Ni_3B directions perpendicular to the foil.

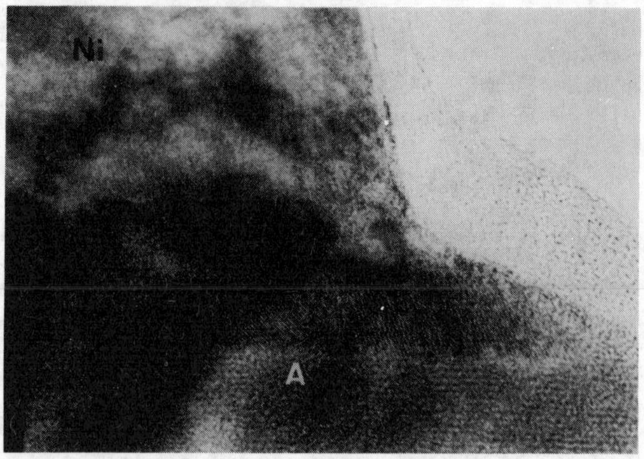

Figure 8. HRE micrograph of a second interface. Orientation : [110]Ni and [631]Ni_3B directions perpendicular to the foil.

EELS analysis

Figure 9 shows typical energy loss spectra recorded both either Ni or Ni_3B in the same region (same thickness). Both elements Ni and B may be observed in the energy range 0–500 eV. The Ni M_{2-3} edge is well observed at 68.1 eV[14] and the K shell ionization edge of Boron appears between 188–215 eV.[14] From Fig. 9, it is shown that the NiM_{2-3} edge is well observed, but the

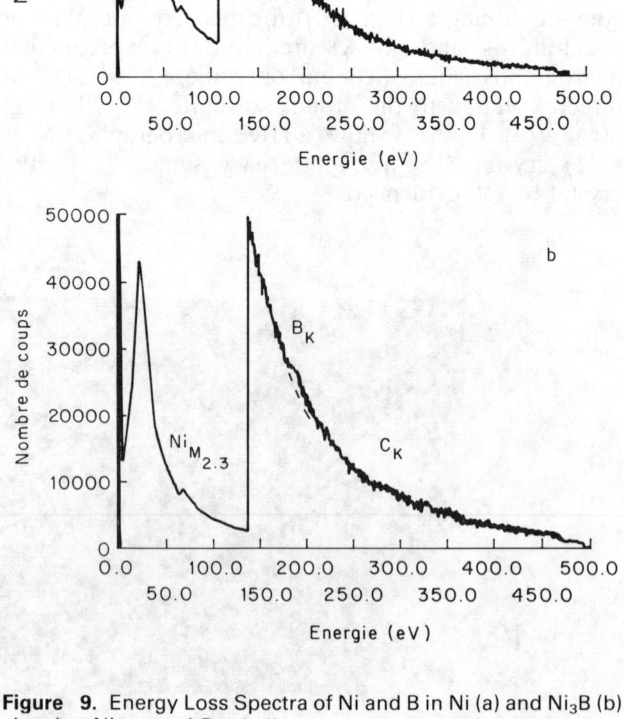

Figure 9. Energy Loss Spectra of Ni and B in Ni (a) and Ni_3B (b) showing $Ni_{M_{2-3}}$ and B_K shells.

BK one is not so clear. It is also noticeable that the K shell ionization edge of carbon at 280 eV due to contamination is often observed but cannot interfere with analysis of BK loss at 188 eV.

In order to measure the edge intensities due to NiM and BK, the spectral background due to the other energy loss processes must be substracted. The background is smoothly decreasing and is fitted with a power law of the form : A ΔE^{-R}. The constraints A and R (with R = >2) are determined by measuring areas under background just before each edge. Near BK the value calculated for R is 2.2.

Fitting the background before NiM_{2-3} with ΔE^{-2}. leads to signal/noise ratio of 0.25 for pure Ni and 0.18 for Ni in Ni_3B. The ratio of these Ni signals leads to Ni concentration of 0.72 in Ni_3B which is in good agreement with the theoretical value of 0.75.

On the contrary, the Boron K loss at 188 eV in Ni_3B is very difficult to observe. The background is very important and a good discrimination between the signal and the noise is necessary to obtain accurate information. The signal/noise ratio is nearly 0.05 (Fig. 9) after substraction of the background.

Moreover, the region of the BK loss at 188 eV is free from characteristic structure due to Ni matrix as observed in some other mono and diborides such as TiB_2[15] or CrB. The CrB compound is presented in hardfacing alloys. For comparison, EEL spectra

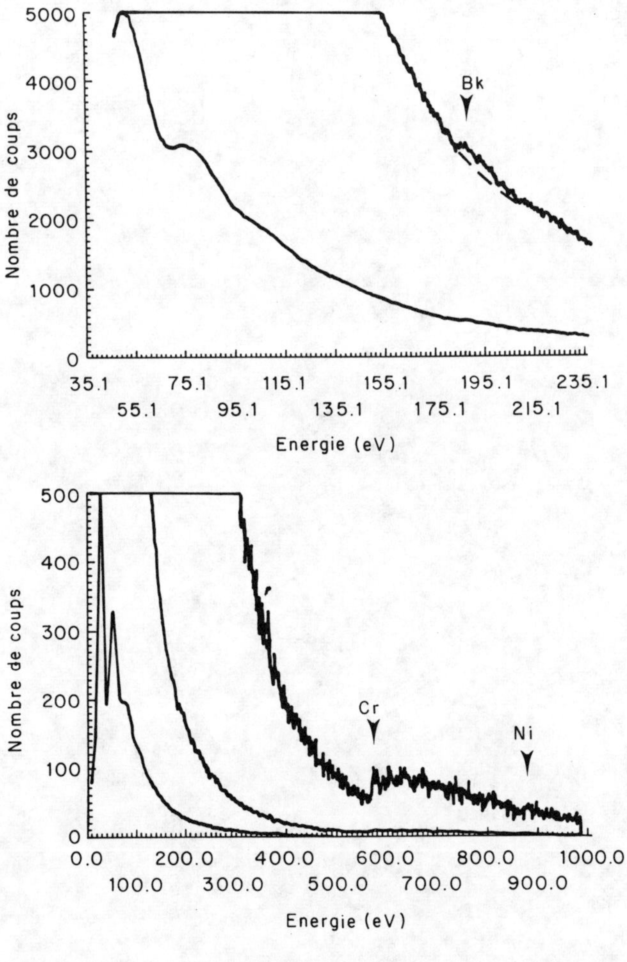

Figure 10. Energy Loss Spectra of CrL and BK in CrB. This compound is present in some Ni hardfacing alloys.

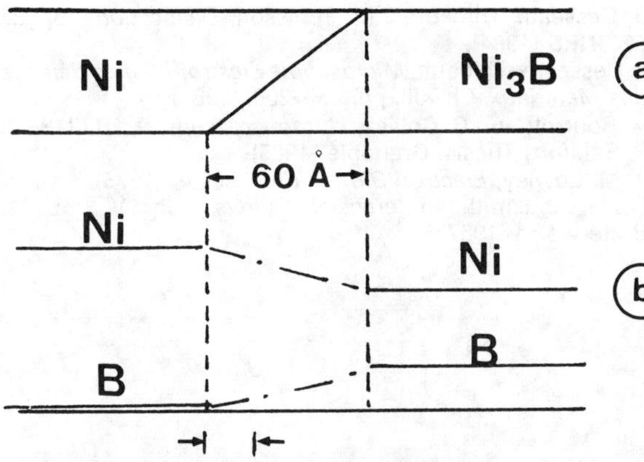

Figure 11. (a) Cross section of the interphase boundary inclined with respect to the foil;
(b) EELS signals of Ni and B. (The scale is the probe size.)

recorded on CrB show that the BK loss presents a fine structure with two large peaks at 187 and 195 eV. (Fig. 10). This lack of fine structure in BK loss in Ni_3B can be interpreted as a very weak interaction between boron atoms due to the presence of a large amount of Ni, whereas the interaction between B atoms in mono and diborides is known to be as much important as B level is higher.

The results obtained by STEM/EELS are summarized on Fig. 11. The EELS signal is reported (Fig. 11(b)) with respect to the sample section and the interphase boundary (Fig. 11(a)). It is to point out that the signal variation is straightforward related to the boundary. The probe size (i.e. the size of sample to be analysed) is about 20 Å. Thus the fact that the boundary is not end-on explained the Ni signal discrepancy and the B signal enhancement through the boundary projection. This result seems to confirm that there is no interdiffusion of Ni and B, detected at our scale.

CONCLUSION

These first results show that both HREM and EELS technique are a helpful combination to study the structure and analyse defects and grain boundaries at high resolution level. From HREM studies, we observed that Ni_3B is well crystallized but further HREM studies are to be performed in order to get better correlation between experimental and calculated images. The boundary between Ni and Ni_3B is well defined and located. At this level, the composition shows also a sharp transition.

REFERENCES

1. O. Knotex, E. Lugscheider and H. Reiman, *Metallurgical Aspects of Wear*, Deutsch Gesell. fur Metalkde (1981).
2. S. Lebaili and S. Hamar-Thibault, *Mem. Sc. Revue Met.* **81**, 519 (1984).
3. R.W. Wickoff, *Crystal Structures*, Wiley Interscience, Ch VB, pp. 114 (1964).
4. R. Gronsky, G. Thomas and K. H. Westmacott, *J. of Metals*, **37**, 36 (1985).

5. J. Desseaux-Thibault, J.M. Penisson, *Dislocations*, pp. 37–52 CNRS (1984).
6. J. Desseaux-Thibault, *Microscopie Electronique en Sciences des Matériaux*, Ch XIII, pp. 257–266, (1981).
7. A. Bourrett and C. Colliex, *Ultramicroscopy*, **9**, 183 (1982).
8. P. Sainfort, Thesis, Grenoble (1985).
9. J. M. Cowley, *Electron Diffraction*, Elsevier (1975).
10. H.J. Goldsmith, *Intersticial Alloys*, Ch V, pp. 215, Butterworth (1967).
11. S. Rundqvist, *Acta Chem. Scand.* **12**, 658 (1958).
12. S. Shapiro, J. A. Ford, *Trans. AIME*, **236**, 536 (1966).
13. J. Ajao, S. Hamar-Thibault, to be published.
14. R. F. Egerton, *Ultramicroscopy*, **6**, 297 (1981).
15. N. Zaluzec, *An Electron Energy Loss library*, Angonne Lab (1984).

SURFACE AND INTERFACE ANALYSIS, VOL. 9, 181–188 (1986), EXTENDED ABSTRACTS

Auger Lines Relative Intensities as a Tool for the Characterizing of Oxides Formed During the Oxidation of Single Crystals of Dilute Cu–Mn Alloys

G. Van Steene, M. Jardinier-Offergeld and F. Bouillon
Laboratoire de Chimie du Solide, Service de Chimie Analytique et Minérale, Faculté des Sciences, CP 160, Université Libre de Bruxelles, 50, av. F. D. Roosevelt, B-1050 Bruxelles, Belgium

A chemical effect affecting the shape of the copper, manganese and oxygen Auger lines can provide a very effective tool for oxides characterization. When several oxides are present at the same time, for example during high temperature and low pressure oxidation of metals and alloys, the usual methods (based on the measure of the Auger transitions peak-to-peak height) are inadequate to establish the surface composition.

During the oxidation process of dilute Cu–Mn alloys, different oxides (MnO, Mn_3O_4, Mn_2O_3, MnO_2, Cu_2O, CuO) may appear. Determination of the nature of these oxides is possible by examination of both oxygen and metallic elements spectra. The oxidation state of these elements changes the electronic distribution in the valence bands, and the relative amplitudes of their Auger transitions (involving one or two valence levels) provide a measurement of this effect.

The ratio $H\mathrm{Cu}_{\mathrm{L3\ M4,5\ M4,5}}/H\mathrm{Cu}_{\mathrm{L3\ M2,3\ M2,3}}$ (H = transition amplitude) is the most sensitive for copper and its oxides, because two valence levels are involved in the L3 M4,5 M4,5 transition.

For Mn and its oxides, we used for a better accuracy the amplitude ratio of the $\mathrm{Mn}_{\mathrm{L3\ M2,3\ M4,5}}$ (with only one valence level) and $\mathrm{Mn}_{\mathrm{L3\ M2,3\ M2,3}}$ transitions. For oxygen, the K$L1$ L1 Auger line also has very poor Auger yield and we preferred to measure the $H\mathrm{O}_{\mathrm{KL2,3\ L2,3}}/H\mathrm{O}_{\mathrm{KL1\ L2,3}}$ ratio for the different oxides.

A statistical treatment of experimental data collected on standard oxides of well known stoichiometry (Cu_2O, CuO, MnO and Mn_3O_4) has given a set of values for the Auger amplitude ratios suitable for determining the compounds formed on the surface during oxidation of dilute 3–10 at% Cu–Mn alloys.

The growth mode of these different oxides is also tentatively presented for the following oxidation conditions: $p = 10^{-4}$ Torr O_2, $T = 700°C$.

Characterization of Buried Interfaces in Metallic Film-structures by SIMS

P. Wilhartitz and M. Grasserbauer
Institute for Analytical Chemistry, Technical University Vienna, A-1060 Wien, Getreidemarkt 9, Austria
H. G. Hiesböck and C. Bartl
Karl Neumayer GmbH., A-2544 Leobersdorf, Günselsdorferstr. 2, Austria

INTRODUCTION

Soldering properties of electroplated coatings are sensitive to contaminations on the interfaces. If such coatings have a thickness of several micrometers, Secondary Ion Mass Spectrometry (SIMS) is one of the best techniques to characterize concealed interfaces. the soldering properties of lead wires for passive electronic components and wires for various connector types are influenced by the existence of an interface zone between base material (copper, bronze, brass, nickel) and electroplated coating (pure tin, tin–lead alloys). One of the elements which are of special interest is carbon. Systematic investigations were performed to study analytical information content of SIMS for such systems.

METHODS

Measurements were performed with a CAMECA IMS 3f ion microprobe mass analyser featuring a double focusing mass spectrometer and direct ion imaging capabilities. A caesium ion source was used producing Cs^+ primary ions with an excitation energy of 15 keV.

The primary ion current was varied between 50 nA and 2 μA. An area of 500 μm in square was scanned with the ion beam but the diameter of the analysed area was 60 μm and 150 μm respectively. The energy slit of the spectrometer has an energy acceptance of 0–130 eV and its position within the spectrometer can be varied by $\pm$ 60 eV. Additionally the energy of the secondary ions can be varied by changing the sample potential by $\pm$ 300 V.

For high mass resolution a maximum resolution $M/\Delta M$ = 20000 can be achieved but for this investigation a mass resolution of 3500 was sufficient.

Samples were prepared by coating a bronze wire with two lead–tin alloyed layers both 4 μm thick. The composition of the two layers was 30% Sn, 70% Pb and 95% Sn, 5% Pb respectively. Between the bronze base material and the first layer intermetallic phases of copper and tin were formed.

Two different types of samples were investigated: (a) high purity material with good soldering properties; (b) base material contaminated with carbon and diminished soldering properties. One sample of the high purity material was allowed to cover with an oxide film before coating but it still exhibited good soldering properties.

RESULTS

The interface between the base material (bronze) and the PbSn-coating could be characterized with SIMS after sputtering layers with a thickness of approximately 8 μm. The high purity sample showed neither oxygen nor carbon enriched at the interface. The sample which was allowed to cover with an oxide layer exhibited a broad oxygen maximum at the interface which had the shape of a diffusion profile. The carbon contaminated sample with diminished soldering properties showed both carbon and oxygen at the interface. Semiquantitative evaluation at the peak maximum with practical sensitivity factors (cts/s.nA) yielded concentrations of $\sim$ 0.1 at% for oxygen and $\sim$ 0.2 at% for carbon. Since the roughness of the crater bottom induced by sputtering was determined to be about 1 μm, local concentrations might be much higher within a thin layer of contaminants. Furthermore the depth profiles showed oxygen precipitates in the base material. They were identified as tin oxide by ion microscopy.

Depth profiling of molecular ion signals was applied to try the characterization of the film-structure and the intermetallic phases at the interface zone. The measured ion intensities were compared to computer simulations. Sn_2 and Cu_3 must have a higher ionization probability since their intensities were increased relative to the monoatomic ion signals. The intensity of Cu_2 was smaller than expected. The mixed molecular ion signals of copper and tin showed a broad maximum at the interface zone towards the base material. Computer simulations taking into account no intermetallic phases resulted in much sharper maxima. Additional broadening of the theoretical values in accordance to the crater bottom roughness gave a difference of more than a factor of five in the width of the maxima. For this reason intermetallic phases must be present.

In order to interpret secondary ion signals encountered at the interface at mass 60, high mass resolution spectra were measured. They offered that the peak at mass 60 consists of at least seven different signals. Since many ion species were possible an identification procedure had to be developed. To distinguish between atomic and molecular ion signals the energy distributions of the two most intensive peaks were measured with high mass resolution. Based on the relationship between the energy of the emitted ion cluster, the sublimation enthalpy, the ionization energy and the number of atoms within the cluster.

$$f(E) \approx \frac{E}{(E + U_0)^n}\ ;\ U_0 = U_s + U_i;\ n = 1 + 2m$$

where U_s = sublimation enthalpy; U_i = ionization energy; m = number of atoms in the cluster.

U. Traxlmayr[1] developed a computer program which allowed to calculate the number of atoms in a cluster from the energy distribution. This program was applied with the modifications necessary. The two most intensive peaks were identified as C_5 and SiO_2. This enabled the interpretation of the other peaks by calculating the mass differences.

CONCLUSION

SIMS was applied to characterize contaminations on concealed interfaces between bronze and lead-tin alloys. Carbon contaminations were found to cause diminished soldering properties whereas oxygen showed no influence. The existence of intermetallic phases at the interface zone was proved by measuring molecular ion signals of copper and tin. Sophisticated identification procedures allowed the interpretation of the complex high resolution mass spectra encountered at mass 60.

REFERENCES

1. U. Traxlmayr, *Untersuchungen zur Diffusion von Chrom in Galliumarsenid mit der Sekundärionen-Massenspektrometrie*, thesis, Technical University Vienna (1984).

The Bonding State of Segregated Elements on α-Iron Surfaces and in their Iron Compounds Studied by Electron Spectroscopy

W. Diekmann, G. Panzner and H. J. Grabke
Max-Planck Institut für Eisenforschung GmbH, D-4000 Dusseldorf 1, FRG

A combined electron spectroscopic study by means of x-ray photoelectron spectroscopy (XPS), ultraviolet photoelectron spectroscopy (UPS) and Auger electron spectroscopy (AES) was performed to characterize the bonding state of the non-metal elements S, C and N segregated on the α-iron(100) surface as a saturated overlayer with c(2 × 2) structure. Solid non-metal iron compounds were prepared for comparative measurements:

FeS	FeS_2		
$(Fe, Mn)_3C$	$(Fe, Cr)_3C$	$(Fe, Cr)_7C_3$	$(Fe, Cr)_{23}C_6$
Fe_2N	Fe_4N		

The segregation of solute impurities in iron and low alloyed steels leads to compositional changes of grain boundaries and free surfaces. Surface equilibrium segregation influences surface properties such as surface reaction kinetics, friction and adhesion, surface diffusion and sintering and corrosion resistance. Grain boundary segregation affects materials performance and may lead to phenomena such as temper embrittlement, creep embrittlement, stress corrosion cracking and intergranular corrosion.

Sulphur is known to cause embrittlement of iron, i.e. a strong tendency to intergranular fracture, whereas carbon and nitrogen will reduce embrittlement. The embrittlement by sulphur has been explained by a weakening of metal–metal bonds in lattice planes adjoining the grain boundary. This hypothesis was supported by cluster calculations,[1] which indicate that sulphur attracts negative charge and will draw electrons from surrounding metal atoms.

The x-ray photoemission spectra of the core lines S 2s, S 2p, C 1s and N 1s from the segregated non-metal exhibit characteristic shifts to lower binding energies in comparison to corresponding lines from appropriate reference samples. The latter samples are solid sulphur S_8, graphite and molecular nitrogen adsorbed on iron at low temperatures. These references represent a homopolar bond of the non-metal under investigation, so that the chemical shift reveals the polarity of the iron–non-metal bond.

FeS and FeS_2 show S 2p binding energies which are 0.3 eV and 0.5 eV below the one of sulphur segregated on Fe(100) of 162.2 eV, i.e. the charge is higher on sulphide sulphur atoms. This value is found also for the sulphur enrichment on the grain boundaries of an *in situ* broken sample. The chemical shift related to the binding energy of 164.0 eV of the reference sample of solid S_8 indicates an almost ionic bond character for the iron-sulphur bond, as it was predicted by cluster calculations.

In the spectra of segregated carbon only two lines appear separated by 2.0 eV. The one with higher binding energy of 284.6 eV corresponds to a graphitic precipitation below the temperature T_S of the solid solubility limit of carbon in α-iron. XPS binding energy and Auger lineshape of this carbon species equal those of graphite. The other line is ascribed to chemisorbed carbon, since it is observable only above T_S, when surface graphite is dissolved into the bulk.

Around T_S a sharp peak arises in the Auger spectrum of the C_{KLL} transition at the high energy side of the $KL_{23}L_{23}$ line. As this peak represents an interatomic transition involving two electrons of the Fe 3d band to neutralize a carbon core hole, the origin of the Auger electron is the very first graphite layer on the iron surface. The C 1s binding energy of the carbides at 283.5 eV indicates a lower bond polarity of the compound in comparison with the strongly heteropolar bonding of chemisorbed carbon with iron atoms.

Fe_4N and Fe_2N were prepared in a gas flow apparatus in thermodynamic equilibrium with a NH_3–H_2 atmosphere. The chemisorbed N 1s binding energy of 397.2 eV lies 0.9 eV below the one of the nitrides and about 3 eV below the binding energy of the hompolar reference bond. According to Siegbahn[2] this shift corresponds to a charge on the chemisorbed nitrogen atom of half the electronic charge.

REFERENCES

1. R. P. Messmer, C. L. Briant, *Acta Met.* **30,** 457 (1982).
2. K. Siegbahn *et al., ESCA, Atomic, Molecular and Solid State Structure Studied by Means of Electron Spectroscopy*, Almquist and Wiksells (1967).

Microchemical Characteristics of Metal Surface Oxidation Related to Quantitative EMPA Profiles in the Depletion Zone of High-temperature Alloys

H. Grübmeier, L. Maldonado, A. Naoumidis and H. Nickel
Institut für Reaktorwerkstoffe, Kernforschungsanlage Jülich GmbH, Postfach 1913, D-5170 Jülich, FRG.

INTRODUCTION

In order to reduce the penetration of carbon or hydrogen into high-temperature alloys under the conditions of process gas generation, especially in view of the application of nuclear heat, the formation of oxide layers as diffusion barrier is taken into consideration. Therefore the surface oxidation of two candidate materials, INCOLOY 800 H and INCONEL 617, was investigated on different samples exposed to a gas-atmosphere relevant to steam reforming process of methane at temperatures from 800°C up to 950°C. The gas composition (H_2O, H_2, CH_4, CO, CO_2) resulted in oxygen partial pressures increasing from 10^{-19} to 10^{-16} bar with temperature.

The microchemical characterization of oxide layers and adjacent metal substrate was intended to be done by electron microprobe analysis (EMPA) including the identification of reaction products and quantitative measurements of concentration profiles in the depletion zone of the metal alloy. The latter method was expected to be sensitive also to minor elements and their oxides forming very thin scales which could not be fully resolved quantitatively in a direct local analysis with EMPA.

METHODS

The oxide scales were fixed by nickel plating and the specimens prepared as cross sections for the analysis. The local arrangement of the multi-layer structure was examined by electronic linescannings which gave the element specific distribution in the transversal direction. If possible, as was the case with the thicker chromium oxide layers, the chemical composition of the major phase was determined by point analyses. Quantitative measurements of concentration profiles in the depletion zone were used to integrate the total depleted mass of oxidized elements corresponding to their amounts in the oxide layers. In combination with the linescannings a partition factor of titanium was estimated for the distinct oxides identified in the duplex layer, so that the titanium contents in these phases could be calculated. As manganese was restricted to the outer oxide phase it was possible to derive from the integrated manganese profile and the titanium mass a Mn : Ti ratio. The complementary chromium concentration could be determined by stoichiometry assuming a solid solution of $Mn^{II}Cr_2^{III}O_4$ — $Mn_2^{II}Ti^{IV}O_4$.

The concentration profiles were recalculated by a diffusion model using a finite difference computer program based on Fick's second law with parabolic time dependent mass flux of cations at the oxide–metal interface and with constant concentration in the unchanged region beyond the depletion depth.

RESULTS

The local distribution of elements represented by x-ray images showed the formation of at least two oxide phases. Linescanning curves of selected elements which formed the oxide scale indicated Ti containing chromium oxide as the main phase adjacent to the metal–oxide interface and additional Mn-Ti-Cr-spinel as a separate phase in the outer layer. The described methods enabled the determination of the chemical composition of both oxide phases from point analyses and profile measurements. The results for INCOLOY 800H after 1000 h oxidation treatment at various temperatures are summarized in Table 1. There was no evidence for manganese solubility in the chromium oxide (detection limit < 250 ppm).

Table 1.

Temperature	Cr-Sesquioxide (at%)						Spinel
°C	Cr	O	Mn	Ti	Ni	Fe	$Mn_{(1+x)}Cr_{2(1-x)}Ti_xO_4$
950	34	65	0.00	0.8	0.09	0.09	$x = 0.23$
900	36	62	0.00	1.0	0.09	0.08	$x = 0.27$
850	33	66	0.00	1.0	0.08	0.09	$x = 0.25$
800	33	65	0.00	1.2	0.45	0.23	$x = 0.27$

The comparison between measured and calculated profiles in the depletion zone of INCOLOY 800 H verified the parabolic oxidation rate of chromium in the temperature range from 800 to 950°C after 1000 h oxidation and at various time periods (600, 1000, 2000, 5000 h) at 900°C. However, it was not possible to explain the manganese profiles with the assumption of a parabolic time dependent mass flux. Manganese seemed to follow a more complex behaviour of oxide growth.

The temperature dependence of the chromium diffusion coefficients showed a linear relationship when plotted in an Arrhenius diagram. The values for INCOLOY 800 H agreed with literature data for Cr in γ-iron, the activation energy derived from the curve slope equals that of INCONEL 617.

Combined use of EPMA, XPS, GDOS for the Characterization of Steel Surfaces

P. De Gregorio, S. Fioravanti, G. M. Ingo
Centro Sperimentale Metallurgico SpA, CP 10747 00100 Rome, Italy

OBJECTIVES

To demonstrate how the combined use of EPMA, XPS and GDOS overcomes the limits of each individual technique, thus providing an analytically integrated system particularly suitable for monitoring surface modifications which take place during industrial processes, such as cold rolling, annealing and pickling.

RESULTS

Surface segregation phenomena were investigated in low carbon steels. More specifically, a morphological and micro-chemical investigation was carried out on the surface layers of decarburized sheets, characterized by a wide range of reactivities to be correlated with the surface and sub-surface chemical composition.

As an example, results are illustrated for two samples, designated HR and LR, having the highest and lowest reactivity respectively. Preliminary information is given by the different SEM morphology. In the case of the low reactivity, small size inclusions are found almost exclusively at grain boundaries and contain P.

In the case of the high reactivity inclusions are scattered everywhere outside grain boundaries and contain the elements Mn, P, Al, Si, Cr in decreasing amounts. Moreover the low reactivity sample submitted to H_2SO_4 pickling shows a predominantly intergranular attack, whereas in the case of the high reactivity the attack is still intergranular, although of much greater scope and characterized by a generalized etching. EPMA was used to investigate the chemical composition and surface distribution of non-metallic inclusions.

Ten representative inclusions were selected and carefully analysed for both HR and LR situations. In the low reactivity sample the chemical composition of the inclusions is basically Fe, P and O with traces of Si. On the contrary the high reactivity sample shows a remarkable Mn enrichment with a smaller amount of P, Cr and Al. These results are visualized by x-ray elemental mappings which add information on topography and relationships among chemical species such as O, P and Mn. To confirm EPMA results, the low reactivity sample was also analysed by STEM + EDS on replicas stripped from the surfaces.

Continuous depth profiles were performed by GDOS. In the low reactivity sample, profiles start by eroding surface inclusions where P predominates and then indicate small P and S enrichments when the matrix begins to be attacked. The high reactivity sample shows a more relevant presence of Mn–rich surface inclusions with lower concentration of Al, S and P. The sub-surface is also enriched in Mn and P to a depth of about one micron, thus demonstrating that the sub-surface composition also contributes to surface reactivity. Cross-sections illustrate morphological differences and the corresponding EDS analysis gives the chemical composition of the oxides and sulphides inclusions at various depths.

XPS depth profiles give further specific details on the top layers by discriminating iron oxide from iron metal, and adding atomic quantification of analytical results according to the first principles model. The agreement between GDOS and XPS results is evident by comparing depth profiles for both high and low reactivity samples.

CONCLUSIONS

This investigation indicates that Mn plays an important role in determining surface reactivity. From a methodological point of view, overall results demonstrate that the combined use of EPMA, XPS and GDOS, with SEM as auxiliary technique, give a set of integrated microanalytical information, supporting each other, fully defining surface and sub-surface microchemistry. In particular, extensive use of GDOS proved to be very efficient to monitor surface composition in industrial situations.

EQUIPMENT USED

XPS	LHS 10 with an EA11 analyser by Leybold & Heraeus
EPMA	Camebax Microbeam
GDOS	Grimm type (RSV) G.D. connected to a Quantovac ARL 31.000
STEM	JEOL CX 200 with microanalysis Link 860/2 system
SEM	Cambridge Stereoscan 250 Mk2 with microanalysis Tracor Northern T200 system

Preliminary Study of the Formation and Characterization of Rh Alloys with an Element of the 1B Group: Copper

M. P. Delplancke, M. Jardinier-Offergeld and F. Bouillon
Laboratoire de Chimie du Solide, Service de Chimie Analytique et Minérale Faculté des Sciences, C.P. 160, Université Libre de Bruxelles 50, av. F.D. Roosevelt, B-1050 Bruxelles, Belgium

The catalytic importance of rhodium in the industry implies an increased knowledge of its surface composition and of the alloying effect on the surface properties of this metal.[1] For this reason, we selected the Cu–Rh system. Massive single crystals of dilute Cu–Rh alloys were prepared by the Bridgman method (fusion and unidirectional solidification). They were characterized by x-ray back reflexion and optical microscopy. All our samples (1.5, 2, 8 and 15 at% Rh) showed cellular-dendritic structures. Their elimination was obtained by heating at 1000°C in hydrogen at atmospheric pressure (to limit the copper loss). The surface composition of the crystals is determined by Auger electron spectroscopy but a quantitative analysis requires a measure of the rhodium relative sensitivity factor. This value was obtained by two methods: (a) the internal standard method by scraping the samples surface;[2,3] (b) the use of elemental sensitivity factors.[4]

SCRAPING

The 1.5, 8, 15 at% Rh alloys were *in situ* scraped with a stainless steel tip. A plot of the heights ratios ($M_5 N_{4,5} N_{4,5}$ transition for Rh and $L_3 M_{4,5} M_{4,5}$ for Cu) versus the rhodium bulk concentration (determined by x-ray fluorescence) is linear. The equation

$$S_{Rh} = \frac{H_{Rh}}{H_{Cu}} \frac{(1-C_{Rh})}{CR_h} \qquad (S_{Cu} = 1)$$

where S_{Rh} represents the rhodium sensitivity factor, H, the absolute height and C_{Rh}, the bulk concentration,[5] is used to obtain $S_{Rh}(S_{Rh}^{CuRh} = 7.1 \pm 0.4)$. This value takes the matrix effects into account.

PURE ELEMENTS

The pure elements spectra were recorded on a 'composite sample': it is a copper single crystal electrolytically polished, face (111), half coated with a thick rhodium layer.

This sample is characterized by a similar roughness for the two elements, identical Auger recording conditions, same position and angle of incidence. The peak heights ratio gives the rhodium sensitivity factor: $S_{Rh}^{Cu} = 7.5 \pm 0.5$. The weakness of the copper peak in the recording conditions reduces the measure accuracy. No correction was made for the matrix effects.

The comparison of the values, obtained by the two methods, shows no important matrix effects in dilute Cu–Rh alloys. They seem to be smaller than the experimental error since the reliability intervals of both methods overlap. This is in agreement with the Hall's prediction.[6]

The calculated value, $S_{Rh}^{Cu} = 7.3 \pm 0.4$, resulting from the Mroczkowski and Lichtman's 'pseudo-first principles' model,[7] is in agreement with ours, while the Handbook sensitivity factor, $S_{Rh}^{Cu} = 2.8$, largely differs. This discrepancy may be explained by the importance of the surface roughness on the Auger detection. We have observed that rhodium is difficult to polish in contrast with copper where a smooth surface is easily obtained by electrolytic polishing. It is possible that the Handbook authors used rhodium and copper samples of different roughness. The rhodium signal is then partly absorbed by the asperities and the resulting sensitivity factor is smaller than the real one. In our experiments, the two elements have always the same roughness and, in their model, Mroczkowski and Lichtman have supposed that the surfaces are ideally smooth. This is probably the reason for the good agreement between the two values.

SURFACE COMPOSITION OF 1.5 at% Rh SPECIMEN

The experimental measurements were made with an Auger/LEED spectrometer (VG Scientific Ltd) equipped with a retarding field analyser (energy resolution of 0.5%). An electron gun provides a beam with an energy of 2.5 keV at a 15° angle of incidence to the target surface. The signal is detected in the $dN(E)/dE$ derivative mode with a 5 V peak-to-peak modulation. The specimen can be cleaned by argon-ion sputtering and by heating in hydrogen atmosphere. The detected impurities are sulphur, boron, carbon and oxygen. Ion sputtering is effective for the sulphur elimination but increases the carbon content. The heat treatment at 600°C in hydrogen removes carbon, reduces the oxygen concentration but brings on the sulphur segregation. Twelve cycles of heating and sputtering are necessary to obtain a low level of surface impurities.

After 90 min at 600°C in vacuum, the rhodium surface concentration is stabilized at a value of 1.6 at%. No solute or solvent enrichment is observable. This result is in agreement with the theoretical predictions of Miedema[8,9] and of Abraham and Brundle.[10]

Acknowledgement

We are grateful to the Institut pour l'Encouragement de la Recherche Scientifique dans l'Industrie et l'Agriculture for the award of a Research Studentship (MPD).

REFERENCES

1. G. A. Somorjai, *Chimia*, **35**, 1 (1981).
2. A. Mogami, M. Uemura, A. Okitsu, N. Takahashi, *Scanning electron Microscopy*, **4**, 1675 (1984).
3. G. Van Steene, thesis, Brussels (1983).
4. L. E. Davis, N. C. MacDonald, P. W. Palmberg, G. E. Riach, R. E. Weber, *Handbook of Auger Electron Spectroscopy*, Physical electronics Industries, Inc. (1976).
5. E. Minni, *Appl. Surf. Sci.* **15**, 270 (1983).
6. P. M. Hall, J. M. Morabito, *Surf. Sci.* **83**, 391 (1979).
7. S. Mroczkowski, D. Lichtman, *Surf. Sci.* **131**, 159 (1983).
8. A. R. Miedema, *Z. Metallkunde*, **69**, 455 (1978).
9. J. R. Chelikowsky, *Surf. Sci.* **139**, L197 (1984).
10. F. F. Abraham, C. R. Brundle, *J. Vac. Sci. Technol.* **18**, 506 (1981).

Automatic Quantitative Measurement of Carbon Profiles in Case-hardened Steels

M. Lamothe and F. Convert
CETIM, 52 avenue Félix Louat, 60304 Senlis, France

Our laboratory's work at CETIM consists of technical assistance in checking surface treatments, or analysing carbon in small samples. Moreover, it consists of studies to improve surface treatments (case hardening, carbonitriding, etc.).

The carbon profile in case-hardened parts determines their mechanical properties. Evaluation of the carbon profile is therefore necessary. Unfortunately few laboratories are able to fully master the technique of quantitative analysis of carbon in steels, using an electron probe. This is due to numerous reasons: sample preparation, decontamination, low emission rate, tedious and long measurements. These kind of measurements are feasible with partial automatization WDS (sample displacement, data processing) using a diamond check sample. This automatization must reproduce the manual sequence: (a) checking for counting stability and (b) optimization of the sample displacements. The purpose of the study undertaken herein is thus to develop a software fulfilling these requirements.

Carbon analysis requires a long counting time (100 to 200 s) because of a low diffraction line to background noise ratio. This long counting time gives rise to two difficulties: (a) surface contamination (by carbon) and (b) tedious manual measurement. We solved the surface contamination problem by use of a liquid nitrogen trap and of a microleak (oxygen or argon) directed to the electron beam impact area.

With partial automatization WDS, these analyses are feasible but it is nevertheless necessary to find an effective test to detect the end of decontamination (decontamination is reached within 1 to 5 min). The test used in our laboratory is the statistical test of X^2 (chi-square test).

We define the counting function $Q(N(i))$:

$$Q = \sum_{i=1}^{n} (N(i) - m)^2/m$$

This function verifies a law of KHI_2 with $(n - 1)$ degrees of freedom if H_1 : the countings $N(i)$ follow a normal law and H_2 : the average (m) is equal to the variance. These hypothesis are verified when the countings are stable and follow a Poisson law ($N(i)$ always sufficiently high).

USE OF TEST

The decontamination is supposed to be finished when

$$Q \leqslant X^2_{1-\alpha}\ (n - 1)$$

where n = counting number used ($n \geqslant 10$) and α = risk with 5% error.

For a good definition of carbon profile in case-hardened parts, near the surface where concentration gradient is high the measurements must be performed very close to each other. Whereas near the core only a limited number of measurements is needed.

An automatic sequence with equal steps implies a greater number of measurements than necessary. So, we chose an automatic sequence with variable steps. The amplitude of the sample displacements is determined during the carbon profile realization, to minimize the number of measurements. The measurement is performed at a particular point only if the concentration difference measured between two points which surround it, is greater than a given value.

The software system developed at CETIM monitors the specimen displacement and checks the counting sequence realized by the task program (TRACOR) which drives our CAMEBAX electron probe. The method developed is easily transposable to any other kind of carbon analysis, and any other element analysis. After the preliminary adjustments, a profile can be automatically obtained in less than two hours (20 measurements) owing to the effective detection of complete decontamination at every measurement point. The precision of the results is similar to that obtained by means of chemical analysis.

SURFACE CHARACTERIZATION FOR THE SOLUTION OF SOME TYPICAL METALLURGICAL PROBLEMS

C. Roques-Carmes
Laboratoire de Microanalyse des Surfaces, ENSMM. 25 030 Besancon Cedex, France

Over the past few years, there has been a sharp interest in solving industrial problems by means of surface techniques. This is particularly true of the general metallurgical field involving above all the phase transformations and their mechanisms. A more comprehensive approach can be extended to the problems of brittle fracture, tribology, wear, lubrication, adhesion, wetting, surface treatment, corrosion and oxidation. There is no ideal technique that would provide an extensive knowledge of the surfaces — their visualization, the quantification of their chemical composition, the local variations in their electronic structure and the imaging of the surface elements with respect to their local concentration and their bonding state. In order to solve these problems, a few complementary techniques are currently used. We will select problems that can be interpreted in the light of physical metallurgy: submicroscopic phase precipitation; segregation in binary and ternary systems; interface transfers in the early stage of wear; lubrification transfer on materials; surface contamination and sheet degreasing; corrosion and pitting corrosion; oxidation of multilayer materials; physico-chemical modifications during work-hardening.

In addition, our aim is to emphasize two artifactual contributions to the analytical results — the surface roughness and the intrinsic non-homogeneity of the metallurgical samples. It will also be shown that we can interpret the results obtained on industrial samples in terms of the physics of the solid state.

SEGREGATION OF TIN AND ANTIMONY ON α–IRON

M. Rüsenberg, H. Viefhaus and H.J. Grabke
Max-Planch-Institut für Eisenforschung GmbH, D–4000 Düsseldorf 1, FRG

AES, ELS, LEED, and XPS investigations of the surface segregation of tin and antimony in a 4wt% Fe–Sn and a 4wt% Fe–Sb alloy were performed for the three low indexed surface orientations. At elevated temperatures from 450 to 650°C, the surface segregation always leads to the same temperature independent coverage.

During tin surface segregation, order-disorder transitions were observed for the (100) and (111) surface orientations. After initial formation of ordered structures (c(2×2) and p(1×1), respectively), a transformation of the ordered structure into a disordered state occurs. For the (110) oriented surface, two different hexagonal surface structures appear successively during tin enrichment. One of them resembles the (111) plane in the diamond structure of pure tin, while the second corresponds to the formation of a thin layer of the intermetallic compound FeSn. Drastic changes of the segregation rate were detected during the transitions.

In contrast, structures of segregated Sb atoms at Fe surfaces with the same orientations do not transform from order to disorder. The c(2×2) and p(1×1) superstructures stay stable as a saturation coverage. Whilst Sb segregates to the Fe (110) surface, one hexagonally ordered structure, probably due to a slightly distorted structure of pure antimony is followed by a dramatic change of the LEED pattern. This phenomenon has been shown to be due to surface faceting.

Small contents of Sn (less than 100 ppm) in polycrystalline Fe–Sn alloys strongly affect surface reaction kinetics. As an example, the rate constants in the carburization-decarburization according to CH_4 = C(dissolved) + $2H_2$ are strongly decreased by the segregated Sn at the surface.

EFFECTS OF THE ALLOYING ELEMENTS Ti, Nb, Mo AND V ON THE GRAIN BOUNDARY SEGREGATION OF P IN STEELS

H.J. Grabke, R. Möller, H. Erhart and S. Brenner
Max-Planck-Institut für Eisenforschung GmbH, Düsseldorf, FRG and University of Pittsburgh, PA, USA

Investigations have been performed and are underway on the effects of the alloying elements Ti, Nb, Mo and V on the grain boundary segregation of phosphorus in iron and steels. The methods are AES-analysis of grain boundaries on intergranular fracture faces and field-ion microscopy and mass spectrometry (atomprobe).

The elements Ti and Nb show strong interaction with P, keeping P in the grain and decreasing its grain boundary segregation. This effect is of importance for the interstitial-free deep drawing steels, where Ti or Nb can help to avoid embrittlement by grain boundary segregation of P. In the alloys and steels containing carbon, Ti and Nb react under formation of carbides, therefore with increasing C–content the grain boundary segregation of P increases. For stoichiometric ratio Ti/C and Nb/C the segregation reaches a maximum, at higher carbon contents C displaces P from the grain boundaries. Some interfacial segregation of P is observed at or near the ferrites/carbide interface in the case of TiC, leading to a decrease of grain boundary segregation. The same effect applies for Mo_2C but not for NbC. In contrast to Ti and Nb there is no direct interaction of Mo and P, which is also in contradiction to previous assumptions on the role of molybdenum in the temper embrittlement of low alloy steels.

SAM AND TEM INVESTIGATIONS OF REVERSIBLE TEMPER EMBRITTLEMENT IN Cr–Ni STEELS

H. Goretzki,[*] A.P. von Rosenstiel,[†] H. Bruis,[‡] M. Weiss[*] and B.H. Kolster[†‡]
[*]Perkin–Elmer Physical Electronics Division, D–8011 Vaterstetten, München, W. Germany
[†]Metaalinstituut TNO, Postbus 541, 7300 AM Apeldoorn, The Netherlands
[‡]Technische Hogeschool Twente, Afd. Werktuigbouwkunde, PO Box 217, 7500 AE Enschede, The Netherlands

Temper embrittlement and de-embrittlement of martensitic Cr–Ni steels were investigated by scanning Auger spectroscopy and transmission electron microscopy. The objective of this work was to determine the structure and chemical composition of embrittled and de-embrittled grain boundaries as a function of thermal and mechanical treatment.

It was confirmed that temper embrittlement of ^{11}Cr0, ^{5}Ni steel is linked to phosphorous segregation and a strong enrichment of Cr and C with subsequent formation of chromium carbides at the grain boundaries.

De-embrittlement yielded restored initial mechanical properties and grain boundary composition. This was obtained already using tempering times of 5-10 min at temperatures above 625°C. The structure and composition of the grain boundaries was studied for several embrittlement and de-embrittlement cycles. Special emphasis is given to the complementary information by Auger spectroscopy and high resolution transmission electron microscopy of this technically important phenomenon.

AN AES STUDY OF THE OXIDATION OF POLYCRYSTALLINE Zr AT ROOM TEMPERATURE AND LOW OXYGEN PRESSURES

J.M. Sanz, C. Palacio, Y. Casas and J.M. Martínez-Duart

Dpto. Física Aplicada. Fac. Ciencias, C–XII, Universidad Autónoma de Madrid, Cantoblanco, 28049–Madrid, Spain

Auger electron spectroscopy has been used to characterize the oxidation of polycrystalline Zr in an oxygen atmosphere at low pressures (10^{-4}Pa$\geqq PO_2 \geqq 10^{-6}$Pa) and at room temperature after cleaning by Ar^+ ion bombardment.

The oxygen and zirconium Auger lines were recorded as a function of the oxygen exposure. The uptake of oxygen has been studied by following the intensity of the oxygen peak as a function of time. A typical adsorption curve, with a linear increase of the intensity followed by a region where the slope decreases continously until the saturation is reached, has been observed.

Changes in the shape of the Zr (MNN) Auger transitions were studied as a function of exposure up to 40 L at 300 K. The Zr($M_{45}N_1N_{23}$) and Zr($M_{45}N_{23}N_{23}$ transitions show typical broadening, reduction of intensity and small energy shifts during the oxidation. The Zr($M_{45}N_{23}N_{45}$) and Zr($M_{45}N_{45}N_{45}$) transitions, which include the valence band, show as expected the most important changes.

Exposure to oxygen reduces both Zr($M_{45}N_{23}N_{45}$) and Zr($M_{45}N_{45}N_{45}$) peak-intensities, accompanied by the appearance of a new peak in the low energy side of the Zr($M_{45}N_{23}N_{45}$) peak, which grows during oxidation. This new peak is interpreted as indicative of the zirconium oxide and probably is due to a Zr(M_{45})Zr(N_{23})O(V) cross transition similar to those observed in other transition metals. The result is a doublet whose components are interpreted as 'metallic' and 'oxide'.

CHARACTERIZATION OF SURFACE LAYERS ON ALUMINIUM AND ALUMINIUM ALLOYS WITH XPS AND SIMS

K.J. Rüegg and M. Textor

Schweizerische Aluminium AG, Forschung und Entwicklung, CH-8212 Neuhausen am Rheinfall, Switzerland

Results obtained by the application of XPS and SIMS for the characterization of surfaces on rolled and extruded products of technically pure aluminium and aluminium alloys are reported. The analytical capability of the two mentioned methods is demonstrated and compared with other surface analytical methods. It is shown how the composition of oxide layers on semi and finished products depends on earlier production stages, particularly rolling and annealing processes. Quantitative data on thermal surface enrichment of alloying and trace elements have been obtained and are, in the case of magnesium, compared with calculated values based on a simple diffusion and surface reaction model which allows a prediction of absolute magnesium surface concentrations as a function of magnesium content of the metal strip gauge annealing temperature and time. Examples are presented which show how surface properties such as corrosion resistance and adhesion in composite aluminium polymer materials depend on oxide layer composition. The combination of XPS and SIMS with sputtering also provides a better understanding of chemical and structural properties of thin conversion layers obtained in chromating, phosphating and anodizing processes. The conclusion is that the analytical characterization of surface layers makes an essential contribution to the optimization and quality control of relevant surface properties of semi and finished aluminium products.

INITIAL STAGES OF GROWTH OF PROTECTIVE OXIDE- AND CARBIDE-LAYERS ON Fe–Al–Ti ALLOYS

H. Viefhaus, J. Peters and H.J. Grabke

Max-Planck-Institut für Eisenforschung GmbH, D–4000 Düsseldorf 1, FRG

Oxide layers protective against carburizing atmospheres are of great importance for high temperature alloys. Very good results were obtained with an alloy Fe–6Al–0.5Ti.

For a single crystalline sample of this alloy the composition and growth of the protective layers were studied by LEED, AES and ELS on a (100) oriented sample. At elevated temperatures (above 400°C) initially Al segregates to the surface within a sequence of two ordered structures. At high temperatures (above 700°C) in oxygen containing atmosphere (5×10^{-8} mbar) epitaxial growth of a titanium oxicarbide layer was observed. This layer growth occurs by surface segregation of Ti and C, which is present as a dissolved impurity. At the same time the segregated Al atoms change their positions to the outer surface of the carbide layer. Reaction with oxygen leads to the formation of an epitaxial Al_20_3 layer on top of the carbide. Both layers are very dense and well adherent, even for high quenching rates.

Experiments in oxidizing and carburizing atmospheres (at 1 bar) have shown that only very small amounts of carbon are able to penetrate through the dense Al_2O_3–layer from the gas phase. The penetrating carbon is captured by the oxicarbide having a wide range of existence with respect to carbon content. Also by further oxicarbide formation the penetrating carbon atoms may be consumed. By this mechanism the protective nature of those layers against carburizing atmospheres is explained.

QUANTITATIVE SURFACE ANALYSIS OF STAINLESS STEEL

C. Jansson,[†] J. Onsgaard,[†] J. Christensen,[*] and K. Aa. Thorsen[‡]

[†] Fysisk Institut, Odense University
[*] Metallurgy Department, Riso National Laboratory
[‡]Department of Metallurgy, The Technical University of Denmark

Surface analysis of stainless steel has been performed by a combination of x-ray photoelectron spectroscopy (XPS) and ion bombardment. The purpose of the study is to characterize stainless steel surfaces under realistic brazing conditions including variations of temperature and vacuum furnace volume. The nominal composition of the selected steel, extra low interstitial 18–2, is Fe (79 wt%), Cr (18 wt%) and Mo (2 wt%) with smaller amounts of Ti and Mn.

The temperature treatments of the stainless steel create interfaces with segregation of Cr to the surface when the annealing temperature is below 700°C, whereas a strong accumulation of Ti at the surface takes place when the annealing temperature is around and above 700°C. Both phases are oxidized due to the oxygen partial pressure. Depth profiling to depths around 500 Å clearly demonstrates a correlation between the segregated metal (Cr or Ti) and oxygen. After passage of the interface, Fe increases to the bulk level, the Cr content assumes its nominal value and O disappears. Chemical binding energy shifts of the metals were followed with XPS. On the basis of line structure analysis and background evaluation, the oxidation numbers of the metal ions could be followed in detail.

Depth profiling of stainless steel samples in the presence of a brazing filler metal demonstrates that the oxygenation is much reduced and that the segregation of Ti to the surface still takes place.

Section 4.

Catalysis

SURFACE AND INTERFACE ANALYSIS, VOL. 9, 195–206 (1986).

Surface Analysis and the Complexity of Real Catalysts

Bernard Delmon
Catalyse et Chimie des Matériaux Divisés, Université Catholique de Louvain, Place Croix due Sud, 1, B-1348 Louvain-la-Neuve, Belgium

This paper is written for specialists of surface techniques not perfectly familiar with catalysis, as well as for scientists working in the area of heterogeneous catalysis. It is aimed at suggesting the problems of characterization that surface analysis is faced with, because of the complexity of real catalysts. We first present the various typical 'architectures' to which catalysts may correspond. We shall then illustrate typical problems, and their solutions, by selecting examples in the field of hydrodesulfurization (HDS) catalysts. We shall show how 10 out of the 12 theories proposed for explaining catalytic activity were eliminated thanks to surface techniques. HDS catalysts are remarkable in several respects, especially because they undergo many 'metamorphoses' (activation, modifications during catalytic work, ageing, regeneration) from the time of their preparation to that of their death by irreversible deactivation. They therefore provide a large body of illustrative examples concerning the characterization of several 'architectures'. In spite of the achievements of surface techniques in clarifying many features of HDS catalysts during their life, investigators are still faced with unsolved problems; many such problems are related to the very fine scale (of the order of 1 nm or less) of the details of importance for catalysis. This might suggest directions for new improvements or the development of new surface techniques. As a conclusion, we shall recall that, because complete characterization needs the joint use of several techniques, a methodology of study had to be developed, now exists, and still continues to improve. The main lines of this general methodology are not restricted to catalysts, but can be used for the study of all complex highly dispersed solids.

INTRODUCTION

Most of the 40 or so[1] techniques used for characterizing surfaces were developed for probing clean and relatively simple surfaces (very flat samples, possibly well defined crystallographic faces, etc . . .). But, they very soon turned to be used also for the study of highly dispersed solids: quite logically, a considerable part of the useful characteristics of the latter, very often the most important ones, are related to surface. Heterogeneus catalysts control partially or totally over 80% of the production of the petroleum, petrochemical and chemical industries together. Because of this key role, they often constitute the category of highly dispersed solids which are subjected in priority to investigations by novel surface techniques. The extent of the surface area and the nature of the surface are of such an importance in catalysts that their determination deserves very costly efforts. But one discerns a clear tendancy to use these techniques also in other areas: pigments, absorbents, adsorbents, various supports, plastic fillers, dispersing aids, raw materials for the manufacture of ceramics, highly dispersed minerals in ores and soils, etc. . . . In most cases, the special way the techniques are used for studying these dispersed materials has been defined in the course of studies on catalysts.

0142–2421/86/120195–12 $06.00

Typical architectures of catalysts

Catalytic technology (and basic research supporting the corresponding innovation and development) has led to solids of especially complex composition and shape. One can conveniently classify catalysts into 8 different basic categories with respect to their 'architecture', namely the distribution, in space, of zones of different composition and structure (Fig. 1). It can be noticed that the other categories of dispersed solids cited above could be classified the same way. One can distinguish unsupported active phases (uns), either homogeneous (uns 1), or with composition gradients, or even constituted of two or several phases (uns 4), and supported catalysts (sup), where the active phase may be atomically dispersed (sup 1), form adhering layers (sup 2) or tiny crystallites (sup 3), or carry one or several well defined phases in the form of relatively large crystals (sup 4). Real catalysts and real dispersed solids may be a combination of the features depicted in Fig. 1.

In the case of catalysts, it is particularly important to know the extent of surface area developed by each phase (or its 'dispersion', defined as the ratio of the number of exposed atoms to the total number of atoms contained in the catalysts) and to describe the porosity, which conditions material transport during catalysis.

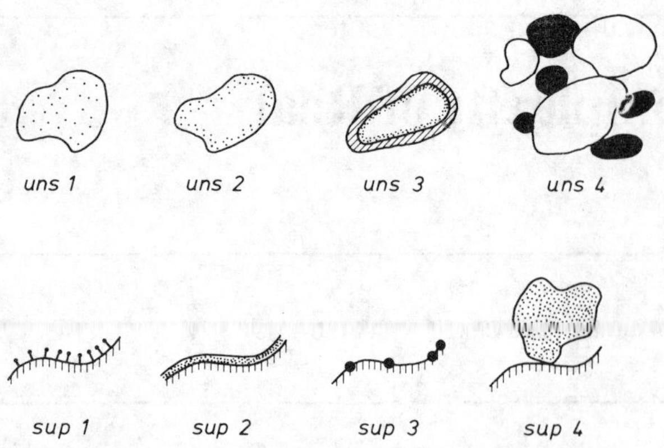

Figure 1. Different types of unsupported (uns 1 through to uns 4) and supported (sup 1 through to sup 4) catalysts.

The study of catalysts from preparation to death

Catalysts are very difficult to prepare. Each step must be controlled carefully. For impregnated catalysts, as those used in hydrotreating (which will constitute the examples cited below) the principal preparation steps are the following:

- impregnation (of the support with solutions of the active elements)
- drying
- calcination (to oxides of the active elements)
- activation (reduction and sulfidation to the really active species, namely sulfides, in the case of hydrotreating)

Catalysts change when they work. They get poisoned, they sinter and eventually deactivate. They may have to be regenerated. The details of their ageing should be described and understood for efficient process operation and for life improvements. Regeneration must be monitored precisely. Surface characterization techniques are necessary all along the life of a catalyst. They help identify the most harmful process in catalyst life.

Examining especially one important characteristic, namely dispersion, one notices that the highest possible dispersion is usually obtained by atomically grafting, or atomically or molecularly adsorbing, the active element on the surface of the support; all successive steps generally tend to diminish dispersion. Chemical techniques, namely selective adsorptions, turn out to be useful only in a limited number of cases. The reason is the lack of molecules adsorbing selectively on each given phase of interest. This was the case, until recently, for sulfided hydrotreating catalysts. One has thus to resort to physical techniques in the vast majority of cases.

Scope of this work

In this work, we shall attempt to suggest how surface analysis can contribute to the understanding of the complexity of real catalysts. A very rich body of illustrating examples can be found in the field of hydrotreating and hydrodesulfurization (HDS). In the course of years, at least 12 theories have been proposed for explaining the conspicuous catalytic synergy between the group VI element (Mo or sometimes W, and potentially U) and the group VIII element (Co, or Ni, or sometimes Fe) which is observed in all corresponding catalysts, in industry and in the laboratory (Table 1). A few words will suffice here to explain these theories or hypotheses, when Table 1 is not self-explanatory. The monolayer model (theory 1) attributed catalytic activity to a molybdenum oxide layer of molecular thickness adhering on the surface of the alumina support; the layer was supposed to be promoted by cobalt ions located below this layer. Intercalation (theory 5) supposed that the Co atoms were situated between layers in the MoS_2 structure, whereas pseudo intercalation (theory 6) restricted this intercalation, at unspecified sites, to places on the edges of the MoS_2 crystallites. Hypothesis theory 9 probably never went beyond the stage of just a possibility;[22] theory 10 assumed the presence of a ternary compound of flexible composition and non specified structure, distinct from $CoMo_2S_4$.

Only theories 11 and 12 remain alive now. Both are based on strong physico–chemical evidences. The Co–Mo–S phase (theory 11) is an association of Co, in a special coordination, with a MoS_2-like structure; Co sits on the edges of the MoS_2 layers. Hypothesis 12 takes into account the presence of MoS_2 and Co_9S_8 crystallites in the cases where crystalline phases and physical properties corresponding to them can be observed. It attributes the synergy to the proximity or contact, between those phases ('contact synergy'). The postulated mechanism involves the creation, on MoS_2, of the active centers, by action of 'spill-over' hydrogen (a surface mobile hydrogen species) emitted by Co_9S_8: thanks to this spill-over species, Co_9S_8 exerts a 'remote control' on the nature and number of active centers.

The existence of these 12 theories is the testimony of a very strange and fascinating evolution of ideas inspired by an incredibly numerous body of experimental facts. Solid state chemistry was extremely important; sometimes it turned out to be misleading, but it very often gave important clues. Surface analysis played a major role in clarifying the situation, as it will in ultimately explaining the synergetic effect. This was made possible because many stages of life of catalysts were investigated. Recent work in our laboratory indicates that still more information can be gained from deeper insight into hitherto little studied steps like

Table 1 Theories proposed for explaining catalytic synergy in hydrodesulfurization

Theory		Reference
1	Monolayer model	2–6
2	Oxide–sulfide layer	3, 7–9
3	Boundary (adlineation)	10
4	2-phase surface complex	10
5	Interlaction	11–13
6	Pseudo-intercalation	14–16
7	Zero-valent cobalt	
8	MoS_2 doping	17–19
9	$CoMo_2S_4$	20
10	$Co_xMo_yS_z$	21
11	Amorphous Co–Mo–S	22–24
12	Contact synergy — remote control	25–29

activation, ageing and regeneration. We shall illustrate some selected aspects of these contributions. Our final purpose will be to show that a methodology exists for combining various techniques in view of more efficient surface analysis.

IMPREGNATED AND CALCINED CATALYSTS

The impregnated HDS catalysts, after gentle drying at moderate temperature, contain cobalt (or nickel) and molybdenum salts (nitrates for the former, ammonium molybdate for the latter) deposited on the support. Under this form, HDS catalysts have been subjected to relatively few surface investigations (but illustrations of the role of surface techniques in clarifying the phenomenon taking place during the impregnation can be found in several other field, e.g. for supported nickel catalysts[30,31]). The scarcity of these studies in just dried HDS catalysts is probably due to the fact that it has been recognized very soon that the calcination stage, namely heating at 400–500°C in air plays the determining role in shaping the catalysts. The results we shall recall hereafter concern the calcined solid, i.e. the precursor of the activated (sulfided) catalyst. The precursor is composed of cobalt and molybdenum in the form of various oxides deposited on alumina.

A first important finding, using essentially XPS, was that cobalt and molybdenum promote their mutual dispersion on γ–Al_2O_3[32–34]. The role of dopes in influencing the dispersion as well as the distribution of the individual elements (namely the variation of active species content between the interior and the near surface part of the support particles) was also explained, thanks to the use of XPS and analytical electron microscopy (AEM).

A further step was to show that the deposited elements were in the form of an intimate association of cobalt and molybdenum i.e. a sort of bidimensional compound, or binary oxide, adhering to the surface of γ–Al_2O_3. AEM[35–36] associated with the above mentioned XPS results, proved this association. The method used was to analyze areas where no crystallites could be detected by electron microscopy, and to show that disperse cobalt was present only where disperse molybdenum was also present. The analogy of this surface compound, or 'bi-layer', with a bi-dimensional 'cobalt molybdate' was further suggested by Raman spectroscopy[37,38] and several techniques. UV diffuse reflectance spectroscopy (DRS)[33] gave proofs that some insertion of cobalt into the alumina support, as a cobalt[41] pseudo aluminate, accompanied the formation of the bi-layer.

Later developments concerned fine details of the structure of the bilayer. ISS indicated that, in certain samples of the bi-dimensional cobalt molybdate, cobalt occupied sites situated below molybdenum.[39,40] It is striking that cobalt[41] as well as nickel, in the case of similar $NiMo/Al_2O_3$ catalyst precursors, may, according to preparation and calcination conditions, lie either above or below molybdenum.[42,43] The interpretation of the ISS results, however, remains difficult[41,44]

In the course of years, it became increasingly obvious that the solids obtained after calcination were only the 'parents' (hence our terminology: precursors) of the active HDS or hydrotreating catalysts, which get completely transformed when they work catalytically. Anticipating on the next section, let us say that physico-chemical techniques proved beyond doubt that the real catalysts are composed of sulfides. It had been shown on model unsupported sulfides that a strong synergy existed between Co and Mo.[45,46] Physico–chemical studies proved that, in practical supported catalysts also, synergy was linked to the simultaneous presence of cobalt and molybdenum associated with sulfur on the surface of the support;[47,48] particularly convincing in this respect were XPS measurements made on catalysts having worked in conditions similar to industrial operation (in particular, under a pressure of 3 MPa) and withdrawn from the reactor without contamination.[49]

Thus, theories 1 and 2 of Table 1 had not to be taken into account for explaining catalytic synergy. The immense merit of these theories has been to point to the crucial role of the precursor in determining the nature and dispersion of the active phase, i.e. ultimately, catalytic activity.

All results concerning this precursor are due to the joint use of many surface techniques. They are typical of the fine details that can be described in complex solids. These results correspond to architecture sup 2 of Fig 1. The essential techniques to be used in such a case are XPS and AEM. They are nicely complemented, for extremely fine details, by ISS. This conclusion would hold also for type uns 2 and uns 3 catalyst geometries (Fig. 1). More generally, these three techniques are, together with Auger spectroscopy (AES) and SIMS, the major techniques for providing essential information, relevant to catalysis, on surface composition, dispersion, distribution and surface structure. These techniques are all relatively widespread and correspond to commercial equipments.

On a more general level, it can be noticed that adequate physico–chemical methods beautifully clarify the vision one has of certain effects in catalysis. The central role of alumina (or carbon) as a support of HDS catalysts is to enable the formation of a highly dispersed precursor, namely the bilayer in the case of alumina. The role of some promoters is not necessarily to exert some complicated effects (e.g. electronic effect). It may simply be to modify the dispersion and distribution, the interaction with the support, and the resistance to sintering of the supported species.

ACTIVATED CATALYSTS

The Co–Mo–S phase

Needless to say that extremely rapid progresses were made when activated, i.e. sulfided, catalysts began to be subjected to physico–chemical studies.

The conclusion that metallic cobalt (hypothesis 7 of Table 1) might be involved turned out to stem from a false effect in some XPS studies, namely their reduction when subjected to erosion by an ion beam. A better understanding of the technique (together with simple thermodynamical considerations), helped eliminate metallic cobalt as active species. The hypothesis that synergy was due to a doping of MoS_2 by Co (in a solid solution, as a defect caused by substitution of Mo by Co) (theory 8 of Table 1) proved unrealistic, in view of the more than 2 orders of magnitude difference between the solubility limit of Co and the concentration giving maximum synergy. $CoMo_2S_4$ (theory 9), as a pure phase, could not be detected in the most active catalysts, only leaving open the possibility of a different, non stoichiometric mixed sulfide (theory 10). It also turned out that the very formal hypotheses 3 and 4 could only be valid if embodied in more precise pictures, as in 11 or 12.

In a lapse of time of about 8 years (1970–1977), many theories were proposed (actually 2 through 11 of Table 1). Some of them rapidly died. Finally, only 5 out of the total of 12 remained worthy of consideration (5, 6, 10, 11 and 12).

A surprising phenomenon occurred during these intense efforts made by several groups for clarifying the structure of the active system. In spite of the use of many already well developed techniques, it turned out that one single method (of much more restricted use than others), namely Mössbauer spectroscopy, was able to give unequivocal evidence of what everybody was looking for eagerly, i.e. an association between cobalt, in a sulfur environment, and MoS_2. This was the discovery of the so dubbed 'Co–Mo–S phase' (theory 11 of Table 1)[22,23,50] The Co–Mo–S phase was obviously an embodiment of one or several of the remaining hypotheses (5, 6 or 11).

Investigation of the Co–Mo–S phase

The Co–Mo–S phase has been subjected to intense investigations, in order to elucidate its structure. This was essentially the work of the group working with Topsoe.[24] But finally XPS, EXAFS and magnetic susceptibility were relatively disappointing. Only IR spectroscopy, using NO as a probe molecule, and AEM gave significant results. The first technique indicated that a special IR band corresponded to NO in interaction with Co, and that the intensity of this band could be correlated to the amount of Co–Mo–S phase present.[51] The conclusion was that Co, in Co–Mo–S, was located on the surface. AEM performed on a recrystallized Co–Mo–S phase indicated that Co sat exclusively on the edges of the MoS_2 crystallites.[51, 52]

Thus, hypothesis 5 and 10 had to be abandoned. The CoMoS phase (hypothesis 11) represented an embodiment of hypothesis 6. Additional information on the structure of the pseudo-intercalated cobalt had been obtained, namely its coordination and the fact that, surprisingly, it only had weak structural relationship with the underlying MoS_2.[53]

For the first time, the real architecture of HDS catalysts could be discerned. Cobalt was detected in three forms, namely as isolated cobalt sulfides, associated with MoS_2 in the Co–Mo–S phase, and trapped in Al_2O_3 as cobalt pseudo-aluminate. Molybdenum was present as MoS_2 or amorphous structures (i.e. Co–Mo–S) in which Mo had the same local environment as in MoS_2, as proven by radial electron distribution measurements[54] or EXAFS.[53]

What follows concerns the hot issue, presently, with respect to the origin of synergy, namely whether all catalytic synergy is due to the Co–Mo–S phase (hypothesis 11 of Table 1), or the presence of two different phases in close contact (e.g. Co_9S_8 and MoS_2), explains the observed effects (hypothesis 12). The problem is to distinguish unequivocally between geometries uns 4 and uns 2 (unsupported model catalysts) or the architectures obtained by supporting these species on Al_2O_3. The task turns out to be arduous. The discussion will allow us to illustrate some important points in the use of surface analysis for understanding the complexity of real catalysts.

Co–Mo–S or $CoS_x + MoS_2$? : modification of architecture as a function of activation

Cobalt atoms in the Co–Mo–S phase are located on the edges of the MoS_2 layers. A major objective is thus to determine the extent of the edge surface area. Detailed theoretical models of the various sites present on the edges and methods to calculate their proportions have been developed.[55] One can thus hope to relate catalytic activity to the geometry of the corresponding catalysts, provided it is possible experimentally to determine this geometry. Unfortunately, XPS and ISS are not very useful in this respect.

At present, the only direct tool for evaluating the edge surface area is high resolution electron microscopy (HREM), Delannay,[56] in our laboratory, measured the number N of layers of the MoS_2 crystallites after full reduction-sulfidation of a commercial, alumina supported, HDS catalyst (HR 306 of Procatalyse). He also measured the 'length' of the corresponding 'slabs' viewed from the side, this corresponding to the diameter D of the MoS_2 crystallites (Fig. 2).

The study was made after sulfidation of the catalysts with a mixture of 15% H_2S in H_2 for 4 h at 10^5 Pa total pressure, and different temperatures. The proportion $n/\Sigma n$ of crystallites with different numbers of layers $N(n_N/\sum_N n_N)$ and the different corresponding sizes (diameters) D_N were measured. The surface area and volume were calculated from n_N and D_N. As an example of the kind of data obtained[56] a plot of $N(n_N/\sum_N n_N$ vs. N is presented in Fig. 3.

This approach made possible to calculate the dispersion of MoS_2, namely the ratio of the total surface area S_T of MoS_2 to volume V. The dispersion relative to edge surface area S_E, namely S_E/V, was also calculated (Fig. 4). In series of experiments made by Prada Silvy with identical quantities of catalysts (V = cst), HDS conversion was determined.[57] It correlates well with S_E/V (Fig. 4). (The ratio of the areas developed by basal (S_B) and edge planes, however, varies little in the series. The obtained result does not indicate unequivocally whether the proportionality of

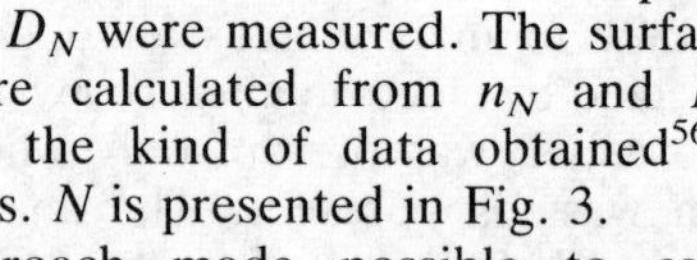

Figure 2.

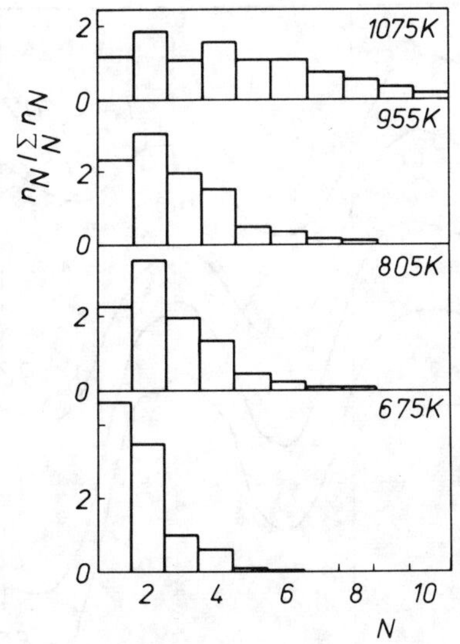

Figure 3. Crystallites with layer spacing of 0.61 nm. Frequency of observation vs. number of layers in a commerical HDS catalyst.

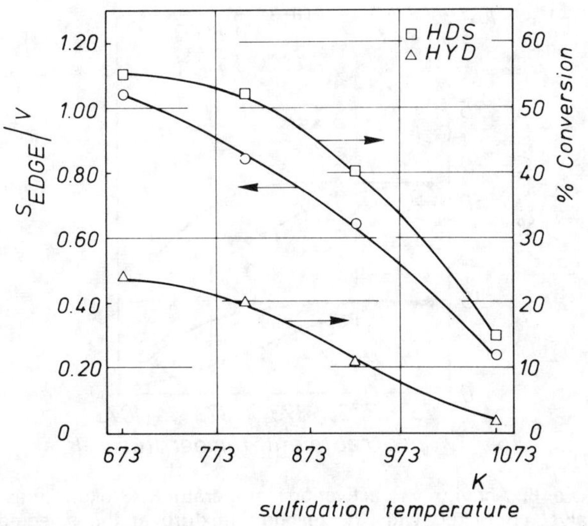

Figure 4. Edge dispersion, activities in HDS and hydrogenation vs. activation temperature.

HDS activity exists with the edge or with the total, i.e. edge and basal planes, surface areas. But this is probably an academic question, as there is little reason to believe that the basal plane, which normally give rise only to van der Waals forces, could be active).

It is worthwhile noticing that there is a satisfactory parallelism between the variations of dispersion indicated by ISS (which correspond to total, i.e. edge and basal planes, dispersion) and those deduced from HREM (Fig. 5). The minor differences observed probably reflect a change in the position of the crystallites inside the pores (due to differences in the extent of sintering of the support).

The results presented so far are perfectly compatible with hypothesis 11, namely that the presence of the Co–Mo–S phase determines catalytic activity. However, hypothesis 12, in addition to the particular supposition that the active centres are controlled by a remote action exerted by Co_9S_8, assumes that the active centres are located on MoS_2. As indicated above. the basal planes cannot be considered as capable of chemisorption or catalytic activity. Hypothesis 12 thus predicts also some positive correlation between surface area of the edge planes and catalytic activity.

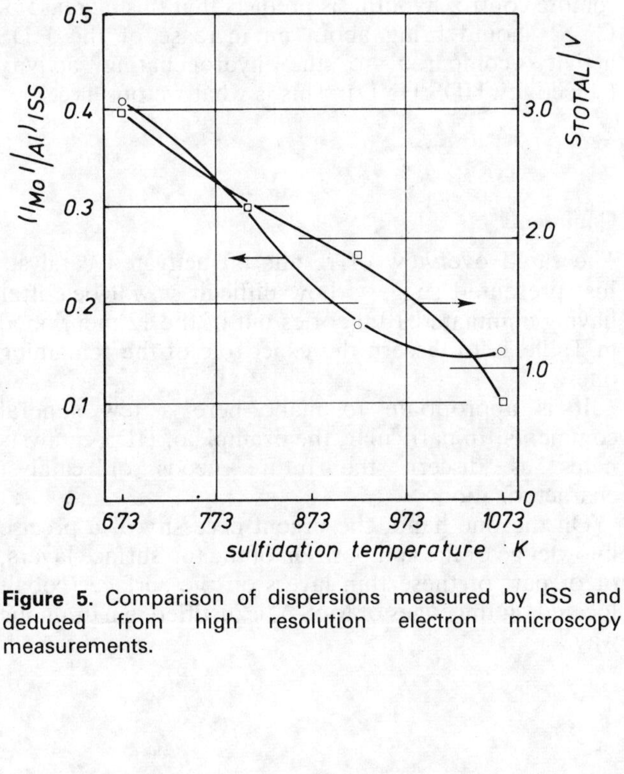

Figure 5. Comparison of dispersions measured by ISS and deduced from high resolution electron microscopy measurements.

Hypothesis 11 suggested that the hydrogenating activity might be due to MoS_2 not decorated by Co. But the quantity of cobalt contained in a commercial catalyst should, according to calculations, permit full decorations after moderate temperature activation (650 K). As a consequence of the diminution of surface area at higher activation temperatures, Co_9S_8 is expelled from the Co–Mo–S phase, but the edges should remain decorated. Selectivity to HDS (compared to the competitive hydrogenation reaction, HYD) should therefore remain unchanged. Figure 4 indicates that, contrary to that expectation, selectivity to HDS increases when more Co_9S_8 gets segregated. This is exactly what the remote control hypothesis 12 implies. Whereas the arguments of the previous paragraph were slightly in favour of hypothesis 11, the present ones would thus give an advantage to hypothesis 12. A decision between these theories is difficult.

We attempted to clarify this ambiguity by preparing a new series of catalysts, with only a change in the nature of the atmosphere during heating, before the temperature of reduction-sulfidation was reached.[58] The reacting mixture, H_2S–H_2, was used instead of Ar. ISS measurements on samples prepared at 923 and 1073 K do not indicate differences between dispersion of Mo in the first and second series. In spite of the extensive diminution of the dispersion of the Mo containing phases between 673 and 923 K, the activity of the second series remains stable (Fig. 6).

It is known that NO adsorbs in large quantities on cobalt sulfides and Co–Mo–S. NO chemisorption remained practically constant for samples prepared between 673 and 803 K. The diminution of the surface area of Co–Mo–S due to sintering should thus be compensated by the development of a new colbalt

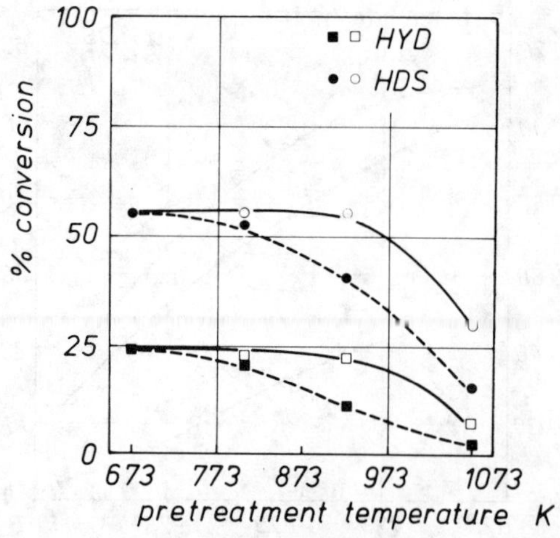

Figure 6. Activity vs. activation temperature. Broken lines: catalyst contacted with the reaction mixture at the selected temperature. Full lines: catalyst heated in the presence of the reaction mixture.

containing phase. Mössbauer emission spectroscopy (MES), in a radioactive catalyst as identical as possible to the industrial ones, indeed indicates the formation of Co_9S_8 and CoS (Fig. 7). Part of this cobalt is expelled from Co–Mo–S, as indicated by the decrease of this phase. But it also turned out[58] that cobalt trapped as cobalt pseudo aluminate became liberated during reduction-sulfidation at the higher temperatures. Figure 7 clearly shows the effect in the case of the laboratory prepared catalyst. In the case of the industrial catalyst, this was shown by reoxidizing at a moderate temperature (673 K) the activated catalysts and investigating the samples by UV diffuse reflectance spectroscopy (UV DRS). A dramatic descrease of the bands at ca.550, 590 and 625 nm due to Co^{2+} (tetr) was observed (Fig. 8). This phenomenon had been suspected for a very long time. It turns out to be extremely intense.

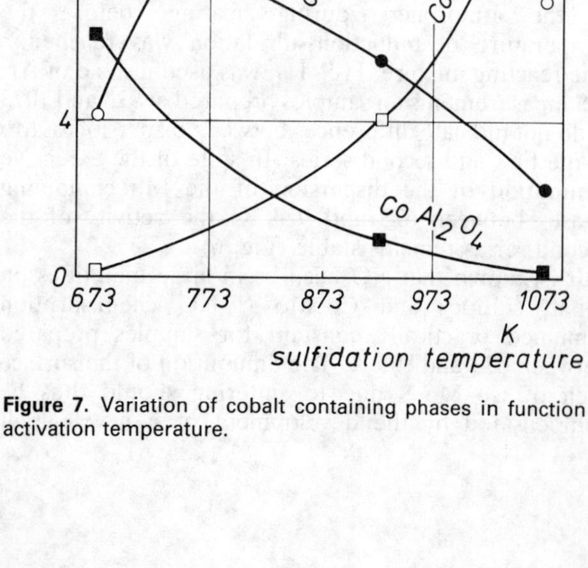

Figure 7. Variation of cobalt containing phases in function of activation temperature.

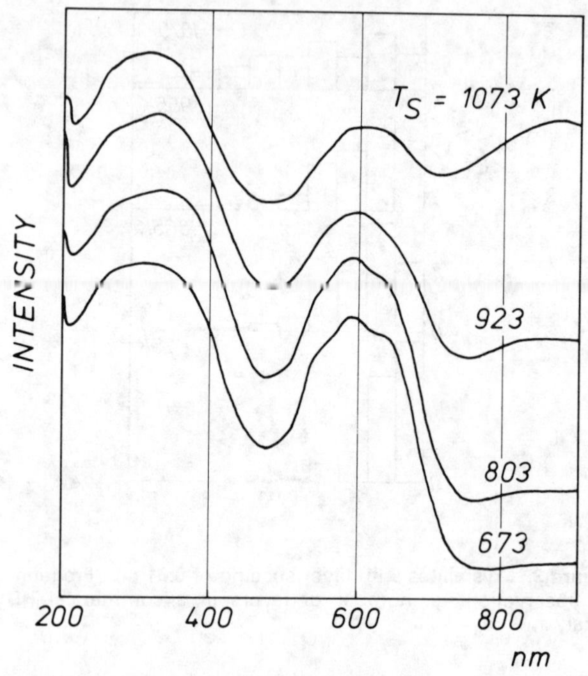

Figure 8. DRS spectra. The catalysts were activated at the temperatures indicated and reoxidized at 673 K.

This second series of results is more readily explained by the remote control theory than by the mere presence of the Co–Mo–S phase. As the number of active centers created on MoS_2 by the remote control mechanism depend on the dispersion of Co_9S_8 as well as MoS_2, and on the quality of the contact between these phases, a nearly constant activity could indeed by achieved with decreasing MoS_2 dispersion, provided more Co_9S_8 (segregated from Co–Mo–S) can, thanks to a more efficient creation of active centres on MoS_2, compensate this descrease. Still more in favour of remote control is the last reported result. At the higher activation temperatures, much cobalt sulfide is formed, not only by segregation from the Co–Mo–S phase, but also from the cobalt pseudo-aluminate present in supported catalysts. We have indicated above that the remote control hypothesis predicts that this increase of Co_9S_8 should bring about an increase of the HDS activity compared to the hydrogenating activity (selectivity HDS/HYD). This is what we observe.

Comments

The rapid overview of results on activated catalysts just presented suggests how difficult it will be, after having eliminated 10 theories out of the 12 mentioned in Table 1, to discern the exact role of the remaining ones.

It is appropriate to make here a few general comments. In particular, the example of HDS catalysts helps us discern the future needs of catalyst characterization.

On the one hand, the urgent necessity of a precise knowledge of the composition of the top surface layers, or of part of these thin layers on the surface (cobalt located on the edges of MoS_2), is justified in a dramatic way.

But the corresponding phases only represent a fraction of the total area exposed. HDS catalysts possess various other types of surfaces: alumina, cobalt pseudo aluminate, cobalt sulfides, etc. . . . Cobalt sulfides may segregate from the Co–Mo–S phase. If the remote control plays a role in HDS catalysis, the proximity, number of contacts, and nature of the latter, between the molybdenum containing phases (MoS_2 and Co–Mo–S) and the sulfides of cobalt must modify the extent of the control phenomenon. This implies that, on the other hand, investigators will demand from surface characterization techniques that they provide them with extremely precise information on the mutual location (proximity relationship) of different types of surfaces or tiny crystallites on the surface.

A high degree of sophistication in the description of catalysts is achieved in certain cases, as platinum monometallic, and, to a lesser extent bimetallic reforming catalysts. HDS and selective oxidation catalysts necessitate still more instrumental capabilities. This is due to the larger number of different phases (or surfaces), the complexity of elemental composition and, in HDS, the very fine scale of the structural features: interlayer spacings of 0.615 nm in MoS_2, making HREM necessary for determining thickness N; lateral dimension of a few nanometers for the MoS_2 slabs; still smaller dimensions for cobalt sulfides.

The HDS castalyst is a combination of architectures uns 1, uns 3 and uns 4 with features of supported catalyst. The essential instruments, for obtaining the basic information with respect to location in space are the analytical electron microscopes. Conventional instruments with analytical attachment (AEM) are extremely useful. The new especially dedicated scanning electron microscopes (STEM) possess still more capabilities in this respect. But as, inherently, AEM is a bulk, no surface technique, it must be complemented by ISS, XPS, and when possible, chemisorption measurements, coupled, if possible, with spectroscopic identification of the chemisorbed species.

On a more philosophical level, a surprising remark was the central role of Mössbauer spectroscopy in detecting one of the essential features of the HDS catalysts, namely the Co–Mo–S phases. Mössbauer spectroscopy is essentially a bulk technique. Paradoxically, it provided, and still provides, the most direct, and nearly only reliable information on very active species in catalysis (obviously surface species). The teaching is that there is no barrier between techniques for the study of surface on the one hand, and bulk, on the other hand. Adaptation of bulk techniques to the study of surfaces becomes widespread. After IR and UV spectroscopies, NMR and ESR are used for investigating surfaces, directly or thanks to the use of probe or probing absorbed molecules.

DEACTIVATED CATALYSTS

The most formidable task in catalysis is not to produce active and selective catalysts, but to make them resistant to deactivation. It is therefore surprising that surface characterization of aged or deactivated catalysts is still relatively rare. Some of the reasons of that situation are the complexity of the phenomena occurring during deactivation and the resulting difficulty of 'disentangling' the various effects by physico–chemical observations.[59–61]

We shall restrict ourselves to a few arguments. They will be selected because, at the same time, they concern physico–chemical techniques and correspond to investigations designed to contribute to the identification of the mechanism of synergy in HDS catalysts, namely: existence of a Co–Mo–S phase or remote control.

There are proofs, due to several authors, that the Co–Mo–S phase has only a limited stability. These proofs essentially involve Mössbauer spectroscopy measurements on unsupported catalysts and correlations with activity.[63–65] Catalysts composed of separate MoS_2 and Co_9S_8 phases, although less active, are more stable. When treated in reducing conditions and at relatively high temperatures, catalysts containing a large quantity of the Co–Mo–S phase suffer segregation of cobalt sulfides while catalytic activity per unit surface area in HDS and hydrogenation increases.

Those data are clear. But it is good to present here another result which, at present, cannot be interpreted unequivocally. This will lead us to raise interesting questions and to speculate on desirable future developments in the field of catalyst characterization.

In principle, the best way to identify an active phase in a catalyst is to study simultaneously how proportion of this phase and catalytic activity change during catalytic work. If a phase appears when the catalyst attains its regime of optimum work, that phase can be supposed to consitute the essential active species. An opposite observation would suggest the contrary. A special experiment was made in the frame of a programme with Góbölös, Wu and Pirotte: catalysts were prepared in such a way as to maximize the Co–Mo–S content. The purpose of this study was to detect the possible segregation of cobalt sulfides after the catalysts had worked catalytically for some time under absolutely normal operating conditions. In principle, this should correspond to a diminution of the XPS signal of cobalt, because an atomically dispersed species on the surface (the cobalt decorating the edges of MoS_2) would form crystallites of non negligible size. XPS measurements were made on two samples of the series also used in the above cited experiments designed to test the stability of the Co–Mo–S phase.[63,64] The intensities of the $Co2p_{3/2}$ (*BE* 779.1 eV) and $Mo3p_{3/2}$ (*BE* 395.3 eV) were measured, and the corresponding relative value I_{Co}/I_{Mo} taken as indicative of the change in dispersion of the cobalt containing phases. The results[66] are reported in Table 2 for two fully sulfided catalysts of different atomic composition Co/(Co+Mo). A hasty conclusion might be that, indeed, the dispersion of cobalt diminishes after the catalysts have worked for some time.

But the kinetic energy of the photoelectrons emitted by cobalt is substantially smaller than that corresponding to molybdenum. Any contamination of the surface would stop more easily the former than the

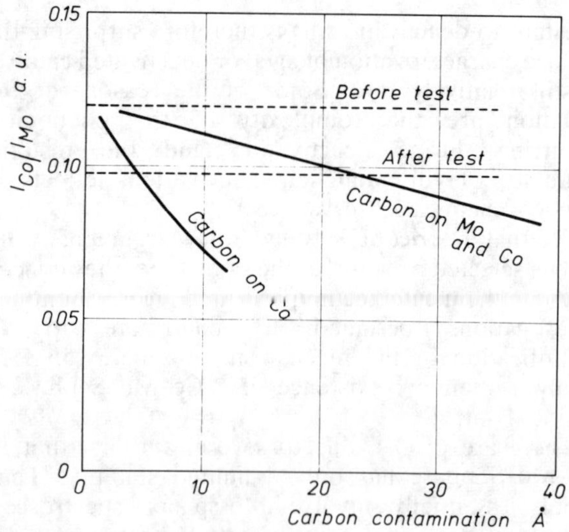

Figure 9. Comparison of XPS I_{Co}/I_{Mo} ratios (before and after test) with calculated changes due to carbon contamination.

latter. The results may thus be simply explained by the deposition of a thin carbon layer during catalysis, a phenomenon which occurs on nearly all catalysts; no change of dispersion of the active species would then be necessary for explaining the results of Table 2. This is illustrated on Fig. 9[66] for the catalyst of composition Co/(Co+Mo) = 0.3. The broken lines correspond to the values of the right-hand column of Table 2, and the curves are calculated from the theoretical fraction of the Co and Mo photoelectrons, respectively, transmitted through carbon layers of different thicknesses. A carbon contamination layer of a fraction of a nanometer, in the case of preferential deposition on cobalt, would suffice to explain the measured effects. The thickness, in the case of uniform deposition, would be larger, but well below those which begin to reduce substantially catalytic activity in many catalysts. Similar problems of contamination have been mentioned in other studies concerning HDS.[44]

The distinction between theories 11 and 12 of Table 1 on the basis of the type of experiments just described raises considerable problems. This conclusion must be generalized. The study of catalysts which have worked for some time and of deactivated catalysts still represents a serious challenge to the techniques of surface analysis. The reason is that many phenomena take place simultaneously: contamination of the surface together with sintering or other transformations of the active phase, sintering of the active phase accompanied by sintering of the support, migrations inside the catalyst pellets, etc. . . . This complexity demands that several techniques are used simultaneously, in order to characterize all relevant changes.

A few techniques begin to contribute significantly. This is, especially, high resolution STEM, for precisely locating harmful deposits,[67] or HREM for analyzing carbon deposits[68] and, now routinely, the sintering of many active phases. Although at unfortunately too large a scale (of the order of a few micrometers), LMMS (laser microprobe mass spectrometry, commercial name LAMMA) and Raman microprobe analysis could also be extremely useful.

But it is necessary to stress the other urgent specific

Table 2.

	I_{Co}/I_{Mo} (a.u.) $\frac{Co}{Co + Mo} = 0.1$	$\frac{Co}{Co + Mo} = 0.3$
Before test	0.087	0.120
After test	0.074	0.097

needs with respect to the analysis of aged catalysts. The first one is to preserve the aged catalyst from contamination when transferred from the reactor to the analysis equipment. Many catalysts used in hydrogenation or hydrotreating can burn spontaneously, or at least oxidize deeply, when contacted with atmosphere. There is no hope to get any valuable information after those uncontrolled chemical alterations. The second need concerns reliable and relatively little harmful erosion techniques, which, in principle, could permit the successive analysis of, first, contaminants (poisons, carbon, other 'fouling' agents), hopefully quantifying their thickness, and, subsequently, of the underlying remaining catalysts. ISS and, more generally, erosion by ions do not match satisfactorily these requirements. There are several reasons for that. The different types of surfaces (e.g. MoS_2, Co_9S_8, carbon) are not eroded at the same rate. A worse inconvenience is the extensive damage, to a depth of several nanometers, brought about by the ions: erosion causes a complete mixing of elements and extensive restructuration. Grazing angle erosion, with potentially less damaging effects, is obviously not relevant to highly dispersed samples.

As a conclusion of this part, one could stress the importance of new developments in surface analysis for the progress of studies on catalysts evolution, ageing, and deactivation. These developments should principally concern the very careful 'peeling off' of the surface. Physicists try to make measurements on absolutely clean and well controlled surfaces. The challenge is to control the state of the surface by perfectly cleaning it. In catalysis, a large part of the interesting information concerns 'the dirt'. It must be carefully preserved and analyzed before removal.

REGENERATED CATALYSTS

Certain very important catalytical processes are economically feasible only because the catalyst can be regenerated. A typical case is catalytic reforming (production of aromatics and high octane gasoline). Still more crucial is regeneration in fluidized bed catalytic cracking (gasoline production) where carbon deposits deactivate catalyst particles in a few seconds and has to be burnt for regeneration: if not, catalyst consumption could be larger than gasoline production!

The regeneration of platinum (or bimetallic) reforming catalysts is exemplative in that the scientific analysis it has been subjected to has permitted the development of fundamental knowledge in several important respects. Summarizing, a first point was that the extremely good dispersion achieved after regeneration and reduction had its origin in a special mechanism: the platinum oxidized compound formed during regeneration undergoes a molecular redispersion on the surface of the alumina. The second point concerned the crucial role of a well controlled amount of an additive, namely chlorine, for achieving this molecular dispersion (the molecularly dispersed platinum species being probably an oxychloride).

Regeneration of HDS catalysts takes place less frequently (every second year, or an average, with large variation in function of the particular type of process). Although this operation is less crucial than for catalytic reforming, it was extremely tempting to investigate the phenomena taking place during regeneration (oxidation of the sulfided aged catalyst) and reactivation (resulfidation). On fundamental grounds, this study could hopefully clarify important details, as it helped in the case of the platinum reforming catalysts. In addition, a strange fact had been reported after regeneration in industrial plants, namely an increase of activity compared to that of the fresh catalyst.[69] In the context of our programme, a reasonable hypothesis was that this kind of study could further help distinguish between theories 11 and 12 of Table 1 (CoMoS and remote control). We present hereafter a few conspicuous results, due to Arteaga and Garcia Fierro.

It was confirmed that a regenerated catalyst could possess higher activities than the fresh one: a redispersion effect similar to that operating in Pt catalytic reforming catalysts takes place. But the part of the study which most directly addressed the problem of the real architecture of active catalysts dealt with a series of catalysts regenerated at different temperatures.[70] The catalysts were first reduced-sulfided (activated), then regeneration was modelled by controlled reoxidation (it turned out to be of secondary importance in many experiments whether carbon had been deposited or not before reoxidation) and finally reduced-sulfided again in standard conditions. Figure 10 reports catalytic activity in standard conditions as a function of the temperature of regeneration. As in the studies concerning activation, several physico–chemical techniques were used for unveiling the architecture of the catalysts. We shall focus here on NO adsorption coupled with IR spectroscopy, which permits to distinguish NO chemisorbed on Co ions (Co–Mo–S or cobalt sulfides).[51] HDS activity increases when the number of superficial Co sites increases; a reverse trend is observed with Mo ions are considered (Fig. 11).

This can be considered to support the Co–Mo–S theory: these opposite effects were indeed taken as an argument in favour of the Co–Mo–S theory.[51] However, one could argue that surface Co ions in segregated cobalt sulfides are also measured by the IR–NO adsorption technique, and that the number of such ions must correlate with the number of active sites created by remote control on MoS_2, thus explaining the results of Fig. 11.

Further questions are raised by the fact that, as in the work on activation, the HDS/HYD selectivity changes (this is easily deduced from the observation of Fig. 10). Still more stimulating to further investigation is the fact[71] that catalysts of identical catalytic activity have extremely different NO adsorption characteristics (Fig. 12).

Again, as in our studies of the influence of activation, the results turn out to be ambiguous. Besides, very few detailed studies on the effect of regeneration or ageing and regeneration, have been reported in literature so far.[72] It is thus not surprising that, based on ageing-regeneration studies, conclusions concerning the respective role of the Co–Mo–S phase and of a remote control effect do not yet emerge. Conspicuous results, as those just reported, should spur a line of investigations in this area.

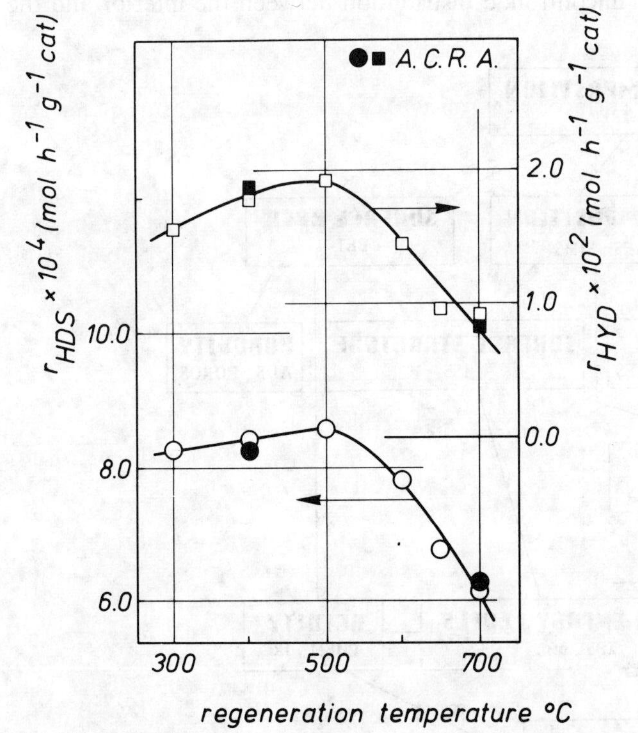

Figure 10. Activity vs. regeneration temperature. Open signs correspond to a simple activation–reoxidation–activation cycle, while full signs (ACRA) corespond to cycles involving carbon deposition.

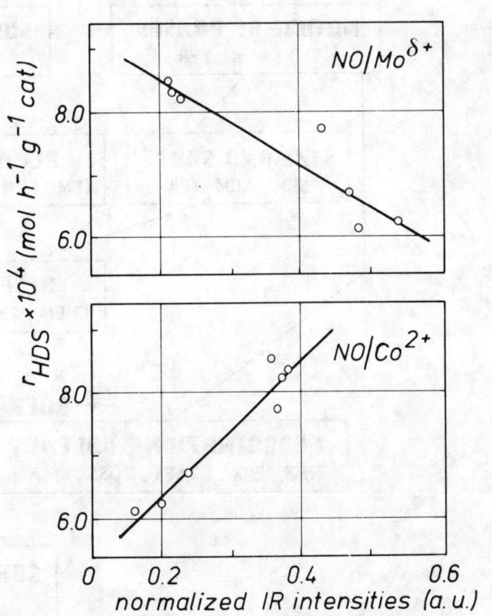

Figure 11. Correlation between catalytic activity and adsorption of NO on, respectively, Mo and Co surface ions.

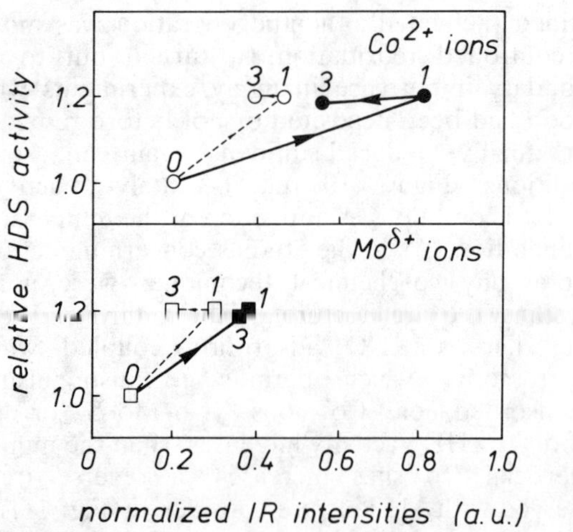

Figure 12. Activities vs. normalized intensities of the IR bands corresponding, to NO adsorbed, respectively, on Co and Mo. Full signs: cycles involving carbon deposition (ACRA, Fig. 10); 1 and 3 refer to the number of regeneration-activation cycles undergone by the sample.

The challenge to physico–chemical techniques will again be to distinguish between highly dispersed phases that contain one or several common elements (namely MoS_2 and Co–Mo–S on the one hand, Co–Mo–S, Co_9S_8, CoS_{1+x} and $CoAl_2O_4$ on the other hand).

These are good reasons to believe that the approach to the study of regeneration will be still more pluridisciplinary and will need the use of still more numerous techniques that in the previously mentioned problems. What might distinguish these studies from those mentioned above is the fact that the solid state chemistry of two phenomena (oxidative regeneration and reactivation by sulfidation) must be understood, and this at a double scale of time, namely during each regeneration and activation, but also in the course of the successive regeneration–reactivation cycles.

CONCLUSION

Catalysts have a complicated life, they undergo many metamorphoses, from preparation to final deactivation and death. Our exercise has been to suggest the role of surface analysis in the understanding of the changes a catalyst undergoes during this complicated life, and the many implications in optimized preparation and use, as well as for identificatin of the active system.

The illustrations chosen, namely those concerning HDS catalysts, are very instructive but leave us still slightly disappointed. The rapid elimination of 10 of the 12 theories of Table 1 took place in parallel with the intense development of the use of surface techniques in catalyst studies and owes much to this development. But it is frustrating to remark that the sensitivity and resolution of these techniques are still too low for giving a decisive advantage to either theory 11 or 12. On the other hand, there is no doubt that the increase in activity and resistance, and the fine adaptation of hydrotreating catalysts to an incredible variety of processes, in the last ten years, owes much to the better knowledge gained on many details of their architecture and, especially, to a very fine monitoring of their preparation: thanks to the corresponding physico–chemical techniques, it has been possible to optimize the characteristics which pertained to both theories (high dispersion, proximity of Co and Mo, minimal trapping of Co in Al_2O_3, etc . . .) and to avoid very harmful accidents (as, at the preparation stage, uncontrolled distribution between the interior and the

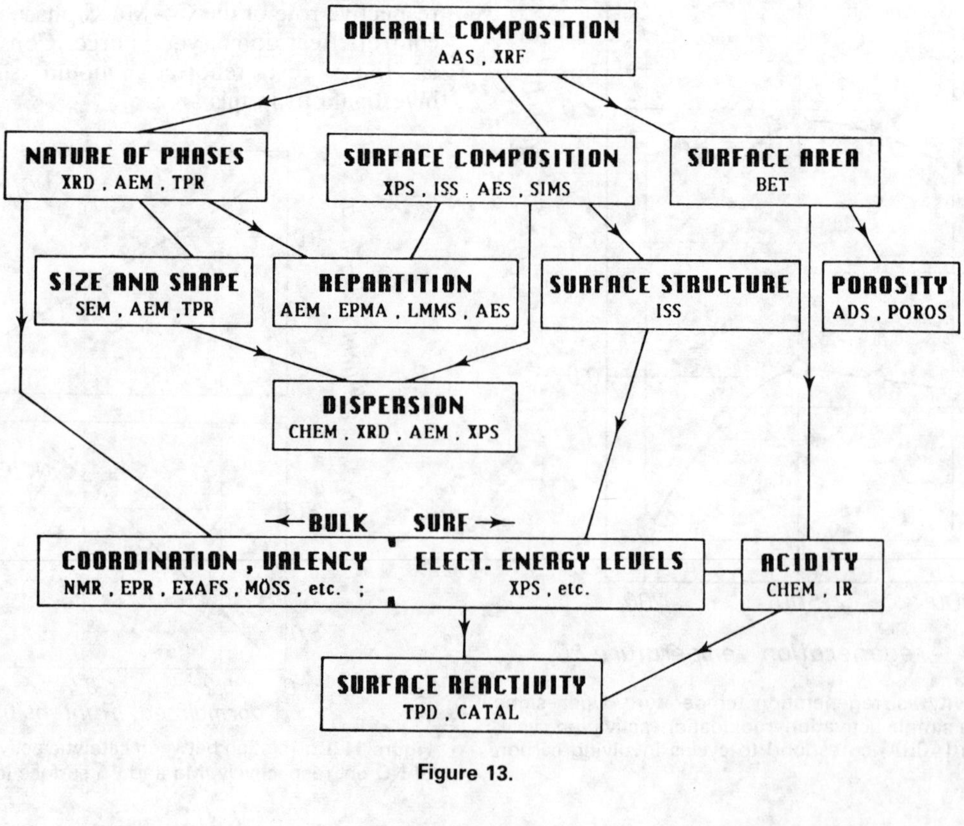

Figure 13.

exterior of the pellets, uneven distribution of both species, etc . . .).

The development of new techniques is continuing very strongly (see, for example, Ref 73 and 74). They become adapted to more difficult tasks. Thanks to much effort from fundamental scientists and equipment manufacturers, these techniques become easier to use, and commercial instruments corresponding to new emerging techniques are built. All signs indicate that heterogeneous catalysis will remain one of the main users of these techniques.

A mundane, but essential, remark is that a meaningful description of catalyst architecture can only be obtained thanks to the joint use of several well selected techniques. If one examines these cases where, as in HDS, platinum reforming, metal catalyzed hydrogenation or ammonia synthesis, a very detailed knowledge of the architecture of the catalyst was obtained, one can easily conclude that a rational approach to characterization had been followed. This rational approach has been described elsewhere.[1] Figure 13 summarizes the lines of a systematic investigation in the case of catalysts. The figure also suggests the techniques which, in most cases, give valuable information, at a reasonable cost (of labour or money).

Acknowledgements

This work stems from the enthusiastic cooperative effort of all the investigators of the Groupe de Physico-chimie Minérale et de Catalyse and of several colleagues (Drs P. Bertrand and J. Ladière particularly), also of the Université Catholique de Louvain. The work of HDS, more particularly, is due to a long standing line of research of the laboratory 'Catalyse et Chimie des Matériaux Divisés'. We wish to thank all investigators for their contribution. Special thanks go to A. Arteaga Berrueta and R. Prada Silvy for their specific help in discussing and preparing the illustrations corresponding to our recent work.

The Groupe de Physico-Chimie Minérale et de Catalyse could develop its expertise in the characterization of highly dispersed solids thanks to the financial support (in equipment and in many other forms) of the Services de Programmation de la Politique Scientifique, whom we gratefully acknowledge.

REFERENCES

1. F. Delannay, B. Delmon, in *Characterization of heterogeneous catalysts* Ch. 1, pp1–28 ed by F. Delannay, M. Dekker (1984).
2. G.C.A. Schuit and B.C. Gates, *A.I.Che.E.J.* **19**, 417 (1973).
3. B.C. Gates, J.R. Katzer and G.C.A. Schuit, *Chemistry of catalytic processes*, Ch. 5, McGraw Hill, (1979).
4. J.M.J.G. Lipsch and G.C.A. Schuit, *J. Catal.* **15**, 163, 174 and 179 (1969).
5. V.H.J. De Beer, T.H.M. van Sint Fiet, J. F. Engelen, A.C. van Haandel, M.V.J. Wolfs, C.H. Amberg and G.C.A. Schuit, *J. Catal.* **27**, 327 (1972).
6. J. Sonnemans and P. Mars, *J. Catal.* **31**, 209 (1973).
7. V.H.J. De Beer and G.C.A. Schuit, in *Preparation of catalysts* p343 ed by B. Delmon, P.A. Jacobs and G. Poncelet, Elsevier, Amsterdam (1976).
8. J. Medena, C. Van Stam, V.H.J. de Beer, A.J.A. Konings and D.C. Koningsberger, *J. Catal.* **53**, 386 (1978).
9. H. Knözinger and H. Jeziorowski, *J. Phys. Chem.* **82**, 2002 (1978).
10. R.W. Phillips, A.A. Fote, *J. Catal.* **41**, 168 (1976).
11. R.J.H. Voorhoeve and J.C.M. Stuiver, *J. Catal.* **23**, 228 (1971).
12. R.J.H. Voorhoeve, *J. Catal.* **23**, 236 (1971).
13. R.J.H. Voorhoeve and J.C.M. Stuiver, *J. Catal.* **23**, (1971).
14. A.L. Farragher and P. Cossee, in *Catalysis: Proc. 5th Int. Cong. Catalysis* p1301 ed by J.W. Hightower, North Holland (1973).
15. A.L. Farragher, *Symposium on the Role of Solid State Chemistry in Catalysis, ACS Meeting, New Orleans, March 20–25, 1977.*
16. A.L. Farragher and P. Cossee, in *Catalysis Proc. 5th Int. Cong. Catalysis* pp1316–1317 ed by J.W. Hightower, North Holland (1973).
17. P.R. Wentrcek and H. Wise, *J. Catal*, **45**, 349 (1976).
18. P.R. Wentrcek and H. Wise, *Preprints, Div. Petrol. Chem. ACS*, **22**, 525 (1977).
19. P.R. Wentrcek and H. Wise, *J. Catal*, **51**, 80 (1978).
20. M. Lo Jacono, J.L. Verbeek, G.C.A. Schuit, *J. Catal.* **29**, 463 (1973).
21. Y. Jacquin, in *Catalyse de Contact* pp495–500 ed by J.F. Le Page *et al.*, Technip Publ (1978).
22. H. Topsøe, B.S. Clausen, N. Burriesci, R. Candia, and S. Mørup, in *Preparation of Catalysts II* pp479–492 ed by B. Delmon, P. Grange, P.A. Jacobs, G. Poncelet Elsevier (1979).
23. H. Topsøe, B.S. Clausen, R. Candia, C. Wiver, S. Mørup, *Bull. Soc. Chim. Belg.* **90**, 1190 (1981).
24. H. Topsøe R. Candia, N.Y. Topsøe, B.S. Calusen, *Bull. Soc. Chim. Belg.* **93**, 783 (1984).
25. G. Hagenbach, Ph. Courty, and B. Delmon, *J. Catal.* **31**, 264 (1973).
26. P. Grange and B. Delmon, *J. Less-Common Met.* **36**, 353 (1974).
27. B. Delmon, *Bull. Soc. Chim. Belg.* **88**, 979 (1979).
28. B. Delmon, *Int. Chem. Eng.* **20**, 639 (1980).
29. D. Pirotte, J.M. Zabala, P. Grange, B. Delmon, *Bull. Soc. Chim. Belg.* **90**, 1239 (1981).
30. M. Houalla, F. Delannay, I. Matsuura, B. Delmon, *J.Chem. Soc., Faraday Trans. 1*, **76**, 2128 (1980).
31. M. Montes, Ch. Penneman de Bosscheyde, B.K. Hodnett, F. Delannay, P. Grange and B. Delmon, *Appl. Catal.* **12**, 309 (1984).
32. B. Delmon, P. Grange, M.A. Apecetche, P. Gajardo and F. Delannay, *C.R. Acad. Sci., Ser. C*, **287**, 401 (1978).
33. P. Gajardo, P. Grange, B. Delmon, *J. Catal.* **63**, 201 (1980).
34. B. Delmon, in *Proc. Climax 3rd Int. Conf. Chemisry and Uses of Molybdenum* pp73–84 ed by H.F. Barry, P.C.H. Mitchell, Climax Molybdenum Cy., Ann Arbor, Mich. (1979).
35. F. Delannay, P. Gajardo, P. Grange, *J. Micros. Spectr. Electron*, **3**, 411 (1980).
36. B. Delmon, *Anal. Quim., Real Soc. Espa. Quimica, Serie A*, **73**, 303 (1983).
37. J. Medena, C. van Stam, V.H.J. de Beer, A.J.A. Konings, D.G. Koningsberger, *J. Catal.* **53**, 386 (1978).
38. C.P. Cheng, G.L. Schrader, *J. Catal*, **60**, 276 (1979).
39. F. Delannay, E.N. Haeussler, B. Delmon, *Bull. Soc. Chim. Belg*, **89**, 255 (1980).
40. F. Delannay, E.N. Haeussler, B. Delmon, *J. Catal.* **66**, 469–472.
41. J.S. Brinen, D.A. D'Avignon, E.A. Meyers, P.T. Teng, D.W. Behnken, *Surf. Interface Anal.* **6**, 295 (1984).
42. H. Knozinger, H. Jeziorowski, E. Taglauer, in *Catalysis, Proc. 7th Int. Congr. Catalysis* pp604–615 ed by T. Seiyama, K. Tanabe, Kodansha, Tokyo, and Elsevier, Amsterdam, (1981).
43. B. Canosa Rodrigo, H. Jeziorowski, H. Knözinger, X, Xh. Wang, *Bull. Soc. Chim. Belg*, **90**, 1339 (1981).
44. S. Kasztelan, J. Grimblot, J.P. Bonnelle, *J. Chim. Phys.* **80**, 793 (1983).
45. G. Hagenbach, Ph. Courty, and B. Delmon, *J. Catal.* **23**, 295 (1971).
46. G. Hagenbach, Ph. Courty, and B. Delmon, *J. Catal.* **31**, 264 (1973).
47. R.I. Declerck-Grimée, P. Canesson, R.M. Friedman, J.J. Fripiat, *J. Phys. Chem.* **82**, 885 and 889 (1978).
48. P. Gajardo, P. Grange, B. Delmon, *J. Chem. Soc., Faraday Trans. I*, **76**, 988 (1980).

49. P. Gajardo, P. Grange, B. Delmon, *Surf. Interface Anal.* **3**, 206 (1981).
50. H. Topsøe, S. Mørup, in *Proceed. Intern. conf. Mössbauer Spectroscopy* Vol.1, p.305, ed by A.Z. Hrynkiewicz, J.A. Sawicki, Cracow, Poland, (1975).
51. N-Y Topsøe, H. Topsøe, O. Sørensen, B.S. Clausen, R. Candia, *Bull. Soc. Chim. Belg.* **93**, 727 (1984).
52. R.R. Chianelli, A.F. Ruppert, S.K. Behal, B.H. Kear, A. Wold, R. Kershaw, *J. Catal.* **92**, 56 (1985).
53. R. Candia, B.S. Clausen, J. Bartholdy, N-Y. Tiosøe, B.L. Engeler, H. Topsøe, in *Proc. 8th Intern. Congr. on Catalysis*, p. II 375, Verlag Chemie, Weinheim, (1984).
54. G. Delvaux, P. Grange, B. Delmon, *Bull. Soc. Chim. Fr.* **I**. 1 (1979).
55. S. Kasztelan, H. Toulhoat, J. Grimblot, J.P. Bonnelle, *Appl. Catal.* **13**, 127 (1985).
56. F. Delannay, *Appl. Catal.* **16**, 135 (1985).
57. R. Prada Silvy, J.M. Beuken, P. Bertrand, B.K. Hodnett, F. Delannay, B. Delmon, *Bull. Soc. Chim. Belg.* **93**, 775 (1984).
58. R. Prada Silvy, F. Delannay, P. Grange, B. Delmon, *Climax 5th Intern. Conf. on Chem. Uses of Molybd., Newcastle upon Tyne, 1–5 July, 1985,* Publisher? Date?
59. B. Delmon, P. Grange, in *Catalyst Deactivation* pp507–543 ed by B. Delmon, G.F. Froment, Elsevier, (1980).
60. B. Delmon, P. Grange, in *Progress in Catalyst Deactivation* pp231–280 ed by J.L. Figuiredo,l NATO advanced Study Institute Series, Series E, no. 54, Martinus Nijhoff, The Hague, (1982).
61. B. Delmon, *Appl. Catal.* **15**, 1–16 (1985).
62. M. Breysse, R. Frety, G. Benaichouba, P. Bussiere, *Radicohem. Radioanal. Letters*, **59**, 265 (1983).
63. M. Vrinat, M. Breysse, R. Frety, *Appl. Catal* . **12**, 151 (1984).
64. M. Breysse, R. Fretty, M. Vrinat, P. Grange, M. Genet, *Appl. Catal.*, **12**, 165 (1984).
65. S. Göbölös, Q. Wu, J. Ladrière, F. Delannay and B. Delmon *Bull. Soc. Chim. Belg.* **93**, 687–696 (1984).
66. D. Pirotte, B. Delmon, unpublished results.
67. J. P. Brunelle, H. Dexpert, J.J. Ehrhardt, E. Freund, N. Pacia, A. Pintero, G. Praline, M. Prigent, in *Catalyst Deactivation* pp233–250 ed by B. Delmon and G.F. Froment, Elsevier, Amsterdam (1980).
68. J.V. Sanders, J.A. Spink, S.S. Pollack, *Appl. Catal.* **5**, 65 (1983).
69. G. Berrebi, *L'Actualité Chimique,* Octobre, 29 (1983).
70. A. Arteaga, J.L. Garcia Fierro, P. Grange, B. Delmon, to be published.
71. A. Arteaga, J.L. Garcia Fierro, F. Delannay, B. Delmon, to be published.
72. A.V. Ramaswamy, L.D. Sharma, A. Singh, M.L. Shingal, S. Sivasanker, *Appl. Catal.* **13**, 311 (1985).
73. Preprints, Petroleum Chem. Div. ACS, *The new surface science in catalysis*, **29**, no.3, pp667–958 ed by M.L. Deviney, J.L. Gland, co-chairmen), Philadephia, Aug. 16–31, Am. Chem. Soc., (1984).
74. *Chem. Engg. News,* March 11, 34, (1985).

SURFACE AND INTERFACE ANALYSIS, VOL. 9, 207–213 (1986)

AEM, XPS and ISS Characterization of Catalyst Modifications During Propene Oxidation Over a Supported Mixed Oxide Catalyst

G. Kremenić,* V. Cortés Corberán* and J.M.D. Tascón†

*Instituto de Catálisis y Petroleoquímica del CSIC, Serrano, 119, 28006 Madrid, Spain

†Groupe de Physico-Chimie Minérale et de Catalyse, Université Catholique de Louvain, Place Croix du Sud 1, B–1348 Louvain-la-Neuve, Belgium

The activity of a (Mo–Fe–Te–Co) O/SiO_2 catalyst for propene partial oxidation increases with time on stream, being accompanied by no noticeable change in selectivity to products. The use of a series of characterization techniques (x-ray diffraction, SEM, analytical electron microscopy, XPS and ISS) allowed us to conclude that the presence of the same phases in the fresh and used catalysts agrees with the constancy of selectivities, while catalytic activity increase is ascribable to a higher degree of dispersion of the active phases. The advantage of the *joint* use of characterization techniques with different degrees of accessibility to the sample in study are emphasized.

INTRODUCTION

It is well known that both the physicochemical and the catalytic properties of a catalyst vary under the action of a reaction medium; this interaction plays a very important role in oxidation processes. No extensive studies have been devoted to the effect of reaction medium on the catalytic properties of mixed oxide catalysts for partial oxidation of olefins. Studies on Bi–Mo–O[1] and Fe–Sb–O[2] catalysts for oxidative dehydrogenation of butenes showed that catalytic activity decreased as catalysts were reduced by the reacting mixture. However, in Te–Mo–O catalysts,[3] the more reduced catalysts formed by interaction with an isobutene–oxygen mixture, were more active for propene and isobutene oxidation than the less reduced ones, formed by interaction with a propene–oxygen mixture.

Also the catalytic properties and phase composition of V–Te–Mo–O catalysts, used in the oxidation of propene to acids, depend on their degree of reduction, which is closely associated with the effect of composition of the reaction medium.[4] More recently, Haber *et al.*[5] have found that in a catalyst of composition $Cu_2Mo_3O_{10}$, depending on the redox properties of the reactant mixture, different surface phases are formed directing the reaction along different reaction paths.

Most of these studies have been performed using a pulse technique. Little work has been done on which transformations take place in the catalyst during continuous operation in a catalytic process. Study of the change in catalytic properties of a surface as caused by the reaction medium in continuous operation is of interest for determining the actual kinetics as well as for providing optimum catalyst utilization conditions.

The aim of the present work was, hence, to study the catalyst modifications during initial stages of continuous operation in partial oxidation of propene. We used a silica supported mixed oxide catalyst known to be very active and selective for acrolein formation.[6,7] In addition to more conventional techniques (XRD, SEM) we tried to show how the *joint* use of three surface or bulk analysis techniques with different degrees of resolution may provide a physicochemical picture on the origin of phenomena involved on catalyst activation during olefin oxidation.

EXPERIMENTAL

Catalyst preparation

The catalyst is composed of oxides of Mo, Fe, Te and Co (at % 4:1.5:0.5:0.2), supported on silica (ca. 20 wt% of active phases). Materials used for the preparation were: silica from BASF (surface area, 172 $m^2\ g^{-1}$), and ammonium heptamolybdate, iron and cobalt (II) nitrates and tellurium oxide, all p.a. from Merck.

The catalyst was prepared by a double impregnation of silica particles (0.42–0.59 mm): first, with an acidic solution of Te, Fe and Co nitrates (tellurium nitrate was previously prepared from the oxide), then evaporated in vacuum and dried at 383 K for 44 h. Then, a second impregnation was carried out with an ammoniacal solution of $(NH_4)_6Mo_7O_{24}\cdot 4H_2O$, followed by drying in an air stream overnight at 383 K and calcination in air at 673 K for 2 h and then at 823 K for 4 h; finally it was cooled slowly to room temperature (RT). The resulting catalyst had a BET surface area of 135 $m^2\ g^{-1}$ (taking a value of 0.162 nm^2 for the cross-sectional area of N_2 molecule).

†Present address: Instituto Nacional del Carbón y sus Derivados, CSIC Apartado 73, 33080 Oviedo, Spain.

0142–2421/86/120207–07 $05.00

Catalytic activity measurements

Oxidation of propene was carried out in a fixed bed Pyrex-glass flow reactor (10 mm internal diameter) provided with a coaxial thermocouple, at 673 K and atmospheric pressure. The catalyst sample (1 g) was diluted with 1 mm-particles of silicon carbide in a 1:6 weight ratio. The reacting mixture, propene–oxygen–steam–helium (1:1:2.5:3.5 molar ratio) was introduced at a $W:F = 15.4$ g cat. h mol^{-1} C_3H_6. Catalytic activities were measured with four different catalyst samples, the results being consistent.

Reactants and effluents were analyzed by gas chromatography using a Zeolite 13X (2.5 m length and 1/8″ outer diameter column for O_2 and CO, and a Porapak Q column (100–120 mesh; 3 m length and 1/8″ outer diameter) for the rest of compounds.

After different times on stream, the reactant feed was cut-off and the catalyst was cooled in helium to RT and separated for its characterization. Characterization data of the catalyst reported below refer to time on steam = 0 ('fresh' catalyst) and 7.5 h ('used' catalyst). The latter is representative for samples at any time on stream ⩾1.5 h.

Catalyst characterization

(a) X-ray diffraction (XRD) X-ray diffractograms were obtained in a Philips-Norelco PW 1010 diffractometer, using Ni–filtered Cu Kα radiation. The samples were finely crushed prior to recording diffractograms.

(b) Scanning Electron Microscopy (SEM) The scanning electron micrographs were obtained with a Jeol–JSM 50 A microscope. For each sample of catalyst, about 30 micrographs were taken so that the one presented in this work are representative of the average aspect observed in each case.

(c) Analytical Electron Microscopy (AEM) After hand grinding in an agate mortar, the samples were dispersed in ethanol and deposited on a thin carbon film supported on a conventional grid for electron microscopy. A JEM TEMSCAN electron microscope equipped with a Kevex x-ray spectrometer was used. In addition to conventional work (bright field and dark field imaging, electron diffraction) this equipment allowed high resolution microanalysis of the electron transparent parts of the specimen, so that the composition of individual particles larger than about 20 nm could be determined with a good quantitative accuracy. On SiO_2 and α–Al_2O_3 supported catalysts most metal oxide particles larger than about 5 nm can be usually distinguished from the carrier itself by electron imaging. The amount of supported phase in a more finely dispersed form on the surface of the support (indistinguishable from the latter on the electron image) can be selectively measured by pointing a finely focused electron probe on areas where no supported particle can be observed. The measurements can be made fairly quantitative for electron transparent speciments by use of proper calibration factors.[8] A detailed account of the use of AEM for the study of complex catalysts can be found elsewhere.[9]

(d) X-Ray Photoelectron Spectroscopy (XPS) The finely ground catalyst was deposited on a double-sided adhesive tape fixed to the sample holder. A Vacuum Generators ESCA–2 equipment used, the exciting radiation being A1 Kα. The x-ray tube operated at 14 kV and 20 mA. Vacuum in the analyzing chamber was better than 1.33×10^{-6} Pa. The recording sequence was maintained the same for the different samples: C 1s, O 1s, Mo 3d, Si 2p, Fe 2p, Co 2p, Te 3d and again O 1s, in order to avoid spurious effects due to progressive contamination of the sample in the course of the analysis. The signal noise ratio was of the order of 20:1 for the C 1s, Mo 3d, Fe 2p and Te 3d lines, and about 40:1 for O 1s and Si 2p. The C 1s line of adventitious carbon was used as a reference for binding energies (BE = 285 eV). The surface composition changes brought about by the activation process were followed by monitoring the Mo 3d, Fe 2p, Te 3d, O 1s intensities; for that purpose, the Si 2p line from the support was chosen as the intensity reference and its value normalized to one.

(e) Ion Scattering Spectroscopy (ISS) The ISS measurements were performed with a Kratos Wg type 515 instrument. The core of the spectrometer consists of a cylindrical mirror analyser with an energy resolution $\Delta E/E = 0.03$ and a predominant scattering angle of 138°. This system is mounted vertically in a UHV chamber in which the base pressure is about 4×10^{-9} mm Hg (1 mm Hg = 133.3 $N.m^{-2}$). During the analysis, the noble gas to be ionized was introduced into the chamber up to a pressure varying from 2×10^{-7} mm Hg to 2×10^{-6} mm Hg.

The catalyst power was compressed in small cups which were supported horizontally in the chamber and were bombarded at normal incidence. $^4He^+$ and $^{20}Ne^+$ ions were used for ISS analysis. The beam energy E_o was 2 keV and the beam current was 15 nA, with a spot size of 350 μm. A hot W filament was used to provide a flow of thermal electrons to compensate the positive charge accumulated on the sample's surface.

During the analysis the ion beam was rastered over a square area of (1×1) mm^2. A spectrum was taken in 6 s and accumulated 60 to 100 times in the memory of a fully programmable multichannel analyser (Lecroy MCA model 3500). The ISS energy spectra were converted into mass spectra by means of the single elastic collision model.

RESULTS

Catalytic activity

The main product of propene oxidation, in the working conditions, is acrolein. In addition, some acetaldehide, carbon oxides and traces of acrylic acid are formed. Typical results of total conversion (x_T%) and selectivities (s_i%) vs. time on stream are given in Fig. 1.

Meanwhile the selectivities to each product have not significant change with time on stream, the total conversion increases to a constant value, which is

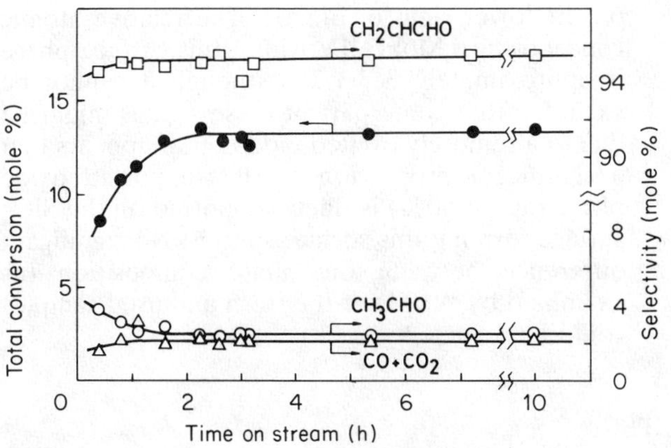

Figure 1. Catalytic activity and selectivities as a function of time on stream. Experimental conditions in text.

reached after about 1.5 h and then remains constant for at least 10 h. Consequently, the data for kinetic studies were taken after 1.5 to 2 h on stream.

XRD

Figure 2 shows the x-ray diffractograms of the fresh and used catalyst. The diffractograms do not vary very much from one sample to the other as concerns the position of the peaks, but the observed intensities for the used catalyst sample are smaller than for the fresh one.

In both samples the lines of orthorhombic MoO_3[10] and of pseudoorthorhombic Fe_2 $(MoO_4)_3$[11] are distinguished. The relative peak intensities seem to be in agreement with peaks reported in the literature which are also drawn for comparison in Fig. 2. The broad band between 15 and 35° (2 θ) is due to the amorphous silica support.

SEM

Representative scanning electron micrographs of the fresh and used catalysts are shown in Fig. 3. On the fresh catalyst, typical microcrystals of orthorhombic molybdite as well as iron molybdate plates and other finely dispersed aggregates can be observed on the whole amorphous silica surface. These materials accumulate preferently at the entrances of the macropores, and also on the fissures and edges of the support particles.

The size of microcrystals has considerably decreased in the used catalyst while dispersion has been clearly improved, this being in agreement with XRD results.

AEM

A large number of aggregate particles are readily distinguished from the SiO_2 carrier by electron imaging. These particles have been analyzed by x-ray microanalysis and, when possible, by electron diffraction. The observations are schematized in Fig. 4,

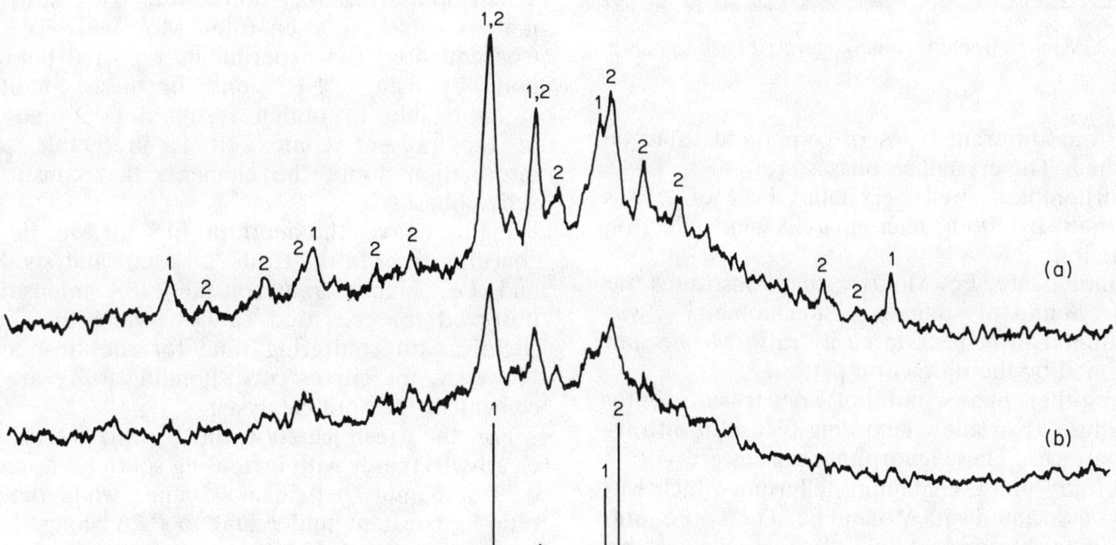

Figure 2. X-ray diffractograms: (a) fresh catalyst; (b) used catalyst; (c) and (d) data from the literature; 1=MoO_3; 2 = Fe_2 $(MoO_4)_3$.

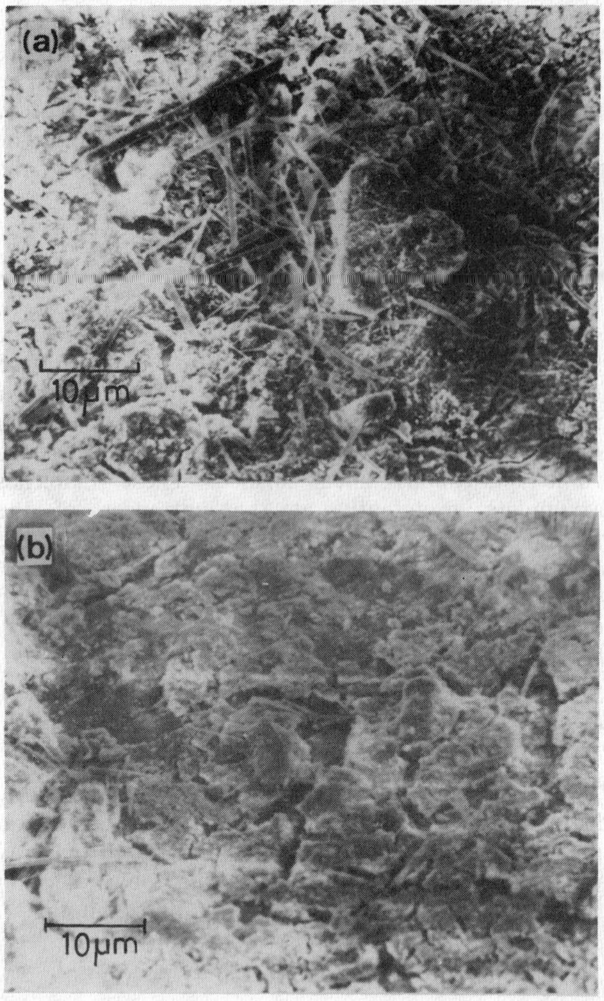

Figure 3. SEM micrographs (a) fresh catalyst; (b) used catalyst.

in which four different types of compounds may be distinguished. The crystalline phases are:

1. Orthorhombic, well crystallized MoO_3, as identified by both microanalysis and electron diffraction.
2. Iron molybdate, $Fe_2(MoO_4)_3$, that constitutes the most abundant phase. Stoichiometry was determined from peak intensity ratio Mo:Fe and confirmed by the diffraction pattern.

   The other phases did not appear sufficiently crystalline as to allow recording of useful diffraction patterns. These amorphous phases are:
3. A ternary phase containing tellurium which was always associated with Mo and Fe. The Te:Fe ratio seemed fairly constant for all the grains of this phase (~0.5), whereas the Mo content varied appreciably. The most frequent composition is shown on the x-ray spectrum, which relative intensities of the peaks suggest the stoichiometry Fe_2 $(TeMoO_8)$.
4. A mixed (Fe – Co) molybdate, also identified from the x-ray microanalysis spectra.

   In addition to these aggregated compounds, the pointing of the electron probe on carrier agglomerates where no supported phase particles are seen reveals the association of some amount of Mo, Te and Fe with the surface of SiO_2 (spectrum on the lower right of Fig. 3). The relative atomic concentration Mo:Te:Fe within this surface phase is approximately 3–4:1:2. Although it cannot be excluded that some part of these signals might be due to a remotely excited radiation, it appears that a significant proportion of the supported oxide phase (about 50%) is finely supported on the SiO_2 surface forming this surface species. No significant difference between the phase composition (as examined by AEM) of the fresh and used samples could be detected.

XPS

The binding energies of the O 1s, Si 2p, Mo 3d and Fe 2p lines did not significantly change upon activation. The essential difference concerned the Te 3d line: for the used catalyst a shoulder towards lower binding energies, with a shift of ca. 3.5 eV from the principal line is observed (Fig. 5).

The peak intensities were ratioed against the Si 2p peak intensity in order to reveal any change in the dispersion of the active phases due to catalytic work. The differences observed between the intensity ratio in the fresh and used catalysts are very small and below the limit of significance, owing to the limited precision of the measurements. Consequently, these data will not be discussed further.

ISS

When sputtering was done with $^4He^+$ ions it was not possible to separate Mo and Te peaks. Consequently, the experiments reported below were done by using $^{20}Ne^+$ ions; in these conditions it is impossible to obtain results for O and Si but in the present context it is preferable to gain information about the elements that constitute the active phases.

Figure 6 gives the depth profiles for Mo, Fe and Te concentrations in the fresh and used catalysts. For Mo and Fe, higher and constant ISS intensities are observed for the used catalyst while they increase slightly with sputtering time for the fresh catalyst. However, the curves corresponding to Te are almost coincident for both catalysts.

For the fresh catalyst the Te:Mo intensity ratio clearly decreases with increasing sputtering time (from 0.99 at 6 min to 0.56 at 41 min) while the Fe:Mo ratio is constant and equal to 0.26 along the same period. However, for the used catalyst the Te:Mo ratio is lower at the beginning (0.60 at 6 min) and decreases much more slowly (0.48 at 41 min); the Fe:Mo ratio is again constant with time and equal to 0.26.

DISCUSSION

The clear increase in catalytic activity with time on stream is accompanied by no significant change in selectivity are generally associated to changes in the

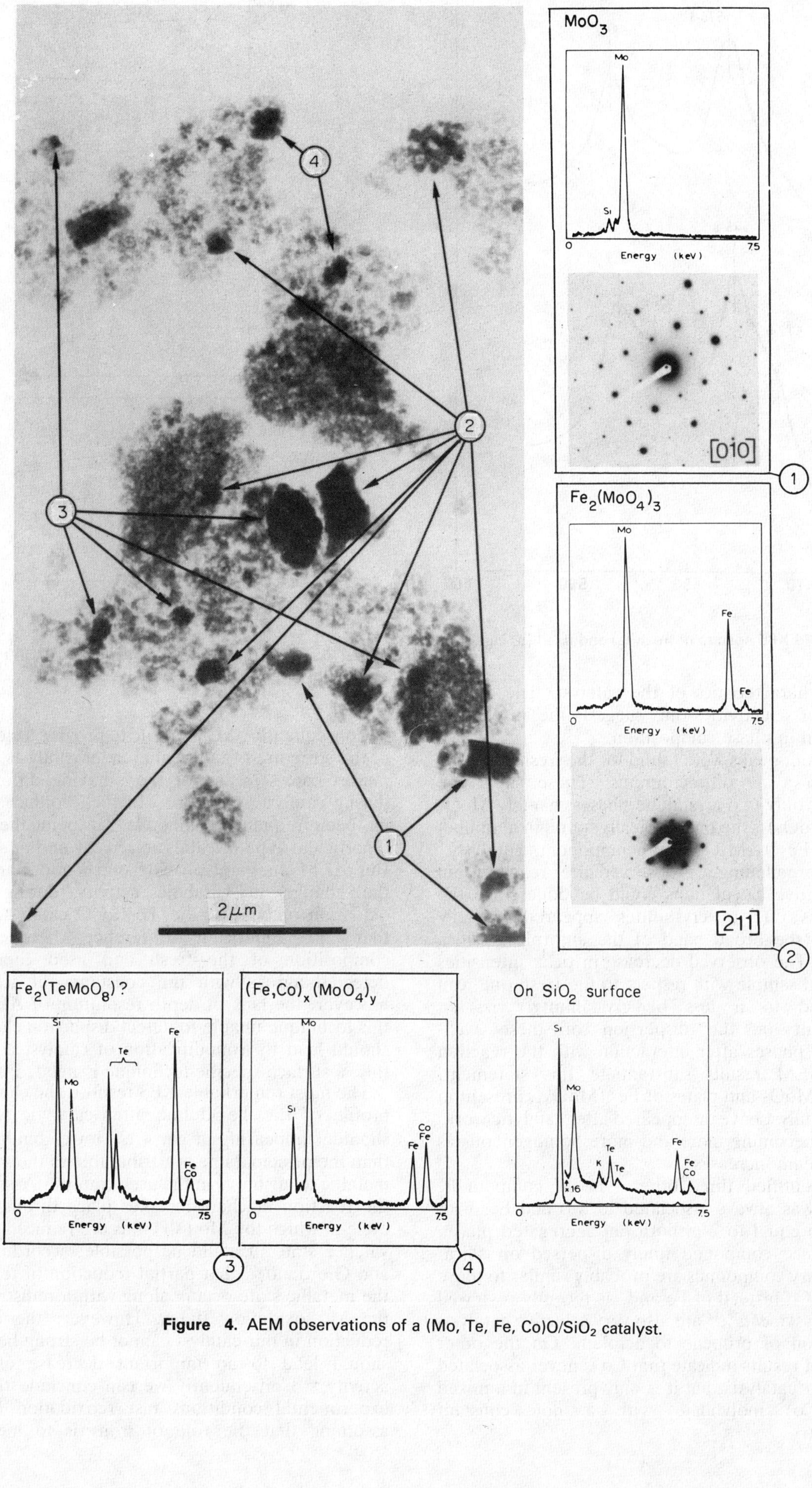

Figure 4. AEM observations of a (Mo, Te, Fe, Co)O/SiO_2 catalyst.

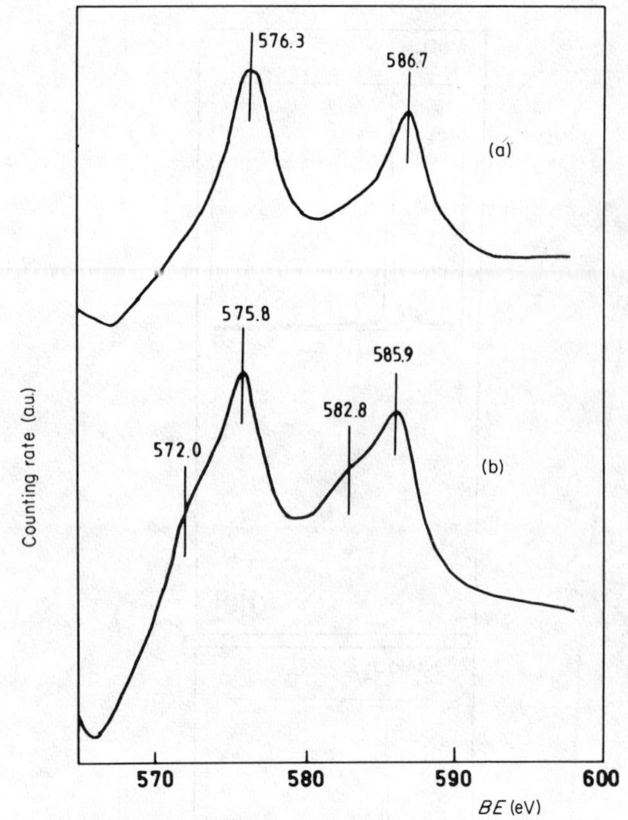

Figure 5. Te 3d XPS spectra of fresh (a) and used (b) catalysts.

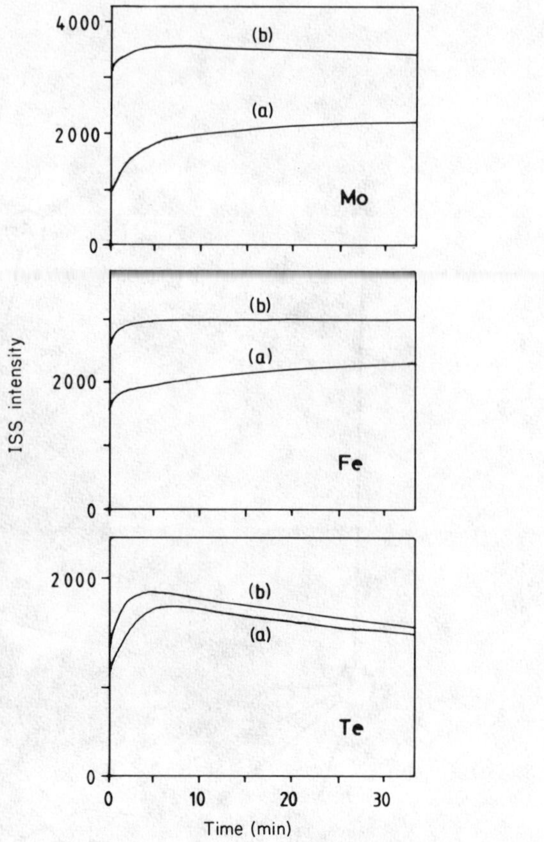

Figure 6. $^{20}Ne^{+}$ ISS profiles of Mo, Fe and Te in the fresh (a) and used (b) catalysts.

structural characteristics of the catalyst. The almost constancy of selectivities thus suggests the absence of modification in phase composition.

No different peaks were found for the fresh and used catalysts in x-ray diffractograms. These show the presence of only two crystalline phases, namely MoO_3 and $Fe_2 (MoO_4)_3$, the ones typically found in a number of MoO_3–Fe_2O_3–M_xO_y multicomponent catalysts.[6] Some line broadening was observed after reaction, but no quantitative calculations could be done since the small peaks due to crystallites appeared partially masked by the broad band of the amorphous silica supported. The observed decrease in peak intensities for the used sample with respect to the fresh one, can be ascribed to a loss of crystallinity, viz. an improvement of the dispersion of these well-crystallized phases after interaction with the reaction mixture. SEM results substantiate this statement: needles of MoO_3 and plates of $Fe_2 (MoO_4)_3$ present in the fresh catalyst have disappeared after catalytic work, the whole becoming more and more homogeneous as time on stream increases.

AEM identified three other type of compounds. Tellurium was always associated to Mo and Fe, with Te:Fe ratio equal to 2 in both, an aggregated phase, and a surface compound finely dispersed on silica. These ternary compounds are probably similar to those containing Co instead of Fe and, as recently reviewed by Grzybowska *et al.*,[12] are also very selective catalysts for oxidation of propene to acrolein. On the other hand, AEM results indicate that Co is never associated to Te in our catalysts, but it is only present in a mixed (Fe + Co) molybdate with a non constant stoichiometry.

Consequently, AEM is able to provide a clear picture of the structure of multicomponent catalysts. Different phases coexist, each of them having different roles during catalytic oxidation. In fact, contact synergism has been frequently claimed [12–14] as being the origin for a particular type of catalytic activity and/or selectivity. But AEM does not provide much information about the change in catalytic activity during propene oxidation on the Mo–Fe–Te–Co–O catalyst. The fact that no significant difference between phase composition of the fresh and used catalysts was detected agrees with the constancy of selectivities; however, the lack of depth resolution of AEM makes this technique unable to detect dispersion changes that should lead to a modification of catalyst activity; for this, a surface-specific technique is needed.

The most remarkable XPS result is the change in the profile of the Te 3d line after catalytic work. The shoulder appearing at ca. 4 eV lower binding energy than the principal line is attributable to the presence of metallic tellurium [15] in the used catalyst. Assuming that the reaction mechanism takes place through a redox cycle in which the Mo (VI) ions are reduced to a lower valence state, it could be possible, according to Bart and Giordano,[16] that partial reduction of tellurium to the metallic state occurs along with a redistribution of the Mo oxidation states. However, the degree of reduction in our catalyst cannot be strong because this should lead to an important decrease of catalytic activity.[16] Consequently we can conclude that in our experimental conditions fast reoxidation can occur assuming that the role of irons is to increase the reoxidation rate, in a similar way as cerium acts in (Te, Mo, Ce) O catalysts.[17]

XPS intensity ratios were not observed to change significantly as a result of catalytic work. This can be due to the fact that the XPS sampling depth corresponds to several atomic layers ca. 5 nm whereas changes taking place in our catalyst could only affect the outmost layer thus resulting in a decrease of XPS sensitivity.

It is readily concluded from ISS results that catalytic work brings about an increase of the dispersion of the phases containing Mo and Fe, while the concentration of Te at the surface of the sample seems unchanged. This suggests that these phases are dispersed on places where no Te is present. The other interesting ISS result is that, for both catalysts, Te concentration diminishes when going from the surface to the bulk thus supporting recent suggestions[12,17] about surface enrichment with tellurium in similar catalysts. This is not surprising since Te is a component with low surface energy. ISS provides a surface-specific insight about phenomena involved in catalyst activation during propene oxidation.

A further proof of the role of phase dispersion in catalyst activation is provided by chemisorption results done with the same catalysts used in this work and reported elsewhere.[18] These indicate that dispersion of the 'active phases' increases considerably with time on stream. So, as the equivalent molybdena area increased from 26.6 $m^2 g^{-1}$ (fresh catalyst, $t = 0$ h) to 37.0 $m^2 g^{-1}$ (used catalyst, $t = 7.2$ h), the total conversion increased from 8.5% at 0.4 h to 13.5% at 7.2 h. This substantiates the correlation between catalytic activity and dispersion of the active phase.

CONCLUDING REMARKS

The main conclusion of this study is that the presence of the same types of phases in fresh and used catalysts agree with the constancy of selectivities, while catalytic activity increase can be attributed to a higher dispersion of the active phases. Due to its inherent ability to probe the outmost layer of a catalyst, ISS is the technique which furnished in this work the most conspicuous results, providing evidence for dispersion of Mo- and Fe-containing phases during catalytic work. Also, a surface enrichment of Te could be detected by ISS.

Owing to the relatively high thickness of the layers analyzed by XPS, this technique was unable to detect changes in dispersion of our catalyst. But it should be kept in mind that XPS furnished a useful complementary information about the redox state of the catalyst surface (reduction to metalic tellurium).

On the other hand, AEM gave a clear picture on the structure of the catalyst, thus providing an explanatin for the constancy of selectivities (presence of the same phases in a fresh and used catalysts). In this sense, AEM compared very favourably with XRD since this last technique only furnished a relatively poor information about crystalline phases.

Consequently, no one of these techniques alone could give enough information for our catalytic problem solving, but, alternatively, the *joint* use of three bulk or surface analysis techniques has proved to be especially advantageous. AEM, XPS and ISS constitute a very powerful set of complementary techniques since they adequately combine depth resolution, lateral resolution and chemical informatin. Furthermore, as suggested by Delmon and coworkers[9,19] they can be considered as the core of the equipment necessary for characterizing catalysts of complex composition and structure, such as the one studied in the present work.

Acknowledgements

The assistance of F. Delannay, M. Genet and B. Doumain from UCL, Belgium, in the experimental part of this work is gratefully acknowledged. We also acknowledge the Secretaría de Estado de Universidades e Investigación and the Consejo Superior de Investigaciones Científicas of Spain for financial assistance.

REFERENCES

1. E.A. Melnik, author's abstract of Candidate's Dissertation, Inst. Phys. Chem., Acad. Sci. Ukran-SSR, Kiev, 1971.
2. V.P. Shchkin, S.A. Ven'yaminov and G.K. Boreskov, *Kinet. Katal.* **12** 621 (1971).
3. V.M. Zhiznevskii, E.V. Fedevich, D.K. Tolopko, M.V. Nikipanchuk, L.F. Yabukovskaya, I.M. Golub and V.Ya. Shipailo, *Kinet. Katal,* **18**, 381 (1974).
4. V.M. Zhiznevskii, L.I. Ivanyk and Huo Hai Sham, *Kinet. Katal,* **18**, 661 (1977).
5. J. Haber, J. Stoch and T. Wiltowski, *React. Kinet. Catal. Lett.* **13**, 161 (1980).
6. G. Kremenić, R. Arriagada Acuña, F. Melo Faus and J.L. Garcia Fierrro, *Proc. 6th Iberoam. Symp. Catal., Rio de Janeiro, Brazil 1978,* Vol. 3, p. 895, ed. by L. Nogueira, M. Schmal, R. Frety and Y.L. Lam (1981).
7. G.Kremenić, F. Melo Faus and R. Arriagada Acuña, *Proc. 7th Iberoam. Symp. Catal., La Plata, Argentina, 1980*. p. 197
8. H.H. Hren, J.I. Goldstein and D.C. Joy (eds), *Introduction to Analytical Electron Microscopy*, Plenum, New York (1979).
9. F. Delannay, *Catal. Rev.-Sci. Eng.* **22**, 141 (1980); F. Delannay and C. Defossé, *Prep. Div. Petrol. Chem. ACS* **26**, 385 (1981); F. Delannay (ed.), *Characterization of Heterogeneous Catalysts,* Chem. Ind. Ser., Vol. 15, Marcel Dekker Inc., New York (1984).
10. ASTM Powder Diffraction File no. 5–508.
11. ASTM Powder Diffraction File no. 20–526.
12. B. Grzybowska, A. Mazurkiewicz and J. Słoczyński, *Appl. Catal.* **13**, 223 (1985).
13. G. Kremenić, V. Cortés Corberán, J.M.D. Tascón and J.L.G. Fierro, in: Mekhanism Kataliza, vol. 2, p. 181, ed. by G.K. Boreskov and T.V. Andrushkevitch, Izd. Nauka, Novosibirsk (1984).
14. J.M.D. Tascón, P. Grange and B. Delmon, *J. Catal.,* in press.
15. W.E. Swartz, Jr., K.J. Wynne and D.M. Hercules, *Anal. Chem.* **43**, 1884 (1971).
16. J.C.J. Bart and N. Giordano, *J. Catal.* **64**, 356 (1980).
17. J.C.J. Bart and No. Giordano, *Ind. Eng. Chem., Prod. Res. Dev.* **23**, 56 (1984).
18. G. Kremenić, V. Cortés Corberán and J.L. García Fierro, *React. Kinet. Catal. Lett.* **17**, 181 (1981).
19. B. Delmon, *An. Quím. A,* **79**, 303 (1983); B. Delmon, *Proc. 8th. Iberoam. Symp. Catal., La Rábida, Spain,* Vol 1, p. 61 (1982). J.M.D. Tascón, P. Bertrand. M. Genet and B. Delmon, *J. Catal.,* in press.

SURFACE AND INTERFACE ANALYSIS, VOL. 9, 215–219 (1986)

AES and SIMS of Carbonaceous Adsorbate Layers on Pt, Rh, and PtRh

A.D. van Langeveld *and J.W. Niemantsverdriet†
*Department of Physics and †Laboratory of Inorganic Chemistry, Eindhoven University of Technology, 5600 MB Eindhoven, The Netherlands

Carbonaceous adsorbate layers on polycrystalline Pt, Rh, and PtRh, formed from C_2H_4 at different temperatures, have been investigated by AES and SIMS. Information on the properties of the carbon has been derived from the fine structure of the carbon Auger peak. SIMS spectra of positive and negative ions in the 12–26 mass range exhibit great sensitivity for the nature of the carbonaceous deposits and reveal the influence of the substrate. Auger spectra indicate that Pt segregates to the surface of the clean PtRh alloy.

INTRODUCTION

Carbonaceous layers form an essential component of working catalysts in hydrocarbon reactions, since they affect catalytic activity, selectivity and stability. These layers are formed immediately when hydrocarbons are passed over the catalyst. The nature of the carbonaceous deposit depends on the substrate and the temperature at which it has been formed.

In the case of Pt, three temperature regimes can be distinguished.[1] At low temperatures, <340 K, hydrocarbons are reversibly adsorbed and can be removed by heating. At intermediate temperatures, 340 – 700 K, the Pt catalyst is readily covered by hydrocarbons which transform into a layer of reactive C_xH_y fragments. These species play an important role in determining the activity and the selectivity of the working Pt catalyst. At high temperatures, above 700 K, the hydrocarbons transform into inactive, graphitic layers which poison the catalyst. AES has successfully been applied to characterize gaseous hydrocarbons[2,3] and carbonaceous adsorbate layers on several metals.[4–8] Previous work on Pt and Rh revealed significant differences in the nature of the carbon species formed on these metals.[8]

SIMS has been applied to study the interaction of hydrocarbons with Ni and Ru.[9–13] To the best of our knowledge, a comparative SIMS investigation of carbonaceous layers on Pt, Rh, and PtRh has not been described in the literature.

In this paper we report an investigation by means of AES and SIMS of carbonaceous layers formed by bringing polycrystalline foils of Pt, Rh, and PtRh in contact with C_2H_4 at three different temperatures, which represent the regimes of hydrocarbon interaction with Pt discussed above. The results show that SIMS spectra of hydrocarbon fragments are sensitive indicators of the state of the carbonaceous deposit.

0142–2421/86/120215–05 $05.00

EXPERIMENTAL

The experiments were carried out in an XPS/AES/SIMS spectrometer (Perkin–Elmer, PHI model 550), equipped with a double-pass cylindrical mirror analyzer and a PHI 2100 specimen introduction/reaction chamber. The base pressure of the UHV chamber is usually 2.10^{-10} Torr. In the present experiments, however, pressures were typically in the low 10^{-9} Torr range, due to the frequent transfer of the sample between the UHV and the reaction chamber. The base pressure of the latter was 1.10^{-7} Torr, obtained with a turbomolucular pump (Pfeiffer, TSU 110).

Auger spectra were measured in differential mode by using a 4.5 μA beam of 2 keV electrons, with a spot diameter of 0.5 mm. All spectra were measured in duplo at a scanning rate of 2 eVs^{-1} and a modulation amplitude of 2 eV. Each spectrum shown is the sum of two scans. Evidence for electronbeam-induced damage of the adsorbate layer has not been observed.

Static SIMS spectra were measured by using a 50 nA beam of 1 keV Ar^+ ions from a differentially pumped ion gun (PHI model 04–30) and a UTI 100C quadrupole mass spectrometer. Each sample was analyzed by measuring a positive SIMS spectrum for 60 s, followed by a negative spectrum also for 60 s. The total exposure of the sample during the SIMS experiments was $3.8.10^{13}$ Ar^+ ions cm^{-2}.

Data acquisition, storage, and instrument control were carried out with a Digital PDP 11/04 computer.

The samples consisted of a polycrystalline foil (specpure quality) of about 15 × 5 × 0.1 mm, spotwelded onto a tantalum hairpin. The temperature was monitored by a chromel–alumel thermocouple spotwelded onto the back of the foil. The cleaning procedure consisted of repeated cycles of sputtering with 5 keV Ar^+ ions and annealing at 1000 K, until no more contaminants could be detected by AES. Finally, the samples were annealed at 1000 K for at least 5 min in order to remove damage caused by sputtering.

The metal foils were treated in the reaction chamber in 0.5 Torr C_2H_4 (Messer Griesheim, research grade purity) for 5 min, at 325, 525 or 775 K. The sample was cooled to room temperature, evacuated to about 10^{-6} Torr, and next transferred into the main chamber for AES and SIMS analysis. Hereafter, the sample was prepared for a subsequent experiment by applying one sputter and anneal cycle as described above.

RESULTS AND DISCUSSION

Surface composition of the PtRh alloy

Surface and bulk composition of the clean, equilibrated PtRh alloy were determined from the intensities of the Pt Auger transitions at 64, 168, and 1967 eV in the spectra of PtRh alloy and Pt metal, by using the Gallon model.[14,15] Backscattering factors for the three Pt transitions in the alloy and in the metal have been calculated with the empirical relation proposed by Reuter,[16] see Table 1.

Application of the Gallon model requires knowledge of the monolayer intensities, defined as the relative contribution of the surface layer to the Auger spectrum of the clean metal. The monolayer intensities (Table 1) were calculated from the empirical expression for the inelastic mean free path of electrons in solids proposed by Seah and Dench.[17] As surfaces of polycrystalline specimens consist mainly of the most densely packed crystallographic planes, the interlayer spacing of Pt (111) planes has been used, the calculations resulted in a bulk content of 62 at % and a content of the uppermost atomic layer of 84 ± 5at% Pt. Hence, the surface of our clean, equilibrated PtRh foil is enriched in Pt.

Surface segregation of Pt in PtRh has been reported by Holloway and Williams[18] and Williams and Nelson[19] for the (111) surfaces of various PtRh single crystals with different bulk compositions, and by van Delft and Nieuwenhuys[20] for the (100) surface of a Pt(0.25)Rh(0.75)single crystal. Ahmad and Tsong,[21] on the other hand, reported surface enrichment of Rh in PtRh alloys, irrespective of the bulk composition. It should be noted, however, that the surface of the PtRh samples used by these authors was contaminated by a

Table 1. Selected parameters of Pt and Rh

	Pt			Rh
(a) Auger transition (eV)	64	168	1967	302
Monolayer intensity (%)	47	43	8	28
Backscattering factor in Pt	2.32	2.28	1.88	
Backscattering factor in Pt	2.07	2.03	1.71	

	Pt	Rh
(b) Additional data		
Matrix density (mole ml^{-1})	0.110	0.121
ΔH^{sub} (kcal mole^{-1} at 298 K)	135.3	133.0
ΔH^{sub} (kcal mole^{-1} at 1000 K)	127.1	124.8
Atomic radius (nm)	0.138	0.134

fraction of a monolayer of sulphur. All results on the surface composition of PtRh alloys are summarized in Fig. 1.

Theory does not provide a clear answer about surface segregation in PtRh. As the thermodynamic data in Table 1 show, the heat of sublimation, ΔH^{sub}, of Rh is 2.3 kcal mole^{-1} less than that of Pt in the temperature range of 298 K to 1000 K, and a slight enrichment of Rh at the surface might be expected. However, as the difference in ΔH^{sub} is small, and, moreover, the difference in atomic size between Rh and Pt is only 3% (Table 1), the driving force for surface segregation due to these effects is small. In such cases other effects, such as differences in surface entropy, or, perhaps, the presence of undetected impurities, may play a decisive role.

In conclusion, the surface of the clean, equilibrated PtRh alloy used in the present study consists of about 84% Pt and 16% Rh. Of course, this composition may be different when the surface is covered by an adsorbate layer.

AES of carbonaceous layers on Pt, Rh, and PtRh.

Auger spectra of the clean, equilibrated Rh, PtRh, and Pt foils, and of the foils after treatment in C_2H_4 at different temperatures are shown in Fig. 2, along with the carbon coverage, θ_c, of the sample. The amount of carbon on the foils has been calculated in the following way. For C_2H_4 on Rh metal, θ_c was estimated from the attenuation of the Rh 302 eV peak with respect to that of the clean metal. In this procedure, the exponential attenuation model, using the effective adsorbate layer thickness and the inelastic mean free path of electrons in the adsorbate layer, was applied. According to the empirical relation of Seah and Dench[17], the inelastic mean free path of electrons in a carbon matrix can be calculated to be 0.65 nm. Previous time dependent experiments indicated the effective monolayer thickness of a carbonaceous adsorbate to be 0.11 nm.[8]

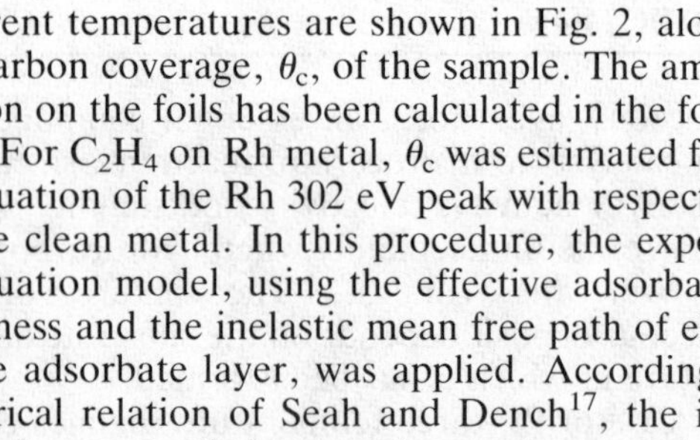

Figure 1. The composition of the uppermost atomic layer of PtRh alloys as a function of their bulk composition for various equilibration temperatures: ○ = present results, 1000 K; △ = Holloway and Williams; 1000 K (AES);[18] + = Williams and Nelson, 1300 K (ISS);[19] □ = Van Delft and Nieuwenhuys, 1200 K (AES/Selective chemisorption;[20] ● = Ahmad and Tsong, 1000 K (Atom probe).[21]

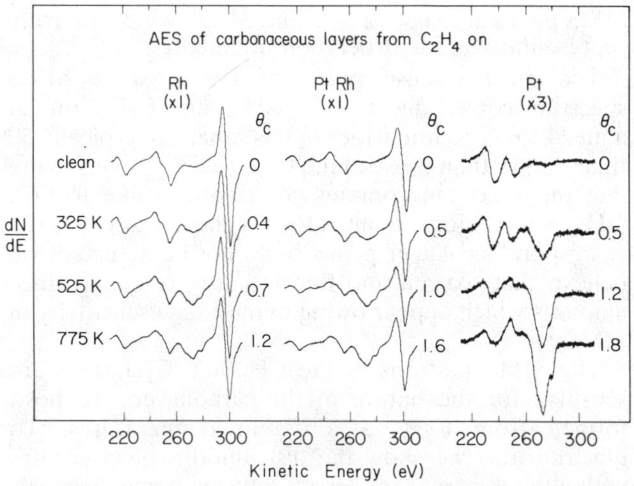

Figure 2. Differential Auger spectra of clean and ethylene-treated Pt, PtRh and Rh. All reactions were carried out in 0.5 Torr C_2H_4 for 5 min at the temperature indicated.

Taking into account the angle of acceptance of the analyzer (45°), and assuming a layer by layer growth of the adsorbate, the attenuation of the intensity of the Rh 302 eV Auger transition due to the presence of one and two monolayers of adsorbed carbon can be calculated; for uncompleted layers a linear interpolation between completed layers was used.

In the case of Pt, θ_c has been derived from the intensity ratios of the C 274 eV and the Pt 237 eV Auger peaks. It was assumed that an intensity ratio C 274/Pt 237 = 2.8 corresponds to monolayer coverage.[1] After corrections for differences in the backscattering factor of both substrates (according to Reuter[16]), the intensities of the various carbon Auger transitions on both pure metals could be compared. These corrected intensities were in good agreement with trends in the θ_c values, calculated. The method of the attenuation of the clean substrate signal and of the ratio of the adsorbate to substrate intensities cannot be used on the PtRh alloy, since the composition of the uppermost atomic layer of the latter may change as soon as hydrocarbons are adsorbed on the surface. Therefore, θ_c values on the alloys were calculated by comparing the corrected intensities of the C 274 eV peak. In these calculations effects due to the presence of different carbon species have been neglected. Of course, this is a serious simplification, and. hence, the θ_c values can only be considered as estimates.

The θ_c values shown in Fig. 2 indicate that for all samples the amount of carbon deposited from 0.5 Torr C_2H_4 increases with increasing temperature. Notice, that the amounts of carbon deposited at temperatures of 525 and 775 are significantly higher for Pt than for Rh, whereas the PtRh alloy takes an intermediate position. On the assumption that the θ_c values of the carbonaceous deposit on PtRh formed at 525 and 775 K reflect the composition of the alloy surface, we find a Pt content of 63 ± 10at%. This is much closer to the bulk composition than the surface composition of the clean alloy. Dynamic chemisorption experiments, which will be reported elsewhere, showed an unexpectedly rapid decrease of the Pt 64 eV Auger peak intensity as soon as the alloy was exposed to CH_4. In contrast to this, the decrease of the Rh 302 eV peak was less pronounced, also after appropriate corrections for the differences in escape depth of the Pt 64 and Rh 302 eV Auger electrons. These results suggest that surface enrichment of Pt in PtRh becomes less pronounced when the alloy is covered by a carbonaceous overlayer. The usual way to derive information on the nature of carbon from its Auger spectrum is to study the fine structure of the carbon spectrum between 240 and 260 eV. Unfortunately, this part of the spectrum is disturbed by the 256 eV transition of Rh and the 237 and 251 eV transitions of Pt. Spectrum subtraction techniques[8] did not yield satisfactory results for the carbon-covered PtRh alloy, most likely due to differences in the surface composition between the clean and the carbon covered alloy, as discussed above.

However, the fine structure in the carbon spectra between 260 and 285 eV contains also information on the nature of the carbon. As Fig. 2 shows, three distinct features appear in the spectrum between 260 and 285 eV. Two of these are located at 268.8 ± 0.4 and 273.6 ± 0.4 eV on all samples. The relative intensity of the 268.8 eV peak decreases with increasing reaction temperature at which the carbonaceous layer was formed. In agreement with previous work[8] we suggest that the 268.8 eV peak is characteristic of carbidic or molecular carbon species which are reactive toward hydrogen. The third feature is located at 277 – 280 eV. The exact position varies somewhat with the reaction temperature and with the substrate. This peak has been observed before in Auger spectra of graphitic overlayers on Ir,[22] Pt,[23,8] Rh,[8] and Fe,[24] but it is absent in the spectrum of pure graphite. It is likely, therefore, that the 277–280 eV peak represents the interaction of graphitic carbon with the substrate, similarly as has been proposed for carbonaceous adsorbate layers on Fe(100).[24]

In conclusion, the Auger spectra of Fig. 2 show that the carbonaceous layer formed on Rh, Pt, and PtRh at lower temperatures consists of carbidic and/or molecular carbon species, whereas at higher temperatures carbon of a more graphitic nature is formed. Pt is more sensitive to formation of graphitic carbon than Rh; the PtRh alloy behaves intermediate. The Auger results provide a useful background for the interpretation of the SIMS results which will be discussed hereafter.

SIMS of carbonaceous layers on Pt, Rh, and PtRh

As a representative example we show in Fig. 3 the positive and negative SIMS spectra of the PtRh foil which has been exposed to C_2H_4 at 325 K for 5 mins. Results of the other experiments are shown in a more condensed form in Fig. 4.

The positive SIMS spectrum in Fig. 3 shows peaks at atomic mass units (amu) 12,13,14, and 15, corresponding to C^+, CH^+, CH_2^+, and CH_3^+, and peaks in the range 24–29, corresponding to C_2^+, $C_2H_2^+$, $C_2H_3^+$ or Al^+, $C_2H_4^+$ or CO^+, and $C_2H_5^+$. Unfortunately, not all peaks in the $C_2H_n^+$ (n = 0,1, ..., 5) can be assigned unambiguously to one species. Both Al^+ and $C_2H_3^+$ appear at 27 amu. As, for example, a chromel–alumel thermocouple has been used, the presence of Al^+ in

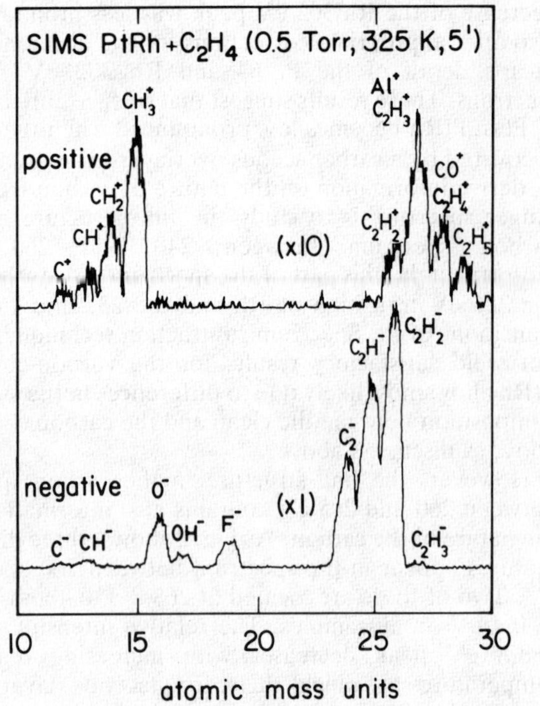

Figure 3. Positive and negative SIMS of PtRh treated in 0.5 Torr C_2H_4 at 325 K for 5 min.

the SIMS spectra cannot be excluded. The peak at 28 amu can be due to either $C_2H_4^+$ or CO^+, the latter being a main constituent of residual gases in vacuum systems. In view of these considerations, we will only use the SIMS peaks at masses 12–15 and 24–26, which are truly representative of hydrocarbon fragments.

The most intense peaks of the negative SIMS spectrum correspond to C_2^-, C_2H^-, and $C_2H_2^-$ ions at amu 24–26. The intensities in this range are typically 50 times larger than those of the CH_n^- ($n = 0$–3) ions. Note that the spectrum contains contributions due to O^-, OH^-, and F^- ions, at amu 16, 17, and 19, also. As the corresponding Auger peaks could not be detected, we believe that oxygen and fluorine are present in trace amounts which appear owing to their high sensitivity in SIMS.[12]

The SIMS patterns of the CH_n and C_2H_n ions are sensitive for the nature of the carbonaceous deposit formed from C_2H_4 at different temperatures. To illustrate this we show the distributions of intensities within the four ranges of the positive and negative CH_n and C_2H_n ions in Fig. 4. Note that the sum of the ion yields within each range has been set equal to 100%, and that, as a consequence, absolute comparisons between different ranges cannot be made.

In order to investigate if the Ar^+ ion current density used in the experiments, 50 nA cm^{-2}, gives rise to damage which may have influenced the results, the yields of selected ions were followed as a function of time during a SIMS measurement. It appeared that, although the yield of almost all ions decreased somewhat with time of exposure to the Ar^+ beam, correction for this result would not alter the picture as presented in Fig. 4 very much. Time-dependent experiments will be presented in a future publication.[25]

The positive SIMS spectra in the 12–15 mass range of the adsorption at 525 K on all substrates indicate the presence of a surface species containing methylgroups,

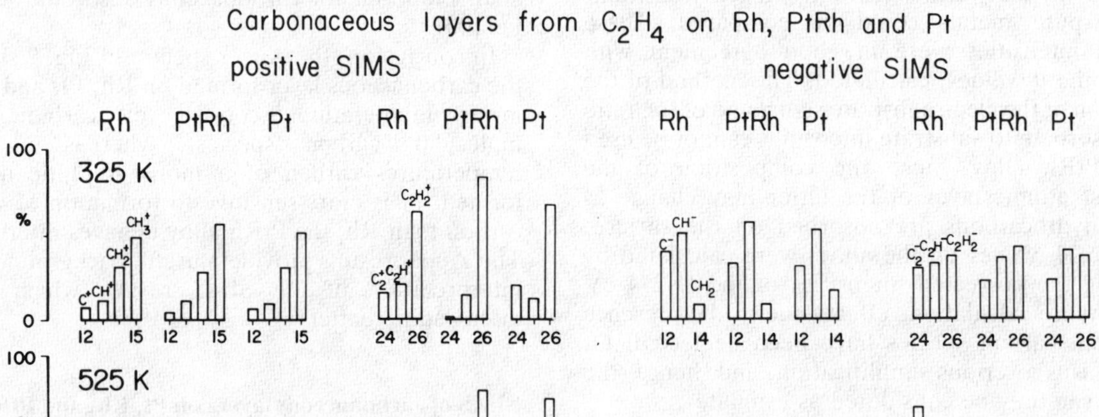

Figure 4. Distribution of intensities of positive and negative ions in the mass ranges 12–15 and 24–26 of Pt, PtRh and Pt after treatment in 0.5 Torr C_2H_4 for 5 min at the temperatures indicated.

such as ethylidyne, in agreement with results of Creighton and White,[26] and Wang *et al.*[27]

The SIMS results in Fig. 4 indicate that the properties of the carbonaceous deposit depends strongly on the temperature of formation and less on the substrate. The distribution of ions in each group of hydrocarbon fragments shows that the hydrogen content of the carbonaceous deposit decreases with increasing temperature of deposition. Comparison of absolute SIMS data showed that secondary ion yields from graphitic carbon are substantially lower than those of adsorbed hydrocarbons or reactive C_xH_y fragments formed at 325 and 525 K, respectively.[25] Hence, the actual hydrogen content of the graphitic carbon will be lower than suggested by the relative intensities of the CH_n and C_2H_n ions.

Since graphitization of carbon occurs through polymerization of smaller hydrocarbon fragments, the graphitic nature of the carbonaceous deposit will be reflected in the intensity ratio of the C_4^- and C_2^- ions (data not shown). In the case of Pt which, as is concluded from the Auger spectra in Fig. 2, is sensitive to formation of graphitic carbon, the C_4^-/C_2^- ratio increases from 0.01 for carbon deposited at 325 K to 0.05 for carbon deposited at 775 K. For Rh, the C_4^-/C_2^- ratio increases from 0.01 to less than 0.02, for carbon deposited at 325 and 775 K, respectively. This difference in graphitization between Pt and Rh is in agreement with a discussion of Beeck.[28] On PtRh the C_4^-/C_2^- ratio increases from 0.01 to 0.04 for C_2H_4 treatments at 325 and 775 K, respectively.

As Fig. 4 shows, the mass distribution of the $C_2H_n^-$ ions is the most sensitive indicator for the state of the carbonaceous layer, followed by CH_n^+ and CH_n^- ions, whereas the distribution of the $C_2H_n^+$ ions between 24 and 26 amu contains the least information. Unfortunately, potentially interesting fragments such as $C_2H_3^+$ and $C_2H_4^+$ cannot be used in the present case, because of possible interference with Al^+ and CO^+, as discussed above.

Comparison of the SIMS patterns of carbonaceous layers formed at a certain temperature on Rh, Pt, and PtRh, reveals the influence of the substrate on the nature of the carbonaceous deposit. The SIMS data in Fig. 4 show that the hydrogen content of carbonaceous deposits formed on Rh at 525 and 775 K is lower than on Pt at these temperatures, suggesting that Rh is more active for C—H bond breaking than Pt, in agreement with catalytic experience.[29] It is also interesting to note that the SIMS patterns of carbon on PtRh differ from those on the constituent metals. The negative SIMS spectra of carbon on PtRh are more characteristic of carbon on Rh than on Pt, in spite of the fact that the alloy contains 62% Pt.

The present work shows that SIMS spectra can be used as fingerprints for the state of carbonaceous layers on metals such as Pt and Rh and their alloys, and that the spectra reveal differences in the properties of the carbonaceous deposits on the different substrates, in agreement with Auger line shape analysis described above. The relation between the SIMS pattern and the type of hydrocarbon species on the surface will be discussed in a future publication.[25]

Acknowledgements

The authors are grateful to Mr. M. Hoddenbagh and Mr. W.C. van der Vleuten for technical support. J.W.N. acknowledges support from a Huygensfellowship granted by the Netherlands Organization for the Advancement of Pure Research (ZWO).

REFERENCES

1. G.A. Somorjai, *Chemistry in Two Dimensions: Surfaces, by.* Somorjai, Cornell Univ. Press, Ithaca (1981).
2. R.R. Rye, T.E. Madey, J.E. Houston and P.H. Holloway, *J. Chem. Phys.* **69,** 1504 (1978).
3. J.E. Houston and R.R. Rye, *J. Chem Phys.* **74,** 71 (1981).
4. M.P. Hooker and J.T. Grant, *Surf. Sci.* **62,** 21 (1977).
5. M.A. Smith, S. Sinharoy and L.L. Levenson, *J. Vac. Sci. Techn.* **16,** 462 (1979).
6. F.P. Netzer, *Appl. Surf. Sci.* **7,** 289 (1981).
7. E.G. Keim. L.L. Labohm, O.L.J. Gijzeman, G.A. Bootsma and J.W. Geus, *Surf. Sci.* **112,** 52 (1981).
8. A.D. van Langeveld, F.C.M.J.M. van Delft and V. Ponec, *Surf. Sci.* **135,** 93 (1983).
9. A. Benninghoven, P. Beckmann, D. Greifendorf and M. Schemmer, *Appl. Surf. Sci.* **6,** 288 (1980).
10. M. Schemmer, P. Beckman, D. Greifendorf and A. Benninghoven, *SIMS III,* ed. by. A Benninghoven, J. Giber, J. Laszlo, M. Riedel and H.W. Werner, Springer Verlag Berlin, Heidelberg (1982).
11. H. Kaarman, B. Leidenberger, H. Hoinkes and H. Wisch, *SIMS III,* ed. by. A. Benninghoven, J. Giber, J. László, M. Riedel and H.W. Werner, Springer Verlag, Berlin, Heidelberg (1982).
12. L.L. Lauderback and W.N. Delgass, *ACS Symp. Series* **248,** 21 (1984).
13. L.L. Lauderback and W.N. Delgass, *ACS Symp. Series* **279,** 339 (1985).
14. T.E. Gallon, *Surf. Sci.* **17,** 486 (1969).
15. A.D. van Langeveld and V. Ponec, *Appl. Surf. Sci.* **16,** 405 (1983).
16. W. Reuter in: *Proc. 6th Intern. Conf. on X-ray Optics and Microanalysis,* ed. by. G. Shinoda, W. Kohra and T. Ichinokawa, Univ. Tokyo Press (1972).
17. M.P. Seah and W.A. Dench, *Surf. Interface Anal.* **1,** 2 (1979).
18. P.H. Holloway and F.L. Williams, *Appl. Surf. Sci.* **10,** 1 (1982).
19. F.L. Williams and G.C. Nelson, *Appl. Surf. Sci* **3,** 409 (1978).
20. F.C.M.J.M. van Delft and B.E. Nieuwenhuys, *Surf. Sci.* **162,** 538 (1985).
21. M. Ahmad and T.T. Tsong, *Surf. Sci.* **149,** L7 (1985).
22. N.A. Kholin, E.V. Rut'kov and A.Y. Tontegode, *Surf. Sci.* **139,** 155 (1984).
23. D.W. Goodman, R.D. Kelley, T.E. Madey and J.T. Yates, *J. Catal.* **63,** 266 (1980).
24. G. Panzer and W. Diekman, *Surf. Sci.* **160,** 253 (1985).
25. J.W. Niemantsverdriet and A.D. van Langeveld, in preparation.
26. J.R. Creighton and J.M. White, *Surf. Sci.* **129,** 327 (1983).
27. P.-K. Wang, C.P. Slichter and J.H. Sinfelt, *J. Phys. Chem.* **89,** 3606 (1985).
28. O. Beeck, *Disc. Farad. Soc.* **8,** 118 (1950).
29. R. Merta and V. Ponec, in *Proc. 4th Int. Congr. Catal. Moscow, 1968,* Vol 2, p. 53, Akadémiai Kiadó, Budapest (1971).

SURFACE AND INTERFACE ANALYSIS, VOL. 9, 221–225 (1986).

Characterization of Supported Bimetallic FeIr/SiO_2 Catalysts by Mössbauer Spectroscopy, Temperature-programmed Reduction and X-ray Photoelectron Spectroscopy

J.W. Niemantsverdriet* and A.M. van der Kraan†
*Laboratory for Inorganic Chemistry and Catalysis, Eindhoven University of Technology, 5600 MB Eindhoven, the Netherlands
†Interuniversitair Reactor Instituut, 2629 JB Delft, the Netherlands

Supported bimetallic FeIr/SiO_2 catalyst have been chacterized by *in situ* Mössbauer spectroscopy, temperature-programmed reduction (TPR) and XPS. The reduction of iron in freshly prepared FeIr/SiO_2 catalysts is greatly facilitated by the noble metal iridium. Reduced catalysts consist of Ir-rich fcc FeIr alloys, possibly some isolated Ir metal, and significant amounts of iron (III) oxide. The Mössbauer parameters of the latter change when the catalyst is exposed to CO gas at room temperature, indicating that the iron (III) oxide is present in a highly dispersed phase. Evacuation of reduced samples at 475 or at 775 K leads to significant oxidation of the iron in the catalyst, most likely by OH groups on the SiO_2 support.

INTRODUCTION

In catalysis it is well known that alloying metals provides a powerful means to optimize the performance of catalysts.[1] As catalysts should prefereably have a large fraction of their atoms at the surface, alloys in the form of small particles stabilized by an inert support, also called supported bimetallic catalysts, are of great practical interest. Supported bimetallic catalysts consisting of iron and one of the more noble group VIII metals have shown attractive improvement of selectivity for reactions of CO and H_2 to light olefins and alcohols.[2,3]

Investigations of the structure of supported bimetallic catalysts are seriously hindered by their small dimensions and by the presence of a nonconducting, porous support. These catalysts are, therefore, not amenable to study by techniques such as x-ray diffraction or Auger electron spectroscopy. In this respect, iron-containing catalysts have the great advantage that Mössbauer spectroscopy can be applied. This technique yields information on oxidation state, lattice symmetry, magnetism and lattice vibrations of the isotope ^{57}Fe (natural abundance 2%) in the catalyst, and can be applied *in situ*.[4] Although in principle a bulk technique, Mössabauer spectroscopy can be made surface sensitive by analyzing the temperature dependence of the lattice vibrations[5] and by studying the changes in the spectra which are the result of exposing the catalyst to a strongly adsorbing gas at room temperature.[6]

Little is known about FeIr catalysts. Preliminary catalytic tests show that methane is by far the dominant product in the CO + H_2 reaction over FeIr/SiO_2 at atmospheric pressure. At higher pressures, however, methanol and ethanol are formed as well.[7] Mössbauer investigations of supported FeIr systems have been reported.[8,9] These studies, however, concern the use of iron as a probe for the noble metal rather than the composition and structure of a supported bimetallic FeIr catalyst.

In this paper we present a characterization of FeIr/SiO_2 catalysts by XPS, temperature-programmed reduction (TPR), and *in situ* Mössbauer spectroscopy. The results illustrate how the latter technique can reveal information on the surface of iron-containing catalysts.

EXPERIMENTAL

The FeIr/SiO_2 catalyst was prepared by adding an aqueous solution of $H_2IrCl_6.6H_2O$ (Merck, PA), $Fe(NO_3)_3.9H_2O$ (Baker JT) and ^{57}Fe (Oak Ridge) in 2N HNO_3, dropwise and under frequent stirring to the SiO_2 support (Cab–O-Sil, EH–5, 310 $m^2\ g^{-1}$), until the incipient wetness point was reached. The catalyst contained 1.1 wt% iron, 10% of which was ^{57}Fe, and 3.9 wt% iridium (atomic ratio Fe:Ir = 1:1). The impregnated sample was dried in air at 295 K for several days, at 325 K for 24 h, and at 400 K for 72 h. About 300 mg of dried catalyst was pressed into a wafer with a diameter of 20 mm. All further treatments were carried out in the Mössbauer *in situ* reactors described elsewhere.[10,11]

Mössbauer spectra were measured with a constant acceleration spectrometer. Isomer shifts are reported with respect to the NBS standard sodium nitroprusside (SNP).

XPS spectra were recorded with a Perkin Elmer PHI 550 spectrometer, equipped with a Mg Ka x-ray source and a double pass cylindrical mirror analyzer operating

0142–2421/86/120221–05 $05.00

at a pass energy of 50 eV. Catalysts were reduced in a pretreatment chamber in flowing H_2. After evacuating the latter to a pressure of 10^{-5} Torr, the sample was transferred to the main chamber. Typical pressures during measurement were about 10^{-8} Torr.

TPR experiments were carried out by measuring the H_2 consumption of a freshly prepared FeIr/SiO catalyst as a function of temperature, which was increased at a rate of 5 K min^{-1}. Further details can be found in Ref. 12.

RESULTS AND DISCUSSION

The freshly prepared FeIr/SiO_2 catalyst, that is, after impregnation and drying but before reduction, has a Mössbauer spectrum which consists of a quadrupole doublet with parameters IS = 0.63 ± 0.03 mm s^{-1} and QS = 0.74 ± 0.05 mm s^{-1}. These values are characteristic of high-spin Fe^{3+} ions in an asymmetric environment as, for example, in highly dispersed iron (III) oxides.[13] The XPS spectra of the fresh catalyst (binding energies in Table 1) indicate that the noble component, iridium, is also present in an unreduced state, as the Ir $4f_{7/2}$ binding energies are characteristic of Ir^{3+} and Ir^{4+} ions.[14]

We will now discuss the reduction of the FeIr/SiO_2 catalyst. Fig. 1 shows the H_2 consumption of fresh FeIr/SiO_2 and Fe/SiO_2 catalysts as a function of reduction temperature in a TPR experiment. Reduction of FeIr/SiO_2 starts at about 350 K and is largely completed at 500 K. As opposed to this, reduction of Fe/SiO_2 starts around 500 K and has not yet been completed at 800 K, as can be inferred from the deviation of the TPR curve from the baseline at 800 K.

Both TPR curves consist of two peaks. TPR profiles of fully oxidized samples consisted of a single peak, for FeIr/SiO_2 at 485 K and for Fe/SiO at 625 K. Hence the peaks at the lower temperatures in the TPR patterns of Fig. 1 can be attributed to reduction of the metal salts and the higher peaks represent the reduction of the metal oxides. Apparently, the drying procedure employed was not sufficiently severe to decompose the metal precursor salts entirely. This in agreement with the presence of an intense Cl 2p signal in the XPS spectrum of the fresh FeIr catalyst.

The Mössbauer spectra in Fig. 2 show how the state of iron changes as function of reduction temperature. A small fraction of the initially present Fe^{3+} has been reduced to Fe^{2+} by H_2 at 375 K, and a larger fraction at 475 K. Formation of zero-valent iron, visible by the single peak in the spectra around 0.35 mm s^{-1} has already occurred at 475 K, but becomes increasingly important at higher temperatures. The maximum extent to which iron in FeIr/SiO_2 is reduced is reached

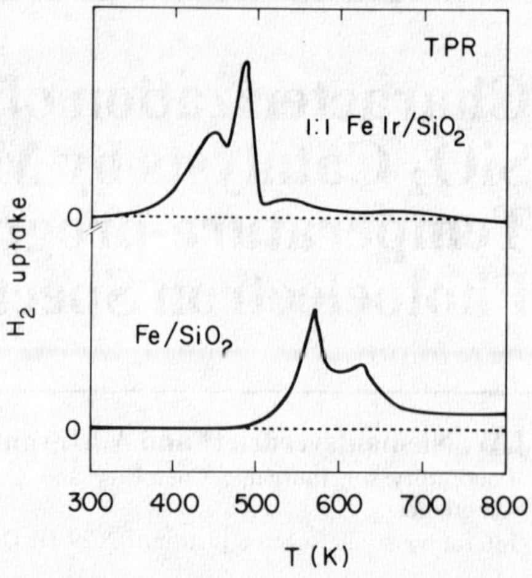

Figure 1. Temperature-programmed reduction of freshly prepared FeIr/SiO_2 and Fe/SiO_2 catalysts.

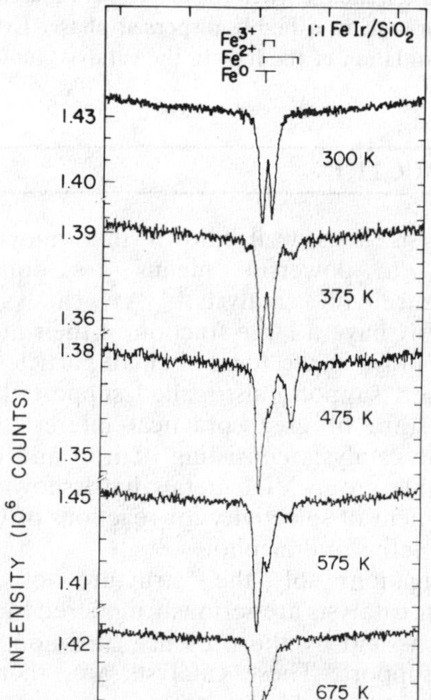

Figure 2. Mössbauer spectra measured in situ at 295 K, of the FeIr/SiO_2 catalyst after reduction in H_2 at the indicated temperatures. Stick diagrams indicate the positions of the peaks for the symmetrical doublets of Fe^{2+} and Fe^{3+} and the singlet of Fe° in fcc FeIr.

Table 1. XPS binding energies of 1:1 FeIr/SiO_2 catalysts

	Fresh	Reduced	Passivated
Si 2p	103.4	103.4	103.4
O 1s	532.6	532.7	532.6
Ir $4f_{7/2}$	62.2	60.7	61.1
Ir $4f_{5/2}$	65.1	63.4	63.9
Cl 2p	199.1	—	—

at 675 – 775 K, and corresponds to a Mössbauer spectrum consisting of a singlet due to fcc FeIr alloy[15] and a doublet of Fe^{3+}. Each component accounts for about 50% of the spectral area.

XPS spectra of the catalyst after reduction at 875 K (Table 1) indicate that Ir is present in the reduced form.[14] The Fe 2p spectra are broad and poorly resolved. Hence, no binding energies for Fe 2p have been included in Table 1. We note, however, that the Fe $2p_{3/2}$ peak extends from about 706 to 712 eV, indicating that iron is present in more than one oxidation state.

The combination TPR, XPS and Mössbauer spectroscopy shows that reduction of Fe and Ir occur in the same temperature region and that iron in FeIr catalysts reduces at significantly lower temperatures than in Fe/SiO_2. Apparently, the presence of the noble metal Ir enhances the reducibility of iron greatly.

In general, reduction of metal oxides takes place in two steps. First, molecular H_2 must react with metal ions to form zerovalent metal, in what is called the nucleation step. Once metal atoms are available, H_2 can easily be dissociated to yield reactive H atoms which take care of further reduction.[16] In the reduction of iron catalyst, nucleation occurs at rather elevated temperatures, around 500 K. In noble metals, on the other hand, nucleation takes place at lower temperatures, generally in the range of 200 – 400 K.[16] Thus, in $FeIr/SiO_2$ catalysts Ir provides the sites where H_2 is dissociated, after which H atoms diffuse to and reduce ionic iron in contact with Ir.

In the discussion of Fig. 2 we have assumed that the Mössbauer spectra of $FeIr/SiO_2$ after reduction at 675 – 875 K consist of a singlet and a doublet, representing Fe° in fcc FeIr alloy and Fe^{3+} in an oxidic environment, respectively. This interpretation, however, needs to be verified as other assignments cannot be excluded based on the room temperature spectra alone. The Mössbauer doublet with IS = 0.75 ± 0.05 mm s^{-1} and QS = 0.75 ± 0.10 mm s^{-1} can in principle be due to high-spin Fe^{3+} ions, low-spin Fe^{2+} ions, and, according to Garten and Sinfelt[8] and Murrell and Garten[9], also to zero-valent iron in the surface of FeIr alloy. Yet another possibility which cannot be discarded without additional information is that the peak in the 295 K spectra at 1 mm s^{-1} represents by itself an unresolved doublet of Fe^{2+}, as in carbon-supported Fe/C catalysts.[17]

Mössbauer spectra at cryogenic temperatures, as shown in Fig. 3, can distinguish between the possibilities mentioned above. The spectrum of reduced $FeIr/SiO_2$ at 77 K is very similar to the one at 295 K and shows in particular that the peak at 1 mms $^{-1}$ has not split at 77 K, as would have been expected for an unresolved Fe^{2+} doublet. Hence, the peak at 1 mms^{-1} is indeed the right half of a doublet. The spectrum of $FeIr/SiO_2$ at 4 K shows that both the singlet and the doublet present at 77 and 295 K have become magnetically split, with a broad distribution in magnetic splittings. As a low-spin Fe^{2+} ion has no magnetic moment, it cannot exhibit magnetic splitting. So the presence of low-spin Fe^{2+} can be discarded.

The maximum values of the magnetic splittings in the 4 K spectrum of $FeIr/SiO_2$ are about 500 kOe. This is much higher than the magnetic fields of fcc FeIr, bcc FeIr, or α–Fe. However, the presence of a broad distribution of magnetic hyperfine fields with values upto 500 kOe is consistent with the assignment of the corresponding phase to ferric iron in highly dispersed iron (III) oxides.[5,11] We therefore conclude that the Mössbauer spectrum at 295 K of a reduced $FeIr/SiO_2$ catalyst consists, as anticipated in the discussion of Fig. 2, of a singlet due to fcc FeIr alloy and a doublet of Fe^{3+}, most likely in the oxidic environment of the SiO_2 support.

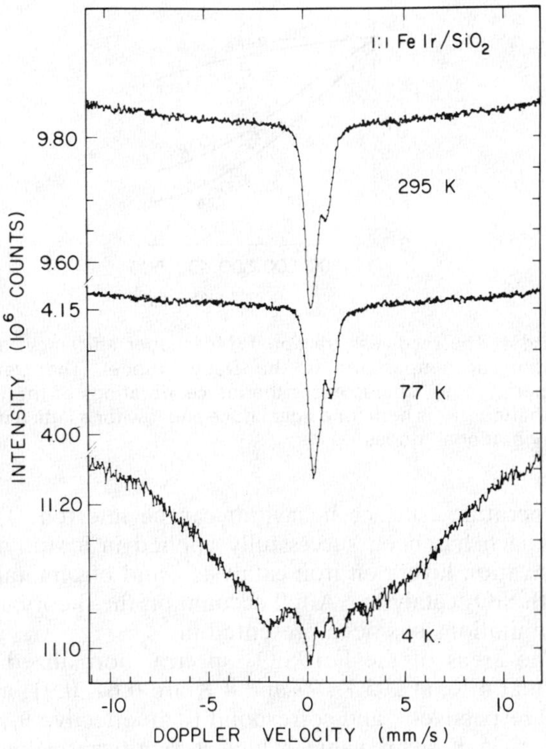

Figure 3. Mössbauer spectra of the reduced $FeIr/SiO_2$ catalyst, measured in situ at the temperatures indicated. The spectra at 77 and 4 K allow for an unequivocal interpretation of the spectrum at 295 K. Analysis of the spectral intensities as a function of temperature yield qualitative information on the dispersion of the sample, see text.

Comparison of Mössbauer spectra at different temperatures provides a way to recognize surface behaviour. The Mössbauer intensity of a certain compound is determined by its concentration times the so-called recoilless fraction, f.[18] This quantity depends on the energy of the lattice vibrations. The temperature dependence of f is satisfactorily described by the Debye model, in which the Debye temperature θ_D characterizes the lattice vibrations such that a high θ_D corresponds to a rigid lattice and a low θ_D to a lattice with soft vibrations. As Fig. 4 shows, the recoilless fraction of typical iron bulk compounds (θ_D 450 – 500 K) does not depend on temperature very much, and hence the intensity of the Mössbauer spectrum decreases only slighly between 4 and 300 K.

According to Somorjai,[19] the θ_D of surfaces is roughly equal to half the value of the corresponding bulk material. The implications of this simple concept for iron atoms at the surface are shown in Fig. 4: the Mössbauer intensity of surface compounds decreases strongly with increasing temperature. Hence, by analyzing Mössbauer intensities as a function of

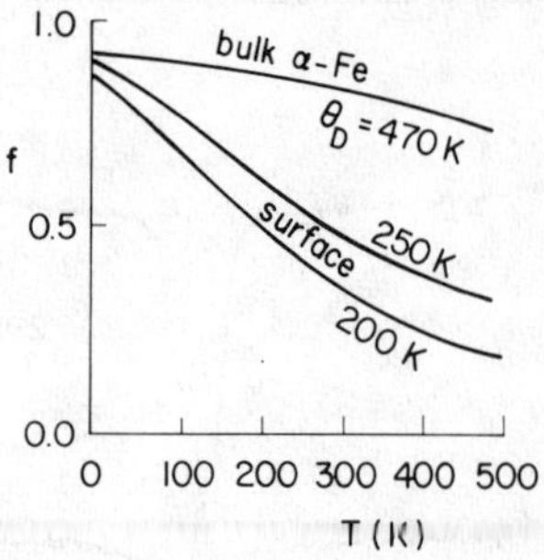

Figure 4. The recoilless fraction *f* (Mössbauer efficiency) as a function of temperature in the Debye model. The Debye temperature, θ_D, characterizes the lattice vibrations of the iron compounds. θ_D is high for a rigid lattice and low for a lattice with soft vibrational modes.

temperature, surface behaviour can be inferred. This approach has been successfully applied in a study of passivation layers on iron catalysts[11] and of bimetallic FeRh/SiO_2 catalysts.[5] A full account on the theory and its limitations has been presented in.[5]

The areas of the FeIr/SiO_2 spectra, normalized to unity at 4 K, at 295 K, 77 and 4 K are 0.64, 0.91, and 1.00 respectively, and correspond to an effective θ_D of 275 ± 25 K. This value, which is significantly lower than the θ_D of iron bulk compounds, indicates that the dispersion of the catalysts is high.

A disadvantage of XPS and other surface techniques is that the sample has to be evacuated prior to measurement. A useful application of *in situ* techniques such as Mössbauer spectroscopy, which can be applied both *in situ* and in vacuum, is to verify if the chemical state of the sample is preserved under evacuation. Fig. 5 shows Mössbauer spectra of a reduced FeIr/SiO_2 catalyst under H_2 and after evacuation of the *in situ* reactor to 10^{-8} Torr at different temperatures. As Fig. 5 shows, the spectra of evacuated samples may differ significantly from the spectrum measured under H_2, particularly when evacuation is carried out at higher temperatures. Evacuation leads to an increase of Fe^{3+} at the expense of Fe°, indicative of oxidation of the iron when H_2 is removed from the gas phase. Comparison with the spectrum of the air-passivated catalyst in Fig. 6 learns that evacuation of reduced FeIr/SiO_2 at 775 K gives rise to the same extent of oxidation as does exposure to air at 295 K. Most likely, the oxidation of iron upon removal of adsorbed hydrogen is due to OH groups on the SiO_2 support, according to a mechanism described by Dutartre *et al.*[20] and Dalmon *et al.*[21] The results imply that care should be taken when XPS is applied to study the state of nonnoble metals such as iron in supported catalysts. We believe that the XPS spectra of Ir reflect correctly the state of the noble metal in the reduced FeIr/SiO_2 catalyst, as the binding energies of the Ir $4f_{7/2}$ are entirely characteristic of Ir metal.

Another way to obtain information on the surface from Mössbauer spectroscopy is to study changes caused by chemisorption of a gas such as CO or NH_3.[6] Figure 6 shows that when a reduced FeIr/SiO_2 catalyst is exposed to CO at 295 K, a considerable fraction of the Fe^{3+}, visible by the peak at 1 mms^{-1}, is converted to Fe^{2+}, corresponding to the new peak at 2 mms^{-1}. The fact that this conversion takes place at room temperature is taken as evidence that the iron ions involved are exposed to the gas phase, and thus are located at the surface. Similar results have been obtained with SiO_2-supported FeRu, FeRh, and FePt catalysts.[22] The results suggest that a supported

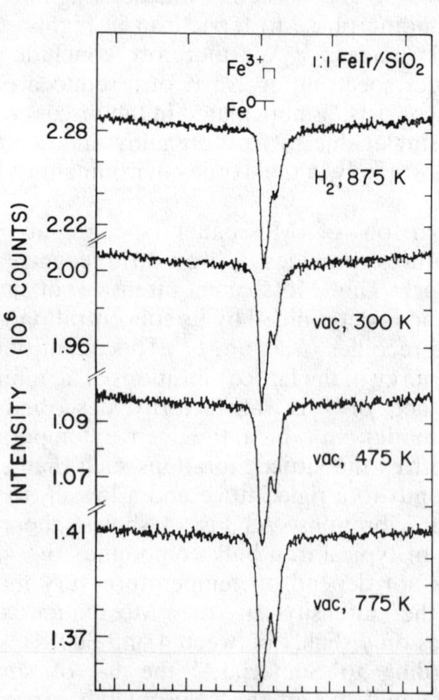

Figure 5. Influence of evacuation on the state of iron in FeIr/SiO_2 catalysts, as reflected in the Mössbauer spectra measured *in situ* at 295 K after the treatments indicated.

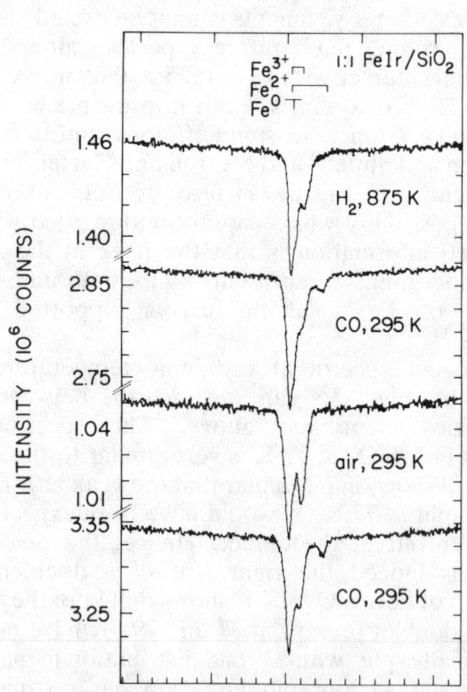

Figure 6. Mössbauer spectra *in situ* at 295 K of a reduced FeIr/SiO_2 catalysts (top) after exposure to CO at 295 K, oxidation in air at 295 K, and again exposure to CO at 295 K.

bimetallic catalyst consists of Ir rich FeIr alloy particles which rest on a thin layer of iron (III) oxide in contact with and stabilized by the SiO_2 support. The alloy particles do not cover the iron oxide completely but leave a fraction of the latter exposed to the gas phase. These are the sites where Fe^{3+} can be converted to Fe^{2+} by CO chemisorption. Minai *et al.*[23] have recently obtained additional support for this model from a combination of Mössbauer spectroscopy and EXAFS (extended x-ray absorption fine structure). We refer to Refs; 22–24 for further details on the structure of a supported bimetallic catalyst.

As Fig. 6 also shows, passivation of the reduced $FeIr/SiO_2$ in air at 295 K gives rise to considerable oxidation of the iron to Fe^{3+}. This is another indication that the alloy particles are small. The XPS results in Table 1 indicate that Ir does not oxidize to a large extent during passivation. Combination of the XPS and Mössbauer results suggest that Fe is partially oxidized upon passivation, whereas Ir remains reduced, which suggests that the passivated alloy particle consists of an Ir rich core and an iron oxide shell around it.

CONCLUDING REMARKS

The present work shows that the combination of TPR, XPS and Mössbauer spectroscopy yields information on the composition, structure and reduction of bimetallic $FeIr/SiO_2$ catalysts. The fact that different gaseous environments of the catalysts are reflected in the ^{57}Fe Mössbauer spectra demonstrates that Mössbauer spectroscopy contains information on the surface of well dispersed samples. A particularly useful application of this technique is that in situ spectra can be compared with spectra obtained under vacuum. In this way Mössbauer spectroscopy may contribute to bridge the gap between UHV investigations of catalysts by surface techniques and the high pressures under which catalysts are active.

Acknowledgement

We thank Mr. J. van Grondelle for measuring TPR profiles, and Mr. E. Gerkema for assistance with the Mössbauer experiments. J.W.N. acknowledges support from a Huygensfellowship of the Netherlands Organization for the Advancement of Pure Research (ZWO).

REFERENCES

1. V. Ponec, *Catal. Rev. Sci. Eng.* **11**, (1975); idem, *Surf. Sci.* **80**, 352 (1979).
2. M.A. Vannice, Y.L. Lam and R.L. Garten, *Adv. Chem.* **178**, 25 (1979).
3. M.M. Bhasin, W.J. Bartley, P.C. Ellgen and T.P. Wilson, *J. Catal.* **54,** 120 (1978).
4. J.A. Dumesic and H. Topsøe, *Adv. Catal.* **26,** 121 (1977).
5. J.W. Niemantsverdriet, A.M. van der Kraan and W. N. Delgass, *J. Catal.* **89** 138 (1984).
6. J.W. Niemantsverdriet, D.P. Aschenbeck, F.A. Fortunato and W.N. Delgass, *J. Mol. Catal.* **25** 285 (1984).
7. F. Stoop, to be published.
8. R.L. Garten and J.H. Sinfelt, *J. Catal.* **62,** 127 (1980).
9. L.L. Murrell and R.L. Garten, *Appl. Surf. Sci.* **19,** 218 (1984).
10. A.M. van der Kraan and J.W. Niemantsverdriet, *Proceedings International Conference on the Applications of the Mössbauer Effect, Alma Ata, 1983,* Gordon and Breach, London, 1985, p. 1795.
11. J.W. Niemantsverdriet, C.F.J. Flipse, A.M. van der Kraan, J.J. van Loef, *Appl. Surf. Sci.* **10,** 302 (1982).
12. H.F.J. van 't Blik and J.W. Niemantsverdriet, *Appl. Catal.* **10,** 155 (1984).
13. H.M. Gager and M.C. Hobson, *Catal. Rev. Sci. Eng. 11,* 117 (1975).
14. C.D. Wagner, W.M. Riggs, L.E. Davis, J.F. Moulder, and G.E. Muilenberg, *Handbook of X-ray Photoelectron Spectroscopy*, Perkin–Elmer, Eden Prairie, Minn (1978).
15. R.L. Mössbauer, M. Lengsfeld, W. von Lieres, W. Potzel, P. Teschner, F.E. Wagner and G. Kaindl, *Z. Naturforsch.* **26,** 343 (1971).
16. N.W. Hurst, S.J. Gentry, A. Jones and B.D. McNicol, *Catal. Rev. Sci. Eng.* **24**, 233 (1982).
17. J.W. Niemant sverdriet, A.M. van der Kraan, W.N. Delgass and M.A. Vannice, *J. Phys. Chem.* **89,** 67 (1985).
18. G.K. Wertheim, *Mössbauer Effect: Principles and Applications,* Academic Press, New York (1964).
19. G.A. Somorjai, *Chemistry in Two Dimensions: Surfaces,* p. 170, Cornell Univ. Press, Ithaca, NY (1981).
20. R. Dutartre, P. Bussiere, J.A. Dalmon, and G.A. Martin, *J. Catal* **59,** 382 (1979).
21. J.A. Dalmon, C. Mirodatos, P. Turlier and G.A. Martin, in *Spillover of Adsorbed Species, p. 169* ed. by G.M. Pajonk, S.J. Teichner and J.E. Germain, Elsevier *Amsterdam* (1983).
22. J.W. Niemantsverdriet, J.A.C. van Kaam, C.F.J. Flipse and A.M. van der Kraan, *J. Catal.* **96,** 58 (1985).
23. Y. Minai, T. Fukushima, M. Ichikawa and T. Tominaga, in *Industrial Applications of the Mössbauer Effect* ed. by. G.J. Long and J.G. Stevens, to be published by Plenum Publishing Company.
24. A.M. van der Kraan and J.W. Niemantsverdriet, in *Industrial Applications of the Mössbauer Effect* ed. by. G.J. Long and J.G. Stevens, to be published by Plenum Publishing Company.

SURFACE AND INTERFACE ANALYSIS, VOL. 9, 227–236 (1986)

XPS and AEM Studies of Carbon-supported HDS Catalysts

J.S. Brinen, L.F. Allard and F.P. Daly
American Cyanamid Company, 1937 West Main Street, Stamford, CT 06904, USA

A series of carbon-supported HDS catalysts has been characterized by XPS and high resolution analytical electron microscopy. the size, distribution and composition of clusters or particles of Co and Mo oxides on the support materials is shown to be related to variations in processing treatments. These results are discussed in terms of catalyst activity measurements and surface analytical information (XPS data).

INTRODUCTION

Commercial hydrodesulfurization (HDS) catalysts currently in use are alumina-supported, molybdenum-based, promoted by cobalt or nickel. In the hydrotreating of residual oils, these catalysts lose activity at a much faster rate than in gas oil processing as a consequence of the feedstock composition. Asphaltenes and higher molecular weight compounds promote coke formation while organometalics, primarily V, Ni, and Fe compounds, lead to deposition of metal sulfides. Although coked catalysts can be regenerated, catalyst deactivation due to metals deposition is irreversible. However, catalyst life can be extended by increasing the pore radius of the support material, thus reducing the rate of pore plugging due to coke and metals deposition.[1]

Carbon-supported HDS catalysts have gained interest in recent years because of ease of metals recovery by burning off the carbon support. Carbon support materials are generally considered in terms of low surface area with essentially no pores (graphitic type) or high surface area with small pores (Darco, product of ICI Americas, Inc). However, a class of carbons having wide pores with intermediate surface areas (WPC, manufactured by American Cyanamid Company) are available and are considered potential support materials for residual oil processing. A series of CoO—MoO_3 catalysts were prepared on these carbon support materials, and have been characterized by XPS, AEM, and by their hydrodesulfurization activity with respect to benzothiophene. Results of activity and XPS analyses have been reported previously,[2] however, the AEM measurements suggest an alternate explanation for the observed XPS results and catalytic behavior. The original results are briefly reviewed, and highlights of the AEM results are described in the present paper.

There is considerable evidence that Mo is highly dispersed on Al_2O_3 in discrete monolayer patches. This is attributed to the strong interaction between the Mo species and the anionic OH groups on the Al_2O_3 surface.[3] High resolution TEM studies generally show an absence of discrete crystallites and a major part of the Mo to be present in the dispersed crystallite-free phase.[4] Silica supports generally have no detectable anionic OH groups[3] and TEM measurements show a large amount of the Mo to be present in the form of Mo crystallites. From our previous experiences with C-supported Rh, Pd, and Pt compounds[5] we expected to observe discrete metallic crystallites, although smaller than in the case of the SiO_2 support. This was not the case, as indicated by the ultrastructural observations reported here.

EXPERIMENTAL

The C-supported CoO—MoO_3 catalysts were prepared by the incipient wetness technique using aqueous solutions of ammonium heptamolybdate and cobalt nitrate sequentially and in combination according to the procedures described in Table 1.

HDS activity was measured with benzothiphene at 270°C and 1 atm. Pressure. Prior to testing, the catalysts were presulfided with 10% H_2S/H_2 mixture at 400°C for 1 hour. The results are reported as relative weight activity (RWA) with American Cyanamid's commercial TRILOBE HDS-20 catalyst (5 wt % CoO-16 wt% MoO_3/Al_2O_3) assigned an RWA value of 100.

XPS measurements were made with an HP 5950A ESCA spectrometer using monochromatic A1 Kα radiation. The spectrometer was equipped with a Surface Sciences Laboratory detector, aperture, and data system. Quantitative surface compositions were determined using Wagner's[6] formalism, and the

Table 1. CoO–MoO_3/Carbon Preparation Procedure

Catalyst[a]	Carbon	Impregnation procedure	Drying conditions	Calcination conditions
A	Darco	One step	120°C, ambient	600°C, 1 hr, N_2
B	WPC	One step	120°C, ambient	600°C, 1 hr, N_2
C	WPC	One step	60°C, vacuum	600°C, 1 hr, N_2
D	WPC	Two step,	60°C, vacuum	250°C, 1hr, air
E	WPC	Two step, Mo first	60°C, vacuum	250°C 1 hr, air

[a]All catalysts contained 5% CoO and 16% MoO_3 by weight.

0142–2421/86/120227–10 $05.00

theoretical cross-sections calculated by Scofield.[7] Samples were examined as powders pressed onto double-sided adhesive tape.

High resolution TEM, STEM and electron diffraction analyses were made using a Philips 420T analytical electron microscope, operated at 120 kV with a LaB_6 emitter. Energy dispersive x-ray fluorescence spectra (EDS) were acquired using a Tracor-Northern 5500 spectrometer with 400 second acquisition times and a nominal 10 nm beam diameter. Samples for AEM analysis were prepared by two different methods. For crystallite size observations at high resolution, catalyst grains were crushed in alcohol and deposited on holey carbon films supported on Be grids. For quantitative EDS measurements and to determine more easily the overall distribution of particles within the bulk support material, microtomed thin sections were prepared. The catalyst powders were embedded in Epon resin (Shell Chemical Company). and 60 nm thick sections were sliced and collected on holey carbon films, again supported on Be grids.

The EDS measurements were made with the specimens held in a single-tilt specimen holder especially designed[8] to minimize 'systems peaks'[9,10] generated in the local specimen region by x-rays and electrons scattered from the point of analysis. With these specimens, 'hole counts'[10] were essentially nil, and did not contribute significant intensity to the acquired spectra. The only spectral contamination noted was a barely discernible Fe peak which was shown to originate from the Fe present in the Be specimen grid.

Quantitative Co and Mo concentrations were determined by the Cliff-Lorimer ratio method.[11] All particles measured were determined to satisfy the thin film criterion for use of this technique without regard to absorption and fluorescence effects. The 'k-factor'[11] for relating Co and Mo concentration ratios to the observed intensity ratios was determined by careful measurements on crushed grains of $CoMoO_4$ as a standard. The spectral intensitites were obtained using the Tracor–Norther metallurgical thin film (MTF) program. Only composition ratios are reported here.

High resolution electron micrographs were recorded at Scherzer focus[12] (95 nm underfocus for the particular objective lens employed in the 420T, having a nominal spherical aberration coefficient of 2 mm), although on crushed grain specimens in particular it was recognized that substantial differences in specimen height might be present in any given field of view. Only the central regions of any image used for focusing were subsequently photographically enlarged and subjected to analysis. An 80 micrometer objective lens aperture was used, giving a collection semiangle of about 15 mrad. These conditions have been shown[13] to give optimum contrast between heavy metal clusters and a light-element support. Clusters of heavy atoms on the order of 0.5 nm in diameter, containing fewer than 10 atoms are easily imaged in bright field with axial illumination under these operating conditions.

Table 2. Catalyst Characterization

Catalyst	Support	XPS[a] Co/C × 100	Mo/C × 100	Co/Mo	HDS activity[b] RWA
A	Darco	0.90	1.88	0.48	48
B	WPC	0.66	0.53	1.25	18
C	WPC	0.48	0.28	1.74	
D	WPC	0.35	0.85	0.41	
E	WPC	0.46	0.96	0.48	36

[a]The Co 2p lines, Mo 3d lines and the C 1s line were used for analysis.
[b]See text for explanation.

RESULTS AND DISCUSSION

Catalysts A and B represent catalysts which were prepared by identical procedures but on carbon supports with different properties. The catalyst prepared on Darco had a hydrodesulfurization RWA value of 18. At first glance, this activity difference might be attributed to surface area differences since the WPC has a surface area of 331 m^2 gm^{-1}, 39% of that of the Darco carbon. However, XPS analyses of the surface of the two catalysts showed the Co/Mo atom ratio of the catalyst supported on Darco was also 39% of the WPC catalyst. Chemical analyses by bulk x-ray fluorescence revealed comparable Co/Mo line intensities for the two catalysts, suggesting that the difference in the observed surface ratios was not a consequence of different bulk ratios. The results of the XPS analyses, given in Table 2, reveal that the higher surface Co/Mo atom ratio for catalyst B is primarily due to the lower intensity of the Mo 3d lines as demonstrated by the lower Mo/C ratio, 0.005 versus 0.019, for catalyst A. These results suggested that the adsorption behavior of Co and Mo were quite different for the two carbons. Note that the Co/C ratios are also different but to a lesser degree. The implication is that for WPC supports, Mo diffuses into the pores to a greater extent than Co when calcined at 600°C. The interaction between Co and Mo is reduced and catalytic activity is reduced.[14,15] By modifying the preparation methodology, most importantly reducing the calcination temperature, the levels of Mo and Co on the external surface can be manipulated to approach both the Co/Mo ratio and the hydrodesulfurization activity of the Darco-supported catalyst. These results are given in Table 2.

A common factor affecting catalyst activity and surface atomic ratios determined by XPS is particle size and dispersion of the active catalytic elements. The conclusions drawn regarding adsoprtion behavior differences of the two carbon supports assumed similar dispersion of Co and Mo species. This question was further addressed by high resolution electron microscopy, which has the potential for providing information regarding both the size and distribution of Mo clusters on catalytic supports. Delannay[16] has recently reviewed the literature and has concluded that for alumina-supported HDS catalysts, high resolution electron microscopy has not been able to detect the presence of Mo in these catalyst precursors (oxidic state) in spite of the high level of Mo and the large atomic weight difference between Mo and Al. However, carbon-supported HDS catalysts do not appear to have been studied to any degree.

Figure 1 shows a comparison of the high resolution TEM images of crushed grain specimens on Darco (a) and of catalyst A which contains Co and Mo on Darco (b). The metal-loaded specimen shows only a slight contrast difference from the blank material, perhaps due to the Co–Mo present on the surface. In contrast, Fig. 2 shows a similar micrograph of Ni–Mo on Al_2O_3, with atom clusters of 0.5 nm diameter easily imaged. Such Co–Mo cluster sizes are not evident to the same degree on the carbon supports. Figures 1(c) and 1(d) show the EDS spectra acquired from the blank material and the loaded specimen, respectively. Mo and Co are clearly detected in the loaded specimen, and the Co/Mo ratio indicated by Fig. 1(d) was essentially constant everywhere in this specimen.

This specimen behaves in a similar manner to the alumina-supported catalysts in that the active catalytic elements are distributed uniformly over the surface, but with cluster sizes too small to clearly detect even under the optimum imaging conditions employed here. The high dispersion and small particle size provide an ideal situation for quantitative XPS analysis. The

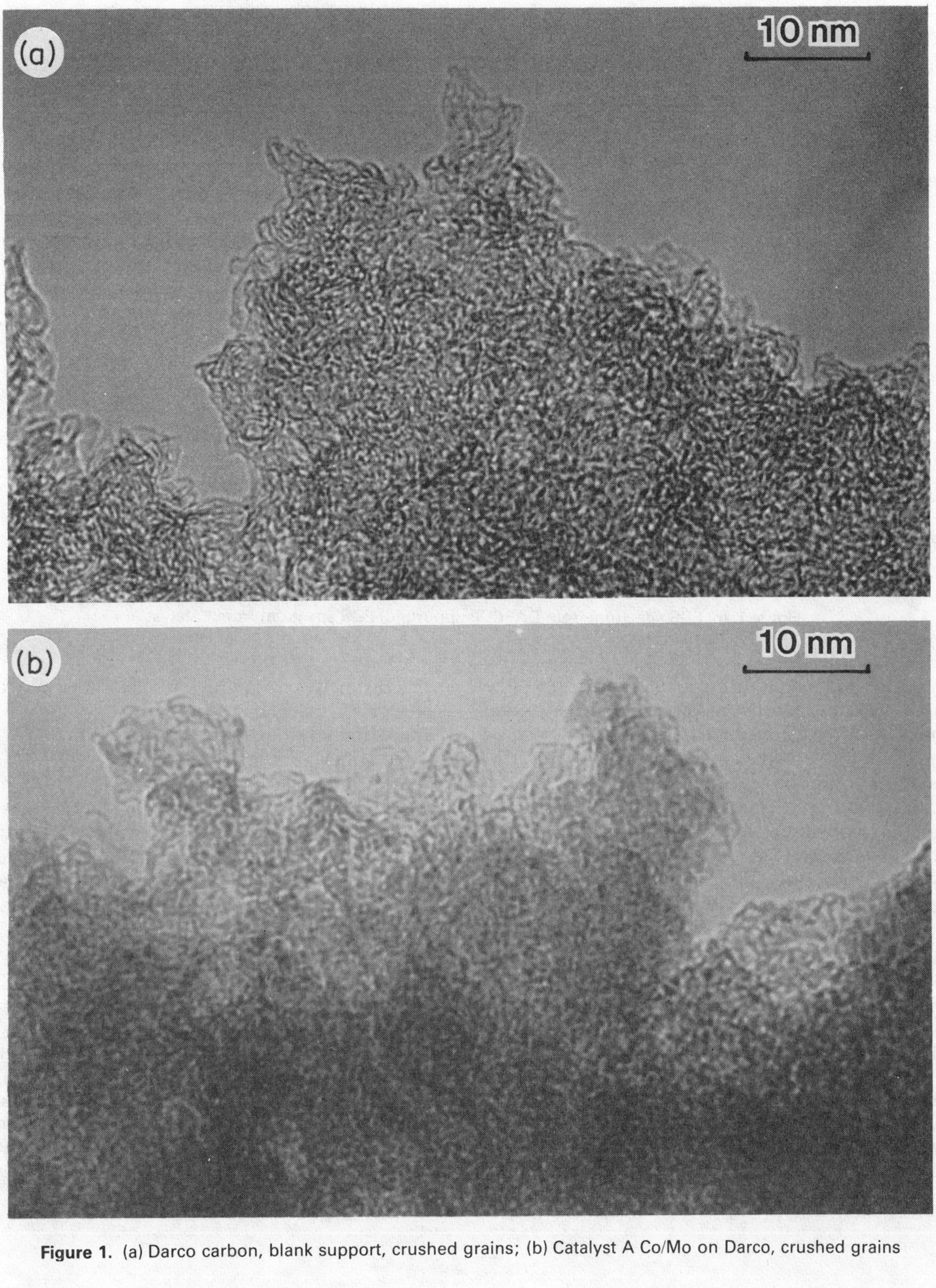

Figure 1. (a) Darco carbon, blank support, crushed grains; (b) Catalyst A Co/Mo on Darco, crushed grains

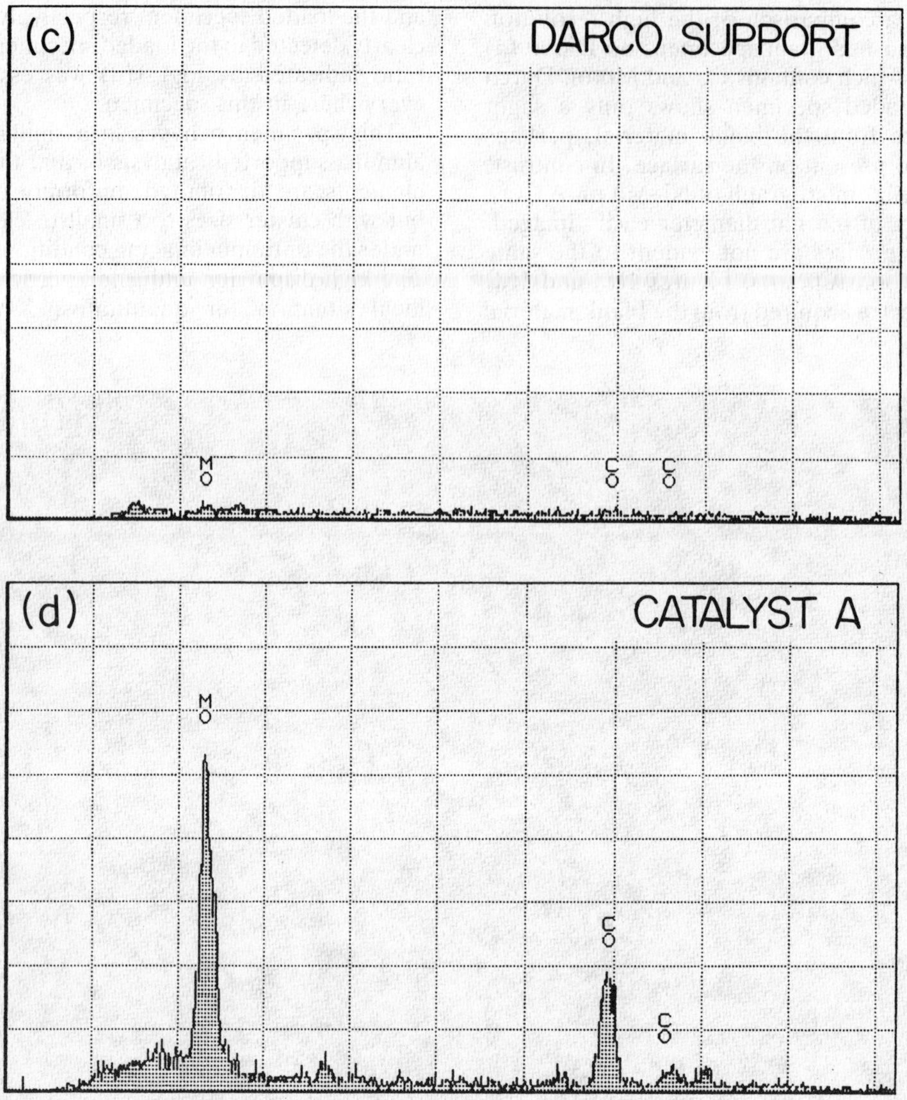

Figure 1. (c) EDS spectrum from DARCO support, (d) EDS spectrum from Catalyst A.

system behaves like a uniform 'solution' as opposed to a layered specimen, resulting in unexpectedly good agreement with the bulk chemical analysis.[17]

Figure 3 shows a comparison of high resolution TEM images of crushed grain specimens of the WPC support and catalyst B. This catalyst was prepared by the identical method used for the catalyst on Darco described above. It is evident from the micrograph that discrete particles are observed for this catalyst preparation. The particle indicated by the arrow is about 5 nm by 15 nm in size. Particles of this size and smaller are observed uniformly distributed throughout the catalyst. Particles were consistently shown to be plate-shaped. The particle arrowed in Fig. 3 is oriented with the beam parallel to the plate surface, and shows distinct 0.4 nm lattice fringes, indicating that it is a single crystal. Fig. 4 shows a bright field TEM image from a microtomed section of Catalyst B. EDS spectra, obtained from the particle arrowed and from a point 20 nm away from the particle, are shown in Fig. 5, to illustrate the typical result obtained from this specimen. The distinctly crystalline particles (note lattice fringes) on this specimen contain Co and Mo with approximately the same Co/Mo ratio (0.5). At point B, 20 nm away from particle A, there is little evidence for the presence of the catalytically active metals. It thus appears that even though identical preparation conditions were employed, the surface chemistry of the WPC and Darco carbon supports are different, resulting in different metal–support interactions and, hence, catalysts with widely different dispersions and crystallite sizes.

In contrast to catalyst B, catalyst C contains much larger crystalline particles, although a wide range of particle sizes was observed. Fig. 6 shows, for comparison, a microtomed section of this catalyst. The particles in catalyst C are on the order of 100 nm in size, an order of magnitude larger than those typically observed in Catalyst B. There are also numerous small particles present, arrowed in the figure, which are of the same general dimension as those in catalyst B. Figure 7 shows EDS spectra of some of the crystalline particles. Particle 1 contains substantial levels of Co and Mo with a ratio of Co/Mo of about 0.9, while particle 2 has a Co/Mo ratio of <0.1. All the particles examined in this specimen had Co/Mo ratios at these two extremes. Regions of the catalyst for which crystallites were not observed did not contain appreciable amounts of Co or Mo. As in catalyst B, the larger crystallites on catalyst C were distinctly plate-shaped, and usually were single crystals. An occasional particle was observed to consist of more than one grain, with each grain often differing appreciably in Co and Mo composition.

Selected area electron diffraction patterns obtained from large areas of catalyst C typically showed spotty diffraction rings which indexed most consistently to the MoO_2 structure (JCPDS #32–671). None of the other catalyst specimens showed measurable intensity in selected area patterns. However, convergent beam electron diffraction patterns were obtained from numerous particles in both catalyst B and catalyst C. The results of these analyses are beyond the scope of the present report, and will be reported in a subsequent paper[18] detailing the AEM results on the present catalysts as well as catalysts which have been sulfided.

Figure 8 shows TEM images of crushed grains of catalysts D and E, where E has been prepared using the two-step impregnation procedure. The two specimens show very similar structures, although the contrast features in catalyst D suggest the presence of small (<1 nm) heavy metal clusters. EDS spectra show Co and Mo to be distributed on the carbon surface with a uniform Co/Mo ratio in areas observed for both specimens.

All of these specimens were highly stable in the electron beam. No evidence for contamination build-up on the specimen, the agglomeration of Co or Mo into distinct clusters, or degradation of the larger particles in the beam was ever observed. In fact, the lattice image of the particle shown in Fig. 4 was recorded after the particle had been subjected to a 400 second microanalysis with the beam in POINT mode on the particle.

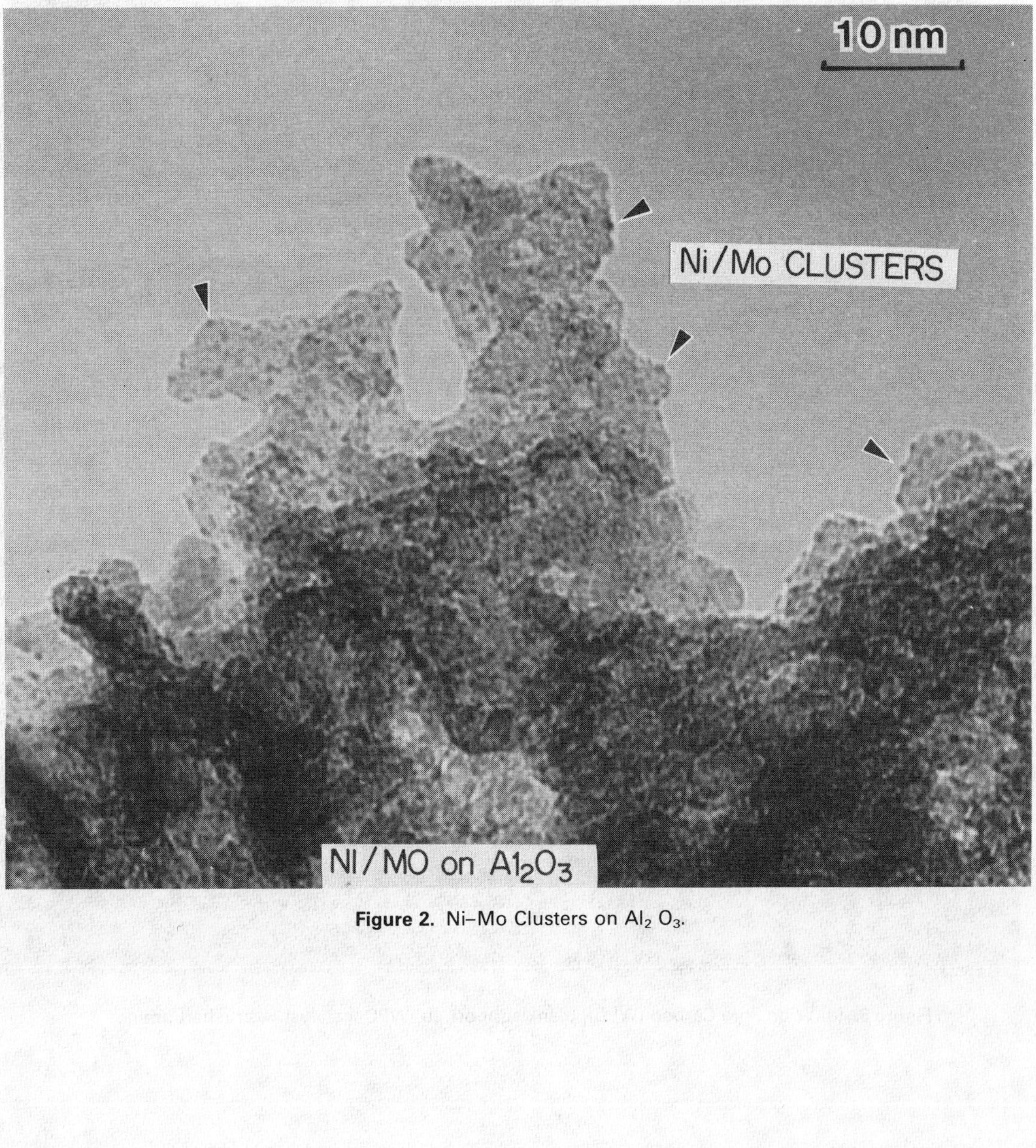

Figure 2. Ni–Mo Clusters on Al_2O_3.

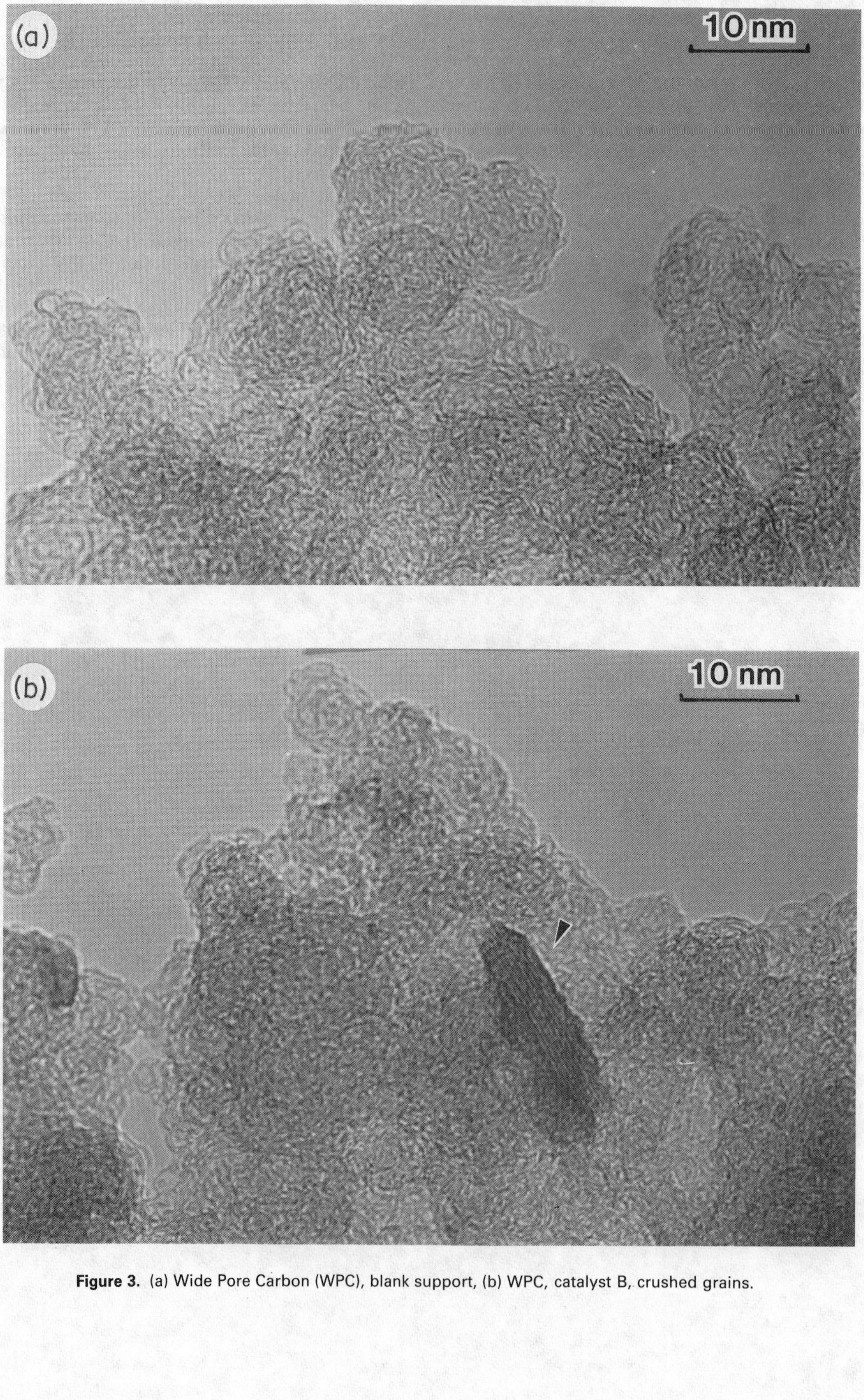

Figure 3. (a) Wide Pore Carbon (WPC), blank support, (b) WPC, catalyst B, crushed grains.

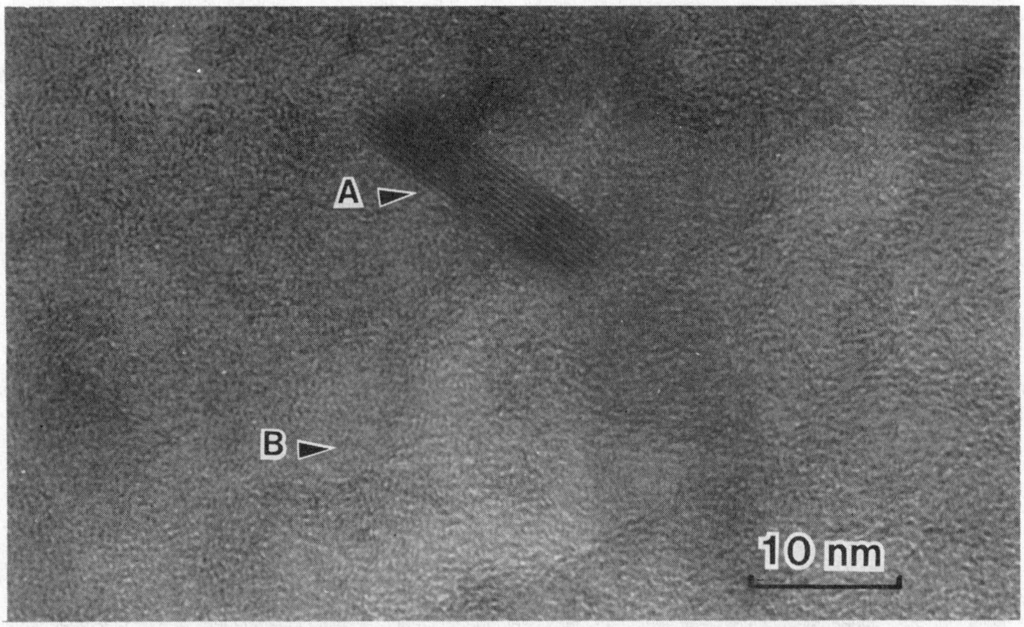

Figure 4. Co–Mo particle on microtomed thin section of catalyst B.

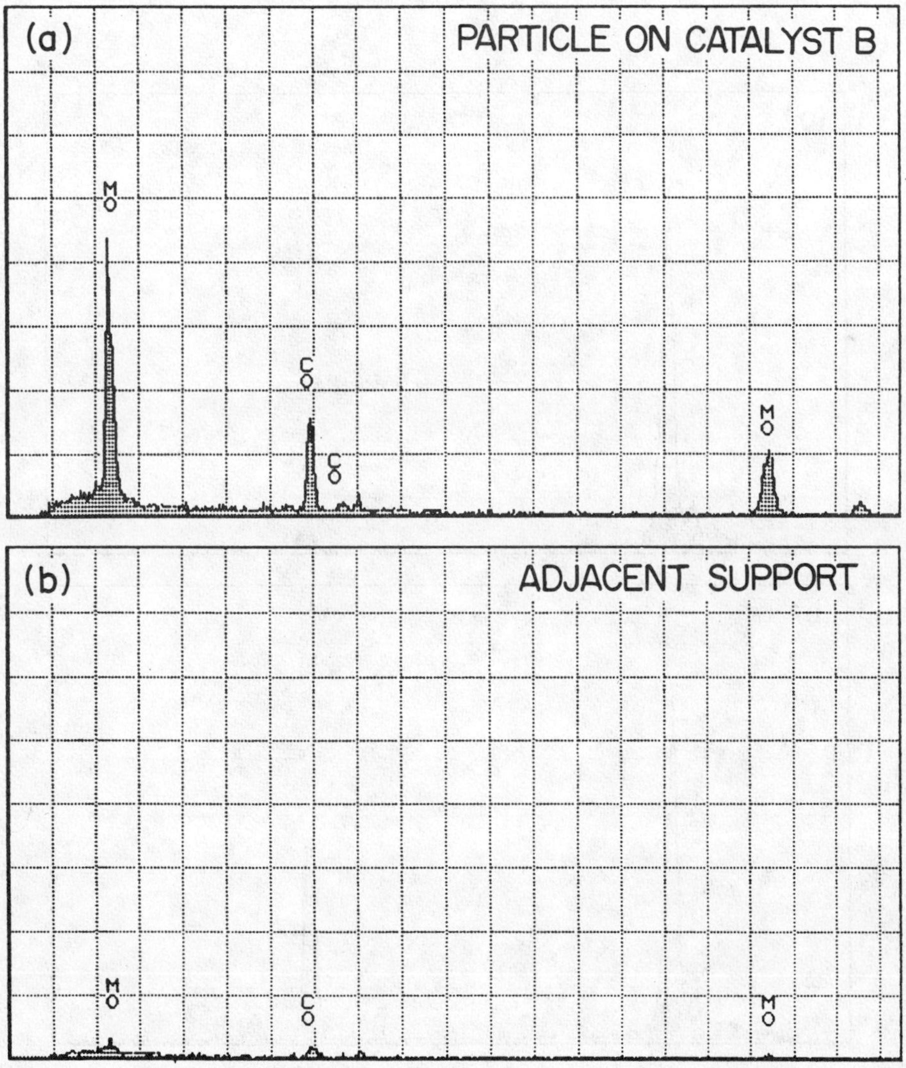

Figure 5. EDS spectra showing typical range of particle compositions on catalyst B, (a) particle on catalyst B, (b) adjacent support.

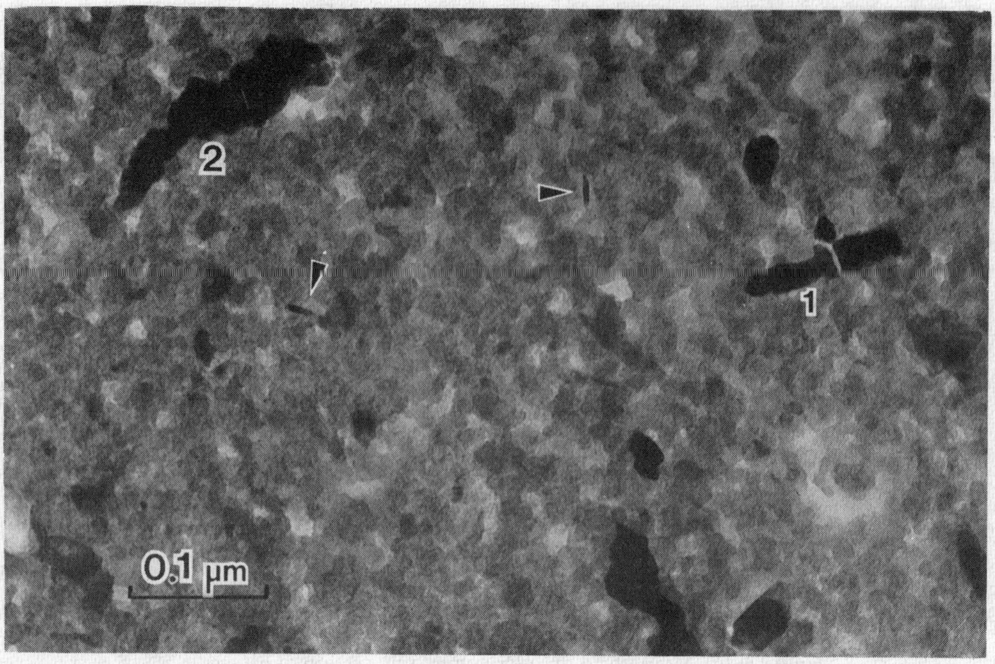

Figure 6. Co–Mo particles on microtomed thin section of catalyst C. Note magnification change relative to Fig. 4.

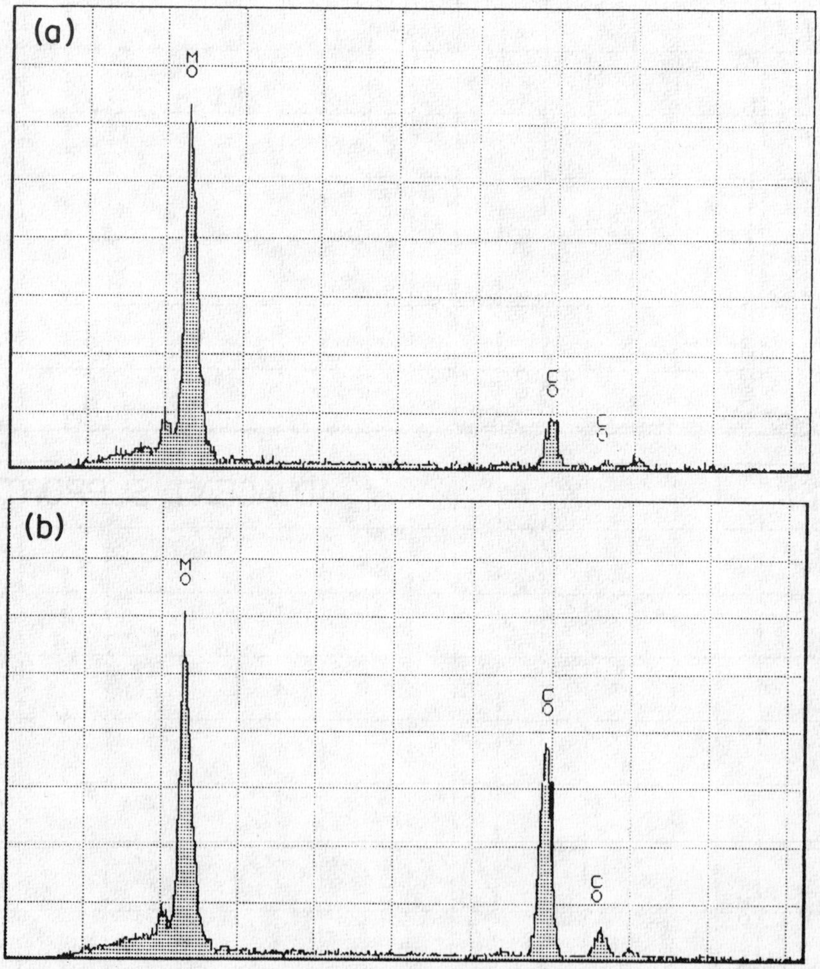

Figure 7. EDS spectra showing typical variation in composition of particles on catalyst C, (a) particle 1, (b) particle 2.

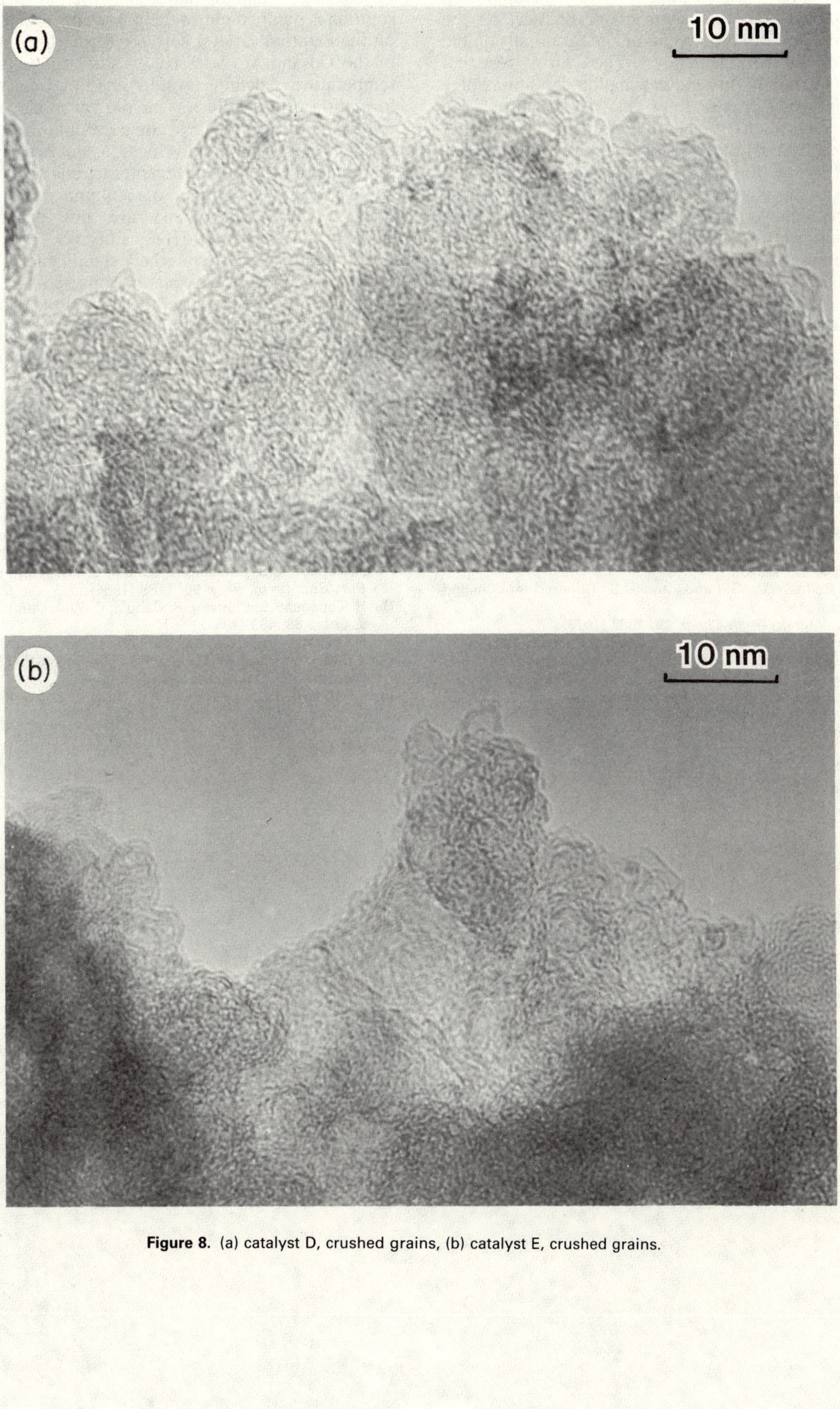

Figure 8. (a) catalyst D, crushed grains, (b) catalyst E, crushed grains.

CONCLUSIONS

High resolution TEM has clearly demonstrated that metal–support interactions and preparation methodologies can have a pronounced effect on the structure of the catalyst precursor and eventually on the activity of the working catalyst. The data suggest that both drying conditions and calcination conditions play significant roles in the final catalyst structure. The difference between drying under ambient conditions at 120°C (catalyst B) and drying under vacuum at 60°C (catalyst c) was the creation of large crystallites with widely varying compositions, a heterogeneous poorly dispersed catalyst. However, if the same drying conditions were employed with a lower calcining temperature, virtually no sintering occurred (catalyst D). (The difference between air drying and N_2 drying was not felt to be critical here.) Drying at 120°C would tend to evaporate the solvent and leave the solute, Co and Mo ions, in the catalyst pores and could result in partial oxidation of the respective salts. Subsequent high temperature calcination would then result in the formation of oxides of Co and Mo with low probability of sintering. When the material is dried in vacuum, the solution is pumped out of the pores and is redistributed on the external catalyst surface. Little or no oxidation of the Co and Mo salts results, and subsequent high temperature calcining results in sintering with the formation of large clusters. If the calcining conditions are milder, the salts are not as mobile and, at low enough temperature, sintering does not occur (catalyst D). Because of the stronger interactions of the surface anionic hydroxyl groups of alumina with the Mo species, these phenomena are not observed on conventional Mo/Al_2O_3 HDS catalysts.

To really ascertain the significance of the measurements on the precursors reported here, similar measurements on the sulfided catalysts should be made. These experiments are in progress and will be reported elsewhere.[18]

REFERENCES

1. E. Newson, *Ind. Engl. Chem. Proc. Div.* **14,** 27 (1985).
2. F.P. Daly, J.S. Brinen, and J.L. Schmit, *App. Cat.* **11,** 161 (1984).
3. G. Muralidar, F.E. Massoth, and J. Shabtai, *JACS Div. Petrol. Chem Prep.* **27(3),** 722 (1984).
4. F. Delannay, P. Garjarde and P. Grange, *J. Microsc. Spect. Elect.* **3,** 411 (1978).
5. J.A. Reffner, C. Garland, and J.S. Brinen (unpublished work).
6. C. D. Wagner, *Anal. Chem.* **49,** 1982 (1977).
7. J.H. Scofield, *J. Elect. Spectros, and Rel. Phen.* **8,** 129 (1976).
8. K.G. Ryan, N.E. Flowers, and M.R. Presland, *J. Micros.* **134(3),** 281 (1984).
9. J. Bentley, N.J. Zaluzec, E.A. Kenick, and R.W. Carpenter, *SEM* Chicago Press (1979).
10. L.F. Allard and D. F. Blake, *Microbeam Analysis 1984,* San Francisco Press.
11. G. Cliff and G.W. Lorimer, *J. Micros.* **103,** 203 (1975).
12. J.C.H. Spence, *Experimental High Resolution Electron Microscopy,* Chap. 6, Oxford University Press (1981).
13. C.R. Hall and R.L. Hines, *Phil. Mag.* **21,** 1175 (1970).
14. H. Topsoe, B.S. Clausen, R. Candia, C. Wivel and S. Morup, *Bull. Soc. Chim. Belg.* **90,** 1189 (1981).
15. H. Topsoe, B.S. Clausen, R. Candia, C. Wivel, and S. Morup, *J. Catal.* **68,** 433 (1981).
16. F. Dellanay *Appl. Cat.* **16,** 135 (1985).
17. J.S. Brinen, *Applied Surface Analysis, ASTM STP699,* T.L. Barr and L.E. Davis, eds., 24–41 (1980).
18. L.F. Allard, J.S. Brinen, and F.P. Daly, to be published.

SURFACE AND INTERFACE ANALYSIS, VOL 9. 237–241 (1986)

Electronic State and Catalytic Activity of SiO_2 Doped γ–Al_2O_3

I. Bertóti, G. Mink, T. Székely, G. Varsányi[†], F. Réti
Research Laboratory for Inorganic Chemistry, Hungarian Academy of Sciences, H–1502 Budapest, Pf. 132, Hungary

Silica–alumina catalysts have been applied for a long time in cracking and reforming reactions. Lewis type surface sites are generally accepted to serve as active centers. In this respect, surface properties of these materials are of main concern. In the present work SiO_2 doped γ–Al_2O_3 with less than a monolayer coverage were prepared by the reaction of $SiCl_4$ (g) with γ–Al_2O_3 at 820–1120 K. Catalytic activity of the samples was tested by a model reaction of *n*–hexane cracking. Surface composition and electronic state were evaluated by XPS investigations performed on a KRATOS XSAM 800 instrument. Catalytic activity (conversion) varies remarkably with increasing Si content showing a sharp maximum at low Si coverage. The observed shift of Al 2p lines with increasing Si content is interpreted as an enhancement of the acceptor or acidic character of the surface Al atoms. The Auger parameter for Si(α'_{Si}), (composed by addition of Si 2p *BE* and Si (KLL) *KE* values) passes also through a maximum at the same Si-content, as does the conversion curve. A close correlation of the Auger parameter and the catalytic activity is established for samples with (Si/Al)$\lesssim$7%. This leads to a conclusion, that the rise in extra-atomic relaxation energy may serve as an additional indication of enhanced catalytic activity.

INTRODUCTION

Silica aluminas, owing to their remarkable properties, have been the subject of extensive investigations for a long time. One of the main fields of their application falls in the catalytic cracking and reforming[1–3], and, in connection with this, attention has turned lately towards the surface properties of this system.[4,5]

Silica–aluminas for catalytic purposes are usually prepared using various solution processes,[3] with preferential formation of the material in amorphous phase.

Gas–solid reactions, however, can also be used for the preparation of mixed oxide phases and among others, they are especially suitable for surface modification of solids.[6,7]

The present work deals with the surface properties and catalytic activity of a series of silica-modified γ–alumina samples prepared by high temperature interaction of $SiCl_4$ vapour with γ–Al_2O_3 surfaces.

EXPERIMENTAL

Materials

In the experiments high purity γ–alumina, produced by Ketjen, The Netherlands, was used: CK–300 type, with impurity level of 90 ppm SiO_2, 100 ppm Fe_2O_3, 10 ppm Na_2O as specified by the manufacturer. The samples were preheated at 1120 K for 30 minutes in Ar flow in order to achieve a standard state with low residual OH content. The specific surface area of the preheated samples was 160 $m^2.g^{-1}$ determined by N_2 adsorption at 77 K. The reagent grade $SiCl_4$ (Schuchard, München) was introduced by a 3N purity Ar steam from a bubbler. The *n*–hexane used for the catalytic test was also of reagent grade (Reanal, Hungary).

Methods

Thermogravimetry was used to follow the sample preparation process. For this purpose a specially designed quartz reactor attached to a Mettler semimicro recording balance was used as described earlier.[8] Atomic absorption spectroscopy, using a Perkin–Elmer AAS 306 instrument was applied for determining the bulk composition after Na_2O_2-flux treatment and dissolution of the samples. GC analysis was applied to follow the catalytic test reactions, using a Chromatron 18.3 type apparatus with a 2 m Porapak R column (ϕ = 4 mm), swept by 30 cm^3 min^{-1} H_2 in temperature-programmed mode.

X-ray photoelectron spectroscopy was used for the analysis of the surface chemical composition and for the evaluation of the surface electronic structure. Spectra were recorded with a KRATOS XSAM800 spectrometer using mainly Mg $K\alpha_{1,2}$ radiation. The instrument was calibrated against the Au $4f_{7/2}$ line fixed at 84.0 eV. The linearity of the energy scale was checked by the dual Al/Mg anode method using the Ag $3d_{5/2}$ line (368.2 eV).[9] The x-ray gun was usually operated at 12 kV and 15 mA. The usual pressure in the sample analysis chamber (SAC) was ca.$3\cdot10^{-7}$ Pa. Spectra were recorded in the fixed retardation ratio (FRR) mode. Data acquisition and data processing were performed using the KRATOS DS300 data system based on a DEC PDP 11/23 LSI processor. Spectra were referenced to the Cls line of adventitious carbon fixed *BE* = 284.6 eV.

[†]Dpt. for Physical Chemistry, Technical University, H-1111 Budapest, Hungary

0142–2421/86/120237–05 $05.00

RESULTS AND DISCUSSION

Preparation of the samples

The interaction of γ-alumina with gaseous $SiCl_4$ was followed by thermogravimetry.[6] The samples were treated in $SiCl_4$/Ar flow under isothermal conditions, in the temperature range of 820–1120 K. The partial pressure of $SiCl_4$ varied between 2 and 20 kPa. A fast initial mass gain is observed, what is followed by a relatively slow mass loss, resulting in a maximum in the TG curves (Fig. 1) The initial mass gain is attributed to the chemisorption of $SiCl_4$ and a stepwise chlorine–oxygen exchange is supposed around the silicon. The mass loss is caused by the volatilization of $AlCl_3$ due to the rearrangement of surface species. The Si content of the sample is only slightly increasing by time, because only the volatilization of $AlCl_3$ releases new reaction sites. The resulting weight gain is attributed entirely to the residual surface chlorine.

Surface and bulk elemental composition

Surface composition of the $SiO_2/A1_2O_3$ samples has been determined by the area intensities (I) of O 1s, C 1s, Cl 2p, Si 2p and Al 2p lines, using the procedure and experimental cross sections of Evans *et al*,[19] as follows;

$$I_i = N_i.\sigma_i{\cdot}E_i^{0.5}{\cdot}E_i.\exp(-\ 14.3{\cdot}E_i^{-0.5})$$

where: N_i is the number of atoms in unit volume; σ_i is the relative differential cross section; E_i is the kinetic energy of the electrons; $E_i^{0.5}$ stands for the escape depth, and E_i stands for the transmission function of the instrument, run in FRR mode; $\exp(-14.3 \cdot E_i^{-05})$: this term accounts for the correction of surface carbon contaminants.

Bulk vs. surface Si/(Si+Al) atomic ratio plot (Fig. 2) suggests, that the Si built into the sample is situated either on the surface or in layers adjacent to the surface. This conclusion has been deduced from a model where four kinds of the distribution of Si were supposed.[10] For these calculations cubic particles with random angular distribution, having an average edge of 10 nm were taken into consideration, as a realistic value for the given γ–$A1_2O_3$ with specific surface area of 160 m^2g^{-1} and mean pore size of 5 nm. λ= 2 nm was taken for the escape depth of the Al 2p and Si 2p electrons. For the case when Si atoms are positioned only in the uppermost layer of the surface a slope of 0.64 is obtained. The best linear fit for the experimental data was obtained with a slope of 0.66 suggesting that in our case the Si atoms are situated in the uppermost layer of the surface.

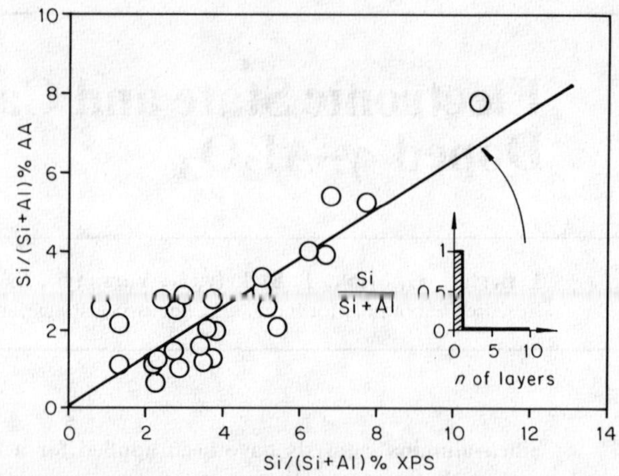

Figure 2. Bulk vs. surface composition of SiO_2/Al_2O_3 samples. Straight line represents the model case, when SiO_2 deposited is located in the uppermost surface layer (see insert).

The relatively high intensity of the O 1s and Al 2p lines offered a problem-free determination of their quantities with the usual accuracy ($\leqslant$ 5 rel%), small quantities of chlorine and silicon, however, can be determined only with greater error. Accurate determination of the Si 2p signal is also affected by the fact, that it coincides in energy with the loss-peak of the Al 2p signal, as demonstrated in Fig. 3. Background substraction, performed arbitrarily as to be linear, contains a great deal of intuitive elements. Substraction of the loss peak from an untreated γ–Al_2O_3 sample is also not satisfactory because of the fact that loss peaks are generally affected by surface treatments and contaminants.[18]

As far as can be detected, there are no other elements beside O, Al, C, Cl and Si, normalization of these can be justified, and elemental composition can

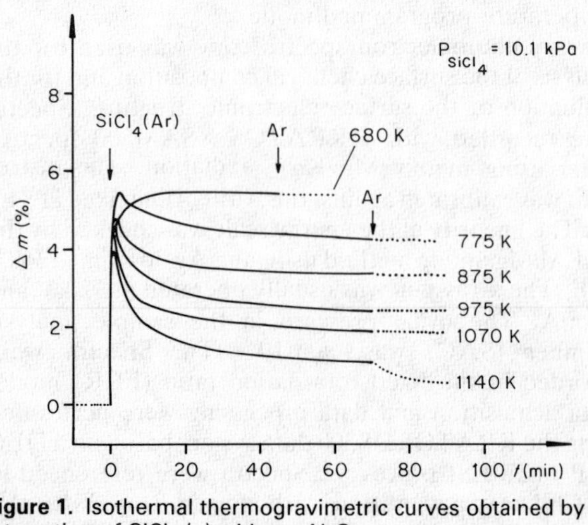

Figure 1. Isothermal thermogravimetric curves obtained by the interaction of $SiCl_4$ (g) with $\gamma-Al_2O_3$.

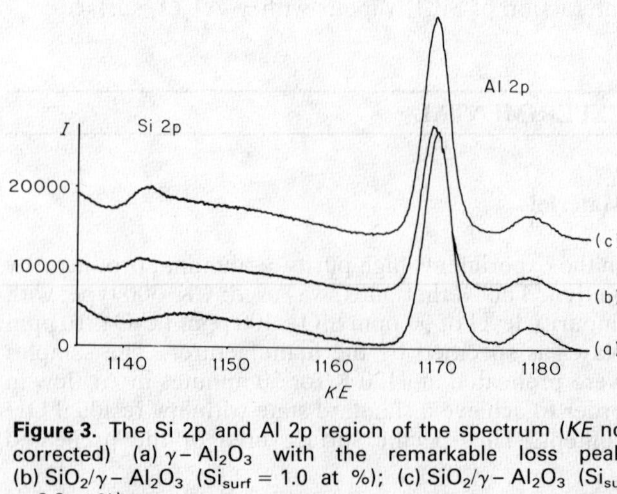

Figure 3. The Si 2p and Al 2p region of the spectrum (*KE* not corrected) (a) $\gamma-Al_2O_3$ with the remarkable loss peak; (b) $SiO_2/\gamma-Al_2O_3$ (Si_{surf} = 1.0 at %); (c) $SiO_2/\gamma-Al_2O_3$ (Si_{surf} = 2.2 at %);

Table 1. Surface composition of some representative samples

Run	Sample	Preparation conditions T K	*t* min	atm.	P kPa	Surface elemental composition (at %) O	C	Cl	Si	Al	(Si/Al) %
QRF	γ–Al_2O_3	850 *in situ*	30	vac.		48.5	16.9	0.1	–	34.5	–
QTE	γ–Al_2O_3+Cl_2	870	45	Cl_2	27	53.3	6.3	2.2	–	38.2	–
TG2089	SiO_2/Al_2O_3	1020	2	$SiCl_4$	2.5	54.3	5.8	2.0	0.5	37.4	1.3
TG2084	SiO_2/Al_2O_3	1070	5	$SiCl_4$	2.5	53.6	7.1	2.3	1.0	36.0	2.8
TG1767	SiO_2/Al_2O_3	820	10	$SiCl_4$	20	55.1	5.3	2.5	1.3	35.8	3.6
TG1892	SiO_2/Al_2O_3	1020	10	$SiCl_4$	10	53.1	11.7	0.9	2.6	31.7	8.3
TG1739	SiO_2/Al_2O_3	920	30	$SiCl_4$	20	54.1	12.5	1.8	3.3_5	28.2	12.0

be given in atomic percents. Some characteristic sample-compositions with low, medium and high silicon-content are given in Table 1.

Looking at the data from more than 20 samples analysed, the amount of chlorine ranges between 1–3 at% and its values are related to the amount of Si. The O/Al ratios are close to 1.5, the stoichiometric composition for alumina. Carbon is always present on the surface of the samples in an amount of 5–17 at%. At this moment it must be mentioned, that C, Cl and Si contents, because of their nonhomogeneous distribution, are a little overestimated.

Catalytic activity of the samples

The catalytic activity of the prepared SiO_2/Al_2O_3 samples has been tested by the reaction of cracking of *n*–hexane in H_2, by pulse technique. Reactions were performed at 820 K, and product fractions were analysed by GC. The C_1, C_2, C_3, C_4, C_5 fractions cover isomers as well as unsaturated hydrocarbons. In the products the C_3 fraction is dominant ($\approx$40%), the C_2 and C_4 fractions represent about 30% each, while the C_1 and C_5 are less than 5%.

Conversions on various samples referenced to the value observed for γ–Al_2O_3 are shown in Fig. 4. The curve goes through a maximum at relatively low Si–content.

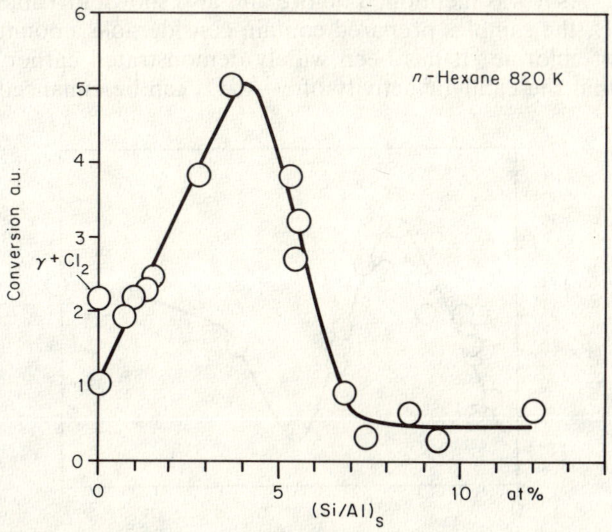

Figure 4. Conversion vs. surface composition curve for the samples.

It is generally accepted that the catalytic activity depends on the number and strength of acidic centers. The amount of acidic centers, in our samples titrated by pyridine adsorption at 470 K, showed a trend similar to that for the catalytic activity, however, the change is less pronounced: a fourfold increase of the conversion is accompanied by a 30% increase of pyridine adsorption.

An increase of surface acidity with increasing SiO_2 content was observed by Murayama and Meguro[5] and the IR data of Defossé *et al.*[4] for adsorbed pyridine also suggest a maximum shape at low SiO_2 concentration. However, the concentration range near to the Al_2O_3, in those works was not explored in such a detail, as it has been done in the present work.

Besides, the preparation conditions were different, so a close comparison of those data with ours is not entirely justified.

Nevertheless, the observed correlation between the catalytic activity and the pyridine chemisorption for our samples shows that acidic centers play a dominant role in the cracking reaction.

Surface electronic structure

In order to characterize the SiO_2/Al_2O_3 samples, beside the above mentioned O 1s, C 1s, Cl 2p, Si 2p and Al 2p lines the O(KVV) region, and the Al($KL_{23}L_{23}$), Si($KL_{23}L_{23}$) lines have also been registered with 0.1 eV steps, using for the latter the bremsstrahlung component of the x-ray radiation.

The change of the *BE* values with Si-content for Si 2p and Al 2p lines is demonstrated in Fig. 5. With increasing Si-content a considerable ($\approx$ 0.8eV) shift in the position of Si 2p signal, and a very similar trend with the Al 2p line can be seen. When Si is built onto the surface of Al_2O_3, the most favourable environment — the tetrahedral coordination with strong covalent bonds to neighbouring oxygens — will not be satisfied, partly because the neighbouring oxygens are bonded to Al atoms. On this basis one can assume that Si atoms might well be in a relatively electron–rich environment, possessing lower *BE* comparing to *BE* values of 103.0 –103.3 eV for various silicas.

On the other hand, the above considerations can be applied also to explain the change of *BE* values of Al 2p lines; the formation of covalent Si–O bonds will lessen the electron density around the Al atoms. This is also predicted by some CNDO–2 calculations, performed for a small Al_2O_3–SiO_2 cluster.[11]

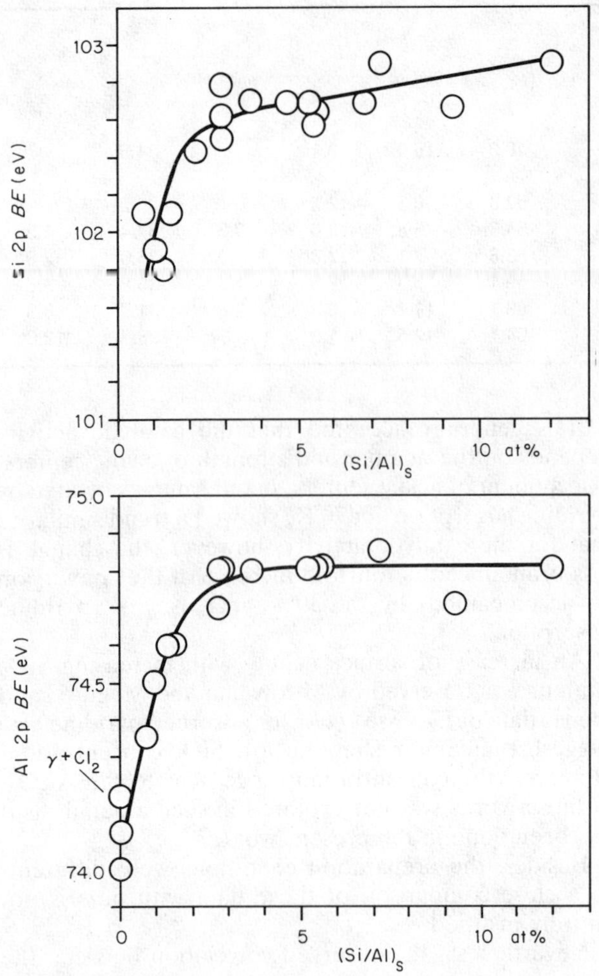

Figure 5. Change of Si 2p and Al 2p *BE* values with composition

Theoretical calculations has been recently performed on γ–Al_2O_3 in order to elucidate the nature of the Lewis acid sites.[12] It has been shown that the valence band consists mainly of the highest occupied oxygen orbitals, the conduction band is composed principally of Al orbitals and that both tetrahedral (Al_{tetr}) and octahedral (Al_{oct}) aluminium result in a level in the forbidden gap, however only the deeper-lying (Lewis-type) acceptor levels originating from Al_{tetr}) atoms seem to be responsible for catalytic activity.[12]

A level detected, and attributed to defects, lying at 1.7 eV below the conduction band in the wide (9.4 eV) forbidden gap γ–Al_2O_3 or amorphous alumina[13] might also be involved in the explanation of our data, as far as, introduction of Si into the lattice of γ-Al_2O_3 can also be considered as a defect. The introduction of a large amount of defect levels would lead also to the pinning up of the Fermi level, or bending down both the conduction and valence bands, as well as inner levels at the interface.[14] An obvious consistancy can also be reached on this basis with the observed shifts of Al 2p and Si 2p lines.

In order to clarify the chemical state of Si in our samples, we composed the Auger parameter (α'_{Si}) by adding Si 2p *BE* values to the Si ($KL_{23}L_{23}$) *KE* values, as proposed by Wagner.[15,16] The data obtained are shown in Fig. 6. The curve goes through a maximum in the same concentration range as did the conversion curve, and the two curves show remarkable correspondence.[17]

As it is stated, the change in the parameter is a measure of the change in the extra-atomic relaxation or screening of the final state ion in the Auger transition by electrons from neighbouring atoms or by conduction electrons.

All the above said about the defected, silica–doped γ–Al_2O_3 surface is in line with this theory, and with the observed, rising part of the α' vs. (Si/Al) curve in Fig. 6.

The descending part of the curve can be interpreted by considering a drastic rearrangement of the structure, with the formation of a compact silicate type network, where Al atoms are preferentially situated in octahedral positions. A clear indication of this tendency is the steep rise of α'_{Si} above Si/Al>6% since, according to Wagner, α'_{Si} is higher in Al_2O_3 with octahedral Al–environment.[15]

In the early study[10] we have developed a model, based on the nonparallel bending of the valence and the conduction bands with growing Si content. As a consequence, the forbidden gap variation goes through a minimum, serving with a straightforward explanation for the observed maximum in the change of Auger parameter. At present we believe, however, that surface states with deep-lying levels in the forbidden gap, causing band-bending variations, should be involved into the explanation of the observed effect.

Recently Bicker and coworkers have published a similar observation, on SiO_2 samples doped with small amounts of alumina, that represent the other end of the concentration scale of Si in the SiO_2–Al_2O_3 system.[18] They also observed higher values of Al-Auger parameter for catalysts possessing higher activity.

Along with our earlier observation,[10,17] this leads to a conclusion that the presence of mobile electrons responsible for higher Auger parameter plays also an important role in the catalytic reactions, facilitating probably the desorption step of the chemisorbed carbocation. Thus the values of Auger parameter may provide an additional indication of the catalytic activity.

As it was mentioned before and also shown in Table 1., the samples prepared contain considerable amount of chlorine. It has been widely demonstrated earlier[3] that the catalytic activity of γ–Al_2O_3 can be enhanced

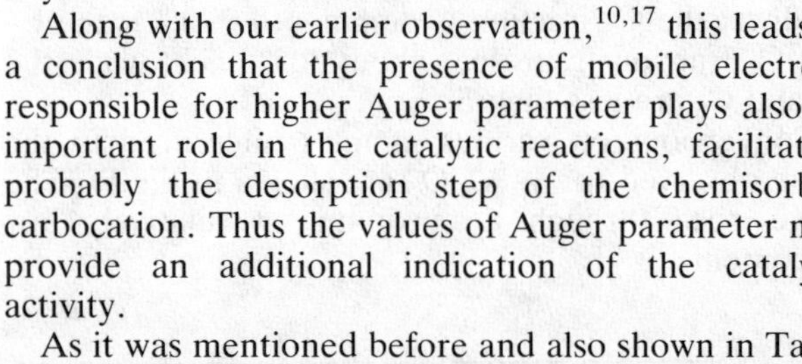

Figure 6. α'_{Si} vs. surface (Si/Al) composition.

by chlorine. In order to estimate the contribution of chlorine we prepared γ–Al_2O_3 samples with surface chlorine content ($\approx$ 2 at%) comparable with the highest values measured in SiO_2/Al_2O_3 samples.

The increase of the catalytic activity (about 2x) was smaller than that for most of the SiO_2/Al_2O_3 samples. The effect of chlorine on surface acidity in pyridine chemisorption and also in alternation of the surface electronic state, turned out to be much smaller compared to SiO_2 doped samples.

On the basis of the obtained results we can conclude that chlorine built in together with Si onto the surface of γ–Al_2O_3 also plays a certain role in the determination of the electronic state of the latter. The scatter of the energies of the O 1s, O(KVV) and also Al($KL_{23}L_{23}$) lines evidences all this. Further detailed studies are needed to elucidate the merit of the influence of the surface chlorine, that is now in progress.

CONCLUSIONS

(a) SiO/Al_2O_3 catalysts possessing high activity in *n*–hexane cracking-reaction has been prepared by the reaction of $SiCl_4$ (g) with γ–Al_2O_3 surface.

(b) The catalytic activity of the samples goes through a maximum at relatively low Si concentrations.

(c) The electronic structure of γ–Al_2O_3 changes considerably even at small doses of Si; the most important features are: (i) the Al 2p lines are shifted to high *BE* which means lower electron density and thus enhanced Lewis-acidity; (ii) the α'_{Si} goes through a maximum with growing Si content.

(d) Close correlation fo the Si-Auger parameter and the catalytic activity is established, and the value of this parameter is supposed to be an additional guide in defining the activity of such types of catalysts.

(e) Further study is necessary to elucidate the role of chlorine both in the catalytic activity and in the surface electronic state of the prepared samples.

Acknowledgment

The authors are indebted to Mrs. Dr. I. S. Pap for her help in the preparation of catalysts, and also to Mrs. E. Ady, Mrs. S. Perlaky, Mr. L. Gulyás and Mr. I. Lengyel for their technical assistance.

REFERENCES

1. H. A. Benesi and B. H. C. Winquist, *Adv. Catal.* **27**, 126 (1978)
2. P. B. Venuto, E. T. Habib, *Catal. Rev.-Sci. Eng.* **18** (1), 1–150 (1978).
3. C. L. Thomas, *Catalytic Processes and Proven Catalysts*, Academic Press, New York (1970).
4. C. Defossé, P. Canesson, P. G. Rouxhet and B. Delmon, *J. Catal.* **51**, 269 (1978).
5. H. Murayama and K. Meguro, *Bul. Chem. Soc. Japan*, **43**, 2386 (1970).
6. I. Bertóti, I. S. Pap, G. Mink, F. Réti, T Székely, *Therm. Anal. Proc.* 8th ICTA, **Suppl**. p. 33 (1985).
7. F. Réti, I. Bertóti. G. Mink, I. S. Pap, T. Székely, *Epitöanyag* (Hungary) **XXXVII** (4), 113 (1985).
8. I. Bertóti, A. Tóth, I. S. Pap and T. Székely, *Thermal Analysis* ICTA 80, Vol. 2, p. 235 Birkhaeuser Verlag, Basel, Boston, Stuttgart, (1980).
9. R. J. Bird and P. Swift, *J. Electron Spectrosc. Relat. Phenom.* **21**, 227 (1980).
10. G. Varsányi. I. Bertóti, G. Mink, F. Réti and M. Révész, Kémiai Közlemények (in press).
11. M. Révész (under preparation).
12. A. O. Litinskii, P. Ya. Hochberg, D. B. Shatkovskaya, A. B. Bolotin, N. N. Litinskaya and G. P. Popov, *Theor. Eksperim. Khim.* **19**, 486 (1983).
13. A. Balzarotti and A. Bianconi, *Phys. Stat. Sol. (B)* **76**, 689 (1976).
14. J. Derrien and F. Ringeisen, *Solid State Commun.* **50**, 627 (1984).
15. C. D. Wagner, D. E. Passoja, H. F. Hillery, T. G. Kinisky, H. A. Six and W. T. Jansen and J. A. Taylor, *J. Vac. Sci. Technol.* **21**, 933 (1982).
16. C. D. Wagner, D. A. Zatko and R. H. Raymond, *Anal. Chem.* **52**, 1445 (1980).
17. Preliminary results were presented at the workshop on "Catalysis and surface chemistry" June, 1984, Budapest, with participation of NSF (USA) and the Hungarian Academy of Sci.
18. R. Bicker, H. Deger, W. Herzog, K. Rieser, H. Pulm, G. Hohlneicher and J. J. Freund, *J. Catal.* **94**, 69 (1985)
19. S. Evans, R. G. Prichard and J. M. Thomas; *J. Electron Spectrosc. Relat. Phenom.* **14**, 341 (1978).

Surface Microstructure Investigation of Calcined θ-Al_2O_3-supported Molybdena Catalyst†

M. I. Zaki, H. M. Ismail and R. B. Fahim
Chemistry Department, Faculty of Science, Minia University, El-Minia, Egypt

INTRODUCTION

The prime industrial importance of the alumina-supported molybdena catalyst[1] has inspired much research on its performance and surface microstructural characteristics.[2,3] In most of these investigations, however, only η– and γ-alumina have been favoured as support materials. θ-alumina has been surprisingly ignored despite of a number of promising properties: (i) comparably high surface area, viz. 121 m^2 g^{-1} compared to 155 m^2 g^{-1} for the γ–modification; (ii) thermally more stable than η– and γ-alumina, and (iii) structurally has a far more ordered spinel lattice.[4] The degree of lattice order/disorder governs the nature of crystallite faces exposed at the surface, which has been found[5] to determine critically the exchangeability of the surface OH groups and, thus, the structure of the supported molybdena species.

The present study investigates the surface microstructural consequences of using θ-Al_2O_3 as a support for the calcined molybdena catalyst, employing a number of physicochemical techniques (XRD, DTA, TGA, IR, SEM and N_2 adsorption). α-Al_2O_3 was also used, but only to provide an alumina reference surface of modest supporting function.

METHODS

Catalyst preparation

The catalyst samples were prepared by impregnation of the support, α- or θ-Al_2O_3, with $(NH_4)_6Mo_7O_{24}$ from aqueous solution (pH 6). After drying, the material was calcined at 550°C for 16 h. The Mo loadings chosen here (0.5–16.0 wt%MoO_3) are swinging about the theoretical monolayer capacity of the respective support, this being *ca.* 10 and 1 wt%MoO_3 for θ- and α-Al_2O_3, respectively.

Characterization

Bulk phases formed at the Mo/support interface were identified by their x-ray diffraction patterns (XRD), infrared absorption behaviour (IR), and by the thermal effects of their formation reactions (DTA).

The surface hydroxylation was quantitatively estimated from the weight loss (TGA) effected by its HMDS-assisted elimination.[6] The nature of the OH groups exposed was revealed by analysing their IR absorption characteristics, exhibited by spectra recorded from self-supporting wafers mounted in a high-temperature IR– cell, on the basis of the alumina surface model proposed elsewhere.[7] Their thermal stabilities were determined by DTA curves and IR hydroxyl spectra.

The nature of the supported Mo-O species was identified by their IR absorption characteristics recorded from thin and light loaded KBr-supported discs.

The S_{BET} and pore structure of the support material and the catalyst were determined by the analysis of N_2 adsorption isotherms (at − 195°C) using known methods (BET-, α_s- and Kelvin-method). A direct texture investigation was done by SEM analysis.

CONCLUSIONS

Support surface

Specific area. α-Al_2O_3 was of a poor surface area (10.7 m^2 g^{-1}) as compared to that of the θ-Al_2O_3 (121 m^2 g^{-1}).

Pore structure. θ-Al_2O_3, exhibiting a type-IV N_2 adsorption isotherm with a high-pressure loop, had a porous surface consisting of micro- and mesopores, whereas α-Al_2O_3, exhibiting an uncommon type-I isotherm with a low-pressure loop, had hardly accessible micropores containing constrictions in between wider cavities. The corresponding SEM micrographs backed evidently this N_2-probed texture of the supports.

Hydroxylation. The surface of θ-Al_2O_3 was found to accommoate a number of OH groups (10.5 OH's/100 Å^2) about three times that measured on α-Al_2O_3 (3.3 OH's/100 Å^2). An analysis of the IR hydroxyl spectra of the supports, based on the data reported elsewhere,[7] suggested that the OH groups exposed belong to various types. The DT curves and IR spectra concluded that the higher the OH stretching frequency (i.e. stronger basic character), the higher the thermal stability of the surface OH group is.

Catalyst surface

Unsupported. Well crystallized MoO_3 was found (XRD, TGA) to be thermally produced by calcination of ammonium paramolybdate (APM) at 300–600°C, and to be sublimed at > 650°C. N_2 adsorption data suggested that MoO_3 obtained at 550°C for 16 h has a uniform surface, in energetic sense (stepwise isotherm), of microporous texture consisting of hardly accessible slit-shaped pores, and of poor surface area (5.6 m^2 g^{-1}).

Supported. The fixation of the impregnated paramolybdate anions was found to lead preferentially to the elimination of the thermally most stable (basic) OH's (DTA) which reside in the interior and/or entrances of the fine pores of the support (i) *On α-Al_2O_3*: Mo loading at 0.5% level, i.e. below the theoretical monolayer capacity of the support (*ca.* 1 wt%MoO_3), was found to lead to the formation of surface molybdates (IR bands at 900, 800 and 350 cm^{-1}). Mo loadings that exceed 1%, i.e. 4–16%, were found to form preferentially $Al_2(MoO_4)_3$, as characterized by thermal (DTA) and IR (bands at 935–860 cm^{-1}) indications, and then MoO_3 bulk phase (XRD proved). The SEM micrograph of the 8% sample demonstrated the existence of MoO_3 aggregations and uncovered domains of the support surface.

(ii) *On θ-Al_2O_3*: Mo loading up to near the monolayer capacity of the support (*ca*. 10 wt%MoO_3) was found to lead to the formation of polymeric molybdates (weak IR bands at 450, 370 and 330 cm^{-1}). At 16% Mo loading level, nucleation of bulk MoO_3 was XRD-verified. Contrary to αAl_2O_3, θ-Al_2O_3 seemed to facilitate good quality spreadings of the supported polymolybdates as indicated by the SEM micrograph of the 8% sample.

Acknowledgement

One of us (MIZ) acknowledges gratefully a grant from the Alexander von Humboldt Stiftung (FRG) which made possible his attendance to the ECASIA 85 meeting.

REFERENCES

1. P. Grange, *Catal. Rev.-Sci. Eng.* **2,** 135 (1980).
2. F. E. Massoth, *Adv. Catal.* **27,** 265 (1978).
3. B. Delmon, *Am. Chem. Soc., Div. Pet. Chem.* **2,** 503 (1977).
4. B. C. Lippens and J. H. de Boer, *Acta Cryst.* **17,** 1312 (1964).
5. H. Jeziorowski and H. Knözinger, *J. Phys. Chem.* **83,** 1166 (1979).
6. A. J. van Roosmalen, M. C. G. Hartmann and J. C. Mol, *J. Catal.* **66,** 112 (1980).
7. H. Knözinger and P. Ratnasamy, *Catal. Rev.-Sci. Eng.* **17,** 31 (1978).

† The full manuscript of this paper has been submitted for publication in Surface and Interface Analysis.

XPS Study of Rhenium on Silica

A. Cimino, D. Gazzoli, M. Inversi and M. Valigi
Centro SACSO (CNR), Dept. of Chemistry, Univ. of Rome, Italy

Supported rhenium catalysts, normally used in the reduced state, are usually prepared from Re(VII) compounds. Deactivated catalysts are subjected to an oxidation step. A chemical interaction between the Re(VII) precursor and the support must take place, in order to have a better dispersion and to avoid volatilization. The extent and the nature of the interaction has therefore been the object of several investigations on different supports (Al_2O_3, TiO_2, MgO, SiO_2). The present study reports on SiO_2 (Cab–O–Sil) as a support, for which a less clear picture could be traced, due to its acidic properties and to its 'relative inertness'. The study was mainly carried out by XPS, paralleled by other techniques (XRD, Chemical analysis and Thermogravimetry).

The specimens, containing rhenium up to 14% by weight, were prepared by impregnation of silica (Cab–O–Sil) with an ammonium perrhenate solution and dried at 110°C. The samples were studied as such ('starting materials') and after heat treatments in air, 3 h, in the temperature range 200–750°C.

The XPS spectrum was analysed for samples of different Re content and after various treatments. The Re 4f feature for the samples dried at 110°C was broader than that of NH_4ReO_4 and the *BE* value of the $4f_{7/2}$ peak, referenced to Si 2p = 103.5 eV, was constantly lower by 0.5–0.6 eV than the values found for NH_4ReO_4. A 'group behaviour' analysis, as separation of *BE* of specific species (O–Re, O–Si, N–Re), indicates that the ammonium perrhenate and Cab–O–Sil show a slightly different charging responsible for the apparent shift of the *BE* values of the Re 4f peak. The starting materials show a very small *I*(Re)/*I*(Si) intensity ratio, in agreement with the morphology deduced by XRD analysis, namely large perrhenate particles dispersed in the small Cab–O–Sil spheres.

Thermal treatments in air in the range 300°–750°C deeply modify the Re 4f feature both in intensity and in spectral shape. The Re 4f peak intensity sharply increases at 300°C and to a lesser extent at 400°C. The intensity decreases at higher temperatures.

The variation of the intensity ratios *I*(Re)/*I*(Si), as a function of the treatment temperatures is reported in Fig. 1. The appearance of a maximum in the curves of all samples is indicative of two factors: one tending to increase the ratio, the other one tending to decrease it. According to XRD analysis and thermogravimetry measurements the first factor can be attributed to the decomposition of NH_4ReO_4 and to the spreading of the resulting rhenium oxides. From TG experiments one can infer that the occurrence of the volatilization of the heptoxide becomes the second prevailing effect.

By curve fitting methods an attempt has been made to analyse the broad and unresolved Re 4f feature developed

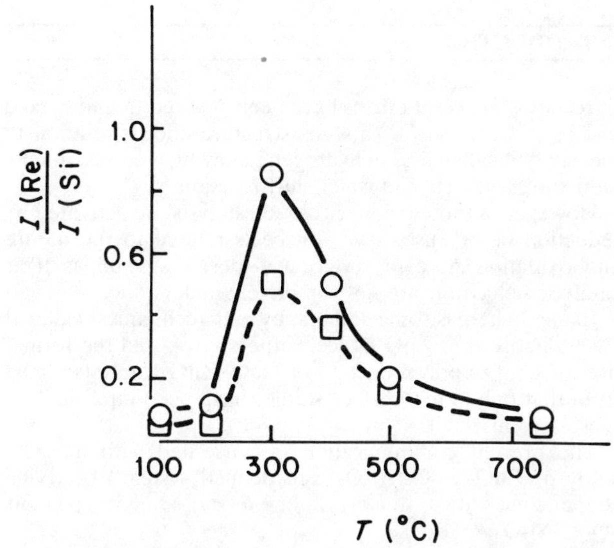

Figure 1. XPS peak intensity ratios, *I*(Re)/*I*(Si), as a function of temperature treatment in air, 3 h; ○ = 14% Re; □ = 8% Re.

upon thermal treatments. The procedures selected were illustrative of two surface situations, the first one with only one species present, the second one with two definite species.

The first way shows the presence of a single species but with a variety of surface situtions, the second one leads to a multiplicity of species but including some with unacceptable *BE* values. The first model is therefore favoured.

It can be concluded that a weak interaction exists between rhenium species formed by decomposition of ammonium perrhenate and the support, as shown by the spreading on the carrier of rhenium oxide and by the high temperature required to cause a complete loss of rhenium.

REFERENCES

1. L. G. Duquette, R. C. Cieslinski, C. W. Jung and P. E. Garrou, *J. Cat.* **90,** 362 (1984).
2. L. Wang and W. K. Hall, *J. Cat.* **82,** 177 (1983).
3. A. Cimino, D. Gazzoli, G. Minelli and M. Valigi, *Z. an. allg. Chem.* **494,** 207 (1982) and refernces therein.
4. P. Arnoldy, E. M. van Oers, O. S. L. Bruinsma, V. H. J. De Beer and J. A. Moulijn, *J. Cat.* **93,** 231 (1985).

Surface Characterization of CuO-ZnO-Al_2O_3 Methanol-Synthesis Catalysts by XPS

G. Moretti. M. Lo Jacono, G. Fierro and G. Minelli
Centro di Studio del CNR SACSO, Chemistry Department, University La Sapienza, P.le Aldo Moro 5, 00100 Rome, Italy

INTRODUCTION

In recent years great efforts have been devoted to understand the ZnO–CuO–Al_2O_3 catalysts structure and its relation to the catalytic behaviour in hydrogen involving rections, mainly methanol synthesis and water-shift reaction.[1]

However, although copper based catalysts are activated by reduction before use, several aspects related to the nature and oxidation state of surface Cu-species as well as their catalytic behaviour are not completely understood.

In fact, whereas some authors, by bulk techniques, claimed Cu^{+1} dissolved in ZnO lattice,[2] others[3] proposed the formation of Cu_2O upon reduction of CuO–ZnO catalysts; apart from that only a few surface studies have been published on these systems.[4–7]

The present communication is concerned with an XPS study of CuO–ZnO–Al_2O_3 well defined system of varying copper–zinc ratios (mainly 2 or 1) and alumina content ($O \leqslant Al_2O_3 < 20\%$).

The aim was: (a) to analyse *in situ* the reduction process related to the copper component, establishing the oxidation states of copper as function of extent of reduction and showing how it is possible to provide a chemical picture of surface states by the analysis of Cu $2p_{3/2}$ photoemission and Cu L_3VV Auger peaks; (b) to assess the mutual interaction between the different and so closely associated phases, as ZnO–CuO or ZnO–Cu; (c) to investigate the role of alumina on the sintering phenomena.

EXPERIMENTAL

All catalysts were obtained by decomposition at 623 K in air of initial precursors, containing one or two phases of basic carbonates. The preparation of carbonate precursors was carried out by following the procedure of the coprecipitation method at constant pH from $NaHCO_3$ and $Me(NO_3)_2$ solutions.[8] The catalysts were characterized by the following techniques: (1) atomic absorption for chemical analysis of zinc and copper; (2) x-ray diffraction for phases identification; (3) diffuse reflectance spectroscopy for Cu^{+2} symmetry; (4) XPS for surface analysis; (5) magnetic susceptibility measurement for Cu^{+2} content; (6) surface area by BET method.

The XPS and Auger spectra of catalysts, in oxidic or reduced state, were performed on a LHS 10 spectrometer with Al K_α radiation. The catalysts were reduced at 408 K and 473 K by flowing H_2 for different intervals of time in a preparation chamber attached to the spectrometer. Data were digitally collected and handled with a 2113 B HP computer interfaced to the spectrometer. Qualitative analysis of copper spectra in terms of Cu^{+2}, Cu^{+1}, Cu^0, was obtained by use of chemical state plot.[9] Quantitative analysis of copper spectra in terms of mole fractions (Cu^{+2}/Cu, Cu^{+1}/Cu, Cu^0/Cu) was tentatively achieved by a careful analysis of Cu $2p_{3/2}$ and Cu L_3VV peaks which respectively yields Cu^{+2}/Cu, $(Cu^{+1} + Cu^0)/Cu$ and $Cu^{+1}/(Cu^0 + Cu^{+1})$. The last one was derived from Cu L_3VV peak heights. The surface Cu/Zn atomic ratios were obtained by the atomic sensitivity factor reported by Wagner *et al.*[10]

RESULTS

The reduction of catalysts in H_2 at 408 K and 473 K leads to a change in the chemical state of Cu^{+2} giving rise to Cu^{+1} and Cu^0. The reduction product of Cu^{+2} (CuO) is mostly Cu^0 and, for a least amount, Cu^{+1} ($Cu^{+1}/Cu < 0.3$). This last one tends to increase with the reduction extent and its amount is not affected by ZnO or Al_2O_3, at least for a small amount of Al_2O_3.

For all Cu–Zn and Cu–Zn–Al catalysts in oxidic state, broadening of the full width at half maximum of Zn $2p_{3/2}$ peak is observed with respect to pure ZnO. Such broadening decreases with increasing reduction extent, until it disappears at higher reduction extent values, indeed when the sample becomes a good conductor. The effect of Zn $2p_{3/2}$ peak broadening, not observable in mechanical mixtures of ZnO–CuO, could be explained by a strong interaction between small particles, by means of a ZnO particle surrounded by CuO particles, with their Fermi levels in equilibrium. In such a case bending of the Zn 3d valence band occurs, leading to higher charging of the ZnO particle surface. The surface charging capacity of a ZnO particle will depend on the surrounding paticles, i.e. CuO, partial reduced CuO and/or Cu^0, and will disappear when the system becomes a good conductor.

According to the preparation method, the nucleation and growth of ZnO and CuO particles occur almost simultaneously by decomposition of the precursor, leading to catalysts with particles well dispersed and fairly uniform in size. that is reflected even in the almost identical composition of the surface and the bulk. However this good initial state of the catalyst can be altered during the exothermic reduction process of copper, with tendency to sintering. the presence of alumina plays a positive role to prevent the sintering of copper.

Acknowledgement

This work was supported by the project 'Chimica Fine e Secondaria' of CNR (Rome, Italy;.

REFERENCES

1. K. Klier, *Adv. Catal.* **31,** 243 (1982).
2. R. G. Herman, K. Klier, G. W. Simmons, B. P. Finn, J. B. Bulko and T. P. Kobilinski, *J. Catal.* **56,** 407 (1979).
3. P. B. Himelfarb, F. E. Wawner, Jr., A. Bieser, Jr., S. N. Vines, *J. Catal.* **83,** 469 (1983).
4. Y. Okamoto, K. Fukino, T. Imanaka and S. Teranishi, *J. Phys. Chem.* **87,** 3740; 3747 (1983).
5. F. Garbassi and G. Petrini, *J. Catal.* **90,** 106; 113 (1984).
6. T. H. Fleish and R. L. Mieville, *J. Catal.* **90,** 165 (1984).
7. F. Boccuzzi, G. Ghiotti and A. Chiorino, *Surf. Sci.* **156,** 933 (1985).
8. F. Trifirò, A. Vaccari, C. Busetto, G. Del Piero and G. Manara, *J. Catal.* **85,** 260 (1984).
9. C. D. Wagner, in *Practical Surface Analysis by Auger and XPS*, ed. by D. Briggs and M. P. Seah, pp. 477–509, John Wiley, Chichester (1983).
10. C. D. Wagner, L. E. Davis, M. W. Zeller, J. A. Taylor and R. H. Raymond, *Surf. Interface Anal.* **3,** 211 (1981).

Surface Investigation, by XPS, ISS and IR Spectroscopy of Adsorbed NO, of the Structural Changes Occurring during the Reduction of Co-Mo/γAl_2O_3 Catalysts

R. Prada Silvy,* J. M. Beuken,† J. L. G. Fierro,* P. Bertrand† and B. Delmon*

* Groupe de Physico-Chimie Minérale et de Catalyse, Unité de Catalyse et Chimie des Matériaux Divisés,
† Unité de Physico-Chimie et Physique des Matériaux,
Université Catholique de Louvain, Place Croix du Sud 1 B-1348 Louvain-la-Neuve, Belgium

The structural changes produced during the reduction by H_2 at different temperatures (300–800 °C range) of a commercial Co–Mo/γAl_2O_3 hydrodesulphurization catalyst have been investigated by: x-ray photoelectron spectroscopy (XPS), ion scattering spectroscopy (ISS) and infrared (IR) of adsorbed NO, as a probe-molecule.

In all samples, different cobalt phases are observed. A fraction of cobalt interacts with the alumina support ($CoAl_2O_4$-like, species which was reducible only at the highest temperatures) and another fraction forms a phase interacting with the molybdenum atoms (species easily reducible at moderate temperatures).

XPS was employed in order to determine the distribution of Mo oxidation states and the nature of surface species; however, this technique is less successful in assessing variations in their dispersion. This limitation was overcome by the use of low energy ion scattering spectroscopy (ISS) which probes only the first atomic layer (sampling depth 0.25 nm). In addition, IR spectroscopy of NO, chosen as a probe molecule adsorbed on the catalyst surface, was used to distinguish between Co and Mo atoms located at the surface of the catalysts and to reveal the changes in the nature of the phases in which these atoms were situated.

The main conclusions of this study are:

(a) XPS, ISS and IR of the NO probe-molecule have been shown to be convenient tools for the physico-chemical characterization of Co–Mo/γAl_2O_3 catalysts. Indeed, this combination gives valuable information on, (i) the oxidation state of the atoms on the surface, (ii) dispersion of the supported phase, and (iii) the nature of the promoter atoms.

(b) With increase of the reduction temperature, several effects have been evidenced on the Co–Mo/γAl_sO_3 catalyst surface: (i) increase of metallic cobalt, (ii) decrease of $CoAl_2O_4$ concentration and simultaneous crystallization of MoO_2 species, and (iii) decrease of molybdenum dispersion.

(c) The increase of C_O^0 exposed on the surface seem to be due to reduction of $CoAl_2O_4$–like species. These species are hardly reducible and only at the highest temperatures.

XPS Study of Irradiated Polycrystalline TiO_2

G. Munuera, A. R. González-Elipe, J. P. Espinós, A. Navío and A. Fernández
Departamento de Química General, Facultad de Química, Universidad de Sevilla, c/o Tramontana s/n, 41012 Sevilla, Spain

TiO_2 is widely used as a photocatalyst, one of its more important applications being the photocleavage of water by irradiating M/TiO_2 powders (M : Pt, Rh, Ni, etc) in aqueous suspensions. In this way, although O_2 is not detected, H_2 is easily produced. To explain the absence of O_2 the formation of peroxide species that would be incorporated into the support has been suggested.

To prove this hypothesis we have studied by XPS a polycrystalline TiO_2 sample (Degussa P-25, 50 m^2g^{-1}) that was suspended in a 1 M in NaOH solution and on which oxygen was photoadsorbed. For comparison an amorphous titanium hydroxiperoxide sample (approximate formula $Ti_2O_{5.4}(OH)_{0.7}$) and an amorphous solid resulting from the chemical attack of a 5 M in H_2O_2 and 1 M in NaOH solution to the TiO_2 have been used. All the samples immersed in aqueous solutions were washed and dried in an Ar flow before their transfer into the spectrometer.

The analysis of the TiO_2 sample showed that some residual Na^+ ions remained on the surface after the washing treatment, while no significant modifications occurred at the O 1s and Ti 2p signals (O 1s BE 529.8 eV, Ti $2p_{3/2}$ *BE* 458.5 eV).

On the contrary, the O 1s spectrum of the titanium hydroxiperoxide presented two almost equivalent peaks at 530.1 and 532.2 eV that can be ascribed to O and O_2^{2-}/OH^- species respectively. Besides that, the O 2s and O 2p levels were slightly broadened if compared with the corresponding peaks of TiO_2. By substracting this region of the spectrum of both samples some remaining features could be ascribed to the M.O. levels of the peroxide species.

The titanium hydroxiperoxide sample decomposed at $T \geqslant 473$ K giving oxygen that was detected by mass spectrometry and a well crystallized Anatase (O 1s) main peak at 529.8 eV).

Since in the irradiated TiO_2 no significant peak at 532.2 eV was recorded for the O 1s level, it might seem unlikely that O_2^{2-} species be incorporated into the TiO_2 support during the water photolysis reaction. However, the H_2O_2 treated TiO_2 had a complicated O 1s spectrum, in that 10 min after the introduction of the sample in the spectrometer, it could be fitted with three curves at 530.1, 531.9 and 532.7 eV (O^{2-}, O_2^{2-}/OH^- and H_2O respectively). This sample was not stable under vacuum at 298 K, giving oxygen and an O 1s spectrum that, thirty minutes later, could be fitted with three curves at the same *BE*, but with much less intensity for the band at 531.9 eV.

These results indicate that O_2^{2-} species can be incorporated into the TiO_2 from a H_2O_2 basic solution forming a kind of titanium hydroxiperoxide that is not stable under vacuum at 298 K. The formation of a similar product is also very likely during the water photocleavage reaction, making its detection by XPS uncertain due to its lack of stability.

REFERENCE

1. G. Mumuera, J. Soria, J. C. Conesa, J. Sanz, A. R. González-Elipe, A. Navío, E. J.López-Molina, A. Muñoz, A. Fernández and J. P. Espinós in *Catalysis on the Energy Scene*, p. 335, ed. by S. Kaliaguine and A. Mahay, Elsevier Amsterdam (1984).

The Influence of Reductive Pretreatments on CuO/MnO_2 Catalytic Activity: an XPS study

V. Di Castro and C. Furlani
Dipartimento di Chimica, Università La Sapienza, 00185 Roma, Italy
C. Fragale, M. Gargano and M. Rossi
Dipartimento di Chimica, Universitá di Bari, 70126 Bari, Italy

Changes in the CuO/MnO_2 system during two different reductive treatments with H_2 have been analysed by XPS. The difference in catalytic activity has been correlated with surface concentration and oxidation state of copper.

INTRODUCTION

Copper supported catalysts are widely used in hydrogenation reactions, and their activity often depends on preparation procedure and reductive pretreatments.[1] In this work we have studied the CuO/MnO_2 system as an example of a system whose catalytic activity is strongly influenced by minor changes in the pretreatments. The aim of this work is to correlate the surface composition of the catalyst and the changes due to the pretreatments measured by XPS with the catalytic activity in a test reaction, that is the hydrogenation of cyclododecatriene.

EXPERIMENTAL

The samples were prepared by impregnation in a $Cu(NH_3)_4^{++}$ solution and heating in air at 350°C for 6 h. XPS spectra were recorded using a VG ESCA 3 spectrometer employing an Al Kα source ($h\nu$ = 1486.6 eV). Samples were dusted as thin films onto a gold probe. Binding energies were corrected with C 1s (285.0 eV) peak for charging effects. All the treatments were performed in the spectrometer preparation chamber.

RESULTS AND DISCUSSION

Two different reductive pretreatments with H_2 have been considered. In the first one the CuO/MnO_2 system is warmed up to 270°C in vacuum and then exposed to H_2 (1 atm) for 30 min, in the second one the sample is warmed up to 270°C slowly in H_2 (1 atm).

This last treatment gives rise to a catalytic activity in the hydrogenation reaction 100 times higher than the first one. We have followed by XPS the changes at the catalyst surface warming up both in vacuum and in H_2 and the main results are listed in Table 1. In both cases there are no significant change up to 150°C, while at higher temperatures the progressive reduction of Mn and Cu is observed. The Mn 2p peak, that is characteristic for MnO_2 species[2] in the untreated sample, is progressively shifted to lower binding energies whilst the XPS peak area ratio shows a monotonic decrease of copper surface concentration.

It is worth noting that the same trend occurs in our catalysts whether they are annealed in H_2 or in vacuum. It suggests that H_2 is not necessary for the reduction of MnO_2 and incorporation of copper and thus that ultra high vacuum is a sufficient reducing environment for this process. On the contrary, copper reduction exhibits a strong dependence on the presence of H_2. In fact, copper is reduced at first to Cu(I) and then almost completely to Cu(O) annealing in H_2 atmosphere, while just a partial reduction to Cu(I) is observed in vacuum. The copper reduction is contemporary to the decrease of copper concentration at the surface, suggesting that copper is incorporated in the substrate as reduced copper in presence of H_2, while an oxidized form diffuse in vacuum. When the sample annealed in vacuum is exposed to H_2 at 270°C a further reduction of copper is observed together with a dramatic decrease of Cu/Mn atomic ratio.

In any case, copper is not completely reduced to Cu(O) as previously reported for other copper based systems,[3,4] and an intense component due to Cu(I) is still observed in the Auger spectra. This difference in the reduction behaviour can be due to a higher stability of the oxidized copper inside the support structure, so that the copper on the surface is reduced to Cu(O) as in the previously studied systems, while copper in the manganese oxide structure remains in a partially oxidized state.

Table 1. Experimental XPS binding energies and Auger kinetic energies

CuO/MnO_2 treatments	Cu $2p_{3/2}$	Satellites ΔE	Satellites %	Cu LMM	Mn $2p_{3/2}$	Cu/Mn atomic ratio	
	934.2 (3.8)[a]	9.2	36	917.4	642.2 (3.8)	0.59	Cu(II)
150°C H_2 1 atm	934.0 (3.7)	9.5	33	917.3	642.0 (4.2)	0.53	Cu(II)
150°C vacuum	934.2 (3.6)	9.2	34	917.2	642.2 (4.0)	0.54	Cu(II)
190°C H_2 1 atm	933.5 (4.0)	10.0	26	917.3	641.8 (4.4)	0.48	Cu(II) ~ Cu(I)
190°C vacuum	933.6 (4.0)	10.2	30	917.2	642.0 (4.6)	0.50	Cu(II)
230°C H_2 1 atm	932.7 (2.0)	—	—	918.2 − 916.2	641.7 (4.4)	0.30	Cu(I) ~ Cu(O)
230°C vacuum	933.4 (3.7)	10.6	25	917.2	641.8 (4.6)	0.31	Cu(II) > Cu(I)
270°C H_2 1 atm	932.8 (2.3)	—	—	918.2	641.4 (4.5)	0.16	Cu(O) >> Cu(I)
270°C vacuum	932.7 (3.0)	11	18	916.5	641.5 (4.5)	0.17	Cu(I) > Cu(II)
270°C vacuum + H_2 1 atm 30 min	932.8 (2.4)	—	—	918.3	641.4 (4.6)	0.09	Cu(O) > Cu(I)

[a] Full widths at half maximum (FWHM) of peaks given in parenthesis.

In conclusion, the XPS data show that the difference in catalytic activity between samples pretreated with the two analysed methods depends on the different copper concentration at the surface and the different degree of copper reduction.

REFERENCES

1. G. C. Bond in *Metal-support and Metal-additive effects in catalysis*, ed. by B. Imelik *et al.* pp 1–10, Elsevier (1982).
2. M. Okv, K. Hirokawa, S. Ideda, *J. Electr. Spectr.* **7**, 465, (1975).
3. F. Capece, V. Di Castro, C. Furlani, G. Mattogno, G. Gargano, C Fragale, M. Rossi, *J. Electr. Spectr.* **27**, 119, (1982).
4. V. Di Castro, C. Furlani, C. Fragale, M. Gargano, M. Rossi, N. Ravasio, *J. Mol. Cat.* (submitted).

Surface Analysis of Pt-black by XPS

L. Kövér, J. Tóth
Institute of Nuclear Research of the Hungarian Acad. of Sci., H-4001 Debrecen, P.O. Box 51, Hungary
Z. Paál
Institute of Isotopes of the Hungarian Acad. of Sci., H-1525 Budapest, P.O Box 77, Hungary

As opposed to the classical concept, regarding metal catalysts as clean crystal planes,[1] recent studies[2–5] have detected many impurities over Pt-black samples. XPS results showed about 11–30% Pt, 7–30% O and 50–75% C in the surface layers of fresh or pretreated samples.[5] The O ls peak was decomposed into two parts: one peak at 530 eV (oxygen attached to Pt)[6], and the other one at 532 eV (oxide of a metalloid, C or Si). *In situ* reduction by H_2 at 573 K removed 'PtO' almost completely. A correlation was found between the peak ratio of the Pt 4f doublet and the amount of oxygen. Thus, the amount of Pt chemically bound to O (and to other impurities as well) could be roughly estimated. Our model suggests that a fraction of Pt is covered by O and by a partly oxidized carbonaceous overlayer (including, perhaps, CO). Only SIMS was able to detect hydrogen.[4]

The present study aims at the quantitative decomposition of various photoelectron peaks observed in spectra of Pt-black surfaces as a function of Ar^+ ion bombardment charge density. Experimental conditions are given below.

Standard thermal cycle (STC) of Pt comprised its heating up in air, followed by 1 h isothermal hydrogen flow.[7]

Al K_α excited XPS spectra of oxygen, carbon and platinum of polycrystalline Pt-black powder samples (5 mm in dia.) were observed as a function of Ar^+ ion sputtering charge density. The parameters of the ion beam used were: 4–5 keV (beam energy), 7–10 μA (beam current), 45° (incidence angle related to the surface). Pt $4f_{7/2}$ photoelectron peaks were decomposed into a metallic (Pt^2) and an 'oxide' (Pt^1) part by computer using an iterative fit of Gaussians (ΔE = 2.6 eV chemical shift[6] was assumed). O 1s peaks were decomposed into two parts one of them is at 530 eV (O^1) and the other one is at 532 eV binding energy (O^2) in a similar way. The same Pt^1/Pt^2 ratio was assumed for the Pt $4p_{3/2}$, Pt $4d_{5/2}$ and Pt $4f_{7/2}$ peaks.

The following statements can be made here: The chemical state of platinum does not change as a function of ion sputtering charge density, after a slight initial change (Table 1). This suggests a picture of non-uniform coverage by impurities: free surfaces between carbonized and/or oxidized islands, in agreement with previous models.[4,8] The 4f doublet ratios were practically constant as the O 1s Pt $4p_{3/2}$ (total) intensity ratios changed as opposed to previous results.[5] The

Table 1. Pt^1 $4f_{7/2}$/Pt^2 $4f_{7/2}$ and O 1s/Pt^1 $4p_{3/2}$ photoelectron intensity ratios as a function of ion bombarding charge density

Sample[a]	Pt^1 $4f_{7/2}$/Pt^2 $4f_{7/2}$						O $1s^{total}$/Pt^1 $4p_{3/2}$				
	Ar^+ ion charge density, μA min cm^{-2}										
	0	35	100	350	670	1100	0	35	100	670	1100
Fresh	0.49		0.48				2.59		0.18		
STC 200°C (1984)	0.48	0.42			0.40		2.10	1.42		1.33	
STC 200°C (1985)	0.18		0.15	0.15		0.16	3.70		2.27	2.77	2.40
STC 360°C	0.42	0.36			0.33		2.89	2.24		2.11	
[b]STC 360°C Ethene	0.15		0.15	0.14		0.15	13.73		6.46	7.51	4.81

[1] = metal
[2] = 'oxide'
[a] Pt-black catalysts were reduced from aqueous solution of H_2PtCl_6 by HCHO in the presence of KOH[5] at room temperature.
[b] Pretreated with ethene.

Table 2. $\rho(C)/\rho(Pt^2)$ atomic densities as a function of Ar^+ ion charge density[a]

Sample[b]	Ar^+ ion charge density, μA mincm^{-2}					
	0	35	100	350	670	1100
Fresh	5.9 (3.9)		2.5			
STC 200°C (1984)	6.8 (6.0)	6.1 (5.51)			5.0 (4.58)	
STC 200°C (1985)	4.9 (4.23)		3.3 (3.05)	3.2 (2.79)		2.9 (2.57)
STC 360°C	8.8 (7.7)	7.6 (6.97)			6.7 (6.17)	
[c]STC 360°C ethene	9.7 (7.52)		6.3 (5.38)	4.4 (3.52)		3.5 (2.81)

[a] In brackets, calculated value of $\rho(C)/\rho(Pt^2)$ corresponding to surface carbon in other chemical form than CO (CO concentration was obtained assuming O^2 1s = O 1s (CO)).
[b] see note a of Table 1.
[c] pretreated with ethene.

ratio of O (total) to the chemically bound fraction of Pt, decreased considerably upon ion bombardment. This means that oxygen, mainly bound to other element as Pt (likely CO or SiO_x)[5] was removed by sputtering. Consequently, ion bombardent is a complementary treatment to high temperature hydrogenation.

The intensity ratio of C 1s to Pt $4d_{5/2}$ and the atomic concentration ratios calculated from these decrease considerably upon bombardment (Table 2). The comparison of O/Pt and C/Pt atomic density ratios indicates that when CO was removed, other type of carbon remained after sputtering. CO is not very strongly bound to platinum and does not represent the main form of surface carbon.

REFERENCES

1. A. A. Balandin, *Adv. Catal.* **19,** 1 (1969).
2. A. Barna, P. B. Barna, L. Tóth, Z. Paál and P. Tétényi, *Appl. Surf. Sci.* **14,** 85 (1982–83).
3. Z. Paál, P. Tétényi, L. Kertész, A. Szász and J. Kojnok, *Appl. Surf. Sci.* **14,** 101 (1982–83).
4. Z. Paál and D. Marton, *Appl. surf. Sci.*, in press.
5. Z. Paál, P. Tétényi, D. Prigge, X.Zh. Wang, G. Ertl, *Appl. Surf. Sci.* **14,** 307 (1982–83).
6. L. Kövér, Cs. Ujhelyi, D. Berényi, D. Varga, I. Kádár, Á. Kövér and J. Miller, *J. Electron Spectrosc. Rel. Phenom.* **14,** 201 (1979).
7. T. Baird, Z. Paál, and S. J. Thomson, *J. Chem. Soc. Faraday I.* **69,** 50 (1973).
8. S. M. Davis and G. A. Somorjai, *J. Catal.* **65,** 78 (1980).

Reactions on Thin Films of Cobalt-Molybdenum Oxides: I. Characterization of Molybdenum and Cobalt Oxide Species by High Resolution XPS

N. S. McIntyre, D. Johnston and L. L. Coatsworth
Surface Science Western, The University of Western Ontario, London, Canada N6A 5B7
J. R. Brown
Energy Research Laboratory, Energy, Mines and Resources Canada, 555 Booth Street, Ottawa, Canada K1A 0R6

Laboratory studies of catalysed hydrodesulphurization reactions on surfaces have usually been studied using the same bulk form of catalyst as is used in the field. Possibly the most heavily investigated system is alumina-supported cobalt oxide molybdenum oxide. The surface of this catalyst has been characterized by many techniques, particularly x-ray photoelectron spectroscopy (XPS), during the reduction, sulphidation and reaction stages.

While a wealth of experimental data has been produced from this approach, there are several compelling reasons for studying the catalyst in a flat, thin film form, rather than a bulk material. Firstly, if the catalyst structure is layered, covering a macroscopic area, it becomes easier to study the interaction between support and reactive layer using techniques such as non-destructive depth profiling. Secondly, the deposition of catalysts in thin films affords a much better control over the surface composition than by the usual bulk precipitation reactions employed. Finally, studies of such films by electron spectroscopic techniques is rendered much easier, since the films are usually deposited on an electrically conducting substrate and are not therefore subject to significant surface charging effects as bulk catalysts usually are. Therefore, photoelectron line shapes and positions become much easier to identify by XPS.

The objective of this particular work therefore, has been to prepare thin film analogues of the oxide precursor of the alumina-supported Co–Mo catalyst used in hydrodesulpherization reactions. Earlier XPS studies of such oxides[1] have suggested that the maximum efficiency of the catalyst, in its reduced and sulphided form, comes from an oxide precursor with a Co/Mo atomic ration of ~1 : 5, with cobalt present as Co_3O_4 and molybdenum oxides present in oxidation states from + 4 to + 6.

This paper reports the fabrication and characterization of thin films of cobalt and molybdenum oxides on an equally thin film of alumina grown on an aluminium metal substrate. The alumina film, 10–20 nm in thickness, was grown by thermal oxidation at 200°C. Subsequently, layers of molybdenum and/or cobalt were deposited onto the alumina substrate using ion beam deposition. These metallic layers were oxidized and calcined under a number of temperature (25°C–500°C) and pressure (760 torr–10^{-4} torr) conditions. Ultimately, several series of thin film specimens were produced:

1. Cobalt oxides on alumina
2. Molybdenum oxides on alumina
3. Mixed cobalt-molybdenum oxides on alumina
4. Molybdenum oxides on cobalt oxides on alumina
5. Cobalt oxides on molybdenum oxides on alumina

The surfaces of some of these films were intended to correspond to the pre-reduced form of the bulk catalyst.

Auger electron spectroscopy and ion bombardment were used to depth profile these films in an effort to measure layer thickness and to determine the degree of interaction of the layers. Prior to calcining, the films were found to be 5–10 nm in thickness while following calcining the distribution of

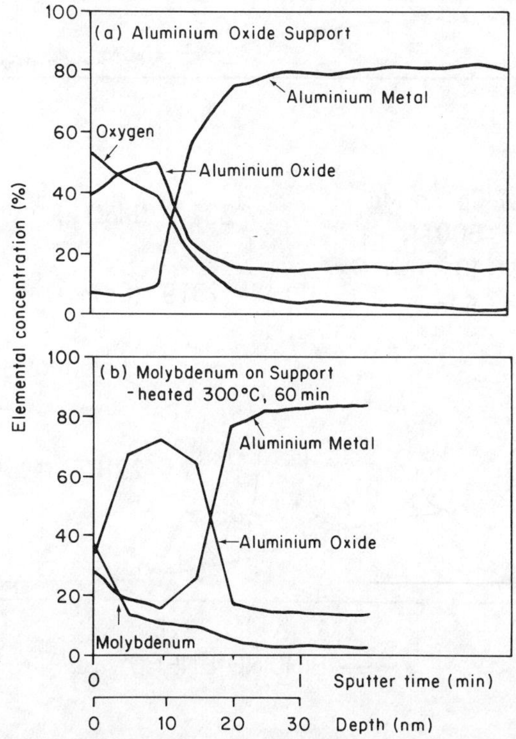

Figure 1. Auger depth profiles through (a) an aluminium oxide substrate (b) a molybdenum layer deposited on the support using ion beam desposition. Both aluminium oxide and aluminium metal LMM Auger line intensities are plotted, in addition to those for the Mo LMM and O KLL lines. The depth scale is calibrated using SiO_2.

cobalt or molybdenum often extended over 20–30 nm. Calcining of the films did not result in significant diffusion of the deposited elements into the supporting oxide or in an increase in the alumina thickness. Figures 1(a) and (b) compare the Auger depth profiles for aluminium oxide, aluminium metal, oxygen and molybdenum concentrations for an uncoated aluminium oxide substrate, and for a molybdenum-deposited film on the substrate calcined at 300°C. Although some penetration of the aluminium oxide by molybdenum is detected, the major part of the molybdenum remains on top of the aluminium oxide.

The surface chemical structures and elemental composition of each film were determined by x-ray photoelectron spectroscopy. A small spot monochromatized x-ray beam was used to produce high resolution spectra of the photoelectron lines, as well as to obtain localized area analysis. Reference spectra of thin films of CoO, Co_3O_4, $CoAl_2O_4$, $CoMoO_4$ and MoO_3 were also obtained to allow a more accurate assessment of the reaction products.

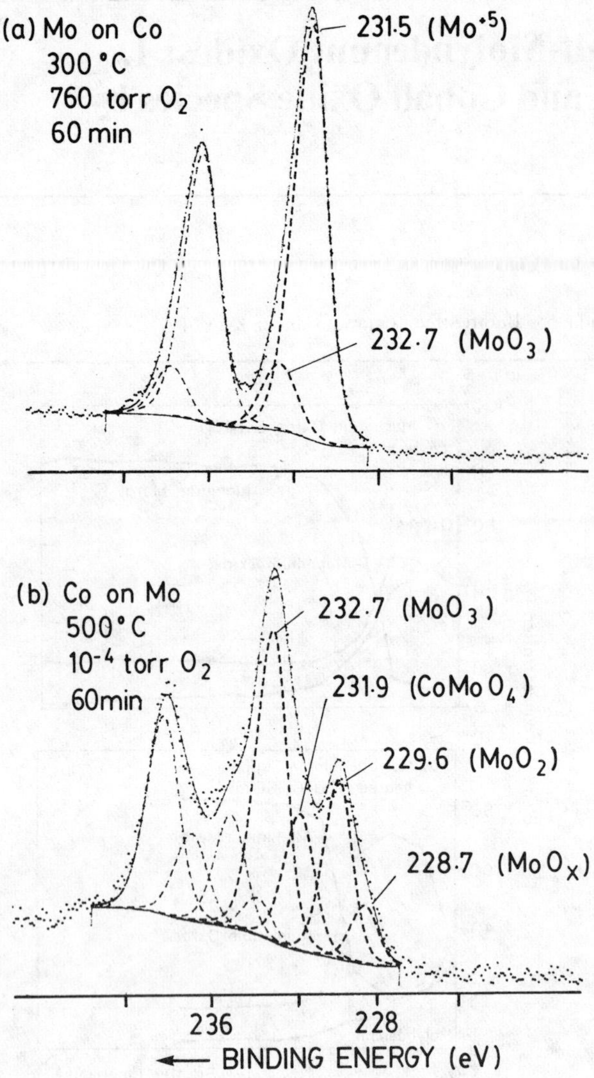

Figure 2. High resolution Mo 3d XPS spectra of 2 calcined bilayer films. The peaks were fitted using a mathematical routine where the spin-orbit doublet splittings and relative intensities are constrained.

The spectra of most of the oxide thin films were free or almost free of charging effects, even without the use of a neutralizing flood gun. Cobalt Co 2p spectra, although broadened by multiplet effects, showed more characteristic line structure than identified in previous work. In the case of the Mo 3d photoelectron spectra, oxide linewidths (FWHM) were 1 eV or less-only slightly broader than the metal line itself. Thus, a number of separate oxide species could be identified. Two examples of the Mo 3d spectra produced are shown in Fig. 2. In Fig. 2(a) the thin film oxide product of a 300°C high-oxygen calcination of a molybdenum on cobalt bilayer film is shown to have most molybdenum in an oxidation state giving a $Mo(3d_{5/2})$ peak centred at 231.5 eV. This is interpreted as a Mo^{+5} oxidation state. Figure 2(b) shows a more complex series of molybdenum oxides which result from oxidation of a cobalt on molybdenum bilayer at 500°C and low oxygen pressure. Certain molybdenum + 6 and + 4 oxidation states are identifiable, while other peaks cannot be assigned at this time.

Many of the surface oxides produced from the metal bilayer films (series 3–5) had compositions and relative elemental concentrations considered desirable for an oxide precursor of the Co–Mo catalyst. In series 5 (cobalt on molybdenum on alumina) the cobalt is detected as an hydroxide under a number of calcining conditions. This suggests the presence of active cobalt sites which react with atmospheric moisture during air transfers of the specimen. Under the same conditions molybdenum oxides in + 5 and + 6 oxidation states predominate. By contrast, in series 3 and 4 specimens, where cobalt is more intimately in contact with molybdenum or aluminium, much of the cobalt appears to have reacted as a molybdate or aluminate while a range of molybdenum oxide states is still detected. Series 2 films are found to consist of Co_3O_4 and $CoAl_2O_4$ in varying concentrations.

A range of thin films with different elemental compositions and structure were thus produced. Some of these have been subsequently subject to reduction, sulphidation and reaction with thiophene, and that is the subject of the following paper.

REFERENCE

1. T. L. Barr, in *Practical Surface Analysis by Auger and Photoelectron Spectroscopy*, p. 337, ed. by D. Briggs and M. P. Seah, Wiley (1984).

Reactions on Thin Films of Cobalt-Molybdenum Oxides: II. Reduction, Sulphidation and Hydrodesulphurization (HDS) Activities

J. R. Brown
Energy Research Laboratory, Energy, Mines and Resources Canada, 555 Booth Street, Ottawa, Canada K1A 0G1

N. S. McIntyre, D. Johnston and L. L. Coatsworth
Surface Science Western, The University of Western Ontario, London, Canada N6A 5B7

Alumina-supported cobalt and molybdenum oxides, after sulphiding, have high hydrogenation and hydrodesulphurization (HDS) activities and are therefore widely used as catalysts in the petroleum refining industry.[1] Such sulphides have recently been prepared in thin film form, and the preparation methods and advantages of such surfaces for catalysis research have been described in a previous paper in these Proceedings.[2] This present paper describes the reaction of some of the prepared films with a model organic sulphide (thiophene) mixed with hydrogen and performed under HDS process conditions.

A micro-activity testing unit (MAT) was used to determine the hydrodesulphurization capability of the five thin film types. The amount of injected thiophene (C_4H_4S) converted to hydrocarbons (butanes and butenes) is monitored by gas chromatography. X-ray photoelectron spectroscopy (XPS) chemically characterized the surface region of the specimens prior to the following reduction (H_2), sulphidation (H_2S/H_2) and thiophene (C_4H_4S/H_2) exposure. A commercial Co–Mo–alumina HDS catalyst was processed identically for reference purposes. Details of the reaction vessel, reaction conditions, specimen transfer procedures and transfer hardware to permit specimen examination by XPS without air exposure will be published elsewhere.

A semi-quantiative surface chemical analysis of each sample is performed using the XPS determined elemental peak intensities and empirically derived elemental sensitivity factors. Peak shape and peak binding energy values are used to determine element oxidation states and, where possible, aid in compound identification.

Reduction, sulphidation and thiophene HDS steps were done sequentially with XPS analysis performed after each step. Each specimen was reduced in flowing hydrogen for 30 min (120 ml min^{-1}) at 623 K, sulphided in flowing 10% H_2S/H_2 for 60 min (120 ml min^{-1}) at 623 K and exposed to thiophene by direct injection of 10 μl of feed into the cool

Table 1. Semi-quantiative XPS analysis of selected Co–Mo–Al oxides prior to and following thiophene HDS

XPS[a] Composition Atomic % (± 15%)	Thin Film Specimen[b] Co-Al-O (Series 1) Fresh	C_4H_4S/H_2[d]	Mo-Al-O (Series 2) Fresh	C_4H_4S/H_2	Mo/Co-Al-O (Series 4) Fresh	C_4H_4S/H_2	Commercial Catalyst[c] (Reference) Co + Mo – Al_2O_3 Fresh	C_4H_4S/H_2
Co	2.0	3.0	N.A.	N.A.	7.0	6.0	0.6	0.4
Mo	N.A.[e]	N.A.	9.0	11.0	7.0	7.0	2.0	1.3
Al	23.0	22.0	10.0	13.0	3.0	6.0	28.4	31.5
O	54.0	48.0	41.0	25.0	42.0	20.0	62.0	59.0
S	N.A.	3.0	N.A.	19.0	N.A.	21.0	N.A.	3.0
C	18.0	20.0	40.0	32.0	42.0	34.0	7.0	4.8
Co/Mo	∞	∞	0	0	1.0	0.9	0.30	0.31
Net C_4H_4S Conversion[f] (vol%)	N.A.	0.0	N.A.	2.0	N.A.	7.0[g]	N.A.	27.0[h]

[a] From XPS peak intensities and elemental sensitivity factors. The initial carbon (hydrocarbon) surface sited on the fresh specimens is greatly decreased after hydrogen and hydrogen sulphide exposures. Most of the carbon present after thiophene exposure is due to thiophene decomposition.

[b] Apparent surface area of each thin film specimen is 1 cm^2.

[c] Bulk composition is reported as 3 wt% CoO, 15 wt% MoO_3 (0.9 at% Co, 2.3 at% Mo) surface area of specimen used was 47 m^2 (0.225 gms) measured by BET N_2 adsorption.

[d] See text for experimental conditions.

[e] N.A. = not applicable.

[f] Net thiophene conversion is calculated using gas chromatography data in the following manner:

$$\% \text{ conversion} = \frac{Ap}{Ap + Bp} \times 100$$

where Ap is the total thiophene content in the gas phase after 1h desulphurization test. Non-catalytic conversion (i.e. reactor vessel walls) ranging from 4 to 6% has been subtracted in this calculation.

[g] Conversion rose to 17% after the 6 h test.

[h] Conversion rose to 52% after the 6 h test.

reactor, pressurizing with hydrogen to 172 kPa while rapidly heating to 623 K. Gas samples were withdrawn for GC analysis after 1 h and, in several cases, 6 h of reaction. Only Matheson ultra-high purity grade gases further purified by moisture and oxygen traps are used.

Series 1 (Co-alumina) specimens containing only Co (2 +) oxide (2 At%) are slow to reduce or sulphide and exhibit no desulphurization activity. Examination of one specimen by XPS (Table 1) after 6 h of thiophene exposure indicates 70% cobalt in the sulphided state (CoSx) [Co $2p_{3/2}$ = 778.8 eV] with the remainder most likely $CoAl_2O_4$ [Co$2p_{3/2}$ = 782.1 eV]. The S 2p spectrum indicates a single peak located at 162.2 eV attributable to a Co sulphide. The surface composition data also indicates a CoS type species (i.e. Co/S $\simeq$ 1). Re-examination by XPS after a 1 min air exposure indicates the sulphided cobalt was re-converted to a Co(2 +) oxide or oxy-sulphate species. The Co 2p 778.8 eV peak intensity decreased significantly and a second peak in the S 2p spectrum at 168.8 eV, attributable to sulphate, is observed.

Examination of series 2 (Mo-alumina) specimen by XPS (Table 1) indicated the Mo present can consist of Mo(6 +) (232.6 eV) and Mo(4 +) (229.6 eV) oxides in varying ratios. Only Mo(4 +) and minor Mo(5 +) oxides are observed by XPS, however, after the reduction step. Hydrogen sulphide and/or thiophene in hydrogen exposures cause 95% Mo (4 + , 5 +) oxide sulphidation to Mo(4 +) (228.8 eV). A trace of residual Mo(5 +) oxide is measured, but the S 2p region indicate only 1 doublet centered at 161.8 eV (MoS_2). Unlike the series 1 case, the sulphided Mo–Al thin films have some HDS activity, converting $\simeq$ 5% of the injected thiophene to hydrocarbons after 1 h (Table 1). Composition data (Table 1) indicates the Mo/S ratio is nearly 0.5, the theoretical value for MoS_2.

Analysis of series 3, 4 and 5 specimens by XPS reveals calcination at a constant set of conditions produces Co–Mo–Al oxides having similar surface chemistries regardless of the metal deposition sequence; surface composition and compounds formed can vary greatly however if the calcination conditions change. A specimen from series 4 (Mo on Co on Al oxide) is discussed here, however the behaviour of a series 3 or 5 specimen calcined identically gives very similar results. Examination of the series 4 specimen by XPS, Table 1, Figure 1, reveals Mo present only as Mo(6 +) oxide (MoO_3 or $Al_2(MoO_4)_3$) while Co sites as Co(2 +) and Co(3 +) (Co_3O_4 and minor $Co(OH)_2$ or $CoAl_2O_4$). After reduction, 70% of the Mo now occurs as Mo(4 +) oxide, the remainder as Mo(5 +) oxide with the mixed Co oxides reduced to metal and to a Co(2 +) species, likely CoO in equal amounts. The sulphiding step readily converts the Mo(4 + , 5 +) to Mo(4 +) sulphide. Only a few at% of the Mo(5 +) and Co(2 +) oxides still remain after sulphidation. This unsulphided Co(2 +) is likely sited as $CoAl_2O_4$. Examination of the surface by XPS after thiophene exposure, (Fig 1, B(a) and (b)), indicates no Co or Mo oxide remain. Note, the S 2s peak located at 225.6 eV in the Mo 3d spectrum (Figure 1. B(a)) is nearly half the Mo 3d peak intensity indicating sulphide anions are associated with the second metal. Note, the XPS Mo 3d spectrum of MoS_2 indicates a S 2s/Mo 3d ratio of 0.1. A high resolution XPS analysis of the S 2p region has revealed the presence of two S(2 −) species, one at 161.8 eV (MoS_2–like) and another at 162.2 eV (CoS–like). Also, the surface composition of this thiophene exposed sample shown in Table 1 indicates a Co + Mo/S ratio of 0.62 in good agreement with a 2 phase mixture of MoS_2 and CoS compounds (0.67). These results however, don't preclude the existence of the 'CoMoS' phase.[3,4]

An examination of the commercial catalyst by XPS, Table 1 and Figure 1C(a) and (b), indicates the presence of Mo sited as Mo(6 +) and Co as Co(2 +). The XPS Co/Mo ratio (0.30) is very similar to the value expected from theory

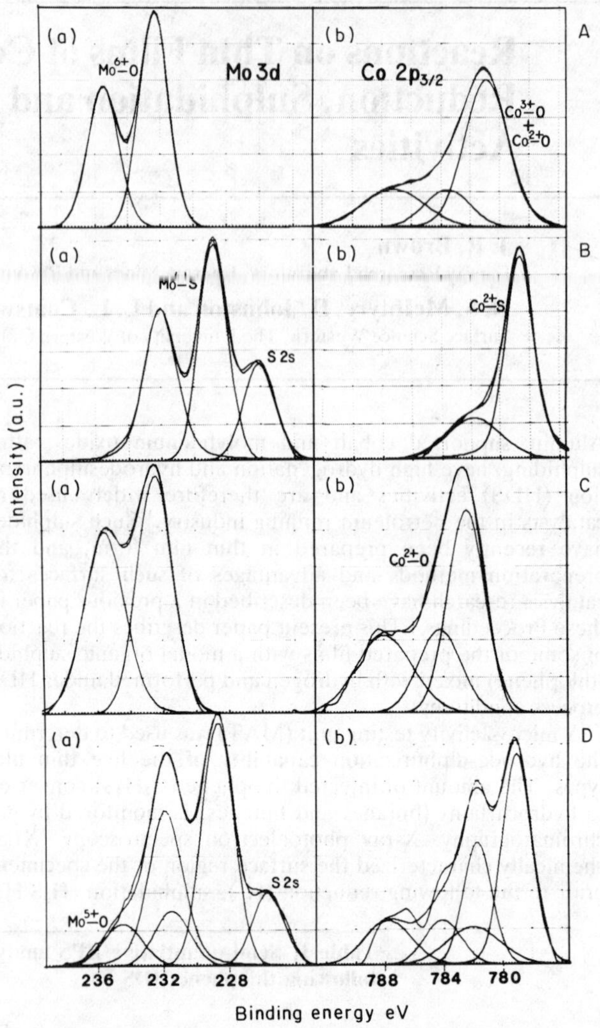

Figure 1. Mo 3d and Co $2p_{3/2}$ XPS spectra of 2 catalyst samples. Spectra labelled A and B refer to the 'series 4' thin film and spectra C and D refer to the commercial catalyst prior to and following HDS testing. The Mo 3d peaks were fitted using a mathematical routine where the spin-orbit doublet splittings and relative intensities are constrained.

(0.39). After the reduction step, 3 surface Mo states are present, Mo(6 +), Mo(5 +) and Mo(4 +) in the ratio 15 : 3 : 2 while only 10% of the Co is reduced to Co metal. Sulphidation further changes these ratios; only Mo(4 +) and Mo(5 +) are now present and in nearly equal amounts while half the Co(2 +) and probably all the Co metal is sulphided. Exposure to thiophene for 1 h results in conversion of half the Mo(5 +) oxide to Mo(4 +) sulphide; after a further 6 h, 90% of the Mo present is sited as a Mo(4 +) sulphide species, Figure 1 D(a). This desulphurization step also increased the CoS–type component to nearly 70% of the total Co present. Note, the Co sited in the commercial catalyst is much more difficult to reduce or sulphide than Co in the series 4 thin film although the Mo components behave similarly. This effect is probably the result of the larger Co/Mo ratio present in the thin film causing less Co-Al-O interaction. The reduction and sulphidation behaviour of several Co-Al oxide films was very similar to this commercial catalyst, specimens all thought to have a large $CoAl_2O_4$–like component.

In conclusion, series 2, 3, 4 and 5 thin films have been found to catalyse the desulphurization of thiophene with conversions ranging from 7 to 22 vol%. The commercial catalysts' HDS activity was much lower than several films (calculated on a unit surface area basis) leading one to speculate inexpensive metallic and bimetallic alumina supported catalysts of high activity might be manufactured by employing modern thin film high technolgy. Even if this goal is never achieved, designer thin film catalysts are providing information highly relevant in catalyst R/D because the physical properties of the layer catalyst (electrical conduction, surface smoothness and surface size, permitting easy handling and mounting, etc.) are ideal for surface spectroscopic investigations.

REFERENCES

1. D.C. McCulloch, in *Applied Industrial Catalysis*, Vol. 1, pp. 70–121, ed. by B. E. Leach, Academic (1983).
2. N.S. McIntyre, D. Johnston, L.L. Coatsworth and J.R. Brown, *Surf. Interface Anal.* **9**, 253 (1986).
3. H. Topsoe, B.S. Clausen, R. Candia, C. Wivel and S. Morup, *J. Catal.* **68**, 433–452 (1981).
4. I. Alstrup, I. Chorkendorff, R. Candia, B.S. Clausen and H. Topsoe, *J. Catal.* **7**, 397–409 (1982).

An ISS Study of Platinized Titania

Gar B. Hoflund, Austin L. Grogan, Jr. and Douglas A. Asbury
Department of Chemical Engineering, University of Florida, Gainesville, Florida 32611, USA

The phenomena of strong-metal-support interactions (SMSI). which was first reported by Tauster *et al.*,[1,2] has been a controversial topic in the field of catalysis during the last seven years. SMSI phenomena is characterized by suppression of the amounts of H_2 and CO which chemisorb on oxide-supported metal crystallites. This occurs after a high-temperature (500°C) reduction in hydrogen and particularly for reducible oxide supports. Annealing in oxygen followed by a low-temperature (200°C) anneal in hydrogen usually restores most of the chemisorption capacity. Numerous explanations of SMSI phenomena have been proposed including suggestions of an electronic interaction between the supported metal and the oxide, i.e. charge transfer,[3–9] changes in metal crystallite morphology.[10] encapsulation of the metal crystallites by support species which form during reduction of the support,[11–22], encapsulation of the metal crystallites by impurity species, creation of special active sites at the metal–support interface, hydrogen spillover and others.

From the studies cited above, it is clear that encapsulation of the metal crystallites does occur and that the electronic effects can be insignificant,[20] small[17] or quite significant.[10] However, the techniques used thus far have not been capable of specifying which species are responsible for the encapsulation. In this study ISS is used to examine platinum crystallites supported on both high purity polycrystalline titania and impure TiO_2 (001) before and after encapsulation is induced. An attempt is made to identify the encapsulating species. The possibility of success using ISS is greater than for other techniques due to its extremely high surface sensitivity. It is essentially outermost layer sensitive because the ions which penetrate below the outermost layer are either neutralized or scattered inelastically thereby not contributing to the ISS peaks. ESCA, AES and SSIMS were also performed, and the results are compared with ISS.

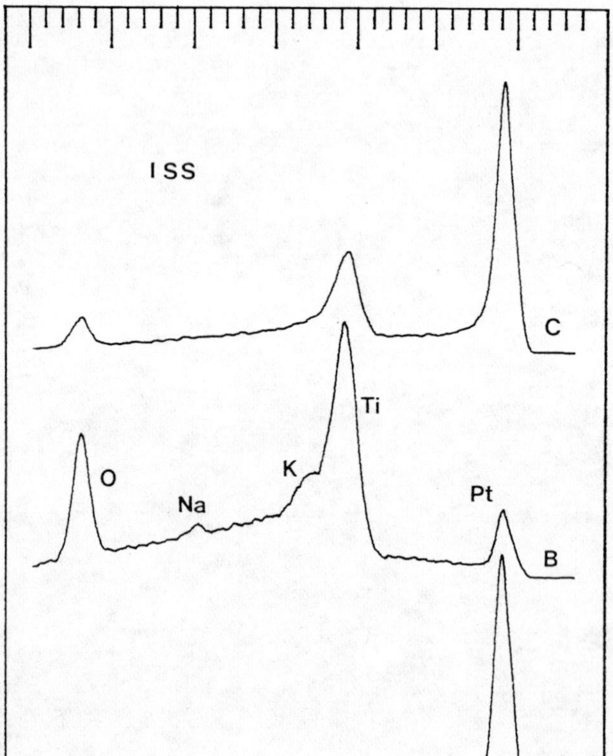

Figure 1. ISS spectra taken from a platinized, polycrystalline TiO_2 surface (A) before heating, (B) after heating at 400°C for 4 hours in 10^{-6} Torr of oxygen and (C) after sputtering.

The samples were heated at different temperatures for varying lengths of time and in different environments to assess the effects of each variable on the encapsulation process. A typical set of ISS spectra are shown in Fig. 1. Spectrum A was taken before heating the sample. The surface species are oxygen, titanium, sodium, potassium and platinum. The platinum peak is much larger than the others, but it does have the largest ISS cross section. Spectrum B shows the same surface after heating at 500°C in 10^{-6} Torr of oxygen for 4 h. Dramatic changes have occurred. Now the platinum peak is much reduced with respect to the titanium and oxygen peaks. Clearly, the platinum crystallites have been encapsulated by titania species. It is possible that the stoichiometry of the encapsulating species corresponds to TiO_2 as estimated from the sizes of the oxygen and titanium peaks and the ISS cross sections. This will be discussed further elsewhere. Spectrum C shows the ISS spectrum taken from the same surface after a light sputter with argon ions. The encapsulating species are removed by the sputtering again exposing a surface which is quite similar to the original surface. However, the sodium and potassium which had segregated to the surface are no longer present.

Encapsulation of the platinum crystallites also occurs on the impure, single crystal Ti (001) surface. However, in this case large amounts of sodium, potassium and sulphur segregate to the surface during the heating, and the oxygen-to-titanium ratio is decreased.

ESCA, AES and SSIMS all show results which are consistent with the ISS studies. However, the results are much less dramatic because AES and ESCA are less surface sensitive than ISS and the SSIMS cross section for platinum is extremely small. ESCA shows that reduction of the titania occurs during annealing and to an even greater extent during sputtering. This is also confirmed by high-resolution AES studies of the most prominent titanium peaks at about 400 eV.

REFERENCES

1. S.J. Tauster, S.C. Fung and R.L. Garten, *J. Am. Chem. Soc.*, **100**, 170 (1978).
2. S.J. Tauster and S.C. Fung, *J. Catal.*, **55**, 29 (1978).
3. C.C. Kao, S.C. Tsai and Y.W. Chung, *J. Catal.*, **73**, 136 (1982).
4. M.K. Bahl, S.C. Tsai and Y.W. Chung, *Phys. Rev. B*, **21**, 1344 (1980).

5. S.C. Fung, *J. Catal.,* **76**, 225 (1982).
6. S.H. Chien, B.N. Shelimov, D.E. Resasco, E.H. Lee and G.L. Haller, *J. Catal.,* **77**, 301 (1982).
7. B.A. Sexton, A.E. Hughes and K.J. Foyer, *J. Catal.,* **77**, 85 (1982).
8. J. Horsley, *J. Am. Chem. Soc.,* **110**, 2870 (1979).
9. S.J. Tauster, S.C. Fung, R.T.K. Baker and J.A. Horsley, *Science,* **211**, 1121 (1981).
10. M.A. Vannice, L.C. Hasselbring and B. Sen, *J. Phys. Chem.,* **89**, 2972 (1985).
11. X-Z Jiang, T.F. Hayden and J.A. Dumesic, *J. Catal.,* **83**, 168 (1983).
12. R.T.K. Baker, E.B. Prestridge and L.L. Murrell, *J. Catal.,* **79**, 348 (1983).
13. P. Meriaudeau, O.H. Ellestad, N. Dufaux and C. Naccache, *J. Catal.,* **75**, 243 (1982).
14. D.E. Resasco and G.L. Haller, *J. Catal.,* **82**, 879 (1983).
15. J.A. Cairns, J.E.E. Baglin, G.J. Clark and J.R. Ziegler, *J. Catal.,* **83**, 301 (1983).
16. H.R. Sadeghi and V.E. Henrich, *J. Catal.,* **87**, 279 (1984).
17. D.N. Belton, Y-M Sun and J.M. White, *J. Phys. Chem.,* **88**, 5172 (1984).
18. S. Taketani and Y-W Chung, *J. Catal.,* **90**, 75 (1984).
19. R.A. Demmin, C.S. Ko and R.J. Gorte, *J. Catal.,* **90**, 59 (1984).
20. J. Santos, J. Phillips and J.A. Dumesic, *J. Catal.,* **81**, 147 (1983).
21. A.J. Simoens, R.T.K. Baker, D.J. Dwyer, C.R. Lund and R.J. Madon, *J. Catal.,* **86**, 359 (1984).
22. Y-W Chung, G. Xiong and C.C. Kao, *J. Catal.,* **85**, 237 (1984).
23. D.J. Dwyer, S.D. Cameron and J. Gland, *Surf. Sci.,* **159**, 430 (1985).

XPS Studies on Doped Vanadia Catalysts in CO Oxidation

G. Mink, I. Bertóti, T. Székely
Res. Lab. Inorganic Chemistry, Hungarian Academy of Sciences, H-1502 Budapest, Pf. 132, Hungary

INTRODUCTION

In oxygen transfer reactions V_2O_5 catalysts are considered as redox systems.[1-3] Transition metal additives generally improve their catalytic activity,[4] however, the probable influence of the transition metal on V_2O_5 is still questionable The catalytic oxidation of CO as a toxic polluting material is of great practical importance. The kinetics of this reaction in the presence of catalysts such as vanadium pentoxide unsupported, supported and even doped with $PdCl_2$ has been studied in a pulse reactor connected to a GC and the chemical state of the catalysts exposed to CO or CO + O_2 has been investigated by *in situ* XPS experiments using a KRATOS XSAM 800 instrument. Although $PdCl_2$ itself, impregnated upon a gamma alumina support, proved to be a good catalyst of the reaction, kinetic studies on V_2O_5/Al_2O_3 catalysts doped with $PdCl_2$, have revealed a synergic effect of Pd resulting in a tremendous increase of the catalytic activity together with the decrease of the apparent activation energy. The observed synergic effect is interpreted in terms of the increased liability of vanadium towards an alteration of its oxidation state in oxygen transfer reactions.

RESULTS AND DISCUSSION

Catalysts were as follows: A: powder like V_2O_5, s^0 = 3 m^2g^{-1}; B: 10% $V_2O_5/\gamma\text{-}Al_2O_3$; C: 7% $PdCl_2/\gamma\text{-}Al_2O_3$; D: 7% $PdCl_2$/10% $V_2O_5/\gamma\text{-}Al_2O_3$ and E: 0.28% $PdCl_2/V_2O_5$ of $s^0 = 4.8$ m^2g^{-1}. The catalysts were preoxidized at 720 K in air for 1 h and in O_2 flow for 0.5 h.

The activity order (per unit weight) was

$$A < B \ll C < D \approx E$$

and the activation energies for catalysts A–B and C–E were 85 ± 15 and 45 ± 7 kJ/mol^{-1}, respectively.

As Pd doped V_2O_5/Al_2O_3 (catalyst D) was far more active than catalysts B and C, we suppose a kind of synergic effect between Pd and V, stimulating the redox process on the V_2O_5 component, too.

The redox nature of catalysts were studied by ESCA, by exposing the samples to CO under mild conditions (p = 10 kPa, T = 320 K, t = 30 min). On catalysts A and B, as well as on single crystalline V_2O_5 no reduction of V_2O_5 (neither broadening, nor shifting or splitting of the V 2p lines) was observed.

On Pd-doped catalysts (D and E), of course, reduction of Pd^{2+} occurs resulting in a splitting of the Pd 3d lines. At the same time, the total amount of the finely dispersed V_2O_5 on catalyst D is reduced to a lower valence state vanadium characterized by a shift of the whole V $2p_{3/2}$ line towards lower *BE*.

In the case of catalyst E, where Pd is deposited on V_2O_5 of considerable particle size, the partial reduction of Pd^{2+} is accompanied by the surface reduction of V_2O_5 resulting in a shoulder of about 10% intensity on the low BE side of the V $2p_{3/2}$ peak. Consequently, Pd seems to stimulate the reduction of the V_2O_5 surface, i.e. enhances the redox nature of V_2O_5.

REFERENCES

1. A. Andersson, *J. Solid State Chem.*, **42**, 263 (1982).
2. A. Andersson, S.T. Lundion, *J. Catal.*, **58**, 383 (1979).
3. Zs. Roder, G. Mink, I. Bertóti, *Supplement to the Proceedings of the 8th ICTA, Bratislava, 1985*, pp. 67–70. Slovak Technical University Ed. 1985; *Thermochim. Acta* (in press).
4. C.L. Thomas, *Catalytic Processes and Proven Catalysts*, p. 110 Academic Press New York and London (1970).

Effect of the Pretreatment on the Surface Structure of Amorphous $Fe_{80-x}Ni_xB_{20}$ Alloys and their Catalytic Activity and Selectivity in the CO + H_2 Reaction

L. Guczi, G. Kisfaludi, Z. Schay and A. Lovas*
Institute of Isotopes of the Hungarian Academy of Sciences, H-1525 Budapest, PO Box 77, Hungary
* Central Research Institute of Physics, H-1525 Budapest, Hungary

Amorphous alloys show superior catalytic activity in the CO + H_2 reaction[1,2] and in the hydrogenation of olefins.[3] It has been shown[2] that the superior activity of the $Fe_{80}B_{20}$ alloy is attributed to small iron clusters stabilized by boron oxide. Crystllization starts from the surface already at 570 K and can be detected only by UPS and conversion electron Mössbauer spectroscopy (CEMS). The accessibility of iron clusters should depend on parameters of the preparation and pretreatment. The effect of these parameters on the catalytic activity has not been studied systematically yet.

In this work the rapid quenching method was used for the preparation of $Fe_{80-x}Ni_xB_{20}$ (x:0,5,10,20,40) ribbons about 10 mm wide and 10–30 μm thick. The catalytic properties of the ribbons in the CO + H_2 reaction were measured in a flow reactor at 520–570 K and at atmospheric pressure. The ribbons were characterized before and after the catalytic reaction by XPS, UPS including depth profiling by Ar ion etching. In some cases electrochemical polarization was also measured.

Among the parameters of the preparation the cooling rate of the melt has the largest effect on the catalytic properties. On increasing the rpm value of the spinning disk from 810 to 2700 the reaction rate decreased from ~ 30 nmol s^{-1} m^{-2} to 250 nmol s^{-1} m^{-2}_{cat} while there was only a modest change in the selectivities. This indicates that only the number but not the structure of the catalytically active sites has been changed. A correlation was observed between the olefine formation and the polarization current. XPS and UPS showed that both iron and boron are in oxidized states while nickel seemed to be in the zero valent state.

The catalytic reaction itself changed the structure of the topmost layer. In every case the thickness of the oxide interface increased and a slight enrichment in boron was observed.

Pretreatments before the catalytic reaction also affected the catalytical properties. While mechanical polishing, anodic polarization and HNO_3 etching proved to be ineffective, cathodic polarization and HCl etching increased the catalytic activity about a hundred fold and decreased significantly the methane selectivity. It was proved by electrochemistry that iron and boron were dissolved preferentially from the alloys by HCl. XPS showed that the oxide layer also became thinner on HCl etching. Some chlorine was left on the surface which was completely removed by the CO + H_2 reaction mixture.

The main conclusions of the present study are:

(a) the catalytic properties of amorphous ribbons are very sensitive to the conditions of preparation (e.g. cooling rate influences the structure of oxide interface) and pretreatment;

(b) the oxide interface belongs to the active components of the catalyst and has a primary effect on the catalytic properties;

(c) pretreatments which change the structure of the oxide interface also affect the catalytic properties;

(d) the oxide interface masks the intrinsic catalytic properties of $Fe_{80-x}Ni_xB_{20}$ alloys;

(e) the oxide interface drastically changes in the CO + H_2 reaction.

REFERENCES

1. A. Yokoyama, H. Komiyama, H. Inoue, T. Masumoto, H. Kimura, *J. Catalysis*, **68**, 355 (1981).
2. G. Kisfaludi, K. Lázár, Z. Schay, L. Guczi, Cs. Fetzer, A. Lovas, G. Konczos, *Appl. Surf. Sci.*, **24**, 225 (1985).
3. S. Yoshida, H. Yamashita, T. Funabiki, T. Yonezawa, *J. Chem. Soc. Chem. Commun.* **655**, (1982).

CHARACTERIZATION OF Fe/Mn FISCHER–TROPSCH CATALYSTS BY CO-CHEMISORPTION, XPS AND AES MEASUREMENTS

H. Papp, D. Hess and T. Grzybek
Lehrstuhl für Technische Chemie, Ruhr-Universität Bochum, W. Germany

Fe/Mn catalysts of different composition have been prepared by coprecipitation followed by calcination at 500°C and reduction at 300°C or 400°C in flowing H_2. Fischer–Tropsch synthesis was performed at 11 bar and a H_2/CO ratio of 2. The catalysts were investigated by XPS, AES and CO-chemisorption.

After calcination an enrichment of Mn is observed in the surface of the Fe rich samples, whereas the surface composition of the Mn rich samples is identical to the bulk.

Reduction leads to a further increase in the Mn content of the surface of the Fe rich samples. The thickness of the enrichment layer is in the order of 200 nm. The irreversibly adsorbed amount of CO on the reduced catalysts is proportional to the Fe concentration on their surface. A correlation between the initial CO conversion during Fischer–Tropsch synthesis and the chemisorbed amount of CO can be established.

During Fischer–Tropsch synthesis the catalysts have been removed after different times on stream and were then analysed by XPS and AES; the Fe/Mn ratio in the surface is changing back to the value of the calcined samples. Two different C 1s peaks can be observed with XPS indicating graphitic and carbidic C on the surface during synthesis. The overall surface carbon concentration depends on the Fe content of the catalysts.

POISONING OF SILVER CATALYSTS FOR PARTIAL OXIDATION OF METHANOL BY IRON AND CHLORINE

R. Jede, E. Manske, O. Ganschow, A. Benninghoven
Physikalisches Institut der Universität Münster, Domagkstr. 75, D–4400 Münster, W-Germany

Formaldehyde formation by partial oxidation of methanol on a technical scale may suffer from iron and chlorine contamination of the catalyst. Using a combined method surface analysis system (SIMS/AES/XPS/ISS/TDMS) coupled to a technical microreactor,[1] we examined the partial oxidation reaction path and its modification by iron and chlorine both on model systems and on technical catalysts. Formaldehyde on the clean catalyst is formed by pairwise oxidative dehydrogenation of methoxy radicals, as indicated by secondary emission characteristics. This step involves only one of two atomic oxygen adsorption states, the other being completely inert with respect to methanol and populating a subsurface adsorption state in the temperature range 320–580 K. Iron coverages even less than 2% of a monolayer change the nature of both oxygen states so that they catalyse total oxidation of methanol instead of partial. Chlorine contaminations on the catalyst reduce the dissociative adsorption of oxygen and thus lead to a decrease in activity. The behaviour of the technical catalyst in model reactions in UHV could be correlated to overall kinetic data under technical conditions. For iron, the poisoning action depends on the amount of iron which can be incorporated in the bulk of the catalyst and its degree of oxidation under technical conditions. Because of this thermal instability on the silver at temperatures above 730 K, chlorine deteriorates the technical reaction only if there is a permanent supply from contaminated reactants.

Reference

1. O.Ganschow *et al.*, *J. Vac. Sci. Technol.* **A1**. 1491 (1983)

THE CHARACTERIZATION OF HETEROGENEOUS CATALYSTS BY QUANTITATIVE XPS

H.P.C.E Kuipers
Koninklijke/Shell-Laboratorium, Amsterdam, Shell Research BV, Badhuisweg 3, 1031 CM Amsterdam, The Netherlands

In order for the preparation and behaviour of a given catalyst to be understood, the catalyst must be characterized. The parameters related to catalyst performance include some geometric ones, which describe the surface areas of supported active phases and the location of these phases with respect to each other. Quantitative XPS can frequently tackle the difficult analytical task of characterization, therefore, because in a geometric sense it 'looks' at catalyst surfaces in a similar way to that in which the reactants do.

A prerequisite of meaningful characterization is the development of physical models which describe signal intensities as measured by XPS. In order to be reliable these models have to take explicitly into account the randomness of supported heterogeneous catalysts. From a study of three ideal cases, namely extended layers of constant thickness, equally-sized spheres and hemispheres on a randomly-oriented support, it is established that the effects of angular and layer-thickness averaging are interconnected in such a way that a general model can be derived. It is found that signal ratios as measured by XPS for convex particles are proportional to surface/volume ratios of the supported phase independent of particle shape. Thanks to the increase in the absolute accuracy of XPS–derived surface quantities which this represents, it becomes possible, by linking the results of XPS to those of other independent techniques, to determine the locations of the various compounds with respect to each other. The usefulness of this approach will be illustrated with an example.

Section 5.

Electronics, Semiconductors and Thin Films

SURFACE AND INTERFACE ANALYSIS, VOL. 9, 265–273 (1986)

Perspectives in Surface and Interface Analysis for Electronic Devices and Circuits

R. Heckingbottom
British Telecom Research Laboratories, Martlesham Heath, Ipswich, IP5 7RE, UK

Over the past two decades a variety of techniques have been developed for the analysis of surfaces and interfaces. Despite the multiplicity of interfaces in electronic devices however, and the need for precise control, surprisingly few of these techniques have had a major impact in the area. Scientific successes such as low energy electron diffraction (LEED) find little application, and a straightforward comparison of the strengths and weaknesses of Auger electron spectroscopy (AES) and x-ray photo-electron spectroscopy (XPS) gives little indication of the massive preference for the former in analysing devices and circuits. A similar disparity in usage has also arisen between dynamic and static SIMS, only the former being significant. It is also clear that several of the more important techniques can be used to analyse thin films as well as surfaces and interfaces. Further, that the techniques are often more quantitative in their application to films. Again it is surprising how little pressure there has been to minimise the shortcomings with respect to interface analysis.

With the benefit of hindsight, some of the features that have led to certain techniques becoming particularly useful in application will be considered. Some new trends in device and circuit research which may increase the emphasis on surface and interface analysis will be identified. In this new area the growing importance of techniques such as transmission electron microscopy, not automatically included as a leading surface and interface analysis technique, will be illustrated.

INTRODUCTION

Over the past two decades surface and interface analysis has grown into a major branch of science. At the same time the electronics industry has provided a major arena for the application of this analysis through its emphasis on precise control of multilayer structures in the devices and circuits used. None the less it is becoming well recognised (by the inauguration of this new conference series and the appearance of several new journals for instance) that the overall scientific field does not cater well for the major applications interest. This review considers, with the benefit of hindsight, some of the major factors responsible for this divergence and some of the implications for the future. The techniques considered are all well established and basic familiarity with them is assumed — many good reviews exist and no attempt is made to repeat them here. Also there is no attempt to be comprehensive. On the contrary, most of the examples are drawn from work with which the author was closely associated, as knowledge of the framework within which the published work occurred is often important in developing the correct perspective.

As with most major developments the origins of surface and interface analysis as recognised today had more than one source. Something is clearly owed to the developments in vacuum and electron beam technology — scanning[1] and transmission[2] electron microscopy, SEM and TEM — during the 1950s. Electron spectroscopy for chemical analysis,[3] ESCA, also played a role in the early 1960s, though at first the surface sensitivity of the technique was largely unexploited. There is little doubt however that the major impetus to the topic was prompted by the relative explosion in interest in low energy electron diffraction (LEED) in the mid 1960s. Again nothing is absolute — LEED studies had continued[4] since their discovery in 1927[5] — but it is clear that the work at Bell Telephone Laboratories in the early 1960s[6] followed by the commercial availability of an appropriate ultra high vacuum system[7] effectively launched the expansion. LEED is therefore the first technique to be considered in more detail.

LOW ENERGY ELECTRON DIFFRACTION (LEED)

There is no doubt that LEED was and remains of major scientific importance. Before its advent the concepts of substantial surface rearrangements and ordered chemisorbed layers were only possibilities. There was simply no data and so detailed consideration of the topic was impossible. It is virtually inconceivable that anyone would have proposed the structure for the silicon (111) surface giving the electron diffraction pattern illustrated in Fig. 1. LEED has in fact lost its dominant position in surface science in general, not only in applied science, but it is important to recognise its importance in scientific terms before considering the reason for its relative unimportance in electronics today.

First of course there are limitations that can frustrate and challenge LEED practitioners themselves: (a) the detailed interpretation of the patterns themselves has proved to be extremely complex; (b) it is a broad area technique and (c) it is chemically ambiguous.

The theoretical difficulties have had major

0142–2421/86/120265–09 $05.00

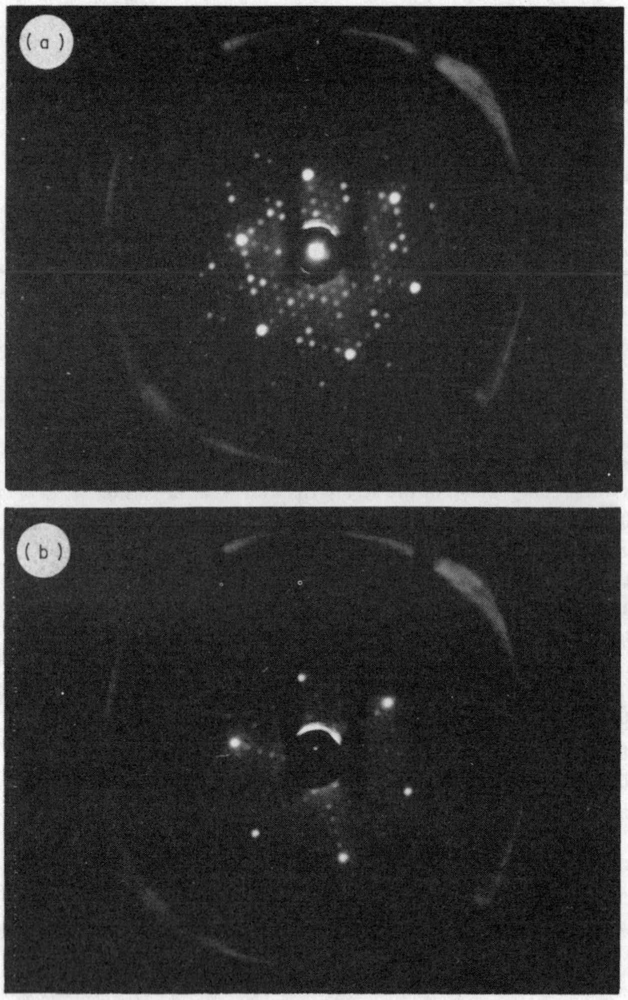

Figure 1. Low energy electron diffraction patterns from the Si(111)7 surface structure of (a) high quality and (b) much poorer quality.

implications for the scientific advance of the topic but this is most unlikely to be the reason for the low level of application. There is no difficulty in monitoring major changes at surfaces as a result of surface treatment either in terms of the quality of a given structure (compare Figs 1(a) and 1(b)) or change of structure (compare Figs 1(a) and 2(a)). The contrast shown in Fig. 1 as a function of time/temperature annealing schedule variations was first stressed by Jona.[8] Changes on chemical treatment as illustrated in Fig 2[9,10] formed the main activity in the late 1960s.

The lack of spatial resolution was less of a handicap for LEED, as its main application lay in surface characterisation and layer growth, than it was for most analytical techniques more appropriate to device and circuit assessment. It was important to recognise the effect of the large differences in electron beam width and coherence width for correct interpretation of diffraction patterns[11,12] but in general the lack of good spatial resolution was far from critical.

The chemical ambiguity was undoubtedly a more serious problem. LEED was in quite a literal sense before its time. In particular its image sufferred seriously because it was applied before the complementary technique of Auger electron

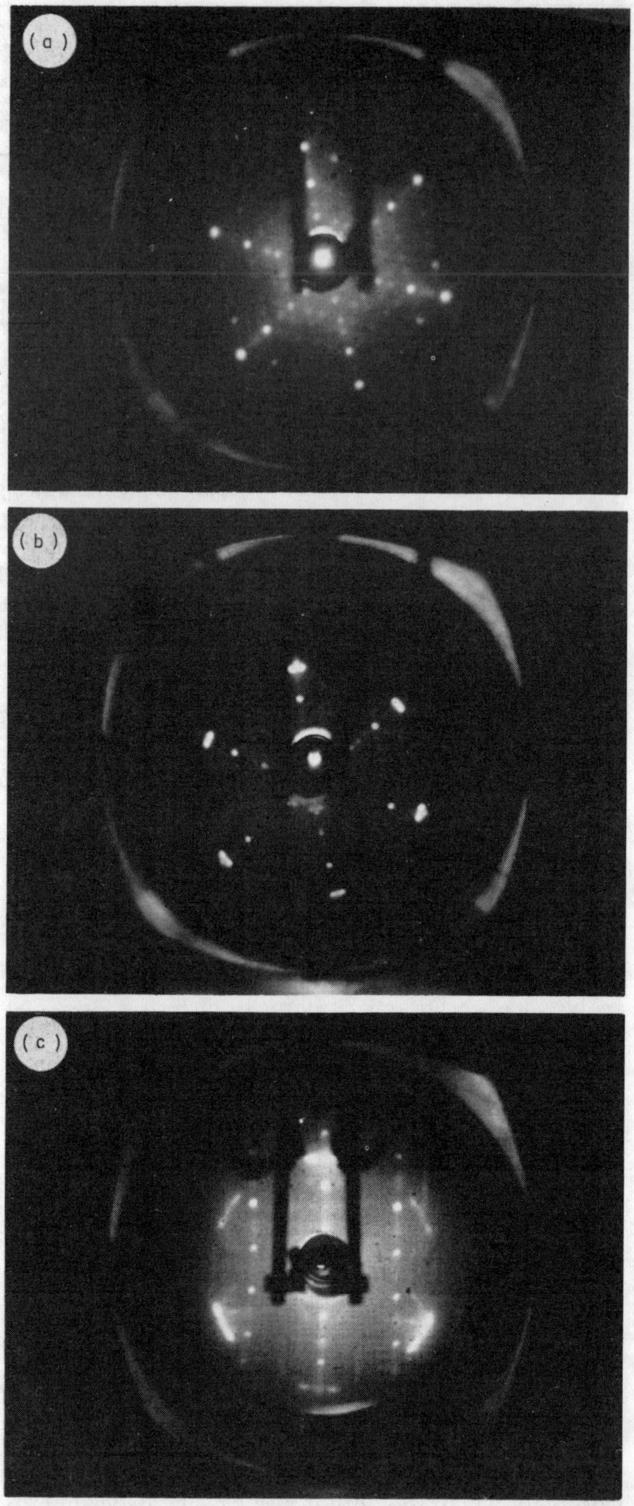

Figure 2. LEED patterns from nitrided Si surfaces, (a) the Si(111)8N pattern, (b) the Si(111) 'doublet' pattern and (c) the Si(311) 'arc' pattern.

spectroscopy (AES) was effectively developed. Figure 3 illustrates the basic confusion that could arise. The identification of Fig. 3(a) as characteristic of the clean surface and Fig. 3(b) as being due to carbon contamination was only possible after months of study using LEED as the primary analytical technique.[13] More complex problems such as that illustrated in Figs 2(b) and 2(c) had to await the application of AES for

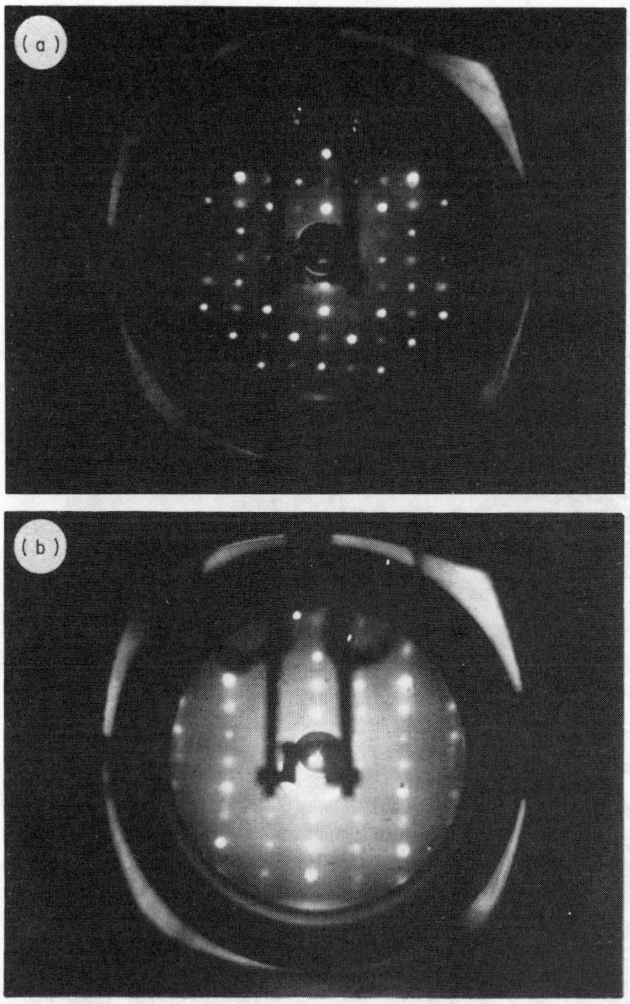

Figure 3. LEED patterns of the Si(311) surface from (a) clean areas and (b) areas contaminated with carbon.

elucidation. Here overlayers of related structure appear on silicon substrates of different symmetry. The overlayer is predominantly silicon nitride as planned but small concentrations of silicon carbide are necessary to nucleate the overlayer structure. LEED alone could not resolve problems like this.

Powerful though this case appears it is not the answer — AES was developed but LEED did not then return to prominence and the main reasons lie not with surface science but with the industry.

The basic problem facing LEED, along with all surface analysis techniques is that for industrial application all ultra clean surfaces (so necessary for definitive scientific research) must be converted into stable interfaces to be useful. Thus in general interface analysis is much more important than surface analysis. The latter is of primary importance as a monitor for surface control during its conversion into an interface. Here again LEED was before its time in terms of industry's needs. The silicon surfaces studied by LEED were incorporated into interfaces with either amorphous silicon dioxide or polycrystalline metals. These were either too accommodating or too imperfect respectively for any detail derived from the silicon surface to have any noticeable effects.

Only now is there scope for surface diffraction techniques to play an important role with increased interest in heteroepitaxial crystal growth. Electron diffraction is an integral part of molecular beam epitaxy which has been the major growth technique in the development of low dimensional structures opening up a new branch of device physics. As it turns out the related technique of reflection electron diffraction (RED) is preferred to LEED. While RED does not give as much information quite so readily its glancing angle configuration allows its application in the growth chamber in a more convenient way. Even then its initial role was largely confined to the determination of suitable substrate treatments prior to growth. Very recently however[14] the oscillation in spot intensity with layer by layer growth is leading to its application as the most precise monitor of layer thickness and interface perfection in III-V heterojunction superlattices typified by GaAs/GaA1As.

AUGER ELECTRON SPECTROSCOPY (AES) AND X-RAY PHOTOELECTRON SPECTROSCOPY (XPS)

By 1970 Auger electron spectroscopy was well established as a surface analysis technique[15] and the potential of XPS for surface analysis was also well recognised.[16] Both techniques represented major advances and during the 1970s comparisons between the two were made frequently. In general there was a slight preference for XPS due to its greater scope for interpretable chemical shift data and to its reduced effect (charging, desorption) on the sample. In applications for electronics however AES was incomparably more useful, with most of the advantages stemming from its much superior spatial resolution. While better lateral resolution is an advantage in many applications, it was the dimensions of the designed-in non-uniformity implicit in semiconductor devices and circuits (5 μm in the mid-seventies and 1 μm later in the decade) that inspired the performance standards of the commercial AES systems. Average values of surface composition are clearly inappropriate in the situation illustrated in Fig. 4. The problem is essentially one of identifying something chemical about the failed device which correlates with its atypical electronic behaviour.

The pressure to exploit this advantage of spatial resolution greatly affected the development of AES. Scanning systems quickly appeared and analysers which gave improved collection from finely focused (and hence lower current) beams — at the expense of optimum energy resolution — were preferred. An added benefit was also quickly recognised. As stressed earlier, the primary emphasis is on interfaces not surfaces. In general surfaces like that shown in Fig. 4 are typically covered with oxide plus some carbon. Even to identify nominally surface contaminants one has to remove the very top layers usually by ion sputtering. Most problems involve designed-in interfaces and their degradation with time or treatment and their analysis demands sputter profiling to depths of up to 1 μm or more. High resolution AES greatly facilitates this depth profiling as analysis can be confined to the central (flat bottomed) region of the sputtered crater and the latter can in turn be smaller in

Figure 4. Optical micrograph of a typical integrated circuit.

extent without 'edge effects' intruding. This feature together with many instrumental technique developments provide the fast effective profiling familiar today. This relative efficiency in use helps make AES the key technique it has become despite the fact that it is destructive in the profiling mode.

Although AES has assumed a dominant position with respect to XPS in electronics applications both are needed and the following comments concerning interface analysis apply to both. The techniques actually apply best to thin film analysis. It is from thin uniform films which are thicker than either the electron escape depths or the uncertainities in the sputter erosion depth that the most quantitative data can be derived. As one passes through the interface region, information about the original interface becomes subject to uncertainties due to the two factors mentioned above plus sputter induced mixing and matrix effects. There have been a range of initiatives — lower ion beam energies, heavier ion beam species and electron escape depth correction — tackling these problems. Figure 5 shows the corrections to a model abrupt interface of silicon nitride on gallium arsenide for electron escape depths (Fig. 5(a)) and sputter ion broadening (Fig. 5(b)). Comparison of the corrected model with the observed profiles gives a clear indication of non-stoichiometry — a silicon rich film — close to the interface. Despite the additional information to be obtained from the more detailed analysis there has been comparatively little work of this kind. The dominant conclusion has to be that for the majority of applications these refinements are not vital.

Most AES and XPS profiles in electronics are of the kind illustrated in Fig. 6 for the gold/titanium metallisation.[17] Comparison of Figs 6(a) and 6(b) shows the extent of the contamination at the interface formed in a conventional vacuum system. Profiles such as those in Figs 6(c) and 6(d) show the degradation of the two interfaces with time/temperature anneals on accelerated life-test. Comparison with electrical data

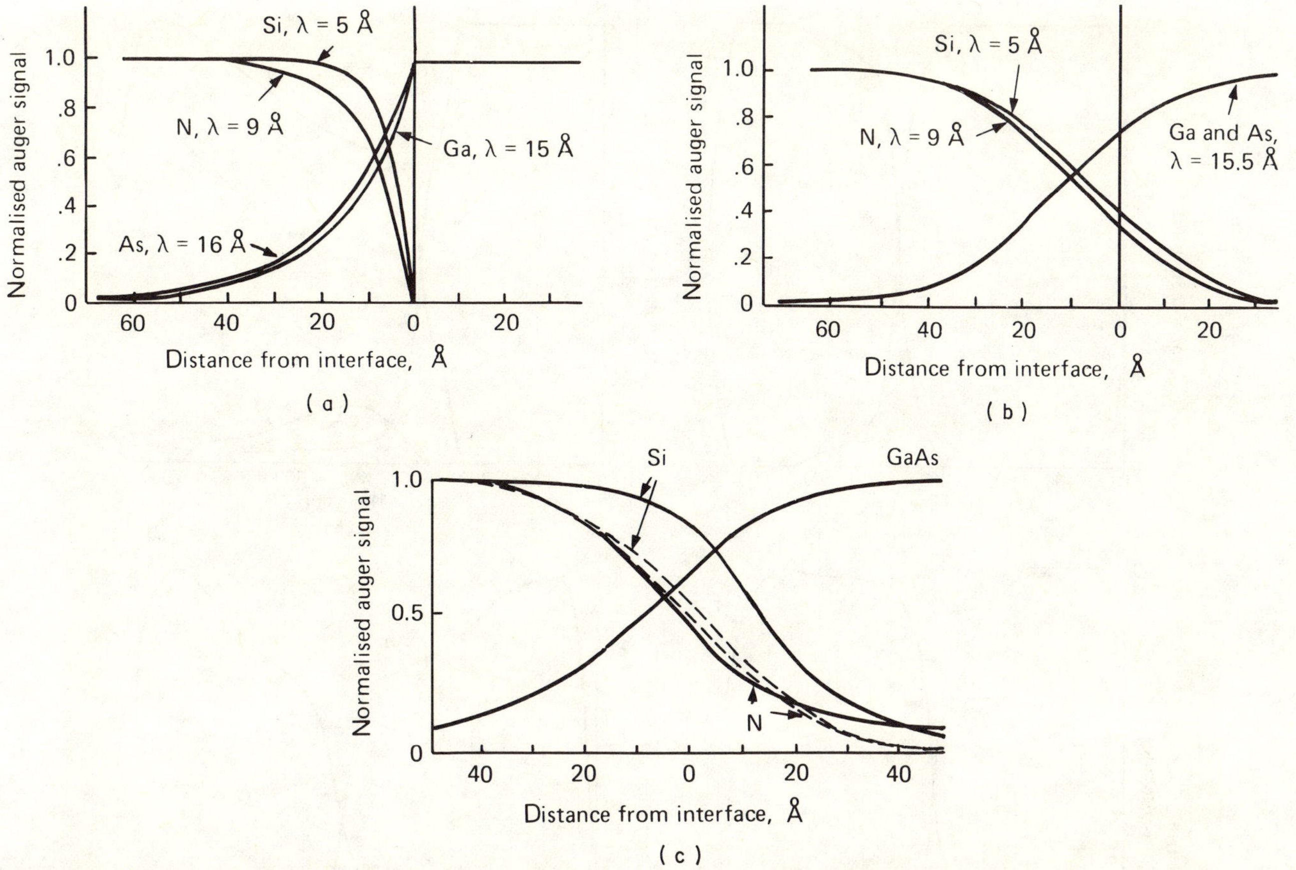

Figure 5. Nonstoichiometry at the Si_3N_4/GaAs interface: (a) model interface with Auger electron escape depth correction, (b) with addition of sputter broadening and (c) comparison of calculated --- and observed — profiles showing silicon rich region near the interface.

established that the observed contamination levels in Fig 6(b) were essential to produce the long operational life needed in highly reliable transistors. Most impact has come from broad conclusions like the above, across a wide range of problems.

SECONDARY ION MASS SPECTROMETRY (SIMS)

The one area of chemical analysis of primary importance to the electronics industry, that neither AES nor XPS can tackle concerns dopants in the semiconductor itself — at concentrations in the part per thousand to part per billion range. SIMS has been able to offer sensitivities in this range since 1970 or thereabouts, for favourable dopant/host matrix combinations, but it was not until the end of the decade that the technique began to make a significant impact in electronics applications.

Basic SIMS is characterised by a very large range in sensitivity, not only between different elements but for the same element in different matrices. While this variability is potentially a source of information, it effectively represented too great a level of uncertainty to be useful for many problems in electronics. Major advances came with the use of primary beams of oxygen and caesium which maximised the yields of positive and negative ions respectively. These were used for depth profiling so that the primary beam ions implanted in the sample dominated the matrix properties. This approach, coupled with the use of ion implants in the matrices as standards, established dynamic SIMS as a powerful technique for thin film analysis.[18] Unfortunately most of these advances are inapplicable to surface analysis and static SIMS as the broad area, almost non-destructive version is termed, has failed to establish a similar role in electronics. Even the destructive dynamic SIMS becomes semiquantitative at interfaces and some of the current research is aimed at reducing the extent of the surface and interface regions. A typical example of dynamic SIMS applied to thin film analysis is shown in Fig. 7. SIMS shows that the unwanted variation in electrical activity is not due to faulty operation of the sulphur doping cell but to a previously unexpected contamination by manganese at a level that does vary through the growth run.[19] Contamination at an interface is shown in Fig. 8. The information is obviously of use as a basis for monitoring surface cleaning procedures but it is uncertain whether the increase in dopant signal at the interface actually represents segregation or merely reflects change in sensitivity with matrix. There is little doubt that it is the usefulness in thin film analysis that has established dynamic SIMS as an important technique in electronics. The base that now exists will however allow remaining problems connected with interface analysis to be tackled with renewed vigour.

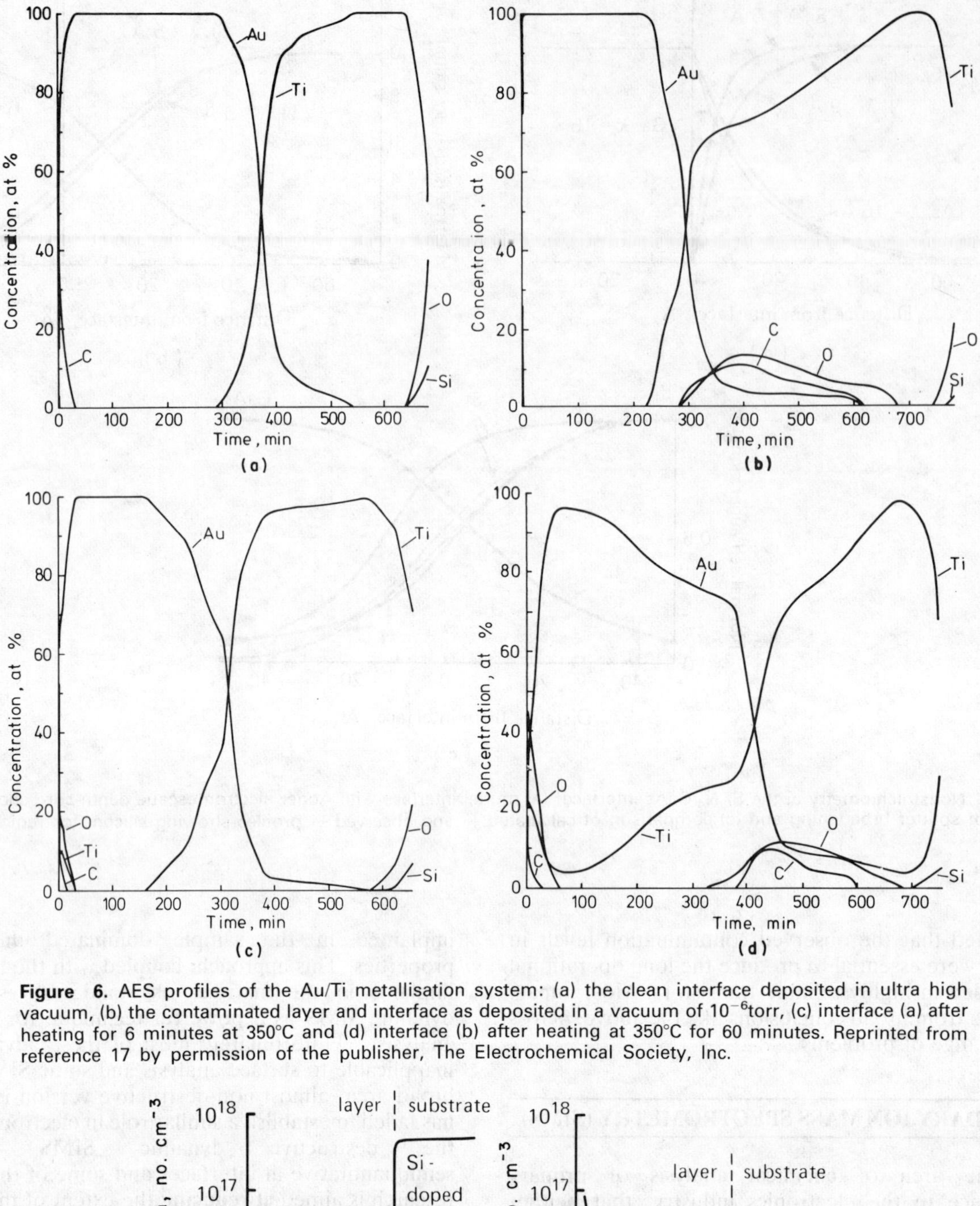

Figure 6. AES profiles of the Au/Ti metallisation system: (a) the clean interface deposited in ultra high vacuum, (b) the contaminated layer and interface as deposited in a vacuum of 10^{-6}torr, (c) interface (a) after heating for 6 minutes at 350°C and (d) interface (b) after heating at 350°C for 60 minutes. Reprinted from reference 17 by permission of the publisher, The Electrochemical Society, Inc.

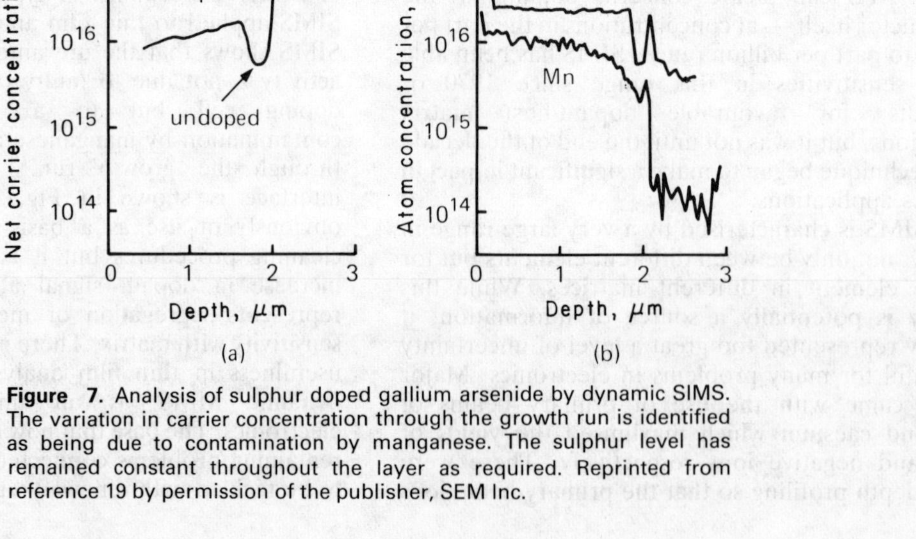

Figure 7. Analysis of sulphur doped gallium arsenide by dynamic SIMS. The variation in carrier concentration through the growth run is identified as being due to contamination by manganese. The sulphur level has remained constant throughout the layer as required. Reprinted from reference 19 by permission of the publisher, SEM Inc.

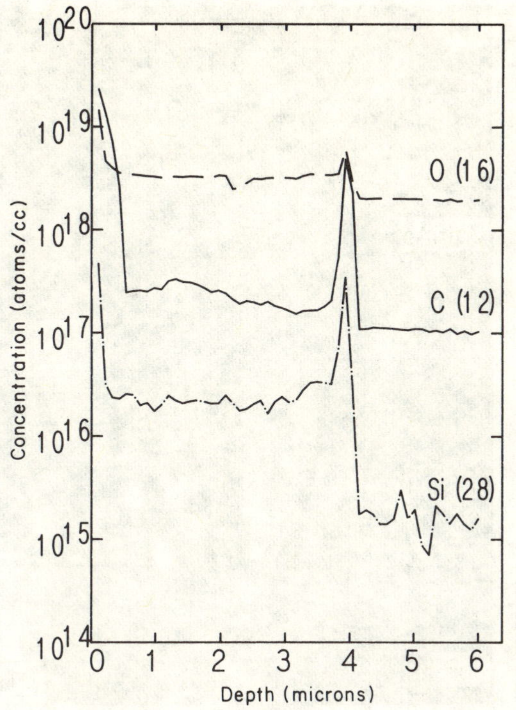

Figure 8. SIMS profiles showing O and C contamination at an interface together with a peak in the Si dopant profile.

RECENT TRENDS

In the examples described so far it is clear that the most useful techniques have been best suited to thin film analysis. Further, that the electronics industry has found the lack of precision in surface and interface analysis at least acceptable even though not ideal. This impression would rightly remain even if the review were extended to include other important techniques such as Rutherford back-scattering. It was commented in the assessment of electron diffraction however that some recent trends do suggest the need for an increased emphasis on surface and interface analysis in the near future. A major research area in electronics at present involves low dimensional structures particularly but not exclusively in compound semiconductors. As is now well known the approach requires the presence of a layer so thin that the bulk properties no longer apply and new energy levels, influenced by the layer thickness, are formed. A single layer bounded by two heterojunctions with material of higher band gap is called a quantum well. Multi quantum well structures result if a superlattice of the two materials is grown. The new structures offer important advances in electronics such as higher speed devices and lower threshold lasers. As the layer thicknesses required are $\leqslant$100 Å the influence of interface perfection is even greater than in conventional devices.

In the majority of cases considered so far the layers in these superlattices have been single crystal with not only the same crystal structure in each layer but the same lattice parameter to within one part per thousand. The single crystal nature, together with the fact that the layer thicknesses are comparable with typical sputter broadening in profiling techniques, has led to a greatly increased reliance on transmission electron microscopy. Samples are thinned for examination perpendicular to the growth direction even though this makes sample preparation even more time consuming. Figure 9 illustrates some of the basic information that can be obtained about interfaces by this approach.[20]

Advances in silicon technology are also generating increasing complexity normal to the plane of the main circuity. A particular example is provided by the variety of approaches being explored to produce silicon on insulator structures. One of these approaches — formation of an insulating oxide layer by oxygen ion plantation — is shown in Fig. 10. Again cross section transmission electron microscopy reveals the layer sequence and the crystal perfection of the silicon layers while Auger electron spectroscopy has been used to determine the oxygen concentration profile.[21]

A final strength of TEM is its ability to provide images with resolution of columns of atoms. A lattice image of a GaAs/GaA1As interface is illustrated in Fig. 11. The power of the approach is obvious even though the interpretation of the images is much more complicated than at first appears.

CONCLUDING REMARKS

Adding the word 'applied' to surface and interface analysis has major implications. A technique for surface and interface analysis is rightly considered powerful if it can be used to generate impressive amounts of new scientific information. For the same technique to be useful to an industry several more criteria must be met. Firstly the industry must have relevant problems, secondly the technique must be able to address the problem in the context posed by the industry and thirdly the technique must be applied to aspects of the problem that the industrialist can act upon. Thus, in the first instance, the electronics industry requires stable interfaces in devices and circuits. A surface technique such as LEED only has an impact if the details of the crystal/vacuum interface carry over and affect the subsequent crystal/solid interface. In the second instance, in the electronics industry, many problems only arise after many stages of processing or during 'life-testing' so that high spatial resolution is a necessity in many cases. In the third instance many interesting conditions are simply intolerable in a device, eg. heavy implantation damage — it is of limited use studying the fine detail of the problem unless it is part of a strategy for solving it.

While it is true that most analytical facilities in electronics companies are established to deal with potential problems that might arise, it is important to note that most of the applications are now likely to be in a related role. Thus it is in the development of new crystal growth sequences or device processing steps that the analytical techniques are most used. It is here that the technologist can best respond to and incorporate the information provided by the analyst. The use of electron diffraction for *in situ* monitoring in MBE is an extreme example of this approach and one that has provided an important role for the technique whereas fault analysis did not.

A third, less glamorous but important application of

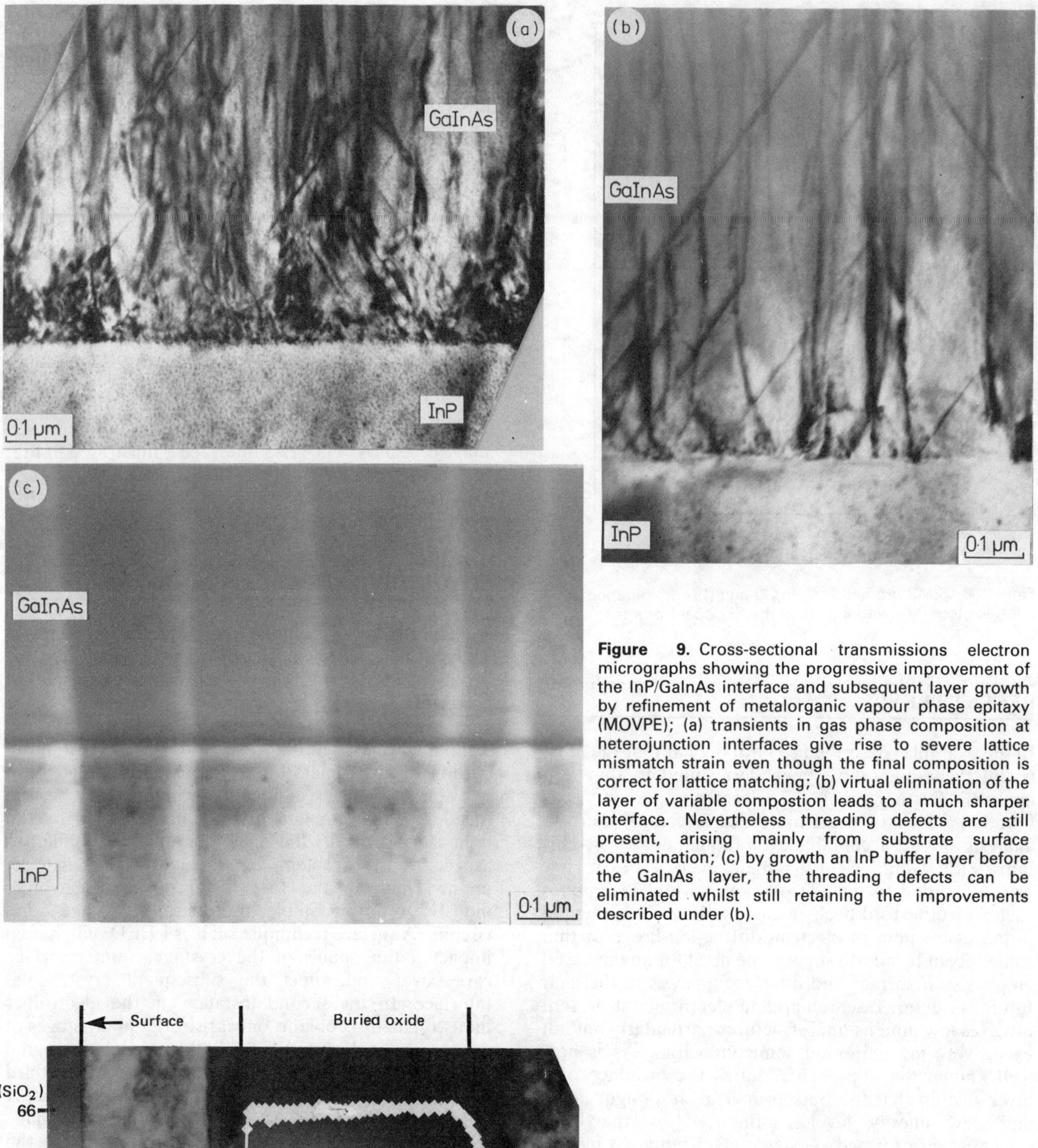

Figure 9. Cross-sectional transmissions electron micrographs showing the progressive improvement of the InP/GaInAs interface and subsequent layer growth by refinement of metalorganic vapour phase epitaxy (MOVPE); (a) transients in gas phase composition at heterojunction interfaces give rise to severe lattice mismatch strain even though the final composition is correct for lattice matching; (b) virtual elimination of the layer of variable compostion leads to a much sharper interface. Nevertheless threading defects are still present, arising mainly from substrate surface contamination; (c) by growth an InP buffer layer before the GaInAs layer, the threading defects can be eliminated whilst still retaining the improvements described under (b).

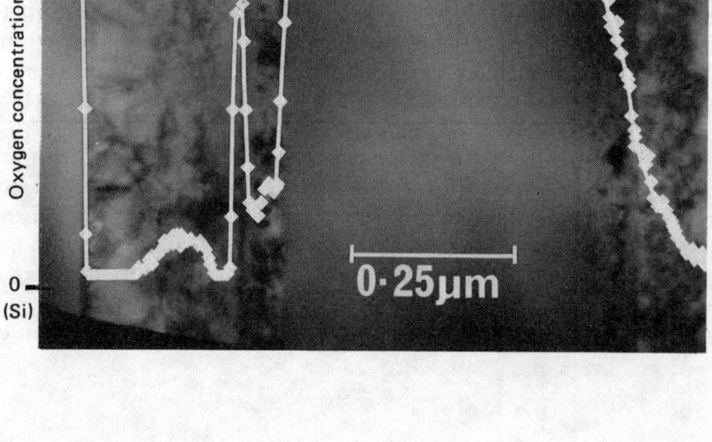

Figure 10. Cross section TEM, with the Auger oxygen depth profile overlayed, for an oxygen implanted silicon sample. An unexpected and important subsidiary feature on the flank of the main silicon dioxide layer is revealed. Reprinted from reference 21 by permission of the American Institute of Physics.

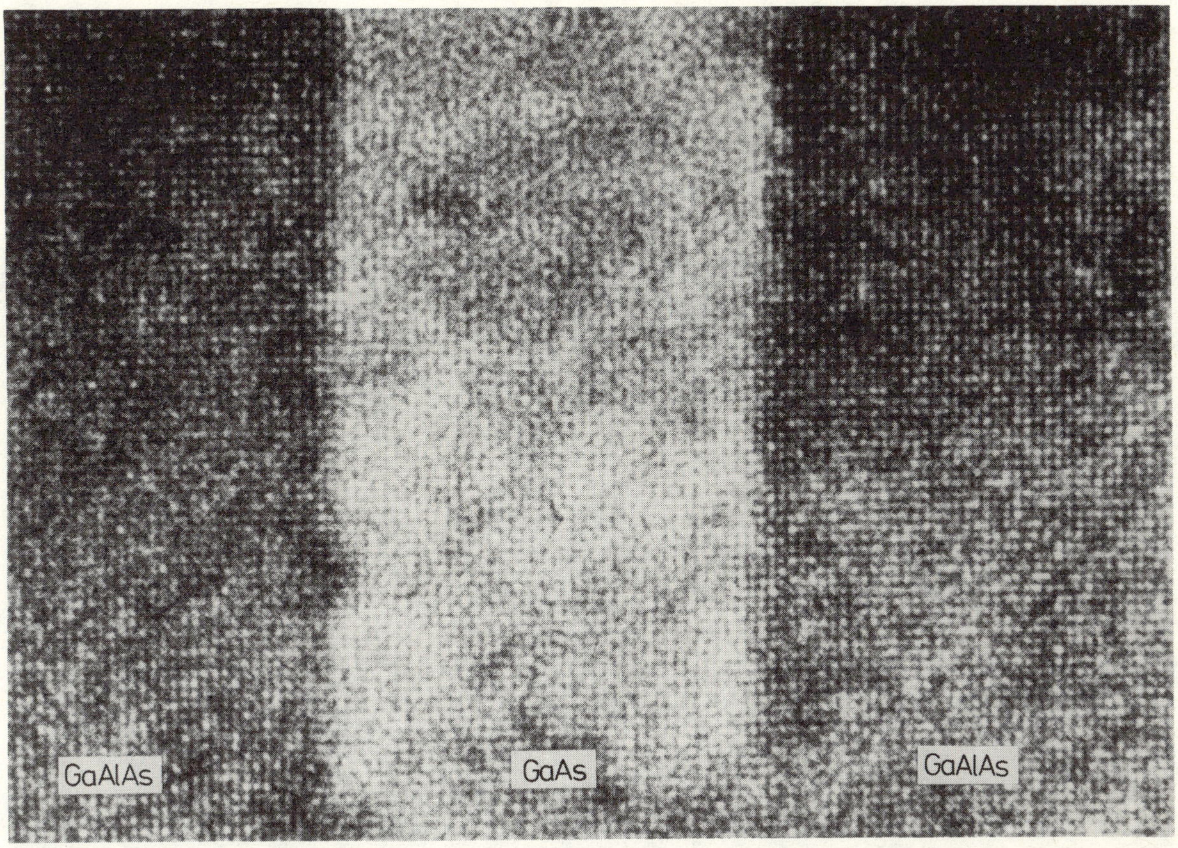

Figure 11. Lattice image of a GaAs/GaAlAs interface — (100) cross section taken with the incident beam parallel to (100).

these techniques is also worth noting. Even when a process is established, the device and circuit fabrication is both expensive and time consuming. It is often worth using surface and interface analysis techniques (even destructively on part of the sample) to determine whether the early processing meets specification before proceeding to the later stages.

Finally one has to recognise that the electrical engineer wishes surfaces and interfaces were just the inactive planes indicated in sketches of devices and circuits. He would prefer not to need analysts. On the other hand, the current trends towards smaller devices and circuit elements and the increasing use of multilayer structures seem likely to guarantee the need for surface and interface analysis in electronics for the foreseeable future.

Acknowledgements

I would like to thank M R Taylor, M Hockly and C J D Hetherington for providing the electron micrographs and GDT Spiller for the SIMS profiles. Acknowledgement is also made to the Director of Research of British Telecom for permission to puolish this paper.

REFERENCES

1. C.W., Oatley, W.C. Nixon and R.F.W. Pease, *Advances in Electronics and Electron Physics,* **21,** 181 (1965).
2. V.E. Coslett, *Practical Electron Microscopy,* Butterworths, London (1951).
3. K. Siegbahn *et al., Electron Spectroscopy for Chemical Analysis,* Nov, *Act, Reg. Soc. Sci. Upsaliencies Ser IV,* **20,** (1967).
4. R.E. Schlier and H.E. Farnsworth, *J. Appl. Phys.* **25,** 1333 (1954).
5. C.J. Davisson and L.H. Germer, *Phys. Rev.* **30,** 705 (1927).
6. J.J. Lander, *Prog. Solid State Chem,* **2,** 26 (1965).
7. Varian Associates Inc, USA.
8. F. Jona, *Surf. Sci.* **8,** 478 (1967).
9. R. Heckingbottom in *The Structure and Chemistry of Solid Surfaces,* ed. by G.A. Somorjai, pp. 78.1 – 78.19, John Wiley, New York (1969).
10. R. Heckingbottom and P.R. Wood, *Surf. Sci.* **36,** 594 (1973).
11. J.E. Houston and R.L. Park, *Surf. Sci.* **21,** 209 (1970).
12. R. Heckingbottom, *Surf. Sci,* **17,** 394 (1969).
13. R. Heckingbottom and P.R. Wood, *Surf. Sci.* **23,** 437 (1970).
14. J.H. Neave, B.A. Joyce, P.J. Dobson and N. Norton, *Appl. Phys.* **A31,** 1 (1983).
15. P.W. Palmberg, G.K. Bohn and J.C. Tracy, *Appl. Phys. Lett.* **15,** 254 (1969).
16. C.R. Brundle in *Surface and Defect Properties of Solids, Vo. 1,* pp. 171 – 204, The Chemical Society, London (1972).
17. G.W.B. Ashwell and R. Heckingbottom, *J. Electrochem. Soc.* **128,** 649 (1981).
18. H.H. Werner, *Surf. Interface. Anal.* **2,** 56 (1980).
19. T. Ambridge, in *Scanning Electron Microscopy,* pp. 31–43. SEM Inc Chicago (1983).
20. M. Hockly and E.A.D. White, *J. Cryst Growth* **68,** 334 (1984).
21. C.G. Tuppen, M.R. Taylor, P.L.F. Hemment and R.P. Arrowsmith, *Appl. Phys. Lett.* **45,** 57 (1984).
22. C.J.D. Hetherington, J.C. Barry, J.M.d Bi, C.J. Humphreys, J. Grange and C.E.C. Wood, *Mat. Res. Soc. Symp. Proc.* **37,** 41 (1985).

SURFACE AND INTERFACE ANALYSIS, VOL. 9, 275–281 (1986)

Study of Near-surface Disorder and Surface Residues after Reactive Ion Etching of Silicon

G.S. Oehrlein, G.J. Coyle,† J.G. Clabes, and Y.H. Lee
IBM Thomas J. Watson Research Center, Yorktown Heights, NY 105988, USA
†IBM East Fishkill Facility, Hopewell Junction NY 12533, USA

Helium ion channelling, nuclear reaction profiling and x-ray photoemission measurements of $CF_4/X\%H_2$ ($0 \leq X \leq 40$) and $CClF_3/50\%H_2$ dry etched Si surfaces are reported. It is shown that plasma exposure of a clean Si surface leads to the deposition of a thin (~30–50 Å thick) C,F-containing film (or C,F, Cl-containing in the case of $CClF_3/H_2$ etching). A subsurface Si carbide layer is formed during reactive ion etching. The near-surface region (~30–50 Å) of the Si substrate is heavily disordered as found by ion channeling. A modified, less damaged layer containing a high concentration of hydrogen is formed in the case of the H_2 containing gases (as shown by nuclear reaction profiling) and extends more than 250 Å from the surface. It is found that heating the plasma exposed substrates to 400°C in dry O_2 for 30 min is efficient in recovering good quality Si surfaces.

INTRODUCTION

Reactive ion etching (RIE) is a plasma-based dry etching technique which is used for fine-line pattern transfer in semiconductor processing.[1,2] Its etching characteristics are anisotropic because of energetic particle (0–1 keV) bombardment of the etched material during processing. Concerns with the use of RIE are that energetic ion exposure of electronic materials can result in displacement damage, and that surface residues might form due to plasma species dependent gas phase-surface reactions.[3,4] The goal of the present work is to establish via utilization of a multi-technique surface characterization approach (x-ray photoelectron emission spectroscopy, He ion channelling/scattering, H profiling) insight as to the major modifications of the near-surface properties of silicon resulting from prototypical dry etching processes.

EXPERIMENTAL

The preparation of the samples used in this work was the same as described in previous publications.[3,4] Briefly, 500 Å layers of thermal oxide on silicon were etched by $CF_4/X\%H_2$ (25 m Torr pressure, 40 sccm total flow) reactive ion etching.[5] The percentage X of H_2 was varied from 0% to 40%. The Si substrates were exposed for 1 min to the plasma after the SiO_2/Si interface was reached during etching (as detected by laser interferometry). A flexible diode reactor made of stainless steel with two 99.999% pure Al electrodes was employed.[1] The reactor was operated in the RIE mode, i.e. the substrate electrode was powered (13.56 MHz, 0.178 W cm^{-2}), while the top electrode was grounded.[1] Considerable ion bombardment of the substrate occurs for this configuration. In our case the maximum ion bombardment energy was ~450 eV (as given by the sum of the plasma potential and the sheath voltage).[1] The etchrate of the SiO_2 is ~500 Å min^{-1} at $100\%CF_4$ and decreases to ~400 Å min^{-1} at $CF_4/40\%/H_2$.[5] The etchrate of the Si decreases from ~400 Å/min at $100\%CF_4$ to ~20 Å/min at $CF_4/40\%H_2$.[5] The $CClF_3/50\%H_2$ reactive ion etched Si samples were prepared in a similar flexible diode reactor. The physically important variables of the etching process, e.g. overetch time (1 min), SiO_2 to Si etchrate ratio (400 Å min^{-1}: 40 Å min^{-1}), were also similar. For control samples the thermal oxide was removed by wet etching. All specimens were air exposed prior to analysis. The air exposure caused the growth of a thin oxide layer at the C,F(C,Cl,F-) film/Si interface.[3]

Helium ion channeling experiments were performed at 1 MeV utilizing the IBM van de Graaf accelerator. The incident beam was aligned with a Si <100> axis, while scattered particles were collected at an angle of 83° from the surface normal to obtain high surface sensitivity.[6] A beam dosage of 20 μC was used for all the experiments reported here (at a beam spot size of ~$(1.5\ mm)^2$). X-ray photoemission spectra (XPS) were obtained either with a Surface Science SSX-100 small-spot XPS instrument or a Hewlett–Packard ESCA 5950B spectrometer. The Si specimens were held against the front side of a Au coated Mo mask via a stainless steel spring mounted on the sample-back. No charging effects were observed. Photoelectrons were collected for various take-off angles.

EXPERIMENTAL RESULTS AND DISCUSSION

Channelling and Nuclear Reaction Profiling

In Fig. 1, He channelling spectra of Si crystals with a <100> axis aligned with the incoming beam and a 7° glancing angle exit geometry are shown. The 500 Å thick SiO_2 layer on top of the Si crystals had been wet etched (BHF control), or removed by $CF_4/40\%H_2$ RIE, $100\%CF_4$ RIE or $CClF_3/50\%H_2$ RIE. The Si substrates were exposed for 1 min to the plasma for

0142–2421/86/120275–07 $05.00

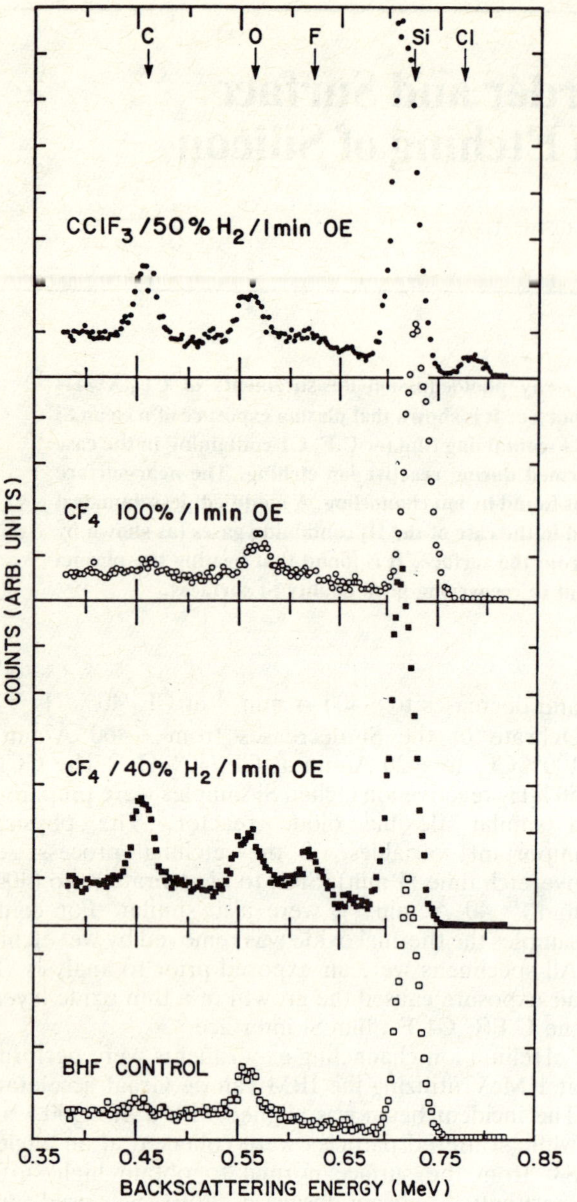

Figure 1. Helium ion channelling spectra of Si specimens plasma exposed for 1 min to a glow discharge of various dry etching gases. A buffered HF etched Si specimen is shown for comparison.

overetching. The control Si sample shows backscattering peaks due to Si atoms near the crystal surface, oxygen from the native oxide layer and carbon due to a hydrocarbon surface contamination. The CF_4/40%H_2 1 min plasma etched Si specimen differs from the control in that the Si surface peak and the C peak are of much greater intensity. Furthermore, an additional F peak is visible, while O-related signal appears similar as for the control. As discussed in Ref. 3, these changes are due to the presence of etching process-related fluorocarbon surface residues and non-lattice site Si atoms in the near-surface region of the substrate. In Ref. 2, the increase in Si surface peak intensity for CF_4/40%H_2 1 min RIE was found to correspond to ~40 Å of disordered Si. For a 100%CF_4 etched Si specimen, the changes are much less pronounced: The Si surface peak intensity is only slightly greater than for the control and the C and O intensities are the same. Also, little F is detected by ion scattering. The top curve in Fig. 1 was obtained with a $CClF_3$/50%H_2 etched Si sample. The RIE-induced changes are, in this case, very similar to the modifications caused by CF_4/40%H_2 etching, e.g. the strong increase in Si and C surface peaks, a similar O peak in both cases and additional surface F (less in the case of $CClF_3$/50%H_2 etching). An extra Cl-related peak is observed for the $CClF_3$/50%H_2 etched Si sample. No evidence of sputtered electrode or vacuum chamber material was seen in the channelling spectra.

The presence of H in the near-surface region of the CF_4/H_2 and $CClF_3/H_2$ dry etched specimens was established by the ^{15}N nucelar reaction profiling technique.[7,3] These results are displayed in Fig. 2, where hydrogen concentration in atomic percent is plotted versus silicon depth. We note that the control for which the SiO_2 was removed by buffered hydrofluoric acid and a 100%CF_4 dry etched sample show, within the experimental uncertainty, the same hydrogen profile. The H concentration close to the surface is greater than 10% due to H_2O in the native oxide (which is present in both cases, see Fig. 1). At a distance greater than 250 Å from the surface, the measured H concentration is equal to the background level. For the Si specimens etched with H_2 containing etching gases, the H concentration is much greater, both at the surface (due to the presence of the C,F- (or C,C1,F-) residue layers) and in the Si subsurface region (due to implanted H). In complementary Raman scattering studies we have found that some of the implanted H is bounded to Si and gives rise to a broad band near a Raman energy shift of 2000 cm^{-1}, due to Si-H stretching vibrations.[8]

X-ray Photoemission Measurements

A detailed XPS analysis of the dry etched Si surfaces was performed. Survey scans produced information

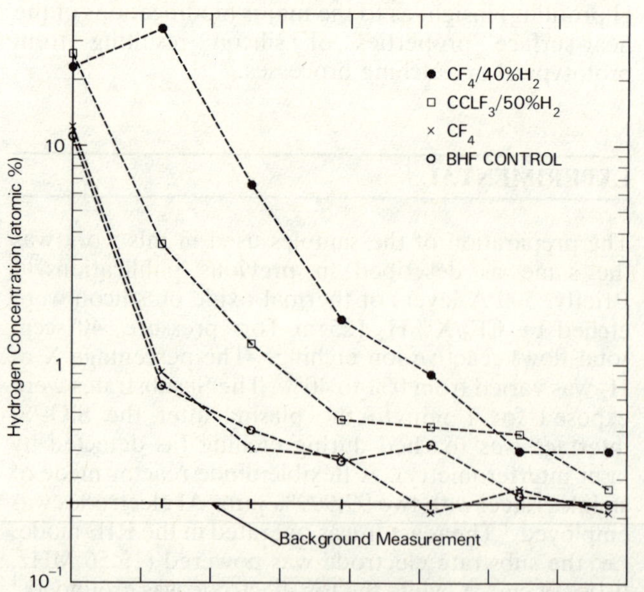

Figure 2. Nuclear reaction (N) hydrogen concentration profiles versus Si depth for reactive ion etched and wet etched contro Si specimens.

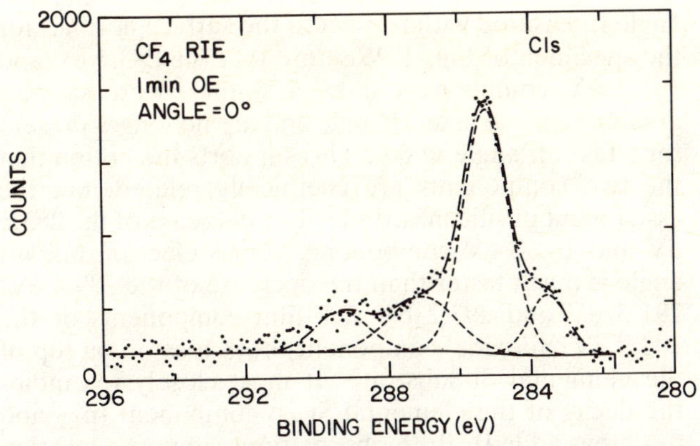

Figure 3. High resolution C 1s photoelectron spectrum measured for CF_4 etched Si specimen. The dashed curves represent the various peak-fitted chemical components. The solid line is the sum of all components.

consistent with the results of the ion channelling measurements of Fig. 1. No Al 2p peak at ~73 eV was observed. Therefore deposition of sputtered electrode material (99.999% pure Al) was negligible in our experiments. High resolution spectra of the carbon 1s and silicon 2p peaks provided additional insight. In the following, we will concentrate on the CF_4 and CF_4/40%H_2 etched Si specimens, since their analysis is more conclusive (due to the absence of surface C1). In Fig. 3 and Fig. 4, high resolution C 1s spectra of CF_4 and CF_4/40%H_2 dry etched samples are shown. The spectral envelope was resolved into its characteristic components by a least-squares fitting technique using Gaussians. Based on the work of others, who studied similar plasma polymerized films,[9,10] the fit was based on the presence and energy position of the following species (as referenced to a C—C and/or C—H binding energy position of the following species (as referenced to a C—C and/or C—H binding energy position of 285.3 eV): *C*—CF_x and/or CO at 287.2 eV (2.3 eV FWHM), CF at 289.4 eV (2.2 eV FWHM), CF_2 at 291.3 eV (1.5 eV FWHM) and CF_3 at 293.2 eV (1.3 eV FWHM), where FWHM denotes the full-width-at-half-maximum of the individual Gaussian. The greater FWHM of the 287.2 eV and 289.4 eV components is explained by the large number of chemical groupings represented by these Gaussians. As the number of bonds to F atoms is increased for a given carbon atom, the number of possible β-substituent patterns[9] decreases. Consequently, the FWHM of the peak representing that

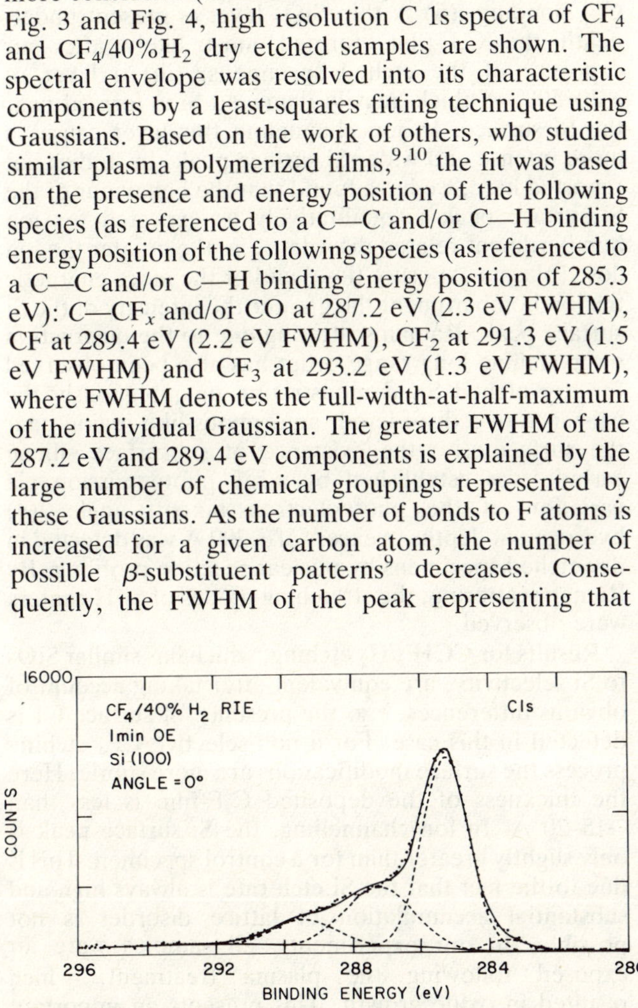

Figure 4. High resolution C 1s XPS spectrum for CF_4/40%H_2 etched Si.

carbon atom is also decreasing. The fits obtained are shown in Figs. 3 and 4. For the CF_4/40%/H_2 1 min overetched Si sample (Fig. 4) the lineshape analysis just described fulfills the following self-consistency check: The C/F-ratio of the C,F-films can be expressed as C/F=(Total C 1s envelope area)/(3xCF_3 component area + 2xCF_2 component area + CF component area). With this expression one finds C/F=4.82. The C/F ratio was directly determined in the ion scattering experiments and found to be C/F=4.92. This agreement supports strongly the soundness of the described deconvolution of the C 1s envelope into characteristic chemical groupings. It also suggests that the fluorocarbon films are rather homogeneous.

An additional peak at 283.5 eV was required to fit the data (see Figs. 3 and 4). In previously published work this peak was related to the presence of siliconcarbide in the Si substrate.[11] The presence of siliconcarbide gives rise also to a Si 2p component at 101.3 eV. We have re-examined the Si 2p structure with greater energy resolution than previously.[11] This resulted in the resolution of the shoulder in the elemental Si component due to the Si $2p_{1/2}$ and $2p_{3/2}$ splitting (see Fig. 5.) The Si 2p spectrum was fitted in the following manner: Two Gaussians were fitted to the principal ~99.6 eV Si peak (elemental Si). Their variability was constrained so as to leave the known photoelectron parameters, such as the branching ratio of 0.51 and energy separation of 0.61 eV,[12] invariant. The 'oxide' Si 2p component was fitted with a broader

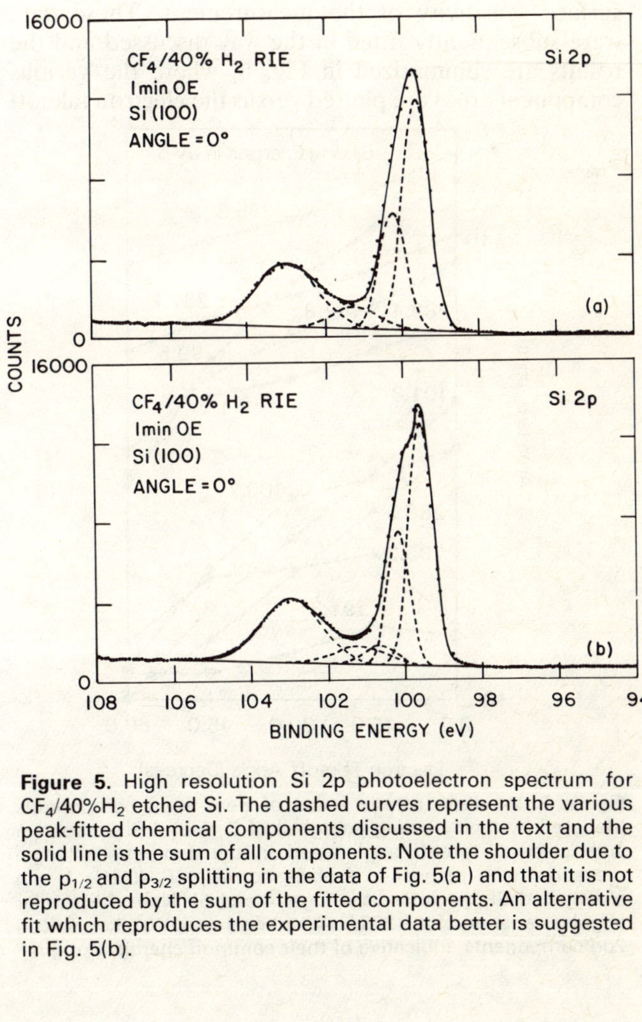

Figure 5. High resolution Si 2p photoelectron spectrum for CF_4/40%H_2 etched Si. The dashed curves represent the various peak-fitted chemical components discussed in the text and the solid line is the sum of all components. Note the shoulder due to the $p_{1/2}$ and $p_{3/2}$ splitting in the data of Fig. 5(a) and that it is not reproduced by the sum of the fitted components. An alternative fit which reproduces the experimental data better is suggested in Fig. 5(b).

peak at 102.8 eV. For the fit shown in Fig. 5(a), an additional silicon-carbide peak at 101.3 eV was used. It is clear from Fig. 5(a) that the overall Si 2p structure can be well represented by the assumption of three chemically distinct species. However, it is apparent from Fig. 5(a) that the asymmetry observed in the elemental Si component is not well reproduced by the sum of the $2p_{1/2}$ and $2p_{3/2}$ peaks. The peak fit can be improved considerably by reducing the FWHM of these physically linked components, as shown in Fig. 5(b). On the other hand, implementation of this procedure caused considerable degradation of the fit in the central region of the spectrum. This discrepancy can be eliminated in two alternative ways: (i) the SiC component at 101.3 eV may be broadened and shifted to lower binding energy. (ii) the inclusion of an additional peak at ~100.7 eV. Procedure (i) is unsatisfactory since the resulting SiC binding energy is 0.7 eV lower than the value measured in our previous model compound study of 101.3 eV.[11] We therefore adopted procedure (ii) and the results are displayed in Fig. 5(b). This fit is clearly superior to that shown in Fig. 5(a).

At this time both the inclusion of the peak at 101.7 eV to fit the data and its physical-chemical significance remain speculative. We note that a chemical species corresponding to SiF has been observed in photoemission measurements at this energy.[12] The presence of SiF in our samples can reasonably be expected. Work is in progress to clarify this question.

Similar spectra as shown in Fig. 4 and Fig. 5 were obtained at other electron takeoff angles to vary the surface sensitivity of the measurement. These data were subsequently fitted in the way discussed and the results are summarized in Fig. 6, where the various component areas are plotted versus the electron takeoff angle (measured with respect to the surface normal) for the specimen of Fig. 4. We note that the 283.5 eV and 101.2 eV components show a similar decrease with greater electron takeoff angle and are no longer present for a takeoff angle of 60°. This supports the notion that the two components are chemically related, i.e. the assignment of silicon carbide. The decrease of the 283.5 eV and 101.2 eV components versus electron takeoff angle is much faster than the decrease of the 289.4 eV, 291.3 eV and 293.2 eV C,F film components of the 102.8 eV 'Si oxide' component, both located on top of the elemental Si substrate. It most closely resembles the decay of the elemental Si 2p component ($p_{1/2}$ and $p_{3/2}$ areas added). Both observations support a location of the Si carbide within the elemental Si matrix.

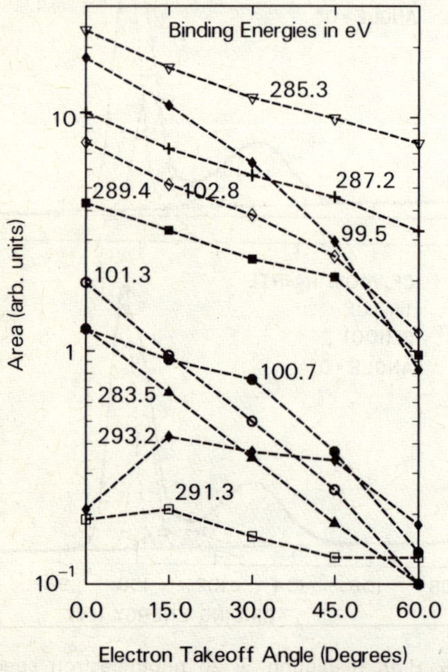

Figure 6. Variation of the areas of peak-fitted components (denoted by their binding energies) of C 1s and Si 2p spectra with electron takeoff angle used in the experiment (angle measured with respect to surface normal, $CF_4/40\%H_2$ 1 min overetched Si specimen, see Figs. 4 and 5). Note the similar angular variation of the 283.5 eV carbon 1s and 101.3 eV silicon 2pd components, indicative of their common chemical origin.

Composite View of Modified Si Near-Surface Region

Our results allow us to construct the following composite view of the modified Si near-surface region caused by SiO_2 to Si *selective* dry etching using the example of CF_4/H_2 (see Fig. 7): (i) in ion channelling and x-ray photoemission studies a C,F-film on the silicon was detected. The thickness of the film is limited to less than ~50 Å. the carbon 1s photoelectron spectrum shows a complicated fine structure, showing C—C and/or C—H, C—CF_x and/or CO, CF, CF_2 and CF_3 bonding, (ii) A thin SiO_2 layer is present underneath the C,F-film, formed most likely after air exposure of the etched specimens, (iii) a layer of ~30–50% Å thick heavily damaged Si is formed near the Si surface. In ion channeling this layer causes a more intense and wider Si surface peak. In studies not reported here, we have found that the behaviour of the Si surface peak is qualitatively as expected for ion bombardment related disorder, e.g. the greater the ion dose (plasma exposure time) and/or the more energetic the ion bombardment the greater the intensity of the Si surface peak. Raman scattering due to the destruction of crystalline long range order has also been observed for dry etched Si, demonstrating unambiguously the existence of a disordered, amorphous-like region near the surface; (iv) the existence of subsurface silicon carbide was established by x-ray photoemission via variation of the photoelectron takeoff angle; (v) hydrogen at depths in excess of ~200 Å was detected in dry etched specimens by nuclear reaction profiling. By Raman scattering, the stretching modes of Si-H centers were observed.

Results for $CClF_3/H_2$ etching, which has similar SiO_2 to Si selectivity, are equivalent after taking account of obvious differences, e.g. the presence of surface Cl is detected in this case. For a non-selective CF_4 etching process the surface modifications are more subtle. Here the thickness of the deposited C,F-film is less than ~15–20 Å. In ion channelling, the Si surface peak is only slightly greater than for a control specimen. This is due to the fact that the Si etch rate is always high and substantial accumulation of lattice disorder is not possible. In our experiments, all samples were air exposed following the plasma treatment, which resulted in oxide growth. This presents an important limitation to our ability to detect displacement damage within the first 10–20 Å of the Si crystal. Some of the

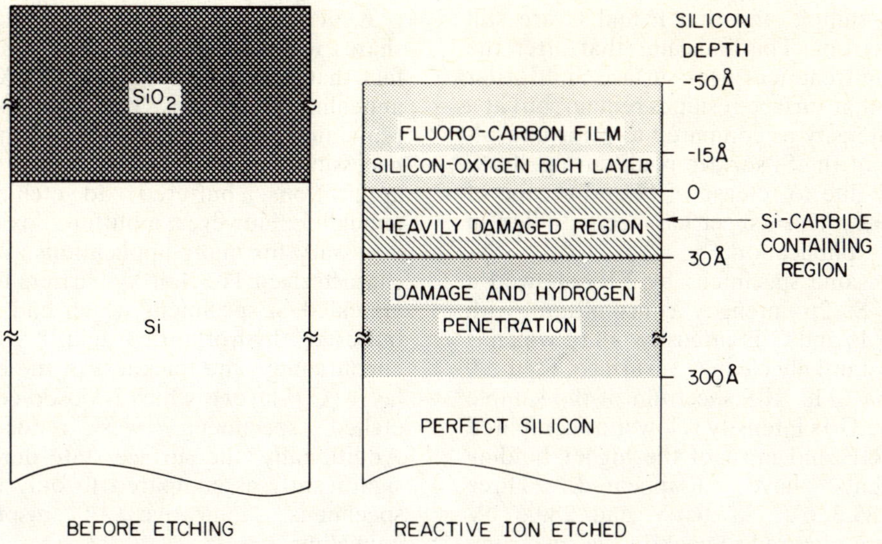

Figure 7. Schematic view of changes in Si near-surface properties caused by SiO_2 to Si selective reactive ion etching via CF_4/40%H_2.

disordered layer will be consumed during oxide growth, making the oxide-converted damaged Si region in ion channelling indistinguishable from oxide-converted undamaged Si. As expected, in the pure CF_4 etching case no H concentration enhancement as compared to the wet etched control was detected. This important difference should serve to reduce etching related Si substrate disorder, as experimentally observed.

Our results can qualitatively be explained by the fluorine-to-carbon ratio concept of Coburn and Winters,[13] which predicts C,F-film deposition for a low F/C ratio and etching for a high F/C ratio. Oxygen addition (from the etching of SiO_2) will increase the F/C ratio, due to formation of CO or CO_2 and etching will proceed. Upon encounter of the SiO_2/Si interface, the F/C ratio will drop due to the absence of O and result in F,C-deposition (or F,C,Cl-deposition in the case of $CClF_3/H_2$) and a small Si etch rate. Hydrogen addition to CF_4 decreases the F/C ratio by formation of HF, leading to greater C,F-film deposition in the case of CF_4/40%H_2 etching as compared to usage of pure CF_4.

Post-RIE Surface Recovery Treatments

The necessity of removing the surface residues and recovering the crystalline quality of the Si substrate prior to post-RIE device processing is obvious. Due to the complexity of the near-surface modifications caused by plasma-exposure of a Si specimen, thermal annealing treatments are ineffective in restoring the surface properties. Two promising approaches are O_2 plasma or O_2 RIE treatments following the standard RIE process. In both treatments the C,F-residue layer is oxidized ('ashed') to form volatile compounds. These treatments cause a thin ≤60 Å thick SiO_2 layer to grow, which can be eliminated by wet oxide etching. Damaged Si at the specimen surface will be consumed during oxide growth and is removed by a buffered oxide dip. A somewhat different approach is described here. It consists of annealing the reactive ion etched Si specimen at 400° for 30 min in dry O_2. During low temperature O_2 annealing, oxidation of the C,F-film occurs and volatile products such as CO, CO_2, H_2O and CF_4 will be formed. This largely results in the removal of the C,F-film. Furthermore some annealing of the damaged Si is expected and outdiffusion of H from the Si should take place. The usefulness of this method is demonstrated here for the case of CF_4/40%H_2 RIE. It can likely be applied also for surface residue removal in the case of other etching gases.

In Fig. 8, the efficacy of the 400°C 30 min O_2 annealing method is demonstrated. There, backscattering spectra of CF_4/40%H_2 dry etched Si wafers which had been subjected to 1 min overetching and were subsequently cleaned/annealed in the stated way are compared to the spectrum of a virgin control Si sample. The degree of F and C contamination of the Si wafer is dramatically reduced by the annealing treatment (compare Fig. 1). Furthermore, the intensity of the Si surface peak is decreased. We note that the Si surface peak has not been reduced to the intensity of

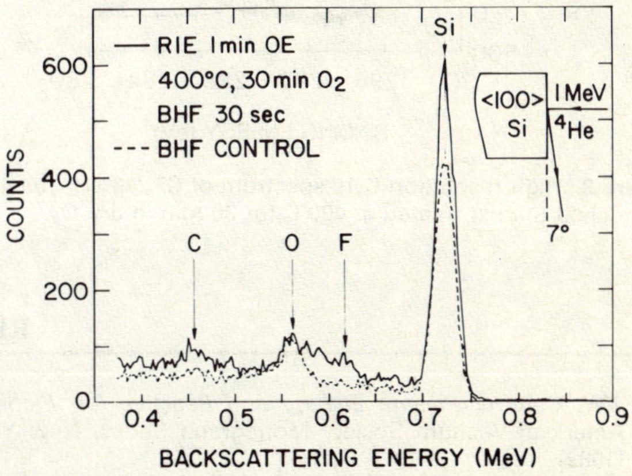

Figure 8. Comparison of He ion channelling spectra of control and reactive ion etched + 400°C O_2 annealed and oxide etched Si samples.

the virgin control sample and that F and C are still visible in the spectrum. This indicates that after the 400°C O_2 annealing treatment near-surface Si disorder is still present and that surface residues remain, but at a greatly reduced intensity as compared to prior annealing. The decrease of the Si surface peak during 400°C annealing is likely due to release of H and escape of trapped SiF_x-species. This aspect has been studied in detail in a separate publication.[14]

Analysis of the same specimens by XPS showed a greatly increased Si 2p intensity concomittant with much decreased F 1s and C 1s intensity, all of which is indicative of a substantially cleaner Si surface. Figure 9 is a high resolution C 1s XPS spectrum of the sample just described. The C 1s intensity is lower as compared to after CF_4/H_2 RIE and most of the higher binding energy components have disappeared. Three components at 285.3 eV, 287.6 eV and 289.5 eV binding energy were required to peak-fit the spectrum. Significantly, the two higher binding energy components have an 'inverted' intensity ratio, i.e. whereas for the peak-fits shown in Fig. 3 and 4, a decrease in intensity was observed for increasing binding energy species, in the present case the 289.5 eV peak has a greater intensity than the 287.6 eV peak after oxidation. Therefore it is likely that the 289.5 eV peak is *in this case* due to carbon-oxygen bonding. For O—C=O bonding the reported C 1s binding energy is 289.0 eV.[15]

No SiC-component was included in the fit of Fig. 9. The mechanism of SiC loss during O_2 annealing is unclear, but may be related to cooxidation of Si and C. A reduced concentration of SiC may also persist after O_2 annealing, which would explain why the fit of Fig. 9 underestimates the C 1s data near 283 eV slightly.

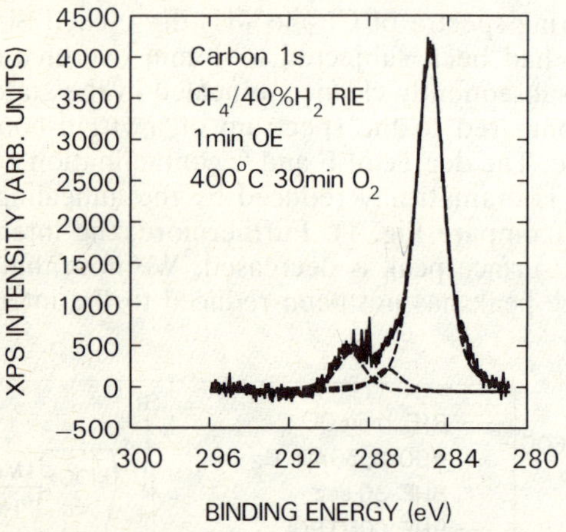

Figure 9. High resolution C 1s spectrum of $CF_4/40\%H_2$ reactive ion etched Si post-treated at 400°C for 30 min in dry O_2.

A disadvantage of the present method, which it shares with O_2 plasma or RIE post-treatments, is the fact that a thin SiO_2 layer (~30 Å) forms during the annealing treatment. This thin oxide layer and the remaining degree of C,F-contamination could necessitate, for best electrical results in certain applications, a buffered oxide etching step following O_2 annealing. However, a buffered oxide dip might not be necessary for many applications. We have electrically characterized Ti-Schottky barriers formed on 400°C O_2 annealed Si specimens which had *not* been etched in buffered hydrofluoric acid.[16] The results were encouraging. The thickness of the electrically blocking layer (C,F-layer), which is observed for as reactive ion etched specimens, was reduced as expected. Additionally, the surface state density was decreased significantly as compared to only reactive ion etched specimens, indicative of displacement damage annealing.

Wet oxide etching of the SiO_2-layer grown during the 400°C treatment reduces the degree of Si contamination further, e.g. the C 1s intensity shown in Fig. 9 is reduced by a factor of greater than two.

SUMMARY

Helium ion channelling, nuclear reaction profiling and x-ray photoemission measurements of $CF_4/X\%H_2$ (X up to 40) and $CClF_3/50\%H_2$ dry etched Si surfaces have shown that plasma exposure of a clean Si surface leads to the deposition of a thin (~30–50 Å thick) C,F-containing film (or C,F, Cl-containing in the case of $CClF_3/H_2$ RIE). A subsurface Si carbide layer is formed during RIE processing. The near-surface region (~30–50 Å) of the Si substrate is heavily disordered as found by ion channelling. A modified, less damaged layer containing a high concentration of hydrogen is formed in the case of the H_2 containing gases (as shown by nuclear reaction profiling) and extends more than 250 Å from the surface. A new method for Si substrate post-RIE cleaning/recovery has been demonstrated which involves heating the plasma exposed substrate to 400°C in dry O_2 for 30 min. The efficacy of the method has been verified by ion channelling and x-ray photoemission measurements.

Acknowledgements

We would like to thank Phil Saunders for assistance in setting up and performing the channelling experiments. Professor W. Lanford of SUNY (Albany, New York) is acknowledged for providing us with time and equipment for the measurement of the H concentration profiles.

REFERENCES

1. J.W. Coburn, *Plasma Etching and Reactive Ion Etching*, American Vacuum Society Monograph Series, New York (1982).
2. D.L. Flamm and V.M. Donnelly, *Plasma Chemistry and Plasma Processing,* **1,** 317 (1981).
3. G.S. Oehrlein, R.M. Tromp, J.C. Tsang, Y.H. Lee, and E.J. Petrillo, *J. Electrochem. Soc,* **132,** 1441 (1985).
4. G.S. Oehrlein, C.M. Ransoms, S.N. Chakravarti, and Y.H. Lee, *Appl. Phys: Lett,* **46,** 686 (1985).
5. L.M. Ephrath and E.J. Petrillo, *J. Electrochem. Soc.* **129,** 2282 (1982).
6. L.C.S. Feldman, J.W. Mayer, and S.T. Picraux, *Materials Analysis by Ion Chanelling,* Academic Press, New York (1982).

7. W.A. Lanford, H.P. Trautvetter, J.F. Ziegler, and J. Keller, *Appl. Phys. Lett,* **28,** 566 (1976).
8. J.C. Tsang, G.S. Oehrlein, I. Haller, and J.S. Custer, *Appl. Phys. Lett.* **46,** 589 (1985).
9. D.W. Rice and D.F. O'Kane, *J. Electrochem. Soc.* **123,** 1308 (1976).
10. A. Dilks and E. Kay, *Macromolecules,* **14,** 855 (1981).
11. G.J. Coyle and G.S. Oehrlein, *Appl. Phys. Lett.* **47,** 604 (1985).
12. F.R. McFeeley, J.F. Morar, N.D. Shinn, G. Landgren, and F.J. Himpsel, *Phys. Rev.* **B30,** 764 (1984).
13. J.W. Coburn and H.F. Winters, *J. Vac. Sci. Technolog* **16(2),** 391 (1979).
14. Gottlieb S. Oehrlein, *J. Appl. Phys.* (submitted).
15. D.T. Clark and A. Dilks, *J. Polymer Sci.: Polymer Chem. Ed. 17,* 957 (1979).
16. P. Spirito, C.M. Ransom, and G.S. Oehrlein, *Solid State Electronics* (accepted for publication).

SURFACE AND INTERFACE ANALYSIS, VOL. 9, 283–291 (1986)

Applications of SIMS, SAES and XPS to Problems in the Semiconductor Industry

P. Alnot, A.M. Huber and J. Olivier
LCR Thomson-CSF BP, 10–91401 Orsay, France

Electronic devices based on III–V compounds can be significantly affected in their behaviour by surface and interface properties. This paper will show the usefulness of surface and interface analytical tools such as SIMS, SAES and XPS in overcoming such problems. Case studies in relation with some of the main problems of III–V compounds technologies will be described.

INTRODUCTION

Silicon is used almost exclusively in the fabrication of semi-conductor devices and modern microcircuits, even though many elements show semiconducting properties. The preponderance and dominance of this elemental semiconductor has resulted from a number of factors, not the least of which has been the excellent interfacial properties of this material when in contact with its thermal oxide SiO_2. This dominance of the interface is not surprising when one realizes that even in large electronic structures most of the carrier transport occurs either across or along an interface and any effect of such surfaces on the motion of the carriers will manifest itself as a modification in device performance.

But silicon is not an optimum choice in some respects; gallium arsenide, and other III-V compounds are thought to be good candidates for future electronic technology. Among the remarkable properties of GaAs are: (a) it moves electrons around three to five times faster than silicon; (b) it emits light, something silicon cannot do; (c) it has a higher resistance to radiation than silicon; (d) it uses less power than silicon.

Although GaAs and other III-V compounds have interesting electrical properties, they are not widely used in the electronic industry. One of the main reasons is that, in contrast to silicon, the instabilities observed concern surface properties. Characterizations of surfaces and interfaces must be performed through technological processes to get enough information in order to solve these problems. In this paper, we shall describe the applications of SIMS, SAES and XPS to the characterization of surfaces and interfaces in III-V compounds.

INSTRUMENTAL

Auger Electron Spectroscopy (AES)

Electron beams provide a convenient probe for the creation of excited states at the solid–vacuum interface. In AES, the energies of these states are studied by analysis of the electrons emitted to conserve energy in the decay back to the ground state. These electrons are known as Auger electrons, and analysis of their energy can result in chemical identification. Today AES and XPS dominate the surface analysis scene because they have been found to show the greatest applicability. They are good surface spectroscopies because the escape depth L of the Auger electrons, which is a function of the electron kinetic energy and of the material being studied, is in the range of 5–30 Å. Combined with ion etching, AES can be used to get elemental concentration profiles within overlayers. Nevertheless several factors affect the resolution depth. We can classify the most important mechanisms which broaden the concentration profiles into two classes: (a) Auger electron escape depth and (b) ion bombardment effects (ion knock-on mixing, preferential sputtering and ion-induced roughness).

(i) In an Auger sputter profiling experiment, the measured elemental signal strength represents an average concentration in the surface region of thickness L. Thus, the depth resolution is limited by the escape depth, and will be optimum if the escape depth is minimized.

(ii) Ions incident on a specimen surface create an atomic collision cascade in the bulk of the material. Some atoms in this cascade reach the surface and are sputtered. Primary recoils are displaced into deeper layers (knock-on effect) and the other atoms are displaced from their original position (atomic mixing), resulting in a broadening of the measured depth profile.

Preferential sputtering from a heterocomposition surface seems to be the rule rather than the exception. Much of it does not influence depth resolution, but due to particle conservation, preferential sputtering is automatically coupled to a surface composition different from that of the bulk. However, the establishment of this different composition takes time, and a sharp change in the bulk composition will induce a transient in the surface composition. Depth resolution will be lost because of these transients.

It is well known that a number of projectile-target combinations leads to a very rough surface during sputtering, and in particular, sharp cones or needles may appear on the surface.[1] It is apparent that the

0142–2421/86/120283–09 $05.00

depth resolution of composition profiling methods will be degraded by the development of topographical structure. Figures 1 and 2 show such structures developed on InP and PbTe samples bombarded by argon ions of energy 2 and 0.6 keV respectively.[2]

As described by Schwartz and Helms,[3] in absence of roughening, the measured interface width can be described as an rms value of the broadening factors:

$$W_m^2 = W_t^2 + 1.65^2 L^2 + X^2$$

Where W_m and W_t are the measured and true interface width, X is the ion knock-on broadening term.

In studying the width of MBE AlGaAs/GaAs interface, Garner *et al.*[4] calculated the ion broadening factor to be about 10, 25 and 40 Å with Ar^+ energies of 250, 500 and 1000 eV respectively. More generally, the influence of atomic mixing on depth resolution is of the order of the projected range of the primary ions.[4] Consequently the disturbing effect of knock-on and atomic mixing can be minimized by using heavy ions (e.g. Xe^+) at low energy ($E \leq 1$ keV) to the detriment of a reasonable sputtering rate.

In order to get round the disadvantages of ion milling, we have developed a method of chemical bevelling coupled with line scan Auger measurements to check abrupt interfaces of epitaxial III-V compounds. For depth epitaxial layers of the order of 1000 Å or more, a current approach involves angle lapping the sample, to reveal the entire depth profile of the film.[7] The angle generally obtained by mechanical polishing, of the order of 0.5°, permits only the

Figure 1. Scanning electron micrograph of an InP(100) surface after Ar^+ ion bombardment (E_p = 2 keV).

Figure 2. Scanning electron micrograph of a PbTe(111) surface after Ar^+ ion bombardment (E_p = 0.6 keV).

thickness measurement of thinner films by SEM or AES.[8–10] In contrast to mechanical bevelling performed with diamond or alumina paste, there is no intermixing of layers. The bevel angle α can be controlled by the speed of the etching solution flow up the sample, and can be as small as 0.05° or less giving a magnification coefficient of $M = (\sin \alpha)^{-1} = 1145$. In contrast to sputtering, the absolute depth resolution R of the bevel Auger profiling method is independent of the film thickness to be examined and is of the order of the electron beam size s plus the beam deflexion resolution d divided by the lateral magnification of the bevel angle:[7]

$$R = (s + d)/M$$

The schematic diagram of the bevel is shown in Fig 3. In the present experiment, the beam diameter was 2000

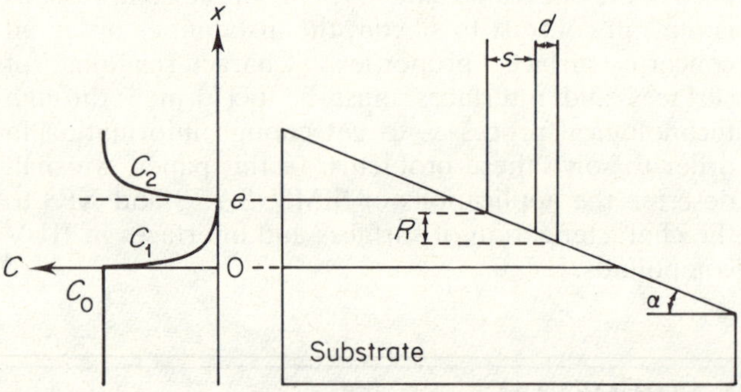

Figure 3. Schematic diagram of the bevel. The mean parameters are the bevel angle α, the electron beam size s, the beam deflexion resolution d, the depth resolution R. The phosphorus exponential composition profiles taken into account are $C_1(x) = C_o \exp(-x/\Lambda_1)$ and $C_2(x) = C_o[1-\exp((-x-e)/\Lambda_2)]$.

Å to 7500 Å (primary energy and current: 10 keV, 1nA and 5 keV, 20 nA respectively) and the deflexion resolution ⩽ 500 Å. With a magnification coefficient $M = 1000$, the absolute resolution R is 8 Å in the worst case. When heterojunctions or quantum wells are studied, the difficulty is to get etching solution with the same etch rate for the two materials of the structure in order to obtain a constant angle along the bevel.

If AES is a useful characterization tool for surface elemental concentration determination, it is sometimes difficult to draw from it chemical bond informations. Concerning this last point, XPS results are easier to handle, these two techniques are complementary.

X-Ray Photoelectron Spectroscopy

X-ray photoelectron spectroscopy (XPS) is based on ionization of a core level of a surface atom by photon. The excitation sources are MgKα and AlKα radiations. This technique gives elemental concentrations in the overlayers, as in AES, and provides chemical information of the specimen analyzed by shifts in the photo-electron binding energy measured. Since depth profiling by ion bombardment will destroy such information, as it was shown in the previous part of this chapter, angular dependent detection of photoelectrons is used to get depth profiles in thin overlayer of the order of 50 Å.

Both AES and XPS have detection limits in between 10^{12} and 10^{13} cm^{-2}. Characterizations of elements having concentrations below these limits can be performed using SIMS.

Secondary Ion Mass Spectroscopy (SIMS)

In secondary ion mass spectroscopy (SIMS), the sample surface under study is bombarded by a beam of energetic ions (O_2^+ or Ar^+ or Cs^+ generally). A fraction of sputtered species are ions. These secondary ions provide the information about the sample composition. The secondary ion intensity measurements are carried out by an electron multiplier.

The analyses in our laboratory are carried out by a CAMECA IMS 3F[11–12] which has the essential characteristics required for semiconductor analysis. These are as follows: (a) high trace sensitivity (ppb range); (b) high mass resolution $M/\Delta M = 10000$; (c) high depth resolution (nm); (d) wide dynamic range in depth profiling (5 in current use, 7 in special conditions); (e) mass separated primary ion beam; (f) simultaneous depth profiling of several elements; (g) Caesium gun equipment.

The target chamber is evacuated by a cryogenic pump to a pressure of 1.10^{-9} Torr. In current analysis of semiconductors, the surface is scanned with focused caesium or a mass filtered oxygen ion beam. (Cs^+, 0.5 μA at 15 keV and 2–3 μ at 10 keV). The scanned area is 250×250 μm^2 and the analyzed region is 60 μm in diameter. Quantitative analysis requires the use of standard samples. Standards for semiconductor analysis by ion implantation were proposed by Werner[13] in 1976 and Leta in 1979.[14] The reliability of this method was verified by systematic study of the effect of matrix composition on ionization probabilities for a wide range of elements. Also, a good correlation was often found between neutron activation and ion implantation calibrated SIMS measurements.

All our standards were prepared by this technique. The calibration factor for each element of each matrix was determined from 9 measurements on three samples, implanted by doses varying from 5×10^{12} to 5×10^{15} at.cm^{-2} at energy of 300 keV.

The statistical results of various experiments show that the quantitative results of SIMS are given with an accuracy of ±20% above a concentration level of 1×10^{16} at.cm^{-3}. Below this level, results are less accurate, for example ±50% at 1.10^{14} at.cm^{-3}. Talysurf measured depth precision is estimated at ±10%. As regards the detection limit of the majority of impurities influencing the electrical properties of III-V compounds, this is situated between 5×10^{12} and 5×10^{13} at.cm^{-3} for Mg, Cr, Fe, Mn, Si, Se, S, Te, B and Be. For Zn, Cd, Sn, O and C this limit is less sensitive, at $5 \times 10^{14} - 5 \times 10^{15}$ at.cm^{-3}.

In current analysis, each sample is analyzed in two different areas, about 5–7 mm. apart. Generally, two recording clean areas give reproducible and representative results for the materials.

One of the problems encountered in the study of the semiconductor materials is the quantitative analysis of doping elements introduced by diffusion, implantation and epitaxy. A further application of SIMS analysis is the determination and localization of undesirable impurities which contaminate the materials during processing and significantly affect device performance.

III-V COMPOUNDS CHARACTERIZED BY SIMS, XPS AND SAES: CASE STUDIES IN SEMICONDUCTOR TECHNOLOGY

We are going to focus our attention of three main problems concerning III-V compounds technologies: (a) thermal stability of semi-insulating SI substrates (Cr doped or undoped, capped or capless annealed); (b) surface preparation by wet etching; (c) interface characterizations.

Thermal Stability of SI Substrate

Direct ion implantation into bulk grown semi-insulating GaAs substrates (Cr doped or undoped) is currently being used as a means of producing electrically active layers for processing MESFET and high speed logic IC. As a necessary part of the preparation, ion implanted wafers must be heated to a high temperature (850°C) in order to remove implantation-induced structural changes and to electrically activate the implanted species. Several heating techniques have been proposed; capped or capless, in controlled atmospheres containing arsenic, to prevent decomposition of the compound by arsenic loss. Unfortunately, at the temperatures normally employed in FET and IC processing, the thermal stability of the semi-insulating substrate has often been a serious problem in the past. In the case of Cr doped Si substrates, n-type layer formation has frequently been observed near the surface, resulting in electrical profiles with tail or poor electrical activity.

The cause of this phenomenom was determined by SIMS about six years ago.[15] An anomalous thermal accumulation of Cr near the surface was established. The existence of an accumulation and depletion zone of the well-known 'Cr profile' explains the change in electrical properties, since this element is responsible for high resistivity in GaAs (Cr is a deep acceptor) – see Fig. 4, Cr profiles in GaAs samples. As a result of this discovery, IC technology was orientated to low Cr 5×10^{15} at.cm^{-3} or undoped SI substrates. The Si content of GaAs was necessarily reduced to about 5×10^{15} at.cm^{-3} (Si is a shallow donor element).

However, thermal stability control of undoped or slightly Cr doped SI GaAs sometimes presents *p*-type layers near the surface. Mn, Fe, Mg outdiffusion during capped or capless annealing from this material has been mentioned by several authors as determined by SIMS.[16–17] (Mg, Mn acceptors, Fe deep acceptor). Contamination from outside, with indiffusion of an electrically active element such as Mn, was also observed.[18]

Figure 5 shows Fe, Mg, Cr and Mn profiles in a slightly Cr doped SI GaAs, annealed by the proximity techniques (885°C 15 min under H_2). No electrical activity at all was obtained in this Si implanted crystal (Si 3×10^{12} cm^{-2}, 150 keV); the presence of considerable deep and shallow acceptor level elements near the surface fully compensates for the electrically active layer.

In our laboratory, reliable annealing (850°C, 15 min) was carried out under sputtering deposited Si_3N_4. Figure 6 presents typical results of this technique of an undoped SI GaAs implanted by ^{29}Si (4×10^{12}cm^{-2}, 150 keV). Mg, Fe, Cr and Mn profiles show only a very thin accumulation zone (100 Å thick) at 1×10^{17} at.cm^{-3} of Si, probably due to chemical adsorption. The electrical concentration profiles were measured by capacitance-

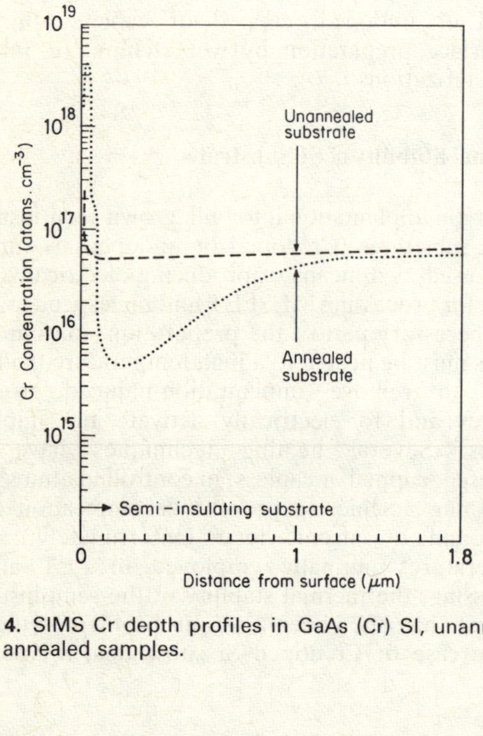

Figure 4. SIMS Cr depth profiles in GaAs (Cr) SI, unannealed and in annealed samples.

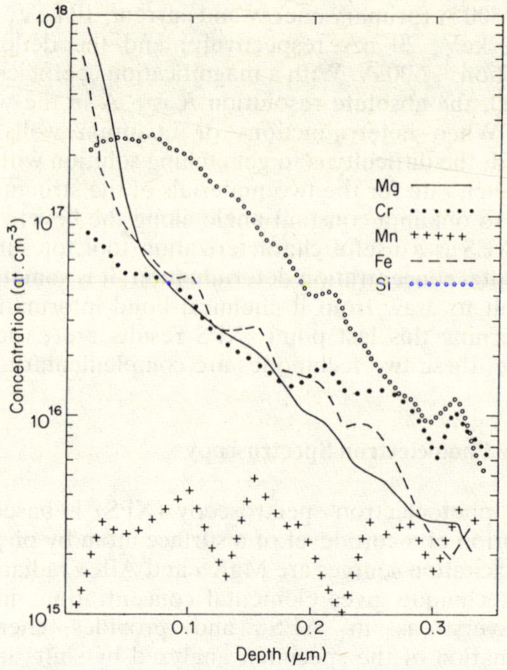

Figure 5. SIMS depth profiles of Mg, Cr, Mn, Fe and Si in GaAs SI Si implemented and annealed with proximity technique; no electrical activity.

voltage technique. A good correlation exists between the chemical Si and electrical profiles. Near ideal electrical activity is achieved currently under these conditions.

This fact could be explained by sputtering effect observed by XPS (see above) in the first step of Si_3N_4 reactive plasma deposition. As we will see later, the source of the GaAs surface contamination could be determined (see above) using SIMS.

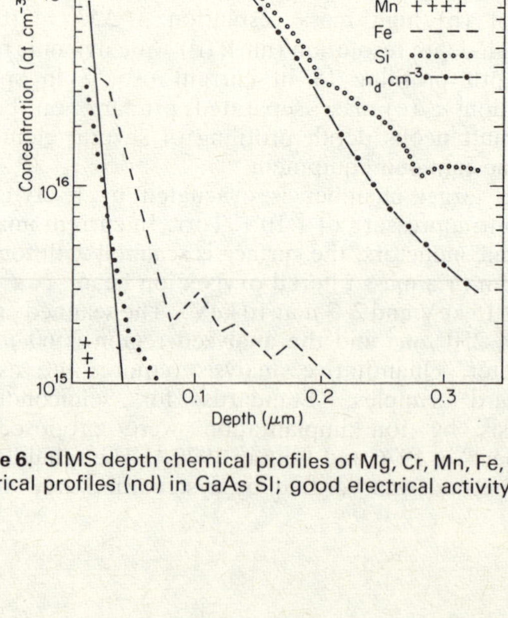

Figure 6. SIMS depth chemical profiles of Mg, Cr, Mn, Fe, Si and electrical profiles (nd) in GaAs SI; good electrical activity.

Surface Preparation by Chemical Etching

XPS characterization. There are different ways to get clean semiconductor surfaces, among them one can distinguish: wet etching or dry etching (plasma etching). Several studies by XPS have been performed in order to determine chemical composition of the surface after wet chemical etching. In general chemical etching solution are composed of one oxidizing agent, one complexing one and one solvent. The effects of several etching baths upon the surface composition have been studied: H_5O_4 : H_2O_2 : H_2O (5:1:1, 3:1:1, 2:1:1),[19,20,21] NaOH + H_2O_2,[22] HCl, HF etc. Based on ellipsometry results,[19] it seems that a general belief is the formation of an oxide layer of around 20 Å thickness after etching. This oxide was thought to be left due to oxidation within the etching solution.[20,22] Recently Massies *et al.*[21] have shown, for ($5H_2SO_4$: H_2O_2 : H_2O) etch, that the main part of the oxide layer was coming from oxidation due to exposure of the fresh etched sample to air.

In the following section, we shall demonstrate the usefulness of XPS in clarifying such problems. Based mainly on chemical shift reported on As 3d and Ga 3d, observed using XPS, compared with tabulated results of Pianetta *et al.*,[23] it was possible to determine the chemical nature of the oxide layers.

As judged by previous works (Ref. 23 and references therein), the As 3d and Ga 3d levels have been extensively observed in order to probe the GaAs surface composition. Figure 7 shows the XPS spectra corresponding to As 3d and Ga 3d of GaAs (100) sample chemically etched with ($5H_2SO_4$: H_2O_2 : H_2O) solution, then rinsed during 1 mn in DI water and exposed to air for different times. The sensitivity of the XPS measurements to the overlayer compound has been enhanced by taking the photoelectron exit angle to 15°.

The binding energies (*BE*) of the Ga 3d and As 3d photolines, corresponding to Ga and As in GaAs, were found to be respectively equal to 19.6 and 41.6 eV. To identify the chemical nature of the surface species it is important to determine their oxidation states. A method for evaluating the effective atomic charges of Ga and As, involved in the surface species, is to

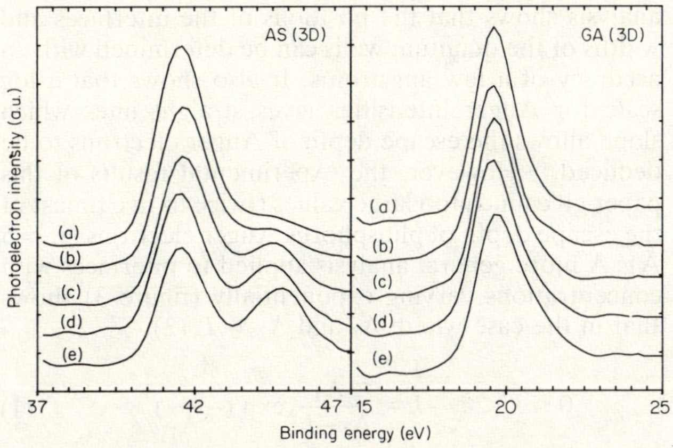

Figure 7. X-ray photoelectron spectra of GaAs(100) etched at T_b=25°C: (a) just after etching; (b) after 6 mn air exposure; (c) 30 mn air exposure; (d) 90 mn air exposure; (e) 160 mn air exposure.

compare their *BE* with those in GaAs and calculate the corresponding chemical shifts. For air exposures in between 0 and 90 mn, one shifted As 3d peak, having a chemical shift of 2.9 eV above the As 3d substrate peak, can be seen in Fig. 7. According to Pianetta *et al.*,[23] such a chemical shift corresponds to oxygen atoms chemisorbed on As atoms. Contrarily to Vasquez *et al.*,[20] no As_2O_5 was observed after etching. This difference could be due to the lower concentration of H_2SO_4 they used in their etching solution ($3H_2SO_4$:H_2O_2:H_2O).

As we can see in Fig. 7, the situation on the Ga 3d is less clear since a resolvable chemical shift is absent, although there is a broadening of this level to the higher *BE* side, as it was previously observed for GaAs(110) oxidation.[24] The amount of broadening of the Ga level was estimated by fitting the asymmetrical peak with two Gaussian components. For all exposures reported in Fig. 7, the energy difference between these two components was found to be equal to 1.4 eV.

This asymmetrical broadening of Ga 3d has been extensively discussed by Chye *et al.*[24] who observed a chemical shift of 0.9 eV, for oxygen adsorption on cleaved GaAs(110). They have interpreted this broadening as being due to direct bonding between oxygen and Ga atoms. Goddard[25] has pointed out that charge transfer between Ga and As following oxygen chemisorption on As atoms, could give a chemical shift of 0.8 eV on Ga 3d. Chemical shhifts of 1.4 eV[26] and 1.3eV[27] have been found on chemically etched GaAs(100) surfaces, and were attributed to the presence of Ga_2O_3.

In Fig. 7, we can see that an air exposure of 160 mn induces a chemical shift of 3.2eV on As 3d. This is indicative of As_2O_3 formation.[23] For this exposure, we did not observe any modification of the *BE* of the Ga 3d shifted component which is still equal to 1.4 eV. Considering the small change of chemical shift on As 3d between oxygen adsorbed on As atoms (2.9 eV) and As_2O_3 (3.2 eV), we did not expect to see a chemical shift modification, as seen on Ga 3d, between oxygen adsorbed on Ga atoms and Ga_2O_2.

We have shown that our particular etching treatment did not produce a thick oxide film oxidation of GaAs surfaces is mainly due to air contamination. This result is in agreement with those of Massies *et al.*[21]

In all this section, we have shown studies concerning the chemistry involved in surface chemical etching in relation with majority of species but one must not forget that in order to create surface state, minorities with concentrations as low as $10^{12}cm^{-3}$ are sufficient to pin Fermi level. Species existing in such low concentrations cannot be detected by XPS and SIMS has to be used.

SIMS characterization. The surface contamination of the III-V compounds by electrical active impurities such as Mg, Fe, Cr, Si, S, C was determined by SIMS. It was studied in the case of impurity redistribution at the metalorganic vapour-phase epitaxy layer/substrate interface.

The origin of these impurities at the interface was determined by study of three possible sources: (a) outdiffusion from the substrate; (b) incorporation from the ambient gas phases during preheating;

(c) adsorption of impurities during chemical surface etching before epitaxy (e.g. GaAs etching by 5:1:1 solution of $H_2SO_4 : H_2O_2 : H_2O$).

It is certain that the phenomenon is partly related to the substrates.[28] However, a simple model of diffusion would not explain the high concentration of impurities. So the second possibility was examined as follows. In one run, 1.5 μm thick GaAs was grown. The wafer was taken out of the reactor and after half an hour in the air was put back for a second layer to grow (about 1 μm thick). Figure 8 shows that, just as if nothing had happened at the first and second epilayer interface, only a very small quantity of Si accumulated there. The peaks M, Cr, Fe, S, Si and O could only be observed at the substrate first epilayer surface.

Examination of the third source involved removal of a small amount of epilayer by means of controlled chemical etching. A second layer was grown on the chemically treated first epilayer. The results are quite different from those obtained previously, as shown by Fig. 9. One can observe two accumulations of peaks of Mg, Cr, Fe, Si, S and O at the first epilayer-second epilayer and first epilayer-substrate interfaces. We can conclude that the majority of impurities results form the chemical etching, despite the fact that the purest existing chemical solution was used. (For example, Fe 10^{-5}, Cr^{-6}, Mg 10^{-5}% in concentration suprapure Merck H_2SO_4) Researchers are now attempting to eliminate completely this chemical adsorption.

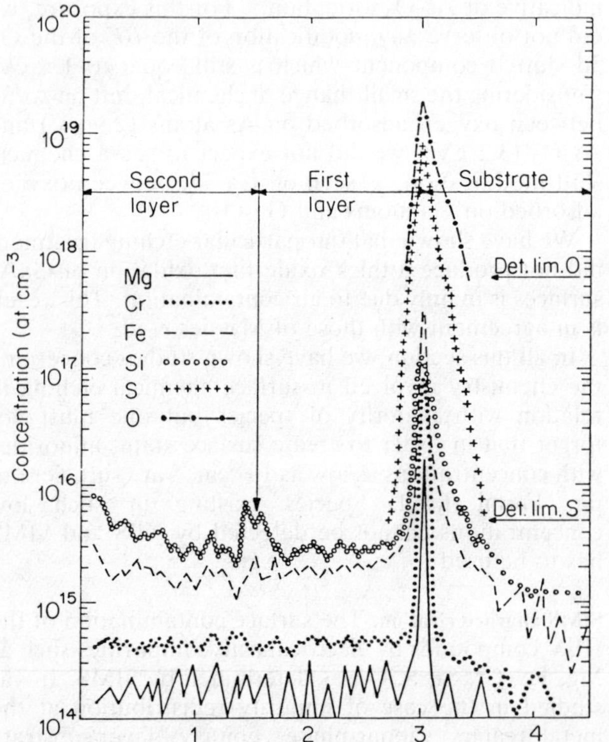

Figure 8. SIMS depth profiles of Mg, Fe, Si, S and O in an MOVPE layer grown on undoped GaAs SI substrate: epitaxy was stopped, the water was taken out of the reactor; a second layer was grown on the first.

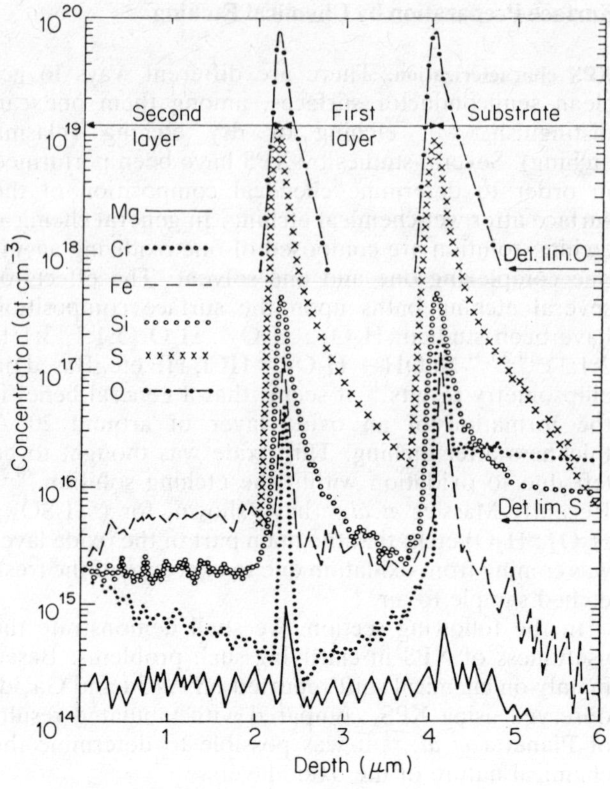

Figure 9. SIMS depth profiles of Mg, Cr, Fe, Si, S and O in an MOVPE layer grown on slightly Cr doped GaAs SI substrate. The first layer was partly removed by controlled chemical etching. A second layer was grown on the chemically treated first epilayer.

Interface Characterizations

III-V LP-MOCVD Heterojunction studies by AES. A precise knowledge of the concentration gradient and position interfaces of III-V compound semiconductor heterojunctions is of tremendous interest for electronic and optoelectronic device technology. One experimental result concerning an InP/GaInAs/InP LP-MOCVD quantum well, and obtained by chemical bevelling and AES is presented. Figure 10 shows the secondary electron image of four InP/GaInAs/InP quantum wells and the Auger line scan.

In the case of abrupt interfaces, the theoretical analysis shows that the positions of the interfaces and widths of the quantum wells can be determined with an accuracy of a few angstroms. It also shows that a log scale for Auger intensities gives straight lines which slope allows the escape depth of Auger electrons to be deduced.[29] However, the experimental results of this paper give much too large values (twice to five times) of the escape depth of phosphorus Auger electrons ($L \simeq 5$ Å). A more general analysis applied to interfaces with concentrations varying exponentially (figure 3) shows that in the case where Λ_1 and $\Lambda_2 \gg L$ (2):

$$0 < x < e \quad I = \frac{\Lambda_1}{\Lambda_1 - L} \exp\left(-\frac{x}{\Lambda_1}\right) \tag{1}$$

$$x > e \quad I = \frac{\Lambda_2}{\Lambda_2 - L} \exp\left(-\frac{x-e}{\Lambda_2}\right) \tag{2}$$

Figure 10. Secondary electron image of four InP/GaInAs/InP quantum wells. The Auger analysis points along the line scan are shown.

Thus in a log-scale, the characteristic widths Λ_1 and Λ_2 govern the slope of the straight lines, when Λ_1 and $\Lambda_2 \gg L$. Figure 11 shows the phosphorus Auger line scan of an InP/$Ga_{0.47}$ $In_{0.53}$As/InP LP-MOCVD quantum well of aimed width e=50 Å, and the theoretical curve calculated using the following parameters e=75 Å, L=4.75 Å, Λ_1 = 11.9 Å and Λ_2 = 21.4 Å. One can notice the good agreement between experimental and theoretical curves. The experimental raising of the Auger signal is less steep than the theoretical one at the two extremities of the well, and can be explained by a slight atomic mixing resulting from a short ion cleaning before the Auger line scan was taken. But, more likely, the explanation comes from the dynamics of the low-pressure MOCVD growth: in the neighbourhood of the growth surface, there is a gas layer whose characteristics change gradually from those of the convective gas phase, to those of the growth surface-gas phase interface.[30] The evolution of that boundary layer when we stop or when we set up again the phosphine flux in the reactor will give the key of the problem.

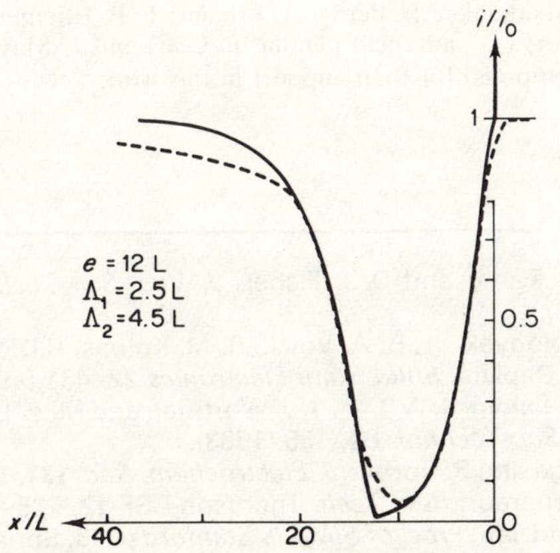

Figure 11. Theoretical Auger line scan of an InP/GaInAs/InP quantum well, the dashed line is the experimental curve.

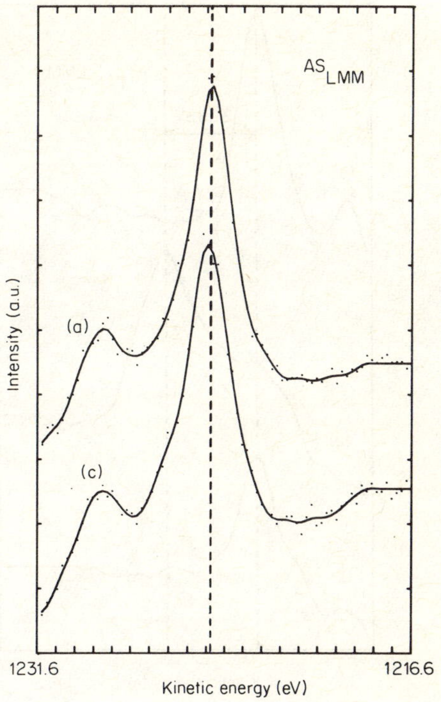

Figure 12. Ga and As Auger lines observed through silicon nitride film (40 Å thick) deposited on: (a) fresh etched (5 H_2SO_4 : H_2O_2 : H_2O) GaAs(100) sample at 13.6 Mhz; (b) etched sample immersed for 60 mn in DI water at 13.6 Mhz; (c) etched sample immersed for 60 mn in DI water at 440 Mhz.

Encapsulation and passivation. Plasma silicon-nitride layers are widely used in the IC technology for passivation of devices. In the case of III-V compounds, it is also used to encapsulate samples to avoid decomposition during the annealing process taking place after ion implantation. Several investigations reported elemental composition of silicon-nitride film[31–34] using techniques such as AES, XPS, RBS. Attempts have been made to correlate physical properties of the silicon nitride films and corresponding electrical properties of devices.[34,35] In doing such correlations, RAO *et al.*[35] have shown the importance of surface preparation prior to plasma silicon-nitride deposition.

The effect of surface preparation upon the chemical composition at the interface silicon-nitride-GaAs, has been studied using XPS. Ga and As Auger lines have been observed through a layer of silicon nitride (40 Å thick) deposited on a GaAs (100) sample. Ga and As Auger lines have been preferred to other transitions (3d or 2p core levels peaks) because they are very sensitive to chemical changes. Silicon-nitride was deposited using CVD RF plasma (N_2, SiH_4, 13.6 Mhz). Since the escape depths of the emitted Auger electrons are equal to about 50–60 Å, Ga and As Auger lines are very characteristic of the chemical composition at the interface silicon-nitride-GaAs.

Figures 12(a) and 13(a) show Auger line corresponding to substrate which has been etched in (5H_2SO_4 : H_2O_2 : H_2O) solution prior to silicon nitride plasma deposition. No features characteristic of oxides can be observed. Figure 13(b) shows Ga (Auger) of an etched GaAs (100) substrate immersed in DI water for 60 mn prior to plasma deposition. One can notice the

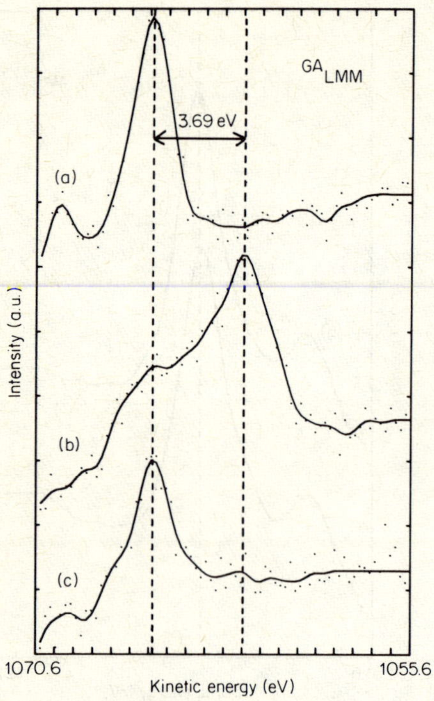

Figure 13. As Fig. 12.

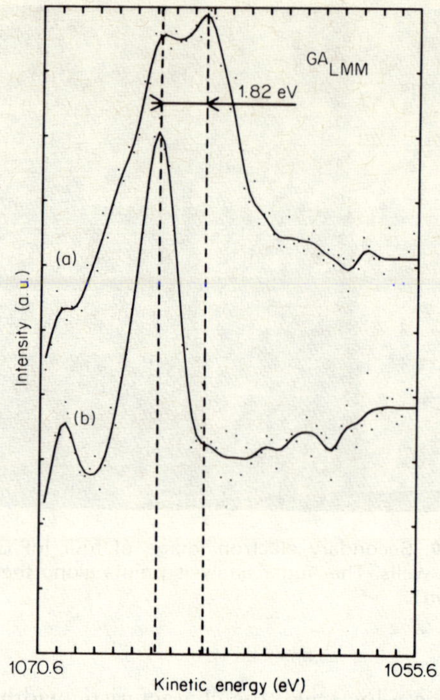

Figure 14. Ga Auger lines observed through silicon nitride film (40 Å thick) (a) deposited on fresh etched (5 H_2SO_4 : H_2O_2 : H_2O) GaAs(100) sample at 13.6 Mhz; (b) deposited on fresh etched sample changing plasma frequency and power as compared with (a).

presence of an oxide layer at the interface. The oxide is mainly composed of gallium oxide (hydroxide) characterized by the chemical shift of 3.69 eV as compared with Ga (Auger) of GaAs. No As (Auger) can be observed for this substrate. This fact could be explained by the preferential dissolution of arsenic oxide into water as compared with gallium oxide, resulting in an oxide rich gallium layer. It was possible to observe the removal of this oxide layer changing deposition plasma frequency from 13.6 Mhz to 440 Mhz (Fig. 12(c) and 13(c)).

These results show the importance of the surface cleaning and plasma characteristics in relation with the deposition of any passivating or encapsulating layer.

GaN is thought to be a good passivation layer[36] due to its higher stability as compared with GaAs and due to its wide gap3.5eV.[37] Through proper plasma changes (flux and power), it was possible to observe GaN formation at silicon-nitride-GaAs interface. This GaN is easily characterized as shown in Fig. 14(a) by a chemical shift of 1.80 eV as compared with Ga (Auger) of GaAs.[38–39]

CONCLUSION

The objective in presenting these typical XPS, SAES and SIMS results in relation to some basic problems of III-V compound technology was, to draw attention to the fact that systematic assessment of materials by these characterization techniques is indispensable for the electronic material improvement. All results presented in this paper are in relation with the chemical characterizations of materials along technological processes. In order to make improvements to the technology, correlations between chemical and electrical characterizations will be needed.

Acknowledgements

The authors thank, J.F. Peray, A. Friederich, P. Etienne, F Wyczisk, G. Morillot, G. Laurencin (Thomson-CSF) and J. Massies (CNRS, Sophia Antipolis) for their support in this work.

REFERENCES

1. J. Olivier, P. Faulconnier and R. Poirier, *JAP* **51**, 4990 (1980).
2. J. Olivier, to be published.
3. S. A. Schwartz, C.R. Helms, W. E. Spicer and N.J. Taylor, *Report NBS.SP.400.67*, Nat. R Stand., Washington DC, USA (1981).
4. C. M. Garner, C. Y. Sú, Y. D. Shen, C. S. Lee, G. L. Person, W. E. Spicer, D. D. Edwall, D. Miller and J. S. Harris, *Jr JAP* **50**, 3383 (1979).
5. R. Bisaro, G. Laurencin, A. Friederich and M. Razeghi, *Appl. Phys. Lett.* **40**, 978 (1982).
6. A. M. Huber, G. Laurencin and M. Razeghi, *Journal de Physique* **C4**, 44, 409 (1983).
7. M. L. Tarng and D.G. Fisher, *J. Vac. Sci. Technol.* **15**, 50 (1978).
8. N. Holonyak, Jr, B. A. Vojak, R. M. Kolbas, R.D. Dupuis and P. D. Dapkus, *Solid State Electronics* **22**, 431 (1979).
9. W. D. Laidig, J. W. Lee, J. J. Wortman and M. A. Littlejohn, *J. Vac. Sci. Technol.* **B1**, 155 (1983).
10. G. Riga and B. Horblit, *J. Electrochem. Soc.* **131**, 1379 (1984).
11. M. Lepareur. *Rev. Tech.* Thomson-CSF **12**, 225 (1980).
12. G. Soldzian, *Proc. of SIMS II, Stanford* p. 18, Springer-Verlag (1979).
13. H. W. Werner. *Acta Electronica* **19**, 53 (1976).

14. D. P. Leta, G. H. Morrison, *Proc. of SIMS II Stanford* p. 61, Springer-Verlag (1979).
15. A. M. Huber, G. Morillot, N. T. Linh, P. N. Favennec, B. Deveaud, B. Toulouse, *Appl. Phys. Lett.* **34**, 858 (1979).
16. P. B. Klein, P. E. Nordquist, P. G. Siebenmann, *J. Appl. Phys.* **51**, 4861 (1980).
17. M. Feng, V.Eú, H. Knaber, W. B. Henderson, *J. of Elect. Mat.* **10**, 973 (1981).
18. J. B. Clegg, G. B. Scott, J. Hallais, A. Mircea-Roussel, *J. Appl. Phys.* **52**, 1110 (1981).
19. A. C. Adams and B. R. Pruniaux, *J. Electrochem. Soc*, **120**, 408 (1983).
20. R. P. Vasquez, B. F. Lewis and F. J. Grunthaner, *Appl. Phys. Lett.* **42**, 293 (1983).
21. J. Massies and J. P. Contour, *J. Appl. Phys.* **58**, 806 (1985).
22. E. Huber and H. L. Hartnagel, *Solid State Electron* **27**, 589 (1984).
23. P. Pianetta, I. Lindau, C. M. Garner and W. E. Spicer, *Phys. Rev.* **B18**, 792 (1978).
24. P. W. Chye, C. Y. Sú, P. Skeath and W. E. Spicer, *J. Vac. Sci. Technol.* **16**, 119 (1979).
25. W. A. Goddard III, *J. Vacuum Sci. Technol.* **16**, 1313 (1979).
26. C. C. Chang, P. H. Citrin and B. Schwartz, *J. Vac. Sci. Technol.* **14**, 943 (1977).
27. P. J. Grunthaner, R. P. Vasquez and F. J. Grunthaner, *J. Vac. Sci. Technol.* **17**, 1045 (1980).
28. B. Tuck. G. A. Adegboyega, P. R. Jay, M. J. Cardwell, *Inst. Phys. Conf. Ser.* **45**, 114 (1979).
29. J. Cazaux, G. Laurencin and J. Olivier, *J. Physique Lett (Paris).* **45**, L999 (1984).
30. M. Razeghi, *Semiconductors and Semimetals Vol. 22*, Part A, p. 305, Academic Press (1985).
31. J. A. Wurzbach and F. J. Grunthaner, *J. Electrochem. Soc.* **130**, 691 (1983).
32. W. A. P. Claassen, W. G. J. N., Valkenburg, M. F. C. Willensen and W. M. V. D. Wijgert, *J. Electrochem. Soc.* **132**, 893 (1985).
33. T. Inada, T. Ohkubo, S. Sawada, T. Hara and M. Nakajima, *J. Electrochem. Soc.* **125**, 1525 (1978).
34. G. M. Samuelson and K. M. Mar, *J. Electrochem. Soc.*, **129**, 1674 (1984).
35. V. B. Rao and P. Y. Koyama, *J. Electrochem. Soc.* **131**, 1674 (1984).
36. J. I. Pankove, J. E. Berkeyheiser, J. J. Kilpatrick and C. W. Magee, *J. Electronic Materials* **12**, 359 (1983).
37. S. Bloom, G. Harbeke, E. Meier and I. B. Ortenburger, *Phys. Status Solidi (B)* **66**, 161 (1974).
38. Y. Mizokawa, H. Iwajaki, P. Jishitani and S. Nakamura, *J. Electron Spectrosc. Relat. Phenom.* **14**, 129 (1978).
39. J. Hedman and N. Martenson, *Physica Scripta* **22**, 176 (1980).

SURFACE AND INTERFACE ANALYSIS, VOL. 9, 293–296 (1986)

Analysis of Plasma Enhanced Chemical Vapour Deposition-Grown Borophosphosilicate Glass with a Combination of Rutherford Backscattering and Elastic Recoil Detection

J.B. Oude Elferink, K. Dunselman, E.J. Evers,* F.H.P.M. Habraken and W.F. van der Weg
Department of Technical Physics,* Department of Nuclear Physics, State University Utrecht, P O Box 80 000, 3508 TA Utrecht, The Netherlands.

J. Holsbrink**
Advanced Semiconductor Materials Europe B V, Jan Steenlaan 9, 3723 BS Bilthoven, The Netherlands

It is shown that with a combination of high energy ion beam methods, Rutherford backscattering spectrometry and elastic recoil detection, the elemental composition of borophosphosilicate glass ($SiO_pP_qB_rN_sH_t$) can be determined without reference to an external standard. Both thin (< 100 nm) and thick films can be analysed, provided the latter are homogeneous over the layer thickness. Some results pertinent to the B and P uptake during the deposition are briefly discussed.

INTRODUCTION

Borophosphosilicate (BPSG) is applied as interlayer dielectric, as final passivation layer and as reflow glass in the production of integrated circuits. If BPSG is flowed after deposition, it will redistribute, thus smoothing down sharp edges of the surface structure. The flow temperature strongly depends on the amount of B and P in the BPSG.[1,2]

Tong *et al.* analysed BPSG by electron microprobe, ion chromatography and x-ray fluorescence. Kern and Schnable mention wet chemical analysis, electron probe microanalysis, secondary ion mass spectroscopy, x-ray fluorescence, infrared spectroscopy, etch-rate measurements and glass fusion tests as method to determine the chemical composition of BPSG.

Disadvantages of these methods are a relatively low accuracy, complexity of sample preparation and/or the necessity of standards for the composition. In this paper we will present a method to obtain the composition of BPSG films with the high energy ion beam methods Rutherford backscattering (RBS) and elastic recoil detection (ERD). We show that, when both techniques are applied to the same sample and the resulting spectra are properly processed, the relative concentration of all elements (H, B, N, O, Si, P) in the films can be determined with a good accuracy, without reference to an external standard.

EXPERIMENTAL SETUP

The samples were grown with plasma enhanced chemical vapour deposition (PECVD) with varying amounts of PH_3 and B_2H_6 diluted in Ar, which is added to a mixture of 70 sccm SiH_4, 2300 sccm N_2O and 10 sccm O_2. The reactor is a commercially available PECVD system, described in more detail elsewhere.[1] It consists of a vertically oriented graphite plate boat assembly with alternate plates connected to a 410 kHz power supply. The pressure in the reactor was 0.8 torr. The temperature was 360°C at the front, 380°C at the center and 415°C at the back. Reactant gases entered the reactor at the front and were pumped off at the back. We measured energy spectra (i.e. the number of particles as a function of the energy) of both backscattered projectiles (RNS) and recoiled target atoms (ERD).

We used a 2 MeV He beam to obtain RBS spectra. The beam was produced with a Van de Graaff accelerator and was incident perpendicular to the sample surface. We measured the energy of ions scattered at 170° with a surface barrier detector.

To obtain ERD spectra we used a 30 MeV Si beam produced with a tandem Van de Graaff accelerator. The beam was incident at 24° to the surface of the sample. The energy of particles recoiled at 32.5° to the beam, was measured with a surface barrier detector. Both the primary beam and the recoiled particles detected moved in a plane perpendicular to the surface of the target. In front of the detector we placed a diaphragm and a 8.89 μm thick mylar foil (density 1.395 g cm^{-3}). The difference in energy loss for various elements in the foil was used to prevent scattered projectiles from reaching the detector and to redistribute the relative position of the contributions of various elements along the energy axis of the ERD spectra. Since the cross section for low angle scattering is much higher than the recoil cross section, the contribution of scattered particles would have obscured the recoil contributions to the spectra.

** Present address: Philips Elcoma.

0142–2421/86/120293–04 $05.00

ANALYSIS

Firstly we will deduce an expression to describe the shape of the energy spectra. With the parameters used in this expression the composition of the films will be calculated.

Shape of an energy spectrum

During collision of a moving particle (the projectile) with an atomic nucleus in e.g. BPSG (the target atom), part of the energy of the projectile is transferred to the target atom. The potential and kinetic energy of the target atom before the collision are negligible compared with the projectile energy. The energy transfer can be described with kinematic factors. The kinematic factors give the fraction of energy retained by the scattered projectile (RBS) and the fraction transferred to the recoiled target atom (ERD).

Kinematic factors for scattered projectiles are tabulated by e.g. Chu.[3] A particle moving through matter gradually loses energy due to interaction with electrons (stopping). As this interaction is a statistical process, the uncertainty in the energy of the particles increased with the distance travelled (range straggling). Stopping and straggling are proportional to the atomic number of the projectile squared and the number of electrons it encounters.[3,4] The probability that a particle scattered at a certain depth in the target is detected with a certain energy can be described with a Gaussian distribution function. The average value of this distribution function is determined by the average energy of the primary beam, and energy loss due to the scattering process and due to stopping. Its width is determined by the energy width of the primary beam, the range straggling and the instrumental resolution. The same description is valid for recoiled particles. However, it should be noted that before the collision energy is lost by the projectile moving in the target material, during the collision energy is transferred to the recoiled particle, while after the collision energy is lost by the recoil particle.

The total number of particles detected with a certain energy due to the collision of projectiles with target atoms of a certain element can be found by multiplying the Gaussian distribution function with the scattering cross section and the number of target atoms at a certain depth and by integrating this product over all depths.

COMPOSITION OF FILM

The composition of a film can be determined without an external standard of reference from ERD and RBS spectra if at least one of the elements is present in both spectra and if all other elements are present in at least one of the spectra. Films can be separated in thin and thick films. For thin films the entire film can be seen in the spectra; for thick films in at least one of the spectra only a surface layer, can be distinguished. Examples of both cases are given in Fig. 1 (thin film BPSG) and Fig.2 (thick film BPSG).

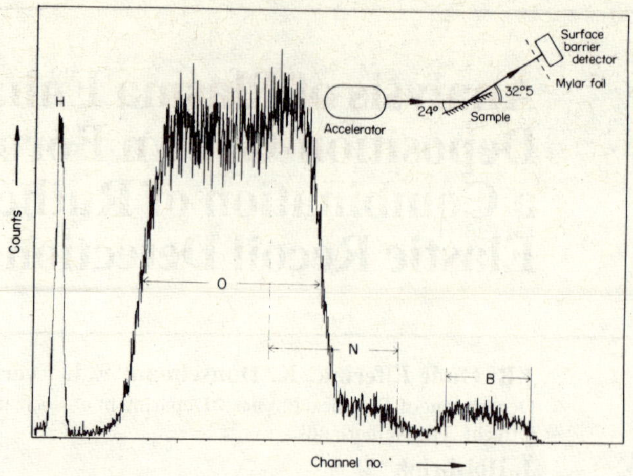

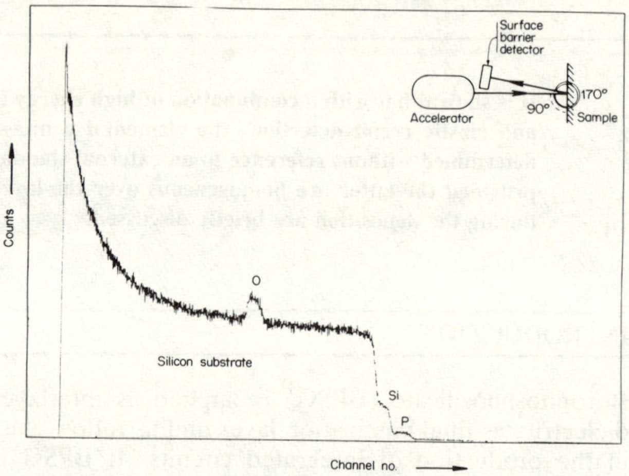

Figure 1. Typical ERD (above) and RBS (below) energy spectra of ~800 Å BPSG. Various elements are indicated. Inserts: sample detector configuration.

For thin films from the spectra the total area of the peaks of the elements can be obtained. The area of peaks with no background (e.g. B), can be calculated directly from the spectra. The area of peaks with a background can be found from fitting to the peak the difference to two cumulative Gaussian distribution functions to which a polynomial of order three is added to describe the background. The fit gives the height of the peak and the position of the steps in the spectra. These can be used to calculate the area P of the peak.

The area of the peak is proportional to the scattering cross section σ and the number of particles N in the film. The results from the ERD and the RBS spectra can be coupled using the element, which can be seen in both spectra, (in our case oxygen) as a mutual reference.

$$\frac{N}{N_O} = \frac{\sigma_O P}{\sigma P_O}$$

In the thin film case we obtained the area of the peaks by multiplication of the height and the width (i.e. the energy lost by particles scattered at the backside of the film) of the peak. However, in the case of thick films, this procedure is not possible, because particles which are recoiled from the interface region do not

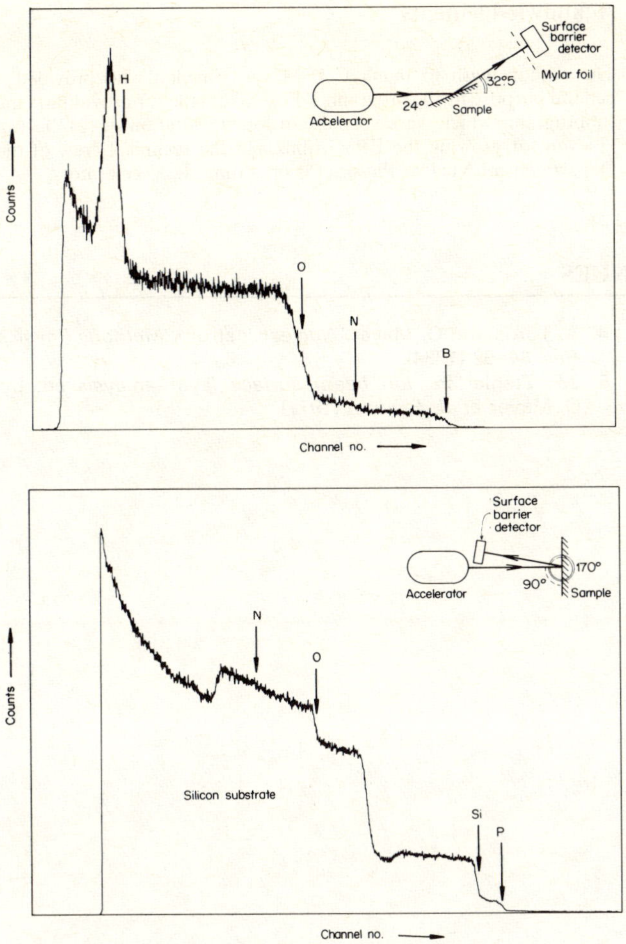

Figure 2. Typical ERD (above) and RBS (below) energy spectra of ~600 Å BPSG. Various elements are indicated. Inserts: sample detector configuration.

have sufficient energy to reach the detector (except hydrogen, which has a very low stopping power). Therefore in this case the energy width in the spectra is not known and only the step heights H of the several elements are available (cf.Fig.2 ERD). To circumvent this problem we proceed as follows: using stopping powers for the elements (given by Ziegler[5]) we calculated the width dE of a peak produced by collisions in a thin layer with thickness y at the surface of the film (surface-energy approximation[3]), converting the thick film into a thin film problem. Then (1) can be rewritten as

$$\frac{N}{N_{\mathrm{O}}} = \frac{\sigma_{\mathrm{O}} H \, \mathrm{d}E}{\sigma H_{\mathrm{O}} \mathrm{d}E_{\mathrm{O}}} \tag{2}$$

It should be noted that for ERD the value of dE is affected by the foil in front of the detector. This effect is taken into account by comparison of the final energy of particles scattered at the surface of the film and at a depth y.

Since the stopping power is determined by the composition of the film, (2) can be solved only in an interative way. We assume that dE is in first approximation the same for all elements. With this assumption we determine an initial value of the composition of the film. Using this result we determine

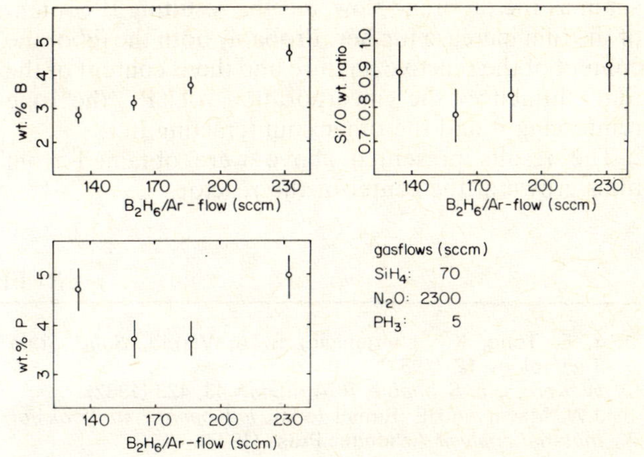

Figure 3. The composition of ~800 A BPSG films as a function of the gas flows in the reactor.

the stopping power for this composition, etc. The iterative process thus obtained converges rapidly (in about 5 steps) to the composition of the film. This method will be further elaborated in a future publication.

RESULTS AND DISCUSSION

Figure 3 shows the composition of thin BPSG films as a function of the diborane flow. Figure 4 shows the same for thick BPSG films. Within experimental error the results obtained for the thick and the thin films are the same. Thus we conclude that the stopping power correction described here is valid.

In pure quartz the Si/O weight ratio is 0.875. For the films presented here we find the same weight ratio. So inclusion of small amounts of B, P, N and H does not affect the relative deposition rates of Si and O. The amount of B and N in the film increases with the diborane flow. So inclusion of N must be reinforced by the presence of diborane in the reactive mixture.

The amount of P in the film initially decreased with an increasing diborane flow. But eventually it increases

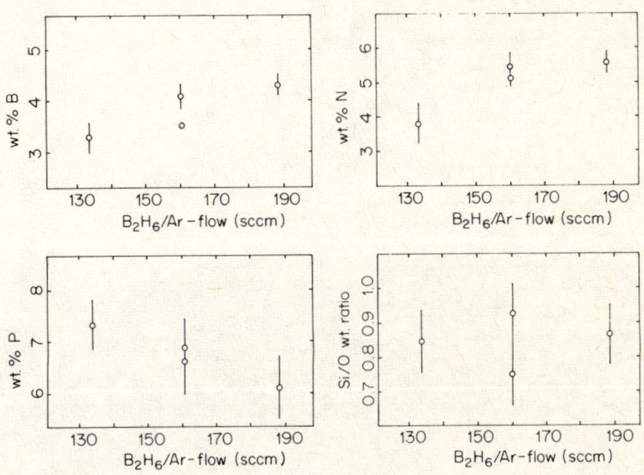

Figure 4. The composition of ~6000 Å BPSG films as a function of the gas flows. o = PH_3 flow is 7 sccm; • = PH_3 flow is 9 sccm.

again as the diborane flow and the resulting B content of the film increase further. Probably both the diborane content of the reactive mixture and the B content of the film influence the incorporation of P, the one reinforcing it and the other counteracting it.

The results presented above were obtained from films grown at the center of the reactor.

Acknowledgements

The authors wish to thank C P M van Engelen who provided a general purpose fitting program, P F A Alkemade for providing some subprograms which were very useful for the solution of (2), T. van Hoeven for growing the BPSG-films and the technical crew of the Department of Nuclear Physics for operating the accelerators.

REFERENCES

1. J. E. Tong, K. Schertenleib, R. A. Varpio, *Solid State Technology* **12**, *(1983).*
2. *W. Kern, G.L. Schnable, RCA Review* **43**, 423 (1982).
3. J.W. Mayer and E. Rimini (eds), *Ion beam handbook for material analysis* Academic Press (1977).
4. A. Turos and O. Mayer, *Nuclear Instrum. Methods Physics Res.* **84**, 92 (1984).
5. J.F. Ziegler in: *Ion beam surface layer analysis* ed. by O. Mayer *et al.*, Plenum (1977).

SURFACE AND INTERFACE ANALYSIS, VOL. 9, 297–301 (1986)

Depth Profiling of Carbonaceous Films, Produced *in situ* in the Tokamak TEXTOR

K.G. Tschersich, U. Littmark and H.P.Fleischhauer
Institut für Grenzflächenforschung und Vakuumphysik, Association EURATOM-KFA, Kernforschungsanlage Jülich. Postfach 1913, D-5170 Jülich, W. Germany

Samples of Si, Ni and Cr exposed to an *in situ* 'carbonization' in the Tokamak TEXTOR have been investigated. Depth profiles have been measured by argon ion sputter etching and AES. From the AES energy spectra (direct counting), the relative AES signals (peak to background ratios) have been derived by properly fitting the background and the peaks. The depth profiles are discussed in light of a ballistic-mixing model for sputter etching and a standard exponential Auger escape probability.

INTRODUCTION

Dense carbonaceous films have attracted increasing interest because of their promising mechanical, chemical and optical properties, such as high hardness, low friction, chemical inertness and infrared transparency. In a recent review, Angus, Koidl and Domitz[1] present a compilation of techniques used to produce such films. An indispensable condition for the growth of dense carbonaceous films seems to be the impingement of 50 eV to 500 eV particles during the growth process. A widely used technique matching this condition is to run DC or RF discharges in a hydrocarbon gas.

A mammoth scale application of this kind of technique, namely to carbon-coat the inside of a Tokamak device, has been tackled for the first time with the Tokamak TEXTOR at KFA Jülich. TEXTOR[2] has a toroidal vacuum vessel with a toroidal liner inside. The first wall facing the plasms is the liner, whose inner surface area is about 35 m^2. Historically, the coating with a carbonaceous film emerged from the standard *in situ* cleaning procedure of the TEXTOR liner by running an RF assisted DC discharge in hydrogen.[2,3,4] Admixing about 1% of methane to the hydrogen and keeping the wall temperature at about 300°C resulted in a 'carbidization' of the liner, characterized by the implantation of carbon into the liner near surface region and by a carbidic peak shape of the AES carbon peak.[5] Recently, the partial pressure of methane was increased to the order of 10% and the liner temperature decreased to about 200°C during the conditioning discharges, enabling the growth of a carbonaceous film which significantly improved the performance of subsequent Tokamak discharges.[6] This type of wall conditioning is called a 'carbonization'.

In order to investigate the composition of the surface films deposited during carbonization, we have placed samples of different materials in the liner surface. Analysis of these samples was performed in our surface analysis station. In this paper, we focus on the acquisition and analysis of depth profiles.

0142–2421/86/120297–05 $05.00

EXPERIMENTAL

The surface analysis system will be described in detail elsewhere.[7] Depth profiles of the samples were obtained by intermittently sputter etching the samples and taking AES spectra at the crater bottom. The ion gun used for sputter etching (Leybold–Heraeus, IQE 12/38) was operated with argon at 2 keV and at a surface current density of 23 $nA \cdot mm^{-2}$. The angle of incidence of the ion beam to the surface normal was 55°. The analysis system is equipped with the slightly modified column of a commercial SEM (Hitachi S–450LB) and a secondary electron detector to allow for physical imaging of the samples. Throughout the analysis, the SEM image was monitored to ascertain that no significant imperfections resided or developed in the area from which AES spectra were taken. For AES the primary electron beam was tuned to 50 nA at 10 keV and rastered over an area of about $2 \times 10^{-2} mm^2$. The electron spectrometer is a 150° spherical sector analyser installed behind an electrostatic lense (Vacuum Generators CLAM 100). The spectrometer entrance slit was set to 4 mm and the retard ratio was held constant at 4. The spectrometer was operated in the pulse counting mode and the direct energy spectra were stored in a multichannel analyser, typically in 1024 channels with a channel width of 0.5 eV. Finally, the spectra were transferred to a PDP 11/23 mini-computer for data analysis and display.

To allow for an *in situ*-like analysis, samples exposed in TEXTOR can be transferred between the analysis station and TEXTOR without being in contact with the ambient atmosphere. This is realized by a vacuum case fitting to manipulators at the analysis station and at TEXTOR.[8]

Three different samples were exposed to the carbonization: a slice of silicon cut from a wafer, a slice of nickel made from sheet nickel, and chromium evaporated to a thickness of about 500 Å onto a metal substrate. The samples were cleaned, locked into the analysis system and pre-characterized. Thereafter, they were moved into the vacuum case, transferred to TEXTOR and inserted by the manipulator at TEXTOR into the liner in such a manner that they made up a part of the inner liner surface.[9] Their spatial separation was less than 50 mm. They were all subjected to the same carbonization and then transferred back into the analysis station. Because the TEXTOR liner material is Inconel 625 it was reasonable to choose Ni and Cr as carbonization collectors. It is well known, however, that carbonaceous films on metals may be long-term unstable and flaking-off may occur. (This was not observed with the present thin films). On Si and Ge this is in general not the case and Si was therefore added as a third substrate.

DATA ANALYSIS

We had to analyse the Auger signals of C in the presence of Si, Ni, Cr and some O. In order to obtain quantitative results it is necessary to introduce a single yet reliable measure for the Auger peaks. We have adopted the so-called peak to background ratio method investigated by Langeron *et al.*[10] This procedure yields relative AES signals. In order to determine the peak maxima and the background at the peak maximum position, separate regions of interest (ROI's) were chosen inside and outside the proper peak region. The ROI's outside the peak were used to identify the background. The experimental data in these ROI's were least-squares fitted to a single low degree (1–3) spline. After substraction of the extrapolated/interpolated background, the peak ROI was fitted in logarithmic values giving a Gaussian-like peak shape. The procedure is illustrated in Fig. 1. The peak extrapolations shown indicate that the high energy branch of the peaks is rather well approximated by this shape.

In most cases, in particular for C(261–269 eV), Cr(525 eV), Ni(844 eV) and O(507–510 eV) on Si and Ni as well as for Si(84–89 eV) at not too low concentrations, we focused on the high energy tail of the peak and its neighbouring high energy background. In these cases the background can be fitted linearly and extrapolated down to the peak energy. This is shown in Fig. 1 for the Cr(525 eV) peak. The peak itself is asymmetric. To take care of this asymmetry, the log of the peak-minus-background data was fitted by a third order spline.

This kind of data processing combines two strong advantages. First, it largely reduces the problems of matrix effects influencing especially the shape of the low energy tail of a peak. Madden[11] has compiled Si peaks taken from different Si compounds showing that the background corrected peak signals are independent of the chemical environment whereas the global shape of the peak varies significantly. Secondly, it is easy to extrapolate the high energy background down to the peak energy, whereas the determination of the background at lower energies is much more tricky. Hesse, *et al.*[12] have used an iterative spline smoothing to determine the background of e.g. a 400 eV scan. It becomes obvious that the background at the peak energy and therefore the background-corrected peak

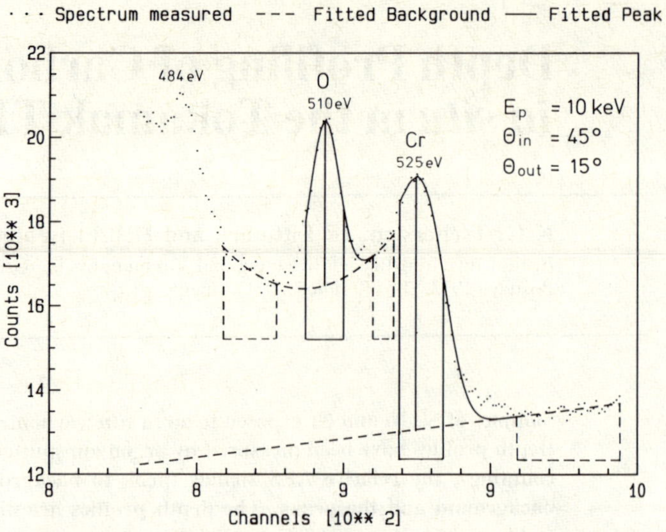

Figure 1. Part of an Auger spectrum obtained from a carbonized Cr-sample, showing the region of the Cr and O Auger peaks. Indicated is the procedure used for the peak and background determination to obtain relative Auger signals as a measure for the Auger activity. To determine the peak maximum and the background below it, one ROI is chosen well inside the peak and one or two ROI's close to, but well free of the peak. From the external ROI's the background is spline fitted (dashed curves). In the peak ROI the background subtracted peak is again spline fitted using logarithmic values yielding a Gaussian-like peak shape (full drawn curves). Finally, the peak value divided by the corresponding background is evaluated and defined as the relative Auger signal.

signal is independent of the degree of iteration whereas the background at lower energies depends strongly on it.

However, the above procedures is not applicable if a peak is superimposed on a strongly varying background. We face this situation for Si (84–89 eV) at low concentrations. At energies around 100 eV the slope of the background changes from being negative (contribution of true secondary electrons) to being positive (increasing transmission; at constant retard ratio the energy window ΔE is proportional to E.) Another example is O(510 eV) on Cr, cf. Fig. 1. In these situations we tried to identify the background by comparing the peak to be analysed with spectra showing the minimum signal of that peak, and determined two background ROI's; one just above and one below that peak. This is again illustrated in Fig. 1 for O on Cr.

Dividing the background-corrected peak value by the background reduces the sensitivity with respect to instrumental parameters.[10] However, one must be aware that the relative AES signal determined in this way depends on the background. It is obvious from Fig. 1 that the background of oxygen includes a contribution due to the presence of Cr. This makes it difficult if not impossible to define a universal sensitivity factor for relative AES signals. Accordingly, the O signals from O on Cr cannot be compared to those from O on Si. The influence of Cr on the relative O signal remains in tolerable limit if the Cr signal does not vary too strongly. In our data the contribution of Cr to the background of O varied from 10 to 20%.

RESULTS AND DISCUSSION

The depth profiles measured on the Si-sample are plotted in Fig. 2. Three observations are conspicious. Firstly, the S-shape of the Si-profile is very steep and typical for erosion-induced ballistic mixing of a sharp interface. Secondly, the C-profile exhibits an extended slow decay followed by an S-shaped profile at the interface. (The initial increase of the C-signal is not yet understood; however, it does not appear to be related to the presence of another component.) Thirdly, the shape of the O-profile is rather symmetric, atypical for the sputter profile of a thin layer.

The thickness of this film has been determined by masking part of the sample with a grid and scanning over the created steps with a Sloan Dektak Stylus instrument. A thickness of 160 Å ± 20 Å was found corresponding to a mean erosion rate of 0.066 $Ås^{-1}$ or a sputtering yield of $Y = 4$ for 2 keV Ar at 55° incidence on the carbonaceous film. Disregarding fluctuations in the erosion rate of the carbon film (if any, they might be mostly due to the presence of O; the H-content in the films is constant laterally as well as in depth as became evident during SIMS depth profiling[13]) we therefore assumed the original C/Si interface to be located at 160 Å.

In order to obtain a more quantitative picture of the original carbonized Si-sample a model calculation was performed with a ballistic-mixing program.[14,15] This traces the dynamically changing internal concentration profiles during sputter erosion. Only purely collisional mixing processes, dominated by cascade recoil relocations, are included. From the instantaneous internal concentration profiles, the Auger signal has been deduced by assuming a standard exponential Auger escape probability determined by the mean free electron path length. (No change in electron reflection was taken into account.) As the program input we used the standard values described in Refs 14 and 15: power potentials for the atomic interactions ($m = 1/3$ for the Ar– and $m = 1/4$ for the recoil-collisions) and an effective displacement energy of 30 eV.

The experimental spectrum (Fig. 2) shows a rather well-behaved Si profile indicative of a step-shaped concentration profile. The first model calculations therefore were on a sharp C/Si interface disregarding the presence of O. These showed that in fact the Si S-shaped AES profile has the theoretically maximum steepness as has the part of the C profile above 130 Å. Two conclusions were drawn: First, no C has diffused into Si and, secondly, no experimentally significant SiO_2 is present at the interface. The latter fact is remarkable since the well known protective SiO_2 film on the virgin Si sample was not purposely removed prior to the film deposition.

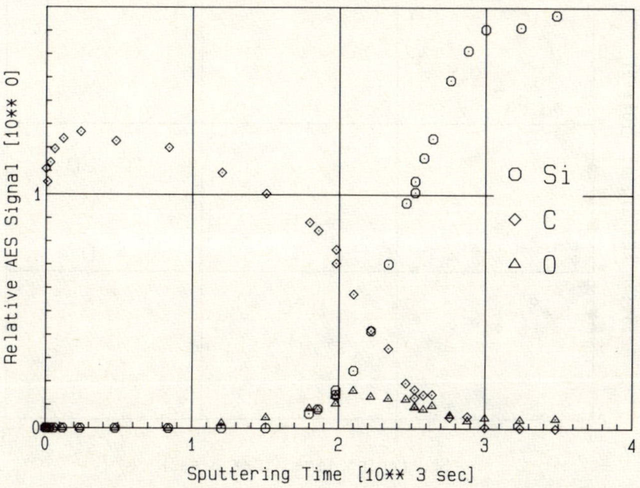

Figure 2. Depth profile of a Si-sample exposed to a carbonization in TEXTOR.

As already mentioned the experimental C depth profile (Fig. 2) shows a characteristic (almost linear) decreasing behaviour in the region below 130 Å indicating a third element here. The only candidate is oxygen and successive model calculations therefore included an O component with the goal to determine an approximate oxygen distribution to account for the decrease in the C profile. Finally, we ended up with a 'triangular' O concentration profile in the region 70–160 Å reaching a maximum of 30 at% at the Si interface.

The calculated O-AES-profile showed up to be pretty symmetric (Gaussian shaped) in accordance with the experimental profile (Fig. 2). (The latter does not show the asymmetric inward-tailing which is characteristic for a thin localized layer.) Obviously the depth resolution and parameter-uncertainty do not allow for determination of the exact shape of the O concentration profile. However, to be in accordance with the model calculation it must be a similarly increasing function in the indicated depth region with a total content of 6 monolayers or 10^{16} O atoms cm^{-2}. The observed oxygen is predominantly incorporated in the deposited carbonaceous film and not to a significant amount part of a Si interface oxide.

The finally adopted model system is depicted in Fig. 3 showing the model concentrations (thick lines), the experimental points from Fig. 2 properly normalized in height and the calculated AES depth profiles (dashed and solid curves). As can be seen it is possible to model the profiles consistently within experimental and theoretical uncertainties.

Most interesting is probably the result for the position of the O. There are two sources for oxygen to be considered: The original SiO_2 passivative layer on the Si substrate and oxygen present in the TEXTOR vessel. The latter was mainly introduced during a previous opening period of the vessel during which in-vessel maintenance was carried out. To account for the measured O profile we put forward the following explanation: The carbonization process includes an ion bombardment (< 500 eV) of the liner surface and thus a sputter erosion process in concurrence with the film growth. It is likely that at the early stage of carbonization, contaminants are brought out from the surface and are slowly redeposited and burried in the excess carbon deposit. This is in accordance with the mass-spectroscopical observation of a relatively high O concentration in the plasma at the beginning, slowly decreasing with time.[16] The effective erosion depth is obviously large enough to remove the surface oxide from a Si sample. Finally, after a film of some 100 Å has grown, a clean carbonaceous film is deposited.

The measured Auger depth profiles for the Ni- and

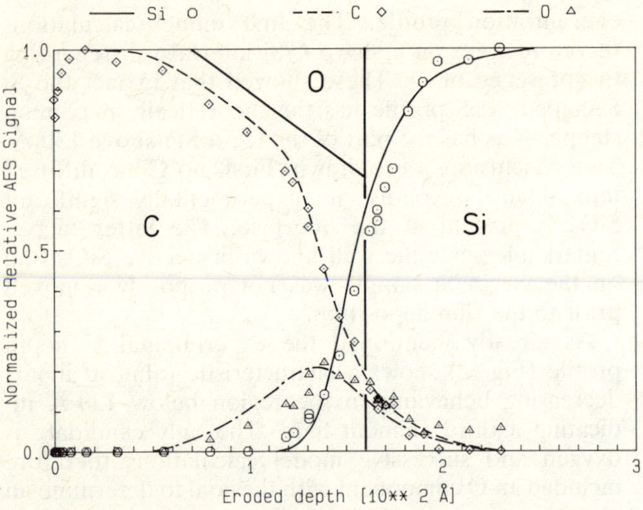

Figure 3. Theoretical modelling of the Auger depth profiles of Fig. 2. The assumed concentration profiles are: C: 100 % in [0, 70 Å], linear decrease to 70 % in [70, 160 Å], Si: 100 % above 160 Å, O: rest. Shown are the correspondingly calculated Auger signals and the experimental points. The theoretical model is based entirely on ballistic mixing due to 2 keV Ar sputter erosion at 55° to the target normal and an exponential Auger escape probability (λ_C(268 eV) = 7.5 Å, λ_O(509 eV) = 11.5 Å, λ_{Si}(89 eV) = 5 Å and θ_{out} = 15°).

Cr-samples, see Fig. 4 and 5, are qualitatively similar to each other but different from the Si profiles. Both C and substrate Auger signals show an exponential behaviour near the surface, typical for a thin layer coverage. We have performed model calculations also for these systems and found that the substrate Auger signal at eroded depth zero is sensitive to and indicative of the thickness of the thin surface film. We estimate the carbon film on Ni to have a thickness of 20 Å and that on Cr to be only 10 Å thick. Again the resolution in the sputter erosion does not allow for determination of the exact shape at a monolayer scale and it is probably more correct to state the 'film thickness' as being $17 \cdot 10^{15}$ and $9 \cdot 10^{15}$ C-atoms cm^{-2}, respectively.

The carbon depth profile from the Ni-sample shows an at least 200 Å long inward tail. This can not be explained by the collisional mixing model and we assume its origin in a diffusion process during either plasma exposure or sputter profiling.

Contrary to Ni, for the Cr-sample, a large content of O is present. We estimated sensitivity factors for Cr and O from a depth profile measured on an area of the Cr-sample that had been covered during the exposure in TEXTOR: C disappeared after a few sputtering steps, the profile measured thereafter being that of a two component system (Cr + 0). From that part of the profile we deduced sensitivity factors for this system. Relying on these sensitivity factors we estimated a constant background of O over the entire measured depth of Fig. 5 to be of the order of 20%. An excess O peak is present in the region 0–80 Å with a content equivalent to 3–4 monolayers of O. This could be brought in accordance with the discussion of the Si-sample indicating the plasma as the source of the excess oxygen.

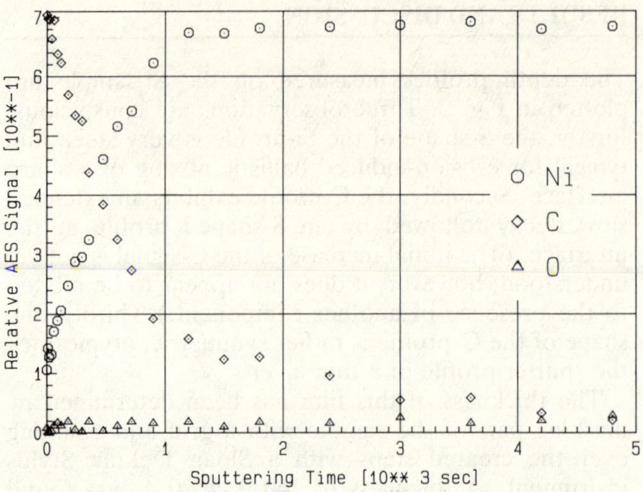

Figure 4. Depth profile of a Ni-sample exposed to the same carbonization in TEXTOR.

The O-background might be due to the preparation of the Cr-sample by evaporation. To further investigate the interface of a carbonaceous film on Cr we have foreseen to deposit a carbonaceous film onto a Cr sample made from a high purity bulk material.

The aim of this paper was not to enter a discussion of the deposition mechanism of carbonaceous films on solid substrates. However, it became obvious that the deposition processes depend strongly on the substrate material. Therefore, even monitoring simple quantities like film thicknesses requires some understanding of the deposition mechanisms. This will be the topic for future works on carbonization of the inner surface of TEXTOR.

CONCLUSIONS

Carbonaceous film deposits from *in situ* carbonization of the 35 m^2 inner surface of the Tokamak TEXTOR have been investigated by means of AES depth

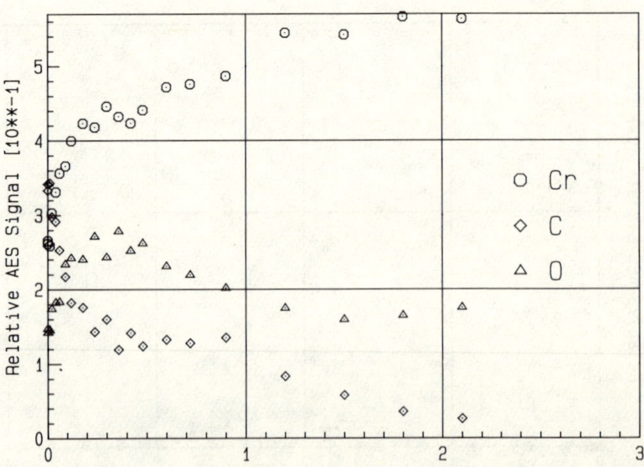

Figure 5. Depth profile of a Cr-sample exposed to the same carbonization in TEXTOR.

profiling of C, O and the sample substrate materials Si, Ni and Cr. 160 Å of carbon were found deposited on the Si-sample, which also showed 6 monolayers of O triangularly distributed in a pre-interfacial region. No implantation or in-diffusion of C or O into Si was observed, rather was removal of original SiO_2 from the Si sample demonstrated. Only 20 Å and 10 Å of carbon were found on the surface of the Ni- and Cr-samples, respectively. Some in-diffusion of C into Ni was observed and a build up of 3–4 monolayers of O in the C–Cr surface was seen. We have demonstrated the suitability of our full vacuum transfer/analysis system and the value of comparing experimental results with mixing theory on the analysis of AES depth profiles.

Acknowledgements

The cooperation of M. Lochter who skilfully operated the demanding manipulator on TEXTOR and J. Winter and his crew who performed the carbonization in TEXTOR is gratefully acknowledged. We thank W. Knörchen for preparing the Cr-sample, G. Flentje for throwing some human style into stiff computer plotting and M. Ferguson for correcting our 'English'.

REFERENCES

1. J. C. Angus, P. Koidl and S. Domitz, in *Plasma Deposition of Thin Films*, CRC Press, in press.
2. J. Winter, F. Waelbroeck, H. G. Esser, L. Könen, T. Banno, E. Rota and R. E. Clausing, *J. Nucl. Mater.* **122** & **123**, 1187 (1984).
3. F. Waelbroeck and the TEXTOR Team, *J. Nucl. Mater.* **121**, 378 (1984).
4. J. Winter, F. Waelbroeck, P. Wienhold, H. G. Esser, L. Könen, M. Braun, B. Emmoth and H. E. Sätherblom, *J. Nucl. Mater.* **128** & **129**, 841 (1984).
5. K. H. Besocke, G. Flentje, J. Winter, P. Wienhold and H. G. Esser, *J. Nucl. Mater.* **136**, 124 (1985).
6. J. Winter, F. Waelbroeck, P. Wienhold, H. G. Esser and the TEXTOR Team, *Verhandl.* **DPG (VI) 20**, 963 (1985).
7. K. G. Tschersich, H. P. Fleischhauer, U. Littmark, H. Gnaser, W. Hofer, J. V. Seggern and K. Besocke, to be published.
8. P. Marmy and U. Stiefel, *Atomkernenergie-Kerntechnik* **44**, 235 (1984).
9. K. Besocke, G. Flentje and K. G. Tschersich, *J. Vac. Sci. Technol.* **A 3**, 2028 (1985).
10. J. P. Langeron, L. Minel, J. L. Vignes, S. Bouquet, F. Pellerin, G. Lorang, P. Alloud and J. Le Héricy, *Surf. Sci.* **138**, 610 (1984).
11. H. H. Madden, *J. Vac. Sci. Technol.* **18**, 677 (1981).
12. R. Hesse, U. Littmark and P. Staib, *Appl. Phys.* **11**, 233 (1976).
13. K. Besocke and G. Flentje, private communication (1985).
14. U. Littmark and W. Hofer, *Nucl. Instrum. Methods*, **168**, 329 (1980).
15. U. Littmark and W. Hofer, *Topics of Current Physics*, vol. 37: Thin Film and Depth Profile Analysis, ed. by H. Oechsner, Springer-Verlag, Berlin (1984).
16. J. Winter, private communication (1985).

SURFACE AND INTERFACE ANALYSIS, VOL. 9, 303–308 (1986)

SIMS Analysis of $Al_xGa_{1-x}As$/GaAs Layered Structures Grown by Metal–Organic Vapour Phase Epitaxy

P. R. Boudewijn, M. R. Leys* and F. Roozeboom
Philips Research Laboratories, PO Box 80.000, 5600 JA Eindhoven, the Netherlands

For $Al_xGa_{1-x}As$/GaAs layered structures, grown by MOVPE, abrupt interfaces in the order of one atomic distance were observed. These structures are well suited to serve as test samples for studying the influence of bombardment conditions on the quality of SIMS depth profiles. The results of such a study are presented and it is shown that depth profiles can be obtained with a depth resolution of about 25 Å. It was found that beyond the interface secondary ion signals decrease exponentially; from the variation of the slope of this exponential decay as a function of primary ion energy it can be concluded that depth resolution is determined by a purely collisional mixing process. From this type of data we can derive a minimal layer thickness in $Al_xGa_{1-x}As$/GaAs quantum well structures for which quantitative SIMS analysis of the Al content is possible.

INTRODUCTION

Devices for opto-electronic applications are often composed of III–V compound semiconductors. An important class of devices is formed by the double heterojunction lasers, of which the laser based on the GaAs/$Al_xGa_{1-x}As$ system and applied in the compact disc player is a good example.[1] These laser devices exhibit a layered structure, i.e. they consist of several layers of material, differing in composition, on top of each other. Current trends in this field, are towards the development of lasers having one or more thin (< 100 Å) active layers (quantum well lasers). Another class of layered structures attracting much attention are the superlattices. These can also be composed of III–V compounds, and consist of alternating thin (< 40 Å) layers of material of different composition.

It has been shown that in the system GaAs/Al_xGa_{1-x} As molecular beam epitaxy (MBE)[2] and metal-organic vapour phase epitaxy (MOVPE)[3,4] can be used for the growth of such thin layer structures. An important aspect in the development and optimization of such growth techniques is the characterization of the grown material by surface analysis techniques. In the present paper the capabilities of secondary ion mass spectrometry (SIMS) for the analysis of GaAs/$Al_xGa_{1-x}As$ layered structures grown by MOVPE will be considered.

An $Al_xGa_{1-x}As$ multilayer structure is shown schematically in Fig. 1. Relevant information to be extracted from a SIMS measurement is indicated, and includes the values of x (Al content), the thickness of the layers, the concentration of intentionally introduced dopant elements in the different layers, analysis of impurities, and finally the gradient in both the Al concentration *and* the dopant element concentration across the interfaces. However, the interpretation of depth profiles in multilayer structures obtained with SIMS is often complicated by a number of disturbing factors, such as the matrix effect (secondary ion yields and sputtering rate are dependent on layer composition) and the finite depth resolution attainable with sputtering methods. Both aspects will be considered in this paper, but emphasis will be placed upon problems related to the depth resolution; in fact the well-characterized interfaces of these MOVPE-grown layers,[4] make them well suited to serve as a model system for studying the influence of various sputtering-related phenomena on the quality of SIMS depth profiles.

* Present address: University of Lund, Sweden

EXPERIMENTAL

SIMS measurements were performed on a Cameca IMS 3F instrument. An O_2^+ primary ion beam, ranging in energy from 1.5 to 12 keV, was rastered across an area

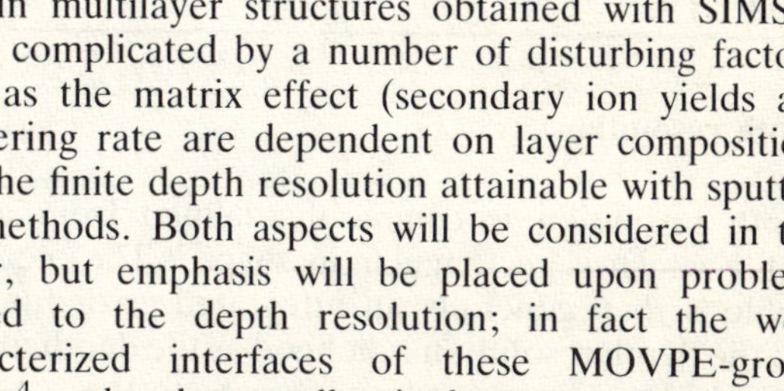

Figure 1. Schematic representation of an $Al_xGa_{1-x}As$ multilayer structure.

0142–2421/86/120303–06 $05.00

of 250 × 250 μm^2 or 500 × 500 μm^2. The maximum beam current was a strong function of the accelerating voltage and varied from 5 μA at 12 keV down to 0.2 μA at 1.5 keV. The primary beam was focused into a spot of 50–100 μm diameter, except at 1.5 keV where proper focusing is very difficult. At this energy an elliptical beam shape was observed with the short axis approximately 150 μm and the long axis exceeding 250 μm. A field aperture, situated at an image plane of the mass spectrometer, was used to select an area to be analysed of 8 or 60 μm diameter.

A Tencor Alphastep stylus instrument was used for determining the sputtered crater depth and estimating the surface roughness of the samples.

All samples were grown in a MOVPE reactor described previously.[4] GaAs(001) substrates were used, either exactly oriented or cut 4° off to [110]; the SIMS depth profiles were not found to be influenced by the orientation.

Samples consisting of thick (> 2 μm) $Al_xGa_{1-x}As$ layers with x values of 0.085, 0.13, 0.27, 0.46, grown on a GaAs substrate, were used as standards for calibrating the $^{27}Al^+$ and $^{69}Ga^+$ intensities. The Al content of these layers was determined with high-resolution x-ray diffraction.[5]

RESULTS

Depth resolution

Depth resolution is commonly defined from a true profile which is rectangular in shape.[6] The measured profile is then generally an integrated error function, and the depth resolution is defined as the depth interval over which the intensity drops from 86 to 16%.[7] Several factors determine the experimental depth resolution:[7] (a) instrumental factors, (b) factors depending on sample constitution, and (c) factors related to the interaction process between the energetic projectiles and the solid (primary recoil mixing, cascade mixing, or radiation-enhanced diffusion). The importance of these various factors for SIMS depth profiling of $Al_xGa_{1-x}As$ multilayer structures will be discussed in the next sections.

Primary ion energy It is well known[7] that the depth resolution of profiles obtained by sputtering methods is strongly dependent on the energy of the bombarding ions. For an $Al_xGa_{1-x}As$ multilayer structure the influence of this parameter on the resultant SIMS depth profiles obtained with O_2^+ bombardment is depicted in Fig. 2. As expected, a strong dependence of the depth resolution on primary ion energy can be observed from this figure. It should be noted in particular that at energies ≥6 keV even a 500 Å thick GaAs layer is incompletely resolved. Furthermore, only at energies below 3 keV the presence of the 40 Å alternating GaAs/AlAs layers does become apparent from the depth profile; this indicates that in this energy region depth resolution and layer thickness are of comparable magnitude. A further point worth attention, apparent from Fig. 2, is the fact that across the first interface of the 500 Å GaAs layer the $^{27}Al^+$ intensity decreases

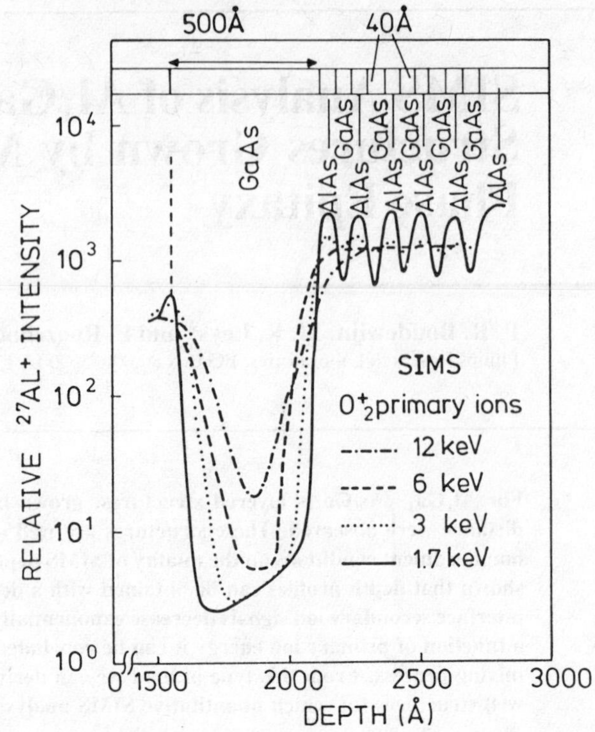

Figure 2. SIMS $^{27}Al^+$ depth profiles of a GaAs/AlAs multilayer structure as a function of the primary ion O_2^+ energy. The depth scale was derived from TEM results.[22] The $^{27}Al^+$ intensity is given relative to the $^{27}Al^+$ background signal obtained from a GaAs substrate.

exponentially with increasing sputtered depth.

A similar behaviour was observed for depth profiles obtained from other $Al_xGa_{1-x}As$/GaAs interfaces.

The increase of the $^{27}Al^+$ signal with increasing depth within the 500 Å GaAs layer is not an artefact of the SIMS measurement, but is due to aluminium actually present in the GaAs layer, which can be attributed to memory effects in the reactor during the growth of this layer.

The results of Fig. 2 demonstrate that low primary ion energy is a prerequisite for sputter depth profiling of this type of layered structures; all results presented in the next sections were thus obtained at the lowest available primary ion energy of 1.5 keV.

Influence of instrumental parameters The most important instrumental factor governing the depth resolution is the homogeneity of the primary beam current density across the analysed area.[7] To ensure sufficient flatness of the sputtered crater bottom, rastering of the bombarding beam is required.[8] This point deserves special attention in view of the poor beam focus at low primary ion energies.

The contribution of such non-homogeneous sputter erosion to the depth resolution can be minimized by decreasing the size of the analysed area and/or increasing the size of the raster. For this reasaon a small analysed area of 8 μm diameter is chosen to obtain the $^{27}Al^+$ and $^{69}Ga^+$ depth profiles. Such a small area, however, does not provide the high sensitivity required for depth profiling of dopant elements in these layered structures. In these cases the size of the analysed area has to be increased towards 60 μm diameter; consequently, the size of the raster has to be adapted to guarantee sufficient depth resolution. The latter is illustrated by the Al profiles given in Fig. 3; these profiles clearly show the improvement in depth resolution when the rastered width is increased.

Influence of sample properties For optimal results in SIMS depth profiling, samples should be selected with a very small initial surface roughness, especially in the case of single crystal samples.[7] We inspected the surface quality of the samples by optical microscopy and by tracing the surface with a stylus-type instrument.

Apart from roughness initially present on the surface, roughness induced by the bombarding ion beam can play an equally important role in determining the depth resolution in sputter depth profiling.[7] The inferior depth resolution of the Al profile in Fig. 4 (a) could be ascribed to such an effect, as is apparent from the stylus surface scans also shown in this figure. We attribute this extreme roughening of the bombarded surface to the presence of a surface layer enriched in aluminium (see Fig. 4 (a)). This layer was not intentionally grown, but is probably due to the desorption of trimethyl aluminium (TMA) from the susceptor, after growth has stopped; under non-ideal growth conditions this TMA reacts at the sample surface, resulting in the formation of a Al-rich layer of poor crystalline quality. Initialized by the crystalline defects in this layer[9,10] a macroscopic surface roughness can be observed after prolonged sputtering.

In these cases removal of the ill-defined surface layer by a non-sputtering method, e.g. chemical etching, is required. Aluminium depth profiles of such an etched sample are given in Fig. 4 (b); the improvement in depth resolution in comparison with the results of Fig. 4 (a) is obvious.

Bombardment-induced roughening and a non-ideal crater bottom (see previous section) give a contribution to the depth resolution which increases with sputtered depth:[7] at large sputtered depths these contributions determine the depth resolution, as is for example apparent from the results shown in fig. 4(b); the decrease in the $^{27}Al^+$ intensity is a factor 7 at the first GaAs well and changes to a factor 4 at the last GaAs well. For optimal depth resolution interfaces situated at a shallow depth (< 1000 Å) beneath the original sample surface are to be preferred.

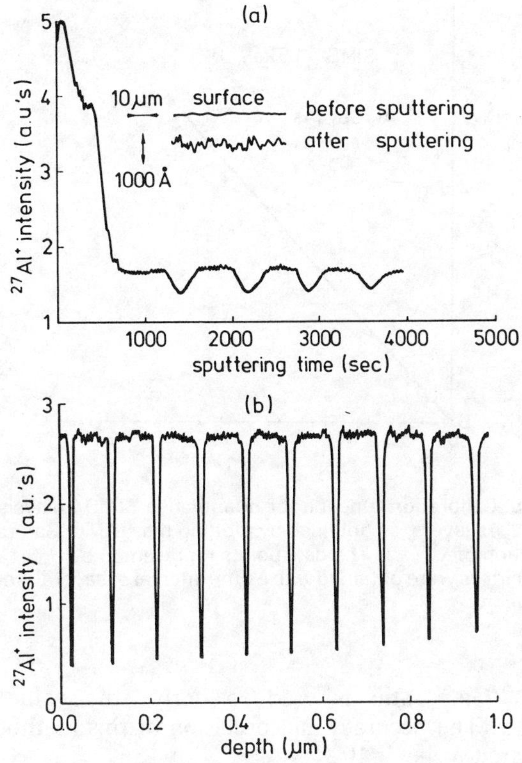

Figure 4. $^{27}Al^+$ depth profiles of a sample consisting of ten 1000 Å thick layers of $Al_{0.5}Ga_{0.5}As$ separated by 80 Å thick GaAs wells. The inferior depth resolution of the profile in (a) is due to bombardment induced surface roughening, as is apparent from the surface scans also shown in (a). After removal of the surface layer by chemical etching the profile shown in (b) was obtained.

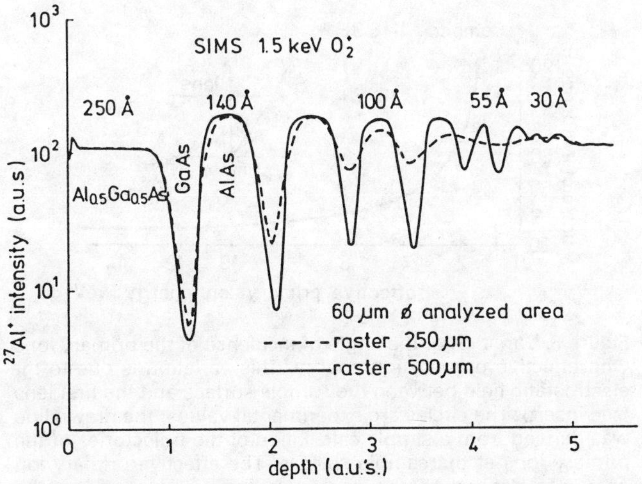

Figure 3. $^{27}Al^+$ depth profiles showing the influence of the raster size on the depth resolution.

Quantitative analysis

Quantitative analysis of the Al content and the dopant element concentration in $Al_xGa_{1-x}As$ layered structures is of vital importance in the assessment of the grown materials. In spite of complications arising from the existence of a matrix effect, it has been shown that in principle quantitative SIMS analysis in any material can be performed using suitable calibration samples.[11]

Using 1.5 keV O_2^+ bombardment SIMS measurements on a series of $Al_xGa_{1-x}As$ samples with known x showed that the $^{27}Al^+$ and $^{69}Ga^+$ intensities depend linearly on Al and Ga concentration, respectively, in the range $x = 0$–0.5. These findings are in agreement with previous results obtained at somewhat different bombardment conditions.[12] Thus by plotting the measured $^{27}Al^+/^{69}Ga^+$ intensity ratio as a function of $x/1 - x$, a straight line is expected as shown in fig. 5. Rationing of secondary ion signals is often preferred in quantitative SIMS analysis since this procedure corrects for small variations in experimental conditions. The validity of this calibration procedure may be judged from the fact that even intensity ratios obtained several weeks later show good reproducibility (see Fig. 5). For quantitative analysis the calibration function of Fig. 5 is used as an analytical function, i.e. the measured

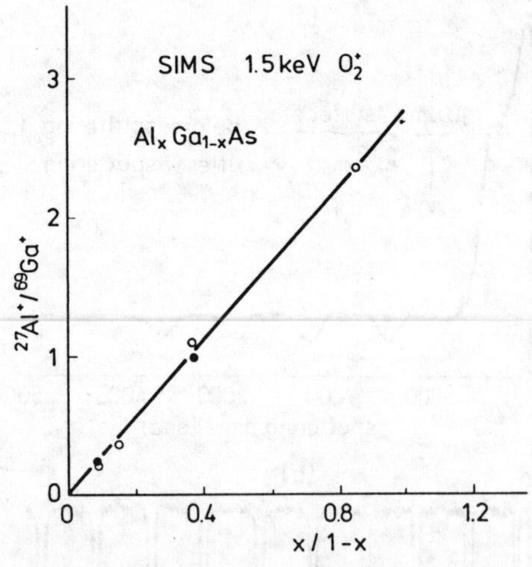

Figure 5. Calibration function for quantitative SIMS analysis of $Al_xGa_{1-x}As$ layers, as obtained by plotting the $^{27}Al^+/^{69}Ga^+$ ratio as function of $x/1-x$. The data points represented by open and closed circles were obtained with a time interval of several weeks in between.

$^{27}Al^+/^{69}Ga^+$ ratio is used to derive the value of $x/1-x$. The accuracy and precision of this method is estimated to be 5–10%.

In principle the same procedure as described above can be applied for quantification of dopant element concentrations; standards prepared by ion implantation can serve as suitable calibration samples.[12]

DISCUSSION

The results described in the previous sections show that by careful optimization of instrumental and sample parameters aluminium depth profiles across an $Al_xGa_{1-x}As$/GaAs interface can be obtained with an estimated depth resolution of 25 Å (see e.g. Figs 2 and 3). An accurate value of the depth resolution can only be obtained if the experimental data are corrected for non-linear intensity and concentration effects and for sputtering yield variations across the interface.[20] Although depth resolution is commonly used to assess the quality of a SIMS depth profile, in practice another property of the profile is often more appropriate for such a characterization: the exponential behaviour of the $^{27}Al^+$ intensity across the $Al_xGa_{1-x}As$ interface. By writing this exponential tail as $I.e^{(-z/\lambda)}$, with I the $^{27}Al^+$ intensity and z the depth coordinate, a decay length λ can be derived.[13] From the type of measurements shown in fig. 2 it was found that λ ranges from 164 Å decade^{-1} at 12 keV O_2^+ to 27 Å decade^{-1} at an O_2^+ energy of 1.5 keV for both $Al_xGa_{1-x}As$/GaAs and GaAs/$Al_xGa_{1-x}As$ interfaces.

A major consequence of such an exponential decay is that it determines the minimum thickness of an $Al_xGa_{1-x}As$ layer for which quantitative analysis of the Al content is possible. As an example let us suppose that a layered structure consisting of a GaAs layer containing small amounts of Al, sandwiched between two $Al_{0.5}Ga_{0.5}As$ layers, has to be analysed. Such an analytical problem is not purely imaginary, since knowledge of the Al content in a GaAs well is sometimes important, e.g. in interpreting the data obtained from spectral photoluminescence.[14] Assuming an Al concentration of 0.5 at % in the GaAs well, the $^{27}Al^+$ intensity will thus change two orders of magnitude across the interfaces. If, in a first approximation, exponential decrease is assumed to start at the interface, it follows from the decay length of 27 Å decade^{-1} at 1.5 keV O_2^+ that a minimum GaAs layer thickness of 108 Å is required in order to obtain one analytical data point from which the Al concentration can be determined quantitatively. Taking into account the fact that for accurate quantitative analysis several data points are needed, and considering the non-exponential behaviour of the $^{27}Al^+$ signal in the near-interfacial region, we arrive at an estimate of the minimum GaAs layer thickness of approximately 150 Å, in order to solve the aforementioned analytical problem.

It can be concluded from this example that knowledge of the characteristic parameter of a depth profile is of great help to the SIMS analyst, since it facilitates the decision whether a certain analytical problem can be solved by a SIMS measurement or not.

Exponential tails in SIMS depth profiles have been observed frequently.[13,15] Theoretical descriptions of such behaviour involve processes such as collisional mixing[16] or a combination of collisional mixing and selective sputtering.[17] It has further been shown that from the energy dependence of the decay length information about the dominant mixing process can be obtained.[18]

Such information is not directly available from our data, since on the particular instrument used a variation in primary ion energy also alters the angle of incidence of the primary ions (Fig. 6). Using the method described by Morgan *et al.*[19] we therefore determined this angle of incidence as a function of primary ion energy. The experimental results are given in Fig. 6, together with the results derived from a simple calculation. It can be seen that at 1.5 keV the incident angle has increased towards nearly 70°. By assuming for the decay length $\lambda(\theta) = \lambda(0)\cos\theta$[20] a set of $\lambda(0)$ values can be obtained, reflecting the influence of incident energy on this parameter.

The energy dependence of the decay length $\lambda(0)$ is

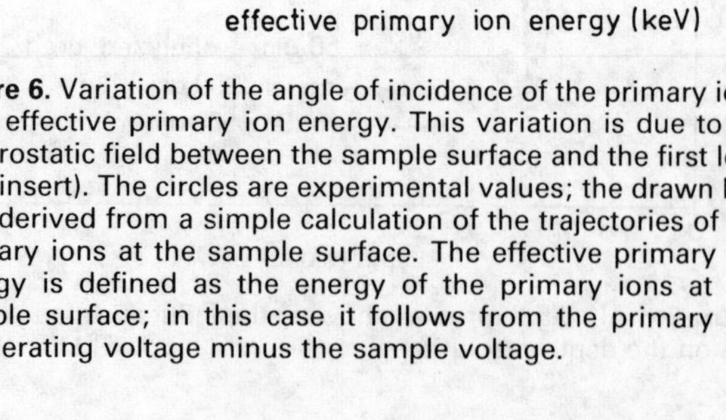

Figure 6. Variation of the angle of incidence of the primary ions with effective primary ion energy. This variation is due to the electrostatic field between the sample surface and the first lens (see insert). The circles are experimental values; the drawn line was derived from a simple calculation of the trajectories of the primary ions at the sample surface. The effective primary ion energy is defined as the energy of the primary ions at the sample surface; in this case it follows from the primary ion accelerating voltage minus the sample voltage.

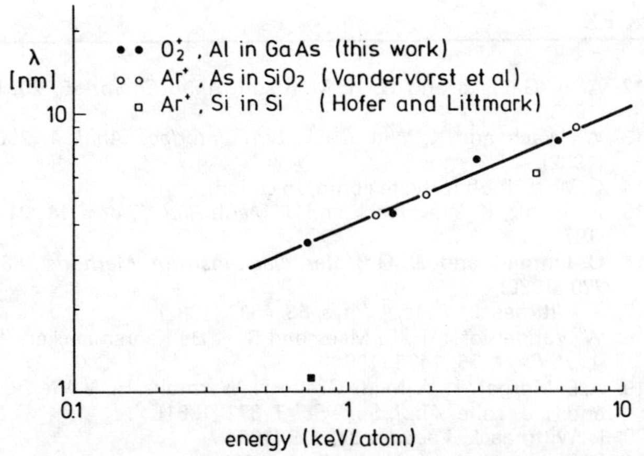

Figure 7. Decay length λ(0) as a function of primary ion energy. The closed circles indicate the results of this work for Al in GaAs at O_2+ bombardment. The open circles are the values obtained by Vandervorst *et al.*[18] for As in SiO_2 (Ar^+ bombardment). The open square gives the theoretical value for Si in Si at Ar^+ bombardment. The closed square gives the actual value obtained at an incident angle of 70° (O_2 bombardment).

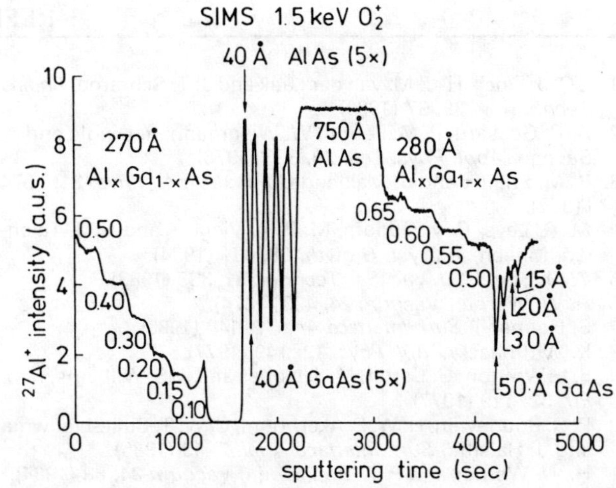

Figure 8. $^{27}Al^+$ depth profile obtained from a complex $Al_xGa_{1-x}As$ multilayer structure. The composition of the various layers is indicated in the figure.

given in Fig. 7. In this figure it is defined as the depth interval over which the signal intensity decreases to $1/e$ of its original value. For comparison the results obtained by Vandervorst *et al.*[18] for a 8 keV arsenic implant in SiO_2 bombarded with Ar^+ are also included in this figure, together with a theoretical result for Si–Si mixing at 5 keV Ar^+ bombardment.[16] This comparison shows that our data are in close agreement with previous results, which is rather surprising considering the differences in system (Al in GaAs versus As in SiO_2, and O_2^+ bombardment versus Ar^+ bombardment). The reasonable agreement of our data with the theoretical value for Si–Si mixing suggests that for Al in GaAs collisional mixing is the dominant process.[16] This is further confirmed by the observed energy dependence of the decay length, which increases approximately in proportion to $E^{1/2}$; such an increase is indeed predicted by the collisional mixing theory.[16,21]

The decay length λ(70) observed at 1.5 keV O_2^+ bombardment, not corrected for angular effects, is also indicated in Fig. 7. This value compares favourably with the decay length expected at perpendicular incidence using the same incident energy. Assuming that extrapolation towards lower energies is valid for the λ(0) values, it follows that the decay length observed at 70° incidence (1.5 keV O_2^+) would correspond to an O_2^+ energy of about 100 eV at perpendicular incidence. This stresses once again the importance of oblique incidence for sputter depth profiling with high depth resolution.

CONCLUSIONS

The present work has indicated the capabilities and limitations of SIMS depth profiling of $Al_xGa_{1-x}As$ multilayer structures. It was shown that at 1.5 keV O_2^+ bombardment the experimentally observed gradient in the Al profile across a $Al_xGa_{1-x}As$/GaAs interface is determined by purely collisional mixing processes. It was further demonstrated that quantitative analysis of the Al content can be performed. The minimum layer thickness required for such an analysis was shown to be dependent on the dynamic range of the analytical signals used for the analysis.

It can be concluded that the present capabilities of SIMS are sufficient to make the technique a valuable tool for the assessment of MOVPE-grown $Al_xGa_{1-x}As$ layered structures. This is exemplified by the SIMS profile shown in Fig. 8, which was obtained from a complex multilayer structure. From a comparison with the growth scheme and cross-sectional TEM results[22] it follows that all the intended features can be observed in the $^{27}Al^+$ profile, even the presence of a few monolayers of GaAs. Furthermore, quantitative analysis showed that the actual Al content of the 270 Å thick $Al_xGa_{1-x}As$ layers agreed within 5% with the intended value. Finally, such SIMS profiles provide important information about the correctness of growth procedures. For example the Al spike observed at the $Al_{0.1}Ga_{0.9}As$/GaAs interface is probably due to inappropriate mass flow control. In summary, SIMS analysis gives a reasonably rapid assessment of overall shape and composition of the layered structure. Although such data can also be obtained from AES depth profiling, SIMS has the additional advantage of providing information on low level Al contents and dopant element concentrations. Layer thickness and interface quality can best be derived from TEM results.[4] This leads to the conclusion that by combining the analytical results from SIMS and TEM optimal assessment of $Al_xGa_{1-x}As$ layered structures is realized.

REFERENCES

1 J. C. J. Finck, H. J. M. van der Laak and J. T. Schrama, *Philips Techn. Rev.* **39,** 37 (1980).
2. A. C. Gossard, P. M. Petroff, W. Wiegmann, R. Dingle and A. Savage, *Appl. Phys. Lett.* **29,** 323 (1976).
3. P. M. Frijlink and J. Maluenda, *Jpn. J. Appl. Phys.* **21,** L574 (1982).
4. M. R. Leys, C. v. Opdorp, M. P. A. Viegers and H. J. Talen-v.d. Mheen, *J. Cryst. Growth*, **68,** 431 (1984).
5. W. J. Bartels, *J. Vac. Sci. Technol.* **B1**, 338 (1983).
6. H. W. Werner, *Vacuum* **24,** 493 (1974).
7. S. Hofmann, *Surf. Interface Anal.* **2**, 148 (1980).
8. K. Wittmaack, *Appl. Phys.* **12,** 149 (1977).
9. J. L. Whitton, G. Carter, M. J. Nobes and J. S. Williams, *Rad. Eff.* **32,** 129 (1977).
10. P. R. Boudewijn, H. W. P. Akerboom, C. W. T. Bulle-Lieuwma and J. Haisma, *Surf. Interface Anal.* **7,** 49 (1985).
11. H. W. Werner and P. R. Boudewijn, *Vacuum* **34,** 83 (1984).
12. A. A. Galuska and G. H. Morrison, *Anal. Chem.* **55,** 2051 (1983).
13. W. Wach and K. Wittmaack, *Surf. Interface Anal.* **4**, 230 (1982).
14. G. W. 't Hooft (private communication).
15. F. Schulz, K. Wittmaack and J. Maul, *Rad. Effects* **18,** 211 (1973).
16. U. Littmark and W. O. Hofer, *Nucl. Instrum. Methods* **168,** 329 (1980).
17. K. Wittmaack, *J. Appl. Phys.* **53**, 4817 (1982).
18. W. Vandervorst, H. E. Maes and R. F. De Keersmaecker, *J. Appl. Phys.* **56,** 1425 (1984).
19. A. E. Morgan, H. A. M. de Grefte, N. Warmoltz, H. W. Werner and H. J. Tolle, *Appl. Surf. Sci.* **7**, 372 (1981).
20. K .Wittmaack, *Vacuum* **34,** 119 (1984).
21. P. Sigmund, *Appl. Phys.* **A30,** 43 (1983).
22. A. F. de Jong (unpublished results).

SURFACE AND INTERFACE ANALYSIS, VOL. 9, 309–317 (1986)

The Application of Secondary Ion Mass Spectrometry to Surface Analysis of Semiconductor Substrates and Devices

A. Brown*, J.A. van den Berg and J.C. Vickerman
Surface Analysis Service, Department of Chemistry, UMIST, PO Box 88, Manchester M60 1QD, UK

The application of static, dynamic and imaging SIMS to semiconductor starting materials and devices is discussed. Static SIMS is used to examine the effect of various chemical preparation methods on the surface of the III–V compounds InAs, InP and GaAs prior to epitaxial growth of device structures. Dynamic SIMS profiles from a superlattice structure involving alternative GaAlAs – GaAs layers are presented along with an examination of the factors affecting depth resolution for low-dimensional devices. Finally the importance of imaging or scanning SIMS for chemical mapping and microanalysis of final device structures is discussed.

INTRODUCTION

The purpose of this paper is to demonstrate the applicability of static, dynamic and imaging secondary ion mass spectrometry (SIMS) to various stages in the production of semiconductor devices.

Static SIMS has been used to examine the effect of cleaning procedures on the semiconductor substrates indium arsenide, gallium arsenide and indium phosphide. These III–V compounds are important materials for the production of semiconductor devices. One of the main pre-requisites for fabrication is a clean, smooth substrate surface prior to epitaxial growth of the actual device structure by LPE, MBE or MOVPE methods.[1] The method of surface preparation should ideally remove both structural damage and electrically active chemical contaminants whilst preserving the stoichiometry of the surface. the general principle employed in practice is to create a surface layer of oxide ('native oxide') which is subsequently removed by chemical etching or *in situ* thermal desorption techniques.[2,3] Chemical preparations of III–V surfaces have previously been assessed for a range of methods, substrates and analytical techniques.[4–10] In this work the effect of degreasing and etching procedures on the surface of InAs, GaAs and InP are monitored by static SIMS. For InAs two published preparation methods were compared directly ie. the first using Br_2/methanol[10] as the etching step, the second using hydrochloric acid.[11] Static SIMS provides highly sensitive elemental and molecular structure information from the uppermost surface layer of any material and has been the subject of a recent review.[12] In this study the positive and negative SIMS spectra permit identification of impurity species and their progressive removal through each preparation stage. The native oxide layers give rise to specific cluster ions eg. GaO_n^-, AsO_n^- whose relative intensity provide information on the success of chemical preparation and the stoichiometry of the surface.

* Author to whom correspondence should be addressed

0142–2421/86/120309–09 $05.00

Dynamic SIMS is the established technique for studying concentration profiles of substrate, dopant or impurity species in semiconductors.[13] This is usually achieved by rastering an energetic (5–10 keV) beam of Ar^+, O_2^+ or Cs^+ across a well defined area of the sample surface and concurrently monitoring the secondary ion signal of the species of interest. Crater edge effects are minimised by only using information from an electronic or optically gated area from the centre of the crater. The demand for faster electronic circuits in the computing industry has led to the development of compound semiconductor heterostructures and in particular modulation-doped GaAs devices. A feature of these devices is their superlattice structure with layer thickness in the range 10 Å to 500 Å. They therefore represent a considerable challenge to the SIMS technique since the effects of recoil mixing in sputter profiling will be a limiting factor in their analysis. In this paper we report on preliminary experiments designed to optimise conditions for routine analysis of these devices.

The development of liquid metal ion sources in recent years has made high spatial resolution SIMS imaging a routine investigative technique in many laboratories.[14,15] The application of a 10 keV gallium ion gun to typical semiconductor device analyses is described along with current developments in computer-controlled SIMS imaging.

EXPERIMENTAL

The following cleaning procedures were applied to cut and polished 1 cm^2 slices of InAs, GaAs and InP.

InAs

Procedure A

Stage 1: Degreasing in boiling trichloroethylene (5 min).
Stage 2: Further degreasing in boiling isopropanol and then methanol (5 min each).

Stage 3: Etch in 2% Br_2/methanol (5 min), rinse in methanol.
Stage 4: Etch in concentrated HF (5 min) rinse in distilled water.

Procedure B

Stages 1 and 2 above then an etch in 150 : 1 : 100 HCl : H_2O_2 : H_2O for 5 min.

GaAs

Stage 1: Degreasing in 'Decon 90' detergent solution (10 min); rinse in water.
Stage 2: Rinse in methanol, degrease by refluxing in isopropanol using Soxhlet extractor (20 min); rinse in methanol.
Stage 3: Etch in 5 : 1 : 1, H_2SO_4 : H_2O : H_2O_2 for 2 min; rinse in water.
Stage 4: Dip in HCl/H_2O (3 : 1); rinse in water until hydrophobic; blow dry in air.

InP

Stages 1 and 2 as for GaAs above then
Stage 3: Etch in 1% w/v solution of bromine/methanol; quench in methanol.
Stage 4: Rinse in water; blow dry in air.

NB. The water used for GaAs and InP was 18 MΩ deionized from a mill Q purification system. 12 MΩ water was used for InAs. All semiconductor substrates were obtained from MCP Electronics Ltd. All reagents were BDH 'Aristar' grade. Prior to analysis the sample slices were clamped onto stainless steel holders.

The equipment used to obtain SIMS spectra is a VG SIMSLAB instrument. The essential components consist of an ion-atom gun designed in this laboratory and described in detail elsewhere,[16] a secondary ion energy analyser after the design of Wittmaack[17] and a quadrupole mass spectrometer (VG MM12–12, 0–1200 Daltons). The analysis chamber is also fitted with a VG MIG100 gallium liquid metal ion gun (minimum spot size 0.1 μm, 0.5–10 keV Ga^+, 10^{-12}–10^{-7}A) and a VG DP50B Duoplasmatron ion gun (O_2^+, Ar^+ 0–10 keV, 0.1nA–2 μA) for SIMS imaging and depth profiling respectively. A precision specimen manipulator permits movement in X, Y and Z directions along with 180° of rotation about the *Z* axis. A railway system and vacuum lock allows rapid transfer of specimens from atmosphere to UHV in approx 3 min. the base pressure in the system is 10^{-10} torr.

Static SIMS spectra were acquired using 2 keV Ar atoms with a flux density of 3×10^9 atoms $cm^{-2}s^{-1}$ measured in the manner described previously.[16] An atom beam is chosen for the analysis of semi or non-conducting materials since charging problems experienced with ion beams are significantly reduced.

For SIMS imaging experiments the gallium ion beam is rastered over the surface using a digital scan unit with the mass spectrometer tuned to the ion of interest. The output of the scan unit *X* and *Y* axes are used to drive a high resolution oscilloscope whose *Z* axis is modulated by the output of the secondary ion pulse counting system. Thus black and white ion images (1056 × 1056 pixels) may be produced by Polaroid or 35 mm photography of the oscilloscope screen. The average acquisition time for one SIMS image is 200 seconds.

Colour imaging is achieved by acquiring the secondary ion information into a DEC Professional 350 computer system with software designed in this laboratory. Images are acquired in approximately 5 min on a 128 × 128 pixel format and the system has the facility for overlaying, additions, ratioing and noise reduction of image data. Depth profiles and mass spectral data are also collected by the computer system.

RESULTS AND DISCUSSION

Static SIMS

InAs

The Stage 1 of Procedure A for InAs, degreasing in trichloroethylene, produces a surprising initial positive ion spectrum (Fig. 1(a)) in that no In^+ signal is observed. The surface is completely covered by a very thin layer of organic residue as indicated by the $C_nH_m^+$ fragments from $n = 1$–10. A very brief period of argon atom etching (10^{12} atoms $cm^{-2}s^{-1}$, 2 keV, 10 min) was sufficient to reduce the intense $C_nH_m^+$ contribution and yield a spectrum dominated by In^+. Other species observed include Ba^+, $BaOH^+$ and mixed cluster ions of the type $BaAsOH^+$ and $InBaO^+$, i.e. mixed oxide ions. The origin of barium in InAs is not clear but introduction may have occurred at the polishing stage. Negative ion spectra (not shown here) contained $C_n(H)^-$ ions, O^-, OH^-, Cl^-, PO_n^-, AsO_n^- and InO_n^-. The origin of phosphate ions is also not clear.

Further degreasing in isopropanol and methanol (stage 2) produced a reduction in the relative intensity of $C_nH_m^+$ ions and with In^+ now visible (Fig. 1(b)).

A dramatic change in the positive ion spectrum is observed after Stage 3 — Br_2/methanol etching (Fig. 1(c)). Much lower relative intensities of Na^+ and $C_nH_m^+$ are noted along with complete removal of Ba^+. Increasing surface cleanliness is shown by the appearance of an intense In_2^+ signal and the substrate cluster ions $InAs^+$, $InAsO^+$, In_2O^+. Residual halide contamination and the formation of surface halide complexes is shown by intense Cl^- and Br^- signals and the cluster ions In_2Cl^+, $AsBr_2^-$ and $InBr_2$.

Na^+ and $C_nH_m^+$ decrease further after the final etch with HF followed by rinsing (stage 4 — Fig. 1(d)). The negative ion spectrum shows however, that F^-, Cl^- and Br^- remain.

Although negative ion spectra are not presented here they convey important information on impurities and the oxide surface. A summary of the trends for some important negative ions is presented in Table 1 with intensities normalised to In^+. As might be expected the halide ions F^-, Cl^-, Br^- increase dramatically after the treatment stages involving their respective acids. This suggests that rinsing with water does not remove them from the surface efficiently due to formation of stable halide complexes. The increase in C_2H^- and AsC^- suggests build up of carbon in the surface. The relative decrease in AsO_2^- after Stage 3 indicates some partial removal of native oxide during etching.

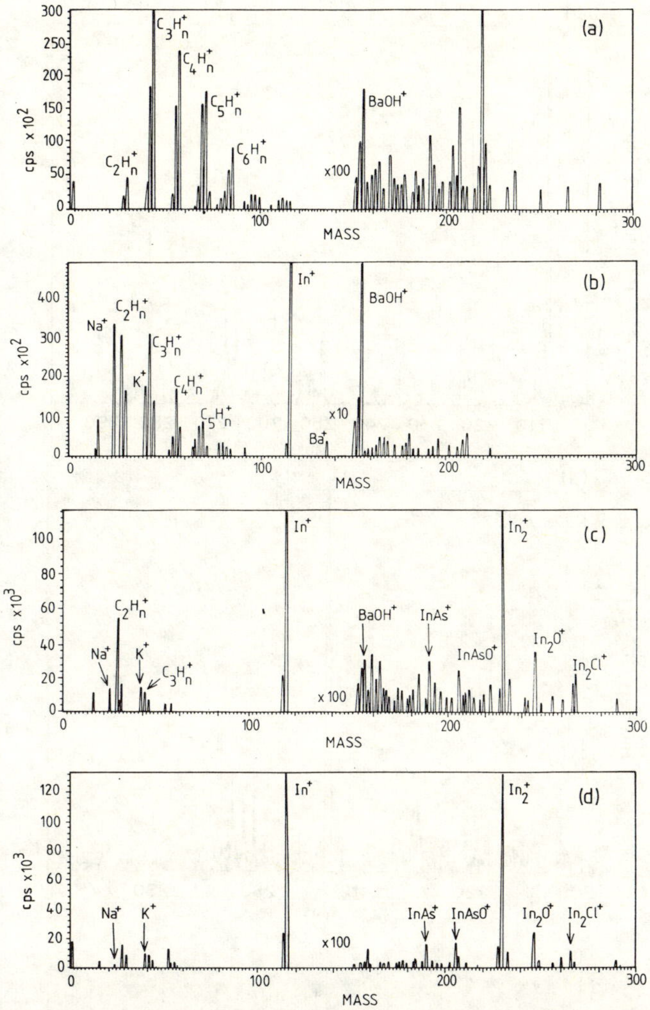

Figure 1. Positive SIMS spectra from successive stages in the chemical preparation of InAs: (a) degreased in trichloroethylene; (b) further degreased in isopropanol/methanol; (c) etched in Br_2/methanol; (d) etched in HF.

Table 1. Variation of impurity and substrate ion intensities, relative to $^{113}In^+$, with sample treatment for InAs

Procedure	Stage	Treatment	F^-	C_2H^-	Cl^-	Br^-	AsO_2^-	AsC^-
A	1	Degreased in trichloroethylene	0.01	0.32	0.63	<0.01	0.22	0.02
	2	Degreased in isopropanol/methanol	0.11	0.77	0.7	0.07	1.62	0.03
	3	2% Br_2/methanol	0.06	1.6	1.2	4.0	0.48	0.24
	4	Conc. HF rinsed in water	0.88	1.2	0.65	0.42	0.17	0.11
B	3	Stages 1, 2 above then HCl/H_2O_2	—	1.3	8.8	1.1	0.71	0.06

Finally the effect of using hydrochloric acid as etchant was to produce an extremely pitted substrate readily visible to the naked eye after only a brief exposure to HCl. Hence this comparative study showed at least that HCl etching is a completely useless treatment for InAs substrates despite producing acceptable levels of impurities such as carbon and sodium. A more comprehensive discussion of the results for InAs will appear elsewhere.[18]

InP

The positive SIMS spectrum of InP, as received, prior to chemical treatment (Fig. 2(a), (b)) is dominated by $C_nH_m^+$ ions and intense ions at 73, 147, 207, 221 and 281. These are characteristic fragments of polydimethyl siloxane and probably originate from the packaging material.[19] The corresponding negative spectrum (Fig. 3(a)) confirms the presence of organic contamination through the C_nH^- ions along with O^-, OH^- from surface oxide and Cl^-. At higher mass $^{59}CH_3SiO^-$, $^{60}SiO_2^-$, $^{75}CH_3SiO_2^-$ confirm the siloxane presence and $^{63}PO_2^-$, $^{79}PO_3^-$ indicate the surface phosphide is actually in the form of phosphate. The degreasing treatment (Stages 1 and 2) results in the complete removal of siloxane contaminations (Fig. 2(c), (d)) although low mass organic species $C_nH_m^+$ (n = 1–6) are still observed. At higher mass the indium oxide ion In_2OH^+ and a small In_2^+ have appeared.

The increasing cleanliness of the substrate is mirrored by the negative spectrum (Fig. 3(b)) where C_nH^- species have decreased relative to O^- and PO_2^-, PO_3^- have increased. The anomalous increase in Cl^- ion intensity may be due to carry over from one of the reagents or vessels used.

Etching in bromine/methanol solution again significantly lowers the intensity of $C_nH_m^+$ clusters (Fig. 2(e), (f)). Likewise In_2^+ and In_2OH^+ have increased relative to In^+. In the negative ions (Fig. 3(c)) PO_2^- and PO_3^- now dominate the spectrum, although C_nH^- and Cl^- have not been completely removed.

GaAs

For GaAs only the negative ion spectra will be discussed since they contain all the essential information on contaminants and surface oxide behaviour. The 'as received' surface (Fig. 4(a)) has intense CH^- and C_2H^- ions from organic contaminants, O^-, OH^-, Cl^- and a Br^- signal of unknown origin. At higher mass AsO_2^- is the most intense oxide species. Degreasing steps (Stages 1 and 2) remove the Br^- ion and reduce the low mass carbon fragments (Fig. (4)). $^{91}AsO^-$, $^{107}AsO_2^-$ and $^{123}AsO_3^-$ increase relatively and weak signals for $^{85,87}GaO^-$ and $^{101,103}GaO_2^-$ are seen.

Etching with sulphuric acid/hydrogen peroxide/water mixture (fig. 4(c)) again produces significant decreases in the contaminant species C_nH^- and Cl^-. However the ions at 80, 96 and 97 Daltons (attributed to SO_3^- and $SO_4(H)^-$) show that surface sulphate species remain even after rigorous rinsing in water.

The final treatment with hydrochloric acid appears to have a less than beneficial effect on the substrate. The overall intensity of oxide species such as O^- and AsO_2^- have decreased (Fig. 4(d)). Most notably a decrease by a factor of 3 is observed for the GaO_n^- ions measured relative to the AsO_2^- ion. Additionally the CH^-, C_2H^- and Cl^- have actually increased relative to the oxide species.

For all the cleaning treatments studied a progressive decrease in organic contamination is generally observed. However the use of halogenated solvents or acids tends to leave halide residues or anionic complexes behind which are not removed easily by water or solvent rinsing. The removal of surface contamination is mirrored in all cases by a concommitant increase in oxide specific ions and clusters from the substrate e.g. AsO_2^- from GaAs and In_2^+ for InP. These ions provide a good indication of relative surface cleanliness.

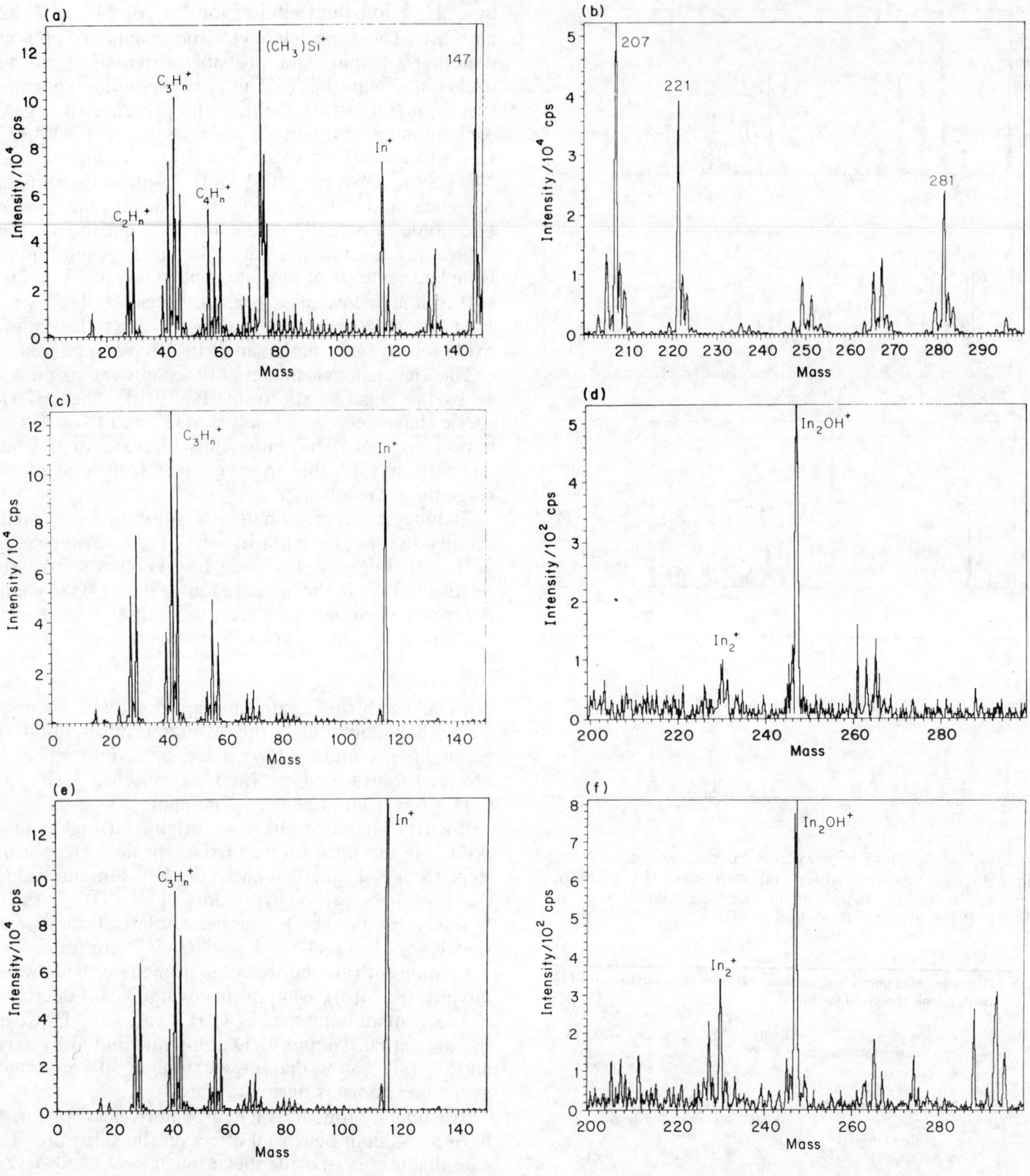

Figure 2. Positive SIMS spectra from successive stages in the chemical preparation of InP: (a), (b) as received; (c), (d) degreased in DECON 90 followed by isopropanol reflux; (e), (f) etched in Br_2/methanol.

For indium arsenide the most notable result of the comparison procedure A is regarded as a successful surface damage incurred during HCl treatment. By comparison procedure A is regarded as a successful preparation method producing a relatively clean oxide surface with no apparent physical damage.

The observation of changes in oxide ion ratios for GaAs upon HCl etching is in good agreement with the literature. It is generally observed by XPS and AES measurements that the oxide layer on GaAs is rich in Ga due to the greater reactivity of Ga towards oxygen.[4,5,8] Consequently the surface which remains after removal of the native oxide layer e.g. by HCl treatment, is As-rich. Hence the observed relative decrease of GaO^- and GaO_2^- species after the HCl stage.

Dynamic SIMS

III–V semiconductor heterostructures are receiving increasing attention since they are interesting substrates for fundamental studies of electron behaviour in solid state physics and from a technological view-

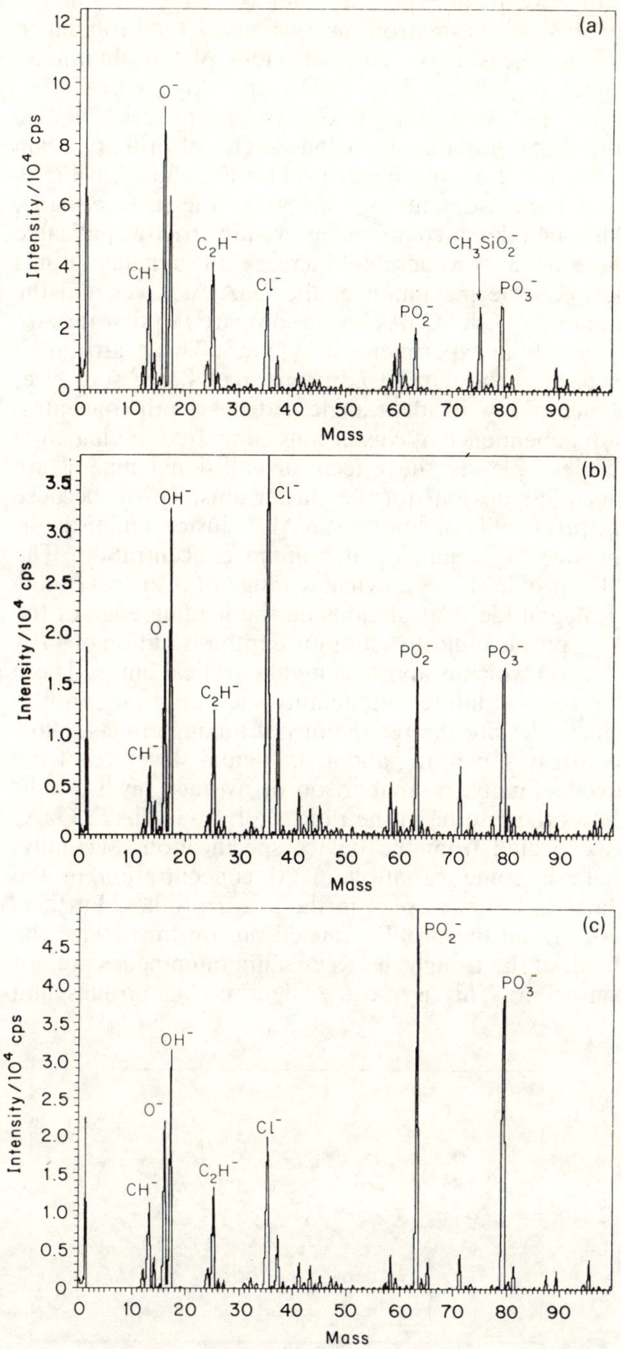

Figure 3. Negative SIMS spectra from stages in the chemical preparation of InP: (a) as received; (b) degreased in DECON 90 followed by isopropanol reflux; (c) etched in Br_2/methanol.

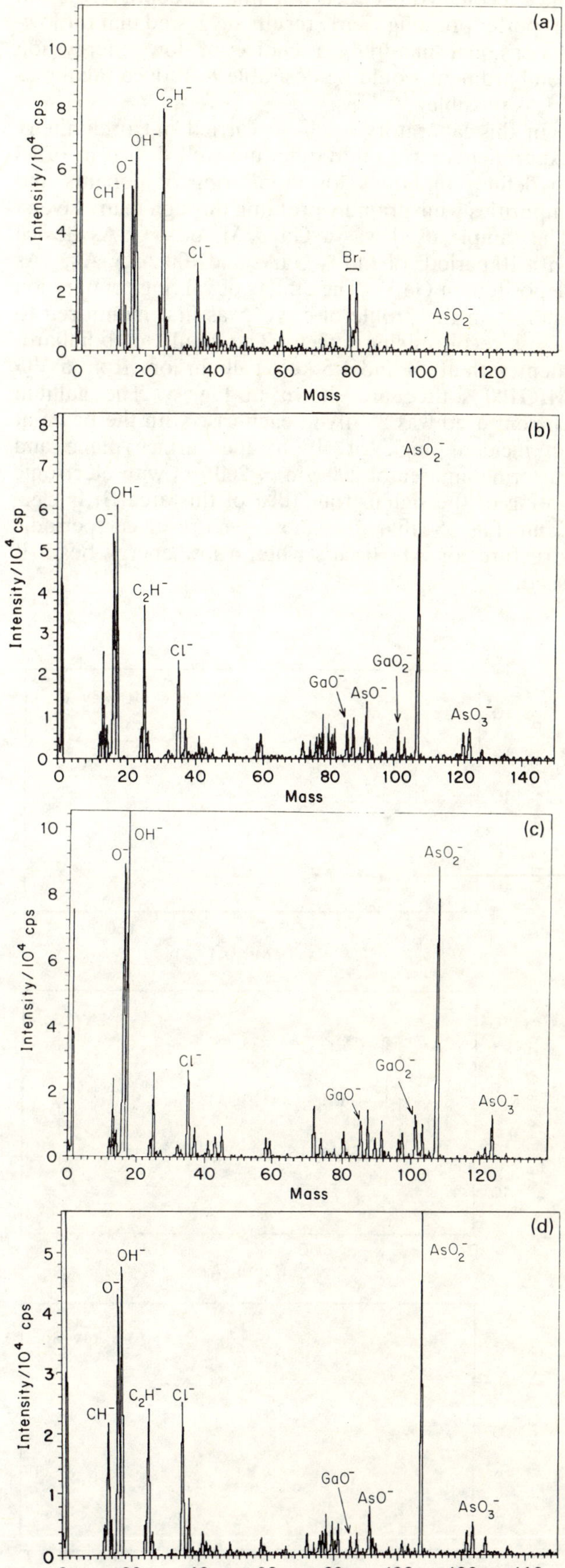

Figure 4. Negative SIMS spectra from stages in the chemical preparation of GaAs: (a) as received; (b) degreased in Decon 90 followed by isopropanol reflux; (c) etched in $H_sSO_4/H_2O_2/H_2O$ (5 : 1 : 1); (d) dipped in HCl/H_2O (3 : 1).

point their high electron mobilities, compared to silicon based materials, make them important in the development of faster switching transistors and other components. The most common quantum well devices reported in the literature are multilayer $Ga_{1-x}Al_xAs$–GaAs structures.[20] They generally consist of repeating layers of GaAlAs and GaAs with thicknesses in the range 10 Å to several hundred Å. The development of epitaxial growth methods such as MBE and MOVPE now means that almost perfect single crystal semiconductor layers can be grown with interface widths approaching one atomic layer. Such devices present an obvious challenge to dynamic SIMS analysis since recoil mixing effects will limit the depth resolution in sputter profiling. Early results suggested that for low-dimensional materials a choice of low energy ion bombardment would be desirable to reduce mixing as far as possible.[21,22]

In this laboratory we have carried out preliminary experiments on a multi-quantum well device in order to define conditions for monitoring of dopants and impurities whilst rapidly profiling through many layers. The sample used was a $Ga_{1-x}Al_xAs$ – GaAs device with 10 periods of 140 Å GaAs and 100 Å $GaAl_{0.24}As$ deposited on GaAs. The effects of varying primary ion energy on the profile of $^{27}Al^+$ was first monitored to assess recoil mixing effects. The results of bombardment with 10, 4 and 1.5 keV gallium ions from a VG MIG100 source are shown in Fig. 5. The gallium current used was 25 nA in each case with the beam at an incident angle of 60° to the surface plane and rastering an area of 200 μm × 200 μm with electronic gating of the signal from 10% of this area. It is clear from Fig. 5 that the most well-resolved periodic structure for Al^+ occurs when a low energy beam is used.

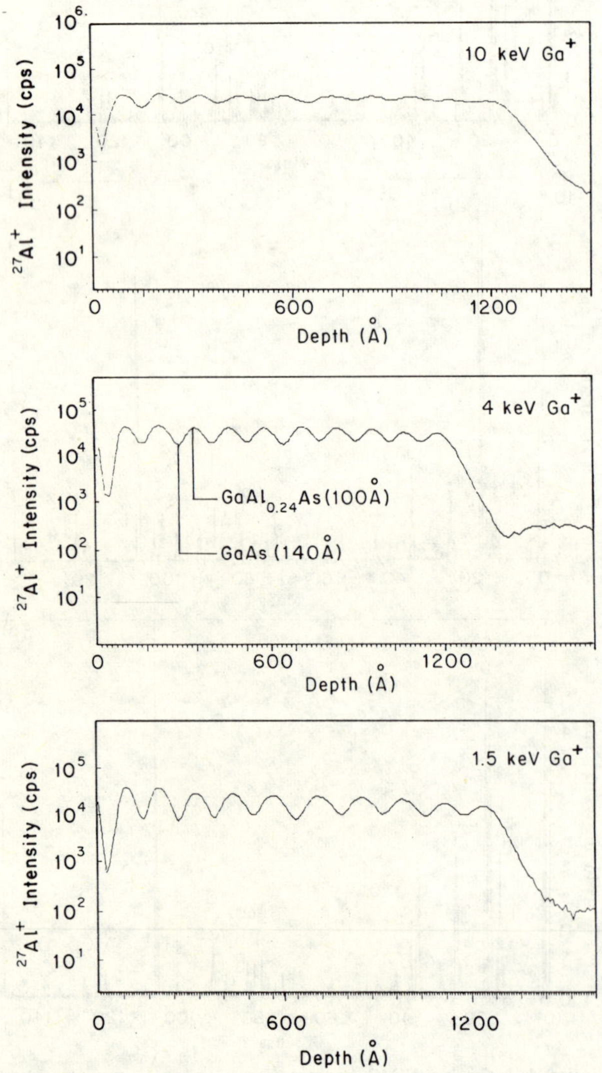

Figure 5. The variation of $^{27}Al^+$ intensity with depth for a GaAlAs–GaAs multi-quantum well device at various primary beam energies; (25 nA Ga^+, 10, 4, and 1.5 keV).

In subsequent experiments an oxygen beam from the DP50B duoplasmatron gun was used for profiling to enhance sensitivity. The behaviour of the aluminium containing cluster ions AlO^+ and Al_2^+ was studied using a 4 keV 25 nA O_2^+ beam rastering a 250 μm × 250 μm area at various angles of primary beam incidence. Log and linear profiles are shown in Fig. 6 for a bombardment angle of 30° to the surface plane. Although direct comparisons are not strictly applicable there is a considerable increase in dynamic range between the maximum of the GaAlAs layer and the minimum of the GaAs layers compared to that for Al^+ in the initial experiments using Ga^+. This is attributed in part to the careful adjustment of beam spot size, choice of low incident angle and the depth concentration dependence of cluster ions compared to elemental species. Clearly the effects of collisional mixing are much less obvious for the cluster ions. This is because the probability of emitting an Al_2^+ cluster diminishes as the inverse square of aluminium concentration. The Al_2^+ profile shows a dynamic range of over two orders of magnitude. Calculations on the leading edge of the Al_2^+ profile indicate a limiting depth resolution of 40 Å (± 5 Å) which is approaching theoretical limits. There are several interesting features about these profiles which, despite the restrictions of mixing effects, allow us to say something about the actual structure of the layers. Firstly, a comparison of average layer widths shows them to be in the ratio 1 : 1.4, GaAlAs : GaAs as expected from the device specification. Secondly, there is some variation in Al concentration in the GaAlAs layers as seen in the linear profiles. Furthermore in all the profiles carried out on this device the depth of the troughs between aluminium peaks are not consistent. This is most evident in log profiles and

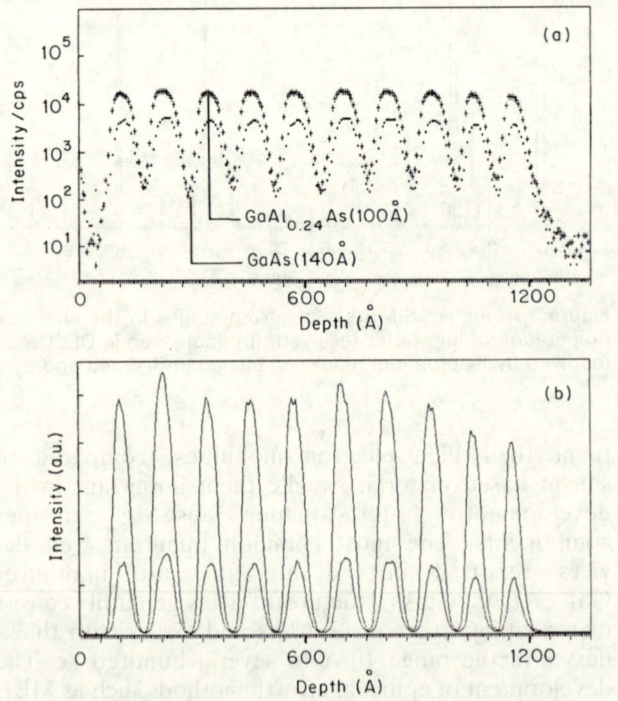

Figure 6. The variation of $^{43}AlO^+$ and $^{54}Al_2^+$ intensity with depth for the $GaAl_{.24}As$–GaAs device; (4 keV O_2^+, 25 nA, 30° incidence to surface plane): (a) logarithmic intensity scale; (b) linear scale; Al_2^+ upper curve, AlO^+ lower curve in (a) and (b).

suggests some minor differences in layer thickness. One of the requirements for quantum well devices will be the ability to analyse any given layer for unknown impurities introduced during fabrication. This is possible using SIMS by carrying out a depth profile to the required layer followed by mass spectral analysis. Positive ion spectra from GaAlAs and GaAs layers in this device show the presence of Na^+ and K^+ as the major impurity species in both layers.

The initial conclusions on quantum well analysis by SIMS are as follows

1. Low energy ion bombardment must be used to reduce recoil mixing effects.
2. Glancing incidence for the primary beam increases depth resolution.
2. The use of very low energies produces the need for careful focussing of the primary beam to yield minimum spot size and hence reduce crater edge effects.
4. Cluster ion detection is a potentially useful method for monitoring matrix element, dopant or impurity levels.
5. The use of heavier bombarding species e.g. Cs or Xe should be beneficial in reducing mixing effects even further.
6. Sample preparation techniques such as microtome slicing should be explored. Samples could be tailor made to the requirements of the analysis beam e.g. by cutting to the dimensions of the rastered area.

SIMS imaging

The last few years have seen a rapid development in liquid metal ion source (LMIS) technology culminating in the routine application of these sources to research and analytical problems in SIMS and scanning ion microscopy.[14,15,23–26] The high brightness of the LMIS (ca $10^6 Acm^{-2}sr^{-1}$) and its small source size (100 nm) permit the formation of a low energy microfocussed beam with enormous potential in surface analysis. Commercially available sources operating at 30 keV are capable of spot sizes below 500 Å and have operating currents in the range 1 pA to several hundred nA. A series of papers by Bayly and co-workers have comprehensively demonstrated the wide range of materials which high resolution scanning SIMS can be used to analyse.[23,24] One of the most demanding areas requiring information on the lateral distribution of chemical species is that of semiconductor devices. The combined requirements of sub-micron spatial resolution and sensitivity to elements in the ppm range make scanning SIMS the most useful of the currently available techniques. There are a number of reasons for this which can be summarised as follows:

1. The common impurity and dopant species in semiconductors can be detected in ppm amounts — particularly ionic contaminants such as halides, sodium and potassium often left behind during processing.
2. Mass spectra can be acquired from a precisely defined area from mm^2 to μm^2 size with a sampling depth of 1–2 layers in the static mode.
3. From 2 above any elemental or molecular species may be mapped by rastering the beam over a given area and using the secondary ion signal intensity to form the image.
4. The use of a secondary electron detector allows high quality SEM images to be generated. Thus local features may be identified and analysed.
5. The development of computer imaging techniques, particularly framestore acquisition, ensure rapid accumulation of data, real time display of the image and considerable manipulation of the stored data.

Several important factors arise from the above. Firstly the microfocussing quality of the LMIS will permit small area depth profiling of sub-micron areas. This will become increasingly important as device structures decrease in size from micron to sub-micron technology. Secondly, computer acquisition allows several species to be imaged simultaneously with total accumulation time per image of less than one minute. Hence ion beam degradation or sputtering of the surface is minimised. This is particularly important for very small features or thin layers.

The most common problems encountered in surface analysis of devices are the chemical identification of microscopic surface features and mapping of the distribution of known contaminants in a specific device area. The application of scanning SIMS to two analytical problems will be used to illustrate several of the points mentioned above.

Semiconductor devices usually consist of extremely complex three dimensional structures. Chemical mapping of impurities or different substrate species can only be analytically meaningful if some method is available whereby overlaying of several images can be achieved. This is demonstrated in Fig. 7 (a)–(c) where

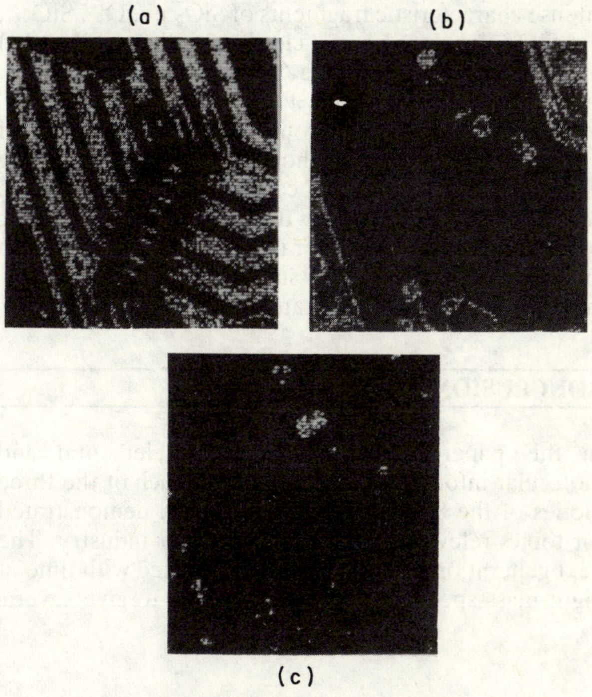

Figure 7. SIMS images of (a) Si^+; (b) Al^+ and (c) Na^+ from a semiconductor device. The field of view for each image is 60 μm. (Images acquired using 10 keV Ga^+, 0.1 nA.)

Si^+, Al^+ and Na^+ images from a 60 μm×60 μm area of a device structure are presented. Si and Al^+ represent the substrate and Na^+ is a localised impurity. The images are intensity scaled from black to white in order of increasing intensity. Obviously precise location of the sodium impurity with the substrate structure is not possible by reference to the individual images. A multi-coloured overlay image provided by computer manipulation techniques provided a readily interpretable solution to the problem but cannot be shown here for technical reasons. The computer system used for acquisition and presentation of this data was developed at UMIST[27] on a DEC 300 series computer fitted with a 10 megabyte Winchester disc, real time interface board and extended bit map in conjunction with a colour monitor.

The high manufacturing tolerances used in device fabrication mean that any production stage can be vulnerable to minute traces of impurity from chemical reagents or processing plant. The presence of stains or surface debris on wafers can cause expensive manufacturing delays. Hence there is a need for rapid analysis of such problems. Indeed, most of the leading semiconductor manufacturers already employ surface analysts on full-time or consulting basis for these reasons. The spectra shown in Fig. 8 were acquired from 2 μm×2 μm areas in and around an oval shaped 'stain' which appeared during fabrication of a silicon based memory device. The stain was clearly visible both optically and in SEM and had a diameter of approximately 50 μm. SIMS imaging with a 10 keV gallium ion beam revealed an excess of O^- in areas of polysilicon tracks which would normally be oxide free. To investigate the problem further secondary ion mass spectra were acquired from 2 μm×2μm areas in and around the stain region. The spectra are shown in Fig. 8 (a–c). A gate oxide region (Fig. 8(a)), which comprises several hundred angstroms of silicon dioxide, shows intense characteristic fragments of SiO_2 i.e. O^-, SiO_2^-, and SiO_3^- whereas normal clean polysilicon (Fig. 8(b)) shows only residual surface O^- and the expected Si^-. the chemical nature of the stain is revealed in Fig. 8(c) with SiO_2^- and SiO_3^- confirming the presence of silicon dioxide on what should be clean Si. Having established the presence of certain impurities it is then possible to carry out more detailed chemical imaging and overlaying e.g. at higher magnifications to establish the exact registry with the substrate and hence determine the origin of the contamination.

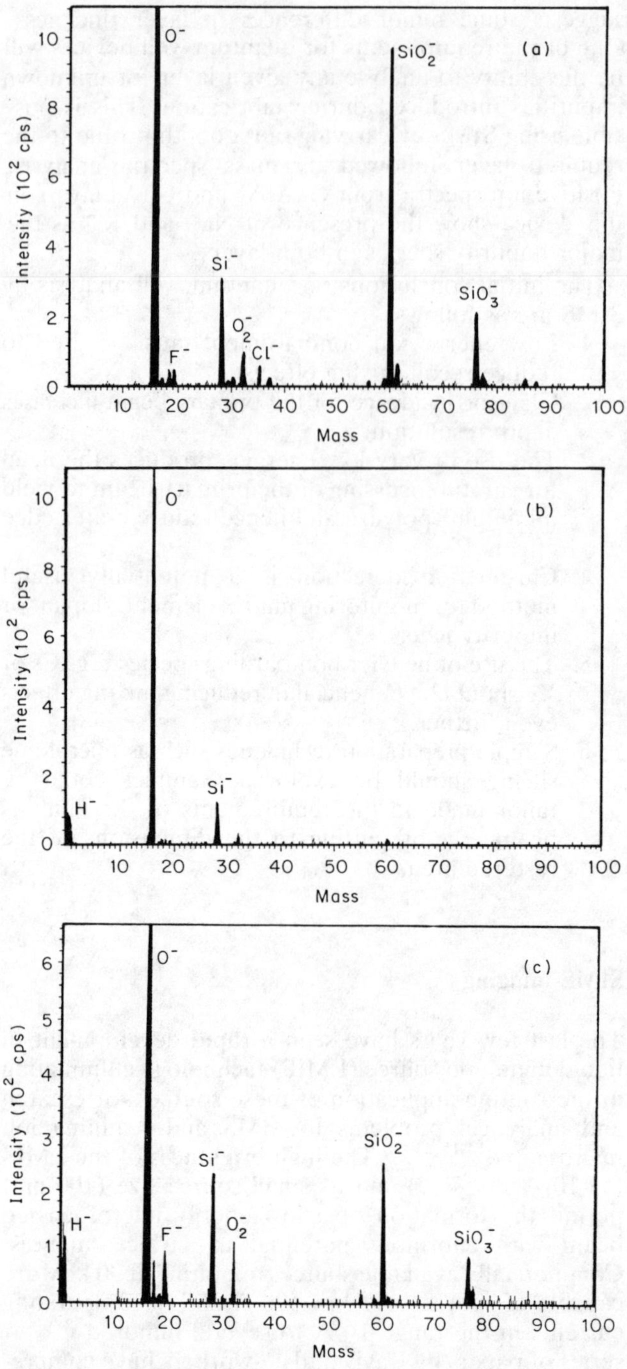

Figure 8. SIMS spectra acquired from 2 μm × 2 μm areas in and around a stain feature on a silicon memory device using 10 keV Ga^+, 0.1 nA; (a) gate oxide (SiO_2); (b) normal polysilicon track; (c) stain area of polysilicon track.

CONCLUSIONS

In this paper the broad range of elemental and molecular information available from each of the three modes of the SIMS technique has been demonstrated for topics relevant to the semiconductor industry. The next generation of SIMS instruments fitted with time of flight mass spectrometers are expected to give several orders of magnitude increase in sensitivity in the static mode compared to quadrupole based instruments. The inclusion of 30 keV liquid metal ion sources will also bring higher sensitivity and spatial resolution to SIMS imaging and small area depth profiling of device structures.

REFERENCES

1. L. Alekzandrov in *Growth of Crystalline Semiconductor Materials on Crystal Surfaces*, Chap. 4, p. 171, Elsevier (1984).
2. C. Ghosh and W. Pothamns, *J. Vac. Sci. Technol.* **B1,** 837 (1983).
3. B. Tuck, *J. Mat. Sci.* **10,** 321 (1975).
4. C. C. Chang, P. H. Citrin and B. Schwartz, *J. Vac. Sci. Technol.* **14,** 4, 943 (1977).
5. T. Oda and T. Sugano, *Japan J. Appl. Physics*, **15,** 1317 (1976).
6. J. L. Zilko and R. S. Williams, *J. Electrochem Soc.* **129,** 406 (1982).
7. J. B. Clegg, *Surf. Interface Anal.* **2,** 91 (1980).
8. E. Huber and H. L. Hartnagel, *Solid State Electronics*, **27,** 589 (1984).
9. D. E. Aspnes and A. A. Studna, *Appl. Phys. Lett.* **39,** 316 (1981).
10. M. G. Astles, O. D. Dosser, A. S. McLean and P. J. Wright, *J. Cryst. Growth*, **54,** 485 (1981).
11. B. R. Hancock and H. Kroemer, *J. Appl. Phys.* **55,** 4239 (1984).
12. A. Brown and J. C. Vickerman, *Surf. Interface Anal.* **6,** 1 (1984).
13. C. W. Magee, R. E. Honig, and C. A. Evans, Jr. in *Secondary Ion Mass Spectrometry — SIMS III*, p. 172, Springer Verlag (1982).
14. P. D. Prewett and D. K. Jefferies, *Appl. Phys.* **13,** 1747, (1980).
15. A. Brown and J. C. Vickerman, *Analyst*, **109,** 851 (1984).
16. A. Brown, J. A. van den Berg and J. C. Vickerman, *Spectrochim. Acta,* **40B,** 871 (1985).
17. K. Wittmaack, J. Maul and F. Schulz, *Int. J. Mass Spectrom. Ion Phys.* **11,** 23, (1973).
18. A. Brown, N. Hunt, A. M. Patterson, J. C. Vickerman and J. O. Williams, *Chemtronics* **1,** 11 (1986).
19. D. Briggs, *Surf. Interface Anal.* **5,** 113, (1983).
20. V. Narayanamurti, *Physics Today*, **Oct** 24 (1984).
21. A. Van Oostrom, *Vacuum*, **34,** 881 (1984).
22. R. J. M. Griffiths, N. G. Chew, A. G. Cullis and G. C. Joyce, *Electronics Lett.* **19,** 998 (1983).
23. A. R. Bayly, A. R. Waugh, and K. Anderson, *Scanning Electron Microscopy*, pp. 23–29, SEM Inc, AMF O'Hare, Chicago (1983).
24. A. R. Bayly, A. R. Waugh and P. Vorhalik, *Spectrochim. Acta*, **40B**, 717, (1985).
25. R. Levi-Setti, P. H. La Marche, K. Lam, T. H. Shields, and Y-L. Wang, *Nuclear Instr. Methods Phys. Res.* **218,** 368 (1983).
26. F. G. Rudenauer, P. Pollinger, H. Studrika, H. Gnaser, W. Steiger and M. Higatsburger in *SIMS III*, p. 43, Springer Verlag (1982).
27. P. Humphrey, unpublished results.

SURFACE AND INTERFACE ANALYSIS, VOL. 9, 319–323 (1986)

A Characterization Study of the Native Oxide Layer Formed on Chemically Etched InP (111)

Gar B. Hoflund and Cheryl F. Corallo
Department of Chemical Engineering, University of Florida, Gainesville, Florida 32611, USA

ISS, AES and angle-resolved ESCA have been used to characterize the native oxide layer formed on the chemically polished and etched InP (111) surface. It is estimated that this oxide layer is about 30 Å thick and forms a sharp interface at the oxide/InP boundary. The oxide layer is nonhomogeneous. ISS shows that the outermost surface layer is In rich, and AES combined with the ISS results shows that the region just below the surface layer is P rich. The In is believed to be present in the oxide layer as InP and $InPO_3$ in approximately equal concentrations. Only a small amount of In_2O_3 is detected. Carbon also comprises the oxide layer, and ESCA shows that it is present as graphite and/or hydrocarbons. The region below the oxide layer is P rich, but depth profiling using AES while sputtering suggests that the composition may approach that of InP with increasing depth.

INTRODUCTION

InP is becoming an important III-V semiconductor material for fabrication of semiconductor devices.[1–4] Small changes in the composition of InP surfaces can significantly influence the electronic structure at the surface.[5] Thus, the properties of epitaxial layers and, consequently, of the final devices are determined by the surface of the InP substrate and any oxide layer which may be present. For this reason it is important to characterize native oxide layers on InP substrates. Often the native oxide layer is completely removed by polishing procedures or is partially removed during chemical etching steps. Characterization of the additional oxide formed upon exposure to air or formed by various methods of oxidation is also important.

Numerous studies of the oxidation of InP have been performed in order to investigate the growth mechanisms, surface compositions and oxidation states of the surface species using various methods of oxidation such as thermal oxidation,[6,7] high-pressure thermal oxidation,[8] photoenhanced thermal oxidation,[9] photoenhanced oxidation,[10] chemical oxidation[11] and electron-stimulated oxidation.[12] The effects of etchants on InP surfaces and oxide layers have also been investigated [13–16] as have the effects of sputtering,[17–19] annealing[20–22] and exposure to vacuum.[5,23]

ESCA, SIMS and x-ray diffraction were used by Nelson *et al.*[6] to study the thermal oxides of p–type InP (100) formed in ~1/3 atm of O_2. They suggest that for a temperature in the range of 400 to 600°C an outer layer of In_2O_3 and $InPO_4$ is formed and an inner layer of primarily $InPO_4$ is formed. Elemental P is observed at the oxide/bulk interface. For $T > 650$°C the oxide is composed of $InPO_4$. Yamaguchi and Ando[7] report that a polycrystalline oxide composed of $InPO_4$ ($In_2O_3+P_2O_5$) forms during a thermal oxidation (450–500°C) of an undoped InP (100) surface in dry oxygen (0.2–2 atm). At higher temperatures the oxide is mostly In_2O_3. Infrared absorption and electron diffraction data suggest that for $T > 620$°C the oxides become In rich and lower oxides (possibly In_2O) form due to the loss of phosphorous. Wilmsen *et al.*[8] find that bulk oxides consisting of $InPO_4$, In_2O_3 and P_2O_5 form at elevated temperature (450°C) and high oxygen pressures (42–500 atm). Faithpour *et al.*[9] used ESCA and SEM to investigate the laser-enhanced thermal oxidation of *n*–type, semi-insulating InP (100) in a N_2O atmosphere. They found that In_2O_3 and a phosphate species form and that the growth rate is a function of the light intensity, N_2O pressure and temperature. The phosphate layer thickness is believed to be limited by the outwardly diffusion of phosphorus, which has also been suggested by Nelson *et al.*[6] The photon-stimulated oxidation of cleaved InP (110) studied by Koenders *et al.*[10] results in the formation of a depletion layer, but the fact that no surface dipole is introduced suggests that oxygen is incorporated into the lattice rather than adsorbed at the surface. Exposure to oxygen with and without photon stimulation results in depletion of phosphorus at the surface in addition to formation of indium oxide. The chemical oxidation of *n*–type InP (100) in hot nitric acid also reveals a depletion of surface phosphorus as discussed by Korotchenkov *et al.*[11] Annealing and exposure to vacuum increases the amount of phosphorus in the surface region. Electron-stimulated oxidation of *n*–type InP (100) was investigated by Oliver *et al.*[12] using AES, LEED and ELS. In_2O_3 and a volatile phosphorus oxide form in the electron irradiated region. Heating above 500°C causes the formation of small droplets of elemental indium. Clark *et al.*[13] used angle-resolved ESCA to investigate the oxide layers and hydrocarbon contamination of InP surfaces after numerous treatments. The as-received samples were found to be phosphorus rich, but various treatments result in substantial differences in the composition of the surface region. Bertrand[15] concludes that the use of different etchants causes the amount of oxidized phosphorus on *n*–type InP (100)

0142–2421/86/120319–05 $05.00

surfaces to vary. However, significant amounts of oxidized indium were found only on a sample which was heavily oxidized in air. Oxides formed in air after chemical etching with Br–MeOH were studied by Wager *et al.*[14] The as-etched surface is only slightly oxidized and is primarily a phosphorus oxide.

Thus far most of the InP studies have focused on the (100) and (110) surfaces with only a few studies of the (111) surface.[7,16,25,26] This may be due to the difficulties encountered in obtaining a mirror finish using conventional etching procedures.[16] In this study angle-resolved ESCA, AES and ISS are used to examine the oxide layer formed on chemically etched InP (111) by exposure to air.

EXPERIMENTAL

An undoped InP (111) face cut to ±1° was furnished by Sumitomo Electric Industries, Ltd. for this study. The sample was grown by the liquid encapsulated Czochralski method and was polished using a mixture of ZrO_2 and NaOCl and then a mixture of SiO_2 and NaOH. The sample was then etched with H_3PO_4: HBr (2:1) to produce a mirror finish. The electrical properties and etch pit density are: carrier concentration $\leqq 10^{16}$ cm^{-3} (*n* type), resistivity at room temperature = 1.8–1.9 × 10^{-1} ohm–cm, mobility = 3.9–4.0 × 10^3 cm^2 V^{-1} s^{-1} and EPD $\leqq 10^5$ cm^{-2}. The as-received sample was cut using a wafer saw to 21 × 15 × 0.35 mm and was ultrasonically cleaned for 5 min in each of the following electronic-grade solvents: trichloroethylene, acetone, methanol and then rinsed in deionized water.

A Perkin–Elmer PHI model 570 analytical system with a base pressure of 2 × 10^{-10} Torr was used in this study. A double-pass cylindrical mirror analyzer (CMA) with an internal electron gun and angularly resolving moveable aperture was used to collect AES, ESCA and ISS data. AES spectra were taken using a primary beam energy of 5 keV and a beam current of 0.3 μA rastered over an area of 2 × 2 mm. ESCA spectra were taken using a Mg anode at a voltage of 14 kV and power of 300 watts. High resolution ESCA spectra were obtained using a pass energy of 25 eV which resulted in a total energy resolution of 0.75 eV. The high resolution ESCA spectra were taken in the angle-integrated mode and angle-resolved mode at take-off angles of 10° and 85° with respect to the surface plane. ISS spectra were obtained using a 1 keV beam of $^3He^+$, a scattering angle of 144° and the moveable aperture for improved resolution. For all spectra taken, the sample was placed at an angle of 50° with respect to the CMA normal.

In this study nondestructive techniques were used before destructive techniques. Thus, the techniques were used in the following order: ESCA, angle-resolved ESCA, AES, ISS and depth profiling using argon-ion sputtering. The results are not presented in this order in the following section.

RESULTS AND DISCUSSION

Figure 1 shows the ISS spectrum taken from the air-exposed InP (111) surface. Three peaks are discerned corresponding to In, P and O, and the In peak is much larger than the other two peaks. ISS is essentially outermost layer sensitive for two reasons. The first is that ions which penetrate beneath the top layer have a very high probability of being neutralized, and the second is that even if they are not neutralized, they usually scatter inelastically during escape from the solid. In either case ions scattering off subsurface atoms do not contribute to the ISS peaks. Inelastically scattered ions comprise the background and, hence, the magnitude of the background depends upon the electron mobility in the near-surface region. Based on previous ISS studies, it appears that electrons in the surface region of this oxide layer are not highly mobile since the background is fairly large. The elastic scattering cross section increases as the mass of the surface atom increases. Even though In is heavier than P, the size of the In peak indicates that the outermost layer consists predominantly (>80%) of In atoms with roughly equal amounts of O and P atoms. No feature due to surface C appears.

Figure 2 shows an angle-integrated AES spectrum of the same surface. An intensity analysis using standard methods[27] yields an In/P atomic concentration ratio

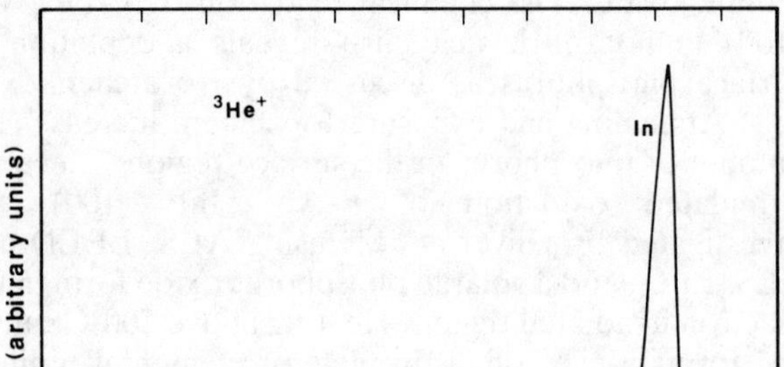

Figure 1. ISS spectrum of air-exposed InP (111).

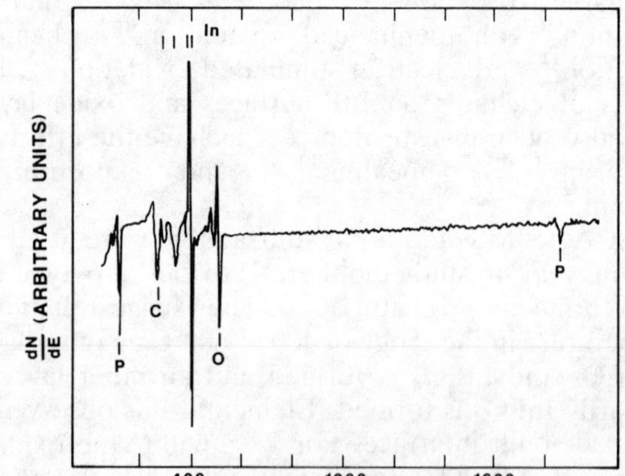

Figure 2. Angle-integrated AES spectrum of the same surface shown in Fig. 1.

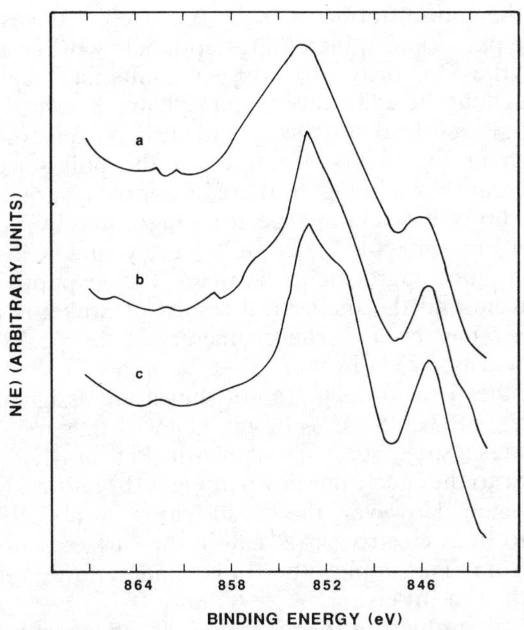

Figure 3. AES spectra of the In $M_{4,5}N_{45}N_{45}$ features taken from the same surface shown in Fig. 2. Spectra (a) and (c) were taken at exit angles of 10° and 85° respectively. Spectrum (b) is angle integrated over all exit angles.

which is slightly greater than 1 but much less than the In/P ratio obtained in ISS. This is due to the fact that AES is much less surface sensitive than ISS. The Auger signal is an exponentially weighted average from electrons originating within about three mean-free paths of the sample surface. Since ISS reveals that the outermost layer is mostly In and AES shows that the outermost few layers contain approximately equal amounts of In and P, the one- or two-layer region below the outermost region must be P rich. Both C and O appear in this spectrum in approximately equal concentrations even though no C is detected by ISS at the surface. High resolution AES spectra of the dominant In peaks at a kinetic energy of about 400 eV are shown in Fig. 3. Spectrum (a) was taken at an exit angle of 10° and spectrum (c) was taken at an exit angle of 85°. Spectrum (b) is angle integrated. These data were taken in the pulse-counting mode and are appropriately referenced so that the energy values are accurate. The spectrum shown in (c) is quite similar in shape to that of In metal[28] and is taken to be representative of InP since most of these electrons are originating deeply below the surface. The kinetic energy of the $M_5N_{45}N_{45}$ transition is 400.6 eV which is about 2.0 eV smaller than values reported for In metal[28,29] This value is also about 0.6 eV larger than values reported for In_2O_3[29] and agrees with the value found by Clark *et al.*[13] for InP. Spectrum (a) is much broader and less well defined in comparison with spectrum (c). Based on mean-free path estimates, most of these electrons originate within 10 Å of the surface since the exit angle is only 10°. It appears that another set of peaks due to another In species is contained in spectrum (a). The kinetic energy of the $M_5N_{45}N_{45}$ transition for this species is about 397.1 eV. It is argued below that this species is a phosphate and probably $InPO_3$ rather than $InPO_4$. Even though the outermost layer is mostly In, it is probably bonded as $InPO_3$ which is present in an amount almost as large as the amount of InP in the near-surface region. Spectrum (a) is similar to the high-resolution AES spectrum of the primary In features of an InP (100) surface which was etched in a 20% solution of $Br_2/HBr/H_2O$ (1:17:35).[13]

Figures 4 and 5 show high resolution ESCA spectra of the P 2p, In 3d, C 1s and O 1s peaks at exist angles of 10° and 85°. A very large difference is seen between spectra taken at 10° and 85° for the P 2p peaks. The major P 2p feature observed at 129 eV in Fig. 4A(b) is assigned as phosphorous in InP. This is in agreement with other studies[6,9,13,14] and the value assigned to P 2p in GaP.[29]. At the surface-sensitive exit angle of 10° shown in fig. 4A(a), another P 2p peak appears at a binding energy of 133.7 eV which is nearly as large as the P peak due to InP. This is assigned as a phosphate species and probably is $InPO_3$ (or $InPO_4$), but it is difficult to rule out the presence of hydrogen-containing phosphate species. The assignment as $InPO_3$ seems somewhat more probable than $InPO_4$ based on an analogy with $NaPO_3$ or $(NaPO_3)_3$ and Na_3PO_4 which have binding energies of 134, 133.5 and 132.2 eV respectively.[29] A study of the In-P-O phase diagram by Schwartz *et al.*[30] also suggests that $InPO_3$ forms at low temperatures while $InPO_4$ is stable at elevated temperatures (>650°C). Based on mean free path arguments, it is estimated that about 95% of the electrons collected are emitted within 50 Å of the

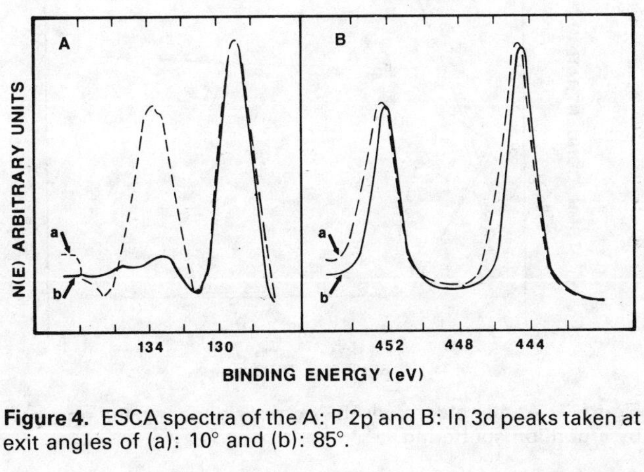

Figure 4. ESCA spectra of the A: P 2p and B: In 3d peaks taken at exit angles of (a): 10° and (b): 85°.

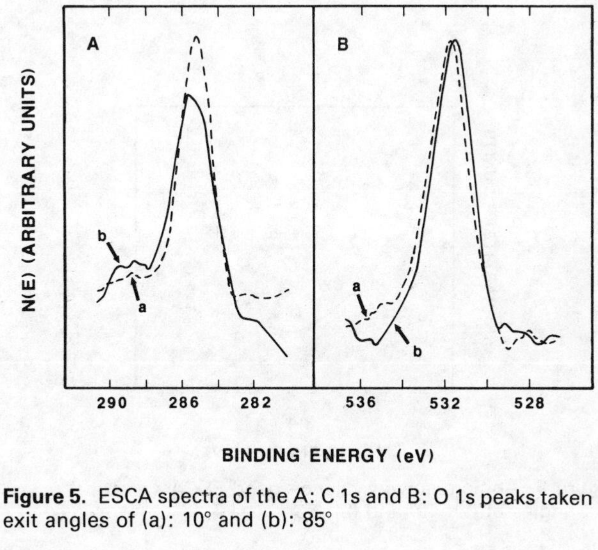

Figure 5. ESCA spectra of the A: C 1s and B: O 1s peaks taken at exit angles of (a): 10° and (b): 85°

surface in the 85° exit-angle spectra and within 10 Å of the surface in the 10° exit-angle spectra. Thus, the spectra shown in Fig. 4A imply that the near-surface region is composed of InP and $InPO_3$ in similar amounts and that the bulk is InP. The In 3d spectra shown in Fig. 4B support this hypothesis. The spectrum shown in (b) is characteristic of In in InP at a binding energy of 444.7 eV. The surface-sensitive spectrum shown in Fig. 4B(a) contains another In species which is assumed to be a phosphate at a binding energy of about 445.5 eV.[9] This is based on the information about P contained in Fig. 4A(a). In in In_2O_3 has a 3d binding energy which is about equal to that of In in InP so it is not possible to determine if In_2O_3 is present from the spectra shown in Fig. 4.

The C 1s and O 1s ESCA peaks are shown in Fig. 5A and B respectively for exit angles of (a): 10° and (b): 85°. The C 1s peak lies at a binding energy of 285.2 eV, in agreement with Hollinger *et al.*,[31] and is most likely in the form of graphite or hydrocarbons. It should be noted that this binding energy is about 0.6 eV higher than the commonly reported reference values for these species.[29] Therefore, selecting carbon as the reference for peak energy assignments at a binding energy of 284.6 eV would shift all the values reported here to lower binding energies by 0.6 eV. The C 1s peak shown in Fig. 5A(b) is broader than that shown in (a) so more than one form of C may be present. It is possible that discrepancies in the C 1s peak position are due to charging effects. However, these are believed to be minimal for the sample used in this study due to its relatively high conductivity and the thinness of the oxide layer.

ESCA O 1s peaks are shown in Fig. 5B. These peaks are discussed assuming that they are composed of a peak due to phosphate at a binding energy of 532.0 eV and a peak due to In_2O_3 at a binding energy of 530.5 eV. This seems reasonable with respect to both the results of this analysis and previous studies.[6,9,14] The surface-sensitive spectrum shown in Fig. 5B(a) is almost entirely due to phosphate. However, the presence of a very small shoulder on the low binding energy side, which is difficult to discern in the figure, may indicate that a very small amount of In_2O_3 is present in the near-surface region. The more bulk-sensitive spectrum shown in Fig. 5B(b) shows a very slight shift to lower binding energy. This may indicate that the concentration of In_2O_3 is greater with respect to the phosphate at increasing depth below the surface, but the majority of oxygen-containing species throughout the oxide layer is phosphate.

Angle-resolved valence band ESCA spectra are shown in Fig. 6. The features of the bulk-sensitive spectrum shown in Fig. 6(b) are in complete agreement with those shown in the spectrum presented by Vesely and Kingston[24] for clean InP. Vesely and Kingston found good agreement between the experimental spectrum and the theoretical results of Stukel *et al.*[32] and Kramer *et al.*[33] The characters of the peaks are assigned as: 3 eV – In 5s, 7 eV – 0 2p, and 9.7 eV – P 3p, and other peaks which are not shown are assigned as: 14 eV – P 3s, 18 eV – In 4d, 23 eV – 0.2s.[29,34] The surface-sensitive spectrum shown in Fig. 6(a) is quite similar to the spectrum shown in Fig. 6(b) in the 5 to 12 eV region. However, the broad feature at about 3 eV due to In 5s electrons is absent in the surface-sensitive spectrum. This implies that In is bonded differently in InP than in $InPO_3$.

Depth profiles for In, P, O and C are shown in Fig. 7. These were obtained by performing angle-integrated AES while sputter etching with argon ions. The atomic concentrations were determined using Auger sensitivity factors. However, due to possible matrix effects and sputtering artifacts, this figure shows only trends in the concentrations. The profiles are shown as a function of sputter time, but from the angle-resolved ESCA results, it is estimated that the sputtering rate was about 50 Å/min. The In/P ratio is greater than 1 at the surface, but the large ratio found using ISS is not observed since angle-integrated AES is less surface sensitive than ISS. The P concentration increases rapidly suggesting that the region below the outermost layer is P rich. The In, P, C and O concentrations all change rapidly in the oxide layer, and a sharp interface is observed at a sputtering time of 0.5 min. Past this interface only In and P appear. The layer beneath the oxide is P rich, but the concentration of In increases while the concentration of P decreases. Although the In and P concentrations change slowly in this region, it appears that a composition corresponding to InP may be approached.

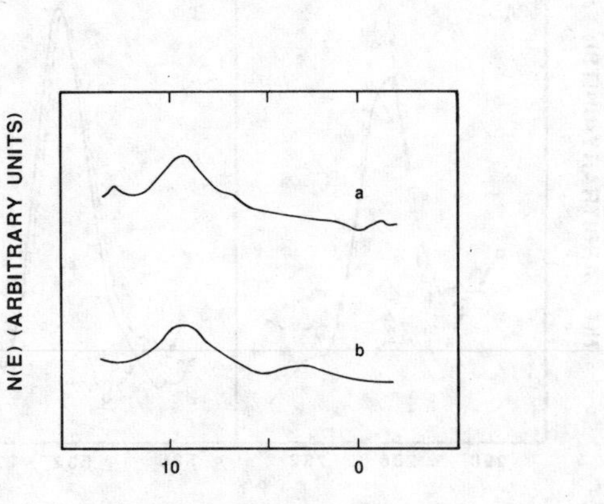

Figure 6. Angle-resolved valence band ESCA spectra taken at exit angles of (a): 10° and (b): 85°.

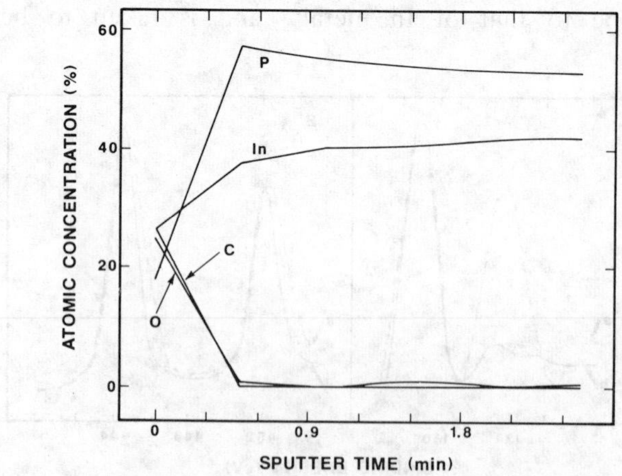

Figure 7. Angle-integrated AES compositional profile obtained by argon-ion sputtering.

CONCLUSION

The oxide layer which forms on a chemically polished and etched InP (111) surface upon exposure to air is very thin (estimated to be about 30 Å thick) and nonhomogeneous. ISS shows that the outermost layer is mostly In and AES combined with the ISS results suggests that the region just below the surface layer is P rich. Indium is believed to exist in the near-surface region as $InPO_3$ and InP in similar concentrations, but a very small amount of In_2O_3 is detected in the subsurface region. Carbon is detected throughout the native oxide layer, and ESCA shows that it is present in graphitic or hydrocarbon forms. Depth profiling using Ar^+ and AES shows that the oxide/InP interface is sharply defined. The region below the oxide layer is P rich, but the composition appears to be approaching that of InP as the sputtering time increases.

Acknowledgements

The data was collected at Perkin–Elmer Physical Electronics. Financial support of this research was received from NSF under grants #CBT-8416380 and #CBT-8506620 and the donors of the Petroleum Research Fund administered under PRF #16995-AC5-C.

REFERENCES

1. L. Messick, *J. Appl. Phys.* **47**, 4949 (1976).
2. D. Fritsche, *Electron Lett.* **14**, 51 (1978).
3. L. Meiners, *J. Vac. Sci. Technol.* **19**, 373 (1981).
4. T. Kawakami and M. Okamura, *Electron Lett.* **15**, 502 (1979).
5. M. G. Dowset and E. H. C. Parker, *J. Vac. Sci. Technol.* **16**, 1207 (1979).
6. A. Nelson, K. Geib and C. W. Wilmsen, *J. Appl. Phys.* **54**, 4134 (1983).
7. M. Yamaguchi and K. Ando, *J. Appl. Phys.* **51**, 5007 (1980).
8. C. W. Wilmsen, K. M. Geib, R. Gann, J. Costello, G. Hrychowian and R. J. Zeto,, *J. Vac. Sci. Technol.* **B3**, 1103 (1985).
9. M. Faithpour, P. K. Boyer, G. J. Collins and C. W. Wilmsen, *J. Appl. Phys.* **57**, 637 (1985).
10. L. Koenders, F. Bartels, H. Ulrich and W. Mönch, *J. Vac. Sci. Technol.* **B3**, 1107 (1985).
11. G. S. Korotchenkov, V. A. Mikhailov and V. I. Tsvitsinsky, *J. Vac. Sci. Technol.* **B3**, 981 (1985).
12. J. Oliver, P. Faulconnier and R. Poirier, *J. Appl. Phys.* **51**, 4990 (1980).
13. D. T. Clark, T. Fok, G. C. Roberts and R. W. Sykes, *Thin Solid Films* **70**, 261 (1980).
14. J. F. Wager, D. L. Ellsworth, S. M. Goodnick and C. W. Wilmsen, *J. Vac. Sci. Technol.* **19**, 513 (1981).
15. P. A. Bertrand, *J. Vac. Sci. Technol.* **18**, 28 (1981).
16. P. A. Kohl, C. Wolowodiuk and F. W. Ostermayer, Jr., *J. Electrochem Soc.* **130**, 2288 (1983).
17. C. W. Tu and A. R. Schlier, *Appl. Surf. Sci.* **11/12**, 355 (1982).
18. O. Wada, *J. Phys. D: Appl. Phys.* **17**, 2429 (1984).
19. B. Achard, B. Guzza and C. Pariset, *Surface Sci.* **160**, L519 (1985).
20. J. Oliver, *J. Appl. Phys.* **51**, 5406 (1980).
21. J. Massies and F. Lamaire-Dezaly, *J. Appl. Phys.* **55**, 3136 (1974).
22. C. R. Bayliss and D. L. Kirk, *Thin Solid Films*, **29**, 97 (1975).
23. J. Massies and F. Lamaire-Dezaly, *J. Appl. Phys.* **57**, 237 (1985).
24. C. J. Vesely and D. L. Kingston, *Phys. Rev, B.* **8**, 2685 (1973).
25. S. N. G. Chu, C. M. Jodlauk and W. D. Johnston, Jr., *J. Electrochem. Soc.* **130**, 2398 (1983).
26. Dia Daoxuan and Zhu Furong, *Chin. J. Semicond.* **6**, 190 (1985).
27. L. E. Davis, N. C. MacDonald, P. W. Palmberg, G. E. Riach and R. E. Weber, *Handbook of Auger Electron Spectroscopy*, Physical Electronics Industries, Inc., Eden Prairie, Minnesota (1976).
28. A. C. Parry-Jones, P. Weightman and P. T. Andrews, *J. Phys. C: Solid State Phys.* **12**, 1587 (1979).
29. C. D. Wagner, W. M. Riggs, L. E. Davis, J. F. Moulder and G. E. Muilenberg (ed.) *Handbook of X-ray Photoelectron Spectroscopy*, Perkin–Elmer Corporation, Eden Prairie, Minn. (1979).
30. G. P. Schwartz, W. A. Sunder and J. E. Griffiths, *J. Electrochem. Soc.* **129**, 1361 (1982).
31. G. Hollinger, E. Bergignat, J. Joseph and Y. Robach, *J. Vac. Sci. Technol. A* **3**, 2082 (1985).
32. D. Stukel, T. Collins and R. Euwema, *Electronic Density of States*, edted by L. Bennett, *Natl. Bur. Std. Spec. Publ. No. 323*, p. 93, US GPO, Washington, DC (1971).
33. B. Kramer, K. Maschke and P. Thomas, *Phys. Status Solidi B* **48**, 635 (1971).
34. R. A. Pollak, S. Kowalczyk, L. Ley and D. A. Shirley, *Phys. Rev. Lett.* **29**, 274 (1972).

SURFACE AND INTERFACE ANALYSIS, VOL. 9, 325–339 (1986). EXTENDED ABSTRACTS

Investigation of Surface Contamination on Silicon Wafers with SIMS

G. Stingeder,* M. Grundner,† M. Grasserbauer*
* Institute of Analytical Chemistry, Technical University Vienaa, A-1060 Wien, Getreidemarkt 9, Austria
† Wacker Chemitronic, D-8263 Burghausen, Postfach 1140, FRG

INTRODUCTION AND DEFINITION OF THE PROBLEM

Silicon wafers generally undergo a cleaning procedure before they are processed by the electronic industry. The most widely used cleaning process are the 'HF-dip' where a hydrophobic surface is formed (SiH) and the RCA cleaning process which results in a hydrophilic surface (SiOH). The cleaning is reported to have an influence on process steps like thermal oxidation or MBE (molecular beam epitaxy).

Characterization of the surface is necessary to achieve a better knowledge of the chemical state of the surface and for control and development of cleaning procedures. Therefore the possibilities for investigation of surface contaminants with SIMS (CAMECA IMS 3f instrument) on silicon wafers which had been treated to get a hydrophilic respectively a hydrophobic surface were studied.

Analytical Goal

Identification and depth distribution analysis of contaminants on silicon wafers in surface zones of approximately 10 nm with SIMS.

The different surface cleaning processes produce typical contaminants. Their lateral distribution is usually homogeneous.

For such a relatively simple system consisting of substrate, native oxide and a hydrocarbon layer the information content of secondary ion signals in surface zones smaller than the implantation depth of the primary ions is investigated empirically.

MEASUREMENT TECHNIQUE

General Problems

Due to the low sputter rates necessary (1–10 monolayers/min) secondary ion intensities are low and the adsorption from the residual vacuum of the sample chamber (esp. hydrocarbons, H_2O and O_2) is an important factor. At nearly every mass number hydrocarbon-species are detected causing severe interferences with the elements to be analysed.

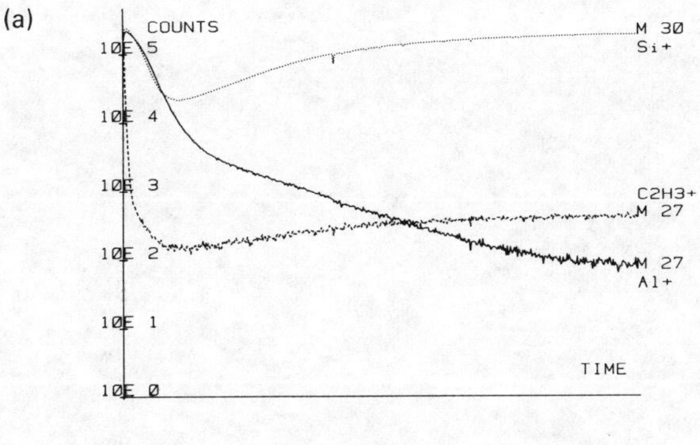

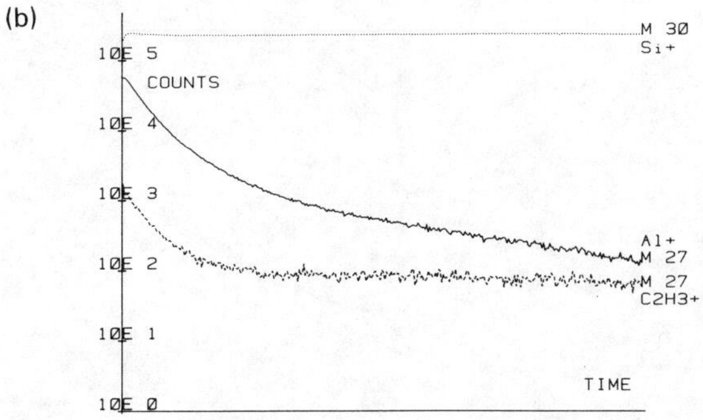

Figure 1. Profiles of $^{27}Al^+$, $^{12}C_2{}^1H_3{}^+$ and $^{30}Si^+$ with high mass resolution ($M/\Delta M$ = 800). Analytical conditions: 5.5 keV $O_2{}^+$, 20 nA, rastering 500 × 500 μm (a) without oxygen leak; (b) with oxygen leak (5 × 10^{-5} mbar O_2).

Chemical matrix-effects and a changing sputter rate are arising from the natural oxide (see Fig. 1) and the dynamic process of saturation with oxygen during sputtering with $O_2{}^+$ primary ions. Minimization of these effects is possible by using an oxygen inlet in the sample chamber (oxygen leak) which however causes an increase in the intensity of various interferences and the interface SiO_2/Si is not visible.

Identification of Secondary Ion Signals

For identification of the secondary ion species in the complex mass spectrum several procedures have to be applied: (a) registration of high resolution mass spectra and calculation of mass differences if at least two species are detectable (see Fig. 2); (b) verification of isotopic ratios by depth profiling of selected masses; (c) evaluation of shape and intensities of depth profiles (for example: hydrocarbons have characteristic profiles because they are only present in the outermost monolayers — see Fig. 1).

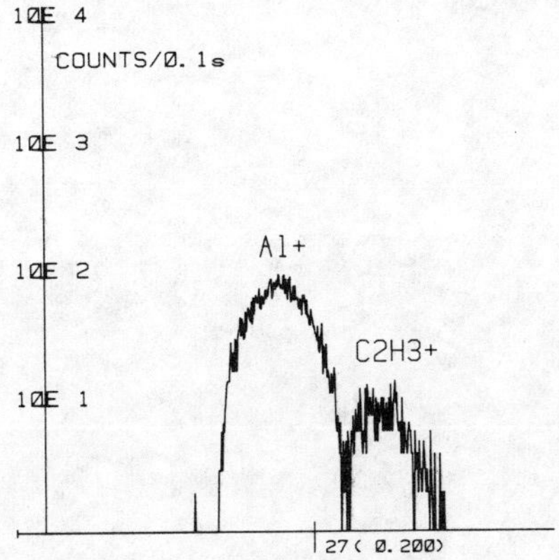

Figure 2. Identification of mass 27 by high resolution mass spectrum $Al^+/C_2H_3{}^+$: ΔM = 0,04192 amu; $M/\Delta M$ > 643.

Depth Distribution Analysis

Depth profiling is absolutely necessary for characterization of the surface because distributions show high gradients and errors due to the presence of different species at one mass line can be eliminated. Distribution analysis was attempted for the mass range 10–12, 19, 23–27 and 39–65 containing the elements B, C, F, Na, Mg, Al, K, Ca, Ti, V, Cr, Mn, Fe, Co, Ni and Cu.

Profiles of up to 29 masses were registered in one measurement to gain information for identification, evaluation of matrix- and sputter-effects and distribution of chemical elements simultaneously. The switching time of the double focusing mass spectrometer was optimized depending on the mass difference and mass resolution.

Distribution information was obtained for the outermost 10 nm of the silicon wafers. It was found that the contaminating trace elements like B, C, Na, Al, K, Ca etc. are enriched in the native oxide (approximately 1.5 nm).

AES and SIMS Investigation of Different Cu_xS/CdS Solar Cells

A. Hubin, H. Terryn and J. Vereecken
Vrije Universiteit Brussels, Depart. Metallurgy, Electrochemistry and Materials Science, Pleinlaan 2, 1050 Brussels, Belgium
M. Moens and F. Adams
Universitaire Instelling Antwerpen, Dept. of Chemistry, Universiteitsplein 1, 2610 Wilrijk, Belgium

Recently much attention has been devoted to the development of solar energy systems. In this regard, the thin film, and in particular the Cu_xS/CdS solar cell appears to become economically interesting. The Cu_xS/CdS heterojunction can be prepared by different Cu_xS formation processes: wet dipping,[1] dry formation[2] and reactive sputtering.[3]

In the present work, the composition of the different Cu_xS/CdS junction types is determined and correlated with the efficiency of the cells by AES and SIMS. Beside their elemental identification, an effort is made to determine the chemical state of the various compounds of the cells.

For each type of Cu_xS formation process, a cell with a typical I–V characteristic is chosen for investigation. For the dry process the influence of an after-treatment (conversion in air at ± 150°C of an evaporated Cu layer of 5 nm thickness) is studied. The properties of the investigated cells are summarized in Table 1.

The samples are analysed with a PHI 545 CMA Auger spectrometer and a CAMECA IMS 300 electrostatic sector analyser

For the identification of the chemical state of Cu and S two methods are applied: (a) intensity ratio of the $L_3M_{45}M_{45}$ (= LVV) and $L_3M_{23}M_{45}$ (= LXV) transition of Cu;[4–6] (b) line shape analysis of the LVV transition of S.[7,8]

Application of the LVV/LXV method on standard Cu specimens indicates the possibility of distinguishing between copper, -oxide, -sulphide. Shape and width of the S transition, determined on Cu_2S and CdS standard specimens, are dissimilar and offer a feasibility for distinguishing between Cu_2S and CdS.

Beside the main components of the surface layer Cu and S, different elements are detected on the four cells (see Table 1). The dry and the wet cells show a considerable amount of Cd and Cl, as contrasted with the sputter cell. O is only observed on the after-treated cell.

Comparison of the CuLVV/LXV intensity ratios of the cells with the reference values shows that Cu is present as sulphide, except for the after-treated cell, where also copper oxide is found.

The depth profiles supply a different picture for the four interface regions between Cu_xS and CdS. The sputter cell shows an abrupt interface, the dry cells show a broad one, but for the wet cell no cross over can be detected.

The observed differences in junction can be explained regarding to the Cu_xS formation process. For the sputter cell, it consists in a deposition of Cu_xS on CdS. For the dry and the wet cells a conversion of the toplayer of CdS must occur (see Table 1).

The significant variation in width and shape of the LVV S peak at the transition from the Cu_xS to the CdS layer offers an alternative method for the localization of the junction. The combination of the results of the surface and depth analysis is shown in Table 1.

The methods, recently developed for chemical identification of compounds, have been applied with success for the characterization of Cu_xS/CdS solar cells.

Acknowledgement

The authors gratefully acknowledge the Laboratorium voor Elektronica en Meettechniek, Rijksuniversiteit Gent for supplying the samples.

REFERENCES

1. L. R. Shiozawa, G. A. Sullivan, F. Augustine, *Conf. Rec. 7th IEEE Photov. Spec. Conf., Pasadena, 1968* (IEEE, New York, 1968), p. 39.
2. T. S. te Velde, J. Dieleman, *Philips Res. Repts.* **28**, 573 (1973).
3. E. Vanhoecke, M. Burgelman, *Thin Solid Films*, **112**, 97 (1984).
4. R. Weissmann, *Solid State Communications*, **31**, 347 (1979).
5. C. N. Rao, D. D. Sarma, M. S. Hegde, *Proc. R. Soc., Lond. A.* **370**, 269 (1980).
6. S. Yashonath, M. S. Hegde, *Proc. Indian Acad. Sci. (Chem. Sci)*, **89–5** 489 (1980).
7. H. Terryn, J. Vereecken, A. Laudet, *Appl. Surf. Sci.*, **24** (1985), 283–286.
8. D. Lichtman, J. H. Craig Jr., V. Sailer, M. Drinkwine, *Appl. Surf. Sci.* **7**, 325 (1981).

Table 1. Characteristics of the analysed Cu_xS/CdS solar cells

Cell	Type	Cu_xS formation process	Efficiency (%)	Characteristics
A	Dry + Treated	$CdS + 2\,CuCl \rightarrow Cu_2S + CdCl_2$	4.9	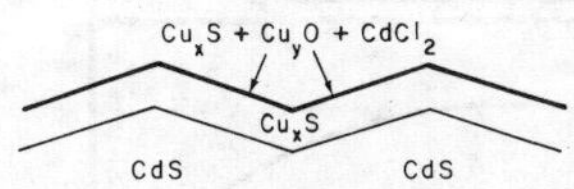
B	Dry		1.7	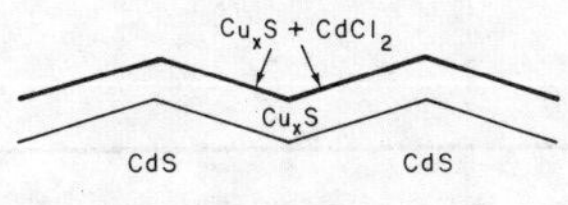
C	Wet	$CdS + 2\,Cu^+ \rightarrow Cu_2S + Cd^{2+}$	2.0	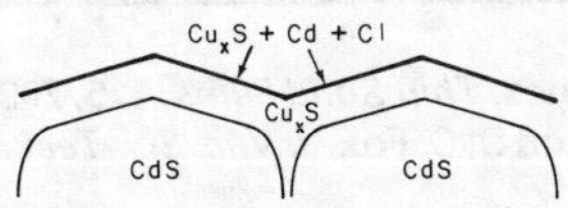
D	Sputter	$2Cu + H_2S \rightarrow Cu_2S$	2.5	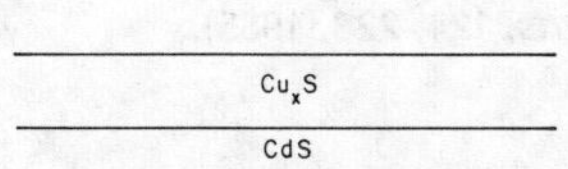

Direct TEM Study of the Roughening of SiO_2/Si Interface Induced during AES Depth Profiling

P. B. Barna, L. Gosztola, A. Zalar* and M. Rasigni**

Research Institute for Technical Physics of Hungarian Academy of Sciences, Budapest PO Box 76, H-1325

* Institute for Electronics and Vacuum Technique, Teslova 31, 61000 Ljubljana, Yugoslavia

**Centre d'Etude's des Couches Minces, Université Aix-Marseille III. 13397 Marseille, Cedex 13, France

The study of the microchemistry of interfaces in layer system by AES is based on ion sputtering. The roughening of the surface induced or enhanced by the sputtering process is now considered as one of the dominant effects determining the evaluated profile of interfaces. Very little direct data is available however on the real surface topography equivalent to the analysed surface.

SEM is suitable only to characterize the rough surfaces developing in polycrystalline materials of large grain sizes.[1] The TEM replica technique[2] has disadvantages in the preparation and also in the conservation of the original structure.

A direct TEM investigation method has been worked out for determining the very fine surface roughness developing in amorphous or fine grained structures during the ion bombardment used for depth profiling.

STEPS OF SAMPLE PREPARATION

1. Samples of 3 mm diameter disks were cut from the Si–SiO_2 wafers. The disks were mechanically thinned from the back side to a thickness of 50 μm (Fig. 1(a)).
2. The depth profiling was carried out either on static disks or or on samples rotated[3] during the sputtering until the oxygen peak disappeared.
3. The sputtered surfaces of disks were shadowed by Pt at an an incidence angle of 65° (Fig. 1(b)).
4. The Pt shadowed rotated disks were thinned by ion milling[4] on the back side to a thickness transparent for 100 keV electrons in the TEM (Fig. 1(c)).

The Pt shadowing of the surface makes the topography visible. By thinning these samples from the back side the topography can be studied in TEM at high lateral and height resolution.

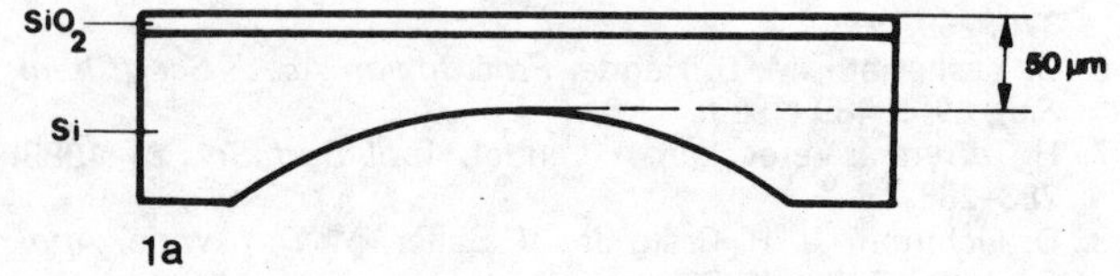

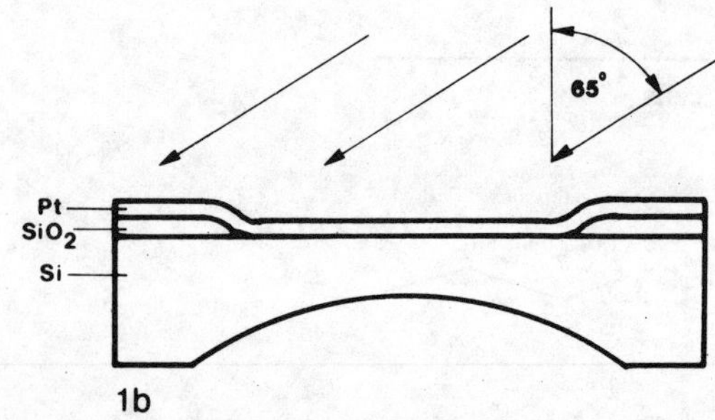

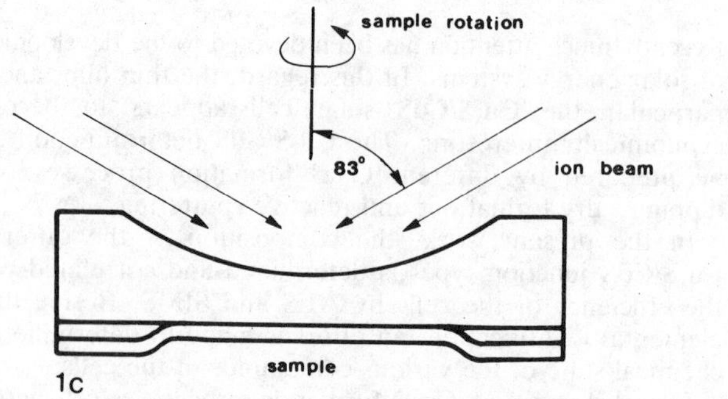

Figure 1. Steps of sample preparation.

The advantages of this method comparing it with the conventional TEM replica techniques are: (a) the interesting area can be localized by high certainty; (b) the real original surface structure can be conserved without any distortion both during the preparation and TEM studies.

As a consequence the TEM images taken from these samples are suitable for quantitative characterization by using the microdensitometer analysis of the electron micrographs.[5] The evaluation of electron micrographs by microdensitometer analysis is in progress.

In Fig. 2. the surface roughness of Si/SiO_2 interface induced during Auger depth profiling on a static sample is shown.

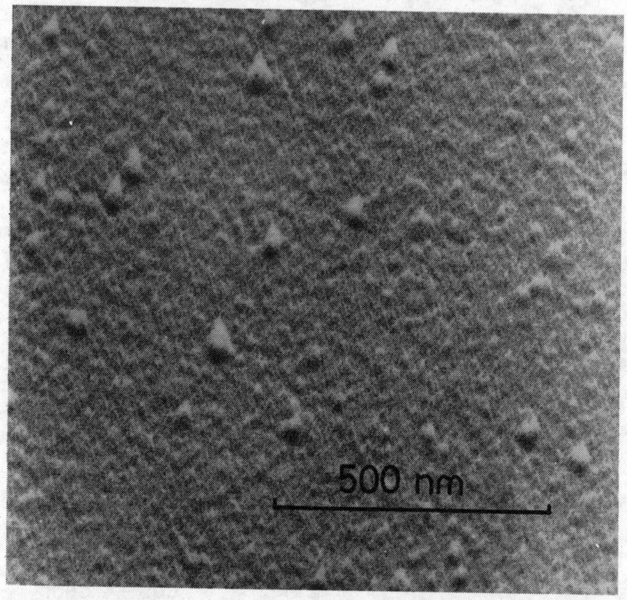

Figure 2. Surface roughness of Si/SiO_2 interface induced during Auger depth profiling/static sample, Ar ion beam at 1 keV energy and 42.6° incidence, thickness of SiO_2 was 84 nm.

REFERENCES

1. M. P. Seah and M. E. Jones, *Thin Solid Films*, **115**, 203 (1984).
2. C F. Cook, C. R. Helms and D. C. Fox, *J. Vac. Sci. Techn.* **17**, 44 (1980).
3. A. Zalar, *Thin Solid Films*, **124**, 223 (1985).
4. A. Barna, *Proc. 8th European Congress on Electron Microscopy*, Ed. Á. Csanády, P. Röhlich and D. Szabó, Publ. Org. Comm., *Budapest, 1984*, p. 107.
5. M. Rasigni, G. Rasigni and J-P. Palmari, *J. Opt. Soc. Am.* **71**, 1124 and 1230 (1981).

Low Temperature Oxidation of Titanium and Titanium Silicide

G. Queirolo and G. De Santi
SGS Microelettronica, SPA 20041- Agrate Brianza, Milan, Italy
G. Ottaviani
Universita' di Modena, Via Campi 213/A- Modena, Italy

In very large scale integration devices (VLSI) the time constant of the doped polycrystalline silicon gate interconnection lines limit the device speed. To overcome this drawback, new high conductivity materials are required. Particularly attractive, due to their compatibility with the microelectronics processes are the refractory and the transition metal silicides. Among these, titanium disilicide ($TiSi_2$) presents the highest electrical conductivity. The very high reactivity of Ti, and in particular its very high oxygen affinity, introduces severe problems when titanium-containing materials are thermal treated in nominally pure nitrogen or in reasonably good vacuo (1 *E*-6 torr).

In this work we have studied the oxidation behaviour of titanium and titanium silicide with Rutherford backscattering and Auger spectrometry, x-ray diffraction and *in situ* electrical resistivity measurements.

In situ electrical measurements show an increase in the sheet resistivity of Ti film deposited on silicon and treated in N_2 + 20% O_2 or in a 1 *E*-6 torr vacuo, up to a temperature of about 500°C, followed by a sharp decrease for higher temperature. At 800°C the sheet resistivity approaches the value found for samples treated in purified helium. The increase in the sheet resistivity is connected to the dissolution of oxygen in the metal film, as shown by the RBS spectra, by AES depth profiles and by the position of the (002) Ti peak in the x-ray diffractograms. The decrease in the sheet resistivity is associated with the silicide formation. During this process the oxygen already present in the Ti film is segregated toward the surface, thereby causing additional problems in the successive growth of the insulating SiO_2 layer necessary in the device processing. If a Ti film is directly heated at high temperature, the film oxidation and the silicide formation compete with each other; these processes stop when the two Ti(O)/Ti and $TiSi_2$/Ti interfaces meet. The same oxidation problems are found when a cosputtered, amorphous $TiSi_2$ film is thermally treated in the same ambients. In fact the oxygen diffuses through the silicide film until the $TiSi_2$ grains are sufficiently grown, forming a $(Ti_xSi_y)\ O_2$ compound at the sample surface. For annealing at higher temperature (1000°C) the surface Ti in the amorphous $TiSi_2$ film is preferentially oxidized. As the thermal treatment proceeds, silicon starts to oxidize, and finally a sharp SiO_2/$TiSi_2$ interface is obtained. Finally, if a well crystallized $TiSi_2$ film is steam oxidized, an uniform layer of SiO_2 film is obtained, without any Ti in it.

In conclusion, we have shown the great importance of the annealing ambient in the thermal treatment of Ti-containing materials, and the absolute necessity of avoiding their oxidation.

AES, SEM and EDX Study of Diffusion Phenomena in Reed Switches

R. Vanden Berghe, R. Vlaeminck, M. Saille and R. Van Paemel
Bell Telephone Manufacturing Company, Gasmeterlaan 106, B9000 Gent, Belgium

The Reed switch is the main switching element in the semi-electronic telephone exchange standard. It consists of two flat, gold plated Fe/Ni cantilever 'Reed' blades, sealed into a nitrogen filled glass capsule so that the blades overlap and are separated by a small gap. Since the blades are ferromagnetic, the switch can be closed by the action of an electromagnetic field; when the electromagnetic field disappears the switch opens again.

The appropriate functioning requires a low contact resistance (CR) and no cold welding phenomena ('no sticking'). Both phenomena can be fulfilled by diffusion-annealing of the gold coated reed blade in an inert atmosphere, or by 'arcing' and 'furbishing'.

In the arcing process the stationary reed blades of the open contact are heated in the overlap area by an electric arc; in the furbishing step, the switch is operated while a high tension is applied to the blades.

A comparison of the diffusion annealing and arcing-furbishing will be made using SEM and SAM methods. The SEM is a JSM scanning electron microscope, equipped with a WDS and a TN5500EDS system, operated at 10^{-3}Pa. The SAM is an Auger electron spectrometer 590 of Physical Electronic Industries.

The secondary electron images of the samples, diffusion annealed in vacuum or in nitrogen, show small uplifted structures scattered over the surface and lined-up around some of the 'grains'.

The SAM in-depth profiles indicate that the layer is homogeneous throughout its entire thickness, its composition depends on the annealing conditions: the Au concentrations (Fe or Ni concentration) decreases (increases) rapidly with annealing time, and eventually reaches a steady state value. This behaviour and the shape of the in-depth profiles can be explained by assuming diffusion both of Fe and Ni along paths of high diffusity through the Au layer towards the surface. From these paths, Fe and Ni diffuse into the surrounding material.[1]

This model corresponds with the boundary conditions set by the diffusion model type II of Wuttig and Birnbaum.[2] Using their formulation, the grainboundary diffusion coefficient, D_p can be estimated as: $9 \times 10^{-11} < D_p < 23 \times 10^{-11}$ cm^2/s^{-1}.

SEM pictures of samples annealed in nitrogen–oxygen mixtures at 550°C and at 700°C for 5 min show grains with disturbed and undisturbed surfaces. The disturbation consists of uplifted structures, scattered all over the surface and of decorated grain boundaries. This decoration is due to a close lining of cavities with concave edges. Line-scans across the surface recorded with SEM show that the strongest Fe signals are found for the grain boundaries, Ni and Au both behave in the opposite way: the most intense signals are found in the grains. In-depth profiles recorded simultaneously on both types of grain share the same general features: an iron oxide layer covering a gold–nickel layer on top of the substrate. This oxide layer is thicker on the disturbed than on the undisturbed grains. the shape of the profiles can be explained by assuming that the Fe atoms reaching the surface are oxidized. This process upsets the mass balance, in such a way that the Fe diffusion towards the surface is accelerated. The formation of nickel oxides is less probable, as can be inferred from the free enthalpy changes for the reactions of Fe and Ni with oxygen.

A reed blade taken from the capsule after arcing and furbishing has a bright spot at the top (the furbished spot) surrounded by a dark area (the arced zone). the furbished spot has a constant Fe, Ni and Au concentration from the surface down to the interface while the arced zone is characterized by a low Au concentration at the surface. the in-depth profile of the arced zone shows that the Fe, Ni and N concentration decrease simultaneously, while the Au concentration increases, reaches a maximum and decreases again in the interface Au/ Fe–Ni substrate.

The shape of the profiles can be explained by assuming that during arcing the blades are heated by the glow discharge and are bombarded by nitrogen ions created in the filling gas. This bombardement forms Ni and Fe nitrides at the surface inducing an accelerated diffusion of Fe and Ni towards the surface. The furbishing increases the homogeneity of the contact points which are sufficiently heated during the furbishing to be melted. This results in a higher superficial Au concentration and in a smaller grain size, which provides for an increased hardness.

The impact of the surface composition of the blades on the CR was investigated: it was found that the concentration of N and C increase with increasing CR.

REFERENCES

1. C. A. Hacque, *IEEE Transactions on components, hybrids and manufacturing technolgy, CHMT-2(1)* March (1971).
2. M. Wuttig and H. K. Birnbaum, *Phys. Rev.* **147**, 495 (1966).

An AES Study of the Electron-induced Dissociation of SiO_2

P. Juliet and B. Blanchard
CEA-CENG/SRCA, 85 X 38041 Grenoble Cedex, France

The damaging effects of the electron beam impact during AES occur via several different processes including beam heating, charging and electronic excitation of both surface and bulk atoms. The latter process may lead to the dissociation of compounds via desorption of ionic or neutral species. The literature reports a number of AES investigations on the different forms of SiO_2 (thin layers, vitreous silica, natural quartz). All the authors have observed electron-induced damages characterized by a charging of the surface and the appearance of the 'free' silicon LVV line.

The aim of this work was to obtain a better understanding of the electronic excitations leading to the oxygen release and to investigate the electronic stucture of the damaged material. In the first part of this study, the derivative mode was used to determine the dissociation cross-section and its behaviour when varying separately the energy and the density of the beam. Afterwards the true secondary electron distribution was qualitatively studied in the $E.N(E)$ mode.

INTENSITY DEPENDENCE OF THE CROSS-SECTION

The cross-section remains nearly constant when the intensity is increased. An increase in intensity induces necessarily an increase in the temperature of the sample. This temperature rise does not contribute to the degradation. Therefore the degradation is exclusively due to electronic excitation.

Carriere *et al.*[1] have proposed that dissociation occurs via a two-stage mechanism involving bond breaking by electronic excitation. They assumed the existence of a mechanism able to stabilize the intrinsic defect resulting from the first bond breakage. Such a stabilization may occur if a strained bond is broken in an amorphous type of SiO_2 as suggested by Lai.[2]

A closer look reveals a slight diminishing of the cross-section for the highest intensity values. This could be explained by a thermal annealing of the intermediate defect. In this way the beam-induced temperature rise reforms the broken bond and inhibits the dissociation.

ENERGY DEPENDENCE OF THE CROSS-SECTION

The cross-section is also a function of the primary beam energy and it is important to note that the line presents a 'shoulder' around 1500 eV. One of the proposed degradation mechanisms is the Knotek–Feibelman mechanism[3] which describes the behaviour of ionically bonded solids in an ionizing environment. The degradation mechanism is initiated by production of a core hole on the Si 2p level, followed by an interatomic Auger process, Auger cascades and desorption of O^+ via the 'Coulomb explosion'. The creation of a core hole on the O 1s level (512 eV) leads also, after intra-atomic Auger decay, to a two-hole final state resulting in the O^+ desorption. This may explain the 'shoulder' in terms of the energy of maximum ionization efficiency of the O 1s level.

THE TRUE SECONDARY ELECTRON DISTRIBUTION OF A DAMAGED SAMPLE

A sharp and narrow peak appears in the true secondary electron spectrum of a damaged sample. It is located at 4 eV energy and its growth law is parallel to the kinetics of the electron-induced dissociation of SiO_2.

The first attempt to correlate Auger spectrometric data to the defect centres was made by Schwidtal.[4] It has been shown that the peak classically attributed to the 'free' silicon resulted from a LVD transition where D is the energy level of the unpaired electron of the E' centre. The D level is assumed to lie 7.2 eV above the valence band, in agreement with the electron-induced increase of the C-band observed in cathodoluminescence. The E' centre is an O^- vacancy with an asymmetric relaxation which localizes the unpaired electron on silicon atom Si I while the other silicon (Si II) has relaxed into the plane defined by its three neighbouring oxygens.[5] The 5.8 eV optical absorption of this center involves electron transfer from Si I to Si II.

Following excitation with ionizing radiation, electrons could be trapped on Si II, the atoms being sufficiently separated to prevent bond formation. The centre is converted into a neutral centre with two dangling bonds. The kinetic energy E_a of an Auger electron resulting from a VSiISiII Auger transition is $E_a = (E_{SiI} - E_V) - (E_{vac} - E_{SiII})$ which, using optical and cathodoluminescence data gives $E_a = 4$ eV. Therefore, the 4 eV peak may be described as an Auger transition involving the localized levels resulting from the oxygen vacancy.

REFERENCES

1. B Carriere and B. Lang, *Surf. Sci.* **64**, 209 (1977).
2. S. K. Lai, *J. Appl. Phys.* **54**, 2540 (1983).
3. M. L. Knotek and P. J. Feibelman, *Phys. Rev. Lett.* **40**, 964 (1978).
4. K. Shwidtal, in *The Physics of SiO_2 and its interfaces*, ed. by S. T. Pantelides, p. 273, Pergamon, New York (1978).
5. K. L. Yip and W. B. Fowler, *Phys. Rev. B.* **11**, 2327 (1975).

AES Study of Thermal Behaviour of Sb-doped Sn-oxide Coatings

P. H. Duvigneaud, P. F. Hennaut and A. Laudet
Laboratoire de Chimie Industrielle et de Chime des Solides, Université Libre de Bruxelles, 30, Avenue F. D. Roosevelt, B-1050 Bruxelles, Belgium

H. Terryn and J. Vereecken
Dienst Metallurgie, Elektrochemie en Materialenkennis, Vrije Universiteit Brussel, Pleinlaan 2, B-1050 Brussel, Belgium

The unique properties of doped tin oxide films (high transparency in the visible region, high reflectance in the infra red region, low emissivity and nearly metallic conductivity) have been extensively studied since the sixties. Their main application consists of transparent and conductive coatings on glass.

The present study deals with thick films prepared by means of spray deposition of $SnCl_4$–$SbCl_3$ solutions at 500°C on alumina substrates. The thickness varied from 1 to 10 μm, so that a sheet resistance of 1 ohm/cm^2 can be obtained. These films permit high-temperature applications, such as heating elements or electrodes for melting.

Previous studies [1,2] of these films aimed to investigate the ageing process of the coatings during thermal treatments, from 1000 to 1300°C by means of x-ray diffraction, x-ray fluorescence and SEM. Significant compositional changes in the bulk of the materials have been observed.

The thick films, prepared at 500°C and grown beyond thickness of 0.5 μm, consisted of strongly oriented grains with [200] axis of rutile perpendicular to the substrate. As the electrical conductivity was independent of the thickness, it was assumed that the distribution of Sb^{5+} is homogeneous.

The increase of carriers, due to the donor level of Sb^{5+} in the SnO_2 conduction band, was already achieved at room temperature and resulted in a nearly constant electrical conductivity of the films after heating up to 1100°C.

Thermal treatments of thick SnO_2 films above 1100°C led to a modification in the microstructure and the composition. The linear weight loss, which appeared during isothermal heating of the film above 1000°C, can be explained by an evaporation process of SnO_2. Microstructural changes occurred together with the evaporation process. Above 1100°C a recrystallization of fine particles took place at the surface of the grains, which became cracked. The preferred [200] orientation decreased as a consequence of the random character of this recrystallization. Sheet resistance tended to increase as a function of time and temperature. However this increase cannot only be explained by the evaporation process.

In order to correlate the ageing of the electrical properties with the compositional change, analyses at different depths have been performed by Auger electron spectroscopy.

The samples were analysed with a PHI-545-CMA Auger spectrometer. Because of the overlapping of the $SnM_2N_1N_1$ and Sb $M_{4,5}N_{4,5}N_{4,5}$ at 451 eV in the Auger spectrum, the following ratio has been used:

$$\frac{M_2N_1N_1\ (Sn) + M_{4,5}N_{4,5}N_{4,5}\ (Sb)}{M_5N_{4,5}N_{4,5}\ (Sn)}$$

Calibration of the Sn $M_2N_1N_1$/Sn $M_5N_{4,5}N_{4,5}$ ratio with pure SnO_2 allowed to measure the Sb/Sn peak ratio from the film.

The surface composition of two layers, which initially contained 3 and 7% antimony, has been studied as a function of the thermal treatment from 500 to 1300°C, during 1 h. The peak ratio, caclulated from the AES spectra for both compositions on the surface and after sputtering of about 50 Å of the SnO_2 layer, was plotted as a function of the temperature. An enrichment in antimony at the surface of the two SnO_2 layers has been observed.

The surface composition of the two layers has also been studied as a function of time of the thermal treatment. The samples were heated at 1200°C from 0–24 hours. In both cases the plot of the peak ratio as a function of time showed the Sb depletion, although an initial enrichment in antimony occured at the beginning of the thermal treatment for the sample containing 3%.

Depth profiling performed on the 3 and 7% Sb doped SnO_2 layers, after 1 h heating at respectively 500 and 1000°C, revealed a maximum amount of Sb on the surface at 1000°C, in contrast with samples heated at 500°C.

AES examinations brought some light on the degradation of the Sb-doped SnO_2 coatings, previously observed above 1100°C. Strong surface segregation of antimony was evidenced, with a maximum at 1000;C. thickness of the Sb enriched layer was not more than 50 Å.

Such a segregation has been previously reported[3,4] for SnO_2–Sb powder catalysts.

According to Cross and Pyke,[4] catalysts, containing more than 4 atomic % Sb, segregate at 1000°C into two phases, with SnO_2 rutile like and Sb_2O_4 structures, respectively.

The highest volatility of the Sb_2O_4 phase can explain the maximum in Sb concentration observed at 1000°C in powder catalyst and in the coatings by means of spray pyrolysis, as well.

AES results thus indicated that the compositional changes, observed at 1200°C using bulk analysis methods, already started at a lower temperature in the first atomic layers.

Furthermore AES emphasized that the contact between the SnO_2–Sb crystals and the ambient atmosphere strongly influenced the degradation of the microstructure and the electrical properties of the coatings.

REFERENCES

1. P. H. Duvigneaud, M. F. Hennaut, *Sprechsaal*, **118**, 402 (1985).
2. M. F. Hennaut, P. H. Duvigneaud, E. Plumat, *Silicates Industriels*, **49**, 171 (1984).
3. J. C. Vedrine, *Analusis*, **9**, 199 (1981).
4. Y. M. Cross, D. R, Pyke, *J. Catal*, **58**, 61 (1979).

An Electron Spectroscopy Study of Sputtered Nickel Ferrite Films

A. G. Fitzgerald and G. Muir
Carnegie Laboratory of Physics, The University, Dundee, Scotland, UK

INTRODUCTION

Ferrimagnetic thin films, such as nickel ferrite, have potential applications in the miniaturization of microwave devices. The structure and compositon of these materials is an important parameter in device efficiency. The aims of this study have been to investigate the structure and composition of sputtered nickel ferrite films and the interaction of these films with a single crystal MgO substrate.

EXPERIMENTAL

The structure and composition of films prepared by dc sputtering from nickel ferrite and invar alloy (Ni 36 wt%, Fe 64 wt%) cathodes in argon and in oxygen has been investigated. The films were prepared in a high vacuum system (ultimate pressure 10^{-4} Pa) and were sputtered at a pressure of 20 Pa onto stainless steel and aluminium stubs and MgO crystals.

RESULTS

With the exception of films sputtered from the invar alloy cathode in argon, where NiO, FeO and nickel–iron alloy were identified, electron diffraction studies indicated that the sputtered films were nickel ferrite.

The surface composition and the oxygen content of these films has been investigated by XPS and AES. In some instances the O 1s photoelectron peak was found to have a shoulder at the position of oxygen in the hydroxyl groups in α-FeO(OH) and $Ni(OH)_2$. This shoulder did not appear to relate to any particular preparation condition and was removed by subjecting the films to a light argon ion etch.

Fe 2p peaks in the x-ray photoelectron spectrum from each type of film have also been studied. In each case the Fe $2p_{3/2}$ peak was found to be broad, suggesting that this peak is the superposition of Fe $2p_{3/2}$ peaks from αFeO(OH), α-Fe_2O_3 and $NiFe_2O_4$. A small satellite peak at 8 eV above the Fe $2p_{3/2}$ position confirmed the presence of α-Fe_2O_3. This satellite peak was removed by a short argon ion etch. After more extensive argon ion etching a small Fe $2p_{3/2}$ peak from unoxidized iron was observed in argon sputtered films from the invar alloy cathode.

The Ni 2p peaks from films sputtered in argon from the nickel ferrite cathode and in oxygen from both the nickel ferrite and invar alloy cathodes were similar in form to the same peaks observed from the surface of a piece of polycrystalline nickel ferrite. The position of the Ni $2p_{3/2}$ peak in x-ray photoelectron spectra indicated the presence of $Ni(OH)_2$ on the surface. When these films were argon-ion etched the structure of the Ni $2p_{3/2}$ peak was modified, with the characteristic satellite shoulder associated with NiO appearing. A similar satellite peak appeared when polycrystalline nickel ferrite was etched. This result suggests that NiO does not exist below the surface of each of these materials. It results from a chemical decomposition that occurs upon argon ion etching.

The structure of Ni $2p_{3/2}$ peak obtained from the surface of films prepared by sputtering from the invar alloy cathode in argon was found to be complex, indicating the existence of Ni $(OH)_2$ and oxides of nickel. On argon ion etching unoxidized nickel was detected below the film surface.

Auger electron spectra have been obtained from the surfaces of films sputtered in argon and oxygen from both cathodes and from the surface of a piece of polycrystalline nickel ferrite. In each spectrum a high oxygen Auger peak is obtained that is reduced upon argon ion etching, indicating that the surfaces are oxygen rich. The ratio of the peak-to-peak amplitude of oxygen and nickel relative to iron has been calculated for each material after etching away this oxygen rich layer. Films sputtered from the nickel ferrite target in argon were found to have identical ratios to those for the piece of polycrystalline nickel ferrite. Films sputtered in oxygen from this cathode and the invar alloy cathode were found to be oxygen rich. Argon ion etching revealed that the argon sputtered films from the invar alloy cathode were nickel rich.

Single crystal magnesium oxide has a suitable crystal structure and lattice parameter for epitaxial growth of ferrite films, and films of these materials have been grown epitaxially on this substrate[1,2] by the chemical vapour deposition technique. Energy dispersive x-ray microanalysis investigations of these epitaxial films have shown that they are invariably nickel deficient.[3]

Auger spectra from freshly prepared films deposited on MgO confirmed earlier surface compositions. The Auger spectra obtained from the surface of each of these films after annealing at 1000°C contained magnesium, oxygen and small iron peaks. It was clear that nickel had diffused preferentially into the MgO surface.

CONCLUSIONS

Films with the nickel ferrite crystal structure can be prepared by sputtering from both polycrystalline nickel ferrite and invar alloy cathodes, although as might be expected, unoxidized iron and nickel are detected in argon sputtered films from the invar alloy cathode. It is likely that the α-Fe_2O_3, α-FeO(OH) and $Ni(OH)_2$ detected on the surface of each film occurred on contact with the atmosphere. The observed preferential diffusion of nickel into the MgO substrate explained the nickel deficiency in epitaxial ferrite films grown upon MgO that had been observed in earlier investigations.

REFERENCES

1. J. E. Mee, G. R. Pulliam, J. L. Archer and P. J. Besser, *IEEE Trans. Magnetics*, **5**, 717 (1969).
2. A. G. Fitzgerald and T. G. May, *Thin Solid Films*, **35**, 201 (1979).
3. A. G. Fitzgerald, *Proceedings of the Twenty-Second Scottish Universities Summer School in Physics (1981)*, ed. by A. P. Cracknell and R. A. Vaughan p. 181, SUSSP Publications (1981).

The Diffusion of Copper and Silver in Amorphous Chalcogenide Thin Films

A. G. Fitzgerald and C. P. McHardy
Carnegie Laboratory of Physics, the University, Dundee DD1 4HN, Scotland, UK

INTRODUCTION

Amorphous chalcogenide films have many technological applications in which their reactions with metals are important. For example, their interaction with silver is potentially useful as a high resolution photoresist and electron resist. Their potential as an electron resist is based on the differing solubilities of irradiated and unirradiated chalcogenide–silver films in alkaline solution. This leads to the possibility of producing high resolution patterns defined by the electron beam size. The aims of this study have been to investigate the extent of metal–chalcogenide interdiffusion in a wide range of chalcogenides, to identify the structural and compositional changes that occur and to examine the effects of electron irradiation over a range of electron intensities.

EXPERIMENTAL

Thin films of As_2X_3, GeX (X = S, Se, Te) and $GeSe_2$, under or overcoated with silver or copper were prepared by vacuum evaporation. Structural and microstructural studies were made in the transmission electron microscope (TEM). Film compositions were determined by energy dispersive x-ray microanalysis and surface compositions were determined by XPS and AES in a UHV SEM.

RESULTS

Structural changes were found to occur at ambient temperature when copper or silver was in contact with each of these amorphous chalcogenides. At less than 30 wt% metal concentrations structural changes occurred in the amorphous phase. At higher concentrations amorphous and polycrystalline material formed. In most cases the polycrystalline material was a silver or copper chalcogenide.

The extent of interdiffusion of both copper and silver in amorphous chalcogenides was examined at 10wt% and 50wt% concentrations. These observations have been made mainly in copper or silver films that have been overcoated with chalcogenide since the reverse preparation procedure, where chalcogenide films are overcoated with one of these metals, leads to films that are sensitive to relatively low electron beam intensities at the typical film thicknesses of specimens examined by electron spectroscopy. XPS and AES from the surface of chalcogenide–metal films show that the diffusion of these metals is most extensive in As_2Se_3, and moderately so in As_2S_3, As_2Te_3, GeTe, GeSe and $GeSe_2$. With GeS neither copper or silver penetrated to the surface on overcoating either of these metals.

The distribution of copper and silver through films of each of these chalcogenides has been investigated by Auger depth profiling using argon ion etching. The most extensive interdiffusion with depth occured for As_2Se_3. the least interdiffusion occurred in GeS. At the 10 wt% metal concentration, in these chalcogenides where the metal penetrates through to the surface, a high concentration of metal has been found at the upper and lower surfaces of the film, with the metal concentration almost dropping to zero at the film centre.

The effects of electron irradiation on chalcogenide–metal films was found to be dependant on the thickness of the composite films, the nature of the supporting substrate, the sequence of layers encountered by the electron beam and the extent of interdifffusion. Thin chalcogenide–copper films were sensitive only to extremely high electron beam intensities however, chalcogenide–silver films were extremely beam sensitive and apart from GeTe–Ag, silver was found to diffuse away from the irradiated area when exposed to a focused electron beam in the TEM.

Observations on beam effects in thicker silver–chalcogenide films have been made with As_2Se_3 in a UHV SEM which has a high intensity field emission source. Films that were overdeposited with metal were found to be much more sensitive and since this film preparation is used in electron resist applications the sensitivity of these films has been studied in detail. At high electron beam intensities, the film temperature is raised sufficiently to cause local melting with the loss of chalcogenide from the irradiated area. the nature of the substrate also modified the behaviour of these films to high electron beam intensities, dendritic growth of silver occurred at the edge of the irradiated area on glass and also on silicon substrates. This behaviour did not occur with a conducting aluminium substrate.

CONCLUSIONS

Contact reactions occur between copper or silver and all of the amorphous chalcogenides of arsenic and germanium. The products of these reactions have been identified by electron diffraction. Electron spectroscopy studies show that interdiffusion is most extensive in As_2Se_3 and least extensive in GeS. The unusual metal distribution through these films at 10 wt% concentrations, when metal diffuses through to the free surface, probably results from the fact that there is a source of vacancies there. When the metal atoms reach this free surface, its properties change and the equilibrium rate at which vacancies can enter and leave alters, leading to increased metal diffusion and a higher metal concentration in this region. The diffusion of silver away from electron irradiated areas in thinner chalcogenide–silver films is probably due to the diffusion of silver under the influence of the temperature gradient that exists between the irradiated and the unirradiated areas. Silver impurities in zinc behave in this way in a temperature gradient. For chalcogenide films overdeposited with silver it is likely, from the evidence of depth profiles, that the upper layers are silver or silver–chalcogenide. The electron beam will raise the lower conductivity chalcogenide layers beneath this to a high enough temperature to cause melting and flow of material away from the area of incidence of the electron beam. This leads to the observed increase in chalcogenide concentration in the neighbouring area of the film. These observations suggest that careful control of the current density of the incident electron beam is required in electron resist applications.

Effects of a Caesium Ion Beam on GaAs

K. Miethe, W. H. Gries and A. Pöcker
Forschungsinstitut der Deutschen Bundespost, Postfach 5000, D-6100 Darmstadt, FRG

Depth profiling of GaAs by use of a caesium ion beam in an ion microprobe mass analyser (ARL-IMMA) has produced prominent topographic features both inside and outside the sputtered areas (craters). After depth profiling of an ion-implanted non-annealed sample of GaAs, a field of 'spots' appears outside the crater. When a crater is sputtered in a virgin GaAs material the spots are located mainly inside the crater.

These spots do not become visible instantaneously, but only after some time (sometimes days) after irradiation and then only if certain Cs doses are exceeded. In all instances the location of spots relative to the crater depends on whether the GaAs has been pre-bombarded or not.

The dose dependence of the formation of spots on virgin conductive GaAs material was investigated with caesium doses ranging from $5 \times 10^{12}cm^{-2}$ to $2 \times 10^{18}cm^{-2}$. For doses higher than $5 \times 10^{15}cm^{-2}$ spots were observed to form inside bombarded areas with a time delay of several days after irradiation. Using SEM, the spots were found to be outgrowths, sometimes with crystal habit.

The distribution of caesium in and around the craters was investigated with SAM and IMMA-SIMS (using O_2^+ primary ions). The spots were found to be enriched with caesium by a factor at least several times the retained dose in the bulk (crater area). In both cases (caesium enriched spots inside or outside craters) measurements by SAM and SIMS have shown caesium to be present over large areas outside the crater.

The effect of electron and ion irradiation on the spots and the Cs distribution was investigated, as well as the influence of an electric field applied to the surface.

From the observations we are led to propose the following model: as the crater develops the concentration of caesium at the crater bottom increases in accordance with the standard model of implantation. At equilibrium the concentration of caesium in the first atomic layer of the sample is estimated to be equivalent to a coverage of less than half a monolayer. For these coverages the caesium on GaAs is of dipole/'ionic' character. The mutual repulsion (and possibly also the interaction with the electric field of the positive primary ion beam) leads to migration of caesium out of the crater (and far afield) at a rate sufficient to save a significant fraction from being sputtered.

We propose that the caesium loses its dipole character when it encounters an ion-bombardement-induced defect (dislocation or dislocation cluster) on the surface. It can then agglomerate to form phase nuclei and grow with time into the outgrowths which have been observed. Simultaneous oxidation could be an essential factor. Formation of outgrowths inside a crater proceeds similarly at defects and/or sputter cones after sputtering has ceased. When the caesium inside the crater has been trapped the caesium dipoles outside the crater observe the crater as a field of sinks into which the caesium re-migrates. If the crater is surrounded by defects at high density, which is the case in pre-implanted non-annealed material, most of the caesium will be trapped there and does not reach the crater. Hence outgrowths in the crater will not reach sizes which are easily visible.

It is concluded that SIMS depth profiling of GaAs by use of a caesium ion beam causes caesium coverage of a large surface area surrounding the sputter crater, and in the presence of ion-bombardement-induced surface defects leads to caesium-rich outgrowths outside and/or inside the crater, which are detrimental to further analysis by SIMS and other surface analysis techniques.

Photoemission Study of Cu/Al Interface Using Synchrotron Radiation

V. Di Castro and G. Polzonetti
Department of chemistry, University La Sapienza, 00185 Roma, Italy

Copper interaction with polycrystalline aluminium in the 1–100 Å range of copper coverage has been studied using synchrotron radiation. For low coverages a strong interaction between the two metals is observed, whilst at higher coverages copper clusters seem to be formed on the surface.

INTRODUCTION

Transition metals have proved to be very suitable for studying epitaxial growth of metals on metals, since shape and position of d bands are very sensitive to changes in aggregation from isolated atoms to bulk materials. Here, we report a photoemission investigation, using synchrotron radiation, of copper growth on polycrystalline aluminium in the 1–100 Å range of Cu coverage. The aim of this study is a better understanding of the behaviour of copper interacting with aluminium both by itself and, previous oxidation, as a possible model of catalysts prepared for coprecipitation.

EXPERIMENTAL

All the experiments were carried out using the vacuum UV beam line at the synchrotron radiation facilities 'PULS' in Frascati described elsewhere.[1] Metals were evaporated from tungsten filaments and monitored by a quartz crystal microbalance. Auger spectroscopy was used to check the surface cleanliness.

RESULTS AND DISCUSSION

Copper was evaporated by step on a film of 300 Å of aluminium at room temperature and the photoemission spectra, measured after each evaporation, are reported in Fig. 1.

At low coverages the Cu 3d band, that is the dominant signal in the valence region, shows a shift of 2 eV to higher BE (binding energies) from the bulk position, similar to that observed in Cu Al_2 alloy.[2] This shift is constant up to 5 Å and could be attributed, as already discussed,[1] to the intermixing of copper and aluminium at the interface. As the coverage increases, the Cu 3d signal progressively shifts to lower BE and, in the meantime, the peak width increases. This behaviour clearly shows that a copper phase is growing up on top of the intermixing region, but the Cu 3d bulk structure does not develop even at 35 Å of Cu coverage in contrast with a layer by layer growth model in which theoretical calculations[3] and experimental measurements for Cu on Ni (111)[4] show convergence to bulk structure occurring around 4 layer thickness. This observation and the presence of multiple components in the 3d band suggest a Valmer–Weber growth mode by clustering on the surface, as confirmed also by the oxidation behaviour of a sample with an intermediate coverage of copper (see Fig. 2).

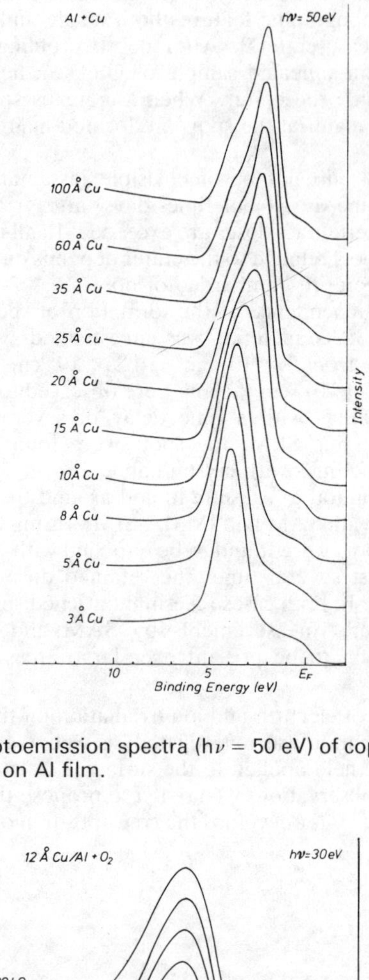

Figure 1. Photoemission spectra (hν = 50 eV) of copper evaporated by step on Al film.

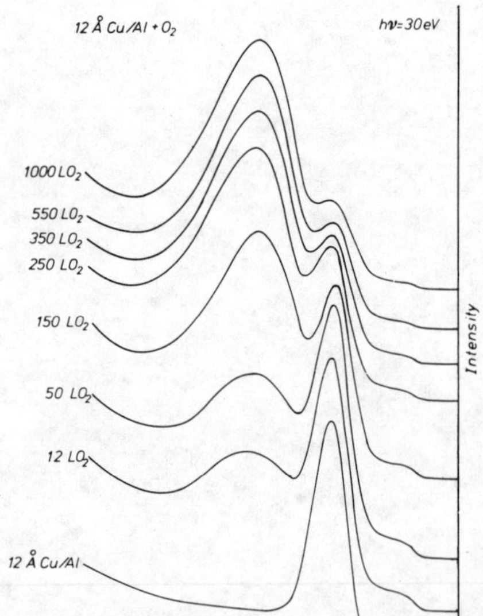

Figure 2. Photoemission spectra (hν = 30 eV) as a function of oxygen exposure for the 12 Å Cu/Al system at room temperature.

During the oxidation process a signal at 6.8 eV, due to O 2p and characteristic of Al_2O_3, progressively increases whilst the Cu3d peak decreases without changing shape or position. Thus, an aluminium oxide phase is formed on top of the copper without any appreciable change in its electronic structure. It follows that the aluminium surface is covered by a discontinuous copper layer that allows oxygen interaction with some bare aluminium surface with Al_2O_3 growth on top of copper clusters. The Cu/Al interface is stable at room temperature but at 150°C, 100 Å of copper diffuse completely in the aluminium substrate in 15 min forming a stable alloy.[5]

REFERENCES

1. V. Di Castro, G. Polzonetti, R. Zanoni, *Surf. Sci.* **162**, 348, (1985).
2. J. C. Fuggle, E. Källne, L. M. Watson, D. J. Fabian, *Phys. Rev. B*, **16**, 750 (1977).
3. J. R. Smith, J. G. Gay, F. J. Arlinghaus, *Phys. Rev. B*, **21**, 2201 (1980).
4. I. Abbati, L. Braicovich, A. Fasana, C. M. Bertoni, F. Manghi, C. Calandra, *Phys. Rev. B*, **23**, 6448 (1981).
5. V. Di Castro, G. Polzonetti, submitted for publication to *Surface Science*.

ISS-study of Dopant Redistribution in $TiSi_2$ Layers during Silicidation by Rapid Thermal Processing

H. Tollet, P. Bielen, C. Creemers, H. Van Hove and A. Neyens
Fysico-chemisch laboratorium, KU Leuven, Celestijnenlaan 200G, 3030 Leuven, Belgium
K. Maex and R. F. De Keersmaecker
ESAT Laboratorium, KU Leuven Kardinaal Mercierlaan 94, 3030 Leuven, Belgium

INTRODUCTION

Transition metal silicides are of great interest in MOS VLSI technology as contact material for shallow junctions. At the same time rapid thermal annealing (RTA) has become of interest as a technique to activate ion implanted dopants with minimal redistribution of dopant atoms. In this contribution the redistribution of *n*-type dopants (As and Sb) upon annealing during simultaneous silicidation using RTA is studied with ion scattering spectroscopy (ISS). The extreme surface sensitivity of ISS together with the intentionally enhanced surface erosion using mixed (He^+ + Ne^+) primary ion beams make this technique ideally suited for the analysis of the $TiSi_2$ formation and the redistribution of the dopant atoms.

EXPERIMENTAL

The analyses were performed with a 3M-ISS spectrometer featuring a 138° CMA analyser and a coaxial, low gas conductance type ion gun. The spectrometer was mounted in a turbo-pumped vacuum chamber equipped with a quick loading system for rapid insertion of samples without breaking the vacuum. The base pressure is about 1×10^{-9} Torr. The samples were analysed using a 200 nA mixed (He^+ + Ne^+) ion beam with a primary energy of 2 keV. The energies of the scattered ions were recorded in scans of 5 s each which are accumulated using multi-channel-scaling (Tracor Northern 1710) and stored on magnetic discs. After retrieval of the spectra data, the peak areas for the main atomic species were determined semi-automatically and plotted as standard depth profiles.

The substrates for the samples were p-type silicon wafers of (100) orientation with a resistivity of 15–30 Ω cm. In a first series of samples the substrate was covered with a 45 nm thick Ti film deposited using DC magnetron sputtering with a base pressure of 10^{-7} Torr and with a deposition rate of about 0.5 nm s^{-1}. These samples were doped with a 5×10^{15} atoms cm^{-2} dose of As or Sb either by ion implantation in the Si substrate before Ti deposition or by implantation through the Ti film (ion beam mixing). In the former case the implantation resulted in a dopant distribution with its maximum at about 200 nm below the original Si surface, while in the latter case the dopant atoms were situated immediately below the Ti/Si interface. Part of the samples of this first series were subjected to rapid thermal annealing in a N_2 atmosphere by a 10 s flash to temperatures above 1150°C to activate the As or Sb implanted profiles and to ensure complete silicidation of the Ti surface film to $TiSi_2$. In a second set of samples the Si substrate was doped with a shallow As-implant of 5×10^{15} atoms cm^{-2} prior to the deposition of a 65 nm thick Ti film. The temperature during the 10s flash was now varied in the range from 600°C to 1100°C.

RESULTS

$TiSi_2$ formation

The depth profiles recorded on the first samples, subjected to RTA at 1150°C or more, showed a complete intermixing of Ti and Si in a surface layer of about 90 nm thick. They also clearly exhibit a zone of quasi constant Ti/Si stoichiometry, which from RBS-evidence[1] was recognised as $TiSi_2$. Analysis of the second set of samples revealed that this silicidation reaction occurs at temperatures from 700–800°C (10 s) on. At higher temperatures the Ti–Si interdiffusion is more complete and the thickness of the $TiSi_2$ layer seems to increase.

Redistribution of the dopant species

It was found that if the As or Sb ions were implanted deeply in the silicon substrate, as in some of the samples of the first series, the distribution of dopant species remains essentially unchanged upon RTA. If on the other hand the dopant element was implanted shallowly in the Si substrate before Ti deposition or was implanted through the Ti layer just below the Ti–Si interface, some redistribution of the impurity atoms does occur on annealing: the As atoms present in the Si that is consumed in the $TiSi_2$ formation, diffuse outwards to the surface and eventually evaporate at the highest temperatures leaving an impurity depleted $TiSi_2$ layer. It is less clear whether the As atoms also diffuse towards the bulk upon heating: although part of the tailing of the signals in the depth profiles must be ascribed to 'crater effects' due to the stationary primary ion beam in the ISS analysis, the increased tailing in the As signal at higher temperatures seems to suggest a minor redistribution towards the bulk.

Nitrogen content in the surface film

Visual inspection of the RTA-treated samples as well as etching tests and recent RBS evidence[2] suggest the presence of a N contaminated surface layer, possibly a Ti–N compound formed between the surface and ambient N_2. However, only small N signals were recorded on the heated samples with a maximum intensity in the $TiSi_2$ layer. Moreover the Ti/Si samples that were not subjected to RTA also show N with a maximum signal at the Ti–Si interface, which evidently cannot result from the N_2 ambient during the heat treatment. The sensitivity of ISS for N was investigated on a pure Si_3N_4 reference sample. Even taking into account the low sensitivity of ISS for the lowest Z elements, the N signal recorded on Si_3N_4 with both He^+ and (He^+ and Ne^+) beams is surprisingly weak. This suggests that even small N signals recorded on the Ti–Si structures can also not be neglected. Further work to elucidate this point is in progress.

CONCLUSIONS

The present experiments have shown that ISS depth profiling over several hundreds of nm below the original surface, using mixed (He^+ + Ne^+) beams, seems to be a promising technique for the investigation of the formation of $TiSi_2$ layers and the redistribution of dopant species. The origin and the observed variation of the N signal are not yet fully understood. Further refinements of the experimental procedure in order to increase the accuracy of the determination of the peak intensities and to eliminate crater effects and the related, artificially induced tailing of the concentration profiles, are presently under investigation.

REFERENCES

1. K. Maex, R.F. De Keersmaecker and P.F.A. Alkemade, in: *Proceedings of the MRS Spring Meeting, San Francisco, April 15–18, 1985*, Vol. 45: 'Ion Beam Processes in Advanced Electronic Materials and Device Technology', pp. 153–158. Publishers: F.H. Eisen, T.W. Sigmon and B.R. Appleton (1985).
2. Private communication.

SIMS-ANALYSIS OF SHALLOW IMPLANTS IN SILICON

D.Avau,* W. Vandervorst, H.E. Maes†
IMEC, Kapeldreef 75, B–3030 Heverlee, Belgium

The increasing use of shallow doping profiles in submicron VLSI-structures poses a serious restraint on the analysis techniques which are used. Especially techniques which employ sputtering to obtain the depth information (like SIMS) are subject to effects induced by the incident ion beam (collisional mixing, preferential sputtering, defect-induced diffusion).

In this work a study was made of the influence of the experimental conditions (primary energy, primary mass, angle of incidence) on the depth profiling of shallow implants in silicon. Ion-implantation profiles of B, P, As and Sb with energies ranging from 4–50 keV were analysed with SIMS under various conditions. For each element the experimental conditions are optimized to reduce the ion beam effects as much as possible without compromising too much the sensitivity and analysis time.

Based on the results of this investigation and especially on the amount of broadening for each element under otherwise identical conditions, a correlation is established between these elements and the broadening mechanism.

Finally from the profiles measured under conditions for minimal broadening, the range parameters (projected range, range straggling, skewness and kurtosis) for the different implant energies are derived by fitting a Pearson-distribution to the measured profiles.

* Fellow IWONL
† Senior Research Associate NFWO

QUANTITATIVE ANALYSIS OF SILICON–SILICON NITRIDE INTERFACIAL REGIONS USING AUGER ELECTRON SPECTROSCOPY

J. Remmerie,* W. Vandervorst and H.E. Maes†
IMEC, Kapeldreef 75, B–3030 Heverlee, Belgium

The increasing application of silicon nitride thin films in VLSI-structures makes it mandatory to better understand the nature and properties of the interfacial region between silicon and silicon nitride. Interface phenomena become indeed increasingly important as device dimensions shrink. Auger electron spectroscopy (AES) combined with Argon ion milling has been the most widely used technique for this kind of analysis. However to obtain reliable information on the interfacial region special care must be taken to account for ion induced artifacts.

In this study, an extensive physical characterization of the silicon nitride–silicon interface and the intermediate oxide (native or intentional) is proposed. In this characterization the depth resolution is optimized by combining low energy Auger sputter profiling with a preceding chemical etch step, merely leaving a shallow nitride covering layer. In order to locate the collision cascade beyond the interfacial region (even during low energy sputter profiling), the thickness of this layer should be less than the mean recoil range of this collision cascade.

A data analysis procedure has been developed which takes into account the dependence of the sputter yield on the instantaneous oxygen surface concentration. This dependence is determined independently by profiling through oxynitride layers of known thickness and varying composition. Electron escape depth and preferential sputtering effects are also accounted for.

Quantitative information on the thickness of the intermediate oxide layer is obtained by comparing these sputter data with results from ellipsometry and infra-red absorption spectroscopy (multiple internal reflection — MIR), and by using nitride–oxide (NO) and oxide–nitride–oxide (ONO) multilayers of known thickness as a reference.

The proposed technique is successfully applied for studying the evolution of the interfacial region during low pressure chemical vapour (LPCVD) silicon nitride deposition on silicon and during thermal nitridation of SiO_2 layers.

*Fellow IWONL
† Senior Research Associate NFWO

APPLICATIONS OF SIMS TO III/V-COMPOUNDS

H. Frenzel
ATOMIKA Technische Physik GmbH, Kuglmüllerstr. 6, 8000 München 19, West Germany
W. Schlaak and A. Steckenborn
Heinrich-Hertz-Institut für Nachrichtentechnik Berlin GmbH, Einsteinufer 37, 1000 Berlin 10, West Germany

The behaviour of dopands in GaAs in respect to SIMS analysis is known. However, there is still a lack of data for applications on InP and combinations like $In_xGa_{1-x}As$. This study was performed to obtain calibration factors and detection limits for these types of semiconductors.

Different samples of InP, GaAs and $In_xGa_{1-x}As$ were implanted with O, C, S, Sn, Si, Mn, Fe at energies up to 400 keV with peak concentrations up to $10E18cm^{-3}$. Depth profiling was performed with an A–DIDA 3000–30 using O_2 and Cs.

Factors for the conversion of secondary ion counts to concentrations were determined and compared to those of GaAs and InP to study the effect of matrix change in epitaxial grown layers. Detection limits are defined as the lowest concentration obtained in the layers under study. We present detection limits down to $10E13$ cm^{-3}.

The effect of different chemical matrices is discussed by LPE-grown $In_xGa_{1-y}As_yP_{1-x}$ on InP.

INVESTIGATIONS OF ELECTRICAL CONTACTS BY COMPLEMENTARY TECHNIQUES OF MICRO- AND SURFACE ANALYSIS

A.P. von Rosenstiel
Metaalinstituut TNO, Postbus 541, 7300 AM Apeldoorn, The Netherlands

Electrical contacts include a broad range of products such as switches, relays, keyboards, terminals, splices, push buttons, foil contacts and connectors. Steadily decreasing voltages and currents, together with decreased dimensions and increased reliability criteria require several complementary techniques for the development, production control and failure analysis of these precision mass products.

Examples are given with particular attention to problems in galvanoplating of intermediate and surface layers with respect to adhesion structure, grainsize, thickness and orientation, surface finish and flatness, inhibitor and trace element analysis, internal stresses, pore reduction, corrosion, erosion, wear and material transfer by a combination of metallographic techniques, XRF, SEM, TEM, EPMA, SIMS, x-ray stress measurements and ultra-microhardness determinations.

CHARACTERIZATION OF CVD-LAYERS BY SCANNING AUGER MICROSCOPY

A. Riahi, M. Puchhammer and H. Störi
Inst. f. Allg. Physik, TU Wien, Karlsplatz 13, A–1040 Vienna. Austria
C. Colombier and B. Lux
Inst. f. Chem. Technologie anorgan, Stoffe, Tu Wien, Getreidemarkt 11, A–1040 Vienna, Austria

Layers of TiC and Al_2O_3 and multilayer structures of these materials on cemented carbide and on pure cobalt substrates have been produced by chemical vapour deposition. The homogeneity of these layers, the distribution of impurities and diffusion effects have been investigated by scanning Auger microscopy. In order to depth profile these comparatively thick layers (several μm) a crater of spherical shape has been produced with a ball cratering device. The composition along the crater walls is investigated by scanning Auger microscopy. Additionally, Auger electron spectra were taken on selected spots along the radius of the crater. Results of the diffusion of cobalt in a TiC layer and on the influence of cobalt on the layer growth have been obtained. It has been demonstrated that a thin Al_2O_3 layer stops Co diffusion. The results demonstrate that the method presented offers for thick layers a depth resolution at least equal to the resolution obtained for sputter depth profiling combined with several advantages such as a short analysis time and absolute depth calibration.

Acknowledgement
This work is supported by the Fonds zur Foerderung der Wissenschaftlichen Forschung, Project No. S43

SCANNING AUGER MICROPROBE INVESTIGATION OF SEGREGATION PHENOMENA IN Sn/Pb ELECTRONIC CONTACTS

W.F.L. Bogaerts, R.F.J. Straetmans and A.A. van Haute
Catholique Universitie of Leuven, Chemical Engineering Dept., de Croylaan 2, B–3030 Leuven, Belgium

Background

Small variations in the production process of specific Sn/Pb microelectronic contacts (overall component width = approx. 0.3 mm) can lead to end products of a very different quality.

In order to trace the cause of these phenomena it is necessary to determine as well the total concentration of Sn and Pb in the 1 micrometer thick toplayer, as their concentration gradient (depth profile) at the small contact area of the device.

Samples produced under different production conditions, each of them known to have a different performance level, have been analysed in a scanning Auger microprobe system with sub-micron spatial resolution. By using some specific data reduction techniques, differential and integral depth profiles have been determined for the various elements at different spots.

Experimental Results

Smaller concentrations of Cl and S are detected in the outer surface layers on both samples. Large amounts of C are present.

In the atomic layers where O is present an important increase of Sn (as an oxide ?) can be noticed. As soon as the O-concentration diminishes, also the Sn concentration drops and the Pb content increases sharply.

At the interface between the surface oxide and the base Sn/Pb metal a nearly identical Sn/Pb ratio is observed on both kinds of samples: 40–45/60–65.

All samples also show the same kind of segregation pattern inside the base Sn/Pb layer, i.e. an increasing Sn/Pb ratio as a function of depth.

Differences between successful and unsuccessful components are only revealed in a relatively narrow zone above the nickel substrate. At this interface a sharp Pb depletion and consequent relative Sn enrichment is observed in one of the sample series. This results in an overall (integral) Sn/Pb concentration ratio which is very different in the different surface coatings. It has been possible to relate these analytical results to specific differences in the processing conditions of the components so that corrective measures could be taken.

Conclusions

This communication illustrates how scanning auger microprobe investigations prove to be a most valuable tool for handling industrial production control problems in a number of critical areas, such as the manufacturing of microelectronic components.

DIFFUSION OF Cr DOPANT FROM SUBSTRATE GaAs INTO LPE GROWN EPILAYERS

C.M. Demanet
NIMR, CSIR, Pretoria, Republic of South Africa

Assessment of the chromium diffusion from semi insulating GaAs substrate into liquid phase grown epitaxial layers is of value for the GaAs device and integrated-circuit industry. It is therefore most important not only to monitor the diffusion phenomenon but also to characterize the process and assess the influence of various growth techniques in this respect. SIMS is a very suitable technique for depth profiling and its sensitivity allows detection of dopant concentration.

In this paper we will report on Cr depth profiles in GaAs liquid phase grown epitaxial layers obtained by SIMS using 0_2^+ as primary ions for optional secondary ions yield. We will compare the various conditions of LPE growth (equilibrium, step and super cooling) in this respect.

Preliminary results tend to show that diffusion in the substrate and in the epilayer could differ. Compared to the value in the epilayer, the higher diffusion coefficient in the substrate could be understood as being due to the doping present there which by increasing the defect concentration facilitates diffusion. Diffusion coefficient will be determined and the profiles for the various growth conditions will be discussed in detail.

ION BEAM ANALYSIS OF ELECTROLESS THIN LAYERS DEPOSITS USED FOR OHMIC CONTACTS ONTO GaAs

D. Lamouche, P. Clechet, J.R. Martin and P. Person
Laboratoire Physicochimie des Interfaces (UA–404), Ecole Centrale de Lyon, BP 163, 69131 Ecully Cedex, France
A. Chevarier, N. Chevarier and M. Stern
Institut De Physique Nucleaire De Lyon (I N2 P3), Université Lyon I, 43, bd du 11 novembre 1918, 69622 Villeurbanne Cedex, France

We have recently demonstrated how to obtain low resistive ohmic contacts on n–GaAs using Au:In/Pd electroless deposits. In particular, concerning contact resistivity, a Au:In/Pd multi-layer has led to superior results when compared to the ones obtained from conventional (i.e. evaporated) Au/GaAs or Pd/GaAs contacts, suggesting important differences in contact's annealing behaviour.

In this work we studied the annealing behaviour of chemical deposition of Pd and Au:In/Pd on GaAs semiconductor. Heavy-ion backscattering analysis, connected with high resolution time of flight spectrometry, has been used to investigate gold and palladium distribution in the alloyed region close to the surface. This method allows the arsenic and gallium contribution to be identified and the ratio between arsenic and gallium on the first 400 Å to be measured and such results to be related to the samples' properties. Investigation of oxygen in these contacts being necessary we used the $^{16}O(\alpha,\alpha')^{16}O$ response at 3.05 MeV which allows an oxygen profiling by alpha Rutherford backscattering technique (RBS).

The RBS analysis on Pd/GaAs shows that the chemical deposition of Pd is connected with a lack of arsenic which is preferentially exchanged. This interface feature, different from the one obtained with evaporated palladium, can at least partly explain the discrepancy measured on the respective barrier heights on Pd/GaAs. After annealing at 360° a slight palladium diffusion is observed and an out diffusion of gallium connected with its surface oxidation takes place. However, a large Pd diffusion is observed in the Au:In/Pd/GaAs case and the presence of Au:In overlayer is shown to prevent from Gallium out diffusion. Reactions mechanism is proposed to illustrate such feature which can partly explain the low resistive ohmic contact obtained on Au:In/Pd/GaAs.

AUGER PROFILE ON Al_xGa_{1-x} As PROTECTED BY As AND GaAs

P. Etienne and P. Alnot
LCR — Thomson-CSF, BP 10, 91401 Orsay, France
J.F. Rochette
DAG — Thomson-CSF, BP 10,91401 Orsay, France
J. Massies
Laboratoire Physique du Solide et Energie Solaire, CNRS, Sophia Antipolis 06560 Valbonne, France

Al_xGa_{1-x} As epilayers are very sensitive to air exposure. It is of prime interest to find a good process to protect this surface. MBE is one of the techniques used to produce such epilayers. Using this technique, it is possible to deposit a protected layer of As or GaAs at the end of the epitaxial growth of Al_xGa_{1-x} As.

In the present work, we intend to examine the protecting capability of this method. Auger electron spectroscopy has been used to obtain concentrations profiles. Annealings under air, molecular and atomic hydrogen atmosphere were studied.

We will show the big differences existing between the protection offered by As or GaAs overlayers. The protection by GaAs has been found to be better than that afforded by As.

AN XPS STUDY OF LAYERED AB2×4 SEMICONDUCTORS

C. Battistoni, L. Gastaldi, G. Mattogno and S. Viticoli
Istituto Teoria e Struttura Elettronica, CNR Area della Ricerca di Roma, Via Salaria Km 29,500, PO Box 10,00016 Monterotondo Stazione, Italy

In recent years a wide interest has been aroused about layered compounds because of their structural, magnetic and electronic properties. Classical studies are particularly inadequate to explain the magnetic behaviour of low-dimensional magnetic systems and the search for a model to test the predictions of statistical thermodynamic theories has made the mono- and two-dimensional systems interesting objects of study among chemists and physicists.

In order to have additional information about the geometries and structural properties of some of these compounds as the solid solutions $Co_xZn_{1-x}In_2S_4$ ($0.00<x<0.60$) an x-ray photoelectron investigation is here reported.

Photoelectron spectra were obtained with a VG ESCA 3MKII spectrometer using the Al K-alpha 1.2 line as the exciting source (H.V. = 1486.6 eV).

It is known that magnetic Co^{2+} ions replaced Zn^{2+} ions in the $ZnIn_2S_4$ structure up to a cobalt concentration ($X = 60\%$) and EPR and optical absorption investigations on this system have not been able to give clear evidence about the redistribution of Co^{2+} ions within octahedral and tetrahedral sites.[2]

The analysis of binding energy values of the Co $2p_{3/2}$ (780.0–780.9 eV) and the presence of the broad satellite peaks in the photoelectronic spectra of these solid solutions are indicative for high-spin Co^{2+} compounds.

In order to know the quantity of Co^{2+} ions replacing tetrahedral or octahedral sites, the analysis of the shape of the photoelectron spectra of the CoS octahedral compound and $CoZn_2S_4$ systems, in which the Co^{2+} ions occupy only tetrahedral sites, are performed.

Two satellites lines at 3.4 and 7.5 eV and 4.5 and 8.3 eV from the main $Co2p_{3/2}$ line are found for CoS and $CoZn_2S_4$ compounds.

By deconvolution of the broad photoelectron spectra of the solid solutions it is possible to give clear evidence that the Co^{2+} ions replace octahedral and tetrahedral coordination sites in $ZnIn_2S_4$ structure.

References

1. L.J. de Jongh, A.R. Miedema, *Adv. Phys.* **23**, (1974).
2. D. Fiorani, L. Gastaldi, S. Viticoli, *Solid State Com.* **48**, 865 (1983).

Section 6.

Corrosion

SURFACE AND INTERFACE ANALYSIS, VOL. 9, 345–356 (1986)

The Role of Electron Spectroscopy in Corrosion Science

J. E. Castle
Dept. of Materials Science, University of Surrey, Guildford, GU2 5XH

The two principal electronspectroscopies – photoelectron and Auger electron – now represent the mature methods of surface analysis and both have been used to further the study of corrosion almost since their inception. This review will examine those areas in which significant developments have been made by their use and will highlight their continuing role in post-corrosion failure analysis.

Specific topics which will be covered include: 1) aqueous corrosion: the formation and composition of passive layers, their interaction with surrounding solutions, and their failure by pitting and stress corrosion; 2) coated surfaces: pretreatment for coating, locus of failure and disbonding under chemical or electrochemical attack; and 3) high temperature oxidation: segregation to the surface, the formation of stable layers and interlayer reaction, and the compostion of barrier layers and of spalling interfaces.

INTRODUCTION

Corrosion has two distinct branches according to whether the environment is aqueous (i.e. supportive of ionic transport) or gaseous. Very different techniques of investigation have developed in the two fields but the respective phenomena still fall within a common thermodynamic and kinetic framework. There is a further common feature which must be fully appreciated if the role of surface analysis is to be understood: Corrosion science concerns the rate of reaction of a solid with its environment and of the prognosis for that rate. Surface analysis, as a mature analytical technique must now be assessed for its value in derivation of this rate or of its future trends. The use of any analytical technique in this special fashion comes high up on the learning curve following its first introduction and it is appropriate that we should now start to look for this attitude in those utilising electron spectroscopy. There have been reviews over the years with titles similar to the present[1–4] and these show an interesting development of the science along the required lines. They also describe the development of the appropriate tools within surface science: recognition of chemical state of an oxide in the presence of its elements by chemical shift, (Fig. 1 (a)), satellites, multiplet splittings or the Auger parameter; the depth distribution, (Fig. 1 (b)), and the spacial distribution. This review while not eschewing analysis for its own sake will attempt to focus on that which has been particularly purposeful. The review is divided for clarity into two parts according to the class of medium used and in each case is introduced by way of the underlying corrosion science. Much material is taken from previously unpublished sources in the authors own laboratory and attention is drawn to the acknowledgement to those kind enough to lend use of their data and examples ahead of publication.

0142–2421/86/120345–12 $06.00

AQUEOUS CORROSION AND ELECTROCHEMISTRY

When metals corrode in an aqueous medium, transport or loss of material in soluble form must be expected and it is impossible to gauge the extent of reaction by the mass of products on the surface. Moreover it is usually difficult to interpret the results of spectroscopy (and also of microscopy) because of the effects of vacuum on the normally immersed surface.[5] Reaction rates must be measured directly or inferred from the electrochemical parameters of exchange current density and of its dependence on electropotential. These measurements give rise to the variety of electrochemical methods for corrosion studies[6] and it is to these that surface analysis will normally be complementary. There are three fundamental relationships:

(a) the Faraday law of electrolysis:

$$m = it/nF$$

relates the number of moles of material, m, undergoing reaction at the electrode to the charge passed, $i{*}t$, and the number of electrons exchanged in the reaction, n. the constant is Faraday's constant, F;

(b) the Nernst relationship:

$$E = E' + (RT/nF)\ \ln(a_{\text{reac}}/a_{\text{prod}})$$

relates the potential of an immersed metal, E, to its free energy in a standard state, E', and the activities, a, of the reactants and the products;

(c) the Tafel relationship:

$$i = i_0\ (\exp(n_A FE/\beta_A RT) - \exp-(n_C FE/\beta_C RT))$$

which shows how the external current through the electrode varies with the exchange current density and

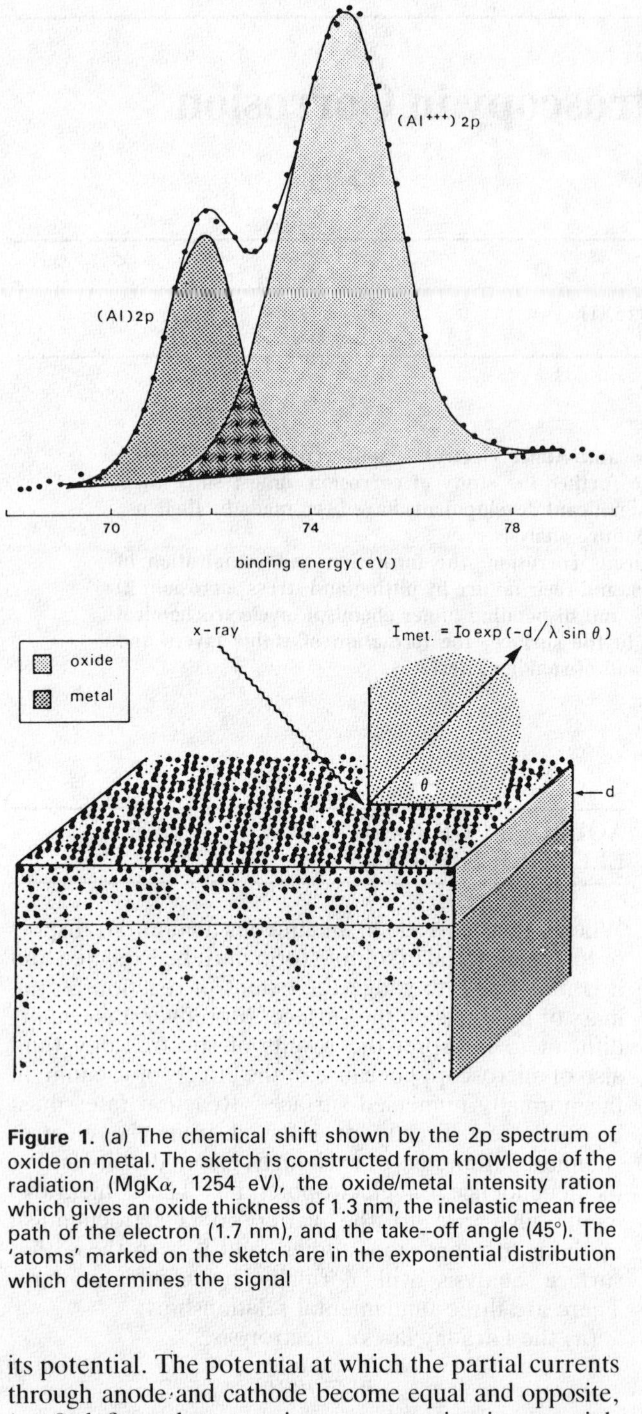

Figure 1. (a) The chemical shift shown by the 2p spectrum of oxide on metal. The sketch is constructed from knowledge of the radiation (MgKα, 1254 eV), the oxide/metal intensity ration which gives an oxide thickness of 1.3 nm, the inelastic mean free path of the electron (1.7 nm), and the take--off angle (45°). The 'atoms' marked on the sketch are in the exponential distribution which determines the signal

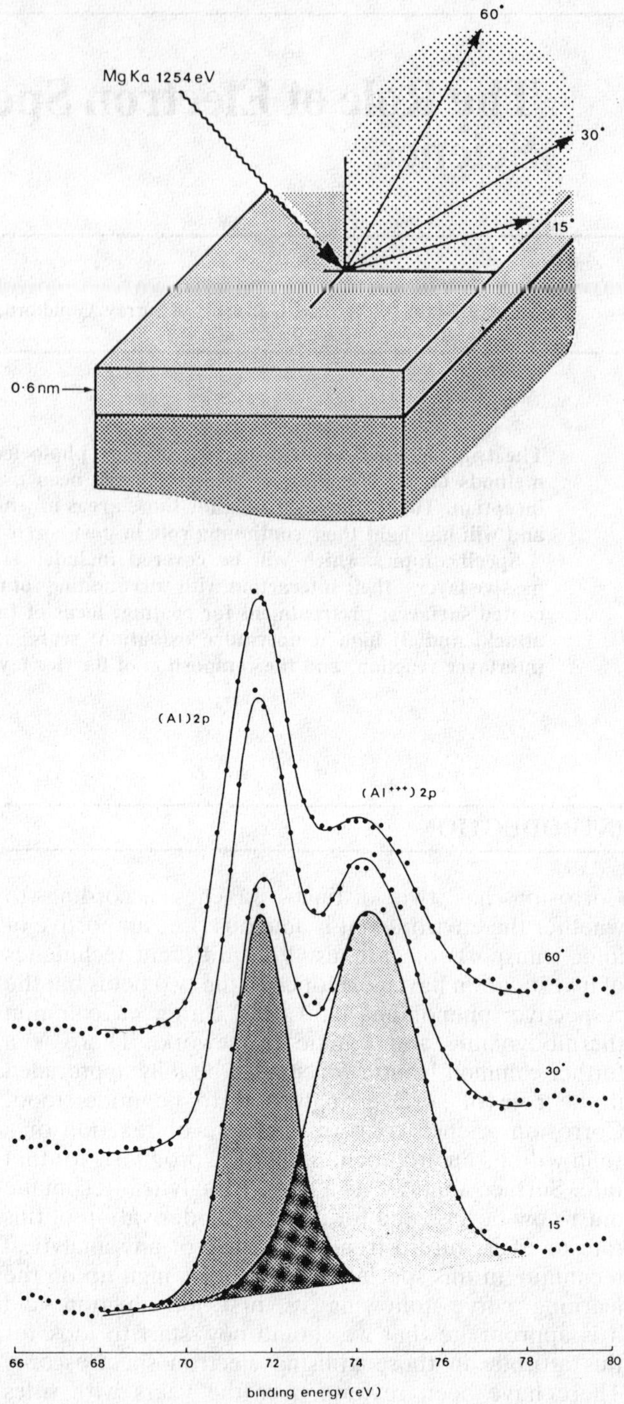

Figure 1. (b) The angular dependence of signal. The oxide/metal intensity ratios vary with the take-off angle in accord with that expected from an oxide thickness of 0.6 nm. Note the strong enhancement of the oxide signal at low take-off angles.

its potential. The potential at which the partial currents through anode and cathode become equal and opposite, $i = 0$ defines the corrosion or open circuit potential, E_{corr}, of a piece of immersed metal. The constants β_A and β_C are known as the Tafel constants and RT has the usual meaning.

If hydrogen or hydroxyl ions are included in the equilibrium at the electrode surface then the electropotential given by the Nernst relation will vary with the pH value of the medium. By determining the potential of metals in equilibrium with their ions Pourbaix[7] was able to plot potential/pH diagrams (now given his name) showing zones in which solid films should precipitate on the surface which would thus tend to become passive. In other zones ions either pass freely into solution giving unacceptable loss of metal or are thermodynamically unable to form, giving a surface which is immune to attack. As the potential of a metal is increased at constant pH value the current through the electrode increases in accord with (c). Then, as the boundary for precipitation is crossed, the current flow becomes impeded by the solid product.

Surface analysis can play a most important role in the interpretation and use of the above equations and diagrams. A recent study[8] carried out on the crevice corrosion of chromium and molybdenum is a good example of the possibilities for comparison of Pourbaix diagrams and the results of surface analysis. Appropriately the work was carried out, in part, in Pourbaix's laboratory by Antoine Pourbaix. This study aided an understanding of changes occuring in crevices during the localised corrosion of alloy steels and showed a good correlation between the thermodynamic predictions and spectroscopy.

It is important to note that these metals become very strongly passivated under the conditions used and thus the thickness of the films is not strongly dependent on time. When examining more active surfaces, continuing reaction at the electrode is likely to give backdeposition of material from solution. This will mask the surface which determines the thermodynamic potential and it follows that thick deposits that can be identified by electron proble x-ray analysis are frequently irrelevant.

The relationship between the thickness of material on the surface and the charge transferred was one of the first topics to be investigated by XPS and the linear relation extended to vanishing point.[1] This ability to analyse a surface after very short reaction times or small charge transfers offers the opportunity to identify the true surface and to confirm the zones on the Pourbaix diagram which has previously been done using methods such as EDXA and x-ray diffraction, requiring much thicker deposits.[9–10] Thus the presence of surface states not predicted by bulk thermodynamics can be allowed for and reprecipitation more accurately distinguished from direct formation of the compound at the electrode. A typical level of charge transfer for this work is 0.5 A s[11] which would disturb about 100 nm according to the Faraday law. Keeping constant the charge transferred through each electrode, exposed at successive potentials in the range of the diagram, ensures that changes in the surface will stem from thermodynamic equilibria or from changes in the rate of delivery of material through the interface rather than from changes in the extent of reaction.

An illustration of work of this type is given in Fig. 2, using the corrosion of copper alloy 706 (copper/nickel 10/iron 1.5%) in 3.4% sodium chloride solution as an

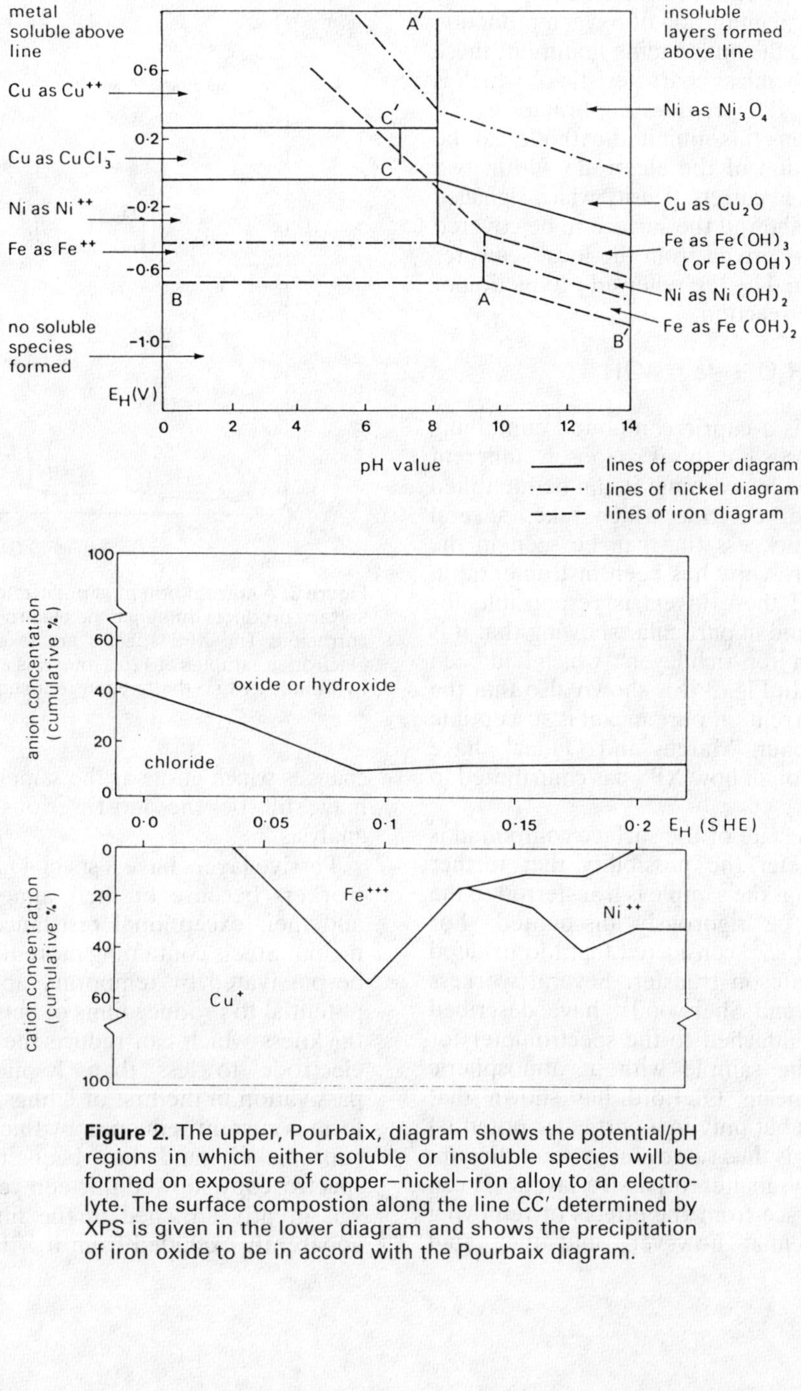

Figure 2. The upper, Pourbaix, diagram shows the potential/pH regions in which either soluble or insoluble species will be formed on exposure of copper–nickel–iron alloy to an electrolyte. The surface compostion along the line CC′ determined by XPS is given in the lower diagram and shows the precipitation of iron oxide to be in accord with the Pourbaix diagram.

example.[12–14] The important features of the relevant Pourbaix diagrams for copper (as in the alloy)[10] nickel and iron (as the pure metals)[7] as determined by conventional methods are included in a composite diagram. At pH values greater than those given by the line AA′ passivating films are predicted whilst below the line BB′ no metal ions are able to form and the alloy is immume from attack. Within the region BAA′ soluble species form and corrosion could therefore occur. In fact tarnish films do form on the surface and the rate of metal dissolution steadily decreases with time. The XPS analysis shows the composition of these films formed at a pH value of 7 as the potential moves from the potential C to C′. At C a film of cuprous chloride (which is predicted by diagrams which include the chloride ions[13]) and perhaps containing cuprous oxide, is formed. As the potential increases this chloride compound is replaced first by insoluble FeOOH and then by nickel(II) compounds, a sequence that suggests the boundary AA′ should lie at lower pH values when thin surface films are used to define regions. The thin film of cuprous chloride is probably responsible for the good inhibition of oxygen reduction at the cathode which is the outstanding feature of these alloys,[14] the electric chemical characteristic of which is illustrated in Fig. 3. XPS provides confirmation that under certain conditions this inhibited cathode can be reactivated. Examination of the electrode within two minutes of an increase in current density which signalled the loss of inhibition, showed the surface to be covered with basic compounds derived from the local seawater environment and formed by the renewed production of hydroxide ions by the reaction:

$$O_2 + 2H_2O + 4e = 4OH^-$$

At higher potentials a cupric compound containing chloride ion also deposits but this is so loosely adherent that it can be removed by washing: it can be identified with the basic copper chloride which takes several days to grow to a thickness that can be seen in the SEM. Electron spectroscopy has been instrumental in determining which of these layers is responsible for corrosion resistance and in particular showing that it is not associated with an iron-rich layer[12] which had been thought responsible: In Fig. 3 it is shown also that the anodic passivation current on pure nickel is susceptible to activation by sulphur. Marcus and Oudar[15] have given a good illustration of how XPS has contributed to this discovery.

Since the oxidation state of the surface compound is an important parameter the possibility that further reaction has occurred as the sample is transferred to the spectrometer, must be rigorously discounted. For example Hashimoto *et al.*[23] considered that Mo oxidised to the hexavalent state on transfer. Several workers including Olefjord[16] and Sherwood[17] have described electrochemical cells attached to the spectrometer for direct transfer of the sample without atmospheric exposure. By this means Olefjord has shown that hexavalent Mo forms but only in a particular potential range. This technique is thus successful in permitting the analysis of apparently unaltered passive layers. It can not preserve the surface from the effects of removing the controlling potential however, and the rapid changes which ensue as the sample leaves the electrolyte may still be the most serious difficulty of *ex situ* analysis.

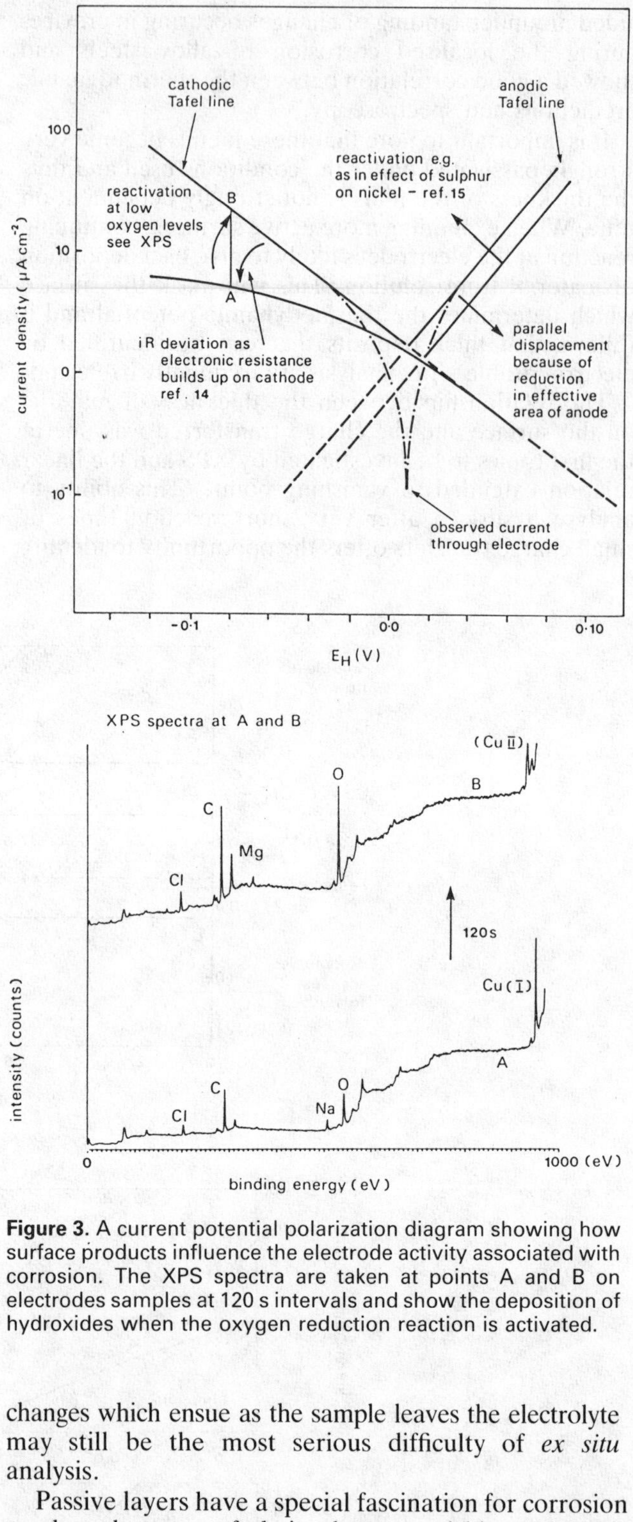

Figure 3. A current potential polarization diagram showing how surface products influence the electrode activity associated with corrosion. The XPS spectra are taken at points A and B on electrodes samples at 120 s intervals and show the deposition of hydroxides when the oxygen reduction reaction is activated.

Passive layers have a special fascination for corrosion workers because of their almost vanishing thickness and their exceptional resistance to corrosion in acid media. Steels containing more than 13% chromium can be passivated by temporary application of an anodic potential to produce films of between 0.5 and 2.5 nm in thickness which can reduce the current density at the electrode to less than 1 microamp. This is true passivation in the first of Uhligs categories[18] (the other form is current reduction by thick films as described for cupronickel) and has been much studied. Auger spectroscopy showed thirteen years ago that chromium was greatly enriched in the film whereas nickel was apparently excluded from it.[19] From the point of view of understanding the role of surface analysis few studies have been as instructive as the series of studies on iron–chromium alloys in sulphuric acid made by Asami, Hashimoto and Shimodaira in Sendai.[20–24] These workers used XPS to measure both the oxide and the metal interface composition, the intensities of which were corrected for the effects of contamination. The work of these authors has been often quoted[1–4] and shows the behaviour of the ferrous ion, chromium (III) and hydroxyl ion contents of the film, formed by passivation in 1 M sulphuric acid, as a function of the chromium content. In each case the concentration of these ions increased relative to total iron, total oxygen and total metal respectively at the critical concentration of chromium necessary for passivation. The composition of metal surface remained identical to that of the bulk metal after passivation and the authors inferred from this that chromium enrichment occurs as a result of the dissolution of iron rather than by selective oxidation of chromium. These characteristic changes at passivation did not take place at potentials below that for passivation. The analysis of the passive film did not correspond to that of a recognisable stoichiometric compound but was nevertheless recognisable by virtue of its high Cr(III), Fe(II) and water contents.[3] A dramatic illustration of the importance of chromium to passivation was given by Hashimoto *et al.* using a series of amorphous alloys.[23] These were based on metalloid additions of P, C, Si, B to an iron 10% chromium alloy and the passive films were formed in = 0.1 N sulphuric acid. The corrosion rate of the films on an alloy containing 13% boron plus 7% of silicon boron carbon or phosphorus fell steadily in that order (Fig 4). XPS analysis showed a steady increase in chromium (III) in the oxide passivating film in the same alloy order and reached 30% on that containing phosphorus. When an alloy containing 13% phosphorus and 7% C was exposed to 1N HCl[24] the chromium content of the film was shown to reach 93% of its cation content and the corrosion rate was found to be immeasurably small. In this work XPS has been vital to understanding the role of the metalloid additions and in recent work phosphorus is shown to oxidize at passivating potentials.[25] XPS has thus shown that instead of forming an important constituent of the film these elements appear to function by stimulating the active dissolution of iron which, as was shown on conventional steels, leads to the enrichment of chromium. Okamoto and Shibata[26] have given a review of this area and further helpful references will be found in their paper.

When iron itself is passivated e.g. in borate buffer solutions the film formed has two layers as indicated by the Auger spectroscopy of Sea, Sato Lumsden and Staehle.[27] Also the outer region differed in iron/oxygen ration and contained elements from the passivating bath whilst the inner layer was free of other ions and corresponded in composition to Fe_2O_3. It was concluded that the inner layer formed by direct oxidation at the metal surface whilst the outer layer formed by solution phase transport and deposition. In accord with this the XPS measurement of Komo and Nagayama[28] show the outer layer to be highly hydrated FeOOH. The degree of hydration decreased with increase in passivation potential. These authors raised the possibility that small shoulders on the iron 2p spectrum could indicate the presence of Fe(VI) ions in the outer film. This has not been confirmed and the shoulders may merely be multiplet splitting satellites associated with the particular sites occupied by Fe(III) ions. The formation of an outer layer of deposited material has recently been confirmed by Graham and co-workers[29] using SIMS and H_2O^{18} as a tracer. They also confirmed the anhydrous character of the inner Fe_2O_3 layer. Their work on the chromium steels however also points to an anhydrous film and this is difficult to reconcile with ESCA work unless back-deposition has a more powerful influence than expected.

One topic which has been investigated by numerous groups but on which no clear consensus has yet been reached is the role of molybdenum in steel. As is well known, molybdenum improves the resistance of austenitic steel to crevice and pitting corrosion. In their pioneering work, Lumsden and Staehle concluded that molybdenum was excluded from the film but enhanced the chromium enrichment and this finding was repeated by other workers using the AES technique. This finding has a parellel in the work of Hashimoto *et al.*[23] on amorphous alloys in which as a result of including 2% Mo in the material the chromium content of the film increased to 18% from the 3% in the bulk alloy and it is frequently suggested that Mo and Cr enriched together on active surfaces. Newman[30] has suggested that they

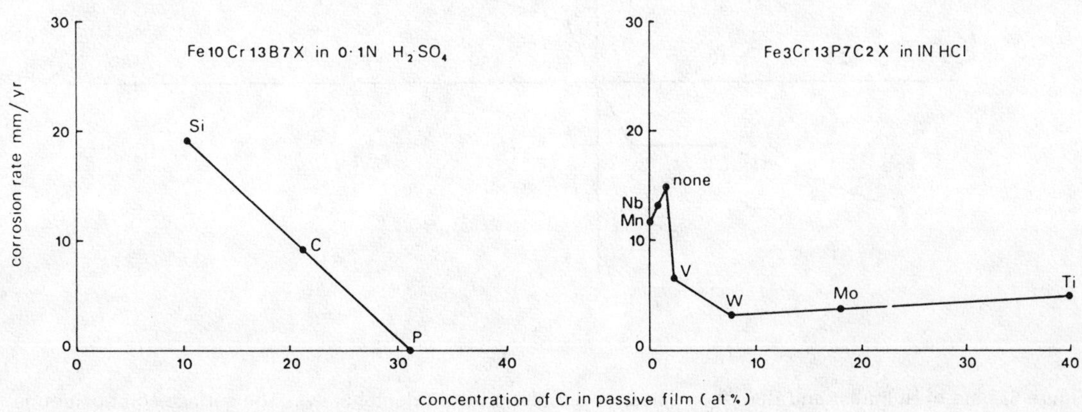

Figure 4. The XPS data of Hashimoto *et al.*,[23] showing how the corrosion rate of an amorphous alloy and the surface concentration of chromium vary with the different elements, *X*, included in the alloy. These elements are not themselves surface specific.

there inhibit dissolution by blocking sites at an atomic level. Other viewpoints are equally strongly expressed[31] and mechanisms suggested include the precipitation of chloride ions as an insoluble Mo (IV) compound and the true inhibition of anodic sites by submonolayer coverage of molybdate ions. The differences found by AES and XPS methods in early work was disconcerting. However, recent work undertaken as part of an EFC round robin shows good agreement between AES and ESCA with regard to Mo distribution.

Sherwood and co-workers[32] have investigated the passivity of molybdenum itself by XPS. Oxidation of this metal could not be detected in air or in deaerated sulphuric acid, but in moist oxygenated conditions both Mo(III) and Mo(IV) are formed, a useful review of the behaviour of molybdenum has been given by Wanklyn.[33] Other metals studied under conditions of anodic oxidation by this group include gold and platinum,[36] chromium,[34] nickel,[35] tin[36] and stainless steels.[37] Selective dissolution from noble alloys has also been studied,[38] in the context of the corrosion of dental alloys. This study illustrates the use of angularly resolved collection of electrons to deduce surface composition. Figure 5 compiled from this paper shows the relation between gold enrichment and rest potential, E_{corr} (see Tafel equation above). The authors make the point that a single potentiodynamic scan, as is usual practice in corrosion studies, does not lead to equilibrium enrichment at the surface and therefore cannot predict long-term corrosion behaviour. This

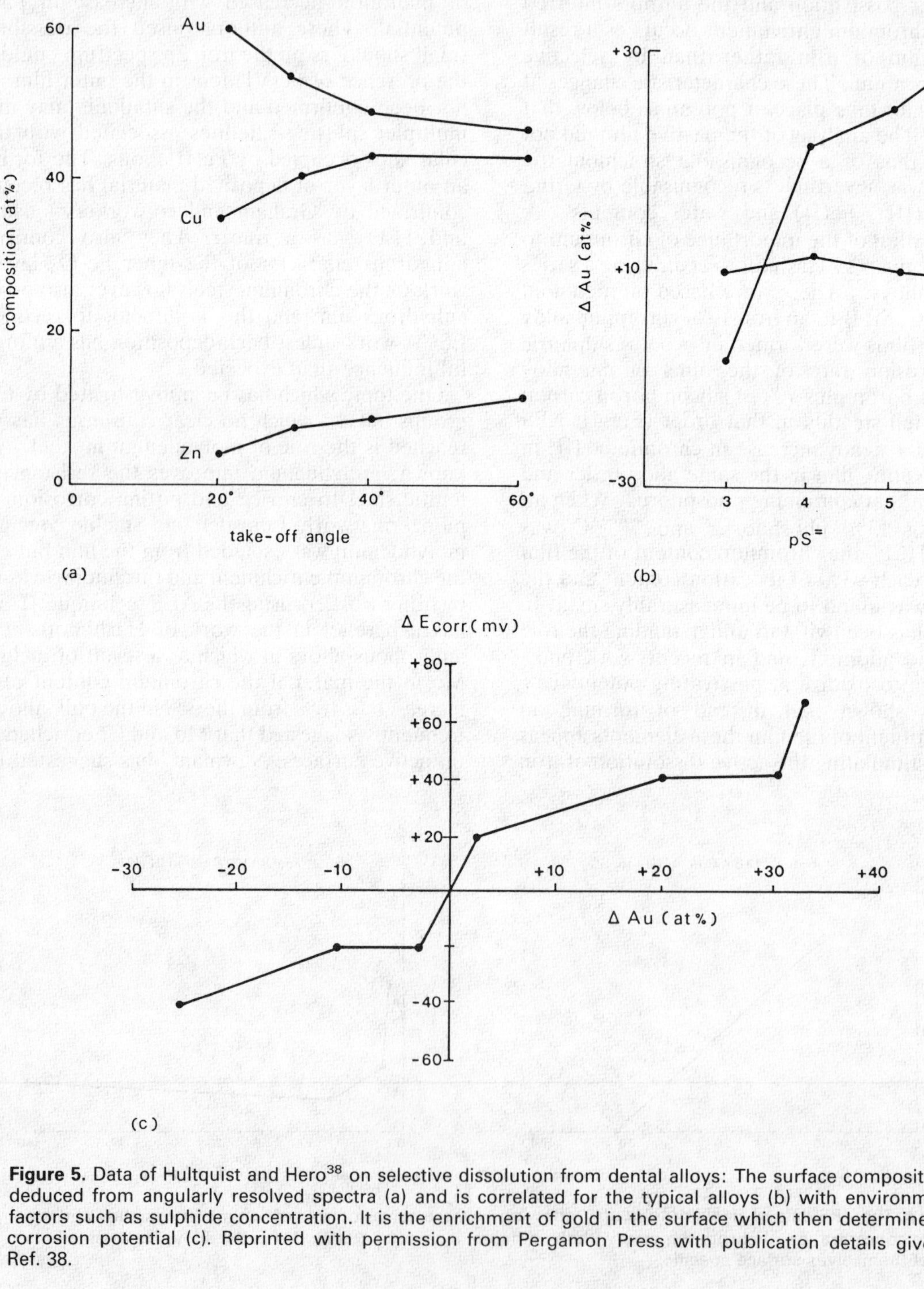

Figure 5. Data of Hultquist and Hero[38] on selective dissolution from dental alloys: The surface composition is deduced from angularly resolved spectra (a) and is correlated for the typical alloys (b) with environmental factors such as sulphide concentration. It is the enrichment of gold in the surface which then determines the corrosion potential (c). Reprinted with permission from Pergamon Press with publication details given in Ref. 38.

paper illustrates well how the depth of altered layers should be cross-correlated between charge transferred through the electrodes and the 'surface' analysis.

Little has been said in this section concerning the role of passivating ions in solution. Chromium, for example, is transferred from passivating solutions into the oxide as has been shown using XPS by Mansfeld *et al.*,[39] and other authors.[40] This general field of study is obviously ideal for XPS analysis in which the oxidation state of the anion can be closely followed during its adsorption and incorporation.[41] Recently, for example, it has been shown that, dichromatic ions, known inhibitors of aluminium corrosion, are reduced to Cr(III) ions on adsorption.[42]

We turn now to those situations where metals, whilst not passive are, in fact, paticularly resistant to attack e.g. the behaviour of stainless steels in pure water. Electrochemistry is difficult in pure water because of its low conductivity and surface analysis can provide primary information on rate of film growth. Castle and Clayton[43] found that a limiting thickness of oxide was reached in under one hour when 18/8 steel was exposed to deaerated water at all temperatures up to 160°C. This limiting thickness increased from around 1 nm at 40°C to 5 nm at 160°C and within this thickness a duplex film structure could be detected. At 200°C film growth did not cease and the film formed after 3 h in some 2.5 times the thickness formed after 1 h. Differentiation into an outer iron-rich layer and an inner chromium rich layer is virtually complete at this temperture. Addition of Co^{++} ions to the water resulted in their incorporation and an enlargement of the outer layer by formation of the iron and cobalt spinel.[44] Olefjord and Fischmiester and Olefjord and Elfstrom[45] have examined chromium steels after exposure to water for much longer periods of time and have concluded that loss of iron compounds continues for up to 50 h. During this time the total oxide film becomes thinner suggesting that chromium (III) rich materials is the passivating phase.

McIntyre and co-workers have studied the corrosion of high nickel alloys such as Stellite 75 and Inconel 600 (77Ni 16Cr 7Fe)[46] monel (66Ni 32Cu)[47] in water and solutions at 280°C. In these alloys, unlike the steels, nickel is oxidised and McIntyre has been able to distinguish NiO and $Ni(OH)_2$ by their multiplet splitting structure. They find evidence for film growth by both solution transport and solid state diffusion, pointing again to the versatility of XPS analysis. The work has been extended to lithiated solutions and this study is a good example of the complimentary use of electron spectroscopy and tracer techniques.[48]

When aluminium brass (76Cu, 22Zn, 2Al), is exposed to marine environments a surface film of the mixed hydroxide, hydrotalcite is formed. This has the composition $(MgZn)_6Al_2(OH)_{18}$ but the divalent ions are readily replaced by others, e.g. Cu(II), or may be lost altogether in acidic environments to leave a residue of Boehmite. The discovery of this unsuspected layer is itself a tribute to surface analysis but its ion exchange behaviour when exposed to seawater of different pH values[49] illustrates another way in which XPS can be used to follow mechanisms of protection; in this case a surface buffering action that was first suggested by Uhlig in a different context.[50] The compositional changes can be followed between different heat exchangers, where it correlates with performance, with depth into a tube, where it relates to the build up of hydrodynamic turbulence, and with the microelectrode system around a small corrosion pit.[51]

Several studies have been made of the interaction between copper and solutions containing benzotriazole. These have confirmed the existance of two states of copper in benzotriazole complexes and have shown that the Cu(I) state is protective.[52] The sequence of spectra obtained by Chadwick and Hashami,[53] showing the interconversion of Cu(I) to the porous Cu(II) complex on exposure to oxygenated water or to air is illustrative of the help which XPS can give.

THE BREAKDOWN OF PASSIVITY

Passivity breakdown is inevitably localised and on the finest scale pitting can only be studied by high resolution Auger microscopy. The analytical resolution of recent instruments is of the order of 0.05 μm and SEM images may be obtained with a resolution of about an order of magnitude better than this. The micrograph in Fig. 6[54] shows the pitting associated with a crevice on 316 steel exposed to magnesium chloride solution, and Fig. 7 illustrates how Auger analysis together with x-ray analysis enables pits to be associated, in some cases, with underlying sulphide inclusions.

In order to demonstrate that such pits are active it is essential to show that electropotential gradients were generated around it. Such work involved the cross-correlation of XPS analyses on whole electrodes as a function of potential, as described above, with Auger analyses as a function of position. The potentials deduced from Auger analyses are in excellent agreement with those calculated by the electrostatic equations for a circular pit.[55] This procedure enables active pits to be distinguished from similar topographic features which are inactive and the sequence of experiments is illustrated in Fig. 8.

GASEOUS CORROSION AND KINETICS OF REACTION

In the laboratory the rate of corrosion or oxidation of a metal or alloy in the gas phase is most simply measured in terms of its weight gain in a given interval of time.[56] Analytical balances with a sensitivity of $\pm$ 0.1 μg are readily available and this corresponds to an oxide thickness of about 1 nm on a sample of 1 cm^2 in area. Thus gravimetric methods can be used throughout the entire thickness range. To avoid unintended thermal cycles it is usual to suspend the sample in the reaction chamber from an external balance so that a continuous readout of sample weight is available under isothermal conditions.

The results of thickness measurements, X, may be extrapolated in time, t, by use of an appropriate 'rate law'.[57] Important examples are (a) parabolic, $X^2 = k_2t$, which applied when diffusion across the corrosion product is rate controlling; (b) linear $X = k_1t$, which applies when the corrosion product offers no im-

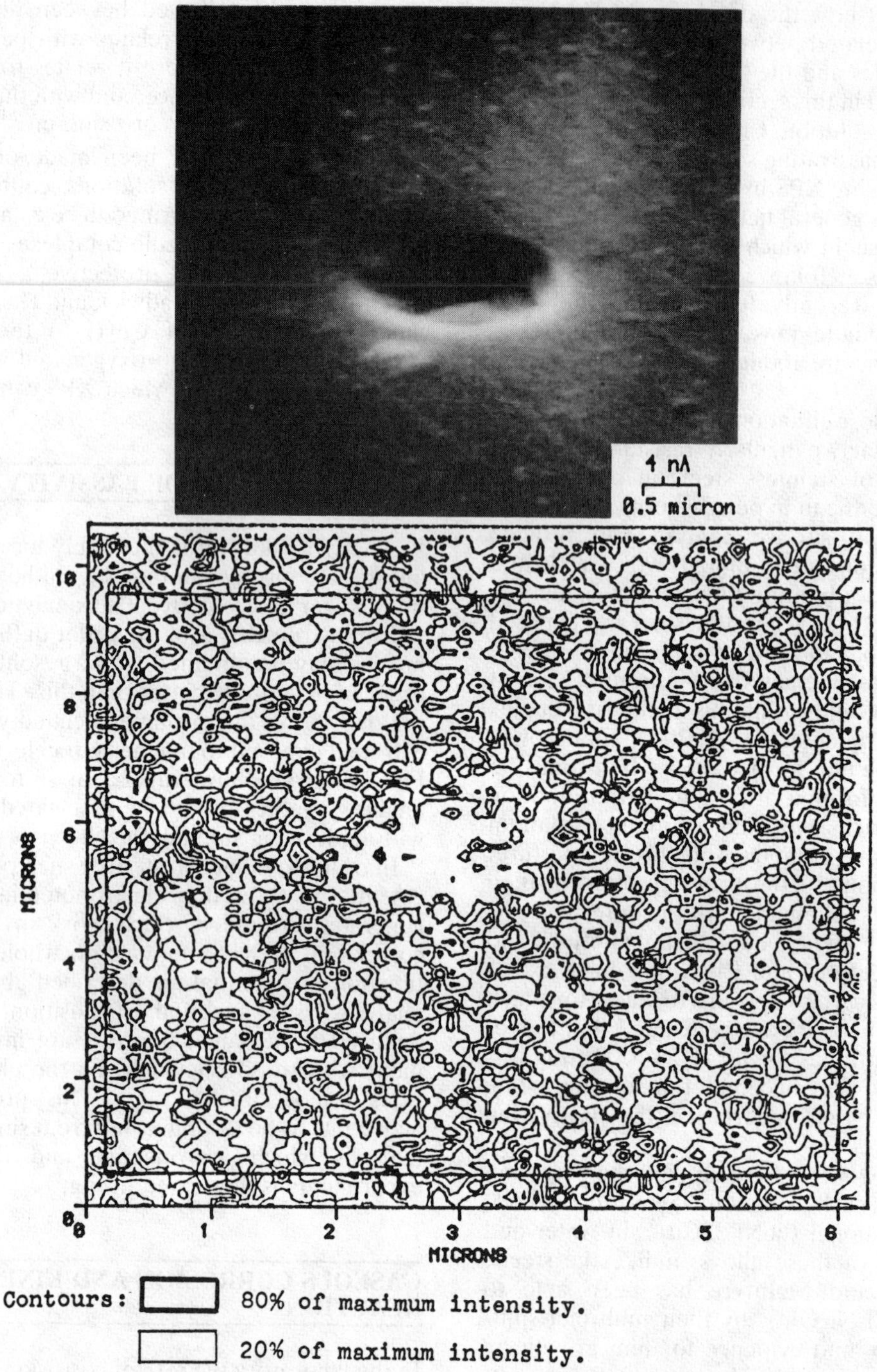

Figure 6. An Auger map of chloride ions associated with a pit on 316 stainless steel formed in $MgCl_2$ solution. Complementary maps show magnesium ions to be concentrated at the annular cathode surrounding the pit.

pedance to reaction; and (c) logarithmic $X = k_e \ln(1 + ct))$ when a passive layer of limiting thickness is reached. k_1, k_2, k_3 and c are all constant at a given temperature. The data can be interpolated and, with care, extrapolated, in temperature by the Arrenhius relationship $k = k_0 e_{(-Q/RT)}$. The nature of the activation energy, Q, may indicate the nature of the controlling transport step i.e. $\simeq 200$ kJ mole^{-1} indicates solid state diffusion. The value is reduced by short circuit diffusion along e.g. grain boundaries, or surface diffusion.

The chemical nature of the corrosion product is fixed by both thermodynamic and kinetic constraints. The composition of the outermost surface will tend towards thermodynamic equilibrium with the gas phase whilst the inner surface will be in equilibrium with elements in the alloy. It is the activity gradients between these equilibrium positions which gives rise to diffusion across the layer and hence to the segregation of distinct phases between them.[58] The relative stabilities of the oxides, sulphides or chlorides commonly of concern in gaseous corrosion are determined by their free energies of formation and may be predicted for a given temperature by the use of Ellingham diagrams,[59] see

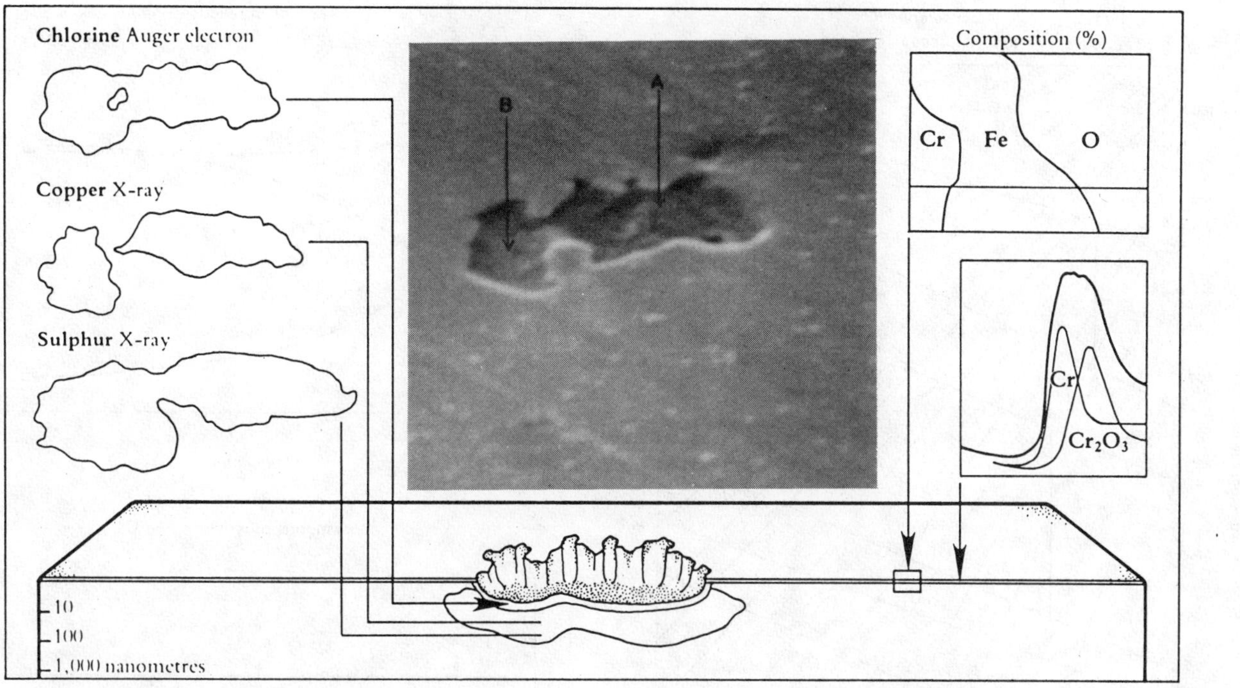

Figure 7. A montage illustrating the complementary information arising from the use of multi-technique instruments to study pitting. Here the Auger and X-ray maps show the relationship between pitting and sulphide inclusions in the steel whilst XPS and Auger depth profiling provide an understanding of the passive film. The regions A and and B on the SEM relate to the manganese distribution which is present in B but not A. The overall length of the pit is 4 μm.

for example Fig. 9. By contrast with the aqueous case, samples may be transferred readily to the electron microscope or spectrometer for identification of composition or morphology.

For films of several micrometers in thickness the X-ray electron probe microanalyser (EPMA) is widely available in the form of an energy dispersive X-ray detector (EDXA) mounted on a SEM and this is much

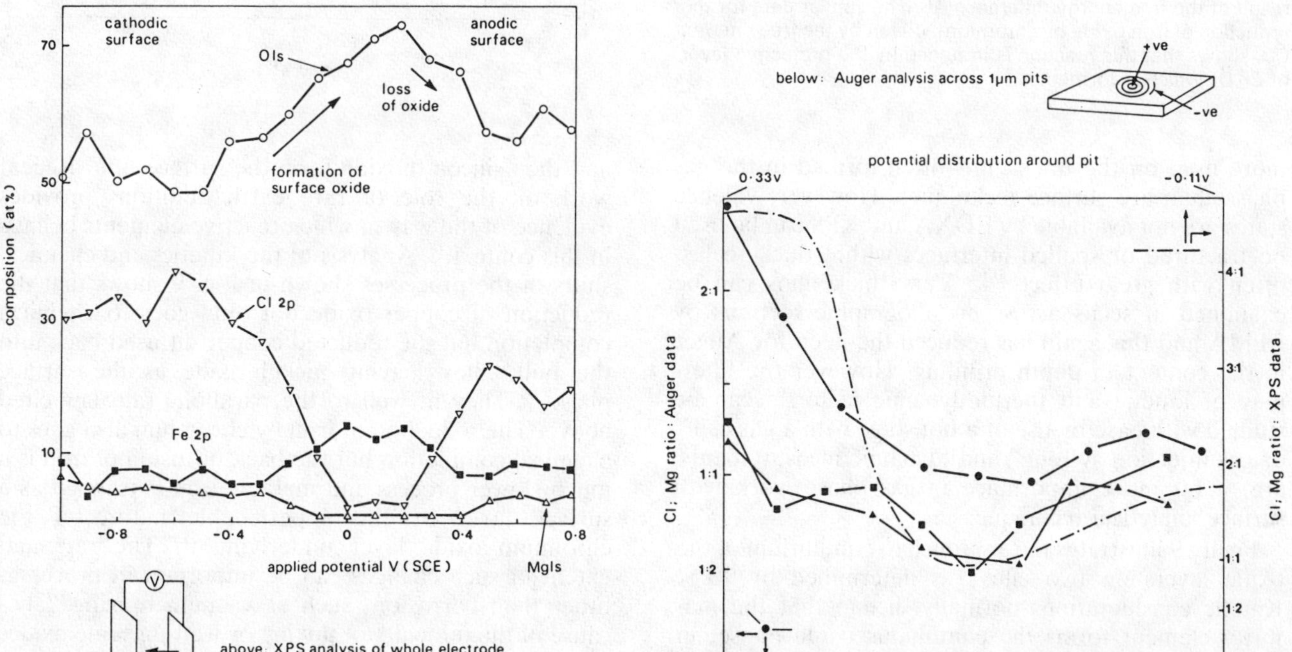

Figure 8. Experiments undertaken in order to relate the Auger maps (e.g. as in Fig. 6) to an electropotential distribution. The surface composition is determined (upper figure) as a function of electropotential by XPS analysis. Then important ratios such as Cl : Mg are transcribed to the radial position around the pit using electrostatic distribution functions to model the potential (broken line, lower figure). These models can thus be tested against the local Cl : Mg ratios observed by SAM. The data in the lower figure is for scans across three pits of 1 μm in radius and show the shape of the distribution to be correctly predicted.

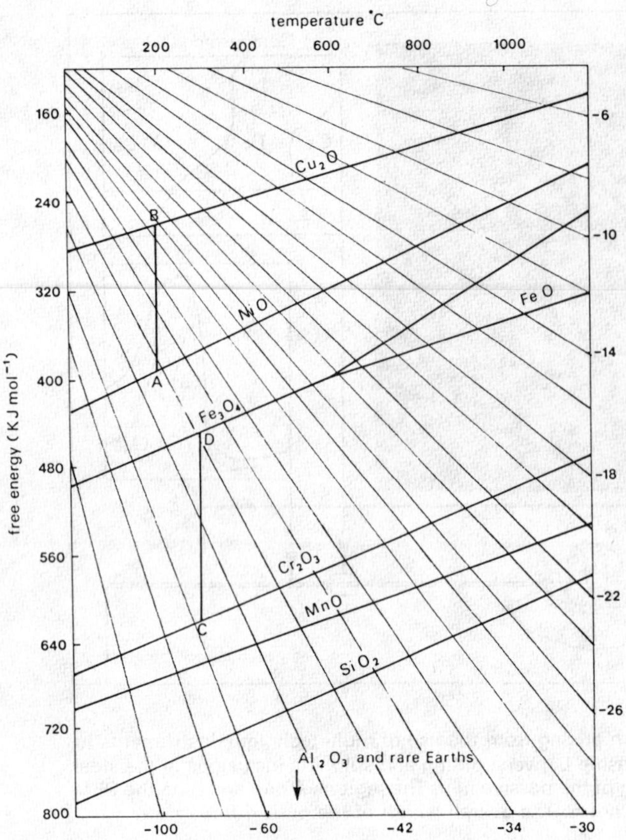

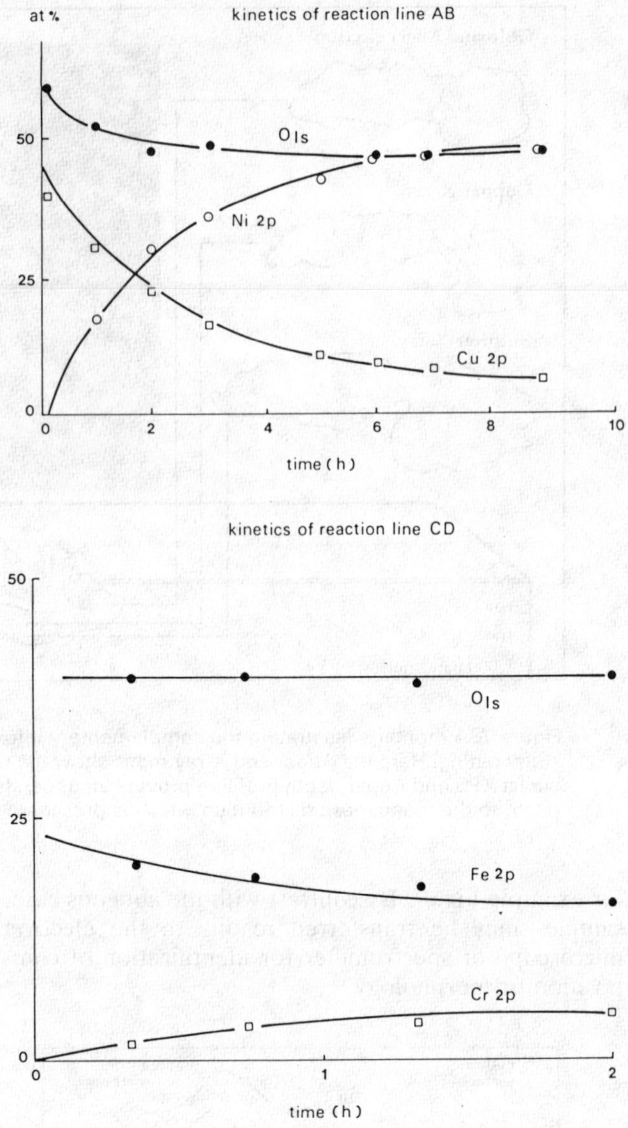

Figure 9. An Ellingham diagram for elements important to the gaseous corrosion of steels. The free energy for the decomposition of the named oxide to form one mole of oxygen is given by the bold lines, the background grid is of the decomposition pressure as a function of temperature. The XPS data shows (upper) how copper oxide is replaced by nickel oxide when a partly oxidized copper–nickel alloy is heated in vacuum, as a result of the free energy difference AB. The similar data for the reduction of iron oxide by chromium, driven by the free energy, CD, shows that this reaction is impeded by the protective layer of Cr_2O_3 which is formed.

more used on the thick films often formed in the gas phase than are surface techniques. However, valence states are not available by EDXA and XPS can be used on fractured or spalled interfaces within thick scales, often with great effect.[60,61] Very thick films can be examined in scans across metallographic sections by EPMA and this again has reduced the need for Auger in the context of depth profiling. However the interplay of kinetic and thermodynamic factors[62] can be studied with ease by use of a hot stage with a gas cell[63] or gas injection system[64] and 'dynamic measurements' are of far more importance in gas phase work with surface analytical techniques.

Figure 9 illustrates the approach to equilibrium of the oxide layers on two alloys as determined by XPS. Kinetic considerations normally dictate that the majority element forms the continuous oxide phase, in these cases, Cu_2O on Cu/Ni and Fe_3O_4 on Fe/Cr alloys. However on continuing to heat, in vacuum after prior oxidation, nickel oxide grows at the expense of Cu_2O, and chromium oxide grows at the expense of the iron oxide, both as predicted by the Ellingham diagram. On a commercial steel the process does not stop at chromium oxide and eventually manganese oxide[65] and then silicon dioxide form the surface film. Recent work on the role of rare earth additions provides evidence of the way in which reactive elements behave in this context.[66] Analysis of the kinetics and chemical shifts of the processes shown in Fig. 9 shows that the reduction of copper oxide not only goes to chemical completion but the reducted copper diffused back into the bulk alloy leaving nickel oxide as the surface phase.[62] They fit well to the parabolic rate law cited above. The reduction of iron by chromium also goes to chemical completion but the back diffusion of iron is a much slower process and metallic iron is trapped as a surface layer by the impermeable nature of the chromium oxide layer underlying it. The rate and extent of such changes can be important in processes other than corrosion, such as vacuum brazing,[67] because of the difficulty of fluxing or wetting some oxides with liquid metals.

Because of the complex interaction of thermodynamics and kinetics the sequence of oxides in the surface hierarchy is important and angular resolution of emitted electrons enables this to be determined.[68] In some cases useful information can also be obtained from the background electrons in the vicinity of the peak because the scattering effect of a surface oxide distorts the spectrum of that which underlies it. Examples can be seen in recent work from this laboratory.[69–70]

Just as analysis has enabled advances in the construction and use of Pourbaix diagrams, so it has been used to improve the accuracy of Ellingham diagrams. An example used for predicting the stability of sulphides relative to oxides is given in Fig. 10.[71] Although sulphur is readily observed to replace oxygen as a monolayer when steel is heated *in vacuo*, this work concerns the behaviour of thick films heated in atmospheres of controlled oxygen and sulphur levels. The transformation of oxide to sulphide implied that the boundary would have to be moved from that predicted by thermodynamics. The need for this was ascribed to activity changes ensuing from the incorporation of iron into the chromium sulphide.

RELATED TOPICS

There are many aspects of materials degradation, related to corrosion which have not been included in this introduction to the subject. Understanding of metallurgical embrittlement has been dramatically advanced by the use of AES to study segregation to grain boundaries[72] and the adhesion of organic polymer paints and coatings has been studies by XPS,[73] to name but two examples of related areas. Some of the guidance offered by papers cited in this review will be of equal value in this much wider context.

CONCLUSION

There are now many examples in the literature of studies in which detailed and intelligent use has been made of electron spectroscopy to observe key aspects of the corrosion process. Relatively few of those published are cited herein but they serve to illustrate the depth of study necessary to advance the very mature science of corrosion. Where properly undertaken, the work has been invaluable. However, in instrument time alone, the work will be expensive. This review will have been worthwhile if it indicates to the would-be users of our techniques the magnitude of their planned investigations and encourages them, by

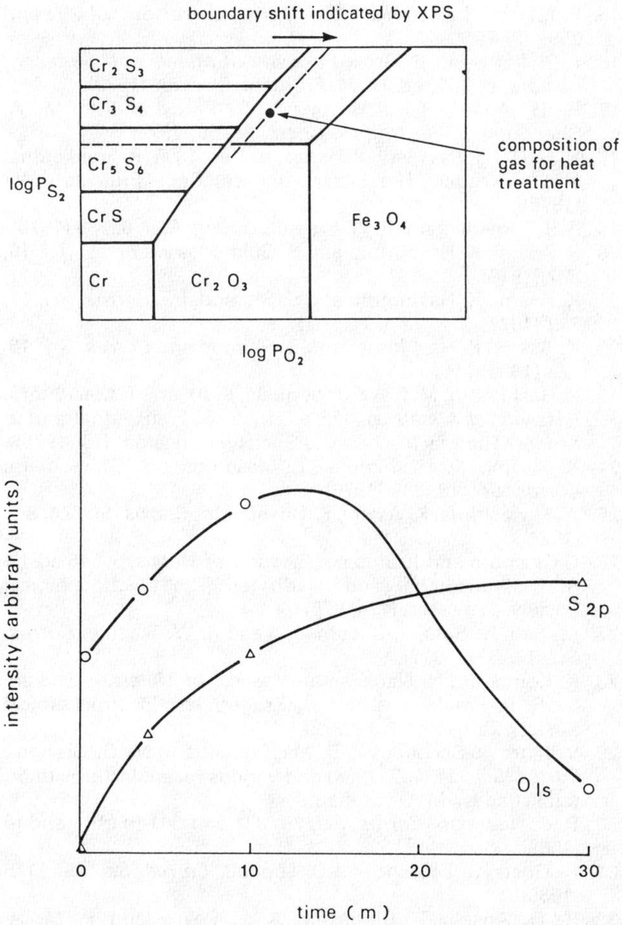

Figure 10. An activity diagram showing the competing formation of sulphide and oxides on chromium. The XPS data (lower) shows how sulphide replaced oxide as the surface anion when oxide samples were heated in the gas composition marked on the O–S diagram, implying that the boundary should be moved. Diagram reprinted with permission from Pergamon Press. Publication detail as Ref. 71.

the examples given, in the close definitions of their objectives.

Acknowledgement

It is a pleasure to thank past and present students and co-workers for their help in the study of corrosion by XPS. Thanks are due to M. S. Parvizi, R. Daud and Ke Ruoru whose unpublished data is used in Fig. 2 and 3; 6, 7 and 8; and 9 respectively. The graphics are by Sheila Rudman from sketches by J.E.C.

REFERENCES

1. J. E. Castle, *Surf. Sci.* **68,** 583–600 (1977).
2. J. E. Castle in *ASTM. STP 699*, p. 182–198 ed. by Barr and Davies, (1980).
3. S. McIntyre, *Practical Surface Analysis*, ed. by Briggs and Seah, Wiley (1983).
4. D. R. Baer and M. T. Thomas, *ACS Symp. Series (82)*, p. 251–282, *Vol. 299*, No. Ind. Appl. Surf. Anal. 1982.
5. J. Kruger and R. P. Frankenthal, *Techniques of Metals Research, IV (2)*, p. 571–667, ed. by R. A. Rapp, Wiley (1970).
6. F. L. Laque and L. L. Shreir, *Corrosion*, p. 20.1–20.108, ed. by L. L. Shreir, Butterworths (1976).
7. M. Pourbaix, *Atlas of Electrochemical Equilibria in Aqueous Solutions*, NACE Houston, Texas (1974).
8. W. Yang, R. Ni, H. Hua and A. Pourbaix, *Corros. Sci.* **24,** 691–709 (1984).
9. D. Verink Jr., M. Pourbaix, *Corrosion — NACE*, **37,** 495 (1971).
10. D. Effird, *Corrosion — NACE*, **31,** 77 (1975).
11. J. E. Castle and D. C. Epler, *Surf. Sci.* **53,** 286 (1976).
12. J. E. Cstle and M. S. Parvizi, *Proc. Sem. Copper Alloy in Marine Environments*, Copper Development Association, Lond., paper 10, (1985).
13. D. D. Macdonald, B. C. Syrett and S. S. Wing, *Corrosion,* **34,** 289 (1978) and **35,** 377 (1979).
14. C. Kato, B. G. Ateya, J. E. Castle, H. W. Pickering, *J. Electrochem. Soc.* **127,** 1980 and 1897 (1980).

15. P. Marcus, I. Olefjord and J. Oudar, *Corros.Sci.* **24** 269 and 259 (1984).
16. I. Olefjord and B. Brox, *Passivity of Metals and Semiconductors*, p. 561, ed. by M. Froment, Elsevier (1983).
17. R. O. Ansell, T. Dickinson, A. R. Porey and R. M. A. Sherwood, *J. Electroanal. Chem.* **98,** 69 (1979).
18. H. H. Uhlig, *Passivity of Metals*, p. 1 ed. by R. P. Frankenthal and J. Kruger, The Electrochemical Soc., Princeton, NJ (1978).
19. J. B. Lumsden and R. W. Staehle, *Scripta Met.* **6,** 286 (1975).
20. K. Asami, K. Hashimoto and S. Shimodaira. *Corros. Sci.* **16,** 387 (1976).
21. K. Asami, K. Hashimoto and S. Shimodaira. *Corros. Sci.* **17,** 713 (1977).
22. K. Asami, K. Hashimoto and S. Shimodaira. *Corros. Sci.* **18,** 125 (1978).
23. K. Hashimoto, M. Naka, J. Noguchi, K. Asami, T. Masumoto, *Passivity of Metals*, p. 156 ed. by R. P. Frankenthal and J. Kruger, The Electrochemical Society, Princeton, NJ (1978).
24. K. Asami, K. Hashimoto, T. Masumoto, S. Shimodaira, *Corros. Sci.* **16,** 909 (1976).
25. A. Kawashima, K. Asami, K. Hashimoto, *Corros. Sci.* **24,** 807 (1984).
26. G Okamoto and T. Shibata, *Passivity of Metals*, p. 646 ed. by R. P. Frankenthal and J. Kruger, The Electrochemical Society, Princeton NJ (1978).
27. M. Seo, N. Sato, J. B. Lumsden and R. W. Staehle, *Corros. Sci,* **17,** 209 (1977).
28. K. Komo and N. Nagayama, *Passivity of Metals*, p. 583 ed. by R. P. Frankenthal and J. Kruger, The Electrochemical Society, Princeton NJ (1978).
29. V. Mitrovic-Scepanovic, B. MacDonald and M. G. Graham, *Corros. Sci.,* **24,** 463 (1984) and various recent lectures to be published by M. G. Graham.
30. R. C. Newman, *Corros. Sci.* **25,** 331 and 341 (parts I and II) (1985).
31. R. Goetz, J. Laurent and D. Laudolt. *Corros. Sci.* **25,** 1115 (1985).
32. R. O. Ansell, T. Dickinson, A. F. Povey and P. M. A. Sherwood. *J. Electro. Anal. Chem.* **98,** 79 (1979).
33. J. N. Wanklyn, *Corros. Sci.* **21,** 211 (1981).
34. T. Dickinson, A. F. Povey, P. M. A. Sherwood. *JCS, Faraday 1* **72,** 626 (1976).
35. T. Dickinson, A. F. Povey, P. M. A. Sherwood. *JCS, Faraday 1* **73,** 327 (1977).
36. R. O. Ansell, T. Dickinson, A. F. Povey, P. M. A. Sherwood, *J. Electrochem. Soc.* **124,** 1360 (1977).
37. R. O. Ansell, T. Dickinson, A. F. Povey, P. M. A. Sherwood *J. Electroanalytical Chem.* **876,** 189 (1978).
38. G. Hultquist and H. Hero, *Corros. Sci.* **24,** 789 (1984).
39. F. Mansfield, J. V. Kenkel and A. B. Harker, *Passivity of Metals*, p. 752 ed. by R. P. Frankenthal and J. Kruger, The Electrochemical Society, Princeton, NJ (1978).
40. W. Meisel, E. Mohs, H-J. Guttman and P. Gutlich, *Corros. Sci.* **23,** 473 (1983).
41. C. Leygraf, G. Hultquist, O. Olefjord, B. O. Elfstrom, V. M. Knajasheva, A. V. Plaskeyer, Y. M. Kolotrykin, *Corros Sci.* **19,** 343 (1979).
42. A. Kh. Bairamow, S. Zakipour and C. Leygraf, *Corros. Sci.* **25,** 69 (1985).
43. J. E. Castle, C. R. Clayton, *The Passivity of Metals*, p. 714 ed. by R. P. Frankenthal and J. Kruger, The Electrochemical Society, Princeton, NJ (1978).
44. S. A. Richardson, J. E. Castle, N. R. Large and A. J. Tench, *BNES, Water Chemistry* **3,** 53 (1983).
45. I. Olefjord and B. O. Elfstrom, *Proc. Eurocorr '77*, p. 21, Society of Chemical Industry (1977).
46. N. S. McIntyre, D. G. Owen and D. Zetasuk, *J. Electrochem. Soc.* **126,** 750 (1979).
47. N. S. McIntyre, T. E. Rummery, M. G. Cook and D. Owen, *J. Electrochem. Soc.* **123,** 1164 (1976).
48. W. H. Hocking, F. W. Stanchell, E. MacAlpine and D. H. Lister, *Corros. Sci.* **25,** 531 (1985).
49. D. C. Epler and J. E. Castle, *Corros.* **35,** 451 (1979).
50. H. H. Uhlig, *Corrosion Handbook*. John Wiley & Sons, N.Y. (1948).
51. J. E. Castle, *J. Vac. Sci. Tech. A.* **1,** 1013 (1983).
52. R. F. Roberts, *J. Electron Spec. Rel. Phenomena* **4,** 273 (1974).
53. D. Chadwick and T. Hashami, *Corros. Sci.* **18,** 39 (1978).
54. R. Daud, PhD Thesis, University of Surrey, 1985.
55. P. H. Melville, *Corros. Sci.* **20,** 633, (1980).
56. S. J. Gregg, *Vacuum Microbalance Techniques,* vol. 7, p. 107, ed. by C. H. Masson and H. J. Van Becham Plenum Press.
57. N. BIrks and G. H. Meier, *Introduction to High Temp. Oxidation of Metals*, Edward Arnold (1983).
58. See e.g. B. D. Barlow and G. R. Wood and D. P. Whittle, *Corros. Sci.* **25,** 253 (1985).
59. I. Barrie, O. Knacke and O. Kubaschewski, *Thermochemical Properties of Inorganic Substances*, Springer Verlag (1977).
60. G. C. Allen, R. K. Wild, *Nature* **307,** 43 (1984).
61. G. C. Allen, R. K. Wild, *Phil. Mag.* **52 A8,** 373 (1985).
62. J. E. Castle and M. Nasserian-Riabi, *Corros. Sci.* **15,** 537 (1975).
63. C-T. Au, S. Sing-Boparai, M. W. Roberts, R. W. Joyner, *J. Chem. Soc. Faraday Trans. 1,* **79,** 1779 (1983).
64. J. E. Castle and M. R. Hunt, *Corros. Sci.* **16,** 137 (1976).
65. J. E. Castle and M. J. Durbin, *Carbon*, **13,** 23 (1975).
66. M. M. El Gomati, C. Walker, D. C. Peacock and M. Prutton, *Corros. Sci.* **25,** 351 (1985).
67. J. M. Cohen, J. E. Castle and M. B. Waldron, *Met. Sci.* **15,** 455 (1981).
68. R. W. Paynter, *Surf. Interface Anal.* **3,** 186 (1981).
69. J. E. Castle, I. Abu-Talib, S. A. Richardson, *Applied Materials Characterization*, ed. by W. Katz, M.R.S. Pittsburgh, USA (1985).
70. S. A. Richardson, I. Abu-Talib and J. E. Castle, *Surf. Interface Anal.* to be published.
71. T. T. Huang, B. Peterson, D. A. Shores and E. Pfender, *Corros. Sci.* **24,** 167 (1984).
72. M. P. Seah, *Vacuum* **34,** 463 (1984).
73. J. E. Castle and J. F. Watts, *I and EC, Produce. R&D,* **24,** 361 (1985).

SURFACE AND INTERFACE ANALYSIS, VOL. 9, 357–365 (1986)

XPS and ISS Examinations of Electrode Surfaces and Passive Layers with a Specimen Transfer in a Closed System

S. Haupt, C. Calinski, U. Collisi, H. W. Hoppe, H.-D. Speckmann and H.-H. Strehblow†
Institut für Physikalische Chemie, Universität Düsseldorf, Universitätsstraße 1, D–4000 Düsseldorf, Federal Republic of Germany.

The application of UHV surface analytical methods to the study of electrodes and surface layers may be affected by changes introduced by contact with the laboratory atmosphere. Therefore, a specimen preparation and transfer in a closed system with its specifications is described. Three examples of its application are discussed: the examination of the electrode – electrolyte interface on Cu and the study of the formation of passive anodic oxides on Fe and on Fe–Cr alloys in 1 M NaOH.

INTRODUCTION

The application of surface analysis to the study of the composition and structure of electrode surfaces is a necessary requirement to achieve a better understanding of electrode processes and corrosion phenomena. Different *in situ* and *ex situ* methods have been applied, especially those working in ultra high vacuum (UHV), such as electron- and ion-spectroscopies for their detailed information concerning the elementary and chemical composition of the surface and the surface near regions of solid materials. However, they require the transfer of a specimen from the electrolyte to the UHV with the risk of possible changes due to loss of potential control, loss of water from the surface and possible air contact. Especially the access of air to the surface might cause oxidation processes which are most critical for specimen preparations at negative electrode potentials and short oxidation times. A specimen transfer in a closed system will help to prevent these changes after its preparation in a protecting clean noble gas atmosphere. The advantages of the use of a closed system for our work are as follows: (a) no oxide formation on semi-noble metals and no oxide growth on reactive metals; (b) no oxidation of components at a surface or of a surface layer; (c) no general contamination by the laboratory atmosphere; (d) the possibility of a well defined specimen pretreatment, as for example sputter cleaning, and a start with an oxide and contamination free surface.

Possible applications are: (a) a more reliable examination of electrosorbed species or even of the double layer under special conditions; (b) the examination of surface layers in the mono- and submono-layer range; (c) examination of chemically unstable surfaces.

In the following section the system is described with its specifications. Three examples are given for its application: the study of the electrode–electrolyte interface of Cu in weak acidic and alkaline solutions and the formation of passivating anodic oxides on Fe and Fe–Cr alloys in 1M NaOH.

† Author to whom correspondence should be addressed.

0142–2421/86/120357–09 $05.00

EXPERIMENTS

Different groups have worked on closed transfer systems and have published their results in the past.[1–4] We used for our work the ESCALAB 5 (VG Instruments). The rack and pinion drives with wobblesticks provides a high flexibility, which is an efficient system for transferring samples to an electrolyte vessel and allows a parallel preparation transfer and analysis of several specimen (Fig. 1). The system contains an analyser and preparation chamber and a fast entry lock for the specimen to be introduced from the laboratory. A fourth chamber is flanged to the fast entry lock containing a small electrolyte vessel for the electrochemical preparation. A rotary feed-through may turn the specimen upside down and its circular front plane is brought into contact with the surface of the electrolyte which is adjusted by the filling level of the 2 cm^3 glass vessel from outside. Argon is purified by an oxisorb system (Messer Griesheim). It serves as a protecting atmosphere within the chamber and is used for the degassing of electrolytes and water in 1 litre glass containers. Before introduction into the system ca. 100 cm^3 of the electrolytes are degassed again in a smaller glass container. Any connections for solutions are made of glass tubes and metal–glass feed–throughs flanged to the system. Gases are introduced by copper or stainless steel tubes with swadge lock fittings. The system was routinely controlled by the examination of sputter cleaned metal surfaces. A semi-noble metal like Cu did not show any oxide formation (< 5% of a monolayer) as checked by its reintroduction and

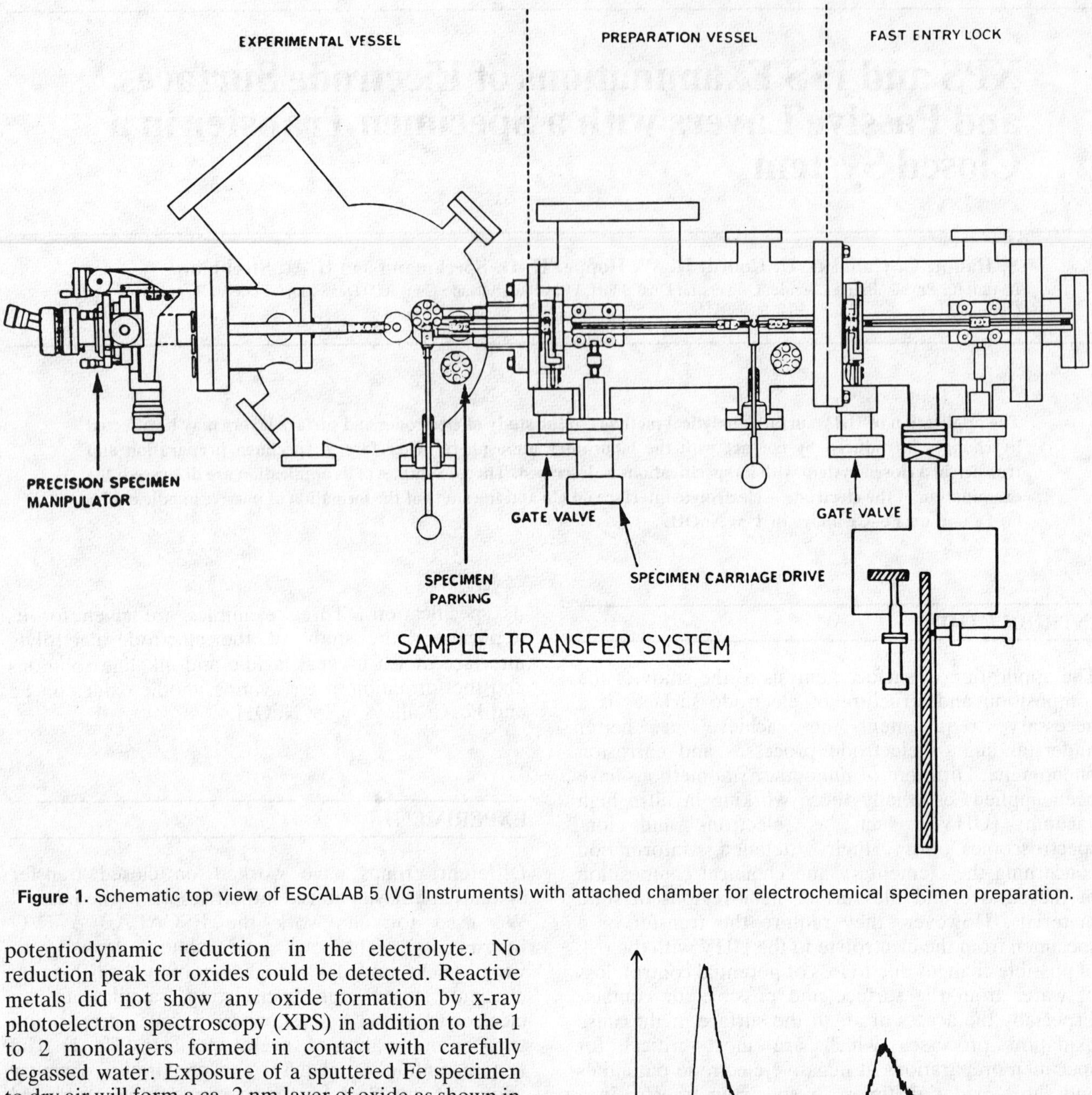

Figure 1. Schematic top view of ESCALAB 5 (VG Instruments) with attached chamber for electrochemical specimen preparation.

potentiodynamic reduction in the electrolyte. No reduction peak for oxides could be detected. Reactive metals did not show any oxide formation by x-ray photoelectron spectroscopy (XPS) in addition to the 1 to 2 monolayers formed in contact with carefully degassed water. Exposure of a sputtered Fe specimen to dry air will form a ca. 2 nm layer of oxide as shown in Fig. 2 by the decrease of the XPS Fe° peak and the increase of the oxide signal. In the following paragraphs the application of the equipment will be described in relation to the examination of metal surfaces and the formation of oxide layers. In most cases the surfaces were argon sputtered to remove any oxides and contamination prior to the electrochemical preparation. For the examination of the copper–electrolyte interface the specimens were only electropolished to preserve their water repelling property, which is needed for this kind of study. For the electrochemical preparation a potentiostat was used with the necessary isolated feed-through to the electrochemical chamber. The working electrode was grounded and the reference electrode Hg/HgO/1M NaOH $E = 0.14$ V and Hg/Hg_2SO_4/1M H_2SO_4 $E = 0.68$ V were used as reference electrodes. All electrode potentials are given relative to the standard hydrogen electrode (SHE) and are corrected for liquid junction potentials if necessary.

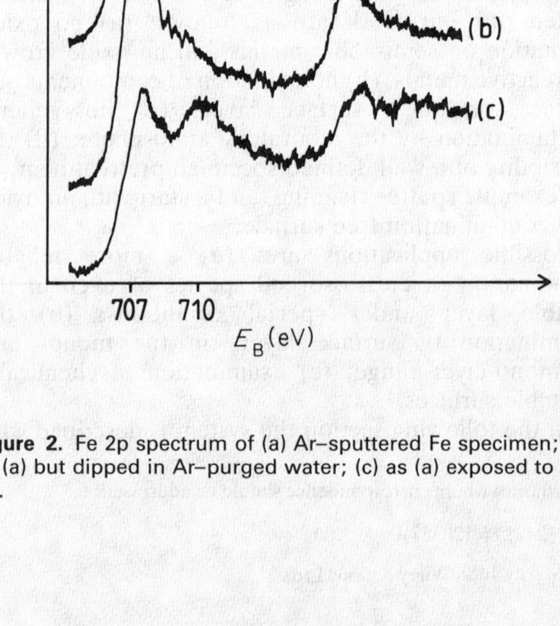

Figure 2. Fe 2p spectrum of (a) Ar–sputtered Fe specimen; (b) as (a) but dipped in Ar–purged water; (c) as (a) exposed to dry air.

RESULTS AND DISCUSSION

The Cu–electrolyte interface

The structure and composition of the electrode–electrolyte interface is of fundamental interest to electrochemistry. If it is possible to transfer a sample from the electrolyte to the UHV without change of this interface, its quantitative study with XPS and other UHV methods can be achieved. For this purpose a hydrophobic surface is required so that no substantial amount of electrolyte may be drawn out of the solution, which would lead to an examination of salt films of the vacuum dried specimen. Rinsing with water is not possible because this might remove the species within the double layer. Hansen and Kolb have shown by XPS studies and other methods [5–7] that the electrolyte strips off at the Helmholz layer on gold electrodes and that the double layer is preserved. These studies could be performed with an air contact of the specimen because of the noble nature of the metal. Studies on copper, however, require a closed system because of oxide formation by air. Electropolished Cu specimen may be extracted from the electrolyte with a hydrophobic surface and introduced into the UHV within ca. 3 mins. The amount of cations and anions (Na^+ and ClO_4^-) was determined by a quantitative evaluation of the Na ls–(E_B = 1072 eV) and the Cl 2p–peak (E_B = 200 eV) including the photoionisation cross-sections of Scofield.[8] The calibration was related to the electrochemical reduction charges of passive thin Cu_2O layers and the area of the O 1s oxides signals (E_B = 530 eV) as described in detail elsewhere.[9,10] The amount of Na^+ ions decreases linearly with increasing electrode potential which has been applied during the extraction of the specimen from a pH 5.0 solution.

Figure 3 shows the difference $n_{ClO_4^-}$ at the interface. It represents the decrease of the positive excess charge within the electrolyte with increasing electrode potential (10^{-9} Mol cm^{-2} equals $F \cdot \Delta n$ = 10^{-4}C cm^{-2}). This linear decrease corresponds to a constant differential electrode capacity found for this potential range. The excess charge is negligible at the potential of zero charge (pzc). The pzc found by Lecoeur and Bellier for Cu(111) and (100) single crystal surfaces is in the vicinity of the value of this study on polycrystalline material (Fig. 3).[11] In more alkaline solution the values follow the same line. However, the formation of anodic oxide at sufficiently positive potentials leads to an increase of Na^+ which is adsorbed at the oxide surface. OH^- adsorption at 100 to 200 mV more negative to the potential of oxide formation causes already an increase of the amount of Na^+ by its co-adsorption with OH^-.

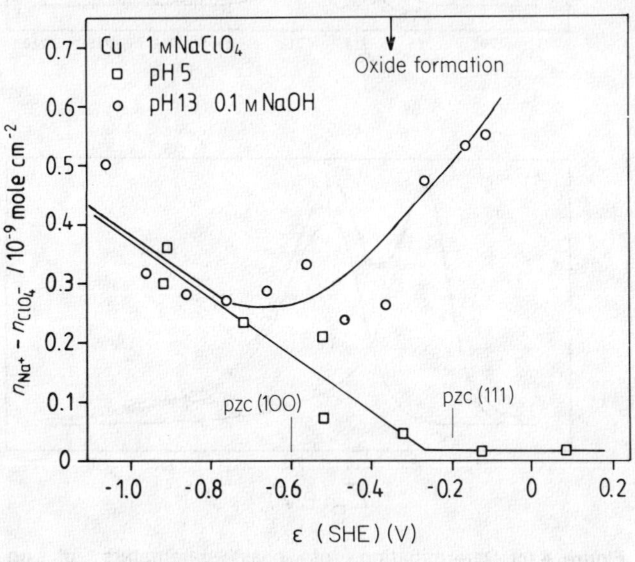

Figure 3. Amount of excess charge $n_{Na^+} - nClO_4^-$ of hydrophobic electropolished Cu electrodes drawn from 1M $NaClO_4$ solution at different potentials pH 5, pH 13 (+0.1 M NaOH). See pzc of (100) and (111) according to ref. 11.

Passive Layers on Fe in Alkaline Solution

The electrochemical preparation of an Fe specimen in a noble gas atmosphere and its following transfer to the

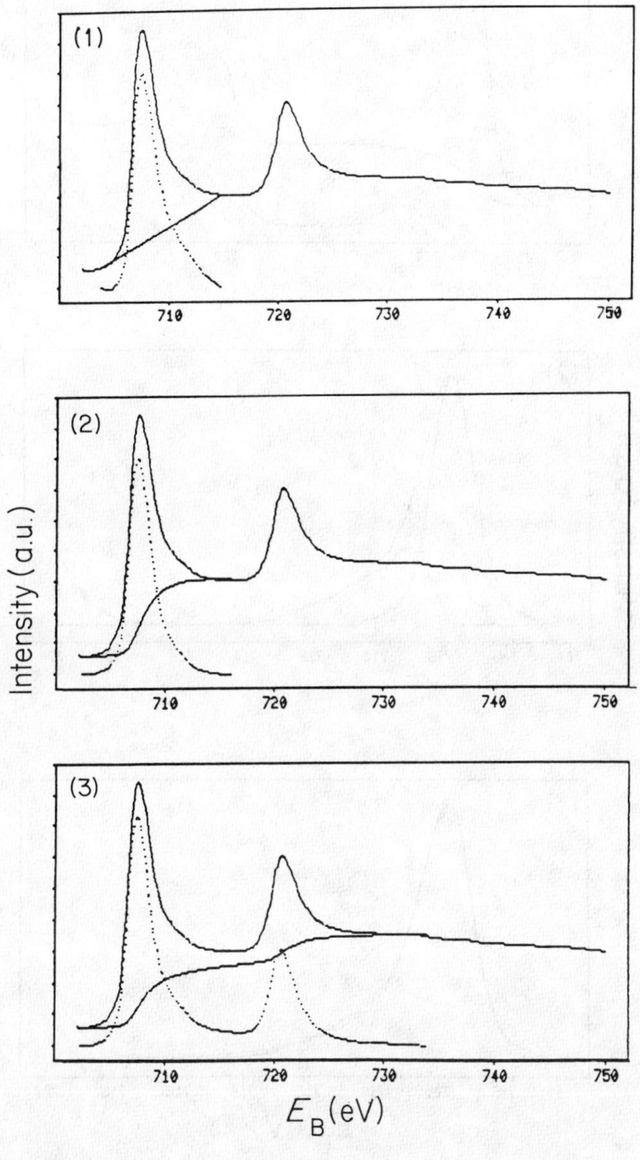

Figure 4.(a) Fe 2p–background subtraction of sputtered Fe–specimens: (1) linear; (2) according to Shirley and Bishop[12,13] with a fit to the Fe $2p_{3/2}$ signal; (3) as 2 with a fit to the Fe$2p_{1/2}$ signal.

UHV allows the study of the passive layer formed at negative potentials and for short oxidation times in the ms range. Fe^{2+} of the film is not oxidised and the passive oxide does not grow after preparation. The samples were sputter cleaned (Ar^+, 5 kV, 40 μA, 10 min) and potentiostatically oxidised in 1 M NaOH for predetermined times.

In these studies we used 1M NaOH to prevent the dissolution of thin oxide films formed at negative potentials and for short times. Strongly acidic electrolytes do not show passive behaviour for these conditions, especially below the Flade potential. After oxide formation the Fe specimens were rinsed four times with three times distilled water by replacing the electrolyte with degassed water within the electrochemical vessel. Short oxidation times were determined by switching the counter electrode on and off via a reed relay. Open circuit conditions do not cause oxide growth over 1 to 2 monolayers as described previously (Fig. 2).

The quantitative evaluation of the amounts of Fe^{2+} and Fe^{3+} within the oxide require the help of a computer (DEC, PDP11). Background corrections were performed according to Shirley and Bishop,[12,13] with a close fit to the Fe $2p_{3/2}$ signal at E_B = 707 eV (Fig. 4(a)). A fit to the Fe $2p_{1/2}$ signal at E_B = 720 eV and a linear background did not cause substantial changes in this case. In order to obtain the Fe°, Fe^{2+} and Fe^{3+} contribution to the Fe $2p_{3/2}$ signal for the peak deconvolution only the peak heights of standard spectra of the three species are varied with fixed peak positions and peak shapes. For the standard signals a 50% Gauss–Lorentz fitting with a tailing function for the asymmetric shape was used (Fig. 4(b), (c)).[14,15] For Fe° a sputtered iron specimen was taken. The Fe^{3+} standard was a passive layer formed on iron at E =

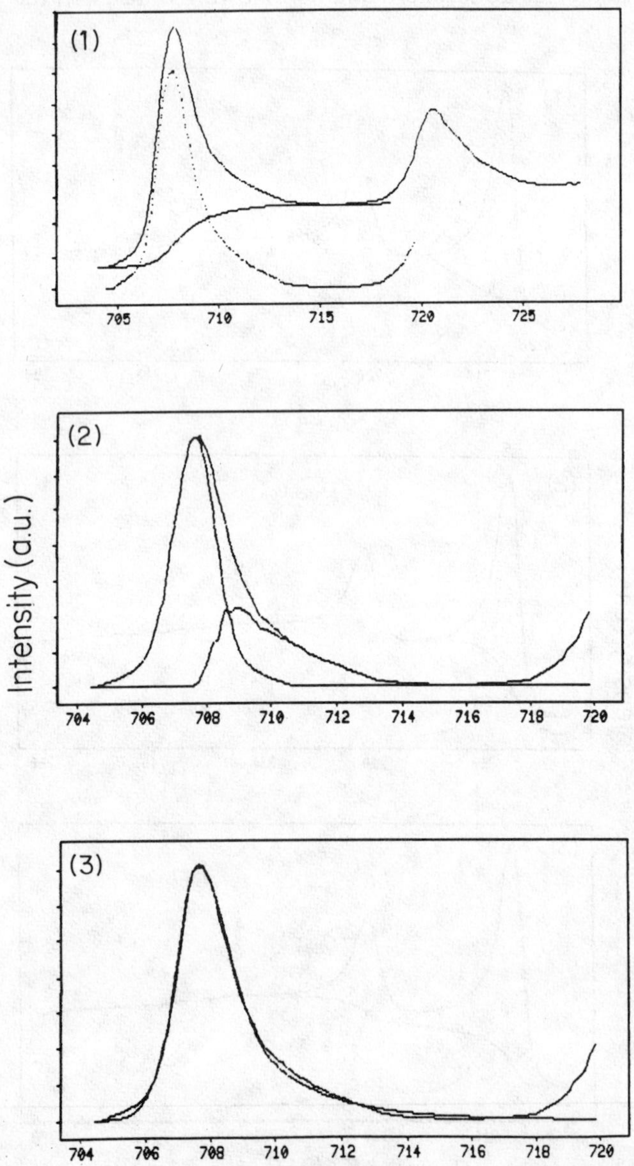

Figure 4.(b) Determination of peak parameters of sputtered Fe (Fe° standard): (1) background subtraction; (2) Gauss–Lorentz fit with tail; (3) close fit of measured and fitted signal.

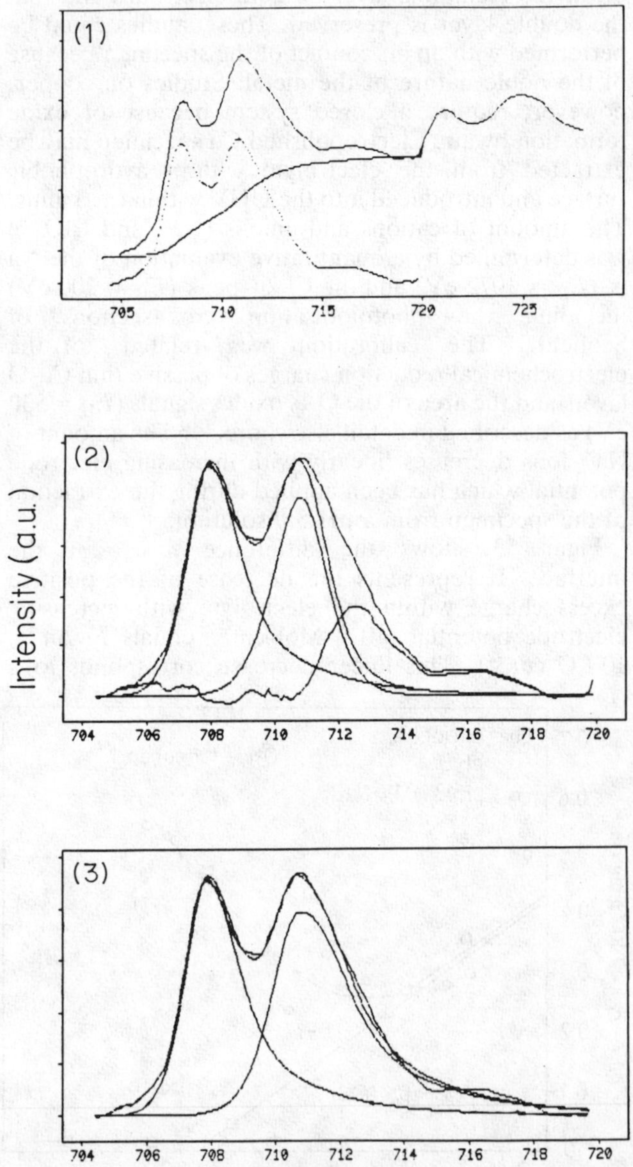

Figure 4.(c) Determination of peak parameters of an FeO, $Fe(OH)_2$ standard, 5 min oxidation of Fe5Cr at E = 0.14 V and subsequent 5 min reduction at E = −0.96 V using the parameter set of the Fe° standard of Fig. 4(b) 1–3 as Fig. 4(b).

0.340 V where no contribution of Fe^{2+} is expected for the positive potentials. For Fe^{2+} a passive layer is formed on a Fe 5% Cr alloy at $E = 0.340$ V for 5 min. and reduced to FeO, $Fe(OH)_2$ at $E = -1.00$ V for 5 min.

Each of the standards provides a set of parameters h, x_o, w, g, a, b, c of (1) which is used for the peak deconvolution of the XPS spectra of the specimen of interest.

$$f(x) = \left\{ \frac{1}{[1 + g(x-x_0)^2/w^2]^* \exp[(1-g)(x-x_0)^2/w^2]} + a^* \exp(b(x-x_o))^* [1-\exp(-c(x-x_o))^2/w^2]^* \delta \right\}^* h$$

$\delta = 1$ for $x > x_0$

$\delta = 0$ for $x < x_0$

h = peak height, x_0 = peak position, w = peak width, g = Gauss/Lorentz percentage, and a, b, c, = tail constants.

The deconvolution involves a linear least squares fitting,[16] whereby the peaks were subdivided into a symmetrical part and a tail, which includes also the 'shake up peaks' of Fe^{2+} and Fe^{3+} (Fig. 4(b), (c)). For our evaluations we used three different methods of background subtraction (Fig. 5), different sets of parameters, and also symmetrical Gauss–Lorentz functions. It can be shown that the main results, as for example linear dependence of the oxide thickness on the potential and the logarithm of time, or an increase and following decrease of the Fe^{2+}/Fe^{3+} ratio with potential and time, do not show a substantial dependence on the method of background subtraction and peak fitting (Fig. 5), and show a close agreement with the electrochemical results known from the application of other methods such as ellipsometry.[17]

After deconvolution the integrated peak areas permit the calculation of oxide thicknesses. For this purpose a simple two layer oxide structure with an inner FeO,$Fe(OH)_2$ and an outer Fe_2O_3 are assumed (Fig. 6). At sufficiently negative potentials the outer Fe_2O_3 should be absent. To compensate for the sensitivity of the electron detection and transmission of the energy analyser and the photoionisaton cross-section, intensity ratios of Fe^{2+} or Fe^{3+} and Fe^0 are

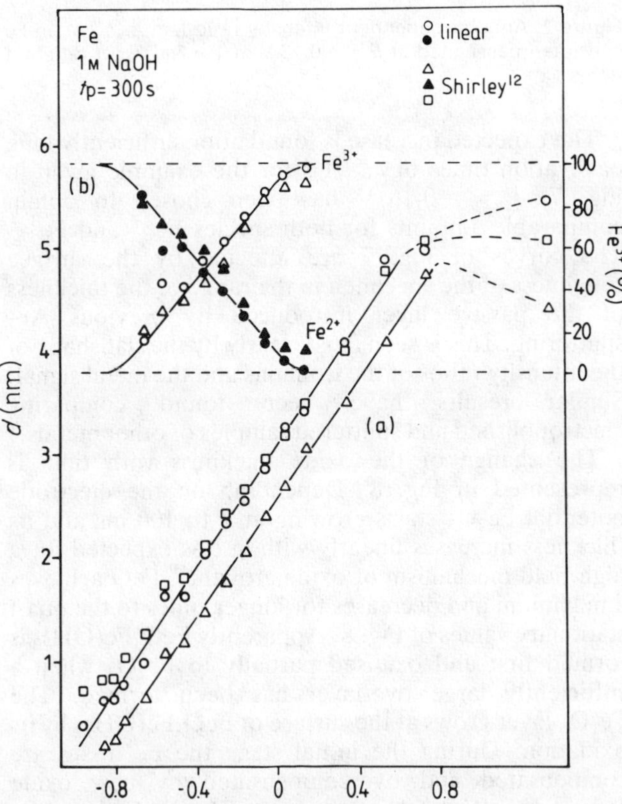

Figure 5. (a) Oxide thickness deduced from Fe $2p_{3/2}$ signal described in the third section for the three methods of background subtraction of Fig 3(a), ○ linear fit, △ fit according to Shirley and Bishop[12,13] at Fe $2p_{3/2}$, □ at Fe $2p_{1/2}$; (b) average composition of the oxide layer deduced from the Fe^{3+} and Fe^{2+} intensities, ○ linear, □ Shirley to Fe $2p_{3/2}$.

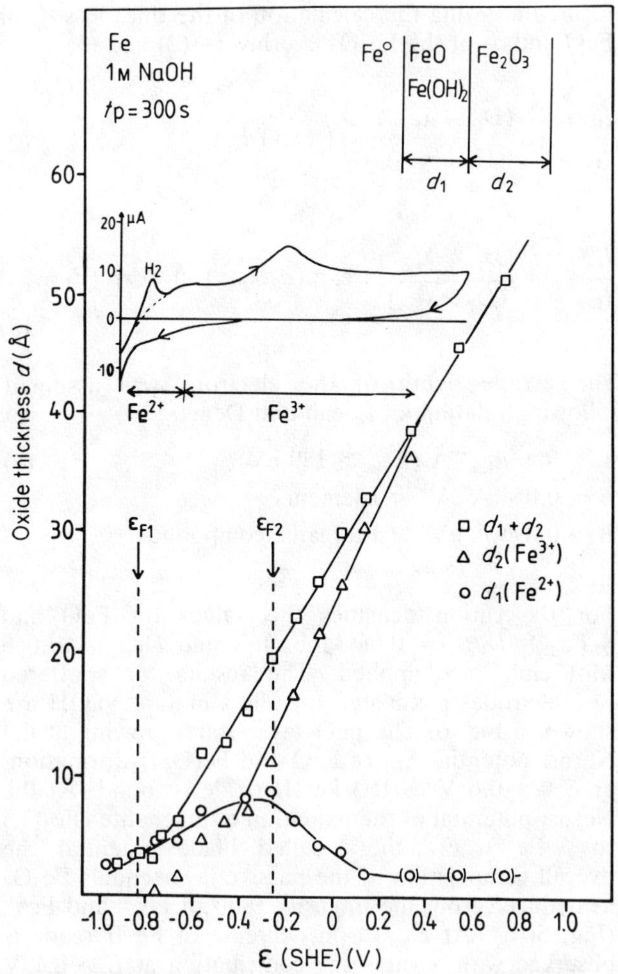

Figure 6. Oxide thickness d_1(FeO) and d_2(Fe_2O_3) in dependence on the oxidation potential, passivation time t_p = 300 s in 1 M NaOH; E_{F1} = potential of oxide formation (Fe_3O_4); E_{F2} = Flade potential (FE_3O_4 Fe_2O_3); insertion: potentiodynamic polarisation curve of reduced Fe and potential domains of formation of soluble Fe^{2+} and Fe^{3+}.

formed allowing the calculation of the thickness d_1 of FeO and d_2 of the Fe_2O_3 overlayer—(2) and (3)

$$\frac{I_{Fe^{2+}}}{I_{Fe^{3+}}} = \frac{(D_{Fe^{2+}}\ \lambda_{Fe^{ox}})}{(D_{Fe^{o}}\ \lambda_{Fe^{o}})} \{ \exp (d_1/\lambda_{Fe^{ox}}) - 1 \} \tag{2}$$

$$\frac{I_{Fe^{3+}}}{I_{Fe^{o}}} = \frac{(D_{Fe^{3+}}\ \lambda_{Fe^{ox}})}{(D_{Fe^{o}}\ \lambda_{Fe^{o}})} \{ \exp (d_2/\lambda_{Fe^{ox}})-1 \} \exp (d_1/\lambda_{Fe^{ox}}) \tag{3}$$

The escape depth of the electrons was assumed following relation 4 of Seah and Dench.[18]

$$\lambda = B\ \sqrt{E_{kin}}\ /\text{Å}\ (E_{kin} > 150\ \text{eV}) \tag{4}$$

$B = 0.054$ Å $eV^{-1/2}$ Elements

$B = 0.096$ Å $eV^{-1/2}$ Inorganic compounds

For the cation densities the values for FeO and γ-Fe_2O_3 $D_{Fe^{2+}} = 0.06$ Mol cm^{-3} and $D_{Fe^{3+}} = 0.066$ Mol cm^{-3} are applied. The results for sputtered Fe–electrodes passivated for 300 s in 1 M NaOH are shown in Fig. 6. The FeO–layer starts growing at the Nernst potential E_{F1} of FeO and $Fe(OH)_2$ formation, at $E > -0.6$ V (SHE) Fe III oxide is found. At the Nernst potential of the oxidation of magnetite (Fe_3O_4) to $\gamma-Fe_2O_3$ E_{F2}, the so-called 'Flade' potential, the overall composition of the passive layer equals Fe_3O_4 as deduced from the intensity ratio of Fe^{3+} and Fe^{2+} (Fig. 5(b)). At $E > E_{F2}$ a decrease of Fe II oxide is observed with a vanishing contribution at $E > 0.1$ V (Fig. 6). Its presence is postulated for thermodynamic reasons but the XPS results are not conclusive, when a large Fe_2O_3 overlayer weakens the Fe^{2+} contribution to the oxide signal. Fe^{3+} is increasing linearly with E demonstrating a high field mechanism of oxide growth with a barrier type film of $\gamma-Fe_2O_3$. Some support by electrochemical methods is indicated in Fig. 6. In the anodic part of the potentiodynamic polarisation curve a peak at E_{F2} indicates the oxidation of Fe^{2+} to Fe^{3+}. Further information may be obtained with a rotating Pt-split-ring-Fe-disc electrode. Hydrodynamic modulation is increasing the sensitivity of the method for the very low corrosion rates for pH 13.[9] For $E<-0.6$ V Fe^{2+} is found as a soluble corrosion product and for $E>-0.6$ V Fe^{3+}. This relatively sharp change within about 100 mV suggests that a thin continuous Fe_2O_3 layer is formed on top of the FeO,$Fe(OH)_2$ film. The composition of the surface of the passive layer determines the nature of the dissolving cations.

Besides these electrochemical studies XPS may give further support for the two layer model of Fig. 6. The increase of the take off angle of the electrons by tilting of the specimen relative to the entrance of the energy analyser yields an increase of intensity of the components of the outer parts of the film relative to the inner ones. The intensity ratio of Fe^{3+} and Fe^{2+} increases with θ after (5).

$$\frac{I_{Fe^{3+}}}{I_{Fe^{2+}}} = \frac{D_{Fe^{3+}}\ \{ \exp (d_2/\lambda_{ox}\cos\theta)-1 \}}{D_{Fe^{2+}}\ \{ 1- \exp (- d_1/\lambda_{ox}\cos\theta) \}} \tag{5}$$

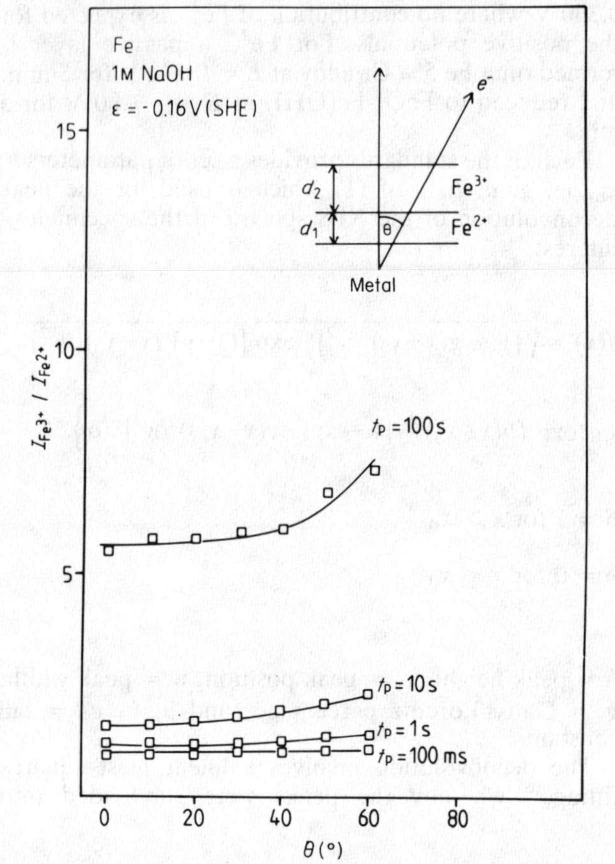

Figure 7. Angular dependent intensity ratio $I_{Fe^{3+}}/I_{Fe^{2+}}$ of an Fe specimen passivated at $E = -0.16$ V in 1 M NaOH for different times t_p.

The expected increase is found after sufficiently long passivation times of $t_p > 1$ s for the example given in Fig. 7. $E = -0.16$ V has been chosen to obtain comparable amounts for both species Fe^{2+} and Fe^{3+}. The curves of Fig. 7 are affected by the surface roughness of the specimen in the range of the thickness of the passive layer introduced by previous Ar–sputtering. This is seen most clearly by the flat shape of the intensity ratios of the iron ions and the metal signal. Similar results have been found comparing electropolished and sputtered samples of other metals.[9]

The change of the oxide thickness with time is represented in Fig. 8. Depending on the electrode potential Fe_2O_3 starts growing at 1 to 100 ms and its thickness increases linearly with lg t as expected for a high field mechanism of oxide growth.[19] FeO achieves a maximum and decreases for longer times to the quasi stationary values of Fig. 8. Apparently FeO,$Fe(OH)_2$ is formed first and oxidised partially to Fe_2O_3 when a sufficiently large overlayer has been formed. The Fe_2O_3 layer grows at the surface of FeO,$Fe(OH)_2$ by its oxidation. During the initial stage the Fe losses are compensated and overcompensated by new oxide formation. With the increasing Fe_2O_3 barrier the FeO,$Fe(OH)_2$ growth is stopped and decreased by a partial net oxidation to Fe_2O_3 and its thickness approaches a final stationary value. With increasing electrode potential the Fe_2O_3 growth is accelerated and stationary conditions are achieved earlier.

The deconvolution of the O ls signal yields the intensity ratio of the OH^- and O^{2-} contribution of the

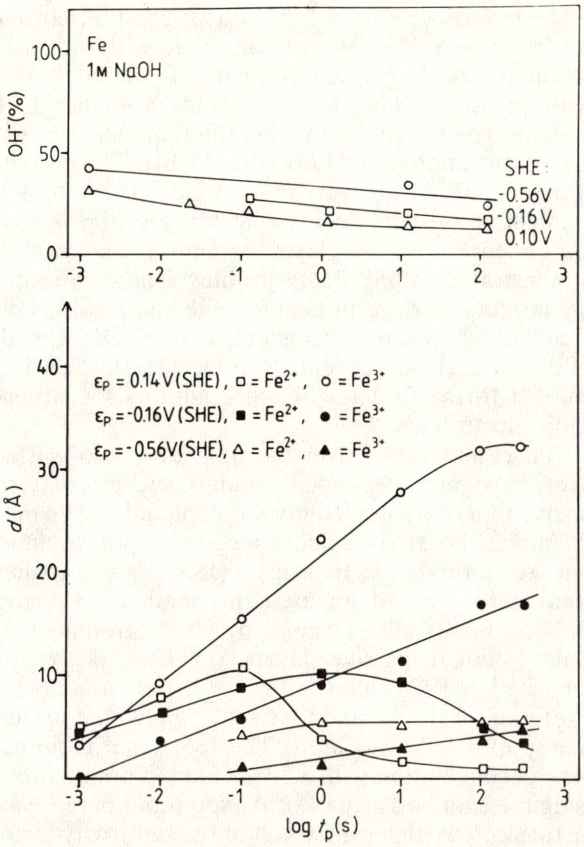

Figure 8. OH^--content and oxide thickness (FeO and Fe_2O_3) of Fe in dependence on passivation time t_p for three different electrode potentials in 1 M NaOH.

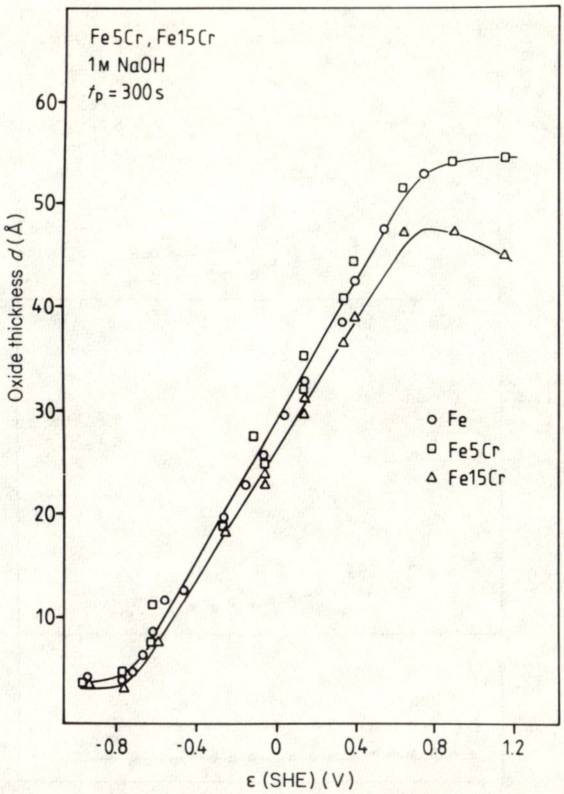

Figure 9. Fe–Oxide thickness of Fe and Fe–Cr alloys in dependence on the electrode potential, passivation time t_p = 300 s in 1 M NaOH.

passive layer. The OH^--content deduced from this ratio decreases with increasing Fe_2O_3 thickness starting with 50% at the beginning. Apparently the Fe II layer contains more water and is more a hydroxide than an oxide, whereas the Fe_2O_3 overlayer has only a minor hydroxide contribution. Similar results are also obtained for long time oxidations for different potentials. A large OH^- content is found only for sufficiently negative electrode potentials.

The reduction of Fe_2O_3 to FeO,$Fe(OH)_2$ causes similarly a reversible change from oxide to major hydroxide contributions. These changes may be followed by the O ls and the Fe $2p_{3/2}$ signal. After oxide formation in 1 M NaOH at 0 V for instance, Fe_2O_3 is reduced to FeO,$Fe(OH)_2$ at $E < -0.6$ V. At $E < -0.9$ V FeO,$Fe(OH)_2$ is further reduced to Fe^o and disappears at $E < -1.0$ V.

Passive layers on Fe–Cr Alloys

The oxide formation on Fe–Cr alloys is complicated by the presence of a second metal. Since both cations are present within the oxide layer and their concentration changes continuously, a multiple layer structure, as in the case of pure iron, is an oversimplification. Fe III oxide grows linearly with the electrode potential up to $E = 0.8$ V, however, with a decrease at $E > 0.8$ V in the potential range of the transpassive increase of the current density with electrode potential. The amount of iron oxide is rather independent of the bulk composition as shown in Fig. 9 for Fe and two Fe–Cr alloys.

The amount of Cr_2O_3 increases linearly with the electrode potential as deduced from the Cr $2p_{3/2}$ signal (Fig. 10). The small step at ca. 0 V may be attributed

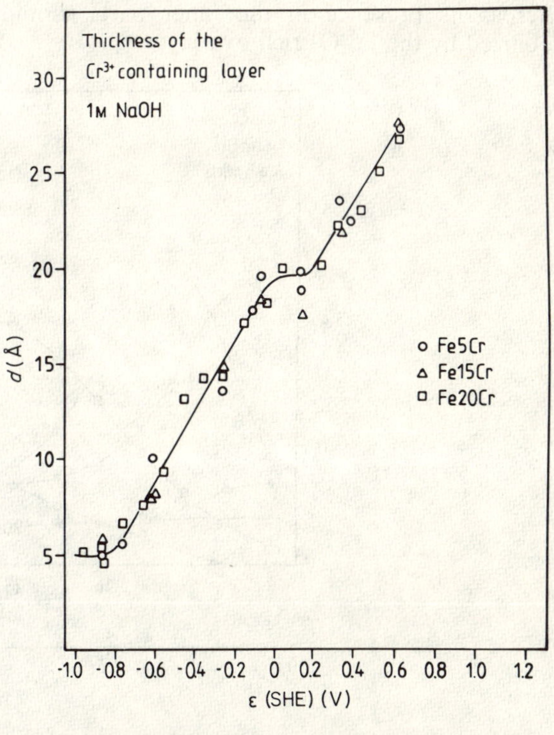

Figure 10. Cr_2O_3 thickness in dependence on the electrode potential as for Fig. 9.

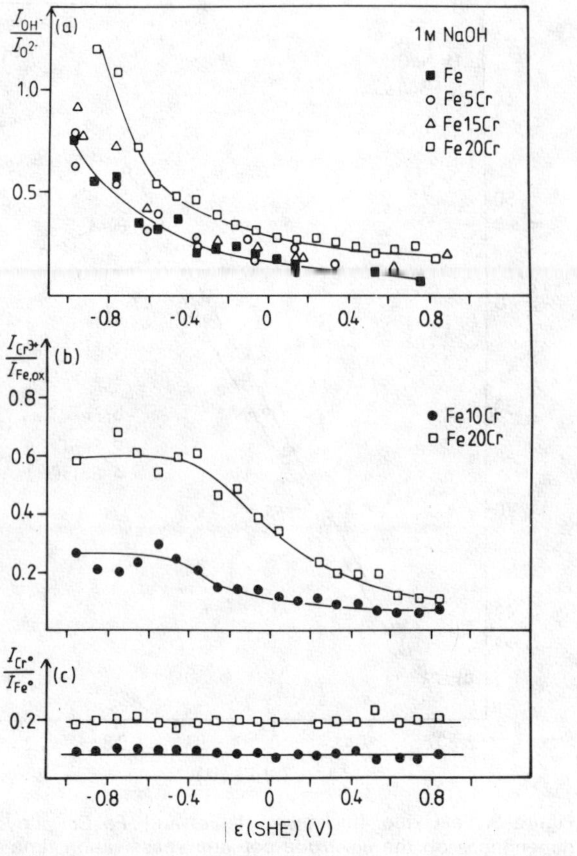

Figure 11. XPS-intensity ratios for the different species in the passive layer (a), (b) and the metal surface (c) in dependence on the electrode potential.

to a release of Cr^{3+} by its oxidation to soluble CrO_4^{2-} which becomes thermodynamically possible at $E > -0.2$ V. This effect yields, at least in the outer parts of the passive film, a reduced Cr – content whereas its presence in the inner parts should be stabilized by the Fe_2O_3 rich overlayer.

The intensity ratios of Fig. 11(a) suggest a change of a large hydroxide content to an oxide with increasing potential. A larger Cr–content of the bulk and consequently within the film yields a higher OH^- content. The Cr–content within the film increases with its concentration in the bulk (Fig. 11(b)). The decrease of the Cr^{3+}/Fe^{3+} intensity ratio at $E = -0.2$ V to +0.2 V might suggest its decrease within the film but is a consequence of the layer structure as may be demonstrated by ISS depth profiles. The composition of the metal surface in contact with the passive layer does not show any change in 1 M NaOH for the different electrode potentials (Fig. 11(c)). This is in contrast to the findings of other authors for strongly acidic electrolytes.[20]

XPS is integrating over the first few nm of surface layers. Angular dependent studies suggest that the passive film contains an outer Cr–depleted and an inner Cr–enriched part. To achieve a better depth resolution ion scattering spectroscopy (ISS) was applied. Frankenthal and Malm used this method to analyse films on Fe–Cr alloys caused by air.[21] Strehblow and Malm studied passive layers on Cu and vapour deposited Al–Cu alloy films.[22,23] For a sufficient resolution of the Fe and Cr signal ^{20}Ne with a primary energy of 3 keV was used. The ISS signal requires a peak deconvolution with a 50% Gauss–Lorentz fitting, using the standard signals of Ar-sputtered pure Fe and Cr surfaces. With a calibration of the sensitivity factors with the signals of the pure elements one obtains the same composition for the oxide free alloys with ISS as the known bulk concentration. Consequently no preferential sputtering for one of the elements is observed. For Fe–Cr oxide standards again no preferential sputtering of one of the cations is detected and the analysis is valid with the same sensitivity factors.[24] Depth profiles were obtained by scanning the focused ^{20}Ne ion beam over an area of 6 × 6 mm with an ion current of 0.3 μA and an ISS analysis with a 3 × 3 mm scan with a reduced current of 0.03 μA. In

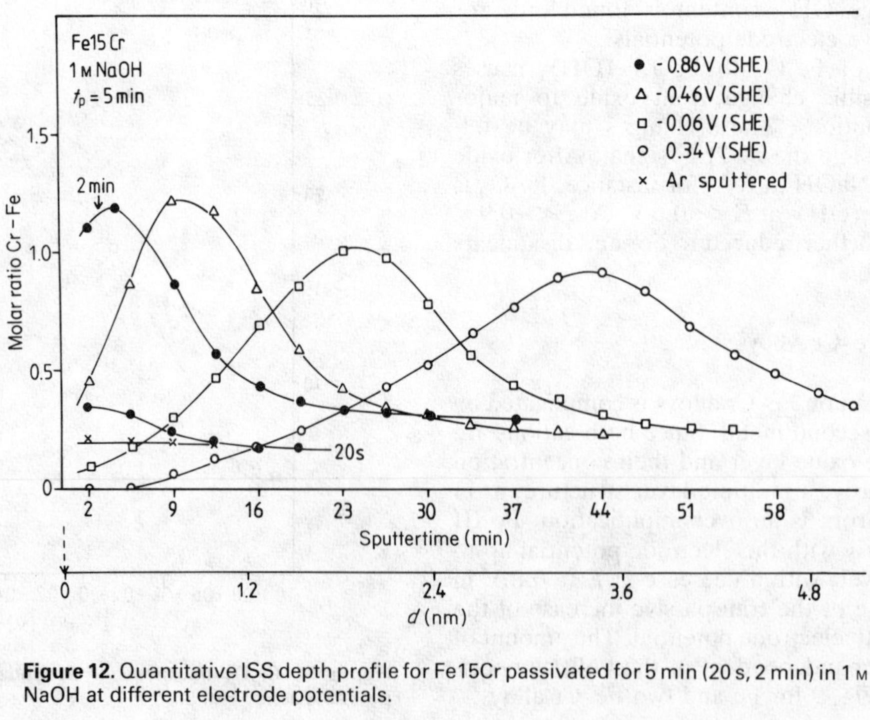

Figure 12. Quantitative ISS depth profile for Fe15Cr passivated for 5 min (20 s, 2 min) in 1 M NaOH at different electrode potentials.

this way only the homogeneously sputtered centre was measured. The sputtering during the ISS measurement was added to the total sputter time taking into account the reduced ion current density.

Figure 12 represents ISS depth profiles of Fe 15 Cr anodically oxidised in 1 M NaOH for 5 min at different electrode potentials in the closed system, starting with sputter cleaned specimen. The molar Cr/Fe ratio has been calculated from the ISS signals with the relative sensitivity factors in dependence on the sputter time. The thickness of the oxide film is added to the abscissa taken from the values of Fig. 6.

At negative potentials ($E = -0.86$ V) a Cr enrichment is formed at the surface during ca. 2 min. Apparently a Cr_2O_3 rich layer forms at the surface, whereas Fe^{2+} is dissolved. At least during rinsing with degassed water the FeO,$Fe(OH)_2$ seems to be dissolved. Anodic oxidation at more positive potentials forms stable Fe_2O_3 which will not dissolve, especially not during rinsing with water. The profiles demonstrate the presence of an outer Cr-depleted oxide film. Especially at $E = 0.34$ V no Cr_2O_3 content is found in the outer parts of the passive layer. The inner parts of the oxide are however enriched with Cr_2O_3 over the bulk composition. A maximum value of Cr:Fe = 1.2 equals 0.55% Cr and consequently a 3.6 fold maximum enrichment of Cr with respect to the bulk material. The increasing width of the depth profile indicates the increase of the oxide thickness with the electrode potential and the incorporation of large amounts of Cr. The ISS depth profile shows directly the outer Cr-depleted parts of the film by the dissolution of CrO_4^{2-} and the increased stability of Fe_2O_3 in contrast to strongly acidic solutions. For low pH values the pronounced depletion in Cr_2O_3 does not occur because of the much higher potential of CrO_4^{2-} -formation.[24] Artifacts for an ISS depth profile might be the possible broadening of the Cr distribution during the sputter process by ion mixing and recoil implantation. However, it gives at least a semiquantitative information which is not available to that extent by XPS. Further studies are necessary to learn about these possible changes.

CONCLUSION

The prevention of air access to the surface of electrodes before and after their electrochemical preparation helps to avoid changes which are not due to the electrode processes themselves. It enables electrochemical experiments and sample preparations to take place with analytically well defined surfaces, i.e. without contamination and any oxide layers formed by air. Surfaces submitted to low electrode potentials or to short electrochemical oxidation times may be studied without major changes. This extension of the potential and time scale may help to get a better understanding of electrode processes, for example the growth of surface films, and may lead to the XPS study for the electrode–electrolyte interface.

Acknowledgement

The financial support given by the Deutsche Forschungsgemeinschaft for projects Str 200/2 and Str 200/3 is gratefully acknowledged.

REFERENCES

1. E. Yeager, W. E. O'Grady, M. Y. C. Woo and P. Aagans, *J. Electrochem. Soc.* **125**, 348 (1978).
2. A. T. Hubbard, R. M. Ishikama and J. Kate Karn, *J. Electroanal. Chem.* **86**, 274 (1978).
3. F. T. Wagner, and P. N. Ross, *J. Electrochem. Soc.* **130**, 1789 (1983).
4. I. Olefjord and B. O. Elfstrom, *Corrosion – NACE*, **38**, 46 (1982).
5. W. N. Hansen, D. M. Kolb, *J. Electroanal. Chem.* **100**, 493 (1979).
6. D. L. Rath, D. M. Kolb, *Surf. Sci.* **109**, 649 (1981).
7. D. M. Kolb, D. L. Rath, R. W. Willie, W. N. Hansen, *Ber. Bunsenges. Phys. Chem.* **87**, 1108 (1983).
8. H. Scofield, *J. Electr. Spectr.* **8,** 129 (1976).
9. To be published by the authors.
10. S. Haupt, U. Collisi, H. D. Speckmann and H. H. Strhblow, *J. Electroanal. Chem.* **194**, 179 (1985).
11. J. Lecoeur, J. P. Bellier, *Electrochimica Acta*, **30**, 1027 (1985).
12. D. A. Shirley, *Phys. Rev* **135**, 4709 (1972).
13. H. E. Bishop, *Surf. Interface Anal.* **3**, 272 (1981).
14. R. O. Ansell, T. Dickinson, A. F. P. Povey, P. M. A. Sherwood, *J. Electroanal. Chem.* **98**, 79 (1979).
15. P. M. A. Sherwood in D. Briggs, M. P. Seah (Ed) *Practical Surface Analysis*, p. 460 Wiley (1983).
16. K. Ebert, A. Ederer, *Computeranwendungen in der Chemie*, VCA Weinheim (1985).
17. N. Sato, in *Passivity of Metals*, p. 46 ed by R. P. Frankenthal, J. Kruger Electrochemical Society (1978).
18. M. P. Seah, W. A. Dench, *Surface Interface Analysis*, **1**, 2 (1979).
19. K. J. Vetter, *Electrochemical Kinetics*, p. 769 Academic Press, New York (1967).
20. I. Olefjord, B. O. Elfstrom, *Corrosion NACE*, **38**, 46 (1982).
21. R. P. Frankenthal, D. L. Malm. *J. Electrochem. Soc.* **123**, 186 (1976).
22. H. H. Strehblow, B. Titze, *Electrochimica Acta*, **25**, 839 (1976).
23. H. H. Strehblow and D. L. Malm, *Corrosion Science*, **19**, 469 (1979).
24. C. Calinski, H. H. Strehblow, *Jahrestagung der Deutschen Bunsen Gesellschaft für Physikalische Chemie*, 16 May 1985.

SURFACE AND INTERFACE ANALYSIS, VOL. 9, 367–370 (1986)

AES/EDX Studies of Reactions in Phosphate Layers During Corrosion

W.J. van Ooij
Department of Materials Engineering, VPI & SU, Blacksburg, VA 24061, USA†

The interface between paint coatings and phosphated automotive steel was investigated using a combined AES/EDX technique. Delamination of the paint during corrosion in salt spray, scab or filiform conditions was found to be controlled by two factors: 1. transformations in the phosphate crystals leading to partial leaching or complete dissolution, 2. wet adhesion retention of the coating in alkaline conditions. The rate of phosphate attack is retarded if the crystals are treated with Cr(III), Cr(VI) or Pb^{2+} ions prior to coating.

INTRODUCTION

The performance of phosphate layers in corrosion control of automotive steel is related to their thickness and porosity.[1] It has also been documented that during corrosion the phosphate crystals are chemically attacked and sometimes completely destroyed.[2,3] Modern conversion layers which are rich in zinc-iron phosphate (phosphophyllite) have been shown to provide a better protection,[1] although no satisfactory explanation is available.

The mechanism of corrosion-induced coating delamination from bare steel has been studied in detail,[4] but very little has been published on delamination in phosphated systems.

The studies summarized here were primarily aimed at the unravelling of mechanisms of the interfacial reactions in phosphated, coated automotive steel during various conditions of corrosion. Some of the questions to be asked are: (a) is delamination corrosion-induced, or is delamination the first step of a sequence of reactions which lead to rust formation? (b) by what mechanism is corrosion performance improved when the phosphate layer is rinsed with a chromate solution prior to electrocoating?

EXPERIMENTAL

Materials, corrosion test conditions (salt spray, scab, filiform) and sample preparation methods have been described.[2,3,5] Standard automotive steel panels activated (Bonder V360 M process) and spray-phosphated (Bonder 130 process) to a zinc-iron phosphate thickness of 2.8–3.3 g/m^2 were used. Chromate/chromium rinse was a commercial process (Bonder 60). Treatment of phosphate layers with H^+ or OH^- was done at ambient temperatures by immersing panels in aerated solutions of H_2SO_4 or NaOH. Treatment with Pb^{2+} ions was done by subjecting panels to 300V for 10 seconds (cathodically) in water to which 100 ppm Pb as acetate had been added. These conditions simulate cathodic electrocoating. The effect of electrocoating on phosphate structure was examined by stripping the electrocoat immediately after deposition by immersion in N-methyl pyrrolidon for 24 hours. Surface and bulk analysis of the phosphate layers was performed by simultaneous AES and EDX in a VG Scientific type MA500 scanning Auger spectrometer equipped with a 30 kV, 50 nm electron source, a Si(Li) LINK EDX detector and facilities for XPS. Quantitative phosphate losses were determined by EDX by comparing the ratios of the P and Zn signals before and after corrosion or other treatment, using a 1×1 mm^2 rastered area. Electrochemical tests were carried out in aerated 5% NaCl solutions using a PAR 350 potentiostat.

RESULTS AND DISCUSSION

In Fig. 1 the phosphate morphology and composition, as per AES and EDX, are shown. According to x-ray diffraction analysis, the layer consists of a mixture of β-hopeite ($Zn_3(PO_4)_2 \cdot 4H_2O$) and phosphophyllite ($Zn_2Fe(PO_4)_2 \cdot 4H_2O$). However, AES indicates that the surface is a mixture of Zn- and Fe-phosphates, regardless of the crystallography. The Zn/Fe surface ratio varies considerably in a lateral sense. The observation that excellent, reproducible Auger spectra can be recorded indicates that the crystals are reasonably conductive. Under high beam currents Fe was seen to diffuse out of the surface layers.

Following cathodic electrocoating no changes in the phosphate layer were detectable in SEM, AES, EDX, XPS or XRD. Anodic electrocoating has a pronounced effect on the phosphate layer (Fig. 2). About 50% of the Zn and the PO_4 has dissolved. Both AES and XRD

†Work done at Alzo Research, Arnhem, The Netherlands.

0142–2421/86/120367–04 $05.00

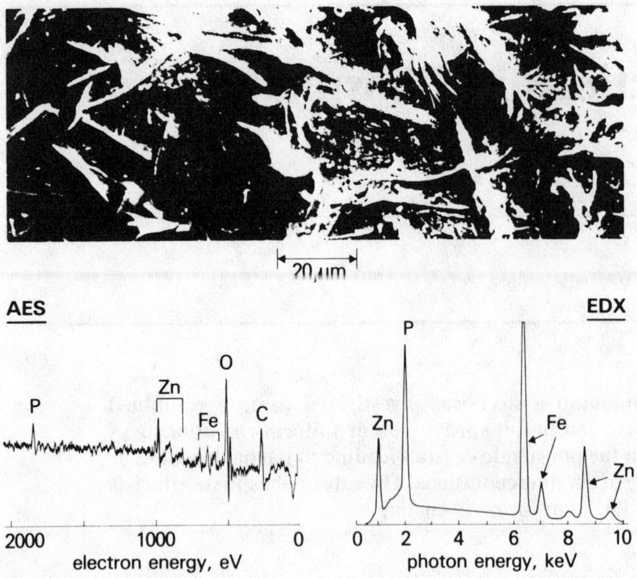

Figure 1. SEM/AES/EDX analysis of Bonder 130 phosphate layer.

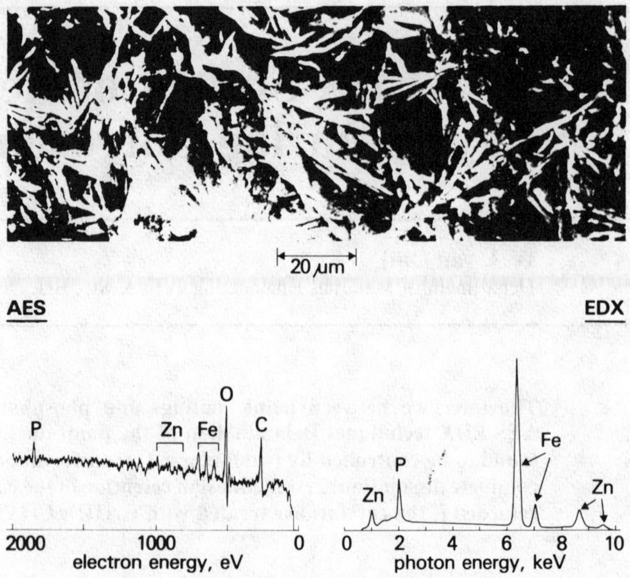

Figure 2. SEM/AES/EDX analysis of phosphate layer following anodic electrocoating for 2 min at 300 V; paint was not cured and stripped in N-methyl pyrrolidon.

show that more zinc phosphate than iron phosphate dissolves, so the layer becomes enriched in phosphophyllite. The surface is pure iron phosphate. This attack during deposition is probably one of the reasons why anodic electrocoat systems are unsatisfactory in certain corrosion tests, especially those which seem to proceed in acidic environments.[2] The pH sensitivity is demonstrated in model experiments with the uncoated phosphated panels immersed in acid or alkali. The residual amounts of P, Zn and Fe after 24 hours are plotted in Fig. 3. At high pH, PO_4 is completely leached out by OH^- leaving a residue of mixed iron and zinc oxides (iron zincate). This has the same morphology as the original phosphate crystals (1). At low pH the entire phosphate dissolves, in moderately acidic conditions $Zn_3(PO_4)_2$ dissolves preferentially and the residual iron phosphate is oxidized and hydrolysed:

$$\text{Zn, Fe(PO}_4)_x + OH^- \longrightarrow Fe_2(ZnO_2)_3 + PO_4^{3-} \quad (1)$$

$$\text{Zn, Fe(PO}_4)_x + H^+ \longrightarrow FePO_4 + Zn^{2+} + HPO_4^{2-} \quad (2)$$

$$FePO_4 + H_2O \longrightarrow Fe(OH)_3 + H_3PO_4 \quad (3)$$

The results of these model studies were used to identify the mechanisms of the reactions in scab, filiform and salt spray tests. Figure 4 shows the simplified mechanism in scab corrosion, outdoor weathering with low concentrations of electrolyte. At both phosphate interfaces reactions proceed simultaneously: at the coating interface the iron–zinc phosphate dissolves and hydrolyses, resulting in delamination, at the other interface a layer of iron oxides is formed. These reactions are inter-related since the phosphate has a fairly good electrical

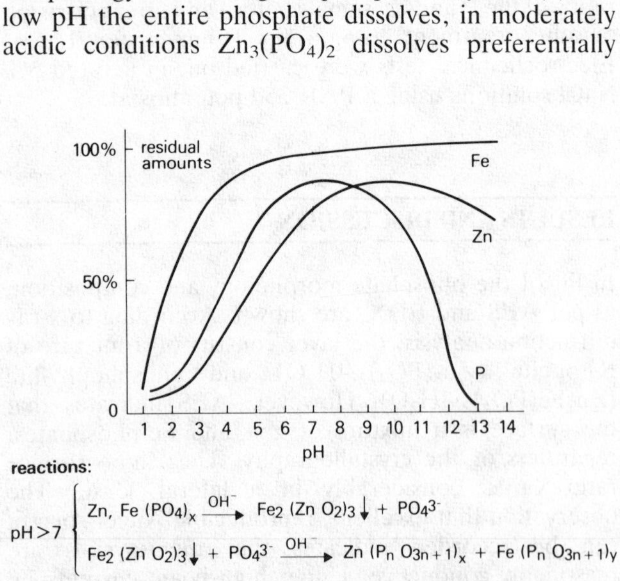

Figure 3. Residual amounts of P, Zn and Fe in phosphate following 24 hours treatment in H_2SO_4 or NaOH at 20°C; analysis by EDX and AES.

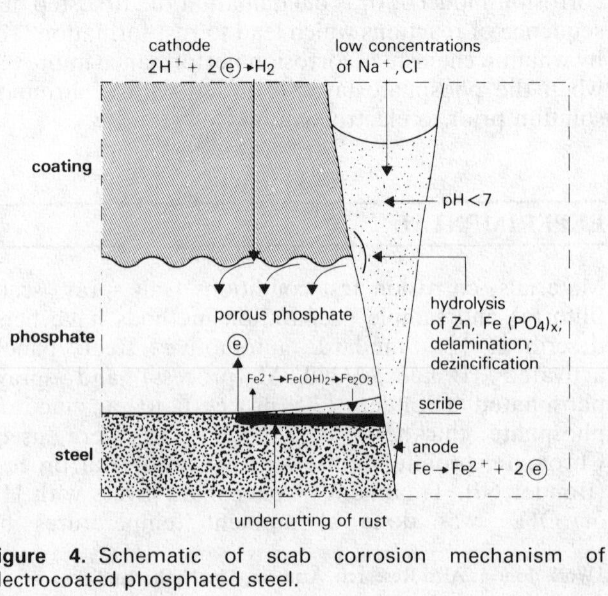

Figure 4. Schematic of scab corrosion mechanism of electrocoated phosphated steel.

conductivity. In this mechanism, thickness of the phosphate layer is important. Hence, anodically electrocoated systems are poor in this test (see above), as has been reported.[2,3] The rate of iron oxide formation is controlled by the rate of delamination at the opposite interface. The latter was found to be dependent on the iron content of the surface layers (see Fig. 2).

Figure 5 shows a schematic of filiform corrosion. Here, delamination is at the phosphate-steel interface, possibly because the iron oxide is attacked by the low pH-liquid. This is an example of anodic delamination. The liquid must be acidic as evidenced by the presence of Cl^- ions and the absence of Na^+ ions (as per AES and EDX), and by the observation that in the head of the filiform there is a region where the phosphate has dissolved completely. The anodic Fe^{2+} ions which oxidize and hydrolise, are, in this model, assumed to be the cause of the strong pH drop. In the tail of the filiform zinc and iron phosphates have reprecipitated but are now mixed with large amounts of iron oxides which is further evidence for a high rate of anodic iron dissolution in the filiform head.

In salt spray corrosion two effects are always observed: formation of microblisters in which the phosphate crystals are completely dissolved but no attack of the steel substrate is observed. Upon drying the phosphate reprecipitates as a mixture of amorphous zinc, iron and sodium phosphates. No chloride ions are detected. The results are consistent with attack at very high pH conditions. Iron zincate (see above) is not formed here, presumably because the phosphate ions remain contained in the blisters. They will polymerize at high pH and then form iron and zinc complexes of the type $(Zn,Fe(OH)_x(P_nO_{3n+1})_y$. In the scratch the anodic area (underrusting) is preceeded by a cathodic region where the phosphate also dissolves entirely (Fig. 6). No attack of the steel occurs here. Consequently, the rate of corrosion (spread of anodic rust) is largely controlled by the ingress of alkaline liquid along the phosphate–polymer interface by an alkaline phosphate attack. The delamination by this mechanism is thus controlled by two factors: the rate of dissolution of the phosphate and the retention of wet adhesion of the

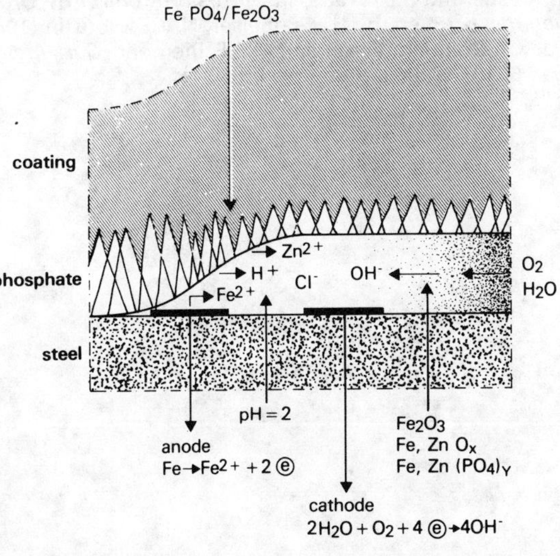

Figure 5. Schematic of filiform corrosion mechanism of electrocoated phosphated steel.

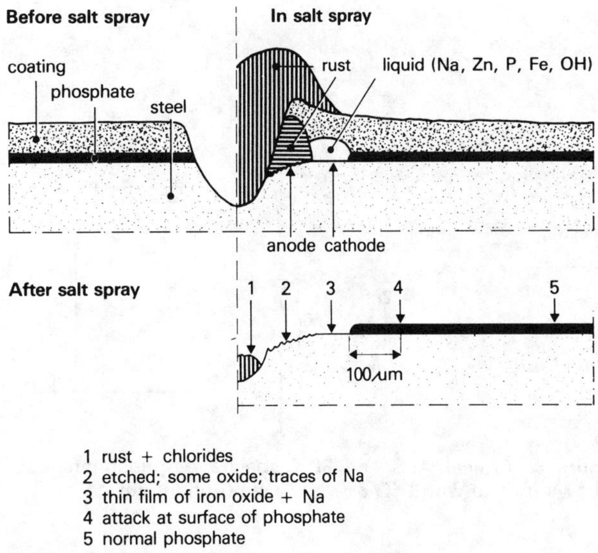

Figure 6. Schematic cross section of coated phosphated panel during salt spray corrosion.

polymer under strongly alkaline conditions. In agreement with this conclusion, AES detected the first stages of phosphate attack relatively far away from the corrosion front (point 4 in Fig. 6). By SEM or EDX no changes could be detected there. A second observation which corroborates on this mechanism is that the ranking of the rate of delamination of a series of punctured coating systems immersed in liquid of pH = 13, i.e. under conditions where iron is passive, agrees well with the performance of the coating in actual salt spray tests and in electrochemical cathodic delamination tests.

Figures 7, 8 and 9 show the effects observed with AES and EDX of treatment of the phosphate layer with Cr(III), Cr(VI) and Pb(II) ions. The conditions were such that electrochemical measurments of the uncoated panels (oxygen reduction current) did not detect any effect of the treatments. Yet, a very pronounced retarding effect of all three treatments was observed in the salt spray tests of the cathodically electrocoated panels. AES/EDX shows that Cr(III) forms a very thin $CrPO_4$ film at the surface of the crystals. Cr(VI) treatment forms a $ZnCrO_4$-rich surface, and Pb is incorporated as a ternary Zn, Fe, Pb

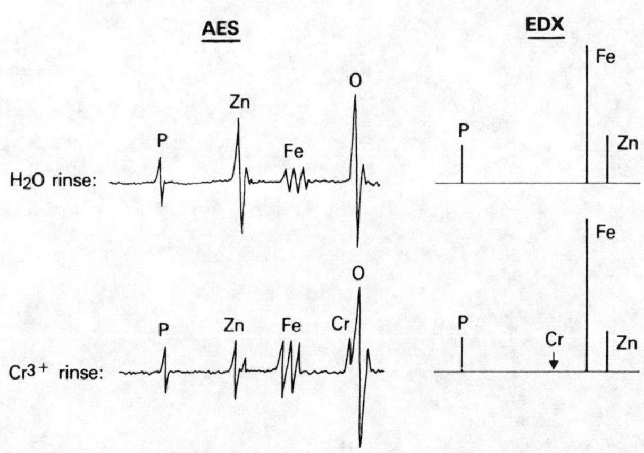

Figure 7. Typical AES and EDX spectra of phosphate layer following rinse with H_2O or with Cr(III) (Bonder 60).

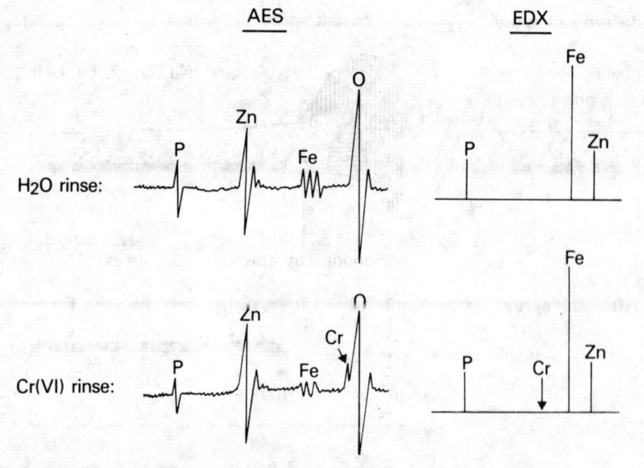

Figure 8. Typical AES and EDX spectra of phosphate layer following rinse with H_2O or with Cr(VI) (Bonder 60).

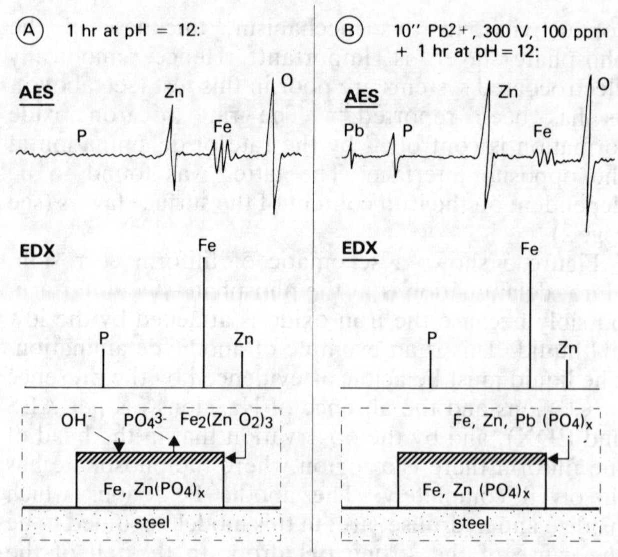

Figure 9. Effect of treatment with Pb^{2+}ions (100ppm, 300 V, 10 sec) on the rate of PO_4 leaching at pH = 12 of zinc–iron phopsphate.

phosphate, down to a depth of 5–10 nm. These modified surfaces are more resistant to alkaline attack, as is shown for Pb in Fig. 9. These observations suggest strongly that, in general, automotive corrosion could be further improved by developing even better treatments than by Cr or Pb, namely treatments which deposit a thin film at the phosphate crystal surface by ion exchange, which stablize phosphate attack both at low and high pH.

CONCLUSIONS

1. Currently used automotive iron-zin phosphates are unstable in water. Depending on the pH, they can dissolve entirely or lose PO_4 or Zn, or both, selectively;
2. Their rate of dissolution in alkali (salt spray conditions) can be retarded by treatment with Cr(III), Cr(VI) or Pb(II) ions which form, by ion exchange, a modified and more resistant surface film;
3. Iron-rich phosphates are relatively poor in corrosion which proceeds at low pH (scab, filiform), zinc-rich phosphates are rather poor at high pH (salt spray);
4. During anodic electrocoating about 50% of the phosphate can dissolve, leaving an iron-rich residue;
5. Scab and salt spray corrosion start at the phosphate-coating interface. The surface composition of the phosphate crystals has an effect on the delamination rate;
6. An important parameter of the polymer in salt spray corrosion is its wet adhesion in strongly alkaline conditions.

REFERENCES

1. L. Wittel, *Ind. Lackierbetr.* **51(5)**, 169 and **51(6)**, 210 (1983).
2. E. Nagel Soepenberg, H.G. Vrijburg, W.J. van Ooij and O.T. de Vries, *Tenth Int. Conf. in Org. Coat. Sci. Technol., Athens (Greece), July 9 – 13, 1984.*
3. E. Nagel Soepenberg, H.G. Vrijburg, J.A.F.M. Schade van Westrum, W.J. van Ooij and O.T. de Vries, *NACE CORROSION 85, Boston, Mass., March 25 – 29, 1985,* paper 388.
4. J.E. Castle and J.F. Watts, in *Corrosion Control by Organic Coatings*, p.78, ed. by H. Leidheiser, NACE, Houstin (1981).
5. W.J. van Ooij, T.H. Visser and M.E.F. Biemond, *Surf. Interface Anal.* **6**, 197 (1984).

SURFACE AND INTERFACE ANALYSIS, VOL. 9, 371–375 (1986)

Application of GDOS (Glow Discharge Optical Spectroscopy) and SIMS to the Analysis of Surface Corrosion Products on Stainless Steel After Exposure to Liquid Alkali Metals

Marten G. Barker
The Department of Chemistry, The University of Nottingham, Nottingham, NG7 2RD, England
Inge E. Schreinlechner
Department of Materials Technology, The Austrian Research Centre Seibersdorf, A–2444, Seibersdorf, Austria

Corrosion of steels by liquid alkali metals involves the transfer of non metals, such as C, N or O, from solution in the alkali metal to the surface of the steel. Surface corrosion products may then be formed. With oxygen in sodium these products may be ternary oxides, with nitrogen in lithium the ternary nitride Li_9CrN_5 is formed and with carbon binary carbides are formed in both alkalies. Along with the surface corrosion products, element depletions and carbon or nitrogen diffusion profiles are created. The analyses of thin corrosion product layers and the determination of the concentration profiles of carbon and nitrogen are important requirements for the understanding of the corrosion process. Lithium penetration into steel is also a major problem in lithium corrosion. Lithium is extremely difficult to determine by conventional analytical methods, due to its low atomic number. In this paper the use of glow discharge optical spectroscopy (GDOS) and SIMS for the study of surface corrosion products, is described, together with a quantitative determination of the carbon-depth profile in type 316 steel exposed to liquid sodium.

INTRODUCTION

The interaction of solid metals and alloys with liquid alkali metals has been the subject of much research over the past years due to the use of liquid Na as the coolant in fast breeder reactors and the potential use of liquid lithium or liquid Li/Pb alloy ($Li_{17}Pb_{83}$) as a coolant and tritium breeder in the fission reactor.

In this paper we will discuss the application of the GDOS technique to the investigation of steel surfaces that have been exposed to the alkali metals Li and Na and to the Li/Pb alloy $Li_{17}Pb_{83}$. Before discussing the problems of applying GDOS to these materials it is necessary to briefly describe the various chemical interactions which take place in liquid alkali metals so that the nature of the surface under investigation may be understood, followed by the principal features of the GDOS technique relevant to the problems we have encountered.

Two main types of interaction are observed when a metal surface comes into contact with a liquid alkali metal at temperatures in the range 400–650°C. The first involves solution processes of particular metals in the liquid metal; in the case of Na and Li a typical solution process involves the dissolution of Ni at high temperatures with the potential precipitation of the Ni at lower temperatures onto the metal surface. The second type of interaction involves the reaction of non-metals always present in solution in the liquid alkali metals, with the solid metal surface. This type of reaction can give rise to surface corrosion products, e.g. the interaction of oxygen dissolved in sodium with Cr on the surface of stainless steel, causes the formation of a layer of the ternary oxide $NaCrO_2$. This material initially grows in grain boundaries in the steel surface but eventually spreads across the grain to form a coherent surface film.[1] In liquid Li surface corrosion processes involving oxygen are less common due to the high stability of Li_2O which ensures that oxygen remains in solution in the liquid metal rather than reacting with elements at the metal surface. Nitrogen and carbon are however more reactive and may form products with elements at the surface of the solid metal.[2] In the case of Li and the Li/Pb alloy, a third form of interaction is observed, namely the penetration of the Li into the surface of the solid metal. Studies of Li penetration into stainless steel are at present in their infancy, and it is therefore worthwhile to investigate the potential of GDOS for such studies. Penetration of the liquid alloy $Li_{17}Pb_{83}$ into stainless steel surfaces has also been reported. The simultaneous measurement of both Li and Pb by GDOS affords a ready means of determining whether both elements penetrate to the same extent or whether separation of the Li and Pb takes place.

Surface techniques capable of detecting light elements such as lithium are not readily available. Only GDOS,[3] SIMS and Auger are possible and in the latter case severe experimental and theoretical problems arise.[4] GDOS and SIMS are extremely sensitive in their detection of Li and are thus the preferred techniques.

In summary, therefore, we are studying many modified steel surfaces which may contain a surface

0142–2421/86/120371–05 $05.00

deposit under which metallic element changes take place to depths of up to 100 μm. Concurrent with these we may have significant concentrations of Li/Pb penetrating into the steel again to depths of 100 μm. Study of the sample surface and depth profiling with GDOS enables us to study up to 19 elements simultaneously and thus build up a comprehensive picture of the chemical changes taking place at the surface. We cannot in this paper interpret these chemical changes, but merely show how the GDOS technique enables us to observe them. We have recently started to use SIMS for a complementary study but so far our experience with this technique is limited. We shall however show how SIMS enables us to examine Li penetration in considerable detail.

EXPERIMENTAL

All the GDOS work described in this paper was carried out using a Siemens Spektraumat 1000 at the Austrian Research Centre Seibersdorf. Initially, the glow discharge was operated at a constant voltage of 1000 V. Several problems were however encountered in using this mode of operation, in particular the effect of compositional changes from sample to sample gave rise to variations in the current flowing from the sample to the anode. We therefore decided to change from constant voltage to constant wattage (120 watts) mode in an effort to maintain constant power of the plasma during depth profiling. The glow discharge is particularly sensitive to surface composition and we soon became aware that the presence of insulating surface layers gave rise to very low wattage conditions. The examination of steel surfaces in which extensive penetration by lithium had taken place, also caused problems with the power levels. Lithium is extremely volatile by comparison with the other elements present in steel and is also highly conductive in the gas phase. To achieve a constant glow discharge condition at 120 watts it was essential that the concentration of lithium metal was not so high that low voltages were produced due to the high current flow through the lithium containing gas phase. A more major problem existed in that the surface corrosion products produced on the steel surface were air sensitive and on reaction with water vapour generated NH_3 gas. The gas evolution prevented the attainment of the vacuum conditions necessary for the generation of the glow discharge. Samples were therefore handled in dry argon filled glove boxes and were transferred to the instrument under cyclohexane. The sample was then quickly placed onto the vacuum port of the instrument and evacuation commenced. The relatively high boiling point of cyclohexane ensured that a film of the liquid covered the sample during transfer but was rapidly removed under the vacuum generated in the sample chamber. By this method, air sensitive samples were easily examined. Samples containing heavy corrosion layers could also be abraded under cyclohexane to leave only a thin surface layer which would not be either highly insulating or, in the case of samples exposed to liquid lithium, contain too high a concentration of lithium at the surface.

The samples employed for much of this study were in the form of flat plates ~2 cms square and 2 mm thick. In some instances where reaction with air was not a problem, circular discs (8 mm Ø) were cut and mounted in aluminium or copper holders. These holders enabled the investigation of depth profiles to several hundred microns by matchining the sample to a flat surface after each profiling sequence. Each sequence being down to ~140 μm. All samples were water cooled on the back face of the sample by a copper cooling block through which water was passed continuously during the glow discharge. This was found to be essential for maintaining consistent readings from sample to sample.

Penetration depths were determined using a randomly selected series of samples which had been studied in the GDOS instrument for different lengths of time using the constant wattage condition. The samples were cut across the crater caused by the glow discharge, mounted in cross section in a conventional metallographic mount and polished. The depth of the crater was then measured using a scanning electron microscope. These measurements indicated a reasonably constant etching rate of 1 μm for every 10 s of glow discharge at 120 watts. Data was therefore collected at 10 s intervals with a continuous glow discharge burn.

RESULTS AND DISCUSSION

In this section we present the results of several different studies of stainless steel (type 316) samples which have been exposed to the liquid metals Li, Na or $Li_{17}Pb_{83}$. The results are presented in several sections each concentrating on a particular aspect of the chemistry of the samples which has been elucidated using either GDOS or SIMS.

Alkali metal penetration studies

A series of flat plate samples exposed to static liquid lithium at 600°C for time periods ranging from 1 week to 1 month were examined by GDOS for penetration by lithium. Both sides of the plate were examined where possible so that some measure of the consistency of the penetration depths could be assessed, it being assumed that both sides of the plate were penetrated equally by the lithium. Figure 1 shows the Li profiles for both sides of a stainless steel exposed to liquid lithium for 2 weeks. It can be seen that there is indeed a very good correlation between the two sets of results for the depth of penetration and for the shape of the profile of the major metallic components of the steel also. Samples were examined from a separate experimental study of material from a thermal gradient flow closed loop in which liquid lithium flows round a circuit across which there is a temperature gradient from 400°C to 600°C. In such a system we would expect to observe higher penetration in the hot section and lower penetration in the cold section. GDOS results are shown in Fig. 2 where the effect of temperature is clear. The penetration values given by the GDOS technique are tabulated in Table 1 for both static and dynamic lithium experiments.

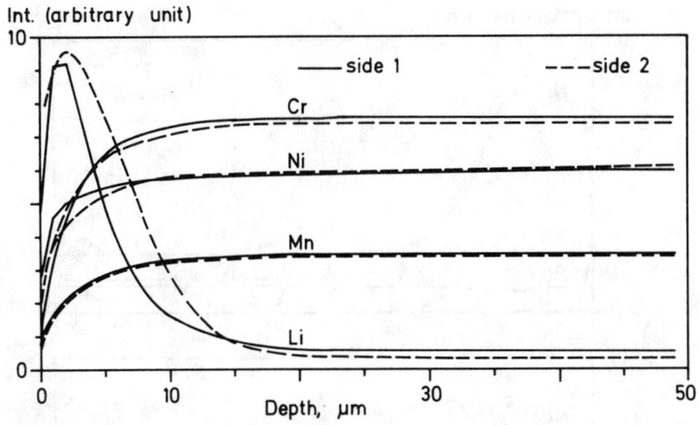

Figure 1. Lithium profiles from stainless steel exposed to liquid lithium at 600°C for 2 weeks (CrNi and Mn also shown). Concentration in the matrix: Cr 16,8 Ni 11,1 Mn 1,5 wt%.

We have also used SIMS to study the penetration of steel by lithium. The SIMS technique clearly showed that lithium penetration was almost exclusively *via* the grain boundaries in the steel[5] and by traversing a cross section of the surface the profile given in Fig. 3 was obtained.

Steels exposed to the liquid alloy $Li_{17}Pb_{83}$ are also expected to show penetration but this time by both lithium and lead. Figure 4 shows the profiles of such a sample in which the excellent correlation between the depths of both elements is obvious. The penetration depth has been varified by microprobe analysis for Pb only, since Li cannot be detected by this method.

Element Depletion Studies

The selective removal of elements from the surface of stainless steels by lithium and lithium lead alloys is ideally studied by GDOS. Figure 4 also shows the profiles of the major elements of the steel as a function of depth. It can be seen that Ni and Mn are selectively

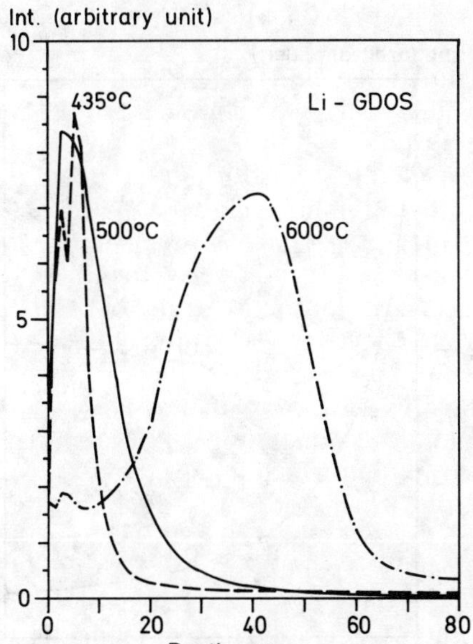

Figure 2. Lithium profiles from loop experiment showing effect of temperature on penetration.

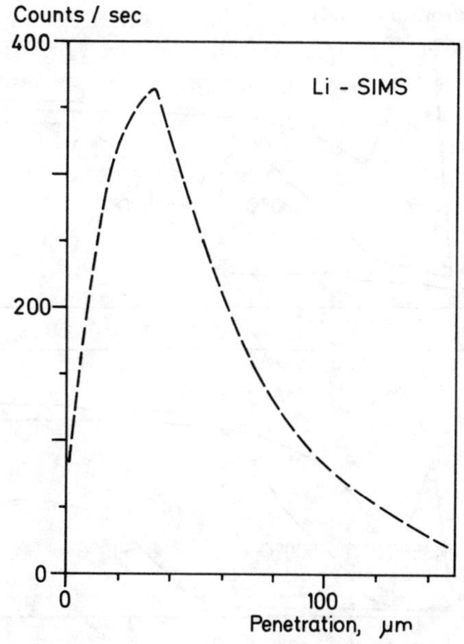

Figure 3. Lithium profile from stainless steel exposed to liquid lithium at 650°C for 1 month. Obtained by SIMS (Cameca IMS 3F).

Table 1. Lithium penetration data for stainless steel type 316

Lithium temperature (°C)	Time of exposure	Penetration (μm)
600	1 week	19
600	2 weeks	27
600	3 weeks	43
600	4 weeks	90
600	4 weeks[a]	80
500	4 weeks[a]	45
435	4 weeks[a]	40
400	4 weeks[a]	19

[a]Circulating lithium loop experiment.

leached, to the same depth as the Li and Pb have penetrated whilst Cr is leached to a much lesser degree and Fe and Mo not at all. (On this sample there is a small enrichment of Mo at the surface since a source of Mo was present in the LiPb alloy.)

The leaching, by dissolution, of certain elements is common in liquid alkali metals at high temperatures. Figure 5 shows the preferential removal of Ni and Mn from the high temperature region of the lithium loop experiment.

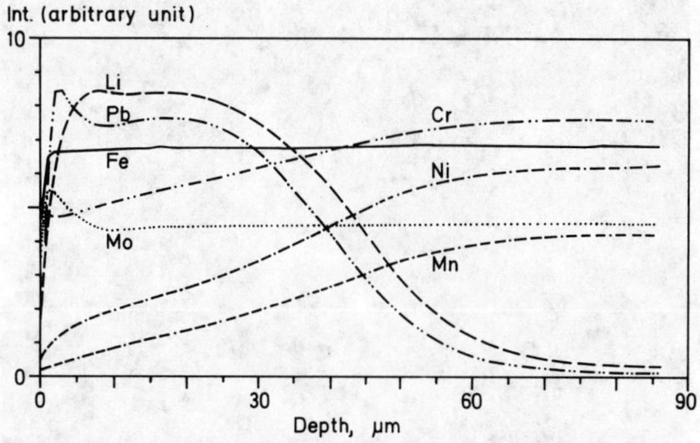

Figure 4. Metallic depletions and penetration by lithium and lead in stainless steel exposed to liquid $Li_{17}Pb_{83}$ at 600°C for 1 month.

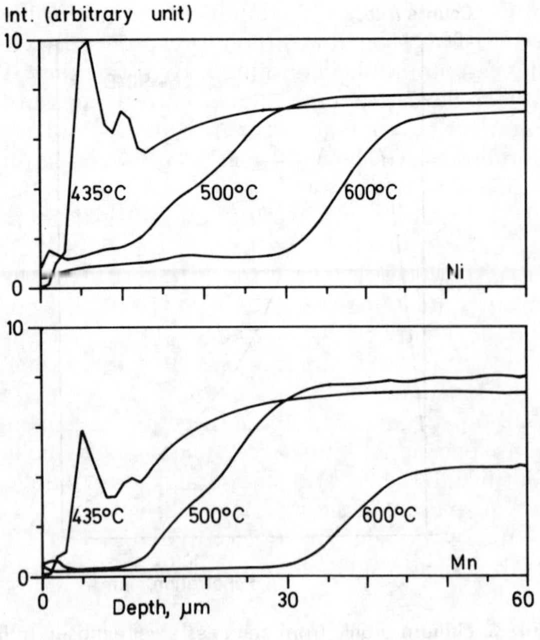

Figure 5. Metallic depletions observed in stainless steel from lithium loop showing effects of temperature.

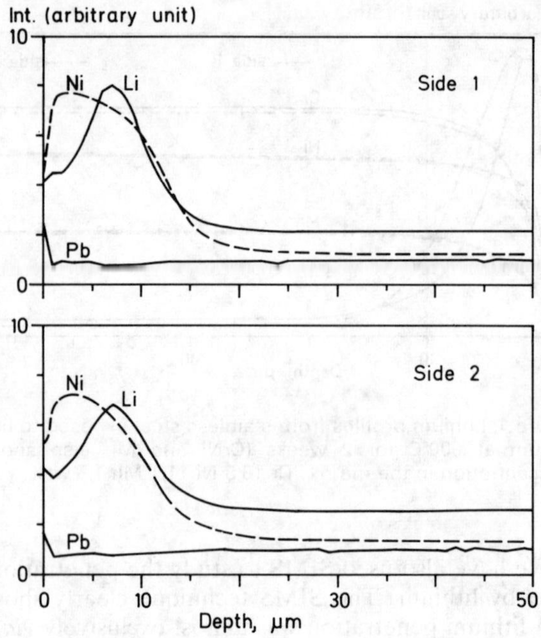

Figure 7. GDOS analysis of Ni deposition on stainless steel exposed to liquid $Li_{17}Pb_{83}$ saturated with nickel.

Element Deposition Studies

Using a liquid metal loop experiment, we can follow material removed in the hot regions to the colder regions, where the liquid metal becomes supersaturated by the dissolved elements leading to the precipitated material being deposited on the steel surfaces.[6] Figure 6 shows Ni deposited on the low temperature steel surfaces of the loop. The GDOS profile for this deposit can be seen in Fig. 5 where the presence of Mn is also observed. Similar deposition of Ni can be observed by placing the steel in lithium lead alloy saturated with Ni. Figure 7 shows the nickel deposition to a depth of 15 μm. A most interesting aspect of this figure is the almost total lack of a lead penetration profile even though there is clearly Li within the nickel deposition zone. This is the only instance we have so far observed of selective penetration of lithium from lithium lead. Both sides of the plate are shown indicating the reproducibility of all the observed effects. High temperature deposition is also possible in loop experiments and Fig. 8 shows a deposit rich in Cr and carbon obtained in liquid Li at 600°C. This layer is thought to be the carbide $Cr_{23}C_6$ formed by the lithium having a high carbon impurity level in this experiment.

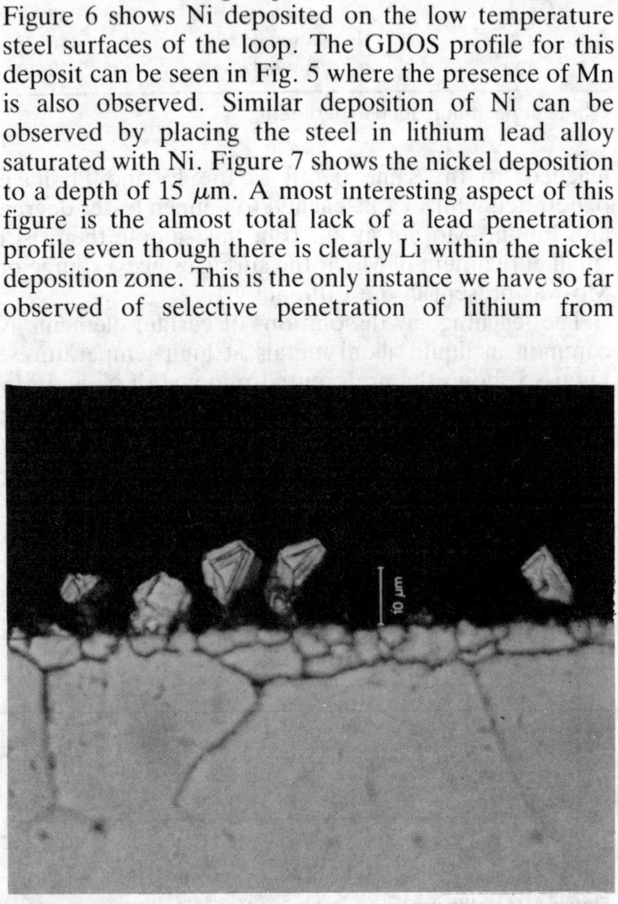

Figure 6. Ni deposit on stainless steel from cold section of loop.

Quantitative GDOS

Of particular importance in the use of liquid sodium as a fast breeder reactor coolant, is the possible transfer of carbon from the sodium to the steel surface.[7] For such studies we require to know the surface carbon level of the steel and the diffusion profile of carbon into the steel. We have therefore attempted to evaluate the

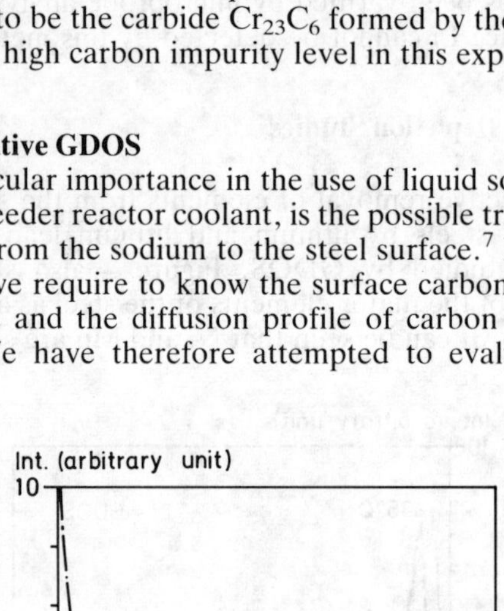

Figure 8. $Cr_{23}C_6$ deposition in hot section of liquid lithium loop.

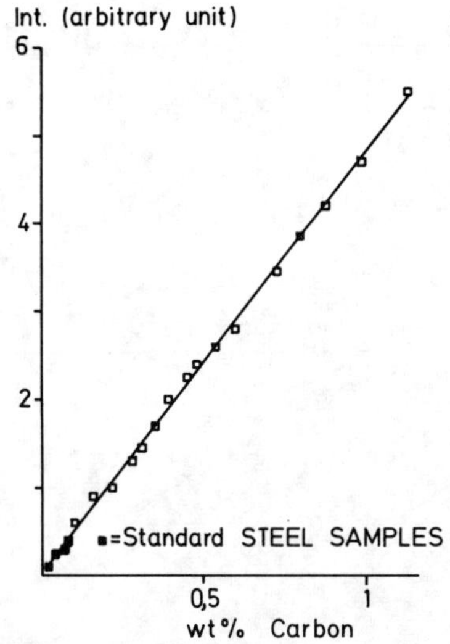

Figure 9. Calibration curve for carbon in stainless steel type 316 obtained using GDOS.

potential of GDOS for the determination of such profiles in a quantitative manner. For the construction of a calibration curve it is usual to use standard materials of known element concentration. This proved to be somewhat difficult in this instance in that the necessity for comparable conductivities between the different standards required them to have very similar compositions, particularly of Fe, Cr and Ni. Only a very limited range of standards could therefore be used and these were deemed to be insufficient for the construction of an accurate calibration curve covering the required carbon concentration range of 0.02 to 1.0 wt%. We were fortunate, however, to have a sample of type 316 steel which had been examined by a variety of techniques (microhardness, SIMS, and nuclear microprobe) and as a result had a known carbon profile extending to a depth of several hundred microns. This sample was therefore studied by GDOS and a carbon profile determined. By fitting the GDOS profile to the carbon profile quantitatively measured by the nuclear microprobe, we were able to construct the calibration curve shown in Fig. 9. We were extremely satisfied to note that the values obtained from the four standard steel samples also fitted onto this calibration line. Values obtained near to the surface of the steel sample showed a deviation from the line giving higher intensities than predicted, but in each case these points could be identified in that the voltage and current readings were not in the same range as those observed for the remainder of the readings.

Carbon appears to be the most suitable element for quantitative application since it appears to be fairly insensitive to the variations in composition of the other elements present in the sample. Clearly, we would wish to extend our quantitative measurements to Fe, Cr and Ni as well as to other light elements. At the present time, it appears that this may be possible for Ni or Cr, but that Fe is very sensitive to compositional changes.

CONCLUSIONS

In this paper we have attempted to show how GDOS has been used to investigate a range of chemical effects involved in the corrosion of steels by liquid alkali metals and alloys. Particular success has been achieved in the analysis of lithium penetration and data has been obtained for the first time using the GDOS technique. Metallic element deposition and depletion are readily studied simultaneously and the comparatively large area of sample examined ensures that consistent results may be obtained.

REFERENCES

1. M. G. Nicholas and I. W. Cavell, *Proc. 2nd Int. Conf. on Liquid Metal Technol. in Energy Prod. Richland (1980)*, Conf. 800401–P1, 3–35.
2. M. G. Barker, S. A. Frankham, D. R. Moore and D. J. Wood, *High Temperatures – High Pressures*, **14**, 523 (1982).
3. R. Berneron and J. C. Charbonnier, *Surface Interface Anal.* **5**, 134 (1981).
4. F. D. Schowengerdt and J. S. Forrest, *Proc. 2nd Int. Conf. on Liquid Metal Technol. in Energy Prod. Richland (1980)*, Conf. 800401–P2, 18–34.
5. M. G. Barker, G. A. Fairhall and S. A. Frankham, *Proc. 2nd Int. Conf. on Liquid Metal Technol. in Energy Prod. Richland (1980)*, Conf. 800401–P2, 18–41,
6. H. R. Konvicka, P. F. Sattler and I. E. Schreinlechner, *Proc. 2nd Int. Conf. on Liquid Metal Technol. in Energy Prod. Richland (1980)*, Conf. 800401–P1, 19–65,
7. M. G. Barker and D. R. Moore, *Liquid Metal Engineering, V and Technology, Proc. 3rd Int. Conf., Oxford (1984)*, Vol. 2, p. 249 BNES, London (1984).

SURFACE AND INTERFACE ANALYSIS, VOL. 9, 377–381 (1986)

ESCA Study of the Oxidation of a Nickel–Molybdenum Alloy

P. Marcus
Laboratoire de Physico-Chimie des Surfaces associé au CNRS, Ecole Nationale Supérieure de Chimie de Paris, 11, Rue Pierre et Marie Curie, 75005 Paris, France.
I. Olefjord
Department of Engineering Metals, Chalmers University of Technology, S–41296 Goteborg, Sweden

The surface composition of Ni–6at% Mo (110) alloy was analysed by ESCA after surface preparation treatments and after oxidation at 25°C and 500°C (p_{O_2} = 10^{-4} Torr). Ion etching causes a marked enrichment of molybdenum on the surface. After annealing at 500°C, the concentration of segregated molybdenum is ~25%. At 800°C there is no segregation of the metal elements, but sulphur segregates to the surface. Oxidation at 25°C gives a thin oxide layer (5–6 Å) with a composition which reflects the composition of the initial surface, so that the oxidation of a surface enriched with molybdenum gives a molybdenum-rich oxide. Oxidation at 500°C gives a molybdenum-enriched oxide. The Mo concentration of the oxide is 12%, i.e. twice that in the bulk. The lateral growth of oxide islands seems to be already complete after a very short oxidation time (2 min) and then a parabolic growth of the oxide layer is observed. The chemical shifts of Ni^{2+} ($2p_{3/2}$) and Mo^{6+} ($3d_{5/2}$) in the oxide formed on the alloy are respectively 2.1 eV and 4.2 eV, while they are 1.6 eV for NiO and 4.8 eV for MoO_3. Adsorbed sulphur has a marked influence on the oxidation. The oxide growth (also parabolic) is faster in the presence of sulphur at 500°C. The oxide film grows on top of the sulphur layer.

INTRODUCTION

Nickel and molybdenum are widely used as base metals or alloying elements in metallurgical and catalysis applications. The aim of this paper is to characterize the surface composition of a nickel–molybdenum alloy during oxidation at low or high temperatures and low oxygen pressure. A forthcoming paper will deal with the oxide produced electrochemically on this alloy. In earlier works we have studied by ESCA the passivation of Ni[1] and of Ni–Fe alloys;[2] and the oxidation of Ni–Fe alloys at 25°C and 500°C.[3] The influence of sulphur on these reactions has been studied in details.[4–8] There are a few reported studies on the passivation of Ni–Mo alloys;[9,10] but there is, to our knowledge, only one reported study on the passivation of Ni–Mo single crystals;[10] and this is the first reported study of the dry oxidation of a Ni–Mo single crystal.

EXPERIMENTAL

Nickel–molybdenum single crystals with a composition Ni–6at%Mo were used. This alloy is f.c.c. The sulphur content of the alloy was ca. 10 ppm. The samples were oriented by x-ray diffraction, using the Laüe method, and spark machining. The crystallographic orientation was (110) within 1°. The samples were mechanically and electrochemically polished and then annealed in hydrogen at 900°C. They were then introduced into the preparation chamber of the ESCA-spectrometer, where argon ion sputtering and annealing in vaccuum could be performed. The effects of these treatments on the alloy surface composition are reported in this paper. The oxidation experiments were also performed in the preparation chamber of the spectrometer.

RESULTS

Preparation of the surface

Figure 1 shows the ESCA spectra recorded after ion etching the alloy and annealing it in UHV at 800°C and 500°C. The main peaks originate from the metallic states of Ni $2p_{3/2}$ and Mo 3d. The binding energies of the peaks are reported in Table 1. The shoulder observed in the Mo region at 226 eV is a S 2s signal due to the presence of sulphur which segregates to the surface during annealing at high temperature. The spectra recorded after annealing at 800°C represent a surface covered by a complete monolayer of sulphur in agreement with other studies.[11] Comparing the two sets of spectra reveals that the surface composition of the alloy is dependent on the annealing temperature. After annealing at 800°C, the ratio of the intensities of the Mo and Ni signals, I_{Mo}/I_{Ni}, is 0.055, while after annealing at 500°C it is 0.103. This shows that the Mo concentration on the surface is much higher at 500°C in agreement with previous findings.[11] The concentrations were calculated from the intensities by using a layer by layer model already described.[2] It shows that at 800°C the Mo and the Ni concentrations on the surface are

0142–2421/86/120377–05 $05.00

Table 1. Binding energies of the elements Ni, Mo, S and chemical shifts of the oxides

Element and valence state	Binding energy (eV)	Chemical shift (eV)	
Ni $2p_{3/2}$	852.8		(measured)
Ni^{2+} (in NiO)	854.4	1.6	(measured)
Ni^{2+} (in the . mixed oxide)	854.9	2.1	(measured)
$Mo3d_{5/2}$	227.5		(measured)
Mo^{6+} (in MoO_3)	232.3	4.8	(measured)
Mo^{6+} (in the mixed oxide)	231.7	4.2	(measured)
Mo^{4+}	229.1	1.6	(best fit in the
	230.4	2.9	deconvoluted spectra)
S 2p (adsorbed S)	162		(measured)
S 2s (adsorbed S)	226		(measured)

identical to the bulk concentrations, while at 500°C Mo is segregated on the surface, with a concentration of ~25%.

A clean surface with a composition identical to the bulk composition can be obtained by annealing at 800°C followed by a very short ion etching to remove the adsorbed sulphur. The ion sputtering is very critical because it causes a strong Mo enrichment on the surface, which is expected when sputtering alloys containing elements of different masses. The surface enrichment of Mo during ion sputtering is characterized in Fig. 2. After 30 min, I_{Mo}/I_{Ni} is very close to the value reported above for the alloy annealed at 500°C.

Oxidation of the alloy at 25°C

Figure 3 shows the ESCA spectra recorded before and after oxidation at 25°C. The initial surface was ion etched for 6 min, and the peak height ratio I_{Mo}/I_{Ni} was 0.075, indicating a molybdenum enrichment on the

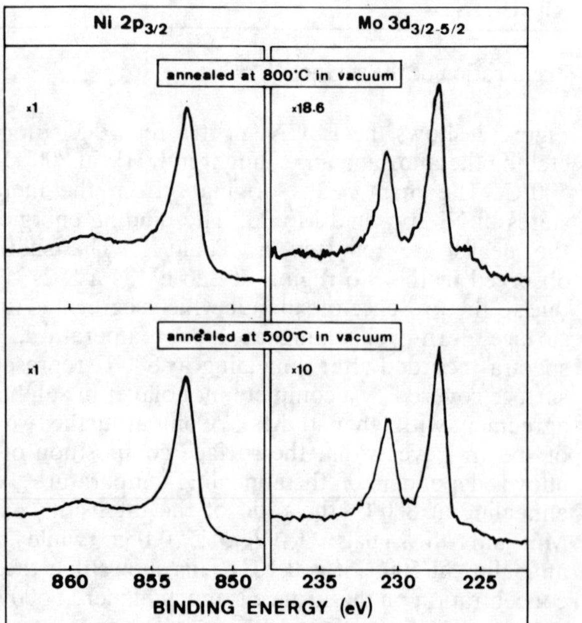

Figure 1. ESCA spectra of Ni–6at%Mo(110) alloy annealed in vacuum at 800°C and at 500°C.

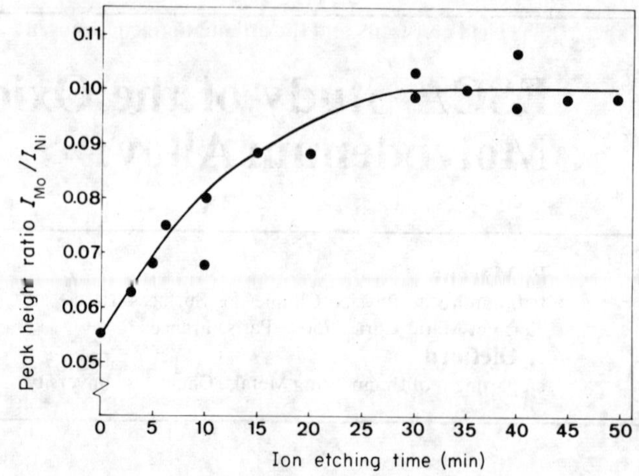

Figure 2. Changes in the peak height ratio I_{Mo}/I_{Ni} during ion etching ($p_{Ar} = 5 \times 10^{-4}$ Torr, 500 V).

surface. Using the layer model and assuming an enriched region of 2 planes, the concentration of Mo in this surface region is found to be ~12%. After oxidation, the spectra in Fig. 3 show that both Ni and Mo are oxidized. However the metallic states are still detected, which shows that the oxide is very thin. The spectra are deconvoluted into elemental peaks representing the chemical states of the elements. The binding energies of the components are given in Table 1. The chemical shift of Ni^{2+} in the oxide formed on the alloy is 1.6 eV, i.e. the same as in pure NiO.[1] The main oxide state of molybdenum in this oxide is Mo^{6+}. The shift is 4.2 eV, which is significantly less than in pure MoO_3, on which we found 4.8 eV, in agreement with reported results.[12] The intensities of the oxide signals of the elements I_x^{ox} (X=Ni, Mo), and the metal signals I_x^{met}, can be expressed in the following way:

$$I_x^{ox} = D_x^{ox} \ \lambda_x^{ox} \ Y_x^{ox} \ 0.62\,(1 - \exp(-a/0.62\ \lambda_x^{ox})) \quad (1)$$

$$I_x^{met} = D_x^{met} \ \lambda_x^{met} \ Y_x^{met} \ 0.62 \exp(-a/0.62\ \lambda_x^{ox}) \quad (2)$$

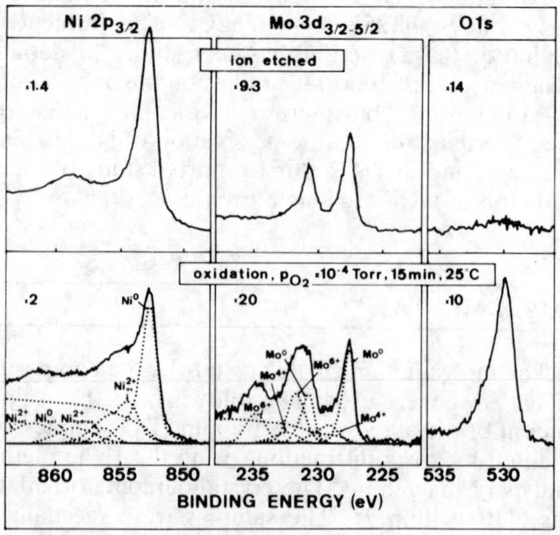

Figure 3. ESCA spectra of Ni–6at%Mo(110) alloy after ion etching and after oxidation at 25°C ($p_{O_2} = 10^{-4}$ Torr).

Table 2. The yield constants and the attenuation lengths used

	$Y \frac{\text{Kc eV cm}^3}{\text{s mole Å}}$	λ (Å)
Mo^{met} 3d	18.5	15
Mo^{ox} 3d	21.5	20
Ni^{met} 2p	36.3	10
Ni^{ox} 2p	34.3	13
O 1s	9.6	17
S 2p	—	21

where D is the density of the element, λ the mean free path of the photoelectrons, Y the experimental yield factor, a the thickness of the oxide layer and 0.62 the correction for the take off angle. The parameters are given in Table 2. Combining (1) and (2) allows us to calculate the oxide thickness and the composition. The thickness is 5.5 Å, and the Mo content is 12%. Thus the composition of the oxide formed at 25°C reflects the composition of the surface region which can be oxidized. Indeed it was shown before that the initial surface was enriched with Mo, and that the Mo concentration was ~12% in a surface region involving two planes. The concentration of metallic Mo below the oxide film is found to be 5.2%, a value very close to the bulk content, which is consistent with the finding that the enrichment of Mo in the oxide results from the enrichment of Mo on the initial metal surface.

Oxidation at 500°C

Samples were ion etched for either a short time to obtain a surface composition close to the bulk composition or for a longer time, so that a Mo-enriched surface was obtained. Then they were oxidized at 500°C (P_{O_2} = 10^{-4} Torr). The oxidation time was ranging from 2 min to 80 min. Figure 4 shows the ESCA spectra recorded after oxidizing the surface without Mo-enrichment for the shortest time (2 min) and the longest time (80 min). After 2 min the signals from the metals dominate while those from the oxide dominate after longer exposure. However, even after oxidation for 80 min the metallic states of Ni and Mo are still detected. This shows that the oxide layer, or a fraction of it, is very thin. The binding energy of the Ni^{2+} peak changes with the oxidation time. The chemical shift after short exposure time is 1.6 eV, as in pure NiO. After longer exposure, it is 2.1 eV. The chemical shift of the Mo^{6+} peak is 4.2 eV, while it is 4.8 eV for MoO_3 formed on pure Mo. The Mo spectra obtained after 2 min of oxidation show that Mo^{4+} and possibly a higher oxide state are also present. The binding energy of Mo^{4+} has been identified before.[12] The concentration of Mo^{4+} in the oxide decreases with exposure time. These results show that the oxide formed in the early stage is different to the mixed oxide which covers the surface at later stages. The influence of the initial surface concentration of molybdenum has been investigated by oxidizing a Mo-enriched surface. The results are summarized in Fig. 5, showing the measured peak height ratio of the Mo^{6+} and Ni^{2+} signals as a function of oxidation time for a normal surface (A) and a Mo-enriched surface (B). It appears that the Mo^{6+}

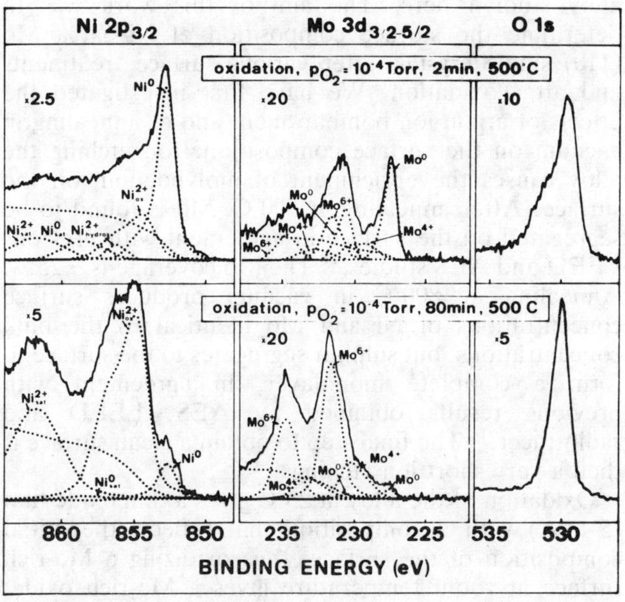

Figure 4. ESCA spectra of Ni–6at%Mo(110) alloy after oxidation at 500°C (p_{O_2} = 10^{-4} Torr).

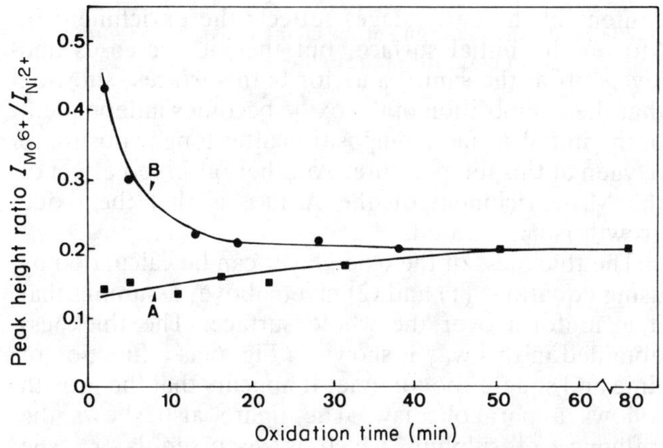

Figure 5. Peak height ratio of Mo^{6+} and Ni^{2+} versus oxidation time (500°C, p_{O_2} = 10^{-4} Torr), (A) without and (B) with Mo enrichment of the initial surface.

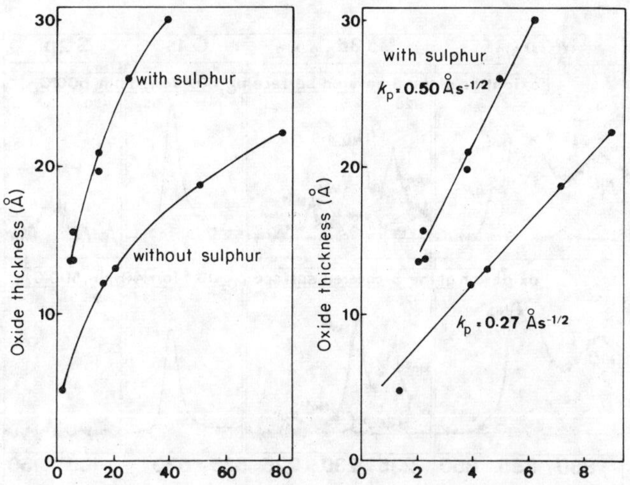

Figure 6. Thickness of the oxide layer growing on a sulphur-free and on a sulphur-covered surface.

content in the early stages reflects the enrichment of Mo on the initial surface, but then it decreases and levels off at the same value for both surfaces, showing that the composition of the oxide becomes independent of the initial surface composition after long exposure to oxygen at this temperature. Another observed effect of the Mo-enrichment on the surface is that the oxide growth is accelerated.

The thickness of the oxide layer can be calculated by using equations (1) and (2) given above, assuming that it is uniform over the whole surface. The thickness obtained in this way is shown in Fig. 6 as a function of time and square root of time. It appears that the growth follows a parabolic law. The figure also shows the influence of sulphur, which is discussed later. The composition of the oxide is derived from the same equations (1) and (2). The obtained concentration of Mo is 12%, i.e. twice as much as in the alloy matrix.

Oxidation of sulphur-covered surfaces

Sulphur was adsorbed by annealing the alloy at 800°C in vacuum. Under these conditions sulphur segregates to the surface and a complete monolayer is formed.[11] The sample was then exposed to $O_2(10^{-4}$ Torr) at 500°C for times ranging from 5 min to 40 min. Fig. 7 shows the ESCA spectra recorded after the shortest and the longest exposure times. The main features of the Ni and Mo regions are similar to those observed before in Fig. 4. The Mo content of the oxide is 7.2%. The main difference concerns sulphur which is still detected after oxidation for 5 min and up to 25 min. After longer exposures a weak sulphur signal is detected. The increase in thickness of the oxide is shown in Fig. 6. It appears that the oxide formed on the S-covered surface is almost twice as thick as the one formed on the S-free surface. Fig. 6 shows that the oxide growth on the S-covered surface also follows a parabolic law; the rate constant is 0.5 Å $s^{-1/2}$ while it is 0.27 Å $s^{-1/2}$ for the S-free surface. Sulphur accelerates the growth of the oxide under the conditions of temperature and oxygen pressure used. The changes in the measured intensity of the S 2p peak during oxidation are shown in Fig. 8. The

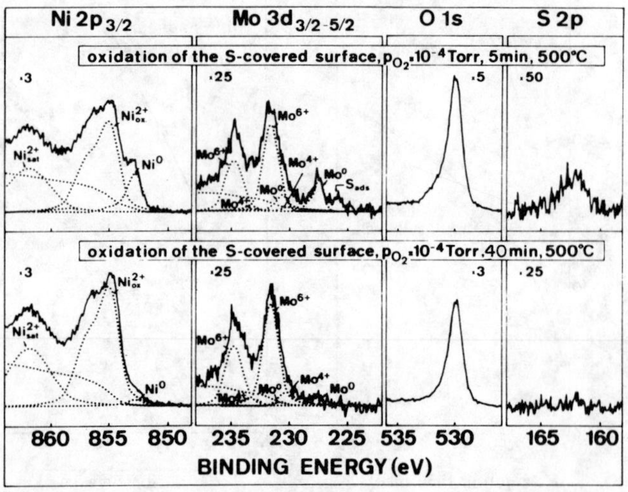

Figure 7. ESCA spectra of Ni–6at%Mo(110) alloy covered with a monolayer of sulphur, after oxidation at 500°C ($p_{O_2} = 10^{-4}$ Torr).

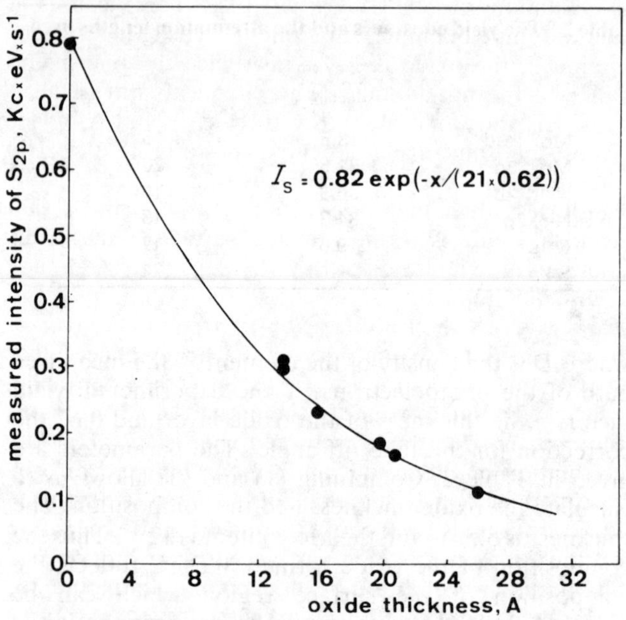

Figure 8. Intensity of the sulphur peak vs. thickness of the oxide.

experimental values are very well fitted by an exponential function for the intensity emitted from a layer of sulphur at the metal/oxide interface. This good agreement shows that sulphur remains at the metal/oxide interface during the growth of the oxide. This finding is also consistent with results obtained for the oxidation of a S-covered surface at room temperature, which shows that the oxide layer is formed on top of the sulphur layer.

DISCUSSION

Earlier studies of the kinetics of oxidation of Ni (110), by RHEED and x-ray emission[13] and by XPS[14] have been reported, while no data were available on the alloy studied here. The aim of this work was to determine the surface composition of Ni–6at% Mo (110) single crystals after various surface treatments and after oxidation. We have first investigated the effects of argon ion bombardment and of annealing in vacuum on the surface composition. Ion etching the alloy causes the enrichment of molybdenum on the surface. After annealing at 500°C, Mo is found to be segregated on the surface in agreement with previous LEED and AES studies.[11] The Mo coverage is ~25%. Annealing at 800°C in vacuum produces surface concentrations of Ni and Mo identical to the bulk concentrations, but sulphur segregates to the surface to form a complete monolayer, in agreement with previous results obtained by AES, LEED and radiotracer.[11] The final step to obtain a clean surface is then a very short ion etching.

Oxidation of the alloy at 25°C gives a thin oxide film (5–6 Å) with a composition that reflects the initial composition of the surface, thus oxidizing a Mo-rich surface at room temperature gives a Mo-rich oxide. Some hydroxide is also present on the oxide film formed at this temperature.

Oxidation at 500°C (p_{O_2} = 10^{-4} Torr) causes a marked molybdenum enrichment. The Mo concentration in the oxide is 12%, i.e. twice the one in the bulk. In the initial stages, the chemical shift of Ni^{2+} ($2p_{3/2}$) is the same as in NiO (1.6 eV), but a longer exposure results in a larger chemical shift (2.1 eV). The chemical shift of Mo^{6+} ($3d_{5/2}$) is 4.2 eV, i.e. significantly less than in MoO_3 (4.8 eV). This shows that a mixed oxide is formed on the surface. Our results show that the lateral growth of oxide islands is already complete after a very short exposure to O_2 (2 min) and the surface is then covered by a continuous layer of oxide which grows in thickness. The evidences that support this statement are: (i) the growth is parabolic, which can only be found for the growth of a continuous film and not for the growth of oxide islands, (ii) the calculation of the cation and anion charges from the recorded spectra, gives a correct balance (within 4%) for a continuous layer model, and (iii) in the oxidation of the sulphur covered surface, the sulphur peak attenuation is well fitted by an exponential function for the growth of a continuous film on the top of the S-layer (and a parabolic growth was also found in this case). The parabolic growth observed here is in agreement with the early theories of Mott for very thin films. However a logarithmic growth has been reported for the oxidation of Ni(110), at lower temperatures (<200°C),[13] while a parabolic growth was reported for the oxidation of Ni(100), also at lower temperatures.[15] The comparison with pevious results on Ni–Fe alloys oxidized under identical conditions[3] shows that the oxide growth is slower on the Ni–Mo alloys.

The influence of sulphur has been investigated. A monolayer of sulphur was preadsorbed by segregation at 800°C. Sulphur has a marked influence on the oxidation of the alloy. The oxide growth is faster in the presence of sulphur. This may be related to the existence of more defects in the oxide formed in the presence of S. We have shown that the oxide grows on top of the sulphur layer. This may have an influence on the number of defects, creating more paths of rapid diffusion. Formation of oxide layer on top of adsorbed sulphur has already been reported for Ni–Fe alloys.[3] On Ni, it was shown that the passive film can be formed on a surface covered by a fraction of a sulphur monolayer.[4] It was also shown by RHEED that adsorbed sulphur can perturb the epitaxial growth of the passive oxide film on nickel.[4] A much smaller Mo enrichment is found in the presence of sulphur. This may be a consequence of the more rapid growth.

CONCLUSION

The conditions to obtain a clean alloy surface with a composition identical to the bulk composition have been determined. The composition and the thickness of the oxides formed at 25°C and 500°C have been measured and the oxide growth mode at 500°C has been investigated. The influence of adsorbed sulphur on the oxidation has been studied. The chemical shifts of Ni^{2+} and Mo^{6+} in the mixed oxide formed on the alloy have been measured.

REFERENCES

1. P. Marcus, J. Oudar and I. Olefjord, *J. Microsc. Spectrosc. Electron.* **4**, 63 (1979).
2. P. Marcus and I. Olefjord, *Surf. Interface Anal.* **4**, 29 (1982).
3. I. Olefjord and P. Marcus, *Surf. Interface Anal.* **4**, 23 (1982).
4. J. Oudar and P. Marcus, *Appl. Surf. Sci* **3**, 48 (1979).
5. P. Marcus, I. Olefjord and J. Oudar, *Mater. Sci. Eng.* **42**, 191 (1980).
6. P. Marcus and J. Oudar, *Passivity of Metals and Semiconductors,* p.119 ed. M. Froment, Elsevier (1983).
7. P. Marcus, A. Teissier and J. Oudar, *Corrosion Sci.* **24**, 259 (1984).
8. P. Marcus, I. Olefjord and J. Oudar, *Corrosion Sci.* **24**, 269 (1984).
9. M.B. Ives, V. Nitrovic-Scepanovic and M. Moriya, *Passivity of Metals and Semiconductors,* p.175 ed. M. Froment, Elsevier (1983).
10. P. Marcus, A.M. Gosnet and J. Oudar, *Mém. Et. Sci. Rev. Met.* **82**, **2**, 83 (1985).
11. P. Marcus, H. Chadli and C. Chandler, to be published.
12. I. Olefjord and B. Brox, *Passivity of Metals and Semiconductors,* p.561 ed. M. Froment, Elsevier (1983).
13. D.F. Mitchell, P.B. Sewell and M. Cohen, *Surf. Sci.* **69**, 310 (1977).
14. P.R. Norton, R.L. Tapping and J.W. Goodale, *Surf. Sci.* **65**, 13 (1977).
15. P.H. Holloway, thesis, Rensselaer Polytechnic Institute, Troy (NY) (1973).

SURFACE AND INTERFACE ANALYSIS, VOL. 9, 383–387 (1986). EXTENDED ABSTRACTS

Surface EXAFS and Magic Angle Spinning NMR Studies of Anodically Formed Oxide Films on Aluminium

A. J. Forty, S. M. El-Mashri*, R. G. Jones†, R. Dupree and I. Farnan
Physics Department, University of Warwick, Coventry CV4 7AL, England
* Present address: Libyan Atomic Energy Establishment, Tripoli, Libya.
† Present address: Department of Chemistry, University of Nottingham, UK

Aluminium oxide (Al_2O_3) films formed by anodic oxidation of aluminium in aqueous electrolytes are of great interest in many technological applications. For example, adhesive bonding of structural components fabricated from aluminium alloys is now used extensively in the aerospace industry.[1] Aluminium is also used in the construction of chemical reactors and water desalination plants and in the construction of mirrors for solar collectors.

When aluminium is anodised in certain electrolytes, an oxide film, consisting mainly of amorphous alumina, is formed on the surface. The nature of the film depends on several factors, the most important of which is the nature of the electrolyte used in forming the film. The strength and chemical stability of these oxides, which depend on the surface preparation, are of prime importance in determining the strength of the bond. A knowledge of the molecular structure of such films should be an important step in reaching an understanding of the structural chemistry of the anodic oxide and, in particular, the hydration in hot water.

The techniques of electron yield-EXAFS and magic angle spinning-NMR have been used to derive information about the different structures of amorphous films, formed anodically on pure aluminium in various electrolyte solutions (tartaric, oxalic, boric, phosphoric and chromic acids), details of which are given elsewhere.[2] The electron yield from the oxidized surfaces were measured to obtain EXAFS spectra over a range of incident x-ray energies extending from 1500 eV (to include the Al K-edge absorption) up to about 2000 eV. This was carried out using the ACO storage ring and double-crystal soft x-ray monochromator operated at the National Scientific Research Centre, Orsay Laboratory (France). Details of the experimental arrangement and method of counting the electron yield are given elsewhere.[3] The analysis of these EXAFS data was carried out by the Fourier transform method. It has been found that the uniform non-porous (barrier type) oxide films which are formed in neutralized sodium tartrate or sodium borate electrolytes are amorphous and have an average Al–O bondlength of 0.19 nm. The amorphous oxide produced in neturalized sodium oxalate gives an average Al–O separation of 0.185 nm, while the porous oxides formed in strong aggressive electrolytes, chromic acid and phosphoric acid, have an average Al–O bondlength of 0.183 nm and 0.180 nm respectively.

The wide variation in measured Al–O bondlength reflects the dependance of molecular structure on the method of preparation. By using a simple relationship between bondlength and the state of co-ordination[3] it is found that the oxide films formed in sodium tartrate and sodium borate have 90% of the aluminium ions in octahedral sites and only about 10% in tetrahedral sites; the aluminium oxide formed in chromic acid appears to have most of the aluminium in tetrahedrally bonded sites and only a small fraction in octahedral sites, with a ratio of 6-fold to 4-fold co-ordination in the range of (10/80) with an uncertainty of $\pm$ 8%. As the Al–O bondlength decreases the average co-ordination number also decreases, which means that more of the Al^{3+} ions are tetrahedrally bonded. In the extreme case of the oxide formed in phosphoric acid where the measured bondlength is 0.180 nm, all the Al^{3+} are tetrahedrally co-ordinated.

A more detailed analysis of the molecular structure can be obtained using magic angle spinning (MAS) NMR. This is an ideal technique for studying the local co-ordination of aluminium ions. For example it reveals well separated peaks for octahedral and tetrahedral aluminium in crystalline aluminas. An initial study[4] of the amorphous alumina films formed in phosphoric acid and sulphuric acid using this technique has revealed octahedral (AlO_6) and tetrahedral (AlO_4) centres and also a new type of centre (AlO_5). The relative proportions of these centres depend on the method of preparation of the film. The 5–co-ordination may be related to the incorporation of anion-species from the electrolyte in the oxide.

REFERENCES

1. J. D. Venables, D. K. McNamara, J. M. Chen, T. S. Sun and R. L. Hopping, *Appl. Surface Science*, **3**, 88 (1979).
2. S. M. El-Mashri, *Scanning electron Microsc.* **I**, 157 (1985).
3. S. M. El-Mashri, R. G. Jones and A. J. Forty, *Phil. Mag. A*, **48**, 665 (1983).
4. R. Dupree, A. J. Forty, S. El-Mashri and I. Farnan, *Proc. 3rd Int. Conf. the Structure of Non-Crystalline Materials*, ed. C. Janot, Collogne C8, *Journal de Physique*, **46**, C8–113 (1985).

Quantitative Analysis of Conversion Layers on Aluminium

H. Bubert,* R. P. H. Garten,† H. Puderbach,‡ and S. Storp§
* Institut für Spektrochemie und angewandte Spektroskopie, Dortmund, FRG
† Max-Planck-Institut für Metallforschung, Laboratorium für Reinststoffanalytik, Bunsen-Kirchhoff-Strasse 13, 4600-Dortmund, FRG
‡ Henkel KGaA, Düsseldorf, FRG
§ Bayer AG, Leverkusen, FRG

Very thin, complex technical layers produced by modern no-rinse processing have been analysed by a combined approach including TEM, AES, XPS and further SIMS, ISS and LAMMA. The morphology, layer thickness, elemental concentrations and in-depth profiles of Al, Cr, C. O. F, Si, P, Mg, as well as some chemical compounds have been sufficiently determined. Problems and limitations are discussed.

INTRODUCTION

Surface and thin-film analytical characterization of the near surface layer has become an important task in product oriented industrial research and development of a wide variety of engineering materials.

We studied the physical and chemical structure of three modern oxide-containing chromate conversion layers on aluminium, produced in an industrial no-rinse process on a large scale for the purpose of corrosion protection and improved adhesion of organic coatings. The most prominent advantage of these 'no-rinse' layers NR 6207, NR 6001, NR 6011 over conventionally produced layers is considerable saving of energy consumption and waste-water treatment.

The purpose of our surface analytical characterization aims at both product oriented improvements (comparison of process variants, failure diagnostics, coating adhesion, layer homogeneity, comparison with conventionally produced oxide/chromate-layers) and the assessment of a multi-method surface analytical approach to characterize technical layers of this type, in a more generalized sense. This includes the comparison of the share of the individual methods applied, in the overall result, the denotion of analytical problems and limitations, when dealing with such types of commercial layers.

RESULTS AND DISCUSSION

We used a combined approach including TEM, AES, XPS and further SIMS, ISS and LAMMA.[1,2] The thickness of layers was estimated to range from 30 to 300 nm — depending on the production process — by means of comparative sputtering measurements on Ta_2O_5, mechanical profilometry and x-ray microanalysis. Based on the information on the composition of the coating bath solutions used, all elements included in these bath solutions, viz. Al, Cr, C, O, F, Si, P, Mg could be identified in the respective coating layers in a characteristic pattern. In-depth profiles of these elements were determined using AES and XPS. Positive LAMMA spectra quickly and easily revealed a survey of the elemental compositions of the layers. Quantitative hydrogen determination has not been thought reasonable at the present state of our investigation due to the large number of hydrogen-containing fragments, combined with the overall difficulties in quantitation of SIMS spectra from complex samples.

Some chemical compounds were identified by means of XPS and LAMMA to be present within these layers, viz. Al_2O_3, CrO_3, $CrPO_4$ and polyacrylic acid. Serious restrictions result from small variations in the XPS-chemical shifts[3,4] of different compounds that can be expected in the layers, according to the preparation process. Thus, the presence of further compounds like AlF_3 and Al-silicate is found to be likely, but difficult to ascertain.

The negative LAMMA spectra showed strong clustering of molecular fragments, that can be attributed to a number of regular cluster series of the types $Al_xO_yF_z \cdot (SiO_2)_n$ and $Cr_xO_y \cdot (SiO_2)_n$. However, comparison with LAMMA spectra from pure powdered compounds Cr_2O_3, CrO_3, $AlF_3 + CrPO_4$ and $AlPO_4 + CrF_3$ did not allow a distinct analytical interpretation of the composition of the original layers or of the oxidation states of Cr in the different layers.

Glow discharge optical emission spectrometry faces prohibitive difficulties to interpret elemental in-depth profiles obtained from such insulating layers.

Stoichiometric balance between electro-positive and electro-negative elements is not obtained following the AES and XPS results; this indicates some systematic errors in quantitation.

As a conclusion, basic problems remain to be solved for thin film analysis of such technical layers, namely: (1) the reliable quantitation of the original elemental in-depth profiles including hydrogen, and (2) the identification of chemical compounds within interface layers, i.e. chemical depth profiling. Even from a combined use of corroborating methods, it remains difficult to obtain a complete characterization of the layers. In the present study, we determined sufficiently the morphology, thickness, elemental composition and in-depth profiles of these layers; also some chemical compounds were reliably identified. Further progress could be made by use of any low-damage thin-film sectioning combined with XPS, and by application of suitable standard reference materials, i.e. the additional application of further analytical methods to similar samples.

REFERENCES

1. H. Puderbach, H. Bubert, and R. P. H. Garten, *Spectrochim. Acta*, **40B**, 817 (1985).
2. H. Puderbach, H. Bubert, A. Quentmeier, R. P. H. Garten, and S. Storp, 12th Kolloqium über Werkstoffanalytik, Vienna 1985, *Mikrochimica Acta*, **Suppl. 11**, 103–112 (1985).
3. R. Holm and S. Storp, in *Analysen- und Meßverfahren*, Ullmanns Encyclopädie der technischem Chemie, Band 5, ed. by H. Kelker, pp. 519–576, Verlag Chemie, Weinheim (1980).
4. D. Briggs, in *Practical Surface Analysis by Auger and X-ray Photoelectron Spectroscopy*, ed. by D. Briggs and M. P. Seah, Chap. 9, pp. 359–396, John Wiley, Chichester (1983).

XPS Study of Thin Oxide Films Formation on Iron Alloys

A. G. Akimov and V. Yu. Demin
Institute of Physical Chemistry, Academy of Science, Leninsky Prospect, Moscow, USSR

The oxidation of alloys leads to the formation of oxide film with the non-uniform compositon depth profile which depends, as a rule, on the oxidation conditions. The numerous studies of alloys oxidation indicate also that the oxide film composition frequently changes strongly with time and the film can be enriched first by one component of the alloy and then by another, though the experimental conditions are the same.[1–3]

In the present study processes of the formation stable protective oxide films on Fe–Cr and Fe–Ni alloys have been investigated. The formation kinetics and the composition of oxide films, formed *in situ* in the analyser chamber of ESCALAB-5 electron spectrometer at 20–600°C and oxygen pressures 10^{-4}–10^{-8} Torr have been studied by XPS depth profiling with Ar^+ ion etching. Fe $2p_{3/2}$, Cr $2p_{3/2}$ and O 1s lines were employed to define the chemical state and for quantitative analysis. The experimental technique has been described in details elsewhere.[2,4] Only the oxide Fe_3O_4, Cr_2O_3 and NiO are themodynamically stable in the temperature range investigated. On Fe–20% Cr alloy at 500°C Cr_2O_3 layer formed at low oxygen pressure ($\leqslant 10^{-7}$ Torr) and higher pressures lead to the formation of scale, consisting at least of two layers: Fe_3O_4 on the gas–oxide interface, and $FeCr_2O_4$ (or Cr_2O_3 solution in Fe_3O_4) on the oxide–alloy interface.[4–6]

The investigations of the initial stages of Fe–Cr alloys oxidation have shown that although Cr plays the determinative role in the formation of the protective oxide film, not Cr oxides as was accepted earlier, Fe oxides are the main components of these films. The composition of oxide films depends upon the alloy composition and oxidation conditions and can be determined by the oxidant concentration on gas–oxide interface, diffusion in oxide film and bulk alloy and also by solid state chemical reactions:

$$2\,Cr^{3+} + Fe_3O_4 \rightarrow FeCr_2O_4 + 2\,Fe^{3+} \qquad (1)$$

in the oxide film.[7] The accumulation of $FeCr_2O_4$ spinel during partial reduction or Cr_2O_3 oxide during total reduction of Fe_3O_4 in oxide film can be the result of this reaction. The greater Cr, compared with Fe, affinity to oxygen is a motive force of the substitution reaction. The reduction of Ni oxide by metallic Fe can proceed in the similar way:

$$2\,Fe^{3+} + 4Ni \rightarrow NiFe_2O_4 + 3\,Ni^{2+} \qquad (2)$$

The reduction in this reaction can also be partial — with the formation of $NiFe_2O_4$ spinel — or total up to the formation of Fe_3O_4.

The kinetics of these reactions is of very great importance as it allows us to find out their limiting stages which is essential to understand the oxidation mechanism. It turned out possible to measure the kinetics of the reaction of total Fe_3O_4 reduction by metallic Cr, which was studied as the variation of thin oxide film composition on Fe–Cr alloy during its vacuum annealing.

The specimen was heated in vacuum up to experiment temperature and cleaned by Ar^+ ion etching. The oxide film formed has the thickness of about 50 Å and consists mainly of Fe_3O_4. The oxygen was pumped away and the oxide film formed was annealed in vacuum at a constant temperature. Cr reduced Fe_3O_4 and accumulated itself in the film as Cr_2O_3. The accumulation kinetics can be approximated by logarithmic function of time. The reduction of Fe_3O_4 can be presented as a two stage process: Cr transport to reaction zone and the reaction of Fe_3O_4 reduction and Cr_2O_3 formation itself. The low value of the reaction activation energy (about 30 kJ mol^{-1}) indicates that the processes of solid state transport are the limiting stage even in the case of very thin films.

The Cr amount, transported through the alloy–oxide interface during oxidation, is limited by the diffusion in alloy and the conditions on this interface. If the interface is in good synch. and Cr concentration is equal to zero on it, then Cr flow reaches maximum and its amount, diffused into the oxide from an alloy in time t, is proportional to $\sqrt{Dt}$ where D = diffusion coefficient of Cr in the alloy.[8]

As was already mentioned the reduction rate of Fe_3O_4 by Cr is limited by solid state transport, i.e. Cr supply to the reaction zone, in a wide temperature range.

At the beginning of the oxidation the oxide growth rate is limited by oxidant adsorption and is close to linear one, while the Cr flow through the oxide–alloy interface is very large and proportional to $1/\sqrt{Dt}$. Consequently, the enrichment of oxide film by Cr_2O_3 is to take place at the initial stages of oxidation. It was confirmed by the variation kinetics of oxide film surface composition for Fe-20% Cr alloy at a different temperature.

REFERENCES

1. C. Leygraf, G. Hultquist, *Surf. Sci.* **61**, 69–84 (1976).
2. A. G. Akimov, I. L. Rozenfeld, L. P. Kazansky, G. V. Machavariany, *Izv. Acad. Nauk SSSR, ser. khim.* **7**, 1482 (1978).
3. S. E. Greco, J. F. Roux, J. M. Blakely, *Surf. Sci.* **120**, 203–222 (1982).
4. A. G. Akimov, V. Yu. Demin, *Poverchnost': Fiz. Khim. Mekh.* **1**, 106–108 (1982).
5. D. R. Baker, *Appl. Surf. Sci.* **7**, 69–82 (1981).
6. C. R. Brundle, *J. Electr. Spectrosc. Relat. Phenom.* **5**, 291–319 (1974).
7. A. G. Akimov, G. V. Machavariany, *Zachita Metallov.* **1**, 32 (1980).
8. B. S. Bokshtein, Diffusion in Metals p. 21, Moscow, Metallurgia (1978).

Surface Layers and Failure Areas on Lacquered Pretreated Steel Sheets

B. Maurickx*, J. M. Anne* and J. Foct†
* DRST, Usinor, B.P. 2508, 59381 Dunkerque, France
† Laboratoire de Métallurgie Physique, Université de Lille 1, 59655 Villeneuve d'Ascq, France

INTRODUCTION

Steel sheets with organic coating are used for food packaging. The multilayer system must be resistant to aqueous food media, drawable and non toxic.

This paper intends to compare the role of two surface pretreatments: chromic passivation and zinc phosphating performed on steel sheets before lacquering. Lacquered samples are submitted to drawing tests and aqueous corrosion media.

Characterization of surface pretreatment layers is obtained by physical techniques (SEM, GDS, XPS). The examination of surface failures provides us with an approach to the damage mechanisms.

EXPERIMENTAL

The steel substrate contains less than 0.09 % C; 0.5 % Mn; 0.02 % P; 0.02 % S; 0.08 % Si and 0.05 % Al.

Steel sheets are coated with commercial epoxy-phenolic or organosol lacquers and heated for about 15 min at 200°C. Thin organic coatings (4–6 μm) are porous, so they do not ensure a full protection of the metallic surface underneath.

The aim of surface pretreatment is to protect steel below lacquer, whilst providing a good basis for adhesion of polymers. The two pretreatments we compare here are:

Chromic passivation

Which is formed by cathodic polarization of steel in sodium dichromate solutions (20–30 g l^{-1}; 50–80°C). The passivation film contains 8 to 12 mg m^{-2} Cr (chemical analysis) which represents a layer thickness of some nm.

Crystalline zinc phosphating

Which is obtained by pulverisation of a dilute solution of phosphoric acid and monozinc phosphates. The deposited layer (1–1.5 gm^{-2}) is mainly composed of hopeite $Zn_3(PO_4)_2$, 4 H_2O and phosphophylite $Zn_2Fe(PO_4)_2$, 4 h_2O. The reactions involved are:

(a) attack of steel by phosphoric acid:

$$Fe + 2\,H_3PO_4 \quad Fe(H_2PO_4)_2 + H_2$$

(b) precipitation of zinc phosphates:

$$3\,Zn(h_2PO_4)_2 \quad Zn_3(PO_4)_2 + 4\,H_3PO_4$$
$$Fe(H_2PO_4)_2 + 2\,Zn(H_2PO_4)_2 \quad Zn_2Fe(PO_4)_2 + 4\,H_3PO_4$$

RESULTS AND DISCUSSION

Characterization of surface pretreatments

The very different nature of the two pretreatments studied imposes the use of different methods of characterization.

Chromic passivation. The thin passivation layer is studied by glow discharge spectrometry (GDS) and x-photoelectron spectrometry (XPS). The unknown erosion rates of passivation compounds make these analyses only qualitative; the GDS profile shows:

(a) Iron is very close to the original surface.

(b) Chromium and oxygen are of course observed in the upper layer. The oxygen signal disappears faster than the chromium one, during abrasion, and this result is consistent with the presence of metallic chromium below chromium oxides.

XPS analyses show:

(a) Presence of chromium, oxygen, iron, carbon at the original surface.

(b) Carbon is due to oil residues.

(c) Presence of iron at the immediate surface confirms that passivation probably does not ensure a total coverage of the steel surface.

(d) Displacements of chromium and oxygen peaks are observed during argon sputtering. The Cr $2p_{3/2}$ peak, observed at the original surface at a binding energy of 577.8 eV is attributed to $Cr(OH)_{3,n}H_2O$. the displacement of 0.5 eV of this peak toward lower binding energy, during abrasion, indicates the presence of increasing amounts of dehydrated oxide approximated to Cr_2O_3. The more significant displacement of the O 1s peak from 531.8 eV to 530.8 eV confirms the change from chromium hydroxides to chromium oxides.

(e) A peak due to metallic chromium appears at the low binding energy side of the $Cr2p_{3/2}$ peak after argon sputtering of 6 min. The peak intensity is very low, but it can be approached by extraction of the chromiumoxide spectrum from the total $Cr2p_{3/2}$ spectrum.[1]

Zinc phosphating. SE micrographs (secondary and retrodiffused electron) show a regular deposit of 'needles and leave' on the steel surface 'needles and leaves' are 10 to 30 μm length crystals, x-ray analyses show that they are mainly composed of zinc and phosphorus, according to the nature of the pretreatment.

Lacquered pretreated steel performances

Drawability. This is evaluated by forming, from lacquered pretreated steel blanks, cups of different depths.

(a) Passivation substrates show a good drawing ability. No lacquer de-adhesion or unusual porosity are detected (lacquer porosity is approached by current density measurements in a NaCl solution).

(b) On the other hand, zinc phosphating substrates show lacquer detachments. SEM and x-ray analyses are then used to characterize the failure surfaces on organosol lacquered substrates:

(i) On the metal surface, x-ray analyses show traces of zinc related to scattered and powdered zinc phosphates crystals, no Cl signal due to organosol lacquer is detected.

(ii) On the organosol surface, cracks appear. Cl and Zn signals are very intensive and give rise to images which are roughly the negative pictures of each other. These observa tions prove that failure probably occurs near the steel and zinc phosphates interface.

Behaviour in aqueous media. Only the corrosion behaviour of the drawable passivated substrate is worth studying. The basic function of organic coatings is to act as barriers to corrosive agents, this requires chemical inertia and adhesion of coating in the presence of water. Usually a good inertia is given by the complete curing of the polymer, but bond durability in a humid environment is much more difficult to obtain. High initial dry adhesion between polar polymers and passivated steel is controlled by peeling tests. This adhesion can be understood in terms of dispersion forces and essentially hydrogen bonds. Hydrogen bonds are presumed to act between the polymer polar groups and the hydrated Fe and Cr oxides forming the metal outermost layer at normal ambient bonding condition.[2] It is well known that a high initial adhesion strength does not necessarily ensure a good wet adhesion: the water molecules can diffuse through all polymers,[3] invade the boundary zone and interrupt or displace hydrogen bonds providing loss of polymer adhesion.[2]

Corrosion exposure of the lacquered passivated steel (citric acid (15 g l^{-1}) + NaCl (15 gl^{-1}), 20°C, 3 weeks) causes homogeneously dispersed corrosion pits surrounded by polymer de-adhesion. SE micrographs show that these pits are localized on pre-existing breaks or pores of the polymer film.

XPS study of the two surfaces created by the polymer de-adhesion near corrosion sites shows: (a) presence of Cr and Fe hydroxides on the metal surface; (b) presence of Fe hydroxides and only traces of Cr on the polymer surface.

This proves Cr and Cr compounds are not involved during the corrosion process and essentially stay on the metal sheet. Fe hydroxides however which are corrosion products do not adhere to the substrate and are observed on the polymer side as well as in the electrolyte (chemical analysis). These observations show that the film detachment in wet environments occurs between polymer and passivated layer. Water molecules and hydroxyl ions are expected to induce polymer disbonding.

The destructive mechanism can then be described by a three steps process: (1) penetration of the electrolyte through pores of polymer film; (2) disruption of the polymer/passivated layer bond around the pore along a circle of radius R; (3) Corrosion at weak passivation sites of which surface density is ρ_c.

Variation of R with time t follows a $K(e,T) \sqrt{t}$ law, K decreases with polymer thickness e and increases with temperature T. The initiation delay (t_i) for corrosion is roughly given by $t_i = K^{-2}(e, T)\rho_c^{-1}$.

CONCLUSION

SEM, XPS and GDS studies of failure surfaces created by polymer detachments show that the composite: 'steel sheet/zinc phosphates/polymer' decays by mechanical phenomena, whilst the composite 'steel sheet/passivation layer/polymer', which is drawable, decays in a wet environment by a three-step mechanism incuding corrosion and polymer disbonding at polymer pores.

Acknowledgements

We thank the Commission of the European Communities for supporting this research

REFERENCES

1. K. Hashimoto and K. Asami, in *Corrosion Engineering*, Vol. 26 p. 375, Boshoku Gijitsu (1977).
2. J. C. Bolger, in *Adhesion Aspects of Polymeric coatings*, p. 3–18, ed. by K. L. Mittal, Plenum Press (1981).
3. M. Svoboda, J. Mleziva, in *Progress in Organic Coating*, **2**, 207–222 (1973/74).

APPLICATION OF XPS ANALYSIS TO HIGH TEMPERATURE CORROSION PHENOMENA

G. Moulin, C. Severac and A.M. Huntz

Lab. Metallurgie, Univ. of Paris XI, Bat. 413, 91405 Orsay, France

Carbon and sulphur impurities diffusing from the metallic substrate during oxidation induce an important modification of the physical and mechanical properties of the oxide scales developed on most of the alloys.

According to the nature of the oxide scale and to the carbon solubility in the oxide, XPS analyses show a large alteration of the carbon chemical state and of its distribution in the thickness of oxides such as Al_2O_3 or Cr_2O_3. In the case of a thermally grown alumina scale formed on the surface of 'pure' FeCrAl alloys, for instance, the chromium and carbon enrichment is large enough to form a new outer scale on the inner alumina scale. Whereas only a sursaturation in carbon engaged in C—C or C—O bonds is noticed in the outer part of the oxide scale on the surface of a nickel chromium base alloy; but, even in this case, such a carbon distribution may lead to an important stress level increase in the oxide scale.

On the contrary the effect of sulphur on nickel chromium base alloy oxidation leads to a strong modification of the nature of the oxide scale, but without modifying notably the accommodation stress level gradient, which only slightly decreases.

The combined effect of carbon and sulphur may alter these processes, for instance by promoting the carbide precipitation in a given part of the oxide scale . . . .

The influence of carbon and sulphur during the oxide scale growth is then discussed on account of their chemical state, distribution and qualitative evolution through the oxide scale.

THE USE OF SIMS WITH A MICROFOCUSED ION BEAM TO STUDY THICK CORROSION SCALES

A.T. Baker,* M.J. Bennett,† A. Brown,‡ D.G. Lees* and D.B. Meadowcroft§

* Joint University and UMIST Department of Metallurgy and Materials Science, Grosvenor Street, Manchester M1 7HS, UK

† Materials Development Division, AERE, Harwell, Oxon OX11 0RA, UK

‡ Surface Analysis Industrial Service, Department of Chemistry, UMIST, PO Box 88, Manchester M60 1QD, UK

§ Central Electricity Research Laboratories, Kelvin Avenue, Leatherhead, Surrey, KT22 7SE, UK

Preliminary SIMS imaging experiments have been carried out on cross-sectional corrosion scales. A Vacuum Generators MIG100 gallium liquid metal ion source (LMIS) was used; this was operated at 10 kV and coupled to a quadrupole mass spectrometer and a digital scan unit. The apparatus generated SIMS images, and hence chemical maps of the corrosion scale, by rastering the Ga^+ beam across the surface.

Two types of information which are of particular importance in high temperature corrosion were obtained: (i) the distribution of cerium which was present in low concentrations and (ii) the distribution of stable isotopes used as tracers. The specimens were mounted and cross-sectioned so that depth-distributions could be determined without sputtering.

(i) Stainless steel to which small amounts of cerium had been added, either by alloying or by ion implantation, were oxidized in CO_2. Cerium greatly improves oxide/metal adhesion and its distribution after oxidation has been studied. When the cerium was present as an alloying addition its distribution in the scale was not uniform and it appeared to be mainly at the outer part of a layer of chromium-rich oxide.

(ii) Pure iron was corroded in SO_2/CO/Ar and then in $^{34}S^{18}O_2$/CO/Ar. The distributions of ^{34}S and ^{18}O in the inner part of the corrosion scale were determined in order to understand the growth-mechanism of the scale. Both these isotopes were found in this region which proves that some scale-growth occurred here and that the sulphur and oxygen were probably transported through the scale in gaseous form.

IN-DEPTH QUANTITATIVE ANALYSIS OF Zn–Fe ALLOY PLATED LAYER ON STEELS BY GLOW DISCHARGE SPECTROMETRY

T. Okano, Y. Furunushi, Y. Matsumura and Y. Ohashi

Iron and Steel Research Laboratories, Kawasaki Steel Corporation, Chiba, Japan

A GDS technique has been developed for quantitative in-depth analysis of Zn–Fe alloy electroplated layer on sheet steels.

For the quantification of Fe concentration and the sputtered depth, the calibration curve method was applied to Zn–Fe alloy deposts. Standard samples were the Zn–Fe alloys electrodeposited homogeneously on Cu sheets. The calibration curve for Fe concentration obtained showed good linearity for the wide range of Fe concentration (9~80wt%).

As the sputter rate of Zn–Fe had strong dependence on the Fe concentration, it is necessary to use the sputter rate-Fe concentration curve for converting the sputter time to the sputtered depth. In the present study, the sputter rate of Zn–Fe alloy with a certain concentration was obtained by measuring the time required to sputter away the standard deposit on Cu sheets.

The in-depth analysis of the practically produced Zn–Fe deposits was carried out by correcting the sputter rate sequentially. The light intensities measured at t_i was converted to the Fe concentration and then the corresponding sputter rate D_i(Fe) was estimated by using the sputter rate–Fe concentration curve. By integrating the depth sputtered during each period from the surface to the interface, the amount of total coated weight was estimated by the equation.

$$Wn = \sum_{i=1}^{n} D_i(\mathrm{Fe}) \cdot \Delta t_i \qquad (i = \text{measuring cycle})$$

Some improvements of depth-resolution in GDS will also be discussed.

Section 7.

Polymers

SURFACE AND INTERFACE ANALYSIS, VOL. 9, 391–404 (1986)

SIMS for the Study of Polymer Surfaces: a Review

D. Briggs
ICI PLC, Petrochemicals and Plastics Division, PO Box 90, Wilton, Middlesbrough, Cleveland TS6 8JE, UK

The following aspects of SIMS application to polymer surface studies are reviewed: experimental procedures; positive and negative ion spectra obtained in the 'static' mode for fingerprinting; identification of impurity molecules on polymer surfaces; microanalysis and chemical imaging and current developments in instrumentation.

INTRODUCTION

The major breakthrough in polymer surface analysis was undoubtedly the development of x-ray photoelectron spectroscopy (XPS or ESCA). The features and applications of XPS in this field have been well documented[1,2] and do not need to be repeated. Over the last ten years or so, however, the challenge of industrial research and problem solving in the field of polymer surfaces and interfaces has ruthlessly exposed current limitations of XPS. These may be summarised briefly as follows:

(1) Binding energy (chemical) shifts of important core levels, eg C, N, O have only a small dynamic range leading to ambiguity in speciation and major peak overlap problems in spectra from multifunctional surfaces.

(2) Poor spatial resolution (conventional instruments analyse regions of mm dimensions, more recent 'selected area' instruments can analyse a region of ~150 μm diameter) and lack of imaging capability.

(3) Limited surface sensitivity (typical sampling depth ~ 50 A and variable depending on kinetic energy of peak studied) especially when electron take-off angle variation cannot be used to minimise the sampling depth, e.g. for rough surfaces or powders.

A technique which has the potential to overcome all of these limitations is secondary ion mass spectrometry (SIMS) used in the 'static' or 'low damage' mode. This review essentially describes work in the author's laboratory over the last five years (1981–5) since other relevant literature is sparse. Early work by Gardella *et al.*[3]. appeared, to suffer from the use of a mass spectrometer of limited transmission, particularly in the very useful region m/z>150 daltons. Subsequently some of their data has been shown to be characteristic of ion-beam damaged polymers,[4] i.e. the SIMS experiments were not 'static'. Some SIMS data from polymers has been reported by Campana *et al.*[5] but these mainly refer to extremely thin films on silver substrates which allowed interference from substrate signals in the spectra. In the work reported herein *all* the samples were typical of real-life materials in that they were highly insulating and required charge neutralisation. Even pure reference materials cast into thin films were studied on glass substrates, thus requiring identical methods for acquiring spectra.

It should be emphasised at the outset, also, that we see SIMS as complimentary to XPS and our technique developments have taken place using UHV systems which combine both techniques.

THE SIMS EXPERIMENT

This is illustrated in Fig. 1 (a). A noble gas ion beam, usually Ar^+ or Xe^+, of $\leqslant$ 4 keV energy is directed onto the sample surface, from which material is sputtered into the vacuum. These fragments can consist of one or more atoms; most are neutral but a few percent carry either positive or negative charge. These are extracted into a mass spectrometer (typically of the quadrupole type) to provide, in sequential experiments, the positive and negative secondary ion mass spectra respectively. The primary ion beam *must* be of low current density (~1 nA cm^{-2}) in order to eliminate sample damage during spectral acquisition (see below) and this is achieved either by using a large diameter defocused static beam or by rastering a focused beam

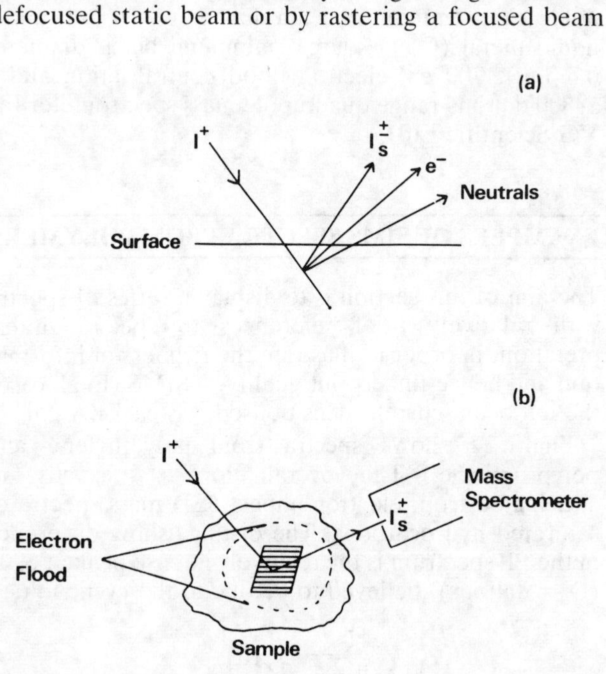

Figure 1. (a) Processes occuring during secondary ion mass spectrometry; (b) schematic of SIMS experiment.

0142–2421/86/120391–14 $07.00

over the surface, as in Fig. 1 (b). The latter mode of operation allows, in principle, SIMS imaging. Here the mass spectrometer is tuned to a chosen mass (strictly m/z) and the signal intensity fed to a CRO scanned in synchronism with the ion beam. This is entirely analogous to SEM, and indeed, since the primary ion beam also gives rise to secondary electron emission an SEM type image can also be obtained from the same field of view.

The major problem in working with polymeric systems is charge build-up since positive ions are being injected into an insulating substrate. This can be overcome by flooding the irradiated area with (relatively) low energy electrons ('electron neutralisation'). In turn this may give rise to electron stimulated ion emission (ESIE, often referred to as 'electron stimulated desorption', ESD) and, by surface potential variation, a variable relative peak intensity situation. By careful attention to experimental detail these effects can be eliminated[6] and, at least in the non-imaging mode, static SIMS data can be obtained as easily and reproducibly as one expects to obtain XPS data from polymer surfaces.

An alternative approach is to use a primary beam of fast atoms so generally minimising the need for neutralisation.[7] This is referred to as 'fast atom bombardment mass spectrometry' (FAB or FABMS). FAB is well established in conventional organic mass spectrometry for bombarding non-volatile organics in a viscous liquid matrix, but these atom sources are too intense for static SIMS. Low current sources are now becoming available for SIMS. The spectra produced using either atoms or ions are, however, essentially the same.[8] It has been assumed that imaging with atom sources would not be possible, but there are now good grounds for believing that imaging at ~10 μm resolution will be possible (see below).

The experimental set up used in our experiments has been described in detail elsewhere.[9] The essential points are the use of two primary ion sources, one for noble gases (minimum beam diameter ~30 μm) and a liquid metal (Ga^+) source (minimum beam diameter 0.5 μm), 700 eV electron flood neutralisation and a 1–800 daltons range quadrupole mass spectrometer (all VG Scientific Ltd).

EXAMPLES OF SIMS SPECTRA FROM POLYMERS

The aim of this section is to display a series of spectra with relatively brief references to spectral interpretation, in order to illustrate the richness of information and hence fingerprint quality of SIMS. In all cases the ion beam current density used was ca 1 nA cm^{-2}.

Figure 2 shows spectra from polyethylene and polypropylene. The overall form is exactly as anticipated from electron-impact (EI) mass spectra of saturated hydrocarbons. The distinguishing difference in the PP spectrum is the relatively intense peak at 69 D (D = daltons), believed to be due to the cyclic ion:

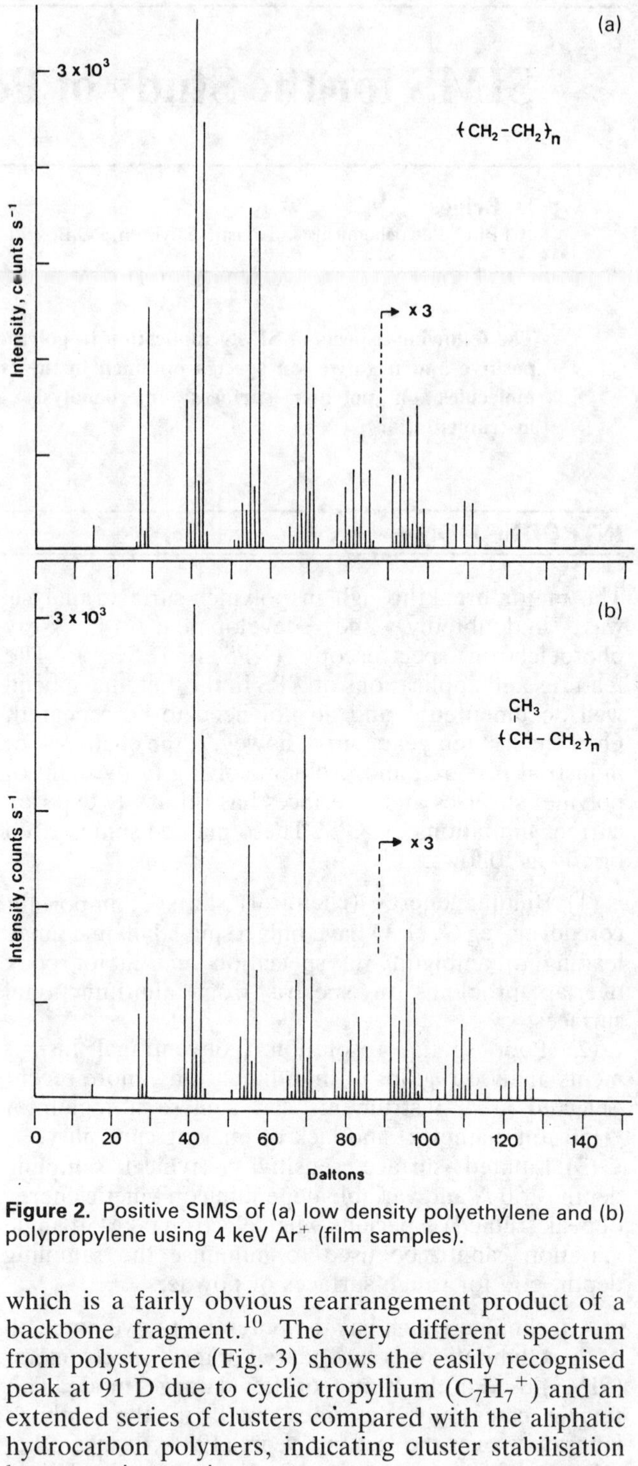

Figure 2. Positive SIMS of (a) low density polyethylene and (b) polypropylene using 4 keV Ar^+ (film samples).

which is a fairly obvious rearrangement product of a backbone fragment.[10] The very different spectrum from polystyrene (Fig. 3) shows the easily recognised peak at 91 D due to cyclic tropyllium ($C_7H_7^+$) and an extended series of clusters compared with the aliphatic hydrocarbon polymers, indicating cluster stabilisation by aromatic moeties.

Figures 4–7 cover a series of polymethacylates in which the backbone of the polymer is the same but which differ in the nature of the pendant alkyl group. Detailed analysis of these spectra shows[4] that the peaks below 100 D are dominated by fragmentation of this alkyl group whereas those above 100 D are dominated by peaks characteristic of the methacrylate backbone. Thus the intense peaks at 15 D (CH_3^+) from PMMA, at 29D ($C_2H_5^+$) from PEMA, at 45 D ($C_2H_4OH^+$) from PHEMA, at 57 D ($C_4H_9^+$) from P^tBMA. The relative intensities of the 29, 43 and 57 D peaks are characteristic of the structure of the isomeric butylmethacrylate polymers, as illustrated by P^tBMA and P^nBMA in Figs 6 and 7.

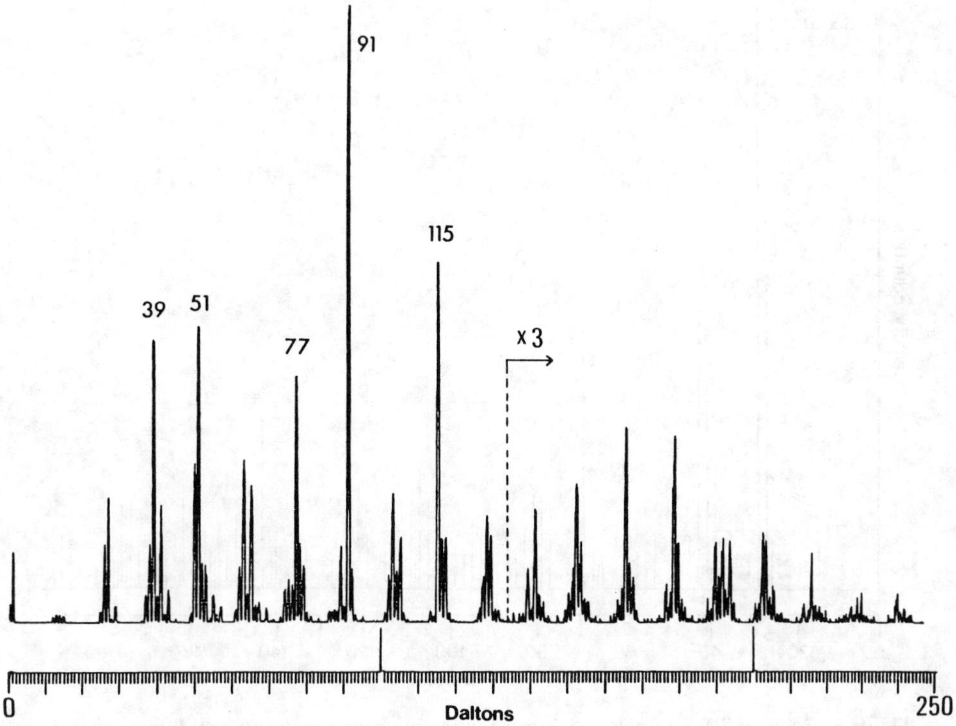

Figure 3. Positive SIMS of polystyrene using 4 keV Ar^+ illustrating the signal: noise characteristics of raw data (film sample). 3×10^3 counts s^{-1} fsd.

Figure 8 is the spectrum from poly (ethylene terephthalate). The protonated repeat unit is seen at 193 D and peaks at lower masses represent logical fragmentation of this cluster with dominant peaks at 149 D (HOOC – C_6H_4 –CO^+) 104 D ($C_6H_4CO^+$) and 77 D ($C_6H_5^+$).

So far these spectra have all been of positive secondary ions. We have previously reported that the negative ion spectra from these molecules contain little *structural* information, being mainly confined to $C_1H_n^-$, O^-, OH^- and C_2 H_m^-. However, this is not always the case. Figures 9 and 10 give the positive and negative ion spectra from poly(ethylene oxide). The ability of oxygen to stabilise negative ion clusters of the type RO^- is clearly seen here. Prominant peaks in both spectra are due to ions with the structures $[(CH_2CH_2O)_nH]^+$, $H(CH_2CH_2O)_nCH_2CH_2O^-$ and HO $(CH_2CH_2O)nCH_2CH_2O^-$ respectively.[11] This

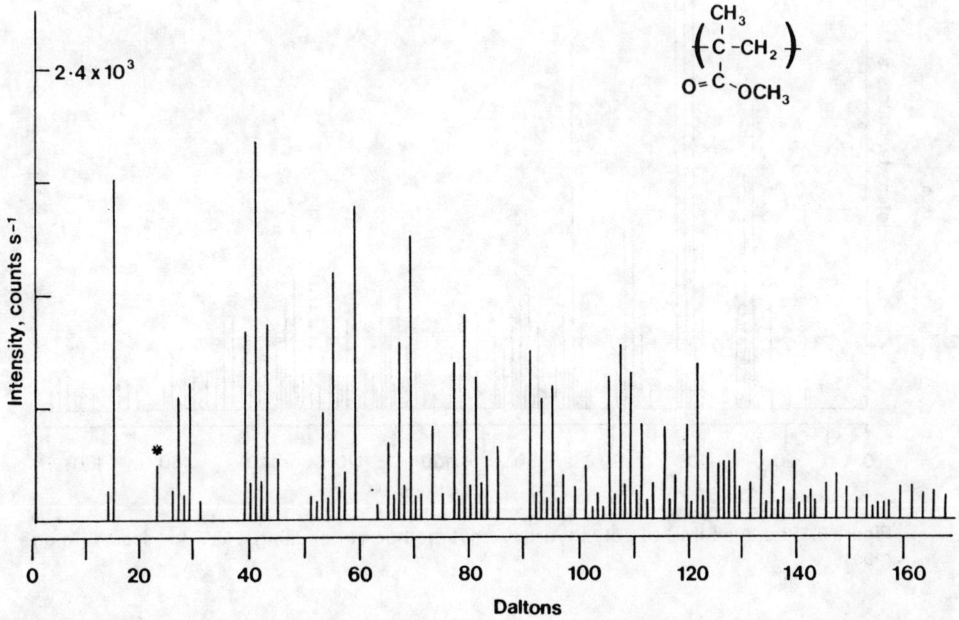

Figure 4. Positive SIMS of poly (methyl methacrylate) using 4 keV Ar^+ (cast film on glass); * = trace Na impurity.

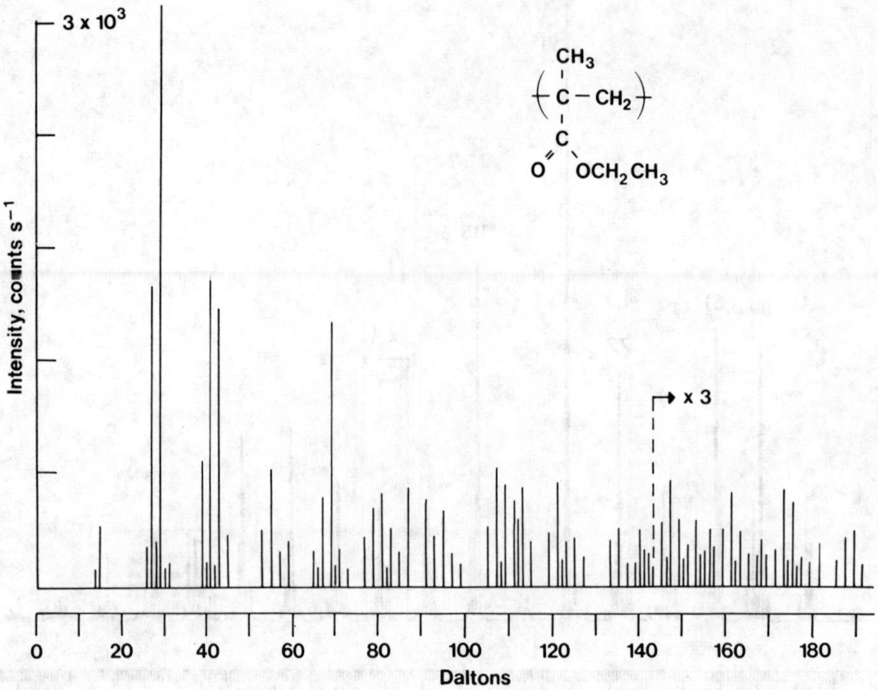

Figure 5. Positive SIMS of poly (ethyl methacrylate) using 4 keV Ar$^+$ (cast film on glass).

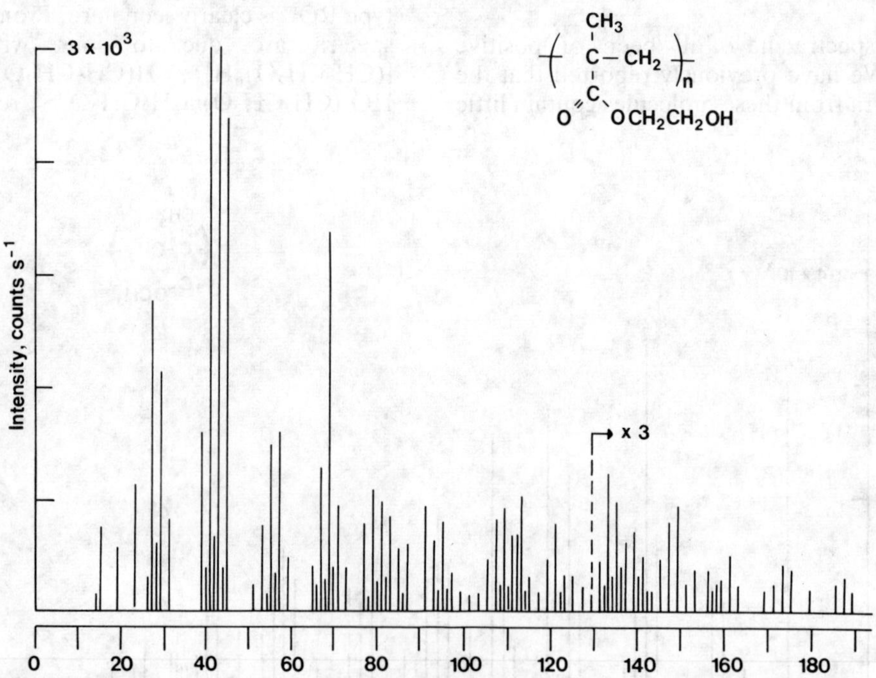

Figure 6. Positive SIMS of poly (hydroxyethyl methacrylate) using 4 keV Ar$^+$ (cast film on glass).

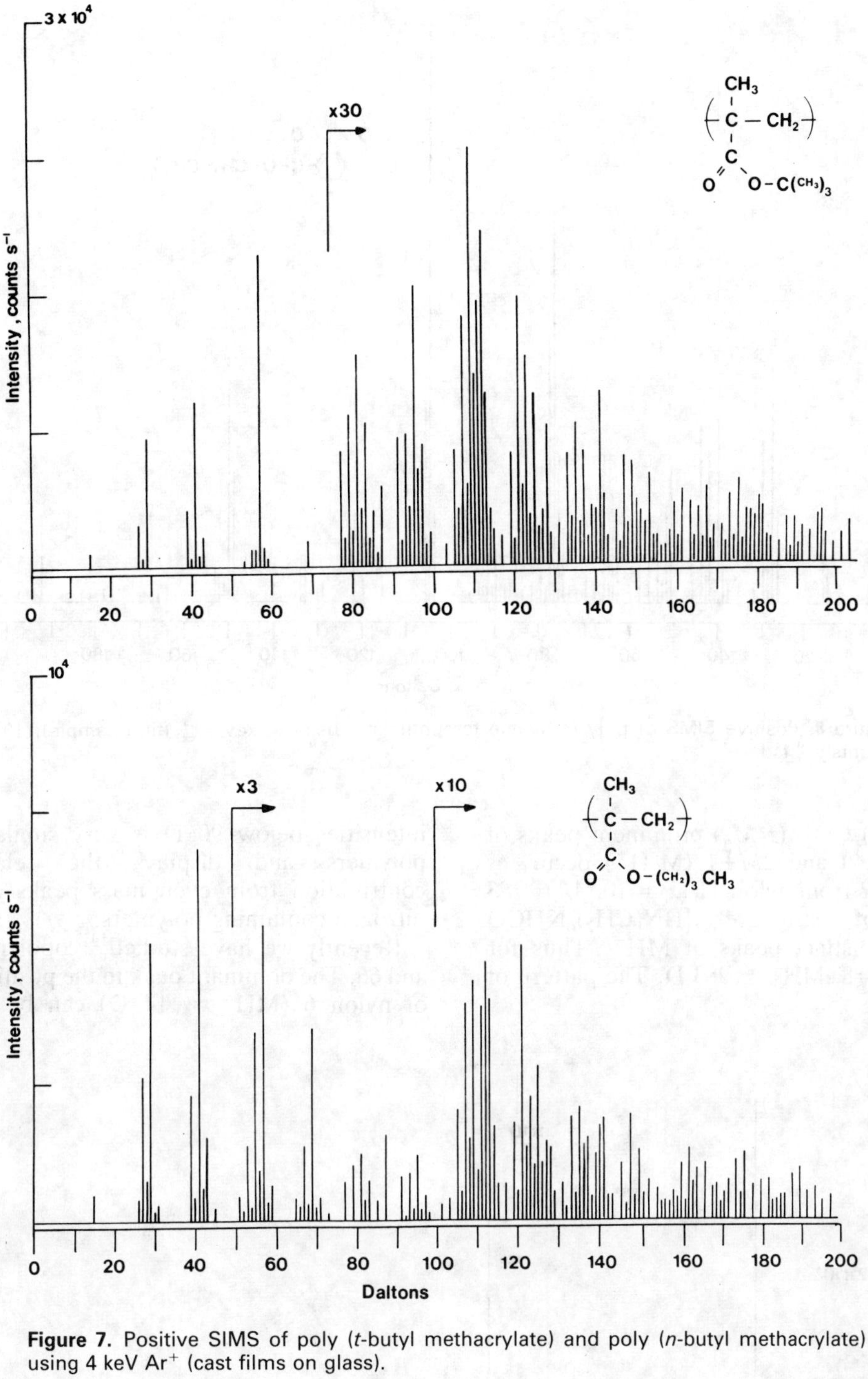

Figure 7. Positive SIMS of poly (*t*-butyl methacrylate) and poly (*n*-butyl methacrylate) using 4 keV Ar^+ (cast films on glass).

effect is also seen in the spectrum from polycarbonate in Fig. 11, in which prominent negative ions probably arise as follows:

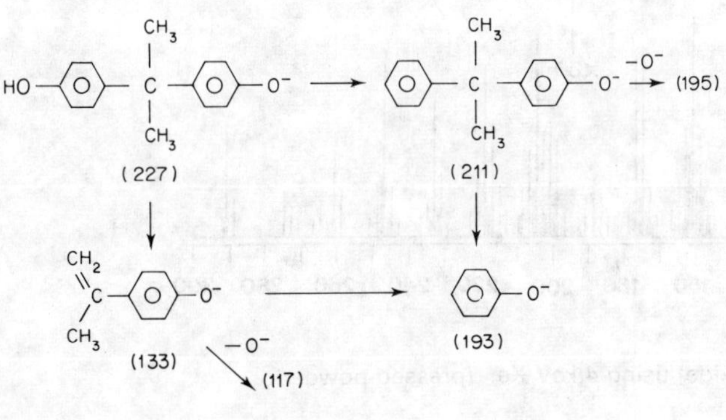

Very recently it has been demonstrated[12] that intense negative ion spectra containing informative peaks from large molecular fragments (like Fig. 11) can be routinely obtained from polymers if the surface potential can be made neutral or slightly negative (optimisation of conditions for positive ion spectra in SIMS gives a positive potential). Ways of achieving this condition were initially found using atom bombardment with an electron flood,[12] but also subsequently under normal SIMS conditions.[13]

The spectra from several polyamides (nylons) have been obtained.[10,14] These are highly characteristic with strong peaks at 26D (CN^-) and 42 D (CNO^-) in the negative ion spectra and fragments based on the repeat unit in the positive ion spectra. Thus for polymers of

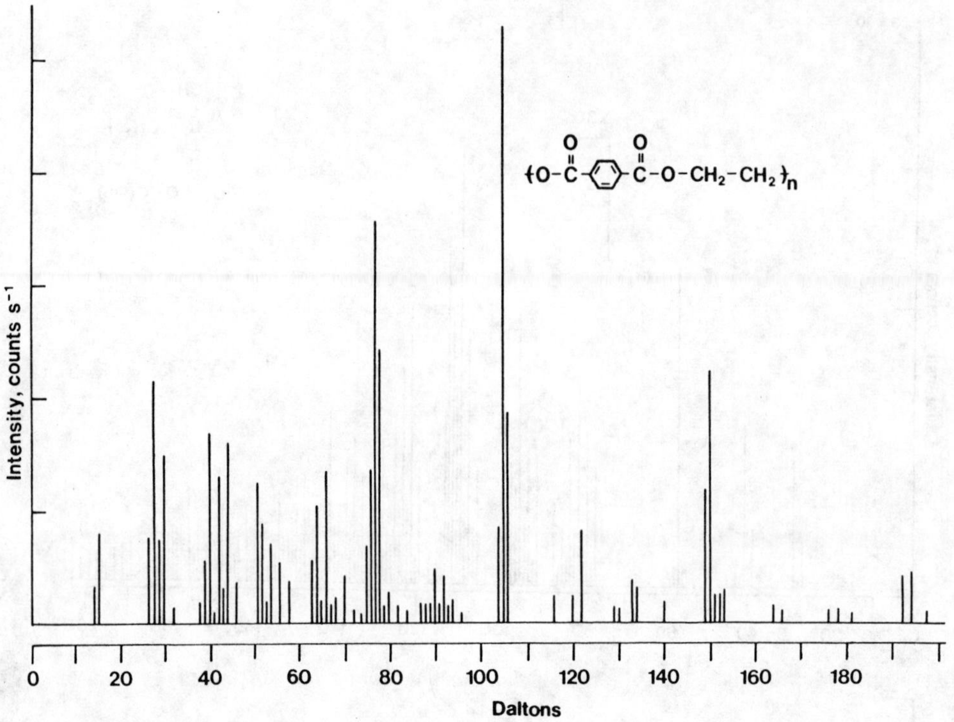

Figure 8. Positive SIMS of poly (ethylene terephthalate) using 4 keV Ar^+ (film sample); 10^3 counts s^{-1} fsd.

structure $[HN(CH_2)_xCO]_n$ $(=M_n)$ prominent peaks of mass $M + 1$ (MH^+) and $2M+1$ (M_2H^+) occur, as illustrated in Fig. 12 from nylon 11 ($x = 10$, $M = 183$ D). Polymers of structure $[HN(CH_2)_6NHCO(CH_2)_yCO]n$ give distinct peaks at MH^+. Thus for nylon 6, 10 when $y=8$, $MH^+ = 283$ D. The pattern of intensities below 90 D is very similar for *all* nylon polymers and displays the relatively intense contribution from even mass peaks characteristic of nitrogen containing polymers.

Recently we have studied[14] copolymers of nylon 6 and 66. The dominant peak in the positive ion spectrum of nylon 6 $(MH^+ = 114$ D) can be used to follow

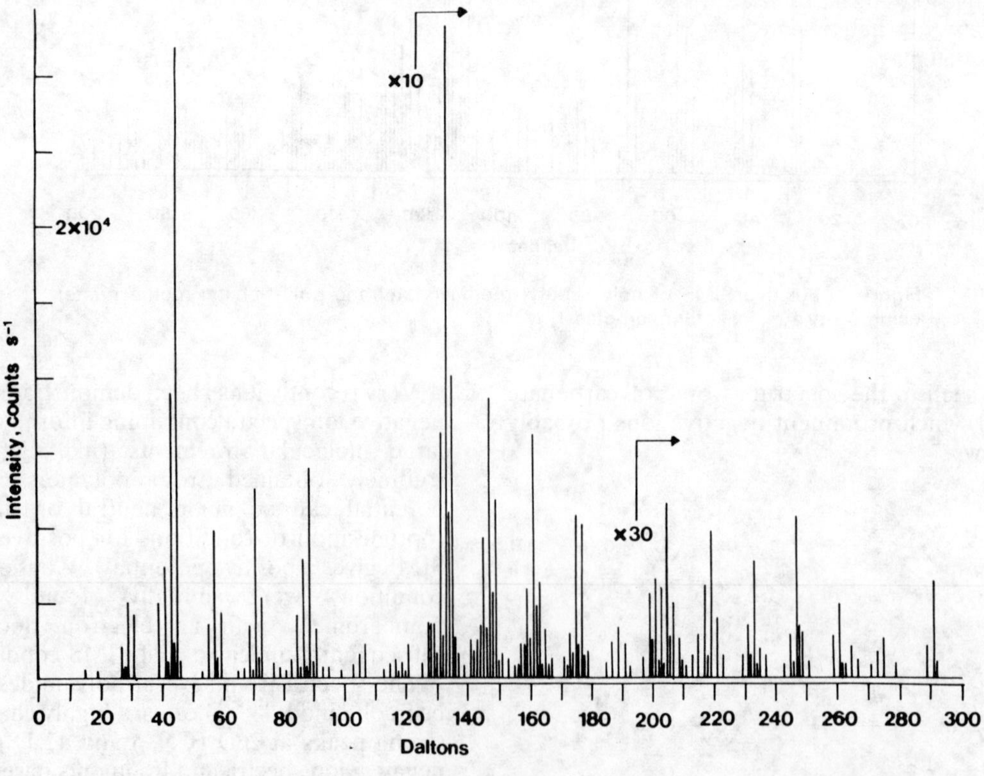

Figure 9. Positive SIMS of poly (ethylene oxide) using 4 keV Xe^+ (pressed powder).

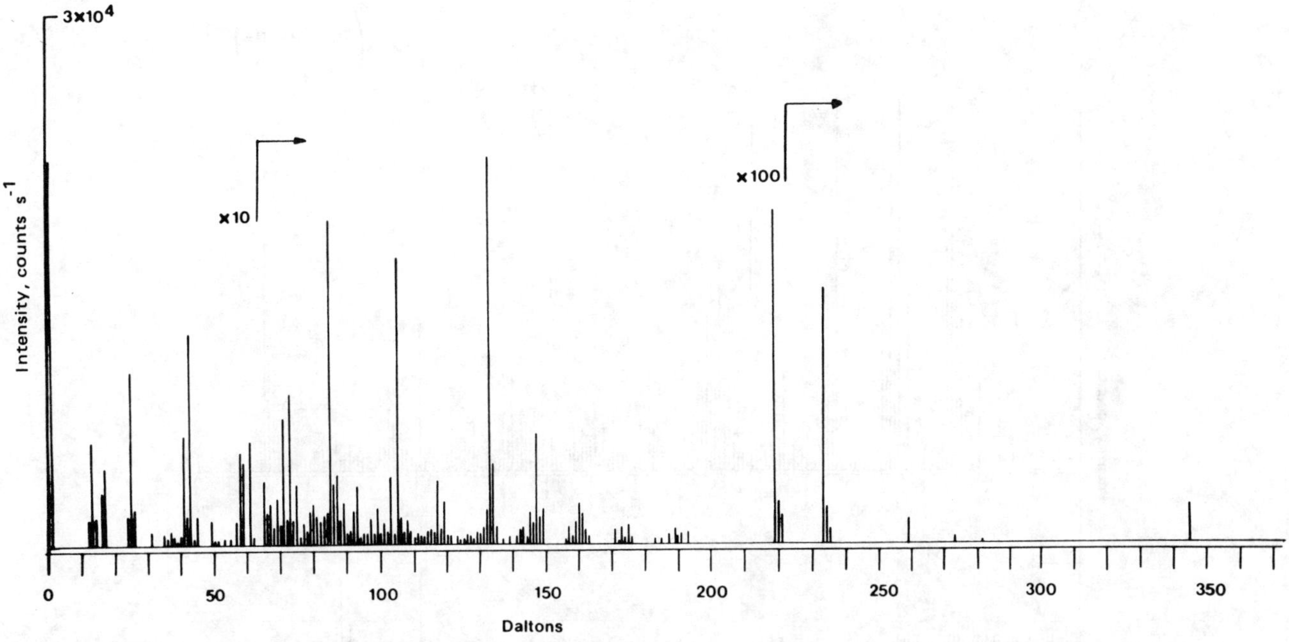

Figure 10. Negative SIMS of poly (ethylene oxide) using 2 keV Xe^+ (pressed powder).

copolymer composition since it is virtually absent in the nylon 66 spectrum. This is shown in Fig. 13. The peak at 55 D is used for normalisation since it is an intense fragment, characteristic of the polyamide backbone, from both homopolymers. Furthermore a peak at 213 D, which is only very weak in the spectra from the homopolymers, is seen to become prominent for the copolymers, and follows composition as also shown in Fig. 13. This peak can realistically be ascribed to an ion involving both monomer units. In a copolymer with statistically distributed monomer units it would be expected to pass through an intensity maximum as the composition changes.

Finally, Figs 14(a) and (b) show the positive ion spectra from two silicone polymers: poly (dimethylsiloxane) and poly(phenylmethylsiloxane).[15] The latter probably contains some oligomeric fractions of the former.

Characteristic peaks are:

$(CH_3)_3Si^+$	$(CH_3)_3Si–O–Si(CH_3)_2{}^+$	$(CH_3)_2PhSi^+$
(73 D)	(147 D)	(135 D)

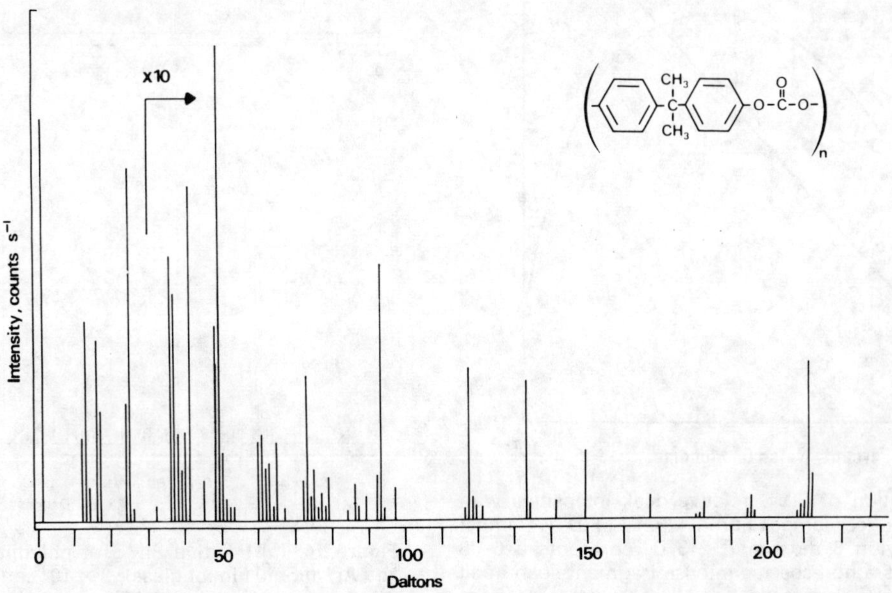

Figure 11. Negative SIMS of polycarbonate using 4 keV Xe^+ (cast film on glass). 10^4 counts s^{-1} fsd.

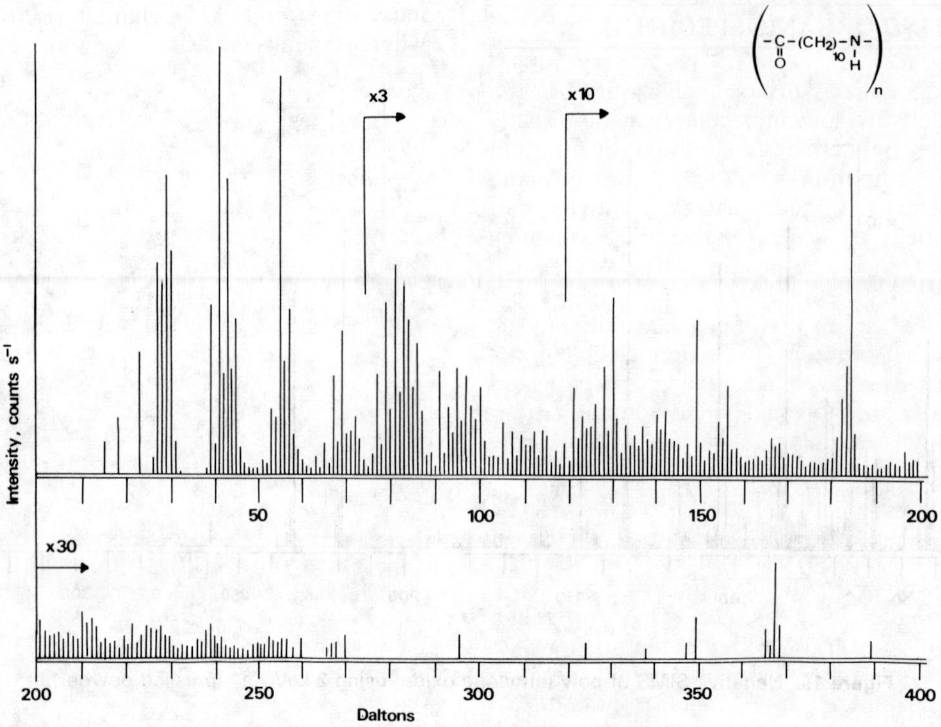

Figure 12. Positive SIMS of nylon 11 using 4 keV Xe^+ (cast film on glass). 3×10^4 counts s^{-1} fsd.

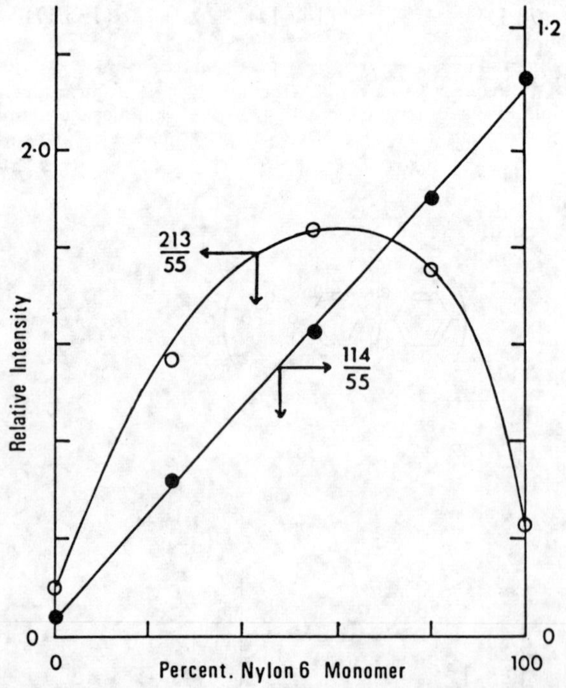

Figure 13. Correlation of SIMS relative peak intensities with nylon 6/66 copolymer composition (wt.%) 114 D is almost entirely due to nylon 6 monomer, 213 D represents a 6–66 linkage and 55 D is a non-specific intense fragment seen in all the spectra useful for normalisation. Films were cast from formic acid solution onto glass.

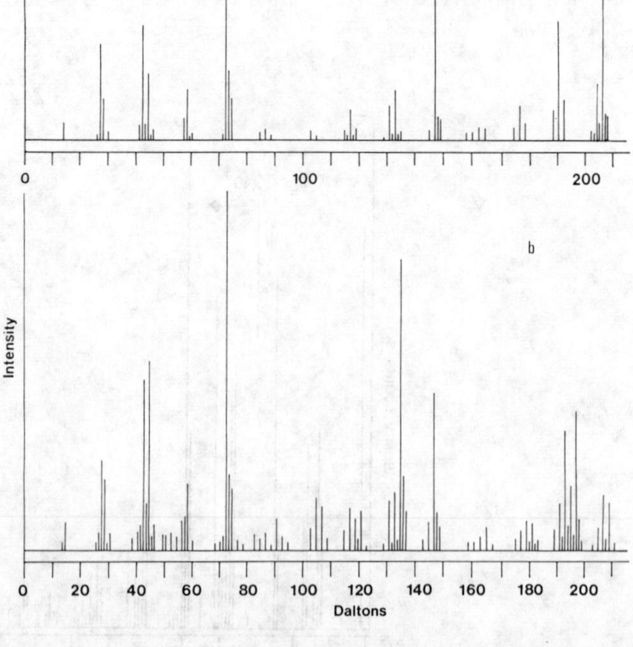

Figure 14. (a) Positive SIMS of polydimethylsiloxane using 4 keV Ar^+ (liquid film of glass); 3×10^3 counts s^{-1} fsd; (b) positive SIMS of polyphenylmethylsiloxane using 4 keV Ar^+ (liquid film on glass); 3×10^3 counts s^{-1} fsd.

FINGERPRINTING ORGANIC SPECIES

Alteration of polymer surface behaviour by the presence of relatively low molecular weight organic species, whether by deliberate addition or through adventitious contamination is a crucial aspect of many technologies. The increased surface sensitivity and molecular specificity of SIMS (relative to XPS) is now providing greater analytical power to investigate such systems, as illustrated below.

Figure 15 shows the positive ion spectrum from the surface of polyethersulphone (PES) film which contact angle measurements showed had a lower than expected surface energy and probably had patchy contamination. XPS indicated an essentially pure surface with a barely detectable Si 2p signal (corresponding to ~0.2 at%). Comparison with Fig. 14(a) clearly proves the presence of dimethylsilicone contamination. Since strong signals from the PES are also seen it is indeed likely that the contamination layer is a patched monolayer.

XPS has been much used to date to characterise the surface composition of 'polyurethane' materials which have potential for many biomedial applications.[16] These are segemented copolymers containing a polyether 'soft' segment and 'hard' segment links based on a di-isocyanate and diamine chain extender. Thus nitrogen characterises the hard segment uniquely, but more detailed structural analysis by XPS requires quantification of weak overlapping high binding energy shoulders on the C 1s peak due to carbamate (–OC(=O)NH–) and urea (–HNC(=O)NH–) units. SIMS spectra of the type discussed above for polyethers and 'hard segment' models allow several characteristic peaks to be identified and used to characterise the copolymers. For example,[17] the same spectral (mass) region from three polyurethanes based on polypropylene glycol (PPG), diphenyl-methylene di-isocyanate (MDI) and ethylene diamine are shown in Fig. 16. Fragments of mass 106 D and 132 D characterise the hard segment, whereas peaks at 101, 115 and 117 D characterise

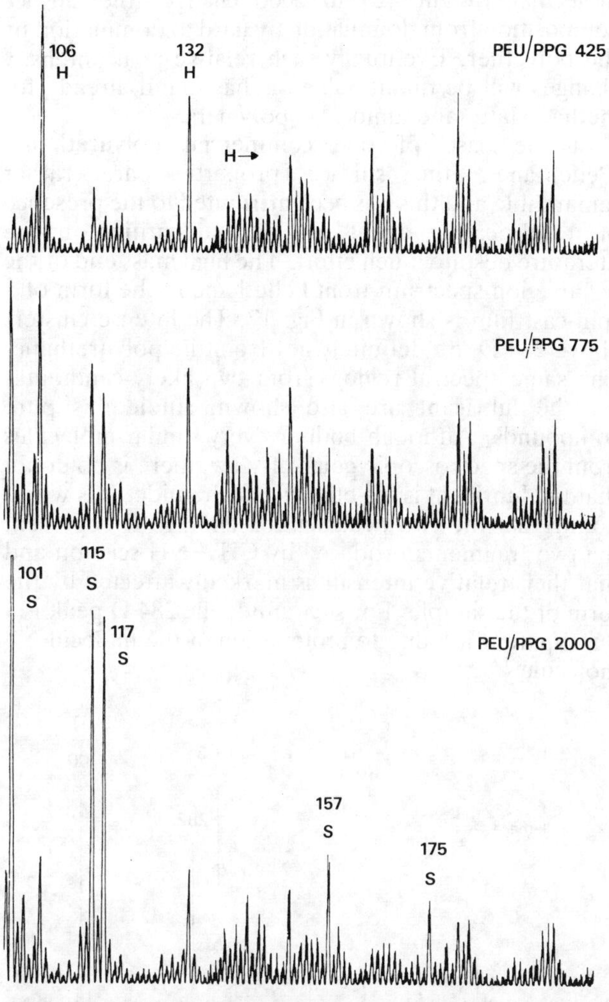

Figure 16. Part of the positive ion spectrum from a series of segmented polyurethanes based on polypropylene glycol (PPG) and methylene di-isocyanate (MDI), with increasing molecular weight PPG segments (see text). 4 keV Xe^+, identical conditions. H, S denote 'hard' and 'soft' segment fragments respectively.

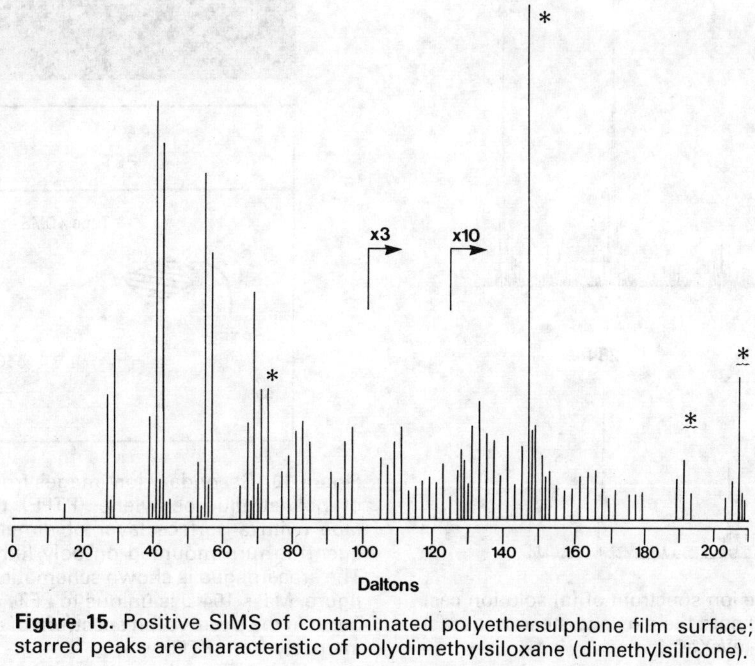

Figure 15. Positive SIMS of contaminated polyethersulphone film surface; starred peaks are characteristic of polydimethylsiloxane (dimethylsilicone).

the soft segment (PPG). The polymers are identical except in the size of the polyether chain, changing from molecular weight 425 to 2000 changes the surface composition from domination by hard to domination by the polyether. Eventually such relative peak intensity changes will be quantifiable, as has found already for methacrylate[4] and amide[14] copolymers.

In the case of one commercial polyurethane, 'Pellethane', the surface properties are rather remarkable and this has been attributed to the presence of a lubricating agent, as yet unidentified in the literature despite much effort. The high mass end of the positive ion spectrum from Pellethane in the form of a spin-cast film is shown in Fig. 17. The intense clusters above 280 D are definitely not from the polyurethane. The same spectral regions from two likely candidates for the lubricant are also shown, studied as pure compounds. Although both are very similar molecules from the spectroscopic point-of-view, there is no doubt that the lubricant is bis-ethylene stearamide. It is worth noting that the two peaks at 282 and 310 D represent the two fragments produced by $CH_2{-}NH$ scission and that their relative intensity is markedly affected by the form of the sample. For stearamide the 284 D peak is a quasi-parent ion due to protonation of the molecule.[17] molecule.[17]

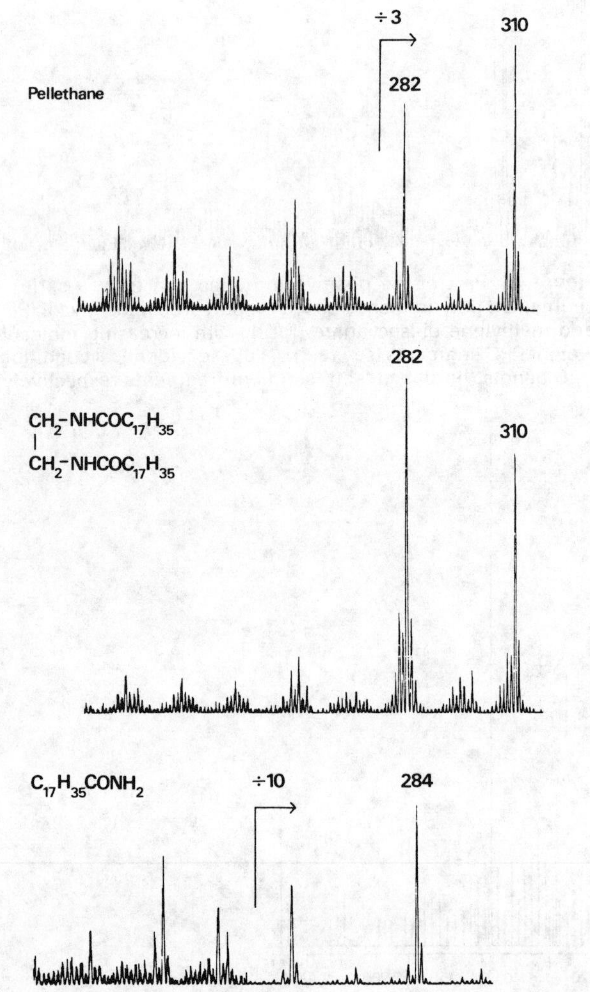

Figure 17. Part of the positive ion spectrum of (a) solution cast 'Pellethane' polyurethane; (b) pure bis-ethylene stearamide; (c) pure stearamide; all using 4 keV Xe^+.

CHEMICAL IMAGING AND MICROANALYSIS

Using the experimental approach outlined at the beginning of the paper, the largest area over which imaging can be achieved is determined by the requirement for uniform charge neutralisation. Figure 18 shows results from a test sample, indicating[18] that

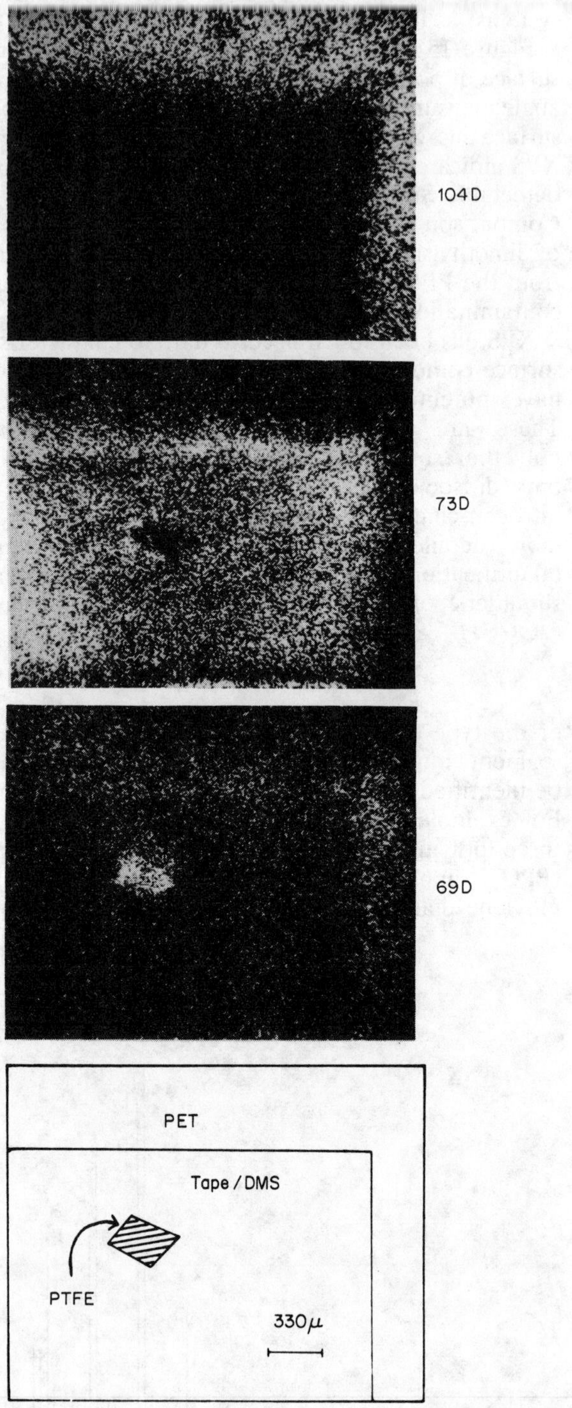

Figure 18. Secondary ion images from a test sample consisting of polytetrafluoroethylene (PTFE) mounted on double-sided tape (with a surface layer of dimethylsilicone (DMS) release agent) in turn mounted on poly (ethyleneterephthalate) (PET). The area imaged is shown schematically in the lower part of the figure. Mass 104 D is unique to PET, 73 D to DMS. Peaks at 69 D occur in the positive ion spectrum of all three materials but most intensely from PTFE.

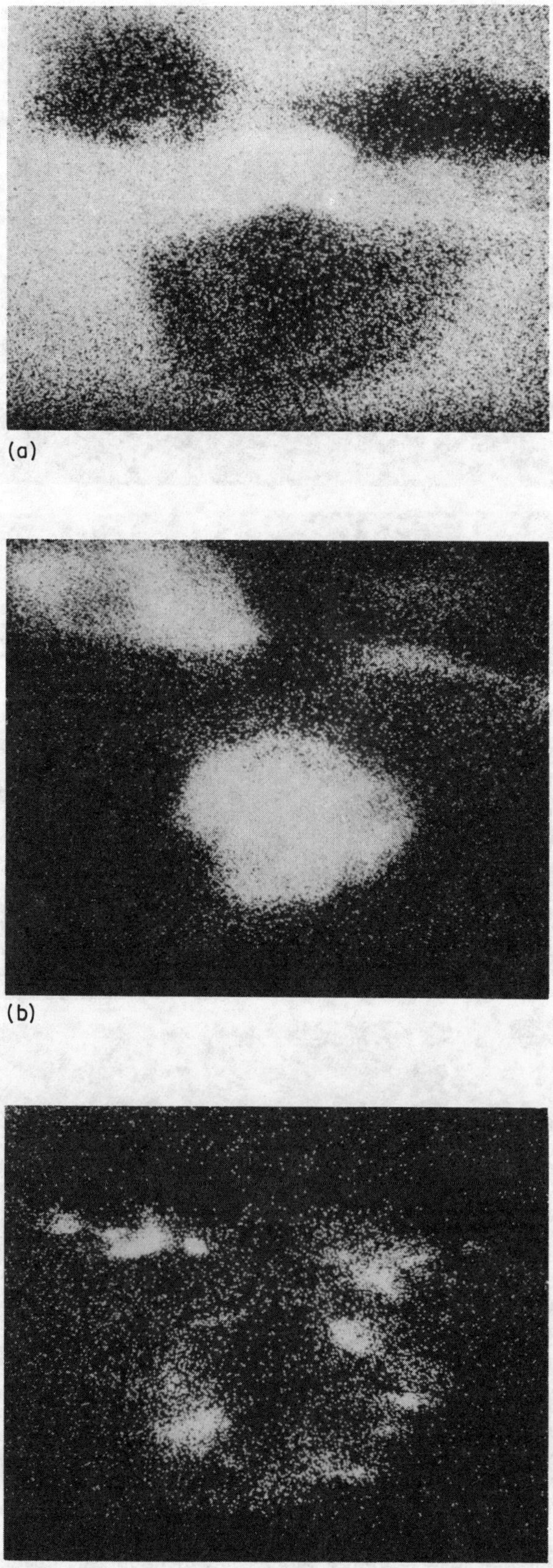

(a)

(b)

(c)

Figure 19. Negative secondary ion images from small (0.5–1 mm) irregular PTFE granules supported on Ag-dag: (a) Cl^- is impurity in the Ag-dag and shows the outline of the particles; (b) F^- illustrates topography effects on secondary ion yields from the PTFE particles. The particles have been in contact with a surfactant, nonyl-phenol ethoxylate which is imaged in (c) by the characteristic molecular fragment $[C_3H_5{-}C_6H_4{-}O]^-$ at 133 D.

uniformity can be achieved with our system over an area ca. 2 mm × 2 mm. At lower magnification part of the image is missing through non-neutralisation in the outer parts of the area scanned. Figure 19 shows images taken from small (0.5–1 mm) irregular PTFE granules which have been in contact with the surfactant nonylphenolethoxylate, mounted on Ag-dag, illustrating the ability to image a molecular species at <100 μm resolution in a non-ideal system. The ability to scan the ion beam within a small area, identified either from a large area image (either from secondary ion of, if possible, secondary electron detection) allows microanalysis. An example is shown in Fig. 20. These three examples made use of a noble gas ion source capable of producing a beam of ~30 μm diameter.

The smallest area which can be studied in principle, is determined by the spot size of the ion beam. In practise limitation comes either from an unacceptable level of ion emission by the neutralising electron beam (the current density of which must increase as the area irradiated by the main ion beam decreases) or from too rapid a rate of sample damage — this effect in particular comes into play when large molecular fragments must be detected to gain the required information. Preliminary work on the ion-beam damage rates of polystyrene and poly (methyl methacrylate) has shown that the threshold dose for damage is $\sim 2 \times 10^{13}$ ions cm^{-2} and that most molecular fragments of analytical value will be lost from the spectra by $\sim 10^{14}$ ions cm^{-2}, depending on the primary ion mass and energy. (Typical undamaged surface

(a)

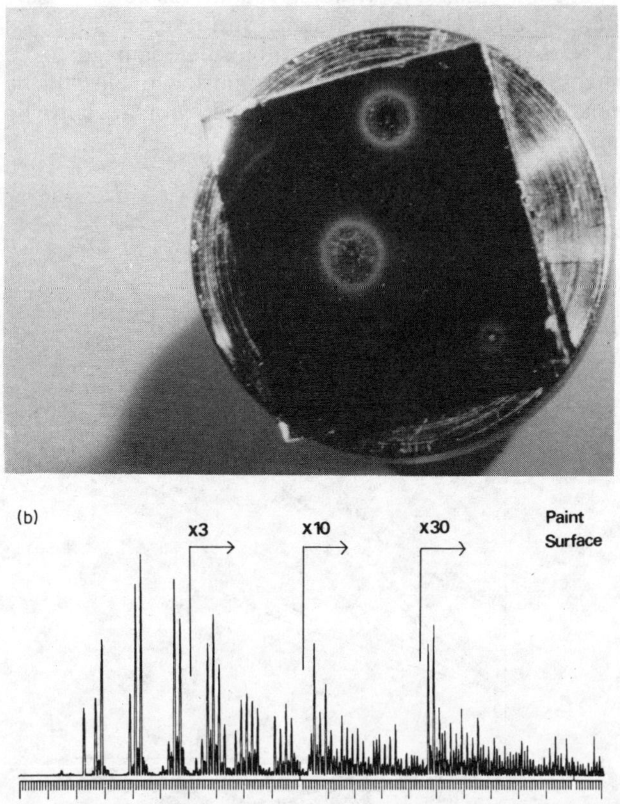

(b)

Figure 20. (a) Paint film on metal sheet showing retraction spots (regions where the paint film has dewetted from the substrate and retracted to leave a bare spot). The sample stub is 1 cm in diameter; (b) positive ion spectrum (2 keV Xe^+) taken from a region coated with paint

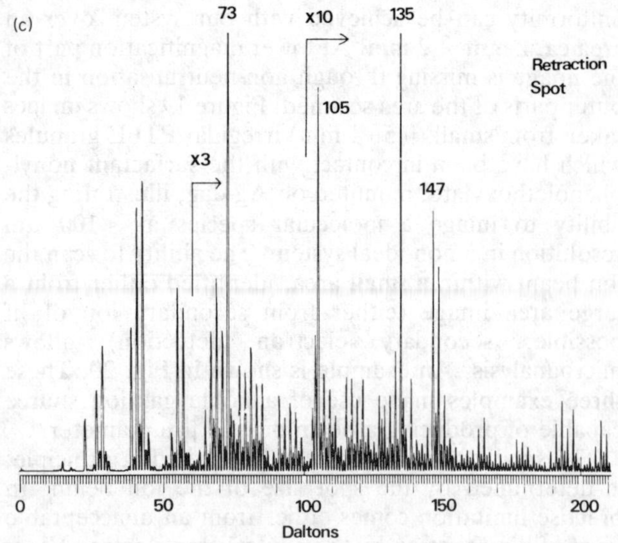

Figure 20. (c) positive ion spectrum (2 keV Xe^+) taken mainly from within a retraction spot. Note the relatively intense peaks at 73 D, 135 D and 147 D characteristic of polyphenylmethylsiloxane (c.f. Fig. 14) and at 105 D and 149 D possibly indicating a phthalate ester.

spectra can be acquired during a dose of $\sim 10^{12}$ ions cm^{-2}).[4,6,16] Figure 21 summarises recent detailed work[19] on damage of PMMA monitored *in situ* by the CH^- (13 D)/O^- (16 D) ratio which correlates with the atomic C/O ratio in the polymer.[8] Damage involves the progressive loss of oxygen and the formation of 'graphitic' carbon.

In Figs 22 and 23 the surfaces of carbon fibres are imaged at increasingly high spatial resolution using liquid-metal (Ga^+) ion sources. In these cases the ion dose is sufficiently high for molecular damage to be severe, but analytical data of importance can still be obtained since molecular species do not need to be probed in these instances.

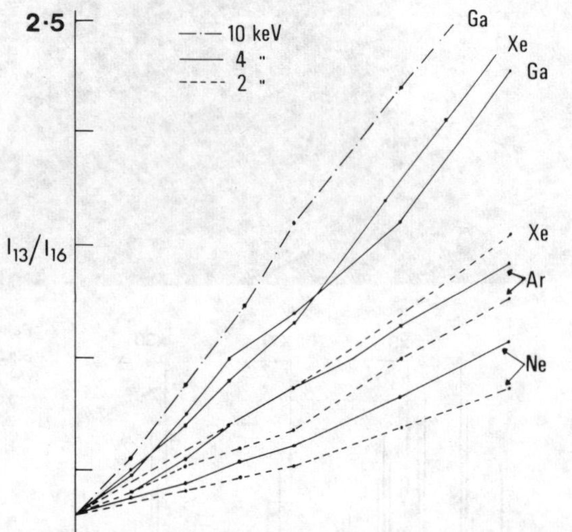

Figure 21. Variation in rate of sample damage of PMMA under various ion beam irradiation conditions in SIMS. The CH^- (13 D):O^- (16 D) ratio, measured directly as a function of dose correlates, with the C:O atomic ratio. Thus the steeper the slope the faster the rate of surface degradation through loss of O.

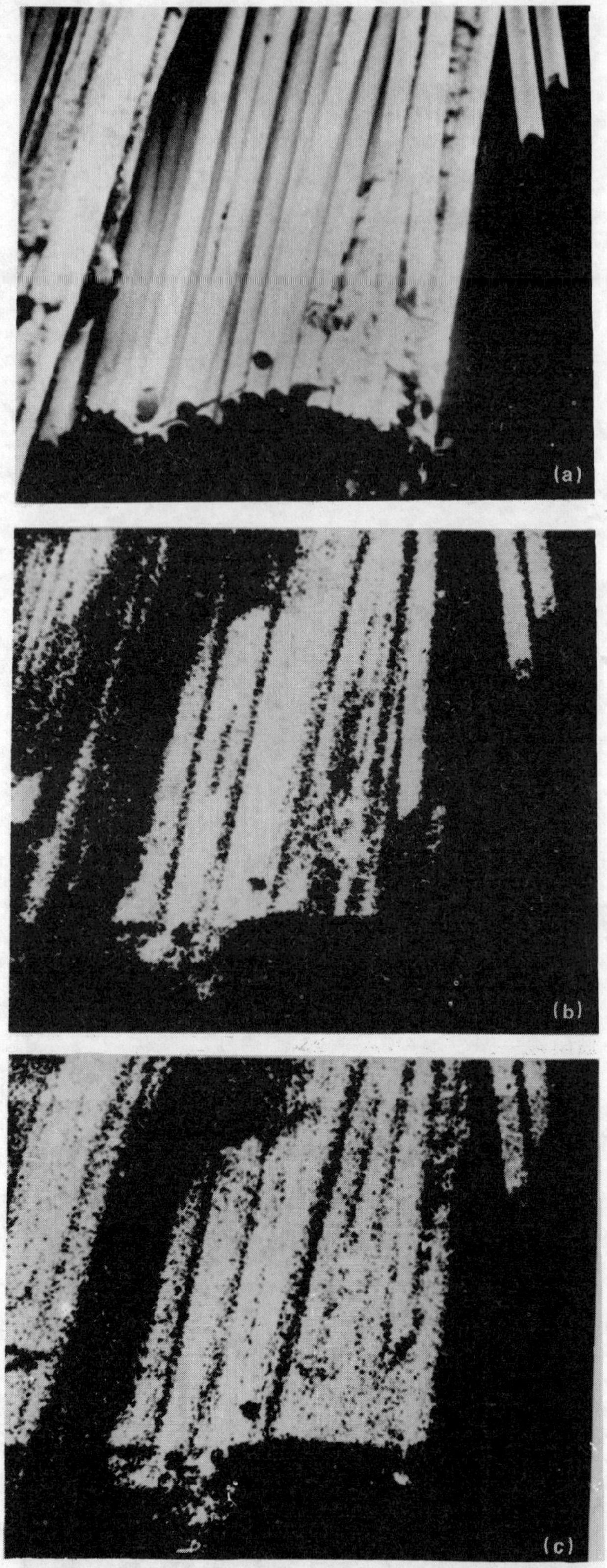

Figure 22. Images from a carbon fibre/thermoplastic fracture surface using 10 keV Ga^+ from a liquid metal ion-source (0.5 μm diameter beam): (a) secondary electrom image; (b) secondary ion image from detection of negative ion mass 26 D ($C_2H_2^-$ and CN^- which is highly characteristic of the carbon fibre surface; (c) image of C_2^- (24 D) which does not discriminate between fibre and polymer. These images confirm the presence of polymer in those regions which are topographically rough in the SEM image, but also demonstrate the absence of polymer on the surface of fibres which appear smooth (fibre diameter ~ 7 μm).

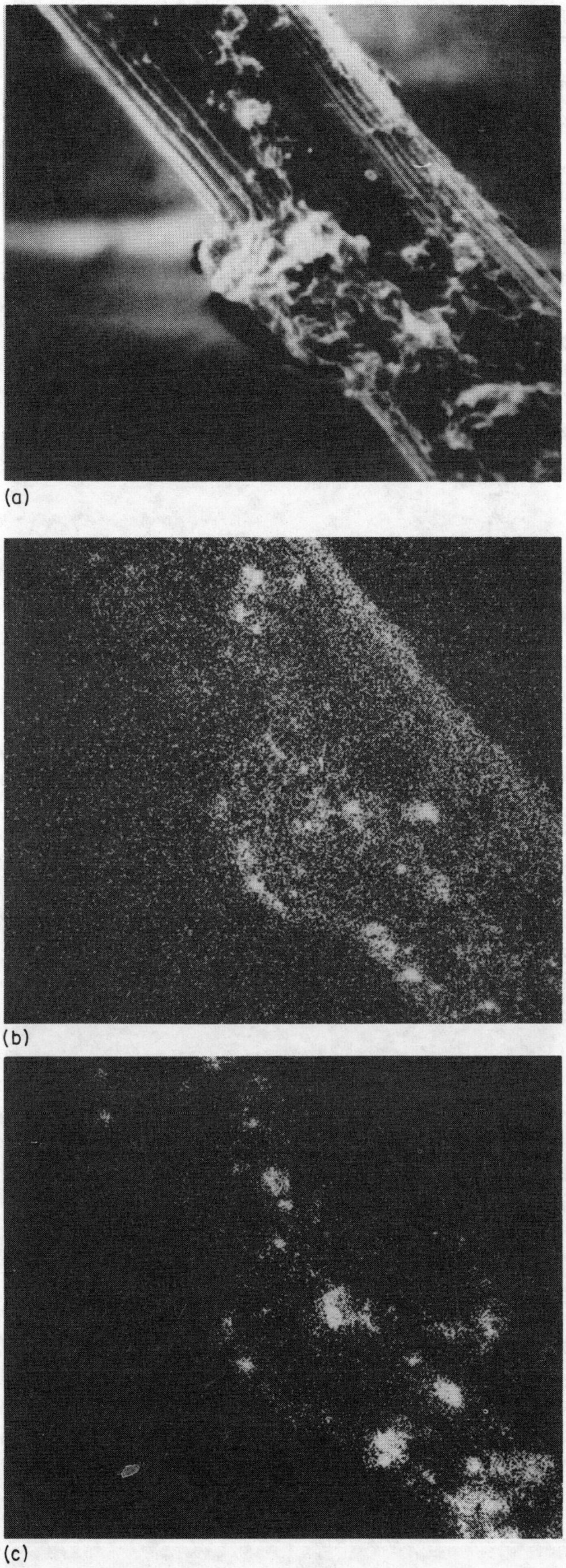

(a)

(b)

(c)

Figure 23. Similar carbon fibre to Fig. 19 from a thermoplastic composite, after extraction of soluble polymer, using a 30 keV Ga^+ beam (diameter ~ 500 A); (a) SEM image showing lumps of adhered polymer; (b) and (c) positive secondary ion images of inorganic dopants in the polymer.

FUTURE DEVELOPMENTS

Major improvements in instrumental performance for SIMS studies of polymer surfaces can be expected within the near future. Secondary ion detection sensitivity using the kind of quadrupole mass spectrometer employed in our studies is about to improve by an order of magnitude through more efficient ion transfer optics. However, a significant drawback of the quadrupole mass spectrometer is its loss of transmission as a function of mass (typically $T \alpha M^{-1} - M^{-2}$), since high mass clusters provide the most unique analytical information for organic species. The time-of-flight (TOF) mass spectrometer, besides having better transmission properties than the quadrupole, does not suffer from mass discrimination and, in principle, has an infinite mass range. Benninghoven has briefly described[20] the performance of a SIMS instrument employing a TOF mass spectrometer. The TOF analyses pulses of ions and therefore requires a pulsed primary ion beam.

This in itself may help avoidance of sample charging since the ion beam will be 'on' for a few nanoseconds and then 'off' during the several microseconds required for mass analysis. With the use of a framestore for digital data aquisition the production of several images from one point-by-point scan of the ion beam over a surface becomes possible. Hence the combination of a small spot pulsed ion-source and TOF detection offers substantial advantages for static SIMS imaging. Such an instrument is currently under construction.[21]

Meanwhile it has also been demonstrated[22] that a scanning atom beam can be produced by charge-exchange of a scanned ion-beam in a collision cell. Atom beam imaging, hopefully with reduced charging problems, and with good resolution may then be possible. Recently, also the use of a collision cell to produce an atom beam in the ion microscope has been described.[23] Imaging at about 1μm resolution is achieved in this case entirely by ion optics. Thus the alternative excitation mode of FAB is undergoing a major advance. Taken together these new developments promise an exciting future for polymer surface characterisation by secondary ion mass spectrometric methods.

REFERENCES

1. D.T. Clark and J.W. Feast, Eds. *Polymer Surfaces* Wiley (1978).
2. D. Briggs in *Practical Surface Analysis by Auger and X-ray Photoelectron Spectroscopy*, Chap. 9, ed. by D. Briggs and M.P. Seah, Wiley (1983).
3. J.A. Gardella, F.P. Novak and D.M. Hercules, *Anal. Chem.* **56**, 1371 (1984) and refs therein.
4. D. Briggs, M.J. Hearn and B.D. Ratner, *Surf. Interface Anal.* **6**, 18 (1984).
5. J.E. Campana and S.L. Rose, *Int. J. Mass Spec. Ion Physics,* **46**, 483 (1983) and refs therein.
6. D. Briggs and A.B. Wootton, *Surf. Interface Anal.* **4**, 109 (1982).
7. A. Brown and J.C. Vickerman, *ibid*, **6**, 1 (1984).
8. D. Briggs, A. Brown, J.A. Van den Berg and J.C. Vickerman in *Ion formation from Organic Solids, p. 162* ed. by A. Benninghoven, Springer-Verlag (1983).
9. D. Briggs, *Polymer*, **25**, 1379 (1984).
10. D. Briggs, *Surf. Interface Anal*, **4**, 151 (1982).
11. D. Briggs and M.J. Hearn, *J. Mass Spec. Ion Processes*, **67**, 47 (1985).
12. A. Brown and J.C. Vickerman, *Surf. Interface Anal*, **8**, 75 (1986).
13. D. Briggs and M.J. Hearn, *Surf. Interface Anal*, to be published.
14. D. Briggs, submitted to *Org. Mass Spectrom.*
15. D. Briggs, *Surf. Interface Anal.* **5**, 113 (1983).
16. See, for example: B.D. Ratner, *Annals Biomed. Eng.* **11**, 313 (1983).
17. M.J. Hearn and D. Briggs and B.D. Ratner, in preparation.
18. D. Briggs and M.J. Hearn, *Spectrochim. Acta.* **40B**, 707 (1985).
19. M.J. Hearn and D. Briggs, this Proceedings.
20. P. Steffens, E. Niehuis, T. Freise, D. Greifendorf and A. Benninghoven, in *Secondary Ion Mass Spectrometry SiMS IV*, p. 404 ed by A. Benninghoven *et al.*, Springer-Verlag (1984).
21. By VG Scientific Ltd, East Grinstead, UK.
22. J.C. Vickerman, personal communication.
23. F. Degreve and J.M. Lang, *Surf. Interface Anal*, **7**, 177 (1985).

SURFACE AND INTERFACE ANALYSIS, VOL. 9, 405–409 (1986)

The π→ π* Shake-up Phenomena in Polyesters Containing Backbone Aromatic Groups

Leonardo C. López and David W. Dwight
Center for Adhesion Science, Virginia Polytechnic Institute and State University, Blacksburg, Virginia 24061, USA
Malcolm B. Polk
School of Textile Engineering, Georgia Tech, Atlanta, Georgia 30332, USA

The only quantitatively different features in the x-ray photoelectron spectra of a series of liquid crystalline homopolyesters and copolyesters (synthesized from substituted aromatic and cycloaliphatic monomers) were found to be shake-up satellites arising from π→ π* transitions. The oxygen shake-up satellite intensities appear to be influenced by a combination of conjugation length and restricted rotation of the backbone aromatic groups. Thus, the three permutations of 1,3– and 1,4–phenylene linkages were distinguished from each other. Angular dependence of the carbon and oxygen shake-up satellites was used to identify preferential enrichment of the surface by the cycloaliphatic blocks.

INTRODUCTION

Several levels of information have been used in applications of x-ray photoelectron spectroscopy (XPS or ESCA).[1] Amongst them, discrete energy loss phenomena due to final state effects have been investigated in some detail. In particular, 'shake-up' transitions have been studied for the analysis of organic systems.[2] Shake-up transitions involve the excitation of a valence band electron from an occupied to an unoccupied molecular orbital coincident with the primary photoionization event, producing a discrete energy loss peak. The energy difference between main line and shake-up peak is equal to the energy of the electronic transition.

In organic molecules, the π→ π* shake-up transition observed in XPS spectra involves a transition from initial to final state molecular orbitals associated with unsaturated carbon groups. This type of transition has been studied in detail for small molecules,[3–10] as well as in macromolecular systems involving aromatic groups.[11–21]

Distefano and co-workers[5] studied the low energy shake-up satellites derived from π→ π* transitions in heterocyclic compounds, such as imidazole, pyrazole, thiazole, isothiazole and isooxazole. In most of these cases, the π→ π* shake-up involved a charge transfer from the core-ionized heteroatom (O, S, N) towards the ring. This kind of observation was also made by Chambers and Thomas[4] on furan, pyrrole and thiophen. Other studies of small molecules involved shake-up features of aromatic compounds containing functional groups bonded to the ring, such as nitro,[6,8] amino[7,10] nitroso,[6] cyano and aldehyde[9] groups. In the study of these *para* substituted aromatic compounds, it was found that in most cases, the intensity of the shake-up peaks corresponding to N 1s and O 1s increased steadily with the electron-releasing ability of the p-substituent. An interesting finding by Distefano *et al.*[8] on nitrobenzenes and nitrothiophens indicated that steric hindrance inhibited resonance and therefore, decreased shake-up intensities.

Several studies of the shake-up features in XPS spectra have been performed in macromolecular systems.[11–21] Clark *et al.*[11–13] Salaneck and Thomas[14] and Duke and co-workers[15] studied the shake-up transitions arising from macromolecular systems having aromatic pendant groups. The detailed work of Clark and co-workers[11,12] focused on the use of the shake-up satellite corresponding to π→ π* transition in aromatic moieties to evaluate the concentration of aromatic species in copolymers of alkanes and styrene. In a different study, Clark and Dilks[13] analysed the effect of para-substituents on the shake-up phenomena of p-substituted polystyrenes. Those authors found that shake-up intensity for the C 1s level decreased with increasing π–electron donating power of the p-substituent.

A few studies of macromolecular systems have focused on polymers containing backbone aromatic groups[16–21] In particular, McGrath *et al.*[16] and Dwight and co-workers[17] studied *bis*-phenol-A-polycarbonate in its copolymers with polysulphone and poly(dimethyl siloxane). Most of this work has focused on the shake-up satellites arising from the C 1s line, as well as the O 1s line when oxygen is bonded directly to the aromatic ring in the backbone. However, Gardella *et al.*[21] reported the observation of shake-up satellites in photoemission from atoms two bond lengths away from the aromatic ring in poly(tetramethyl-*p*-silphenylene siloxane), in which there is no strong chemical evidence for delocalization of molecular orbitals across the siloxane bond.

The present study reports on the effects of isomerism on the shake-up satellite intensities from C 1s and O 1s lines of some aromatic polyesters. Also, the use of carbon and oxygen shake-up satellites with angular-dependent XPS is demonstrated for the determination of surface selective enrichment in block copolyesters containing aromatic and cycloaliphatic groups.

0142–2421/86/120405–05 $05.00

EXPERIMENTAL

The polymers investigated in this study were synthesized in the laboratory of Prof.M.L. Polk. The synthesis routes and characterization data have been published.[22–24] Tables 1 and 2 show the polymer structures and nomenclature utilized.

Table 1. Homopolyesters

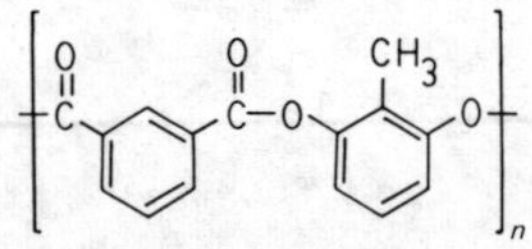

IAR. Poly (oxy (2-methyl-1, 3-phenylene) oxy isophthaloyl); η_{inh}=0.21 dL g^{-1} in *m*-cresol at 30°C

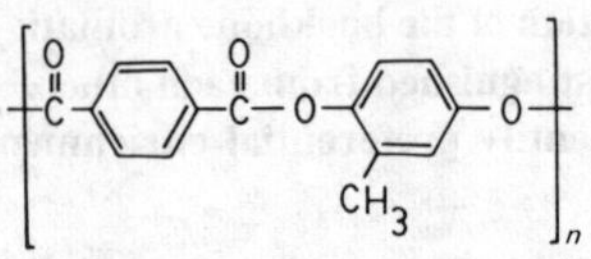

IIIAR. Poly (oxy (2-methyl-1, 4-phenylene) oxy terephthaloyl); η_{inh}=0.58 dL g^{-1} in 3:1 (v/v) $CHCl_3/CF_3COOH$ at 30°C

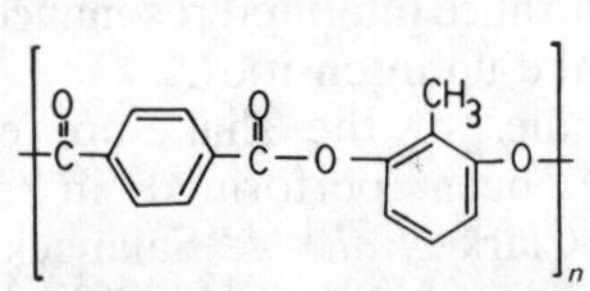

VIIAR. Poly (oxy (2-methyl-1, 3-phenylene) oxy terephthaloyl); η_{inh}=0.52 dL g^{-1} in *o*-chlorophenol at 30°C

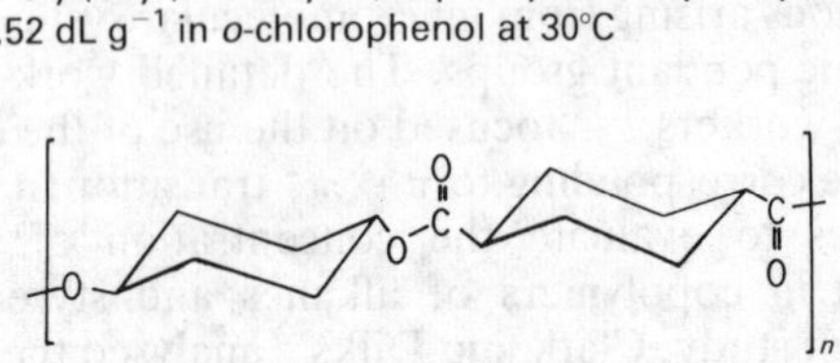

IIIAL. Poly (oxy-trans-1 4-cyclohexylene oxy carbonyl-trans-1, 4-cyclohexylene carbonyl); η_{inh} = 0.55 dL g^{-1} in 3:1 (v/v) $CHCl_3/CF_3COOH$

Table 2. Copolyesters

$(IIIAL)_x$—$(IIIAR)_y$

IIB. Poly (oxy-trans-1, 4-cyclohexylene oxycarbonyl-trans-1, 4-cyclohexylene carbonyl-b-oxy (2-methyl-1,4-phenylene) oxy terephthaloyl); η_{inh}=0.85 dL g^{-1} in 3:1 (v/v) CH_2Cl_2/CF_3COOH at 30°C

$(IIIAL)_x$—$(VIIAR)_y$

VIIB. Poly (oxy-trans-1, 4-cyclohexylene-oxy-carbonyl-trans-1, 4-cyclohexylene carbonyl-b-oxy (2-methyl-1, 3-phenylene) oxy terephthaloyl); η_{inh}=0.74 dL g^{-1} in *o*-chlorophenol at 30°C

$(IAR)_x$—$(VIIAR)_y$

XB. Poly (oxy (2-methyl-1, 3-phenylene) oxy isophthaloyl-b-oxy (2-methyl-1, 3-phenylene) oxy terephthaloyl); η_{inh}=0.60 dL g^{-1} in 3:1 (v/v) of $CHCl_3/CF_3COOH$ at 30°C

An XSAM–800 x-ray photoelectron spectrometer, manufactured by Kratos Ltd., England, was utilized to obtain the spectra. This instrument is equipped with a dual anode x-ray source and a hemispherical analyser. The analyser was operated in the Fixed Retardation Ratio (FRR) mode. All spectra were collected using the Mg $K\alpha_{1,2}$ x-ray source operated at 13 kV and 20 ma. Angular-dependent depth profiling was performed by varying the electron take-off angle between the analyser and the sample surface.[25,26] Electron exit angles of 90°, 30° and 10° were utilized. The C 1s photoelectron line corresponding to hydrocarbon at *BE*=285.0 eV. was used to calibrate the binding energy scale.[27] Curve-fitting was performed utilizing Gaussian line shapes for the component peaks in the Kratos standard software. The homopolyesters were analysed in the powder form mounted on the sample probe by means of double-sided adhesive tape. The copolyesters were solution cast from chloroform onto the sample probe.

RESULTS AND DISCUSSION

Homopolyesters

Figures 1 and 2 depict the core level spectra corresponding to the homopolyesters IAR and IIIAR, VIIAR and IIIAL, and a typical computer curve-fitted peak. Table 3 shows the binding energies of the core level spectra of the homopolyesters. The multicomponent character of the C 1s spectra is readily observed. The computer curve-fitted peak indicates

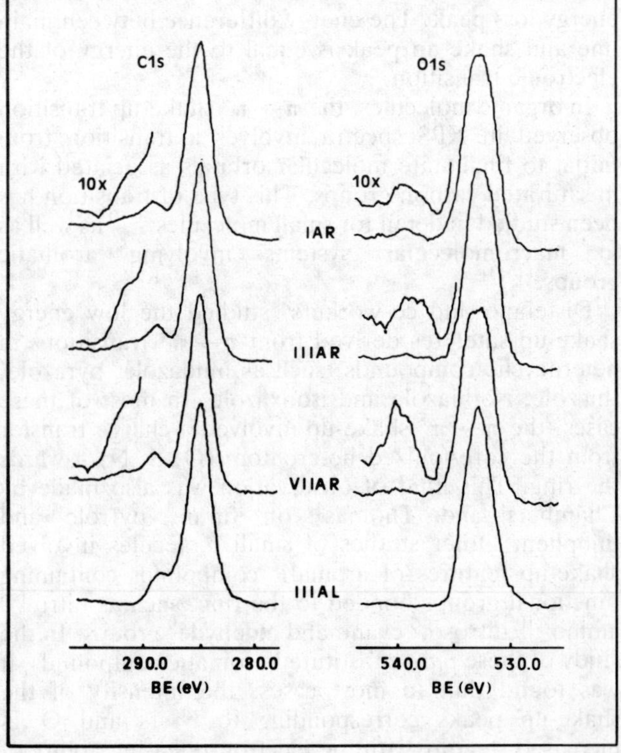

Figure 1. Carbon 1s and oxygen 1s high resolution spectra of the series of homopolyesters.

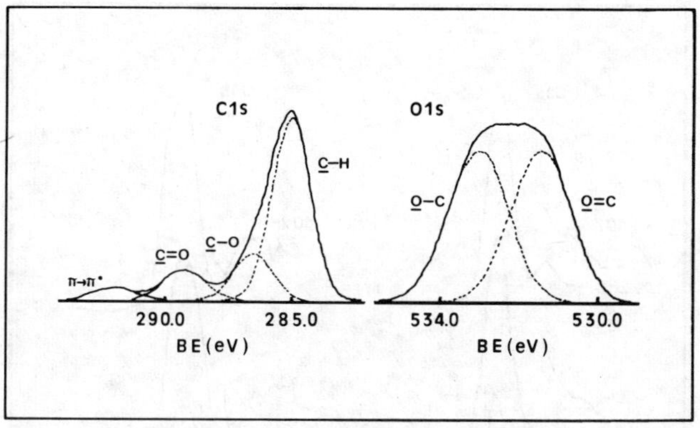

Figure 2. Typical carbon 1s and oxygen 1s computer curve-fitted spectra of homopolyester VIIAR.

three components: hydrocarbon linkages at $BE = 285.0$ eV., singly bonded carbon-to-oxygen at $BE = 286.6$ eV., and doubly bonded carbon-to-oxygen at 289.0 eV. The binding energies of the core levels of the different homopolymers are in agreement within ±0.2 eV, as shown in Table 3. The presence of a shake-up satellite 7.1 eV. from the hydrocarbon component of the C 1s peak is observed in the spectra of the aromatic homopolyesters. Given the energy shift, the satellites can be assigned to $\pi \rightarrow \pi^*$ transitions occurring in the aromatic groups. The C 1s shake-up satellite line shapes are overlapped with the carbonyl, but do appear to be similar for all the homopolyesters. This indicates that these isomers can not be distinguished by means of C 1s shake-up satellites as has been done in other systems.[12,13]

The O 1s spectra of the homopolyesters present two component peaks that correspond to oxygen doubly bonded to carbon at $BE = 532.1$ eV. and to oxygen singly bonded to carbon at $BE = 533.7$ eV. The O 1s spectra pertaining the aromatic polyesters also show a shake-up satellite centered at approximately 7.1 eV. from the doubly bonded oxygen-to-carbon component. The energy of the transition is the same of the $\pi \rightarrow \pi^*$ transition observed in the C 1s spectra. Therefore, the O 1s shake up satellites suggest some degree of conjugation between the benzene rings and the

Table 3. Core level binding energies (eV) for the homopolyesters

Homopolymer	C 1s[a]	O 1s[a]	C 1s shake-up[a]	O 1s shake-up[a]
IAR	285.0	532.3	292.1	539.4
	286.5	533.9		
	289.1			
IIIAR	285.0	532.1	292.1	539.2
	286.5	533.7		
	289.1			
VIIAR	285.0	532.2	292.1	539.3
	286.6	533.8		
	289.1			
IIIAL	285.0	532.0	—	—
	286.6	533.5		
	288.9			

[a] $\Delta E = \pm 0.2$ eV.

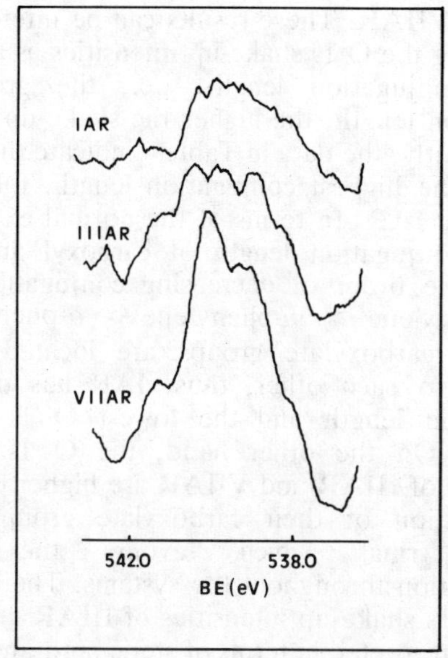

Figure 3. Oxygen shake-up satellites corresponding to the homopolyesters.

carboxylate groups bonded to the aromatic groups. Although of similar origin, the satellite line shapes are characteristic of the specific aromatic polyester as shown in Fig 1 and 3. Those characteristic line profiles can be attributed only to the isomerism of the three aromatic homopolyesters. IIIAR and VIIAR differ in the phenylene groups linked to the terephthaloyl moiety: IIIAR has a 2–methyl–1, 4–phenylene group linked to the terephthaloyl moiety, whereas VIIAR presents a 2-methyl-1, 3–phenylene group (Table 1). On the other hand, IAR differs from VIIAR in the isophthaloyl moiety. That is, in the case of IAR, the carboxylate groups are in *meta* position to each other, whereas in VIIAR the carboxylate groups are in *para* position to each other.

Table 4 presents the shake-up intensities expressed as a percentage of the intensity of the main photoelectron peak. The values presented are averages of three measurements performed on different spectra. The intensity of the C 1s shake-up satellite is about 14% for all the aromatic polyesters. This value is higher than previously reported in the studies of Clark *et al.*[11–13] However, those studies were performed on polymers containing *pendant* aromatic groups, whereas in the present case the aromatic groups are in the polymer backbone. Therefore, there may be extension of π-electron delocalization through the bonds of the carboxylate group, creating a highly conjugated system.

The O 1s shake-up intensities are different for each of the polyesters, 0.6% for IAR, 2.7% for IIIAR, and

Table 4. C 1s shake-up and O 1s shake-up intensities for the homopolyesters

Homopolymer	C 1s shake-up[a] Intensity	O 1s shake-up[a] Intensity
IAR	13.3	0.6
IIIAR	14.0	2.7
VIIAR	14.0	4.4

[a]The estimated error is less than ±10%

4.4% for VIIAR. These results can be interpreted by considering the O 1s shake-up intensities as a measure of the conjugation length, i.e., the greater the conjugation length, the higher the shake-up intensity. Consequently, the data in Table 4 indicate that VIIAR presents the highest conjugation length, followed by IIIAR and IAR. In terms of the normal expectations for the conjugation length of carboxyl and phenyl groups, the order of decreasing conjugation should be *p*-phenylene > *m*-phenylene > *o* phenylene. In IAR the carboxylate groups are located at meta positions to each other, thus, IAR has the lowest conjugation length and the lowest O 1s shake-up intensity. On the other hand, the O 1s shake-up intensities of IIIAR and VIIAR are higher due to the para location of their carboxylate groups on the aromatic rings which favours the electron delocalization throughout the systems. The difference in the O 1s shake-up intensities of IIIAR and VIIAR may be interpreted in terms of steric hindrance and the position in which the phenylene moiety of the polymer is linked in the backbone. In VIIAR, the phenylene group is bonded in a 1,3– position. However, in IIIAR, the phenylene group is bonded in a 1,4– position, therefore allowing a rotation to occur as shown below:

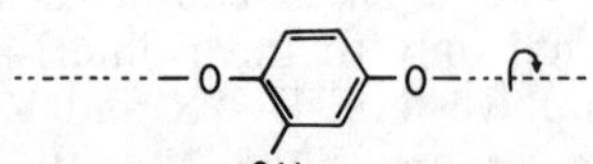

If rotation takes place, the (2-methyl-1, 4-phenylene) group may come out of the plane that contains the terephthaloyl moiety. This can affect the conjugation in two ways. First, the loss of planarity between the two aromatic rings does not allow the symmetry necessary to extend delocalization of π-electrons to both aromatic systems through the carboxyl functionality. Second, when the (2-methyl–1, 4–phenylene) group comes out of plane, the methyl group can sterically interfere with the p-orbitals of the carbonyl functionality, decreasing the extent of conjugation between the carbonyl and aromatic groups of the terephthaloyl moiety. That rotation is not possible in VIIAR due to a 1,3-linkage, consequently, IIIAR has a lower O 1s shake-up intensity than VIIAR.

The analysis of the data for the aromatic homopolyesters indicates that 1,3– and 1,4-phenylene linkages can be distinguished by means of the O 1s shake-up satellite intensities. In the next section, angular-dependent shake-up satellite intensity data are used to investigate preferential enrichment of the surface with a specific block in a series of block copolyesters.

Copolyesters

Figure 4 depicts the core level spectra of IIIB, VIIB and XB copolyesters. Table 5 presents the core level binding energies found for those polymers. The C 1s level spectra show components at BE = 285.0 eV. (C–H), BE = 286.5 eV. (C–O) and BE = 289.1 eV. (C=O). Shake up satellites are observed located at 7.1 eV. from the hydrocarbon component of the C 1s spectra. The C 1s line shape is similar for the three polymers as shown in Fig. 4. However, the C 1s shake up satellite of XB is slightly different in shape and more intense than those corresponding to IIIB and VIIB, which are similar to each other. This is expected since XB is an all aromatic copolymer and VIIB and IIIB have a saturated block in their structure.

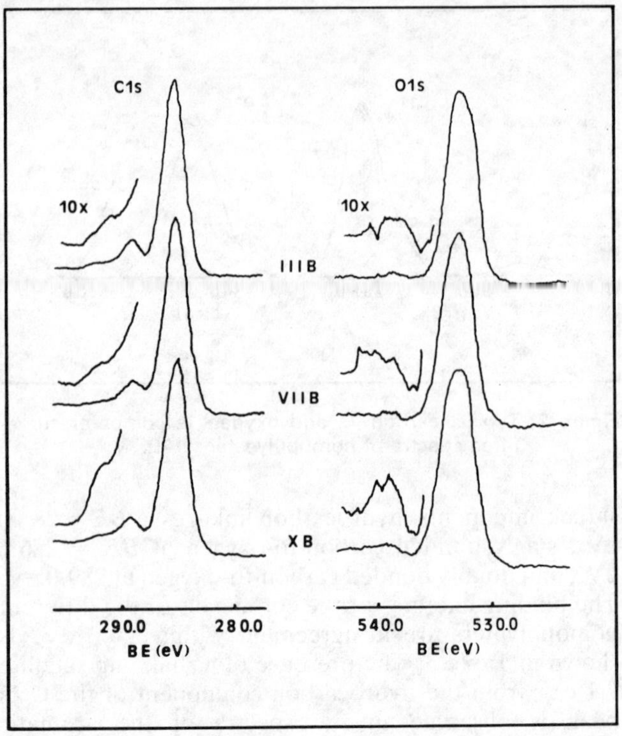

Figure 4. Carbon 1s and oxygen 1s high resolution spectra of the series of copolyesters

The O 1s spectra present two components at BE = 532.1 eV. (O=C) and BE = 533.7 eV. (O—C). Similar to the homopolyesters, a shake up satellite located at 7.1 eV. from the O=C component of the O 1s spectra is observed. The shake up satellite line shapes are characteristic for each block copolyester, indicating their 'finger-print' character. The O 1s shake-up line shape of XB does not resemble those of the homopolyesters. However, this is not surprising considering that XB contains blocks of VIIAR and IAR. Also, the copolyesters were analysed as thin films.

Table 5. Core level binding energies (eV) for the copolyesters

Polymer	C 1s[a]	O 1s[a]	C 1s shake-up[a]	O 1s shake-up[a]
IIIB	285.0	532.1	292.1	539.2
	286.5	533.7		
	289.0			
VIIB	285.0	532.1	292.2	539.3
	286.5	533.7		
	289.1			
XB	285.0	532.2	292.1	539.2
	286.6	533.8		
	289.2			

[a]$\Delta E = \pm 0.2$ eV.

Table 6 presents the angle-dependent XPS data corresponding to the three copolyesters in terms of C 1s and O 1s shake-up intensities. IIIB shows a decrease of the C 1s and O 1s shake-up intensities as the electron exit angle decreases (the surface of the polymer is emphasized). Since only the aromatic block of IIIB can present shake up satellites, the results imply that the surface of IIIB is enriched with the cycloaliphatic block IIIAL. In a similar fashion, VIIB also shows a decrease in the C 1s and O 1s shake-up intensities as the electron exit angle decreased. The gradient in VIIB is steeper than in IIIB. Therefore, the enrichment of the surface with the saturated block (IIIAL) seems to take place to a greater extent in VIIB than in IIIB. The C 1s shake-up intensity of XB was constant as the electron exit angle decreased, consistent with an all aromatic copolymer. However, the O 1s shake-up intensity increased strongly as the electron exit angle decreased. An increase in the O 1s shake-up intensity indicates an increase in the extent of conjugation between the π orbitals of the carboxylate group and aromatic group. The blocks that comprise XB have the chemical structure of IAR and VIIAR and the characterization data for IAR and VIIAR (Table 4) suggested a greater conjugation length in VIIAR. Therefore, the increase of the O 1s shake-up intensity of XB suggests the enrichment of the surface with the VIIAR block.

Table 6. Angle-dependent XPS data for the copolyesters

Polymer	Exit angle	C 1s shake-up Intensity[a]	O 1s shake-up Intensity[a]
IIIB	10°	8.3	1.2
	30°	8.3	2.0
	90°	10.0	2.0
VIIB	10°	7.1	0.5
	30°	7.1	0.4
	90°C	10.0	2.6
XB	10°	10.0	2.4
	30°	10.0	0.9
	90°	10.0	0.5

[a]The estimated error is less than ±10%

CONCLUSIONS

The shake-up satellites arising from $\pi \rightarrow \pi^*$ transitions were utilized to characterize a series of homopolyesters and copolyesters. O 1s shake-up satellites revealed conjugation of benzene rings and carboxylate functionalities bonded to the aromatic groups. Also, steric effects were found to have an influence on the shake-up satellite line shapes and intensities. The shake-up transitions proved to be an effective level of information to distinguish between 1,3- and 1,4-phenylene linkages.

The angular-dependence of the shake-up intensities was demonstrated to be effective in the determination of surface enrichment with a specific block in the copolyesters. The all aromatic copolymer showed enrichment of the surface by the block with the highest conjugation length. The mixed copolymers were enriched at the surface by the cycloaliphatic blocks.

Acknowledgement

The financial support of the Office of Naval Research (ONR) is gratefully acknowledged.

REFERENCES

1. D.T. Clark, in *Handbook of X-ray Photoelectron Spectroscopy, p. 211,* ed. by D. Briggs, Heyden (1978).
2. A. Dilks, In *Electron Spectroscopy: Theory, Techniques and Applications* Vol. 4, p. 278, ed. by C.C. Brundle and A.D. Baker, Academic (1981).
3. J. Riga, J.J. Pireaux, and J.J. Verbist, *Molec. Phys.* **34,** 131 (1977).
4. S.A. Chambers and T.D. Thomas, *J. Chem. Phys.* **67,** 2596 (1977).
5. G. Distefano, M. Guerra, D. Jones and A. Modelli, *Spectrochimica Acta* **38A,** 85 (1982).
6. G. Distefano, M. Guerra, D. Jones, A. Modelli and F.P. Colonna, *Chem. Phys.* **52,** 389 (1980).
7. R. Nakagaki, D.C. Frost and C.A. McDowell, *J. Elect. Spectros. Relat. Phenom.* **27,** 69 (1982).
8. G. Distefano, M. Guerra, F.P. Collona, D. Jones, G. Consiglio and D. Spinelli, *Chem. Phys.* **72,** 267 (1982).
9. G. Distefano, M. Guerra, D. Jones and A. Modelli, *Chem. Phys.* **71,** 169 (1981).
10. F.P. Colonna, G. Distefano, M. Guerra and D. Jones, *J. Elect. Spectrosc. Relat. Phenom* **18,** 309 (1980).
11. D.T. Clark, B.D. Adams, A. Dilks, J. Peeling and H.R. Thomas, *J. Elect. Spectrosc. Relat. Phenom.* **8,** 51 (1976).
12. D.T. Clark and A. Dilks, *J. Polym. Sci.: Polym. Chem. Ed.* **14,** 533 (1976).
13. D.T. Clark and A. Dilks, *J. Polym. Sci.: Polym. Chem. Ed.* **15,** 15 (1977).
14. W.R. Salaneck and H.R. Thomas, *Solid State Comm.* **27,** 685 (1978).
15. C.B. Duke, W.R. Salaneck, T.J. Fabish, J.J. Ritsko, H.R. Thomas and A. Paton, *Physical Review B* **18,** 5717 (1978).
16. J.E. McGrath, D.W. Dwight, J.S. Riffle, T.F. Davidson, D.C. Webster, and R. Viswanathan, *Am. Chem. Soc., Div. Polym. Chem., Polym. Preprints* **20(2),** 528 (1979).
17. D.W. Dwight, J.E. McGrath, A.R. Beck and J.S. Riffle, *Am. Chem. Soc., Div. Polym. Chem., Polym. Preprints* **20(1),** 702 (1979).
18. J.S. Riffle, *PhD Dissertation,* Department of Chemistry, VPI & SU (1980).
19. A.K. Sha'aban, *M.S. Thesis,* Department of Chemistry, VPI & SU (1983).
20. N.M. Patel, *M.S. Thesis,* Department of Chemical Engineering, VPI & SU (1984).
21. J.A. Gardella, R.L. Chin, S.A. Ferguson and M.M. Farrow, *J. Elect. Spectrosc. Relat. Phenom.* **34,** 97 (1984).
22. M.B. Polk, K.B. Bota, N. Nandu, M. Phingbodhipakkiya and C. Edougu, *Macromolecules* **17,** 129 (1984).
23. M.B. Polk, K.B. Bota, E.C. Akubuiro and M. Phingbodhipakkika, *Macromolecules* **14,** 1626 (1981).
24. M.B. Polk and F.C. Onwumere, *J. Macromol. Sci. A, Chem Ed.,* in press.
25. (a) C.S. Fadley and S.A. L. Bergstrom, *Phys. Letters* **35A,** 375 (1971); (b) C.S. Fadley and S.A.L. Bergstrom, in *Electron Spectrscopy, p. 233,* ed. by A.A. Shirley, North Holland (1972).
26. W.A. Fraser, J.V. Florio and W.D. Robertson, *Surf. Sci.* **36,** 661 (1973).
27. P.Swift, *Surf. Interface. Anal.* **4,** 47 (1982).

SURFACE AND INTERFACE ANALYSIS, VOL. 9, 411–417 (1986). EXTENDED ABSTRACTS

Ion Beam Damage Effects in Poly(Methyl Methacrylate)

M. J. Hearn and D. Briggs
ICI PLC, Petrochemicals and Plastics Division, PO Box 90, Wilton, Cleveland, TS6 8JE, UK

INTRODUCTION

For the surface characterization of polymeric materials by SIMS there are two essential elements which should be considered in any optimization of experimental parameters. Both are a function of the various ion beam parameters and are mutually exclusive. Firstly, a high signal : noise ratio is required where spectral intensity and contribution of structurally informative ions is maximized. This issue has recently been addressed.[1] Secondly, spectral stability is important and spectral time dependence resulting from changes in the structural composition of the surface due to damage effects must be minimized. In principle, this can be achieved by utilizing total ion doses $< 10^{13}$ ions cm^{-2}. However, damage effects associated with SIMS analysis of polymers have only been poorly characterized, so detailed knowledge of polymer damage rates is required to define more exactly at which point damage effects prove prohibitive. This is especially true for SIMS imaging at high spatial resolution using liquid metal ion sources. Early studies of ion bombardment of polystyrene and PMMA, which relied solely on XPS for monitoring damage showed that these organic polymers are significantly damaged at ion doses $> 10^{14}$ ions cm^{-2}. In this study both XPS and SIMS have been used to investigate various aspects of ion beam damage in PMMA for dose regimes extending from 'static' (10^{13} ions cm^{-2}) to near dynamic (10^{15} ions cm^{-2}) SIMS conditions.

METHOD

The XPS/SIMS instrument based on the VG scientific ESCALAB LAB Mk 1 UHV system has been previously described.[2] In particular this utilizes an AG61 noble gas ion source and a MIG 100 Ga^+ surce (beam angles 0° and 35° wrt sample normal respectively). In all experiments charge neutralization was by a 700 eV electron flood. Static SIMS measurements use ion currents of 0.7 nA cm^{-2} whilst damage was induced by increasing the ion current to 42 nA cm^{-2}. XPS measurements used Mg Kα at 300 W. PMMA samples were always freshly prepared by spin casting from dilute solution in chloroform onto cleaned glass cover slips.

RESULTS

In the XPS studies damage rates were monitored using C 1s and O 1s core level intensities and are expressed in terms of the change in C : O atomic ratio as a function of total ion dose. During ion bombardment the oxygen content of the PMMA surface rapidly decreases. With 4KeV xenon structural damage is evident at 5×10^{13} ions cm^{-2} and is well pronounced after 2×10^{14} ions cm^{-2}. Damage is lessened to some extent by reducing primary ion mass and there is an apparent marked decrease in damage at lower primary ion energies. However the lack of surface sensitivity of XPS may distort data if in damage terms there is a significant damage gradient within the XPS sampling depth. Hence negative ion SIMS has also been utilized to measure surface damage rates.

High resolution XPS with component analysis of the C 1s peak envelopes reveals major losses of intensity in the chemically shifted

$$-\mathrm{C}-\mathrm{O} \text{ and } -\overset{\overset{\displaystyle \mathrm{O}}{\|}}{\mathrm{C}}-\mathrm{O}$$

components due to a loss of mainly oxygen species from the surface regions of the polymer.

This disruption of the molecular structure of PMMA results from direct loss or reaction of alkyl ester side chains besides breaking of main chain bonds. At higher fluences ($> 10^{15}$ ions cm^{-2}) a surface rich in carbon exists (oxygen content $< 10\%$) which is highly crosslinked and contains some degree of unsaturation.

For a series of homogeneous polyalkylmethacrylates it has previously been shown[3] that the relative intensity ratio of the negative ion peaks due to CH^- and O^- (13D and 16D respectively) closely reflects the C : O atomic ratio of the bulk polymer. The I13D/I16D peak intensity ratio regarded as being representative of the average oxygen content of the uppermost molecular layers was used as a sensitive probe for measuring surface damage. Results are summarized in Fig 21, Ref 4 which plots I13D/I16D ratio as a function of ion dose for various combinations of primary ion mass and energy. The general trend is quite clear. Reducing either the primary ion mass, at constant energy, or, the primary ion energy itself, significantly reduces the rate of polymer damage. Damage rates were more or less independent of rate of ion dose (i.e. current density) over the 10–100 nA cm^{-2} range.

Positive ion SIMS spectra from PMMA are particularly sensitive to ion beam damage effects. Processes induced by ion-polymer interactions cause rapid changes in the molecular composition of the surface region from where the secondary ions are ejected, thus having a pronounced effect on the SIMS fragmentation pattern. To a first approximation those ions invoked as containing oxygen suffer the major intensity decays, this decay occurring at more or less the same rate for all primary ion masses at 4 KeV. The rate is somewhat lower at 2 KeV. Results from dose dependent studies can be summarized as follows:

1. In the dose regime $10^{12} - 10^{13}$ ions cm^{-2} spectral intensities remain unchanged and relate to undamaged polymer.

2. For doses between $10^{13} - 10^{14}$ ions cm^{-2} progressive degradation of the PMMA structure is observed, as reflected by major changes in both absolute and relative secondary ion intensities.

3. Above 10^{14} ions cm^{-2} molecular information is severely diminished and identification of PMMA structural features is impossible. The positive ion spectrum tends to a pattern of hydrocarbon clusters ($C_nH_m{}^+$) common to many similarly damaged polymers.

CONCLUSIONS

Static SIMS spectra from undamaged polymer surfaces can be obtained for total ion doses of $< 10^{13}$ ions cm^{-2}. Damage effects, to which SIMS is more sensitive than XPS, manifest themselves at higher doses. Sample damage rates are best minimized by lowering primary ion energies. Reducing primary ion mass reduces damage but at the expense of having good high mass cluster intensities which is important for SIMS analysis of polymers.

REFERENCES

1. D. Briggs and M. J. Hearn, *J. Mass Spec. Ion Processes*, **67**, 47 (1985).
2. D. Briggs, *Polymer*. **25**, 1379 (1984) and refs therein.
3. D. Briggs, M. J. Hearn and B. D. Ratner, *Surf. Interface Anal.* **6**, 184 (1984).
4. D. Briggs, *Surf. Interface Anal.* **9**, 391 (1986). (this Proceedings)

Application of XPS to the Characterization of Polymer-loaded Drugs

Fabio Garbassi
Istituto Guido Donegani SpA, Via Fauser, 4 28100 Novara, Italy
Fabio Carli
Farmitalia Carlo Erba SpA, Via Imbonati, 24 20123 Milano, Italy

The practice of loading drugs on suitable polymeric materials is assuming an increasing importance in pharmaceutical technology. It is for example connected to the low physiological solubility of several active molecules, and to the tendency to decrease the collateral effects by reducing the dosage.

However the characterization of the drug/polymer system by traditional techniques suffers from a lack of information mainly on the reciprocal location of the two moieties.

This work deals with the investigation by XPS of some drug–polymer systems. The aim is to contribute to solving problems like the mechanism of loading, the effectiveness of loading methods and the protection again humidity using suitable packaging techniques.

The work concerns mainly the griseofulvin/crospovidone system. Griseofulvin is taken as a model of a drug scarcely soluble in water (in the physiologic liquors) and crospovidone is cross-linked polyvinylpyrrolidone. Also another active principle, here named AP, was taken into account.

The griseovulvin loading was made by swelling the polymer powder with a solution of the drug in dimethylformamide under continuous mixing in a mortar. A solution volume smaller than the maximum capacity of the polymer was always used. After swelling, samples were dried at 120°C under vacuum, gently milled to 60 mesh and homogeneously mixed.

Loading of AP on crospovidone was made using three different methods, here indicated as I, II and III. All samples were examined by XRD and XPS and sometimes observed at SEM. The elemental sensitivity factor method was used for drawing surface compositions from photoemission peak intensity values. Measurements on pure substance indicate that experimental compositions from XPS data differ from theoretical ones within 1–2%.

Table 1. Griseofulvin/crospovidone system

Polymer	Drug/polymer w : w	Surface composition (at%) O	C	N	Cl	Cl/N exp.	Cl/N theor.
A	—	11.5	76.6	11.9	—	—	—
	1 : 5	13.0	76.0	10.2	0.5	0.049	0.063
	1 : 7	13.3	75.9	10.3	0.5	0.049	0.044
	1 : 10	12.3	77.4	8.6	0.5	0.058	0.032
D	—	17.7	70.8	11.5	—	—	—
	1 : 5	17.3	70.4	11.3	0.5	0.055	0.063
	1 : 7	16.8	73.9	8.7	0.3	0.036	0.044
	1 : 10	18.9	72.3	8.2	0.3	0.038	0.032

Table 2. Griseofulvin/crospovidone system

Residual crystallinity vs. Cl/N atomic ratio

%cryst.	Cl/N
5	0.074
25	0.055
35	0.049
45	0.041

Table 3. AP/crospovidone system

Comparison of loading techniques and humidity effect

Loading technique	N/O theor.	N/O exp.	N/O exp.[a]
I	0.74	0.63	0.38
II	0.82	0.65	0.65
III	0.82	1.10	0.44

[a] After exposure to 90% rh for 3 months.

THE MECHANISM OF DRUG LOADING

Two possible mechanisms can be envisaged for the griseofulvin loading on crospovidone: *surface deposition* and *intramacromolecular diffusion*. In the first case, taking into account the specific surface area of the polymer powder and the molecular dimensions of griseofulvin, the polymer should be fully covered by a layer of drug about 10 nm thick. As a consequence, a strong decrease of the N 1s peak intensity could be forecast (in fact, only crospovidone contains N, while only griseofulvin contains C1). Since, on the contrary, the nitrogen concentration is always very near to that of unloaded crospovidone, intramacromolecular diffusion is likely to occur. Some XPS composition data are reported in Table 1. The Cl/N atomic ratios are compared to the theoretical values calculated considering a homogeneous molecular mixing. XPS results appear able to distinguish behaviour differences among the various polymers used in loading experiments. This argument is fully explored elsewhere.[1]

Care must be taken in evaluating quantitative results when griseofulvin shows some residual crystallinity (from XRD and DSC data). In Table 2, a linear inverse relationship is reported between such a parameter and the surface Cl/N ratio: the larger are the crystals, the weaker are the signals from the drug, and as a consequence, the Cl/N values.

THE EVALUATION OF THE LOADING TECHNIQUE AND THE ACTION OF HUMIDITY

In Table 3, the different techniques used for AP loading are compared, giving rise to rather large differences of N/O ratios (AP contains only C, O and H atoms). Thus, XPS results allow a specific loading technique to be chosen, whether a larger or smaller amount of drug is requested on the support surface.

In the same Table, data are reported after exposure of the

same samples for three months to 90% of relative humidity (rh). A different stability is observed, depending on the loading technique.

THE EFFECTIVENESS OF THE PACKAGING SYSTEM

Some blisters of griseofulvin/crospovidone samples underwent a long-period exposure (up to 6 months) to 80% rh. The Cl/N ratio does not show variations. By comparison, in unprotected powder samples exposed to 70% rh, Cl/N varied from 0.04 to 0.24. It can be concluded that the blister packaging is suitable for the storage of such systems

REFERENCE

1. F. Carli and F. Garbassi, *J. Pharmaceut. Sci*, **79**, 963 (1985).

XPS Investigation on Fluoropolymers Thin Films Deposited in Glow Discharges

E. Desimoni, C. Malitesta, L. Sabbatini, P. G. Zambonin
Laboratorio di Chimica Analitica, Dipartimento di Chimica dell Universita di Bari, Via Amendola, 173–70126 Bari, Italy

INTRODUCTION

Particular attention has been devoted in the last years to thin film deposition and surface treatments of polymers generated in glow discharges. Fluoropolymers thin films deposited in a radiofrequency discharge fed with C_2F_6–H_2 mixtures have been investigated by XPS in the present study and the influence of some plasma parameters on the polymer chemical structure has been evaluated. The results indicate that the overall F/C stoichiometry and the relative abundance of fluorinated carbon groups depend on the electrical characteristics of discharge, its duration and feed composition.

EXPERIMENTAL

The plasma apparatus was fed by a mixture of C_2F_6–H_2 at a pressure of 300 mTorr; hydrogen composition was varied in the range of 20–70%. The experiments have been performed at a fixed power (about 40 W) with two peak voltages: 1200 V (HV) and 850 V (LV). Gold substrates were utilized after cleaning with organic solvents in an ultrasonic bath and after drying at 110°C.

XPS spectra of the deposited films were recorded by a Leybold–Heraeus LHS 10 Spectrometer, operated at 280 W with Mg Kα as the exciting radiation ($h\nu$ = 1253.6 eV). Wide scan spectra were recorded in the FRR mode with a retarding ratio B = 3; detailed spectra were recorded in the FAT mode with a pass energy E_o = 50 eV. Spectra were calibrated against the Au $4f_{7/2}$ signal (84.0 eV) arising from a little portion of the gold substrate not covered by the polymer since, when the polymer films were sufficiently thin (i.e. less than one micron), the sample charging was negligible. Typical pressure in the analysis chamber was around 5×10^{-9}mbar.

Spectra were collected as series of datafiles on floppy disk by using an Apple II microcomputer interfaced[1] to the spectrometer; processing of the spectra was performed by an IBM 4341 by using an interative non-linear least squares curve fitting program based on a Gaussian–Lorentzian product function.

RESULTS AND DISCUSSION

A systematic XPS investigation was performed on plasma-polymerized films deposited under HV and LV electrical conditions from different C_2F_6–H_2 mixtures in order to obtain information on the polymer structure and stoichiometry. For each sample detail spectra were recorded of carbon 1s, oxygen 1s, fluorine 1s and fluorine 2s core level signals.

The C 1s spectra could be fitted by using a series of peaks attributed to $-CF_3$, $-CF_2$, $-CF$, $-C-CF$ and $-CH$ groups on the basis of literature data.[2–4] The percent distribution of carbon components as determined by XPS analysis of samples obtained under the same electrical conditions on varying the feed composition or, vice versa, at the same feed composition on varying the electrical parameters of discharge, is reported in Table 1, beside the overall F/C stoichiometry of the films. A typical C 1s spectrum is reported in Fig. 1.

No appreciable signal due to C = O or C − OH species could be detected in any samples; at the same time the O 1s signal was always very weak and attributable to adsorbed molecular oxygen.

The data reported in Table 1 clearly indicate that a correlation exists between polymer structure, feed composition and electrical conditions: the degree of fluorination increases on decreasing the hydrogen percent and on increasing the peak-to-peak voltage.

XPS analyses were also performed on films deposited under identical experimental conditions but at different growth stages. The results indicate that the primary precursor

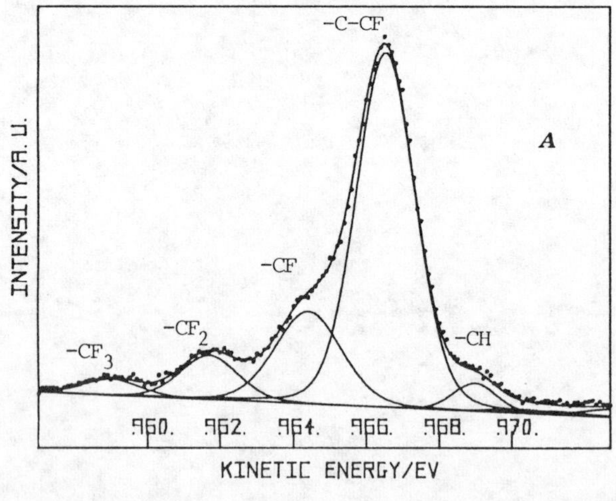

Figure 1. A typical C 1s XPS spectrum recorded on a thin polymer film deposited under HV conditions and 70% hydrogen in the feed mixture.

Table 1.

Sample	%H_2	Electrical Conditions	Total Stoich.	% − CF_3	% − CF_2	% − CF	% − C − CF	% − CH
A	20	HV	$CF_{1.6}$	13	53	19	11	4
B	20	LV	$CF_{0.8}$	10	14	25	49	2
C	50	HV	$CF_{0.6}$	5	11	25	59	< 1
D	50	LV	$CF_{0.5}$	4	7	20	67	2
E	70	HV	$CF_{0.4}$	3	8	19	66	4

of the film turns to a polymer with a different overall stoichiometry at longer exposure in plasma.

A deeper investigation on this matter is still in progress; the influence of other plasma parameters (i.e. the deposition temperture) on both polymer chemical structure and its growth mechanism will be evaluated and bulk and surface homogeneity of the deposited polymers will be checked.

Acknowledgements

This work is part of a joint research program with the Centro di Studio per la Chimica dei Plasmi of the Italian National Council of Research, which is strongly acknowledged also for samples preparation.

REFERENCES

1. E. Desimoni and C. Malitesta, *Computer Enhanced Spectroscopy*, accepted, (1986).
2. A. Dilks in *Electron Spectroscopy*, vol. 4, chap 5, ed. by C. R. Brundle, A. D. Baker, Academic Press, London (1981).
3. D. T. Clark, in *Handbook of X-Ray and Ultraviolet Photoelectron Spectroscopy*, Chap 6, ed. by D. Briggs, Heyden, London (1978).
4. A. Dilks and E. Kay, *Macromolecules*, **14**, 855 (1980).

An ESCA Study of a Contact Lens Co-polymer RF Plasma-discharge Treated at Various Power Levels

D. Fakes,*§ M. Newton,† J. F. Watts,‡ and M. J. Edgell‡

* Department of Pharmacy, Kings College, London, UK
† Department of Pharmacy, School of Pharmacy, University of London, UK
‡ Department of Materials Science and Engineering, University of Surrey, UK
§ Present address: SK & F Laboratories Ltd., Welwyn Garden City, Herts

INTRODUCTION

The contact lens co-polymer BX3 includes an organosiloxane component which renders the lens gas-permeable. there is, as a result of the presence of this silicon-bearing material, a reduction in lens wettability, which may restrict its utility in a physiological environment. This trend can be reversed by means of a RF oxygen plasma treatment. This paper describes an investigation into the changes that occur in the surface chemistry of the BX3 co-polymer, as a function of the RF plasma-treatment process variables.

EXPERIMENTAL

RF treatments were carried out at power levels of 10, 30, 50 and 100 W for time periods in the range 0–160 s. Quantitative surface analyses were obtained by XPS in both MgKα and AlKα x-rays. The data obtained in AlKα radiations show a strong dependence on treatment energy (power × time). The surface oxygen assay being enhanced as treatment progresses at the expense of the carbon concentration. A complementary enhancement of silicon is also observed. Selected specimens were analysed in monochromated AgLα (hμ = 2984.3 eV), and the additional probing depth indicates that although sub-surface carbon is increased it is still not comparable with the corresponding level in the untreated specimen.

The use of these photon sources provides a non destructive depth-profiling capability. The sampling depths, taken as 3 λ (where λ = electron inelastic mean free path) have been calculated using the formalism of Seah and Dench; for the carbon 1s electron the analysis depth is approximately 6.0, 7.0, and 10.0 nm in MgKα, AlKα, and AgLα respectively. The elemental composition data suggest subtle near surface concentration gradients with the most consistent observation being the surface depletion of carbon and enhancement of silicon. Oxygen is also enhanced during the RF treatment. This suggests the depth of the plasma treatment zone is greater than the AgLα analysis depth.

The data for the 10 W treatment is presented graphically in Fig. 1 and it can be seen that the C/Si ratio increases with treatment time indicating the surface enhancement of silicon. This is further confirmed by the offset of the MgKα results; the more surface sensitive analysis (by approximately 1.0 nm) consistently recording a lower C/Si value.

By taking energy as power x treatment time, 'universal curves' were produced for all power and treatment times. The C/O and C/Si ratios exhibited a linear relationship with log energy.

By examination of the high resolution C 1s spectra it is possible to gain an insight into changes in surface functionality that occur on RF treatment. The changes that occur on treatment are seen as an increase in both C − O and C = O species. this was observed on all specimens; the relative increase in the oxygenated carbon moieties being essentially independent of treatment power and time.

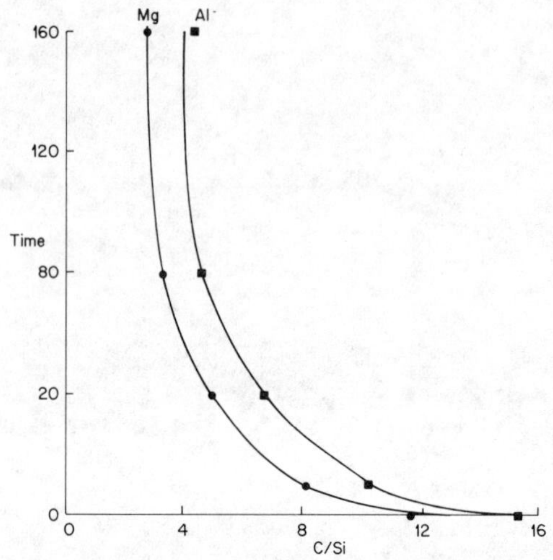

Figure 1. C/Si ratio vs. treatment time (s) for BX3 treated at 10 W, 0.1 mbar.

The changes in silicon chemistry were more difficult to elucidate. Progressive plasma treatments caused a shift in the Si 2p binding energy, computer curve-fitting showed initially a dominant component at approximately 101.8 eV and a very small (5%) contribution at 103.8 eV. On treatment at 10 W for 320 s, the situation has reversed, the dominant component is now that at 103.8 eV. This most likely reflects the replacement of Si–C groups by inorganic Si–O. With more vigorous treatments the Si 2p spectrum FWHM is reduced suggesting total replacement of Si–C by Si–O.

CONCLUSIONS

The present study has shown, at the elemental level, that plasma-discharge treatment of the contact lens co-polymer BX3 results in a surface grossly depleted in carbon but enriched with silicon and oxygen. Clearly the oxygen displaces carbon from the surface but chemically bonds to that remaining; the extent of this interaction is not accounted for by any increase in oxygenated carbon species and clearly the oxygen has an important role to play in the modification of the polysiloxane component. The silicone of the polysiloxane is not displaced from the surface on treatment, there is, however, evidence of additional oxygenation of the silicon bearing molecules during the RF treatment. The changes in surface composition (Ci/Si and C/O) show a logarithmic dependence on energy.

Acknowledgement

DF would like to acknowledge the support of the SERC and the assistance of Cambridge Precision Instruments Ltd.

XPS Study on the Thermal Degradation of Poly-N,N′/4,4′-diphenylether/pyromellitimide

A. Tóth, I. Bertóti, T. Székely, T. A. Antonova* and J. N. Sazanov*
Research Laboratory for Inorganic Chemistry, Hungarian Academy of Sciences, H-1502 Budapest, Pf. 132, Hungary
* Research Institute for Macromolecular Compounds, Academy of Sciences of the Soviet Union, Leningrad, USSR

Polyimides offer certain desirable traits like high temperature stability and high radiation resistance, which render them reliable materials for a variety of applications in a number of differing advanced technologies, including aerospace application, microelectronics, etc.[1,2] XPS has successfully been applied in characterization of surfaces of polyimide samples.[3–5] In this work the surface changes of the title polyimide caused by high temperature heat treatment were followed by XPS.

Samples of a poly-N,N′/4,4′-diphenylether/pyromellitimide foil, produced in the Soviet Union under the trade name 'PM', were preheated in vacuo (*ca.* 10^{-1} Pa) with a heating rate of 7.5 K min^{-1} up to different temperatures with the highest one reaching 1198 K. X-ray photoelectron spectra were recorded in FRR mode with a KRATOS XSAM 800 spectrometer by using Mg $K\alpha_{1,2}$ radiation. The instrument was calibrated against the Au $4f_{7/2}$ line fixed at 84.0 eV. Spectra were referenced by the 'gold decoration' method. Data acquisition and processing were performed using the KRATOS DS 300 data system. For quantification the experimentally based cross-sections of Evans *et al.*[6] were adapted.

Quantification of the detailed spectra of the O 1s, N 1s and C 1s regions of the starting polyimide foil cured at 653 K showed a non-stoichiometric elemental composition: an excess of carbon content and a lack of nitrogen and oxygen contents were found. Results of the peak synthesis for the C 1s peak gave a deficiency for the carbonyl type carbon atoms (BE = 288.8 eV), with respect to the stoichiometric value.

After heating the samples in vacuo drastic changes in peak intensities, positions and shapes were observed. On heat treatment the oxygen content decreased montonously from 14 at% to 4 at%. Ether type oxygen atoms (BE = 533.8 eV) proved to be more stable than carbonyl type ones (BE = 532.1 eV): their ratio changed from ca. 1 : 3.8 to 1 : 0.9. The relative content of nitrogen increased from 4.6 at% up to a maximum value of 6.1 at% at 823 K and then decreased. It was possible to decompose the N 1s peak to at least 4 components. Based on literature data and measurements performed on bicyclohexane-carbodiimide as a model compound, the peak components at 401.4 eV, 400.8 eV, 399.3 eV, and 398.6 eV were tentatively assigned to amide functionalities,[7] imide,[3,4] isoimide[4] and/or carbodiimide,[8] and pyridine type[9] structural features, respectively.

The surface concentration of carbon increased monotonously from 81.5 % to 93 %. The intensities of the original components of the C 1s peak (at 288.8 eV, 286.1 eV and 285 eV) decreased, while a new component appeared at *ca.* 291 eV), evidence about the graphitization of the surface structure. In agreement with this observation a gradual increase in electrical conductivity was found as evidenced by the gradual decrease in sample charging with increasing pretreatment temperature.

REFERENCES

1. M. I. Bessonov (ed.), *Poliimidy — klass termostoikich polimerov*, Nauka, Leningrad (1983).
2. T. O. Herndon, R. L. Burke and J. A. Yasaitis, *Solid State Techn.* **27,** 179 (1984).
3. H. J. Leary Jr and D. S. Campbell, *Surf. Interface Anal.* **1,** 75 (1979).
4. P. L. Buchwalter and A. I. Baise, in *Polyimides, Synthesis, Characterization and Applications*, Vol. 1, p. 537, ed by K. L. Mittal, Plenum Press (1984).
5. T. E. Svarney and J. F. Turner, *Solid State Techn.* **28,** 171 (1985).
6. S. Evans, R. G. Pritchard and J. M. Thomas, *J. Electron Spectrosc. Relat. Phenom.* **14,** 341 (1978).
7. M. Barber and D. T. Clark, *Chem. Comm.* **24,** (1970).
8. A. Tóth, I. Bertóti and T. Székely, unpublished results.
9. R. B. Jones, C. B. McCourt and P. Swift in *Proceedings of the International Conference on Coal Science, Düsseldorf*, p. 657
Verlag Glückauf, Essen, (1981).
10. R. Schlögl and H. P. Boehm, *Carbon*, **21,** 345 (1983).

ANALYSIS OF COPOLYMER SURFACES BY SIMS

S. Nagaoka, M. Koyama, S. Iwamaru and T. Shima
R&D Lab. for New Materials, Konishiroku Photo Ind. Co. Ltd., No.1 Sakura-machi, Hino-shi, Tokyo, Japan

Secondary ions emitted from styrene (St) methylmethacrylate (MMA) copolymer surfaces were detected using SIMS. We measured the positive finger print spectra with different St / MMA samples of which the surface compositions were derived from XPS. Those spectra contain the same characteristic fragment peaks as found in both polystyrene and polymethylmethacrylate. And the intensities of those peaks sensitively reflect the surface composition. We will discuss the advantage of SIMS from a quantitative aspect, and the problems in depth-profiling applications.

PROBING POLYMER SURFACE DYNAMICS BY CONTACT ANGLE HYSTERESIS AND ESCA

J.D. Andrade and W-Y Chen
College of Engineering, University of Utah, Salt Lake City, Utah 84112, USA

Polymeric materials generally exhibit various molecular motions and relaxations. Such relaxation processes, which include the glass transition temperature, have significant effects on physical and mechanical behavior. Polymer molecules and segments at surfaces and interfaces also exhibit motions and relaxations. In air or vacuum, such motions 'permit' the surface to structure to minimize the surface free energy. In aqueous solution, the polymer surface structures and orients to optimally interact with the aqueous solvent, thereby minimizing the interfacial free energy. XPS and related high vacuum techniques probe the vacuum-equilibrated surface. The best way to probe the polymer–liquid interface is via dynamic contact angle or wetting methods. We describe the Wilhelmy Plate method for dynamic contact angle measurement, permitting both advancing and receding angles to be measured continuously. Case studies using various methacrylate copolymer and polyethylene oxide surfaces will be presented and discussed.

SURFACE FREE ENERGIES AND SURFACE COMPOSITIONS OF PVC/ PMMA POLYMER BLENDS

H.J. Busscher,[*] W. Hoogsteen,[†] L. Dijkema,[†] G.A. Sawatsky,[‡] A.W.J. van Pelt,[*] H.P. de Jong,[*] G. Challa[†] and J. Arends[*]
[*] University of Groningen, Laboratory for Materia Technica, Ant. Deusinglaan 1, 9713 AV Groningen, The Netherlands
[†] Laboratory of Polymer Chemistry
[‡] Laboratory of Physical Chemistry, Nijenborg 16, 9747 AG Groningen, The Netherlands

The surface free energies of PVC/PMMA polymer blends have been determined by contact angle measurements employing the dispersion and polar component concept. A comparision of measured surface free energies with surface free energies calculated from the parachor, yields the conclusion that the surfaces of the blends are enriched by the low energy component. This conclusion has been confirmed by ESCA data, which showed reasonable correspondence with surface chemical compositions calculated from contact angles. The differences can be explained from the fact that surface free energies were determined in an aqueous environment, while the electron spectroscopy was carried out in vacuum. Phase separation in the blends at 60 wt% PMMA became obvious from a stronger presence of PVC at the surface above 60 wt% PMMA.

A COMPARISON OF POSITIVE AND NEGATIVE CLUSTER ION EMISSION FROM POLYMERS BY STATIC SIMS

A. Brown and J.C. Vickerman
Surface Analysis Service, Department of Chemistry, UMIST, Manchester, UK

The high sensitivity and molecular specificity of static SIMS make it an important technique for surface analysis of polymers. In general it has been found that polymers, in common with all other materials, yield SIMS spectra which are uniquely characteristic of their surface chemical structure and are interpretable by the established rules of electron impact mass spectrometry. This has been demonstrated for a wide range of polymers providing that surface chemical damage is minimised by usnig total primary beam doses well below 10^{13} particles cm^{-2}. Recent FABMS studies on polystyrene have demonstrated additional advantages in using neutral primary beams. Firstly, surface potential drift is minimised which leads to more stable, reproducible spectra and a consequent reduction in analysis times. Secondly, there is evidence for a charge induced damage effect in insulators which makes neutral primary beams preferable to ions for SIMS analysis of polymers.

However, in most of the published static SIMS data on polymers all of the useful molecular structure information has been derived from *positive ion spectra*. The negative ion spectra have usually been restricted to low mass fragments of the type C^-, CH^-, O^-, OH^-, C_2^-, C_2H^- and X^- for halogen containing polymers. Clearly from a chemical point of view there is no reason why polymers containing electronegative elements such as O and F eg. polyethyleneterephthalate, polytetrafluoroethylene and polyalkylmethacrylates should not emit stable, negative molecular cluster ions under ion or atom bombardment which are typical of the polymer structure. Indeed many examples exist in the literature for the presence of negative molecular ions in SIMS spectra of, for example, monomolecular organic thin films on metallic substrates.

In this paper negative ion SIMS spectra showing a high degree of molecular structure information will be presented for a range of thick film polymers. The structural content of these spectra will be discussed in relation to the previously characterised positive ion spectra of the same materials. The polymers studied include polyethyleneterephthalate, polyalkylmethacrylates and polydimethylsiloxane. The important role of surface potential and its influence of spectral content and stability will be discussed.

Section 8.

Adhesion

SURFACE AND INTERFACE ANALYSIS, VOL. 9, 421–429 (1986)

Use of Surface Behavior Diagrams to Study Surface and Interfacial Reactions

G.D. Davis
Martin Marietta Laboratories, 1450 S. Rolling Road, Baltimore, Maryland 21227, USA

The concept of surface behavior diagrams (SBD's) is presented and their construction from x-ray photoelectron spectroscopy or Auger electron spectroscopy measurements is described. Their use in investigating changes in composition of a sample's surface as a function of a reaction parameter (such as time, solution concentration, or applied voltage) or a depth parameter (such as deposited film thickness, sputter time, or distance along a crater wall) is illustrated with several examples from the literature. These include: hydration of phosphoric-acid-anodized aluminum, adsorption of an organic hydration inhibitor from an aqueous solution onto aluminum surfaces, electrochemical polarization of an aluminum surface, deposition of aluminum onto (HgCd)Te surfaces, and depth profiling of an anodized semiconductor and a multilayer engine bearing. In each example, the SBD's proved very beneficial in describing the surface or interfacial reactions occurring and, in many cases, in establishing the mechanisms involved.

INTRODUCTION

Surface behavior diagrams (SBD's) are a recently developed means of display and analysis of quantitative surface-sensitive data, such as those provided by x-ray photoelectron spectroscopy (XPS) or Auger electron spectroscopy (AES). The purpose of this paper is to describe what SBD's are, the conditions in which they can be useful, and how to construct them. In part, this will be done by using several examples from the literature.

Surface behavior diagrams resemble ternary or quaternary phase diagrams in that they represent a surface composition as a weighted sum of three or four basis compounds. However, they differ from phase diagrams in that surface compositional information rather than bulk phase information is represented and in that the equilibrium condition is relaxed so that changes in the surface composition during nonequilibrium reactions can be traced.

With an SBD, changes in the surface composition can be traced as a function of some parameter of interest, such as reaction time, solution concentration, applied potential, temperature, overlayer thickness, or depth into the sample. The path along which the composition evolves can then be compared to those predicted by proposed models to help to establish the mechanisms involved during a reaction. In other applications, the dependence with depth of the composition can be measured. Interfacial reactions between a substrate and an overlayer can then be detected and identified and, in some cases, material properties can be predicted based on the chemistry of a thin film coating. Often, insights into the surface chemistry can be obtained that would be less apparent by other means.

We have used SBD's to study the evolution of the surface composition under a variety of conditions, including hydration of aluminum oxide surfaces,[1–5] adsorption of hydration inhibitors from aqueous solutions,[5] depth profiling of oxidized semiconductors,[6,7] deposition of thin metallic overlayers onto semiconductor surfaces,[8] and corrosion of aluminum.[9] Several of these studies will be used in the third section to illustrate different types of SBD's. Other uses of SBD-like diagrams have also appeared in the literature, e.g., for characterizing the diffusion behavior and intermetallic compound formation at diffusion barriers in engine bearings[10,11] and for examining the redistribution of chromium and the segregation of phosphorus to the grain boundaries of Ni–Cr–Fe alloys.[12]

CONSTRUCTION OF SURFACE BEHAVIOR DIAGRAMS

To construct an SBD, good quantitative measurements of the surface composition are needed. In most cases, we have preferred to use high-resolution XPS spectra rather than AES spectra because of the better possible quantification and chemical state identification. For each measurement, elemental sensitivity factors were derived from spectra of standards obtained with the particular instrument used. For a compound SBD, the next step is to identify a set of basis compounds with which the data can be expressed in the form of linear combinations. These compounds normally are selected based on the prior knowledge or expectations of the surfaces. The number of basis species can be as many as the number of distinct chemical entities provided by the data, i.e., the number of elements or chemical states detected. However, in most cases, the chemistry is such that three or four basis compounds will characterize the surface. In other cases, simplifications can often be made to limit the basis species to three or four for ease of presentation.

In our study of the hydration of aluminum anodized

0142–2421/86/120421–09 $05.00

in phosphoric acid,[1,2] XPS measurements indicated the presence of Al, P, and O with the Al and P being in the +3 and +5 oxidation states, respectively. Accordingly, we chose as our basis compounds Al_2O_3, H_2O, and $AlPO_4$. In this system, the compositions of the hydroxides of aluminum, AlOOH and $Al(OH)_3$, which might be present, can be represented as $Al_2O_3+H_2O$ and $Al_2O_3+3H_2O$, respectively, even though the hydroxides are distinct chemical compounds and not mixtures of Al_2O_3 and H_2O.

Once the basis compounds are chosen, the surface composition is converted from atomic to molar concentrations of the basis compounds. (This step is omitted during construction of an elemental SBD). The conversion is achieved by use of a linear transformation. Assuming the surface composition of the anodized aluminum can be written as a sum of the molar concentrations of Al_2O_3, H_2O, and $AlPO_4$, the elemental concentrations can be derived from

$$\begin{aligned}
&[P]\ \alpha\ [AlPO_4] \\
&[Al]\ \alpha\ [AlPO_4] + 2[Al_2O_3] \qquad (1)\\
&[O]\ \alpha\ 4[AlPO_4] + 3[Al_2O_3] + [H_2O].
\end{aligned}$$

Inverting these relationships, we obtain

$$\begin{aligned}
&[AlPO_4] = 2k[P] \\
&[Al_2O_3] = k([Al] - [P]) \qquad (2)\\
&[H_2O] = k(2[O] - 3[Al] - 5[P])
\end{aligned}$$

where $k=[2([O]-[Al]-2[P])]^{-1}$ is a normalization factor so that the molar concentrations provided by (2) sum to unity. The linear transformation given by (2) can be represented in matrix form as

$$\begin{bmatrix} AlPO_4 \\ Al_2O_3 \\ H_2O \end{bmatrix} = k \begin{bmatrix} 2 & 0 & 0 \\ -1 & 1 & 0 \\ -5 & -3 & 2 \end{bmatrix} \begin{bmatrix} P \\ Al \\ O \end{bmatrix} \qquad (3)$$

If the basis compounds chosen were inadequate to describe the surface, at least one of their molar concentrations would be negative. For example, if the surface had a film of $Al(H_2PO_4)_3$, the composition coordinates in {[$AlPO_4$], [Al_2O_3],[H_2O]} space would be {0.600,−0.200,0.600}, indicating that this particular set of basis compounds is invalid. Consequently, another set, e.g., {[P_2O_5],[Al_2O_3],[H_2O]}, would have to be chosen. However, as we will see in the next section, the {[$AlPO_4$], [Al_2O_3],[H_2O]} space adequately describes the data.

Another example, in which the XPS data provide more information than necessary to construct an SBD, involves a depth profile of anodized (HgCd)Te where the XPS spectra indicated the presence of O, Cd, Hg, Te^{+4}, and Te^{-2}.[6,7] Possible constituents of the oxide included TeO_2, CdO, HgO, $CdTeO_3$, $HgTeO_3$, $CdTe_2O_5$, and $HgTe_2O_5$. To construct a ternary (two-dimensional) SBD, we chose as our basis compounds (HgCd)O, TeO_2, and (HgCd)Te. The compositions of the oxide complexes could then be written as $(HgCd)TeO_3 = (HgCd)O+TeO_2$ and $(HgCd)Te_2O_5 = (HgCd)O+2TeO_2$. The SBD can be constructed using

$$\begin{bmatrix} (HgCd)O \\ TeO_2 \\ (HgCd)Te \end{bmatrix} = k' \begin{bmatrix} 1 & 1 & -1 & 0 & 0 \\ 0 & 0 & 0 & 1 & 0 \\ 0 & 0 & 1 & 0 & 0 \end{bmatrix} \begin{bmatrix} Hg \\ Cd \\ Te^{-2} \\ Te^{+4} \\ O \end{bmatrix} \qquad (4)$$

where the normalization factor $k'=([Hg]+[Cd]+[Te^{+4}])^{-1}$. In this case, the concentration of oxygen was not needed to generate the SBD. However, it was used to confirm that the assignments indicated by the SBD analysis were consistent.[6]

In this particular SBD we were unable to display the two cation species separately; however, by expanding to three dimensions, we constructed a Hg–Cd–Te^{-2}–Te^{+4} elemental SBD where we chose the elemental chemical states themselves as the basis species.[6] This representation has the advantage of not requiring prior information about the surface in order to select compounds as the basis states. On the other hand, the elemental SBD may not display the chemistry of the surface as readily as the compound SBD.

The coordinates obtained from the linear transformations are then plotted onto an equilateral triangle or a tetrahedron with each vertex representing the composition of a basis compound. The evolution of the surface composition can be traced as a function of some parameter of interest, which is usually expressed as a label (either explicitly or implicitly) beside the points on the SBD or occasionally as a separate axis normal to the plane of the ternary SBD. Examples of each means of display are provided in the next section.

APPLICATIONS

The best way to demonstrate the usefulness of the SBD's is to present examples where they have been applied to actual problems. The examples chosen are representative of the different SBD types.

Use of an SBD to study a surface reaction: hydration of phosphoric-acid-anodized aluminum

Phosphoric acid anodization (PAA) is one method of surface preparation of aluminum adherends prior to adhesive bonding for aerospace applications. PAA surfaces are more resistant to hydration than Al surfaces prepared by other treatments, and this is a major reason for the superior durability of bonds formed with PAA adherends over bonds formed with adherends treated with other processes. To understand why PAA surfaces resist hydration, we examined, with

XPS and high-resolution scanning electron microscopy, fresh PAA surfaces and surfaces at various stages of hydration.[1–3] The surface compositions as a function of time in a humidity chamber are presented in Fig. 1. In the course of the experiment, the P level decreases to zero while the O and Al levels change only slightly with most of the difference occurring at the end of the exposure. Little insight into the hydration chemistry can be derived from this representation. On the other hand, the Al_2O_3–H_2O–$AlPO_4$ SBD (Fig. 2) provides a clearer description of the initial surface composition and the hydration process. The fresh surfaces (solid dots) are clustered around line I, which corresponds to the equivalent of approximately one molecular layer of phosphate on top of Al_2O_3 (based on photoelectron mean-free-path considerations). Variations off the line (corresponding to differences in the phosphate concentration) are believed to be caused by small variations in the rinsing conditions. Variations along the line (corresponding to differences in the water concentration) are believed to be caused by variations in the storage conditions of the sample prior to analysis.

Hydration is seen to be a three-step process. The first step (line I) can occur prior to insertion into the humidity chamber and involves the adsorption of water. The adsorption is reversible as evidenced by points b and b′ which represent the same sample shortly after being inserted into the ultrahigh vacuum system (b) and after several days in that dehydrating environment (b′). More extensive hydration occurs once the sample is exposed to the high humidity (line II). At this stage, dramatic changes in the surface morphology are observed as the surface is converted to boehmite, AlOOH. Later, further hydration ensues with the growth of bayerite [$Al(OH)_3$] crystallites on the surface (line III). Because it is during the second step that the adhesive bond fails,[13] we examined that stage more closely. A comparison with various possible mechanisms showed that only two processes were compatible with the evolution of the surface composition directly to that of boehmite (a) growth of

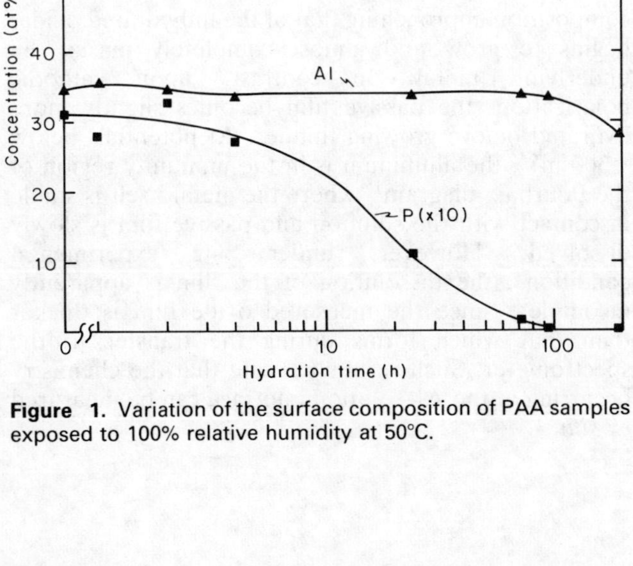

Figure 1. Variation of the surface composition of PAA samples exposed to 100% relative humidity at 50°C.

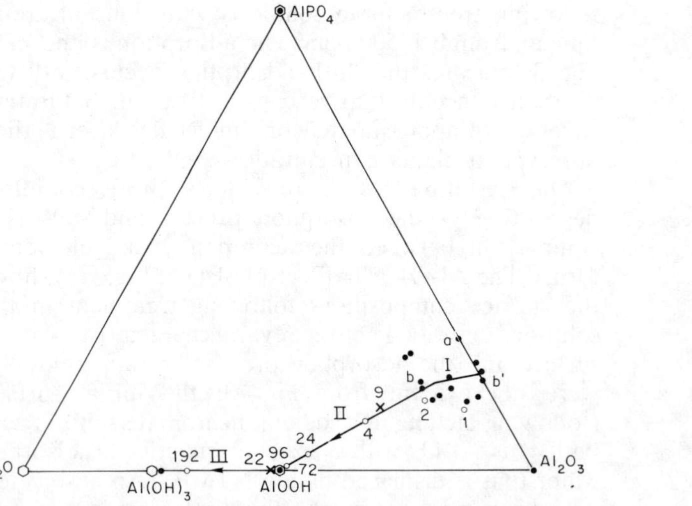

Figure 2. Al_2O_3–$AlPO_4$–H_2O SBD showing the three steps of hydration of PAA oxides in 100% relative humidity at 50°C (circles) and 60°C (crosses). Hexagons represent calculated compositions; solid circles represent compositions of standards and of PAA samples prior to hydration; open circles and crosses represent compositions of samples at various stages of hydration. The numbers beside the points correspond to the exposure time, in hours, to the humid environment. The circles represent the same data shown in Fig. 1.[1–3]

boehmite on top of the original surface and (b) slow dissolution of the surface phosphate followed by rapid hydration of the newly exposed amorphous aluminum oxide. Because the dissolution process is expected to be much slower than the hydration process, we could not differentiate between the two cases with the SBD alone. However, in the first case, the original phosphorus-containing surface would be buried below the hydration product so that AES sputter depth profiles should allow the distinction. Such profiles of the hydrated surfaces revealed that very little of the initial phosphorus was present, indicating that the second possibility was the hydration mechanism and that the limiting factor in the hydration resistance was the dissolution of the inhibiting phosphate.

Comparison of compound and elemental forms of SBD's: adsorption of nitrilotris methylene phosphonic acid on aluminum

Based, in part, on the results discussed above, we decided to investigate the use of several organic hydration inhibitors on aluminum adherend surfaces that are more susceptible to hydration.[4,5] One of these that is very successful is nitrilotris methylene phosphonic acid {NTMP, $N[CH_2P(O)(OH)_2]_3$}. It provides hydration resistance in a manner similar to anodization in phosphoric acid, i.e., it forms a protective molecular layer that does not allow the aluminum oxide substrate to hydrate until the inhibitor dissolves.

Treatment with NTMP consists of immersing the adherend into a dilute aqueous solution (generally 100–300 ppm) of inhibitor and rinsing. To gain an understanding of the processes involved during the adorption of NTMP, we examined aluminum surfaces following treatment in solutions with concentrations ranging from 0.1–500 ppm. The adsorption isotherm in Fig. 3 indicates that little adsorption occurs until the solution concentration reaches 1–10 ppm. Saturation coverage of approximately one molecular layer is then observed at higher concentrations.

The resulting SBD's provide a more complete description of the adsorption process and well as a comparison between their compound and elemental forms. The Al_2O_3–NTMP–H_2O SBD of Fig. 4(a) shows the surface compositions following treatment in the solutions described above. Several characteristics of the surface and the adsorption process are apparent that were not evident from Fig. 3: the initial surface (following etching in a sodium dichromate/sulfuric acid bath)[14] is Al_2O_3 with a sizeable quantity of adsorbed water that is displaced by the NTMP. (An adsorption process that did not involve the displacement of water, e.g., if the initial surface was AlOOH and not Al_2O_3 with adsorbed water, would follow line b toward the NTMP vertex). Upon the adherend's treatment in 100–500 ppm concentration solutions, the NTMP has displaced all the water initially present and formed approximately one molecular layer on the Al_2O_3 surface. The curvature of path a, more clearly shown in the elemental SBD of Fig. 4(b), indicates the adsorption to be a two-step process. At low solution concentrations, the NTMP competes less successfully with water for the available adsorption sites; as a result, only one inhibitor–oxide bond is formed, with the resulting release of a molecule of water. In this stage, the P concentration (and hence that of the inhibitor) increases rapidly to its saturation value, but the O concentration (a fraction of which represents adsorbed water) remains higher than expected from a single-stage adsorption process. At the higher solution concentrations, all three methylene phosphonic acid legs of the NTMP form bonds to the oxide, thus displacing more water. In this stage, there is relatively little increase in the number of adsorbed inhibitor molecules, but a distinct decrease in the oxygen coming from the absorbed water. This displacement of water is so efficient that the surfaces immersed in the highest concentration solutions exhibit no residual water.

For comparison, the adsorption characteristics of amino methlyene phosphonic acid [AMP, $H_2NCH_2P(O)(OH)_2$] are also shown (Fig. 4(b). This inhibitor has only one methylene phosphonic acid leg so that it should exhibit a single-step adsorption behavior. The SBD shows this to be the case; there is only a slight curvature of the adsorption isotherm for AMP. In addition, the SBD indicates that AMP does not displace all the water initially present. Because the remainder may act as nucleation sites for boehmite formation, these results suggest that AMP may not be as good a hydration inhibitor as NTMP. Mechanical testing of structures treated with the two inhibitors prior to bonding confirmed this prediction.

Use of a prismatic SBD: electrochemical polarization of aluminum

In an ongoing study of corrosion of pure aluminum, we immersed specimens into dilute solutions of Na_2SO_4 at

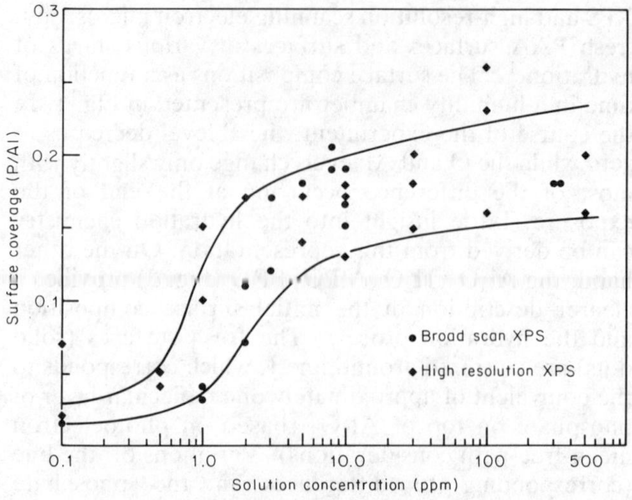

Figure 3. Adsorption of NTMP onto etched aluminum surfaces as a function of solution concentration. Samples were immersed for 30 min. The high resolution measurements are also plotted in Fig. 4.

pH=7 and polarized them to a given potential.[9] They were then removed from the solution, rinsed, and inserted into the analytical system. XPS measurements detected O, C, S, Al^{+3} and metallic Al, indicating the presence of a passive film on the surface thin enough to allow the underlying metal to also be observed. To simplify the presentation of the results, the small amount of S detected is not shown on the SBD. To avoid assigning too much O to the Al, we reduced the O concentration by four times the S concentration to compensate for the O bound to the S as a sulfate. The surface compositions, then, were plotted on a prismatic Al_2O_3–H_2O–Al:ΔV SBD (Fig. 5) where the applied potential (relative to the open-circuit potential) is the z axis. Here the SBD can be envisaged as a stack of ternary SBD's with the surface compositions represented by the crosses in the interior of the prism. Their projections onto the Al_2O_3–H_2O–Al and Al_2O_3–H_2O:ΔV planes are also given. The distance of the crosses from the $Al_2O_3-H_2O$:ΔV plane is shown by the line segments connecting the dots and the crosses; they correspond to the amount of metal detected in the XPS spectra.

The data show a passive film of AlOOH with a slowly varying thickness for anodic potentials up to 1.5–2 V. As the potential is increased, an anodic film, having a composition approaching that of the unhydrated oxide, begins to grow and almost completely masks the underlying metal. In contrast, upon cathodic polarization, the passive film becomes slightly more hydrated before growing thinner. At potentials below ~300 mV, the aluminum is in the immunity region of the Pourbaix diagram[15] where the metal itself is stable in contact with the solution and passive film is slowly dissolved. However, under our experimental conditions, the dissolution of the film is apparently incomplete since the measured oxide film is thicker than that which forms during the transfer to the spectrometer. Such results indicate that the chemistry occurring at the Al–solution interface can be measured *ex situ*.

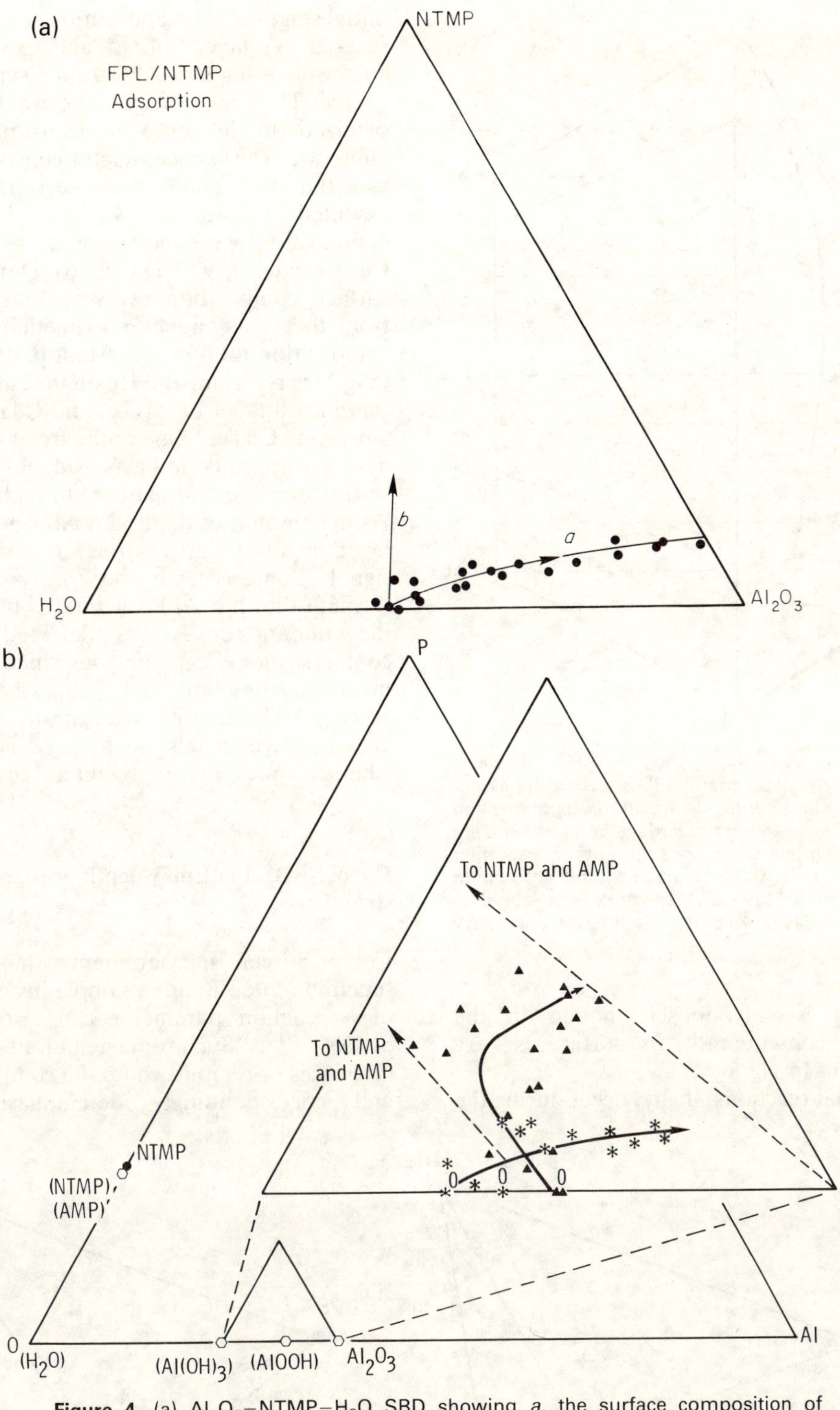

Figure 4. (a) Al_2O_3–NTMP–H_2O SBD showing *a*, the surface composition of aluminum surfaces after 30 min immersion in aqueous solutions of NTMP at concentrations ranging from 0.1 to 500 ppm, with increases going from left to right, and *b*, path representing no displacement of water; (b) Al–P–O elemental SBD showing the surface composition of aluminum surfaces after immersion of NTMP (triangles) and AMP (stars) at various concentrations. Open hexagons are calculated compositions. Compositions denoted by O represent surfaces not immersed in NTMP solutions.[5]

Use of a quaternary SBD: initial metallization of (HgCd)Te

As an example of a quaternary SBD, we examine the deposition of ultra-thin aluminum layers onto (HgCd)Te,[8] a material used for infrared detectors because its bandgap can be tuned for a given wavelength by varying the ratio of Cd and Hg. However, the technology used in processing (HgCd)Te and in fabricating devices on it is much less advanced than the corresponding technology for Si or even GaAs. In part, this is a result of its material properties.

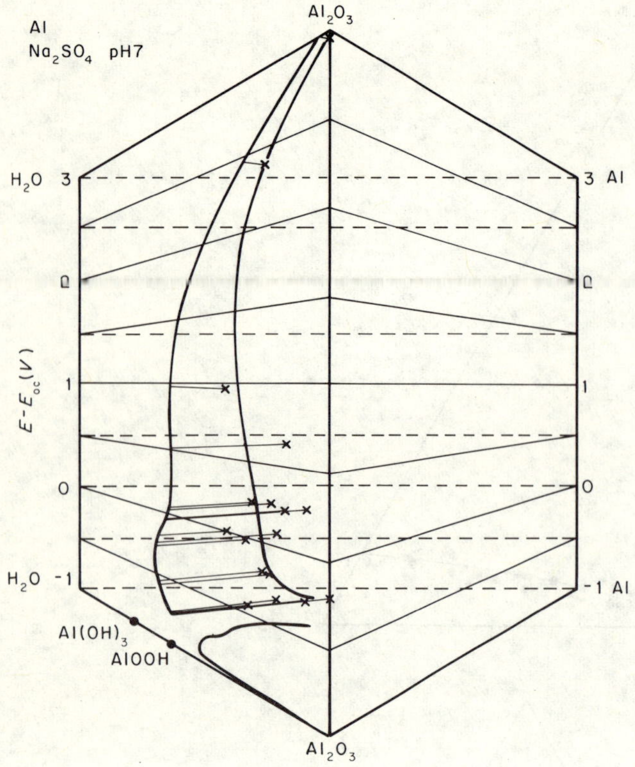

Figure 5. $Al_2O_3-H_2O-Al{:}\Delta V$ prismatic SBD showing the evolution of an aluminum surface during electrochemical polarization (*z* axis). The crosses, representing the surface composition, are in the interior of the prism; the short line segments indicate their distance from the $Al_2O_3-H_2O{:}\Delta V$ plane (i.e., the amount of underlying metal seen in the XPS spectra). The curve is projected onto the $Al_2O_3-H_2O-Al$ plane and the $Al_2O_3-H_2O{:}\Delta V$ plane.

For example, Hg is only loosely bound in the (HgCd)Te lattice; consequently, its surface is very reactive with respect to Hg loss.

To determine what reactions, if any, occur during the initial stages of Al metallization, we deposited very thin (0.2–20 Å) layers of Al and examined the surface chemistry using soft XPS from a synchrotron radiation source. The series of spectra showed both metallic and oxidized (in the sense of electro-positive) Al present along with the semiconductor constituents. Because Te was the only anion detected, the oxidized Al was assumed bound as Al_2Te_3. Thus, the surface compositions were plotted on an $Al^0-Al_2Te_3-HgTe-CdTe$ quaternary SBD (Fig. 6). During deposition, the surface composition evolves along a C-shaped path from the semiconductor to metallic Al. Initially, the composition follows a constant [CdTe] line, suggesting that Al is replacing the Hg on the surface and forming a thin mixed layer of Al_2Te_3 and CdTe without masking any of the CdTe. Some of the freed Hg diffuses into the near-surface bulk and acts as donors, as indicated by a shift of the spectral features to higher binding energy. As the amount of deposited Al increases, a metallic Al layer begins to grow, thus attenuating the substrate signal. Some Te, in the form of Al_2Te_3, diffuses through the thin Al layer but, as the overlayer grows, the amount of Al_2Te_3 detected diminishes. The contact, then, can be described schematically as metallic Al/metallic Al with Al_2Te_3/Al_2Te_3+CdTe/n^+(HgCd)Te/unaffected substrate. Such results suggest that reactive metals, such as Al and In, would form ohmic contacts to *n*-type, but not to *p*-type, (HgCd)Te.

Use of SBD's to display depth profiles of multilayer structures

Compositional measurements that are taken as a function of depth into a sample instead of as a function of a reaction parameter can also be plotted onto SBD's.[6,7,10,11] Such representations tend to accentuate interfaces or regions where the composition is changing and correspondingly de-emphasize homogeneous

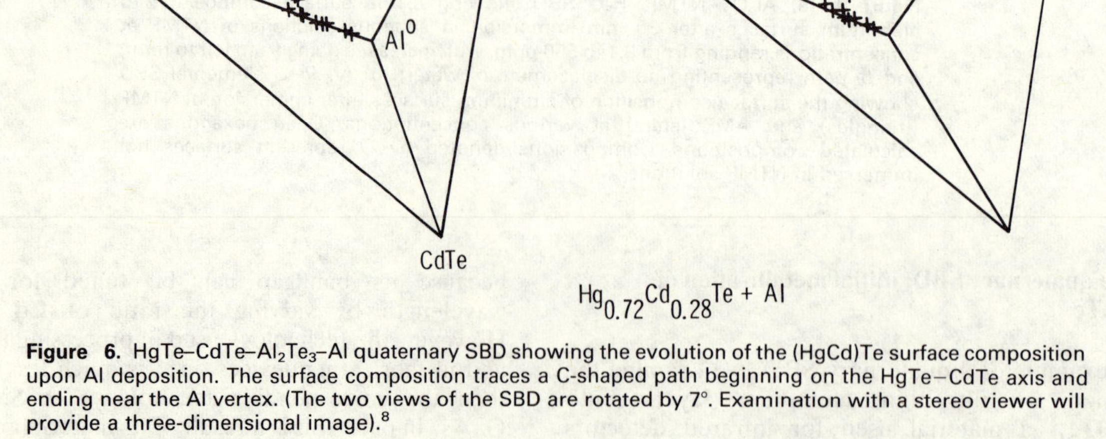

Figure 6. HgTe–CdTe–Al_2Te_3–Al quaternary SBD showing the evolution of the (HgCd)Te surface composition upon Al deposition. The surface composition traces a C-shaped path beginning on the HgTe–CdTe axis and ending near the Al vertex. (The two views of the SBD are rotated by 7°. Examination with a stereo viewer will provide a three-dimensional image).[8]

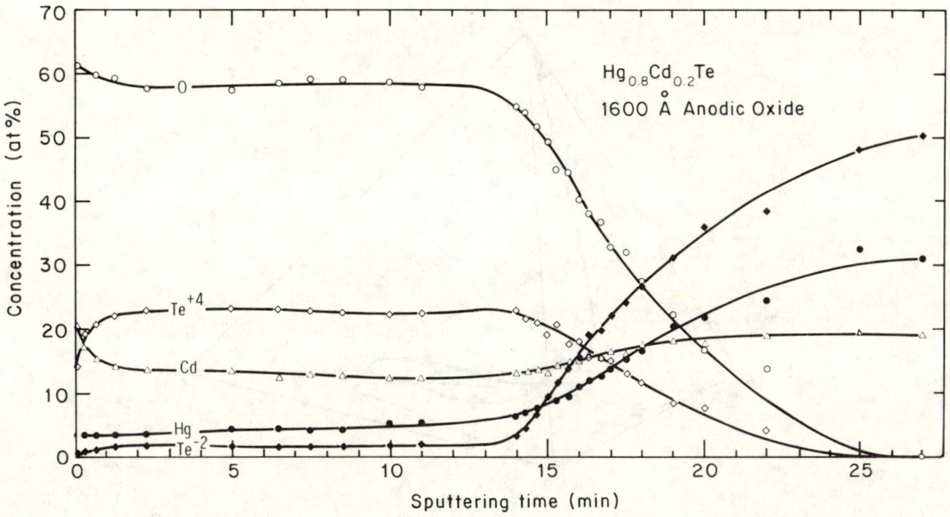

Figure 7. XPS sputter depth profile of an anodic oxide on (HgCd)Te.[6,7]

layers. Two examples have been chosen to illustrate this application — one where the interfacial chemistry was not apparent in the conventional display and one in which material properties could be correlated with the compositional information.

Analysis of interfacial reactions: anodic oxide of (HgCd)Te. Anodization of (HgCd)Te is one method of surface passivation during fabrication of infrared detectors. To investigate the chemistry of the anodic oxide and its interactions with the semiconductor substrate, we used XPS and ion bombardment to obtain a sputter depth profile.[6,7] The conventional presentation of the depth profile is given in Fig. 7. It shows the oxide composition to be uniform beyond the initial surface with a fairly abrupt interface, allowing for the depth resolution of the measurement. No insights about the interfacial chemistry are apparent with this representation. The (HgCd)O–TeO_2–(HgCd)Te SBD (Fig. 8) also indicates a uniform oxide composition, as indicated by the small scatter of nine points. However, the SBD suggests that the interfacial chemistry is more complex than expected. If the compositions of the oxide and semiconductor were uniform up to the interface (which need not be abrupt), the composition should evolve along the dashed line from the bulk oxide composition to the bulk semiconductor composition. Instead, the measurements consistently fall below this line, indicating an apparent cation deficiency at effective oxide thicknesses of 30 –10 Å. Further analysis of these and other data shows that Hg is the depleted species and that the deficiency can be seen for other samples having different anodic oxide thicknesses, but not for samples oxidized using other procedures.[6,7] Although other measurements have also indicated an unusual chemistry[16] or poor electrical properties[17] at the interface, no agreement has yet been established concerning the mechanisms involved. The various models that have been proposed include a field-assisted dissolution of the Hg during anodization,[6,7,18] a Hg–Cd exchange reaction between the oxide and semiconductor,[19] and a preferential sputtering of Hg.[20] Regardless of the cause of the observed properties, we could more completely describe the interface from the depth profile in the SBD format than from the same data in the conventional format.

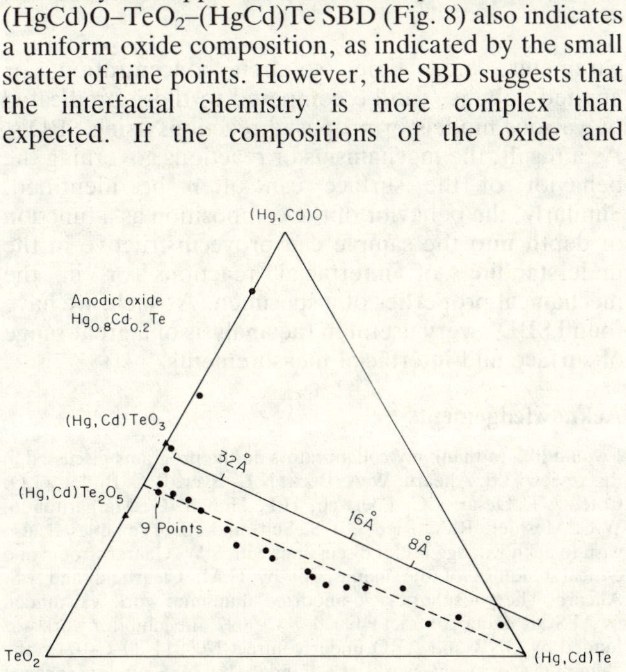

Figure 8. (HgCd)O–TeO_2–(HgCd)Te SBD showing the depth profile of Fig. 7. The superimposed depth scale gives the thickness of an atomically abrupt oxide.[6,7]

Prediction of material properties: multilayer engine bearings. For our final example, we discuss the work of Bierlein and coworkers[10,11] who examined diffusion occurring during the use of engine bearings and correlated the chemistry of the multilayer structure with its mechanical properties. Figure 9(a) shows a schematic representation of a depth profile, plotted on a Pb–Ni–Cu–Sn quaternary elemental SBD, of a bearing coating consisting of babbitt metal (Pb, 10 wt% Sn, 2 wt % Cu) on top of a nickel diffusion barrier on top of a copper-lead alloy (24 wt% Cu). Because Pb is immiscible in Cu and Ni, its concentration was factored out for ease of analysis and presentation and the depth profiles replotted on Ni–Cu–Sn ternary SBD's (e.g., Figs 9(b) and 9(c), before and after a heat treatment to simulate use). The results from two methods of depth profiling — ion sputtering and ball cratering where the electron beam is scanned across the wall of the crater — are presented. The poorer depth resolution of the sputtering procedure when profiling relatively thick

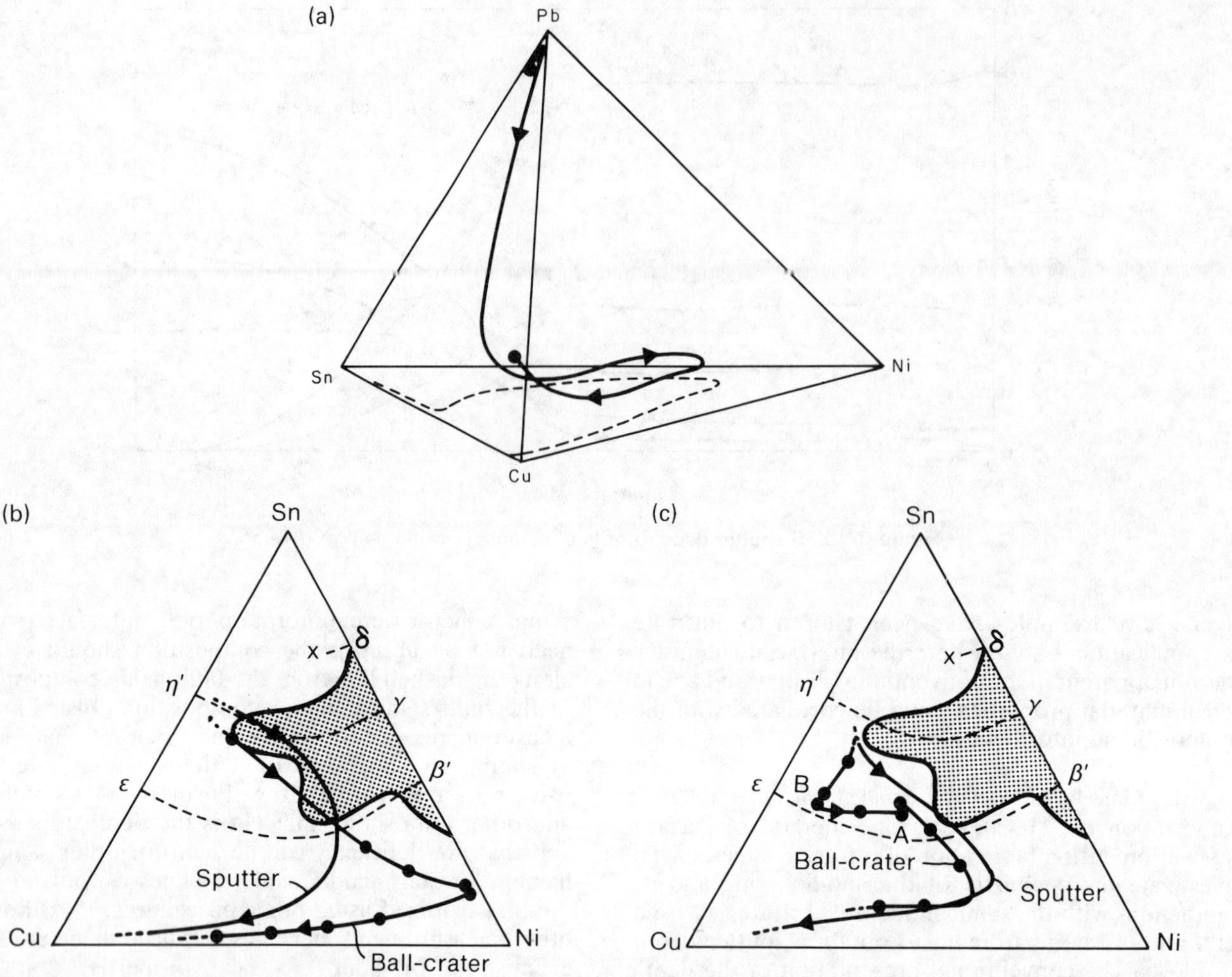

Figure 9. (a) Pb–Ni–Cu–Sn quaternary SBD showing a schematic depth profile through an engine bearing. Ni–Cu–Sn ternary SBD's showing depth profiles through (b) a new engine bearing and (c) a heat-treated bearing. The Pb concentration has been suppressed, i.e., the ternary SBD is the projection of quaternary SBD's similar to that in (a). The dashed lines correspond to proposed intermetallic compounds. The cross-hatched region corresponds to compositions exhibiting poor score resistance. Two different procedures were used to obtain the depth profiles: ion sputtering and ball cratering. In these SBD's the compositions are shown in weight percents and not atomic percents, as are all others presented (from refs. 10 and 11).

films (due to ion-induced roughness) is demonstrated by its inability to resolve the thin Ni layer in Fig. 9(b). The ball-cratering profile of the heated sample suggests the formation of two intermetallic compounds as represented by the slow change of composition at point A near the center of the SBD and the sharp deviation of the profile toward the ϵ binary compound at point B. Also shown in the SBD's is the range of compositions that exhibit poor score-resistance (the cross-hatched region). The depth profiles indicate that such compositions occur within the new bearing and are close to those measured in the heat-treated bearing. As the bearing wears, layers having these compositions are exposed, thus allowing scoring distress. To prevent this occurrence, they proposed and tested a new trimental (Ni/Sn/Cu) barrier layer. Corresponding depth profiles indicated no region of poor score-resistant compositions.

SUMMARY

As these examples have shown, the evolution of the surface composition as a function of a reaction parameter, e.g., time, solution concentration, or applied voltage, can be compared to those predicated by various models or proposed reactions using SBD's. As a result, the mechanisms or reactions governing the behavior of the surface can often be identified. Similarly, the behavior of the composition as a function of depth into the sample can prove instructive in the understanding of interfacial reactions or in the mechanical properties of a specimen. As such, we have found SBD's very useful in the analysis of a great range of surface and interfacial measurements.

Acknowledgements

I would like to thank my collaborators on the programs discussed in this review: J.S. Ahearn, W.A. Beck, N.E. Byer, R.C. Butler, G.O. Cote, A.I. Desai, J.C. Foertsch, H.F. Hough, G. Margaritondo, W.C. Moshier, R.A. Riedel, T.S. Sun, and J.D. Venables. I also wish to acknowledge useful discussions with S.W. Gaarenstroom and critical readings of the manuscript by H.M. Clearfield and J.S. Ahearn. The phosphoric-acid-anodized aluminum work was funded by AFSOR under contract F49620–78C–0097; the inhibitor work was funded by ONR and ARO under contract N00014–80–C–0718; the aluminum corrosion work was funded by NSF under contract DMR–8401401; and the (HgCd)Te experiments were performed at the University of Wisconsin Synchrotron Radiation Center which is supported by NSF grant DMR83–13523.

REFERENCES

1. G. D. Davis, T. S. Sun, J. S. Ahearn, and J. D. Venables, *J. Mater. Sci.* **17**, 1807 (1982).
2. J. S. Ahearn, G. D. Davis, T. S. Sun, and J. D. Venables, in *Adhension Aspects of Polymer Coatings* p. 281 ed. by K. L. Mittal, Plenum Press, New York (1983).
3. G. D. Davis, J. S. Ahearn, and J. D. Venables, in *Proc. 15th Nat. SAMPE Tech. Conf.*, Cincinnati, OH (1983) P. 202, SAMPE, Azusa, CA (1983).
4. G. D. Davis, J. S. Ahearn, and J. D. Venables, *J. Vac. Sci. Technol. A* **2**, 546 (1984).
5. G. D. Davis, J. S. Ahearn, L. J. Matienzo, and J. D. Venables, *J. Mater. Sci.* **20**, 975 (1985).
6. G. D. Davis, S. P. Buchner, W. A. Beck, and N. E. Byer, *Appl. Surf. Sci.* **15**, 238 (1983).
7. G. D. Davis, S. P. Buchner, J. S. Ahearn, and N. E. Byer, in *Technical Issues in Focal Plane Development, Proc. Soc. Photo-Opt. Instrum. Eng.*, Vol. 409, p. 35, Arlington, VA (1983)
8. G. D. Davis, N. E. Byer, R. A. Riedel, and G. Margaritondo, *J. Appl. Phys.* **57**, 1915 (1985).
9. G. D. Davis, W. C. Moshier, J. S. Ahearn, and J. C. Foertsch (to be published).
10. J. C. Bierlein, S. W. Gaarenstroom, R. A. Waldo, and A. C. Ottolini, *J. Vac. Sci. Technol. A* **2** 1102 (1984).
11. J. C. Bierlein and A. O. DeHart, *SAE Technical Paper 830064*, 1983.
12. G. S. Was and J. R. Martin, *Met. Trans.* **16A**, 349 (1985).
13. J. D. Venables, D. K. McNamara, J. M. Chen, B. M. Ditchek, T. I. Morgenthaler, T. S. Sun, and R. L. Hopping, *Proc. 12th Nat. SAMPE Tech. Conf., Seattle, WA, October 1980*, p. 909, SAMPE, Azusa, CA (1980).
14. H. W. Eichner and W. E. Schowalter, *Report No. 1813* Forrest Products Laboratory, Madison, WI, (1950).
15. M. J. N. Pourbaix, *Atlas of Electrochemical Equilibria in Aqueous Solutions*, Pergamon Press, Oxford (1966).
16. U. Kaiser, O. Ganschow, J. Neelson, H. M. Nitz, L. Wiedmann, and A. Benninghoven, in *Proc. 3rd Intl. Conf. SIMS*, ed. by A. Benninghoven, J. Gibbs, J. Laszlo, M. Riedel, and H. W. Weimer, *Springer Series in Chemical Physics* 19, 365 (1982).
17. B. K. Janousek and R. C. Carscallen, *J. Appl. Phys.* **53**, 1720 (1982).
18. G. D. Davis, T. S. Sun, S. P. Buchner, and N. E. Byer, *J. Vac. Sci. Technol.* **19**, 472 (1981).
19. D. R. Rhiger and R. E Kvaas, *J. Vac. Sci. Technol. A* **1**, 1712 (1983).
20. M. Seelmann-Eggebert, G. Brandt, and H. J. Richter, *J. Vac. Sci. Technol. A* **2**, 11 (1984).

SURFACE AND INTERFACE ANALYSIS, VOL. 9, 431–435 (1986)

XPS Analysis of Barium Labelled Carbon Fibres and Interfacial Effects in Fibre–Epoxy Composites

P. Denison,* F. R. Jones* and J. F. Watts
Department of Materials Science and Engineering, University of Surrey, Guildford, Surrey GU2 5XH, UK

A barium labelling and XPS technique has been applied to a range of Type I carbon fibre samples, with differing degrees of surface oxidative treatment. This has yielded a quantitative analysis of the concentration and distribution of fibre surface oxygen, acid groups and chemisorptive sites. The latter two species could play a major role in the formation of the fibre surface–resin interface in a compositve material. Additionally, it is shown that a relatively high degree of treatment does not result in a correspondingly high surface concentration of either oxygen containing species or acidic groups, as both these quantities reach maxima at medium levels of treatment.

INTRODUCTION

The nature of the fibre surface–resin interface in carbon fibre–epoxy composites is still a subject of extreme controversy, since it has been asserted by a number of authors, eg,[1] that the interface may have both chemical and physical features which play an important role in bonding. A chemical bond could arise from the interaction between mutually reactive functional groups on the fibre surface, and in the resin matrix. Ehrburger *et al.*[2] observed an apparent correlation between the concentration of fibre surface acid groups and the interlaminar shear strength (ILSS) of composites. However, Harvey[3] found that the removal of surface functional groups, by heating fibres to 1000°C in a vacuum, resulted in no significant decrease in fibre composite ILSS, thus suggesting that physical effects, such as fibre surface topography, could play an important role in interface formation.

As part of an overall programme initially aimed at investigating the effect of fibre surface acidity on the interfacial bond strength, the authors have developed a barium labelling and XPS technique for fibre surfaces, using commercially treated Type II fibre samples.[4] Carbon fibres are generally available in two types, I and II; the former is of high modulus (≈400 GPa) and low strength (≈2 GPa), whereas the latter is of low modulus (≈200 GPa), but high strength (2.5 GPa).

Subsequently, as presented in this paper, a slightly modified form of the technique has been applied to a set of Type I fibre samples, with a range of different oxidative treatment levels, since it has been known for some time that composite interfacial bond strength is a function of the level of fibre surface oxidation.[5]

MATERIALS

Fibre samples in the form of ≈1 m lengths of 'tow' (6000 × 7 μm diameter fibres), were supplied by Dr J. Harvey, of the Royal Aircraft Establishment, Farnborough, which had been subjected to a range of degrees of surface oxidative treatment. The treatment process in which active oxygen is evolved at the fibre surface, is patented[6] and the degrees of fibre treatment, DFT, are quantified from 0 to 100%, as given in Table 1.

EXPERIMENTAL

At each level of treatment, 'control' and 'Ba-labelled' specimens were preparing for XPS analysis. The former were analysed as received, and the latter given a labelling treatment.

Labelling Process

A ≈10 cm length of tow was washed in water at room temperature for 1 min, then immersed in a saturated solution of $BaCl_2$, containing 20 g l^{-1} $Ba(OH)_2$, whilst

Table 1. Degrees of surface oxidative treatment applied to fibre samples

Parameter $N(0^*)$	$\log_{10}[1 + N(0^*)]$	Degrees of fibre treatment (DFT/%)
0	0	0
2.2	0.51	23
7.7	0.94	43
11.1	1.08	49
40.0	1.61	73
158.0	2.2	100

N.B. $N(0^*)$ is a treatment parameter which is directly proportional to the number of μmoles of active oxygen evolved at the fibre surface per cm of tow length.

$$DFT[N(0^*)] = \frac{\log_{10}[1 + N(0^*)]}{\log_{10}[1 + 158]} \times 100$$

* Current Address: Department of Ceramics, Glasses and Polymers, University of Sheffield, Elmfield, Northumberland Road, Sheffield S10 2TZ, UK.

0142–2421/86/120431–05 $05.00

simmering for 1 h. The tow was then rinsed in five separate portions of water for 1 min to remove excess Ba salts, before being dried in a stream of cold air and a dessicator, prior to mounting for XPS.

In the development of this technique on commercial fibres,[4] it was observed that with three 1 min post labelling rinsings in H_2O, to remove excess Ba salts there was a small quantity of $BaCl_2$ remaining on the surface of one highly oxidised fibre specimen. Whilst this may have been due to the presence of a relatively high degree of surface microporosity in this case, to ensure that inadequate post labelling rinsing could not alone cause this effect, with subsequent work on Type I fibres it was decided to increase the number of such rinsings to five, instead of the original three. Since Bradley and Czuha[7] in their work on labelling COOH with Ba on polymers, observed that the Ba signal from the sample surface, remained, even after a 1 h long post labelling rinse in water, our 5 × 1 min rinsing process should not result in any loss of Ba bound to surface acid groups on fibres.

XPS analysis of Fibre Samples

Samples were prepared for XPS by mounting lengths of fibre tow across gaps in metal holders. Spectra were acquired on a VG Scientific ESCA 3 Mk II, at an operating vacuum of $\lesssim 10^{-9}$ torr, using Al(Kα) radiation ($h\nu$ = 1,486.6 eV) and an analyser pass energy of 50 eV, with 4 mm entrance and exit slits.

The spectrometer was interfaced to a VG 3040 data system, based on a PDP8 computer. Each fibre sample was analysed by a combination of 1000 eV survey and 20 eV high resolution scans, for all the relevant elements that could be present. Elemental surface atomic percentages were calculated from peak areas, after linear background subtraction, using published sensitivity factors.[8]

RESULTS AND DISCUSSION

As a test for reproducibility of both the oxidative treatment process and the analytical technique, two or three analyses were initially performed at each level of treatment, on both 'control' and 'labelled' specimens.

However, as the work progressed, further analyses were possible, which were conducted at the 49% DFT level, since it became apparent that at this treatment level fibre surface acidity undergoes a significant change, as a function of DFT.

The calculated surface atomic percentages are presented in Table 2. In interpreting the data, a model graphitic unit segment of material at the fibre surface is considered, which is 10 nm × 10 nm in area and 3.5 nm in depth. As we have described previously[4] an inelastic mean free path of 3.5 nm is adopted for C 1s photoelectrons, which have a kinetic energy 1200 eV. It is appreciated that the intensity of photoemission from layers below the segment surface plane decays

Table 2. Calculated surface atomic percentages and number of oxygen and barium atoms present, per unit surface segment

DFT %	Control specimens					Ba-Labelled specimens					
	No	C %	O %	N %	No. of O atoms	No	C %	O %	N %	Ba %	No. of Ba atoms
0	3	96	4.1	<.3[a]	1700	2	93	6.1	0.9	0.06	26
		94	6.4	<.3[a]	2700		91	7.7	1.5	0.12	53
		96	3.8	<.3[a]	1600						
23	2	90	7.7	2.5	3400	2	87	10.0	2.4	0.15	69
		91	7.5	1.8	3300		86	11.5	2.2	0.18	84
43	2	90	7.6	1.9	3400	3	90	8.6	0.9	0.26	120
		90	8.1	1.9	3600		86	11.5	1.9	0.33	150
							85	13.4	1.0	0.29	140
49	4	87	9.8	2.9	4500	4	86	10.6	2.8	0.42	200
		87	9.5	3.7	4400		84	12.6	3.3	0.49	220
		86	10.1	3.6	4700		80	15.9	3.1	0.78	390
		87	9.2	3.8	4300		83	14.1	2.2	0.65	310
73	2	87	10.5	3.0	4900	2	81	15.6	3.3	0.39	192
		83	13.1	4.0	6300		80	16.5	3.0	0.42	210
100	2	88	8.4	3.9	3800	3	81	14.7	3.6	0.24	120
		87	8.9	4.1	4000		83	13.8	3.2	0.44	210
							83	14.0	2.7	0.60	290

N.B. No = Number of specimens analysed at a given DFT; C% is rounded to the nearest integer.

[a] 0.3% N is the minimum level that can be detected.

Cl was not detected on any labelled specimen.

according to the Beer–Lambert law. However assuming a graphitic structure to a depth of at least 10 nm, integration of the decay function gives a result which is equivalent to all the atoms detected by the XPS technique being within the analysis area, and a depth of 3.5 nm. Hence this effective analysis depth is used.

As the unit segment contains a total of 40 000 C atoms, from the density of graphite, the 'surface' atomic percentages in Table 2 which effectively apply to the top 3.5 nm of material, can also reveal the true number of oxygen atoms on the segment surface, eg at zero DFT, 3.8% 0 per 96% C implies the presence of ≈(3.8/96 × 40 000 = 1,600 oxygen atoms. The dependence of the number of oxygen atoms present on DFT is given in Table 2 and Fig. 1.

Similarly, the calculated number of Ba atoms present on the labelled specimens together with the number, n, and distribution of labelled acid groups, on the model segment, is given in Tables 2 and 3.

With n COOH groups, distributed randomly over an area of 100 nm^2, each of those groups can be considered to occupy a sub-sector of the surface, bounded by nearest neighbour COOH groups, which is of area ~ $(100/n)$ nm^2 on average. We therefore define an average separation to be $\sqrt{100/n} = 10/\sqrt{n}$ nm, between adjacent COOH groups. At zero DFT, for example, this is >1 nm, which implies that a significant number of the acid groups present exist in single isolation, in this case, and could not be labelled by Ba^{2+}, as the maximum separation between adjacent groups for dibasic labelling to occur is ≈ 0.4 nm, from relevant bond lengths and angles.[4]

Therefore a statistical approach is required to reveal the true number of acid groups present on the segment, x, given the number which exist as adjacent pairs, and labelled by Ba, n. Dividing the (10 nm × 10 nm) segment surface into 625 (.4 nm × .4 nm) sectors, with a random distribution of x groups over the segment, the number of adjacent sectors occupied by groups, n, is simply $x(x/625)^2$, i.e. $x = \sqrt[3]{(625^2 n)}$. As a test for self consistency, with every sector containing one group, x and n both equal 625. The calculated values of x, over the whole DFT range, are given in Table 3 and in graphical form, in Fig. 2.

Table 3. Derived number of acid groups present per unit surface segment

DFT %	No. of labelled groups (n) (labelled specimens)	Separation scale of labelled groups $= \frac{10\text{ nm}}{\sqrt{(n)}}$	True no. of groups = $\sqrt[3]{(n)} \times 625^2$	Fibre acidity $\frac{\mu\text{eq.}}{\text{gram}}$
0	52	1.4	270	3.1
	110	0.95	350	±0.4
	140	0.85	380	3.9
23	170	0.77	400	±0.1
	240	0.65	450	4.7
43	300	0.58	490	±0.2
	280	0.60	480	
	400	0.50	540	6.3
	440	0.48	560	±0.8
49	780	0.36	780	
	620	0.40	620	
73	380	0.51	530	5.4
	420	0.49	550	±0.1
	240	0.65	450	5.4
100	420	0.49	550	±0.9
	580	0.42	610	

[a] Except where separation scale <0.40 nm, when this equals n.

Referring to Figs 1 and 2 and Tables 2 and 3, it is apparent that at a given DFT, there is for example, some scatter in the surface oxygen concentrations on the control specimens and the corresponding barium, and hence derived acid group concentrations, on labelled specimens.

Whilst this in part probably reflects the reproducibility of our analytical technique, on different specimens with nominally the same levels of surface treatment, DFT, there is a further possible, contributing

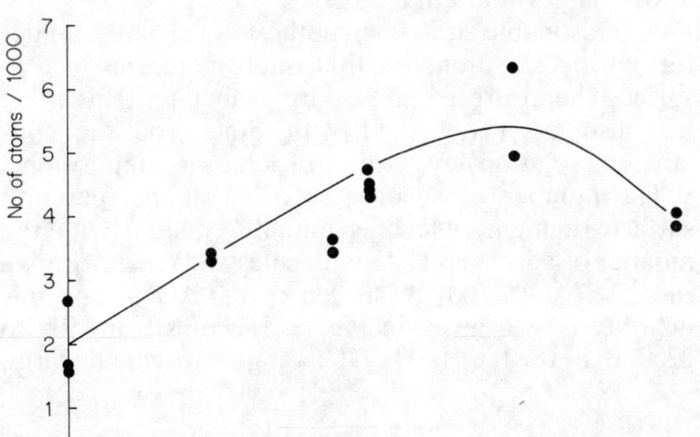

Figure 1. Number of oxygen atoms present, per unit surface segment on control specimens versus degree of fibre treatment (DFT).

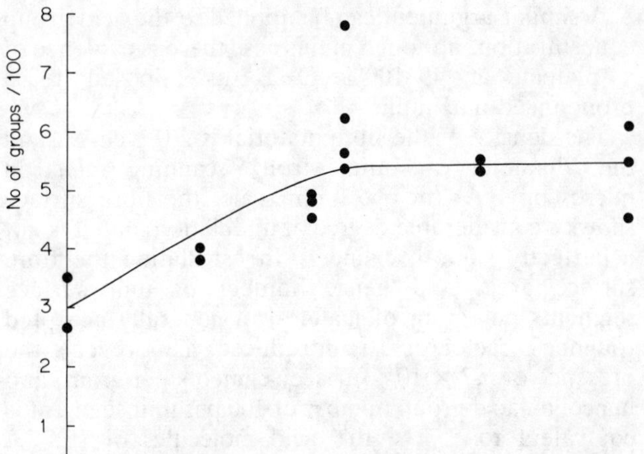

Figure 2. Number of acid groups present, per unit surface segment on labelled specimens versus degree of fibre treatment (DFT).

factor. As described above, ~1 m lengths of fibre tow, at each value of DFT, were available for analysis. From these, ~10 cm sections were cut, from which specimens were prepared and mounted for XPS.

In the spectrometer chamber, only a few mm of specimen length is in the analysis area, and hence an actual statistical variation in surface chemical composition, at different, points along a ~1 m tow length, could also partly account for the observed scatter in the XPS data, at a given value of DFT.

Specifically, for DFT $\geqslant$ 49%, there are nine values of Ba%, all but one of which are $\geqslant$ 0.39. This suggests that the value of 0.24% Ba at 100% DFT, is probably statistically unrepresentative at this particular level of treatment.

Returning to Figs 1 and 2, at lower levels of treatment, the surface concentration of oxygen and acid groups increases with DFT, an obviously expected trend. However, the concentrations reach maxima at DFT = 73 and 49% respectively, and as DFT is raised to 100%, the oxygen level decreases somewhat, although still remaining at a higher level than on the unoxidised specimens, at zero DFT.

This initially surprising result can be understood by considering the mechanism of the oxidative treatment process, in which active oxygen is evolved at the fibre surface, and a number of different reactions involving surface carbon atoms are likely: (a) $C_s + O + H_2O \rightarrow$ C/O/H groups; (b) $C_s + 2O \rightarrow CO_2 \uparrow$ evolved gas.

The evolution of CO_2 during oxidative treatment of fibres has been reported in the literature.[9] Whilst it is not intended to analyse the reaction scheme in detail, it can be appreciated that after treatment, the total amount of oxygen remaining as functional groups on the fibre, would be determined by the nett competition between reaction paths (a) and (b), each of which would probably depend upon DFT.

Therefore, the increase in O% with DFT at values < 73% suggests that reaction (a) increases in relation to (b), over this treatment range, however, for 73% < DFT < 100%, the CO_2 evolution reaction becomes relatively more dominant, thus accounting for the decrease in O% at the highest treatment level.

A similar argument can be applied to the acid group concentration, although in this case there is evidence of a plateau at 49–100% DFT, as opposed to a pronounced maximum.

The density of the fibre material is 2.0 g cm^{-3}, and the diameter = 7 μm, from scanning electron microscopy. At the ≈0.1 μm scale, the fibre surface shows a considerable degree of undulation, and it is not a perfectly smooth cylinder. In calculating the fibre surface area, and hence number of unit surface segments per gram of material, a generally accepted roughness factor of 2 is introduced. This reveals the presence of 5.7×10^{15} surface segments per gram, and hence, an acid group number of 100 per unit segment is equivalent to 5.7×10^{17} acid molecules or ($5.7 \times 10^{17}/6.02 \times 10^{23}$) = 0.95 μeq. of acid groups, per gram of fibre. Effectively, 100 acid groups per segment is equivalent to 1 μeq. gram^{-1}, within the limits of experimental error, and this latter unit is generally used in expressing the degree of fibre surface acidity, obtained by titration, for example, as presented in the literature.[2]

Referring to Table 3, at a given DFT, the average and standard deviation in the number of acid groups, yields the degree of acidity, and its uncertainty; as presented in column 5, and it is apparent that the uncertainty is always $\leqslant 1$ μ eq. gram^{-1}.

In titration work on oxidised and unoxidised Type I fibres, Ehrburger *et al.*[2] quotes uncertainties of ≈±2 μeq. gram^{-1}, which even bearing in mind the assumptions made in calculating acidity levels from raw XPS data, suggests that the XPS and Ba labelling technique provides a relatively accurate method of determining surface acid group distributions and concentrations.

On untreated Type I samples, our acidity levels of ~3.1 ± 0.4 μeq. g^{-1} (Table 3), compare well with the values of 3 and 2 μeq. g^{-1}, obtained by Ehrburger *et al.*,[2] using NaOH and $NaOC_2H_5$ as respective alkalis. However, it is not possible to make such a comparison between the two sets of results on oxidised samples, in view of the different oxidation conditions used for our fibres, and those analysed by titration.[2]

A survey of existing literature on carbon and fibre surfaces[10,11] reveals that a number of different forms of carbon–oxygen functional group, including —COOH, —OH C=O, and lactonic ring structures, probably exist on the surfaces of oxidised fibres, and the results presented in Table 3 are consistent with this type of surface chemistry. For example, at DFT = 23%, the ~390 COOH groups account for only 780 out of the total of ~3400 oxygen atoms on the surface segment. Hence, most of the oxygen is present as a non-carboxylic acid form, which is not labelled by barium. (It has been pointed out[4] that in view of the relatively much weaker acidity of OH, in comparison to COOH, pairwise labelling of the former, by Ba^{2+}, is unlikely to occur to any significant degree, during the labelling process.) The possible role of groups ofther than COOH in composite interface formation must therefore be considered.

A further interesting observation from this work is the apparent uptake of oxygen, by the fibre surface during labelling, defined as ΔO% = average O% (labelled specimens) – average O% (control specimens), at a given DFT, and this uptake is shown in Fig. 3, over the whole DFT range.

A reasonable initial hypothesis, currently under further investigation,[12] is that water molecules bind to surface chemisorptive sites, during labelling. This effect has also been observed with other trial aqueous labelling solutions, regardless of solution composition.[4]

The number of adsorbed oxygen atoms, per unit surface segment, which is initially equated with the number of adsorbed H_2O molecules and chemisorptive sites, $\simeq (\Delta O\%/100) \times 40\,000 = 400\,\Delta O\%$, and this quantity is also given in Fig. 3. It varies from 800 to 2200, over the 0 to 100% DFT range, virtually linearly.

CONCLUSION

The XPS and Ba labelling technique has yielded a full quantification of the fibre surface concentration and distribution of acid groups, and chemisorptive sites, as a function of oxidative treatment level, either or both

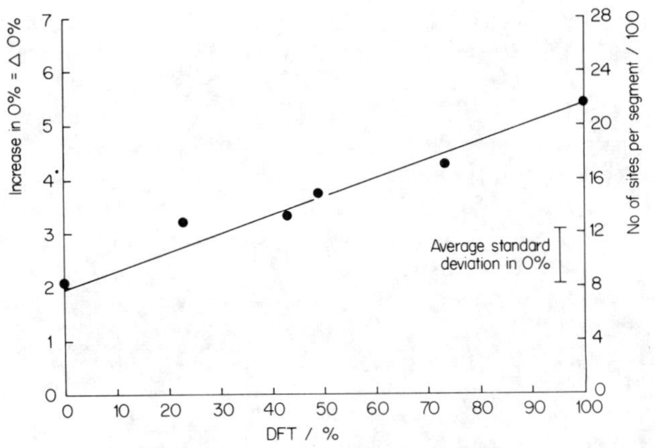

Figure 3. Increase in the surface atomic percentage of oxygen, during labelling, and the derived number of chemisorptive sites, per unit surface segment, versus degree of fibre treatment (DFT).

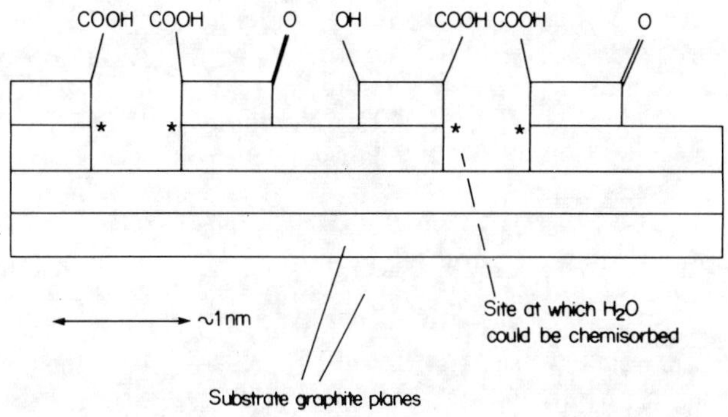

Figure 4. Schematic diagram of the fibre surface structure.

of which could play a major role in composite interface formation. This is a significant step forward in understanding the true and probably complex nature of the interface at the molecular level, and taking into account our results, it is possible to propose a structure of the fibre surface as shown in Fig. 4.

Furthermore, some light has been thrown upon the relationship between the conditions of oxidative treatment, and the resulting chemical composition of the fibre surface. It has been shown that there is nothing to be gained, with regard to surface oxygen and acid groups, by resorting to relatively high degrees of treatment, DFT $\geqslant$ 73%, whereas the chemisorptive power of the surface continues to increase at these higher levels. It is intended, in future work, to measure interfacial bond strengths in composite specimens directly, by loading them in a straining stage in a scanning electron microscope. Ultimately, we hope to relate interfacial bond strength, and composite mechanical properties in general to the conditions of fibre surface oxidative treatment, thus permitting optimisation of the latter, which would be of obvious interest to manufacturers of fibres and composites.

Acknowledgements

This work was carried out with financial support from the Procurement Executive, Ministry of Defence. We also thank the following for advice and discussion during the project: Professor J. E. Castle, University of Surrey, Professor J. E. Bailey, University of Sheffield and Dr. G. Dorey and Dr. J. Harvey of the Royal Aircraft Establishment, Farnborough.

REFERENCES

1. B. Rand and R. Robinson, *Carbon*, **15**, 257 (1977).
2. P. Ehrburger, J. J. Herque and J. B. Donnett, *Proc. 5th London Int. Carbon and Graphite Conf.* Vol 1, p. 398, Society of Chemical Industry, London (1978).
3. J. Harvey, private communication, (1984).
4. P. Denison, F. R. Jones and J. F. Watts, *J. Mat. Sci.*, **20** 4647 (1985).
5. D. Clarke, N. J. Wadsworth and W. Watt, *Proc. 2nd Int. Carbon Fibres Conf., London, The Plastics Inst., London, Paper 7*, p. 44, (1974).
6. J. Harvey, private communication (1985).
7. A. Bradley and M. Czuha, *Anal. Chem.* **47**, 1838 (1975).
8. C. K. Jørgensen and H. Bērthou, *Discuss. Faraday Soc.* **54**, 269 (1972).
9. P. Ehrburger, J. J. Herque and J. B. Donnett, *Proc. 4th London Int. Carbon and Graphite Conf.* p. 201, Society of Chemical Industry, London (1974).
10. D. M. Brewis, J. Comyn, J. R. Fowler, D. Briggs and V. A. Gibson, *Fibre Sci. and Tech.* **12**, 41 (1979).
11. J. B. Donnet, *Carbon*, **6**, 161 (1968).
12. P. Denison, F. R. Jones and J. F. Watts, to be published.

SURFACE AND INTERFACE ANALYSIS, VOL. 9, 437–440 (1986)

The Adhesion of Evaporated Metal Films to Polyethylene Substrates

P. Bodö and J.-E. Sundgren
Department of Physics and Measurement Technology, Linköping University, S–581 83 Linköping, Sweden

Various metal films (Ti, Al, Cr, Ni, Cu, Ag and Au) were deposited onto high density polyethylene (HDPE) substrates by electron beam evaporation. The adhesion of these films was determined from pull strength measurements. The results show that the adhesion to as-prepared HDPE samples was relatively high for Ti, Ni and Cr films while it was low for Al, Cu, Ag and Au films. As the HDPE samples were *in situ* Ar ion bombarded prior to film deposition the adhesion for all the metals, except for Ag and Au, increased to values exceeding the cohesive strength of the HDPE bulk. Photoelectron spectroscopy, XPS, studies show that this high adhesion is due to a metal–polymer interaction as well as to an increased strength of the substrate surface itself. The metal–carbon bonding observed, require a dehydration of the HDPE surface. On the basis of the adhesion results obtained on asprepared samples this C—H bond breaking is suggested to be activated by the heat released during condensation of the films. Ion bombardment also results in C—H bond breaking and thus an activated and also a cross-linked HDPE sample surface. For prolonged ion bombardment a graphite-like surface was obtained as determined from XPS valence band studies.

INTRODUCTION

Metallization of polymer substrates is a topic of high technological importance and is today, for example, extensively being used in multilayer structures and packaging in the microelectronic area. New fields such as various medical applications are also currently being exploited. The main focus of the research in all these areas is the metal–polymer interactions and the resulting film–substrate adhesion.

Several studies of metals film adhesion to different polymers have also been reported during the last decade.[1–7] Vogel and Scoonhorn[1] have, for example, shown that the adhesion of evaporated metal films to polyethylene and polytetrafluorothylene substrates was dependent on the type of metal deposited. During film growth the polyethylene surfaces were cross-linked to depths that were different for the various metals used. They concluded that a strong film adhesion was due to an increased strength of the polymer surface region as it was cross-linked during the metal condensation process. Depending on the metal–polymer interaction different degrees of cross-linking were achieved. Metals forming the strongest joints with the substrates yield the greatest amounts of cross-linking. Cross-linking of a substrate surface can of course also be achieved prior to metal deposition by means of various pretreatments such as, for example, plasma treatments. These methods have been extensively reviewed by Brewis and Briggs.[8] However, a high metal film adhesion not only requires a high substrate surface strength but also a strong interaction between the metal and the polymer. X-ray photoelectron spectroscopy (XPS or ESCA) combined with *in situ* metal film growth provides a powerful method for identifying chemical bonds at the interfaces and in the past few years a number of studies on this topic have also been reported.[2–9] Burkstrand[5] has reported on XPS results showing formation of metal–oxygen–polymer complexes at the interface between Cr, Ni and Cu films and oxygen plasma treated polymer samples and correlated their presence with an increase in the adhesion of the metal films. Chou and Tang[4] have also studied metal–oxygen reactions at the interface in relation to film adhesion during metallization of polymide. They observed that Cr and Ni reacted with the oxygen in the polyimide whereas no reaction was obtained for Cu and Ag. The differences in the reactions and resulting film adhesion between the metals were discussed in terms of changes in the free energies of the reactions involved.

In previous papers[2,3] we have reported on metal–polymer interfacial interactions between evaporated Ti films and high density polyethylene (HDPE) and polydimethylsiloxane (PDMS) substrates. XPS studies of *in situ* grown Ti films on HDPE showed that Ti—C bonds were formed at the interface during film growth.[2,10] In the study using PDMS substrates the Ti—C formation was again observed but also interactions with the oxygen and silicon were observed in these samples.[3] It was also shown, in agreement with Vogel and Schonhorn,[1] that the chemical bonding is not enough to achieve a strong film adhesion since eventually surface layers of low strength has to be eliminated by cross-linking and/or by physical sputtering. This can be done by, for example, a plasma or an argon ion bombardment pretreatment.

In this paper we report on adhesion measurements of Ag, Al, Au, Cr, Cu, Ni and Ti films deposited through evaporation onto HDPE substrates and how argon ion bombardment influence the HDPE surface chemistry and the adhesion of the various metal films.

0142–2421/86/120437–04 $05.00

EXPERIMENTAL

Substrates (10×10×1 mm) were prepared by pressing HDPE granula (Hostalen GA7260, Hoeschst) between clean Al foils in an Ar atmosphere at a temperature of 150°C. This has previously[10] been shown to result in samples with a high crystallinity containing a rather small amount of low weight molecular products at the surface. Substrates prepared at a higher temperature of 200°C were less crystalline and did not have as good surface properties for adhesion as HDPE substrates prepared at 150°C. Ti, Ni, Cr, Al, Cu, Au and Ag films were deposited onto these substrates by electron beam evaporation to a thickness of 300 nm. The cryo-pumped vacuum system used had an ulitmate pressure of 10^{-7} torr ($\approx 10^{-5}$ Pa), and was equipped with a four crucible electron gun, a crystal quartz monitor for deposition rate and film thickness control and a sample load lock for sample exchange. The process conditions used for each metal are given in Table 1. In a set of experiments argon ion bombardment pretreatments were also performed *in situ* in the evaporation chamber immediately prior to film deposition. The ions were created using a Physical Electronics (model 4161) sputter ion gun operated at an Ar pressure of 5×10^{-5} torr. The ion energy was 2 keV and the ion dose approximately 4×10^{14} ions cm^{-2}.

The metal film adhesion was measured with a Sebastian I adherence tester by which the films are pulled off the substrate and the force at film failure is measured, as described in more detail in Refs. 2,3, and 10.

As-prepared and Ar ion bombarded HDPE samples were studied with XPS. Photoelectron valence band spectra were recorded with a VG ESCA3 spectrometer using unmonochromatized Al Kα radiation (1487 eV). *In situ* ion bombardments were performed using the same conditions as described above.

RESULTS AND DISCUSSION

The adhesion of the various metals both to as-prepared HDPE samples and to ion bombarded samples is given in Table 1. For as-prepared samples the adhesion is low for Ag, Al, Au and Cu films and then gradually increases for Cr, Ni and Ti films. For ion bombarded substrates the adhesion is higher than the cohesive strength of the HDPE substrates for all the metals except Ag and Au. For Ag and Au no significant difference in the adhesion between films deposited on as-prepared and ion bombarded substrates could be observed. A high adhesion of thin metal films to polymer substrates can never be established unless there is an interfacial interaction between the deposit and the substrate. In the case of Ti deposited onto HDPE substrates we have, based on XPS analysis, previously observed a Ti—C formation at the film–substrate interface. In order for such an interfacial reaction to occur a C—H bond breaking is required at the substrate surface. For as-prepared HDPE samples this has to occur during the initial stages of growth. One possible source which could supply the energy necessary for this C—H bond breaking is the heat

Table 1. Deposition conditions and results of adhesion measurements for Ag, Al, Au, Cr, Cu, Ni and Ti films deposited onto as-prepared and Ar ion bombarded HDPE substrates

	Deposition conditions		Adhesion measurements (Pull strength in MPa)	
Metal	Deposition rate (nm min^{-1})	Power applied to e-gun (KeV) (mA) (kW)	As-prepared HDPE samples	Ar ion bombarded HDPE samples (4×10^{14}ions cm^{-2})
Ag	250	10 100 1.0	0	0
Al	300	10 300 3.0	0	20 (cohesive)
Au	300	10 150 1.5	0	0
Cr	150	10 140 1.4	4	22 (cohesive)
Cu	150	10 240 2.4	1	19 (cohesive)
Ni	90	10 240 2.4	10	20 (cohesive)
Ti	300	10 240 2.4	13	22 (cohesive)

released during the metal condensation process. The heat of condensation ΔH_c can to a first approximation be set equal to the heat required for vaporization, which can be found tabulated in the literature.[11] In Fig. 1 the adhesion of Ag, Al, Au, Cr, Cu, Ni and Ti films to as-prepared HDPE samples is shown as a function of their ΔH_c. For ΔH_c values lower than the C—H bond strength a very low adhesion is obtained. However, when ΔH_c becomes of the same order as the C—H bond strength an increased adhesion is observed as H_c is increased further. Energy could of course also be transferred to the substrate surfaces through direct radiant heating from the evaporation source or through a flux of secondary electrons or reflected primary electrons from the evaporation source. Especially high energy electrons have been shown to be effective in cross-linking HDPE surfaces.[10] Both the radiant heat and the amount of electrons from the electron beam heated source is proportional to the input beam power. However, no systematic variations in the metal–film adhesion with electron beam power have been observed, as also can be seen in Table 1.

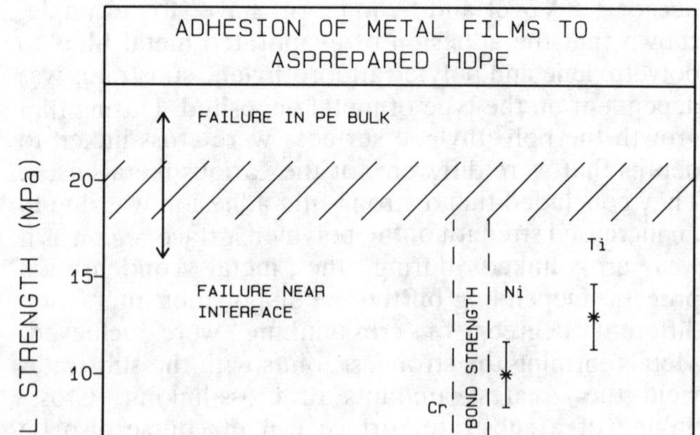

Figure 1. Adhesion of evaporated Ag, Al, Au, Cr, Cu, Ni and Ti Films to as-prepared HDPE samples as a function of their heat condensation, ΔH_c. The values of ΔH_c were taken from Ref. 11.

Vogel and Schonhorn[1] also observed that evaporated Ti and Ni films had a higher adhesion to as-prepared polyethylene substrates compared to Al and Cu films. They also concluded that the adhesion of the various metals films did not reflect the evaporation source temperature but were rather inherent properties of the metal–polymer interfaces themselves.

Vogel and Schonhorn also found that the metal depositions resulted in a cross-linking of the outermost polymer surface region. The thickness of this cross-linked surface region was different for the various metals used. Those metals with the highest adhesion also showed polymer surfaces with the thickest cross-linked regions. Chou and Tan,[4] who observed metal–oxygen–polymer interfacial interactions, have correlated the adhesion of Al, Cu, Cr and Ni films to polyimide subtrates to the sum of the heat of condensation and the heat of oxidation. However, it is questionable whether the heat of formation, ΔH_f, for oxides or in our case for carbides can be used to predict the interfacial reactions that occur at metal–polymer interfaces. In our case, the use of ΔH_f values for carbides gives no correlation to the measured adhesion values. For example, ΔH_f for Cr_7C_3 and TiC are -44 and -54 kcal/mol^{-1} respectively while the carbide for Ni (Ni_3C) is unstable ($\Delta H_f = +11$kcal mol^{-1}).[12] Still Ni films show a higher adhesion compared to Cr films. It is thus more probably that the interfacial reactions lead to the formation of more complicated organometallic complexes. However, the energies involved in these metal -polymer reactions are difficult to calculate especially when detailed chemical and structural information about the interfaces are missing.

The adhesion of all evaporated metals except Ag and Au to HDPE substrates is, as already has been mentioned, enhanced by an argon ion bombardment pretreatment. The pull strength required for film failure of Ti films as a function of ion dose, as has been determined previously,[10] is shown in Fig.2. The film adhesion is relatively high on as-prepared samples but as the HDPE sample surface is ion bombarded prior to film deposition the adhesion becomes high enough to cause a cohesive failure in the HDPE bulk. The ion dose required for this enhancement is as low as approximately 1×10^{14} ions cm^{-2}. At the backside of Ti films detached from as-prepared samples a thin HDPE residual has been observed by XPS analysis, indicating that a thin surface layer of low strength is present on as-prepared sample surfaces. This weak layer which limits the film adhesion is eliminated by the ion bombardment through cross-linking and for higher ion doses also through a significant amount of physical sputtering. This cross-linking has also been observed from the changes in the XPS C 1s spectra after ion bombardment of HDPE samples.[10] The integrated C 1s peak intensity increased as a result of the increased density within the cross-linked region. The cross-linking ocurring during ion bombardment is activated by a dehydration process which is reflected in a small shift of the C 1s level towards a lower binding energy and in a slight asymmetry of the peak shape. The asymmetry is due to a change in the loss characteristics of dehydrated HDPE.[13] The dehydration induced by the ion bombardment has also been detected in the XPS valence band spectra as shown in Fig. 3. The valence band spectrum of as-prepared HDPE have two main peaks at approximately 19 and 14 eV corresponding to C—C bands and a small broad peak at approximately 10 eV corresponding to a C—H band.[14] These characteristic features are successively lost as the surface is ion bombarded. The valence band spectrum taken after a prolonged ion bombardment of HDPE, is shown in Fig. 3, together with a spectrum of highly oriented pyrolytic graphite, HOPG. As can be seen the valence region of ion bombarded HDPE samples resemble the spectrum of graphite. However, the modified surface layer formed on HDPE being graphitic in nature must have an amorphous structure.

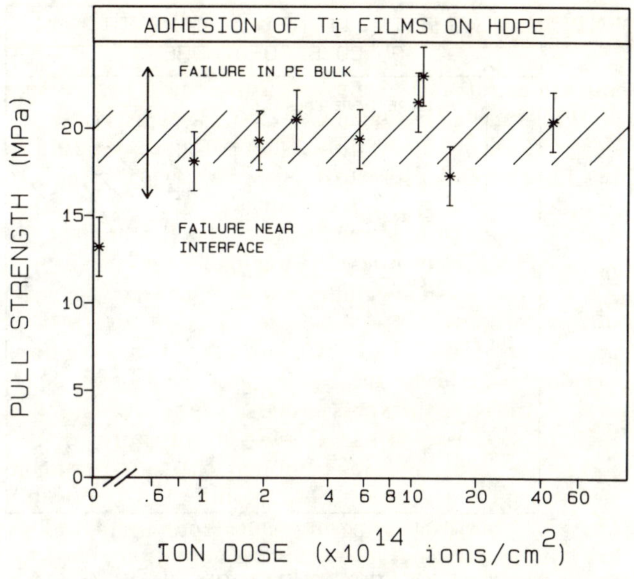

Figure 2. Adhesion of evaporated Ti films to HDPE substrates as a function of Ar ion dose. The energy of the Ar ion was 2 keV.

As the HDPE samples are ion bombarded the surface is activated, through dehydration as has been shown by the XPS core level and valence band studies, and any metal that forms a stable Me—C bond or organometallic complexes with the substrate will adhere strongly to its surface. Figure 4 show the adhesion

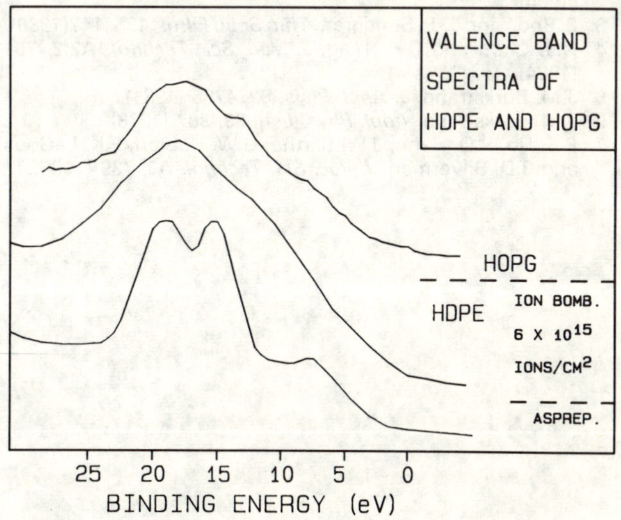

Figure 3. XPS valence band spectra of as-prepared and ion bombarded HDPE as well as of highly oriented pyrolytic graphite (HOPG).

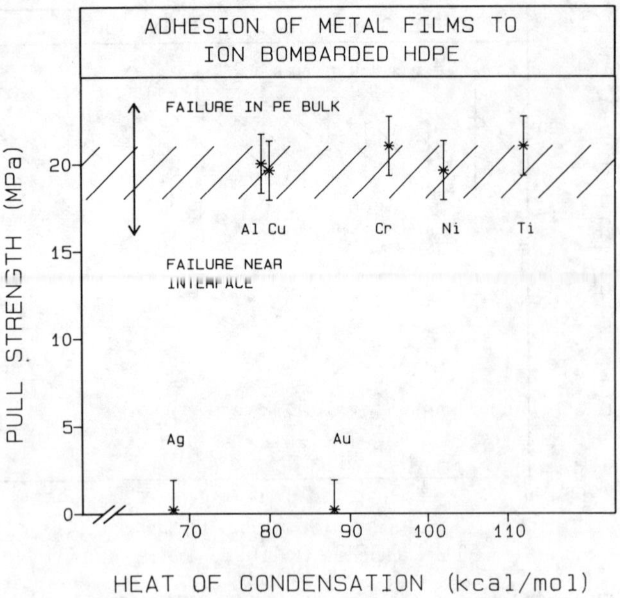

Figure 4. Adhesion of evaporated Ag, Al, Au, Cr, Cu, Ni and Ti film to HDPE substrates as a function of their heat of condensastion.

of the various metals to HDPE that has been exposed to an ion dose of approximately 4×10^{14} ions cm^{-2} prior to film deposition.

All the metals, except Ag and Au, have an adhesion high enough to cause a cohesive failure in the HDPE bulk. Since no C—H bond breaking is required on ion bombarded samples there is no dependence on the heat of condensation for the various metals as in the case of as-prepared samples. All the metals, except Ag and Au, must form a stable organometallic Me—C complex. It is not possible to exactly predict the metals that will form strong Me—C complexes, but it seems reasonable that noble metals such as Au and Ag will form less stable organometallic compounds. The complexity of the situation will of course increase as other polymers such as polyimide containing both O and N in addition to C and H are used as substrates for metal–film deposition. More fundamental research in order to determine the chemical and structural nature of the metal–polymer interfaces and the bond strength of various possible organometallic complexes are thus needed in order to make quantitative predictions on metal–polymer interfacial strengths possible.

CONCLUSIONS

The adhesion of various metal films to high density polyethylene samples has been studied. Both substrates in an as-prepared condition and argon ion bombarded prior to metal deposition have been studied. The results show that a high adhesion to as-prepared samples is only obtained as the heat of condensation of the deposited metal is of the same order or higher than the C—H bond strength. A high adhesion can, as C—H bonds have been broken, only be obtained in those cases where stable organometallic complexes are formed at the interface. The ion bombardment pretreatments dehydrates and cross-links the sample surface thereby eliminating the criterium of C—H bond breaking during the initial stages of film growth.

Acknowledgement

The financial support of the Swedish Board for Technical Development, (STU), is gratefully acknowledged.

REFERENCES

1. S.L. Vogel and H. Schonhorn, *J. Appl. Polym. Sci.* **23,** 495. (1979).
2. P. Bodö and J.-E. Sundgren, *J. Vac. Sci. Technol.* **A2,4,** 1498 (1984).
3. P. Bodö and J.-E. Sundgren, *Thin Solid Films,* **136,** 147(1986).
4. N.J. Chou and C.H. Tang, *J. Vac. Sci. Technol.* **A2,2,** 751 (1984).
5. J.M. Burkstrand, *J. Appl. Phys.* **52,** 4795 (1981).
6. J.M. Burkstrand, *Appl. Phys. Lett.* **33,** 387 (1978).
7. P.S. Ho, P.O. Hahn, J.W. Bartha, G.W. Rubloff, F.K. LeGous and B.D. Silverman, *J. Vac. Sci. Technol.* **A3,** 739 (1985).
8. D.M. Brewis and D. Briggs, *Polymer* **22,** 7 (1981).
9. D.T. Clark in *Polymer Surfaces,* p.309 ed. by D.T. Clark and W.J. Feast, Wiley (1978).
10. P. Bodö and J.-E. Sundgren, submitted for publication.
11. *Handbook of Chemistry and Physics,* 56th Ed., p. F–223, CRC Press, (1975).
12. *ibid* p. D–73.
13. The interaction of plasmas and ion beams with organic materials as studied by ESCA, p. 186, D.R. Hutton, Ph.D., thesis, University of Durham, (1983).
14. J.J. Pireaux and R. Caudano, *Phys. Rev.* **15,** 2242(1977).

SURFACE AND INTERFACE ANALYSIS, VOL. 9, 441–444 (1986)

Kinetics of Bromine Migration and Copper Reactions at Copper–Epoxy Resin Interfaces

Alberto Torrisi and Salvatore Pignataro
Dipartimento di Scienze Chimiche dell'Universita', Viale Andrea Doria, 8–95125 CATANIA, Italy

Two chemically different copper surfaces have been moulded to a commercial epoxy resin of Novolac type. The obtained samples have been stored at high temperatures in the range 150–200°C. These systems have been mechanically fractured at given time intervals and the surfaces obtained in the fracture have been analysed by ESCA. The kinetics of migration of Br to the interfacial zone have been evaluated and connected to the performances of the finite electronic device. Reaction kinetics involving the copper substrate have also been followed. These given evidence of the strong dependence of the interfacial chemistry on the physico-chemical nature of the involved surfaces.

INTRODUCTION

In a recent series of papers from this laboratory, ESCA was extensively used in connection with several problems encountered in the fabrication of semiconductor devices.[1–6] Nickel diffusion through gold barriers, bad performance of soft solder alloy, organic contamination on the pads of the silicon chip leading to failures are some of these problems. Very recently some investigations have been made on the effect of organic contaminants on the oxidation kinetic of silicon,[6] on the influence of surface chemical composition on the reliability of the Al/Cu bond[5] and on the degradation of the dielectric properties of a Pb glass.[7]

In addition to those studies a great effort has been made in order to have a better knowledge of the interfacial chemistry involved in metal–polymer interfaces. This has been done by considering commercial resins and metals and therfore 'real' multicomponent systems used in the production lines of a semiconductor device. In these studies[1,2,4] the resins were moulded to various metal surfaces under plant conditions, the joint was mechanically broken and the two faces obtained after fracture analysed by ESCA. Several observations could be made. In particular, some of the resin constituents were found to remain confined in the bulk of the moulded resin. Others were found to be homogeneously distributed between the bulk of the resin and the interface. Another group of constituents of the polymeric material were found to segregate to the interfacial zone, often in a chemical state different from that found in the bulk. The migration phenomena were found to selectively depend on the chemical nature of the metal and resin respectively.

Dramatic chemical composition differences were found between the two faces obtained in the fracture. The adhesion of the resin to the metal face was found to be influenced by these migration phenomena. Proof of various chemical reactions taking place at the interface with time of storage or moulding temperature could be obtained. In particular, the possibility to better quantify the observed phenomena by following the kinetics of migration or reaction at the metal polymer interfaces was shown.[5] This paper reports some further experiments along these lines.

In this regard it has to be noted that, while surface kinetics have been frequently reported for a variety of systems, to the best of our knowledge no examples of interface kinetics are available in the literature. This is connected to the fact that it is in general very difficult to have: (a) a reproducible fracture locus; (b) a fracture locus which does not change with the time and therefore with the evolution of the reaction. These two conditions are satisfied for systems which do not react vigorously (like those discussed in this paper) so that the chemical modifications remain within the sampling depth of the surface technique used.

EXPERIMENTAL

The resin used was a Novolac B type amine cured commercial epoxy resin from Hysol Corporation. All the Cu (OFC) frames before their use were chemically etched with an acid treatment followed by a decreasing treatment and an electropolishing process. Some samples were prepared moulding the resin on the Cu frames immediately after the electropolishing process ('etched' samples). Other samples were moulded 15–20 days after the polishing procedure in order to simulate plant conditions ('unetched' samples). The moulding process was done in the plant apparatus. The moulding temperature was 175°C; the moulding pressure was about 70 kg cm^{-2}, the post curing temperature was also 175°C.

After the moulding step, the resin moulded on the back of the metal frame — in order to ensure a mechanical graft between the metal and the resin itself — was taken off.[3,4] The samples were then fractured in air. The interfaces obtained in the rupture of the nominal metal–resin bond were analysed by using an ESCA ES 300 (KRATOS) electron spectrometer

0142–2421/86/120441–04 $05.00

equipped with an Ar^+ ion gun for the sputtering process. The radiation used was the Al K_α. The calibration of the spectrometer was achieved according to recently proposed procedures.[5]

The metal side of the joint was in general analysed immediately after the fracture. The time needed for the complete analysis was about four hours. The corresponding resin face was in general analysed after the analysis of the metal face. No appreciable variations were found in the ESCA spectra of the resin faces during this four hours time. The relative intensity for the various ESCA peaks were found to be reproducible within 20% in two consecutive experiments.

The calculations of Cu_2O thickness on the metal side of the fracture were made using the approximation and the formulas reported in Ref. 5. The evaluations of the Cu_xO_y thickness on the resin side of the fracture were made using equation (1) reported in Ref. 5 substituting I_m with $I_{C\ 1s}$, assuming $\lambda_{C\ 1s} = 12$ Å and a value of $\sim 10^{23}$ atoms cm^{-3} for the atomic density (graphite).

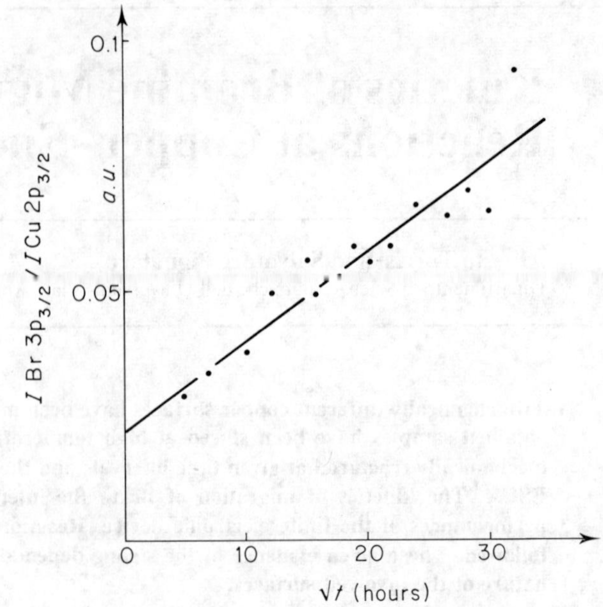

Figure 1. Ratio of the Br $2p_{3/2}$/Cu $2p_{3/2}$ ESCA peak area on the metal side of the fracture of a metal–resin joint vs. the square root of the time of storage at 150°C.

RESULTS AND DISCUSSION

Bromine migration

A commercial resin used to encapsulate a semiconductor device is a multicomponent system. Its formulation is made by several different chemical constituents. Those generally found are the organic macromolecule (mainly epoxide or silicone); the polymerization catalyst; the filler (often SiO_2); an adhesion promoter e.g. an epoxysilane; some light stabilizers; a 'curing' agent; a dye; a fluidifier (generally a low molecular weight organic material); some flame retardants (usually a bromine organic derivative in conjunction with antimony oxide).

In connection to these last components, it was shown in a previous study[4] that while antimony remains mainly confined into the bulk of the resin, the bromine migrates with time to the resin interface. Also the increase of the moulding temperature from 160°C to 190°C was found to cause an increase of the bromine in the interfacial zone.

Bromine is a potentially dangerous element because reaction with water from the environment of the finite device or from the encapsulating resin itself, gives HBr which could corrode the metal contacts. It was therefore highly desirable to study in greater detail this phenomenon.

For this reason several samples prepared by moulding a commercial epoxy resin to a copper surface ('etched' as reported in the experimental section above) were stored at 150°C. At given time intervals one of these samples was taken and mechanically fractured. The metal face of the fracture was analysed by ESCA in order to obtain the ratio between the Br $3p_{3/2}$ and the Cu $2p_{3/2}$ peak areas.

Figure 1 shows the plot of this ratio vs. the square root of the time of storage at 150°C after moulding. The migration of bromine to the metal surface is clear and can be quantified as depending on a simple rate law of the form:

$$Y = a\sqrt{X} + b$$

where X is the time of storage, Y is the Br $3p_{3/2}$/Cu $2p_{3/2}$ peak areas ratio a is the slope and b is the value of Y when the time is equal to 0. In a very complex system, like that here studied, it is very hard to discriminate among the possible driving forces for the observed migration. However the observed time dependence of the phenomenon suggests in any case a diffusion controlled migration.

In order to gain more information on the phenomenon the experiment was repeated by using various storage temperatures and a different type of copper surface. In particular the copper surface was an 'unetched' copper surface (see experimental section above) and therefore containing the usual carbonaceous contaminants and the well known[5] mixture of copper oxides and hydroxides.

Figure 2 shows the same type of plot described for Fig. 1. The importance of the chemical nature of the 'copper surface' is immediately evident. For the 'etched' copper surface the bromine growth at the interface is faster than that observed for the 'unetched' copper surface at the same temperature (150°C). This finding may support the hypothesis that the phenomenon is determined by the interfacial free energy.

A dramatic increase of the bromine content in the interfacial zone is found with time, for the 'unetched' samples, when the experiment is made at 200°C. The phenomenon appear to be almost as a threshold effect since, at least within the time of observation ($\sim$1000 h), not much difference in the Br$3p_{3/2}$/Cu$2p_{3/2}$ peak area ratio is found on going from 20°C to 175°C for the storage temperature.

However, the effect is not surprising assuming that it is due to the melting and/or the decomposition of the flame retardant aromatic bromine derivative that should occur just around 175°C. A support to this hypothesis comes from the fact that a new resin having a bromine containing flame retardant with a higher

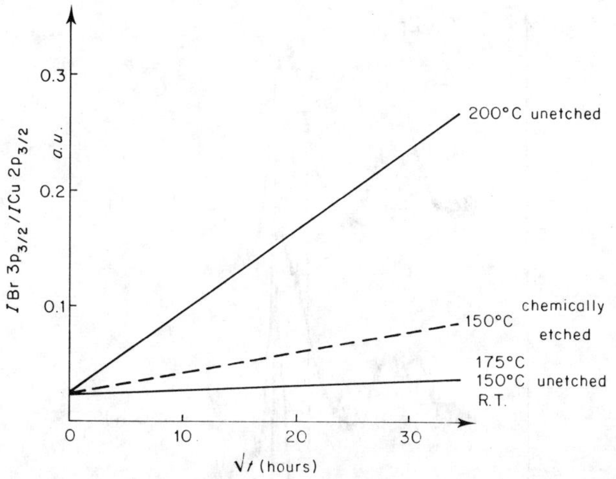

Figure 2. Regression lines of the Br $3p_{3/2}$/Cu $2p_{3/2}$ ESCA peak area ratios on the metal side of the fracture of a metal–resin joint vs. the square root of the time of storage at high temperature.

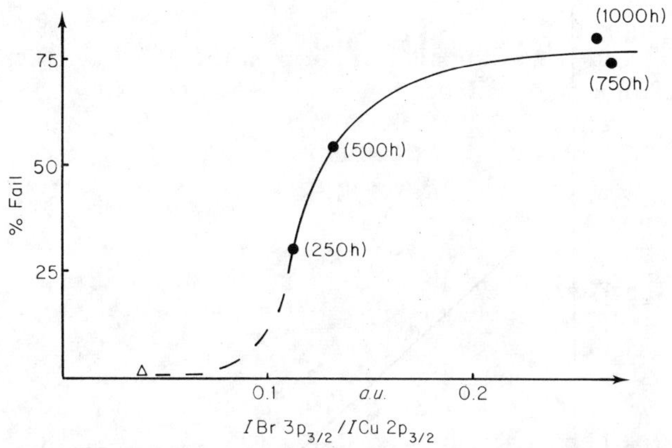

Figure 3. Percent of failure of a finite electronic device vs. the Br $3p_{3/2}$/Cu $2p_{3/2}$ ESCA peak area ratio on the metal side of the resin–metal joint interface(●). The reported figure in brackets refer to hours of storage of the finite devices at 200°C/(Δ): this point refers to a storage temperature of 175°C and a time of storage of 1000 hours. The Cu copper surface used was the 'unetched' one.

melting point shows a much better thermal stability (Hysol Co., private communication). Also an incipient thermal decomposition of the resin network may concur to facilitate the bromine migration to the interface.

The above quantitative results become of great technological interest when they are compared with the performance of a finite electronic device using the materials above considered.

Figure 3 shows the bromine relative content at the metal side of the resin-metal interface vs. the percentage of failure of the finite device. The copper frames used were 'unetched' ones and the storage temperature was 200°C. If this content is lower than about 0.1 (in our arbitrary scale) the percentage of failure is very low, while it increases abruptly just above this figure.

This threshold behaviour may be connected to the corrosion of the metallic interconnections caused by the bromine migration. An independent failure mechanism might also be that mentioned above, involving the beginning at 200°C of a decomposition of the resin network with consequent mechanical stresses in the finite device.

Copper reactions

In the previous paper of this series[4] other relevant observables were the Cu signals at the interface. In particular, the following points on the interface reactions of the Cu-resin system were observed:

(a) $Cu(OH)_2$ was found to grow at the interface on leaving the sample in air[2] at room temperature for six months.

(b) The copper oxides were found to interpenetrate into the 'resin phase' with time.

(c) This phenomenon seemed to be dependent on the chemical nature of the resin.

(d) The relative amount of copper derivatives as well as the nature of the various oxides changed at the fracture locus with the moulding temperature.

In summary, the Cu ESCA signals were found to be good observables to follow interfacial reaction kinetics. These signals were then recorded and worked up during the experiment above described for the bromine migration.

In particular, the average thickness of the Cu_2O layer, found after the fracture of the samples on the metal side of the fracture itself, is plotted in Fig. 4 vs. the log. of the storage time at 150°C. The copper surface used in these samples was immediately moulded after the electrochemical polishing (see also experimental secion above). The thickness of Cu_2O decreases with time according to a simple law of the type:

$$A_t = m \log(t + 1) + A_o$$

where A_t was the thickness of the Cu_2O in Å; m was the slope of the line; t was the storage time at 150°C and A_o was the Cu_2O thickness at time zero.

Considering that the Cu content on the resin side is very low and probably also decreasing with time, it may be concluded that the copper oxides disappear from the interfacial zone with time. This finding can be interpreted in at least two different ways. The first is that the oxides (or the anion) migrate with time in the bulk of the resin. The second and, perhaps most appealing, is that the oxides are reduced at the expense of some chemical compound present at the interface. One possible candidate for this reduction process is just the brominated flame retardant derivative considered above which should contain phenolic moieties which could be oxidized to quinonic units by the copper oxides. Unfortunately we do not have for the moment any evidence supporting either mechanism.

A different behaviour is found for the samples made by using the 'unetched' copper surface. Figure 5 shows the $Cu2p_{3/2}$ peak shapes for the metal side of the fractures observed for the samples stored at 150°C for different times. The Cu $2p_{3/2}$ peak at time zero is very similar to that reported for 250 hours of storage.

The Cu LMM Auger peak shape (not shown) indicates this surface is mainly Cu_2O having a thickness greater than 40 Å (no Cu^o signal is found in the LMM Auger peak). The spectra reported in the figure show

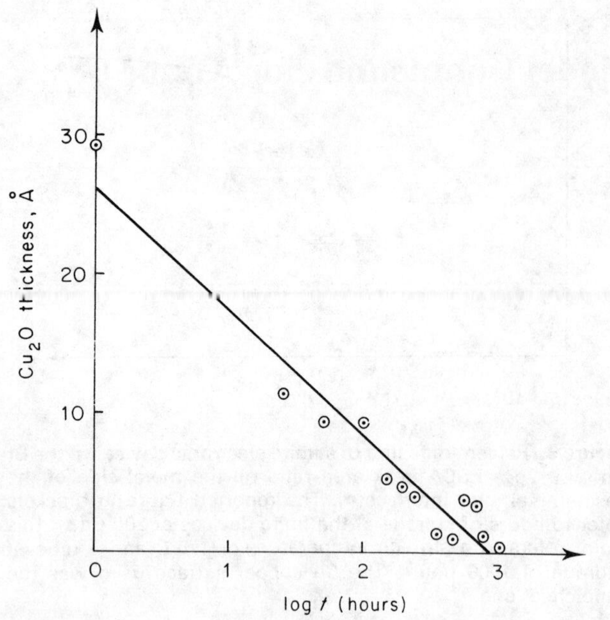

Figure 4. Cu_2O average thickness on the metal side of the fracture of the resin–metal joint vs. the log of storage time at 150°C. The samples were prepared by using an 'etched' surface.

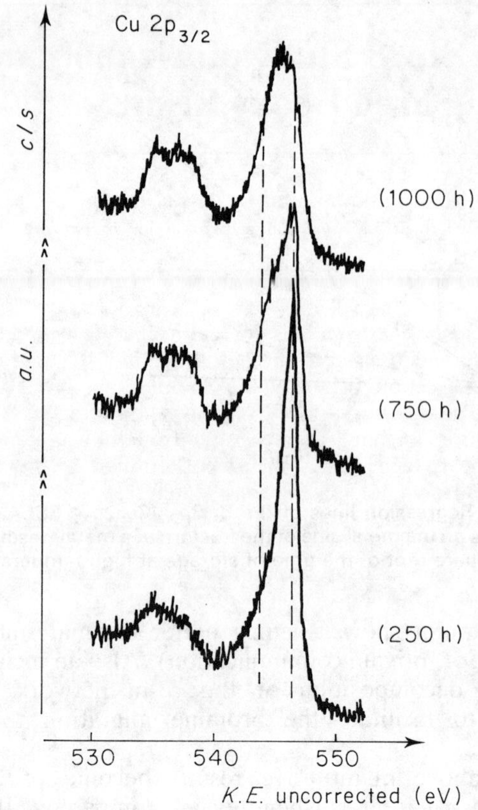

Figure 5. ESCA Cu 2p 3/2 signals for the metal side of the fracture of the resin–metal joint. The samples were prepared by using an 'unetched' copper surface. The time of storage is reported in brackets. The temperature of storage was 150°C.

therefore that at the observed fracture locus Cu_2O oxidizes to Cu (II) (probably a mixture of $Cu(OH)_2$ and CuO) in agreement with the previous finding.[4] The amount of total copper observed on the resin side of the fracture remains instead constant and at a low level even if it is greater than for the sample 'etched'.

The different behaviour of these kind of samples with respect to those obtained using 'etched' copper surface is again hard to be interpret in a unique way. One may think that it is connected to the different thickness of the Cu_2O and to a different type of fracture locus which for the 'unetched' copper might mainly pass through this Cu_2O layer. However further work is needed in order to prove a particular mechanism. In any case, whatever is the explanation of this behaviour, it confirms the strong dependence of the interfacial chemistry on the physico-chemical nature of the surfaces involved.

Acknowledgements

Dr. Angelo Cavallaro (SGS Microelettronica, Catania) is gratefully acknowledged for his help in the experimental part of the work. Ing. Giuseppe Ferla (SGS Microelettronica, Catania) is gratefully acknowledged for his helpful suggestions. The work has been partially supported by CNR Rome (Progetto Finalizzato Chimica Fine e Secondaria, Progetto Finalizzato Materiali e Dispositivi per l'Elettronica a Stato Solido).

REFERENCES

1. A. Torrisi, S. Pignataro, G. Nocerino, *Appl. Surf. Sci.* **13**, 389 (1982).
2. A. Licciardello, S. Pignataro, M. Panczel, *Surf. Technology* **18**, 189 (1983).
3. A. Torrisi, G. Marletta, O. Puglisi, S. Pignataro, *Surf. Interface Anal.* **5**, 61 (1983).
4. S. Pignataro, A. Torrisi, G. Ferla, *Surf. Interface Anal.* **7**, 129 (1985).
5. S. Pignataro, A. Torrisi, O. Puglisi, A. Cavallaro, A. Perniciaro, G. Ferla, *Appl. Surf. Sci.*, in press, and references therein.
6. A. Licciardello, O. Puglisi, S. Pignataro, *J. Appl. Phys. Letters*, in press.
7. A. Puglisi *et al.*, in preparation.

SURFACE AND INTERFACE ANALYSIS, VOL. 9, 445–448 (1986). EXTENDED ABSTRACTS

SIMS Study of Alcohol-amines: model Compounds for Amine Cured Epoxy Resins

S. Affrossman and R. A. Pethrick
Chemistry Department, University of Strathclyde, Glasgow G1 1XL, Scotland

Amine-alcohols are models for amine cured epoxy resins, which are widely used in the aerospace industry for joining aluminium sections. The model compounds are applied to the prepared aluminium surface from the vapour or from solution to produce a thin overlayer of, preferably, monolayer coverage. Evidence that the deposited film is the model compound and not adventitious material would be useful. SIMS is capable of providing fragmentation patterns which, at least, may be used as a fingerprint for identification of chemical compounds in trace amounts, and at best, may give identifiable fragments from which the structure of the compound can be deduced. A model compound which closely resembles the cured resin is the adduct of the diglycidyl ether of bisphenol A and diethylamine:

$$(Et)_2 - C - C(OH) - C - O - Ph - C(Me)_2 - Ph - O - C - C(OH) - C - N(Et)_2$$

The compound was deposited from c–hexane solution on to an aluminium substrate for analysis. The SIMS spectrum of the compound was obtained with a Vacuum Science Workshop 600 mass quadrupole and mass filtered ion gun using an argon ion beam at 2 kV and 10^{-9} amp. and is shown in Fig. 1. The parent peak at *m/z* 486 is observed, but the most prominent feature in the spectrum is the large signal at *m/z* 86. This fragment arises from breakage of the R′.C(OH) − C(NR_2) bond, which would be expected, from data from conventional mass spectroscopy studies, to be susceptible to irradiation. Signals will generally decrease in intensity with increase in *m/z* because of the characteristics of the analyser, but the marked decrease in signal intensities for *m/z* above 130 is very evident, and is related to the structure of the molecule. The signals at *m/z* 130 and lower values correspond to successive breaking down of the aliphatic chain. The small signals at *m/z* 135 and 177 are probably phenolic fragments arising from disintegration of the aromatic centre portion of the molecule. The signals at *m/z* 400 and 356 are the parent minus *m/z* 86 and 130 respectively. The absence of fragments corresponding to the parent minus two times 86 or 130, suggests that the energy of excitation of this large molecule tends to be restricted to one end. r if not, leads to a severe disintegration of the molecule. It is apparent that the SIMS spectrum of the model compound may be directly related to its structure.

The SIMS spectrum of the compound deposited from a very dilute hexafluorobenzene solution on to an aluminium substrate is composed of the same signals observed in the above, but with additional features. These are, principally, new signals at *m/z* 468, 338 and (enhanced) 112. We note that these correspond to (486–18), (356–18) and (130–18) respectively. This suggests that the new signals are the result of loss of water from part of the sample. Examination of the formula shows that loss of water would only be observed from certain fragments i.e. the new signals, noted above, were observed but there was no (86–18) signal, consistant with the deduction of dehydration of the molecule. The absence of a (486 − 2 × 18) signal suggested only one water is lost per molecule.

The apparent lack of similar dehydration when c-hexane was used as solvent may be a consequence of the difficulty of depositing less than monolayer amounts of a relatively high molecular weight species from very dilute solution on to a contaminant free surface. However, hexafluorobenzene, is known to complex strongly with amine groups, so an alternative possibility is that the dehydration may be entirely due to the solvent.

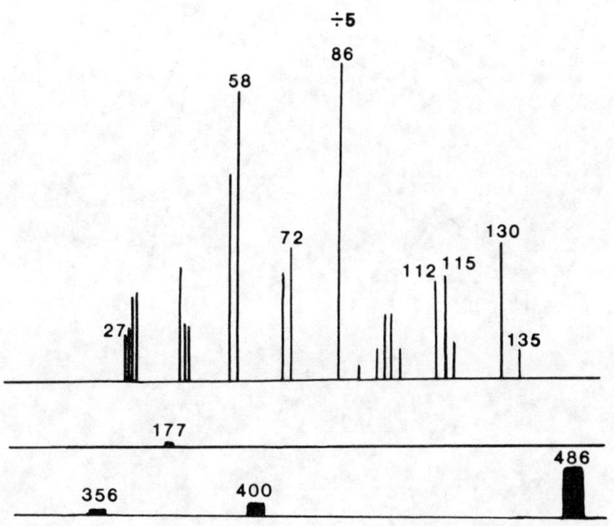

Figure 1. SIMS spectrum of adduct of diglycidyl ether of bisphenol A and diethylamine (model compound deposited from cyclo-hexane).

Surface Chemistry of Aminoalkylated Polystyrene Resins and Adhesion of Yeast Cells

V. Dourtoglou, W. E. E. Stone*, M. Genet and P. G. Rouxhet
Université Catholique de Louvain, Unité de Chimie des Interfaces, Place Croix du Sud, 1, B-1348 Louvain-la Neuve, Belgium
* Section de Physico-Chimie Minérale (MRAC)

Adhesion of micro-organisms to a support allows them to be confined in a reactor and to improve their performance as biocatalysts. The ability of a support to retain micro-organisms is related to its surface charge and to its surface energy (hydrophobic–hydrophilic balance). This work is dedicated to a modification of the surface properties of a polystyrene resin, in order to promote the adhesion of the yeast *Saccharomyces cerevisiae*.

Polystyrene was chloromethylated and then reacted with triethylamine or triethanolamine to obtain various products differing according to the degree of aminoalkylation: $-C_6H_4-CH_2Cl + R_3N \rightarrow C_6H_4-CH_2-N^+-R_3 + Cl^-$. Representative samples have been examined by x-ray photoelectron spectroscopy (XPS) (Vacuum Generator ESCA 3 MkII with unmonochromatized Mgk radiation).

Alteration of the grafted polymers under the x-ray beam is marked by a modification of the N 1s peak which indicates the transformation of quaternary ammonium into amine groups. Therefore, the analysis has to be performed within 60 min.

The Cl 2p peak shows two components, attributed respectively to the chloride counterions and $-CH_2-Cl$ groups. The N 1s peak also shows two distinct components due to quaternary ammonium and amine, indicating that amine non covalently bound is strongly held by the polymer. This has been confirmed by contacting non modified polystyrene with triethylamine. A comparison, for the various samples, of the relative intensity of the two Cl components, and of the N/C ratios measured by XPS and by chemical analysis indicates that the surface concentration of polystyrene grafted by triethylamine is about the same as the bulk concentration. On the contrary, grafting with triethanolamine is enhanced on the surface.

Zeta potential measurements (Lazer Zee Meter 500, Pen Kem) show that the chloromethylated polystyrene is negatively charged above pH 3.5. A small degree of aminoalkylation is sufficient to make the surface positive up to pH 10.

The adhesion of yeast has been examined by sedimenting a suspension of cells in water over a film obtained by pressing the polymer under 10 tons cm^{-1}; after washing by demineralized water, the samples were examined by scanning electron microscopy (JEOL JSM-35). Yeast adhesion occurs only when the polymer surface is positively charged.

The Catalysis of Electroless Copper Deposition on Fibreglass–Epoxy Laminates

C. Lea, M. T. Anthony and G. C. Smith
National Physical Laboratory, Teddington, Middlesex, UK

INTRODUCTION

The electronics assembly industry relies heavily on fibreglass–epoxy laminate sheet for the production of high quality double-sided printed circuit boards. The base laminate usually comprises eight partially cured epoxy-impregnated glassfibre cloths pressed together at the curing temperature, between copper sheets. During PCB manufacture the copper outer surfaces are imaged and chemically etched to produce the electrical tracks. Holes are drilled through the board for component insertion, but these also act, after being copper plated, as electrical pathways between the tracks in the two copper surfaces on each side of the board. The plating of the drilled holes is a crucial step in PCB manufacture, and its success depends on the hole-wall preparation and the deposition of a catalytic precursor for electroless copper which, in turn, is a necessary substrate for the final electroplated copper layer.

THE INDUSTRIAL REQUIREMENT

During automated soldering of the finished board the laminate outgasses. If the copper barrel is weak or porous, this gas blows through into the liquid solder to produce cavities and blowholes in the fillets; conditions which are unacceptable to quality control and require expensive manual touch-up. Any pores in the copper barrel arise directly from poor coverage of the hole walls by the under-lying electroless copper. The electroless deposit is covered with 20–30 μm of electroplated copper. Voids in the electroless copper which have dimensions greater than about 20 μm are not successfully bridged over, leaving an escape path for gas evolving from the laminate during soldering. A typical void in the electroless deposit is shown in Fig. 1: a glass fibre has not been adequately catalysed for the copper to nucleate.

THE CATALYSIS OF ELECTROLESS COPPER

The most common catalyst for the deposition of electroless copper is colloidal palladium. The surface is first sensitized, usually with stannous ions, before the activation by Pd ions. In the PCB industry generally a mixed catalyst is used of strongly acidic stannous and palladium chlorides. This approach is favoured because, whereas separate solutions produce a palladium film over the entire panel, the mixed solutions leave the existing copper foil uncoated, thus not compromising the adhesion between the base copper and the plated copper.

After the adsorption of the catalyst an accelerator is used to decompose the colloidal film and remove unwanted tin complexes; this increases the initial speed of copper deposition. Changes in the surface coverage of Pd and Sn, during the accelerator process, have been monitored using XPS. Copper will nucleate on surfaces for which the accelerator stage has been omitted but the deposition is slower, more grainy and uneven. XPS has also been used to measure the Pd coverage for a variety of different chemical suppliers and ages of the solutions. From these measurements it turns out that the quality of the electroless copper deposit is strongly correlated with the coverage by the palladium.

QUALITY OF THE ELECTROLESS DEPOSIT

The quality of the electroless copper is defined in terms of its coverage, i.e. its opacity using a visual backlighting inspection method. The deposit is generally 2–3 μm thick. The smaller and more evenly dispersed are the initial copper nuclei, the more opaque will be the electroless deposit. The size of the nuclei is controlled by the coverage of the catalytic particles, but also by the rate of deposition of the copper. Artificial growth inhibitors can be incorporated to reduce the growth rate to produce a finer grained deposit. The uniformity of the deposit can be improved by homogenising the surface of the drilled hole. Several methods have been tried. A chemical etch using H_2SO_4 and HF removes the wetting agents present on the glass fibre surfaces to aid their wetting by the epoxy during fabrication, and also removes loose epoxy swarf and cusps left by the drilling, from the surface. The situation is then further improved using an alkaline permanganate oxidation treatment to remove sulphurated compounds from the epoxy surface. A second, simpler treatment is to remove the unwanted swarf with an ultrasonic water bath, and to homogenise by a bake in air; 1 h at 130°C is suitable.

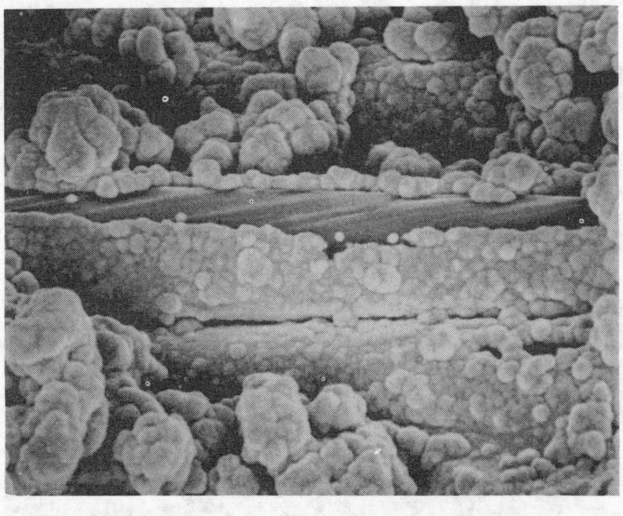

Figure 1. A scanning electron micrograph of electroless copper deposited on a drilled surface of fibreglass-epoxy laminate, showing poor coverage of an exposed glass fibre. Imaged area about 50 μm × 40 μm.

CONCLUSIONS

The quality of an electroless copper deposit on the drilled surface of a hole in fibreglass–epoxy laminate during PCB manufacture is critical in determining the subsequent soldering quality of the board. In turn, the quality of the electroless copper is defined by the effectiveness of the adsorption of the Pd catalytic precursor as measured by XPS. This adsorption can be improved by a prior chemical etch or a bake, most importantly to remove the wetting agents present on the exposed glass fibres.

APPLICATIONS OF SURFACE ANALYSIS TECHNIQUES IN ADHESION SCIENCE

W.J. van Ooij
Department of Materials Engineering, VPI & SU, Blacksburg, Va. 24061, USA

An overview is given of the usefulness of surface analysis techniques such as XPS, AES and SIMS, for the study of adhesion problems. Four major areas are distinguished, namely:

1. interfaces between organic coatings and substrates, such as paints on metals, adhesives on glass or metals;
2. interfaces in composite materials, e.g. fibers-polymers, tire cords;
3. polymer surfaces, e.g. in metallized plastics.
4. adsorption of coupling agents, inhibitors or adhesion promoters on surfaces to be bonded.

Examples are given of surface or interface characterization studies in which adhesion or failure mechanism could be elucidated conclusively. These studies encompass a number of fundamentally different mechanisms.

The special requirements and adaptations of surface techniques to the study of adhesion problems, the synergism of combination instruments, sample preparation techniques and the usefulness and limitations of model systems to study adhesion phenomena are discussed.

Section 9.

Ceramics

SURFACE AND INTERFACE ANALYSIS, VOL. 9, 453–462 (1986)

Results of Microanalytical Investigations on SiC-Ceramics and their Impact on Materials Development

F. Thümmler, R. Hamminger, G. Grathwohl
University of Nuclear Research Center Karlsruhe, Postfach 3640, D-7500 Karlsruhe, Federal Republic of Germany

Sintered and hot isostatically pressed SiC-materials are leading candidates in the group of non-oxide ceramics for high performance applications. The physical and mechanical properties are largely controlled by interfacial features such as grain boundaries, grain boundary segregation and the presence of second phases. This concerns the sintering additives, i.e. B and C or Al and C, as well as various impurity elements, incorporated by the original powders and during processing. Detailed microstructural analyses of interfaces are required in order to understand the behaviour of these elements during processing and to promote the development of advanced materials with improved properties. Elucidating results were achieved by Auger electron spectroscopy using high resolution equipment in combination with an ultra high vacuum fracture device. The analyses of interfaces pointed out characteristic differences between the various types of sintered and hot isostatically pressed SiC. Neither B nor impurity elements, including O, could be detected in the grain boundaries of (B,C)-doped materials. In the case of (Al,C)-doping, the grain boundaries exhibit extremely thin segregation layers. Hot isostatically pressed SiC, in contrast, exhibitis grain boundary films enriched with O. Furthermore, a general characteristic of SiC-ceramics was found in the existence of inclusions of free C. The relevance of these findings is discussed with respect to the manufacturing process as well as to important high temperature properties.

INTRODUCTION

The application of ceramics for engineering purposes, e.g. Diesel engines, gas turbines, cutting tools and other wear resistant components is well in progress or under consideration. For some applications, like several types of tools, wear and sealing parts, some mass productions have already been established. Advanced ceramics are here in a strong competition with other more conventional materials, like cemented carbides and metallic alloys. In more difficult applications, considerable progress has also been achieved during recent years and some ceramic components are just being introduced in Diesel car engines by Japanese producers. Although ceramics are generally brittle, a technically acceptable reliability has been achieved for several applications in engineering. Nevertheless, the level of strength and toughness, the behaviour under subcritical crack growth and, generally, the long-term behaviour at high temperatures have to be improved for new and more severe applications of these structural ceramics. This is one of the main tasks for advanced materials development in ceramics.

Today, non-oxide ceramics like SiC and Si_3N_4 are considered as leading candidates for high performance applications at low as well as at high temperatures. The bonding of SiC and Si_3N_4 is mainly covalent, resulting in very high lattice strengths and low diffusivities. Consequently, their sinterability is bad and sintering to high densities can only be achieved by using certain additives. In the case of Si_3N_4 additives like MgO, Y_2O_3 are used which essentially form a silicate liquid phase. In SiC, the additives like B, C, Al provide deoxidizing effects and alter the relevant surface energies; they also may form small amounts of transient liquid phases. Thus, the additives together with the impurities of the raw materials, lead to multi-phase rather than single-phase ceramic compounds, most obvious in the case of Si_3N_4. However, multi-phase microstructures are not only a result of sintering additives, but can be used in some cases for improving mechanical and physical properties.

Among non-oxide ceramics, the SiC-based materials and especially sintered silicon carbide (SSiC) belong to the most interesting high-performance ceramics. Since the pioneer work of Prochazka[1] on SSiC, numerous investigations have been published on its manufacturing and properties. It exhibits excellent oxidation and creep resistance up to 1500°C, nearly temperature-independent strength up to 1300°C and low subcritical crack growth. However, the strength level and the Weibull modulus show only moderate values. Generally, the physical and mechanical properties are influenced by some microstructural features such as presence and nature of second phases and by the state of the grain boundaries. The microstructures of many available SSiC-materials are still heterogeneous, in spite of the efforts to develop a monophase and homogeneous material.

The aim of this paper is to summarize the microstructural and microchemical analytical work done in Karlsruhe, to correlate these results with some properties of the materials and to give comments for the improvement of these properties and the materials processing.

0142–2421/86/120453–10 $05.00

Table 1. Characteristics of various SSiC and HIPSiC

Material	Elements [wt.%]									
	B	Al	N	0	Mg	Ca	Fe	$C_{(tot)}$	$C_{(free)}$	$Si_{(tot)}$
SSiC I	<0.1	0.40	0.07	0.10	<0.01	<0.05	<0.1	29.5	0.5	69.2
SSiC II	<0.1	0.58	0.29	0.07	<0.01	<0.05	<0.1	31.1	0.6	67.0
SSiC III	0.25	0.10	0.03	0.08	<0.01	<0.05	<0.1	30.9	0.7	67.8
SSiC IV	0.34	0.05	0.06	<0.01	<0.01	<0.05	0.05	30.4	0.5	68.6
SSiC V	0.23	0.02	0.01	0.02	<0.01	<0.02	0.07	29.9	0.8	68.9
SSiC VI	0.55	0.02	0.01	0.01	<0.01	<0.01	0.02	30.6	1.1	67.7
HIPSiC I	0.06	0.03	0.01	0.01	<0.01	<0.01	0.26	29.0	0.6	70.0
HIPSiC II	0.05	0.02	0.02	0.02	<0.01	<0.01	0.07	28.9	<0.3	70.7

Material	Density [g/cm^3]	T.D. [%]	Grain size,av. [μm]
SSiC I	3.11	97.1	3.4
SSiC II	3.09	96.5	3.5
SSiC III	3.13	97.7	3.7
SSiC IV	3.15	98.3	5.6
SSiC V	3.10	96.4	4.0
SSiC VI	3.00	93.3	3.7
HIPSiC I	3.20	99.6	2.3
HIPSiC II	3.21	99.9	2.5

MATERIALS

Different high-quality α-SSiC-materials as delivered from international suppliers were investigated as listed in Table 1. Additionally, two qualities of undoped hot isostatically pressed silicon carbide (HIPSiC) were analysed. The concentrations of the intentional sintering additives Al and B were determined to be 0.40 – 0.58 wt% and 0.22 – 0.55 wt%, respectively. While SSiC I and SSiC II are (Al_4C_3C) and (AlN,C)-doped, respectively, SSiC III – VI are (B,C)-doped materials. However, as shown in Table 1, the (Al,C)-doped qualities are not really free of Al and B. The content of impurities is low and at a similar level for all materials. The content of N is relatively high in material II, referring to AlN added as a sintering aid.

Except of SSiC VI all materials have high densities, as shown in Table 1. The grain growth during sintering was controlled; exaggerated grain growth was not detected but some individual grains up to 12 μm were found. The average grain sizes differ by a factor of less than 3 for these materials.

EXPERIMENTAL

In order to get knowledge of the interesting microstructural and microchemical features, several complementary investigation methods have been used. In the case of SSiC containing (Al,C) or (B,C), analytical problems arise from the detectability of the small concentrations of these sintering additives, thus requiring high sensitivities and both high lateral and spatial resolution. Concerning the sintering aids, they are elements of low atomic numbers. The analytical work of this study was performed applying the following methods: (a) Wavelength dispersive analysis of x-rays (WDX); (b) Auger electron spectroscopy (AES) and high resolution AES (HRAES); (c) α-microautoradiography.

WDX, AES and α-microautoradiography were used to identify the inhomogeneities in the form of inclusions and precipitations, whereas high resolution Auger electron spectroscopy (HRAES) was found to be a powerful tool to examine the UHV-exposed grain boundaries. The degree of uniformity of the B-distribution of (B,C)-doped materials was determined by α-microautoradiography. Conventional ceramography together with optical microscopy have been used to reveal the microstructure and microstructural defects; the SiC-polytypes were determined by x-ray diffraction.

RESULTS

Microstructure

In all SSiC-materials typical microstructural defects such as irregular or crack-like pores (Fig. 1 (a) – (d)), zones of variable densification (c), and inclusions (d) have been observed on polished specimens. In the two HIPSiC-materials, however, these microstructural features could not be observed which was mainly due to the small grain sizes, the high purity and the density near the theoretical value.

The (Al,C)-doped SSiC- and both HIPSiC-materials were found to consist mainly of the 4 H-polytype, the (B,C)-doped SSiC-qualities of the 6 H-polytype.

Inhomogeneities

The results give clear evidence of the heterogeneous distribution of the sintering additives Al respectively B and C in both types of SSiC-materials. A WDX-line profile of B is shown in Fig. 2. In these (B,C)-doped

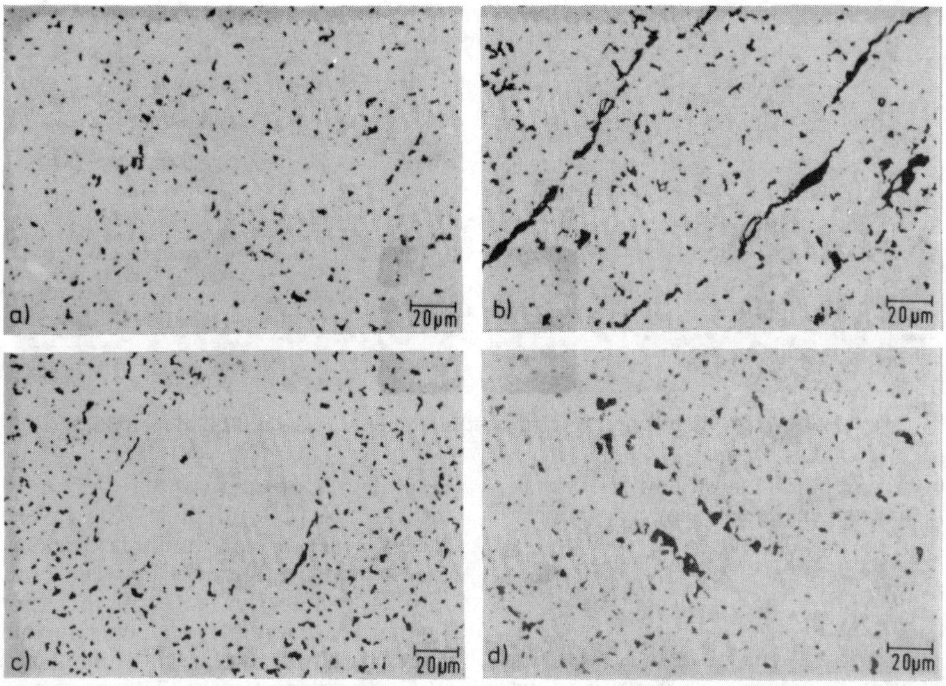

Figure 1. Typical microstructural faults on optical micrographs:(a) pores in SSiC I;(b) orientated pores in SSiC IV;(c) inhomogeneities in SSiC III;(d) inclusions in SSiC V.

materials, no B-enrichment at the grain boundaries could be detected, but higher B-contents in some of the grains. The inhomogeneous B-distribution is demonstrated in Fig. 3, where α-microautoradiograms are shown for two materials.

Considerable amounts of the impurity elements are concentrated within differently sized inclusions which were found in all materials. Many inclusions with very different chemical compositions have been analysed by WDX and AES as well. Some of these inclusions, where extremely high concentrations of O have been measured, did not contain Si and/or C. In Fig. 4 and 5 such high-oxygen containing inclusions, namely SiO_2 and CaO are presented. Figure 4 gives the WDX-line profiles of Si and O together with the absorbed electron imaging revealing the existence of small SiO_2-particles which were analysed only in the (B,C)-doped materials by the application of WDX and AES. CaO (Fig. 5) was found in all materials. This was measured by element distribution maps (WDX) and point analysis (AES). Other O-containing inclusions were found to exist as silicates in the system Si-O-Al, Si-O-K, Si-O-Ca and Si-O-Na.

A general characteristic of SSiC seems to be the existence of particles of free C. The dark inclusions found in all materials were investigated by HRAES-point analyses and x-ray diffraction, they consist of excess or residual graphite, as shown for (B,C)-doped SSiC in Fig. 6 and 7.

Excess C is considered to be a remainder of the production process, which involves the usage of C-additions. Characteristic differences were elaborated with respect to the accompanying elements of these C-particles in the various materials. In one case B (Fig. 8), in another case N was detected in these particles. Quantitative measurements gave further evidence for stoichiometric BN being present. The atomic concentrations were calculated to be 4.9 at % B and 5.3 at % N. Thus, the material contains BN inside its graphitic inclusions.

HIPSiC-materials, on the other side, only rarely revealed microchemical inhomogeneities. Small sized

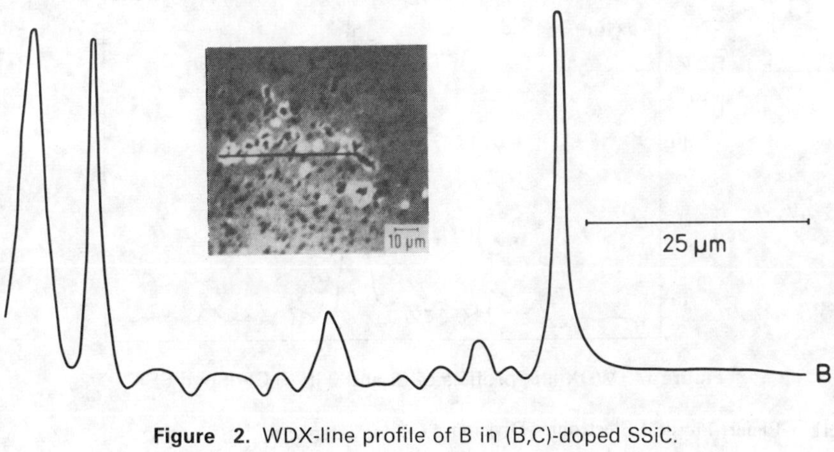

Figure 2. WDX-line profile of B in (B,C)-doped SSiC.

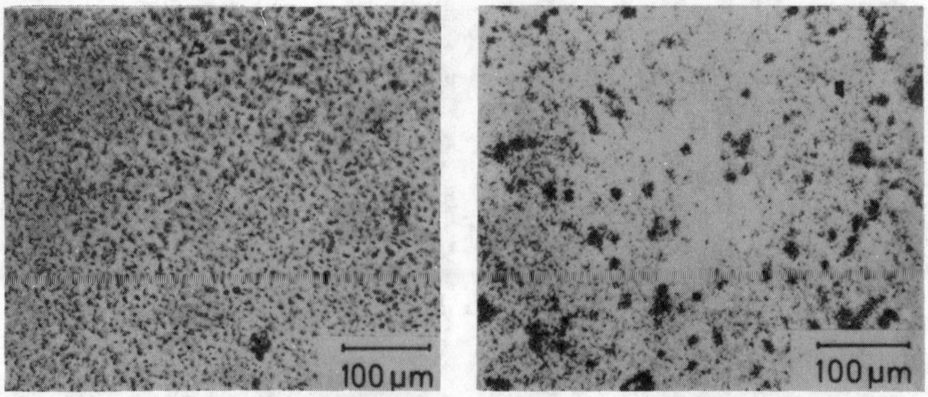

Figure 3. Microautoradiograms of polished specimen surfaces of (B,C)-doped SSiC.

inclusions of free C have been analysed in HIPSiC I, while this type of inhomogeneity was not observed in case of HIPSiC II.

Grain boundaries

The analysis of the grain boundaries was performed by the application of HRAES† in combination with an UHV specimen fracture attachement. The lateral resolution offered here was as high as 50 nm in contrast to the value of conventional AES being much lower. Thus, the main advantage of using HRAES can be seen in its excellent resolution limit which enables illustration and analysis of single grains and grain boundaries, as demonstrated earlier by the authors.[2] The fracture modes of SSiC have been found to be mixed, transgranular and intergranular. Generally, a small degree of intergranular fracture was observed in the case of (B,C)-doping. The investigations of the freshly generated intergranular regions pointed out to characteristic differences between both types of SSiC with respect to the status of their grain boundaries. Neither B nor impurity elements, including O, could be detected along the grain boundaries of different (B,C)-doped materials. These grain boundaries are absolutely free of any intergranular layer within the monolayer-sensitive depth resolution of the method. B could only be measured in the matrices of (B,C)-doped specimens and in enriched spots as mentioned before. Fig. 9 shows the typical secondary electron image of an intergranular fractured region of (B,C)-doped SSiC. The Auger spectrum of this figure was taken by a point analysis from the marked intergranular area. Besides the Si- and C-peaks there is no further element detectable.

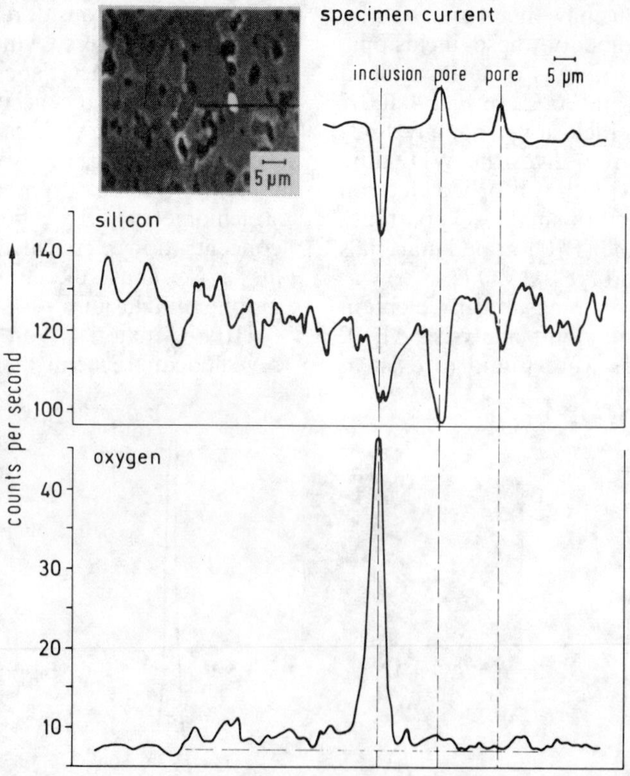

Figure 4. WDX-line profiles of Si and 0 in (B,C)-doped SSiC.

†Equipment PHI 600, Perkin–Elmer, Physical Electronics Division, D–8011 Vaterstetten.

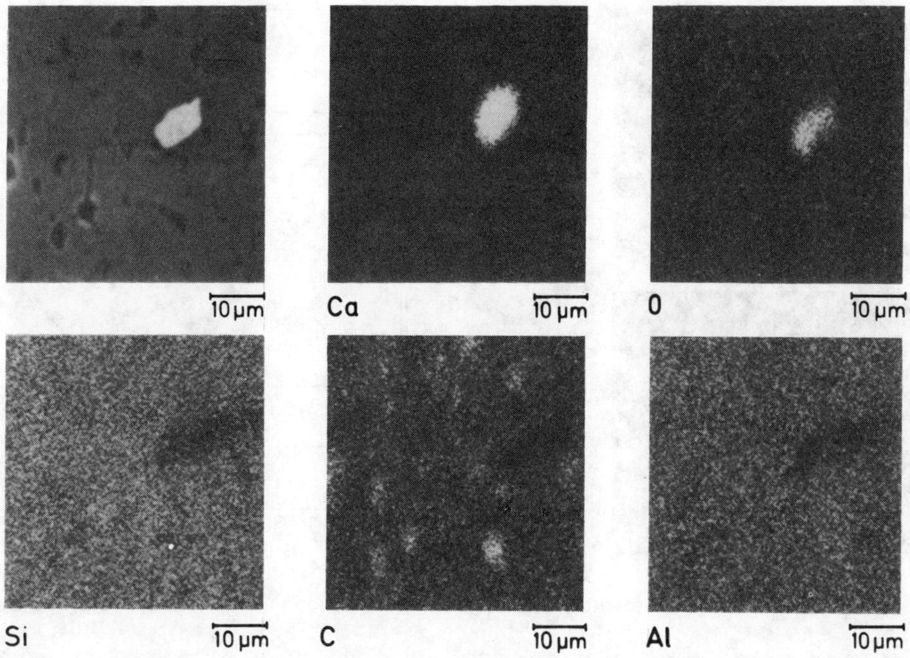

Figure 5. WDX-investigation of an inclusion of CaO in (B,C)-doped SSiC; absorbed electron image and distribution maps.

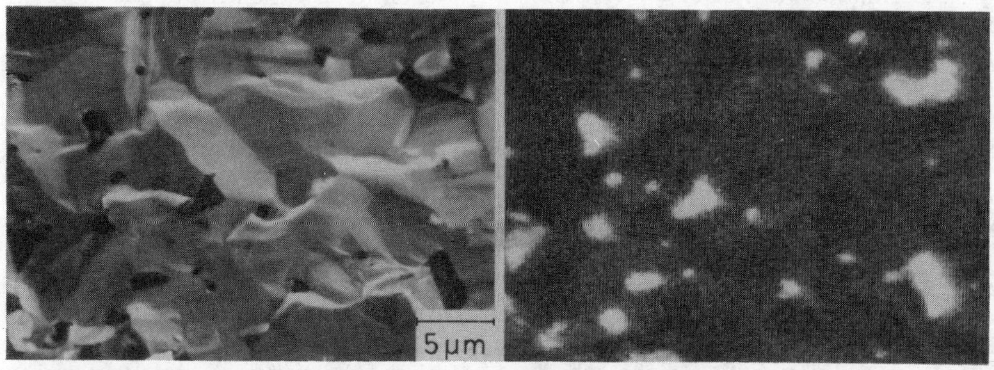

Figure 6. HRAES-distribution map of free C on a fracture surface of (B,C)-doped SSiC.

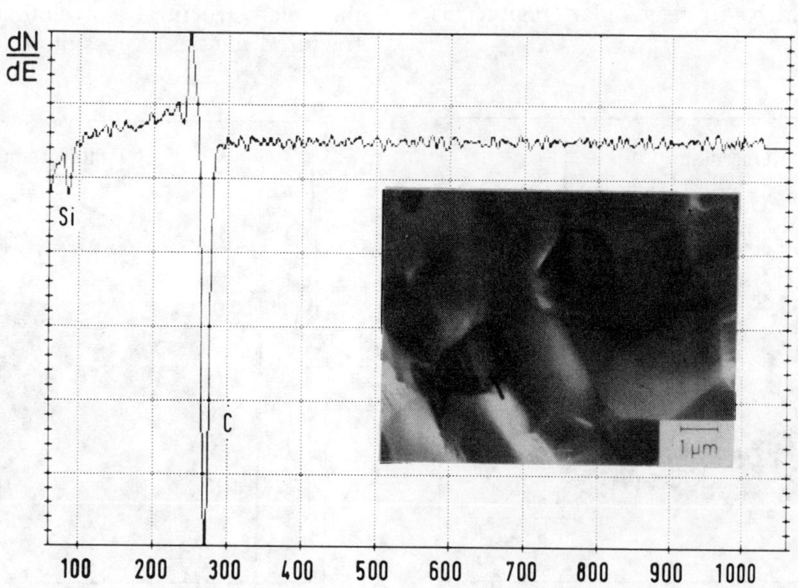

Figure 7. HRAES-point analysis of (B,C)-doped SSiC, indicating pure graphite.

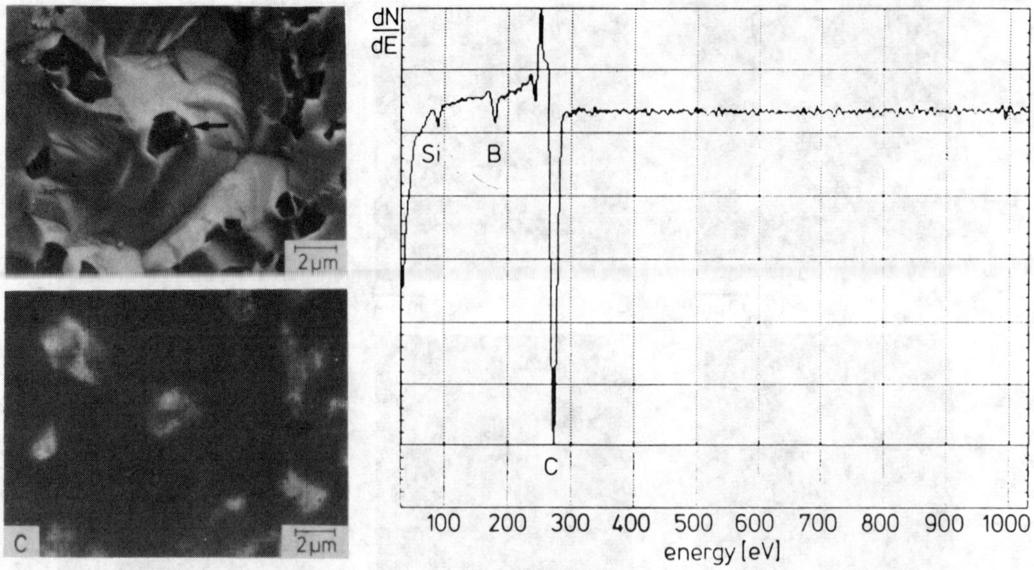

Figure 8. HRAES-point analysis of (B,C)-doped SSiC, indicating enrichments of B beside graphite.

In contrast to the results received from the (B,C)-doped materials, the grain boundaries of the (Al,C)-doped materials were found to be enriched with Al. These enrichments of the sintering additive Al are generally accompanied by B. The total B-content of some 100 ppm was also measured by chemical analysis. The grain boundary analysis of (Al_4C_3,C)-doped SSiC I is shown in Fig. 10. The spectrum was taken from the intergranular region at point 1 of the appropriate secondary electron image while point 2 marks a transgranular region.

Slight sputtering treatments gave evidence about the extremely small spatial extension of the indicated grain boundary film. This is shown in Fig. 11 by expanded spectra before and after sputtering by Ar-ions, which removed a film of about 1 nm. The Al-, B-depressed low energy Si-peak increases and the stoichiometric C/Si-ratio of the matrix (transgranular region) is re-installed.

In addition to the Al- and B-content the grain boundaries of (AlN,C)-doped SSiC II contain also some N which is referable to the sintering additive AlN. The grain boundary analysis of this material is presented in Fig. 12, where the enrichment of Al, B and N is manifested. The expanded spectra of Fig. 13 before and after sputtering of about 1 nm gives the evidence for the extremely small thickness of the enriched grain boundary film.

Similar results as those received from the (Al,C)-doped SSiC were found in case of HIPSiC. The enrichment of Al and B at the grain boundaries of HIPSiC I (containing excess C) is revealed in Fig. 14, the total contents of some hundred ppm have been analysed chemically. Additional enrichments of O have been measured at the grain boundaries of HIPSiC II (without excess C). The distribution analysis of O and the microstructural correlation with intergranularly fractured regions is possible in spite of a high level of

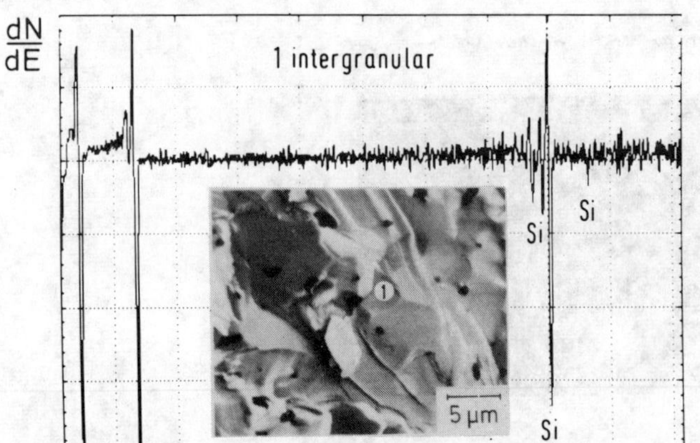

Figure 9. Grain boundary spectrum of (B,C)-doped SSiC.

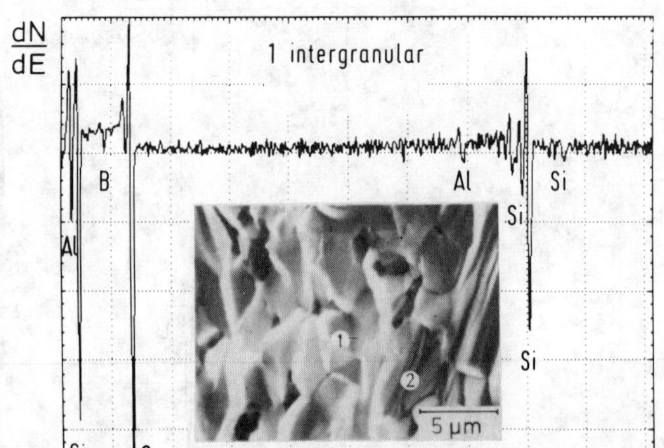

Figure 10. Grain boundary spectrum of (Al_4C_3,C)-doped SSiC.

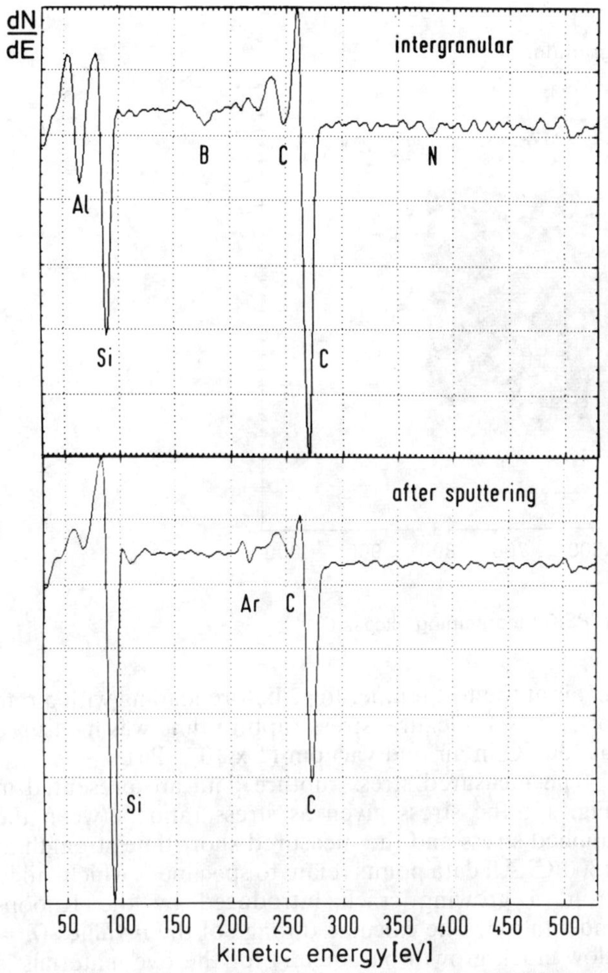

Figure 11. Expanded grain boundary spectra (Al_4C_3,C)-doped SSiC.

noise and the very small grain sizes. Both the grain boundary spectrum and the distribution of O are shown in Fig. 15.

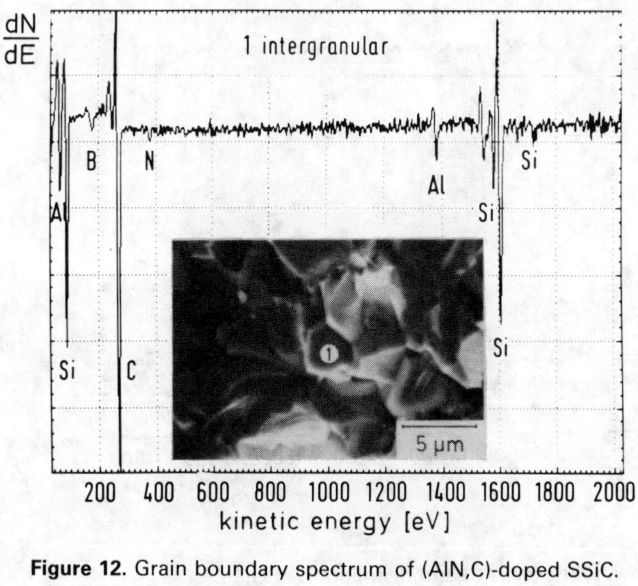

Figure 12. Grain boundary spectrum of (AlN,C)-doped SSiC.

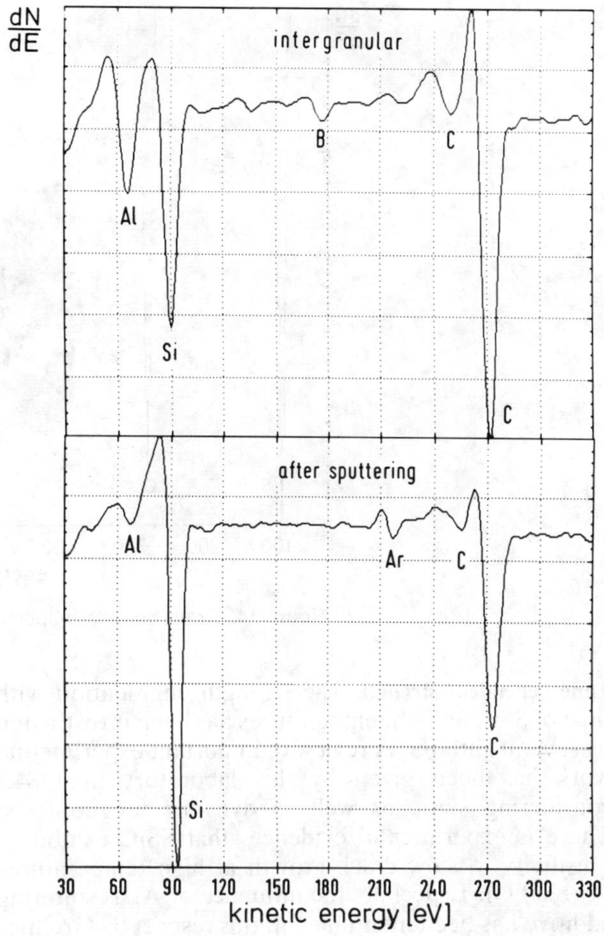

Figure 13. Expanded grain boundary spectra of (AlN,C)-doped SSiC.

High temperature properties

Creep The creep behaviour was investigated using specimens of 45 × 4.5 × 3.5 mm in a four-point bending mode with 20 mm inner span. Isothermal tests as well as load changing and temperature changing experiments were performed up to 1600°C.

α-SSiC is generally a very creep resistant material. However, remarkable differences were found in the creep rates of the materials with different dopings. As shown in Fig. 16 the (Al,C)-doped materials generally tend to exhibit higher steady state creep rates compared with the (B,C)-doped materials. It became further evident from stress changing experiments (Fig. 16) that the stress sensitivity of the creep rates can be described under these conditions by a linear relationship. This result can be used for the interpretation of the controlling creep mechanism as stress induced diffusion processes: in the (Al,C)-doped materials grain boundary diffusion processes are suggested as creep controlling mechanism, while in the (B,C)-doped materials volume diffusion processes become more important.[3] The creep behaviour of HIPSiC has not been investigated in detail. However, as for β-SSiC, it seems to be controlled by more complicated mechanisms, with an extended transient creep range as described in ref. 4.

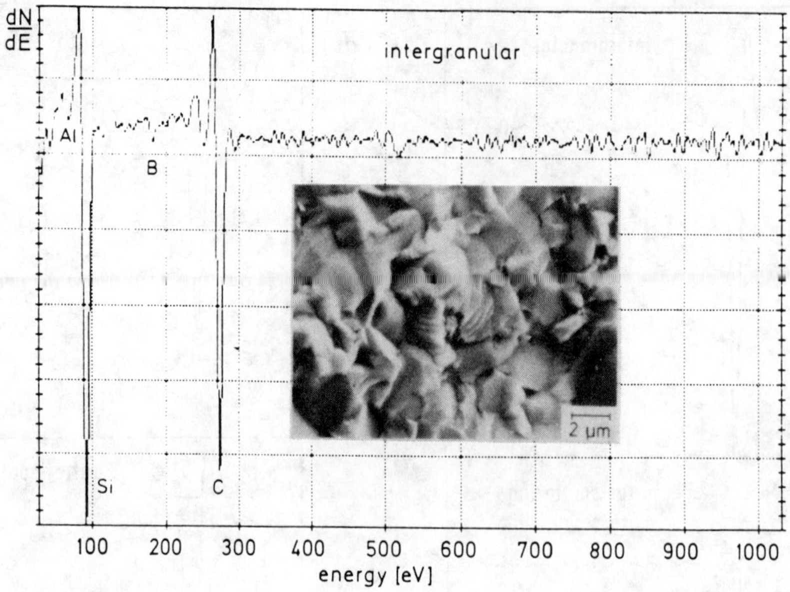

Figure 14. Grain boundary spectrum of HIPSiC I (containing excess C).

Time dependent strength The strength degradation with time of SSiC at high temperatures has been investigated by several authors. A review of important experimental work has been given by this laboratory in 1984,[5] considering static as well as dynamic fatigue tests. There is experimental evidence, that SSiC exhibits a sensitivity to slow crack growth at high temperatures (≥ 1200°C). Especially the influence of Al as sintering additive has been mentioned in this respect by Grellner Schwetz and Lipp.[6] The state of the grain boundaries (e.g. their strength) should exert a decisive influence since slow crack growth normally occurs intergranularly.

The materials investigated in this work were machined to bending bars (see creep specimens), preflawed by Knoop indentation (50 N) and heated 2 h in air at the test temperature before loading with a rate of 1.25 N/s. Flexural stress rupture time was measured at 1300°C in air and vacuum (1×10^{-3} Pa).

The measured stress rupture data are presented in Fig. 17, the stress given as stress ratio between the applied stress and the measured short time strength at 1300°C. All data points relate to specimens which failed from a growing crack introduced by the Knoop-indentation. The calculated range of the ***n***-values (***n*** = slow crack growth parameter) for the five materials is from ***n*** = 64 to ***n*** = 23. In vacuum, SSiC showed a high resistance against slow crack growth. The room temperature strength of 'vacuum-survivors' was considerably higher than the strength of the preflawed specimens and increased with increasing load they had sustained.

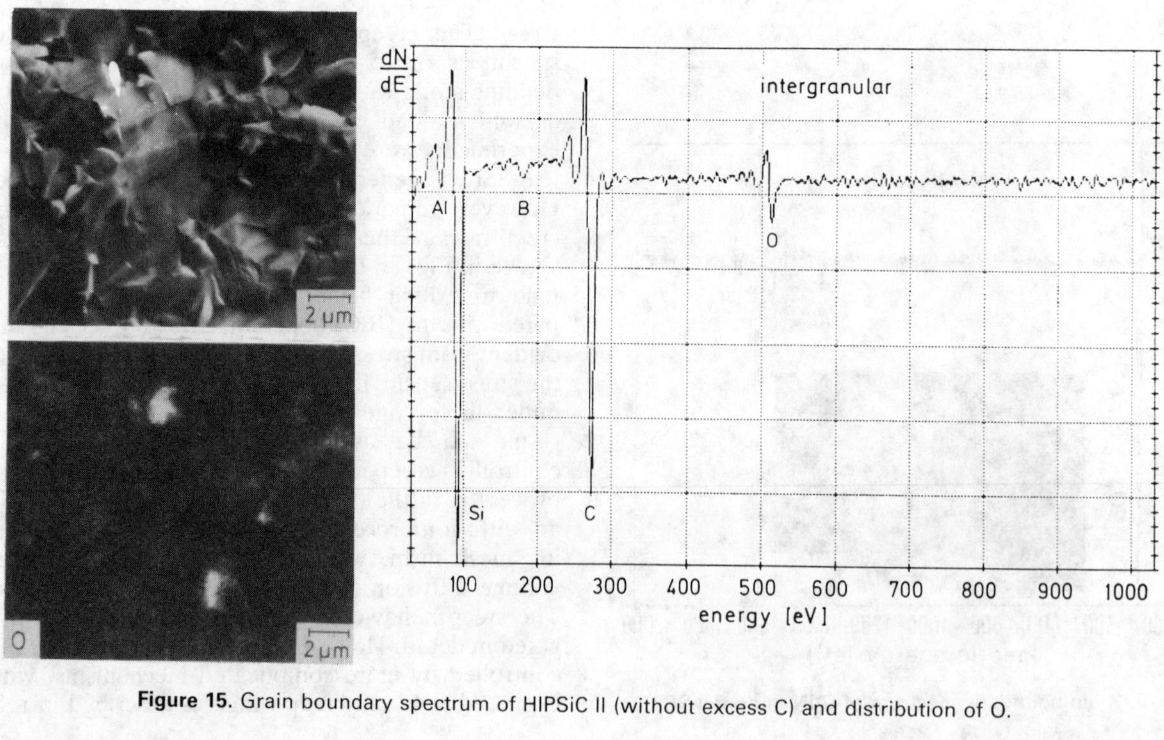

Figure 15. Grain boundary spectrum of HIPSiC II (without excess C) and distribution of O.

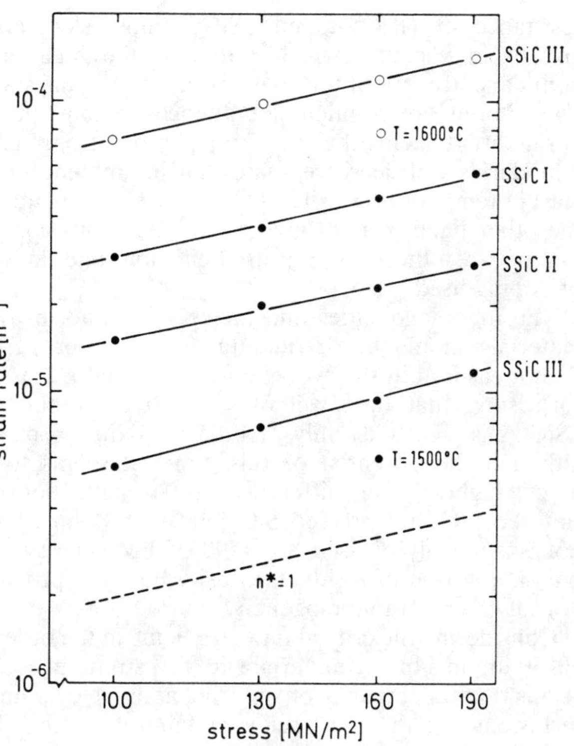

Figure 16. Log strain rate – log stress curves of SSiC in vacuum.

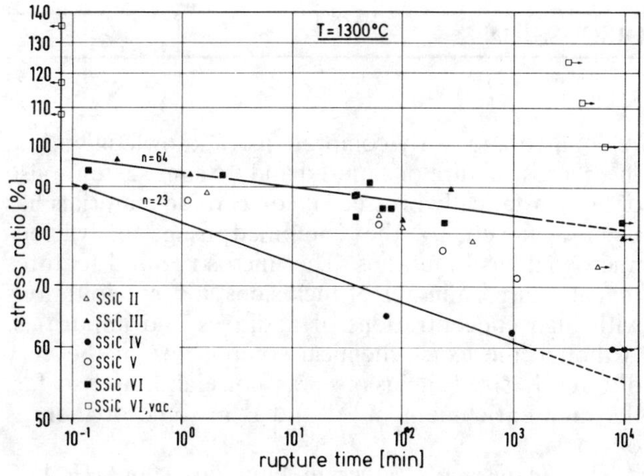

Figure 17. Results of SSiC-rupture tests in air and in vacuum. Fitting lines through extreme slope values of SSiC III (n = 64) and SSiC IV (n = 23).

The SEM-micrographs in Fig. 18 illustrate the fracture surfaces of (Al,C)-doped SSiC II and (B,C)-doped SSiC VI. The semicircular Knoop-flaw in Fig. 18(a) is surrounded by a ring-like region of slow crack growth. A clear distinction between this region and the fracture surface of the catastrophic failure can be made for the (B,C)-doped material. As can be seen from Fig. 18(b), the fracture mode in the region of slow crack growth is intergranular, while the fast fracture occurs mainly transgranular (Fig. 18(c)). In contrast, the (Al-C)-doped SSiC does not allow to identify the transition between the slow and the fast crack growth regions, because the fracture mode is intergranular in both regions. This is illustrated by the fracture surface of SSiC II in Fig. 18(d).

Other properties

The four-point bend strengths of the materials investigated are for the SSiC-qualities 400-450 MNm^{-2} and for HIPSiC 500-550 MNm^{-2}. The oxidation rates of all SSiC-materials investigated at 1300°C and 1500°C are extremely small, as described in.[7] Substantial differences between (B,C)- and (Al,C)-doped materials were not detected. The oxidation behaviour of the HIPSiC-materials has not been investigated so far.

Figure 18. SEM-micrographs of (B,C)-doped SSiC VI (a)–(c) and (Al,C)-doped SSiC II (d).

DISCUSSION

A detailed and fairly comprehensive knowledge about the microstructure of sintered and, to some extent, also of hot isostatically pressed silicon carbide materials has been achieved by the combined usage of various microanalytical methods. The microstructural features reveal a large number of inclusions and precipitations with high concentrations of additives and impurities. Although the exact chemical composition of the very differently sized inclusions was not elaborated so far, the high enrichment of Al and B in some of them is obvious.

The analysis of the grain boundaries by HRAES offering a lateral resolution of 50 nm in combination with a UHV fracture device provides an excellent tool for investigating uncontaminated fracture surfaces including grain boundaries and single grains. The difference in the grain boundary state of the (B,C)- and (Al,C)-doped qualities is evident: the grain boundaries of the former are free of B or any other foreign element in the sintered state. The origin of O at the grain boundaries of a HIPSiC-material in the case, when no free C is present in the microstructure, is understood as a consequence of the insufficient deoxidation of the powder surface during sintering. However, the occurrence of B and Al at the grain boundaries of HIPSiC cannot be explained as clearly at the moment.

With respect to creep, it seems that the highest creep strength can be reached in the material with extremely pure grain boundaries, i.e. (B,C)-doped SSiC, in which a classic diffusion creep process can be active. Pure grain boundaries in a covalently bonded ceramic seem to be generally favourable with respect to high temperature long-term properties. Materials, whose grain boundaries are impure, i.e. (Al,C)-doped SSiC, behave in a more complex manner and show a somewhat lower creep strength. Although small differences in grain size, porosity and content of free C are neglected in this interpretation, the conclusion seems to be obvious.

A similar statement is possible with respect to the strength degradation effects. An important fact at high temperatures is the different fracture mode of both types of materials: The (B,C)-doped SSiC shows intergranular fracture in the region of slow crack growth and mainly transgranular fracture during catastrophic failure, while (Al,C)-doped SSiC is fracturing intergranular in both regions. This again is a consequence of the different grain boundary states in both types of materials. Another point of interest is the very high resistance of (B,C)-doped SSiC against slow crack growth in vacuum, which refers to the high grain boundary strength of this material in its 'original state', i.e. without environmental influences during the experiment. It seems that this property of 'strong' grain boundaries will decrease somewhat in ambient air, at least at temperatures ≥ 1200°C. These experiments, on the other hand, cannot give an unambiguous correlation between the stress-rupture behaviour and the type of doping used.

The inhomogeneities and inclusions found in these materials should be detrimental for other properties than measured in this work, e.g. for strength. It is not surprising, that the level of strength of commercial SSiC has reached only 400–450 MNm^{-2} so far, although the potential of this material seems to be much higher. The difference in strength between sintered SiC, hot pressed SiC (600–650 MNm^{-2}) and hot isostatically pressed SiC (500–550 MNm^{-2}) is not only a question of residual porosity but very probably also of chemical inhomogeneity.

Considering the critical flaw size being in the order of 30–50 μm in a brittle material with this strength level, it seems that the impurity or sintering additive containing inclusions are the most relevant failure sources. The free C inclusions are generally smaller; also the grain size of these materials seems not to be a strength limiting factor at this impurity level.

Finally, these investigations demonstrate that the manufacturing process of SSiC and HIPSiC has not been fully optimized so far. The inhomogeneities lead to the question whether the amounts of the sintering additives are chosen and admixed properly with respect to their optimal effectivity. Sintering additives, which are concentrated in certain spots of the microstructure of the sintered body, could not have had their highest efficiency during sintering. Although the microstructural inhomogeneity detected in the sintered samples may not be identical with the state in the green compacts before sintering, the presence of inhomogeneities also in the green compacts is obvious. Consequently, the processing should be improved, mainly in the different steps of the powder technology. These are e.g. purer starting powders, lower content of O or improved deoxidation, improved homogeneity and adjusted state and amount of additives. Microstructural and interfacial analysis, elaborated in this paper, are qualified tools for supervision of materials development and can help to promote advanced ceramics with improved properties.

REFERENCES

1. S. Prochazka, *Ceramics for High Performance Applications*, 239, ed. by J. J. Burke, Brook Hill, Chestnut Hill (1974).
2. R. Hamminger, G. Grathwohl, F. Thümmler, *J. Mat. Sci.* **19** 3154 (1983).
3. G. Grathwohl, Th Reetz, F. Thümmler, *Sci. of Ceramics* **11** 425 (1981).
4. H. Cohrt, F. Thümmler, *Int. Journ. High Technol. Ceremics* (accepted).
5. G. Grathwohl, R. Hamminger, H. Iwanek, F. Thümmler, *Sci. of Ceramics* **12** 583. (1983).
6. W. Grellner, K. A. Schwetz, A. Lipp, *Proc. Brit. Ceram. Soc.* **31** 27 (1981).
7. F. Thümmler, G. Grathwohl, R. Hamminger, *Ceramic Components for Engine*, 548, ed. by S. Sōmiya, E. Kanai and K. Ando, KTK, Tokyo (1984).

SURFACE AND INTERFACE ANALYSIS, VOL. 9, 463–466 (1986)

Combined TEM and AES Study of CVD TiC and Al_2O_3/TiC Coatings on Cemented Carbides

E. Minni
Turku Regional Institute of Occupational Health, Hämeenkatu 10, SF–20500 Turku, Finland, and Department of Physical Sciences, University of Turku, SF–20500 Turku, Finland

S. Vuorinen
Helsinki University of Technology, Laboratory of Physical Metallurgy, Vuorimiehentie 2 A, SF–02150 Espoo, Finland

The diffusion processes occuring during chemical vapour deposition of single coating TiC and multicoating Al_2O_3/TiC on cemented carbides were studied. Diffusion of cobalt in addition to tungsten diffusion was found. Cobalt was present on the outer surface of both coatings and at the Al_2O_3/TiC interface. The anomalously high concentration of cobalt at Al_2O_3/TiC interface cannot be fully explained by diffusion, a probable reason is chemical vapour transport.

INTRODUCTION

Vapour deposited coatings produced by chemical vapour deposition (CVD) are successfully used to reduce friction and wear. One of the most important applications of the CVD technique is a CVD coated cemented carbide cutting insert. The first industrial process using this technique started in the beginning of the 1970's. A 3–6 μm thick TiC coating on cemented carbide cutting tools produced a substantial increase in cutting performance.

The multicoating is a later development. For example, an extra layer of aluminum oxide (Al_2O_3) can be applied on previously TiC coated cemented carbide tools. Al_2O_3 is not as hard as TiC, but is chemically inert and hence increases the resistance against diffusion wear. In the Al_2O_3/TiC double coating, the diffusion wear resistance of Al_2O_3 is thus combined with the abrasion wear resistance of TiC. The Al_2O_3/TiC double coating, where deposition and compatibility problems between alumina and cemented carbide substrate have been avoided by using the intermediate TiC layer, shows considerably better performance in high speed machining than the straight TiC coating.

The industrial deposition processes for TiC and Al_2O_3 are well established. The formation of these coatings can be described by the following overall reactions[1,2]

$$TiCl_4(g) + CH_4(g) \rightarrow TiC(s) + 4HCl(g) \quad (1)$$

$$2AlCl_3(g) + 3CO_2(g) + 3H_2(g) \rightarrow Al_2O_3(s) + 3CO(g) + 6HCl(g) \quad (2)$$

Cemented carbides are composite materials combining the hardness of hexagonal tungsten carbide (WC) and the ductility of the cobalt binder phase. Cubic transition metal carbides (TaC, TiC, NbC, etc), referred to as γ-carbides, are often added in order to improve wear resistance. During CVD deposition the cemented carbide substrate, described above, is decarburized. This results in the formation of a brittle carbide of tungsten and cobalt, usually referred to as η-carbide, between the coating and the substrate. Since the η-carbide formation can be limited by improved coating and precarburization techniques to small interfacial islands, the TiC cemented carbide interface can be described as a three dimensional combination of seven sub-interfaces, namely the interfaces of WC–TiC, Co–TiC, γ–TiC, η–TiC, η–WC, η–Co and η–γ.[3] Direct contacts thus occur between TiC and the substrate constituents (WC, Co-binder, γ-carbide) and between TiC and η-carbide. Consequently, the heterogeneous structure of the cemented carbide substrate is clearly reflected in the structure of the interfacial TiC.[3]

The microstructure of a CVD coating, formed by a heterogeneous nucleation and growth process, is important for the performance of the coating. In previous studies[3] the single TiC coating has been described as being composed of two regions with different grain morphologies: a fine grained region, referred to as interfacial carbide layer, next to the cemented carbide substrate and an upper large grained region. The interfacial carbide layer extends to a distance of 1.5–2μm from the TiC/substrate interface and is composed of fine (grain size 0.1μm), equi-axed, randomly-oriented TiC grains. The uppermost part of the TiC coating is distinguished from the interfacial region by virtue of grain size only, which is larger by an order of magnitude in the former region. This upper layer is typically 2–4 μm thick with a grain size reaching 1 μm at the top of the coating layer. While the CVD coating can be characterized to be almost dislocation free, the CVD Al_2O_3 layer shows a fairly high dislocation density.[4]

In this paper diffusion processes occuring during CVD of TiC and Al_2O_3/TiC have been studied using scanning transmission electron microscopy with an energy dispersive x-ray spectrometer and Auger electron spectroscopy. Especially the behaviour of cobalt during CVD will be dealt with.

0142–2421/86/120463–04 $05.00

EXPERIMENTAL

Commercially available TiC and Al_2O_3/TiC coatings (Sandvik, Coromant GC 1025 and GC 015) have been studied. AES measurements were carried out using VG ESCA3 MkII electron spectrometer equipped with a scanning Auger facility for elemental mapping. The instrument was calibrated using the XPS-lines Au $4f_{7/2}$ and Cu $2p_{3/2}$. The Auger-spectra were recorded in the derivative mode with a plotter. A large area sputtering gun (VG AG2) was used for cleaning and depth profiling (DP) of the specimens. The base pressure of the instrument was 10^{-8} Pa.

By combining AES and sputtering it is possible to study the changes in concentrations in the direction perpendicular to the surface. The sputtering process increases the roughness of the surface. The results obtained using too large sputtering doses are therefore unreliable. Consequently, the depth profiling curves were measured only for outer layers, and chemical analysis of bulk coatings were performed using a scanning transmission electron microscope (JEOL 200 CX) with an energy dispersive x-ray spectrometer (STEM/EDX). Specimens for TEM were first mechanically thinned to obtain wafers approximately 50 μm thick. Electron transparent specimens were then obtained by ion-beam milling. Ion-beam milling was predominantly carried out from the polished wafer surface opposite to that covered by the coating, using argon ions at a voltage of 5 kV and a current of 30–40 μA.

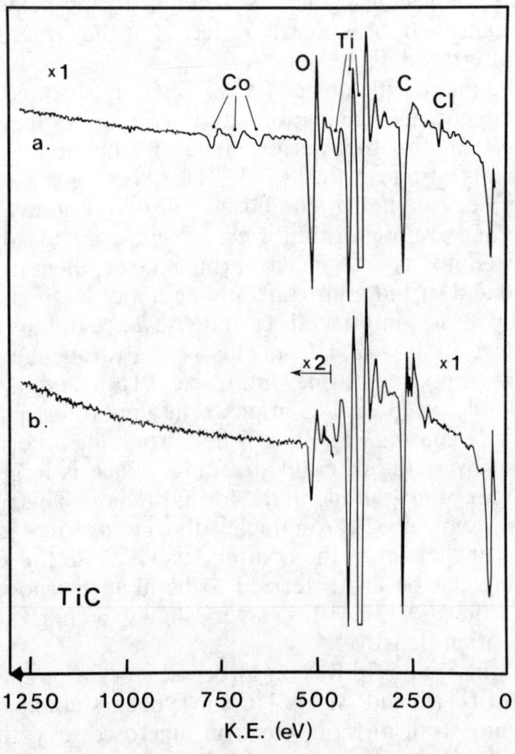

Figure 1. Derivative Auger-spectra of TiC coating after different sputtering times (5 kV, 20 μA) (a) 10 min; (b) 190 min.

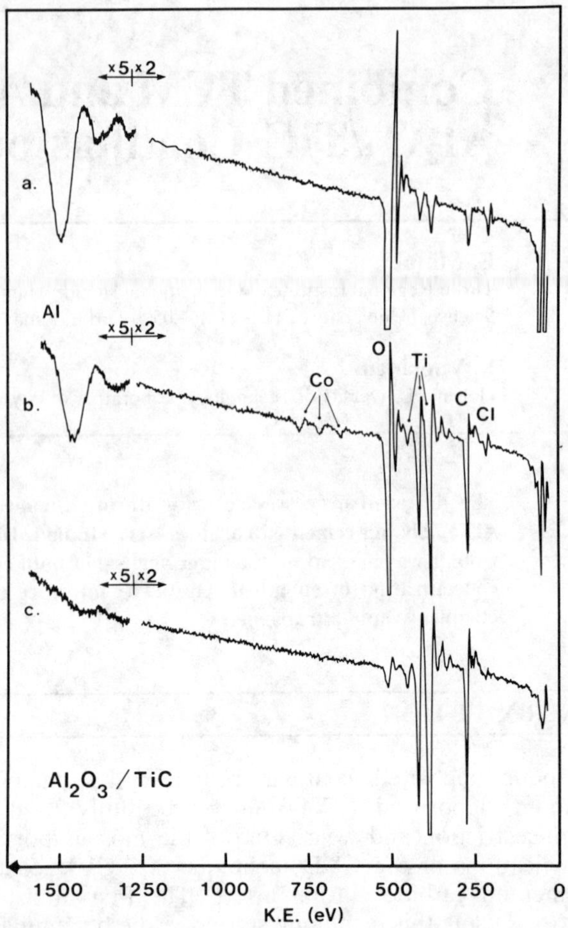

Figure 2. Derivative Auger-spectra of Al_2O_3/TiC coating after different sputtering times (5 kV, 20 μA) (a) 90 min; (b) 190 min; (c) 330 min.

RESULTS AND DISCUSSION

An Auger-spectrum from the single TiC coating after cleaning of the surface is presented in Fig. 1(a). The elements found are Co, O, Ti, C and Cl. During sputtering the cobalt signal gradually disappears and the oxygen signal decreases with progressive sputtering. A typical spectrum is presented in Fig. 1(b).

An Auger-spectrum measured from the Al_2O_3/TiC double coating is presented in Fig. 2(a). The elements found were Al, O, Ti, C and Cl. Before sputtering, a weak signal of cobalt was also found on the surface of this specimen. This signal disappeared during the first light sputtering, indicating that the Co-layer was very thin. After extended sputtering, the signal of cobalt reappeared (Fig. 2(b)), and the maximum of the cobalt content was obtained at the Al_2O_3/TiC interface. After the interface the Co-content decreased (Fig. 2(c)). The depth profiling curve is presented in Fig. 3.

The STEM/EDX studies of bulk coatings confirmed diffusion of tungsten and cobalt along TiC/TiC grain boundaries into the fine grained interfacial carbide layers in both the single TiC and intermediate TiC (Al_2O_3/TiC) layers (Fig. 4). Diffusion of tungsten and cobalt was found to be clearly limited in the interfacial part of the coatings and consequently in the outer,

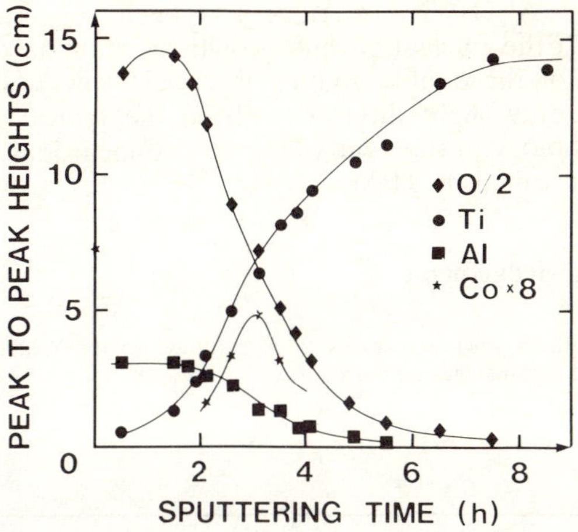

Figure 3. AES depth profiling curves of Al_2O_3/TiC coating (5 kV, 20 μA).

larger-grained (Fig. 5) TiC coating layer the tungsten and cobalt signals were found to have disappeared.[3]

TEM studies of Al_2O_3/TiC interface reveal (Fig. 6) that cobalt, the existance of which was confirmed by AES, does not occur as a continuous film at this interface, but is rather found as inclusions at the Al_2O_3/TiC interface. TEM studies also show that the grain size of Al_2O_3 in the immediate vicinity of cobalt inclusions is considerably smaller (0.1 μm) than the average grain size of the order of 1.5 μm reported for the Al_2O_3 layer.[4]

Existence of cobalt on the coating surfaces as well as at the Al_2O_3/TiC interface could be due to grain boundary diffusion in the coating layers, surface diffusion along the cracks developed in the coatings, or vapour phase transport.[5,6] Cobalt, transported through the intermediate TiC layer to the Al_2O_3/TiC interface, has been suggested to be responsible for disturbances found in the nucleation and growth of Al_2O_3.[7] The DP curves (Fig. 3) also suggest that oxidation of the intermediate TiC can be avoided during CVD of Al_2O_3.[8,9]

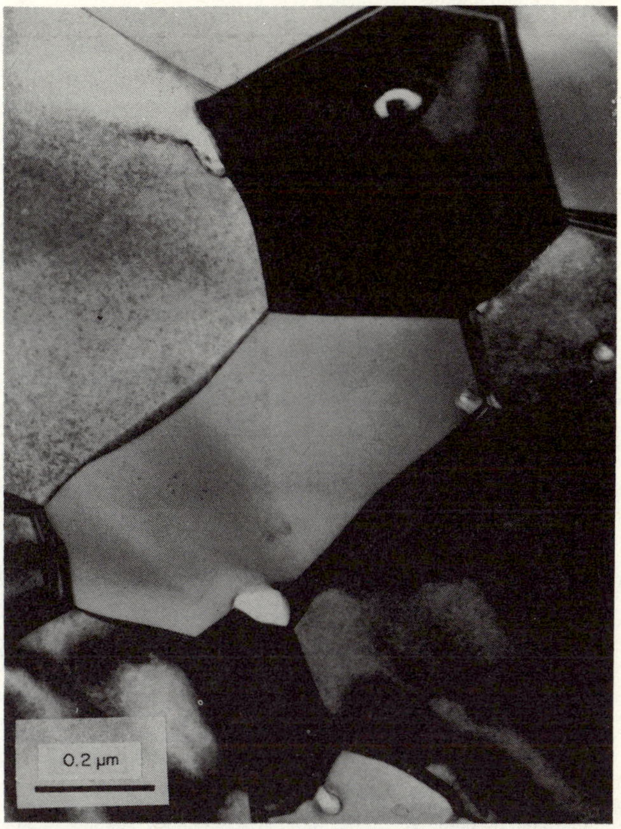

Figure 5. Structure of the intermediate TiC layer near the Al_2O_3/TiC interface.

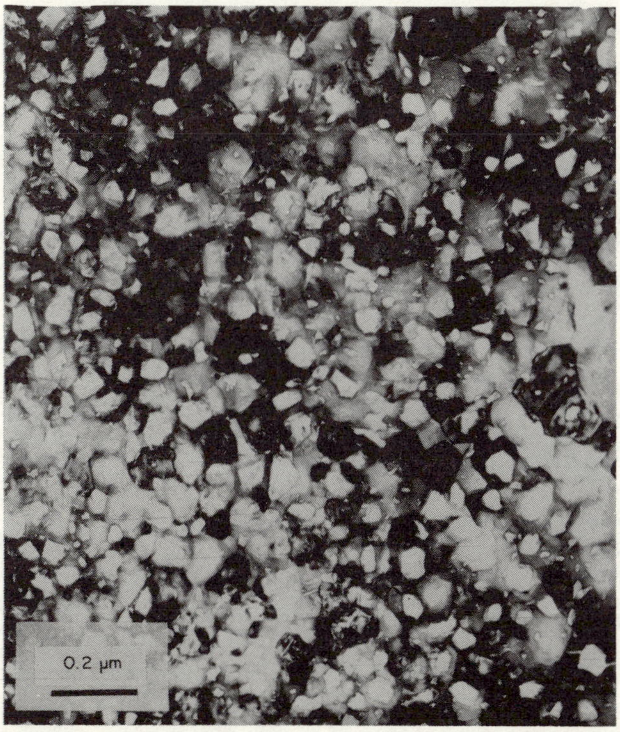

Figure 4. Structure of the interfacial TiC layer near the cemented carbide substrate.

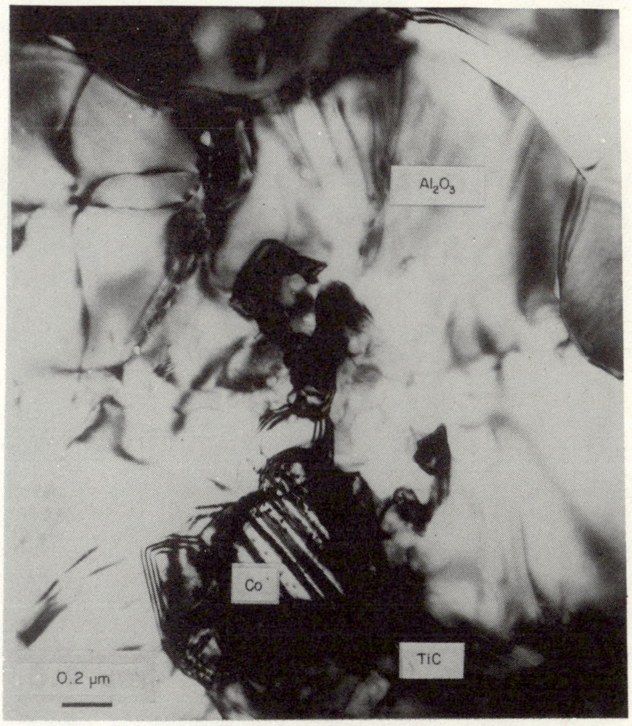

Figure 6. Cobalt inclusion found in Al_2O_3 at the Al_2O_3/TiC interface.

CONCLUSIONS

The relatively thick TiC layer cannot prevent transport of cobalt through it. Although the existence of cobalt in the interfacial carbide layer is obviously mainly due to grain boundary diffusion, the anomalously high concentration of cobalt at the Al_2O_3/TiC interface cannot be fully explained by diffusion processes. The possible explanation is that cobalt is transported through the TiC coating via chemical vapour transport. Cobalt transported through the intermediate TiC layer to the Al_2O_3/TiC interface has been suggested to disturb the nucleation and growth of alumina during CVD of the double coating layer.[7] However, in this study only slight differences from the typical Al_2O_3 morphology in the near vicinity of Co-inclusions could be confirmed by TEM.

Acknowledgements

The authors wish to express their gratitude to the Academy of Finland for financial support.

REFERENCES

1. H. E. Hintermann, *VDI-Berichte* **333**, 53 (1979).
2. M. Kornmann, H. Schachner, R. Funk and B. Lux, *J. Crystal Growth* **28**, 259 (1975).
3. S. Vuorinen and A. Horsewell, *J. Mat. Sci* **17**, 589 (1982).
4. S. Vuorinen, in *Proc. 9th International Conference on Chemical Vapour Deposition (CVD)*, ed. by McD. Robinson, G. W. Cullen, C. H. J. van den Brekel, J. M. Blocher, Jr. and P. Rai-Choudhury, pp. 719–717, The Electrochemical Society, Pennington (1984).
5. J.-O. Carlsson and M. Boman, in *Proc 9th International Conference on Chemical Vapour Deposition (CVD)*, ed. by McD. Robinson, G. W. Cullen, C. H. J. van den Brekel, J. M. Blocher Jr. and P. Rai-Choudhury, pp. 129–149, The Electrochemical Society, Pennington (1984).
6. L. Kjellsson, Thesis, Chalmers University of Technology, Göteborg (1982).
7. H. Altenna, Ch. Colombier, A. Lebl, J. Lindström and B. Lux, in *Proc 4th European Conference on CVD*, ed. by J. Bloem, G. Verspui and L. R. Wolff, pp. 428–443, Philips Centre for Manufacturing Technology (1983).
8. J. N. Lindström M. Collin and A. Thelin, in *Proc 9th International Conference on Chemical Vapour Deposition (CVD)*, ed. by McD. Robinson, G. W. Cullen, C. H. J. van den Brekel, J. M. Blocher, Jr. and P. Rai-Choudhury, pp. 689–708, The Electrochemical Society, Pennington (1984).
9. J. N. Lindström and H. Schachner, in *Proc 3rd European Conference on Chemical Vapour Deposition*, ed. by H. E. Hintermann, pp. 208–217. Laborataire Suisse de Recherches Horlogères, Neuchâtel (1980).

SURFACE AND INTERFACE ANALYSIS, VOL. 9, 467–471 (1986).

Study of the Thermal Behaviour of $^{56}Fe^+$ Implanted Yttria Stabilized Zirconia

D.Scholten and A.J. Burggraaf
Twente University of Technology, Department of Chemical Engineering, Laboratory of Inorganic Chemistry and Materials Science, PO Box 217, 7500 AE Enschede, the Netherlands

By ^{56}Fe implantations (15 to 110 keV) in the oxygen ion conducting material, 0.86 ZrO_2–0.14 $YO_{1.5}$ (ZY_{14}) concentration profiles were obtained with maxima exceeding the percolation limit for electronic conduction. In this way a mixed conductive layer in the surface of the oxidic electrolyte has been prepared. As this material finds application under elevated temperatures (range 450–1000°C), the thermal stability of the prepared surface layer was investigated. Profile analyses were obtained using RBS and AES, which methods give complementary information. The introduced distributions remain unchanged at temperatures up to 600°C for at least 100 h and up to 800°C for at least 24 h, although the profile maxima exceed the thermodynamically stable Fe concentration limits. At 900°C the 15 keV, 2 × 10^{16} Fe cm^{-2} profile shows some Fe diffusion after 1 h treatment, while for the 110 keV 8 × 10^{16} Fe cm^{-6} profile with the same maximum concentration (≈12 × 10^{21} Fe cm^{-3}) no rearrangement is observed yet. This is ascribed to a higher damage concentration and/or concentration gradient for the low energy implantation. At 1000°C all profiles change rapidly and at 1550°C the implanted iron dissolves completely in the ZY_{14} matrix. The Fe mobility behaviour of the implanted material is compared with the behaviour of a chemically prepared (metastable) ZY_{14} with 10 mol% Fe_2O_3. In this material measurable Fe mobility is found at temperatures exceeding 1000°C.

INTRODUCTION

Zirconium dioxide can be stabilized in the fluorite structure by addition of different cations. This oxide doped with cations of lower valency than Zr^{4+} (e.g. Y^{3+}), is an oxygen–anion conductor and an electronic insulator. This material is used as a solid electrolyte in sensors and other devices[1,2] and for its outstanding mechanical properties in numerous other applications. For electrical applications, however, also electronic conductivity and/or catalytic activity in a surface layer are needed in order to achieve electrode properties. This can be introduced by doping with lanthanide or transition elements or metallic elements. Iron has been chosen because it is reasonably soluable in the stabilized oxide and while iron oxides and solid solutions of ZY containing 6 mol% Fe_2O_3 or more show electronic conductivity[3] and interesting properties as oxidation catalysts. To obtain electronic conductivity, relatively high concentrations exceeding the percolation limit for conductivity in the surface are needed.

Ion implantation is used as a technique for modification of the surface. With this technique dopant surface concentrations can be obtained beyond the limits of thermodynamical stability and metastable compounds or solid solutions can be investigated. Different profile shapes can be introduced with surface concentrations exceeding the percolation limit for electronic conduction.

The stability of a dopant profile has to be examined as well as the annealing of damage, introduced during the implantation process. Of specific interest is the Fe profile stability in the working temperature range of sensors and electrodes made of this material. From literature some information on the thermal behaviour of implanted oxygen conductive materials of this kind is known.

Cohen[4] found a full crystalline recovery after low dose implantation of Yb in YSZ (1.2 × 10^{16} Yb cm^{-2}, 150–210 keV). Temperatures of 1000°C for 48 h were used and no change in the Yb implanted profile was detected. For high dose implants damage remains. Only after 1300°C anneal some broadening of the Yb profile and recovery of crystallinity occurred. Schnell[5] worked with thin layers of ZrO_2 implanted (60 and 90 keV) with Ca (up to 22 mol% CaO) followed by air annealing. High electronic conductivity was found in first measurement runs up to 200°C. This was induced by defects 'introduced by ion implantation' according to the author. Temperature treatments in the range of 200–300°C caused oxidation of the metallic excess. After 830°C anneal the ionic conductivity in the Ca stabilized zirconia layer was somewhat higher than the conductivity of chemically prepared CSZ. As is shown by the above mentioned authors as well as the authors[6] of this article, implantation permits one to obtain a large range of dopant/Zr ratio's in the implanted layer and thus to study the effect of chemical composition on conductivity even in metastable solid solutions. However, knowledge of the thermal behaviour of the implanted layer is important. For the investigation of dopant mobility good surface analysing techniques are indispensable. These techniques therefore are an important part of the presented paper.

0142–2421/86/120467–05 $05.00

EXPERIMENTAL

The substrate material with compositon 86 mol% ZrO_2–14 mol% $YO_{1.5}$ [ZY_{14}] was prepared following the procedure as described in Ref. 7. The grain size of the sintered material was kept small (≈ 20 μm) relative to beam diameters of the analysing techniques. In this way the crystal orientation of the individual grains does not influence the reproducibility of the measurements.

The metastable standard material ZY_{14} containing 10 mol% Fe_2O_3 was obtained in accordance with the method described in Ref. 8, and using sintering temperatures of 1450°C in air for 10 h followed by quenching to keep the Fe dissolved in the ZY lattice. After slowly cooling down to room temperatures (in 8 h) exsolution takes place. The standard material having a Fe–Zr composition exceeding the percolation limit for electronic conduction shows electrical conductivity. For the $^{56}Fe^+$ implantations a 120 keV ion implantation machine has been used with $FeCl_2$ as feed material in the hot cathode ion source. Beam current densities were kept constant at about 1 μA cm^{-2} to avoid strong heating and charging up of the substrate. The composition of the implanted layer was determined by Rutherford backscattering spectroscopy (RBS) with a Van de Graaff generator producing a 2 MeV ^4He beam. The detection angle was 170°. Auger electron spectroscopy (AES) has been used as well. The AES results have been made (semi-)quantitative using results of measurements on the standard material of known composition. Fe distribution measurements were achieved by AES in combination with depth profiling with an Ar^+ beam of 2 keV, 36 μm cm^{-2} at about normal incidence.

RBS is a good and quick technique for the analyses of the implanted profiles giving direct information on dopent concentration in depth taking system resolution in account (Mayer and Chu Ref. 9 pp. 120, 140). However, no high depth resolution is achieved but as this technique does not involve sputtering no complications arise due to compositional changes during the measurements. If the currents are kept low the sample heating does not exceed 300°C. This was verified using measurements on ZY_{14} containing freshly implanted material. As is shown further on, this material changes colour after treatment at about 300°C. No such effects have been found near the ^4He beam spot.

For measurements of the (Zr+Y)/Fe ratio with higher resolution and for accurate investigations of the first 50 Å below the surface, AES has been used in combination with depth profiling. Good similarity of the results from both techniques has been found.[10] The absolute Fe_2O_3 contents in the standard material was verified with XRF. The Fe contents in the ZY lattice was calculated from XRD data. All heat treatments were carried out in air with rapid heating up and quenching after given duration of the thermal treatment.

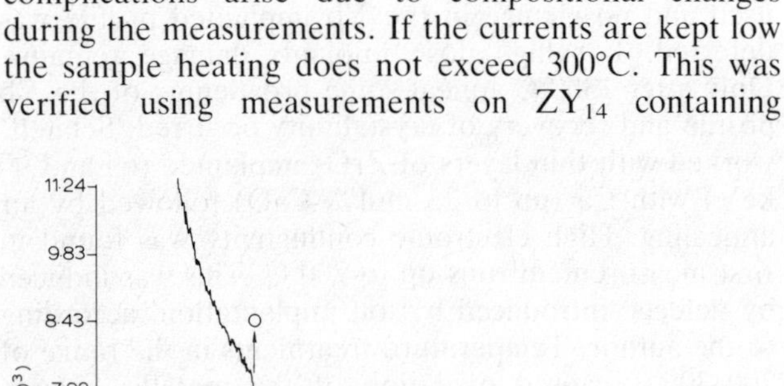

Figure 1. 2 MeV ^{4}He RBS spectrum of ZY14. Indicated are the energy positions for the different surface elements present.

RESULTS AND DISCUSSION

Analysing techniques

A typical 2 MeV ^4He RBS spectrum of ZY_{14} has been given in Fig. 1. For the interpretation of this spectrum the method and values for kinematic factors, stopping cross sections and scattering cross sections as described by Mayer and Chu (Ref. 9) have been used. In the figure energy positions for backscattering after collisions at the surface with O, Zr–Y and Hf have been marked. The Hf is an impurity, substitional for Zr and amounts about 1.5 wt%. The mass difference of Zr and Y is too small to separate these elements. The way the concentration profiles have been calculated from the RBS and AES diagrams will be described below. ^4He backscattered from Zr ions situated below the surface have lost more energy due to stopping in the ZY matrix. The deeper the collision the lower is the energy of the outcoming He particle. The 'shoulder' in the figure represents the Zr and Y cations in the matrix, showing the normal increasing yield with higher depth as caused by the energy loss of the ^4He. ^4He backscattered after collision with Fe at the surface, results in an energy decrease of the ^4He from 2 MeV to 1.5 MeV. So the implanted iron appears in the spectrum on top of the Zr/Y 'shoulder' (Fig. 2). The electronic energy loss of the ^4He in the ZY matrix and after Zr collision is about 106 eV $Å^{-1}$ (using $\Delta E_{ZY} = \Delta E_{Zr} + \Delta E_Y + \Delta E_O$; ΔE_x is the energy loss due to element X in the ZY matrix calculated separately for the incoming and outgoing particle in the same way as given in Ref. 9). This energy loss in case of a He collision with Fe ion amounts 109 eV $Å^{-1}$ ZY matrix. With this value and the measured energy position of the Fe peak, a peak depth can be calculated. For 3.3×10^{16} Fe cm^{-2} implanted with 110 keV the peakdepth amounts about 40 nm (Fig. 2). This is in agreement with the expected value of 40.3 nm as calculated for Fe implantation in ZY_{14} using computer program DIMUS (Ref. 12). In the spectrum of Fig. 2 (insert) a step in the Zr-Y signal at the edge is found. This is caused by an extra electronic stopping effect in the ZY material due to the incorporation of Fe of maximal 8% (for 110 kV, 3.3×10^{16} Fe cm^{-2}). This means that about 8% less matrix material is measured per channel (1 channel ≈ 2.6 keV) and the yield (Zr–Y) decreases with about 8% as well. The Fe concentration is quantified by using the

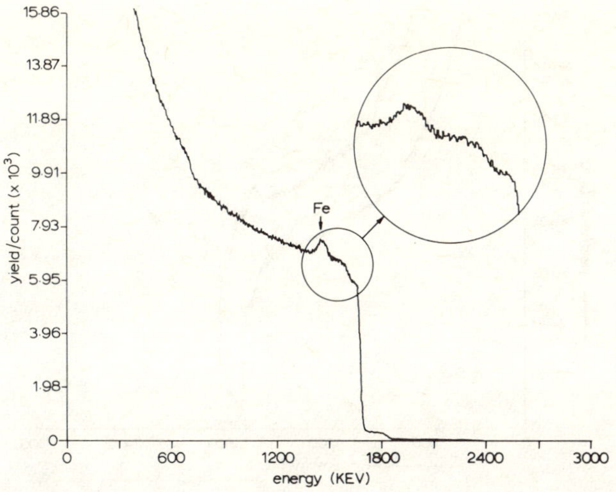

Figure 2. 2 MeV ^4He RBS spectrum of ZY14 containing implanted Fe. The arrow in the insert indicates the step in the Zr–Y shoulder.

yield at the Fe peak relative to the yield at the Zr–Y step, corrected for difference in scattering cross section. The Zr–Y is used as the internal standard representing a concentration of 2.8×10^{22} Zr+Y(cm^{-3}ZY). The Fe top concentration amounts 6×10^{21} Fe cm^{-3} ZY (in the case of Fig. 2).

An Auger spectrum of the surface of Fe implanted ZY_{14} is shown in Fig. 3. A high carbon peak is found at 230 eV. Y and Zr are measured at 120 and 140 eV respectively. Fe appears in the spectrum at three positions ≈615, 670 and 685 eV. After Ar bombardment in the Auger chamber for 30 sec. the carbon disappears. The high surface contents has to be ascribed to surface pollution as caused by the vacuum in the target chamber of the implantation machine. Cleaning at elevated temperature of the surface before measuring is impossible as this migh have an influence on the implanted material which is part of the investigation. For this reason analyses of the surface composition have not been used in measurements of the implanted profile. The Auger depth profiling analyses of the layers under the surface have been made semiquantitative using sensitivity factors for Zr, Y and Fe as calculated from analyses on chemically prepared standard material ZY/Fe_{10}. No preferential cation sputtering has been found. Also no radiation enhanced diffusion during depth profiling of the ZY_{14} + 10 mol% Fe_2O_3 (ZYFe10) standard material was experienced by comparison surface concentrations and bulk concentrations after sputtering. As this material has a metastable Fe concentration as for the implanted material, we assumed no strong radiation enhanced diffusion to occur. Absolute values of Fe (cm^{-3}ZY) result by assuming the Zr–Y concentration as being constant 2.8×10^{22} at cm^{-3}ZY. In this way iron concentrations were calculated from the Auger spectra after each Ar sputter interval, resulting in a concentration profile as shown for a 3.3×10^{16} Fe cm^{-2}, 110 keV in Fig. 4. The maximum in the distribution amounts 5×10^{21} Fe cm^{-3}. These values agree well with a value of 6×10^{21} as found by RBS analyses. The sputtertime abcissa in this figure can be converted into a depth scale by comparison with the peakdepth as measured with RBS. Twenty minutes Ar bombardment under a vertical incident beam of 30 μA cm^{-2} will be equivalent to about 400 Å, which means a sputter rate of 20 Å per minute. This same sputter rate was found also in a different way. A rf sputtered thin layer of ZY_{14} of 1100 Å on Al_2O_3 as measured with talysurf and XRF, sputters off during Auger depth profiling in 54 mins. Finally with SIMS depth profiling the sputter rate was found to be of the same order of magnitude[6] as expected.

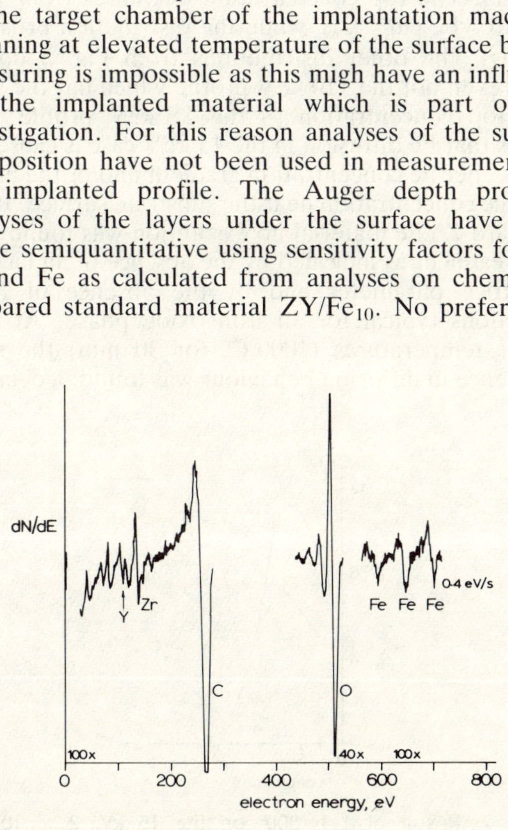

Figure 3. Auger profile of the surface of Fe implanted ZY_{14}.

The composition of chemically prepared standard material ZYFe was measured by XRF. From XRD data the lattice parameter of this cubic standard material can be calculated. Because the ionic radius of Fe^{3+} is smaller than that of Zr^{4+}, the parameter of the host lattice decreases linearly with increasing substitutional Fe content in the lattice as described in Ref. 8. For the chemically prepared standard $ZYFe_{10}$ the lattice parameter was 5.100 Å, indicating about 10.5 mol% Fe_2O_3.

Stability of implanted Fe profiles

The Fe implantations with 15, 55 and 110 keV and doses in the range from 2 to 8 Fe cm^{-2} ZY resulted in Auger profiles as shown in Fig. 5. The distributions were calculated from Auger data and were similar to what was found with RBS. The maxima of the 2×10^{16} Fe cm^{-2}, 15 keV and of the 8×10^{16} Fe cm^{-2}, 110 keV profiles are comparable ≈12×10^{21} Fe cm $^{-3}$) and about twice as high as the 2.76×10^{16} Fe cm^{-2}, 55 keV and the 3.3×10^{16} Fe cm^{-2}, 110 keV profiles.

In this way a comparison of the temperature behaviour of profiles with the same maxima but different damage concentration and different

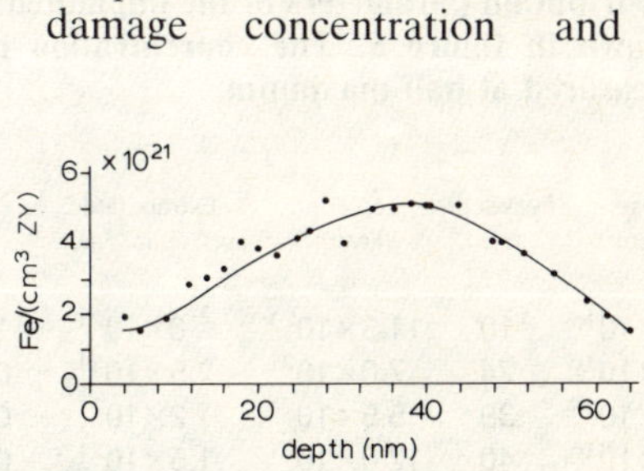

Figure 4. Fe distribution after implantation of 3.3×10^{16} Fe cm^{-2}, 110 kV, as measured with AES depth profiling.

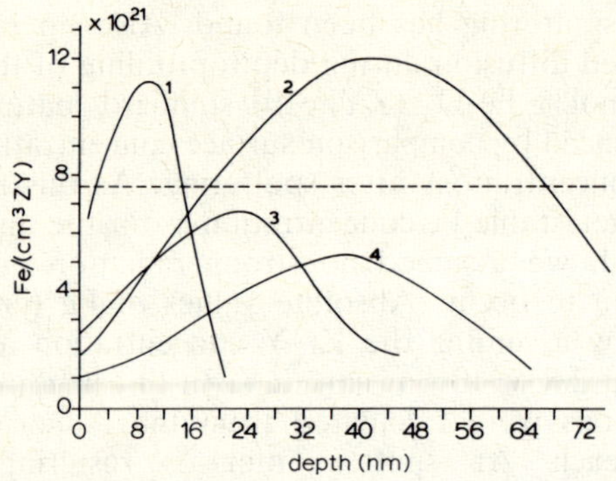

Figure 5. Fe distributions in ZY_{14} as implanted (AES): 1: 2×10^{16} Fe cm^{-2}, 15 kV; 2: 8×10^{16} Fe cm^{-2}, 110 kV; 3: 2.76×10^{16} Fe cm^{-2}, 55 kV; 4: 3.3×10^{16} Fe cm^{-2}, 110 kV;---- : standard ZY14 + 10 mol% Fe_2O_3.

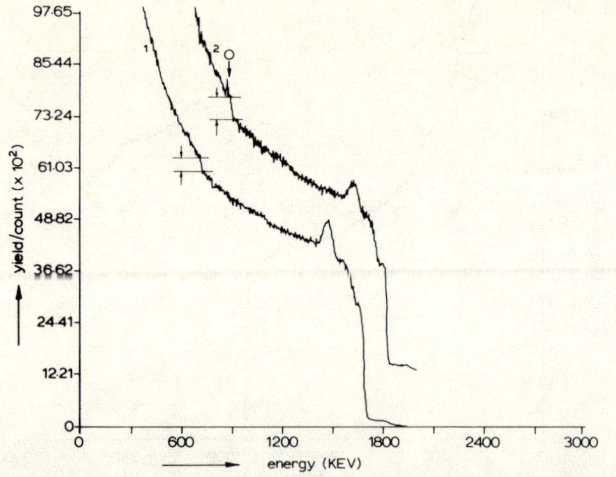

Figure 6. (a) RBS spectrum of 8×10^{16} Fe cm^{-2}, 110 kV compared with the spectrum after annealing at 600°C for 100 h. Surface enrichment in oxygen is shown by the increase in stepheight at the O edge in curve 2. The spectra are given in the same figure for comparison convenience. The curves have the same energy axes but are shifted with respect to each other arbitrarily to show them separately. For specification of the edges in both curves see Fig. 1.

concentration gradients can be made. The iron contents in the standard amounts to about 8×10^{21} Fe cm^{-3} and lies in the range of the implanted iron concentrations. The major part of the concentration profiles discussed above exceeds the value for which electronic conductivity was found ($\approx 4 \times 10^{21}$ Fe cm^{-3}) in chemically prepared material. Distribution peak depths, top concentrations, surface concentrations as well as gradients of the distribution slope directed away from the surface are given separately in Table 1.

A change of colour of the implanted material from silvery to orange red is observed after all temperature treatments exceeding 300°C. This rapid change (in a few seconds) does not involve any rearrangement of the profile. It represents a change in oxidation state of the implanted material and not annealing of radiation damage, because after annealing >300°C of argon implanted material a change of colour from purple to plain white (which is the colour of the unimplanted material) was observed. We found that this effect corresponds with a rapid change in surface resistance[10] in analogy with phenomena reported by Perez[11] for Fe implanted MgO and by J.P. Schnell[5] for Ca implanted ZrO_2 layers.

The high and low concentration profiles as described (Fig. 5) remain stable at temperatures of 600°C up to 100 h and 800°C for 24 h. Only some increase of the oxygen concentration was detected by comparison of RBS spectra of fresh and annealed material (e.g. Fig. 6). Comparison with literature[4] indicates that nonstoichiometry of the oxygen lattice induced by ion implantation is removed at temperatures higher than 500°C. Also the metastable 10 mol% Fe_2O_3 standard material did not show any exsolution of second (iron oxide) phase as measured with XRD. The lattice parameter remains 5.100 Å and obviously the diffusion coefficient of iron is too low for rearrangement across larger distances.

At a temperature of 900°C for 1 h for the 2×10^{16} Fe cm^{-2}, 15 KeV, a change in the distribution is detected (Fig. 7). The top concentration decreases from 11 to 7×10^{21} Fe cm^{-3} ZY while the distribution broadens slightly. The other distributions from Fig. 5 do not alter, even not the 110 keV profile which has the same high top concentration as the 15 keV profile. This means that Fe diffusion in the 15 keV case is caused by the steeper Fe concentration gradient and/or the higher damage concentration near the substrate surface. In the standard ZYFe material no exsolution was found after this treatment as indicated by the absence of an increase in lattice parameter and by the absence of XRD reflections typical for an iron oxide phase. At even higher temperatures (1000°C, for 30 min) the same difference in diffusion behaviour was found: a decrease

Table 1. **Distribution parameters of the implanted profiles as shown in figure 5. The concentration gradient is measured at half maximum.**

Acc. volt (keV)	Dose (Fe/cm^2)	Peakdepth (nm)	Top conc. (Fe/cm^3)	Extrapolated surf. conc. Fe/cm^3	gradient in Fe/cm^4
15	2×10^{16}	10	11.3×10^{21}	6×10^{21}	1.25×10^{28}
55	2.76×10^{16}	24	7.0×10^{21}	2.5×10^{21}	0.27×10^{28}
110	3.3×10^{16}	39	5.5×10^{21}	2×10^{21}	0.13×10^{28}
110	8×10^{16}	40	12.0×10^{21}	1.5×10^{21}	0.29×10^{28}

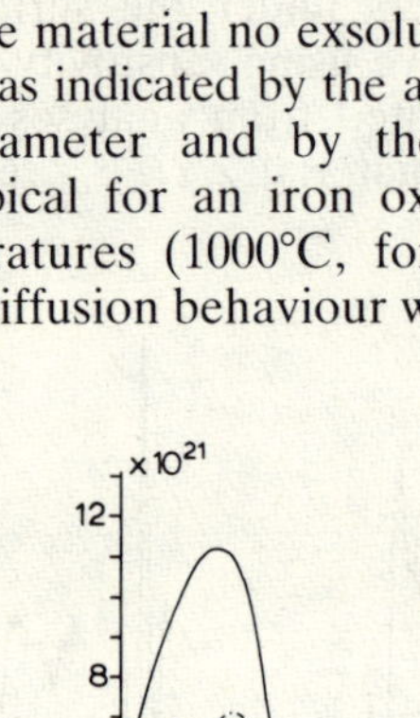

Figure 7. Result of 1 h 900° on the 15 kV, 2×10^{16} Fe cm^2; fresh; after annealing (-·-).

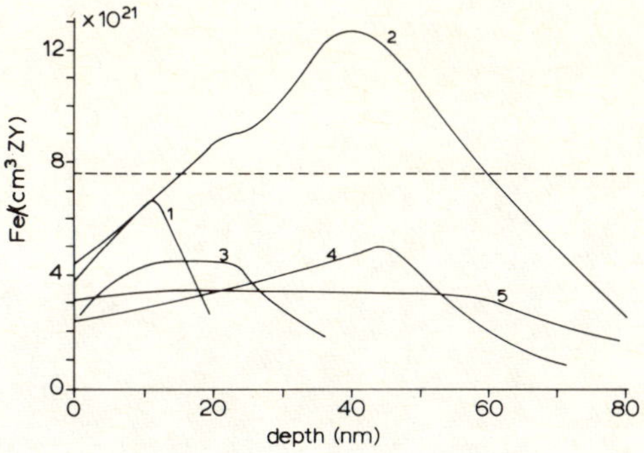

Figure 8. Influence of temperature treatment at 1000°C (compare with Figure 5):

1: 2×10^{16} Fe cm^{-2}, 15 kV, 30 min; 2: 8×10^{16} Fe cm^{-2}, 110 kV, 30 min; 3: 2.76×10^{16} Fe cm^{-2}, 55 kV, 1 h; 4: 3.3×10^{16} Fe cm^{-2}, 110 kV, 1h; 5: 2×10^{16} Fe cm^{-2}, 15 kV, 1 h; ----- : standard 10 mol % Fe_2O_3.

in top concentration of the 15 keV and no changes for the other profiles.

However, after prolonged treatments at 1000°C, diffusion was found for all distributions with concentrations exceeding about 4×10^{21} Fe cm^{-3}. The low doses all show a decrease of top concentration towards this value (Fig. 8) except for the 8×10^{16} Fe cm^{-2}, 110 keV which decreases in top concentration after 1 h to 8×10^{21} Fe cm^{-3}.

No iron mobility (e.g. precipitation) has been observed in the metastable standard material at 1000°C (even after treatments up to 20 h). This means that Fe mobility is only found in implanted material and is caused by the (steep) concentration gradients and/or the presence of radiation induced defects. After treatment at a temperature of 1450°C all implanted Fe is assumed to be dissolved in the ZY lattice substitutionally (XRD) in concentrations of about 1 mol% in less than 30 min, because no precipitation was found.

CONCLUSIONS

– Auger depth profiling in combination with RBS makes calculation of the concentration profile of implanted Fe ions possible. On ZY_{14} the argon sputter rate used is about 20 Å/min.
– No radiation enhanced diffusion for metastable Fe concentrations up to about 75×10^{21} Fe/cm^3 is found.
– Profile stability for maximum concentrations up to 12×10^{21} Fe/cm^3 was found after treatment at 600°C for at least 100 hours and 800°C for at least 24 hours. An increase in oxygen concentration compared with implanted specimens was found (RBS)
– A difference in stability at 900°C for 1 hour was found for profiles with high top concentration (12×10^{21} Fe/cm^3) implanted with 15 and 110 keV respectively. This is caused by a difference in concentration gradient and/or damage concentration.
– At temperatures of 1000°C for more than 1 hour a measurable Fe mobility was found for all profiles implanted with 15–110 keV and doses of $2–8 \times 10^{16}$.
– The mobility of implanted iron (110 keV) at concentrations up to 12×10^{21} Fe/cm^3 is comparable with the behaviour at temperatures up to 900°C of a chemically prepared specimen ZY (metastable) containing 7.5×10^{21} Fe/cm^3.
– At temperatures of 1450°C full dissolution of the implanted Fe in the ZY matrix is achieved.

Acknowledgements

The authors would like to thank Dr. E.H. du Marchie van Voorthuysen and Mr. J.J. Smit of the State University of Groningen for the ion implanter facilities, Drs. E.W.A. Young of the State University of Utrecht for the R.B.S. facilities. Philips Elcoma Eindhoven is thanked for the Auger facilities. The investigations were supported by the Netherlands Foundation for Chemical Research (S.O.N.) with financial aid from Z.W.O.

REFERENCES

1. J. Olmer, J.C. Vigue, E.J. Schouler, *Solid State Ionics,* **7,** 23 (1982).
2. M. Stoukides, C.G. Vayenas, *J. Catal.* **70,** 137 (1981).
3. A.J.A. Winnubst, D. Scholten, A.J. Burggraaf, PhD Thesis Twente Univ. of Techn. Enschede, The Netherlands, (1987).
4. C. Cohen, J. Siejka, M. Bertie, A.V. Drigo, H. Croset, M.M. Tosic, Radiat. Eff. **64,** 221 (1982).
5. J.P. Schnell, M. Croset, D. Dieumegard, G. Velasco, *Nucl. Instr. and Meth.* **209/210,** 1187 (1983).
6. D. Scholten, A.J. Burggraaf, *Proceedings of the Conference on Radiation Effects in Insulators 3,* Guildford (UK), (July 1985), special issue of Radiat. Effects.
7. M.A.C.G. van de Graaf, A.J. Burggraaf, in *Advances in Ceramics,* **12** ed. by N. Claussen, M. Ruhle, and A.H. Heuer, p 744, The American Ceramic Society, Columbus, O.H. (1984).
8. M.J. Verkerk, A.J.A. Winnubst, A.J. Burggraaf, *J. Mat. Sci.,* **17,** 3113 (1982).
9. Wei-Kan Chu, J.W. Mayer, M.A. Nicolet, *Backscattering Spectrometry,* Academic Press, New York, San Francisco, London (1978).
10. D. Scholten, A.J. Burggraaf, *Solid State Ionics* **16,** 147 (1985).
11. A. Perez, J. Bert, G. Marest, B. Sawicka, J. Sawicki, *Nucl. Instr. and Meth.* **209/210,** 281 (1983).
12. *Ion Implantation Techniques,* eds. H. Rijssel, H.G. Glawischnig, Springer Verlag, Berlin Heidelberg, New York (1982).

SIMS Analysis of Garnet Epilayers

H. J. Tolle and P. Willich
Philips GmbH, Forschungslaboratorium Hamburg, 2000 Hamburg 54, W. Germany

Magnetic garnet layers have a wide variety of different physical properties. Modern applications are in the field of printing devices and displays[1] and they are the basic elements for components in magneto-optic waveguiding.[2] The film thickness ranges up to 15 μm. The resistivity depends on the grade of substitution but shows values of $\approx 10^{10}$ Ωcm. The nominal composition of yttrium iron garnet (YIG) is $Y_3Fe_5O_{12}$, substituted garnets are more complex and contain up to 10 elements.[3] Different layers grown from a solution of $PbO - B_2O_3$ by the technique of liquid phase epitaxy (LPE) are analysed by SIMS (secondary ion mass spectrometry).[4]

The physical properties (i.e. absorption, refractive index) depend on the incorporation of melt components into the layer. The incorporation of lead is strongly influenced by the rotation of the substrate during the growth process. Thus growth induced variations of the composition of epitaxial garnet layers are studied by SIMS. In order to give quantitative results the SIMS counting rates were translated in atomic concentrations (garnet formula units) by suitable calibration standards, the composition of which was defined by EPMA (electron probe microanalysis).[5] In some cases the film to be analysed was used as a standard itself. For example a 15 μm thick multi layer of YIG, consisting of 4 single layers each with a different substitution (lead concentration) but with thicknesses in the range of 2.8 to 4.6 μm. For the EPMA measurements craters of various depths are sputtered, each crater corresponding to a different lead concentration. Within these sections the incorporation of the elements to be analysed (i.e. lead) is homogeneous over a depth of more than 1 μm. A quantitative analysis is performed by EPMA within such a crater. A linear calibration function is obtained when plotting the SIMS counting rates versus the concentrations measured by EPMA. The described method of calibrating SIMS counting rates by EPMA can be also applied to other elements and analytical problems. The minimum concentration should be about 500 ppm over a homogeneous depth section of 0.5 μm. The SIMS depth profiles are carried out by bombarding the sample with 10 keV O_2^+ ions (Atomika ion microprobe). For charge compensation electrons with 1.5 keV energy are used. The instrument was adjusted in respect to get high sputtering rates ($\approx 4 \mu m\ h^{-1}$) combined with sufficient depth resolution ($\Delta z/z \approx 2\%$).

REFERENCES

1. P. Hansen, B. Hill and W. Tolksdorf, *Philips Tech. Rev.* **41**, 33 (1983).
2. H. Dammann, E. Pross, W. Tolksdorf, M. Zinke, *Proc. 5th. Intern. Conf. on Integrated Optics and Optical Fibre Communic. (IOOC–ECOC), Venice 1985*, p. 249.
3. C.-P. Klages and W. Tolksdorf, *J. Crystal Growth*, **64**, 275 (1983).
4. H. J. Tolle, *Surf. Interface Anal.* **4**, 14 (1982).
5. P. Willich, W. Tolksdorf and D. Obertop, *J. Crystal Growth*, **53**, 483 (1981).

ELIMINATION OF ION BOMBARDMENT INDUCED ARTEFACTS IN COMPOUND IDENTIFICATION DURING THIN FILM ANALYSIS: DETECTION OF INTERFACE CARBIDES IN 'DIAMOND-LIKE' CARBON FILMS ON SILICON

P. Sander, U. Kaiser, O. Ganschow and A. Benninghoven
Physikalisches Institut der Universität Münster Domagkstr.75, D-4400 Münster, W-Germany
P. Koidl and R.E. Sah
Fraunhofer-Institut für Angewandte Festkörperforschung, Eckerstr.4, D-7800 Freiburg, W-Germany

Proving positively the *a priori* presence of compounds in thin film analysis involving sputter depth profiling is usually prevented by artefacts like bombardment induced reactions and preferential sputtering. Looking for interfacial carbides in diamond-like carbon layers on silicon deposited by plasma deposition of hydrocarbon vapour, we tried to overcome these problems. As a first step, by combined SIMS, AES and XPS analysis we inspected possible signature data indicative of SiC, the compound to be looked at, using stoichiometric bulk samples. Neither changes of the C KLL lineshape[1] nor the Si 2p plasmon loss structure[2] turned out to be useful, the first one disappearing under ion bombardment, the second one being observable from other carbon binding states as well.[3] The only decisive electron spectroscopic indication of SiC was a 2 eV shift and 2 eV broadening of the Si LVV Auger line taken in the $E{\cdot}dN/dE$ mode of the electron spectrometer. In the corresponding SIMS data, $SiC^{\pm}$ secondary ions could of course be observed. However, SIMS/AES data from hydrocarbons adsorbed on Si revealed the existence of these signals with any binding state of carbon. As a second step, we looked for preferential sputtering under inert gas ion bombardment. From SiC, with increasing ion energy carbon is sputtered more preferentially. As a third step, we looked for SiO formation under simultaneous adsorption and various ion bombardment conditions. Inert gas ion as well as H_2^+ and O_2^+ bombardment always leads to SiC formation over a suitable range of hydrocarbon partial pressures and ion energies. In SIMS, the transformation of hydrocarbons into carbide is indicated by an increase of the $SiC^{\pm}$ yields by a factor 5 without changes of the carbon coverage. Finally, we compared depth profiles from films deposited at different substrate bias. At low bias, film adhesion is known to become poor. In this case, no SiC was detected, whereas in films deposited at higher bias considerable amounts of interfacial SiC were detected. This proves that the interfacial carbide is related to the adhesive strength of the films. At the same time, its formation can be related to the occurrence of bombardment induced reactions at sufficiently high substrate bias.

References

1. T.W. Haas, J.T. Grant, G.J. Dooley III *J. Appl. Phys.* **43** 1853 (1972).
2. F. Bozso, L. Muehlhoff, M. Trenary, W.J. Choyke, J.T. Yates Jr., *J. Vac. Sci. Technol.* **A2**, 1271 (1984).
3. T.J. Morarec, T.W. Orent, *J. Vac. Sci. Technol.*, **18**, 226 (1981).

Section 10.

Tribology and Environmental

SURFACE AND INTERFACE ANALYSIS, VOL. 9, 477–492 (1986).

Electron Spectroscopy of Reaction Layers Formed Under Different Wear Test Conditions

H.J. Mathieu, D. Landolt
Materials Department, Swiss Federal Institute of Technology (EPFL) CH-1007 Lausanne, Switzerland
R. Schumacher
Ciba-Geigy SA, Plastics and Additive Division, CH-1701 Fribourg-Marly, Switzerland

Auger electron spectroscopy (AES) depth profile analysis and x-ray electron spectroscopy (XPS) have been applied to the study of reaction layers formed during different wear tests (Reichert, Four ball, FZG, SRV and thermal immersion). Under heavy load conditions with normal pressure between 150–2000 N mm^{-2} reaction layers of 10–20 nm thickness are obtained in the presence of extreme pressure (AW/EP) additives. Isogeometrical phosphorous/sulphur compounds (thiophosphate structure) were used as model additives to investigate and compare their role under increasingly severe lubrication conditions in the temperature range 40–120°C. Results of the different methods are presented using the same additive in commercial synthetic hydrocarbon and conventional mineral oils. Preparation of the samples, transfer from oil bath to UHV and reproducibility of the mechanical and analytical characterisation are discussed. Correlations between composition of the reaction layer after tribofragmentation with the coefficient of friction and wear scar area are obtained. It is shown that application of AES depth profile analysis to the study of worn surfaces gives additional information towards a better understanding of AW/EP additives mechanisms.

INTRODUCTION

Tribology deals with the interaction of surfaces in relative motion. It is the study of friction, lubrication and wear of real world surfaces to reduce wear in given applications. Wear is not an intrinsic material property, but rather one of the characteristics of any engineering system.

Bearings, gears, pistons and many other moving parts of a machine are subjected to wear. However, there exist different types of wear. The four most common types of wear are (a) adhesive, (b) abrasive, (c) corrosive, (d) fatigue.[1,2] The present paper will treat mainly adhesive and abrasive wear problems in the presence of an additional containing lubricating oil. Figure 1 shows a plot of the friction coefficient as a function of viscosity times speed divided by load.[3] When two lubricated surfaces move under a light load relative to each other, they are separated by an elasto-hydrodynamic oil film (zone III). Such fully hydrodynamic lubrication presumes the presence of a fluid completely separating the rubbing surfaces.

Under increasingly severe loading conditions the oil film providing hydrodynamic lubrication may break down locally allowing occasional direct metal–metal contact at asperities to occur. This is the so called mixed regime corresponding to zone II in Fig. 1. If loading conditions become even more severe seizure and welding between the two sliding surfaces may take place leading to rapid destruction of the workpiece (zone I).

The so called EP (extreme pressure) or AW (anti-wear) additives are added to the base oil to improve friction and wear behaviour in the mixed regime. They act by forming a thin surface film which effectively prevents direct metal–metal contact at asperities of the sliding surfaces. The formation of surface films involves chemical reactions of the additives with the metallic surfaces under the combined action of shear forces and temperature spikes. This process is called tribofragmentation.[20] Additive reactions in presence and absence of shear forces have been extensively studied[4–16] but their mechanism is still insufficiently understood. A basic understanding of the principal reaction mechanism involved in surface layer formation, however, is a prerequisite for the rational development of improved AW/EP additives.

The application of surface analysis methods like Auger electron spectroscopy (AES) and photoelectron-

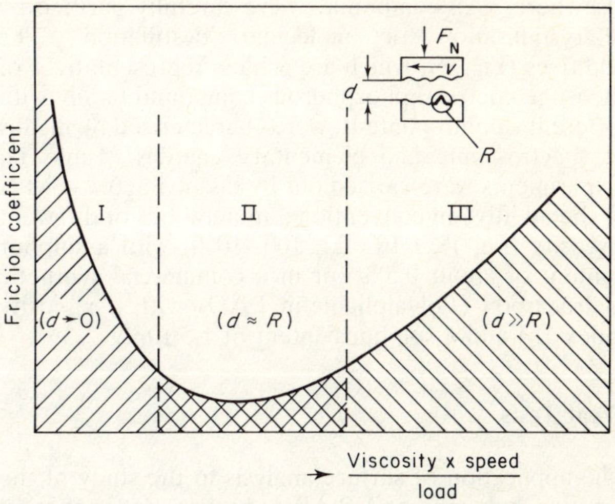

Figure 1. Lubrication regimes for sliding surface pairs.[3]

0142–2421/86/120477–16 $08.00

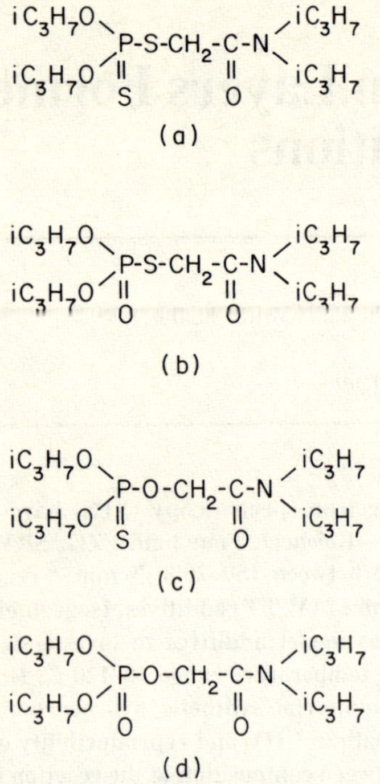

Figure 2. Model compound formulae.

spectroscopy (ESCA-XPS) has greatly expanded the possibilities for the study of AW/EP-additive reaction mechanisms.[2] However, despite many publications found in the literature, in which these methods are employed, only a few systematic studies relating additive reactivity and efficiency to molecular structural parameters have been published.[4,5,11] It is the aim of the present paper to review our own studies of reaction layers formed under different loading conditions in presence of isogeometrical phosphorus/sulphur compounds in the temperature range 40–120°C.

EXPERIMENTAL

Model compounds were prepared by general procedure according to the synthetic reactions described elsewhere.[4] All compounds were carefully purified by re-crystallisation or molecular destillation. The additives (Fig. 2), which are ashless representatives of the isogeometrical phosphorous compound family with different sulphur content, were characterized by means of spectroscopic and elementary analysis (Table 1). Experiments were carried out by dissolving 0.5–4 wt% of the additive in conventional mineral oils of different viscosity (i.e. ISO VG 22, 100, 1000) with a sulphur content of about 0.5%[4] or in a commercial synthetic hydrocarbon (Polyalphaolefin PAO) with a viscosity index 124 and a sulphur content of 1.5 μm g^{-1}.[4]

Wear Tests

The application of surface analysis to the study of the reaction behavior of EP/AW additives under friction and wear conditions puts special demands on the mechanical wear test. The most important requirements are listed in Table 2. Only few of the many different tests used in practice for testing EP/AW-additives, therefore, are suited for mechanistic studies. In the following, four commonly used wear tests will be considered in more detail from this point of view.

Table 1. Analytical Data

Compound	Molecular weight	MP (°C)	Acid number (mg K OH/g	% P	% S	^{31}P–NMR (ppm/H_3PO_4)
A	355.5	30–32	<0.04	8.7	18.0	90.6
B	339.4	28–30	<0.04	9.1	9.5	24.0
C	339.4	39–41	<0.04	9.1	9.5	65.4
D	323.4	38–40	<0.04	9.6	—	2.9

The Four-ball lubricant testing apparatus (VKA) represents a standard test for AW/EP-additives.[7,11,17] It consists of 4 steel spheres of 12.5 mm diameter. One top sphere revolves at a constant speed of 25 rpm under variable load for 1 minute upon 3 identical stationary balls clamped in a ball pot and immersed in the test oil. Typical wear scars are represented in Fig. 3 for two different loads. It shows that with increasing load the area of the wear scar increases. Characteristic data for this test are the intitial seizure load, the weld load (WL) and the wear scar diameter (WSD). The latter is determined at a load at 400 N after 1 h test duration.

The Reichert test is a cylinder-on-disc type wear test.[15] Its experimental arrangement is shown in Fig. 4(a). A test 100 Cr 6 steel cylinder of 2cm length and 1 cm diameter is pressed with a certain force on a rotating disc turning at constant velocity in a bath containing 25 ml of lubricant.

The test results in severe wear of the cylinder (cf. Fig. 4(b)). The diameter of the wear scar is used as a measure of wear and hence of the quality of the additive. The method is illustrated by the data of Fig. 5 obtained for two different additives under otherwise identical conditions.

Table 2. Wear test requirements for mechanistic studies of EP/AW additives

1. Well defined experimental variables

   Load/surface area
   Sliding velocity
   Oil viscosity
   Friction path
   Additive concentration

2. Reproducible determination of friction and wear

   Friction coefficient
   Wear rate

3. Amendable to surface analysis

   Cleanliness
   Small sample size
   Reproducible surface conditions

4. Economically acceptable

   Simple sample shape
   Limited quantity of oil and additive
   Short test duration

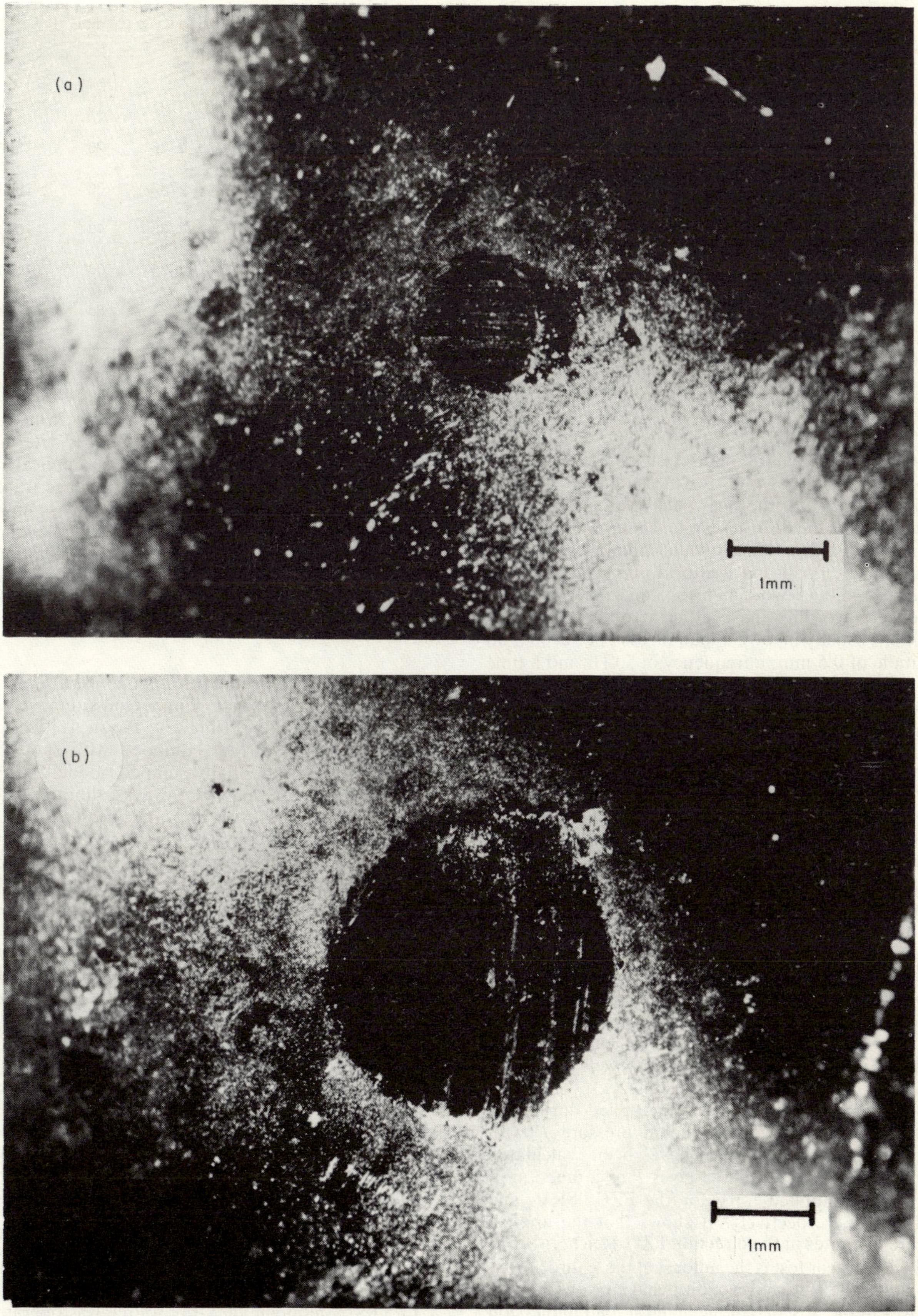

Figure 3. VKA-worn surfaces for (a) F_N = 220 kg, (b) F_N = 260 kg.

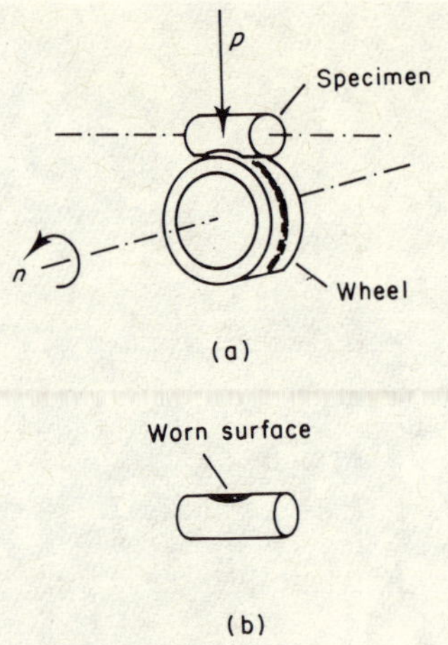

Figure 4. Schematic representation of Reichert wear test.[6]

The modified ball-on-plate test device (Optimol SRV-system) allows one to measure the friction coefficient as well as the wear behaviour. A schematic arrangement of the SRV-test system is shown in Fig. 6.[4] A small steel ball oscillates while in contact with the test specimen. Load and friction force are controlled and measured by a two-dimensional piezoelectric force transducer recording the coefficient of friction. Test conditions usually involve a load of 100 N or 200 N, an amplitude of 0.5 mm, a frequency of 50 Hz and a time of 2 h in the temperature range 40–120°C. A typical friction curve is shown in Fig. 7. After the 'running-in' phase the coefficient of friction practically constant. It is that friction coefficient which is determined after a test duration of 2 h.

The FZG–Spur Gear machine is another standard test for lubricants.[18] In this test representing test conditions close to application a variable load is applied on the tooth flank of a gear. Figure 8 shows such a tooth flank together with a picture of the surface in contact with the opposing gear flank. As a measure of the quality of the lubricant one defines 12 different load stages. The last stage before welding of the surface is determined. The whole gear is immersed in a temperature controlled oil bath and new gears are needed for each test. Duration of the test can be up to 24 h. The FZG and test is very time consuming and expensive and test specimen are less well suited for surface analysis.

For the comparison of the load applied during the different wear tests, the maximum pressure P with respect to surface normal have been calculated according to Hertz.[19] Table 3 gives P as a function of load stage or applied force for the FZG, SRV- and Reichert-test, respectively. It shows that the applied pressure increases in the direction FZG-Reichert-SRV, i.e. the FZG-gear test is the mildest of these three tests.

To study the surface reaction behaviour of EP/AW-additives in the absence of mechanical load immersion tests can be used. In our laboratory we have developed the so-called single pulse heating [SPH] method.[20] A schematic diagram of the experimental set up is given in Fig. 9. It consists of a Teflon body (1) with a rectangular cavity (2) into which a heated metal sheet (3) is placed. The cell cavity (2) is filled with the lubricating oil. The metal sheet consists of Armco iron. Temperature of the heated metal sheet is determined from the voltage transient obtained on application of the constant current pulse. A typical voltage transient corresponding to a pulse length of 12 ms is shown in Fig. 10. From the measured voltage the sample temperature can be calculated. For data correlation the temperature at the end of a pulse is used.

Table 3. Maximum normal pressure p (N mm^{-2})[17]

FZG		SRV		Reichert	
LS	P	$F_{(N)}$	P	$F_{(N)}$	P
1	146	50	1727	10	668
2	295				
3	474	100	2176	20	842
4	621				
5	773	200	2742	30	963
6	927				
7	1080	300	3138	40	1060
8	1232				
9	1386	400	3454	50	1140
10	1538				
11	1691	500	3721	60	1210
12	1841				

Surface Analysis

In our studies surface and thin film analysis by AES or XPS was performed using commercial scanning Auger microprobes (Perking Elmer – Physical Electronics Model 545A–550/590 A) equipped with sputter ion guns that were differentially pumped (model 550/590 A only). The sputter rate was always calibrated against Ta_2O_5 anodic films, so that thicknesses (in Ta_2O_5 equivalents) can be compared easily for measurements performed at different times. For AES, primary beam current density were varied between 2–10 mA cm^{-2} at primary energies of 2 keV. The electron beam was always scanned to reduce current density and to obtain data information of an area approx. 200 × 200 μm^2. Both electron and ion beam were centered by visualisation on a TV monitor using the absorbed current image.

Before and after each test the samples of the different wear tests were carefully cleaned in an ultrasonic bath containing chemically pure acetone and white spirit in a 1/1 ratio. After cleaning, samples were stored over silica gel under an argon atmosphere.

Typical AES depth profiles of specimen subjected to the SRV wear test in the presence and absence of additive A are shown in Fig. 11. In case of pure base oil (Fig. 11(a)) the oxide film hardly exceeds 3–4 nm, whereas in the presence of additive A (Fig. 11(b)) a reaction layer containing O, P and S with a maximum at a depth of approx. 2 nm is observed.

In XPS, a monoanode x-ray source was operated at 10 keV energy and 40 mA emission current using Mg Kα (1253.4 eV) radiation. Binding energies were based on an instrumental calibration of the Au 4f$_{7/2}$ at 84.0 eV and Cu 2p$_{3/2}$ at 932.6 eV corresponding to C 1s at 83.8 eV.[21] Further experimental details are found elsewhere.[20,22]

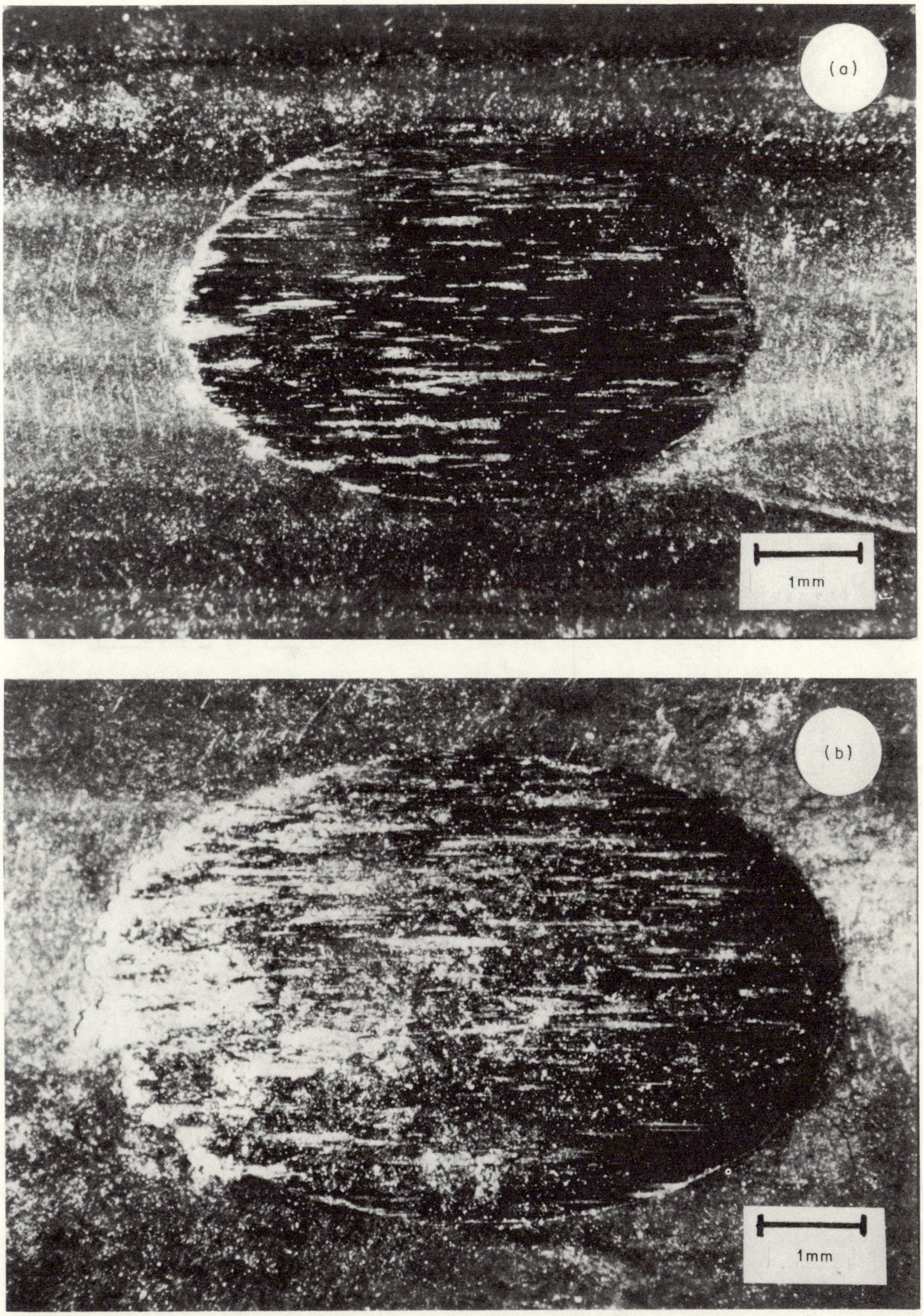

Figure 5. Reichert wear scars for additive (a) A, (b) A′ after a path length of 400 m. Additive A and A′ have a rest-groupe C_3H_7 and C_8H_{17}, respectively (cf. Table 4).

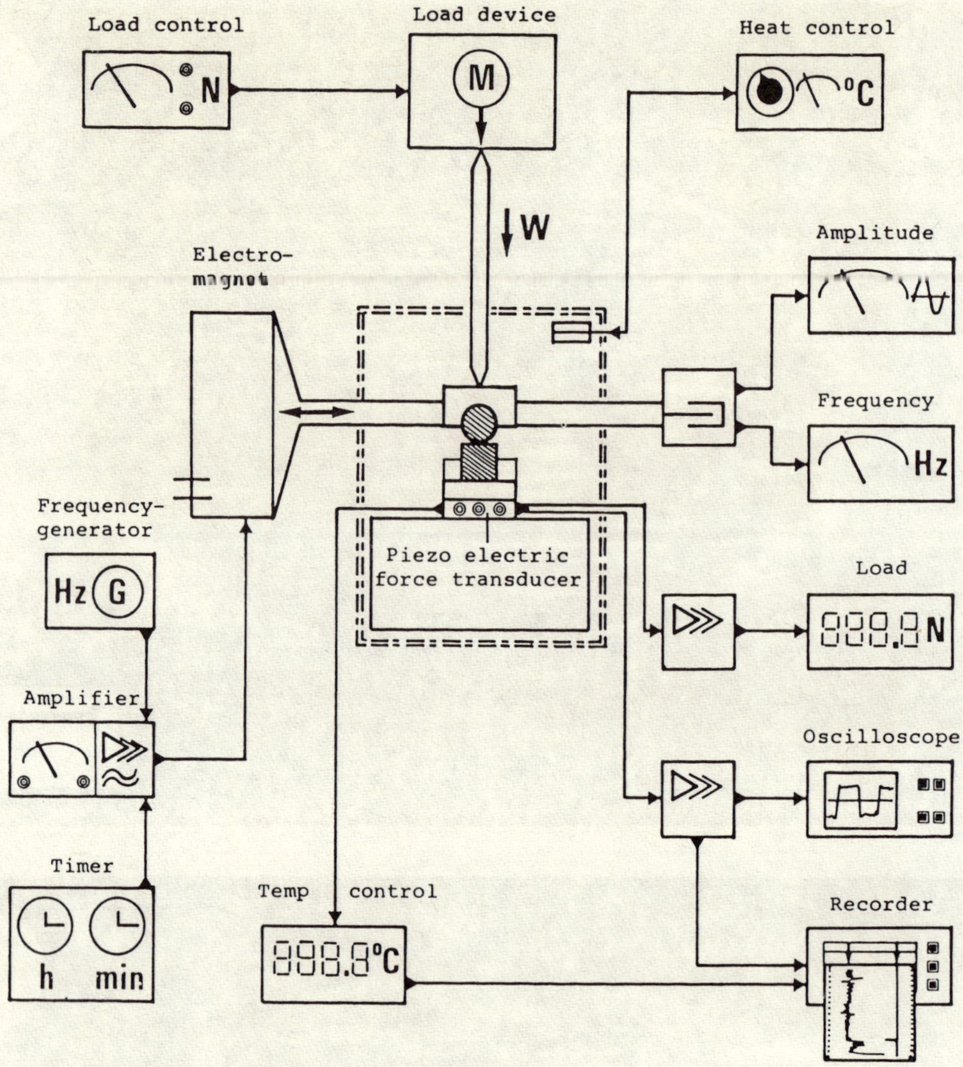

Figure 6. Schematic diagram of SRV-tester.[4]

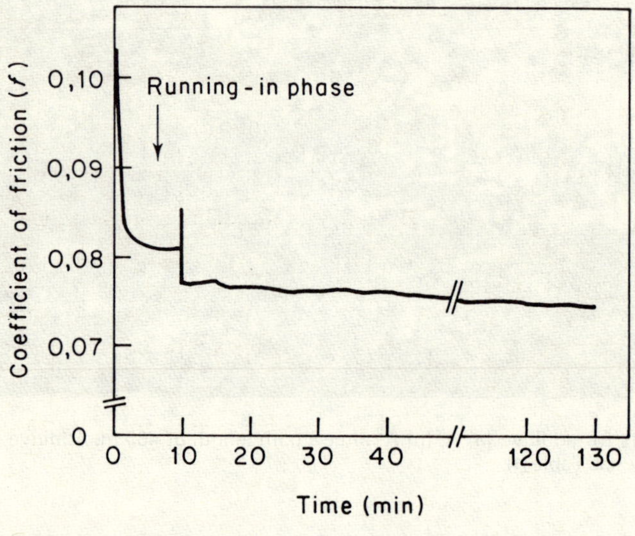

Figure 7. Typical friction curve under standard SRV test conditions for additive A.

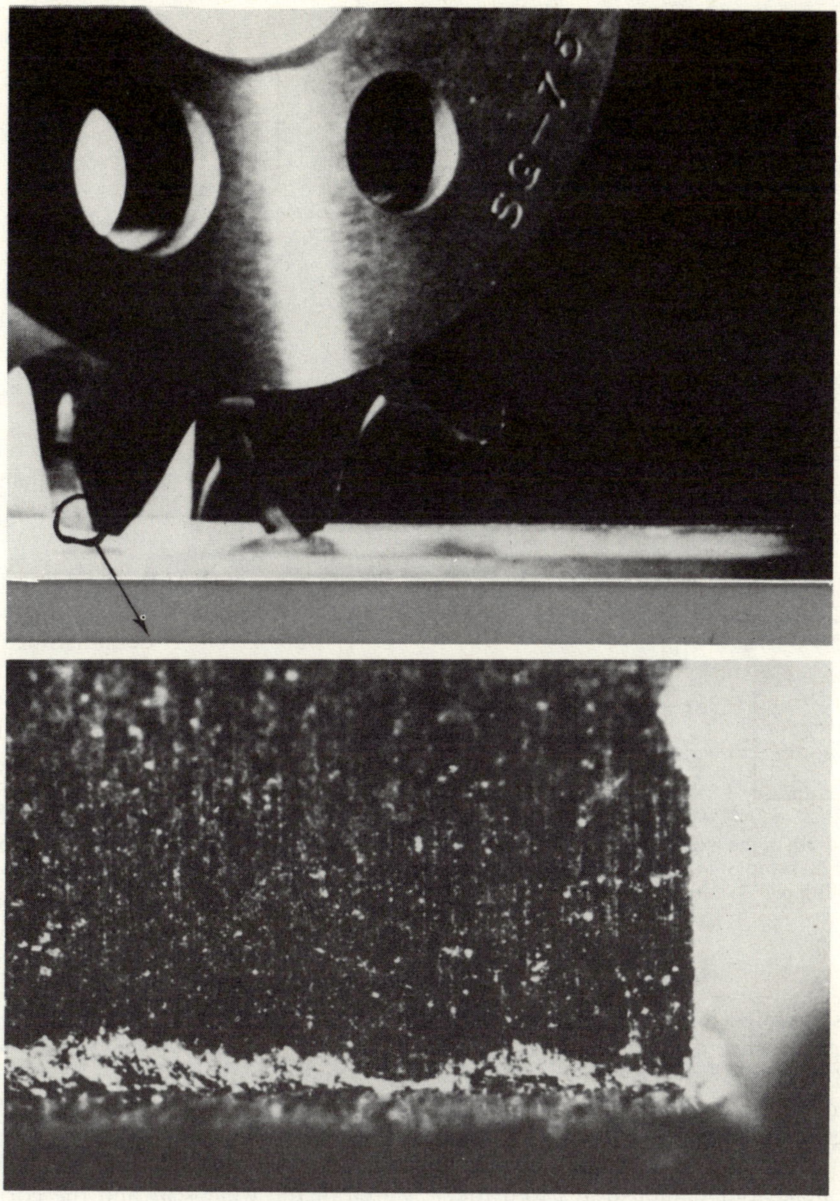

Figure 8 FZG spur gear machine — surface of tooth flank.

RESULTS

Figure 12 shows the wear scar area as a function of friction path of samples subjected to the Reichert wear test at a load of 15 N in the presence and absence of AW/EP-addititives. When the additives A–D are present, the wear scar area reaches a steady value after a short time corresponding to a friction path of 20–50 m. In the absence of additives the wear scar area is larger and increases continously. The sulphur containing additives A–C are significantly more effective in this test than additive D containing only phosphorous.

The influence of applied load on wear was investigated for a constant friction path of 100 m with the Reichert test for additive A and B. As shown in Fig. 13, increasing load leads to more wear. The wear scar area increases approximately linearly with applied load.

The additive concentration in the lubricant also influences the wear behaviour as shown in Fig. 14 for additives A and B. One observes an initial decrease in wear scar area with increasing additive concentration, then the curves seem to level out. For a fixed concentration, additive A is more effective than additive B. The same general behaviour is observed, when the wear scar area is plotted against the molar additive concentration.

Fig. 15 shows for additive A that in the SRV test both friction coefficient and wear scar area vary in a similar manner with solvent temperature indicating a parallelism between friction and wear. The variation of the friction coefficient with solvent temperature for different additives (A–D) is shown in Fig. 16 together with the P and S content of the reaction layer measured by AES depth profiling. Up to approx. 80°C the friction coefficient, μ, remains almost constant, but it increases abruptly when the so-called high wear region is reached. The increase of the friction coefficient in the

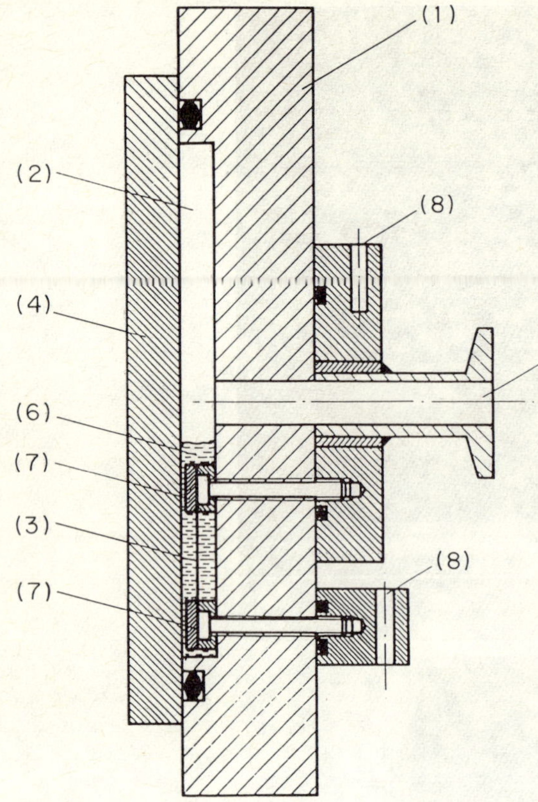

Figure 9. Schematic diagram of the experimental cell: (1) Teflon cell body; (2) cell cavity; (3) sample; (4) glass plate; (5) inert gas inlet; (6) level of lubricating oil; (7) sample holder; (8) electrical contacts.[20]

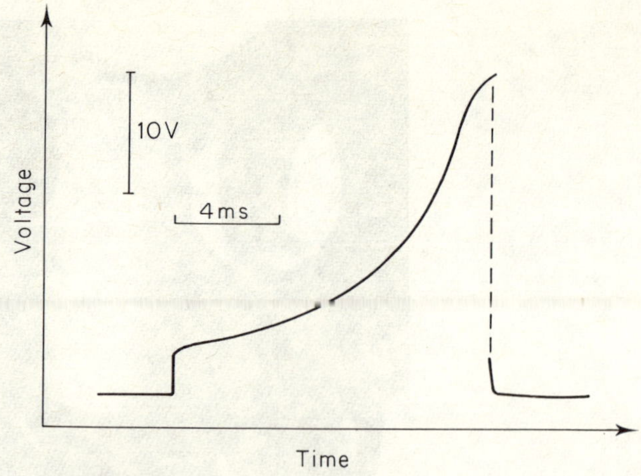

Figure 10. Shape of the voltage transient measured at the application of a constant-current pulse.[20]

high wear region is coupled to a decrease of the phosphorous and sulphur concentration in the reaction layers. The defected sulphur is that of the additives, since these measurements were performed in the pure base oil (PAO). Figure 16(d) gives also the friction coefficient of pure PAO-base oil (triangles). The completely different behaviour demonstrates well the beneficial role of the additives. The steady decrease of the friction coefficient of the base oil is explained by the decreasing viscosity with temperature.

Table 4 presents performance data of additives A–D in the different wear tests. The effectiveness of the additives may be judged by comparison with the data for the base oil (PAO). In addition, data of additive A′ having a different C—H group than A (A: CH_7; A′: C_8H_{17}) are given. It demonstrates the steric influence of the molecule, which is important during the first stage of reaction, i.e. the adsorption of the additive molecule to the surface.[4,23] Comparison of the different data of the additives reveals that the sulphur-free additive D, containing only phosphorous, is practically ineffective in the Reichert wear test but active in the four ball test and particularly so in the FZG gear test. The performance in the SRV test is poorer than additive A and B, but comparable to the phosphorous-free additive C. The latter, however, containing only sulphur is effective in the Reichert and four ball tests but relatively inactive in the FZG gear test.

Table 4. Wear test results

Compound	R_1	Reichert	VKA ISL	VKA WL	VKA WSD	FZG LS	SRV
		mm²	N	N	mm²		mm²
PAO base oil	—	36–38	400	1600	0.8	5	0.77
A	$i-C_3H_7$	17.5	1100	2300	0.5	12	0.50
A′	$i-C_8H_{17}$	22.5	600	1600	1.0	7	—
B	$i-C_3H_7$	22.5	600	1600	1.0	7	0.58
C	$i-C_3H_7$	14.5	1100	2400	0.7	7	0.70
D	$i-C_3H_7$	33	1200	2100	0.4	11	0.66

Reichert: 1% in ISO VG 22.
VKA: 1% in ISO VG 22; WSD: 400 N during 1 hour.
FZG: 0.05% in ISO VG 32.
SRV: 2% in PAO ISO VG 100; 100°C.

Repeatability of wear test and surface analysis data is very important. Using the Reichert test, the repeatability of wear scar area determination was checked by subjecting a series of 10 samples to the same standard test conditions in the presence of additive A and B. The measured average values of the area of the wear scar and the corresponding standard deviations are listed in Table 5 showing good repeatability. Therefore, all wear test data reported here were repeated 2–4 times and only average values used.

Surface and Depth Profile Analysis

Reproducibility. To test repeatability .of AES depth profile data a series of samples were subjected to standard Reichert test conditions in the presence of additives A and B in a base oil containing 0.5 wt% S.

Table 5. Repeatability of reichert wear test data[5]

Compound	Number of measurements	Area of wear scar [mm²] X	S
A	10	18.2	1.0
B	10	24.6	1.9
No additive	20	35.3	1.7

X = average S = standard deviation

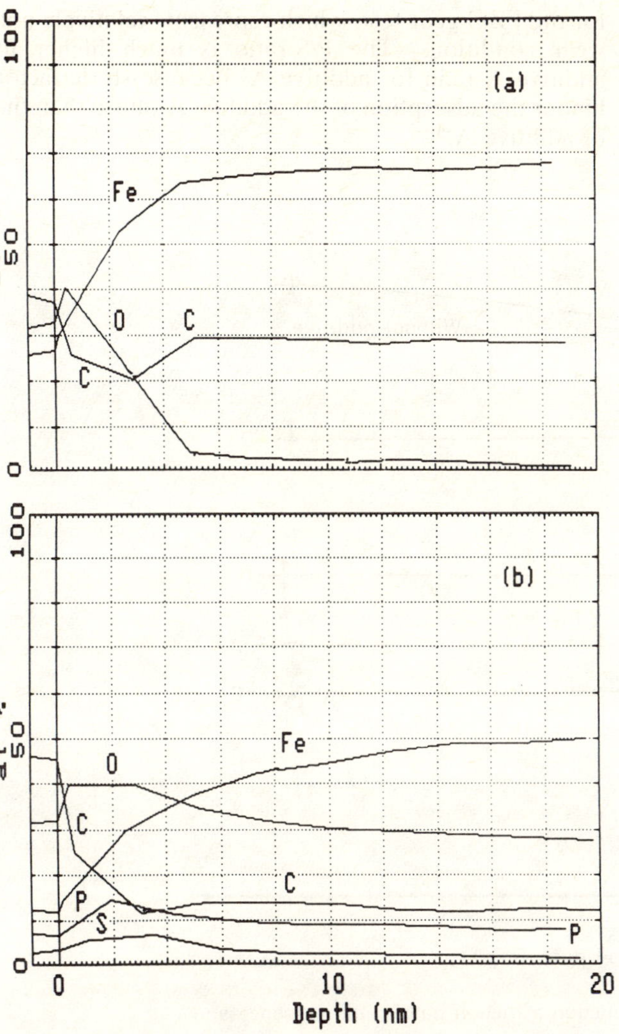

Figure 11. Typical AES depth profiles after SRV tests for (a) pure PAO base oil (b) additive A in PAO, T = 100°C, F_N = 200 N.

Table 6 shows average values and the standard deviation at three different depths of the AES profile of 10 independent samples. Reproducible data within approx. 10% can, therefore, be obtained if several independent measurements are performed. Contamination effects do not influence considerably the results, if cleaning procedures and transfer conditions are maintained constant. However, different artefacts, if not avoided may falsify the interpretation of AES/XPS data. The preparation of test samples, i.e. their mechanical polishing has to be controlled to avoid contamination or modification of the surface layers before testing. Sulphur containing polishing agents are often applied which in turn can modify the concentrations during wear testing. This was the case in particular for VKA-test spheres, where the mechanically polished outer surface contained considerable amounts of calcium oxide, obviously used with the polishing agent. Another source of error is sulphur present in the base oil. It is often assumed that such sulphur is chemically not active. However, sulphur compounds present in the base oil can become active at elevated temperatures or due to tribofragmentation. The behaviour is illustrated in the following data. Additive A was added to different base oils and submitted to the SRV-test. Results of the

Table 6. Repeatability of AES Data[5]

Additive	Number of	sulphur content [wt%]					
		18Å		43Å		170Å	
		X	S	X	S	X	S
A	10	7.3	1.2	4.9	1.4	2.6	1.0
B	10	4.6	0.8	2.4	0.47	1.0	0.2
no additive	4	1.8	0.4	0.7	0.1	0.2	0.03

X = average S = standard deviation

wear tests and the corresponding AES depth profiles are shown in Fig. 17. Inspection reveals that in different viscous base oils the behaviour of the friction coefficient μ can be quite different as well as the P and S concentrations in the reaction layer. In particular, with the base oil BP solvent N 100 (a) and Vitrea 41 (b) sulphur is found enriched in the reaction layer at temperatures above 100°C. The other base oils, Shell Valvata 1000 (c) and the low sulphur containing PAO ISO VG 100 (d) show only a slightly rising sulphur content or even a decrease with increasing temperature. To investigate the role of the base oil sulphur an additional experiment was carried out using the sulphur free additive D. Figure 18 shows that sulphur is found in the reaction formed during an SRV test. The S-concentration as well as the coefficient of friction (μ) is shown as a function of oil temperature. The low sulphur oil (PAO−1.5 ppm S) exhibits almost no sulphur at all temperatures and a phosphorous concentration maximum around 80°C. Above this temperature μ rises to the high wear region. In case of the mineral oil containing 0.46 % S the behaviour of the friction coefficient is different and large quantities of sulphur are found above 80°C obviously due to the activation of the base oil sulphur. Therefore, it is important that wear tests also involve characterisation of the base oil. Surface analytical experiments are preferably performed with sulphur free solvents to avoid artefacts.

AES Depth Profiling of Worn Surfaces. AES depth profiles were measured up to a depth of approx. 20 nm. Profiles of the different wear test are similar to that shown in Fig. 11, exhibiting a maximum of S and P just below the surface. Fig. 19 shows the variation in sulphur and phosphorous contents, respectively, in the reaction layer of samples subjected to Reichert test conditions in presence of additive A. The concentration at a depth of 1.8 nm is plotted as a function of friction path. The depth of 1.8 nm was chosen to avoid influence of surface contamination due to cleaning and transfer procedures. Figure 19 shows that a steady state sulphur concentration is reached rapidly, whereas, the phosphorous content increases steadily and a steady state is apparently not reached even after a friction path of 400 nm.[5]

For the SRV test, the sulphur and phosphorous-concentrations determined from AES-profiles are shown in Fig. 16. Under mild temperature conditions (40–60°C) a relatively high phosphorous content is found. With increasing temperature (80–100°C), phosphorous passes through a maximum. Above 100°C, phosphorous and sulphur decrease. This leads to higher friction and wear. Under more severe conditions, only compounds A and B show activity. At

120°C none of the additives tested is active as is indicated by the low phosphorous content of the reaction layer. Steric parameters such as the size of the C_XH_Y-rest (compare Table 4, additives A and A′) also influence the amount of phosphorous and sulphur found in the reaction layer. This is illustrated by Fig. 20 for the FZG gear test, which represents relatively mild wear conditions. The P/S-ratio is much higher for additive A than for additive A′ because steric factors hinder the adsorption of the additive A′ more than that of additive A.[4]

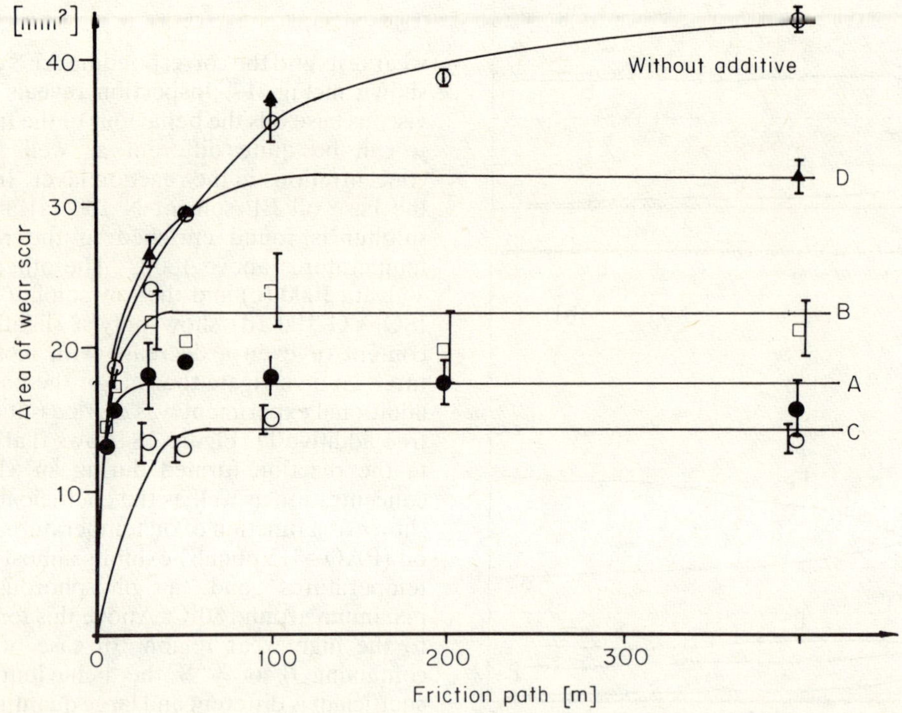

Figure 12. Reichert wear test curves as a function a friction path in the presence and absence of EP additives.[5]

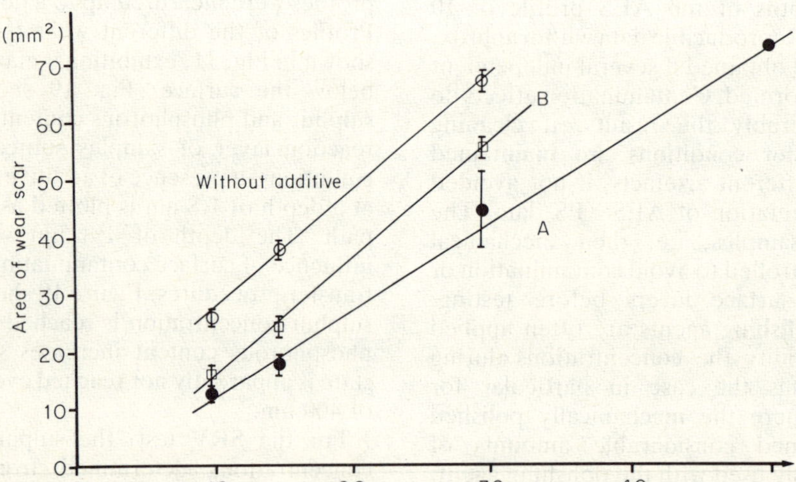

Figure 13. Reichert wear scars under increasing load conditions in the presence and absence of additives A and B.[5]

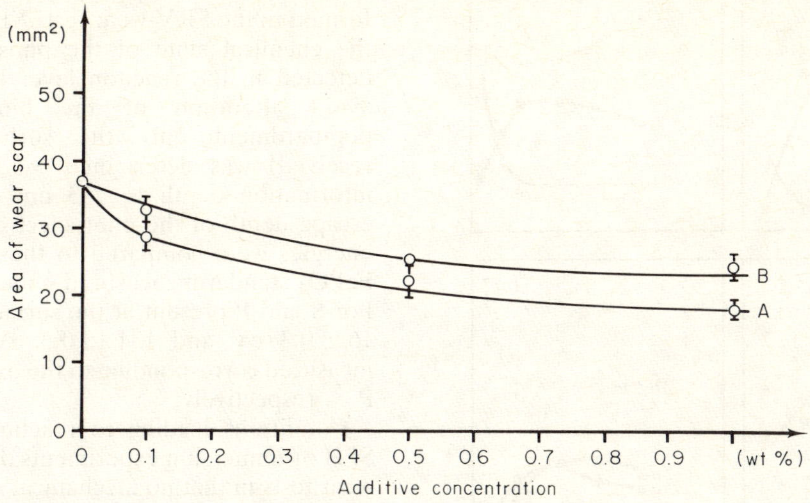

Figure 14. Reichert wear scar area as a function of additive concentrations A and B.

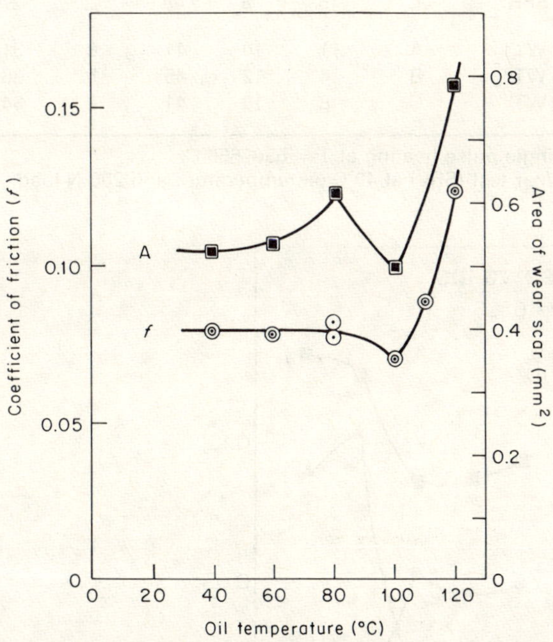

Figure 15. Correlation between friction and wear for additive A (SRV-test).[4]

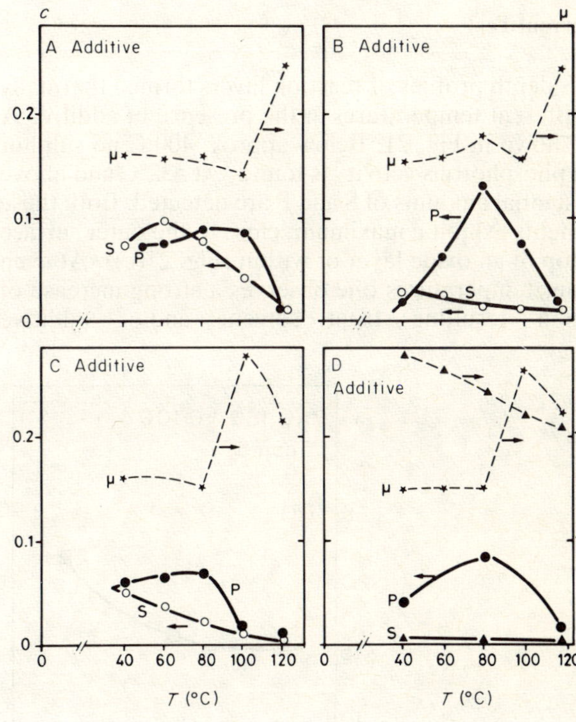

Figure 16. Tribochemical fragmentation in PAO ISO VG 100 base oil containing additives A–D. Concentration, c (wt%×0.01), and friction coefficient, μ, are shown as a function of temperature T (SRV-test). Filled traingles give the friction coefficient of the pure PAO base oil.

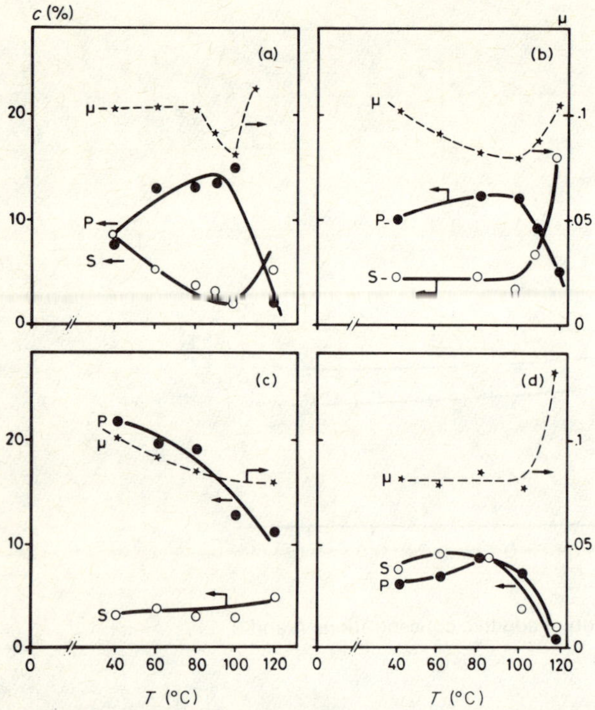

Figure 17. Tribochemical fragmentation in (a) BP solvent N 100, (b) ISO VG 100 (Vitrea 41), (c) Shell Vavlata 1000, (d) PAO ISO VG 100 containing 2% additive A. Concentrations of P and S (wt%) were determined at 4.3 nm.

Thermal Test

AES depth profiles of reaction layers formed thermally at different temperatures in the presence of additive A are shown in Fig. 21. Below approx. 400°C no sulphur and phosphorous activity is found. At 532°C and above substantial amounts of S and P are detected. Both these elements exhibit a maximum, close to the outer surface on top of an oxide layer or within (Fig. 21(c)). At even higher temperatures one observes a strong increase of carbon resulting from solvent and/or additive decomposition. One notices that the depth profile of the reaction layer formed during thermal tests with additive A is definitely different from that of Fig. 11(b) formed in the SRV-wear test. XPS was used to identify the chemical state of the phosphorous and sulphur detected in the reaciton layer by AES profiling. To avoid alteration of the binding state by ion bombardment only the surface composition (as received) was determined by XPS. This limits the information depth to 2 3 nm corresponding to the escape depth of the photoelectrons. The XPS binding energies were compared to those of FeS, $FeSO_4$ and $FePO_4$ standards measured under identical conditions. For S and P present at the surface binding energies of 162±0.3 eV and 134.1±0.3 eV for S and P were measured corresponding to the oxidation states S^{2-} and P^{+5}, respectively.

Conditions leading to reaction layer formation in SPH or immersion experiments differ from those of the wear tests in that no mechanical removal of the surface layer occurs. The chemical composition of the reaction layer, therefore, is not the same. This is evidenced by Table 7 in which elemental concentrations measured by AES are compared for SPH- and SRV-wear test experiments for the additives A–C in the same solvent (PAO). The S/P and the O/P ratios in the SPH

Table 7. Comparison of sulphur and phosphorous concentration in reaction layers at 1.8 nm depth (at%)

Method	Additive	S	P	O	C	Fe
SPH[a]	A	33	6	18	12	31
SPH[a]	B	8	3	48	4	37
SPH[a]	C	15	8	48	2	27
WT[b]	A	11	10	41	8	31
WT[b]	B	4	12	45	5	36
WT[b]	C	8	10	41	7	34

[a] Single pulse heating at T = 530–550°C.
[b] Wear test (SRV) at 40°C oil temperature and 200 N load.

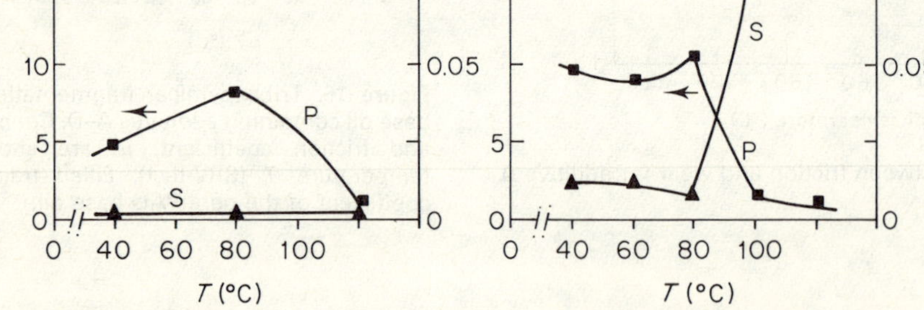

Figure 18. Wear behaviour of 2% additive D in (a) PAO ISO VG 100 (s=1.5 ppm) and (b) ISO VG 100 (S=0.46 %). Concentrations of P and S (wt%) were determined at 4.3 nm.

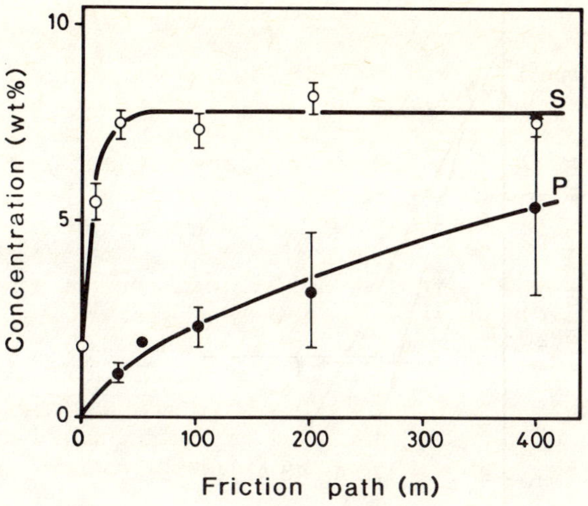

Figure 19. Concentrations of S and P determined by AES depth profile analysis for different path lengths of Reichert test (additive A).

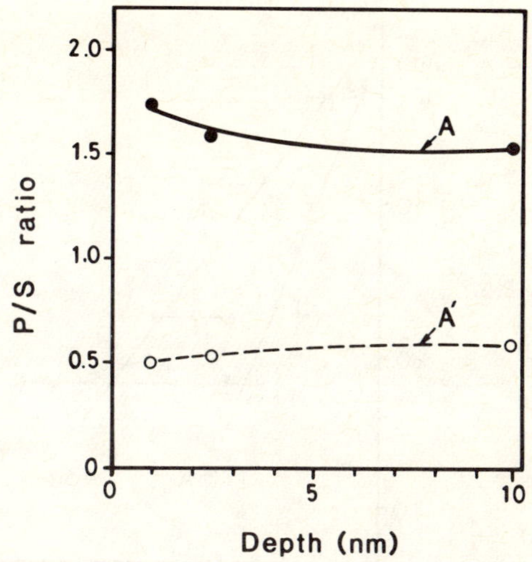

Figure 20. AES depth profile concentration ratio (P/S) after FZG test in ISO VG 32 base oil containing additives A and A′.

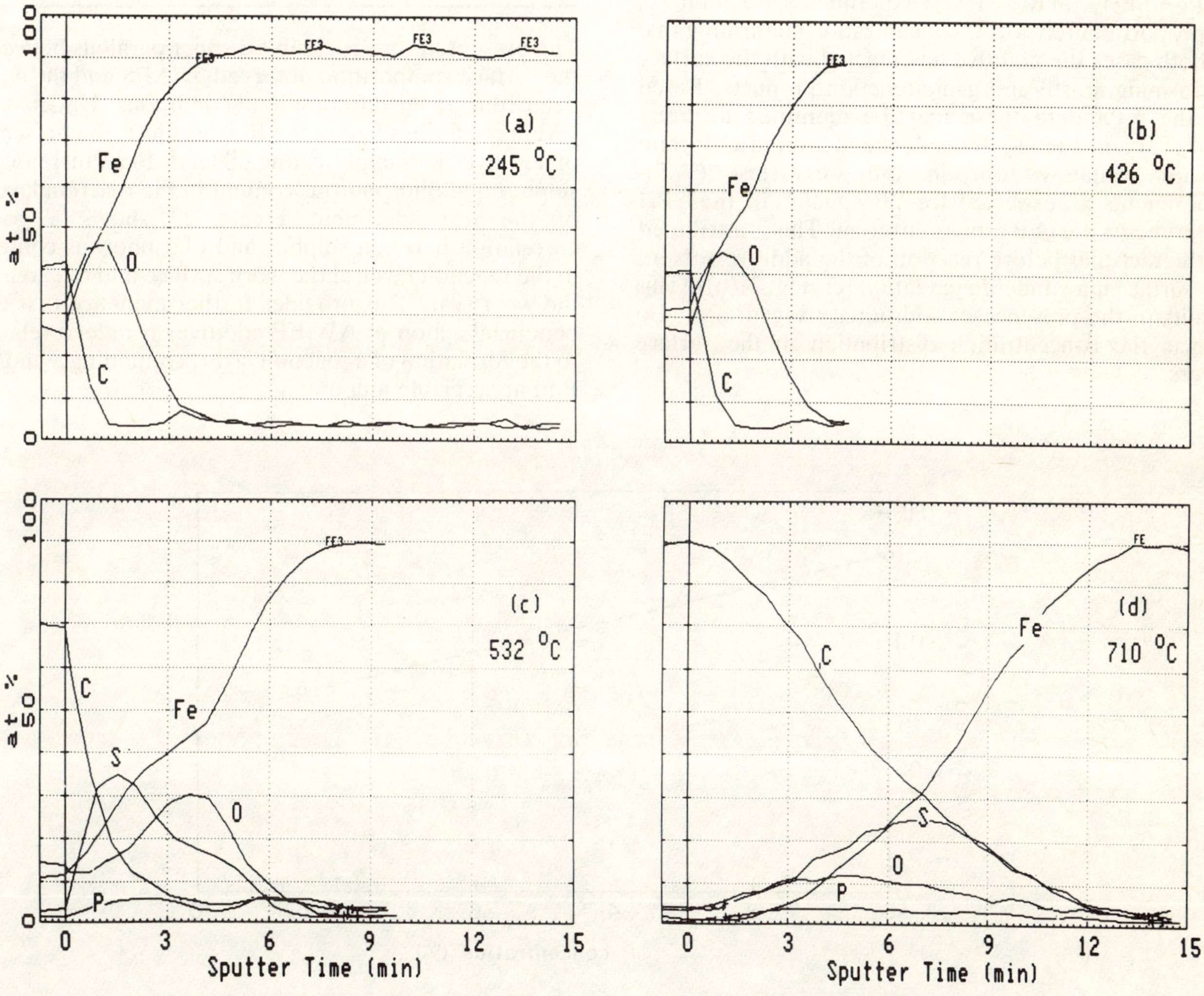

Figure 21. AES depth profiles of thermally heated samples (SPH-method at (a) 245°C, (b) 426°C, (c) 532°C, (d) 710°C for additive A in PAO,[22] sputter rate was 1 nm min^{-1}.

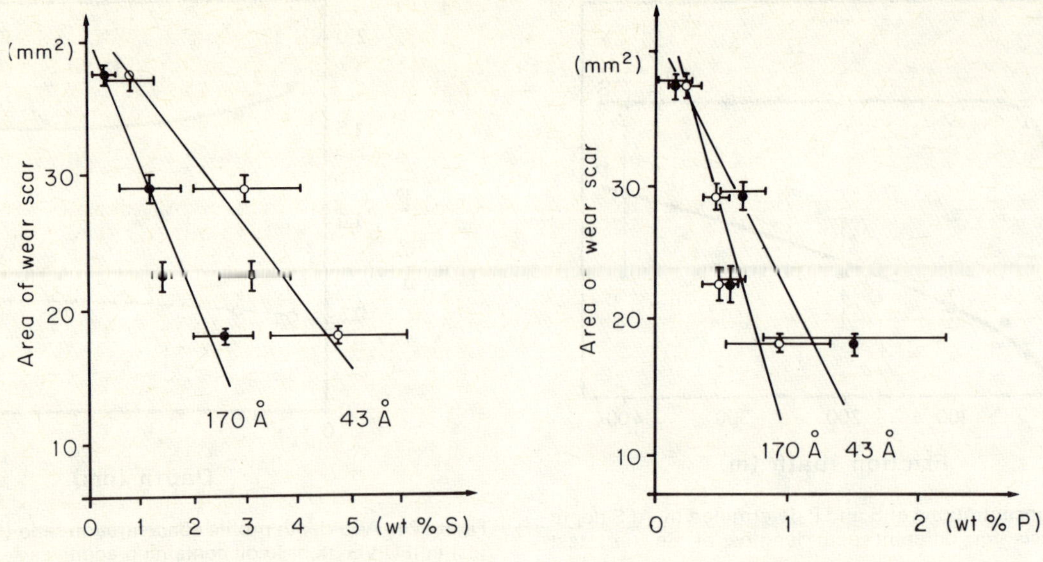

Figure 22. Worn area produced by Reichert wear test as a function of the sulphur concentration determined by AES of different depths of the reaction layer in the presence of additive A.[5]

experiments are much higher than in the SRV tests. Both SPH and SRV experiments exhibit a disproportionately low C concentration confirming that in both cases the additive has reacted with the surface by forming mostly inorganic reaction products. Based on the XPS data these may be identified as FeS_X (X=1–2) and $FePO_4$. Indeed, the measured O/P-ratio is approximately four in the wear test (WT)-experiments as expected for phosphate. In the SPH experiments the O/P ratio is different. This is attributed to the fact that before reaction of the additive sets in, the surface may undergo oxidation (cf. Fig. 21(c)). This results in the presence of additional oxygen, and also affects the concentration distribution in the surface layers.

DISCUSSION

The presented results exhibit distinct parallels between the surface composition observed by AES and the wear behaviour under different load conditions. Under very severe conditions like in the Reichert test, the presence of additives in the lubricating oil leads to an increase of sulphur and phosphorous content in the reaction layers of the test specimen. Figure 22 shows a good correlation between sulphur and phosphorous content in the reaction layer of the worn surface and the area of the wear scar. This provides further evidence that the beneficial action of AW/EP-additives is indeed related to the formation of a reaction layer, containing S and/or P in appreciable amounts.

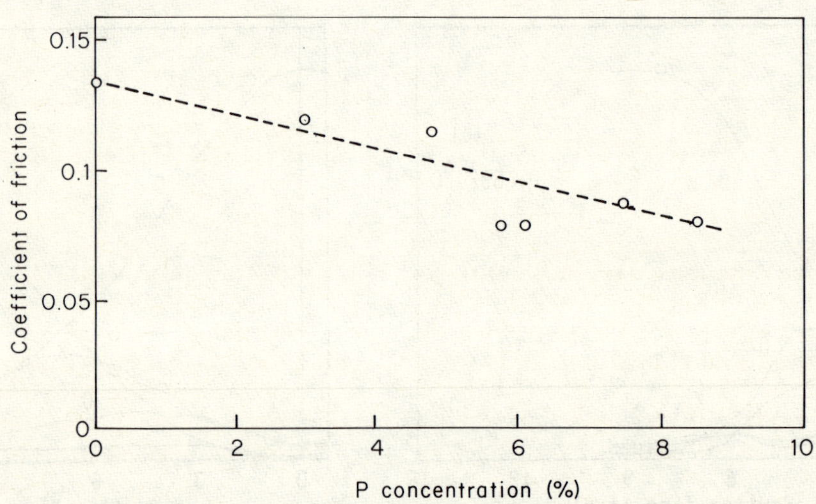

Figure 23. Coefficient of friction determined by SRV test as a function of P-concentration determined by AES depth profile analysis (PAO, additive A, T = 80°C, depth 4.3 nm).[23]

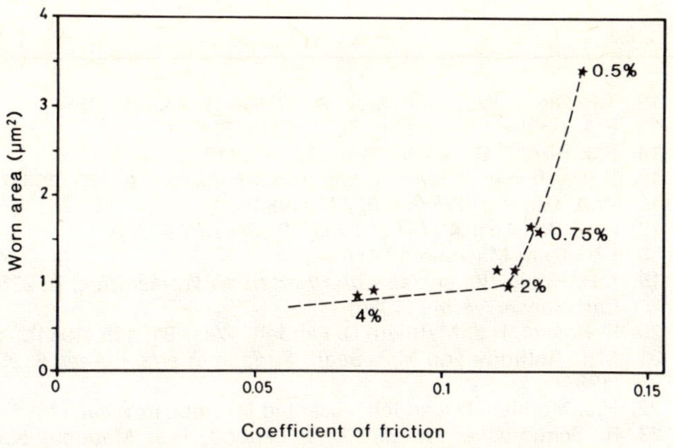

Figure 24. Worn area as a function of friction coefficient for various concentrations (wt%) of additive A in PAO (SRV-test).[23]

Figure 16, which represents wear test and AES depth profile data for the milder SRV-test exhibits an anologous result. The friction coefficient increases steeply above 80°C for all additives. A–D coinciding with a decrease of the phosphorous concentration in the reaction layer. The influence of sulphur under SRV-test conditions correlates less well. This is best seen if one compares the behaviour of additive B and C (Fig. 16 B and C) which both contain sulphur, although in a different binding state. Whereas phosphourous passes through a maximum when increasing temperature the concentration of sulphur varies only slightly. This leads to the conclusion that under milder wear test conditions (SRV) mainly phosphorous is responsible for the wear preventing action of the additives. Figure 23 shows that the friction coefficient is directly related to the phosphorous concentration. With higher P-concentration in the reaction layers the coefficient of friction is reduced considerably. This in turn reduces wear as illustrated by Fig. 24 where the wear scar area is plotted as a function of the friction coefficient for different additive concentrations. The data also demonstrate that wear can be reduced considerably provided the additive is present in sufficiently high concentration to build up a continuous reaction layer.

CONCLUDING REMARKS

Results and discussion have evidenced that many aspects have to be considered when investigating the role of a wear preventing additive. The type of solvent and the wear test have to be chosen as a function of the goals which are persued. Small wear test samples are particularly suited for AES and XPS analysis on the worn surfaces as demonstrated here for the Reichert and SRV test. FZG-test gears, on the other hand, cannot be entered into a UHV system without cutting them first into smaller samples. This is time consuming and may create problems of contamination which may negatively affect the accuracy of the surface analysis. In general, for surface analysis it is extremly important that transfer conditions from the oil bath to the UHV analysis chamber are well controlled. Reproducible preparation of the test specimen surface is essential. For example, in the present study AES depth profile analysis on VKA spheres gave unsatisfactory results because different mechanical polishing procedures have been used by the manufacturer.

To study the influence of the molecular structure of AW-additives using surface analysis the SRV-test was found to be best suited. It yields the friction coefficient as well as the area of the wear scar. It allows to vary easily such important parameters as additive concentration, load and temperature. The size of the test specimen is well suited for surface and thin film analysis. Furthermore, only a small volume of lubricant is needed allowing one to use highly purified, expensive solvents. While for mechanistic studies the SRV-test, therefore, is particularly well suited it is understood that application oriented additive screening must still involve a variety of different tests simulating as closely as possible the multitude of solicitations encountered in practice.

AES depth profile analysis has turned out to give more valuable information on reaction layers than non destructive XPS analysis which gives information on the very surface composition only. Because of unavoidable reactions at the surface during cleaning and transfer the surfaces analyzed showed always a superficial layer containing C and O (~1–2 nm), independent of the wear test. Therefore, the composition of the first few monolayers is usually not significant for establishing wear test correlations. The composition of deeper layers (>1.5 nm), however, can be reproducibly measured and represents adequately the chemical reactions taking place during the wear test. Depth profile analysis of reaction layers, therefore, allows correlations to be formulated between layer composition, the nature of the additive and the applied wear test conditions:

This leads to the following conclusions:

(a) Application of AES depth analysis complemented by XPS leads to better understanding of the wear preventing role of AW/EP-additives. Under severe lubrication conditions it reveals the formation of reaction layers of a thickness of about 10–20 nm containing fragments of the additive molecule.
(b) Reaction layer composition can be correlated to wear parameters like wear scar area or coefficient of friction. This allows one to study the influence of structure and composition of the additive molecule on its reactions at the surface.
(c) For model studies involving surface analysis the SRV-test allowing for control of the influence of load, friction path and temperature was found to be best suited.
(d) Unexpected side reactions due to tribochemical activation of base oil sulphur can lead to artefacts. Therefore, the characterisation of the base oil is an essential step in any friction and wear study.

REFERENCES

1. E. Rabinowicz *Friction and Wear of Materials,* Wiley (1965).
2. D.M. Buckley, *Surface in Adhesion, Friction, Wear and Lubrication,* Elsevier (1981).
3. D.F. Moore, *Principles and Applications of Tribology,* p. 134, Pergamon (1975).
4. R. Schumacher, D. Landolt and H.J. Mathieu, H. Zinke, *ASLE Trans,* **26,** 94 (1983).
5. H.J. Mathieu, D. Landolt, R. Schumacher, *Wear* **66,** 87 (1981).
6. R. Schumacher, E. Gegner, A. Schmidt, H.J. Mathieu, D. Landolt, *Tribology* **13**, 313 (1980).
7. E.S. Forbes, *Wear* **15,** 87 (1970).
8. T. Sakurai, *J. Lubrication Technology* **103,** 473 (1981).
9. K.H. Gaukler, G. Möllensted, P. Stöhr, *Motorentechnische Zeitschr.* **38,** 34 (1977).
10. J.J. McCarrol, R.W. Mould, H.B. Silver, M.I. Sims, *Nature* 266, 518 (1977).
11. M.R. Phillips, M. Dewey, D. Hahl, T.F. Quinn, H.N. Southworth, *Vakuum* **26,** 451 (1977).
12. T.P. Davis, W.G. Johnston, *ASLE Trans.* **23,** 289 (1980).
13. B.A. Baldwin, *Lub. Eng.* **32,** 125 (1975).
14. R.J. Bird, G.D. Galvin, *Wear* **37,** 143 (1976).
15. D.R. Wheeler, *Scanning Electron Microscopy* **II,** 589 (1984).
16. W.A. Glaeser, *Wear* **100,** 477 (1984).
17. Four Ball Test *ASTM D–2783, IP 239, DIN 51350.*
18. FZG Gear Machine, IP 166
19. K.H. Habig, *Verschleiss und Härte von Werkstoffen,* p. 22ff Carl Hanser Verlag (1980).
20. E. Rosset, H.J. Mathieu, D. Landolt, *Wear* **94,** 125 (1984).
21. M.T. Anthony and M.P. Seah, *Surface Interface Anal.* **6,** 95 (1984).
22. H.J. Mathieu, D. Landolt, accepted for publ. in *Wear,* (1986).
23. R. Schumacher, H. Zinke, D. Landolt, H.J. Mathieu, *5th Intern. Kolloquium Additive für Schmierstoffe und Arbeitsflüssigkeiten, 14–16. Jan. 1986, Techn. Akademie Esslingen,* vol I, 3.9–4 (1986)

SURFACE AND INTERFACE ANALYSIS, VOL. 9, 493–499 (1986)

The Influence of Surface Films on the Protection of Metal Surfaces Under Boundary Lubrication Conditions

J. L. Sullivan and L. F. Wong
Surface Analysis and Tribology Group, University of Aston, Birmingham, England

Present methods of refining aviation kerosenes produce the fuels of high thermal stability required by modern jet aircraft, but remove many of the polar impurities present in fuels refined by more traditional techniques. This leads to a reduction in the fuels boundary lubricating ability and such fuels have in the past produced high wear rates and seizure in some aircraft fuel systems. The majority of such problems have been overcome by the addition of boundary lubricants at concentrations of about 12 to 15 ppm to the refined kerosenes, but the mechanisms of protection of these additives are not understood. This investigation was conducted to examine the surfaces of aluminium bronze and tool steel (a metallurgical combination used in the fuel pumps of the majority of British jet aircraft engines) worn under boundary lubrication conditions in the presence of kerosenes both with and without additives. The object was to isolate the mechanisms of boundary film formation and protection of the surfaces through extensive analyses of surfaces and debris, using techniques such as Auger electron spectroscopy, electron probe microanalysis ellipsometry and x-ray diffraction. A model is proposed to explain these mechanisms in terms of material transfer and subsequent competing reactions between dissolved oxygen and fatty acid molecules with the metal surfaces.

INTRODUCTION

Jet aircraft fuel systems, and in particular fuel pumps, must operate without external lubrication under relatively severe conditions, where for large parts of their operating cycle metal to metal contact is prevented only by absorbed hydrocarbon boundary films. Few problems were experienced with kerosenes produced by traditional chemical refining processes, but when hydrofining techniques were introduced to produce the fuels of high thermal stability required by current generation aircraft engines, high wear rates and seizure were experienced in some fuel systems. This was thought to be due to the more complete removal by the hydrofining process of the hetrocyclic and other polar impurities required for boundary film formation.

The majority of problems associated with wear and seizure were initially overcome by the introduction to hydrofined kerosene of a boundary lubricant additive consisting of 45% dimeric acid plus 5% phosphate ester in a fuel oil base at a concentration of about 12 ppm. The action of this additive in producing protective boundary films with an aluminium bronze on tool steel conjunction (a metallurgical combination used in the majority of British aircraft fuel pumps) has been studied by one of the authors and the results are reported elsewhere.[1–3] From these investigations models were proposed to explain the mechanisms of protection in terms of aluminium depletion, aluminium phosphate production and metal oxide – fatty acid surface film formation.

Although present at a concentration of only 0.6 ppm, these initial investigations indicated that the phosphate ester had a profound effect on the wear process. However, objections were raised to the inclusion of the ester due to possible damage to turbine blades and new generation additives were introduced consisting solely of dimeric acids in a pure base. In the practical situation these were found to protect surfaces against high wear and seizure at least as well as the early formulations which indicates that although the phosphate esters interact with the aluminium bronze surfaces this interaction is not essential for boundary protection. Thus in order to extend the work, eliminate the effects due to phosphate ester interactions and isolate the dominant mechanisms of protection this investigation was undertaken using a similar metal combination, but a new generation additive. An understanding of the mechanisms was sought through measurements of wear rates and by extensive use of surface sensitive techniques such as Auger electron spectroscopy, electron probe microanalysis, scanning electron microscopy and x-ray debris analysis.

It is hoped that the results of this investigation will have wider implications in explaining general modes of protection by boundary additive films.

EXPERIMENTAL DETAILS

The wear test rig used for the generation of samples has been described elsewhere,[1] but consists essentially of a 2 mm diameter vertical aluminium bronze pin on a 100 mm diameter horizontal rotating steel disk. Speed could be varied from 0 to 600 rpm and load from 0.5 to 200 N. Earlier work[4] has shown that surfaces generated

0142–2421/86/120493–07 $05.00

on this rig were almost identical to those developed on pump simulation rigs and those taken from in service components. A peristaltic pump with flow adjustment reservoir was used to pump fluid to the conjunction using a single pass non-recovery system in order to avoid debris contamination. The rate of fluid flow could be varied in order to maintain boundary lubrication conditions. Friction and wear were continously monitored using a strain gauge load cell and linear displacement transducer respectively, the amplified outputs of which were displayed on a chart recorder.

The materials used in the investigation were DTD 197A aluminium bronze and KE 180 tool steel, the composition of which are:

DTD 197A: 11%Al, 4%Fe, 4%Ni, 2%Mn, remainder Cu

KE 180: 13%Cr, 1.5%C, 0.5%Mn, 5%Si, remainder Fe

The KE 180 steel disks were of hardness 700 HV and were randomly ground and lapped to a surface finish of 0.02 μm cla. The pins of hardness 240 HV were produced from rod and surfaces finished to 1 μm cla. Prior to each test all components were cleaned in a petroleum vapour bath at 80°C for a period of one hour.

The fluid used in the investigation was an additive free hydrotreated aviation kerosene. Prior to each experiment this was percolated through a one metre column of alumina catalyst to remove any polar impurities formed or taken up during storage. The additive used in the investigation consisted of dimers of ethoxylated alkyl phenol acids in a process oil and kerosene base and this was added to some of the kerosene samples at a volume concentration of 12 ppm immediately prior to the wear tests. Fluid flow rates were adjusted in the experiments to maintain a coefficient of friction of 0.1 and hence ensure boundary lubrication conditions. Tests were then carried out at room temperature with a new pin on new disk track for each experiment. Equilibrium wear rates of aluminium bronze against steel were then determined as a function of load for velocities of 0.6, 2 and 4 ms^{-1} with loads increasing from 24.5 N in increments of 24.5 N to a maximum of 196 N or until seizure had occurred. The wear tests usually lasted for periods between 90 and 120 min, all measurements being taken after an initial 'running-in' period when equilibrium conditions had been established. The tests were conducted in the presence of percolated hydrofined kerosene and then repeated for kerosene plus additive. An average of five such wear tests were conducted for each load and speed combination.

Fluids used in the experiment were collected after each wear test and filtered to extract wear debris which was analysed using powder x-ray diffraction with Cu K_α radiation at 35 kV and 30 μA. Samples were exposed for periods of 40 minutes and measured d values were compared with standard powder diffraction files in order to identify debris constituents.

Auger electron spectroscopy was employed to identify surface compositional changes. During the analysis a primary electron beam of 3 keV was used given a specimen current of 2 μA into a 20 μm diameter spot. In order to gain information on changes in the concentration of elements of interest (C, O, S, Al, Cu and Fe) with depth into the surface, successive layers of the surface were removed by periods of 0.5 keV Argon ion bombardment, the Auger spectra being recorded between each successive period. Relative atomic concentrations were calculated from spectral peak to peak heights and published sensitivity factors[5] for each spectrum and the resulting depth profiles plotted.

Selected samples of pin and disk surfaces were analysed using electron probe microanalysis (15 keV beam energy, 80 μA beam current and 3000 Å spot size), where spot analyses and large area elemental distribution maps of elements of interest in the surface and on some 11° taper sections were recorded. Further examination of surfaces, sub-surfaces and debris was carried out using scanning electron microscopy and x-ray energy dispersive analysis.

RESULTS

Figure 1 shows wear rate as a function of load for speeds of 0.6, 2 and 4 ms^{-1} for experiments conducted with kerosene and kerosene plus additive. The errors in measurement of wear rates were always negligible compared with the systematic scatter of points. All points lie about three straight lines. Wear rates without additive show no systematic variation for speeds of 0.6 and 2 ms^{-1}, Fig. 1 (a), but a significant increase was observed, Fig. 1 (b), at 4 ms^{-1} at all loads. No variation in wear rate with speed was observed when the additive was present. The additive was pro-wear at the lower speeds. Seizure occurred at 196 N at 2 ms^{-1} and 122.5 N at 4 ms^{-1} in pure kerosene. No seizure occurred at any load or speed when the additive was present or for speeds of 0.6 ms^{-1} in pure kerosene.

X-ray powder diffraction analysis showed the debris generated in kerosene plus additive and that in pure kerosene up to the point of seizure to be very similar. The majority constituents were Cu_2O and Cu. No

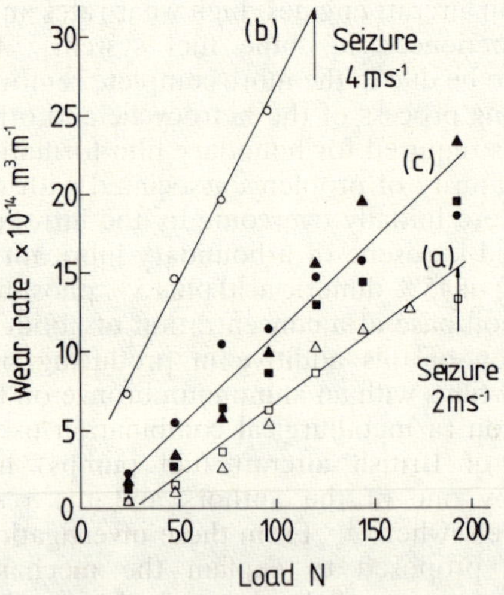

Figure 1. Wear rate versus load (a) 0.6 and 2 ms^{-1}, pure kerosene; (b) 4 ms^{-1}, pure kerosene; (c) 0.6, 2 and 4 ms^{-1}, kerosene plus additive; △ 0.6 ms^{-1}, □ 2 ms^{-1}, ○ 4 ms^{-1} = pure kerosene; ▲ 0.6 ms^{-1}, ■ 2 ms^{-1}, ● 4 ms^{-1} = kerosene plus additive.

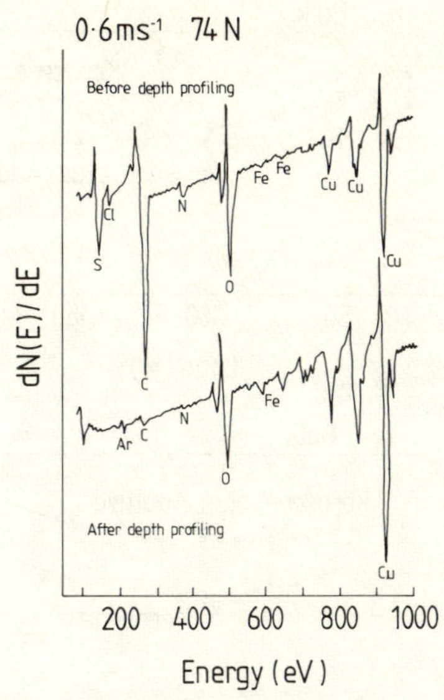

Figure 2. Auger electron spectra from an aluminium bronze pin worn under a load of 74 N and speed 0.6 ms^{-1} in pure kerosene.

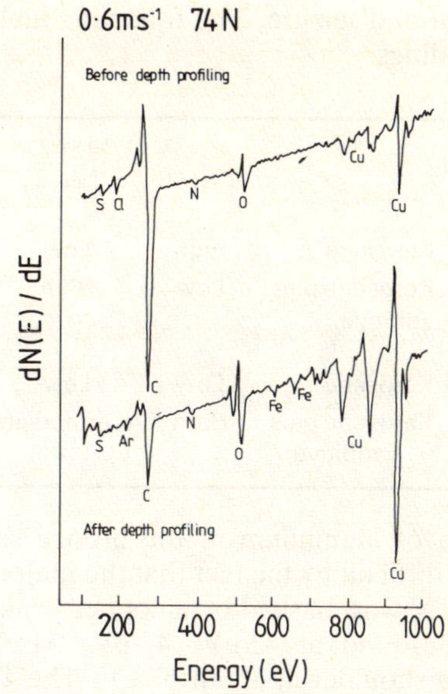

Figure 3. Auger electron spectra for an aluminium bronze pin worn under a load of 74 N and speed 0.6 ms^{-1} in kerosene plus additive.

evidence of either Al or Al_2O_3 was found using this technique, although x-ray energy dispersive analysis showed aluminium or its oxide to be present at concentrations of less than 4%, the detection limit of the powder XRD technique.

Figures 2 and 3 show typical Auger spectra from aluminium bronze surfaces generated under relatively mild conditions, in this case a load of 73.5 N and speeds of 0.6 ms^{-1}.

Comparing the two figures the spectra for experiments without the additive show that the surface contaminant carbon layer is rapidly removed by argon ion bombardment leaving a mainly copper oxide film. In contrast to this the spectra for experiments in kerosene plus additive show less oxygen (by implication less oxide) and a considerable amount of hydrocarbon film remaining on the surface after 80 minutes of argon ion bombardment. The sputtering rate for such a complex surface matrix is difficult to accurately assess, but calculation from the data of Carter and Colligon[6] would suggest that 80 μA minutes of argon ion bombardment under the conditions of the experiments corresponds to the removal of about 150 to 200 A° of the surface. Spectra recorded under more severe conditions at speeds of 4 ms^{-1} conversely show more oxide on the surface of pins worn in the presence of the additive. Depth profiles further illustrate these points. In the interest of clarity the profiles shown here compare carbon and oxygen concentrations only, although concentration for all other elements of interest were recorded. Figure 4 shows a profile for a surface generated at a load of 74.5 N and speed 2 ms^{-1} and is typical for low load, low speed conditions where the additive exhibits pro-wear effects, the profiles show that there is more carbon than oxygen in the surface films from pins worn in kerosene plus additive while the converse is true for surfaces worn in pure kerosene. The profiles produced from surfaces run at higher speeds, Fig. 5 is the profile of a pin worn at 98 N load and 4 ms^{-1}, show that the results are reversed and that more oxygen than carbon is present on surfaces generated in the presence of the additive. Without the additive, particularly at higher loads, there is very little oxygen or, after the removal of the initial contaminant layer, little carbon. These results suggest that under relatively mild conditions the additive inhibits oxidation resulting in a thinner or more patchy film, but that under severe conditions the additive gives sufficient protection to the surface to allow oxidation and prevent

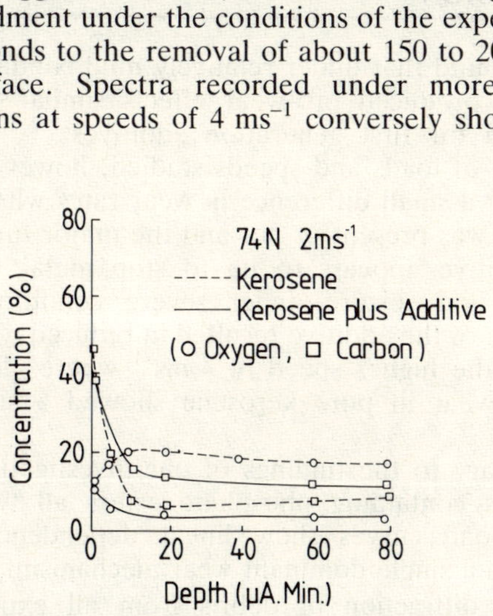

Figure 4. Oxygen and carbon depth profiles for an aluminium bronze pin surface worn under a load of 74 N at 2 ms^{-1}.

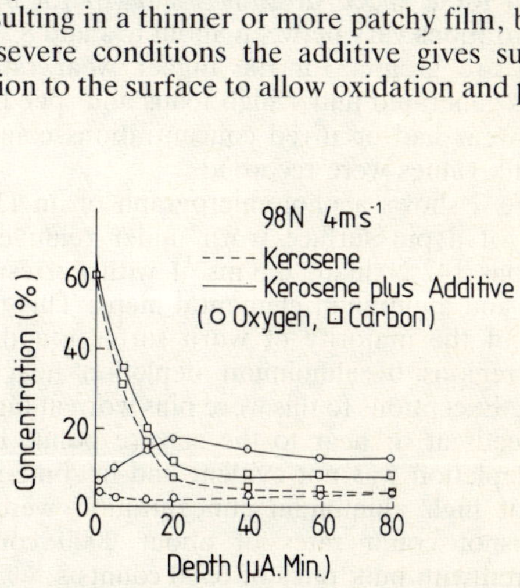

Figure 5. Oxygen and carbon depth profiles for an aluminium bronze pin surface worn under a load of 98 N at 4 ms^{-1}.

metallic wear and seizure. The following table summarizes the findings.

Load/speed	Fluid	Surface concentrations		
		Oxygen	Carbon	Copper oxide
Low	Kerosene	High	Low	High
	Kerosene plus additive	Low	High	Low
High	Kerosene	Low	Low	Low
	Kerosene plus additive	High	Intermediate	High

Detection of aluminium in the bronze surface was made more difficult by the fact that the major Al Auger peak at 67 eV was masked by a copper peak at 60 eV. Peaks were found at 42 eV, 47 eV, 54 eV, 58 eV, 60 eV and a shoulder at about 67 eV. The 47 eV peak can be assigned to iron and the 58 and 60 eV peaks are due to copper. This leaves peaks at 42, 54 and 67 eV. The 67 eV shoulder is a result of the main aluminium peak and those at 42 and 54 eV would appear to be aluminium oxide plasmon loss peaks. Thus there appears to be some aluminium or Al_2O_3 in the surface, but the fact that the peak at 1396 eV was generally not visible above noise levels indicates that aluminium concentration was low and that the oxide formed was primarily copper oxide. Aluminium at concentration close to bulk values was detected through the higher energy peaks under severe conditions close to seizure point for samples worn in pure kerosene.

The low concentration of aluminium and aluminium oxide in wear debris and on the surface suggests that there is preferential depletion of that element from the bronze. This was confirmed by electron probe microanalysis on surfaces generated over the whole load range and for speeds of 0.6, 2 and 4 ms^{-1}. Concentrations were calculated from the average of a number of electron probe spot analyses on a given pin surface after corrections for atomic number, adsorption and fluorescence. For the conditions of our experiment the effective sampling depth was 1 μm. Figure 6 shows typical curves of mean aluminium concentration versus load (a) for a speed of 2 ms^{-1} and (b) for 0.6 ms^{-1}. Concentrations vary between about 6% and 8% and in general are greater for the higher wear rates. For samples generated under high loads and speeds where severe wear had occurred concentrations comparable with bulk values were recorded.

Figure 7 shows a photomicrograph of an 11° taper section of a pin surface worn under relatively mild conditions (47 N load, 0.6 ms^{-1}) with corresponding copper and aluminium elemental maps. This figure is typical of the majority of worn surfaces and clearly shows regions of aluminium depletion near to the surface. Exceptions to this were pins worn at high loads and speeds at or near to the seizure point. In these cases depletion was not evident and in some samples areas of high aluminium concentration were found giving spot count rates of about 9000 counts s^{-1} compared with bulk rates of 6500 counts s^{-1}.

Figure 8 shows a photomicrograph of a section of wear track from the steel counterface from an experiment conducted under similar conditions of load and speed to that of Fig. 7. Corresponding maps of copper aluminium and iron are also included in this figure. These photomicrographs are typical of surfaces worn in the presence of pure kerosene and kerosene plus additive and clearly show that transfer from pin to disk had occurred. EPMA spot analyses showed that the aluminium content of this transfer film on the steel surface was always higher than in the bulk bronze, usually ranging from 20 to 80% depending on position.

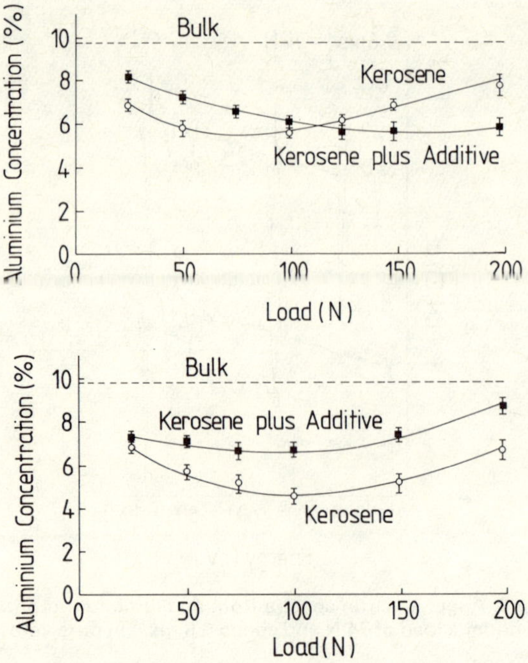

Figure 6. Aluminium concentration (weight %) versus load (a) 2 ms^{-1}; (b) 0.6 ms^{-1}.

Scanning electron micrographs taken from pin surfaces worn with and without additive show little apparent difference under relatively mild conditions and each appears smooth and oxidized.

DISCUSSION

It was found that under relatively mild conditions the additive produced pro-wear effects similar to those observed for first generation additives.[1,2,7] For the majority of loads and speeds studied, however, there was only a small difference in wear rates whether the additive was present or not and the major function of the additive appears to be to stop metal to metal contact and seizure under severe conditions. The inclusion of the additive resulted in reduced wear rates only at the higher speed of 4 ms^{-1} where aluminium bronze wear in pure kerosene showed a significant increase.

Contrary to the findings of our investigations with additives containing phosphate esters all wear rate versus load curves show linear dependency which indicates a single dominant wear mechanism.

X-ray diffraction of debris from all experiments where seizure had not occurred showed it to consist of Cu_2O and Cu, the major constituent being the oxide.

Electron probe microanalysis showed that there was transfer of material from the bronze pin to the steel counterface and that this film was rich in aluminium. EPMA further showed that at equilibrium, under relatively mild conditions, considerable depletion of aluminium from the first μm of surface had occurred and this was confirmed by AES analysis. The fact that aluminium concentration in the transfer layer was high, but concentration in the debris low suggests that depletion and preferential transfer occurs only during the initial stages of wear and subsequently during film breakdown. This process leaves a copper rich surface on the bronze which oxidizes to form Cu_2O and an extremely adherent aluminium rich film on the steel surface. Aluminium is known to readily diffuse into steel to produce a hard intermetallic compound.[8] Wear then occurs almost entirely at the bronze surface.

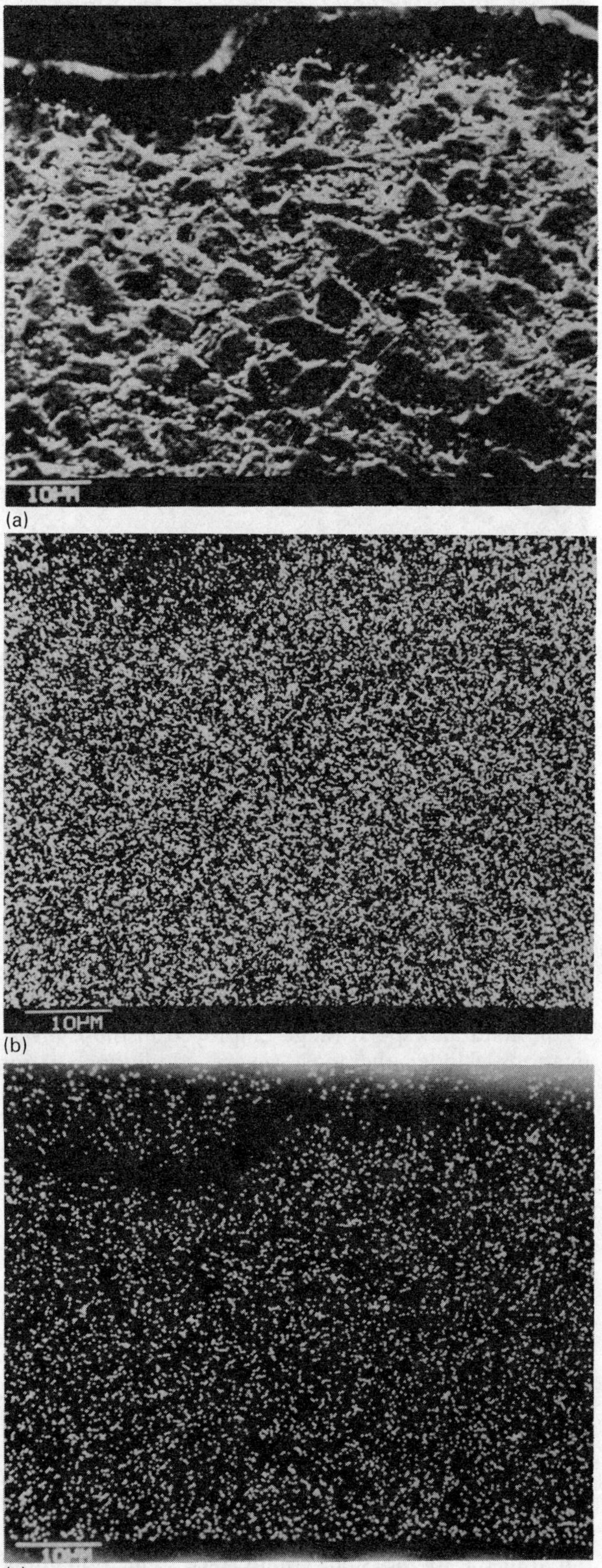

Figure 7. Photomicrographs of 11° taper section of a worn pin surface (a) reflected electron image; (b) copper x-ray distribution; (c) aluminium x-ray distribution.

The results for surface and debris analysis discussed above show little difference whether the additive was present in the kerosene or not. Surface topographies revealed by SEM were also similar, both types of fluid producing smooth oxidized surfaces for the majority of the load and speed range. The only surfaces which differed were those generated in pure kerosene at high loads and speeds, where considerable surface distress was observed. The results therefore indicate that both with and without additive present in the kerosene an oxidational wear mechanism is dominant.[9] It is only when the outermost layers of the surface are anlysed using AES that differences of degree become apparent. The Auger depth profiles in Fig. 4 show that surfaces generated in the presence of the additive under mild boundary conditions have less oxide and more hydrocarbon than those generated in pure kerosene. For surfaces at higher speeds and loads when wear rate tends are reversed, the profiles, Fig. 5, show a very sparse coverage of both oxide and hydrocarbon for surfaces generated in pure kerosene, but a relatively thick oxide film for those worn in the presence of the additive. Narrow scans of the major carbon peak also show fine detail in the surfaces generated in the presence of the additive which are not present on surfaces generated in additive free kerosene, indicating different adsorbed hydrocarbon species.

The results then show that the mechanism of boundary film formation and subsequent protection depends on the preferential surface depletion of aluminium from the bronze surface and transfer to the steel counterface to produce a stable adherent aluminium rich layer. Subsequently the copper rich surface of the bronze oxidizes to form a stable protective film of Cu_2O, this process being aided by the fact that the elastic moduli of Cu and Cu_2O are similar. The extent of coverage of the oxide Cu_2O and the degree of protection afforded depends on whether the additive is present or not and on the severity of the conditions. Under relatively mild conditions with the additive present in the kerosene there will be a competing reaction between the dimer acid and entrapped oxygen in the fluid with the surface. A typical example for heat of adsorption of such an acid is 42 kJ $mole^{-1}$[10] compared to 80 kJ $mole^{-1}$[11] for the activation energy for the oxidation of copper. Thus after initial oxidation a stable conventional hydrocarbon boundary film will be preferentially formed on the bronze surface inhibiting further oxidation. Asperity contacts will produce sharp increases in temperature and this will lead to desorption of polar molecules from contact sites and allow limited oxidation to take place. The extent of the oxide coverage, however, may be expected to be less

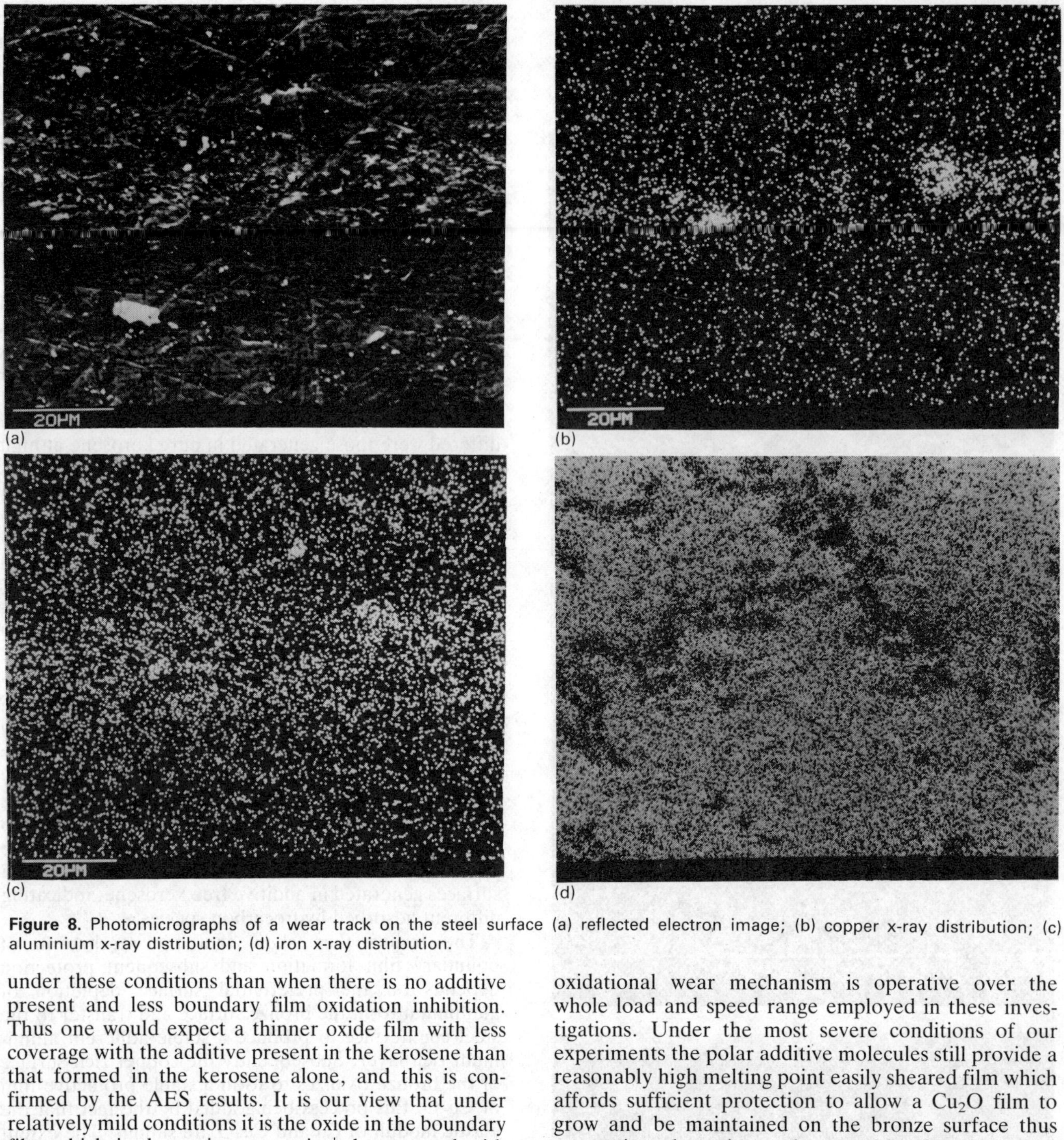

Figure 8. Photomicrographs of a wear track on the steel surface (a) reflected electron image; (b) copper x-ray distribution; (c) aluminium x-ray distribution; (d) iron x-ray distribution.

under these conditions than when there is no additive present and less boundary film oxidation inhibition. Thus one would expect a thinner oxide film with less coverage with the additive present in the kerosene than that formed in the kerosene alone, and this is confirmed by the AES results. It is our view that under relatively mild conditions it is the oxide in the boundary film which is the major protecting element and with limited oxide coverage one might expect wear to be greater than when a thicker stable oxide film is present on the surface. Hence under these conditions the additive exhibits pro-wear effects.

When conditions become more severe and load and speed increase, without the additive and the protection afforded by the boundary polar film, the rate of removal of the oxide becomes greater than its rate of formation and catastrophic metallic wear and seizure ensue (Fig. 1). Auger profiles (Fig. 5), under these conditions show little oxide or hydrocarbon coverage. When the additive is present, however, our results show that wear rates do not increase with speed and increase only linearly with load in exactly the same manner as for the mild boundary conditions. This shows that when the additive is present the same mild oxidational wear mechanism is operative over the whole load and speed range employed in these investigations. Under the most severe conditions of our experiments the polar additive molecules still provide a reasonably high melting point easily sheared film which affords sufficient protection to allow a Cu_2O film to grow and be maintained on the bronze surface thus preventing the seizure that would otherwise occur under these conditions.

CONCLUSIONS

These investigations show that oxidational wear is the single dominant mode for aluminium bronze on tool steel under boundary lubrication conditions in the presence of aviation kerosene.

In the early stages of contact aluminium depletion from the bronze surface regions occurs with preferential transfer to the steel counterface. The remaining copper rich bronze surface then oxidizes to form a Cu_2O protective layer.

With the dimeric acid additive present in the

kerosene there is a competing reaction between acid and oxygen molecules from the fluid and the surface and this leads to a thinner or more patchy oxide film than that developed in additive free kerosene. Hence the slight pro-wear effect of the additive under relatively mild conditions.

Under more severe conditions when oxide film destruction and metal to metal contact would normally occur, the dynamically absorbed fatty acid surface film affords sufficient protection to allow oxide films to be maintained and hence prevents seizure.

In this application, therefore, the major function of the additive is to preserve and maintain a protective oxide film.

Acknowledgements

The authors wish to thank J. Hadley and J. Hutton of Shell Research, Thornton, for their helpful discussions and for the provision of materials and analytical facilities during this study. L. F. Wong would further like to thank Shell and the Science and Engineering Research Council for financial assistance.

REFERENCES

1. Poole W. and Sullivan J. L. *ASLE Trans.* **22**, 154–169 (1979).
2. Sullivan J. L. and Poole W. *Thermal Effects in Tribology*, 316–324 I.Mech.E. (1980).
3. Poole W. and Sullivan J. L. *ASLE Trans.* **23**, 401–409 (1980).
4. Poole W. Ph.D. Thesis, University of Aston (1980).
5. Davis L. E. *Handbook of Auger Electron Spectroscopy*, Physical Electronics Industries (1978).
6. Carter G. and Colligon J. S. *Ion Bombardment of Surfaces*, Heineman (1968).
7. Vere R. A., Askwith T. C. and Hardy P. J. *M.O.D. Report AX/395/014* (1974).
8. Dzhavaga I. I. and Lebeder Y. M. *Aut. Svarka*, **8**, 11–14 (1970).
9. Quinn T. F. J. and Sullivan J. L. *Wear*, **94**, 175–195 (1984).
10. Allen C. M. and Drauglis E. *Wear*, **14**, 363–383 (1969).
11. Tylecote R. F. J. Inst. Met. **78**, 259–300 (1951).

SURFACE AND INTERFACE ANALYSIS, VOL. 9, 501–505 (1986)

XPS Studies of Xanthate Adsorption on Metals and Sulphides

L.-S. Johansson, J. Juhanoja, K. Laajalehto and E. Suoninen
Department of Physical Sciences, University of Turku, SF-20500 Turku, Finland

J. Mielczarski
Department of Chemistry, Silesian Technical University, 44–100 Gliwice, Poland

The influence of experimental factors in XPS studies of adsorption of xanthates (dithiocarbonates) on metal or mineral surfaces has been studied in order to explain discrepancies in the results previously published by different authors. Serious decomposition of the adsorption layer on copper substrate was observed as a result of prolonged exposure to non-monochromatized radiation from a magnesium anode tube, whereas monochromatized AlKα radiation did not cause any noticeable layer damage. The most probable explanation for the damage is heating caused by the x-ray flux absorbed by the sample. Analysis of new XPS measurements of ethyl xanthate adsorption layers on copper substrates confirms earlier interpretation by Mielczarski *et al.* Comparison with results for lead sulphide substrate (galena) by Pillai *et al.* indicates a dominating role of preadsorption contaminants in their measurements, which explains the discrepancies between the interpretations presented by the two groups for the layer structure. The effect of first monolayer orientation on the spectral intensities is emphasized. This effect is already present in the first spectra of adsorbed layers of thio compounds reported by Clifford *et al.*

INTRODUCTION

The selective adsorption of so-called collector molecules on mineral surfaces forms the basis of the industrial process of flotation[1] used for separating different minerals from each other. The basic aspects of the adsorption process have been studied by infrared methods.[2] In 1974, Clifford *et al.*[3] demonstrated the use of XPS for studies of the adsorption of dithio compounds commonly used in flotation of sulphides. The method has since been used by several groups.[4–6] These studies have shown that XPS methods can easily be used for qualitative observations of the adsorption process, but quantitative interpretation is complicated by the anomalous intensity relationships already reported by Clifford *et al.*[3] and confirmed by later work using both metal[7] and sulphide[8,9] substrates. The original explanation of these anomalies based on huge variations in the ionization cross sections of the core orbitals[3] is implausible and has been disproved by gas-state XPS measurements of xanthate (dithiocarbonate) derivatives.[10] Selective attenuation of the XPS signals from different depths of origin in the adsorption layer because of inelastic scattering was suggested on the basis of an analysis of the intensity relationships as functions of the adsorption time.[7] These relationships, both between the different components of a particular line and between different lines, are for different collectors and substrates[7–9] consistent with this model. Strong intensity anomalies have also been reported by Pillai *et al.*[5] between the components of the emission lines emitted by an ethyl xanthate layer on galena (PbS), but their interpretation of the observed line shapes differs essentially from Ref. 7.

We give in the following a preliminary report of our recent work on this problem. The aim of the work is twofold: to study the possible effect of some experimental factors on the XPS measurements of adsorbed ethyl xanthate layers, and to resolve and explain the discrepancies between the interpretations.[5,7]

EXPERIMENTAL

Elemental polycrystalline copper sheet was used as the substrate. The surface was ground to 1/4 μm smoothness with diamond paste, washed with soap solution, rinsed with distilled water and dried. In order to minimize the contamination, the substrate was subsequently coated in the spectrometer vacuum with a thin layer of evaporated copper. The adsorption layer was made by dipping the substrate in a 8.0×10^{-5} mol l^{-1} solution of potassium ethyl xanthate (KEX) for varying times. No attempt was made to exclude atmospheric oxygen during the adsorption treatment. The resulting coverage of KEX molecules could be estimated from previous IR[11] and XPS[7] measurements. The estimates were, however, only approximate because of some differences in the method of preparing the substrate surface.

The XPS spectra were obtained as in Ref. 7 with a VG ESCA3 MkII spectrometer using monochromatized AlKα excitation. The pressure in the analyser chamber was about 10^{-7} Pa.

A curve fitting program allowing description of the components by any one of several model functions (Gaussian, Lorentzian or Voigt) was developed[12] for optimal separation of the different components in the emission lines.

0142–2421/86/120501–05 $05.00

RESULTS AND DISCUSSION

Effect of the measurement technique on the adsorption layer

Xanthate compounds have been found to be under certain conditions relatively unstable. In XPS measurements of gaseous xanthate derivates,[10] the combined effect of the exciting radiation and slight heating (to about 50°C) of the sample, needed to vaporize the compound, was found to cause partial disintegration of the sample molecules. In order to establish whether similar effects could influence the XPS measurements of adsorbed xanthate layers, it was decided to study the effect of sample irradiation and the resulting heating of the sample. This was considered particularly important in the present measurements, since relatively long measuring times are often needed for sufficient statistical accuracy. Since the monochromatization used in the present work on the other hand greatly reduces the radiation doses, it was considered interesting to compare the effects of non-monochromatized and monochromatized radiation.

Figure 1 shows the curve fitting analysis performed on S 2p spectra from samples exposed before the measurement for increasing times to non-monochromatized radiation from a magnesium anode x-ray tube. A two-doublet fit is seen to give components corresponding to two different binding states with a binding energy (*BE*) difference of 0.8 eV. Comparison with previous measurements[7] indicates that the higher BE state corresponds to an adsorbed EX^- group remaining intact after the irradiation, whereas the lower *BE* component corresponds to sulphide-like bonding produced by radiation damage of the EX^- groups. The relative strength of the latter contribution can therefore be used as a measure of the radiation damage of the sample. The contributions to the other measured lines can be traced in a similar manner either to the EX^- groups, copper substrate or oxidation products.

Figure 2 shows the application of this method to a series of samples with different radiation histories. The results show that extensive non-monochromatized radiation causes considerable decrease in the intensity ratio S 2p/Cu $2p_{3/2}$, indicating both decrease in the S 2p intensity due to a gradual decomposition of the xanthate layer, and simultaneous increase in the Cu 2p intensity from the increasingly uncovered substrate. On the other hand, damage due to monochromatized Al K excitation is during a typical measuring time of a few hours not serious. Exposure to the ambient ultrahigh vacuum without simultaneous radiation was found to cause slow desorption of the EX^- groups, as reflected by attenuation of the emission lines and by some pressure increase in the chamber when the sample was introduced. However, no changes in the line shapes were found.

When a thin vacuum deposited layer of copper on a glass plate was used as a substrate, some increase in the amount of radiation damage was observed for the same radiation intensity level. Since the most important difference between the bulk copper substrate and a thin copper layer on the poorly heat-conducting glass would seem to be quicker heating of the latter substrate, this observation seems to support the temperature rise, caused by the exposure, as the probable reason for the disintegration of the layer. This result agrees with Ref. 10, in which the exact amount of sample heating was also found to control the disintegration.

Slight cooling of the sample holder during the

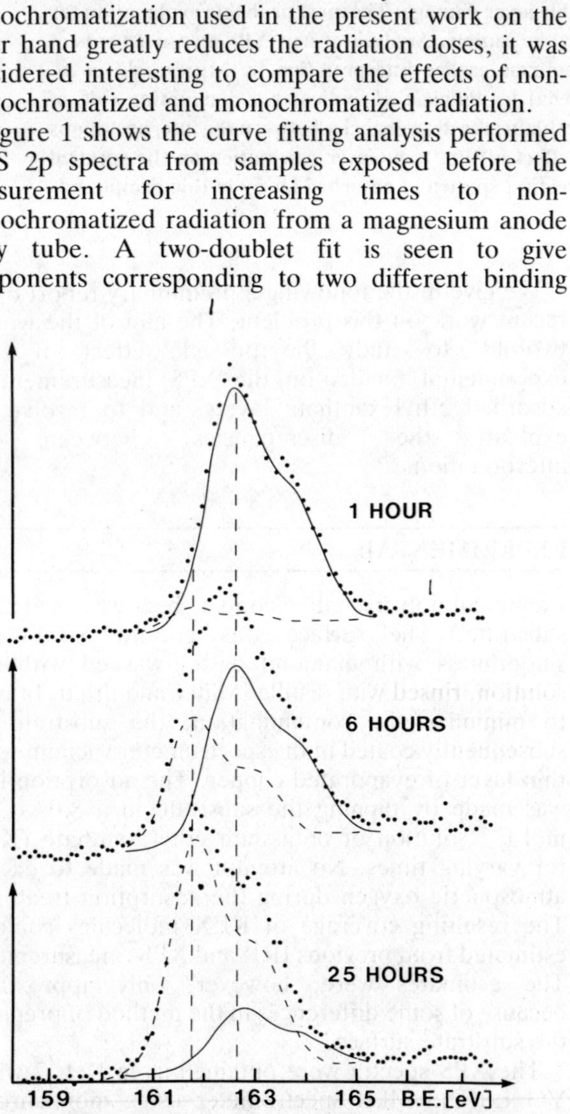

Figure 1. Change in the shape of the S 2p emission from a sample treated in xanthate solution for 10 min with increasing exposure to non-monochromatized radiation from a magnesium anode x-ray tube (13 kV, 10 mA) before the measurement. The radiation causes gradual increase in the relative intensity of the sulphide component (dashed curve) as compared to the intensity of the xanthate component (solid curve).

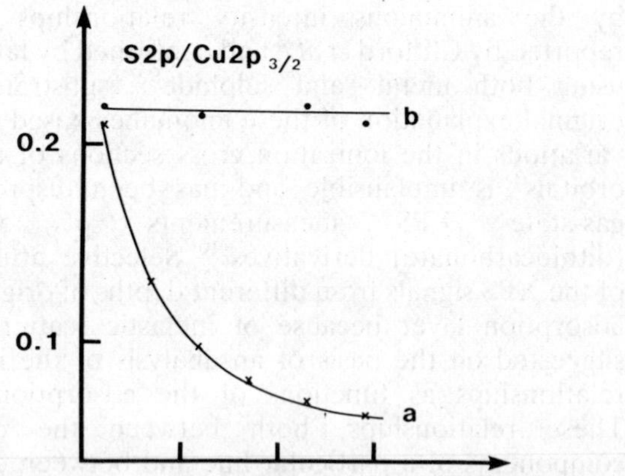

Figure 2. The S 2p/Cu $2p_{3/2}$ intensity ratio for samples irradiated for increasing times as in Fig. 1 (curve (a)) or with monochromatized AlKα radiation from an aluminium anode tube (13 kV, 10 mA; curve (b)). The samples were treated before irradiation for 10 min in xanthate solution. The intensities include in each case all components contributing to the measured lines.

irradition was found to eliminate any noticeable changes in the layer structure. This observation further confirms the decisive role of sample heating in the decomposition process. Another contributing factor could be bombardment of the sample by secondary electrons created by x-rays in the tube window adjacent to the sample surface in the case of non-monochromatic excitation. Although it cannot be completely disregarded, this mechanism does not seem probable as the main reason for the observed damage.

The above experiments show that the effect of the excitation on the structure of the adsorption layer was in the present studies obviously rather small and can be ruled out as a significant source of error in the interpretation of the spectra.

C 1s emission

In the measurements by Pillai *et al.*[5] of EX^- layers on galena, the sample thicknesses were probably several monolayers because of the long adsorption treatments (2–3 h). A three-component fit of the C 1s spectra showed a predominant contribution at the approximate position of the 'normal' hydrocarbon $(CH_2)_n$ contamination (284.8 eV) and another component at 286.3 eV. A third, quite weak component was placed at 288.0 eV. In spite of the authors' opinion that all these components were dominated by contributions from various contaminations, these three components were also assumed to correspond, respectively, to the three differently bonded carbon atoms I-III in the EX^- ion: $(C^IH_3C^{II}H_2OC^{III}S_2)^-$. The very different relative component intensities and their observed variations with the surface treatment were explained by the authors as being caused by the unequal and varying contributions of the contamination components. They also maintained that the binding of the carbon in the CS_2 group is not influenced by the interaction of the ion with the substrate but is essentially the same as in the bulk KEX phase.

Mielczarski *et al.*[7] interpreted the shape of the C 1s emission observed by them for EX^- layers of different thicknesses on copper as being mainly due to a component (C^I) at 284.8 eV and another component ($C^{II} + C^{III}$) at 286.5 eV. The only contamination component observed was the 'normal' one at about 285.0 eV. The observed pronounced changes in the relative component intensities were explained by effects of selective attenuation due to orientation of the adsorbed groups in the first monolayer. Essentially the same results were also obtained using a chalcocite (Cu_2S) substrate.[8]

Because of the contradicting results in Refs. 5 and 7, the C 1s emission from EX^- layers on copper for samples with different times of xanthate treatment was remeasured and analyzed by curve fitting. Two examples of the analysis are shown in Fig. 3(a). The obtained *BE* values are given in Table 1 together with the results reported in Refs. 5 and 7. The integrated intensities, normalized to the integrated S 2p intensities, are given in Fig. 4(a). Except for a somewhat larger initial contamination of the present samples, the results are compatible with the layer model presented in Ref. 7, if the $(CH_2)_n$ contribution is assumed to gradually disappear with increasing xanthate coverage. Hence, the present work supports the results and interpretation presented in Ref. 7. The *BE* values of the C atoms I-III seem to be confirmed. Since the intensity of the component marginally found in the region 288 eV is very small and decreases somewhat with increasing coverage, assigning it at least partly to the C^{III} as suggested in Ref. 5 seems doubtful. Presence of some carbonate groups on the surface is a plausible and sufficient explanation of this component.

It should be pointed out that the predominant part of

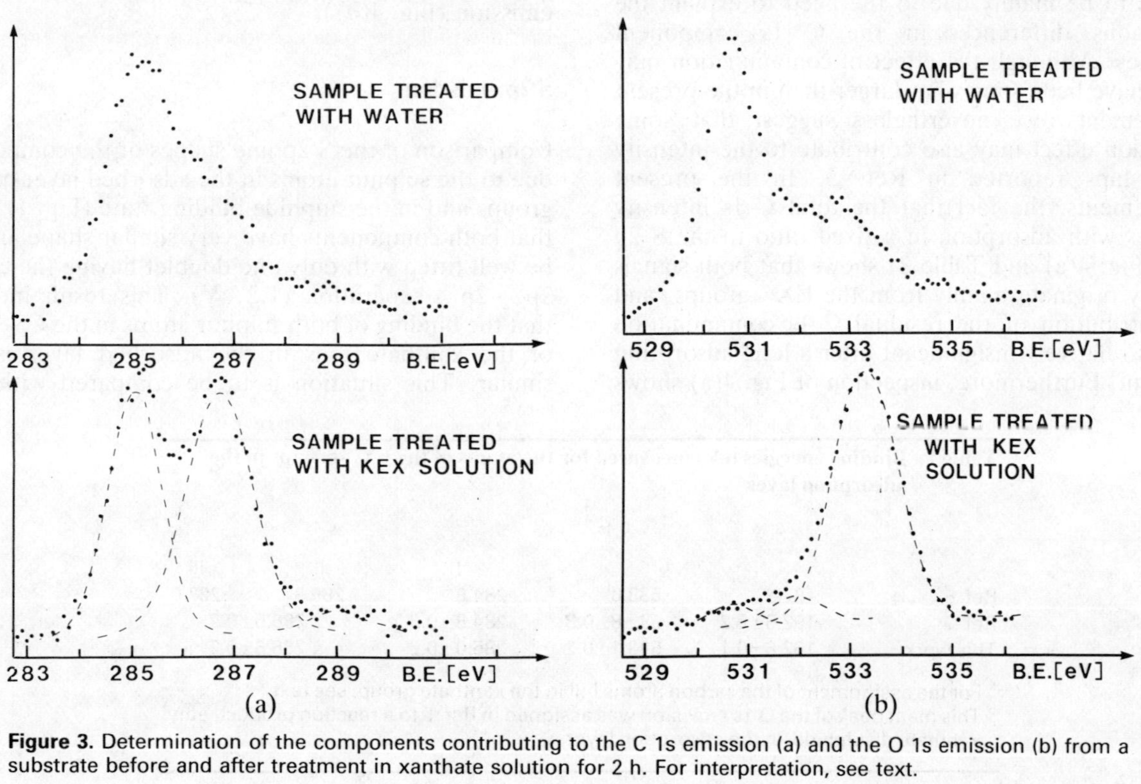

Figure 3. Determination of the components contributing to the C 1s emission (a) and the O 1s emission (b) from a substrate before and after treatment in xanthate solution for 2 h. For interpretation, see text.

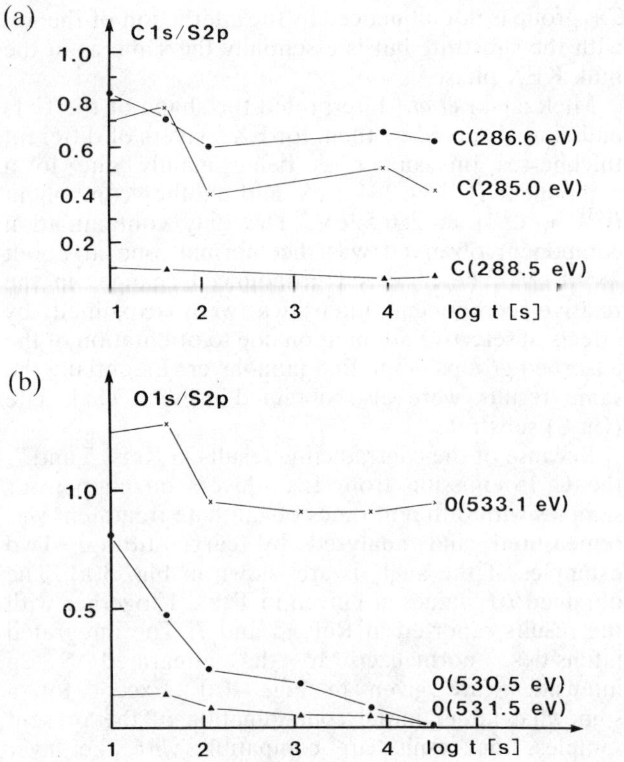

Figure 4. (a) The ratios of the integrated intensities of the different components of the C 1s spectra (Fig. 3(a)) to the S 2p intensities for samples treated for increasing times in xanthate solution; (b) the same results for the components of the O 1s spectra (Fig. 3(b)).

the contamination in all observed C 1s components claimed in Ref. 5 would seem to make any conclusions on the *BE*s of the carbon atoms in the EX^- group difficult. Although some contamination of the substrate is inevitable, the predominant role assigned to it in Ref. 5 seems to be mainly due to the need to explain the conspicuous differences in the C 1s component intensities. Although the effect of contamination may indeed have been somewhat larger than in the present measurements, we nevertheless suggest that some orientation effect may also contribute to the intensity relationships reported in Ref. 5. In the present measurements, the fact that the total C 1s intensity increases with adsorption in a fixed ratio to the S 2p signal (Fig. 4(a) and Table 2) shows that both signals probably originate mainly from the EX^- groups, and the contribution of the residual C 1s contamination signal is relatively insignificant after a long adsorption treatment. Furthermore, inspection of Fig. 4(a) shows

Table 2. Relative values of measured and calculated integrated intensities from a sample treated for 10 h in xanthate solution (cf. Fig 4). The calculation was based on the stoichiometric composition at the EX^- group and the Scofield atomic ionization cross sections (Ref. 15).

	S 2p	O 1s	C 1s	
			I	II
Calculated	1.0	0.87	0.30	0.60
Measured	1.0	0.95	0.41	0.62

that the ratio of the two components assigned to C^I and $(C^{II} + C^{III})$, respectively, is after long treatment for the present samples about 0.6 which is in reasonable agreement with the value 0.5 expected if the C 1s emission is assumed to be from EX^- groups only.

O 1s emission

Analysis of the O 1s emission from a xanthate treated sample (Fig. 3(b) and Table 1) indicates, besides the component due to the oxygen atom in the EX^- group *BE* 533.1 eV), also a component due to oxide (530.5 eV) and another one probably due to OH^- groups (531.4 eV). These assignments can be made by comparison with the O 1s emission from an untreated substrate and previous work by, e.g., Barr[13] and Hirokawa and Danzaki.[14] Again, our interpretation differs from that in Ref. 5 which assigns the component at about 533 eV to 'adsorbed oxygen containing species' originating from the EX^- groups (see below). The relative intensity of the O 1s component assigned to the EX^- group in this work is again reasonable compared with the intensities of the other lines (Table 2) and grows roughly as the intensity of the S 2p emission (Fig. 4(b)).

S 2p emission

Comparison of the S 2p line shapes of the components due to the sulphur atoms in the adsorbed layer of EX^- groups and in the sulphide binding state (Fig. 1) shows that both components have very similar shape and can be well fitted with only one doublet having the correct $2p_{3/2}$–$2p_{1/2}$ separation (1.2 eV). This result indicates that the binding of both sulphur atoms in the CS_2 group of the xanthate ions in the adsorbed layer is very similar. This situation is to be compared with their

Table 1. Binding energies (eV) measured for the atoms of the EX^--groups in the adsorption layer

	S 2p	O 1s	C 1s[a]		
			I	II	III
Ref. 5	162.3	533.0[b]	284.8	286.3	287.0
Ref. 7	162.5±0.2	532.8±0.3	284.8±0.3	286.5±0.2	
This work	162.6±0.1	533.1±0.2	285.0±0.2	286.6±0.2	

[a] For the assignment of the carbon atoms I-III in the xanthate group, see text.
[b] This main peak of the O 1s emission was assigned in Ref. 5 to a reaction product, ethyl xanthate disulphide, in the adsorption layer.

clearly different bonding in gaseous xanthate derivatives.[10] A careful analysis shows that the half-widths of the S 2p components due to the xanthate groups are slightly larger than those obtained for the sulphide-bonded contribution. A possible explanation for this broadening may, however, be a small difference between the binding energies of the sulphur atoms in the first monolayer, consisting of chemisorbed EX^- groups, and those in the rest of the layer, consisting of CuEX molecules.[7] Within the limitations of the measurements, no difference in the bonding of the two atoms can therefore be detected.

Interpreting their results for a galena substrate, Pillai *et al.*[5] suggest that ethyl xanthate groups form upon adsorption either ethyl dixanthogen ($(C_2H_5OCS_2)_2$) or carbonate disulphide ($(C_2H_5O_2CS)_2$). In the former case, two clearly different S 2p doublets would be expected for the differently bonded sulphur atoms on the basis of the results for gaseous methyl dixanthogen.[10] In the latter case, the intensity of the S2p emission should be substantially reduced because of the different stoichiometry of the adsorbed groups from that of the EX^- groups, and a new signal due to the elemental sulphur produced by the reaction should be observed. The present results do not support the presence of either of the above species, at least on the copper substrate.

CONCLUSIONS

The present results are with respect to the interpretation of the C 1s and O 1s spectra essentially the same as those previously reported in Ref. 7. The interpretation differs considerably from the one presented by Pillai *et al.*[5] for xanthate treated galena surfaces. The main reason for the difference seems to be the predominant role of contamination in the measurements reported in Ref. 5. An additional contributing factor could be radiation damage during the measurements reported in Ref. 5.

Analysis of the S 2p spectra from adsorbed xanthate layers reveals no difference between the bonding of the two sulphur atoms forming the adsorption bond.

Acknowledgements

This work has been supported by research grants from the Center for Advancement of Technology (Finland) and the Academy of Finland.

REFERENCES

1. R. Woods in *Flotation*, A.M. Gaudin Memorial Volume 1, ed by M.C. Fuerstenau, American Institute of Mining, Metallurgical, and Petroleum Engineers, Baltimore (1976).
2. J. Leja, L. H. Little and G. W. Poling, *Trans. Inst. Mineral. Met.* **72**, 407 (1968).
3. R. K. Clifford, K. L. Purdy and J. D. Miller, *AIChE Symposium Series* **71**, 138 (1975).
4. L. Ranta, E. Minni, E. Suoninen, S. Heimala, V. Hintikka, M. Saari and J. Rastas, *Appl. Surf. Sci.* **7**, 393 (1981).
5. K. C. Pillai, V. Y. Young and J. O'M. Bockris, *Appl. Surf. Sci.* **16**, 322 (1983).
6. M. D. Veremeenko, V. A. Makeev, V. I. Nefedov and V. I. Zelenov, *Poverkhnost'. Fizika, Khimiya, Mekhanika*, **6**, 121 (1983).
7. J. Mielczarski, F. Werfel and E. Suoninen, *Appl. Surf. Sci.* **17**, 160 (1983).
8. J. Mielczarski and E. Suoninen, *Surf. Interface Anal.* **6**, 34 (1984).
9. J. Mielczarski and E. Minni, *Surf. Interface Anal.* **6**, 221 (1984).
10. E. J. Suoninen, T. D. Thomas, S. E. Anderson, M. T. Runyan and L. Ungier, *J. Electron Spectrosc. Relat. Phenom.* **35**, 259 (1985).
11. J. Mielczarski, P. Nowak and J. W. Strojek, *Roczniki Chemii (Ann. Soc. Chim. Polonorum)* **50**, 917 (1976).
12. L.-S. Johansson, thesis, Univ. of Turku 1985, (in Finnish).
13. T. L. Barr, *Surf. Interface Anal.* **4**, 185 (1982).
14. K. Hirokawa and Y. Danzaki, *Surf. Interface Anal.* **4**, 63 (1982).
15. J. H. Scofield, *J. Electron Spectrosc. Relat. Phenom.* **8**, 129 (1976).

SURFACE AND INTERFACE ANALYSIS, VOL. 9, 507–513 (1986)

An AES Depth Profiling Study of the Deposits Formed on Aluminium During the Jet Fuel Thermal Oxidation Test

L.B. Hazell,* C. Baker, P. David and A.D. Fackerell
BP Research Centre, Chertsey Road, Sunbury-on-Thames, Middx, TW16 7LN, UK

The jet fuel thermal oxidation tester (JFTOT) is the primary method of assessing the thermal stability of aviation kerosene. It involves flowing a measured quantity of aerated fuel over a heated aluminium tube at 260°C for 150 minutes under 34 Bar of nitrogen. To research fuel degradation phenomena and identify the limitations of the test, we have used temperatures up to 365°C. Auger electron spectroscopy (AES) combined with argon ion bombardment (AIB) has been used to determine the thickness and chemical composition of the carbonaceous deposits formed. Segregation of magnesium from the bulk of the aluminium tube to its surface oxide and the flow dynamics of the test can combine to reduce dramatically the thickness of deposit. However, the low optical reflectivity of the magnesium rich oxide influences the standard optical methods of assessment. This can indicate a thick deposit has been formed where AES depth profiling shows it to be extremely thin. Properties of the tube/deposit interface region, as revealed by AES depth profiling, are described and the implications of these results for the JFTOT are discussed.

INTRODUCTION

The jet fuel thermal oxidation tester (JFTOT) is one of the industry standard tests[1] for thermal stability of aviation kerosene. Thermal degradation of jet fuel is invariably accompanied by the formation of carbonaceous deposits or debris. In the 200°C–400°C temperature range, oxidation of hydrocarbons produces condensed or polymeric solid deposits.[2] Sulphur and nitrogen compounds and trace metal contaminants accelerate this process.[3] There are three main areas where these deposits can cause problems in aircraft. Firstly, since the fuel is used as a coolant throughout the aircraft, a reduction of the efficiency of heat exchangers can occur. Secondly, malfunction of the fuel metering and control units can be caused by deposits formed in fine tolerance valves and actuators. Thirdly, fuel trapped in the burner arm pipework can be subjected to severe local heating, particularly during periods of low flow rate such as on deceleration. Any deposits formed here might flake off and block the injector, ultimately leading to damage of the turbine blades. As higher performance, energy efficient engines are developed, the demands on fuel quality are increased and the need for a reliable test of thermal stability becomes paramount.

With these future requirements in mind, we have studied the fundamental operation of the current standard JFTOT test to determine its limitations. This paper describes how this has been accomplished using combined Auger electron spectroscopy (AES) and argon ion bombardment (AIB) depth profiling to investigate the composition and thickness of the deposits produced.

*Author to whom correspondence should be addressed

0142–2421/86/120507–07 $05.00

Two types of fuel have been used. Hydrofined fuel is 'desulphurised' using hydrogen to convert sulphur compounds to H_2S which is removed by scrubbing.[4] Depending on the severity of the treatment it is possible to decrease the sulphur content of the fuel to, typically, less than 0.1 wt%. Other species detrimental to thermal stability, eg. nitrogen compounds and olefins, are also removed. Copper sweetening is used for low sulphur feedstocks. This involves heating the oxygenated feedstock in the presence of a copper chloride catalyst. In this process oxidation of mercaptans to innocuous disulphides occurs. These remain in the fuel.

EXPERIMENTAL

The JFTOT

The JFTOT procedure[1] requires a measured quantity, 450 ml, of aerated fuel to flow over a directly heated standard aluminium JFTOT tube at 260°C in 150 minutes under 34 bar of nitrogen (see Fig. 1). The fuel passes over the tube once at a rate which given laminar flow conditions over the tube surface. Any deposit formed in the test is assessed in two standard ways. Initially, a visual comparison is made against ASTM standard coloured panels. This provides rating of 0 (unused tube) to 4. Ratings of 3 and 4, together with unusual deposits not conforming to the ASTM standard, are considered failures for current specification purposes. A more objective assessment can be made by placing the tube in an instrument designed to measure the local light reflectivity of the tube surface relative to

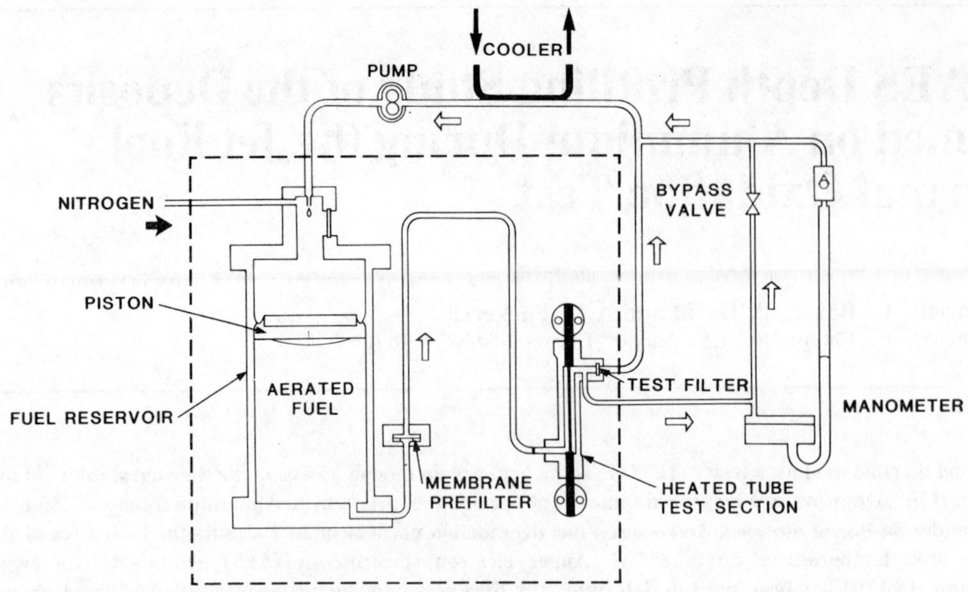

Figure 1. Schematic diagram of JFTOT apparatus.

an unused tube. This tube desposit rater (TDR) allows the tube to be rated either at specific spots, or, by spinning, an average value at each circumference along the length of the tube can be determined and a trace plotted. This trace is expected to be centred about the hottest point of the tube which, for the operating temperature profile obtained, is close to 39 mm from the fuel inlet. However, the deposit is not always central, as the results described in this paper show, especially when high temperature tests are performed.

AES/AIB depth profile analysis

AES analysis in conjunction with AIB depth profiling was used as a potentially better method of assessing deposit thickness at specific points on the JFTOT tubes. This technique has the added advantage of being able to give information on the deposit composition and how this varies with fuel type, temperature and time. In this work a Varian AES spectrometer with an Apple-based datasystem[5] designed for depth profiling was used. The tubes were cut as required and mounted on a carousel for insertion into the spectrometer. The system was baked to attain background pressures of better than $\sim 2.7 \times 10^{-10}$ mbar before back-filling with argon to a pressure of $\sim 2.7 \times 10^{-5}$ mbar. This was necessary in order to ensure that the spectrum of AIB cleaned aluminium was oxygen free and remained so for many hours. Under these conditions, low levels of oxygen species could be detected and reliably assigned as a component of the desposit rather than a contaminant. An argon beam energy of 3 keV was used at a current density if 71.3 μA cm^{-2}. This gives an etch rate of 3.1 Å s^{-1} calibrated using the NPL Ta_2O_5 standard. Calibration of the etch rate appropriate to the material of the deposit itself is uncertain, particularly as the hydrogen content has a major influence and this is not detectable using AES. Our best estimate of the etch rate for the deposits is 0.56 Å s^{-1} and is based on fitting the profile shape with a model[6] which assumes a value for the inelastic mean free path for the carbon Auger peak. This value is 6 Å and is based on the compilation of Seah and Dench.[7]

The profiles were obtained using sequential bombardment and AES analysis. Quantification of the spectra at each depth was carried out using the accepted relationship based on sensitivity factors[8] after correction for background and noise thresholds as well as overlapping peaks in the manner described in Ref. 5. No attempt was made to calibrate the sensitivity factors for our particular instrument.

RESULTS AND DISCUSSION

Results using copper sweetened fuel with a standard JFTOT tube

The profiles shown in Figs 2 and 3 were obtained at two different points on a tube tested in copper sweetened Jet A–1 fuel at 365°C. This fuel degrades rapidly at this higher temperature and so the test duration was restricted to 45 minutes, by which time sufficient deposit had formed to represent a grade 3 failure. In these tests the TDR traces along the tube length, shown inset in the figures, were doublets in which each peak was of approximately equal height.

The profiles obtained at the 26 mm position show a carbonaceous deposit of thickness equivalent to ~1500 s argon ion etching. There is ~3 at% of oxygen throughout the bulk of the deposit and ~0.4 at% sulphur, probably present in organic compounds as opposed to elemental sulphur. Both appear relatively enriched at the surface and at the deposit/tube interface. The surface enrichment is a common feature in this work which may be contamination or a beam induced effect that is not yet fully understood. The peak in oxygen concentration at the interface is due to the oxide present on the aluminium tube surface. Close

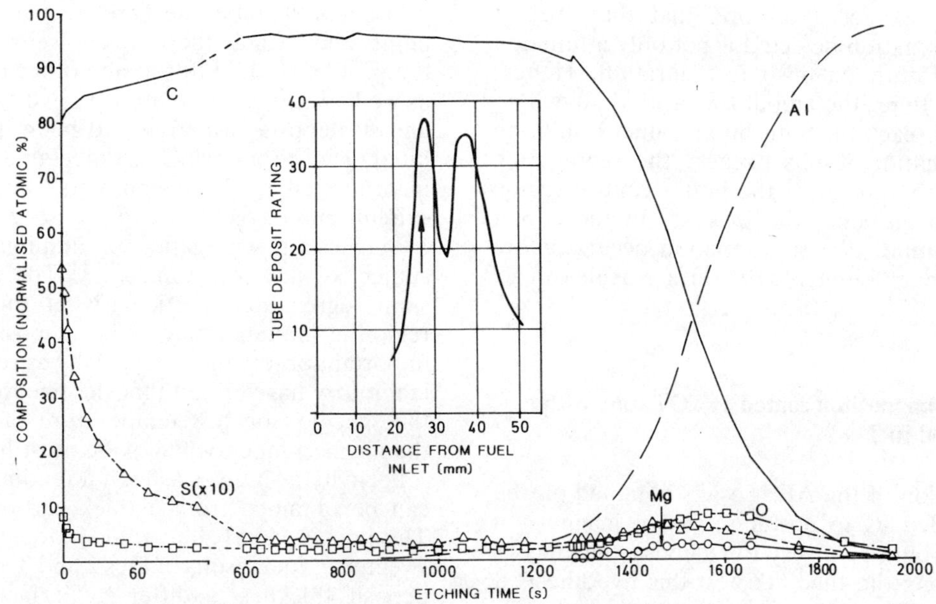

Figure 2. AES depth profile of deposit formed in 45 minutes at 26 mm position (~330°C) with copper sweetened fuel on an aluminium JFTOT tube.

inspection of the interface region reveals that magnesium has segregated into the surface oxide to a maximum detected level of ~2.5 at %. X-ray fluorescence analysis indicates ~1 at% of magnesium impurity present in the bulk.

The profile of Fig. 3. was obtained on the same tube but at the 36 mm position which is close to the hottest region, at ~39 mm. The TDR trace suggests a similar deposit to that at the 26 mm position but, although the composition is qualitatively similar, the actual deposit thickness is much reduced and equivalent to only ~210 s argon ion etching. Additionally, the peak level of magnesium in the oxide has apparently increased to ~10 at% which, in this case, may be accounted for by the improved interfacial resolution. In the many tests carried out in this laboratory we have found the deposit thickness is inversely related to the magnesium content of the surface oxide. However, the magnesium composition in the oxide reaches a maximum of ~20 at%. Presumably this is related to the ability of magnesium to react with the aluminium oxide on the tube surface, probably forming a spinel phase,[9] and the rate at which

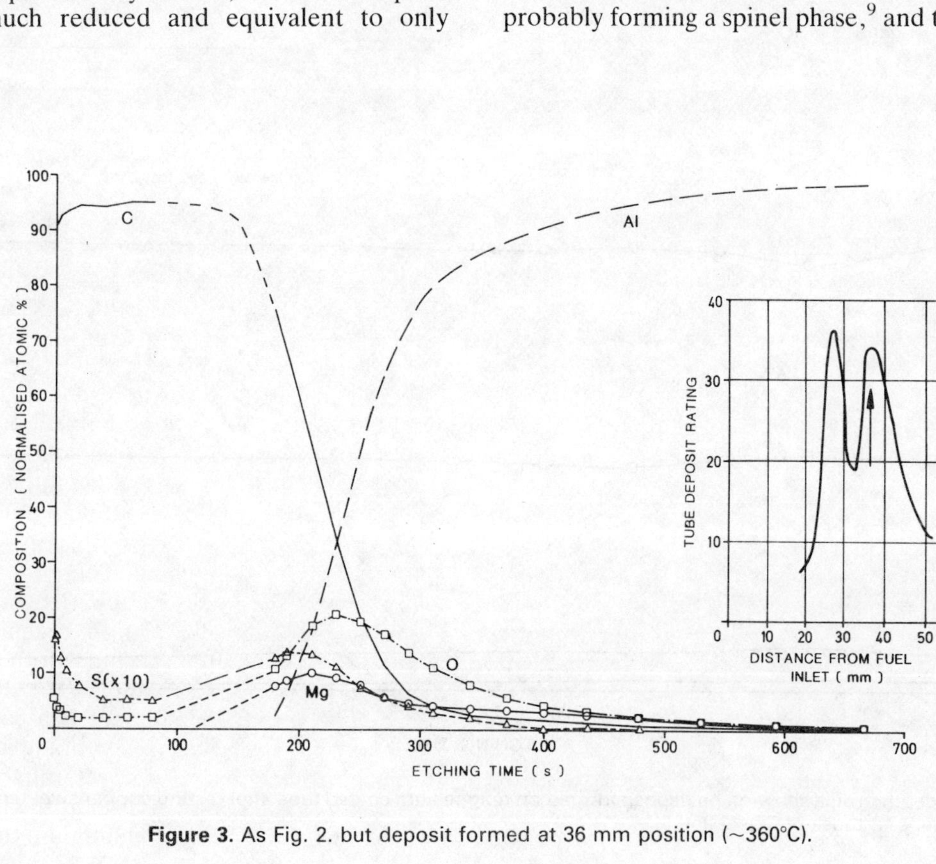

Figure 3. As Fig. 2. but deposit formed at 36 mm position (~360°C).

it does so. We expect, therefore, that the level of magnesium segregation detected is not only a function of tube temperature but also test duration. Hence, given sufficient time, the amount measured anywhere on the tube will reach the 'saturation' value. Inhibition of deposit formation results because the segregation occurs sufficiently quickly at the hotter central region of the tubes to influence the process. In the cooler regions of the tube, deposit formation occurs before segregation of magnesium into the surface oxide can act to prevent it.

Results using a magnesium coated JFTOT tube with copper sweetened fuel

Our interpretation of the AES results obtained on the standard tube led us to postulate that a magnesium oxide surface would give very low deposit thicknesses at all points along the tube. To test this hypothesis a tube was deliberately pre-coated with magnesium prior to testing. Coating was achieved by repeated vacuum evaporation of magnesium turnings from a heated molybdenum boat. As a relatively poor vacuum was used, $\sim 10^{-4}$ mbar, the magnesium was oxidised during coating and afterwards by atmospheric oxidation.

The profile shown in Fig. 4. was obtained on the magnesium coated JFTOT tube used in a test at 365°C with the same copper sweetened fuel for 20 minutes. Although the profile was taken at the 39 mm hottest position of the tube, the TDR trace indicated zero at all points and visually there was no reason to believe that the profile would differ at any other position. Clearly, as we had suspected, there is very little carbon present and so deposit formation had been dramatically inhibited. The carbon which is detectable could have been incorporated in the evaporated film, along with the sulphur and oxygen, due to the poor vacuum used.

Magnesium segregation in aluminium has been the subject of detailed studies.[10,11] The fact that magnesium segregation, particularly at the hottest central region of the tube, may inhibit deposit formation is an important observation, especially as recent work in this laboratory has verified that the phenomenon occurs at the specification test temperature of 260°C. Further, the magnesium enriched oxide must have a low optical reflectivity as it gives as high a reading on the TDR as can be obtained from a thick carbonaceous deposit. This point is clearly illustrated in the case of the copper sweetened fuel results of Figs 2 and 3. The AES derived deposit thicknesses differ by a factor of ~7.5 but magnesium segregation at the 36 mm position has 'darkened' the tube resulting in approximately equal readings on the TDR.

Results using hydrofined fuel with a gold coated JFTOT tube

Since the temperature profile of the tube is roughly symmetrical about the hottest point, the question arises

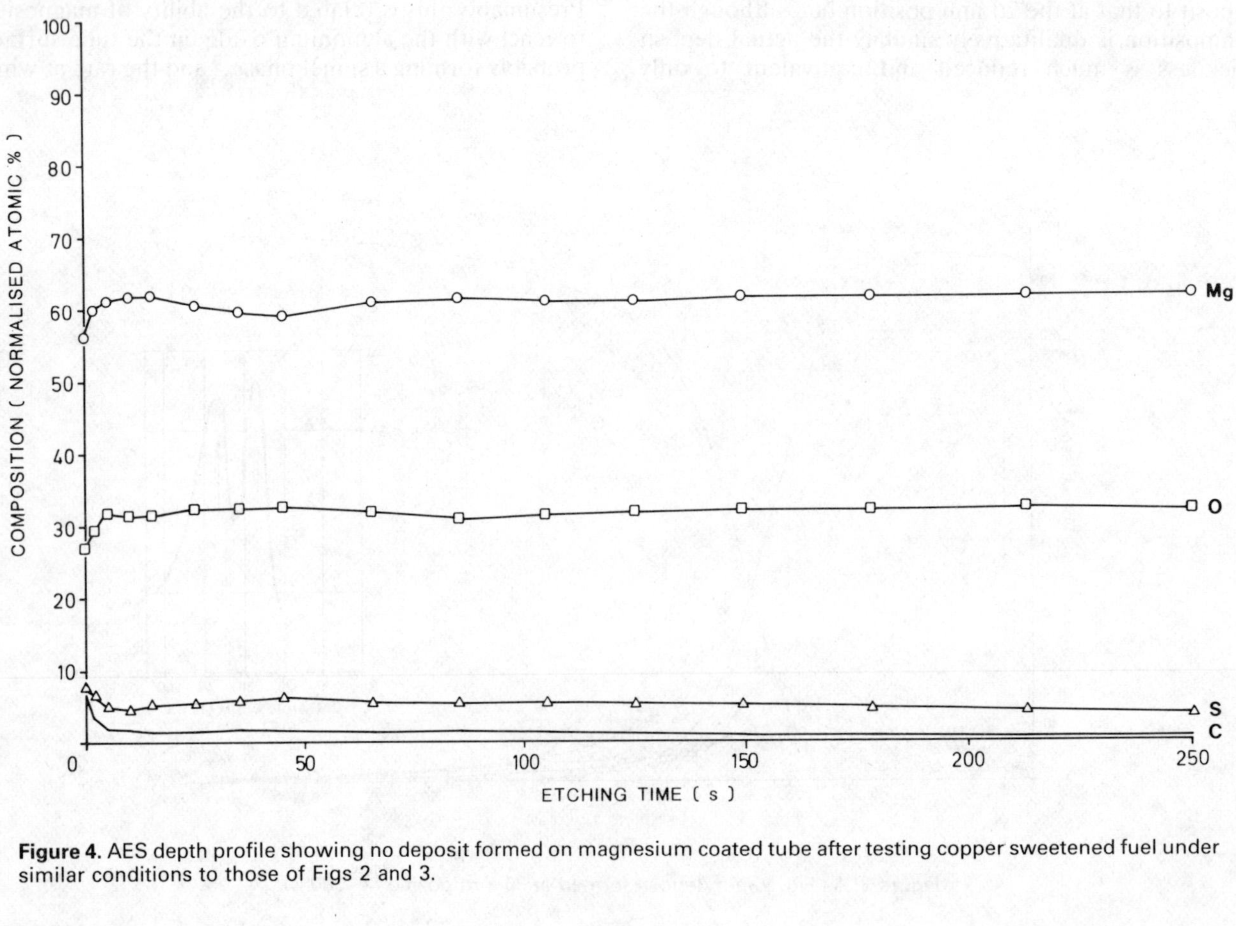

Figure 4. AES depth profile showing no deposit formed on magnesium coated tube after testing copper sweetened fuel under similar conditions to those of Figs 2 and 3.

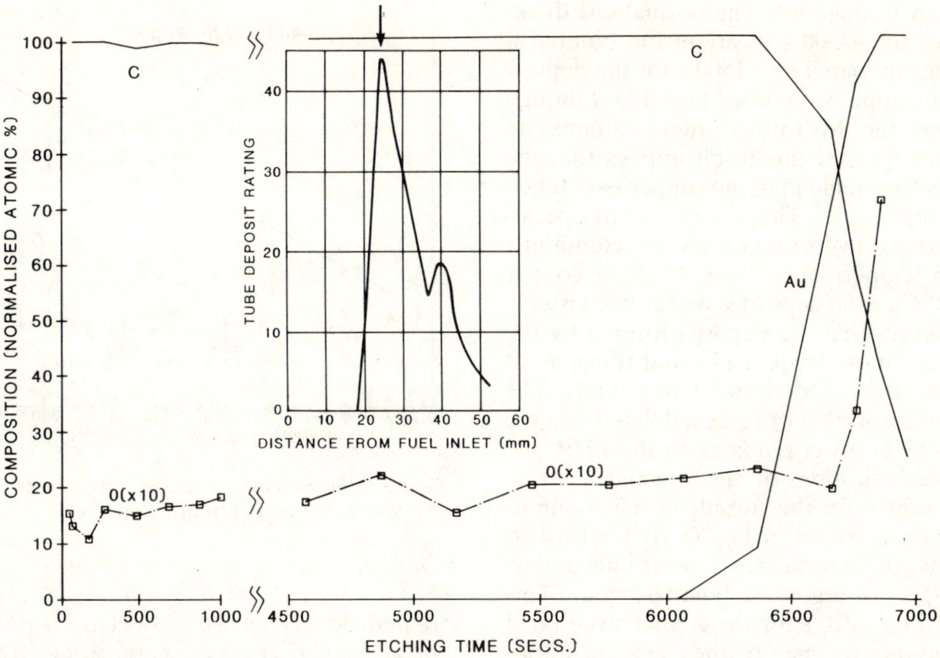

Figure 5. AES depth profile deposit formed in 90 minutes at 24 mm position (~330°C) with hydrofined fuel on a gold coated tube.

as to why there are not three peaks in the TDR trace. To establish whether this is a further consequence of magnesium segregation or is a separate problem, a gold coated JFTOT tube was manufactured using a plasma sputter coating unit. The intention of this modification was to suppress the effect of magnesium segregation. It also enabled oxygen detected near the interface region to be associated, unequivocally, with the deposit rather than the tube surface oxide. The results reported here were obtained using a hydrofined fuel with a very low sulphur content, 0.0001 wt%. This is generally believed to have a higher thermal stability than copper sweetened fuel and so, to obtain comparable visual ratings, the test duration was increased from 45 minutes to 90 minutes. In the following discussion, therefore, all thickness values have been halved to correlate with the earlier results.

The profiles shown in Figs. 5 and 6 are from the two maximum positions in the TDR trace of the gold coated JFTOT tube used at 365°C. The profile obtained at the 24 mm position, Fig. 5 shows a surprisingly thick deposit of a carbonaceous compound containing

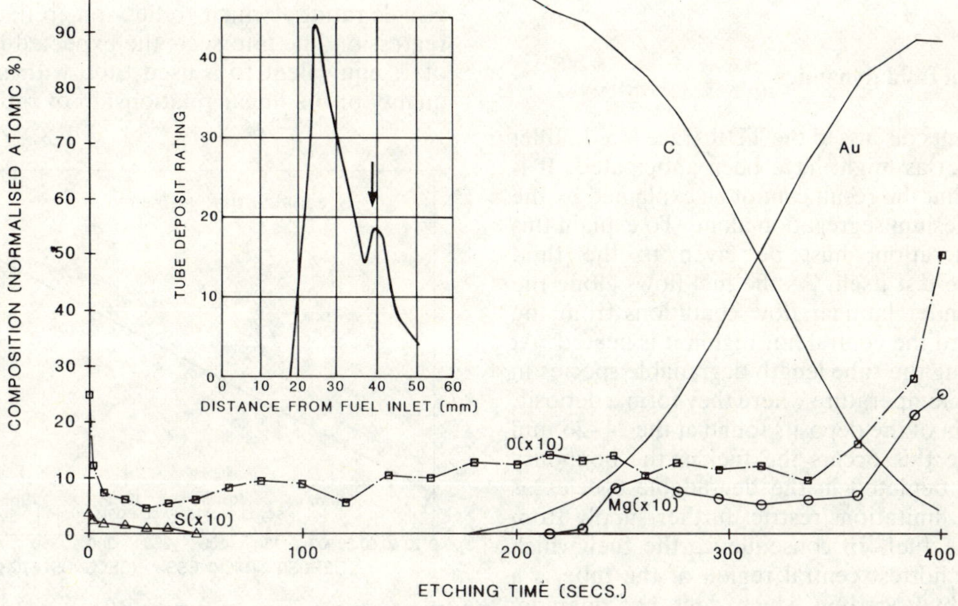

Figure 6. As Fig. 5, but deposit formed at 39 mm position (~365°C).

~2 at% of oxygen throughout. The normalised thickness is equivalent to ~3300 s of argon ion bombardment. This can be compared to ~1500 s for the deposit obtained with the copper sweetened fuel. Even though the deposit from the hydrofined fuel contains no detectable sulphur species, this result implies that the hydrofined fuel is less stable than the copper sweetened variety at this temperature. This is contrary to operating experience where hydrofined fuels are commonly tested at higher temperatures, ~330°C, than copper sweetened, ~290°C, as no deposit is detectable visually at 290°C. We suspect that the deposits formed by the hydrofined fuel are more lacquer-like and transparent than those from the copper sweetened fuel. The consequent lower absorbtion of reflected light from the aluminium tubes gives lower readings on the TDR and leads to deceptive indications of higher stability.

The profile obtained for the subsidiary maximum at the 39 mm position is shown in Fig. 6. By contrast to the 26 mm results, the normalised deposit thickness is equivalent to ~155 s of argon ion bombardment. This may be compared to ~210 s for the copper sweetened fuel measured at approximately the same position. Although some sulphur is detectable on the surface at the 39 mm position, any present in the bulk of the film is below detection limits. Note also that the gold coating is quite thin and the temperature is sufficient to cause magnesium segregation not only to the aluminium oxide surface but into the gold coating itself. However, the level is low and we do not believe it to be a significant influence in deposit formation. Unfortunately, it does impart a purple colouration to the gold coating. This has a low optical reflectivity and causes a peak in the TDR trace, the height of which is unrelated to the amount of deposit present. In fact, the ratio of the thicknesses of the deposit as registered using AES depth profiling at the two maxima is a factor of ~21 whereas the TDR rating differs by a factor of only ~2.4. This shows the greater degree of discrimination available using AES depth profiling. However, at present, this is not a commercially feasible approach for routine assessment of the deposits.

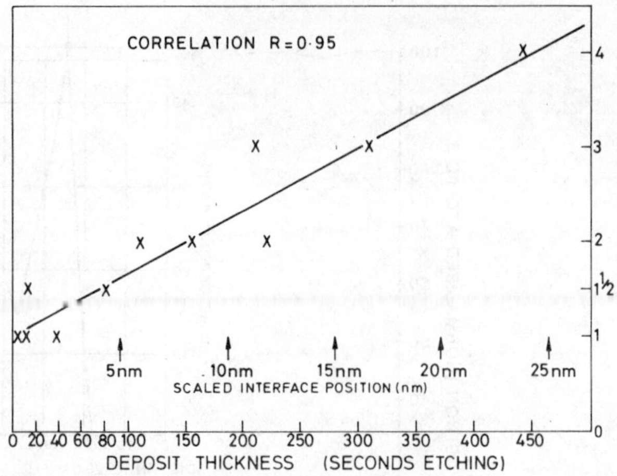

Figure 7. Correlation of visual rating with deposit thickness measured by AES depth profiling.

Consideration of fluid dynamics

In both the results discussed the TDR trace is a doublet and not a triplet as might have been anticipated. It is also apparent that the result cannot be explained by the effects of magnesium segregation alone. To explain this feature consideration must be given to the fluid dynamics of the test itself. As the fuel flows along the tube surface under laminar flow conditions from the cool inlet toward the central hot region it is heated. At some point along the tube length degradable species in the fuel reach a temperature where they form a deposit. This is the origin of the deposits found at the 24–26 mm positions. Once this occurs the fuel in the boundary layer becomes depleted in the degradable species as mass transport limitations restrict further supply from the bulk of the fuel. In consequence the fuel which flows over the hottest central region of the tube is a more refined hydrocarbon which does not degrade. This has been verified in our laboratory for alumina treated dodecane. The effect is to produce a deposit thickness variation along the tube. This is a skewed singlet function which increases rapidly at the inlet end and tails off towards the centre in some, as yet, undefined way. A doublet in the TDR trace is formed when this is combined with the low reflectivity caused by magnesium segregation, as in the results reported here. The expected symmetrical triplet cannot arise when there is boundary layer depletion.

Relationship of deposit thickness to standard rating measurements

The features of the JFTOT test procedure discussed in the previous sections indicate potential problems in obtaining a reliable assessment of the deposit thickness. Figures 7 and 8 show the correlation of thickness, derived from AES depth profiling, with visual and TDR ratings, respectively. A range of non-standard temperatures and conditions have been used. For the visual rating, which relies on reflected light, the regression line intersects the expected minimum rating of 1; equivalent to a used tube with no deposit. The quality of the linear relationship of rating with deposit

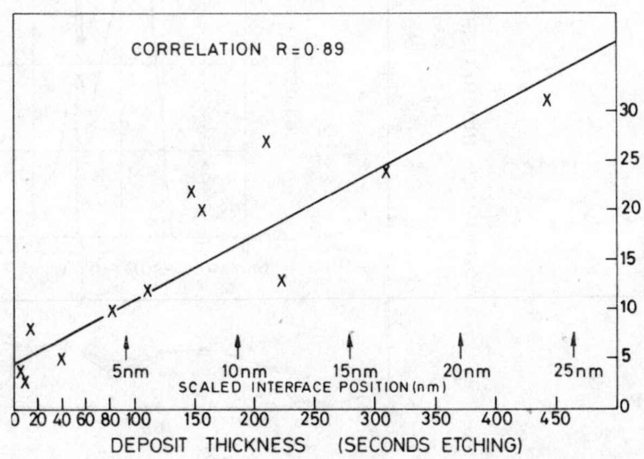

Figure 8. Correlation of TDR value with deposit thickness measured by AES depth profiling.

thickness is surprisingly good considering the relatively crude nature of the method. By contrast, even though the TDR is calibrated for a used tube with no deposit, the regression line does not pass through the origin. Presumably, a non-linear relationship between rating value and thickness may be expected since the TDR method involves a combination of specular reflection from the deposit and the tube surface. Although this could form the basis of a calibration curve, the effects discovered in this work influence deposit formation in an undefined way and would render this, and the test, unreliable. The value of this work is in defining those features of the test which should be considered of primary importance in specifying an improved test procedure.

CONCLUSION

AES/AIB depth profiling analysis of the deposits formed in the jet fuel thermal oxidation test at high temperatures has revealed the inhibiting effect of magnesium segregation into the tube surface oxide. Boundary layer depletion of degradable species at the tube surface under the laminar flow conditions of the test can give rise to skewed tube deposit rater traces with the maximum displaced from the expected hottest central region of the tube. At the higher than standard temperatures used here, the TDR values cannot be correlated with deposit thickness reliably. For high temperature tests, the most appropriate method of assessing the deposits is using the ASTM visual comparison standards.

Acknowledgements

The work was part sponsored by the Ministry of Defence and the authors would like to thank them and BP International plc for permission to publish the results.

REFERENCES

1. *ASTM Standard D 3241–82*, IP No. 323/84.
2. F. R. Mayo, N. Kirsher, H. Richardson and R. Stringham, *Stanford Research Institute Report No. PYU — 1681* (1972).
3. F. R. Mayo, H. Richardson and G. D. Mayorga, *Stanford Research Institute Report No. PYU–2361* (1973).
4. S. A. Berridge in *Modern Petroleum Technology*, 5th edition, ed. by G. D. Hobson, Wiley (1984).
5. L. B. Hazell, C. Baker and D. P. Dearden, *Surf. Interface Anal.* **7** 150 (1985).
6. J. E. Castle and L. B. Hazell, *J. Elect. Spec. Relat. Phen.* **12** 195 (1977).
7. M. P. Seah and W. Dench, *Surf. Interface Anal.* **1** 2 (1979).
8. L. E. Davis, N. C. MacDonald, P. W. Palmberg, G. E. Riach and R. E. Weber, *Handbook of Auger Electron Spectroscopy*, Physical Electronics Industries Inc., 6509 Flying Cloud Drive, Eden Prairie, Minnesota 55343, USA.
9. P. Ascarelli and M. Cini, in *Emmission and Scattering Techniques*, 353, ed. by P. Day, Reidel (1981).
10. C. Lea and C. Molinari, *J. Mat. Sci.* **19** 2336 (1984).
11. C. Lea and J. Ball, *Appl. Surf. Sci.* **17** 344 (1984).

SURFACE AND INTERFACE ANALYSIS, VOL. 9, 515–518 (1986). EXTENDED ABSTRACTS

Studies of Ion Exchange Reactions on Naturally Weathered Medieval Glass Surfaces

M. Schreiner
Academy of Fine Arts, Inst.f.Farbenchemie, Schillerplatz 3, A-1010 Wien, Austria

G. Stingeder and M. Grasserbauer
Technical University, Inst.f.Analytical Chemistry, Getetreidemarkt 9, A-1060 Wien, Austria

OBJECTIVES

The degradation of a silicate glass surface due to interaction with the ambient atmosphere is referred to as 'weathering'. It occurs when sufficient quantities of water or moisture are deposited on the glass surface. Depending on the conditions of exposure of the glass surface corrosion of glass by weathering has been classified into dynamic and static. It can be a serious problem for storage, transport or optical and HLW (high level waste) glass as well as glass paintings used as window panes at historical buildings.

Similar to the corrosion of silicate glass in aqueous solutions an ion exchange process can be assumed,[1] when water from precipitation or condensation is in contact with the glass surface. The extraction of the alkali components of the glass and the incorporation of protons and/or water into the silicate structure leads to the formation of a surface layer called 'gel zone' or 'leached layer'. The pH of the attacking solution will increase also but under dynamic weathering conditions the attacking medium will be replenished before the pH will rise into alkaline range. Under static conditions a small volume of condensate will contact a large surface area long enough to reach alkaline conditions. Consequently an attack on the siloxane bonds will cause breakdown of the silicate network and severe morphological damages occur. Additionally, reaction products of the leached elements of the glass with components of the ambient atmosphere (e.g. CO_2, SO_2) can be formed and weathering crusts consisting of hydrated silica, carbonates and sulphates are observed on the surface of naturally weathered glass panes.

ANALYTICAL PROCEDURE

Studying the weathering mechanism of medieval stained glasses used as window panes in glass paintings and exposed to the ambient atmosphere for nearly 600 years, SEM and SIMS were used to obtain surface analysis of the morphological and chemical changes which take place during weathering. Medieval stained glass is characterized by a high sensitivity towards atmospheric attack due to its high content of modifier ions such as potassium, calcium and magnesium and due to a much lower silica content than it has been used for common modern glass.[2] Nevertheless, a remarkable difference in the chemical durability of the various window panes in a medieval glass painting can be observed. Depending on their chemical composition, some medieval glass decays predominantly by pitting, some by crusting and some look nearly unweathered showing a smooth and bright surface. Therefore, two types of glasses have been investigated: (1) glass samples containing 15–20 wt% K_2O, 15–25 wt% CaO and 40–50 wt% SiO_2 as main components and showing thick laminated crusts and (2) K_2O-CaO-SiO_2 samples with a similar SiO_2 content but also with large amounts of PbO, ranging from 15 to nearly 30 wt%. This type of glass is rather unaffected by the long term exposure.

The SIMS measurements were carried out by a CAMECA IMS 3f instrument using O^- as primary ions with an energy of 14.5 keV for positive charged and 5.5 keV for negative charged secondary ions. The beam current was 300 or 500 nA, the beam diameter about 60 μm and the raster size 225 × 225 μm or 500 × 500 μm. In order to avoid surface charging during analysis a gold layer (50 nm) was deposited on the specimens and the accelerating voltage for the secondary ions (4500 V) was increased during measurements (20 to 60 V depending on analysing time). The depths of the sputter craters were obtained by stylistic measurements (Dektak).

QUANTIFICATION OF THE SIMS DEPTH PROFILES

In order to convert the raw depth profiles obtained by SIMS, where the secondary ion intensity versus analysing time is recorded, into concentration versus depth profiles[3] relative sensitivity factors were used defined as

$$S_{X/M} = (I_X/c_X \cdot f_X)/(I_M/c_M \cdot f_M)$$

where I is the secondary ion signal, c the atomic concentration, which was obtained with XRF, EPMA and ICP-AES, and f the isotopic abundance of the analytical (X) and reference (M = Si) element.[4]

The relative sensitivity factors were obtained from the secondary ion intensities measured at the cross sections of the glass specimens representing the bulk concentrations. The sputter rate of the several layers was determined by shorter runs to significant depths and subsequent crater depth measurements by Dektak.[5] During quantification the main elements of the glass samples were considered except hydrogen incorporated in the leached layer.

CONCLUSION

The surface analytical investigations carried out by means of SEM and SIMS reveal that weathering of both glass types is governed by an ion exchange process. The depletion of the alkaline but also earth alkaline modifier ions and of Pb for panes of glass type 2 could be proved as well as the incorporation of H in the leached surface layer. Further the formation of an outermost surface layer consisting of the leached elements and components of the ambient atmosphere was determined at the Pb-containing window panes. SEM investigation carried out at cross sections of medieval glass specimens reveal that the weathering process leads to an increase in the effective surface area of the glass in contact with the attacking medium leaving micropores, where the leaching process takes place predominantly.

REFERENCES

1. D. E. Clark, C. G. Pantano and L. L. Hench, *Corrosion of Glass*, Books for Industry and the Glass Industry (1979).
2. R. G. Newton, *Glass Techn.*, **26/1**, 21–38 (1985).
3. M. Schreiner, G. Stingeder and M. Grasserbauer, *Fres. Z. Anal. Chem.* **319**, 600–605.
4. D. E. Newburry, *Scanning*, **3**, 110–118 (1980).
5. A. Lodding and H. Odelius, *Microchim. Acta Suppl.* **10**, 21 (1983).

Trace Element Determination by a Combined AES–Chemical Deposition Procedure

K. Macho*†, R. P. H. Garten‡, H. Bubert†, J. Auffarth*, D. Klockow*, G. Tölg†‡

* Fachbereich Chemie der Universität Dortmund, FRG

† Institut für Spektrochemie und angewandte Spektroskopie, Dortmund, FRG

‡ Max-Planck-Institut für Metallforschung, Laboratorium für Reinststoffanalytik, Bunsen-Kirchhoff-Strasse 13, 4600-Dortmund, FRG

Quantitative determinations of elemental traces of sulphur and fluorine were performed by means of a combined procedure of chemical deposition of the analyte onto a metallic receptor and thin film analysis by Auger electron spectrometry (AES) of the layer thus formed. Deposition of the analyte as a thin homogeneous layer was achieved, yielding calibration graphs ranging from 7 nmol to 200 nmol for the complete procedure.

INTRODUCTION

The quantitative analysis of trace elements in engineering and environmental materials represents a very important section of analytical chemistry. Very often trace analyses have to be carried out using only small sample amounts, e.g. in environmental research or in materials characterization, resulting in extremely small absolute amounts of analyte to be determined.

Surface analytical methods typically exhibit a power of detection in absolute terms in the 10^{-15} g range. The aim of this contribution is to profit from this advantage for the purpose of determining trace elements.[1,2,3] Such an approach should be of special interest for elements like fluorine, that are difficult to determine at the nanogram range by conventional methods.[4]

In order to make proper use of the high power of detection inherent in surface analytical methods, the analyte has to be separated from the sample matrix and to be deposited in a very thin layer of a surface area of some mm^2 or less (for SIMS and AES) on a suitable getter material. In former studies similar depositions were used for ESCA-studies by means of transfer to a solid receptor from gas[5] or liquid phase.[1,2]

EXPERIMENTAL

The following analyte-getter systems were investigated: fluorine–aluminium, fluorine–magnesium, fluorine–zirconium and sulphur–silver. Targets were prepared by chemical deposition of trace amounts (0.2 to 500 nmol) of sulphide or fluoride either directly from aqueous solutions or via a volatile compound of the analyte onto a suitable metallic getter (receptor) in a closed cell. Sulphide was volatilized, e.g. as H_2S. In the case of fluorides, it has been shown that the analyte can be selectively separated from the matrix and can be enriched on alkaline receptors.[6] The separation is accomplished by the formation of the volatile product $(CH_3)_3$ SiF that diffuses via the gas phase (microdiffusion)[8] to the receptor mounted at the top of the cell.

Aluminium and silver sheets (2×5 mm^2 size, 0.2 mm thickness) were carefully cleaned and etched prior to use as receptors. Alternatively spot type receptors of Al, Mg, Zr and Ag (400 nm thickness, 0.5, 1 or 5 mm diameter) were prepared by vacuum vapour deposition onto the centre of a polished disc of either spectral grade graphite or glassy carbon (6 mm diameter, 1 mm thickness). In this way a receptor of a well defined surface area is obtained on a smooth, sturdy, conductive, and chemically inert base material of high purity. Quantitative determination of the collected analyte is completed by AES/Argon-sputtering.

RESULTS AND CONCLUSIONS

Quantitative determination of traces of sulphur and fluorine was successfully achieved using a combination of analyte deposition in the form of a thin layer and thin film in-depth profiling of this layer by means of AES/sputtering. Through the deposition process and the inherent transport reaction (microdiffusion) high enrichment factors at the order of $f \approx 10^4$ to 10^5 were obtained. This enrichment was made selective by optimizing carrier reagent as well as receptor material. Up to now calibration graphs ranging from 7 nmol up to 200 nmol of the respective trace elements were obtained and determinations in this range have been carried out. The reproducibility, however, of the overall procedure is not yet satisfactory. (Estimated uncertainty of a single measurement $\pm$ 7 to 10 % for the Ag–S system, coefficients of correlation $r^2 = 0.997$; $n = 7$ (Ag-S), $r^2 = 0.94$; $n = 5$ (Al–F), total yield of the enrichment procedure was estimated from relative sensitivity factors and sputter rate to range in the order of 0.1 to 0.5 for the systems investigated.)

Stringent restrictions of the surface analytical parameters had to be taken into account, namely the primary electron energy ($E_e = 3.0$ keV), the ion energy (E_i 1.5 keV) and the beam current densities ($j_e = 0.4$ μAcm^{-2} [fluorine], $j_e = 6$ μAcm^{-2} [sulphur]; $j_i = 30$ μAcm^{-2}). This is due to the appearance of sulphur-segregation[9] in the Ag–S system and fluorine-desorption[10] in the Al–F, Mg–F, Zr–F systems.

The performance of the getter materials Al, Mg, Zr is insufficient as yet, due to incomplete deposition at the subnanogramme-scale. Although the procedure requires relatively high expenditure in terms of working (0.5 h/sample) and instrumental (ca. 1h/sample) time, we see good prospects for application to a restricted number of elements where selective carrier reagents are available.

REFERENCES

1. J. S. Brinen and J. E. McClure, *Anal. Lett.* **5**, 737–743 (1972).
2. D. M. Hercules, L. E. Cox, S. Onisick, G. D. Nichols and J. C. Carver, *Anal. Chem.* **45**, 1973–1975 (1973).
3. D. Briggs, V. A. Gibson and J. K. Becconsall, *J. Electr. Spectr. Relat. Phenom.* **11**, 343–347 (1977).
4. J. Auffarth, dissertation, Universität Dortmund (1979).
5. M. B. Carvalho and D. M. Hercules, *Anal. Chem.* **50**. 2030–2034 (1978).
6. J. Auffarth, D. Klockow, *Anal. Chim. Acta*, **111**, 99–102 (1979).
7. D. R. Taves, *Talanta*, **15**, 969–74 (1968).
8. E. J. Conway, *Microdiffusion Analysis and Volumetric Error*, 4th ed. C. Lockwood & Son Ltd., London, (1957).
9. D. P. Griffis and R. W. Linton, *Surf. Interface Anal.* **4**, 197–203 (1982).
10. C. G. Pantano and T. E. Madey, *Appl. Surf. Sci.* **7**, 115–141 (1981).

Rutherford Backscattering and Channelling Studies of Laser Irradiated Stainless Steel

T. Laursen and J. L. Whitton
Department of Physics, Queen's University, Kingston, Ontario, Canada K7L 3N6
J. Nilson and Wing Nip
Lumonics Inc., 105 Schneider Road, Kanata, Ontario, Canada K2K 1Y3
I. V. Mitchell, H. H. Plattner and M. L. Swanson
Atomic Energy of Canada Limited Research Company, Chalk River Nuclear Laboratories, Chalk River, Ontario, Canada KOJ 1JO

Rutherford backscattering in combination with ion channelling is one of the preferred techniques for characterizing surface alloys produced by directed energy processing (laser and electron irradiation and ion implantation). Rutherford backscattering spectrometry (RBS) in itself measures the elemental composition as a function of depth and, in combination with channelling, microstructural information related to the lattice location of alloying elements and crystal damage is obtained. Because of the numerous possible applications there is a particular interest in surface alloys of stainless steel. In a recent study, we examined the effects of pulsed laser annealing in nitrogen implanted stainless steel.[1] We are now extending our study to include surface alloys produced by laser alloying thin evaporated films of gold and molybdenum on stainless steel.

The single crystals of fcc austenitic stainless steel have a quoted Fe : Cr : Ni composition of 70 : 17 : 13. Metal films of gold or molybdenum were vacuum deposited by electron beam evaporation on to separate specimens. The laser irradiation was made using a XeCl excimer laser $\lambda = 308$ nm, Lumonix Model 292). The energy densities were 2.0, 3.2, and 8.4 $J.cm^{-2}$ for a pulse duration of 40 ns. In most cases a cover gas of flowing helium was used to reduce oxidation, but irradiations in air were also made. Each laser irradiated spot and the unirradiated part were analysed by RBS and channelling using a 2 MeV He^+ beam.

The effect of laser irradiation on the channelling spectrum of the gold coated <100> single crystal is shown for two different energy densities in Fig. 1. The spectra are shown together with the unirradiated aligned and random spectra. The relatively high χ_{min} of 0.2 in the unirradiated case is due to the 15 nm thick gold film. The laser treatment increases the χ_{min} value even though the gold peak is seen to diminish. At the same time, the stainless steel edge (a combination of Fe, Cr and Ni which differ in mass by only a few mass units) is seen to have moved up in energy while the gold peak forms a tail towards lower energies indicating an interface region of mixed composition. The depth distributions of gold can be derived from the RBS spectra obtained in a random direction.

The impurity profiles show that along with the inward diffusion there is a gradual loss of impurity with increasing laser power.

The diffusion constants, which are needed to explain the impurity redistribution, are several orders of magnitude higher than the diffusion constants in the solid state. It is, therefore, concluded that the diffusion has taken place in the liquid phase, and thus the laser irradiations have involved surface melting.

The channelling results show that substitutional surface alloys are formed with gold and molybdenum. They are formed beneath a narrow surface layer (~ 20 nm), which has a high gold or molybdenum content. For this surface layer, these impurities are located in mainly non-substitutional positions.

The quality of the epitaxial regrowth, as seen in the channelling spectra, range from good to very poor. The dechannelling in the alloyed surface has a complex behaviour, and is very dependent on the energy density of the irradiation. It is also important whether or not the irradiation is carried out in air or flowing helium. Laser irradiation in air results in channelling spectra with a higher χ_{min} than seen in Fig. 1. The reason is likely to be that oxidation takes place and obstructs the epitaxial regrowth.

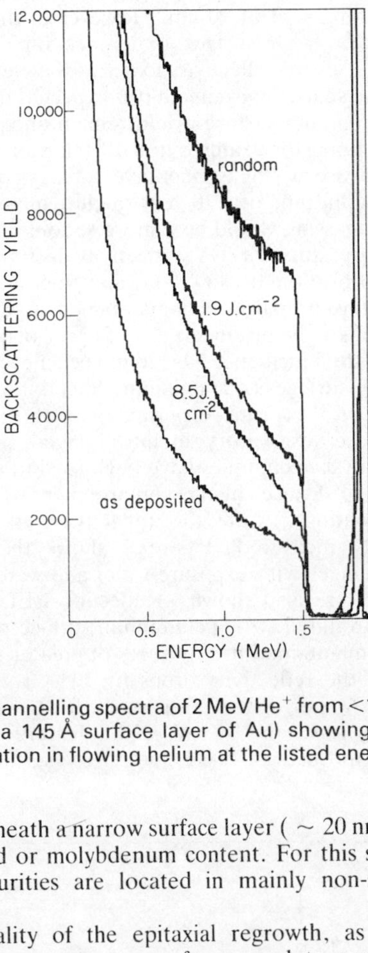

Figure 1. Channelling spectra of 2 MeV He^+ from <100> stainless steel (with a 145 Å surface layer of Au) showing the effect of laser irradiation in flowing helium at the listed energy densities.

REFERENCE

1. J. L. Whitton, T. Laursen, J. A. Nilson, Wing Nip, I. V. Mitchell, H. H. Plattner, and M. L. Swanson, *Mat. Res. Soc. Symp. Proc.* **41**, 313 (1985).

Aluminium Mirror Coatings in Space: a Study of the Decrease in Ultraviolet Normal Incidence Reflectance Produced by Controlled Oxidation of Evaporated Aluminium Mirror Surfaces

M. Grande and W. M. Burton
Space & Astrophysics Division, Rutherford Appleton Lab., Chilton, Didcot, Oxfordshire OX11 0QX, UK

Pure aluminium is an excellent normal-incidence reflector at wavelengths longward of 90 nm. However in the terrestrial environment, a fresh surface undergoes rapid oxidation, which leads to severe reflectivity losses at wavelengths below 180 nm. In the space environment it is expected that a freshly deposited aluminium surface would remain unoxidized for a considerable time; for example in a 1000 km orbit, the time needed to absorb a single monolayer of oxygen would be more than the lifetime of a 10 year satellite mission. In fact, spacecraft outgassing would be a more serious constraint.

A laboratory study of UV reflectivity loss produced by controlled absorption of air and water vapour onto fresh aluminium films prepared and maintained in UHV (10^{-10} torr) conditions is being undertaken as part of a feasability study for a normal incidence 90–120 nm echelle spectrograph with reflecting surfaces of aluminium, coated in orbit, which is proposed for the prime experiment on th ESA–NASA ultraviolet space observatory satellite 'Lyman'.

Surface analysis techniques have been used to monitor the evolution of the surface chemical environment of evaporated aluminium coatings, while the reflectivity at 120 nm is simultaneously measured. Figure 1 shows the measured reflectivity changes with exposure to air and water; previous measurements are also shown.[1] Reflection EELS measurements show an initial strong aluminium surface plasmon (10 eV) which is maintained to exposures of greater than 100 L, during which the reflectivity drops by 10% from its initial value of 90%. Further exposure of 10^4 L reduced the reflectivity to 70%; the 10 eV EELS feature is lost and replaced at exposures of 200 L by a 7 eV surface plasmon

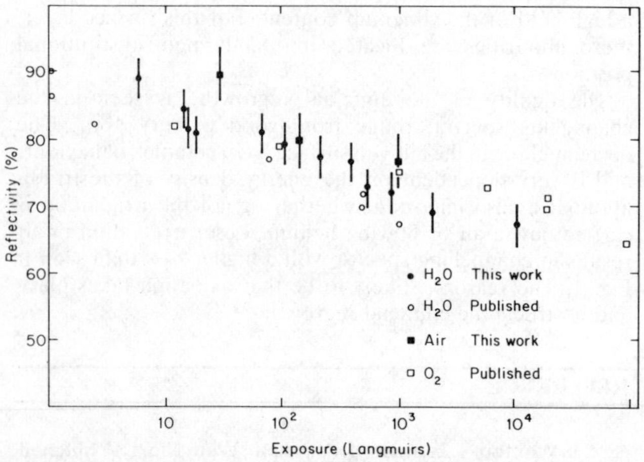

Figure 1. Reflectivity of freshly deposited aluminium surfaces at 120 nm versus exposure in Langmuirs.

modified by oxygen uptake. With higher exposure a 20 eV feature, generally attributed to excitation of incorporated oxygen, is seen. The low energy AES spectrum shows the replacement of the 63.5 eV aluminium metal $L_{23}VV$ feature by the 50 eV Al(L_{23}) O(L_{23}) O(L_{23}) feature which becomes visible at about 100 L and dominates at exposures greater than 500 L. This behaviour is seen for both H_2O vapour and air exposure, although apparently the initial reflectivity loss is slightly more rapid for air.

The uptake is interpreted as a two stage process, with a rapid reflectivity loss associated with the initial oxygen adsorbition in the first 100 L, followed by a slower drop as the first monolayers of stable Al_2O_3 are formed by activated place exchange.

We have calculated the expected reflectivity at 121.6 nm of aluminium with an oxide overlayer, using optical constants from Bunner 1983,[2] which indicate a somewhat steeper fall in reflectivity with oxide thickness than those of Hass and Hunter.[3] It appears clear, however, that the reflectivity change between 10^2 and 10^4 L from 80% to 70% is attributable to at most two monolayers of Al_2O_3. It is uncertain what optical constants should be used for the topmost discrete O and Al layers (see e.g. Ref 4) which account for the drop from 90% to 80% in the first 100 L; using values for Al_2O_3 produces too great a drop, while those for O_2, appropriately corrected for a condensed monolayer, produce far too small a loss. Since the reflectivity loss in this regime should be linear with coverage, with more accurate reflectivity measurements this technique can be an accurate measure of the initial rate of coverage.

In conclusion, it appears that in a spacecraft spectrometer constructed to conventional UHV standards, we anticipate an exposure of 10^4 L in one year, which would reduce the 121.6 nm reflectivity to 70% and the spectrometer throughput to 50%. The loss at 90 nm would be more severe. This represents a performance within the design requirements of the mission, but one which could clearly be improved considerably if the overall vacuum environment could be improved, for example by using large gettering areas on the walls of the spectrometer enclosure.

REFERENCES

1. W. M. Burton, M. Grande, M. C. W. Sandford, S. J. B. Corrigan and G. E. Holmes, *Rutherford Appleton Laboratory Report No. RAL-85-041*, (1985) and references therein.
2. A. Bunner, *Perkin–Elmer report no ER-591*, (1983) and references therein.
3. G. Hass and W. R. Hunter, *Physics of thin films*, **12**, 2–51 (1982) and references therein.
4. W. H. Krueger and S. R. Pollack, *Surf. Sci*, **30** 263–279 (1972).

QUANTITATIVE AES ANALYSIS OF CARBIDES, NITRIDES AND CARBONITRIDES

M. Puchhammer, A. Riahi and H. Störi
Inst.f. Allg. Physik, TU Wien, Karlsplatz 13, A–1040 Vienna, Austria

Films of different refractory transition metal carbides, nitrides and carbonitrides have been analysed by Auger electron spectroscopy. As the physical properties of these films depend critically on their elemental composition, a precise method of quantitative analysis is required. A new method for evaluation of the AES spectra, based on the calculation of peak areas has therefore been developed. This method is comparatively insensitive to chemical effects in the Auger spectrum which introduce errors of several ten percent into the composition data if the classical evaluation method based on peak to peak amplitudes is applied.

The new method has been implemented on a commercial PHI surface analysis system with a double pass cylindrical mirror analyser coupled to a PDP–11/23 Plus microcomputer system. A special software package for recording, archiving and evaluation of spectra has been developed.

Acknowledgement

This work was supported by the Jubiläumsfond der Österreichischen Nationalbank, Project No. 2309.

SELECTIVE OXIDATION SURFACE ANALYSIS (SOSA) ON METALS AND MINERALS

R.J. Pugh* and K. Husby†
* Institute for Surface Chemistry, Box 560, 114 86 Stockholm, Sweden
† Swedish Insitute for Metal Research, Drottning Kristinas väg 48, 114 28 Stockholm, Sweden

Selective oxidation surface analysis (SOSA) studies are reported for (a) metal surfaces containing carbonaceous impurities and (b) powders with adsorbed hydrocarbon additives.

The principle of the method consisted of heating the sample in a stream of high purity oxygen. The surface carbon was selectively oxidized over characteristic temperature ranges (between 50 and 800°C) and the effluent gas fed over a catalyst. The resulting carbon dioxide was continously analysed by an IR detection system.

The method was shown to be capable of differentiating between carbon and various carbonaceous compounds (i.e. hydrocarbons, carbides, allotropic forms of carbon, etc). It was therefore found possible to quantify the amounts of high molecular weight carbonaceous impurities (i.e. oils, greases etc) in low carbon steel sheets and copper tubes.

In the case of mineral powders such as quartz, fluorite and galena the amount of long chain fatty acid, amine or xanthate adsorbed during processing of the material was determined. In addition, the degree of surface carbonation of the fluorite and galena could be estimated.

SOSA could prove useful in the areas of mineral processing and in evaluating the effectiveness of degreasing agents in the metallurgical industry.

Author Index to Volume 9

*The Authors of Short Abstracts are not included in this index.

Adams, F. 327
Affrossman, S. 445
Akimov, A. G. 385
Allard, L. F. 227
Alnot, P. 283
Andreoni, I. 157
Anne, J. M. 386
Anthony, M. T. 447
Antonova, T. A. 417
Asbury, D. A. 169, 258
Auffarth, J. 516

Balk, L. J. 47
Baker, C. 507
Barker, M. G. 371
Barna, P. B. 328
Bartl, C. 182
Berneron, R. 21
Bertóti, I. 237, 260, 417
Bertrand, P. 247
Beuken, J. M. 247
Beyer, W. 131
Bielen, P. 338
Bishop, H. E. 105, 163
Blanchard, B. 331
Bodö, P. 437
Boudewijn, P. R. 303
Bouillon, F. 181, 187
Briggs, D. 70, 391, 411
Brinen, J. S. 227
Brown, A. 70, 309
Brown, J. R. 253, 255
Brunel, M. 139
Bubert, H. 384, 516
Burggraaf, A. J. 467
Burton, W.M. 518

Calinski, C. 357
Carli, F. 412
Castle, J. E. 345
Cazaux, J. 133
Cazet, J. P. 21
Charbonnier, J. C. 21
Chen, J.-R. 131
Cimino, A. 245
Clabes, J. G. 275
Clark, E. A. 68
Clarke, N. S. 31
Coatsworth, L. L. 253, 255
Collisi, U. 357
Combarieu, R. 125
Convert, F. 188
Corallo, C. F. 319
Cortés Corberán, V. 207
Coyle, G. J. 275

Creemers, C. 338
Cserny, I. 69

Daly, F. P. 227
Darque-Ceretti, E. 125
Davey, A. R. 31
David, P. 507
Davis, G. D. 421
De Gregorio, P. 186
De Keersmaecker, R. F. 338
Delamare, F. 125
Delmon, B. 195, 247
Delplancke, M. P. 187
Demin, V. Yu. 385
Denison, P. 431
De Santi, G. 329
Desimoni, E. 414
Desseaux-Thibault, J. 175
De Wit, J. H. W. 163
Deydier, P. 139
Di Castro, V. 249, 336
Diekmann, W. 184
Dingle, T. 65
Dourtoglou, V. 446
Dunselman, K. 293
Dupree, R. 383
Duvigneaud, P. H. 332
Dwight, D. W. 405

Edge, G. J. 68
Edgell, M. J. 416
El Gomati, M. E. 99
El-Mashri, S. M. 383
Erickson, N. E. 111
Espinós, J. P. 248
Evers, E. J. 293

Fackerell, A. D. 507
Fahim, R. B. 243
Fakes, D. 416
Farnan, I. 383
Fernández, A. 248
Fierro, G. 246
Fierro, J. L. G. 247
Fioravanti, S. 186
Fitzgerald, A. G. 333, 334
Fleischhauer, H. P. 297
Foct, J. 386
Forty, A. J. 383
Fragale, C. 249
Furlani, C. 249

Garbassi, F. 412
Gargano, M. 249
Garten, R. P. H. 384, 516

Gazzoli, D. 245
Gee, M. G. 55
Genet, M. 446
González-Elipe, A. R. 248
Gosztola, L. 328
Grabke, H. J. 184
Grande, M. 518
Grant, J. T. 132
Grasserbauer, M. 182, 325, 515
Grathwohl, G. 453
Gregory, G. 157
Gries, W. H. 335
Griffiths, B. W. 65
Grogan, Jr, A. L. 258
Grübmeier, H. 185
Grundner, M. 325
Guczi, L. 261

Habraken, F. H. P. M. 293
Hamar-Thibault, S. 175
Hamminger, R. 453
Harris, A. 67
Häupl, K. 27
Haupt, S. 357
Hazell, L. B. 507
Hearn, M. J. 411
Heckingbottom, R. 265
Hennaut, P. F. 332
Herion, J. 71, 131
Hiesböck, H. G. 182
Hilgers, W. 71
Hoflund, G. B. 169, 258, 319
Hofmann, S. 3
Holsbrink, J. 293
Hoppe, H. W. 357
Huang, W. 119
Huber, A. M. 283
Hubin, A. 327

Ingo, G. M. 186
Inversi, M. 245
Ismail, H. M. 243

Jardinier-Offergeld, M. 181, 187
Johansson, L.-S. 501
Johnston, D. 253, 255
Jones, F. R. 431
Jones, R. G. 383
Juhanoja, J. 501
Juliet, P. 331

Kirszensztejn, P. 169
Kisfaludi, G. 261
Klockow, D. 516
Koenig, M. F. 132

Kohler, V. 67
Kövér, Á. 69
Kövér, L. 69, 251
Kremeníc, G. 207
Kühlein, M. 66

Laajalehto, K. 501
Lai, S. Y. 70
Laitinen, H. A. 169
Lamothe, M. 188
Landolt, D. 477
Lang, M. 27
Laudet, A. 332
Laursen, T. 517
Lea, C. 145, 447
Lee, Y. H. 275
Leys, M. R. 303
Littmark, U. 297
Lo Jacono, M. 246
Lopez, L. C. 405
Lovas, A. 261

Macho, K. 516
Maex, K. 338
Maldonado, L. 185
Malitesta, C. 414
Marcus, P. 377
Mathieu, H. J. 477
Maurickx, B. 386
McHardy, C. P. 334
McIntyre, N. S. 253, 255
Meda, L. 129
Mielczarski, J. 501
Miethe, K. 335
Minelli, G. 246
Mink, G. 237, 260
Minni, E. 463
Mitchell, I. V. 517
Moens, M. 327
Moncoffre, N. 139
Moreau, J. P. 21
Moretti, G. 246
Muir, G. 333
Mullock, S. J. 67
Munuera, G. 248

Naoumidis, A. 185
Navío, A. 248
Newton, M. 416
Neyens, A. 338
Nickel, H. 185
Niemantsverdriet, J. W. 215, 221
Nilson, J. 517
Nip, W. 517

Oehrlein, G. S. 275
Olefjord, I. 377
Olivier, J. 283
Onsgaard, J. 119
Ottaviani, G. 329
Oude Elferink, J. B. 293

Paál, Z. 251
Panzner, G. 184
Pethrick, R. A. 445
Pignataro, S. 441
Plattner, H. H. 517
Pöcker, A. 335
Polak, M. 151
Polk, M. B. 405
Polzonetti, G. 336
Pons-Corbeau, J. 21
Powell, C. J. 79, 111
Prada Silvy, R. 247
Prutton, M. 99
Puderbach, H. 384

Quantin, J. C. 125
Queirolo, G. 129, 329

Rasigni, M. 328
Réti, F. 237
Roozeboom, F. 303
Rossi, M. 249
Rouxhet, P. G. 446
Ruckman, J. C. 31

Sabbatini, L. 414
Saille, M. 330
Sazanov, J. N. 417
Schay, Z. 261
Schiffmann, B. 151
Scholten, D. 467
Schreiner, M. 515
Schreinlechner, I. E. 371
Schumacher, R. 477
Seah, M. P. 79, 85
Sigmund, P. 130
Smith, G. C. 55, 145, 447
Southon, M. J. 67
Speckmann, H.-D. 357
Stingeder, G. 325, 515
Stone, W. E. E. 446
Storp, S. 384
Strehblow, H.-H. 357
Stritzker, B. 131
Sullivan, J. L. 493
Sundgren, J.-E. 437

Suoninen, E. 501
Swanson, M. L. 517
Székely, T. 237, 260, 417

Taglauer, E.
Tascón, J. M. D. 207
Terryn, H. 327, 332
Thummler, F. 453
Toftorup, A. L. 130
Tölg, G. 516
Tolle, H. J. 473
Tollet, H. 338
Torrisi, A. 441
Tóth, A. 417
Tóth, J. 251
Tougaard, S. 130
Tousset, J. 139
Tschersich, K. G. 297

Valigi, M. 245
van den Berg, J. A. 309
Vanden Berghe, R. 330
van der Kraan, A. M. 221
van der Weg, W. F. 293
Van Hove, H. 338
van Langeveld, A. D. 215
van Ooij, W. J. 367
Van Paemel, R. 330
Van Steene, G. 181
Varsányi, G. 237
Vereecken, J. 327, 332
Veszprémi, T. 69
Vickerman, J. C. 70, 309
Vlaeminck, R. 330
Vuorinen, S. 463

Wagner, H. 131
Wallach, E. R. 67
Watts, J. F. 416, 431
Whitton, J. L. 517
Wilhartitz, P. 182
Willich, P. 473
Wirth, A. 157
Wissman, P. 27
Wong, L. F. 493

Young, E. W. A. 163

Zaki, M. I. 243
Zalar, A. 41, 328
Zambonin, P. G. 414
Zsombok, Gy. 69

Information for Authors

General

Surface and Interface Analysis is devoted to the publication of papers dealing with the development and application of techniques for the characterization of surfaces, interfaces and thin films. Papers dealing with standardization and quantification are particularly welcome, and also those which deal with the application of these techniques to industrial problems. Papers dealing with the purely theoretical aspects of the technique will also be considered. Review articles will be published; prior consultation with Martin Seah (Review Editor) is advised in these cases. Papers must clearly be of scientific value in the field and will be submitted to two independent referees. Contributions must be in English and must not have been published elsewhere, and authors must agree not to communicate the same material for publication to any other journal. Authors are invited to submit their papers for publication to David Briggs or John T. Grant (papers for the American continents and Japan only). In every case, receipt of the paper will be acknowledged by the receiving Editor. Authors are advised to contact the appropriate Editor if notice of receipt is **not** received within one month. It is in the interest of an author to ensure accurate and consistent presentation and thus avoid publication delays. There are no page charges.

Manuscripts

Manuscripts should be typed with double line spacing and margins of at least $1\frac{1}{2}$ inches must be left at the top, bottom and left-hand side of each page. The name and full address of the author to whom page proofs should be despatched must be typed at the top of the first page. Denote with an asterisk (*) the author to whom enquiries regarding the work should be directed. All manuscripts should be submitted in triplicate and a set of copies or prints of the diagrams and spectra must accompany each copy of the manuscript. This will enable papers to be sent to two referees simultaneously and so assist rapid publication. The title of the paper should describe the contents fully but concisely and should include keywords suitable for retrieval systems. Compounds and figures should be numbered in separate sequences using Arabic numerals. All figures and tables must have captions. Footnotes should be avoided. Greek letters used in the text should be defined in the margin. A running head of up to 70 characters should be provided.

Abstract

Every paper for publication must be preceded by an abstract of not more than 250 words setting out the essential contents and typed on a separate sheet. This should be sufficiently comprehensive to allow effective use by abstracting services. It must be self-contained and compound numbers must not be used unless they are defined with the abstract. A list of 'keywords' should be included with the abstract.

Nomenclature

Authors should conform to nomenclature, symbols, abbreviations and procedures adopted by the International Union of Pure and Applied Chemistry (IUPAC). SI units are preferred; if more commonly used units are adopted, conversion factors should be given at their first occurrence.

Description of Experimental Methods

The Experimental section should be precise and give all details necessary for repeating the work. Particular attention should be given to methods used for calibration of instruments and spectra when electron spectroscopic data are presented.

Abbreviations

All abbreviations should be defined the first time they are used. The following terms are acceptable without definition: XPS, UPS, ESCA, SIMS, AES, ISS, SEM, RBS.

Illustrations

Illustrations on tracing cloth, tracing paper or white card should be supplied separately from the text and must bear the names of the authors, the title and the figure number. All formulae and figures should be carefully drafted and never drawn freehand. The accuracy of the printed version depends upon the author's accuracy.

References

This section should be used only for references to previous work. Additional information, such as experimental details, etc. should be incorporated into the text. References should be numbered in the sequence in which they occur in the text, cited by superior numbers and listed at the end of the paper. Authors should as far as possible cite available published work. If unpublished or personal work must be mentioned, then these citations should be included in sequence with the normal literature references. If the material cited is not readily accessible, then authors should also give, for example, the Chemical Abstracts reference.

References should be listed in the following style:

1. D. C. Jackson, T. E. Gallon and A. Chambers, *Surf. Sci.* **36**, 381 (1973).
2. T. L. Barr, in *Practical Surface Analysis by Auger and X-ray Photoelectron Spectroscopy*, ed. by D. Briggs and M. P. Seah, Chapt. 8, pp. 283–358. John Wiley, Chichester (1983).

Journal titles must be abbreviated according to the system co-sponsored by the American Chemical Society and listed in *Bibliographic Guide for Editors & Authors.*

Acknowledgements

These are placed at the end of the text preceding the references and should be brief.

Proofs

Page proofs for the correction of printers' errors only, will be despatched to the author indicated on the first page of the manuscript. Any further alteration at this proof stage may have to be charged to the author and may seriously delay publication.

Reprints

Fifty reprints of each paper and one copy of the journal issue in which it appears will be supplied free of charge. Additional copies may be ordered if required on a form supplied with the proofs.

Short Communications

Communications which are not intended for publication as full papers may be published in this section. These are to be presented in the same style as a full paper.

Letters Section

Letters will be published in SIA at the discretion of the Editor-in-Chief. Contributions may be addressed to either David Briggs (Editor-in-Chief) or John T. Grant (North American Editor).

Copyright

Upon acceptance of an article by the journal, the author(s) will be asked to transfer copyright of the article to the publisher. This transfer will ensure the widest possible dissemination of information under applicable Copyright law. A copy of the Publishing Agreement to be used is reproduced in each volume and additional copies are available from the Editors or the publisher.